PRINCIPLES OF NEUROBIOLOGY

PRINCIPLES OF NEUROBIOLOGY

SECOND EDITION

LIQUN LUO

CRC Press
Taylor & Francis Group

A GARLAND SCIENCE BOOK

Second edition published 2021
by CRC Press
6000 Broken Sound Parkway NW, Suite 300, Boca Raton, FL 33487-2742
and by CRC Press
2 Park Square, Milton Park, Abingdon, Oxon, OX14 4RN

© 2021 Taylor & Francis Group, LLC
First edition published by Garland Science (Taylor & Francis Group, LLC) 2016
CRC Press is an imprint of Taylor & Francis Group, LLC

Library of Congress Cataloging-in-Publication Data
Names: Luo, Liqun, 1966– author.
Title: Principles of neurobiology / Liqun Luo.
Description: Second edition. | Boca Raton : Garland Science, 2020. | Includes bibliographical references and index. | Summary: "Principles of Neurobiology, Second Edition presents the major concepts of neuroscience with an emphasis on how we know what we know. The text is organized around a series of key experiments to illustrate how scientific progress is made and helps upper-level undergraduate and graduate students discover the relevant primary literature. Written by a single author in a clear and consistent writing style, each topic builds in complexity from electrophysiology to molecular genetics to systems level in a highly integrative approach. Students can fully engage with the content via thematically linked chapters and will be able to read the book in its entirety in a semester-long course. Principles of Neurobiology is accompanied by a rich package of online student and instructor resources including animations, figures in PowerPoint, and a Question Bank for adopting instructors"— Provided by publisher.
Identifiers: LCCN 2020023964 (print) | LCCN 2020023965 (ebook) | ISBN 9780815346050 (paperback) | ISBN 9780367514716 (hardback) | ISBN 9781003053972 (ebook)
Subjects: LCSH: Neurobiology.
Classification: LCC QP355.2 .L86 2020 (print) | LCC QP355.2 (ebook) | DDC 612.8—dc23
LC record available at https://lccn.loc.gov/2020023964
LC ebook record available at https://lccn.loc.gov/2020023965

ISBN: 9780367514716 (hbk)
ISBN: 9780815346050 (pbk)
ISBN: 9781003053972 (ebk)

Typeset in Utopia Std and Avenir LT Std
by Carol Pierson, Chernow Editorial Services, Inc.

Visit the companion website: www.crcpress.com/cw/luo

To Lubert Stryer—my mentor, colleague, and dear friend.

PREFACE
TO THE SECOND EDITION

Neurobiology has witnessed rapid advances in the past five years, thanks in part to the support from the U.S. National Institutes of Health's BRAIN Initiative and similar initiatives internationally. To give a few examples: deciphering single-cell transcriptomes across the nervous system has produced valuable information regarding the development and function of specific cell types and has shed light on what constitutes a cell type in complex brain regions. Technological advances in neural circuit dissection, from genetics to anatomy and neurophysiology, have enabled better understanding of many neurobiological processes, from sensation of internal organs to the organization of memory systems in the brain. Breakthrough nucleic acid–based therapies have enabled treatment of devastating neurodegenerative disorders.

The second edition of *Principles of Neurobiology* intends to capture these and many other new advances while maintaining its discovery–based approach: to teach students how knowledge is obtained. This new edition has also added or strengthened many features, thanks to feedback from students and instructors around the globe who have used the first edition in their courses. Major changes include:

- New sections on theory and modeling in Chapter 14 to reflect an increasingly important role theory and modeling play in modern neurobiology. These new sections encompass a wide range of topics from neuronal encoding and decoding to neural circuit architectures and learning algorithms, further expanding the horizon of students of neurobiology.
- Expanded coverage of motor and regulatory systems in separate chapters. The new motor systems chapter has more in-depth discussions of brainstem, cerebellum, basal ganglia, and parietal and frontal cortex in motor coordination, planning, and sensorimotor integration. The new regulatory systems chapter includes new advances on the interoceptive system and the links between homeostatic need and motivated behavior.
- Open questions at the end of each chapter to stimulate students and researchers to explore new terrains.

I would also take this opportunity to highlight several features for students and instructors:

- The current sequence of chapters reflects the course I have been teaching at Stanford, but no single linear sequence can capture the rich interconnections in neurobiology. Embedded in each chapter are many references to other sections and chapters to enable students to make such links. In the electronic version of the textbook, such connections are just one click away.
- Subsets of chapters can be reorganized to cover a variety of courses. For example, Chapters 5, 7, and 11 can be used for a *developmental neurobiology* course. Relevant sections in Chapter 4, 6, 8, 9, 10, and 11 can be used for a *systems neurobiology* course. Both courses can benefit from the basic foundations in Chapters 1–3, the disease connections in Chapter 12, and the evolutionary perspective in Chapter 13. Students can benefit from connections with the rest of the neurobiology. Finally, relevant sections of the entire textbook can be used in a *molecular and cellular neurobiology* course.
- Chapter 14 contains systematic descriptions of major techniques used in neurobiology, from molecular genetics to circuit and behavioral analyses,

and now theory and modeling, and are frequently referred to throughout the text. Students should study the relevant sections in Chapter 14 as often as needed to enhance their understanding of earlier chapters.

- Material in "Boxes" are just as important as the main text. Boxes are created so important materials can be discussed in more depth, with additional examples, or from a different perspective, without interrupting the storylines of the main text.
- Students interested in finding out more about how discoveries are made are highly encouraged to study the primary literature on subjects of interest. These are cited in the figure legends and in "Further Reading" at the end of each chapter (often complementary).

I would like to extend my gratitude to numerous students and instructors who have used first edition of *Principles of Neurobiology* for their feedback and encouragement. I thank the previous Garland Science team, in particular Denise Schanck, for encouraging me to work on this new edition. I am grateful to Chuck Crumly, my editor from CRC Press, whose unwavering support and sage advice have guided me throughout the journey. I am continually indebted to Nigel Orme, whose expert illustrations have made the textbook vivid; working with Nigel on the figures added much fun. I thank many colleagues (see the Acknowledgments) for their expert review and critiques. I owe much gratitude to my PhD student Andrew Shuster, who carefully edited the entire textbook and substantially improved its clarity and accuracy. I thank Jordan Wearing, whose remarkable organization skills and attention to details have enabled smooth transition from manuscripts to final production. I also thank Barbara Chernow and her team for the superb production of the final pages. Finally, I am very grateful to the continuous support of my wife, Charlene Liao, and our two daughters, Connie and Jessica.

Liqun Luo
April 2020

PREFACE
TO THE FIRST EDITION

Neurobiology has never seen a more exciting time. As the most complex organ of our body, the brain endows us the ability to sense, think, remember, and act. Thanks to the conceptual and technical advances in recent years, the pace of discovery in neurobiology is continuously accelerating. New and exciting findings are reported every month. Traditional boundaries between molecular, cellular, systems, and behavioral neurobiology have been broken. The integration of developmental and functional studies of the nervous system has never been stronger. Physical scientists and engineers increasingly contribute to fundamental discoveries in neurobiology. Yet we are still far from a satisfying understanding of how the brain works, and from converting this understanding into effective treatment of brain disorders. I hope to convey the excitement of neurobiology to students, to lay the foundation for their appreciation of this discipline, and to inspire them to make exciting new discoveries in the coming decades.

This book is a reflection of my teaching at Stanford during the past 18 years. My students—and the intended audience of this book—include upper division undergraduates and beginning graduate students who wish to acquire an in-depth knowledge and command of neurobiology. While most students reading this book may have a biology background, some may come from physical sciences and engineering. I have discovered that regardless of a student's background, it is much more effective—and much more interesting—to teach students how knowledge has been obtained than the current state of knowledge. That is why I have taken this discovery-based teaching approach from lecture hall to textbook.

Each chapter follows a main storyline or several sequential storylines. These storylines are divided by large section headings usually titled with questions that are then answered by a series of summarizing subheadings with explanatory text and figures. Key terms are highlighted in bold and are further explained in an expanded glossary. The text is organized around a series of key original experiments, from classic to modern, to illustrate how we have arrived at our current state of understanding. The majority of the figures are based on those from original papers, thereby introducing students to the primary literature. Instead of just covering the vast number of facts that make up neurobiology in this day and age, this book concentrates on the in-depth study of a subset of carefully chosen topics that illustrate the discovery process and resulting principles. The selected topics span the entire spectrum of neurobiology, from molecular and cellular to systems and behavioral. Given the relatively small size of the book, students will be able to study much or all of the book in a semester, allowing them to gain a broad grasp of modern neurobiology.

This book intentionally breaks from the traditional division of neuroscience into molecular, cellular, systems, and developmental sections. Instead, most chapters integrate these approaches. For example, the chapter on 'Vision' starts with a human psychophysics experiment demonstrating that our rod photoreceptors can detect a single photon, as well as a physiology experiment showing the electrical response of the rod to a single photon. Subsequent topics include molecular events in photoreceptors, cellular and circuit properties of the retina and the visual cortex, and systems approaches to understanding visual perception. Likewise, 'Memory, Learning, and Synaptic Plasticity' integrates molecular, cellular, circuit, systems, behavioral, and theoretical approaches with the common goal of understanding what memory is and how it relates to synaptic plasticity. The two chapters on development intertwine with three chapters on sensory and motor systems to help students appreciate the rich connections between the development and function of the nervous system. All chapters are further linked by abundant cross-referencing through the text. These links reinforce the notion that topics in neurobiology form highly interconnected networks rather than a linear sequence. Finally and importantly, Chapter 13 ('Ways of Exploring') is dedicated to key methods in neurobiology research and is extensively referenced in all preceding chapters. Students are encouraged to study the relevant methods in Chapter 13 when they first encounter them in Chapters 1–12.

This book would not have been possible without the help of Lubert Stryer, my mentor, colleague, and dear friend. From inception to completion, Lubert has provided invaluable support and advice. He has read every single chapter (often more than once) and has always provided a balanced dose of encouragement and criticism, from strategic planning to word choice. Lubert's classic *Biochemistry* textbook was a highlight in my own undergraduate education and has continued to inspire me throughout this project.

I thank Howard Schulman, Kang Shen, and Tom Clandinin, who, along with Lubert, have been my co-instructors for neurobiology courses at Stanford and from whom I have learned a tremendous amount about science and teaching. Students in my classes have offered valuable feedback that has improved my teaching and has been incorporated into the book. I am highly appreciative of the past and current members of my lab, who have taught me more than I have taught them and whose discoveries have been constant sources of inspiration and joy. I gratefully acknowledge the National Institutes of Health and the Howard Hughes Medical Institute for generously supporting the research of my lab.

Although this book has a single author, it is truly the product of teamwork with Garland Science. Denise Schanck has provided wise leadership throughout the journey. Janet Foltin in the initial phase and Monica Toledo through most of the project have provided much support and guidance, from obtaining highly informative reviews of early drafts to organizing teaching and learning resources. I am indebted to Kathleen Vickers for expert editing; her attention to detail and demand for clarity have greatly improved my original text. I owe the illustrations to Nigel Orme, whose combined artistic talent and scientific understanding brought to life concepts from the text. Georgina Lucas's expert page layout has seamlessly integrated the text and figures. I also thank Michael Morales for producing the enriching videos, and Adam Sendroff and his staff for reaching out to the readers. Working with Garland has been a wonderful experience, and I thank Bruce Alberts for introducing Garland to me.

Finally, I am very grateful for the support and love from my wife, Charlene Liao, and our two daughters, Connie and Jessica. Writing this textbook has consumed a large portion of my time in the past few years; indeed, the textbook has been a significant part of our family life and has been a frequent topic of dinner table conversation. Jessica has been my frequent sounding board for new ideas and storylines, and I am glad that she has not minded an extra dose of neurobiology on top of her demanding high-school courses and extracurricular activities.

I welcome feedback and critiques from students and readers!

Liqun Luo
April 2015

NOTE ON GENE AND PROTEIN NOMENCLATURE

This book mostly follows the unified convention of *Molecular Biology of the Cell* 6th Edition by Alberts et al. (Garland Science, 2015) for naming genes. Regardless of species, gene names and their abbreviations are all in italics, with the first letter in upper case and the rest of the letters in lower case. All protein names are in roman, and their cases follow the consensus in the literature. Proteins identified by biochemical means are usually all in lower case; proteins identified by genetic means or by homology with other genes usually have the first letter in upper case; protein acronyms usually are all in upper case. The space that separates a letter and a number in full names includes a hyphen, and in abbreviated names is omitted entirely.

The table below summarizes the official conventions for individual species and the unified conventions that we shall use in this book.

Organism	Species-Specific Convention		Unified Convention Used in this Book	
	Gene	Protein	Gene	Protein
Mouse	*Syt1*	synaptotagmin I	*Syt1*	Synaptotagmin-1
	Mecp2	MeCP2	*Mecp2*	MeCP2
Human	*MECP2*	MeCP2	*Mecp2*	MeCP2
Caenorhabditis	*unc-6*	UNC-6	*Unc6*	Unc6
Drosophila	*sevenless* (named after recessive phenotype)	Sevenless	*Sevenless*	Sevenless
	Notch (named after dominant mutant phenotype)	Notch	*Notch*	Notch
Other organisms (e.g. jellyfish)		Green fluorescent protein (GFP)	*Gfp*	GFP

RESOURCES FOR INSTRUCTORS AND STUDENTS

The teaching and learning resources for instructors and students are available online. We hope these resources will enhance student learning and make it easier for instructors to prepare dynamic lectures and activities for the classroom.

Instructor Resources

Instructor Resources are available on the Instructor Resources Download Hub, located at www.routledgetextbooks.com/textbooks/instructor_downloads/. These resources are password-protected and available only to instructors adopting the book.

Art of Principles of Neurobiology
All figures from the book are available in two convenient formats: PowerPoint® and PDF. They have been optimized for display on a computer.

Figure-Integrated Lecture Outlines
The section headings, concept headings, and figures from the text have been integrated into PowerPoint presentations. These will be useful for instructors who would like a head start creating lectures for their course. Like all of our PowerPoint presentations, the lecture outlines can be customized. For example, the content of

these presentations can be combined with videos and questions from the book or Question Bank, in order to create unique lectures that facilitate interactive learning.

Animations and Videos

All animations and videos that are available to students are also available to instructors. They can be downloaded from the Instructor Hub in MP4 format. The movies are related to specific chapters, and callouts to the movies are highlighted in green throughout the textbook.

Question Bank

Written by Elizabeth Marin (University of Cambridge), and Melissa Coleman (Claremont McKenna, Pitzer, and Scripps Colleges), the Question Bank includes a variety of question formats: multiple choice, fill-in-the-blank, true-false, matching, essay, and challenging 'thought' questions. There are approximately 40–50 questions per chapter, and a large number of the multiple-choice questions will be suitable for use with personal response systems (that is, clickers). The Question Bank provides a comprehensive sampling of questions that require the student to reflect upon and integrate information, and can be used either directly or as inspiration for instructors to write their own test questions.

Student Resources

Resources for students are available on the books Companion Website, located at www.crcpress.com/cw/luo.

Art of Principles of Neurobiology

All figures from the book are available in two convenient formats: PowerPoint® and PDF. They have been optimized for display on a computer.

Animations and Videos

There are over 40 narrated movies, covering a range of neurobiology topics, which review key concepts and illuminate the experimental process.

Flashcards

Each chapter contains flashcards, built into the student website, that allow students to review key terms from the text.

Glossary

The comprehensive glossary of key terms from the book is online.

Blog

A blog associated with Principles of Neurobiology companion website has monthly new entries, which introduce students to the latest discoveries in research and extend the concepts discussed in the textbook.

ADDITIONAL NOTES ON HOW TO USE THIS BOOK

- Key terms in the text are highlighted in bold font, with glossary entries.
- Extensive cross-references of sections and figures help strengthen the connections between different parts of neurobiology. In the e-book, hyperlinks have been created for these cross-references so students can click the link to study a related figure or a section in a different part of the book, and click again to return to the original page.
- Students are particularly encouraged to study the relevant sections in Chapter 14 when referenced in earlier chapters.
- To emphasize the discovery–based approach, most figures have been adapted from the original literature. For simplicity, error bars and statistics have been omitted for most figures. Interested students can find such details by following the citations in figure legends.

ACKNOWLEDGMENTS

The author and publisher of *Principles of Neurobiology* specially thank Andrew Shuster (Stanford University) for editing the entire textbook, and Melissa Coleman (Claremont McKenna, Pitzer and Scripps Colleges) and Lisa Marin (University of Cambridge) for creating the Question Bank.

The author and publisher of *Principles of Neurobiology* gratefully acknowledge the contributions of the following scientists and instructors for their advice and critique in the development of the second edition of this book:

Chapter 1: Eric Knudsen (Stanford University), Doris Tsao (California Institute of Technology).

Chapter 2: Josh Huang (Cold Spring Harbor Laboratory/ Duke University), John Huguenard (Stanford University), Lily Jan (University of California, San Francisco), Yulong Li (Peking University), Kang Shen (Stanford University), Gina Turrigiano (Brandeis University), Nieng Yan (Princeton University).

Chapter 3: Josh Huang (Cold Spring Harbor Laboratory), Lily Jan (University of California, San Francisco), Erik Jorgensen (University of Utah), Yulong Li (Peking University), Kang Shen (Stanford University), Tom Südhof (Stanford University), Rachel Wilson (Harvard University).

Chapter 4: Tom Clandinin (Stanford University), E. J. Chichilnisky (Stanford University), Tirin Moore (Stanford University), Bill Newsome (Stanford University), Massimo Scanzioni (University of California, San Francisco), Lubert Stryer (Stanford University), Doris Tsao (California Institute of Technology), Wei Wei (University of Chicago).

Chapter 5: Tom Clandinin (Stanford University), Marla Feller (University of California, Berkeley), Andy Huberman (Stanford University), Alex Kolodkin (Johns Hopkins University), Susan McConnell (Stanford University), Carla Shatz (Stanford University), Larry Zipursky (University of California, Los Angeles).

Chapter 6: Diana Bautista (University of California, Berkeley), Xiaoke Chen (Stanford University), Xintong Dong (Johns Hopkins University), Xinzhong Dong (Johns Hopkins University), David Ginty (Harvard University), Eric Knudsen (Stanford University), Shan Meltzer (Harvard University), Adi Mizrahi (Hebrew University), John Ngai (University of California, Berkeley), Greg Scherrer (University of North Carolina).

Chapter 7: Yuh-Nung Jan (University of California, San Francisco), Alex Kolodkin (Johns Hopkins University), Susan McConnell (Stanford University), Sergiu Paşca (Stanford University), Larry Zipursky (University of California, Los Angeles).

Chapter 8: Silvia Arber (University of Basel), Rui Costa (Columbia University), Josh Huang (Cold Spring Harbor), Eve Marder (Brandeis University), Krishna Shenoy (Stanford University), Mark Wagner (Stanford University), Kevin Yackle (University of California, San Francisco).

Chapter 9: Will Allen (Harvard University), Xiaoke Chen (Stanford University), Yang Dan (University of California, Berkeley), Steve Liberles (Harvard University), Brad Lowell (Harvard University), Ruslan Medzhitov (Yale University), Louis Ptáček (University of California, San Francisco), Chen Ran (Harvard University), Bill Snider (University of North Carolina).

Chapter 10: Barry Dickson (Howard Hughes Medical Institute Janelia Research Campus), Catherine Dulac (Harvard University), Weizhe Hong (University of California, Los Angeles), Mala Murthy (Princeton University), Nirao Shah (Stanford University).

Chapter 11: Lu Chen (Stanford University), Edvard Moser (Norwegian University of Science and Technology), Roger Nicoll (University of California, San Francisco), Bill Snider (University of North Carolina), Gerry Rubin (Howard Hughes Medical Institute Janelia Research Campus), Mark Schnitzer (Stanford University), Liz Steinberg (Stanford University), Gina Turrigiano (Brandeis University), Ilana Witten (Princeton University).

Chapter 12: Xiaoke Chen (Stanford University), Lief Fenno (Stanford University), Anirvan Ghosh (Unity Biotechnology), Aaron Gitler (Stanford University), Wei-Hsiang Huang (McGill University), Bill Snider (University of North Carolina), Ryan Watts (Denali Therapeutics).

Chapter 13: Tom Clandinin (Stanford University), Chuck Crumly (CRC Press), Ruslan Medzhitov (Yale University), Alex Pollen (University of California, San Francisco), David Stern (Howard Hughes Medical Institute Janelia Research Campus), Lubert Stryer (Stanford University).

Chapter 14: Will Allen (Harvard University), Tom Clandinin (Stanford University), Claire Cui (Google), Shaul Druckmann (Stanford University), Catherine Dulac (Harvard University), Surya Ganguli (Stanford University), Scott Linderman (Stanford University), Ken Miller (Columbia University), Bill Snider (University of North Carolina), Lubert Stryer (Stanford University), Alice Ting (Stanford University), Mark Wagner (Stanford University), Rachel Wilson (Harvard University), Dan Yamins (Stanford University), Zheng Zhang (New York University, Shanghai).

The author and publisher of *Principles of Neurobiology* gratefully acknowledge the contributions of the following scientists and instructors for their advice on and critiques of the first edition of this book:

Will Allen, Silvia Arber, Steve Baccus, Bruce Baker, Ben Barres, Michael Baum, Richard Benton, Peter Bergold, Nic Berns, Tobias Bonhoeffer, Katja Brose, Linda Buck, John Carlson, Sidi Chen, Xiaoke Chen, Tom Clandinin, Melissa Coleman, Yang Dan, Karl Deisseroth, Laura DeNardo Wilke, Claude Desplan, Hongwei Dong, Xinzhong Dong, Serena Dudek, Catherine Dulac, Dave Feldheim, Marla Feller, Guoping Feng, Russ Fernald, Joe Fetcho, Gord Fishell, Hunter Fraser, Sam Gandy, Surya Ganguli, Xiaojing Gao, David Ginty, Lisa Giocomo, Aaron Gitler, Casey Guenthner, Joachim Hallmayer, Craig Heller, Shaul Hestrin, Simon Hippenmeyer, Weizhe Hong, Hadley Wilson Horch, Mark Horowitz, Josh Huang, Wei-Hsiang Huang, Andy Huberman, Steve Hyman, Lily Jan, Yuh-Nung Jan, Patricia Janak, Greg Jefferis, William Joo, David Julius, Haig Keshishian, Eric Knudsen, Alex Kolodkin, Takaki Komiyama, Richard Levine, Yulong Li, Charlene Liao, Jeff Lichtman, Manyuan Long, Chris Lowe, Jan Lui, Rob Malenka, Dev Manoli, Eve Marder, Lisa Marin, Mike McCloskey, Susan McConnell, Emmanuel Mignot, Kazunari Miyamichi, Adi Mizrahi, Bill Mobley, Tim Mosca, Jeremy Nathans, Bill Newsome, Lisa Olson, Karen Parfitt, Josef Parvizi, Ardem Patapoutian, Dmitri Petrov, John Pizzey, Mu-ming Poo, Chris Potter, David Prince, Martin Raff, Geert Ramakers, Jennifer Raymond, Jing Ren, Michael Rosbash, Botond Roska, Ed Ruthazer, Greg Scherrer, Mark Schnitzer, Tom Schwarz, Matthew Scott, Idan Segev, Nirao Shah, Mehrdad Shamloo, Carla Shatz, Kang Shen, Krishna Shenoy, Annemarie Shibata, Alcino Silva, Malathi Srivatsan, Scott Sternson, Chuck Stevens, Tom Südhof, Karel Svoboda, Larry Swanson, Bosiljka Tasic, Karl Wah Keung Tsim, Mark Wagner, Joy Wan, Fan Wang, Xinnan Wang, Eric Warrant, Ryan Watts, Brady Weissbourd, Marius Wernig, Rachel Wilson, Boon-Seng Wong, Daisuke Yamamoto, Jian Yang, Charles Yanofsky, Larry Young, Haiqing Zhao, Weimin Zhong, Huda Zoghbi.

SPECIAL FEATURES

CONTENTS

Chapter 8
Motor Systems   335

Chapter 9
Regulatory Systems   375

Chapter 10
Sexual Behavior   411

Chapter 11
Memory, Learning, and Synaptic Plasticity **445**

Chapter 12
Brain Disorders **499**

Chapter 13
Evolution of the Nervous System   547

CHAPTER 1

An Invitation to Neurobiology

How does the nervous system control behavior? How do we sense the environment? How does the brain create a representation of the world out of the sensations? How much of our brain function and behavior is shaped by our genes, and how much reflects the environment in which we grew up? How is the brain wired up during development? What changes occur in the brain when we learn something new? How have nervous systems evolved? What goes wrong in brain disorders?

We are about to embark on a journey to explore these questions, which have fascinated humanity for thousands of years. Our ability to address these questions *experimentally* has greatly expanded in recent years. What we currently know about the answers to these questions comes mostly from findings made in the past 50 years; in the next 50 years, we will likely learn more about the brain and its control of behavior than in all of prior human history. We are at an exciting time as students of neurobiology, and it is my hope that many readers of this book will be at the forefront of groundbreaking discoveries.

PRELUDE: NATURE AND NURTURE IN BRAIN FUNCTION AND BEHAVIOR

As we begin this journey, let's discuss one of the questions we raised regarding the contributions of genes and environment to our brain function and behavior. We know from experience that both genetic inheritance (**nature**) and environmental factors (**nurture**) make important contributions, but how much does each contribute? How do we begin to tackle such a complex question? In scientific research, asking the right questions is often a critical step toward obtaining the right answers. As evolutionary geneticist Theodosius Dobzhansky put it, "The question about the roles of the genotype and the environment in human development must be posed thus: To what extent are the *differences* observed among people conditioned by the differences of their genotypes and by the differences between the environments in which people were born, grew and were brought up?"

1.1 Human twin studies can reveal the contributions of nature and nurture

Francis Galton first coined the phrase *nature versus nurture* in the nineteenth century. He also introduced a powerful method for studying this conundrum: statistical analysis of human twins. Identical twins (**Figure 1-1**), or **monozygotic twins**, share 100% of their genes in almost all cells, as they are products of the same fertilized egg, or **zygote**. One can compare specific traits among thousands of pairs of identical twins to see how correlated they are within each pair. For example, if we compare the intelligence quotients (IQs)—an estimate of general intelligence—of any two random people in the population, the correlation is 0. (Correlation is a statistic of resemblance that ranges from 0, indicating no resemblance, to 1, indicating perfect resemblance.) This correlation is 0.86 for identical twins (**Figure 1-2**), a striking similarity. However, identical twins also usually grow

Figure 1-1 Identical (monozygotic) twins. Identical twins develop from a single fertilized egg and therefore share 100% of their genes in almost all cells (some lymphocytes are an exception due to stochasticity in DNA recombination). Most identical twins also share similar childhood environments. (Courtesy of Christopher J. Potter.)

up in the same environment, so this correlation alone does not help us distinguish between the contributions of genes and the environment.

Fortunately, human populations provide a second group that allows researchers to tease apart the influence of genetic and environmental factors. Nonidentical (fraternal) twins occur more often than identical twins in most human populations. These are called **dizygotic twins** because they originate from two independent eggs fertilized by two independent sperm. As full siblings, dizygotic twins are 50% identical in their genes according to Mendel's laws of inheritance. However, like monozygotic twins, dizygotic twins usually share very similar prenatal and postnatal environments. Thus, the differences between traits exhibited by monozygotic and dizygotic twins should result from the differences in 50% of their genes. In our example, the correlation of IQ scores between dizygotic twins is 0.60 (Figure 1-2).

Behavioral geneticists use the term **heritability** to describe the contribution of genetic differences to trait differences. Heritability is defined as the difference between the correlations of monozygotic and dizygotic twins multiplied by 2 (because the genetic difference is 50% between monozygotic and dizygotic twins). Thus, the heritability of IQ is (0.86 − 0.60) × 2 = 0.52. Roughly speaking, then, genetic differences account for about half of the *variation* in IQ scores within human populations. Traditionally, the non-nature component has been presumed to come from environmental factors. However, "environmental factors" as calculated in twin studies include *all* factors not inherited from the parents' DNA. These include the postnatal environment, which is what we typically think of as nurture, but also prenatal environment, stochasticity in developmental processes, somatic mutations (alterations in DNA sequences in somatic cells after fertilization), and gene expression changes due to **epigenetic modifications**. Epigenetic

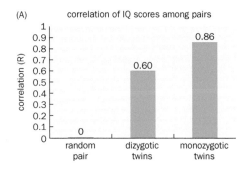

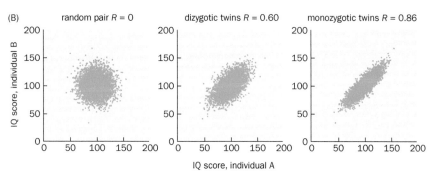

Figure 1-2 Twin studies for determining genetic and environmental contributions to intelligence quotient (IQ). (A) Correlation, or *R* value, of IQ scores for 4672 pairs of monozygotic twins and 5546 pairs of dizygotic twins. The correlation between the IQ scores of randomly selected pairs of individuals is zero. The difference in correlation between monozygotic and dizygotic twins can be used to calculate the heritability of traits. The large sample size makes these estimates highly accurate. **(B)** Simulation of IQ score correlation plots for 5000 pairs of unrelated individuals (*R* = 0), 5000 pairs of dizygotic twins (*R* = 0.60), and 5000 pairs of monozygotic twins (*R* = 0.86). The *x* and *y* axes of a given dot represent the IQ scores of one pair. The simulations assume a normal distribution of IQ scores (mean = 100, standard deviation = 15). (A, based on Bouchard TJ & McGue M [1981] *Science* 212:1055–1059.)

modifications refer to changes made to DNA and chromatin that do not modify DNA sequences but can alter gene expression—these include DNA methylation and various modifications of histones, the protein component of chromatin. As we will learn later, all of these factors contribute to nervous system development, function, and behavior.

Twin studies have been used to estimate the heritability of many human traits, ranging from height (~90%) to the chance of developing schizophrenia (60–80%). An important caveat regarding these estimates is that most human traits result from complex interactions between genes and the environment, and heritability itself can change with the environment. Still, twin studies offer valuable insights into the relative contributions of genes and nongenetic factors to many aspects of brain function and dysfunction in a given environment. The completion of the Human Genome Project and the development of tools permitting detailed examination of the genome sequence data, combined with a long history of medical and psychological studies of human subjects, have made our own species the subject of a growing body of neurobiological research (Section 14.5). However, mechanistic understanding of how genes and the environment influence brain development, function, and behavior requires experimental manipulations that often can be carried out only in animal models. The use of vertebrate and invertebrate model species (Sections 14.1–14.4) has yielded much of what we have learned about the brain and behavior. Many principles of neurobiology revealed by experiments on specific model species have turned out to operate in a wide variety of organisms, including humans.

Figure 1-3 **Penguin feeding.** The instinctive behaviors of an adult penguin and its offspring photographed in Antarctica, 2009. Top, the young penguin asks for food by bumping its beak against its parent's beak. Bottom, the parent releases the food into the young penguin's mouth. (Courtesy of Lubert Stryer.)

1.2 Examples of *nature*: animals exhibit instinctive behaviors

Animals exhibit remarkable instinctive behaviors that help them find food, avoid danger, seek mates, and nurture their progeny. For example, a baby penguin, directed by its food-seeking instinct, bumps its beak against its parent's beak to remind its parent to feed it; in response, the parent instinctively releases the food it has foraged from the sea to feed its baby (**Figure 1-3**).

Instinctive behaviors can be elicited by very specific sensory stimuli. For instance, experimenters have tested the responses of young chicks to an object resembling a bird in flight, with wings placed close to either end of the head–tail axis. When moved in one direction, the object looks like a short-necked, long-tailed hawk; when moved in the other direction, the object looks like a long-necked, short-tailed goose. Seeing the object overhead, a young chick produces different responses depending on the direction in which the object moves, running away when the object resembles a hawk but making no effort to escape when the object resembles a goose (**Figure 1-4**). This escape behavior is **innate**: it is with the chick from birth and is likely genetically programmed. The behavior is also stereotypic: different chicks exhibit the same escape behavior, with similar stimulus specificity. Once the behavior is triggered, it runs to completion without

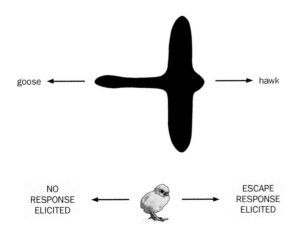

goose ← → hawk

NO RESPONSE ELICITED ← → ESCAPE RESPONSE ELICITED

Figure 1-4 **Innate escape response of a chick to a hawk.** A young chick exhibits instinctive escape behavior in response to an object moving overhead that resembles a short-necked, hawk-like bird; moving the pictured object from left to right triggers this instinctive behavior. Moving the object from right to left so that it resembles a long-necked goose does not elicit the chick's escape behavior. (Adapted from Tinbergen N [1951] The Study of Instinct. Oxford University Press.)

Figure 1-5 Barn owls use their auditory system to locate prey in complete darkness. The photograph was taken in the dark with infrared light flashed periodically while the camera shutter remained open. (Courtesy of Masakazu Konishi.)

further sensory feedback. **Neuroethology**, a field of study that emphasizes observing animal behavior in natural environments, refers to such instinctive behaviors as following **fixed action patterns**. The essential features of the stimulus that activates the fixed action pattern are referred to as **releasers**.

How do genes and developmental programs specify such specific instinctive behaviors? In Chapter 10, we will explore this question using sexual behavior as an example. We will learn about how a single gene in the fruit fly named *fruitless* can exert profound control over many aspects of fruit fly mating behavior.

1.3   An example of *nurture*: barn owls adjust their auditory maps to match altered visual maps

Animals also exhibit a remarkable capacity for learning as they adapt to a changing world. We use the ability of barn owls to adjust their auditory maps to changes in their vision to illustrate this capacity.

Barn owls have superb visual and auditory systems that help them catch prey at night when nocturnal rodents are active. In fact, owls can catch prey even in complete darkness (**Figure 1-5**), relying entirely on their auditory system. They can accurately locate the source of sounds made by prey, based on the small difference in the time it takes for a sound to reach their left and right ears. The owl's brain creates a map of space using these time differences, such that activation of individual nerve cells at specific positions in this brain map informs the owl of the physical position of its prey.

Experiments in which prisms were attached over a juvenile barn owl's eyes (**Figure 1-6**A) revealed how the owl responds when its auditory and visual maps provide conflicting information. Normally, the owl's auditory map matches its visual map, such that perceptions of sight and sound direct the owl to the same location (Figure 1-6B). The prisms shift the owl's visual map 23° to the right. The owl rapidly learns to adjust its motor responses to restore its reaching accuracy on visual targets. However, a mismatch occurs between the owl's visual and auditory maps on the first day after the prisms are placed (Figure 1-6C): sight and sound indicate different locations to the owl, causing confusion about the prey's location. The juvenile owl copes with this situation by adjusting its auditory map to

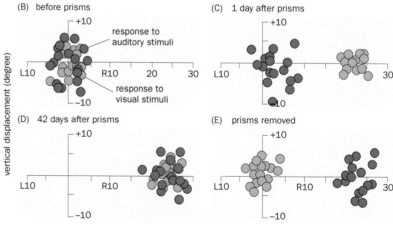

Figure 1-6 Juvenile barn owls adjust their auditory map to match a displaced visual map after wearing prisms. (A) A barn owl fitted with prisms that shift its visual map. **(B)** Before the prisms are attached, the owl's visual map (blue dots) and auditory map (red dots) are matched near 0°. Each dot represents an experimental measurement of an owl's head orientation in response to an auditory or visual stimulus presented in the dark. **(C)** One day after the prisms were fitted, the visual map is displaced 23° to the right of the auditory map.

(D) After a juvenile owl has worn the prisms for 42 days, its auditory map has adjusted to match its shifted visual map. **(E)** The visual map shifts back immediately after the prisms are removed, causing a temporary mismatch. This mismatch is corrected as the auditory map shifts back soon after (not shown). (A, courtesy of Eric Knudsen. B–E, from Knudsen EI [2002] *Nature* 417:322–328. With permission from Springer Nature.)

match its altered visual map within 42 days after starting to wear the prisms (Figure 1-6D), eliminating the positional conflict between sight and sound. The owl adjusts its strike behavior to accurately target a single location. When the prisms are removed, a mismatch recurs (Figure 1-6E), but the owl adjusts its auditory map and strike behavior back to their native states shortly afterward.

The story of the barn owl is an example of how the nervous system learns to cope with a changing world. Neurobiologists use the term **neural plasticity** to refer to changes in the nervous system in response to experience and learning. But the story does not end here. Studies have shown that plasticity declines with age: juvenile owls have the plasticity required to adjust their auditory map to match a visual map displaced by 23°, but owls will have lost this ability by the time they reach sexual maturity (Figure 1-7A). Some human learning capabilities, such as the ability to learn foreign languages, likewise decline with age. Thus, experiments targeted toward improving the plasticity of adult owls may reveal strategies for improving the learning abilities of adult humans as well.

Several ways have been found for adult owls to overcome their limited plasticity in shifting their auditory maps. If an owl experiences adjusting to a 23°-prism shift as a juvenile, it can readily readjust to the same prisms as an adult (Figure 1-7B). Alternatively, even adult owls that cannot adjust to a 23° shift all at once can learn to shift their auditory maps if the visual field displacement is applied in small increments. Thus, by taking baby steps, adult owls can eventually reach nearly the same shift magnitude as young owls. Once they have learned to shift via gradual increments, adult owls can subsequently shift in a single, large step when tested several months after returning to normal conditions (Figure 1-7C).

What are the neurobiological mechanisms underlying these fascinating plasticity phenomena? In Chapters 4 and 6, we will explore the nature of the visual and auditory maps. In Chapters 5 and 7, we will study how neural maps are formed during development and modified by experience. And in Section 11.25, we will address the mechanism of owls' map adjustment in the context of memory and learning. Before studying these topics, however, we need to learn more basics about the brain and its building blocks. We devote the rest of this chapter to providing an overview of the nervous system and introducing how key historical discoveries helped build the conceptual framework of modern neuroscience.

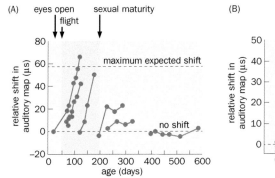
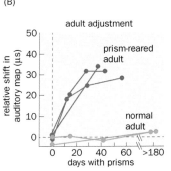
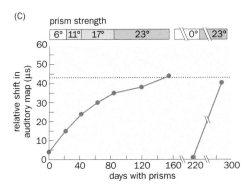

Figure 1-7 Ways to improve the ability of adult barn owls to adjust their auditory maps. (A) Owls' ability to adjust their auditory maps to match displaced visual maps declines with age. The y axis quantifies this ability to shift the auditory map, measured by the difference in time (μs, or microseconds) it takes for sounds to reach the left and right ears, which the owl uses to locate objects. Each trace represents a single owl, and each dot represents the average of auditory map shift measured at a specific time after the prisms were applied. The shaded zone indicates a sensitive period, during which owls can easily adjust their auditory maps in response to visual map displacement. Owls older than 200 days have a limited ability to shift their auditory maps. **(B)** Three owls that had learned to adjust their auditory maps in response to prism attachment as juveniles also shifted their auditory maps as adults (red traces). Two owls with no juvenile experience could not shift their maps as adults (blue traces). **(C)** Adult owls could learn to shift their auditory maps if given small prisms in incremental steps, as shown on the left side of the graph. This incremental training enabled adult owls to accommodate a sudden shift to the maximal visual displacement of 23° after a period without prisms, as shown on the right side of the graph. The dotted line at y = 43 μs represents the median shift in juvenile owls in response to a single 23°-prism step. (A & B, after Knudsen EI [2002] *Nature* 417:322–328. With permission from Springer Nature. C, after Linkenhoker BA & Knudsen EI [2002] *Nature* 419:293–296. With permission from Springer Nature.)

HOW IS THE NERVOUS SYSTEM ORGANIZED?

For all vertebrate and many invertebrate animals, the nervous system can be divided into the **central nervous system** (**CNS**) and **peripheral nervous system** (**PNS**). The vertebrate CNS consists of the **brain** and the **spinal cord** (**Figure 1-8**A,B). Both structures are bilaterally symmetric; the two sides of the brain are referred as **hemispheres**. The mammalian brain consists of morphologically and

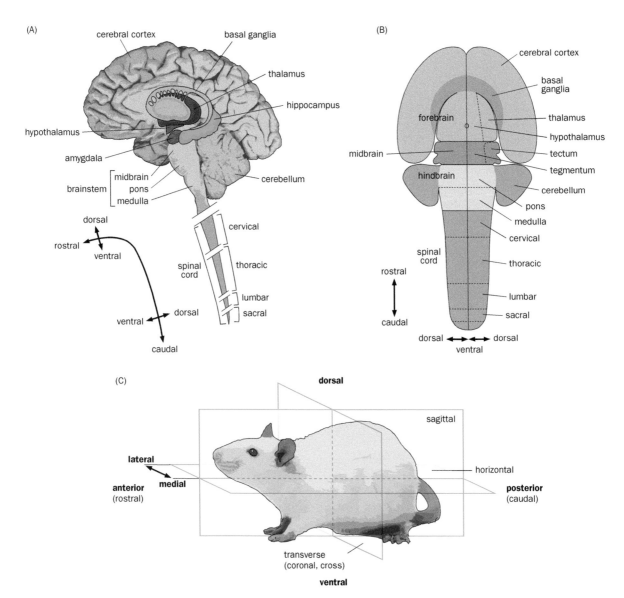

Figure 1–8 The organization of the mammalian central nervous system (CNS). (A) A sagittal (side) view of the human CNS. The basal ganglia (orange), thalamus (purple), hypothalamus (dark blue), hippocampus (light blue), and amygdala (red) from the left hemisphere are superimposed onto a midsagittal section of the CNS (tan background), the left half of which has been cut away to reveal right hemisphere structures (see Panel C for more explanation of the section plane). Major brain structures are indicated and will be studied in greater detail later in the book. From rostral to caudal, the brainstem is divided into midbrain, pons, and medulla. Spinal cord segments are divided into cervical, thoracic, lumbar, and sacral groups. Bottom left, illustration of the rostral–caudal neuraxis (CNS axis). At any given position along the neuraxis in a sagittal plane, the dorsal–ventral axis is perpendicular to the rostral–caudal axis. **(B)** A flatmap of the rat CNS reveals the internal divisions of major brain structures. The flatmap is a two-dimensional representation based on a developmental stage when progenitor cells of the nervous system are arranged as a two-dimensional sheet. It can be approximated by cutting the CNS along the midsagittal plane from the dorsal side and opening the cut surface using the ventral midline as the axis; the ventral-most structures are at the center and the dorsal-most structures are at the sides. (Imagine a book opened to display its pages; the spine of the book—the ventral midline—lays face down.) The left half of the flatmap indicates the major CNS divisions; the right side indicates major subdivisions. **(C)** Schematic illustration of the three principal section planes defined by the body axes. Transverse sections are perpendicular to the rostral–caudal axis, sagittal sections are perpendicular to the medial–lateral axis, and horizontal sections are perpendicular to the dorsal–ventral axis. (B, adapted from Swanson LW [2012] Brain Architecture. 2nd ed. Oxford University Press.)

functionally distinct structures, including the **cerebral cortex, basal ganglia, hippocampus, amygdala, thalamus, hypothalamus, cerebellum, midbrain, pons,** and **medulla**; the last three structures are collectively called the **brainstem.** The brain can also be divided into **forebrain, midbrain,** and **hindbrain,** according to the developmental origins of each region (Figure 7-3A). The spinal cord consists of repeated structures called segments, which are divided into cervical, thoracic, lumbar, and sacral groups. Each segment gives off a pair of spinal nerves. The PNS is made up of **nerves** (discrete bundles of axons) connecting the brainstem and spinal cord with the body and internal organs as well as isolated **ganglia** (clusters of cell bodies of nerve cells) outside the brain and spinal cord. We will study the organization and function of all of these neural structures in subsequent chapters.

The internal structure of the nervous system has traditionally been examined in histological sections. Three types of sections are commonly used and are named following the conventions of histology. In **transverse sections**, also called cross or **coronal sections**, section planes are perpendicular to the long, **anterior–posterior** axis of the animal (also termed the **rostral–caudal** axis, meaning snout to tail). In **sagittal sections**, section planes are perpendicular to the **medial–lateral** axis (midline to side) of the animal. In **horizontal sections**, section planes are perpendicular to the **dorsal–ventral** (back to belly) axis (Figure 1-8C). Note that in humans and other primates, which have a curved CNS, some of the anatomical terms may differ from these definitions. For uniformity, the definition of the rostral–caudal axis in this book always follows the **neuraxis** (axis of the CNS; bottom left of Figure 1-8A) rather than the body axis. Transverse or coronal sections are perpendicular to the neuraxis while horizontal sections are in parallel with the neuraxis. The neuraxis is defined by the curvature of the embryonic **neural tube**, from which the vertebrate nervous system derives, as we will learn in Chapter 7.

1.4 The nervous system consists of neurons and glia

The nervous system is made up two major categories of cells: **neurons** (nerve cells) and **glia**. A typical neuron has two kinds of **neuronal processes** (cytoplasmic extensions): a long, thin process called the **axon**, which often extends far beyond the cell body (**soma**), and thick, bushy processes called **dendrites**, which are usually close to the soma (**Figure 1-9**A). At the ends of the axons are **presynaptic terminals**, specialized structures that participate in the transfer of information between neurons. Dendrites of many vertebrate neurons are decorated with

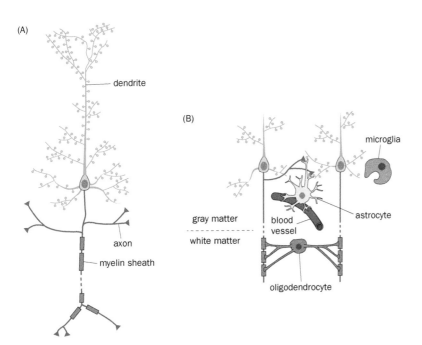

Figure 1-9 Neurons and glia.
(A) Schematic drawing of a typical neuron in the mammalian CNS. Dendrites are in blue; the axon is in red. The dashed break in the axon indicates that it can extend a long distance from the cell body. The brown structures surrounding the axon are myelin sheaths made by glia. The triangles at the ends of the axonal branches represent presynaptic terminals and the protrusions along the dendritic tree are dendritic spines. **(B)** Schematic drawing of glia in the CNS. Oligodendrocytes form myelin sheaths to wrap the axons of CNS neurons. (Schwann cells, not shown here, play a similar role in the PNS.) Astrocyte end feet wrap around connections between neurons (or synapses, which will be introduced later) in addition to blood vessels. Microglia are immune cells that engulf damaged cells and debris upon activation by injury and during developmental remodeling. (B, based on Allen NJ & Barres BA [2009] *Nature* 457:675–677.)

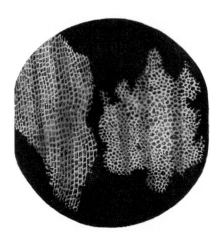

Figure 1-10 The first image of cells. A drawing by Robert Hooke illustrates the repeating units visible in thin sections of cork under a primitive microscope. Hooke thought the units resembled small rooms and coined the term *cells* to describe them. (From Hooke R [1665] Micrographia. J. Martyn and J. Allestry.)

small protrusions called **dendritic spines,** which likewise function in intercellular information transfer. Over the course of this book, we will encounter many neuronal types with distinct morphologies. Most of them have well-differentiated axons and dendrites serving distinct functions, as will be discussed in Section 1.7.

There are four major types of glia in vertebrate nervous systems: **oligodendrocytes, Schwann cells, astrocytes,** and **microglia** (Figure 1-9B). Oligodendrocytes and Schwann cells play analogous functions in the CNS and PNS, respectively: they wrap axons with their cytoplasmic extensions, called **myelin sheath,** which increases the speed at which information propagates along axons. Oligodendrocytes and myelinated axons constitute **white matter** in the CNS because myelin is rich in lipids and thus appears white. Astrocytes play many roles in neural development and regulation of neuronal communication; they are present in the **gray matter** of the CNS, which is enriched in neuronal cell bodies, dendrites, axon terminals, and connections between neurons. Microglia are the resident immune cells of the nervous system: they engulf damaged cells and debris and help reorganize neuronal connections during development and in response to experience. Invertebrate nervous systems have a similar division of labor for different glial types.

1.5 Individual neurons were first visualized by the Golgi stain in the late nineteenth century

Contemporary students of neurobiology may be surprised to learn that the cellular organization of the nervous system was not uniformly accepted at the beginning of the twentieth century, well after biologists in other fields had embraced the cell as the fundamental unit of life. Robert Hooke first used the term *cell* in 1665 to describe the repeating units he observed in thin slices of cork (**Figure 1-10**) when using a newly invented piece of equipment—the microscope. Scientists subsequently used microscopes to observe many biological samples and found cells to be ubiquitous structures. In 1839, Matthias Schleiden and Theodor Schwann formally proposed the **cell theory**: all living organisms are composed of cells as their basic units. The cell theory was widely accepted in almost every discipline of biology by the second half of the nineteenth century, except among researchers studying the nervous system. Although cell bodies had been observed in nervous tissues, many histologists of that era believed that nerve cells were linked together by their elaborate processes to form a giant net, or reticulum, of nerves. Proponents of this **reticular theory** believed that the reticulum as a whole, rather than its individual cells, constituted the unit of the nervous system.

Among the histologists who supported the reticular theory of the nervous system was Camillo Golgi, who made many important contributions to science, including the discovery of the Golgi apparatus, an intracellular organelle responsible for processing proteins in the secretory pathway (Figure 2-1). Golgi's greatest contribution, however, was the invention of the **Golgi stain**. When a piece of neural tissue is soaked in a solution of silver nitrate and potassium dichromate in the dark for several weeks, black precipitates (microcrystals of silver chromate) stochastically form in a small fraction of nerve cells, rendering these cells visible against an unstained background. Importantly, once black precipitates form within a cell, an autocatalytic reaction occurs such that the entire cell, including most or all of the elaborate extensions, can be visualized in its native tissue (**Figure 1-11**). Golgi stain thus enabled visualization of the entire morphology of individual neurons for the first time. Despite inventing this key method for neuronal visualization, however, Golgi remained a believer in the reticular theory (**Box 1-1**).

It took another great histologist, Santiago Ramón y Cajal, to effectively refute the reticular theory. The work of Ramón y Cajal and several contemporaries instead supported the **neuron doctrine**, which postulated that neuronal processes do not fuse to form a continuous reticulum. Instead, neurons intimately contact each other, with communication between distinct neurons occurring at these contact sites (Box 1-1). The term **synapse** was later coined by Charles

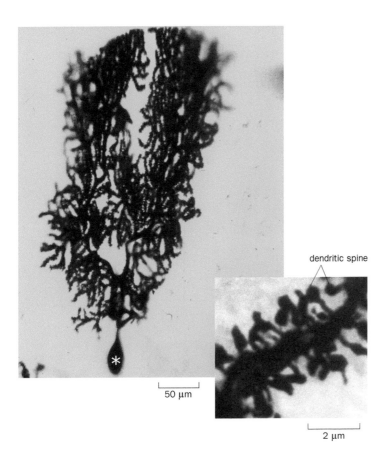

50 µm

dendritic spine

2 µm

Figure 1-11 Golgi stain. An individual Purkinje cell in the mouse cerebellum is stained black by the formation of silver chromate precipitate, allowing visualization of its complex dendritic tree. The axon, which is not included in this image, projects downward from the cell body indicated by an asterisk. The inset shows a higher magnification of a dendritic segment, highlighting protruding structures called dendritic spines. (Adapted from Luo L, Hensch TK, Ackerman L, et al. [1996] *Nature* 379:837–840. With permission from Springer Nature.)

Sherrington to describe these sites, at which signals flow from one neuron to another. After systematically applying the Golgi stain to study tissues in many parts of the nervous systems of many organisms, ranging from insects to humans, and at many developmental stages, Ramón y Cajal concluded that individual neurons are embryologically, structurally, and functionally independent units of the nervous system.

Box 1-1:   The debate between Ramón y Cajal and Golgi: why do scientists make mistakes?

Camillo Golgi and Santiago Ramón y Cajal were the most influential neurobiologists of their time. They shared the 1906 Nobel Prize for Physiology or Medicine, the first to be awarded for findings in the nervous system. However, their debates on how nerve cells constitute the nervous system—via a reticular network or as individual neurons communicating with each other through synaptic contacts—continued during their Nobel lectures (**Figure 1-12**A,B). We now know that Ramón y Cajal's view was correct and Golgi's view was largely incorrect. For example, utilizing the brainbow method (Section 14.18), individual neurons, their dendritic trees, and even their axon terminals can be visualized in and distinguished by distinct colors (Figure 1-12C). Interestingly, Ramón y Cajal used the Golgi stain to refute Golgi's theory. Why didn't Golgi reach the correct conclusion using his own method? Was he not a careful observer? After all, he made many great discoveries, including those describing the Golgi

apparatus. According to Ramón y Cajal's analysis, "Golgi arrived at this conclusion by an unusual blend of accurate observations and preconceived ideas. . . . Golgi's work actually consists of two separate parts. On the one hand, there is his method, which has generated a prodigious number of observations that have been enthusiastically confirmed. But on the other, there are his interpretations, which have been questioned and rejected."

Before the invention of the Golgi stain, histologists could not resolve processes of individual nerve cells and therefore believed that nerve processes were fused together in a giant net. Golgi was trained in a scientific environment in which this reticular theory was the dominant interpretation of nervous system organization and so tried to fit his observations into existing theory. For example, even though Golgi was the first to discover, using his staining method, that dendritic

(Continued)

Box 1-1: continued

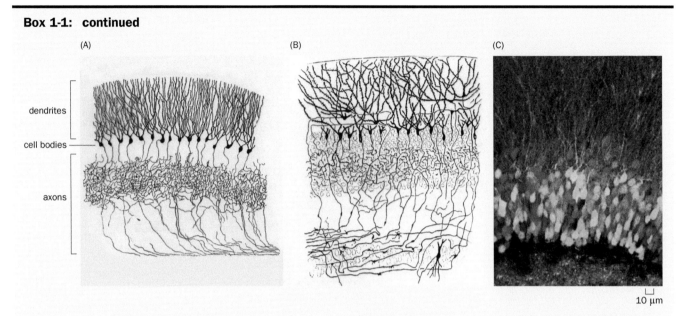

(A)
dendrites
cell bodies
axons

(B)

(C)
⊔ 10 μm

Figure 1-12 Three different views of hippocampal granule cells.
(A) Golgi's drawing of granule cells of the hippocampus. The dendritic, cell body, and axonal layers are indicated on the left. In Golgi's drawing, all axons are fused together to form a giant reticulum. **(B)** Ramón y Cajal's depiction of the same hippocampal granule cells. Note that axons below the cell bodies have definitive endings. **(C)** Hippocampal granule cells labeled by the brainbow technique, which allows the spectral separation of individual neurons expressing different mixtures of cyan, yellow, and red fluorescent proteins. Not only cell bodies but also some dendrites above and axon terminals below can be resolved by different colors. (A, after Golgi C [1906] Nobel Lecture. B, after Ramón y Cajal S [1911] Histology of the Nervous System of Man and Vertebrates. Oxford University Press. C, after Livet J, Weissman TA, Kang H, et al. [2007] *Nature* 450:56–62. With permission from Springer Nature.)

trees have free endings (Figure 1-12A, top), he thought that dendrites were used to collect nutrients for nerve cells. He believed that it was their axons, which formed an inseparable giant net as he viewed them (Figure 1-12A, bottom), that performed all the special functions of the nervous system. This story teaches an important lesson: scientists need to be observant, but they also need to be as *objective and unbiased* as possible when interpreting their own observations.

1.6 Twentieth-century technology confirmed the neuron doctrine

Ramón y Cajal could not convince Golgi to abandon the reticular theory, but many lines of evidence since the Golgi–Ramón y Cajal debate (Box 1-1) have provided strong support for the neuron doctrine. For example, during development, neurons begin with only cell bodies. Axons then grow out from the cell bodies toward their final destinations. This was demonstrated by observing axon growth *in vitro* via experiments made possible by tissue culture techniques, which were initially developed for the purpose of visualizing neuronal process growth (**Figure 1-13**). Axons are led by a structure called the **growth cone**, which changes its shape dynamically as axons extend. We will learn more about the function of the growth cone in axon guidance in Chapter 5.

The final pieces of evidence that neuronal processes are not fused with each other came from observations made possible by the development of **electron microscopy**, a technique allowing visualization of structures at nanometer (nm) resolution. (Conventional **light microscopy**, which scientists since Hooke have used to observe biological samples, cannot resolve structures less than 200 nm apart because of the physical properties of light.) The use of electron microscopy to examine **chemical synapses** (so named because communication between cells is mediated by release of chemicals called **neurotransmitters**) revealed that the **synaptic cleft**, a 20–100 nm gap, separates a neuron from its target, which can be another neuron or a muscle cell (**Figure 1-14**A). Synaptic partners are not symmetric: presynaptic terminals of neurons contain small **synaptic vesicles** filled with neurotransmitters, which, upon stimulation, fuse with the plasma membrane and release neurotransmitters into the synaptic cleft. Postsynaptic target cells have **postsynaptic specializations** (also called **postsynaptic densities**) enriched in neurotransmitter receptors on their plasma membrane surfaces. Chemical synapses are the predominant type of synapse allowing neurons to communicate with each other and with muscle cells. We will study them in greater detail in Chapter 3.

Neurons can also communicate with each other by **electrical synapses** mediated by **gap junctions** (Figure 1-14B). Here, each partner neuron contributes protein subunits to form gap junction channels that directly link the cytoplasms of two adjacent neurons, allowing ions and small molecules to travel between them. These gap junctions come closest to what the reticular theory would imagine as a fusion between different neurons. However, macromolecules cannot pass between gap junctions, and the neurons remain distinct cells with highly regulated communication. The existence of gap junctions, therefore, does not violate the premise that *individual neurons are the building blocks of the nervous system*.

1.7 In vertebrate neurons, information generally flows from dendrites to cell bodies to axons

As introduced in Section 1.4, neurons have two kinds of processes: dendrites and axons. The dendritic morphologies and axonal projection patterns of specific types of neurons are characteristic and are often used for classification. For example, the most frequently encountered type of neuron in the mammalian cerebral cortex and hippocampus, the **pyramidal neuron**, has a pyramid-shaped cell body with an apical dendrite and several basal dendrites that branch extensively (Figure 1-15A). Much of the dendritic tree sprouts dendritic spines (Figure 1-11 inset), which contain postsynaptic specializations in close contact with presynaptic terminals of partner neurons. Another widely encountered neuronal type, **basket cells** (Figure 1-15B), wrap their axon terminals around the cell bodies of pyramidal cells in the cerebral cortex or **Purkinje cells** (Figure 1-11) in the cerebellum.

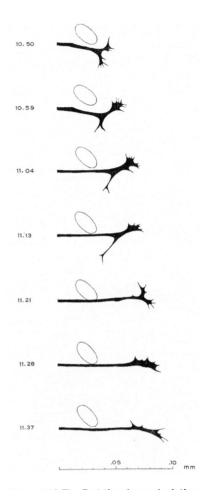

Figure 1-13 The first time-lapse depiction of a growing axon. Frog embryonic spinal cord tissue was cultured *in vitro*. Growth of an individual axon was sketched with the aid of a camera lucida at the time indicated on the left (hour.minute). The stationary blood vessel (oval) provided a landmark for the growing tips of the axon, called growth cones, which undergo dynamic changes in shape, including both extensions and retractions. A distance scale is at the bottom of the figure. (From Harrison RG [1910] *J Exp Zool* 9:787–846.)

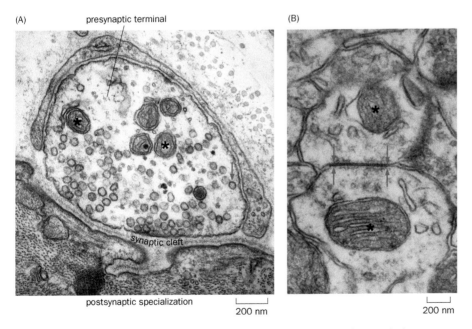

(A) presynaptic terminal

postsynaptic specialization 200 nm

(B) 200 nm

Figure 1-14 Chemical and electrical synapses. (A) Electron micrograph of a chemical synapse between the presynaptic terminal of a motor neuron and the postsynaptic specialization of its target muscle cell. A synaptic cleft separates the two cells. The arrow points to a synaptic vesicle. **(B)** Electron micrograph of an electrical synapse (gap junction) between two dendrites of mouse cerebral cortical neurons. Two opposing pairs of arrows mark the border of the electrical synapse. Asterisks indicate mitochondria in both micrographs. (A, courtesy of Jack McMahan. B, courtesy of Josef Spacek & Kristen M. Harris, SynapseWeb.)

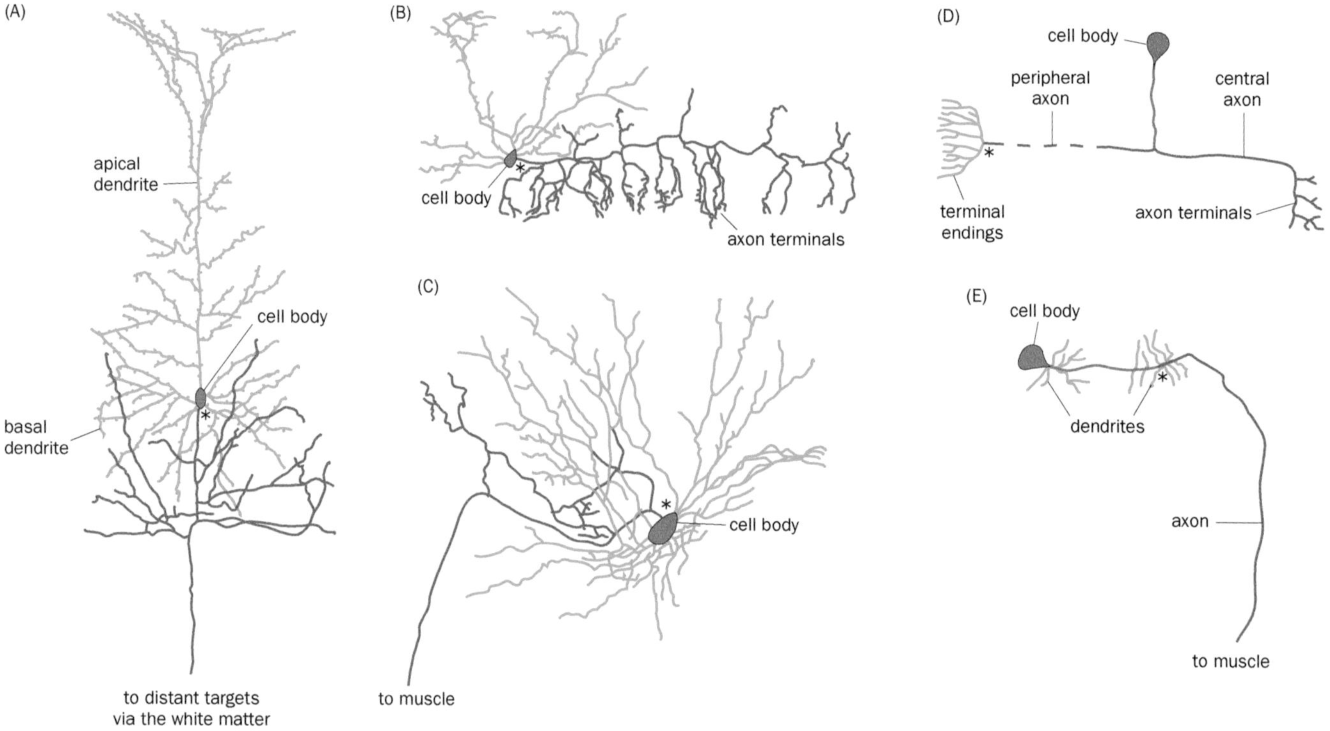

(A)

apical
dendrite

cell body

basal
dendrite

to distant targets
via the white matter

(B)

cell body

axon terminals

(C)

cell body

to muscle

(D)

cell body

peripheral
axon

central
axon

terminal
endings

axon terminals

(E)

cell body

dendrites

axon

to muscle

Figure 1-15 Morphological diversity of neurons. (A) A pyramidal cell from rabbit cerebral cortex. A typical pyramidal cell has an apical dendrite (blue) that gives off branches as it ascends, several basal dendrites (blue) that emerge from the cell body, and an axon (red) that branches locally and projects to distant targets. **(B)** A basket cell from mouse cerebellum. The basket cell axon (red) forms a series of "basket" terminals that wrap around Purkinje cell bodies (not drawn). **(C)** A motor neuron from cat spinal cord. Its bushy dendrites (blue) receive input within the spinal cord, and its axon (red) projects outside the spinal cord to muscle, while also leaving behind local branches. **(D)** A mammalian sensory neuron from a dorsal root ganglion. A single process from the cell body bifurcates into a peripheral axon (dashed to indicate the long distance) with terminal endings in the skin (equivalent of dendrites for collecting sensory information) and a central axon that projects into the spinal cord. **(E)** A motor neuron from the fruit fly ventral nerve cord (equivalent to the vertebrate spinal cord). Most invertebrate central neurons are unipolar: a single process extends out of the cell body, giving rise to dendritic branches (blue) and an axon (red). In all panels, asterisks denote axon initiation segments; as will be discussed in Section 1.8, action potentials are usually initiated at these sites. (A–D, adapted from Ramón y Cajal S [1911] Histology of the Nervous System of Man and Vertebrates. Oxford University Press. E, based on Lee T & Luo L [1999] Neuron 22:451–461.)

The spinal cord **motor neuron** extends bushy dendrites within the spinal cord (Figure 1-15C) and projects its axon out of the spinal cord and into muscle. Located in the **dorsal root ganglion** just outside the spinal cord, a **sensory neuron** of the **somatosensory system** (which processes bodily sensation) extends a single process that bifurcates, forming a peripheral axon that gives rise to branched terminal endings and a central axon that projects into the spinal cord (Figure 1-15D). Most vertebrate neurons have both dendrites and an axon leaving the cell body, and hence are called **multipolar** (or **bipolar** if there is only a single dendrite); somatosensory neurons are *pseudounipolar* because, although there is just one process leaving the cell body, it gives rise to both peripheral and central branches.

What is the direction of information flow within individual neurons? After systematically observing different types of neurons in various parts of the nervous system, Ramón y Cajal proposed the **theory of dynamic polarization**: transmission of neuronal signals proceeds from dendrites and cell bodies to axons. Therefore, every neuron has (1) a receptive component, the cell body and dendrites; (2) a transmission component, the axon; and (3) an effector component, the axon terminals. With few exceptions (the somatosensory neuron being one), this important principle has been validated by numerous observations and experiments since it was proposed a century ago and has been used extensively to deduce the direction of information flow in the vertebrate CNS. We will study the cell biological basis of neuronal polarization in Chapter 2.

How did observing the morphologies of individual neurons lead to the discovery of this rule? Ramón y Cajal took advantage of the fact that, in sensory systems, information must generally flow from sensory organs to the brain. By examining different neurons along the visual pathway (**Figure 1-16**), for example, one can see that at each connection, dendrites are at the receiving end, facing the external world, while axons are oriented so as to deliver such information to more central targets, sometimes at a great distance from the cell body where the axon originates. This applies to neurons in other sensory systems as well. Conversely, in motor systems, information must generally flow from the CNS to the periphery. The morphology of the motor neuron indeed supports the notion that its bushy dendrites receive input within the spinal cord, and its long axon, projecting to muscle, provides output (Figure 1-15C).

Neuronal processes in invertebrates can also be defined as dendrites and axons according to their *functions*, with dendrites positioned to receive information and axons to send it. However, the morphological differentiation of most invertebrate axons and dendrites, especially in the CNS, is not as clear-cut as it is for vertebrate neurons. Most often, invertebrate neurons are **unipolar**, extending a single process giving rise to both dendritic and axonal branches (Figure 1-15E). Dendritic branches are often, but not always, closer to the cell body. In many cases, the same branches can both receive and send information; this occurs in some vertebrate neurons as well, as we will learn in Chapters 4 and 6. Thus, in the "simpler" invertebrate nervous systems, it is more difficult to deduce the direction of information flow by examining the morphology of individual neurons.

1.8   Neurons use changes in membrane potential and neurotransmitter release to transmit information

What is the physical basis of information flow *within* neurons? We now know that the nervous system uses electrical signals to propagate information. The first evidence of this came from Luigi Galvani's discovery, in the late eighteenth century, that application of an electric current could generate muscle twitches in frogs. It was known by the beginning of the twentieth century that electrical signals were spread in neurons via transient changes in **membrane potential**, the electrical potential difference across the neuronal membrane. As we will learn in more detail in Chapter 2, neurons at the resting state are more negatively charged inside the cells compared to outside the cells. When neurons are excited, their membrane potentials change transiently, creating **nerve impulses** that propagate along their axons. But how is information relayed through nerve impulses? Quantitative studies of how sensory stimuli of different magnitudes induce nerve impulses provided important clues.

Studies of muscle contraction in response to electrical stimulation of motor nerves suggested that an elementary nerve impulse underlies different stimulus strengths. An all-or-none conduction principle became evident when amplifiers for electrical signals built in the 1920s made it possible to record nerve impulses from single axon fibers in response to sensory stimulation. Edgar Adrian and co-workers systematically measured nerve impulses from somatosensory neurons (Figure 1-15D) that convey information about touch, pressure, and pain to the spinal cord. They found that individual nerve impulses were of a uniform size and shape, whether they were elicited by weak or strong sensory stimuli; stronger stimuli increased the frequency of such impulses but not the properties of each impulse (**Figure 1-17**).

These experiments led to two important concepts in modern neuroscience. The first concept is the presence of an elementary unit of nerve impulses that axons use to convey information across long distances; we now call this elementary unit an **action potential**. In Chapter 2, we will study in greater detail the molecular basis of action potentials, including why they exhibit the all-or-none property. The second concept is that neurons use the frequency of action potentials to convey the intensity of signals. Whereas the frequency of action potentials is the most widely used means to convey signal intensity throughout the nervous system, the timing of action potentials can also convey important information.

In addition to action potentials, another important form of communication within neurons are **graded potentials**—membrane potentials that vary continuously

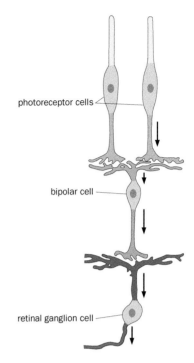

photoreceptor cells

bipolar cell

retinal ganglion cell

Figure 1–16 Neurons and information flow in the vertebrate retina. Visual information is collected by photoreceptor cells in the retina, communicated to the bipolar cell, and then to the retinal ganglion cell, which projects a long-distance axon into the brain. Note that for both the bipolar cell and the retinal ganglion cell, information is received by their dendrites (blue) and sent via their axons (red). The photoreceptor processes can also be divided into a dendrite equivalent that detects light (blue) and an axon that sends output to the bipolar cell. Arrows indicate the direction of information flow. We will learn more about these cells and connections in Chapter 4. (Adapted from Ramón y Cajal S [1911] Histology of the Nervous System of Man and Vertebrates. Oxford University Press.)

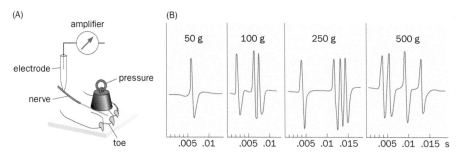

Figure 1-17 **Stimulus strength is encoded by the frequency of uniformly sized nerve impulses.** **(A)** Experimental setup for applying a specified amount of pressure to the toe of a cat, while recording nerve impulses (action potentials) from an associated sensory nerve. **(B)** With increasing pressure applied to a cat's toe, the frequency of action potentials measured at the sensory nerve also increases, but the size and shape of each action potential remain mostly the same. The *x* axis shows the time scale in units of seconds (s). (Adapted from Adrian ED & Zotterman Y [1926] *J Physiol* 61:465–483.)

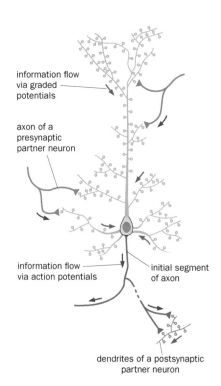

Figure 1-18 **The fundamental steps of neuronal communication.** A typical neuron in the mammalian CNS receives thousands of inputs at dendritic spines (blue) distributed along its dendritic tree. Inputs are collected in the form of synaptic potentials, which travel toward the cell body (blue arrows) and are integrated at the axon initial segment (red) to produce action potentials. Action potentials propagate to axon terminals (red arrows) and trigger neurotransmitter release, thus conveying information to postsynaptic partner neurons.

in magnitude. One type of graded potential, called **synaptic potentials**, is produced at postsynaptic sites in response to neurotransmitter release from presynaptic partners. Graded potentials can also be induced at peripheral endings of sensory neurons by sensory stimuli, such as the pressure on the toe in Adrian's experiment mentioned earlier; these are called **receptor potentials**. Unlike action potentials, the sizes of graded potentials vary depending on the strength of the input stimuli and the sensitivity of postsynaptic or sensory neurons to those stimuli. Some neurons, including most neurons in the vertebrate retina, do not fire action potentials at all. These **non-spiking neurons** use graded potentials to transmit information, even in their axons.

Synaptic potentials are usually produced at dendritic spines, along the dendrite tree, and at the soma of a neuron. A typical mammalian neuron contains thousands of postsynaptic sites on its dendritic tree, allowing it to collect input from many individual presynaptic partners (**Figure 1-18**). As we will learn later, there are two kinds of inputs: excitatory inputs facilitate action potential production in the postsynaptic neuron, whereas inhibitory inputs impede action potential production. In most neurons, the purpose of these synaptic potentials is to determine whether, when, and how frequently the neuron should fire action potentials so that information can propagate along its axon to its own postsynaptic target neurons. The site of action potential initiation is typically the **axon initial segment** (or the **axon hillock**) adjacent to the soma (Figure 1-15A–C). Thus, synaptic potentials generated in dendrites must travel through the soma to the axon initial segment to contribute to action potential generation.

The rule of action potential initiation near the soma has notable exceptions. For example, in the sensory neuron in Figure 1-15D, action potentials are initiated at the junction between terminal endings and the peripheral axon of the sensory neuron such that sensory information can be transmitted by the peripheral and central axon to the spinal cord across a long distance. In invertebrate neurons, which are mostly unipolar, action potential initiation likely occurs at the junction between the dendritic and axonal compartments (Figure 1-15E).

How is information transmitted *between* neurons? At electrical synapses, membrane potential changes are directly transmitted from one neuron to the next by ion flow across gap junctions (Figure 1-14B). At chemical synapses, the arrival of action potentials (or graded potentials in non-spiking neurons) at presynaptic terminals triggers neurotransmitter release. Neurotransmitters diffuse across the synaptic cleft and bind to their receptors on postsynaptic neurons to produce synaptic potentials (Figure 1-18; Figure 1-14A). The process of neurotransmitter release from the presynaptic neuron and neurotransmitter reception by the postsynaptic neuron is collectively referred to as **synaptic transmission**. Thus, whereas *intraneuronal* communication is achieved by membrane potential changes in the form of graded potentials and action potentials, *interneuronal* communication at chemical synapses relies on neurotransmitter release and reception. We will study these fundamental steps of neuronal communication in greater detail in Chapters 2 and 3.

1.9 Neurons function in the context of specialized neural circuits

Neurons perform their functions in the context of **neural circuits**—ensembles of interconnected neurons that act together to perform specific functions. The simplest circuits in vertebrates, those that mediate the spinal reflexes, comprise as few as two interconnected neurons: a sensory neuron that receives external stimuli and a motor neuron that controls muscle contraction. Many fundamental neurobiological principles have been derived from studying these simple circuits.

When a neurologist's hammer hits the knee of a subject during a neurological exam, the lower leg kicks forward involuntarily (**Figure 1-19**). The underlying circuit mechanism for this **knee-jerk reflex** has been identified: sensory neurons embed their endings in specialized apparatus called **muscle spindles** in an extensor muscle whose contraction extends the knee joint. These sensory neurons detect stretching of the muscle spindles caused by the physical impact of the hammer and convert this stimulus into electrical signals—namely, receptor potentials—at the sensory endings. Next, the peripheral and central axons of the sensory neurons propagate these electrical signals to the spinal cord as action potentials. There, central axon terminals of the sensory neurons release neurotransmitters directly onto the dendrites of their partner motor neurons. These motor neurons extend their own axons outward from the spinal cord and terminate in the same extensor muscle in which the sensory neurons embed their endings. Sensory axons are also called **afferents**, referring to axons projecting from peripheral tissues to the CNS, whereas motor axons are called **efferents**, referring to axons that project from the CNS to peripheral targets. Both the sensory and motor neurons in this circuit are **excitatory neurons**. When excitatory neurons are activated—that is, when they fire action potentials and release neurotransmitters—they make their postsynaptic target cells more likely to fire action potentials. Therefore, mechanical stimulation activates sensory neurons. This in turn activates the postsynaptic motor neurons. Neurotransmitter release at motor axon terminals leads to contraction of the extensor muscle.

The knee-jerk reflex involves coordination of more than one muscle. The flexor muscle, which is antagonistic to the extensor muscle, must *not* contract at the same time in order for the knee-jerk reflex to occur. (As we will learn in Chapter 8, contraction of extensor muscles increases the angle of a joint, while contraction of flexor muscles decreases the angle of a joint.) Therefore, the sensory axons must *inhibit* contraction of the corresponding flexor muscle in addition to causing contraction of the extensor muscle. This inhibition is mediated by **inhibitory interneurons** in the spinal cord, a second type of postsynaptic neuron targeted by the sensory axons. Note that neurobiologists use the term *interneuron* in two different contexts. In a broad context, all neurons that are not sensory or motor neurons are interneurons. But in most contexts, the term *interneuron* refers to neurons that confine their axons within a specific region, in contrast to **projection neurons**, whose axons link different regions of the nervous system. The spinal inhibitory interneurons fit both criteria.

In this reflex circuit, activation of sensory neurons causes excitation of these inhibitory interneurons, which in turn inhibit the motor neurons innervating the flexor muscle. This inhibition makes it more difficult for the flexor motor neurons to fire action potentials, causing the flexor muscle to relax. Thus, coordinated contraction of the extensor muscle and relaxation of the flexor muscle brings the lower leg forward. First analyzed in studies of spinal reflexes by Charles Sherrington in the 1890s, the role of inhibition is crucial in coordinating neuronal function throughout the nervous system.

In summary, the knee-jerk reflex involves one of the simplest neural circuits: coordinated excitation and inhibition is executed by *monosynaptic* connections between sensory neurons and motor neurons and *disynaptic* connections between sensory neurons and a different group of motor neurons via inhibitory interneuron intermediates. Nervous system functions rely on establishing proper connections

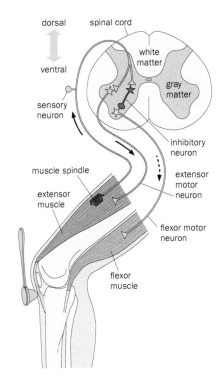

Figure 1-19 The neural circuit underlying the knee-jerk reflex. A simple neural circuit is responsible for the involuntary jerk that results when the front of the knee is hit with a hammer. In this simplified scheme, a single neuron represents a population of neurons performing the same function. The sensory neuron extends its peripheral axon to the muscle spindle of the extensor muscle and its central axon to the spinal cord. In the spinal cord, the sensory neuron has two postsynaptic targets: the green motor neuron that innervates the extensor muscle, and the red inhibitory interneuron that synapses with the yellow motor neuron innervating the flexor muscle. When the knee is hit, mechanical force activates the sensory neuron, resulting in excitation of the extensor motor neuron, which causes contraction of the extensor muscle (following solid arrows). At the same time, sensory neuron activation causes inhibition of the flexor motor neuron, which relaxes the flexor muscle (dashed arrow). The spinal cord is drawn as a cross section. The gray matter at the center contains cell bodies, dendrites, and the synaptic connections of spinal cord neurons; the white matter at the periphery consists of axons of projection neurons. The sensory neuron cell body is located in a dorsal root ganglion adjacent to the spinal cord.

between neurons in numerous neural circuits like the knee-jerk reflex circuit; we will study how the nervous system wires up precisely during development in Chapters 5 and 7.

Most neural circuits are much more complex than the spinal cord reflex circuit. Box 1-2 discusses commonly used circuit motifs we will encounter in this

---

Box 1-2: Common neural circuit motifs

The simplest circuit consists of two synaptically connected neurons, such as the sensory neuron–extensor motor neuron circuit in the knee-jerk reflex. In circuits containing more than two neurons, individual neurons can receive input from and send output to more than one partner. Further complexity arises when some neurons in a circuit are excitatory and others are inhibitory. The nervous system employs many **circuit motifs**—common configurations of neural circuits that allow the connection patterns of individual neurons to execute specific functions. Here, we introduce the most common circuit motifs (**Figure 1-20**).

Let's first consider circuits containing only excitatory neurons. **Convergent excitation** (Figure 1-20A) refers to a circuit motif wherein several neurons synapse onto the same postsynaptic neuron. Conversely, **divergent excitation** (Figure 1-20B) refers to a motif wherein a single neuron synapses onto multiple postsynaptic targets via branched axons (axonal branches are also called **collaterals**). Convergent and divergent connections allow individual neurons to integrate input from multiple presynaptic neurons and to send output to multiple postsynaptic targets, respectively. Serially connected excitatory neurons constitute a **feedforward excitation** motif (Figure 1-20C) for propagating information across multiple brain regions, as in the relay of somatosensory stimuli to the primary somatosensory cortex (Figure 1-21). When a postsynaptic neuron synapses onto its own presynaptic partner, this motif is called **feedback excitation** (Figure 1-20D). Neurons that transmit parallel streams of information can also excite each other, forming a **recurrent (lateral) excitation** motif (Figure 1-20E).

When excitatory and inhibitory neurons interact in the same circuit, as is most often the case, many interesting circuit motifs with diverse functionalities can be constructed. The names of motifs involving inhibitory neurons usually emphasize the nature of the inhibition. In **feedforward inhibition** (Figure 1-20F), an excitatory neuron synapses onto both an excitatory neuron and an inhibitory neuron, and the inhibitory neuron further synapses onto the excitatory postsynaptic neuron. In **feedback inhibition** (Figure 1-20G), the postsynaptic excitatory neuron synapses onto an inhibitory neuron, which synapses back onto the postsynaptic excitatory neuron. In both cases, inhibition can control the duration and magnitude of the

excitation of the target neuron. In **recurrent (cross) inhibition** (Figure 1-20H), two parallel excitatory pathways cross-inhibit each other via inhibitory neuron intermediates; the inhibition of the flexor motor neuron in the knee-jerk reflex discussed in Section 1.9 is an example of recurrent inhibition. In **lateral inhibition** (Figure 1-20I), an inhibitory neuron receives excitatory input from one or several parallel streams of excitatory neurons and sends inhibitory output to many postsynaptic targets of these excitatory neurons. Lateral inhibition is widely used in processing sensory information, as we will study in greater detail in Chapters 4 and 6. Finally, when an inhibitory neuron synapses onto another inhibitory neuron, the excitation of the first inhibitory neuron reduces the inhibitory output of the second inhibitory neuron, causing **disinhibition** of the final target neuron (Figure 1-20J).

The circuit motifs discussed here are often used in combinations, giving rise to many different ways of processing information. In Chapter 3, we will encounter another group of neurons, the **modulatory neurons**, which can act on both excitatory and inhibitory neurons to up- or downregulate their excitability or synaptic transmission, adding further richness to the information processing functions of neural circuits. In Chapter 14, we will examine circuit architecture from theoretical and computational perspectives. We will see that excitatory and inhibitory neurons can be connected in specific ways to produce logic gates for computation; these logic gates are also the bases of all operations in modern computers (Section 14.32).

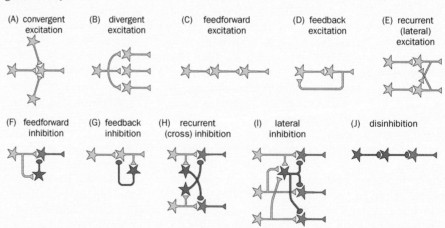

Figure 1-20 **Common circuit motifs.** In all panels, the general information flow is from left to right. Green, excitatory neuron; red, inhibitory neuron; gray, any neuron. (**A–E**) Circuit motifs consisting of only excitatory neurons. (**F–J**) Circuit motifs that include inhibitory neurons. See text for more details. For recurrent inhibition (H) and lateral inhibition (I), only the feedforward modes are depicted; the feedback modes of these motifs may also be used (not shown), in which case the inhibitory neuron(s) receive input(s) from postsynaptic excitatory neurons, as in Panel G.

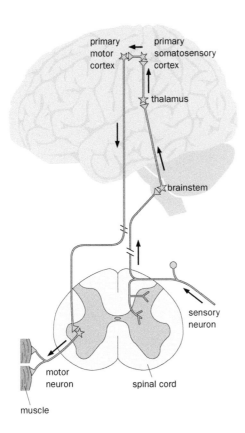

Figure 1-21 Sensory and motor pathways between the spinal cord and cerebral cortex. Some sensory neurons, in addition to participating in the spinal cord reflex circuit, send an ascending branch that connects with relay neurons in the brainstem, which deliver information to neurons in the primary somatosensory cortex via intermediate neurons in the thalamus. Through intercortical connections, information is delivered to neurons in the primary motor cortex, which send descending output directly and indirectly to spinal cord motor neurons for voluntary control of muscles. Shown here are the most direct routes for these ascending and descending pathways. The spinal cord is represented in cross section. The brain is shown from a sagittal view (not at the same scale as the spinal cord). Arrows indicate the direction of information flow.

book. For example, a subject becomes aware that a hammer has hit her knee because sensory neurons also send axonal branches that ascend along the spinal cord. After passing through relay neurons in the brainstem and thalamus, sensory information eventually reaches the **primary somatosensory cortex**, the part of the cerebral cortex that first receives somatosensory input from the body (**Figure 1-21**). Cortical processing of such sensory input generates the perception that her knee has been hit. Such information also propagates to other cortical areas, including the **primary motor cortex**. The primary motor cortex sends descending output directly and indirectly to spinal cord motor neurons to control muscle contraction (Figure 1-21), in case we want to move our leg voluntarily (in contrast to the knee-jerk reflex, which is involuntary). We will study these sensory and motor pathways in greater detail in Chapters 6 and 8, but in general we know far less about the underlying mechanisms of these ascending, cortical, and descending circuits than we do about the spinal cord reflex circuit. Elucidating the principles of information processing in complex neural circuits that mediate sensory perception and motor action is one of the most exciting and challenging goals of modern neuroscience.

1.10 Specific brain regions perform specialized functions

It is well established today that specialized functions of the nervous system are mostly performed by specific parts of the brain. However, throughout prior centuries, philosophers argued about whether brain functions underlie mind, let alone whether specific brain regions are responsible for specific mental activities. Even in the early twentieth century, a prevalent view was that any specific mental function is carried out by neurons across many areas of the cerebral cortex.

Franz Joseph Gall developed a discipline called **phrenology** in the early nineteenth century. Gall supposed that all behavior emanates from the brain, with specific brain regions controlling specific functions. The centers for each mental function, he reasoned, grow with use, creating bumps and ridges on the skull. Based on this reasoning, Gall and his followers attempted to map human mental function to specific parts of the cortex, correlating the size and shape of the bumps

Figure 1-22 Phrenologists' depiction of the brain's functional organization. According to phrenology, the brain is divided into individual areas specialized for defined mental functions. The size of each area is modified by use. For example, a cautious person would have an enlarged area corresponding to cautiousness.

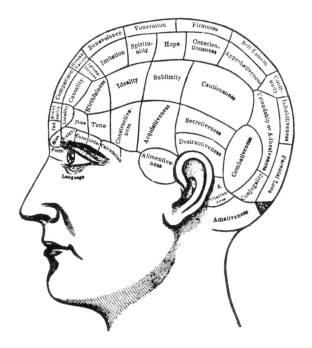

and ridges on individuals' skulls with their talents and character traits (**Figure 1-22**). While we now know that these maps are false, Gall's thinking about brain specialization was actually quite advanced for his time.

Brain lesions provided the first instances of scientific evidence that specialized regions of the human cerebral cortex perform specific functions. Each hemisphere of the cerebral cortex is divided into four lobes, the **frontal**, **parietal**, **temporal**, and **occipital lobes**, based on the major folds (called **fissures**) separating the lobes (**Figure 1-23**A). In the 1860s, Paul Broca discovered lesions in a specific area of the human left frontal lobe (Figure 1-23B) in patients who could not speak. This area was subsequently named **Broca's area** (Figure 1-23A). Carl Wernicke subsequently found that lesions in a distinct area in the left temporal lobe, now named **Wernicke's area** (Figure 1-23A), were also associated with defects in language. Interestingly, lesions in Broca's area and Wernicke's area give distinct symptoms. Patients with lesions in Broca's area have great difficulty producing language, whether in speech or writing, but their understanding of language is largely intact. By contrast, patients with lesions in Wernicke's area have great difficulty understanding language, but they can speak fluently, although often unintelligibly and incoherently. These findings led to the proposal that Broca's and Wernicke's areas are responsible for language production and comprehension,

Figure 1-23 Language centers in the human brain were originally defined by lesions. (A) Major fissures divide each cerebral cortex hemisphere into frontal, parietal, temporal, and occipital lobes. Broca's area is located in the left frontal lobe adjacent to the part of the primary motor cortex that controls movement of the mouth and lips (Figure 1-25). Wernicke's area is located in the left temporal lobe adjacent to the auditory cortex. **(B)** Photograph of the brain of one of Broca's patients, Leborgne, who could speak only a single syllable, "tan." The lesion site is circled. Observation of similar lesions in language-deficient patients led Broca to propose that the area is essential for language production. (B, from Rorden C & Karnath H [2004] *Nat Rev Neurosci* 5:813–819. With permission from Springer Nature.)

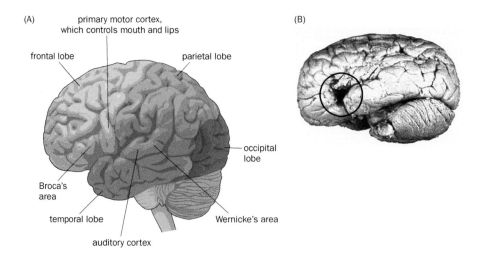

respectively. These distinct functions are consistent with the locations of Broca's and Wernicke's areas being close to the motor cortex and the **auditory cortex** (the part of the cortex that analyzes auditory signals), respectively (Figure 1-23A).

In the twentieth century, two important techniques—brain stimulation and brain imaging—confirmed and extended findings from lesion studies, revealing in greater detail specific brain regions that perform distinct functions. Brain stimulation is a standard procedure for mapping specific brain regions to guide brain surgeries, such as severing axonal pathways to treat intractable **epilepsy**. (Epilepsy is a medical condition characterized by recurrent seizures—strong surges of abnormal electrical activity that affect part or all of the brain; Box 12-4.) Such surgeries are often performed without general anesthesia (the brain does not contain pain receptors) so that patients' responses to brain stimulation can be observed. Stimulation of Broca's area, for instance, causes a transient arrest of speech in patients. These brain stimulation studies have identified additional areas involved in language production.

One of the most remarkable methods developed in the late twentieth century is the noninvasive functional brain imaging of healthy human subjects as they perform specific tasks. The most widely used technique is **functional magnetic resonance imaging (fMRI)**, which monitors signals originating from changes in blood flow that results from local neuronal activity. By allowing researchers to observe whole-brain activity without bias while subjects perform specific tasks, fMRI has revolutionized our understanding of brain regions implicated in specific functions. Such studies have confirmed that Broca's and Wernicke's areas are involved in language production and comprehension, respectively.

Because fMRI offers higher spatial resolution than do lesion studies, it has enabled researchers to ask more specific questions. For example, do bilingual speakers use the same cortical areas for their native and second languages? The answer depends on the cortical area in question and the age at which an individual acquires the second language. In late bilinguals who were first exposed to the second language after 10 years of age, representations of the native and second languages in Broca's area map to adjacent but distinct loci (**Figure 1-24**A). In early bilinguals who learned both languages as infants, the two languages map to the same locus in Broca's area (Figure 1-24B). Thus, the age of language acquisition appears to determine how the language is represented in Broca's area. It is possible that after a critical period during development (we will study this important concept in Chapter 5), native language has consolidated a space in Broca's area, such that a second language acquired later must utilize other (adjacent) cortical areas. By contrast, the loci in Wernicke's area that represent the two languages are inseparable by fMRI even in late bilinguals (Figure 1-24C).

1.11 The brain uses maps to organize information

Thanks to a combination of anatomical, physiological, functional, and pathological studies on human subjects, we now have a detailed understanding of the gross organization of the human nervous system (Figure 1-8A). Experimental studies of mammalian model organisms, which share this gross organization (Figure 1-8B), complement our understanding. We will study the organization and function of many nervous system regions in detail in subsequent chapters.

An important organizational principle worth emphasizing now is that the nervous system uses maps to represent information. We have already seen this phenomenon in our earlier discussion of the auditory and visual maps that barn owls use to target their prey. Two striking examples of maps in the human brain are the **motor homunculus** and the **sensory homunculus** (**Figure 1-25**). These homunculi ("little men") were discovered through the use of electrical stimulation during brain surgeries to treat epilepsy, as discussed in Section 1.10. For example, stimulation of cortical neurons in specific parts of the primary motor cortex elicits movement of specific body parts on the contralateral side. (Movement of the left side of the body is controlled by the right side of the brain and vice versa.) Systematic studies revealed a cortical **topographic map** corresponding to movement of specific body parts: nearby neurons in the motor homunculus control the

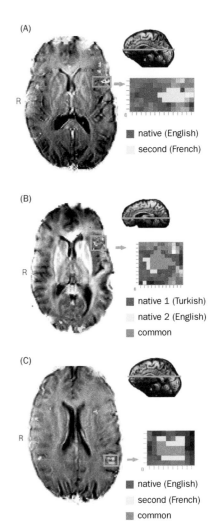

Figure 1-24 Representations of native and second languages as revealed by functional magnetic resonance imaging (fMRI). The detection of blood-flow signals associated with brain activity by fMRI provides a means for imaging the brain loci where native and second languages are processed. In the brain scans on the left side of the figure, green rectangles highlight language-processing areas in the left hemisphere; the highlighted areas are magnified on the right. In the miniature brain profiles located at top right of each panel, the green lines represent the section plane visualized in the scanned images. R, right hemisphere. **(A)** In a late bilingual, the two languages are represented in separate, adjacent loci within Broca's area. **(B)** In an early bilingual who learned both languages from infancy, the language representations in Broca's area overlap. **(C)** In Wernicke's area, the representations of native and second languages overlap regardless of when the second language was acquired; Panels A and C came from the same late bilingual subject. (From Kim KH, Relkin NR, Lee KM, et al. [1997] *Nature* 388:171–174. With permission from Springer Nature.)

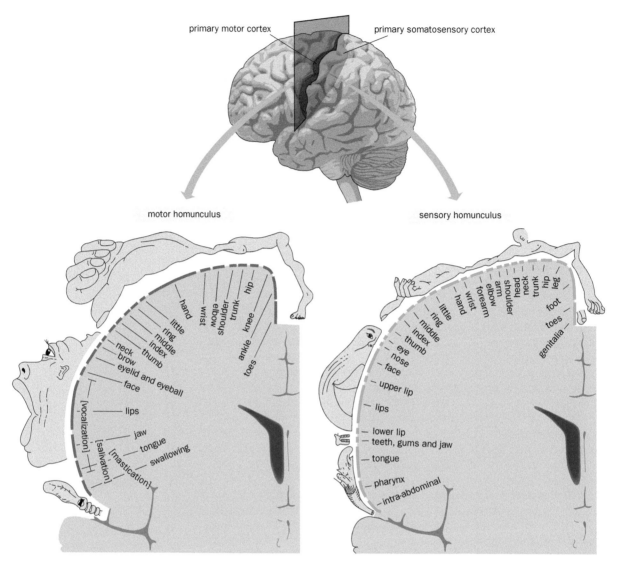

Figure 1-25 Sensory and motor homunculi. Top, the locations in the brain of the primary motor and primary somatosensory cortices. Arrows indicate that sections along the plane with a 90° turn would produce the homunculi in the bottom panels. Bottom left, cortical neurons in the primary motor cortex control movement of specific body parts according to a topographic map. For example, neurons that control movement of the lips and jaw are close together but are distant from neurons that control finger movement. Bottom right, cortical neurons in the primary somatosensory cortex represent a topographic map of the body. For example, neurons that represent touch stimuli on the lips, jaw, and tongue are in adjacent areas but are distant from neurons that represent touch stimuli on the fingers. (From Penfield W & Rasmussen T [1950] The Cerebral Cortex of Man. Macmillan.)

movements of nearby body parts. This map is distorted in its proportions: the hand, and in particular, the thumb, is highly overrepresented, as are the muscles surrounding the mouth that enable us to eat and speak (Figure 1-25, bottom left). These distortions reflect disproportional use of different muscles. As we will learn in Chapter 8, the motor homunculus is a simplified representation of a more complex organization of the motor cortex for movement control.

In the adjacent primary somatosensory cortex, there is a corresponding sensory homunculus. Stimulation of specific areas in the primary somatosensory cortex elicits sensations in specific body parts on the contralateral side (Figure 1-25, bottom right). Again, cortical neurons in adjacent areas represent adjacent body parts, forming a topographic map in the primary somatosensory cortex. This preservation of spatial information is all the more striking, considering that these cortical neurons are at least three synaptic connections away from the sensory world they represent (Figure 1-21). Obvious distortions are also evident: some body parts (for example, the hand and especially the thumb) are overrepresented compared to others (for example, the trunk). These distortions reflect differential sensitivities of different body parts to sensory stimuli such as touch. Interestingly, cortical neurons in the sensory and motor homunculi that represent the same body part are physically near each other, reflecting a close link between the two cortical areas in coordinating sensation and movement.

Neural maps are widespread throughout the brain. We will learn more about maps in the visual system (Chapter 4); the olfactory, taste, auditory, and somato-sensory systems (Chapter 6); the motor system (Chapter 8); and the hippocampus and **entorhinal cortex** (part of the temporal cortex overlying the hippocampus), where maps represent spatial information of the outside world (Chapter 11). We will also study in detail how neural maps are established during development (Chapters 5 and 7).

1.12 The brain is a massively parallel computational device

The brain is often compared to the computer, another complex system with enormous problem-solving power. Both the brain and the computer contain a large number of elementary units—neurons and transistors, respectively—wired into complex circuits to process information conveyed by electrical signals. At a global level, the architectures of the brain and the computer resemble each other, consisting of input, output, central processing, and memory (Figure 1-26). Indeed, the comparison between the brain and the computer has been instructive to both neuroscientists and computer engineers.

The computer has huge advantages over the brain in speed and precision of basic operations (Table 1-1). Personal computers nowadays can perform elementary arithmetic operations such as addition at a rate of 10^{10} operations per second. However, the rate of elementary operations in the brain, whether measured by action potential frequency or by the speed of synaptic transmission across a chemical synapse, is at best 10^3 per second. Furthermore, the computer can represent quantities (numbers) with any desired precision according to the *bits* (binary digits, or 0s and 1s) assigned to each number. For instance, a 32-bit number has a precision of 1 in 2^{32} or $\sim4.3 \times 10^9$. Empirical evidence suggests that most quantities in the nervous system have variability of at least a few percent due to biological noise, or a precision of 1 in 10^2 at best. However, calculations performed by the brain are neither slow nor imprecise. For example, a professional tennis player can follow the trajectory of a tennis ball after it is served at a speed of 150 miles per hour, move to the optimal spot on the court, position her arm, and swing the racket to return the ball to the opponent's court—and all of this within a few hundred milliseconds. Moreover, the brain can accomplish all of these tasks (with the help of the body it controls) with a power consumption about 10-fold less than a personal computer. How does the brain achieve that?

A notable difference between the brain and the computer is the methods by which information is processed within each system. Computer tasks are largely performed in **serial processing** steps; this can be seen in the way engineers program computers by creating a sequential flow of instructions and by the fact that the operation of each basic unit, the transistor, has only three nodes for input and output altogether. For this sequential cascade of operations, high precision

Figure 1-26 Architectures of the computer and the nervous system. **(A)** Schematic of the five classic components of a computer: input (such as a keyboard or a mouse), output (such a screen or a printer), memory (where data and programs are kept when programs are running), datapath (which performs arithmetic operations), and control (which tells datapath, memory, and input/output devices what to do according to the instructions of the program). Control and datapath together are also called the processor. **(B)** The nervous system can be partitioned in several different ways, one of which is shown here. In this four-system model, the motor system controls the output of the nervous system (behavior). It is in turn controlled by three other systems: the sensory system, which receives input from external environment and the body; the cognitive system, which mediates voluntary behavior; and the behavioral state system (such as wake/sleep), which influences the performance of all other systems. Arrows indicate extensive and often bidirectional connections between the four systems. As we will learn in Chapter 11, memory is primarily stored in the form of synaptic connection strengths in neural circuits in all of these systems. (A, after Patterson DA & Hennessy JL [2012] Computer Organization and Design. 4th ed. Elsevier. B, after Swanson LW [2012] Brain Architecture. 2nd ed. Oxford University Press.)

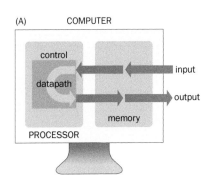

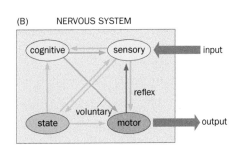

Table 1-1: Comparing the computer and the brain

Properties	Computer[a]	Human brain
Number of basic units	~10^9 transistors[b]	~10^{11} neurons; >10^{14} synapses
Speed of basic operation	10^{10}/s	<10^3/s
Precision	1 in 4×10^9 for a 32-bit number	~1 in 10^2
Power consumption	10^2 watts	~10 watts
Processing method	Mostly serial	Serial and massively parallel
Input/output for each unit	1–3	~10^3
Signaling mode	Digital	Digital and analog

[a] Based on personal computers in 2008.

[b] The number of transistors per integrative circuit has doubled every 18–24 months in the past few decades; in recent years the performance gains from this transistor growth have slowed, limited by energy consumption and heat dissipation.

(Data from von Neumann [1958] The Computer & the Brain. Yale University Press; Patterson & Hennessy [2012] Computer Organization and Design. 4th ed. Elsevier.)

is necessary at each step because errors accumulate and amplify in successive steps. The brain also uses serial steps for information processing; in the tennis return example, information flows from the eye to the brain and then to the spinal cord to control contraction of leg, trunk, arm, and wrist muscles. However, the nervous system also employs **massively parallel processing**, taking advantage of the large number of neurons and large number of connections each neuron makes. For instance, the moving tennis ball activates many retinal photoreceptors, which transmit information to different kinds of bipolar and retinal ganglion cells (Figure 1-16) that we will learn about in Chapter 4. Information regarding the location, direction, and speed of the ball is extracted by parallel circuits within two to three synaptic connections and transmitted in parallel by different kinds of retinal ganglion cells to the brain. Likewise, the motor cortex sends commands in parallel to control contraction of leg, trunk, arm, and wrist muscles, such that the body and arms are simultaneously well positioned to return the incoming ball.

This massively parallel strategy is possible because each neuron collects inputs from and sends outputs to many other neurons—an average of 10^3 for both inputs and outputs for a mammalian neuron. Using the divergent projection motif (Figure 1-20B), information from one neural center can be delivered to many parallel downstream pathways. The convergent projection motif (Figure 1-20A) allows many neurons that process the same information to send their inputs to the same postsynaptic neuron. While information represented by individual neurons may be noisy, by pooling inputs from many presynaptic neurons, a postsynaptic neuron can represent the same information with much higher precision.

The brain and the computer also have similarities and differences in the signaling modes of their elementary units. Transistors employ **digital** signaling, which uses discrete values (0 or 1) to represent information. Action potentials in neuronal axons are also all-or-none digital signals, enabling reliable long-distance information propagation. However, neurons also utilize **analog** signaling, which uses continuous values to represent information. In non-spiking neurons, output is transmitted by graded potentials that can transmit more information than can action potentials (we will discuss this in more detail in Chapter 4). Neuronal dendrites also use analog signaling to integrate up to thousands of inputs. Finally, signals for interneuronal communication are mostly analog, as synaptic strength is a continuous variable.

Another salient property of the brain clearly at play in the tennis return example is that the connection strengths between neurons can be modified by experience, as we will see in greater detail in Chapter 11. Repetitive training enables the circuits to become better configured for the tasks being performed, resulting in greatly improved speed and precision.

Over the past decades, engineers have taken inspiration from the brain to improve computer design. The principles of parallel processing and use-dependent modification of circuits have both been incorporated into modern computers. For example, increased parallelism, such as the use of multiple processors (cores) in a single computer, is a current trend in computer design. As another example, "deep learning" (a branch of machine learning and artificial intelligence) has enjoyed great success in recent years and accounts for rapid advances in object and speech recognition by computers and mobile devices (Box 14-6); these advances were inspired by findings from the mammalian visual system. At the same time, neurobiologists can enhance their understanding of the nervous system and the potential strategies it employs to solve complex problems by looking at the brain from an engineering and computational perspective, a subject we will expand on in the final part of Chapter 14.

GENERAL METHODOLOGY

The development and utilization of scientific methodology is essential to advancing our knowledge of neurobiology. We devote the last chapter of this book (Chapter 14) to discussing important methods for exploring the brain. *The relevant sections of Chapter 14 should be frequently consulted when these methods are introduced in Chapters 1–13.* We conclude this chapter by highlighting a few general methodological principles that will be encountered throughout the book.

1.13 Observation and measurement are the foundations of discovery

At the beginning of this chapter, we noted that asking the right question is often a crucial first step in making important discoveries. A good question is usually specific enough to be answered with clarity in the framework of existing knowledge. At the same time, the question's answer should have broad significance.

Careful observation is usually the first step in answering questions. Observations can be made with increasing resolution by using improving technology. Our discussion in this chapter about the organization of the nervous system provides good examples. Cells were discovered because of the invention of the light microscope. The elaborate shapes of neurons were first observed because of the invention of the Golgi staining method. The debate between the neuron doctrine and the reticular theory was finally settled with electron microscopy. Inventing new ways of observing can revolutionize our understanding of the nervous system.

While observations can give us a qualitative impression, some questions can be answered only with quantitative measurements. For instance, to find out how sensory stimuli are encoded by nerve signals, researchers needed to measure the size, shape, and frequency of action potentials induced by stimuli of varying strengths. This led to the discovery that stimulus strength is encoded by the frequency of action potentials, but not by their size or duration. The development of new measurement tools often precedes great discoveries.

Observation and measurement go hand in hand. Observations can be quantitative and often form the basis of a measurement. For example, electron microscopy first enabled visualization of the synaptic cleft. At the same time, it also permitted researchers to measure the approximate distance a neurotransmitter must travel across a chemical synapse and to estimate the physical size of the membrane proteins needed to bridge the two sides of a chemical synapse.

1.14 Perturbation experiments establish causes and mechanisms

While observation and measurement can lead to discovery of interesting phenomena, they are often inadequate for investigating the underlying mechanisms. Further insight can be obtained by altering key parameters in a biological system and studying the consequences. We call such studies **perturbation experiments**. Putting prisms on a barn owl is an example of a perturbation experiment. Artificial displacement of the visual map allowed researchers to measure the owl's ability to adjust its auditory map to match an altered visual map. We will encounter numerous perturbation experiments throughout this textbook.

Most perturbation experiments can be categorized as loss-of-function or gain-of-function. In **loss-of-function experiments**, a specific component is removed from the system. This type of experiment tests whether the missing component is *necessary* for the system to function. As an example, specific brain lesions in Broca's patients caused loss of speech, suggesting that Broca's area is *necessary* for speech production. In **gain-of-function experiments**, a specific component is added to the system. Gain-of-function experiments can test whether a component is *sufficient* for the system to function in a specific context. As an example, electrical stimulation (in epileptic patients) indicated that activation of specific motor cortical neurons is *sufficient* to produce twitches of specific muscles. Both loss- and gain-of-function experiments can be used to deduce causal relationships between specific components in a given biological process.

Originating from genetics, the terms *loss-of-function* and *gain-of-function* refer to the deletion and misexpression, respectively, of a gene—a unit of operation for many biological processes. These perturbations allow researchers to test the function of a gene in a biological process. Powerful genetic perturbations can be performed with high precision in many model organisms (Sections 14.6–14.11). Indeed, we will encounter many examples of gene perturbation experiments that have revealed the mechanisms underlying myriad neurobiological processes.

As the lesion and electrical stimulation exemplify, loss- and gain-of-function perturbations do not refer only to experiments that manipulate genes. In contemporary neuroscience, a central issue is the analysis of neural circuit function in perception and behavior, and here single neurons or populations of neurons of a particular type have been conceptualized as the organizational and operational units. To assess the function of specific neurons or neuronal populations in the operation of a circuit, tools have been developed to conditionally silence their activity (loss-of-function) or artificially activate them (gain-of-function) with high spatiotemporal precision (Sections 14.11 and 14.23–14.25). Given that neurons participate in neural circuits in many different ways (Box 1-2), precise perturbation experiments are crucial in revealing the mechanisms by which neural circuits control neurobiological processes. These experiments also help establish causal relationships between the activity of specific neurons and the processes they control.

With these basic concepts and general methodological frameworks in hand, let us begin our journey!

SUMMARY

In this chapter, we introduced the general organization of the nervous system and some fundamental concepts in neurobiology, framing these topics from a historical perspective. Neurons are the basic building blocks of the nervous system. Within most vertebrate neurons, information, in the form of membrane potential changes, flows from dendrites to cell bodies to axons. Graded potentials in dendrites are summed at the junction between the cell body and the axon to produce all-or-none action potentials that propagate to axon terminals. Neurons communicate with each other through synapses. At chemical synapses, presynaptic neurons release neurotransmitters in response to the arrival of action potentials, and postsynaptic neurons change their membrane potential in response to neurotransmitters binding to their receptors. At electrical synapses, ions directly flow from one neuron to another through gap junctions to propagate membrane potential changes. Neurons act in the context of neural circuits and form precise connections with their synaptic partners to process and propagate information within circuits. Neural circuits in different parts of the brain perform distinct functions, ranging from sensory perception to motor control. The nervous system employs a massively parallel computational strategy to enhance the speed and precision of information processing. In the rest of this book, we will expand our studies of these fundamental concepts in the organization and operation of the nervous system.

FURTHER READING

Books and reviews

Adrian ED (1947) Physical Background of Perception. Clarendon.

Allen NJ & Lyons DA (2018) Glia as architects of central nervous system formation and function. *Science* 362:181–185.

Knudsen EI (2002) Instructed learning in the auditory localization pathway of the barn owl. *Nature* 417:322–328.

Ramón y Cajal S (1995) Histology of the Nervous System of Man and Vertebrates. Oxford University Press. (Original French version 1911)

Swanson LW (2012) Brain Architecture: Understanding the Basic Plan, 2nd ed. Oxford University Press.

Tinbergen N (1951) The Study of Instinct. Oxford University Press.

von Neumann J (1958) The Computer & the Brain. Yale University Press.

CHAPTER 2
Signaling within Neurons

Nervous systems can react rapidly to sensory stimuli. For example, a 5-day-old zebrafish larva begins responding to a pulse of water representing a potential threat within 3 milliseconds, and completely changes its direction of movement within 12 milliseconds, propelling itself away from a potential threat (**Figure 2-1**). This escape behavior relies on detection of mechanical force by sensory neurons, transmission of this sensory information through interneurons to motor neurons, and coordinated contraction or relaxation of appropriate muscles, all of which happens within several milliseconds. Because these escape behaviors, such as avoiding a predator (**Movie 2-1**), are crucial for animal survival, the speed at which neurons communicate has been subject to strong evolutionary selection.

As introduced in Chapter 1, the nervous system uses electrical signals to transmit information within a neuron. Individual neurons are the basic units of the nervous system, receiving, integrating, propagating, and transmitting signals based on changes in the membrane potential (Figure 1-18). A typical vertebrate neuron receives inputs from its presynaptic partners in the form of synaptic potentials at its dendrites and cell body. The neuron integrates these synaptic potentials along their paths toward the axon initial segment, where action potentials are generated. Action potentials propagate along the axon to the neuron's presynaptic terminals, where they cause neurotransmitter release. Neurotransmitters then bind to receptors on the postsynaptic target neurons, producing synaptic potentials and thus completing a full round of neuronal communication.

In this chapter and in Chapter 3, we will discuss the fundamental mechanisms of neuronal communication, focusing on three key steps: (1) the generation and propagation of action potentials, (2) the release of neurotransmitters by presynaptic neurons, and (3) the reception of neurotransmitters by postsynaptic neurons. Before we delve into these key steps, we will first study the special cellular properties of neurons as large cells with elaborate cytoplasmic extensions and as conductors of electrical signals. Understanding these properties is essential for our study of neuronal communication. The first four sections of this chapter also serve as an introduction to or a refresher of basic concepts and terms in molecular and cell biology used throughout this textbook.

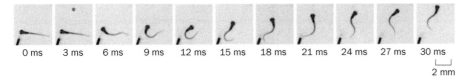

0 ms   3 ms   6 ms   9 ms   12 ms   15 ms   18 ms   21 ms   24 ms   27 ms   30 ms

2 mm

Figure 2-1 Rapid escape response of a zebrafish larva. Time-lapse images of the escape behavior of a zebrafish larva in response to a water pulse from the tube at bottom left. The asterisk (*) indicates the first detectable response at 3 milliseconds (ms) after stimulus onset, which is difficult to see in the frames shown here but is evident when observed in video clips. (From Liu KS & Fetcho JR [1999] *Neuron* 23:325–335. With permission from Elsevier Inc.)

CELL BIOLOGICAL AND ELECTRICAL PROPERTIES OF NEURONS

Neurons are the largest cells in animals. For instance, the cell body of a sensory neuron that innervates the toe is located in a dorsal root ganglion at about the level of the waist, but its peripheral branch extends down to the toe, and its central axon extends up to the brainstem (Figure 1-21); thus this sensory neuron spans about 2 m for a tall person and 5 m for a giraffe. Many neurons have complex dendritic trees. For example, the dendritic tree of the cerebellar Purkinje cell (Figure 1-11) has hundreds of branches and receives synaptic inputs from up to 200,000 presynaptic partners. The surface area and volume of axons or dendrites usually exceed those of the cell bodies by several orders of magnitude. This unique architecture, along with signaling molecules that decorate this architecture in specific patterns, enables rapid electrical signaling across long distances and allows individual neurons to integrate information from many cells.

In order for neurons to initiate, integrate, propagate, and transmit electrical signals, they must continuously synthesize proteins and deliver them to the appropriate subcellular compartments. Thus, there are two ways in which communication happens within the neuron: by the transport of RNA, proteins, and organelles along their long processes to get those components to the right part of the cell and by electrical signals moving along those processes. These two might, respectively, be compared to sending a package through the mail, whereby a physical object is delivered, and to sending a text or email, whereby only information is conveyed. Each is critical to nervous system function but relies on very different mechanisms. In the following sections, we will study these mechanisms, beginning with the basic molecular and cell biology of the neuron.

2.1 Neurons obey the central dogma of molecular biology and rules of intracellular vesicle trafficking

Macromolecule synthesis in neurons obeys the **central dogma** of molecular biology, which states that information flows from DNA → RNA → protein (**Figure 2-2**, left). **Genes**, the genetic substrates that carry instructions for how and when to make specific RNAs and proteins, are located in the nucleus on **DNA** molecules. DNAs are long double-stranded chains of nucleotides containing the sugar deoxyribose, a phosphate group, and one of four nitrogenous bases: adenine (A), cytosine (C), guanine (G), or thymidine (T). **Transcription** is the process by which RNA polymerase uses DNA as a template to synthesize single-stranded **RNA** (a chain of ribose-containing nucleotides, in which uracil [U] replaces T); the part of the gene that serves as a template for RNA synthesis is called the gene's **transcription unit**. The premessenger RNAs (pre-mRNAs) produced during gene transcription carry information in their specific ribonucleotide sequence that corresponds to the deoxyribonucleotide sequence of the transcription unit. Pre-mRNAs undergo a series of RNA processing steps. These include capping (adding a modified guanosine nucleotide to the 5′ end of the RNA), **RNA splicing** (removing RNA sequences that don't code for protein, called **introns**, and joining together the remaining sequences, called **exons**), polyadenylation (adding a long sequence of adenosine nucleotides to the RNA's 3′ end), and occasionally **RNA editing** (a regulated event that changes specific RNA sequences). The resulting mature **messenger RNAs** (**mRNAs**) are exported from the nucleus to the cytoplasm, where they are decoded by ribosomes during protein synthesis (**translation**). The information in the mRNA sequence dictates the amino acid sequence of the newly synthesized polypeptide (**protein**).

In addition to mRNAs, cells also transcribe a variety of RNAs that do not encode proteins. These include ribosome RNAs (rRNAs) and transfer RNAs (tRNAs) essential for the translation process as well as a variety of other RNAs collectively called noncoding RNAs. One class of noncoding RNAs that function in the cytoplasm is the **microRNAs**; these short (21–26 nucleotides) RNAs regulate protein production by triggering the degradation and inhibiting the translation of mRNAs

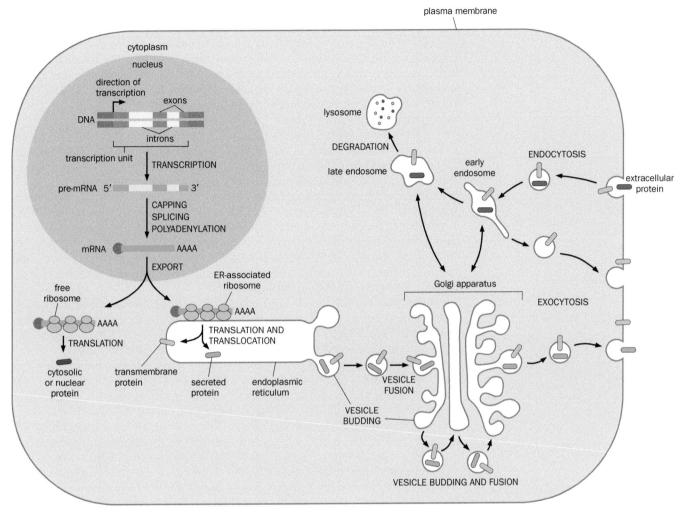

Figure 2-2 Schematic summary of the central dogma of molecular biology and intracellular vesicle trafficking. Left, in the nucleus, double-stranded DNA serves as a template for transcription to produce a pre-mRNA, which grows longer as nucleotides are added to the 3' end. Pre-mRNA is processed by capping the 5' end, splicing to remove introns and join exons, and polyadenylation at the 3' end to produce mature mRNA, which is exported to the cytoplasm. mRNAs encoding cytosolic and nuclear proteins (purple) are translated on free ribosomes in the cytosol (left branch). mRNAs encoding secreted (blue) or transmembrane (green) proteins are translated on ribosomes associated with the endoplasmic reticulum (ER). Bottom right, after synthesis and translocation across the ER membrane, transmembrane and secreted proteins exit the ER through vesicle budding, pass through the Golgi apparatus via a series of vesicle fusion and budding steps, and are transported to the plasma membrane. Fusion of a vesicle with the plasma membrane (exocytosis) leads to the release of secreted proteins into the extracellular space and delivery of the transmembrane proteins to the plasma membrane. Top right, extracellular proteins (red) or transmembrane proteins on the plasma membrane can be internalized through vesicle budding from the plasma membrane (endocytosis) into early endosomes. The content can be recycled back to the plasma membrane through exocytosis or can be delivered to late endosomes and lysosomes for degradation. There is also bidirectional trafficking between the endosomes and the Golgi apparatus. Much of the intracellular vesicle transport is along the microtubule cytoskeleton using molecular motors; we will discuss these in Section 2.3.

with complementary sequences. Other noncoding RNAs play a variety of roles in the nucleus to regulate gene expression.

Translation occurs in one of two distinct locations, depending on the destination of the protein products in eukaryotic cells. Proteins localized in the cytosol and nucleus are synthesized on free ribosomes in the cytoplasm, whereas proteins destined for export from the cell (**secreted proteins**) or that span the lipid bilayer of a membrane (**transmembrane proteins**) are synthesized on ribosomes associated with the **endoplasmic reticulum (ER)**, a network of membrane-enclosed compartments (Figure 2-2, left).

For most secreted proteins and transmembrane proteins destined for the plasma membrane, all or part of their sequence is translocated across the ER membrane as they undergo translation. Fully translated proteins then undergo a series of trafficking steps via **intracellular vesicles**, which are small, membrane-enclosed organelles in the cytoplasm. Secreted and transmembrane proteins exit the ER via budding of vesicles from the ER membrane and transit through the Golgi apparatus via a series of vesicle fusion and budding events. Eventually, vesicles that carry these proteins fuse with the plasma membrane in a step called **exocytosis**, so that secreted proteins are released into the extracellular space and transmembrane proteins are delivered to the plasma membrane (Figure 2-2, bottom right).

In addition to exocytosis, neurons (like many other cells) employ a process called **endocytosis**, which allows cells to retrieve fluid and proteins from the extracellular space or transmembrane proteins from the cell's plasma membrane. Endocytosis products are first delivered to early **endosomes**, which are membrane-enclosed organelles that carry newly ingested materials and newly internalized transmembrane proteins. Proteins from early endosomes can either cycle back to the plasma membrane through exocytosis or be transported to late endosomes and **lysosomes**, which contain enzymes for protein degradation (Figure 2-2, top right). In Chapter 3, we will study specific examples of exocytosis and endocytosis in the context of presynaptic neurotransmitter release and postsynaptic neurotransmitter receptor regulation.

While obeying the central dogma and rules of intracellular vesicle trafficking, neurons also have special properties to accommodate their large size and the great distance between the tip of their axonal or dendritic extensions and the cell body (soma). We can ask a simple question: how does a specific protein, such as a neurotransmitter receptor or a protein associated with the presynaptic membrane, get to the dendritic tip or axon terminal? The answer to this "simple" question is quite complex, and we are far from having complete answers for it. In principle, the corresponding mRNAs can be transported to the final destination before directing protein synthesis there. Alternatively, the protein can be synthesized at the soma and can either diffuse passively or be actively delivered to its final destination. For a transmembrane protein, delivery can take one of the following routes: (1) the intracellular vesicle that carries the protein can fuse with the plasma membrane at the soma, and the transmembrane protein can then diffuse over the plasma membrane to its destination; (2) the vesicle can be transported within dendrites or axons and can then fuse with the plasma membrane at its final destination; or (3) the protein can first be targeted to the plasma membrane of one compartment (axon or dendrite) and then endocytosed and trafficked to the final destination, a process called **transcytosis**. Each synthesis and transport mechanism has been observed. Their relative prevalence depends both on the type of protein and the type of neuronal compartment to which that protein is targeted. We will study some of these mechanisms in the next two sections.

2.2 While some dendritic and axonal proteins are synthesized from mRNAs locally, most are actively transported from the soma

Substantial evidence has demonstrated that mRNAs encoding a subset of proteins are targeted to dendritic processes, where they direct **local protein synthesis**. Electron microscopic studies revealed the presence of polyribosomes (clusters of ribosomes) in the dendrites, suggesting the translation of mRNA (**Figure 2-3A**). *In situ* **hybridization** studies, which can determine mRNA distribution in native tissues (Section 14.12), have identified specific mRNAs present in dendrites (Figure 2-3B). These dendritically localized mRNAs encode a variety of proteins known to function in dendrites and postsynaptic sites, such as the α subunit of the Ca^{2+}/calmodulin-dependent protein kinase II (CaMKII; Figure 2-3B, left panel), cytoskeletal elements such as actin and microtubule-associated protein 2 (MAP2), and neurotransmitter receptors. We will revisit many of these proteins later in this chapter and in Chapter 3. The list of dendritically localized mRNA has greatly expanded in recent years, thanks to high-throughput methods such as isolating mRNA from dendritic compartments for RNA sequencing (Section 14.12). In

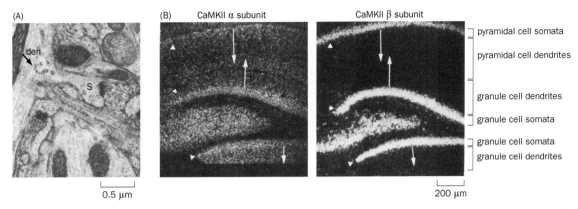

Figure 2-3 **Dendritic protein synthesis. (A)** Electron micrograph of part of a rat hippocampal granule cell (Figure 1-12). At the left of the micrograph is a segment of a dendrite (den), with a dendritic spine (S) branching to the right. A cluster of ribosomes (arrow) is seen at the junction between the dendritic trunk and the spine. **(B)** Localization of mRNA encoding the α and β subunits of the Ca²⁺/calmodulin-dependent protein kinase II (CaMKII) in sections of rat hippocampus, detected by *in situ* hybridization using probes specific to each gene. mRNAs for the CaMKII α subunit localize to cell bodies (somata; arrowheads) as well as dendrites of these cells (indicated by the arrows leaving the cell body layers), whereas mRNAs for the CaMKII β subunit are restricted to layers containing somata of both granule cells and pyramidal cells. See Figure 11-5 for a schematic of the hippocampus. (A, from Steward O & Levy WB [1982] *J Neurosci* 2:284–291. Copyright ©1982 Society for Neuroscience. B, from Burgin KE, Waxham MN, Rickling S, et al. [1990] *J Neurosci* 10: 1788–1798. Copyright ©1990 Society for Neuroscience.)

addition to polyribosomes and mRNAs, ER and Golgi apparatus–like membrane organelles have also been observed in distal dendrites, enabling locally synthesized transmembrane and secreted proteins to go through the secretory pathway just as in the soma (Figure 2-2). Finally, local translation in dendrites has been directly demonstrated using a number of *in vitro* preparations.

Local protein synthesis in dendrites solves several problems: it ameliorates the costs of long-distance transport, spatially focuses production of specific proteins where they are needed, and, perhaps most interestingly, allows protein production to be regulated in a specific compartment of the dendritic tree. As discussed in later chapters, such local protein synthesis in dendrites enables rapid changes of protein translation in response to synaptic signaling; these locally synthesized proteins in turn help modify synaptic signals, which may result in regional remodeling of dendrites and synapses in response to synaptic activity. Although much less studied compared to dendritic protein synthesis, local protein synthesis has also been observed in developing and mature axons. The products of local protein synthesis in developing axons may play important roles in axon guidance, a process we will study in detail in Chapters 5 and 7.

Even for proteins known to be synthesized in dendrites, the mRNA is usually more abundant in the soma (Figure 2-3B, left panel), suggesting that they are also synthesized in the soma. For most proteins, mRNAs appear predominantly in the soma (Figure 2-3B, right panel). How do these proteins get to their final destinations in dendrites and axons? This question has been explored primarily in axons because of the relative experimental ease of isolating distal axons from cell bodies. For example, radioactively labeled amino acids can be injected into regions that house cell bodies of sensory or motor neurons, which extend long axons to distant sites. Newly synthesized proteins that incorporate these radioactively labeled amino acids can be isolated from their axons at different times after injection and at different distances from the cell bodies, and analyzed by biochemical methods such as gel electrophoresis to determine their identities (**Figure 2-4**). These studies have identified two major groups of proteins based on the speeds of their appearances in axons. The fast component travels at 50–400 mm per day (0.6–5 μm/s); this includes mostly transmembrane and secreted proteins. The slow component travels at 0.2–8 mm per day; this includes mostly cytosolic proteins and cytoskeletal components. These two modes are termed **fast axonal transport** and **slow axonal transport**, respectively. In addition to **anterograde** transport from the cell body to the axon terminal, some proteins, such as those taken up via endocytosis, travel in the **retrograde** direction from axon terminals back to the cell body;

Figure 2-4 Studying axonal transport by isolating radioactively labeled proteins. Top, radioactively labeled amino acids are injected near the neuronal cell body and are either incorporated into newly synthesized proteins shortly after injection or metabolized. Middle and Bottom, at two time periods after the initial injection, proteins are isolated at specific segments of the axons (blue boxes), analyzed by gel electrophoresis, and visualized by autoradiography. (Adapted from Roy S [2014] *Neuroscientist* 20:71–81. With permission from SAGE.)

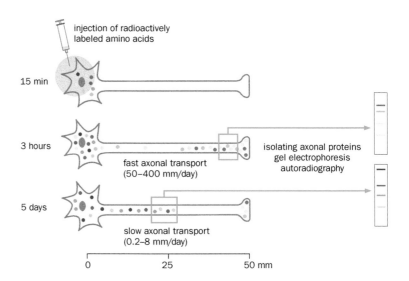

the speed of retrograde transport is similar to that of the fast anterograde axonal transport.

What mechanisms account for different modes of axonal transport? Theoretical studies indicate that passive diffusion within axons is too slow to account for even slow axonal transport, suggesting that all of these transport modes are active processes. Although less well studied, protein and mRNA transport into dendrites likely utilizes similar active processes. To understand the mechanisms underlying active transport and to appreciate why some proteins are transported to dendrites and others to axons, we need to examine the cytoskeletal organization of neurons.

2.3   The cytoskeleton forms the basis of neuronal polarity and directs intracellular trafficking

Like all eukaryotic cells, neurons rely on two major cytoskeletal elements for structural integrity and motility—**filamentous actin** (**F-actin**, also called microfilaments) and **microtubules**. F-actin is composed of two parallel helical strands of actin polymers, whereas microtubules are hollow cylinders consisting of 13 parallel protofilaments made of α- and β-tubulin subunits (**Figure 2-5**). Most cells also have intermediate filaments, cytoskeletal polymers with diameters between those of F-actin (~7 nm) and microtubules (~25 nm). The most prominent intermediate filaments in vertebrate neurons are the **neurofilaments**, which are concentrated in axons where they promote cytoskeletal stability.

F-actin and microtubules are both polar filaments that have *plus* and *minus* ends with distinct properties. As in all cells, F-actin is mostly concentrated near the plasma membrane. These include the plasma membrane along axonal and dendritic processes, at presynaptic terminals and postsynaptic dendritic spines in mature neurons, and at growth cones of axons and dendrites in developing neurons. Actin subunits are added to the plus ends of F-actin, which localizes close to the plasma membrane, such that actin polymerization can cause membrane protrusions responsible for morphological dynamics and motility of cells. F-actin does not mediate long-distance transport along dendritic or axonal processes, however. Microtubules occupy the center of axonal and dendritic processes, and are thus the cytoskeletal element along which long-distance transport in neuronal processes occurs. In most nonneuronal cells, the more dynamic plus ends of microtubules point toward the periphery, whereas the more stable minus ends are at the centers of cells, anchored in microtubule organizing centers. Microtubule orientation is more complicated in neurons and contributes to the distinction between dendrites and axons.

As introduced in Chapter 1, information generally flows from dendrites to axons. More than a century after Ramón y Cajal's original proposal, there is now a

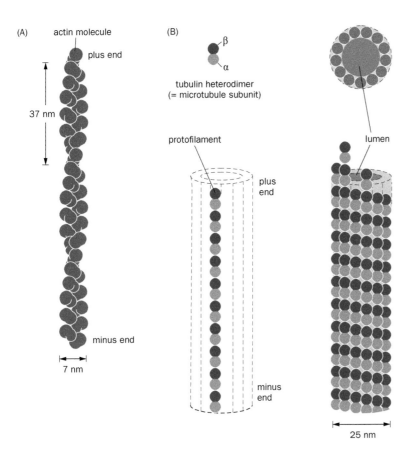

Figure 2-5 Filamentous actin (F-actin) and microtubules are two major cytoskeletal elements. (A) Schematic of F-actin, composed of two helical strands with a repeating unit of 37 nm. **(B)** Schematic of the microtubules. Left, α- and β-tubulin heterodimers assemble into longitudinal protofilaments. Right, the microtubule is made of 13 parallel protofilaments that form a tube with a hollow lumen (cross section is shown above). (Adapted from Alberts B, Johnson A, Lewis J, et al. [2015] Molecular Biology of the Cell, 6th ed. Garland Science.)

good understanding of the cell biological basis of **neuronal polarity**—the distinction between axons and dendrites. Subcellular structures and proteins related to reception of information, such as neurotransmitter receptors, are generally targeted to dendrites. Conversely, structures and proteins related to transmission of information, such as synaptic vesicles, are targeted to axons. These are largely achieved by the asymmetric cytoskeletal organization of the neuron, which enables **motor proteins**—which convert energy from ATP hydrolysis into movement along the cytoskeletal polymers—to transport specific cargos to specific destinations.

The orientation of microtubules differs in axons and dendrites. In axons, parallel arrays of microtubules are oriented with their plus ends pointing away from the cell body and toward the axon terminal, following the plus-end-out rule. However, dendrites have a mixed population of microtubules: plus-end-out and minus-end-out populations are about equally prevalent in proximal dendrites (**Figure 2-6A**). To date, this rule has applied to all vertebrate neurons examined in culture and *in vivo*. As discussed in Section 1.7, most invertebrate neurons have a single neurite that exits the cell body, which then gives rise to both dendritic and axonal branches. Some sensory neurons in the nematode *C. elegans* and the fruit fly *Drosophila* have distinct axonal and dendritic processes (similar to bipolar neurons in vertebrates). In invertebrate bipolar neurons thus far examined, dendrites appear to have mostly minus-end-out microtubules, whereas axons have plus-end-out microtubules. Thus, although details may differ, both vertebrate and invertebrate neurons share the principle that dendrites and axons differ in their microtubule orientations.

Two types of motor proteins move cargos along microtubules: (1) the cytoplasmic **dynein** and (2) a large family of proteins called **kinesins**. Dynein is a minus-end-directed motor; as such, it transports cargos from axon terminals back to the cell body. Most kinesins are plus-end-directed motors and therefore transport cargos from the cell body to axon terminals (Figure 2-6B). In vertebrate dendrites, both dyneins and kinesins can mediate bidirectional transport because of

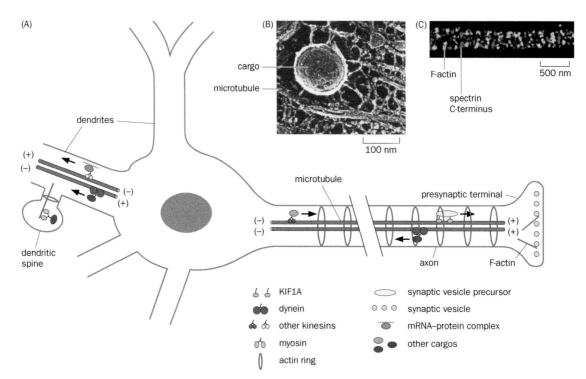

Figure 2-6 Cytoskeletal organization and motor proteins in axons and dendrites. (A) In the axon of a typical vertebrate neuron, microtubules (green) are oriented with the plus end (+) pointing toward the axon terminal, or plus end out. By contrast, dendrites contain microtubules with both plus- and minus-end-out orientations. Cargos destined for axon terminals, such as synaptic vesicle precursors (cyan), are preferentially transported toward axon terminals by plus-end-directed kinesins such as KIF1A (orange). Other cargos are transported by dynein (which moves toward the minus end of microtubules) or other kinesins in the axon and dendrites. For example, an mRNA (cyan line) with an associated protein complex (light brown) is transported by a kinesin into dendrites. Dynein and most kinesins are dimers with two heads (motor domains), whereas KIF1A acts as a monomer. F-actins (red) are distributed near the plasma membrane within axons (where they form rings) and dendrites (not shown) and are particularly enriched in dendritic spines and presynaptic terminals. After leaving microtubules, cargos may be further transported to their local destination by myosin-based movement along F-actin. **(B)** Quick-freeze deep-etch electron microscopy reveals the structure of the axonal cytoskeleton. The arrow points to a structure consistent with kinesin protein moving a cargo along the microtubule. **(C)** Super-resolution fluorescence microscopy image of an axon segment of a cultured rat hippocampal neuron stained with phalloidin (green), which labels F-actin, and an antibody against the C-terminus of βII-spectrin (magenta). Note the periodic distribution of both proteins, with the period being ~190 nm. (B, from Hirokawa N, Niwa S, & Tanaka Y [2010] *Neuron* 68:610–638. With permission from Elsevier Inc. C, from Xu K, Zhing G, & Zhuang X [2013] *Science* 339:452–456.)

mixed polarity microtubules. mRNAs used in dendrites for local protein synthesis are transported by dynein and several kinesins on microtubules in the form of mRNA–protein complexes (Figure 2-6A).

Dynein and kinesins have specific proteins that link them to specific cargos, and some kinesins may bind directly to cargos. For example, synaptic vesicle precursors are transported from the cell body to axon terminals by binding directly to a specific kinesin called KIF1A (Figure 2-6A). Certain types of kinesins are highly enriched in axons whereas others can transport cargoes into both axons and dendrites, increasing the specificity with which cargo is delivered to defined neuronal compartments. The asymmetric organization of the microtubule cytoskeleton and specific motor–cargo interactions together establish and maintain neuronal polarity. Other factors contributing to neuronal polarity include diffusion barriers at the axon initial segment for both cytosolic and membrane proteins. In Chapter 7, we will explore how polarity is initiated in developing neurons.

In vitro motility studies indicate that kinesins and the cytoplasmic dynein mediate fast axonal transport. Kinesins can move along microtubules at about 2 μm/s (**Box 2-1**; **Movie 2-2**), in the same range as fast anterograde axonal transport (Figure 2-4). Recent studies indicate that kinesins also mediate slow anterograde axonal transport. However, slow transport is characterized by much longer pauses between runs (periods when cargos are being transported), whereas fast axonal transport features longer runs and shorter pauses. During its brief runs, slow transport achieves speeds comparable to runs of fast axonal transport.

Box 2-1:   How were kinesins discovered?

Breakthroughs in biology often result from utilization of new techniques and appropriate experimental preparations to address important, unsolved questions. The identification of kinesins illustrates how the combination of these ingredients drives new discoveries. In the early 1980s, the invention of video-enhanced differential interference contrast (VE-DIC) microscopy enabled visualization of subcellular organelles in unstained live tissues. When applied to the squid giant axon, a specialized axon whose diameter can reach 1 mm (we will encounter this axon again in our studies of the action potential later in this chapter), VE-DIC microscopy revealed that organelles move along filament-

like structures running along the axon's length inside the plasma membrane (Figure 2-7A). Furthermore, following extrusion of the axonal cytoplasm (axoplasm), organelle movement along filaments in the axoplasm, free of axonal membrane, was similarly observed.

The ability of the extruded axoplasm to support organelle movement opened the door to many experimental manipulations. For example, researchers could dilute the axoplasm to track organelle movement along a single filament over time and measure its speed (Movie 2-2). The speed was found to be around 2 μm/s for transportation of small

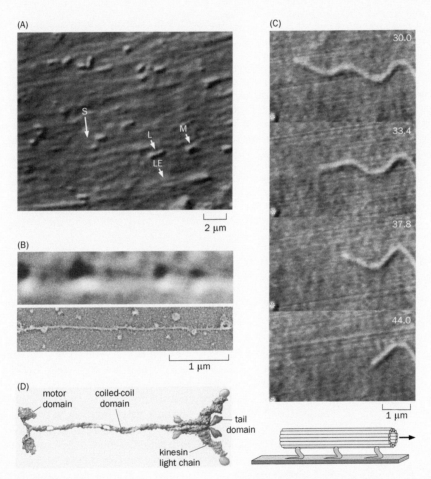

Figure 2-7 Discovery of the first kinesin. (A) An image of a segment from an intact squid giant axon, taken using video-enhanced differential interference contrast (VE-DIC) microscopy, shows horizontal linear elements (LE, which are microtubules) running in parallel with the axon. In video records, many small (S), medium (M), and large (L) organelles can be seen moving along the linear elements. **(B)** Top, VE-DIC image of organelles moving horizontally along what appears to be a single transport filament from extruded squid axoplasm; bottom, the same field of view in electron microscopy, taken after VE-DIC study, confirming the presence of a single microtubule. The apparent diameter of the transport filament in the light microscope is inflated by diffraction to about 10 times its true diameter (25 nm). **(C)** Top, time-lapse movie (time at upper right indicates seconds) of a single microtubule moving rightward on a glass slide to which a soluble fraction purified

from squid axoplasm had been immobilized, in the presence of ATP. The object marked with an asterisk at the bottom left serves as a stationary marker. Bottom, a schematic of putative motor proteins from the squid axoplasm attached to the glass and to the microtubule. ATP hydrolysis by multiple motor proteins oriented in the same direction causes microtubule movement relative to the glass. **(D)** Molecular structure the kinesin from the squid giant axon. (A, from Allen RD, Metuzals J, Tasaki I, et al. [1982] *Science* 218:1127–1129. With permission from AAAS. B, adapted from Schnapp BJ, Vale RD, Sheetz MP, et al. [1985] *Cell* 40:455–462. With permission from Elsevier Inc. C, adapted from Vale RD, Schnapp BJ, Reese TS, et al. [1985] *Cell* 40:559–569. With permission from Elsevier Inc. D, from Vale RD [2003] *Cell* 112: 467–480. With permission from Elsevier Inc.)

(Continued)

Box 2-1: continued

organelles, in the same range as the fast axonal transport speed determined by tracking radioactively labeled proteins in vertebrate neurons *in vivo* (Figure 2-4). Chemicals or drugs could be added to the axoplasm to study their effects on organelle motility. It was already known then that the actin-based motor myosin utilized ATP hydrolysis to power movement along F-actin, so researchers tested whether ATP hydrolysis is also required for axonal transport. They found that motility was blocked when ATP was depleted from the axoplasm or when a nonhydrolyzable ATP analog was added, indicating that organelle movement along axoplasmic filaments indeed depends on ATP hydrolysis. Finally, following motility studies along a single transport filament using VE-DIC microscopy, electron microscopic analysis of the same filament (Figure 2-7B) provided unequivocal evidence that the individual filaments supporting organelle movement are individual microtubules.

These studies suggested the presence of motor proteins that utilize energy from ATP hydrolysis to move organelles in the squid axoplasm. Indeed, just as microtubules can support organelle movement, a soluble fraction of axoplasm from the squid giant axon containing the putative motor proteins, when immobilized on glass, could also cause individual microtubules to move in the presence of ATP (Figure 2-7C). Using this functional assay, biochemical purification

of squid axoplasm led to the identification of a protein complex that could support microtubule movement on glass. Similar protein complexes purified from bovine and chick brains exhibit similar capabilities. Members of this protein family were named kinesins (from the Greek *kinein*, "to move").

We now know that kinesins are evolutionarily conserved molecular motors found in all eukaryotes. The kinesin complex originally purified from the squid axoplasm belongs to a specific kinesin subfamily, consisting of two heavy chains and two light chains. Each heavy chain has an N-terminal globular domain that contains the microtubule-binding site and an ATPase, a long coiled-coil domain that mediates the dimerization of two heavy chains, and a C-terminal domain that binds to the light chain and to cargo (Figure 2-7D). Biochemical and biophysical studies have revealed detailed mechanisms of how kinesins move along microtubules (Movie 2-3). Each mammalian genome has about 45 genes encoding different kinesins, many of which are expressed in neurons and are responsible for carrying different cargos to specific subcellular compartments of neurons (Figure 2-6). Mutations in kinesins and proteins associated with kinesins (and dynein) in humans underlie a variety of neurological disorders, highlighting their importance to human health.

Microtubules are integral structural components of dendritic trunks and axons and can be considered the highways that mediate long-distance transport in neurons. However, microtubules are usually absent from dendritic spines and presynaptic terminals. After cargos get off the microtubule highway at their approximate destinations, such as distal segments of dendrites, F-actins direct local traffic utilizing a large family of **myosin** proteins as molecular motors (Figure 2-6). We will study the mechanism by which myosin–actin interactions produce motility in the context of muscle contraction in Section 8.1.

Examination of the actin cytoskeleton at the axonal membrane using super-resolution fluorescence microscopy (see Section 14.17 for details) revealed that F-actin is distributed in periodic rings associated with a network of spectrin, a plasma membrane-associated cytoskeletal element (Figure 2-6C). The period of these actin rings, at ~190 nm, is just below the resolution of conventional diffraction-limited light microscopy, preventing its discovery until recently. Actin rings are also found in dendrites and at the necks of dendritic spines, although less commonly than in axons. These actin rings and their associated spectrin networks may provide stability to long axons and regulate axonal diameters.

In summary, membrane proteins (associated with intracellular vesicles) and cytosolic proteins destined for dendrites or axons are delivered to their destinations via interactions with specific motor proteins, which enable their transport along microtubules for long distances and sometimes along actin filaments for local movements. Although we have an outline of the trafficking rules, we still lack complete answers to many questions: How is motor–cargo selection achieved? How is cargo loading and unloading regulated? What regulates the transition between pauses and runs? Why do certain motors prefer axons? Answers to these questions will reveal how each neuronal compartment acquires a unique assortment of specialized proteins to carry out its functions, thus enriching our understanding not only of neuronal cell biology but also of neuronal physiology and function.

2.4   Channels and transporters move solutes passively or actively across neuronal membranes

The mechanisms we have studied thus far are concerned with how proteins and organelles inside the cell move around, but haven't addressed how extracellular molecules or ions enter the cell via the plasma membrane. This requires a different type of transport: across the lipid bilayer. The lipid bilayer of the plasma membrane and membranes of intracellular vesicles is impermeable to most ions and polar molecules that are soluble in aqueous environments such as the cytosol or extracellular milieu. The lipid bilayers serve as essential compartmental boundaries, delineating cells and intracellular organelles, such as the ER, Golgi apparatus, and synaptic vesicles. Inorganic ions and water-soluble molecules—including nutrients, metabolites, and neurotransmitters, collectively referred to as **solutes**—cannot easily diffuse across the lipid bilayer and thus require specific transport mechanisms to move across. Transport across the lipid bilayer is essential for many neuronal functions, such as electrical signaling.

Specialized transmembrane proteins transport solutes across the membranes of neurons and other cells. These membrane transport proteins can be divided into two broad classes: **channels** and **transporters**. Channels have an aqueous pore that allows specific solutes to pass directly through when they are open. In later sections of this chapter we will study **ion channels**, each of which allows selective passage of one or more specific ions. Transporters have two separate gates that open and close alternately, allowing movement of solutes from one side of the membrane to the other side (**Figure 2-8**A). In general, solutes move through open channels much more rapidly than they do through transporters.

Channels and transporters usually support *net movement* of solutes in one direction across the membrane under a given condition. Uncharged solutes move from the side with higher concentration to the side with lower concentration, or down their **chemical gradient**. Movement across the membrane of charged solutes, such as ions, creates an **electrical gradient**, a difference in the electrical potential across the membrane. The **electrochemical gradient**, which combines the chemical and electrical gradients, determines the direction and magnitude of net solute movement (**Figure 2-9**). When the electrical and chemical gradients are in the same direction, they enhance each other in driving solute movement (Figure 2-9, middle). When these gradients are in opposing directions, they partially (Figure 2-9, right) or sometimes fully cancel out each other's effects. Transport of solutes down their electrochemical gradients does not require external energy and is called **passive transport** (Figure 2-8A).

Some transporters can move a solute across the membrane against its electrochemical gradient (from low to high) using external energy; this process is called **active transport** (Figure 2-8B). Energy for active transport can come from the following sources. (1) Chemical reactions: the most frequent form is ATP hydrolysis, in which the transporter is an ATPase, and chemical energy from ATP hydrolysis

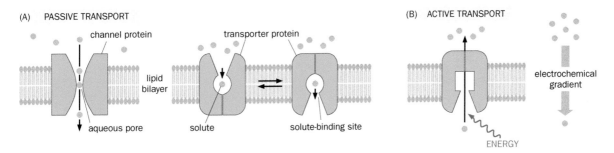

Figure 2-8 Channels and transporters mediate passive and active transport. (A) A channel protein has an aqueous pore that allows solutes to pass through directly when the channel is open, whereas a transporter protein moves the solute across the membrane through sequential opening and closing of at least two gates. In the absence of energy input, solutes move down their electrochemical gradients (Figure 2-9) through channels and transporters; this is called passive transport. **(B)** Transporter proteins can also utilize energy to mediate active transport to move solutes up their electrochemical gradients. (Adapted from Alberts B, Johnson A, Lewis J, et al. [2015] Molecular Biology of the Cell, 6th ed. Garland Science.)

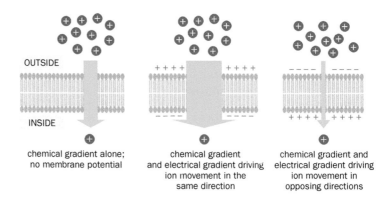

Figure 2-9 Electrochemical gradients of charged solutes such as ions. When there is no electrical potential difference across a membrane, the chemical gradient alone determines the direction of ion movement, from high to low concentrations (left). When there is an electrical potential difference across a membrane, the chemical and electrical gradients act together to determine the direction and magnitude of the force governing ion movement. These two components of the electrochemical gradient may work in the same direction (middle) or in opposing directions (right). The thickness of the arrows symbolizes the magnitude of the force driving ion movement. (Adapted from Alberts B, Johnson A, Lewis J, et al. [2015] Molecular Biology of the Cell, 6th ed. Garland Science.)

is used to drive conformational change of the protein. (2) Light: energy is derived from photon absorption. Transporters driven by chemical reactions or by light are also called **pumps**. We will discuss light-driven pumps in the context of the evolution of vision in Chapter 13. (3) Coupled transport: when a transporter moves multiple solute species together, energy gained from transporting some species down a gradient can be used to transport other species up another gradient. Coupled transporters (also called cotransporters) can be divided into two types: those that move solutes in the same direction are called **symporters**, and those that move solutes in opposite directions are called **antiporters** or **exchangers** (**Figure 2-10**; **Movie 2-4**). (In analogy to cotransporters, transporters that transport a single species of cargo are also called uniporters.) In the next section, we will encounter specific examples of an ATP-driven pump and cotransporters that play crucial roles in establishing electrochemical gradients of different ions across the plasma membrane, which forms the basis of electrical signaling in neurons.

2.5 Neurons are electrically polarized at rest because of ion gradients across the plasma membrane and differential ion permeability

Electrical signals in neurons, as well as in other **excitable cells** (cells that produce action potentials, such as muscle cells), rely on a difference in electrical potential across the plasma membrane. This electrical potential difference between the inside of the cell and the extracellular environment is called the **membrane potential** of the cell. We can measure the membrane potential directly by inserting a microelectrode into the cell, a procedure called **intracellular recording**. A microelectrode for intracellular recording is usually made of glass with a very fine tip filled with a conducting salt solution, so that it makes electrical contact with the inside of a cell; at the other end, the electrode is connected via a wire to an amplifier and an oscilloscope (see Figure 14-34 and Sections 14.20–14.21 for different

Figure 2-10 Three types of active transport. An active transporter can be an ATP-driven pump, which utilizes energy from ATP hydrolysis to move solutes against their electrochemical gradients (left); a light-driven pump, which derives its energy from photon absorption (middle); or a coupled transporter, which derives its energy from transporting a second species of solute down its electrochemical gradient. Coupled transporters can be symporters or antiporters, depending on whether the two solutes move in the same direction or in opposite directions. The schematic at the far right summarizes the electrochemical gradients of different solutes, with the larger number of a given symbol indicating the high end of an electrochemical gradient. The downward gray arrow shows the electrochemical gradient of the solute represented by yellow circles. (Adapted from Alberts B, Johnson A, Lewis J, et al. [2015] Molecular Biology of the Cell, 6th ed. Garland Science.)

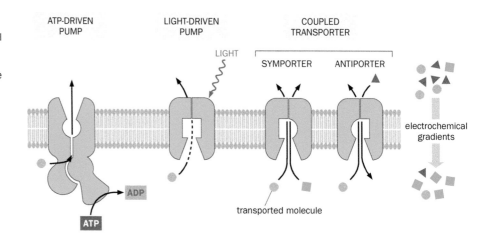

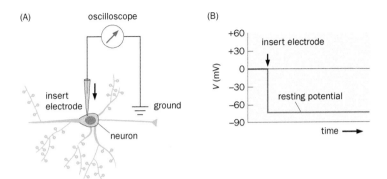

Figure 2-11 Measuring the resting potential of a neuron. (A) Schematic of recording the membrane potential of a neuron at rest. The oscilloscope measures the voltage difference (*V*, in millivolts) between the ground (the electrical potential in the extracellular environment) and the tip of the electrode. **(B)** Before the insertion of the electrode tip into the cell, *V* is zero. After insertion, *V* drops to –75 mV, the resting potential of our model neuron.

methods for recording neuronal activity). The membrane potential of a neuron at rest (the **resting potential**) is typically between –50 and –80 millivolts (mV), depending on the cell type (**Figure 2-11**). Thus, neuronal membranes are electrically polarized. A change in the electrical potential inside the cell toward a less negative value is termed **depolarization**. A change in the electrical potential inside the cell toward a more negative value is called **hyperpolarization**.

Neurons are electrically polarized because (1) ion concentrations differ between the intracellular and extracellular environments; (2) the **permeability** of the plasma membrane to each of the three major ions is different (as we will learn later, permeability is determined by the number of open ion channels that conduct specific ions; when an ion channel is open, it becomes permeable to one or more ions; ions can flow in either direction, with the net flux determined by the ion's electrochemical gradient). For a typical neuron, the concentrations of sodium ions (Na^+) and chloride ions (Cl^-) are 10- to 20-fold higher in the extracellular space than in the cytosol, whereas the intracellular potassium ion (K^+) concentration is 30-fold higher than in the extracellular environment (**Figure 2-12**A), although these values vary with types of neurons and animals. At a neuron's resting state, the plasma membrane is not very permeable to Na^+ and Cl^- but is relatively permeable to K^+. (That is, K^+ crosses the neuronal membrane more readily than Na^+ or Cl^-.) In addition to these two **cations** (positively charged ions such as K^+ and Na^+) and one **anion** (a negatively charged ion such as Cl^-), some organic

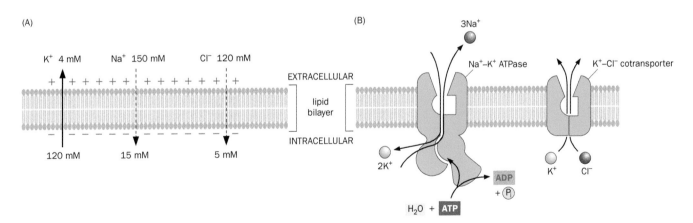

Figure 2-12 Ionic basis of the resting potential. (A) Numbers indicate typical concentrations of K^+, Na^+, and Cl^- inside and outside of a mammalian neuron, in millimoles per liter (mM). At the resting state, the membrane is permeable to K^+ (arrow indicating the direction down the electrochemical gradient) but less permeable to Na^+ and Cl^- (dashed arrows). The resting membrane potential is largely determined by the balance between the chemical gradient that drives K^+ outward, and the electrical gradient that drives K^+ inward. The + and – signs on opposite sides of the membrane indicate that the intracellular electrical potential is more negative than that of the extracellular

environment. **(B)** Left, the Na^+–K^+ ATPase uses energy from ATP hydrolysis to transport three Na^+ ions outward and two K^+ inward, against their electrochemical gradients, each cycle. The activity of the Na^+–K^+ ATPase maintains the intracellular concentrations of Na^+ and K^+ and thus the resting potential by counteracting the leak of Na^+ and K^+ across the resting membrane. Right, the K^+–Cl^- cotransporter utilizes energy released as K^+ moves down its electrochemical gradient to move Cl^- up its electrochemical gradient, helping maintain a concentration gradient of Cl^- across the membrane.

anions are enriched intracellularly, but neuronal membranes are usually not permeable to these organic anions.

The ionic concentrations across the neuronal membrane are maintained by active transport, principally by a transporter called the **Na$^+$–K$^+$ ATPases**. This ion pump uses energy derived from ATP hydrolysis to pump Na$^+$ outward and K$^+$ inward, against their respective electrochemical gradients (Figure 2-12B; Movie 2-5), thus maintaining the concentration differences of these two important ions across the resting membrane. It has been estimated that one third or more of neurons' energy is used by the Na$^+$–K$^+$ ATPase, highlighting both the importance of this particular pump and the importance of maintaining Na$^+$ and K$^+$ concentration gradients. The Cl$^-$ gradient is maintained by several cotransporters, such as the K$^+$–Cl$^-$ symporter that couples K$^+$ and Cl$^-$ export (Figure 2-12B).

To understand how chemical gradients and membrane potentials influence ion movement, let's first consider a hypothetical situation in which the membrane is permeable only to K$^+$. (This situation applies to glia quite well.) The concentration difference across the membrane causes K$^+$ to diffuse outward down its chemical gradient. As K$^+$ flows outward, however, the intracellular compartment becomes more negatively charged, thus increasing the electrical potential difference across the membrane. This deters further outward K$^+$ diffusion, because K$^+$ is positively charged. Eventually, the chemical and electrical forces will reach the **equilibrium potential** of K$^+$, E_K, the membrane potential at which the electrical and chemical forces balance each other out, such that there is no net K$^+$ flow. E_K follows the **Nernst equation**:

$$E_K = \frac{RT}{zF} \ln \frac{[K^+]_o}{[K^+]_i}$$

where $[K^+]_o$ and $[K^+]_i$ are the extracellular and intracellular K$^+$ concentrations, R and F are the gas and Faraday constants, respectively, T is the absolute temperature (in Kelvin), z is the valence of the ion (+1 for K$^+$), and ln is the natural logarithm function. At room temperature, the expression RT/F is about 25 mV. Using the K$^+$ concentration difference of our model neuron (Figure 2-12A), we can calculate the E_K to be about –85 mV, which is slightly hyperpolarized relative to the resting potential.

The Nernst equation can also be used to determine the equilibrium potentials of Cl$^-$ and Na$^+$. Using the concentration differences across our model neurons (Figure 2-12A), we can determine that E_{Cl} = –79 mV and E_{Na} = +58 mV. Note that despite having a chemical gradient opposite to that of K$^+$, Cl$^-$ has a similar equilibrium potential to K$^+$ because it has a negative charge (z = –1). A positive equilibrium potential for Na$^+$ means that if the membrane were permeable only to Na$^+$, the intracellular membrane potential would be positive relative to the extracellular environment. We will see the significance of this when we study the ionic basis of the action potential later in this chapter.

In reality, the resting potentials of most neurons are slightly less negative than the K$^+$ equilibrium potential, because the membrane is also somewhat permeable to Na$^+$ and Cl$^-$. When the membrane is simultaneously permeable to multiple ions, the resting membrane potential V_m at equilibrium (that is, when there is no net ion flow) can be calculated using the **Goldman–Hodgkin–Katz (GHK) equation**:

$$V_m = \frac{RT}{F} \ln \frac{P_K[K^+]_o + P_{Na}[Na^+]_o + P_{Cl}[Cl^-]_i}{P_K[K^+]_i + P_{Na}[Na^+]_i + P_{Cl}[Cl^-]_o}$$

where P_K, P_{Na}, and P_{Cl} are the permeabilities for K$^+$, Na$^+$, and Cl$^-$, respectively. In essence, the GHK equation states that each ion makes an independent contribution to the resting potential that is weighted according to the permeability of the resting membrane to that ion.

Because the membrane potential is generally not identical to the equilibrium potential for any single ion, there is a force tending to push each ion into or out of the cell. This is called the **driving force** for that ion, and it is equal to the difference between the membrane potential and the equilibrium potential for that ion. If the membrane were permeable only to K^+ (that is, $P_{Na} = 0$, $P_{Cl} = 0$), then $V_m = E_K$ according to the GHK equation; the driving force for K^+ would be 0, and the net current would be zero, despite a high permeability for K^+. However, neither P_{Na} nor P_{Cl} is actually 0. As the equilibrium potential of Cl^- is generally very close to the resting potential, Cl^- flow is small at rest because of the small driving force. On the other hand, the driving force for Na^+ is very large as both the chemical and electrical gradients favor its entry into the cell. Na^+ will leak inward, down its electrochemical gradient, despite a small P_{Na}. This will raise the membrane potential slightly above E_K, creating a driving force and causing K^+ to leak outward. If unopposed, these leak currents would steadily decrease the intracellular K^+ concentration and increase the intracellular Na^+ concentration; however, they are counterbalanced by active transport via the Na^+–K^+ ATPase, which pumps Na^+ out and K^+ in (Figure 2-12B). The Na^+–K^+ ATPase thus maintains the intracellular K^+ and Na^+ concentrations, thereby stabilizing the resting potentials of neurons.

2.6 The neuronal plasma membrane can be described in terms of electrical circuits

While ions cannot cross the lipid bilayer of the neuronal plasma membrane by themselves, the intracellular and extracellular environments are both aqueous solutions that support ion movement. We can model how a neuron functions using terminology developed to describe **electrical circuits**—interconnections of electrical elements containing at least one closed current path. In this section, we introduce the basic components and rules of electrical circuits, which will be instrumental to our discussion of neuronal signaling in subsequent sections.

The simplest electrical circuit consists of two electrical elements: a battery and a resistor. The **battery** maintains a constant voltage (electrical potential difference) across its two terminals, thus providing an energy source. The **resistor** implements electrical resistance (that is, opposes the passage of electric current) and produces a voltage across its two terminals when current flows through it (**Figure 2-13**A, left). The electric current (I, the flow of electric charge per unit time) that passes through the resistor follows **Ohm's law**:

$$I = \frac{V}{R}$$

in which V is the voltage across the resistor and R is the **resistance** of the resistor. As the electrical wires connecting the battery and the resistor are assumed to have zero resistance, the voltage across the resistor is the same as the voltage across the battery. The units of I, V, and R are the ampere (A), volt (V), and ohm (Ω), respectively.

When two resistors are connected in series (Figure 2-13A, middle), the current passing through each resistor is the same; the voltage across both resistors is the sum of voltages across each resistor, or $V = V_1 + V_2$; and the combined resistance is $R = R_1 + R_2$. An equivalent but more widely used measure of a resistor in electrophysiology is **conductance** (g), which is the inverse of resistance: $g = 1/R$.

Figure 2-13 Electrical circuits with only resistors or capacitors. (A) Left, an electrical circuit consisting of two elements, a battery (E) with a voltage V across its two terminals and a resistor with a resistance R. The current (I, arrow) flows outside the battery from the positive terminal (+, represented by the longer line of the battery) to the negative (–) terminal. Middle, two resistors (with resistance R_1 and R_2) connected in series. The current flowing through them is the same. The sum of voltages across each resistor ($V_1 + V_2$) equals the voltage across the battery. Right, two resistors connected in parallel. The voltage across each is the same as the voltage across the battery. The total current (I_T) equals the sum of the currents passing through each path ($I_1 + I_2$). **(B)** A circuit consisting of a battery (E), a capacitor (with a capacitance C), and a switch (s). When the switch is turned on, a transient current (dashed arrow) charges the capacitor until the voltage across it equals the voltage across the battery.

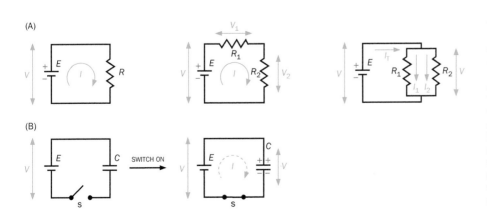

Thus, when two resistors are connected in series, $1/g = 1/g_1 + 1/g_2$; in other words, the conductance *decreases*. When the two resistors are connected in parallel (Figure 2-13A, right), the voltages across each resistor are the same; the total current is the sum of the currents passing through each resistor, or $I = I_1 + I_2$; the combined resistance follows the formula $1/R = 1/R_1 + 1/R_2$, and the combined conductance can be calculated as $g = g_1 + g_2$. In other words, the conductance *increases* for resistors in parallel. The unit of conductance is the siemens (S). It follows from the definition of conductance that Ohm's law can also be expressed as:

$$I = gV$$

Note that a resistor is at the same time a **conductor** of electric current, and these two terms are used interchangeably, depending on the context; a resistor with high resistance is a poor conductor, and a resistor with low resistance is a good conductor.

We can now relate these simple electrical circuits to what we have learned so far about the neuron. The lipid bilayer is an **insulator**, a resistor with infinite resistance, as it does not allow electric current to pass through. As noted in Section 2.5, the plasma membrane is not a perfect insulator—even at the resting state, ions can leak through the membrane via specific channels. These ion channels can be modeled as parallel current paths, each consisting of a resistor with a specific resistance and a battery with a specific voltage equivalent to the equilibrium potential of the ion, interposed between the electric potential across the membrane. We will discuss this model in detail in Section 2.7. We will also encounter resistors connected in series when we study propagation of electrical signals along neuronal fibers (dendrites and axons) in Section 2.8.

Another important electrical element is a **capacitor**, consisting of two parallel conductors separated by an insulator. A capacitor is a charge-storing device, as it does not allow current pass through the insulator. The lipid bilayer of the plasma membrane, along with the extracellular and intracellular milieu, is an excellent example of a capacitor. In a simple circuit consisting of a battery and a capacitor (Figure 2-13B), when the switch is turned on, current flows from the battery to the capacitor until the capacitor is charged to a voltage equal to that of the battery. Positive charge accumulates on one conductor, while negative charge accumulates on the other; this is how charge is stored. The **capacitance** (C), the ability of a capacitor to store charge, is defined as $C = Q/V$, where Q is the electric charge stored when the voltage across the capacitor is V. The unit of capacitance is the farad (F), and the unit of charge is the coulomb (C). When two capacitors are connected in series, the combined capacitor (C) follows the formula $1/C = 1/C_1 + 1/C_2$. When two capacitors are connected in parallel, $C = C_1 + C_2$.

In theory, when a circuit has no resistance (Figure 2-13B), the capacitor is charged instantaneously when the switch is turned on. In practice, circuits always have some resistance. In a circuit containing both resistors and capacitors (an **R-C circuit**), the current flowing through the resistor and the capacitor changes over time after the switch is turned on. The product of resistance and capacitance has the unit of *time* and is called the **time constant** (designated as τ). The time constant defines how quickly capacitors (such as the plasma membrane) charge or discharge over time in response to external signals, such as a sudden change of current flow (as would result from the opening of channels). The larger the time constant, the longer it takes to charge a capacitor and the more an electrical signal is spread out over time.

Let's examine a parallel *R-C* circuit to help clarify the important concept of a time constant (see **Box 2-2** for a quantitative treatment of this subject). In this circuit, a resistor and a capacitor are connected in parallel to a constant current source (**Figure 2-14**, left). Once the switch is turned on, while the sum of the currents (I_T) remains constant, the current flowing through the resistor (I_R) and the voltage across the resistor ($V = I_R R$) increases over time (Figure 2-14, middle), while the current flowing across the capacitor (I_C) decreases over time (Figure 2-14, right). Intuitively, this is because the role of the capacitor is to store charge and the voltage across it cannot change until charge builds up, which takes time, and

---

Box 2-2: A deeper look at the parallel *R-C* circuit

In Section 2.6 we encountered a parallel *R-C* circuit (Figure 2-14). For students with a background in differential equations, we discuss here how the temporal dynamics of these circuits are derived.

When the parallel *R-C* circuit is connected to a constant current source (after the switch is turned on in Figure 2-14, left), there is a redistribution of current over time from the capacitor path (I_C) to the resistor path (I_R), but their sum (I_T) is constant. Also, the voltage (V) across the resistor and the capacitor both equal $I_R R$. Thus, we have $I_R(t) = I_T - I_C(t) = I_T - dQ/dt = I_T - C\,dV/dt = I_T - RC\,dI_R(t)/dt$. Solving this differential equation gives:

$$I_R(t) = I_T\left(1 - e^{-\frac{t}{RC}}\right)$$

$$V(t) = I_T R\left(1 - e^{-\frac{t}{RC}}\right)$$

$$I_C(t) = I_T e^{-\frac{t}{RC}}$$

The first and third functions are graphically represented in the middle and right panels of Figure 2-14. For example, $I_R(t)$ increases to 63% $(1 - e^{-1})$, 86% $(1 - e^{-2})$, and 95% $(1 - e^{-3})$ of I_T at times equivalent to one, two, and three time constants.

Note that we connect the parallel *R-C* circuit to a constant current source rather than a battery because this is a better model of what happens to neuronal membrane during electrical signaling: inputs can be modeled as a transient supply of a constant current source (opening and closing of ion channels), whereas the membrane potential *continuously* changes with the opening and closing of ion channels. If we were to connect a parallel *R-C* circuit to a battery with a constant voltage, the capacitor would be charged instantaneously, reducing the circuit to one with only a resistor, as in Figure 2-13A.

---

the voltage across R and C must be the same in the parallel paths. The time constant τ serves as the unit on the time (x) axis and defines how rapidly electrical signals change over time. The voltage reaches 63%, 86%, and 95% of the peak value at times equal to one, two, or three time constants (Figure 2-14, middle; see Box 2-2 for how these values are derived).

The parallel *R-C* circuit is widely used in neurobiology, as it is an excellent description of the neuronal plasma membrane. The ion channels function as resistors, and the lipid bilayer together with the extracellular and intracellular environments act as a capacitor, storing electrical charge in the form of ions accumulating near the surface of the membrane. Whereas the membrane capacitance per unit area is mostly constant ($\sim 1\ \mu F/cm^2$), the membrane resistance can change substantially with membrane potential and time. We will now apply these concepts to examine ion flow across the neuronal plasma membrane.

2.7 Electrical circuit models can be used to analyze ion flow across the glial and neuronal plasma membranes

Having introduced the basics of electrical circuits, we discuss two examples to illustrate how these electrical circuit models can help us understand current flow across the plasma membrane.

Let's first suppose that the membrane is permeable only to K^+, a good approximation for glia. This is equivalent to a circuit consisting of two parallel paths, a K^+ conducting path and a membrane capacitance path (**Figure 2-15A**). The membrane capacitance path is symbolized by a capacitor (C_m). The K^+ conducting path

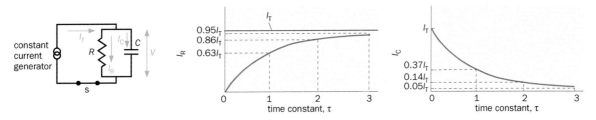

Figure 2-14 Temporal dynamics of a parallel *R-C* circuit. Left, a circuit with a resistor and a capacitor connected in parallel, supplied with a constant current source (with a constant total current I_T). Middle and Right, after the switch (s) is connected, I_R gradually increases to approach its maximal value of I_T (middle), whereas I_C exhibits exponential decay, with a time constant τ equal to the product RC (right). I_R and I_C values are indicated at $t = \tau$, 2τ, and 3τ. V changes similarly to I_R (not shown).

Figure 2-15 **Electrical circuit models of the neuronal plasma membrane. (A)** In a simplified model in which the membrane is permeable only to K$^+$, the plasma membrane consists of two parallel paths, a membrane capacitance (C_m) path and a K$^+$ path. The resistance of the resistor is indicated by the inverse of the conductance, $1/g_K$. The K$^+$ path also has a battery corresponding to the equilibrium potential of K$^+$ (E_K). The battery is positive on the extracellular side because the equilibrium potential of K$^+$ is negative intracellularly. V_m is the membrane potential. At rest, since there is no net current flow, $V_m = E_K$. **(B)** In a more realistic model for neurons, the plasma membrane can be modeled as four parallel paths: one path for membrane capacitance plus one path each for K$^+$, Cl$^-$, and Na$^+$. Arrows indicate the directions of currents within each path according to the equilibrium potential of each ion and the resting potential of our model neuron in Figure 2-12. Note that while Cl$^-$ ions flow inward (as the resting potential is less negative than the equilibrium potential of Cl$^-$ in our model neuron), the current carried by Cl$^-$ is shown flowing outward; this is because electrical circuit diagrams conventionally indicate current as the flow of positive charge. As a result, current carried by negatively charged ion (like Cl$^-$) is indicated as occurring in the direction opposite to the actual direction of ion movement. (B, adapted from Hodgkin AL & Huxley AF [1952] *J Physiol* 117:500–544.)

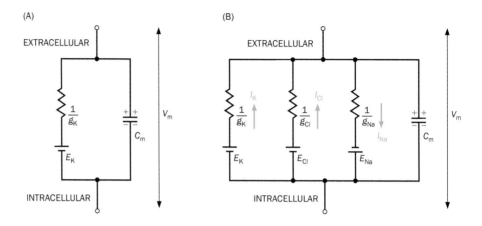

has two electrical elements, a resistor and a battery representing the equilibrium potential of K$^+$. Since resistance is the inverse of conductance, the resistor is symbolized by $1/g_K$, where g_K is the conductance of the K$^+$ path. The battery in series with the membrane potential symbolizes the electrochemical gradient driving K$^+$ movement along this path. We can determine the membrane potential, V_m, from the K$^+$ path as the sum of the voltage across the resistor (I_K/g_K according to Ohm's law) and E_K. At the resting state, the influx and efflux of K$^+$ balances out, or $I_K = 0$. Thus, $V_m = E_K$; that is, the resting membrane potential is the same as the equilibrium potential of K$^+$.

Now let's consider a more realistic situation for neurons, in which the membrane is permeable to Cl$^-$ and Na$^+$, in addition to K$^+$. We can add two parallel paths to Figure 2-15A, one for Cl$^-$ and one for Na$^+$ (Figure 2-15B). In what follows, we demonstrate how V_m can be determined based on the conductance and equilibrium potentials for each ion. In this parallel circuit, the voltage across each path is V_m, so we have three equations:

$$V_m = \frac{I_K}{g_K} + E_K \tag{1}$$

$$V_m = \frac{I_{Cl}}{g_{Cl}} + E_{Cl} \tag{2}$$

$$V_m = \frac{I_{Na}}{g_{Na}} + E_{Na} \tag{3}$$

From the three equations, we have $V_m(g_K + g_{Cl} + g_{Na}) = E_K g_K + E_{Cl} g_{Cl} + E_{Na} g_{Na} + I_K + I_{Cl} + I_{Na}$. At rest, the net current that flows across the membrane should be zero. As the membrane potential is constant, the current flow in the capacitance branch is also zero. Thus,

$$I_K + I_{Cl} + I_{Na} = 0 \tag{4}$$

Accordingly, V_m can be derived as

$$V_m = \frac{E_K g_K + E_{Cl} g_{Cl} + E_{Na} g_{Na}}{g_K + g_{Cl} + g_{Na}}$$

where g and E are the conductance and the equilibrium potential for each ion, respectively. This is in fact the circuit model equivalent of the Goldman–Hodgkin–Katz equation introduced in Section 2.5. This, however, is a more useful formula because conductance and equilibrium potential are easier to determine experimentally than permeability and the absolute ionic concentrations used in the formula in Section 2.5. Note that conductance and permeability are both used to describe how easy it is for an ion to flow across the plasma membrane and are often used interchangeably. But there is a subtle difference. Permeability is an intrinsic property of the membrane reflecting the number of opened channels, as we will learn later, and does not vary whether the ions to be conducted are present or not, whereas conductance depends not only on the permeability but also on the presence of ions.

Once we have determined V_m, we can also determine the currents within each parallel path:

$$I_K = g_K(V_m - E_K)$$
$$I_{Cl} = g_{Cl}(V_m - E_{Cl})$$
$$I_{Na} = g_{Na}(V_m - E_{Na})$$

Note that the values in parentheses represent the driving force for each ion as defined in Section 2.5. Thus, *the current each ion carries is the product of the conductance and the driving force for that ion.* As we will learn later in this chapter, g_{Na} and g_K change as a function of membrane potential, and this voltage dependence underlies the production of action potentials. We will also learn in Chapter 3 that synaptic transmission is mediated by a change in the postsynaptic membrane conductance in response to neurotransmitter release from the presynaptic terminal.

2.8 Passive electrical properties of neurons: electrical signals evolve over time and decay over distance

Having introduced the ionic basis of resting potentials and the electrical circuit model of the neuronal plasma membrane, we are now ready to address two key topics in electrical signaling: how neurons respond to electrical stimulation and how electrical signals propagate within neurons. We start with observations from an idealized experiment on a neuronal fiber (a dendrite or an axon), which summarizes results from real experiments across different preparations and approximates the properties of neuronal dendrites. In this experiment, an electrode connected to a current source is inserted into the neuronal fiber, such that it can "inject" electric current into the neuronal fiber at the command of the experimenter. We call this electrode a **stimulating electrode**. Inserted into the membrane right next to the stimulating electrode is a **recording electrode** (*a*), which is connected to an amplifier and oscilloscope so it can record the membrane potential change in response to the current injection from the stimulating electrode. Two additional recording electrodes (*b* and *c*) are inserted at different distances along the fiber from the stimulating electrode to record membrane potentials at distant sites (**Figure 2-16**A; Movie 2-6).

We start by injecting a small depolarizing current (that is, injecting positive charge into the neuron) in the form of a rectangular pulse (Figure 2-16B, left). The injected current will flow through the electrode across the membrane and along the inside of the fiber. We observe that the membrane potential at recording electrode *a* becomes *gradually* depolarized at the beginning of the current pulse (t_1) and returns *gradually* to the resting potential at the end of the current pulse (t_2). At the more distant sites, membrane potentials recorded by electrodes *b* and *c* exhibit similar gradual changes. However, the magnitudes of the membrane potential changes are much diminished (Figure 2-16B, middle and right).

We use an electrical circuit model of the fiber (Figure 2-16C) to explain these results. Each segment of the membrane can be modeled as a parallel *R-C* circuit (Figure 2-15B), with membrane conductances of all ions in Figure 2-15 combined as R_m, the resting membrane potential as a battery E_r, and the membrane capacitance as C_m. Different segments of the process can be modeled as parallel *R-C* circuits, linked by internal (or axial) resistances (R_i) of ion movement along the longitudinal axis in the fiber interior. Ions flow more freely in the large extracellular environment; therefore, the extracellular resistance is often approximated as zero.

Let's consider our first observation: membrane potentials change gradually in response to a step current pulse, already evident from the recording at the current injection site. In addition to causing ion flow across the membrane, represented by I_R in the circuit diagram, part of the injected current flows across the capacitance path to charge the membrane capacitor (I_C). As discussed in Section 2.6,

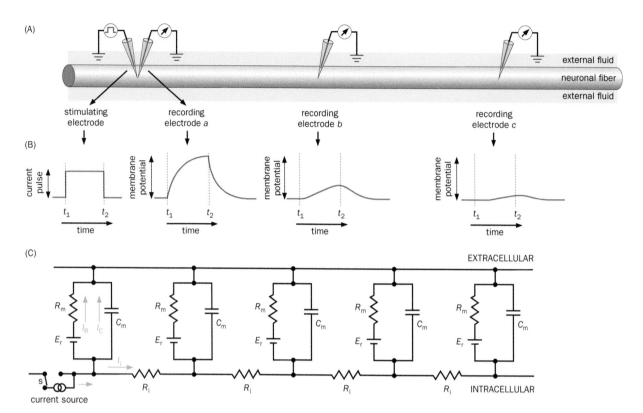

Figure 2-16 Passive electrical properties of neurons observed in an idealized experiment. (A) Illustration of the experimental preparation. A stimulating electrode provides a source of electrical signal in the form of a rectangular current pulse. Three recording electrodes are inserted into the neuronal fiber at different distances from the stimulating electrode. **(B)** When the stimulating electrode delivers a rectangular depolarizing pulse (left), the membrane potential changes at the sites of the three recording electrodes as illustrated (right). Dashed lines represent times (t_1 and t_2) aligned to the onset and offset of the current pulse. The y axes of the three membrane potential plots have the same scale. Two properties are evident: (1) membrane potential changes are gradual in response to the square current pulse, and (2) the magnitudes and speeds of the membrane potential changes decay across distance. **(C)** An electrical circuit model of the neuronal fiber. Each membrane segment is approximated as a parallel R-C circuit, with a membrane capacitance (C_m); a membrane resistance (R_m) that integrates K^+, Cl^-, and Na^+ conductances; and a battery representing the resting membrane potential (E_r). These segments are joined internally by resistors (with resistance R_i), reflecting internal or axial resistance within the neuronal fiber; the resistance of the external fluid is approximated as 0. Current injection is modeled by transiently connecting the intracellular side with a constant current source through the switch (s). Arrows indicate the direction of current flow caused by injecting positive charges from the current source, including current into the resistor branch (I_R) and capacitance branch (I_C) across the membrane, as well as current that flows within the neuronal fiber (I_i). (Adapted from Katz B [1966] Nerve, Muscle, and Synapse. McGraw Hill.)

I_R follows an exponential curve with the product $R_m C_m$ as the time constant (Figure 2-14), as does the change in membrane potential. The smaller the time constant, the faster the membrane potential changes in response to current injection. This experiment reveals a general property of electrical signaling in neurons. Because of the membrane capacitance, electrical signals evolve over time even when current injection is constant, with the product $R_m C_m$ as a key parameter of these temporal dynamics. This property limits the temporal resolution of electrical signals but also provides opportunities for temporal integration: when two individual signals are delivered within a small time interval, they may not be resolved as individual signals, but rather detected as an integrated signal. We will discuss this further in Chapter 3 in the context of dendritic integration of synaptic inputs.

Let's now turn to our second observation: the magnitude of the change in membrane potential decreases as the distance from the site of current injection increases. Suppose the greatest change in membrane potential at electrode *a* in response to the current injection is V_0. The spread of this change in membrane potential along the fiber is carried by the axial current, which diminishes due to the continual leak of current via membrane conductance along the way. An easy way to visualize this is to simplify the circuit in Figure 2-16C further by considering only the conductance path (**Figure 2-17A**). (This simplification is equivalent

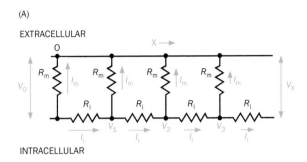

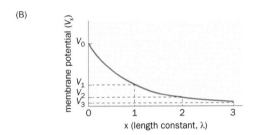

(A) EXTRACELLULAR

(B) membrane potential (V_x) — x (length constant, λ)

Figure 2-17 **A circuit model illustrating electrical signal decay along a neuronal fiber. (A)** In this simplified model of Figure 2-16C, batteries symbolizing the resting potential are omitted because only the *change* in the membrane potential from the resting potential is considered here; capacitors are also omitted because only the *peak changes* in membrane potential are considered. The current injection at position 0 causes a peak membrane potential change of V_0. In addition to passing through the membrane (I_m), part of the injected current also spreads in both directions along the interior of the neuronal fiber as an internal current (I_i) (only the rightward spread is shown here). Along the way, the magnitude of I_i diminishes because of leaky transmembrane current (I_m), symbolized by the decreasing lengths of the arrows. **(B)** The changes in membrane potential decay across the fiber, following an exponential decay curve. The x axis unit is the length constant of the neuronal fiber. The values for membrane potentials at 1×, 2×, and 3× the length constant equal 0.37 V_0, 0.14 V_0, and 0.05 V_0, respectively. (Adapted from Katz B [1966] Nerve, Muscle, and Synapse. McGraw Hill.)

to considering the peak magnitude of changes in membrane potential after the membrane capacitance is charged.) The axial current (I_i) gradually diminishes in magnitude because part of the current leaks away to the outside due to membrane conductance. The membrane potential change $V(x)$ at distance x from electrode a is given by the following formula:

$$V(x) = V_0 e^{-\dfrac{x}{\sqrt{dR_m/4R_i}}}$$

where R_m is the membrane resistance per unit area of membrane surface, R_i is the internal (axial) resistance per unit volume of the neuronal cytoplasm, and d is the diameter of the fiber. This equation represents an exponential decay of electrical signal across distance (Figure 2-17B), in analogy to the exponential decay of current over time in an R-C circuit we discussed in Section 2.6 and Box 2-2.

The term $\sqrt{dR_m/4R_i}$ is called the **length constant** or **space constant** (designated as λ). It is expressed in units of length; one length constant corresponds to the distance at which the peak magnitude of the membrane potential change has attenuated to $1/e$, or about 37%, of the original peak magnitude. As specific examples, the positions of the electrodes b and c in our idealized experiment (Figure 2-16A) were chosen to be 1.5 and 3 length constants away from electrode a. The longer the length constant, the further electrical signals can be transmitted before they decay to a given fraction of their original value. As indicated by the formula, the length constant increases with increasing membrane resistance or neuronal fiber diameter. This is because if R_m increases, then more current will flow down the axial path (as current is inversely proportional to resistance) and λ will increase; similarly if R_i is reduced by increasing fiber diameter (essentially adding more resistance in parallel, which decreases the total resistance), then more current will flow down R_i, thus increasing λ. Indeed, animals have evolved various strategies to increase the distance across which electrical signals spread, such as enlarging fiber diameter, as in the squid giant axon, or increasing unitary membrane resistance, as in myelination; we will discuss these strategies in more detail in Section 2.13.

The temporal spread of electrical signals and their attenuation across distance are often referred to as **passive electrical properties** of neurons (as opposed to active properties, which we will begin to study in the next section). They are also called the **cable properties** of neuronal fibers, in analogy to the transocean cables that transmit electrical signals using insulators to separate the interior conductor from the exterior conducting seawater. The time constant, which characterizes the

Table 2-1:   Time and length constants of axons, dendrites, and muscle cells

Fiber	Diameter (μm)	Length constant (mm)	Time constant (ms)
Squid giant axon[a]	500	5	0.7
Lobster nerve[a]	75	2.5	2
Frog muscle[a]	75	2	24
Apical dendrite of mammalian cortical pyramidal neuron[b]	3	1	~20

[a] Data from Katz B (1966) Nerve, Muscle, and Synapse. McGraw-Hill. Length constants were measured in large extracellular volume.

[b] Data from Stuart G, Spruston N, & Häusser M (1999) Dendrites. Oxford University Press.

temporal spread of electrical signals, and length constant, which characterizes the attenuation across distance, are the two key passive electrical properties of neurons. Table 2-1 lists experimentally determined time and length constants of neurons and muscles in various experimental preparations that have played important roles in the history of neurophysiology. We will learn about these specific preparations in later sections and chapters.

2.9   Active electrical properties of neurons: depolarization above a threshold produces action potentials

As is evident from Table 2-1, electrical signals would decay considerably across distance if neurons had only passive properties. Even in fibers with very large diameters, length constants are at most a few millimeters. This means that signals would decay to 37% of their original magnitude across just a few millimeters. How do electrical signals propagate faithfully over much greater distances? To state the problem concretely, how do signals propagate reliably through human motor neurons across distances of approximately a meter in order to control muscles in the toe?

To answer this question, let's continue our idealized experiment using the setup in Figure 2-16A. Through the stimulating electrode, we inject step current pulses with varying magnitudes and directions into a neuronal fiber, this time an axon (**Figure 2-18**, top). We begin by injecting negative current, which hyperpolarizes the membrane potential. The magnitude of changes in membrane potential, as measured by electrode *a* (Figure 2-18, bottom), is proportional to the magnitude of injected negative current. If we reverse the sign and inject positive current into the axon, we see that the membrane potential is depolarized rather than hyperpolarized. The magnitude of depolarization is also proportional to the magnitude of injected positive current, provided that only a small amount of positive current is injected. However, when the injected current exceeds a certain magnitude, a much larger and transient elevation of the membrane potential is

Figure 2-18 Depolarization exceeding a threshold results in action potentials. Using the experimental preparation diagrammed in Figure 2-16A, a series of rectangular current pulses were applied through the stimulating electrode (top). The corresponding changes in membrane potential recorded by electrode *a* are shown at the bottom. The unit of the *x* axis is 1 ms. For both hyperpolarization pulses and the first two depolarization pulses (current pulses 1–4), the membrane potential changes follow the sign of the current pulses, and their magnitudes are proportional to the magnitudes of the current pulses. In response to the fifth current pulse, the membrane potential change becomes unstable and varies across different trials (as illustrated by multiple curves). Occasionally the stimulation results in a very large depolarization—the action potential. Action potentials of the same magnitude are always produced in response to the sixth current pulse. (Adapted from Katz B [1966] Nerve, Muscle, and Synapse. McGraw Hill.)

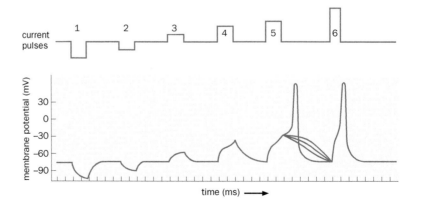

produced in some fraction of trials. Above that magnitude of current injection, each current pulse invariably produces a large and transient elevation of the membrane potential. This is called an **action potential** or a **spike**. It is caused by depolarization of the membrane potential above a specific level, the **threshold**, in response to an injection of positive current of a certain magnitude (Movie 2-6). A stimulus that can cause the neuron to generate an action potential is called a **suprathreshold stimulus**, and a stimulus that cannot is called a **subthreshold stimulus**.

Note that the size of the action potential does not change with the magnitude of depolarization once the threshold is reached. Furthermore, if an action potential is recorded by electrode *a*, an action potential of similar magnitude and waveform will also be recorded by electrodes *b* and *c* in Figure 2-16A. In other words, action potentials propagate with little or no decay. As opposed to the passive spread of electrical signals discussed in Section 2.8, action potentials are an **active electrical property** of neurons. As we will learn later in this chapter, active electrical properties of neurons are a result of voltage-dependent changes in ion conductance.

Not all neurons fire action potentials: some neurons use only graded potentials to transmit electrical signals even in their axons. From the earlier discussion of length constants, these must necessarily be neurons with short axons; we will see examples of such neurons in the vertebrate retina in Chapter 4. It should also be noted that active electrical properties are not exclusive to axons: some neurons exhibit active properties in dendrites as well; we will discuss these properties in Chapter 3.

How are action potentials produced? How do they propagate along the axon? We devote the next part of this chapter to addressing these fundamental questions in neuronal signaling.

HOW DO ELECTRICAL SIGNALS PROPAGATE FROM THE NEURONAL CELL BODY TO ITS AXON TERMINALS?

In this part of the chapter, we follow the discovery path that has led to our current understanding of mechanisms by which action potentials are produced and propagate. In addition to answering a key question in neuronal communication—how electrical signals propagate from the neuronal cell body to its axon terminals— these studies also established the concept of ion channels and highlighted the mechanisms by which ion channels function.

2.10 Action potentials are initiated by depolarization-induced inward flow of Na⁺

The discovery of the ionic basis of the action potential is an excellent example of how scientific breakthroughs can result from the introduction of new methods, model organisms, and analytic tools. Squid of the genus *Loligo* have a giant axon whose diameter reaches up to 1 mm, many times larger than nearby axons (**Figure 2-19A**) or the axons of typical mammalian neurons. The giant axon conducts action potentials very rapidly and controls the squid's jet propulsion system, allowing it to quickly escape danger. The giant axon's large diameter also enabled researchers to insert electrodes and measure action potentials more accurately than before (Figure 2-19B). During such measurements, it was discovered that the membrane potential during the rising phase of the action potential far exceeded zero (Figure 2-19C), indicating that the action potential is not caused by a transient breakdown of the membrane that allows the membrane potential to become zero, a prevalent view held before these measurements.

At the peak of the action potential, the membrane potential was observed to approach the Na⁺ equilibrium potential. (Recall that E_{Na} = +58 mV in our model neuron in Figure 2-12.) This finding suggested that the membrane is preferentially

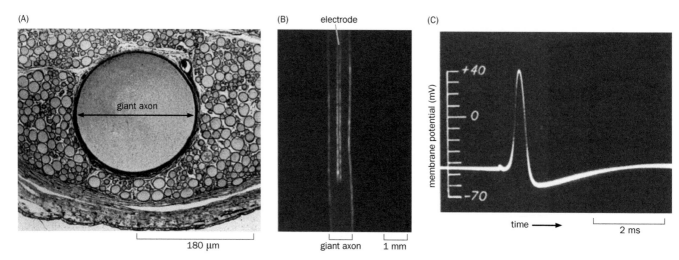

Figure 2-19 Studying action potentials using the squid giant axon. **(A)** Electron micrograph of a cross section of a squid giant axon showcasing its large diameter (~180 μm for this sample), in contrast to neighboring axons (for example, the axon indicated by *). **(B)** Photograph of an electrode inserted into a squid giant axon with a diameter close to 1 mm. **(C)** An action potential recorded from the squid giant axon by an intracellular electrode. (A, courtesy of Kay Cooper and Roger Hanlon; B, from Hodgkin AL & Keyes RD [1956] *J Physiol* 131:592–616. C, from Hodgkin AL & Huxley AF [1939] *Nature* 144:710–711. With permission from Springer Nature.)

permeable to Na^+ at the peak of the action potential and that the inward Na^+ flow is responsible for the rising phase of the action potential. To test this hypothesis, the extracellular Na^+ concentration was systematically reduced. If Na^+ were responsible for the rising phase of the action potential, one would predict from the Nernst equation that the magnitude of the action potential would decrease with lower concentrations of extracellular Na^+. This was indeed the case (**Figure 2-20**).

But how does the membrane become permeable to Na^+? An important conceptual breakthrough was the realization that depolarization could induce an increase in membrane permeability to Na^+, with the influx of Na^+ resulting in further depolarization. Such a self-reinforcing process (positive feedback loop) could account for the rapid change in membrane potential observed during the rising phase of the action potential.

2.11 Sequential, voltage-dependent changes in Na^+ and K^+ conductances account for action potentials

To test whether depolarization could render axonal membranes more permeable to Na^+, it was important to quantitatively measure ion flow across the membrane under conditions mimicking the action potential. However, ion flow across the membrane changes the membrane potential, which in turn can affect the permeability of ions, thus complicating the measurement of ion flow. A new method called the **voltage clamp** was introduced to simplify the measurements of ion flow in response to voltage changes (**Figure 2-21**). The voltage clamp compares the intracellular membrane potential with a command voltage set by the experimenter. Differences between the two voltages automatically produce a feedback current that is injected back into the cell, which rapidly changes the intracellular membrane potential to the value of the command voltage. After the initial stimulation, which is usually in the form of a step change in the command voltage, ion flow across the membrane as a consequence of the membrane potential change can be measured by recording how much current must be injected into the cell in order to maintain the membrane potential at a specified value.

Using the voltage clamp technique, Alan Hodgkin and Andrew Huxley carried out a series of classic experiments around 1950 to determine the ionic basis of the action potential. By subjecting the squid giant axon to depolarizing voltages, they were able to dissect the composition of ionic flows that underlie an action potential. Importantly, holding the membrane potential at a constant value eliminated the capacitive current (current that charges the membrane in response to

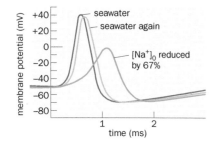

Figure 2-20 Testing the hypothesis that the rising phase of the action potential is caused by Na^+ influx. The magnitude and speed of the action potential are diminished when the normal extracellular solution (seawater, red trace) was replaced by a solution of 33% seawater and 67% isotonic dextrose (hence the extracellular Na^+ concentration, or $[Na^+]_o$, was reduced by 67%; blue trace). Reapplication of seawater (green trace) restored the magnitude and speed of the action potential. (Adapted from Hodgkin AL & Katz B [1949] *J Physiol* 108:37–77.)

voltage change, equivalent to I_c in Figure 2-16C) so that they could measure ionic currents across the membrane (equivalent to I_R in Figure 2-16C) and observe how they changed over time. For example, a 56 mV depolarizing step produced an initial inward current, followed by an outward current (**Figure 2-22**, green trace). (According to electrophysiology convention, inward current is net flow of cations into the cell or anions out of the cell; vice versa for outward current.) The inward current was abolished when extracellular Na^+ was replaced by choline, an organic ion that carries a +1 charge similar to Na^+ but is unable to permeate the membrane (Figure 2-22, blue trace). This finding indicated that the initial inward current is indeed caused by Na^+ influx. Thus, the Na^+ current could be calculated by comparing the difference between the two conditions (Figure 2-22, red trace). Other evidence suggested that in the case of choline replacement, the remaining current is caused by outward K^+ flux.

Voltage clamps also allowed systematic measurement of Na^+ and K^+ conductances over a series of different voltages. As introduced in Section 2.7, conductance is the ratio of the current that passes through the membrane and the driving force, which is the difference between the membrane potential and the equilibrium potential. Thus, the conductances for Na^+ and K^+ are given by

$$g_{Na} = \frac{I_{Na}}{V_m - E_{Na}}$$

$$g_K = \frac{I_K}{V_m - E_K}$$

where I_{Na} and I_K are the Na^+ and K^+ currents as measured in Figure 2-22, V_m is the membrane potential, and E_{Na} and E_K are the equilibrium potentials of Na^+ and K^+. By changing V_m (which equals V_{CMD}) and measuring the currents in the voltage clamp experiments, Hodgkin and Huxley could experimentally determine g_{Na} and g_K at different membrane potentials. They found that both Na^+ and K^+ conductances increased when the intracellular membrane potential became more depolarized and that these conductance changes evolve over time (**Figure 2-23**).

Hodgkin and Huxley made several key discoveries from these experiments. First, they confirmed that the rising phase of the action potential results from an influx of Na^+ and determined that the Na^+ influx is caused by a rapid increase in Na^+ conductance as a consequence of membrane depolarization. Second, after the initial depolarization-induced increase, Na^+ conductance would invariably decrease despite continued depolarization (Figure 2-23A), accounting for the falling phase of the Na^+ current (Figure 2-22, red trace). This was termed **inactivation** of the Na^+ conductance. Third, depolarization also caused an increase in K^+ conductance, resulting in K^+ efflux. Importantly, the change in K^+ conductance lagged behind the change in Na^+ conductance (Figure 2-23B). Fourth, the Na^+ and K^+ conductances appeared to be independent of each other, but both depended on the membrane potential. Well before the molecular mechanisms of membrane transport became known, these findings paved the way for the modern concept of ion channels as transmembrane proteins selectively permeable to specific ions (Section 2.4). In particular, in the squid giant axon, channels selectively permeable

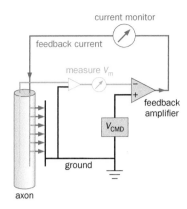

Figure 2-21 Illustration of the voltage clamp technique. The membrane potential (V_m) is measured by the blue wire inserted into the squid giant axon with respect to the ground wire outside the axon. It is then compared to a command voltage set by the experimenter (V_{CMD}) as two different inputs to the voltage clamp feedback amplifier (large triangle). The difference between the two voltages ($V_{CMD} - V_m$) produces a feedback current as the output of the amplifier, which is injected by a second inserted wire (red) into the axon. When $V_{CMD} = V_m$, there is no feedback current. Upon a step change of V_{CMD}, the feedback current rapidly changes V_m to the new V_{CMD} (within microseconds). Thus, the voltage clamp enables experimenters to control V_m of the axon being studied and, at the same time, to measure the amount of feedback current needed to hold V_m to the value of V_{CMD}; this quantity of feedback current equals the current flowing across the axon membrane (parallel red arrows). The feedback current can flow in either direction (that is, the red arrows can reverse) depending on the relative values of V_m and V_{CMD}.

Figure 2-22 Dissociation of Na⁺ and K⁺ currents via voltage clamp experiments. Top, a voltage step increase of 56 mV was applied to the squid giant axon. Middle, ion flow across a unit area of the axonal membrane in response to the depolarizing voltage step was measured by determining how much current was injected into the axon in order to maintain the axon's membrane potential at the command voltage established by the experimenter (the +56 mV step). The green trace shows the current flow under physiological conditions; an initial inward current (the downward portion of the trace) is followed by an outward current (the upward portion). The dashed line demarcates zero net current. The blue trace shows the current flow under conditions in which the external Na⁺ was mostly replaced by choline⁺, which cannot cross the membrane; this trace illustrates the K⁺ current only. Bottom, the red trace represents the deduced Na⁺ current, which is the difference between the green and blue traces. (Adapted from Hodgkin AL & Huxley AF [1952] *J Physiol* 116:449–472.)

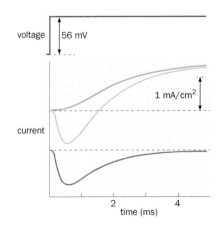

Figure 2-23 Voltage-dependent changes in Na⁺ and K⁺ conductances. Na⁺ **(A)** and K⁺ **(B)** conductances (*y* axis, in millisiemens per square centimeter) change over time following a depolarizing voltage step from the resting potential. A series of measurements were performed, each after a voltage step of a different magnitude (indicated on each trace). Over time, the Na⁺ conductance first increases then decreases, whereas the K⁺ conductance only increases, but more slowly than the initial rise of the Na⁺ conductance. Larger voltage steps cause more rapid rises in both conductances. (Adapted from Hodgkin AL & Huxley AF [1952] *J Physiol* 116:449–472.)

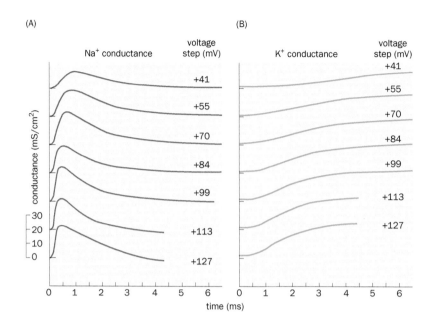

to Na⁺ or K⁺ allow these ions to flow through the membrane. The conductance of these ion channels increases when the axon is depolarized. These channels are now called **voltage-gated ion channels** because their conductances change as a function of the membrane potential.

In summary, the action potential can be accounted for by sequential changes in the Na⁺ and K⁺ conductances (**Figure 2-24**; **Movie 2-7**), which we now know are caused by the opening and closing of voltage-gated Na⁺ and K⁺ channels. At the resting state, voltage-gated Na⁺ and K⁺ channels are both closed. (A different set of K⁺ channels accounts for K⁺ permeability at rest.) During the rising phase of the action potential, when the membrane is depolarized, opening of voltage-gated Na⁺ channels allows Na⁺ to flow into the cell down its electrochemical gradient. Depolarization also causes an increase in K⁺ efflux (via resting K⁺ channels) because the force produced by the new, smaller electrical gradient is less effective in countering the force produced by the chemical gradient. When the Na⁺ influx exceeds the K⁺ efflux, the neuron passes the threshold for firing an action potential. (For most excitable cells the threshold is 10–20 mV above the resting potential.) More depolarization causes opening of more voltage-gated Na⁺ channels, which causes further depolarization. This positive feedback loop generates the rapid rising phase of the action potential.

During the falling phase, Na⁺ channels are inactivated after the initial opening, preventing further Na⁺ influx. At the same time, voltage-gated K⁺ channels open, allowing more K⁺ efflux. These two events together account for the falling phase of the action potential, allowing neurons to repolarize to the resting potential and prepare for the next action potential (Figure 2-24). Indeed, based on the

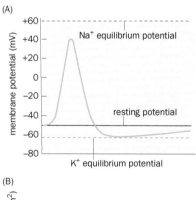
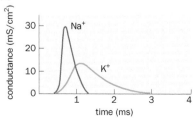

Figure 2-24 A summary of the ionic basis of the action potential. (A) Schematic of an action potential, with reference to the resting potential and the equilibrium potentials of Na⁺ and K⁺. (The K⁺ equilibrium potential and the resting potential of the squid giant axon are more depolarized than for our model neuron in Figure 2-12.) **(B)** Sequential changes in Na⁺ and K⁺ conductance (calculated according to data in Figure 2-23) during the action potential. Both graphs share the same *x* axis. The rising phase is caused by an increase in Na⁺ conductance, leading to Na⁺ influx. The falling phase is accounted for by both the inactivation of the Na⁺ conductance, which stops the Na⁺ influx, and an increase in the K⁺ conductance, leading to K⁺ efflux. The transition between the rising and falling phase occurs before the rising phase reaches the Na⁺ equilibrium potential. The falling phase overshoots the resting potential and approaches the K⁺ equilibrium potential, before the membrane potential gradually returns to the resting potential, which is slightly above the K⁺ equilibrium potential (Section 2.5). (Adapted from Hodgkin AL & Huxley AF [1952] *J Physiol* 117:500–544.)

measurements in voltage clamp experiments, Hodgkin and Huxley established a model that *quantitatively recapitulates* the dynamics of action potentials and many of their properties (see Section 14.30 for details). Importantly, the ionic basis of the action potential, originally discovered in the squid giant axon, applies to neurons and other excitable cells across most of the animal kingdom, including humans.

2.12 Action potentials are all or none, are regenerative, and propagate unidirectionally along the axon

The Hodgkin–Huxley model satisfactorily explains several properties of the action potential that ensure faithful transmission of information from the cell body to axon terminals.

First, action potentials are **all or none**. When a stimulus-induced neuronal membrane depolarization is below the threshold, the action potential does not occur. When depolarization exceeds the threshold, the waveform of the action potential is determined by the timing of Na^+ and K^+ conductance changes and the relative concentrations of Na^+ and K^+ inside and outside the cell, which remain mostly constant for any given neuron. (The Na^+ influx and K^+ efflux during an action potential cause very small changes in intracellular, and even smaller changes in extracellular, Na^+ and K^+ concentrations.) To a first approximation, action potentials assume the same form in response to any suprathreshold stimulus.

Second, action potentials are **regenerative**—they propagate without attenuation in amplitude. Suppose that an action potential occurs at a particular site on the axon. The rising phase creates a substantial membrane depolarization, which spreads down the axon and brings an adjacent region to threshold, which in turn does so for its adjacent downstream region, and so on (**Figure 2-25**). In this way, the action potential propagates in a similar form continuously and faithfully down the axon toward its terminals.

Third, action potentials propagate *unidirectionally* in the axon, from the cell body to the axon terminals. When an action potential occurs at a given site on the axon (for example, site A in Figure 2-25), in principle depolarization should also spread up the axon toward the cell body (site Z) in addition to spreading down the axon toward the axon terminals (site B). However, the delayed activation of the K^+ channels and the inactivation of the Na^+ channels combine to create a **refractory period** after an action potential has just occurred, during which time another action potential cannot be reinitiated. Because the action potential normally initiates at the axon initial segment and passes through Z before reaching A, another action potential cannot immediately back-propagate from A to Z. This refractory period ensures that the action potential normally propagates only from the cell body down the axon to its terminals, not in the reverse direction.

In most projection neurons, whose axons form synapses on distant target neurons, the action potential first arises at the axon initial segment, where voltage-gated Na^+ channel density per unit membrane area is the highest; this high channel density lowers the threshold for action potential initiation. The axon initial segment is a critical site for the integration of depolarizing and hyperpolarizing

Figure 2-25 Action potential propagation. This schematic of an action potential as it sweeps across an axon provides a snapshot of electrical signaling events in the axon. The wave front is at site A, where voltage-gated Na^+ channels open and depolarize the membrane. Positive charges at site A within the axon spread to site B, where they will cause depolarization above the threshold. Thus, at the next moment, the wave front will reach site B. Site Z, where the wave front has just passed, is experiencing a refractory period during which the delayed activation of K^+ channels and the inactivation of Na^+ channels prevent the action potential from propagating backward from A to Z. Red and blue arrows indicate Na^+ influx and K^+ efflux, respectively. Curved arrows represent current flow completing the left and right circuits as a result of Na^+ influx. The charges below represent the membrane potentials at different segments of the axon.

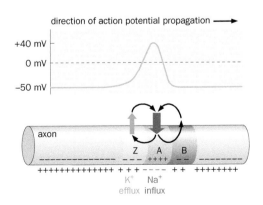

synaptic potentials from the dendrites and the cell body; this integrative process is discussed further in Section 3.24. After initiation, action potentials travel unidirectionally along the axon toward its terminals. At the initiation site, however, action potentials can in principle travel in both directions; indeed, in some mammalian neurons, action potentials can back-propagate to dendrites, which, like axons, contain voltage-gated Na^+ and K^+ channels. In artificial situations where experimenters electrically stimulate the axon or its terminals, action potentials can propagate in a retrograde direction from axon terminals to the cell body, producing so-called **antidromic spikes**, which can be recorded from the cell body. However, antidromic spikes have not been found to occur under physiological conditions *in vivo*.

Altogether, these three properties make the action potential an ideal means to transmit information faithfully from neuronal cell bodies across long distances to their axon terminals. But since action potentials are all or none, the size of action potentials cannot encode information about the stimulus. Rather, the information is usually encoded by the rate (number of action potentials per unit time) or the timing of action potentials in response to a stimulus (Section 1.8). The spike rate is limited by the refractory period. Some neurons, such as fast-spiking inhibitory neurons in the mammalian cortex, can fire up to 1000 Hz, or one action potential per millisecond; the interval between these action potentials is shorter than the refractory period of many neurons. This requires specializing the ion channels so the action potential is repolarized quickly and the refractory period is complete in time for the next action potential. Thus, ion channel properties (such as Na^+ channel inactivation and the delayed opening of K^+ channels) have been selected during evolution to ensure unidirectional propagation of action potentials, and neurons with high spike rates use specialized ion channels with fast kinetics. The broad range of possible spike rates expands the information-coding capacity of individual neurons.

2.13 Action potentials propagate more rapidly in axons with larger diameters and in myelinated axons

The speed at which action potentials propagate is not the same for all neuronal types but instead depends on axonal properties. This is because the speed at which depolarization spreads down the axon is determined by the cable properties of the axon. Returning to our circuit model of neuronal membranes (Figures 2-16C and 2-17A), we see that this speed is determined by how quickly a change in membrane potential charges the membrane capacitor as well as the relative distribution of current flowing forward versus leaking out into the extracellular environment. If all other properties are equal, these values would be a function of axon diameter: the larger the axon diameter, the lower the axial resistance of the axon and the larger the proportion of current flowing forward. You can deduce this from the equation for the length constant (λ), which is equal to $\sqrt{dR_m/4R_i}$ (Section 2.8). Hence, the larger the diameter (d), the larger the length constant and the further depolarization can spread at a suprathreshold value to produce the next action potential at more distant sites. This is why the squid and some other animals have evolved giant axons with very large diameters, which implement rapid escape behaviors.

In principle, increasing R_m (the membrane resistance per unit area) can also increase the length constant. However, increasing R_m also increases the time constant (which is equal to $R_m C_m$; Section 2.8) required to charge the membrane along the way, which slows down the propagation of action potentials. One way to compensate for this is to also reduce C_m, the membrane capacitance. Indeed, this is how **axon myelination** works. Many vertebrate axons are wrapped in a **myelin sheath**, formed by layers of cytoplasmic extensions of glia—Schwann cells in the PNS and oligodendrocytes in the CNS. Some axons in large invertebrates are also myelinated (Section 13.7). The cytoplasmic extensions of glial cells wrap around the axon many times, with most of the cytoplasm compressed out of the extensions toward the soma to form **compact myelin** consisting of closely packed glial plasma

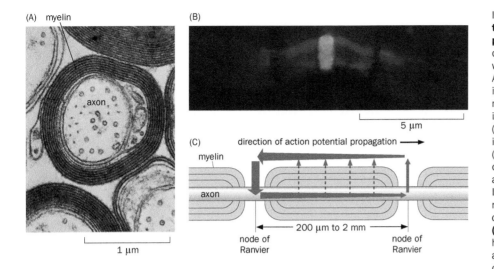

Figure 2-26 Axon myelination increases the speed of action potential propagation. (A) An electron micrograph of a cross section of spinal cord axons wrapped by oligodendrocyte membranes. At the center is a single axon wrapped in myelin sheath. **(B)** A fluorescence microscopic image of the rat optic nerve immunostained to visualize three proteins (see Section 14.12 for more details of the immunostaining method). Na^+ channels (green) are highly clustered at the center of the node of Ranvier. K^+ channels (blue) are distributed peripherally at the node. In between are transmembrane proteins named Caspr (red) that help organize channel distribution at the node. **(C)** Schematic of an action potential hopping between nodes of Ranvier. After an action potential occurs at the left node of Ranvier, positive charges rapidly flow to the next node to the right. This is because, as a consequence of myelin wrapping, the internodal membranes have low capacitance, which requires less charging, and high resistance, which allows only a small amount of current to leak through (dashed arrows). The arrival of positive charges at the right node causes rapid depolarization above threshold to regenerate the action potential there. Red arrows indicate the direction of current flow that completes the circuit as a result of Na^+ influx at the left node. For simplicity, the circuit resulting from depolarization spreading leftward from the left node is omitted. (A, courtesy of Cedric Raine. B, adapted from Rasband MN & Shrager P [2000] *J Physiol* 525:63–73.)

membranes (Figure 2-26A). From an electrical circuit perspective, compact myelin is equivalent to having many resistors connected in series, such that the total membrane resistance R_T is equivalent to nR_m, where n is the number of layers of glial membrane. However, in serial connections, the total capacitance C_T is equivalent to C_m/n (recall from Section 2.6 that the total capacitance of two capacitors connecting in series follows $1/C_T = 1/C_1 + 1/C_2$; the combined capacitance of n identical capacitors connected in series follows $1/C_T = n/C$, or $C_T = C/n$). Thus, myelination greatly increases membrane resistance, and hence the length constant, without increasing the time constant (as the product $R_T C_T$ remains unchanged compared to the original $R_m C_m$); this means that once an action potential is produced at a specific site on the axon, depolarization spreads across a large distance to cause distant regeneration of the action potential.

Despite the increased membrane resistance, small amounts of current still leak out of myelinated axons. Therefore, it is important to have **nodes of Ranvier**, occurring at regular intervals (usually 200 μm to 2 mm apart), where the axon surfaces are exposed to the extracellular ionic environment with highly concentrated voltage-gated Na^+ and K^+ channels (Figure 2-26B). As a result of these channel activities, depolarizing current spreads rapidly in between the nodes, and action potentials are "renewed" only at the nodes of Ranvier (Figure 2-26C). Thus, action potentials in a myelinated axon hop from node to node. This is termed **saltatory conduction** (from the Latin *saltare*, "to jump"), as opposed to continuous propagation in unmyelinated axons (Figure 2-25).

Myelination greatly increases the conduction speed of action potentials and the capacity for high-frequency firing. Action potentials can travel at speeds of up to 120 m/s in myelinated axons, as compared to <2 m/s in unmyelinated axons. Although unmyelinated axons usually have smaller diameters than myelinated axons (Box 2-3), the diameter difference alone does not account for the large difference in propagation speeds. Saltatory conduction also saves energy; there is a reduced demand for the Na^+–K^+ ATPase to pump Na^+ outward and K^+ inward, as there is little transmembrane current except at the nodes of Ranvier. Because myelin is so important for proper conduction in the axons of vertebrates, including humans, improper myelination is responsible for several major neurological disorders, including multiple sclerosis and Charcot–Marie–Tooth disease (Box 2-3).

2.14 Patch clamp recording enables the study of current flow through individual ion channels

Studies of the action potential in the squid giant axon suggested the existence of dedicated ion channels for Na^+ and K^+. This idea was later supported by the characterization of toxins that specifically block Na^+ or K^+ channels. The most famous

Box 2-3: Axon-glia interactions in health and disease

Axon myelination provides a striking example of the intimate interactions between glia and neurons. In the white matter of the central nervous system, each oligodendrocyte typically extends several processes that myelinate multiple axons (Figure 1-9). In the peripheral nervous system, each Schwann cell is usually dedicated to wrapping a segment of a single axon. During development and remyelination in adults, oligodendrocyte and Schwann cell extensions wrap the axon many times like a spiral and compress the cytoplasm in between layers of the extensions to form the myelin sheath (**Figure 2-27**A).

Axons and glia interact in diverse ways. For example, as will be discussed in Chapter 6, the somatosensory system contains distinct types of sensory neurons with characteristic axon diameter, degree of myelination, and action potential conduction speed. The thickness of myelin matches the size of the axon. Sensory neurons that innervate muscle and provide rapid feedback regulation of movement have large-diameter axons, thicker myelin sheaths, and conduct action potentials more rapidly. Sensory neurons that sense touch have intermediate axon diameters and conduction speeds. Many sensory neurons that sense temperature and pain are unmyelinated and conduct action potentials more slowly. These unmyelinated axons are nevertheless associated with **Remak Schwann cells**, whose cytoplasm extends in between individual axons to form a Remak bundle (Figure 2-27B). Here the glia's role is simply to segregate individual axons rather than to support saltatory conduction.

What determines whether an axon should be myelinated or not, and if so, to what degree? These questions have been answered in the PNS. An axonal cell-surface protein called

type III neuregulin-1 (**Nrg1-III**) plays a key role: axons expressing high levels of Nrg1-III are associated with thick myelin sheaths, axons expressing intermediate levels of Nrg1-III are thinly myelinated, and axons expressing low levels of Nrg1-III are associated with a Remak bundle (Figure 2-27B). Nrg1-III acts on the erbB receptor complex on Schwann cells to direct their differentiation, including the expression of myelin-associated proteins and the spiral wrapping of axons. Nrg1/erbB signaling is not required for myelination by oligodendrocytes, suggesting alternative axon-glia signals in the CNS. Schwann cells and oligodendrocytes also signal back to axons to provide long-term support to their health and integrity.

The importance of myelination to human health is highlighted by the plethora of **demyelinating diseases**, in which damage to the myelin sheath decreases the resistance between nodes of Ranvier and disrupts the organization of ion channels in the nodal region (Figure 2-26). This slows down or even stops action potential conduction, causing deficits in sensation, movement, and cognition. Demyelinating diseases can be caused by several factors, including autoimmune responses that attack glial cells and mutations in proteins necessary for myelin function.

The most common CNS demyelinating disease is **multiple sclerosis** (**MS**), an adult-onset inflammation-mediated disease that affects 1 in every 3000 people globally. The hallmarks of MS include the formation of inflammatory plaques in white matter caused by destruction of myelin by immune cells. Most MS patients begin with a phase of relapsing-remitting MS, during which patients cycle between inflammatory demyelination with neurological symptoms, and

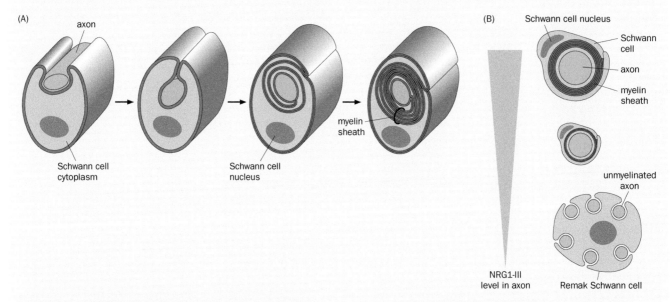

Figure 2-27 Schwann cell wrapping of axons and its regulation by neuregulin signaling. (A) Schematic of sequential steps illustrating a Schwann cell wrapping an axon, forming spiral extensions, and compressing its cytoplasm between the layers of plasma membrane to form myelin sheath. **(B)** Top, large-diameter axons express the highest levels of type III neuregulin-1 (NRG1-III), which direct thicker myelination. Middle, intermediate-diameter axons express

intermediate levels of NRG1-III, which direct thinner myelination. Bottom, small-diameter unmyelinated axons express the lowest levels of NRG1-III, which direct their interaction with Remak Schwann cells, forming a Remak bundle. (Adapted from Nave KA & Salzer JL [2006] *Curr Opin Neurobiol* 16:492–500. With permission from Elsevier Inc.)

Box 2-3: continued

remyelination and recovery. The next phase is characterized by continual neurological symptoms and progressive deterioration, which is often irreversible. Although abnormal immune responses clearly play a major role, the causes of MS remain mostly unknown. Variants of certain genes such as the major histocompatibility loci confer risks, but environmental factors appear to play a major role. Thanks to the recent development of drugs that inhibit immune destruction of CNS myelin, the life prognosis for a first diagnosis of MS today is far better than it was several decades ago.

Compared to MS, much more is known about the mechanisms of demyelinating diseases in the PNS because many are caused by inherited mutations in specific genes. **Charcot-Marie–Tooth (CMT) disease** (first described by J. M. Charcot, P. Marie, and H. H. Tooth in 1886) is the most common inherited disorder of the PNS, affecting 1 in 2500 individuals. CMT patients exhibit age-progressive deficits in sensation and/or movement in a length-dependent manner (that is, distal limbs exhibit the most severe deficits). Genetic alterations in several dozen different genes underlie various forms of the CMT disease that display similar symptoms. Some CMT genes act in Schwann cells. For example, the

most common cause of CMT (CMT1A) results from a duplication of a chromosome segment, causing overexpression of the peripheral myelin protein 22 (PMP22). CMT1B is caused by mutations in a gene encoding myelin protein zero (MPZ). Both PMP22 and MPZ are transmembrane proteins expressed abundantly in Schwann cells and involved in the formation of compact myelin. CMT1X is caused by mutations in a gene encoding a gap junction channel (Box 3-5) expressed in Schwann cells, which can facilitate transport of molecules from the area around the nucleus to inner layers of the myelin sheath by introducing shortcuts across the myelin membranes. Some CMT genes act in axons and encode a variety of proteins, such as neurofilaments and those involved in mitochondria fusion and protein translation. Studies of these CMT genes can help us understand both the normal biology of myelination and the pathogenesis of demyelination diseases.

Fundamental studies in neurobiology and research on disorders of the nervous system can greatly benefit each other. We will encounter examples of this synergy throughout the book and will address many in Chapter 12, which focuses on brain disorders.

---

toxin is puffer fish **tetrodotoxin (TTX)** (Figure 2-28), which potently blocks voltage-gated Na⁺ channels of many animal species and is widely used to experimentally silence neuronal firing. In recordings from the squid giant axon, for example, TTX application mimics the replacement of Na⁺ with choline⁺ in the original Hodgkin–Huxley experiment (Figure 2-22). Other drugs, such as **tetraethylammonium (TEA)**, selectively block voltage-gated K⁺ channels. Experiments using these drugs provided evidence for the existence of ion channels selectively permeable to specific ions.

Direct support for the existence of ion channels and characterization of individual channel properties came as a result of an important technical innovation in the late 1970s called **patch clamp recording** (see Section 14.21 and Box 14-3 for more details). In its original form, now called a **cell-attached patch**, a **patch pipette** (also called a patch electrode, a glass electrode with a small opening at its tip) forms a high-resistance seal with a small patch of the plasma membrane of an intact cell. This high-resistance seal is called a giga seal because the resistance exceeds 10^9 ohms (giga = 10^9), thus preventing ion flow between the pipette and extracellular solutions. When performing a patch clamp recording, the experimenter can "clamp" the voltage in the patch pipette, which corresponds to the extracellular potential for the small patch of the membrane under the pipette. Ion flow through the small membrane patch, which sometimes contains only a single ion channel, can be resolved and studied in isolation (Figure 2-29A).

When a cell-attached patch clamp was applied to cultured rat muscle cells, for example, the opening and closing of a single channel could be detected as discrete events. When a single channel opened in response to a depolarization of the patch, it produced a unitary inward current of ~1.6 picoamperes (circled in Figure 2-29B). When hundreds of these single channel recording traces were summed and averaged over time, a "macroscopic" Na⁺ current ensemble much like the Hodgkin–Huxley voltage clamp recording was reconstituted, with characteristic voltage-dependent opening and subsequent inactivation (Figure 2-30B, bottom; compare with the bottom trace of Figure 2-22). This experiment thus revealed the biophysical basis of voltage-dependent changes in Na⁺ conductance at the level of single molecules: each voltage-gated Na⁺ channel exists in discrete states, open or closed, with stochastic transitions; the proportion of time an individual channel

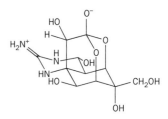

Figure 2-28 Tetrodotoxin (TTX) from the puffer fish. Puffer fish (above), whose resident symbiotic bacteria produce tetrodotoxin (TTX, below), is a delicacy in Japanese cuisine that must be carefully prepared. TTX is a potent blocker of voltage-gated Na⁺ channels in many species. (Image courtesy of Brocken Inaglory/Wikipedia.)

Figure 2-29 Studying ion flow across individual Na⁺ channels using patch clamp. (A) A cell-attached patch pipette can record ion flow within a small patch of muscle membrane. The configuration is analogous to that of the voltage clamp schematic shown in Figure 2-21. The patch pipette serves two functions: to measure the voltage of the pipette V_p, and to inject current through a feedback circuit such that V_p matches that of the command voltage (V_{CMD}). The current that needs to be injected in order for V_p to match V_{CMD}, which can be measured with the current monitor, is equivalent to the current that flows through the ion channel(s) under the patch pipette (I_p). **(B)** In response to a depolarization step (V_p) applied to the patch pipette (top), current flows across the patch (I_p, middle). Traces are shown for nine individual current measurements. Na⁺ channel openings can be seen as downward, rectangular steps (for example, circles on the first, second, and fourth traces), as positively charged Na⁺ ions leave the recording pipette and flow into the cell. The average of 300 I_p traces (bottom) resembles the macroscopic Na⁺ current measured by conventional voltage clamp, including voltage-dependent activation and inactivation. In these recordings, a K⁺ channel blocker was included in the pipette solution to block ion flow through possible K⁺ channel(s) in the membrane patch. (Adapted from Sigworth FJ & Neher E [1980] *Nature* 287:447–449. With permission from Springer Nature.)

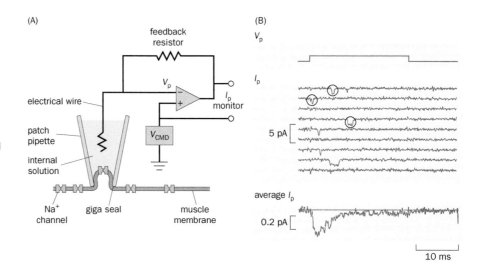

is open and able to conduct current—that is, the channel's **open probability**—is temporarily increased by depolarization.

Note that in the experiment each channel opening takes a square-like form, which means that the channel typically transitions between a state that is non-conducting (closed) to one that is conducting (open) without sliding through intermediates. Also note that even though an individual channel makes the closed-to-open transition abruptly, channels do not all open immediately after the membrane potential changes, and they close soon thereafter through an apparent inactivation process. Indeed, a difference in the delay in depolarization-induced open probability between Na⁺ and K⁺ channels accounts for the temporal difference in Na⁺ and K⁺ conductance rises during the action potential (Figure 2-24B).

In general, the current (I) carried by a particular ion species across a piece of neuronal membrane can be determined from single channel properties by the following formula:

$$I = N P_o \gamma (V_m - E)$$

where N is the total number of channels present, P_o is the open probability of an individual channel, V_m is the membrane potential, E is the equilibrium potential of that ion (hence $V_m - E$ is the driving force), and γ is the **single channel conductance**. Compared to the relationship between current and driving force we learned in Section 2.7, we see that the product $N P_o \gamma$ is equivalent to the macroscopic conductance, g. Thus, the ion conductance across a neuronal membrane is the product of (1) the number of channels present on the membrane, (2) the open probability of each channel, and (3) the single channel conductance. As discussed earlier, the open probability P_o is a function of both membrane potential and time, whereas the single channel conductance γ is a physical property of the channel protein that can vary with changes in its ionic milieu.

2.15 Cloning of genes encoding ion channels allows studies of their structure–function relationship

The molecular structures of ion channels as individual proteins were determined after the cloning of genes encoding specific ion channels, as the revolution in molecular biology spread to neuroscience in the 1980s. Being able to clone a gene requires one or more of the following approaches: (1) purifying the corresponding protein and using its amino acid sequence to deduce its nucleotide sequence and design a probe to screen a **cDNA library** (consisting of cloned cDNAs, or complementary DNAs, synthesized from mRNA templates derived from a specific tissue), (2) identifying a mutant defective in the gene product and using molecular genetic techniques to trace the causal gene, or (3) expressing the candidate gene product (by partitioning of a cDNA library) in a host cell and using a functional assay to identify the presence of the gene product. If there is a rich source of the protein

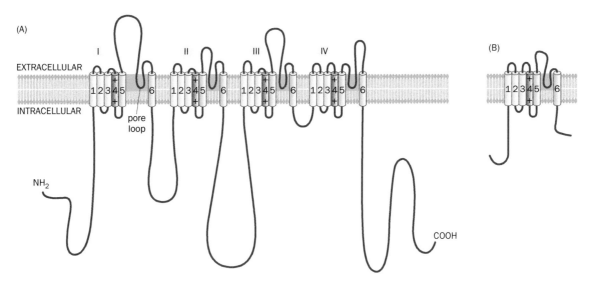

Figure 2-30 Primary structure of voltage-gated Na⁺ and K⁺ channels.
(A) A voltage-gated Na⁺ channel comprises four repeating modules. Each module consists of six helical transmembrane segments (TMs), with the fourth TM (red) containing positively charged amino acids that play a key role in voltage sensing. The pore loop and the adjacent fifth and sixth TMs together constitute the ion conduction pore (green). This structure was originally derived from the voltage-gated Na⁺ channel from electric eel (see Noda M, Shimizu S, Tanabe T et al. [1984] *Nature* 312:121–127). **(B)** A voltage-gated K⁺ channel protein resembles one of the four repeating modules of a voltage-gated Na⁺ channel; note the positively charged amino acids in the fourth TM and the pore loop between the fifth and sixth TMs. Four such subunits constitute a functional channel. The structure was originally derived from a K⁺ channel identified in *Drosophila* after the positional cloning of the *Shaker* gene (see Papazian DM, Schwarz TL, Tempel BL et al. [1987] *Science* 237:749–753 and Tempel BL, Papazian DM, Schwarz TL et al. [1987] *Science* 237:770–775). (From Yu FH & Catterall WA [2004] *Science STKE* 253:re15. With permission from AAAS; adapted from Sato et al. [2001] *Nature* 409:1047.)

and a functional assay (such as a high-affinity ligand) with which to look for the presence of the protein in biochemical fractions, the protein purification route is available; indeed, this route led to the cloning of the first Na⁺ channels.

Voltage-gated Na⁺ channel proteins were first purified from the electric eel *Electrophorus electricus,* whose electric organ is densely packed with Na⁺ channels. (Electric eels use their electric organ to shock their prey with large currents.) Peptide sequences from purified electric eel Na⁺ channel proteins were used to identify cDNAs encoding these proteins, and the electric eel cDNAs were then used to identify homologous genes in other organisms, including mammals, leading to the determination of their complete amino acid sequences. These studies revealed a highly conserved primary structure (**Figure 2-30**A): Animals from invertebrates to mammals have voltage-gated Na⁺ channels consisting of four repeating modules, each containing six transmembrane segments. This structural conservation explains why toxins such as TTX block Na⁺ channels across the animal kingdom. The fourth transmembrane segment, termed S4, contains many positively charged amino acids and was hypothesized to be the sensor that detects voltage changes for channel gating. A hydrophobic stretch of amino acids between the fifth and sixth transmembrane segments (S5 and S6) forms an extra pore loop within the membrane. As we will learn in more detail in Section 2.16, the pore loop, S5, and S6 together form the central pore for ion conduction.

While studies on voltage-gated Na⁺ channels benefited from the electric organ, the lack of a similarly enriched source of K⁺ channels and the overall heterogeneity of K⁺ channels made them resistant to similar biochemical approaches. Fortunately, genetic studies in the fruit fly *Drosophila melanogaster* provided an alternative strategy for cloning genes encoding K⁺ channels. A *Drosophila* mutant named **Shaker**, so-called because the mutant flies shake their legs under ether anesthesia, exhibited defects in a fast and transient K⁺ current in muscles and neurons, as well as defects in action potential repolarization. These findings led to the hypothesis that a K⁺ channel was disrupted in *Shaker* mutant flies. **Positional cloning** of the DNA corresponding to the *Shaker* locus (see Section 14.6 for details) identified the first voltage-gated K⁺ channel (Figure 2-30B). Interestingly, this K⁺ channel protein resembles one of the four repeating modules of the Na⁺ channel; like each of these modules, it has six transmembrane segments, including the positively charged S4 and the pore loop. Subsequent work showed that four such polypeptides (subunits) constitute one functional K⁺ channel.

Cloning of ion channels enabled structure–function studies investigating the molecular mechanisms underlying channel properties. One such property is the inactivation of voltage-gated Na⁺ channels, which contributes to the repolarization phase of the action potential. This enables action potentials to travel unidirectionally by enforcing a refractory period (Sections 2.10 and 2.11). Biophysical studies in the 1970s led to a **ball-and-chain** inactivation model positing that a cytoplasmic portion of the channel protein (the ball), connected to the rest of the channel by a polypeptide chain, blocks the channel pore after the ion channel opens. It was further hypothesized that depolarization not only opens the channel but also causes movement of charged amino acids to create a negatively charged inner channel pore, which would facilitate binding of a positively charged ball (**Figure 2-31**A).

This ball-and-chain model was elegantly validated in the voltage-gated Shaker K⁺ channel, which undergoes inactivation similarly to the voltage-gated Na⁺ channel. By using a molecular biology technique called *in vitro* mutagenesis to alter the DNA sequence of the cloned *Shaker* gene, researchers could express Shaker proteins in which selected stretches of amino acids were deleted, inserted, or replaced. This work revealed that the cytoplasmic N-terminal domain of the Shaker K⁺ channel is necessary for fast inactivation. Deleting a stretch of amino acids in this domain generated a mutant K⁺ channel that could not undergo fast inactivation after depolarization-induced channel opening. Supplying a peptide containing the first 20 amino acids of the cytoplasmic domain was sufficient to

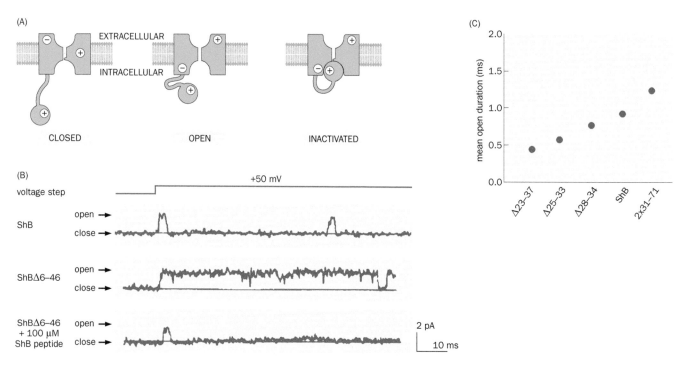

Figure 2-31 Molecular mechanisms of voltage-gated ion channel inactivation. (A) The ball-and-chain model of voltage-gated Na⁺ channel inactivation. Depolarization causes the channel to open and, at the same time, moves charged amino acids such that the inner pore of the channel becomes more negatively charged, creating a binding site for a portion of the channel's cytoplasmic domain (the ball), which is positively charged. The binding of the ball to the inner pore inactivates the channel. **(B)** Mutagenesis studies of the Shaker K⁺ channel (the ShB isoform) support the ball-and-chain model. In patch clamp experiments, the wild-type ShB channel (top) inactivates after initial depolarization-induced opening, and remains closed despite continued depolarization at +50 mV. When amino acids 6–46 are deleted from ShB (middle), the channel exhibits defective inactivation, as the channel remains open after depolarization. The inactivation defect of the mutant channel is corrected by supplying a soluble ball peptide made up of the first 20 amino acids of the cytoplasmic domain (bottom). **(C)** Compared to the wild-type channel (fourth column), the mean channel open duration is shortened by deleting 15, 9, and 7 amino acids from the chain connecting the ball to the rest of the channel (first three columns, respectively). The mean channel open duration is lengthened by adding 41 amino acids to the chain (fifth column, duplicating amino acids 31–71). (A, adapted from Armstrong CM & Bezanilla F [1977] *J Gen Physiol* 70:567–590. B & C, adapted from Hoshi T, Zagotta WN, & Aldrich RW [1990] *Science* 250:533–538 and Zagotta WN, Hoshi T, & Aldrich RW [1990] *Science* 250:568–571.)

restore the inactivation of this mutant K$^+$ channel (Figure 2-31B). Thus, the first 20 amino acids correspond to the ball. As predicted by the ball-and-chain model, several positively charged amino acids within the first 20 amino acids were found to be crucial for inactivation. Furthermore, decreasing the length of the chain—the intervening polypeptide between the first 20 amino acids and the rest of the channel—yielded a channel open for shorter periods before inactivation, whereas lengthening the chain increased the duration that the channel remained open (Figure 2-31C). These data suggest that the length of the chain dictates the "search" time for the ball to find the open channel. This example illustrates the power of combining molecular biology and electrophysiology to understand the mechanisms of ion channel function.

2.16 Structural studies reveal the atomic bases of ion channel properties

Ion channels are remarkable molecular machines. A voltage-gated K$^+$ channel can conduct up to 10^8 K$^+$ ions per second, which is near the diffusion rate of K$^+$, while also maintaining a high selectivity for K$^+$: the channel conducts ~10,000 times fewer Na$^+$ ions than K$^+$ ions. Central to the channel's conduction and ion selectivity is the pore loop (Figure 2-31B). *In vitro* mutagenesis studies have shown that mutations in the pore loop alter ion selectivity. Subsequent structural studies at atomic resolution using X-ray crystallography have provided detailed mechanisms of ion conduction and selectivity.

The amino acid residues of the K$^+$ channel pore loop are highly conserved from bacteria to humans. Thus, the crystal structure of the bacterial K$^+$ channel KcsA, the first ion channel whose structure was determined at atomic resolution, revealed mechanisms of K$^+$ conduction and selectivity that are likely universal. The KcsA channel is not voltage gated but nevertheless resembles part of the voltage-gated K$^+$ channel: each of the four KcsA subunits contains only two transmembrane helices (equivalent to transmembrane segments S5 and S6 of a voltage-gated K$^+$ channel subunit) and a pore loop in between, part of which forms a pore helix (**Figure 2-32**A). From the side, the channel looks like a conical

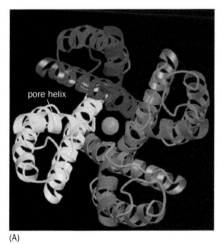
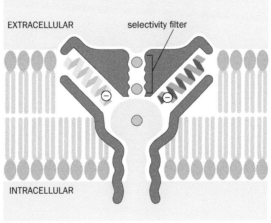
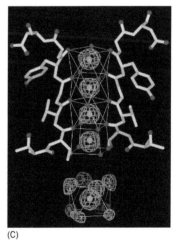

(A) (B) (C)

Figure 2-32 Atomic structure of KcsA, a bacterial K$^+$ channel.
(A) The KcsA channel viewed from the extracellular side. Each of the four subunits (differently colored) consists of two transmembrane helices and a shorter pore helix (indicated for the yellow subunit). A green K$^+$ ion is passing through the central pore. **(B)** From a side view, the KcsA channel resembles a conical funnel. An aqueous passage extends from the intracellular side to a central cavity in the middle of the lipid bilayer. Three K$^+$ ions are shown, one in the cavity and two in the selectivity filter above. The gray shading represents the channel protein viewed from the side. Two of the four pore helices are shown in this side view, with their electronegative carboxyl ends facing the cavity to stabilize the positively charged K$^+$ ion in the cavity. **(C)** Atomic structure of the selectivity filter. Oxygen atoms from carbonyl groups of the main polypeptide chain (red dots) create four K$^+$ binding sites within the selectivity filter. At each binding site, the K$^+$ ion is surrounded by eight oxygen atoms, just as the hydrated K$^+$ ion is surrounded by eight oxygen atoms from water in the cavity shown at the bottom. This mimicry renders conduction of K$^+$ (but not the smaller Na$^+$) more energetically favorable, providing a mechanism for ion selectivity. (A & B, from Doyle DA, Cabral JM, Pfuetzner RA, et al. [1998] *Science* 280:69–77. With permission from AAAS. C, adapted from Zhou Y, Morais-Cabral JH, Kaufman A, et al. [2001] *Nature* 414:43–48. With permission from Springer Nature.)

funnel (Figure 2-32B), with a cavity at the center of the channel to accommodate a hydrated K$^+$ ion. (Ions in solution, including K$^+$ and Na$^+$, are normally present in hydrated form.) Electronegative carboxyl ends of the four pore helices face the cavity and stabilize the K$^+$ ion as it travels through. When the channel is open, intracellular hydrated K$^+$ has free access to the cavity. Between the cavity and the extracellular side is the **selectivity filter**, through which K$^+$ ions pass in dehydrated form. This is accomplished via the interaction of K$^+$ with the electronegative carbonyl groups from the main polypeptide backbone corresponding to the most highly conserved amino acids of K$^+$ channels. These close carbonyl interactions, which mimic and replace the water molecules surrounding the K$^+$ ion in solution, perfectly match the size of the K$^+$ ion but not the smaller Na$^+$ ion. This accounts for the K$^+$ channel's high degree of ion selectivity and the favorable energetics of K$^+$ conduction through an open channel (Figure 2-32C; Movie 2-8). There are four possible positions for K$^+$ at the selectivity filter, which is usually occupied by two K$^+$ ions at alternate positions. The repulsion between the two K$^+$ ions forces K$^+$ to flow rapidly through the selectivity filter when the channel is open and when there is a driving force (an electrochemical gradient).

Following the pioneering studies on the KcsA K$^+$ channel, the structures of many ion channels (Box 2-4), including eukaryotic voltage-gated K$^+$ and Na$^+$ channels (Figure 2-33) have been solved by X-ray crystallography and, more recently, cryogenic electron microscopy (**cryo-EM**). The overall architectures of voltage-gated K$^+$ and Na$^+$ channels resemble each other, reflecting the similarities in their primary structures (Figure 2-30). Each of the four subunits of the K$^+$ channel and the repeating units of the Na$^+$ channel has a voltage-sensing domain (consisting of the transmembrane segments S1–S4) at its periphery and a pore domain (consisting of S5, the pore loop, and S6) at the center of the channel. In both cases, the voltage-sensing domain of one unit latches onto the pore domain of an adjacent unit (Figure 2-33A, B). Depolarization sensed by the S4 segments, which are enriched in positively charged amino acids (Figure 2-30), moves the voltage-sensing domain within the lipid bilayer. This causes conformational

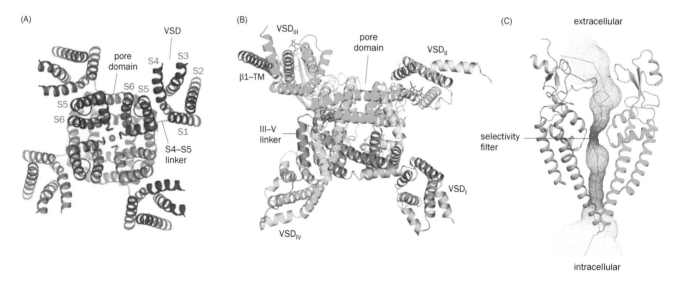

Figure 2-33 Structures of voltage-gated K$^+$ and Na$^+$ channels.
(A) Organization of transmembrane segments of a rat voltage-gated K$^+$ channel determined by X-ray crystallography, viewed from the extracellular side. There is a fourfold symmetry reflecting the four identical subunits constituting the channel. Each subunit (with a unique color) has six transmembrane segments. The S1–S4 segments constitute the voltage-sensing domain (VSD), whereas the S5, pore loop, and S6 segments constitute the pore domain. These two domains are connected by the S4-S5 linker. Note that the VSD of one subunit latches onto the pore domain of an adjacent subunit—for example, S4 of the red subunit is closest to S5 of the blue subunit. **(B)** Structure of a human skeletal muscle-specific voltage-gated Na$^+$ channel in complex with the auxiliary subunit β1 determined by cryo-EM. The four repeating units from the N- to C-terminus are colored in gray, yellow, green, and cyan, respectively, with the transmembrane segment of β1 in gold and the linker between the third and fourth repeats in orange. The organization of transmembrane segments resembles that of the voltage-gated K$^+$ channel. **(C)** Side view of the same structure as in Panel B, showing only the pore segments (S5, pore loop, S6) of two repeating units. The asymmetrical permeation path is illustrated by purple dots. (A, from Long SB, Campbell EB, & MacKinnon R (2005) *Science* 309:903–908. With permission from AAAS. B & C, from Pan X, Li Z, Zhou Q, et al. (2018) *Science* 362:eaau2486. With permission from AAAS.)

changes of the pore domain, leading to the opening of the ion conduction pore. The close interactions between the voltage-sensing domains and the pore domains of different units contribute to the coordinated conformational changes of the pore domains in response to membrane potential changes.

Unlike the fourfold symmetry of the voltage-gated K^+ channels consisting of four identical subunits, the arrangement of the voltage-sensing domains and the ion permeation path in the voltage-gated Na^+ channel are asymmetric (Figure 2-33B, C), reflecting the fact that the channel is made of four homologous but nonidentical repeating units. Furthermore, while K^+ ions pass through the selectivity filter in dehydrated form in K^+ channels, in Na^+ channels (as well as structurally similar Ca^{2+} channels; Box 2-4), ions pass through the selectivity filter in hydrated form. Ion selectivity is determined by specific amino acid side chains in the selectivity filters of the Na^+ and Ca^{2+} channels in addition to the carbonyl groups of the backbone, rather than entirely by the latter as in K^+ channels.

Since the ionic basis of the action potential was discovered in the 1950s, researchers have come a long way in elucidating the mechanisms underlying this most fundamental form of neuronal communication.

Box 2-4: Diverse ion channels with diverse functions

The voltage-gated Na^+ and K^+ channels we discussed in the context of the action potential are just two of many kinds of ion channels. In the human genome, more than 230 genes encode ion channels (Table 2-2). Ion channels are usually classified by the ions they conduct and the mechanisms by which they are gated. Many ion channels share sequence similarities, reflecting their shared evolutionary history (see Section 13.6 for details). Figure 2-34 depicts a phylogenetic tree for 143 structurally related ion channels in the same

Table 2-2:   Number of genes encoding ion channels in the human genome

Channel type	Gene number
K^+ channels	78
Voltage-gated K^+ channels	40
Inward-rectifier K^+ channels	15
Two-pore domain K^+ channels	15
Ca^{2+}/Na^+-activated K^+ channels	8
Na^+/Ca^{2+} channels	29
Voltage-gated Na^+ channels	9
Voltage-gated Ca^{2+} channels	10
Other Ca^{2+} and Na^+ channels	10
TRP channels	28
Cyclic-nucleotide-gated and HCN channels	10
Cl^- channels	10+
Neurotransmitter-gated channels	70
Other ligand-gated channels	11
Piezos	2

Summarized from the Guide to Pharmacology database (www.guideto pharmacology.org). See Figure 2-34 for a phylogenetic tree of channels listed above the row of Cl^- channels. The human genome also encodes 11 channels that conduct water (aquaporins) and 24 channels that form gap junctions between adjacent cells (Box 3-5), which are not included here. Abbreviations: TRP, transient receptor potential; HCN, hyperpolarization-activated cyclic-nucleotide-gated.

superfamily to which voltage-gated Na^+ and K^+ channels belong. All 143 channels share a common pore structure with two transmembrane helices (2TMs) and a pore loop. All except two subfamilies of K^+ channels use a 6TM-unit similar to voltage-gated Na^+ and K^+ channels (Figure 2-30).

K^+ channels make up the most diverse channel family and are encoded by at least 78 genes, many of which have alternatively spliced isoforms. The mix and match of different channel subunits to form heteromeric channels also contributes to channel diversity. K^+ channels play important roles in diverse functions of excitable cells, establishing properties such as the resting potential, the kinetics of repolarization after action potential initiation, and the spontaneous rhythmic firing of **pacemaker cells** (Figure 2-35A). Some K^+ channels are activated by depolarization, with diverse activation and inactivation kinetics adapted to neuron types with different firing frequencies. Some K^+ channels are activated by a rise in intracellular Ca^{2+} or a drop in ATP concentration, allowing cells to alter membrane potentials in response to changes in $[Ca^{2+}]_i$ or energy levels. Members of the **inward-rectifier K^+ channel** subfamily preferentially pass inward current at membrane potentials more hyperpolarized than E_K, and allow minimal outward current at membrane potentials more depolarized than E_K. This is because under depolarizing conditions, the inward-rectifier K^+ channels are blocked from the intracellular side by positively charged polyamines and Mg^{2+}. Among other functions, inward-rectifier K^+ channels help maintain the resting potential near E_K and are a major substrate for modulation by metabotropic neurotransmitter receptors (discussed in detail in Chapter 3). K^+ channels are also present in many nonexcitable cells. For example, K^+ channels are the predominant ion channels in glia.

Like K^+ channels, **Cl^- channels** generally stabilize the resting membrane potential, as E_K and E_{Cl} are both near the resting potential. Cl^- channels also play diverse roles in different cell types, often involving intracellular vesicles. Cl^-

(Continued)

Box 2-4: continued

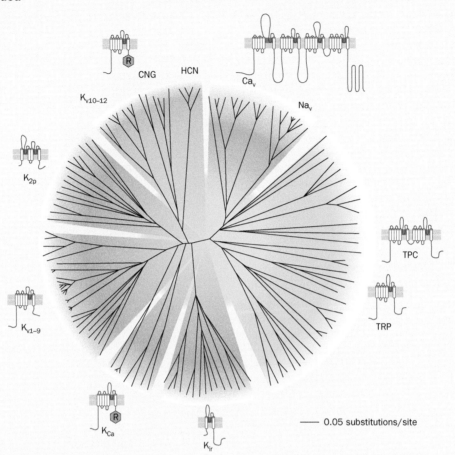

— 0.05 substitutions/site

Figure 2-34 Phylogenetic tree of 143 ion channels. This tree is constructed according to similarities in the amino acid sequences of the conserved pore regions of 143 structurally related ion channels, which correspond to a subset of the ion channels listed in Table 2-2; channels that do not show sequence similarity to this superfamily, such as Cl⁻ channels and neurotransmitter-gated channels, are not included here. The scale bar represents a distance on the tree corresponding to 0.05 amino acid substitutions per site in the sequence. Background shades separate ion channels into families: blue, voltage-gated Na⁺ (Na$_v$) and Ca²⁺ (Ca$_v$) channels; green, transient receptor potential (TRP) and related channels, including two-pore channels (TPC); red, most K⁺ channels, including inward-rectifier K⁺ channels (K$_{ir}$), Ca²⁺-dependent K⁺ channels (K$_{Ca}$), the first nine subfamilies of voltage-gated K⁺ channels (K$_{v1-9}$), and two-pore domain K⁺ channels (K$_{2p}$); orange, cyclic nucleotide-gated (CNG) channels, including the structurally related hyperpolarization-activated cyclic nucleotide-gated (HCN) channels and the last three subfamilies of voltage-gated K⁺ channels (K$_{v10-12}$), which contain a cyclic-nucleotide-binding domain. The schematics surrounding the tree illustrate the membrane topologies of the channel proteins, with the pore loops shaded in dark gray. R within a red hexagon represents cytoplasmic domains that possess cyclic nucleotide- or Ca²⁺-binding domains. (From Yu FH & Catterall WA [2004] *Science STKE* 253:re15. With permission from AAAS.)

channels consist of heterogeneous families of proteins with structures distinct from cation channels. One particular family consists of channels containing two subunits, each of which has 18 membrane-embedded helices and an ion conduction pore. In fact, these Cl⁻ channels are in the same family as bacterial Cl⁻/H⁺ exchangers, suggesting that ion channels and transporters can share structural similarities. New Cl⁻ channels are still being identified.

Voltage-gated Ca²⁺ channels constitute another important family of proteins in excitable cells. Their primary structure resembles those of voltage-gated Na⁺ channels, with four repeating modules each containing six transmembrane helices. Different voltage-gated Ca²⁺ channels differ in their activation thresholds, single channel conductances, and inactivation speeds. Neurons usually maintain a very low intracellular Ca²⁺ concentration of ~0.1 μM, which is >10,000-fold

lower than the extracellular Ca²⁺ concentration (~1.2 mM); this produces a very high E_{Ca} of approximately +120 mV. Opening of voltage-gated Ca²⁺ channels thus leads to depolarization due to Ca²⁺ influx driven by both chemical and electrical gradients. Indeed, action potentials in some neurons and cardiac myocytes are mediated by voltage-gated Ca²⁺ channels instead of voltage-gated Na⁺ channels. As we will learn in later chapters, voltage-gated Ca²⁺ channels play important roles in regulating neurotransmitter release at axon terminals. They are also essential for excitation–contraction coupling of muscles, for dendritic integration in some mammalian neurons, for shaping spiking patterns such as rhythmic burst firing (Figure 2-35B), and for regulating gene expression and neuronal differentiation in response to neuronal activity. Other Ca²⁺ channels are present on the membrane of internal Ca²⁺ stores and are gated by intracellular messengers.

Box 2-4: continued

(A)

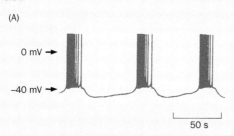

(B)

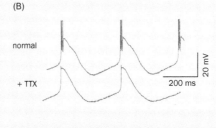

Figure 2-35 Ion channels that contribute to rhythmic burst firing of neurons. (A) Intracellular recording of a marine mollusk pacemaker cell (a neuron that can produce rhythmic output in the absence of input), showing periodic bursts of action potentials. Repeated action potentials result in a cumulative rise in intracellular $[Ca^{2+}]$, which activates Ca^{2+}-dependent K^+ channels that contribute to the hyperpolarization after the burst. As $[Ca^{2+}]_i$ gradually decreases during the interburst intervals, more Ca^{2+}-dependent K^+ channels close, raising the membrane potential until it reaches the threshold for another burst of action potentials. **(B)** Intracellular recording from a cat thalamic neuron in slice in the absence (top) and presence (bottom) of TTX. TTX blocks the action potential firing but not the rhythmic membrane potential change, indicating that voltage-gated Na^+ channels do not contribute to the latter. Further analysis revealed that the rhythmic membrane potential change is mainly a result of interactions between low-threshold voltage-gated Ca^{2+} channels, voltage-gated K^+ channels, and HCN channels. Depolarization opens the Ca^{2+} channels and produces a Ca^{2+} spike. Inactivation of Ca^{2+} channels and opening of K^+ channels cause hyperpolarization, which then activates HCN channels. HCN channel opening causes depolarization until reaching the threshold of the Ca^{2+} channels for the next Ca^{2+} spike. Without TTX, Ca^{2+} spikes depolarize the neuron above the threshold of voltage-gated Na^+ channels, resulting in burst firing. (A, from Smith SJ & Thompson SH [1987] *J Physiol* 382:425–428. B, from McCormick DA & Huguenard JR [1992] *J Neurophysiol* 68:1384–1400.)

Whereas K^+, Cl^-, Na^+, and Ca^{2+} channels are so named because of their ion selectivity, some channels are not as selective for the ions they conduct. For example, most **TRP channels** (named after the founding member, transient receptor potential, a *Drosophila* protein essential for visual transduction) and **CNG channels** (for cyclic nucleotide-gated channels) are *nonselective cation channels*, which means that they are permeable to Na^+, K^+, and sometimes Ca^{2+}. Because the driving force for Na^+ is typically larger than that for K^+ at the resting state, opening of nonselective cation channels causes more Na^+ influx than K^+ efflux and therefore produces a net depolarization. As we will learn in Chapters 4 and 6, CNG channels and TRP channels play important roles in sensory neurons to convert environmental stimuli—including light, odorants, pheromones, temperature, and noxious chemicals—into membrane potential changes. TRP channels also contribute to mechanosensation in *Drosophila* and *C. elegans*. **HCN channels** (for hyperpolarization-activated cyclic nucleotide-gated channels) are activated by hyperpolarization (usually below –55 mV) and cyclic nucleotides and are structurally related to CNG channels. Because HCN channels conduct cations and thus depolarize cells in response to hyperpolarization, they are particularly important for rhythmic neuronal firing and heart beating (Figure 2-35B).

At least 70 genes in the human genome encode ion channels gated by neurotransmitters (Table 2-2); these channels belong to different gene families from those depicted in Figure 2-34. Many neurotransmitter-gated channels are nonselective cation channels and thus their opening depolarizes and causes excitation of postsynaptic neurons. Some neurotransmitter-gated channels are selective for Cl^-; their opening usually mediates inhibition of postsynaptic neurons. We will study the structure and function of these neurotransmitter-gated ion channels in greater detail in Chapter 3.

Since the sequencing of the human genome in early 2000s, we are still discovering new ion channels. For instance, as we will discuss in Chapter 6, ion channels gated by mechanical forces mediate hearing and touch sensation. In mammals, the molecular nature of mechanosensitive channels is still being intensely investigated, as they do not appear to belong to the ion channel families discussed here. For example, a subset of mechanosensitive channels that mediate touch belong to an evolutionarily conserved family of proteins called the **Piezos**, which contain >30 transmembrane segments per subunit with no sequence resemblance to other known ion channels. We expect further additions to the ion channel list (Table 2-2) in the future.

Finally, mutations in many human ion channels cause or increase susceptibility to a variety of nervous system disorders, including epilepsy (Box 12-4), schizophrenia, autism (Section 12.24), migraine, and abnormal pain sensitivity. More than 1000 pathogenic mutations have been identified in voltage-gated Na^+ channels alone. These findings highlight the importance of ion channels in human health.

SUMMARY

Neurons are extraordinarily large cells. They adopt specialized cell biological properties to support their dendrites and axons, whose surface areas and volumes often exceed those of their cell bodies by orders of magnitude. mRNAs, ribosomes, and secretory pathway components are present in dendrites (and axons to a limited extent) so that cytosolic and membrane proteins can be synthesized and processed locally. Organelles and soma-synthesized proteins are actively transported to axons and dendrites by specific microtubule motors. Axonal microtubules are oriented uniformly with their plus ends facing outward. In vertebrate neurons, dendrites possess both plus-end-out and minus-end-out microtubules. The microtubule polarity difference in axons and dendrites is critical for directing specific cargos to appropriate subcellular compartments. Kinesins are mostly plus-end-directed microtubule motors and mediate axonal transport to deliver membrane proteins (via intracellular vesicles) and cytosolic proteins from the soma to axon terminals. Dynein and minus-end-directed kinesins mediate retrograde transport from axon terminals back to the soma. Kinesins and dynein also transport cargos within dendrites. Whereas microtubules run along the centers of dendritic and axonal processes, F-actin is enriched at the peripheries and can help cargos reach their final destination via myosin-based transport after they leave the microtubule highway.

Electrical signaling in excitable cells is enabled by the properties of the lipid bilayer and the activities of transporters and ion channels on the plasma membrane. Active transporters, such as the Na^+–K^+ ATPase, use energy to transport ions across the membrane against their electrochemical gradients; these transporters maintain ionic concentration differences across the plasma membrane. In most neurons and muscles, the intracellular compartment is high in K^+ but low in Na^+, Ca^{2+}, and Cl^- compared to the extracellular environment. Because the membrane at rest is more permeable to K^+ than to any other ions, the resting membrane potential is close to the K^+ equilibrium potential.

The neuronal plasma membrane can be effectively described as a parallel *R-C* circuit, with conductance paths for each ion representing ion flow through specific channels and a capacitance path representing the lipid bilayer. Electrical signals, such as the change in membrane potential in response to current injection, evolve over time. When electrical signals passively propagate along neuronal fibers, leaky membrane conductance along the way causes the signals to decay across distance. To propagate electrical signals reliably across a long distance, axons employ active properties such as the action potential.

Action potentials are produced by suprathreshold depolarization. Depolarization first opens voltage-gated Na^+ channels, leading to further depolarization and accounting for the rapid rising phase. The falling phase of action potentials is caused by inactivation of Na^+ channels and delayed opening of voltage-gated K^+ channels. This sequence ensures that action potentials are all or none, regenerative events that propagate unidirectionally along the axon from the cell body to the axon terminals. Studies utilizing important techniques developed in the past decades, such as patch clamp recording, molecular cloning, and atomic structural analysis, have revealed the molecular and mechanistic bases of how ion channels conduct ions with exquisite selectivity, how channel opening is controlled by voltage, how inactivation occurs, and how properties of individual ion channels account for macroscopic current in response to membrane potential changes.

Ion channels serve diverse functions. We will study these functions in greater detail in subsequent chapters, starting with the central subjects of the next chapter: neurotransmitter release at the presynaptic terminal and neurotransmitter reception at the postsynaptic specialization.

OPEN QUESTIONS

- How can we predict the neuronal compartment (axon, dendrite, soma) to which a protein is localized based on its amino acid and/or mRNA sequences?

- What mechanisms regulate the loading and unloading of cargo from molecular motors?

- What mechanisms control timing, extent, and neuronal type specificity of myelination of CNS axons?

- How would you discover previously unknown ion channels?

FURTHER READING

Books and reviews

Alberts B, Johnson A, Lewis J, Morgan D, Raff M, Roberts K, & Walter P (2015). Molecular Biology of the Cell, 6th ed. Garland Science.

Catterall WA, Wisedchaisri G, & Zheng N (2017). The chemical basis for electrical signaling. *Nat Chem Biol* 13:455–463.

Hille B (2001). Ion Channels of Excitable Membranes, 3rd ed. Sinauer.

Holt CE, Martin KC, & Schuman EM (2019). Local translation in neurons: visualization and function. *Nat Struct Mol Biol* 26:557–566.

Katz B (1966). Nerve, Muscle, and Synapse. McGraw-Hill.

Miller C (2006). ClC chloride channels viewed through a transporter lens. *Nature* 440:484–489.

Nirschl JJ, Ghiretti AE, & Holzbaur ELF (2017). The impact of cytoskeletal organization on the local regulation of neuronal transport. *Nat Rev Neurosci* 18:585–597.

Cell biological properties of neurons

Allen RD, Metuzals J, Tasaki I, Brady ST, & Gilbert SP (1982). Fast axonal transport in squid giant axon. *Science* 218:1127–1129.

Lasek R (1968). Axoplasmic transport in cat dorsal root ganglion cells: as studied with [3-H]-L-leucine. *Brain Res* 7:360–377.

Park HY, Lim H, Yoon YJ, Follenzi A, Nwokafor C, Lopez-Jones M, Meng X, & Singer RH (2014). Visualization of dynamics of single endogenous mRNA labeled in live mouse. *Science* 343:422–424.

Vale RD, Reese TS, & Sheetz MP (1985). Identification of a novel force-generating protein, kinesin, involved in microtubule-based motility. *Cell* 42:39–50.

Xu K, Zhong G, & Zhuang X (2013). Actin, spectrin, and associated proteins form a periodic cytoskeletal structure in axons. *Science* 339:452–456.

Electrical properties of neurons, action potentials, and ion channels

Armstrong CM & Bezanilla F (1977). Inactivation of the sodium channel. II. Gating current experiments. *J Gen Physiol* 70:567–590.

Doyle DA, Morais Cabral J, Pfuetzner RA, Kuo A, Gulbis JM, Cohen SL, Chait BT, & MacKinnon R (1998). The structure of the potassium channel: molecular basis of K+ conduction and selectivity. *Science* 280:69–77.

Hodgkin AL & Huxley AF (1952a). Currents carried by sodium and potassium ions through the membrane of the giant axon of Loligo. *J Physiol* 116:449–472.

Hodgkin AL & Huxley AF (1952b). A quantitative description of membrane current and its application to conduction and excitation in nerve. *J Physiol* 117:500–544.

Noda M, Shimizu S, Tanabe T, Takai T, Kayano T, Ikeda T, Takahashi H, Nakayama H, Kanaoka Y, Minamino N, et al. (1984). Primary structure of *Electrophorus electricus* sodium channel deduced from cDNA sequence. *Nature* 312:121–127.

Pan X, Li Z, Zhou Q, Shen H, Wu K, Huang X, Chen J, Zhang J, Zhu X, Lei J, et al. (2018). Structure of the human voltage-gated sodium channel Nav1.4 in complex with beta1. *Science* 362:eaau2486.

Schwarz TL, Tempel BL, Papazian DM, Jan YN & Jan LY (1988). Multiple potassium-channel components are produced by alternative splicing at the *Shaker* locus in *Drosophila*. *Nature* 331:137–142.

Sigworth FJ & Neher E (1980). Single Na+ channel currents observed in cultured rat muscle cells. *Nature* 287:447–449.

Taveggia C, Zanazzi G, Petrylak A, Yano H, Rosenbluth J, Einheber S, Xu X, Esper RM, Loeb JA, Shrager P, et al. (2005). Neuregulin-1 type III determines the ensheathment fate of axons. *Neuron* 47:681–694.

Zagotta WN, Hoshi T & Aldrich RW (1990). Restoration of inactivation in mutants of Shaker potassium channels by a peptide derived from ShB. *Science* 250:568–571.

CHAPTER 3
Signaling across Synapses

In this chapter, we continue to explore neuronal communication. We discuss first how arrival of an action potential at the presynaptic terminal triggers neurotransmitter release and then how neurotransmitters affect postsynaptic cells. This process, called **synaptic transmission**, results in information transmission from the presynaptic cell to the postsynaptic cell across the chemical synapse. In the context of studying postsynaptic reception, we will also introduce the fundamentals of signal transduction and study how synaptic inputs are integrated in postsynaptic neurons. Finally, we will discuss the electrical synapse, an alternative to the chemical synapse. Intercellular communication mediated by chemical and electrical synapses is the foundation of all nervous system functions.

HOW DOES NEUROTRANSMITTER RELEASE AT THE PRESYNAPTIC TERMINAL OCCUR?

In Chapter 2, we addressed the basic cell biological and electrical properties of neurons that are required to understand how molecules, organelles, and action potentials get to axon terminals. We now address the main purpose of these movements: to transmit information across synapses to postsynaptic targets, which can be other neurons or muscle cells. To illustrate general principles, we focus first on model synapses and neurotransmitter systems and then expand our discussion to other neurotransmitter systems.

3.1 Arrival of the action potential at the presynaptic terminal triggers neurotransmitter release

Neurotransmitters are molecules released by presynaptic neurons that diffuse across the synaptic cleft and act on postsynaptic target cells. The vertebrate **neuromuscular junction (NMJ)**, the synapse between the motor neuron axon terminals and skeletal muscle, is a model synapse that has been used to understand basic properties of synaptic transmission, many of which were later found to apply to other synapses. The neurotransmitter at the vertebrate NMJ was identified in the 1930s to be **acetylcholine (ACh)** (Figure 3-1A). An important advantage of studying the neuromuscular synapse is that the postsynaptic muscle cell (also called a muscle fiber) is a giant cell that can easily be impaled by a microelectrode for intracellular recording (see Section 14.21 for details); sensitive and quantitative measurement of synaptic transmission can be achieved by recording the resulting current or membrane potential changes in the muscle fiber. The NMJ is also an unusual synapse in that the motor axon spreads out to form many terminal branches that harbor hundreds of sites releasing ACh onto the target muscle. This property makes the NMJ a strong synapse that reliably converts action potentials in the motor neurons into muscle contraction (to be discussed in more detail in Section 8.1). In the experiments described in this chapter, researchers typically adjusted experimental conditions to prevent muscle contraction, so as to avoid recording artifacts induced by movement.

Figure 3-1 Studying synaptic transmission at the vertebrate neuromuscular junction (NMJ). **(A)** Structure of acetylcholine (ACh), the first identified neurotransmitter. **(B)** Measuring depolarization of a muscle fiber in response to motor axon stimulation or ACh iontophoresis in an NMJ *ex vivo*. The intracellular electrode is inserted into the muscle fiber close to the NMJ to record the end-plate potential (EPP). The square wave on the motor axon represents an application of current through a stimulating electrode that depolarizes the motor axon, causing it to fire an action potential. The square wave attached to the ACh pipette represents application of positive current that drives positively charged ACh out of the micropipette and onto the surface of the muscle close to the NMJ. **(C)** EPPs in the muscle fiber in response to motor axon stimulation (top) or focal ACh application (bottom) are similar in waveform. The first downward dip in the top trace indicates the time of axon stimulation. (C, adapted from Krnjevic K & Miledi R [1958] *Nature* 182:805–806. With permission from Springer Nature.)

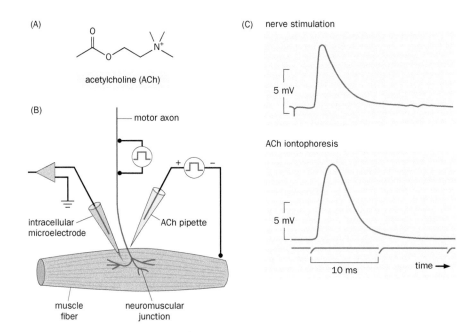

In a typical setup for studying synaptic transmission across the NMJ, an *ex vivo* preparation containing the muscle and its attached motor nerve is bathed in a solution mimicking physiological conditions. The motor nerve is then stimulated with an electrode to produce an action potential, and the membrane potential of the muscle fiber is recorded with an intracellular electrode at the NMJ (Figure 3-1B). Motor nerve stimulation was found to induce a transient depolarization in the muscle fiber within a few milliseconds (Figure 3-1C, top panel). This transient depolarization is called an **end-plate potential**, or **EPP**, as the postsynaptic area of the muscle fiber is flattened like an end plate. We will study the postsynaptic mechanisms that produce the EPP in greater detail in the second part of this chapter. For now, we simply use the EPP as a readout as we investigate the mechanisms of neurotransmitter release.

How does motor nerve stimulation evoke an EPP? Researchers found that motor nerve stimulation can be mimicked by ACh application through a micropipette at the contact site between motor axon terminals and the muscle (Figure 3-1C, bottom panel). (This method is called **iontophoresis**; here, positively charged ACh is driven out of a micropipette by applying positive current.) Subsequent experiments revealed that application of the tetrodotoxin (TTX; Figure 2-29), which blocks voltage-gated Na⁺ channels and thus prevents action potential propagation in motor axons, blocks the muscle EPP in response to motor nerve stimulation. However, ACh iontophoresis could evoke an EPP even when action potentials were blocked or when the motor axon was removed altogether. These results suggested that the ultimate effect of action potentials in the motor axon is to trigger ACh release at the axon terminals, and binding of ACh to the muscle membrane triggers depolarization of the muscle fiber in the form of an EPP.

We now know that fusion of synaptic vesicles with the presynaptic plasma membrane causes release of discrete packets of ACh molecules into the synaptic cleft. As we shall soon see, however, the release of neurotransmitters in discrete packets was deduced before the actual discovery of synaptic vesicles.

3.2 Neurotransmitters are released in discrete packets

In the early 1950s, Bernard Katz and colleagues applied intracellular recording techniques, then newly invented, to muscle cells to study the mechanisms of neuromuscular synaptic transmission. While studying muscle EPPs evoked by nerve stimulation in the frog NMJ, they observed that muscle fibers also exhibited small

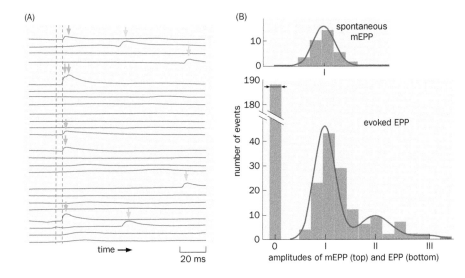

Figure 3-2 **Miniature end-plate potentials (mEPPs) and a statistical test of quantal neurotransmitter release. (A)** At low extracellular Ca²⁺ concentrations, nerve stimulation (at the time indicated by the first dashed vertical line) infrequently evokes EPPs, each of which follows the nerve stimulus with a specific latency (second dashed vertical line). In the 24 trials shown here (each represented by a horizontal trace), five EPPs (yellow arrows) were evoked. In one case the action potential evoked two quanta. Note also the presence of four depolarization events not linked with nerve stimulation (cyan arrows). **(B)** Using spontaneous mEPPs (top) as the unitary size, the frequency distribution of evoked EPPs (bottom) was predicted by a Poisson distribution (red line). This fits with the experimental data plotted as a histogram showing the number of EPPs (*y* axis) whose amplitudes fell within a certain bin (*x* axis). Note that the mEPP amplitude (top) varies; the sizes of individual neurotransmitter packets are not exactly the same. The evoked EPP amplitude (bottom) is 1×, 2×, or 3× the average mEPP amplitude. The frequency of synaptic failure (trials with an EPP amplitude of 0) also matches well the prediction from the Poisson distribution (red line flanked by two arrows). (Adapted from Del Castillo J & Katz B [1954] *J Physiol* 124:560–573.)

EPPs in the *absence* of any nerve stimulation; these were termed **miniature end-plate potentials**, or **mEPPs**. mEPPs had an intriguing property: for a given neuromuscular preparation, they seemed to have a defined unitary size. The amplitude of mEPPs, hypothesized to be due to spontaneous release of ACh from motor axon terminals, was usually two orders of magnitude lower than EPPs evoked by nerve stimulation. To understand if mEPPs and EPPs are related to each other, Katz and colleagues lowered the concentration of extracellular Ca²⁺, a condition known then to diminish neurotransmitter release. They reached a condition in which most nerve stimulations did not evoke any EPPs. When stimuli did trigger EPPs under these conditions, the amplitude of those evoked EPPs were usually the same size as mEPPs, and occasionally two or three times the unit size (**Figure 3-2**A). Further reduction of the Ca²⁺ concentration reduced the *frequency* of observing any evoked EPP, but it did not further diminish the EPP *amplitude*. These observations suggested that mEPPs reflected the basic unit of synaptic transmission. EPPs evoked by nerve stimulation under normal conditions were caused by the simultaneous occurrence of hundreds of mEPPs. These results led to the **quantal hypothesis of neurotransmitter release**, which posited that neurotransmitters are released in discrete packets (quanta) of relatively uniform size.

To further test the quantal hypothesis, Katz and his colleagues used statistical methods (see **Box 3-1** for details) to predict the frequencies of releasing no quanta, a single quantum, or multiple quanta in response to nerve stimulation. If the larger evoked potentials are caused by release of multiple quanta, they should occur at frequencies related to the frequency of release of a single quantum. When Ca²⁺ concentration is low and the release probability is small, the frequency (f) that k quanta are released during each nerve stimulation follows the **Poisson distribution**:

$$f = \frac{m^k}{k!} e^{-m}$$

where m is the mean number of units (quanta) that respond to an individual stimulus. Since mEPPs following each spontaneous release correspond to one unit, m can be experimentally determined as the mean EPP amplitude divided by the mean mEPP amplitude. Indeed, the frequency distributions of EPPs calculated earlier and those observed experimentally were an excellent fit: the frequency of **synaptic failures**, cases when nerve stimulation did not cause an EPP, matched precisely with the statistical prediction; there was a prominent peak at around the size of the unitary mEPP and a small peak at twice the mEPP amplitude (Figure 3-2B). Thus, this statistical analysis provided strong support for the notion that neurotransmitters are released in discrete packets.

Box 3-1:   Binomial distribution, Poisson distribution, and neurotransmitter release probability

The Poisson distribution and the related **binomial distribution** are both probability distributions describing the frequency of discrete events that occur independently. Let's start our discussion with the binomial distribution. Imagine that we are tossing a coin and that the probability that the coin lands with heads facing up is p. A coin toss is an example of a Bernoulli trial (binomial trial), a trial with exactly two outcomes. The binomial distribution describes the frequency (f) in which k events occur after n trials (that is, k times heads facing up after n coin tosses):

$$f(k; n, p) = \frac{n!}{k!(n-k)!} p^k (1-p)^{n-k}$$

where $k = 0, 1, 2, \ldots n$, ! is factorial (for example, $4! = 4 \times 3 \times 2 \times 1 = 24$), and $n!/k!(n-k)!$ is the binomial coefficient. Suppose you want to know the likelihood of tossing a coin four times and having the heads face up only once. The probability for heads, p, is 0.5 for any given toss of a fair coin. According to the formula, the binomial coefficient for $k = 0, 1, 2, 3, 4$ is respectively 1, 4, 6, 4, 1 (note that $0! = 1$), and the frequency of occurrence (f) for the five k values are calculated to be 0.0625, 0.25, 0.375, 0.25, 0.0625, respectively. In other words, out of four coin tosses, the probability that heads faces up only once (or three times) is 25%; the probability that heads faces up twice is 37.5%, and the probability that heads faces up four times (or never) is 6.25%.

If neurotransmitter release occurs in discrete quanta and if the release of each quantum occurs at a probability of p and if each quantum is independent of the other quanta, then we can think of each spike as initiating a set of simultaneous Bernoulli trials (equal to the number of quanta available for release), and we can calculate the frequency that k quanta out of the total n quanta are released using the binomial formula. However, researchers did not know the actual values for n (how many quanta are available to be released) or for p (how likely is any individual quantum to be released), so it was not possible to apply a binomial distribution. Fortunately, according to probability theory, when n is large (>20) and p is small (<0.05), the binomial distribution can be approximated by the Poisson distribution, in which the frequency (f) that k events occur can be determined by a single parameter λ (which equals the product of n and p in the binomial distribution) according to the following formula:

$$f(k; \lambda) = \frac{\lambda^k}{k!} e^{-\lambda}$$

One can experimentally estimate λ (same as m in Section 3.2) because, as the product of n and p, it equals the mean number of quanta that are released in response to a stimulus and thus is equivalent to the ratio of the mean EPP amplitude and the mEPP amplitude (assumed to be the quantal unit). Thus, researchers can calculate the probability of release in response to nerve stimulation—estimating the likelihood that no release occurs ($k = 0$), that a single quantum is released ($k = 1$), that two quanta are released ($k = 2$), and so on—and can then compare these calculations with the actual experimental data, as shown in Figure 3-2B.

Note that to apply the Poisson distribution, the release probability (p) must be small and the number of available quanta (n) must be large so that p does not change during the measurement of λ. Researchers cannot control n, but it turns out that n is very large in the vertebrate NMJ, because there are typically hundreds of neurotransmitter release sites between a motor axon and its muscle target. Researchers can experimentally reduce p by studying neurotransmitter release in low-Ca^{2+} extracellular solutions. Synaptic transmission at the NMJ also follows closely other assumptions required for the Poisson distribution: independent release of each quantum (because of the large number of release sites), the uniformity of the population (p is the same for all quanta), and the relative uniformity of their size (each vesicle contains a similar amount of neurotransmitter molecules). In many CNS synapses these assumptions either fail (for example, n is often too small) or cannot be tested adequately. Thus, the probability of neurotransmitter release in CNS synapses may not follow the Poisson distribution. Interestingly, recent studies suggest that release probability at some CNS synapses containing a single presynaptic active zone can be fit by binomial distribution with n ranging between 1 and 10, likely representing the number of docking sites for readily releasable synaptic vesicles.

3.3   Neurotransmitters are released when synaptic vesicles fuse with the presynaptic plasma membrane

Physiological and anatomical studies often complement each other in driving neuroscience discoveries. A strong candidate for the physical substrate of quantal neurotransmitter release became evident when electron microscopy (EM) was first applied to the nervous system in the mid-1950s. Thin sections across the nerve terminals revealed that they contain abundant vesicles that are ~40 nm in diameter. At the NMJ, many such vesicles appear stacked near the presynaptic membrane juxtaposed to the muscle membrane (**Figure 3-3**A). These **synaptic vesicles** were immediately hypothesized to be vesicles filled with neurotransmitters. The relatively uniform size of synaptic vesicles could explain why neurotransmitters are released in packets with a uniform quantal size. (The quantal size at the frog NMJ has been estimated to be about 7000 ACh molecules.) The unitary release of neurotransmitters occurs when a single synaptic vesicle fuses with the plasma

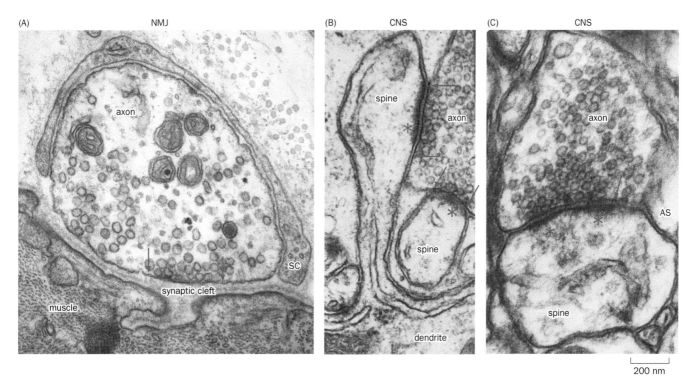

Figure 3-3 **Structures of synapses revealed by electron microscopy.** Pairs of arrows define the extents of active zones in the presynaptic terminals. Asterisks indicate postsynaptic densities. Note the abundance of ~40 nm diameter synaptic vesicles in each presynaptic terminal; some of these vesicles are docked at the active zone, ready for release. **(A)** A frog NMJ. The synaptic cleft is considerably wider than at the CNS synapses shown in the other two panels. SC indicates a Schwann cell process that wraps around the motor axon terminal.

A typical motor axon forms hundreds of such presynaptic terminals onto a muscle fiber. **(B)** Two synapses formed between a single axon and two dendritic spines in rat cerebellar cortex. **(C)** A synapse from human cerebral cortex. AS indicates an astrocyte process wrapped around many CNS synapses. All images share the scale bar. (A, courtesy of Jack McMahan. B & C, courtesy of Josef Spacek and Kristen M. Harris, SynapseWeb.)

membrane, dumping its neurotransmitter content into the synaptic cleft and producing a miniature depolarization in the muscle cell. Nerve stimulation under normal conditions (ordinary external Ca^{2+} rather than low Ca^{2+}) causes hundreds of these vesicle fusion events at a given NMJ, thereby producing EPPs two orders of magnitude larger than mEPPs. Thus, the NMJ has a high **quantal content** (releasing transmitters from several hundred synaptic vesicles per action potential). By contrast, many synapses in the CNS have much lower quantal contents (releasing transmitters from only one or a few synaptic vesicles per action potential).

The basic structural elements of chemical synapses are highly similar across the entire nervous system and in different animal species (Figure 3-3). In all cases presynaptic terminals have clusters of synaptic vesicles "docked" at the presynaptic membrane ready for release at the **active zone**. Across the synaptic cleft from the active zone and at the postsynaptic membrane is an electron-dense structure called **postsynaptic density**. We will study the molecular composition of the active zone and the postsynaptic density later in this chapter.

Although EM studies revealed many vesicles in presynaptic terminals, observing a fusion event in response to stimulation was necessary to confirm the hypothesis that neurotransmitter release is due to fusion of synaptic vesicles and the presynaptic plasma membrane. These events occur very transiently and are thus difficult to detect in typical electron microscopic preparations. To visualize fusion events, researchers stimulated a neuromuscular preparation while the entire sample was falling toward a copper block cooled to 4 Kelvin that froze the tissue immediately upon contact. Fusion events between synaptic vesicles and the presynaptic plasma membrane were indeed caught in action (**Figure 3-4**). Such studies provided definitive evidence that synaptic vesicle fusion with the presynaptic plasma membrane causes neurotransmitter release.

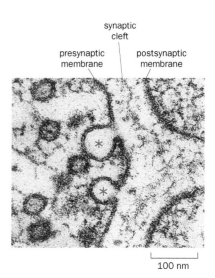

Figure 3-4 **Synaptic vesicle fusion caught in action.** This electron micrograph was taken from a frog NMJ preserved 3–5 ms after nerve stimulation, revealing the fusion of two synaptic vesicles (asterisks) with the presynaptic plasma membrane. (Courtesy of John Heuser. See also Heuser JE & Reese TS [1981] *J Cell Biol* 88:564–580.)

3.4 Neurotransmitter release is controlled by Ca²⁺ entry into the presynaptic terminal

How does action potential arrival cause synaptic vesicle fusion in the presynaptic terminal? As noted in Section 3.2, external Ca²⁺ facilitates action potential–triggered neurotransmitter release: bathing NMJ preparations in solutions with progressively lower concentrations of Ca²⁺ rendered the stimulation of motor axons increasingly ineffective at generating EPPs in muscle cells. Supplying Ca²⁺ locally at the NMJ in preparations with very low extracellular Ca²⁺ through iontophoresis (Figure 3-1) provided a means to determine when Ca²⁺ is required during action potential–induced synaptic transmission. A brief application of Ca²⁺ enabled neurotransmitter release if it occurred immediately before the depolarization pulse, but not if it occurred after the depolarization pulse. Thus, extracellular Ca²⁺ is required during the brief period when depolarization occurs and preceding transmitter release itself.

How does external Ca²⁺ contribute to neurotransmitter release? Key insights were obtained from studies of the squid giant synapse, whose presynaptic as well as postsynaptic terminals are so large that researchers can insert electrodes into both compartments for intracellular recordings. (One of the postsynaptic target cells is the neuron that extends the giant axon featured in Chapter 2.) It was found that action potentials could be replaced by depolarization, which opens voltage-gated Ca²⁺ channels (Box 2-4) in the presynaptic plasma membrane, causing an inward flux of Ca²⁺ that triggers neurotransmitter release.

Let's study in detail one specific experiment (**Figure 3-5**), which showcased the Ca²⁺ dependence of neurotransmitter release and provided information about the timing of different steps of the process. In this experiment, the voltage clamp technique was applied to both the presynaptic terminal and postsynaptic target of the squid giant synapse in the presence of Na⁺ and K⁺ channel blockers, such that the only cation that could cross the presynaptic membrane was Ca²⁺. From a resting potential at –70 mV, a depolarizing voltage step to –25 mV applied to the presynaptic terminal (Figure 3-5A, top) triggered Ca²⁺ influx, as measured by presynaptic current (Figure 3-5A, middle). This resulted in synaptic transmission, as measured by an inward postsynaptic current (Figure 3-5A, bottom; we will study the nature of such postsynaptic currents in later sections).

However, a voltage step to +50 mV applied to the presynaptic terminal did not trigger presynaptic Ca²⁺ influx or postsynaptic currents (Figure 3-5B, left portion). At this potential, voltage-gated Ca²⁺ channels are open, but because +50 mV is close to the equilibrium potential of Ca²⁺ in the presynaptic terminal under the experimental conditions used, there was little driving force for Ca²⁺ influx (Section 2.5). Nonetheless, returning the presynaptic membrane potential from +50 mV

Figure 3-5 Voltage clamp studies of Ca²⁺ entry into the presynaptic terminal of the squid giant synapse. Voltage steps were applied to the presynaptic terminal (top traces) using the voltage clamp technique (Figure 2-21). The current injected into the presynaptic terminal to maintain the clamped voltage is equivalent to the Ca²⁺ current across the presynaptic membrane (middle traces), as Na⁺ and K⁺ channel blockers were applied in these experiments. Postsynaptic current was simultaneously recorded in a voltage clamp setting as an assay for neurotransmitter release (bottom traces). **(A)** A depolarizing step in the presynaptic terminal triggered the opening of voltage-gated Ca²⁺ channels, which caused Ca²⁺ influx and a subsequent postsynaptic response. **(B)** A larger depolarization step of the presynaptic membrane potential (bringing the membrane close to the Ca²⁺ equilibrium potential) prevented Ca²⁺ entry due to lack of a driving force; no postsynaptic response occurs. A tail current representing Ca²⁺ influx was produced when the presynaptic membrane potential returned to –70 mV, which triggered a postsynaptic response. The pairs of dashed lines represent the presynaptic voltage step (left) and the onset of the postsynaptic response (right). The delay in Panel B includes the time between Ca²⁺ entry and the postsynaptic response. In Panel A, the interval is lengthened by the time required to open Ca²⁺ channels. (Adapted from Augustine GJ, Charlton MP, & Smith SJ [1985] *J Physiol* 367:163–181. See also Llinás RR [1982] *Sci Am* 247(4):56–65.)

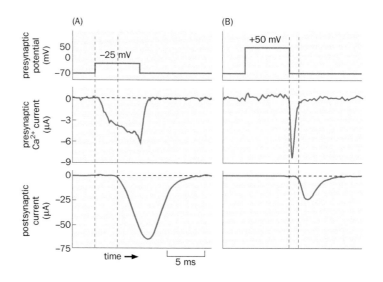

to –70 mV produced a presynaptic "tail current" (Figure 3-5B, middle). This is because the membrane potential change was faster than the closure of voltage-gated Ca^{2+} channels; thus there was a transient period with a strong driving force for Ca^{2+} influx while Ca^{2+} channels remained open. Ca^{2+} influx produced a corresponding postsynaptic current response (Figure 3-5B, bottom). Interestingly, the Ca^{2+} tail current after presynaptic depolarization to +50 mV triggered a postsynaptic response more rapidly than did the presynaptic depolarization to –25 mV (compare the time interval between the two dashed lines in the two panels). This suggests that the normal synaptic delay between presynaptic depolarization and postsynaptic response in Panel A consists of two components: a delay due to the time it takes to open voltage-gated Ca^{2+} channels (which was bypassed in the tail current condition, as the channels were already open) and a delay between Ca^{2+} entry and the neurotransmitter-triggered postsynaptic response.

The hypothesis that Ca^{2+} entry triggers neurotransmitter release was further validated by other techniques. In one type of experiment, a chemical dye used to indicate changes in Ca^{2+} concentration (see Section 14.22 for more details) was injected into the presynaptic terminal of the squid giant synapse. Nerve stimulation was found to increase the intracellular Ca^{2+} concentration at the presynaptic terminal (Figure 3-6). The Ca^{2+} concentration was highest in specific regions of the presynaptic terminal. As will be discussed in Section 3.7, this is because voltage-gated Ca^{2+} channels are highly concentrated at the active zone, where synaptic vesicles dock and fuse with presynaptic membrane. Another type of experiment featured chemical compounds that "cage" Ca^{2+}, preventing it from interacting with Ca^{2+}-binding proteins, and can be triggered by light to release Ca^{2+}. When caged Ca^{2+} was introduced into the presynaptic terminal of the squid giant axon, light could trigger neurotransmitter release in the absence of action potentials or Ca^{2+} entry from the extracellular solution. This experiment demonstrated that vesicle fusion and neurotransmitter release do not require depolarization or Ca^{2+} entry from the extracellular solution; rather, they are caused by an instantaneous increase in Ca^{2+} concentration in the presynaptic terminal.

Together, these experiments firmly established a sequence of events from action potential to neurotransmitter release:

> *Action potential from the axon → Depolarization of the presynaptic terminal → Opening of voltage-gated Ca^{2+} channels → Ca^{2+} entry into the presynaptic terminal → Fusion of synaptic vesicles with presynaptic plasma membrane → Neurotransmitter release*

This sequence of events, originally worked out in the frog NMJ and the squid giant synapse, has been found to apply to all chemical synapses across the animal kingdom, regardless of the type of synapse and neurotransmitter used.

The short latency between Ca^{2+} entry into the presynaptic terminal and postsynaptic events (~2 ms in Figure 3-5B, and often shorter) indicates that there must be a pool of synaptic vesicles that are ready to fuse with the presynaptic plasma membrane immediately upon a rise in intracellular Ca^{2+} concentration. This is consistent with observations in electron microscopy (Figure 3-3). Furthermore, membrane fusion is energetically unfavorable, as breaking two membranes and resealing them necessitates exposing hydrophobic surfaces to water and thus requires external energy such as ATP hydrolysis. However, the final step of synaptic vesicle fusion is so fast that it is unlikely to involve an ATP hydrolysis-dependent catalytic process. Instead, as we will soon learn, synaptic vesicles are primed for fusion by a specialized protein complex already existing in a high-energy configuration, simply waiting for Ca^{2+} to trigger the sudden conformational change that permits fusion.

3.5 SNARE proteins mediate synaptic vesicle fusion

We now turn to the molecular mechanisms mediating the fusion of synaptic vesicles with the presynaptic plasma membrane (a process also called neurotransmitter

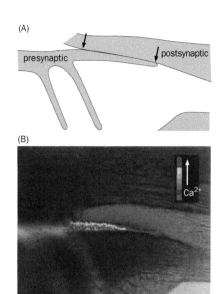

(A)

presynaptic   postsynaptic

(B)

Ca^{2+}

500 μm

Figure 3-6 Nerve stimulation triggers Ca^{2+} entry into the presynaptic terminal of the squid giant synapse. (A) Schematic of the squid giant synapse image in Panel B. The presynaptic terminals resemble finger-like extensions that contact postsynaptic neurons (one of which is shown). The two arrows indicate the extent of synaptic contact between the two neurons. **(B)** A brief train of presynaptic action potentials caused the Ca^{2+} concentration to increase in the presynaptic terminal, as reported by fluorescence changes of microinjected fura-2, a Ca^{2+} indicator (see Section 14.22 for details). The Ca^{2+} increase is seen as a shift from cool colors to warm ones. (From Smith SJ, Buchanan J, Osses LR, et al. [1993] *J Physiol* 472:573–593. With permission from the Physiological Society.)

exocytosis). Our current understanding of these mechanisms comes from a convergence of multiple experimental approaches:

- *Biochemical purification* to identify presynaptic protein components. Because of the uniform size and buoyancy of synaptic vesicles and their abundance, researchers can purify them to a high degree, which permits the identification of their key components. Indeed, the synaptic vesicle is one of the best-characterized organelles in the cell, with quantitative information about its protein and lipid composition (**Figure 3-7**; **Movie 3-1**). We will study key synaptic vesicle proteins in this and subsequent sections.
- *Genetic studies in yeast* that identified genes required for membrane fusion in the secretion pathway (Figure 2-2).
- *Biochemical reconstitution* of mammalian vesicle fusion reactions *in vitro*, which identified the minimal components sufficient for membrane fusion. Strikingly, these and yeast genetic studies identified a common set of evolutionarily conserved proteins (which we will soon discuss) and demonstrated that neurotransmitter exocytosis is a specialized form of membrane fusion that occurs in all eukaryotic cells and in many parts of the cell (Figure 2-2; see also Figure 13-13).
- *Genetic tests for the necessity* of these evolutionarily conserved proteins in synaptic transmission in *C. elegans, Drosophila,* and mice *in vivo*.
- *Studies of toxins* that block specific steps of neurotransmitter release and identification of their protein targets.

Together, these approaches have given rise to our current understanding of the neurotransmitter release mechanisms summarized in the following.

At the core of vesicle fusion are three **SNARE** proteins and **SM proteins**. (We will introduce the origin of the SNARE acronym later. SM stands for Sec1/Munc18-like proteins—Sec1 was originally identified in yeast for its requirement in secretion; Munc18 is the mammalian homolog of Unc18, originally identified in *C. elegans* as mutants that exhibit an uncoordinated phenotype). The first SNARE is a transmembrane protein on the synaptic vesicle called **synaptobrevin** (also called VAMP for vesicle-associated membrane protein), which is the most abundant synaptic vesicle protein. Since synaptobrevin is associated with the synaptic vesicle, it is designated as a **v-SNARE**. The second SNARE is a transmembrane protein on the plasma membrane called **syntaxin**. Owing to its location on the target membrane for vesicle fusion, syntaxin is called a **t-SNARE**. The third SNARE, named **SNAP-25** (synaptosomal-associated protein with a molecular weight of 25 kDa), is a t-SNARE anchored to the cytoplasmic face of the plasma membrane via a lipid modification. Once the synaptic vesicle is in the vicinity of the presynaptic plasma membrane, the cytoplasmic domains of synaptobrevin, syntaxin, and SNAP-25 assemble into a very tight complex. Current data indicate that the assembly of the SNARE complex proceeds from binding of v- and t-SNAREs from the ends farthest from the membrane to the ends closest to the membrane, like a zipper that zips up. The force generated by the assembly of the SNARE complex pulls the synaptic vesicle membrane even closer to the plasma membrane, leading the lipid bilayers to fuse, such that the contents of the synaptic vesicle are exposed to the extracellular space (**Figure 3-8**A; **Movie 3-2**).

The structure of the SNARE complex has been determined at atomic resolution by X-ray crystallography. Three SNARE proteins form a four-helix bundle, with synaptobrevin and syntaxin each contributing one helix and SNAP-25 contributing two helices (Figure 3-8B). Many naturally occurring protease toxins that inhibit neuronal communication target these three SNARE proteins at specific amino acid residues (**Box 3-2**). Proteolytic cleavage by these proteases is predicted to inhibit the attachment of the four-helix bundle to the membrane, thereby blocking neurotransmitter release.

The SNARE-based mechanism of membrane fusion applies to many fusion reactions in intracellular vesicle trafficking. The v- and t-SNAREs for other specific fusion events (for example, fusion of ER-derived vesicles with the Golgi mem-

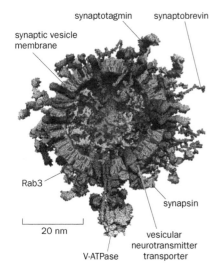

Figure 3-7 The molecular anatomy of a synaptic vesicle. This model is based on quantitative determination of the protein components associated with synaptic vesicles. Each colored structure in this cross section of the synaptic vesicle represents a synaptic vesicle protein. The synaptic vesicle membrane and six synaptic vesicle proteins whose functions are discussed in this and subsequent sections are indicated (see also Table 3-1). (Adapted from Takamori S, Holt M, Stenius K, et al. [2006] *Cell* 127:831–846. With permission from Elsevier Inc.)

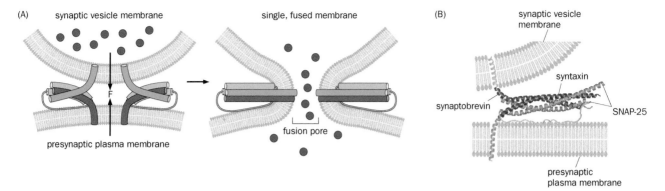

Figure 3-8 Model and structural basis of synaptic vesicle fusion.
(A) Schematic models of SNARE complexes before and after membrane fusion. Before fusion (left), the vesicular and target SNAREs are on separate membranes (the synaptic vesicle membrane and the presynaptic plasma membrane, respectively). The strong binding of their cytoplasmic domains, in a zipper-like fashion starting from the two sides and progressing toward the center, produces a force (F) that brings the vesicle and target membranes together, causing them to fuse (right). Colored rods represent helices from the SNARE proteins detailed in Panel B. **(B)** Structure of the SNARE complex for synaptic vesicle fusion determined by X-ray crystallography.

Blue, red, and green represent α helices from the cytoplasmic domains of synaptobrevin, syntaxin, and SNAP-25, respectively. Faded blue and red represent transmembrane domains of synaptobrevin and syntaxin, respectively, which were not part of the solved crystal structure. The orange strand links the two SNAP-25 helices and is attached to the presynaptic plasma membrane through a lipid modification, which was also not part of the crystal structure. (A, adapted from Südhof TC & Rothman JE [2009] *Science* 323:474–477. B, from Sutton, RB, Fasshauer D, Jahn R, et al. [1998] *Nature* 395:347–353. With permission from Springer Nature.)

brane; Figure 2-2) resemble the v- and t-SNAREs for synaptic vesicle exocytosis. These findings suggest that the mechanism of synaptic vesicle exocytosis was co-opted from general vesicle trafficking (see Section 13.9 for more discussions).

In all of these reactions, including synaptic vesicle fusion, however, SNARE proteins were found to be insufficient to mediate fusion. A partner for SNARE proteins in all fusion reactions is an SM protein called Munc18 in mammals. Munc18

Box 3-2: From toxins to medicines

Research in neurobiology has greatly benefited from naturally occurring toxins that have evolved to block specific steps of neuronal communication. These toxins are produced by organisms from a wide range of phylogenetic groups, including bacteria, protists, plants, fungi, and animals. Despite the energetic costs of producing toxins, they offer adaptive advantages such as deterring herbivores, fending off predators, and immobilizing prey. Scientists have used these toxins to study the biological functions and mechanisms of action of their target proteins. Some of these toxins have even been developed into medicines.

Virtually all steps of neuronal communication are targets for toxins. Action potentials are potently blocked by tetrodotoxin (Figure 2-29), an inhibitor of voltage-gated Na⁺ channels produced by symbiotic bacteria in puffer fish, rough-skinned newts, and certain octopi. The presynaptic voltage-gated Ca²⁺ channels essential for neurotransmitter release are specifically blocked by a small peptide, ω-conotoxin, from marine snails. Synaptic vesicle fusion is blocked by a number of proteases produced by the bacteria *Clostridium tetani* and *Clostridium botulinum*. **Tetanus** and **botulinum toxins** specifically cleave SNARE proteins, with each toxin cleaving a specific SNARE at a specific amino acid residue, thereby preventing synaptic vesicle fusion with presynaptic membrane (Figure 3-8). Indeed, identification of the protein targets of tetanus and botulinum toxins was

instrumental in establishing that SNARE proteins play a central role in synaptic vesicle fusion. Toxins that target neurotransmitter receptors will be discussed later in this chapter. For instance, **curare**, a plant toxin used by Native Americans on poisonous arrows, and **α-bungarotoxin** and cobratoxin from snakes are all potent competitive inhibitors of acetylcholine receptors at the vertebrate NMJ and thereby block motor neuron–triggered muscle contraction. **Picrotoxin**, another plant toxin, is a potent blocker of the GABA$_A$ receptors that mediate fast inhibition in vertebrates and invertebrates alike. **Muscimol**, produced by toxic mushrooms, is a potent activator of these GABA$_A$ receptors. The venoms of predators such as snakes, scorpions, cone snails, and spiders have been a rich source of tools for investigating neuronal communication. The fact that most toxins affect many different animal species also indicates that the molecular machinery of neuronal communication is highly conserved across the animal kingdom.

Natural toxins and their derivatives have also been used extensively in medicine. Channel blockers are used to treat epilepsy and intractable pain. Synaptic transmission blockers are used as muscle relaxants. For example, botulinum toxin A, commonly known as Botox, can be injected into specific eye muscles to treat strabismus (misaligned eyes). Botox injections have also become a popular cosmetic procedure to temporarily remove wrinkles.

binds to SNAREs throughout the fusion reaction and is essential for fusion, although its exact role in mediating vesicle fusion remains unclear.

3.6 Synaptotagmin acts as a Ca²⁺ sensor to trigger synaptic vesicle fusion

How does Ca^{2+} entry regulate neurotransmitter exocytosis? A prime candidate linking these two events is a class of transmembrane proteins on the synaptic vesicle called **synaptotagmins** (Figure 3-7), which possess five or six Ca^{2+} binding sites on their cytoplasmic domains. To test the function of synaptotagmin in synaptic transmission, **knockout** mice were created in which synaptotagmin-1, the predominant form of synaptotagmin expressed in forebrain neurons, was disrupted using the gene targeting method (see Section 14.7 for details). To assay for synaptic transmission, embryonic hippocampal neurons from control or knockout mice were dissociated and cultured *in vitro* to allow synapse formation and subjected to a variation of the patch clamp technique called **whole-cell patch recording** (**whole-cell recording** in short). (In whole-cell recordings, the membrane underneath the patch pipette is ruptured such that the patch pipette is connected to the entire neuron; see Section 14.21 and Box 14-3 for details.) Pairs of cultured neurons were recorded simultaneously to test for synaptic connections between each pair. In a connected wild-type pair, depolarization of the presynaptic neuron, which caused it to fire action potentials, resulted in inward currents in the postsynaptic neuron, indicating successful synaptic transmission. By contrast, in knockout pairs, depolarization of the presynaptic neuron elicited much smaller postsynaptic responses, indicating that synaptotagmin-1 is required for normal synaptic transmission (**Figure 3-9**A). Parallel studies in *Drosophila* and *C. elegans* indicated that disruption of synaptotagmin homologs in these invertebrate model organisms (see Section 14.2) also impaired synaptic transmission.

The knockout experiment did not prove that synaptotagmin acts as a Ca^{2+} sensor, as disrupting other genes encoding proteins essential for synaptic transmission, such as the v-SNARE synaptobrevin, similarly blocked synaptic transmission. Subsequent experiments have provided strong evidence that synaptotagmin is a major Ca^{2+} sensor that regulates neurotransmitter release. For example, a mutant synaptotagmin-1 with a single amino acid change that reduces Ca^{2+} binding by 50% in an *in vitro* biochemical assay was identified. When researchers replaced the endogenous synaptotagmin-1 with this mutant synaptotagmin-1 in a varia-

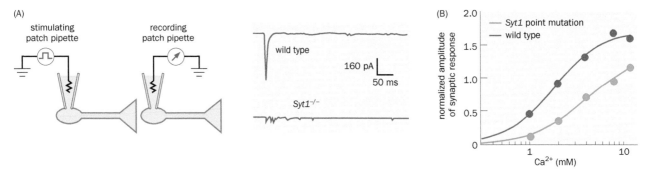

Figure 3-9 Synaptotagmin, a Ca²⁺ sensor that functions in synaptic transmission. (A) Left, schematic of experimental preparation to examine the role of synaptotagmin-1 (Syt1) in synaptic transmission. Pairs of cultured hippocampal neurons were subjected to simultaneous whole-cell recording. Depolarizing currents were injected into presynaptic neurons to cause them to fire action potentials, and postsynaptic responses were recorded as inward currents when the membrane potential was clamped at −70 mV. Right, compared with the inward current triggered by a presynaptic action potential between a pair of wild-type neurons (top trace), the synaptic response between a pair of neurons from *Syt1* knockout mice (lacking both copies of the *Syt1* gene) was greatly diminished (bottom trace). **(B)** A point mutation

in *Syt1* that reduced Ca^{2+} binding by 50% also reduced the sensitivity of neurotransmitter release in cultured hippocampal neurons to Ca^{2+} by about 50%, as indicated by the downward shift of the mutant curve relative to the wild-type curve; each curve depicts normalized synaptic transmission amplitude against Ca^{2+} concentration. This finding suggests that synaptotagmin-1 acts as a Ca^{2+} sensor for synaptic vesicle fusion in hippocampal neurons. (A, adapted from Geppert M, Goda Y, Hammer RE, et al. [1994] *Cell* 79:717–727. With permission from Elsevier Inc. B, adapted from Fernández-Chacón R, Königstorfer A, Gerber SH, et al. [2001] *Nature* 410:41–49. With permission from Springer Nature.)

tion of the knockout procedure called **knock-in** (Section 14.7), neurons derived from the knock-in mice exhibited a corresponding 50% reduction in the Ca^{2+} sensitivity of neurotransmitter release (Figure 3-9B), consistent with synaptotagmin-1 acting as a Ca^{2+} sensor. Another protein involved in neurotransmitter release is complexin, which has a complex role, both activating the SNARE complex and blocking it in an intermediate step. One current model is that synaptotagmin releases the inhibitory block of complexin in a Ca^{2+}-dependent manner, thus allowing SNAREs to complete the vesicle fusion reaction in response to a rise in intracellular Ca^{2+} concentration.

At fast mammalian CNS synapses at physiological temperatures, action potential arrival can cause postsynaptic depolarization within as little as 150 μs. This interval includes about 90 μs for the opening of voltage-gated Ca^{2+} channels during the action potential upstroke, allowing Ca^{2+} influx, and 60 μs in total for Ca^{2+}-triggered vesicle fusion, neurotransmitter diffusion across the synaptic cleft, and the postsynaptic physiological response. To enable this rapid action, synaptic vesicles are docked at the active zone ready for release (Figure 3-3), with their SNARE proteins partially preassembled in high-energy configurations, yet blocked, waiting for the action of a Ca^{2+} sensor to release the break and complete SNARE assembly, which drives membrane fusion.

The rapidity of neurotransmitter release following Ca^{2+} entry described earlier, as well as its transience, allows the presynaptic terminal to respond to future action potentials with more neurotransmitter release. This requires the presynaptic concentration of free Ca^{2+} to be rapidly reduced after Ca^{2+} entry stops; this is accomplished by proteins that rapidly sequester Ca^{2+} or pump it out of the cytoplasm. It is also important that the presynaptic Ca^{2+} sensor has a low binding affinity for Ca^{2+}, as a high-affinity sensor would be bound for a longer time period. Indeed, synaptotagmin employs multiple low-affinity Ca^{2+}-binding sites that bind Ca^{2+} cooperatively (that is, binding of one Ca^{2+} facilitates binding of a second); only when multiple sites bind to Ca^{2+} is it able to trigger neurotransmitter release. Together, these mechanisms ensure that neurotransmitter release is triggered only transiently and locally at the site of Ca^{2+} entry.

3.7 The presynaptic active zone is a highly organized structure

The rapidity and transience of Ca^{2+}-induced neurotransmitter release relies on the proximity of voltage-gated Ca^{2+} channels and docked synaptic vesicles in the active zone. Indeed, Ca^{2+} imaging of presynaptic terminals (e.g., Figure 3-6) suggested that the rise in intracellular Ca^{2+} concentration in response to depolarization is highly restricted to microdomains near the active zone. Although the intracellular Ca^{2+} concentration is normally very low (~0.1 μM), it can shoot up transiently several orders of magnitude in the microdomain; this facilitates cooperative binding of Ca^{2+} to multiple Ca^{2+}-binding sites of synaptotagmin, allowing it to achieve the conformational change necessary for triggering vesicle fusion.

The molecular machinery that organizes the active zone has been extensively characterized (**Figure 3-10**). A central player is a protein called Unc13. Unc13 binds and activates the t-SNAREs and also reaches up and tethers the v-SNARE synaptobrevin and synaptic vesicle to the release site. Voltage-gated Ca^{2+} channels are recruited by two other active zone core components, RIM (Rab3-interacting molecule) and RIM-BP (RIM-binding protein). RIM binds Rab3, a synaptic vesicle–associated small GTPase, and thus brings synaptic vesicles close to voltage-gated Ca^{2+} channels. RIM and RIM-BP also interact with other active zone proteins, which in turn associate with the actin cytoskeleton that supports the structural integrity of the presynaptic terminal and transports molecules into the presynaptic terminal (Figure 2-6).

The speed of neurotransmission requires neurotransmitters to diffuse only a short distance to neurotransmitter receptors on the postsynaptic cell. The active zone has many synaptic adhesion molecules linking it to the postsynaptic membrane rich for neurotransmitter receptors. These include the presynaptic **neurexins**, which bind postsynaptic **neuroligins** (called **heterophilic binding**), and

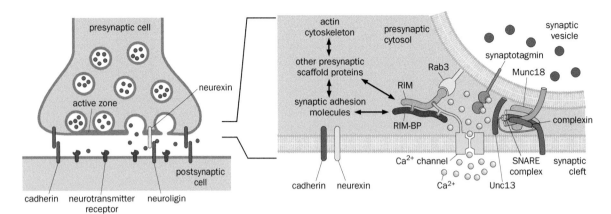

Figure 3-10 Molecular organization of the presynaptic terminal.
Left, a lower-magnification model of a chemical synapse showing presynaptic and postsynaptic cells. Transsynaptic adhesion molecules, such as neurexin (yellow), neuroligin (red), and cadherin (blue), align the active zone with a postsynaptic density enriched in neurotransmitter receptors, facilitating the rapid action of neurotransmitters. Right, a magnified model of the presynaptic active zone. Unc13 binds both t- and v-SNAREs and thus brings the synaptic vesicle close to the presynaptic plasma membrane. The RIM/RIM-BP protein complex binds to voltage-gated Ca^{2+} channels directly and to

synaptic vesicles via the Rab3 protein; this allows Ca^{2+} entry to activate synaptotagmin with minimal diffusion, which in turn releases the complexin inhibitory block on the SNARE/SM complex and causes neurotransmitter release (the SNARE complex is represented as in Figure 3-8A; Munc18 is the SM protein in mammalian synapses). RIM and RIM-BP are also associated with other presynaptic scaffold proteins, which are in turn associated with the actin cytoskeleton and with synaptic adhesion molecules. (Adapted from Südhof TC [2012] *Neuron* 75:11–25. With permission from Elsevier Inc.)

cadherins (Ca^{2+}-dependent cell adhesion proteins), present both pre- and postsynaptically, which bind each other (called **homophilic binding** in the case that the pre- and postsynaptic cadherins are the same isoform). These and other cell-adhesion molecules bring the presynaptic and postsynaptic plasma membranes within 20 nm of each other (Figure 3-10) and align the active zone with the portion of the postsynaptic membrane rich in neurotransmitter receptors.

Recent studies using super-resolution fluorescent microscopy (see Section 14.17 for more details) have begun to reveal how pre- and postsynaptic units are organized to form a functional synapse. For example, super-resolution localization of molecules in the *Drosophila* NMJ (**Figure 3-11**) indicated that RIM-BP proteins form a ring around a cluster of voltage-gated Ca^{2+} channels at the active zone presynaptic membrane. An active zone scaffold protein called Bruchpilot (similar to a mammalian protein called ELKS) extends from the center of the active zone to the periphery. Glutamate receptors are enriched in the postsynaptic density, which is aligned with the presynaptic active zone (glutamate is used as a neurotransmitter in the *Drosophila* NMJ; see Section 3.11 for details). A similar subsynaptic molecular architecture aligns presynaptic vesicle fusion with postsynaptic neurotransmitter receptors in vertebrates.

Figure 3-11 The organization of selected proteins in the *Drosophila* NMJ. This model is based on two-color labeling of different pairs of proteins (shown at right) and measurement of their distances using a technique called stimulated emission depletion microscopy, which can resolve structures ~50 nm apart (see Section 14.17 for details). For instance, the distance between the C-terminal of Bruchpilot and postsynaptic glutamate receptors was estimated by measuring the distance between the fluorescence signal from an antibody against the C-terminus of Bruchpilot and that from an antibody against the glutamate receptor (151 ± 24 nm apart). RIM-BP, Rab3-interacting-molecule binding protein; VGCC, voltage-gated Ca^{2+} channel; GluR, glutamate receptor; N, amino terminus; C, carboxy terminus; SEM, standard error of the mean. (From Liu KSY, Siebert M, Mertel S, et al. [2011] *Science* 334: 1565–1569. With permission from AAAS.)

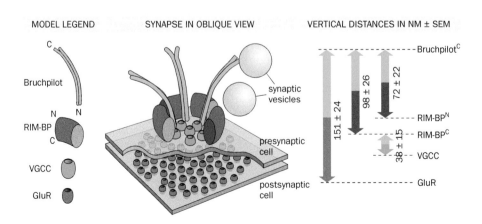

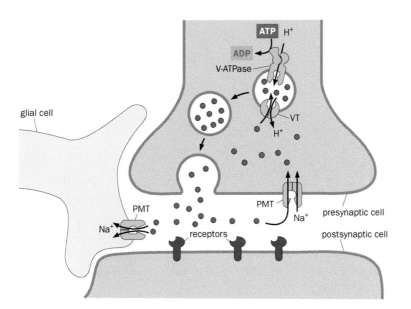

Figure 3-12 Clearance and recycling of neurotransmitters. After being released into the synaptic cleft, excess transmitters are taken up by plasma membrane transporters (PMTs) on the presynaptic or nearby glial plasma membranes; both are symporters that utilize energy from Na$^+$ entry. Within the presynaptic cytosol, neurotransmitters are transported into synaptic vesicles by vesicular neurotransmitter transporters (VTs), antiporters that use energy from transport of protons (H$^+$) out of the synaptic vesicle down their electrochemical gradient. The V-ATPase on the synaptic vesicle membrane establishes the H$^+$ gradient across the vesicular membrane using energy from ATP hydrolysis. (Based on Blakely RD & Edwards RH [2012] *Cold Spring Harb Perspect Biol* 4:a005595.)

3.8 Neurotransmitters are cleared from the synaptic cleft by degradation or transport into cells

For postsynaptic neurons to continually respond to firing of presynaptic neurons, neurotransmitters released in response to each presynaptic action potential must be efficiently cleared from the synaptic cleft. Although neurotransmitters can diffuse out of the synaptic cleft, additional mechanisms are employed to speed up neurotransmitter clearance.

Acetylcholine at the NMJ is rapidly degraded by **acetylcholinesterase**, an enzyme enriched in the synaptic cleft. Indeed, this enzyme is so active that most acetylcholine molecules released by motor axon terminals are degraded while diffusing across the short distance of the synaptic cleft. Some of the physiology experiments described in earlier sections involving mEPP measurements actually included acetylcholinesterase inhibitors in the saline to boost the mEPP amplitude.

In most other neurotransmitter systems, transmitter molecules are recycled after they are released, rather than being degraded. In a process called **neurotransmitter reuptake**, excess neurotransmitters are first taken back into the presynaptic cytosol via **plasma membrane neurotransmitter transporters**. These are symporters that move Na$^+$ into the cell along with the neurotransmitter itself; because Na$^+$ is moving down its electrochemical gradient as it enters the cell, the energy stored in the Na$^+$ gradient is harnessed to drive neurotransmitter reuptake (**Figure 3-12**; Movie 3-1). Once in the cytosol, neurotransmitters refill new and recycled synaptic vesicles (see Section 3.9) via a second transporter, the **vesicular neurotransmitter transporter** on the synaptic vesicle (Figure 3-7). The energy for vesicular transporters derives from transport of protons out of the synaptic vesicle down their electrochemical gradient. The electrochemical gradient of protons (high in vesicles and low in the cytosol) is created by the **V-ATPase**, which hydrolyzes ATP to pump protons (H$^+$) into the synaptic vesicle. In some neurotransmitter systems, including glutamate, secreted transmitters are cleared from the synaptic cleft by neurotransmitter transporters on the plasma membrane of glial cells, which wrap around many synapses (Figure 3-3). In Chapter 12, we will learn more about neurotransmitter reuptake mechanisms, as drugs altering these mechanisms are widely used to treat psychiatric disorders.

3.9 Synaptic vesicle recycling by endocytosis is essential for continual synaptic transmission

To maintain the ability to respond to sustained neuronal firing, presynaptic terminals must be able to replenish the stockpile of synaptic vesicles filled with

Figure 3-13 The synaptic vesicle cycle. After membrane fusion between the synaptic vesicle and the presynaptic membrane and release of neurotransmitters into the synaptic cleft, synaptic vesicles can be recycled via three pathways. In the kiss-and-run pathway, synaptic vesicles reform after a very transient fusion with limited exchange of proteins and lipids with the presynaptic plasma membrane (1a). In clathrin-mediated endocytosis, the synaptic vesicle membrane fuses fully with the presynaptic plasma membrane and is then retrieved via clathrin-mediated endocytosis (1b). In ultrafast endocytosis, the presynaptic plasma membrane is rapidly endocytosed, forming vesicles larger than synaptic vesicles; synaptic vesicles are produced subsequently from endosomes via a clathrin-mediated process (1c). The interior of vesicles is acidified by pumping protons (H⁺) inside using the V-ATPase on the synaptic vesicle membrane; the synaptic vesicle is then ready to be filled with neurotransmitter by the proton export-coupled vesicular transporter (2; Figure 3-12). Synaptic vesicles filled with neurotransmitters join the pool of vesicles in the presynaptic terminal (3). Some vesicles transit into the readily releasable pool and are docked at the active zone (4) ready for exocytosis. Finally, Ca²⁺ entry through voltage-gated Ca²⁺ channels at the active zone triggers vesicle fusion (5). (Based on Südhof TC [2004] *Ann Rev Neurosci* 27:509–547; Watanabe S, Trimbuch T, Camecho-Pérez M, et al. [2014] *Nature* 515:228–233.)

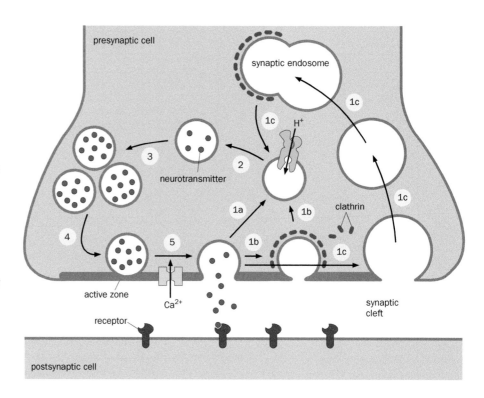

neurotransmitters. The synaptic vesicle membrane and proteins are mostly synthesized in the soma. Because the distance between synaptic terminals and the soma can be up to 1 meter, synaptic vesicles must be rapidly recycled at the synapse to support future rounds of synaptic transmission. Two components of synaptic vesicles must be recycled and sorted into new vesicles: membranes and proteins.

Several mechanisms have been proposed for retrieving synaptic vesicle membranes (**Figure 3-13**). The first mechanism, "kiss and run," involves a very transient fusion of the synaptic vesicle with the presynaptic plasma membrane for neurotransmitter release, followed by closure of the pore so that mixing of the vesicle's protein and lipid content with the presynaptic plasma membrane is limited. In the second mechanism, the synaptic vesicle collapses into the presynaptic plasma membrane after fusion and is retrieved in the presynaptic terminal by clathrin-mediated endocytosis. (Clathrin is a protein that assembles into a cage on the cytoplasmic side of a membrane to form a coated pit, which buds off to form a clathrin-coated vesicle.) More recently, a third mechanism has been described, in which endocytic vesicles larger than synaptic vesicles are produced from the presynaptic membrane surrounding the active zone; these large endocytic vesicles then become part of synaptic endosomes, from which synaptic vesicles are subsequently produced by a clathrin-mediated process. Time-resolved electron microscopy revealed that large endocytic vesicles form within 100 ms after vesicle fusion; this is much faster than clathrin-mediated endocytosis, which usually takes seconds, potentially allowing more rapid production of synaptic vesicles for repeated use. Which of these mechanisms predominates at any given synapse is not known.

Proteins must be properly sorted to regenerate functional synaptic vesicles. Most importantly, the SNARE proteins must be pulled apart and sorted to the plasma and vesicle membranes. The SNARE complexes are disassembled by NSF in an ATP-dependent manner. (NSF stands for <u>N</u>-ethylmaleimide-<u>s</u>ensitive <u>f</u>usion protein, named after a chemical inhibitor that blocks vesicle fusion reactions *in vitro*; SNARE stands for <u>s</u>oluble <u>NSF</u>-<u>a</u>ttachment protein <u>re</u>ceptor.) Syntaxin and SNAP-25 remain in the presynaptic plasma membrane, whereas synaptobrevin is returned to the synaptic vesicle. Synaptobrevin and other vesicle proteins, such as synaptotagmin and the vesicular transporters, are recruited to the site of vesicle formation by specialized adaptor proteins that interact with the clathrin machinery. Once assembly of the vesicle is complete, the neck attaching it to the membrane narrows and is cleaved by a protein called **dynamin**. The clathrin coat is then stripped off the vesicle.

Fully regenerated vesicles are then acidified by the proton pump V-ATPase and refilled with neurotransmitters (Figure 3-12). Filled vesicles join the synaptic vesicles in the presynaptic terminal. A synaptic vesicle-associated protein called synapsin, a marker widely used to identify synapses, links vesicles together at the synapse to maintain their proximity to the presynaptic zone. A small subset of synaptic vesicles constitutes the **readily releasable pool**, operationally defined as the pool of vesicles most easily released. Readily releasable vesicles may be the vesicles docked at active zones in the high-energy configuration of a preassembled SNARE complex and readied for further rounds of neurotransmitter release in response to depolarization-induced Ca^{2+} entry (Figure 3-13).

We use a specific example to illustrate the importance of synaptic vesicle retrieval for continued synaptic transmission and neuronal communication. To identify genes necessary for neuronal communication, forward genetic screens (see Section 14.6 for details) were carried out in the fruit fly *Drosophila* to isolate mutations causing paralysis in flies kept at high temperatures. This led to the discovery of a temperature-sensitive mutation called *Shibire^ts^*. Mutant flies behave normally at room temperature (~20°C) but are paralyzed shortly after shifting to elevated temperatures (>29°C); their motility returns within a few minutes after flies are returned to 20°C. Molecular-genetic analysis revealed that the *Shibire* gene encodes dynamin, the previously mentioned protein that cleaves vesicles from the membrane. Dynamin is essential for both clathrin-mediated endocytosis of synaptic vesicles and ultrafast endocytosis. The *Shibire^ts^* mutation causes the dynamin collar to become locked and unable to cleave vesicles from the membrane at elevated temperatures. Without vesicle recycling, presynaptic terminals are rapidly deprived of synaptic vesicles (**Figure 3-14**) and became unable to release neurotransmitters in response to further action potentials, thus causing paralysis. The *Shibire^ts^* mutation has proven to be a useful tool for rapidly and reversibly silencing specific neurons *in vivo* to analyze their function in information processing within neural circuits (Section 14.23).

(A)

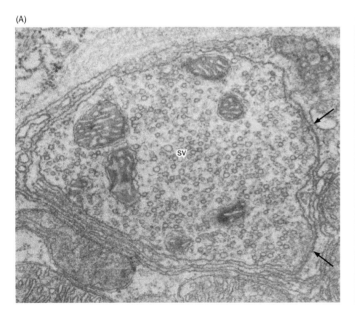

(B)

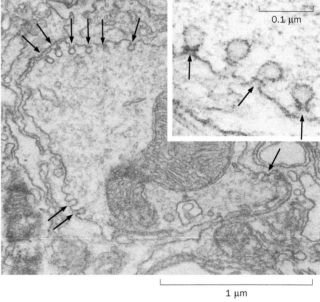

Figure 3-14 Electron micrographs of synapses in temperature-sensitive *Shibire^ts^* mutant fruit flies. (A) A neuromuscular junction from a *Shibire^ts^* mutant fly fixed at 19°C. The presynaptic terminal is abundant in synaptic vesicles (sv). Arrows indicate active zones. **(B)** An NMJ fixed 8 minutes after raising the temperature to 29°C.

Note the reduced number of synaptic vesicles in the presynaptic terminal compared with Panel A and the presence of "collared" vesicles (arrows, higher magnification in inset), indicating a block of the last step of endocytosis. (From Koenig JH & Ikeda K [1989] *J Neurosci* 9:3844–3860. Copyright ©1989 Society for Neuroscience.)

Table 3-1:   A molecular cast for neurotransmitter release

Molecule	Location	Functions
Synaptic vesicle fusion with presynaptic membrane		
Synaptobrevin/VAMP	Synaptic vesicle	Mediates vesicle fusion (v-SNARE)
Syntaxin	Presynaptic plasma membrane	Mediates vesicle fusion (t-SNARE)
SNAP-25	Presynaptic plasma membrane	Mediates vesicle fusion (t-SNARE)
Sec1/Munc18 (SM)	Presynaptic cytosol	Binds to SNARE complex and essential for vesicle fusion
Unc13	Active zone	Promotes assembly of SNARE complex
Ca^{2+} regulation of synaptic transmission		
Voltage-gated Ca^{2+} channel	Active zone of presynaptic membrane	Allows Ca^{2+} entry in response to action potential-triggered depolarization
Ca^{2+}	Entering from extracellular space to presynaptic cytosol	Triggers synaptic vesicle fusion
Synaptotagmin	Synaptic vesicle	Senses Ca^{2+} to trigger vesicle fusion
Complexin	Presynaptic cytosol	Binds and regulates SNARE-mediated vesicle fusion
Organization of presynaptic terminal (and alignment with postsynaptic density)		
RIM	Active zone	Organizes presynaptic scaffold
RIM-BP	Active zone	Organizes presynaptic scaffold
ELKS/Bruchpilot	Active zone	Organizes presynaptic scaffold
Rab3	Synaptic vesicle	Interacts with active zone components
Cadherin	Presynaptic and postsynaptic plasma membranes	Trans-synaptic adhesion
Neurexin	Presynaptic plasma membrane	Trans-synaptic adhesion
Neuroligin	Postsynaptic plasma membrane	Trans-synaptic adhesion
Neurotransmitter and vesicle recycling		
Acetylcholinesterase	Synaptic cleft	Degrades neurotransmitter acetylcholine
Plasma membrane neurotransmitter transporter (PMT)	Presynaptic plasma membrane, glial plasma membrane	Transports excess neurotransmitter molecules back to presynaptic cytosol or to nearby glia
Vesicular neurotransmitter transporter (VT)	Synaptic vesicle	Transports neurotransmitters from presynaptic cytosol to the synaptic vesicle
V-ATPase	Synaptic vesicle	Establishes proton gradient within the synaptic vesicle
Synapsin	Synaptic vesicle	Maintain vesicles in the vicinity of the active zone
Clathrin	Presynaptic cytosol	Shapes recycled vesicles from presynaptic plasma membrane or endosome membrane during endocytosis
Shibire/dynamin	Presynaptic cytosol	Retrieves vesicles from presynaptic plasma membrane via endocytosis
NSF	Presynaptic cytosol	Disassembles SNARE complex after fusion

As a summary of what we have learned so far, Table 3-1 provides a list of molecules that play key roles in mediating and regulating the sequence of events essential for neurotransmitter release.

3.10 Synapses can be facilitating or depressing

Because synaptic transmission is the key mode of interneuronal communication, the **efficacy of synaptic transmission**, measured by the magnitude of the postsynaptic response to a presynaptic stimulus, is regulated in many ways. **Synaptic plasticity**, the ability to change the efficacy of synaptic transmission, is an extremely important property of the nervous system. Synaptic plasticity is usually

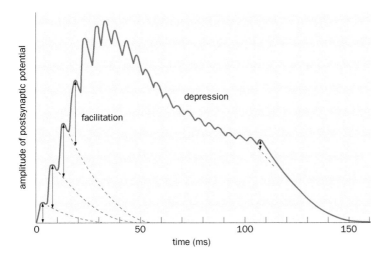

Figure 3-15 Synaptic facilitation and depression. In this schematic, the amplitudes of postsynaptic potentials, indicated by the lengths of the double arrows parallel to the y axis, change in response to a train of action potentials. The first series exhibits facilitation, as each successive action potential produces a larger response; the latter series exhibits depression as responses become smaller and smaller for each successive action potential. The dashed lines represent the natural decay of postsynaptic potentials, had there not been follow-up action potentials, and were used to determine the amplitude of postsynaptic potentials in response to successive action potentials. (From Katz B [1966] Nerve, Muscle, and Synapse. With permission from McGraw Hill.)

divided into **short-term synaptic plasticity**, which occurs within milliseconds to minutes, and **long-term synaptic plasticity**, which can extend from hours to the lifetime of an animal. We discuss in the following the two simplest forms of short-term plasticity, which involve changes in neurotransmitter release probability. Long-term synaptic plasticity will be examined in Chapter 11 in the context of memory and learning.

Although Ca^{2+}-dependent synaptic vesicle fusion provides an essential link between action potential arrival and neurotransmitter release, not every action potential produces the same neurotransmitter release probability. As discussed earlier, the quantal content of CNS synapses is much lower than that of the NMJ, as presynaptic axons typically form at most a few active zones onto any given postsynaptic partner neuron. In most mammalian CNS synapses *in vivo*, the average **release probability** (defined as the probability that an active zone of a presynaptic terminal releases the transmitter contents of one or more synaptic vesicles following an action potential) is estimated to be far smaller than 1. If many active zones exist between a presynaptic cell and a postsynaptic cell, as is the case at the vertebrate NMJ, the probability that at least one active zone releases a vesicle approaches 1; however, the magnitude of the postsynaptic response still depends on the release probability of each active zone.

The release probability can be affected by prior synaptic activity. At **facilitating synapses**, successive action potentials trigger larger and larger postsynaptic responses. By contrast, at **depressing synapses**, successive action potentials result in smaller and smaller postsynaptic responses (**Figure 3-15**). Fast facilitation and depression are most often due to presynaptic factors, such as altered amounts of neurotransmitter release. The same synapse can be facilitating or depressing, depending on its intrinsic properties and recent activity.

In the simplest case, facilitating synapses have low starting release probabilities. The amount of release increases during repeated action potentials as active zone Ca^{2+} builds up. Depressing synapses, on the other hand, are usually characterized by high starting release probabilities that result in substantial release at the beginning of a stimulus train; this exhausts the pool of readily releasable vesicles, leading to a decline in the amount of release per spike as the stimulus train proceeds. Typically, large numbers of vesicles in the presynaptic terminal can replenish depleted vesicles in the readily releasable pool, so this sort of depression can recover in seconds. In the course of this book, we will encounter additional mechanisms for adjusting synaptic strength via distinct processes and at different temporal scales.

3.11 Nervous systems use many neurotransmitters

To illustrate the basic principles of synaptic transmission, we have focused primarily on the vertebrate NMJ, which employs the neurotransmitter acetylcholine.

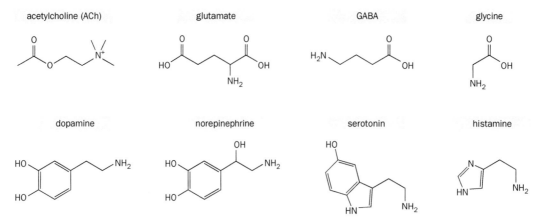

Figure 3-16 Structures of widely used small-molecule neurotransmitters. Acetylcholine is produced from choline and acetyl-CoA. Glutamate and glycine are natural amino acids. GABA (γ-amino butyric acid) is produced from glutamate. Dopamine is derived from the amino acid tyrosine. Norepinephrine is produced from dopamine and is a precursor for the hormone epinephrine. Serotonin is derived from the amino acid tryptophan. Histamine is derived from the amino acid histidine. See Figure 12-20 for the biosynthetic pathway that produces dopamine, norepinephrine, and epinephrine.

The principles learned thus far apply to virtually all chemical synapses, regardless of the neurotransmitter they use (**Figure 3-16**; **Table 3-2**). Two major neurotransmitters used in the vertebrate CNS are **glutamate** (glutamic acid), a natural amino acid, and **GABA** (γ-amino butyric acid), which is converted from glutamate by the enzyme **glutamic acid decarboxylase** (**GAD**). Glutamate is the predominant **excitatory neurotransmitter** in the vertebrate nervous system because its release depolarizes postsynaptic neurons, making them more likely to fire action potentials. GABA is an **inhibitory neurotransmitter** because its release usually renders postsynaptic neurons less likely to fire action potentials. The amino acid **glycine**, another inhibitory neurotransmitter, is used by a subset of inhibitory neurons in the vertebrate brainstem and spinal cord.

GABA is the major inhibitory neurotransmitter in diverse species, including invertebrates such as the nematode *C. elegans*, the fruit fly *Drosophila melanogaster*, and crustaceans. Indeed, GABA's inhibitory action was first established in the crab. Like vertebrates, *C. elegans* also uses ACh as the excitatory transmitter at its NMJ and glutamate as the major neurotransmitter in its CNS. Curiously,

Table 3-2: Widely used neurotransmitters

Neurotransmitter	Major uses in the vertebrate nervous system[a]
Acetylcholine	Motor neurons that excite muscle; ANS[b] neurons; CNS excitatory and modulatory neurons
Glutamate	Most CNS excitatory neurons; most sensory neurons
GABA	Most CNS inhibitory neurons
Glycine	Some CNS inhibitory neurons (mostly in the brainstem and spinal cord)
Serotonin (5-HT)	CNS modulatory neurons; neurons in the gastrointestinal tract
Dopamine	CNS modulatory neurons
Norepinephrine	CNS modulatory neurons; ANS[b] neurons
Histamine	CNS modulatory neurons
ATP, adenosine	Some sensory and CNS neurons
Neuropeptides	Usually co-released from excitatory, inhibitory, or modulatory neurons; neurosecretory cells

[a] See text for variations in invertebrate nervous systems.

[b] ANS, autonomic nervous system; as will be discussed in more detail in Chapter 9, acetylcholine and norepinephrine are used in different types of ANS neurons.

Drosophila utilizes ACh as the major excitatory neurotransmitter in its CNS and glutamate as the transmitter at its NMJ (Figure 3-11).

Although we often identify neurotransmitters as excitatory or inhibitory, it is important to note that neurotransmitters do not possess such qualities intrinsically—they are just molecules. The excitatory or inhibitory quality arises from the ion conductances these molecules activate in the postsynaptic neuron. Thus, we consider glutamate and acetylcholine excitatory neurotransmitters because of their actions on most postsynaptic neurons, but in some cases they can be inhibitory, as we will see in the next part of this chapter.

Another important class of neurotransmitters plays a predominantly modulatory role. **Modulatory neurotransmitters** (also called **neuromodulators**) typically act on longer time scales and/or larger spatial scales than the classical neurotransmitters described so far. Neuromodulators can shift the membrane potential up or down, thereby influencing neuronal **excitability** (how readily a neuron fires an action potential). Classic neuromodulators include **serotonin** (also called **5-HT** for 5-hydroxytryptamine), **dopamine**, **norepinephrine** (also called noradrenaline), and **histamine** (Figure 3-16); these are all derived from aromatic amino acids and are collectively called **monoamine neurotransmitters**. In addition to being released into the synaptic cleft, these neurotransmitters can be released into the extracellular space outside of morphologically defined synapses to affect nearby cells; this process is called **volume transmission**. In vertebrates, the cell bodies of neurons that synthesize monoamine neurotransmitters are mostly clustered in discrete nuclei in the brainstem or hypothalamus. They send profuse axons that collectively innervate a large fraction of the CNS (see Box 9-1 for details). Dopamine and serotonin act as neuromodulators throughout the animal kingdom. In place of norepinephrine, a chemically similar molecule called **octopamine** is used in some invertebrate nervous systems. Finally, ATP and its derivative adenosine can act as neurotransmitters in some sensory neurons and in the CNS.

Some neurotransmitters have different roles in different parts of the nervous system (Table 3-2). In vertebrates, ACh is used as an excitatory neurotransmitter by motor neurons to control skeletal muscle contraction at the NMJ. It is also one of two neurotransmitters employed by the **autonomic nervous system** for neural control of visceral processes such as heartbeat, respiration, and digestion. In the brain, ACh can act as both an excitatory neurotransmitter and a neuromodulator. Likewise, norepinephrine functions as the autonomic nervous system's other neurotransmitter, but acts as a neuromodulator in the brain.

The type of neurotransmitter a neuron releases is often used as a major criterion for neuronal classification: glutamatergic, GABAergic, cholinergic, etc. Neurons of a given neurotransmitter type express a specific set of genes associated with that type, including enzyme(s) that synthesize the neurotransmitter, a vesicular transporter that pumps the neurotransmitter into synaptic vesicles, and in many cases a plasma membrane transporter that retrieves the neurotransmitter from the synaptic cleft after release (Figure 3-12). However, recent studies have identified an increasing number of cases in which the same neuron releases more than one neurotransmitter. For example, some serotonin or dopamine neurons co-release glutamate or GABA. Some cholinergic neurons co-release GABA. These findings muddy the definition of neuronal types by the neurotransmitter they release. In most cases, it remains unclear whether two transmitters are released from the same synaptic vesicle or distinct vesicles, and whether they are released from the same presynaptic terminal or distinct terminals. These are important questions to be addressed in the future.

In addition to the small-molecule neurotransmitters we have discussed thus far, some neurons also secrete **neuropeptides** that act as neurotransmitters. The mammalian nervous system utilizes dozens of neuropeptides, with lengths ranging from a few amino acids to several dozen. As we will learn in Chapters 9 and 10, neuropeptides regulate diverse and vital physiological functions such as eating, sleeping, and sexual behaviors. Neuropeptides are usually produced by proteolytic cleavage of precursor proteins in the secretory pathway (Figure 2-2). They are packaged into **large dense-core vesicles** (which are larger than synaptic vesicles and contain electron-dense materials) that bud off from the Golgi apparatus and are delivered to axons and presynaptic terminals via fast axonal transport. Neuropeptides cannot be locally synthesized or recovered after release and must be transported across long distances from the soma to axon terminals. This may account for their more sparing use: the probability of neuropeptide release seems to be much lower than that of small-molecule neurotransmitters in the same terminals. We know far less about the mechanisms controlling neuropeptide release from large dense-core vesicles than we do about the mechanisms controlling the release of small-molecule neurotransmitters from synaptic vesicles. In most cases, neuropeptides play modulatory roles and are released by neurons that also use a small-molecule neurotransmitter. Indeed, large dense-core vesicles can contain not only peptides but also small-molecule neuromodulators like monoamines. As we will learn in Chapter 9, some neurons and neuroendocrine cells secrete neuropeptides into the bloodstream; in these cases, neuropeptides act as **hormones** to influence the physiology of remote recipient cells.

The reason different neurotransmitters have different effects is that their postsynaptic receptors have different properties. We now turn to the next step of neuronal communication: mechanisms by which neurotransmitters influence postsynaptic neurons.

HOW DO NEUROTRANSMITTERS ACT ON POSTSYNAPTIC NEURONS?

In the first part of the chapter, we used postsynaptic responses, such as the end-plate potential (Figure 3-1) and postsynaptic inward current (Figures 3-5 and 3-9), as assays to investigate the mechanisms of presynaptic neurotransmitter release. In the following sections, we study the mechanisms by which postsynaptic neurons produce these responses. We first discuss rapid responses, occurring within milliseconds, caused by direct changes in ionic conductances. We then study responses occurring in tens of milliseconds to seconds mediated by intracellular signaling pathways. We further highlight responses occurring in hours to days and involving expression of new genes. Finally, we summarize how postsynaptic neurons integrate different inputs to determine their own firing patterns and neurotransmitter release properties, thus completing a full round of neuronal communication.

3.12 Acetylcholine opens a nonselective cation channel at the neuromuscular junction

We begin our journey across the synaptic cleft by returning to the vertebrate neuromuscular junction (NMJ). In Section 3.1, we learned that acetylcholine (ACh) released from motor axon terminals depolarizes the muscle membrane and that iontophoretic application of ACh to muscle mimics ACh release from presynaptic terminals (Figure 3-1). How does ACh accomplish this? By locally applying ACh to different regions of muscle fibers, researchers found that exogenous ACh produced the most effective depolarization near motor axon terminals. These experiments implied that receptors for ACh must be present on the muscle membrane and concentrated at the NMJ. Upon ACh binding, ACh receptors trigger a change in the muscle membrane's ionic conductances within a few milliseconds.

To explore the underlying mechanisms, voltage clamp experiments analogous to those carried out on squid giant axons (Section 2.10) were performed on muscle fibers. These experiments enabled researchers to test how ACh release induced by motor axon stimulation changes ion flow across the muscle membrane (**Figure 3-17**A). In these experiments, two electrodes were inserted into the muscle cell, one to measure the membrane potential (V_m) and compare it to a desired command voltage (V_CMD) and the other to pass feedback current into the muscle to maintain V_m at the same value as V_CMD. The current injected into the muscle, which can be experimentally measured, equals the current that passes through the muscle membrane in response to ACh release, or the **end-plate cur-**

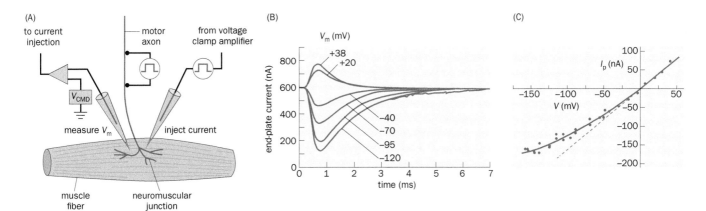

Figure 3-17 Properties of an acetylcholine-induced current studied by voltage clamp. (A) Experimental setup. Two intracellular electrodes were inserted into a muscle cell at the frog NMJ. The first (left) was to record the membrane potential (V_m), which was compared with an experimenter-determined command potential (V_{CMD}). The second electrode injected feedback current into the muscle to maintain V_m at V_{CMD}. The end-plate current in response to ACh release caused by motor axon stimulation was determined from the feedback current injected into the muscle cell to hold V_m at V_{CMD}. See Figure 2-21 for more details of voltage clamp. **(B)** The end-plate current elicited by single motor axon stimulation was measured at the six different membrane potentials indicated. At negative potentials the end-plate current was inward (positive ions flowing into the muscle cell), whereas at positive potentials the end-plate current was outward. **(C)** Peak end-plate current (I_p, y axis) as a function of the muscle membrane potential (V, x axis). Experimental data (represented as dots) fell on a curve (the I–V curve) that is close to linear (dashed line), indicating that the conductance (represented by the slope of the I–V curve) is mostly unaffected by voltage. The current switches signs between negative (inward) and positive (outward) at 0 mV, the reversal potential of the channel opened by ACh. (B & C, adapted from Magleby KL & Stevens CF [1972] *J Physiol* 223:173–197.)

rent. ACh release was found to cause an inward current at negative membrane potentials, but an outward current at positive membrane potentials (Figure 3-17B). The current–voltage relationship (plotted on an *I–V* **curve**) was nearly linear, and the membrane potential at which the current flow reversed direction (called the **reversal potential**) was approximately 0 mV (Figure 3-17C).

If the ACh-induced current were carried by a single ion, the reversal potential would equal the equilibrium potential of that ion, as both reversal potential and equilibrium potential define a state in which net current is zero. However, the reversal potential of the ACh-induced current is unlike the Na⁺, K⁺, or Cl⁻ currents we discussed in Section 2.5. These three ions have equilibrium potentials around +58 mV, –85 mV, and –79 mV, respectively. Thus, we can infer that the ACh-induced current is carried by more than one ion. Indeed, when researchers performed experiments measuring the reversal potential of the ACh-induced current while varying extracellular ion concentrations, they found that ACh opens a channel permeable to K⁺, Na⁺, and other cations but not to anions such as Cl⁻. Further evidence indicated that ACh acts on a single channel permeable to both Na⁺ and K⁺. At positive membrane potentials, the driving force for K⁺ efflux exceeds that for Na⁺ influx (because V_m is further from E_K than from E_{Na}), so K⁺ efflux exceeds Na⁺ influx, causing a net outward current. At negative membrane potentials, the driving force for Na⁺ influx exceeds that for K⁺ efflux, causing a net inward current. Ca²⁺ influx also makes a small contribution to the inward current. Importantly, since the reversal potential of 0 mV is far above the muscle membrane's resting potential (around –75 mV) and the threshold for action potential production (usually 10–20 mV more depolarized than the resting potential), the end-plate current under physiological conditions is always inward, carried by more Na⁺ influx than K⁺ efflux (**Figure 3-18A**). This depolarizes the muscle membrane, resulting in the end-plate potential (EPP) we introduced in Section 3.1.

The action of the ACh-induced current can be represented by an electrical circuit model of the muscle membrane, in which the ACh-induced current is represented in parallel with the resting current (Figure 3-18B). Immediately after the switch is on (representing ACh release), $I_{Na} = g_{Na}(V_m - E_{Na})$, and $I_K = g_K(V_m - E_K)$. Because at rest V_m is around –75 mV, the absolute value $|V_m - E_{Na}|$ far exceeds $|V_m - E_K|$. Assuming the ACh-activated channel has similar conductances for Na⁺ and K⁺ (see the following), the inward current on the Na⁺ branch will far exceed

Figure 3-18 ACh opens a nonselective cation channel on the muscle membrane. (A) Schematic of how ACh release causes depolarization of the muscle membrane. At rest (left), the membrane potential of the muscle cell is around –75 mV, similar to the resting membrane potential of many neurons, with higher K⁺ concentration inside the cell and higher Na⁺ concentration outside. ACh binding opens a cation channel on the muscle membrane permeable to both Na⁺ and K⁺. This produces more Na⁺ influx than K⁺ efflux because of the larger driving force on Na⁺, and so the muscle membrane is depolarized. **(B)** An electrical circuit model. The left part represents the resting muscle membrane, which includes a membrane capacitance branch (C_m) and a membrane resistance branch (R_m) with a battery representing the resting potential (E_r) (Sections 2.7 and 2.8). The right part represents the ACh-induced current, with the K⁺ and Na⁺ paths (with resistances of $1/g_K$ and $1/g_{Na}$, respectively) in parallel. After the switch is turned on (green arrow) by ACh release, the current passing through the Na⁺ path is much larger than the current passing through the K⁺ path because the driving force for Na⁺ (= E_{Na} – V_m; Section 2.5) is far greater than the driving force for K⁺ (= V_m – E_K). A net inward current discharges the membrane capacitance, and depolarizes the membrane potential.

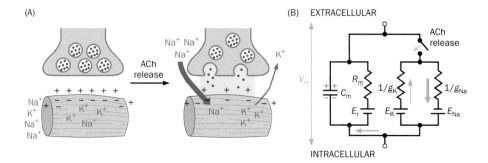

the outward current on the K⁺ branch. Thus, ACh release will activate a net inward current.

The reversal potential, designated E_{rev}, is an important property of ion channels permeable to multiple ions and is determined by the relative conductances and equilibrium potentials of each ion. Using the electrical circuit model in Figure 3-18B, we can determine their relationship as follows: at the reversal potential ($V_m = E_{rev}$), Na⁺ influx equals K⁺ efflux, thus $I_K = -I_{Na}$. Since $I_K = g_K(V_m - E_K)$ and $I_{Na} = g_{Na}(V_m - E_{Na})$, we have

$$E_{rev} = \frac{g_{Na}E_{Na} + g_K E_K}{g_{Na} + g_K} = \frac{\dfrac{g_{Na}}{g_K}E_{Na} + E_K}{\dfrac{g_{Na}}{g_K} + 1}$$

We can see from this formula that if the conductances for Na⁺ and K⁺ were equal ($g_{Na}/g_K = 1$), E_{rev} would simply be an average of E_{Na} and E_K. If the ionic concentrations across the muscle membrane were the same as our model neuron in Figure 2-12A, with $E_K = -85$ mV and $E_{Na} = +58$ mV, then E_{rev} would be –13.5 mV. However, since $E_{rev} = 0$ mV, as shown in Figure 3-17C, we can calculate that g_{Na}/g_K is approximately 1.5 (i.e., 85/58); in other words, the channel opened upon ACh binding has a higher conductance for Na⁺ than for K⁺.

3.13 The skeletal muscle acetylcholine receptor is a ligand-gated ion channel

A deeper understanding of the ACh-induced conductance change required identification of the postsynaptic **acetylcholine receptor (AChR)** and the ion channel whose conductance is coupled to ACh binding. Further studies indicated that the muscle AChR is itself the ion channel. Just as the NMJ served as a model synapse because of its experimental accessibility, the AChR served as a model neurotransmitter receptor because of its abundance, particularly in the electric organ of the *Torpedo* ray, which is highly enriched for an AChR similar to that from the skeletal muscle. Biochemical purification and subsequent cloning of the *Torpedo* AChR revealed that it consists of five subunits: two α, one β, one γ, and one δ (**Figure 3-19**A). Each AChR contains two ACh binding sites, located at the α–γ and α–δ subunit interfaces, respectively. Both sites need to bind ACh for the channel to open. Evidence that this heteropentameric receptor was the ACh-activated channel came from a reconstitution experiment: co-injection of mRNAs encoding all four AChR subunits into the *Xenopus* oocyte caused the oocyte, which normally does not respond to ACh, to produce an inward current in response to ACh iontophoresis in voltage clamp experiments. This ACh-induced inward current was reversibly blocked by curare (Box 3-2), an AChR **antagonist** (agent that acts to counter the action of an endogenous molecule); washing out the curare restored the inward current (Figure 3-19B). Omitting an mRNA of any of the AChR subunits abolished the ACh-induced inward current in the oocyte expression system.

The three-dimensional structure of the *Torpedo* AChR has been determined by high-resolution electron microscopy (**Figure 3-20**). All AChR subunits contain four transmembrane helices, with the M2 helices from all subunits lining the ion

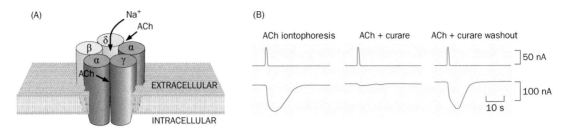

Figure 3-19 Composition of the acetylcholine receptor (AChR).
(A) Schematic illustrating the subunit composition of the AChR.
The two ACh binding sites are at the α–γ and α–δ subunit interfaces.
(B) Functional expression of AChR was achieved by injecting mRNAs
encoding the four AChR subunits into *Xenopus* oocytes. Top traces,
current used for iontophoresis of ACh. Bottom traces, inward current in
response to ACh application measured in a voltage clamp setup. ACh
application led to an inward current (left) blocked by curare, an AChR
inhibitor (middle), that was restored after curare was washed out
(right). The membrane potential was held at –60 mV. (B, adapted
from Mishina M, Kurosaki T, Tobimatsu T, et al. [1984] *Nature*
307:604–608. With permission from Springer Nature.)

conduction pore. The transmembrane helices form a hydrophobic barrier or "gate"
when AChR is closed, preventing ion flow. ACh binding causes rotation of the
α subunits, inducing an alternative conformation of the M2 helices and opening
the gate to allow cation passage.

To summarize synaptic transmission at the vertebrate NMJ: action potentials
trigger ACh release from motor axon terminals; ACh molecules diffuse across the
synaptic cleft and bind to postsynaptic AChRs, which are highly concentrated
on the muscle membrane directly apposing the motor axon terminal; upon ACh
binding, muscle AChRs open a nonselective cation channel that allows more Na⁺
influx than K⁺ efflux, thus depolarizing the muscle cell in the form of an EPP; when
this depolarization reaches threshold, the muscle cell fires action potentials,
resulting in muscle contraction. We will study the mechanisms of muscle contraction in Section 8.1.

Whereas the open probability of the voltage-gated Na⁺ and K⁺ channels we
studied in Chapter 2 is increased by depolarization, the open probability of muscle AChR channels is not affected by membrane potential changes; its conductance (I/V) is mostly constant across different voltages, as can be seen by the near

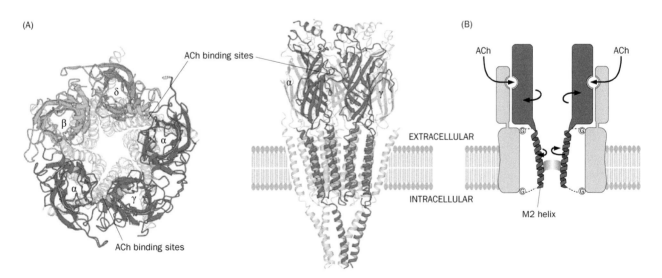

Figure 3-20 AChR structure and gating model. (A) Structure of
Torpedo AChR in a closed state at a resolution of 4 Å by electron
microscopy. Left, a surface view from the extracellular side. The
tryptophan in the α subunit implicated in ACh binding is highlighted
in gold. Only the extracellular portions are colored. Right, a side view
showing the transmembrane helices. The front α and γ subunits are
highlighted in color. **(B)** A model of AChR activation. ACh binding
induces a rotation of part of the extracellular domain of the α subunit
(red). This rotation triggers a conformational change in the
transmembrane helix M2 that lines the ion conduction pore, leading
to the opening of the ion gate. Dotted lines with circled Gs (for glycine
residues) denote the flexible loops connecting M2 to the rest of the
protein. (A, from Unwin N [2005] *J Mol Biol* 346:967–989. With
permission from Elsevier Inc. B, adapted from Miyazawa A Fujiyoshi Y,
& Unwin N [2003] *Nature* 423:949–955. With permission from
Springer Nature.)

linear *I–V* curve in Figure 3-17C. Rather, the open probability of muscle AChR channels is increased by ACh binding. The muscle AChR is therefore called a **ligand-gated ion channel** and is the prototype of a large family of such channels (Table 2-2). Most ligands of these channels are extracellular neurotransmitters such as ACh; however, some channels are gated by intracellular signaling molecules.

3.14 Neurotransmitter receptors are either ionotropic or metabotropic

Following the pioneering work on vertebrate skeletal muscle AChRs, receptors for several other neurotransmitters were found to be ion channels. All neurotransmitter-gated ion channels in vertebrates belong to one of three families. GABA-, glycine-, and serotonin-gated ion channels are in the same family as muscle AChRs (**Figure 3-21**, left), with five subunits each possessing four transmembrane segments. Glutamate-gated ion channels constitute a second family, with four subunits each possessing three transmembrane segments (Figure 3-21, middle). Finally, some neurons use ATP as a neurotransmitter, and ATP-gated ion channels are trimers, with each subunit having just two transmembrane segments (Figure 3-21, right).

Neurotransmitter receptors that function as ion channels, allowing rapid communication across the synapse, are also called **ionotropic receptors** (**Figure 3-22**A). For example, the direct gating of muscle AChR channels by ACh allows transmission of electrical signals from presynaptic neuron to postsynaptic muscle within a few milliseconds (Figure 3-1C). Ionotropic receptors are synonymous with the ligand-gated ion channels introduced in the previous section. Both terms encompass receptors that are gated by ligands (neurotransmitters) and conduct ions across the membrane; the decision of which term to use depends on whether channel- or receptor-relevant properties are under discussion.

In contrast to fast-acting ionotropic receptors, **metabotropic receptors** (Figure 3-22B), when activated by neurotransmitter binding, modulate the membrane potential indirectly by triggering intracellular signaling cascades that regulate ion channel conductances. (The intracellular signaling molecules are often referred to as second messengers, as opposed to the first messengers—the extracellular ligands.) Accordingly, they operate over a longer time scale, ranging from tens of milliseconds to seconds. In addition, unlike ionotropic receptors, which are mostly concentrated in the postsynaptic density across the synaptic cleft from the

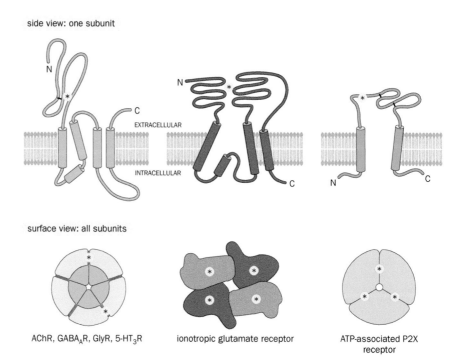

Figure 3-21 Three families of ionotropic receptors in vertebrates. Left, like subunits of the ionotropic AChR (Figure 3-20), subunits of the ionotropic GABA, glycine, and serotonin receptors span the membrane four times. Five subunits constitute a functional receptor with two neurotransmitter-binding sites (stars). Middle, an ionotropic glutamate receptor has four subunits and four neurotransmitter/agonist-binding sites; each subunit spans the membrane three times. Right, an ionotropic P2X receptor consists of three subunits, each spanning the membrane twice. (From Hille [2001] Ion Channels of Excitable Membranes. With permission from Sinauer.)

side view: one subunit

EXTRACELLULAR

INTRACELLULAR

surface view: all subunits

AChR, GABA$_A$R, GlyR, 5-HT$_3$R   ionotropic glutamate receptor   ATP-associated P2X receptor

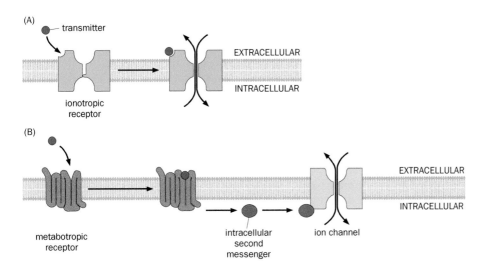

Figure 3-22 Ionotropic and metabotropic neurotransmitter receptors. (A) Ionotropic receptors are ion channels gated by neurotransmitters. Neurotransmitter binding causes membrane potential changes within a few milliseconds. **(B)** Metabotropic receptors act through intracellular second messenger systems to regulate ion channel conductance. Neurotransmitter binding causes membrane potential changes in tens of milliseconds to seconds.

presynaptic active zone, metabotropic receptors are typically not concentrated at the postsynaptic membrane apposing the presynaptic active zone and therefore are termed *extrasynaptic.*

Many neurotransmitters have both ionotropic and metabotropic receptors (Table 3-3). For example, ACh can act on metabotropic receptors in addition to the ionotropic AChR we just studied. To distinguish between the two receptor types, we refer to them according to their specific **agonists** (an agonist is an agent that activates a biological process by interacting with a receptor, often mimicking the action of an endogenous molecule). Hence, ionotropic AChRs are also called **nicotinic AChRs** because they are potently activated by nicotine. Nicotinic AChRs are expressed not only in muscles but also in many neurons in the brain, where nicotine acts as an addictive stimulant. Metabotropic AChRs are also called **muscarinic AChRs** because they are activated by muscarine, a compound enriched in certain mushrooms.

In the following sections, we highlight the actions of key ionotropic and metabotropic receptors for major neurotransmitters in the CNS (Table 3-3).

Table 3-3:   Ionotropic and metabotropic neurotransmitter receptors encoded by the human genome

Neurotransmitter	Ionotropic		Metabotropic	
	Name	Number of genes	Name	Number of genes
Acetylcholine	Nicotinic ACh receptor	16	Muscarinic ACh receptor	5
Glutamate	NMDA receptor	7	Metabotropic glutamate receptor (mGluR)	8
	AMPA receptor	4		
	Others	7		
GABA	GABA$_A$ receptor	19	GABA$_B$ receptor	2
Glycine	Glycine receptor	5		
ATP	P2X receptor	7	P2Y receptor	8
Serotonin (5-HT)	5-HT$_3$ receptor	5	5-HT$_{1,2,4,6,7}$ receptors	13
Dopamine			Dopamine receptor	5
Norepinephrine (epinephrine)			α-adrenergic receptor	6
			β-adrenergic receptor	3
Histamine			Histamine receptor	4
Adenosine			Adenosine receptor	4
Neuropeptides			Neuropeptide receptors	Dozens

Based on the Guide to Pharmacology database (www.guidetopharmacology.org). Abbreviations: GABA, γ-aminobutyric acid; P2X receptor, ATP-gated ionotropic receptor; P2Y, ATP-gated metabotropic receptor; 5-HT# receptor, serotonin (5-hydroxytryptamine) receptor subtype #; ACh, acetylcholine; NMDA, N-methyl-d-aspartate; AMPA, 2-amino-3-hydroxy-5-methylisoxazol-4-propanoic acid.

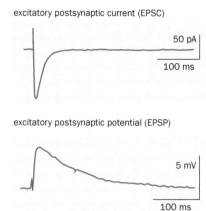

excitatory postsynaptic current (EPSC)

50 pA

100 ms

excitatory postsynaptic potential (EPSP)

5 mV

100 ms

Figure 3-23 Excitatory postsynaptic current and excitatory postsynaptic potential at a glutamatergic synapse. Representative EPSC (top) and EPSP (bottom) recorded using whole-cell patch clamping from hippocampal pyramidal neurons in an *in vitro* slice preparation, in response to electrical stimulation of glutamatergic input axons. The EPSC was recorded in a voltage clamp mode with the membrane potential held at –90 mV, and the EPSP was recorded in a current clamp mode (see Section 14-21 for details). The vertical ticks before the EPSC and EPSP are electrical stimulation artifacts. (Adapted from Hestrin S, Nicoll RA, Perkel DJ, et al. [1990] *J Physiol* 422:203–225.)

3.15 AMPA and NMDA glutamate receptors are activated by glutamate under different conditions

Ionotropic glutamate receptors mediate the fast actions of glutamate, the major excitatory neurotransmitter in the vertebrate CNS. Indeed, glutamatergic excitatory synapses account for the vast majority of synapses in the vertebrate CNS: virtually all neurons—whether they are themselves excitatory, inhibitory, or modulatory—express ionotropic glutamate receptors and are thus excited by glutamate.

Like muscle AChRs, ionotropic glutamate receptors are cation channels permeable to both Na^+ and K^+, with a reversal potential near 0 mV. Under physiological conditions, glutamate binding to ionotropic glutamate receptors produces an inward current called the **excitatory postsynaptic current** (**EPSC**) (**Figure 3-23**, top), as more positively charged ions flow into the cell than out of it. This is analogous to the end-plate current we saw at the NMJ (Figure 3-17). The inward current produces a transient depolarization in the postsynaptic neuron called the **excitatory postsynaptic potential** (**EPSP**) (Figure 3-23, bottom), analogous to the EPP at the NMJ. The recordings shown in Figure 3-23 were made in acutely prepared **brain slices** (fresh sections of brain tissue a few hundred micrometers thick) that preserve local three-dimensional architecture and neuronal connections while allowing experimental access, such as whole-cell recording of individual neurons and control of extracellular solutions.

Historically, ionotropic glutamate receptors have been divided into three subtypes named for their selective responses to three agonists: AMPA (2-amino-3-hydroxy-5-methylisoxazol-4-propanoic acid), kainate (kainic acid), and NMDA (*N*-methyl-D-aspartate). Molecular cloning of these receptors revealed that they are encoded by distinct gene subfamilies of ionotropic glutamate receptors (Table 3-3). Because the properties of **AMPA receptors** and **kainate receptors** are more similar to each other, they are collectively called non-NMDA receptors; **NMDA receptors** have distinctive properties. In the following we use the AMPA and NMDA receptors to illustrate these differences.

AMPA receptors are fast glutamate-gated ion channels that conduct Na^+ and K^+; depending on subunit composition, some AMPA receptors are also permeable to Ca^{2+} (see the next section). They mediate synaptic transmission at most glutamatergic synapses when the postsynaptic neuron is near the resting potential. Because the driving force of Na^+ is much greater than that of K^+ near the resting potential, AMPA receptor opening causes a net influx of positively charged ions, resulting in depolarization of postsynaptic neurons (**Figure 3-24**A).

NMDA receptors have two unusual properties. First, they must be bound to two distinct ligands (glutamate and glycine) to open. However, their glycine binding site is a high-affinity site and so may already be bound at ambient extracellular glycine levels. Second, NMDA receptors do not open unless the postsynaptic membrane is depolarized. The mechanism of voltage dependence is different from that of the voltage-gated Na^+ and K^+ channels discussed in Chapter 2. At the extracellular face of the membrane, the mouth of the NMDA receptor is blocked by Mg^{2+} at negative membrane potentials, such that the channel remains closed despite glutamate binding (Figure 3-24A). However, depolarization of the postsynaptic membrane relieves the Mg^{2+} block (Figure 3-24B). In the absence of external Mg^{2+}, the NMDA receptor conductance is not affected by the membrane potential, as can be seen by the nearly linear *I–V* curve (Figure 3-24C, blue line), similar to the *I–V* curve for nicotinic AChR (Figure 3-17C). By contrast, under physiological external Mg^{2+} conditions, the conductance is greatly reduced when the membrane potential is negative (Figure 3-24C, red line). Thus, the NMDA receptor acts as a **coincidence detector**, opening only in response to concurrent presynaptic glutamate release *and* postsynaptic depolarization. This property is very important for synaptic plasticity and learning, as well as for activity-dependent wiring of the nervous system, as we will learn in Chapters 5 and 11. Once opened, NMDA receptors have high Ca^{2+} conductance. While AMPA receptors provide initial depolarizations to release the Mg^{2+} block of nearby NMDA receptors—these

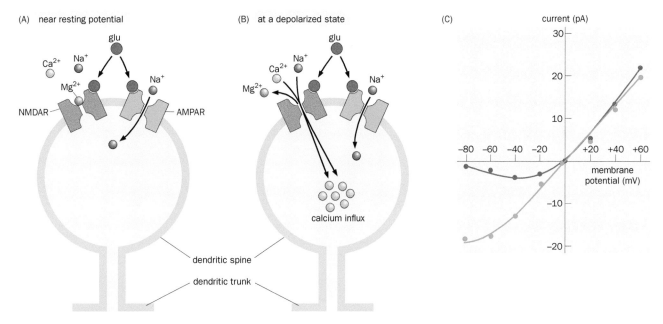

Figure 3-24 Properties of AMPA and NMDA glutamate receptors.
(A) When the postsynaptic neuron (represented by a dendritic spine) is near the resting potential, glutamate (glu) released from the presynaptic neuron opens only the AMPA receptor (AMPAR), causing Na^+ entry and producing excitatory postsynaptic potentials (EPSPs). The NMDA receptor (NMDAR) is blocked by external Mg^{2+} and therefore cannot be opened by glutamate binding alone. **(B)** When the postsynaptic neuron is depolarized, the Mg^{2+} block is relieved. Both NMDAR and AMPAR can now be opened by glutamate binding. The NMDAR is highly permeable to Ca^{2+}. For simplicity, the small K^+ efflux through open AMPARs and NMDARs is omitted. **(C)** Current–voltage relationship of the NMDAR in the presence or absence of external

Mg^{2+}. Blue curve, the nearly linear slope of the *I–V* curve indicates that the conductance of the NMDAR in Mg^{2+}-free media is nearly constant between –60 mV and +60 mV, revealing that the NMDAR is not gated by voltage per se. Red curve, physiological concentrations of extracellular Mg^{2+} markedly diminish the inward current at negative membrane potentials, as Mg^{2+} blocks cation influx. Data were obtained using whole-cell recording of cultured mouse embryonic neurons. (A & B, adapted from Cowan WM, Südhof TC, & Stevens CF [2001] Synapses. Johns Hopkins University Press. C, adapted from Nowak L Bregestovski P, & Ascher P [1984] *Nature* 307:462–465. With permission from Springer Nature. See also Mayer ML, Westbrook GL, & Guthrie PB [1984] *Nature* 309:261–263.)

two glutamate receptors are often co-localized to the same postsynaptic site—NMDA receptors contribute additional depolarization alongside AMPA receptors. Furthermore, Ca^{2+} influx via NMDA receptors contributes to many biochemical changes in postsynaptic cells, as will be discussed later in the chapter.

3.16 Properties of individual ionotropic glutamate receptors are specified by their subunit compositions

All ionotropic glutamate receptors have four subunits (Figure 3-21). Each subunit consists of several modular domains (**Figure 3-25A**): an amino terminal domain, a ligand-binding domain, a transmembrane domain comprising three membrane-spanning helices (M1, M3, and M4) and an additional pore loop (M2), and a carboxy-terminal intracellular domain. AMPA receptors can form functional homotetramers (composed of four identical subunits) although they are usually found *in vivo* as heterotetramers of two or more of the four variants, GluA1, GluA2, GluA3, and GluA4. Comparisons of cryo-EM structures of GluA2 homotetramers in closed and open states revealed that glutamate binding results in a large conformational change featuring closing of the clamshell-like structure of the ligand-binding domain. This change leads the opening of two gates in the ion conductance pore formed by the M3 and M2 helices, respectively (Figure 3-25B).

NMDA receptors are obligatory heterotetramers composed of two **GluN1** subunits and two **GluN2** subunits. (Some neurons express NMDA receptors containing GluN3 subunits instead of GluN2 subunits.) GluN1 is encoded by a single gene, whereas GluN2 has four variants, GluN2A, GluN2B, GluN2C, and GluN2D, encoded by four separate genes. Glutamate and glycine bind to GluN2 and GluN1 subunits, respectively, which triggers the opening of the ion channel.

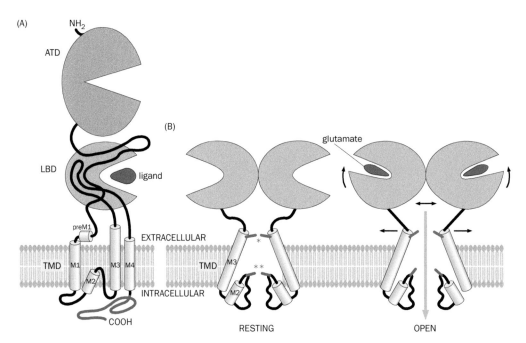

Figure 3-25 AMPA receptor structure and activation mechanism.
(A) All ionotropic glutamate receptor subunits comprise an amino-terminal domain (ATD), a ligand-binding domain (LBD), a transmembrane domain (TMD), and a carboxy-terminal intracellular domain (red). The line represents the polypeptide chain from the extracellular amino terminus (NH_2) to the intracellular carboxy-terminus (COOH). The M1–M4 cylinders represent helices that span across (M1, M3, M4) or loop into (M2) the plasma membrane. The schematic is based on the crystal structure of a homotetramer of GluA2.
(B) Schematic summary of the AMPA receptor activation mechanism, based on comparisons of cryo-EM structures of homotetrameric GluA2 structures with the ion conduction pore in the closed and open states (these states were trapped when GluA2 was in complex with different auxiliary proteins in the absence or presence of agonists). Glutamate binding induces closure of the LBD clamshells, leading to conformational changes (indicated by arrows) transduced to the M3 and M2 membrane helices and opening of the upper (*) and lower (**) gates of the ion conduction pore. Only two subunits are shown; ATDs are omitted from this schematic. (A, adapted from Sobolevsky AI, Rosconi MP, & Gouaux E [2009] *Nature* 462:745–756. With permission from Springer Nature. B, adapted from Twomey EC, Yelshanskaya MV, Grassucci RA, et al. [2017] *Nature* 549:60–65. With permission from Springer Nature. See also Chen S, Zhao Y, Wang Y, et al. [2017] *Cell* 170:1234–1246.)

The subunit composition of both AMPA and NMDA receptors has important functional consequences. For example, most AMPA receptors contain the GluA2 subunit; most GluA2-containing AMPA receptors are impermeable to Ca^{2+} due to a posttranscriptional modification called **RNA editing**, which changes the mRNA sequence encoding a key residue in GluA2's channel pore (see Figure 3-26A). AMPA receptors lacking GluA2 or containing unedited GluA2 subunits are permeable to Ca^{2+} (though not as permeable as NMDA receptors). AMPA receptors lacking GluA2 are also susceptible to voltage-dependent block by intracellular polyamines, preventing Na^+ influx when the neuron becomes depolarized. These AMPA receptors are thus inwardly rectified, like the inward-rectifier K^+ channels we discussed in Box 2-4. NMDA receptors containing different GluN2 variants also have distinct channel conductances and cytoplasmic signaling properties, and bind differentially to postsynaptic scaffold proteins (see the next section). Combining different subunits thus allows both AMPA and NMDA receptors to exhibit a rich repertoire of functional and regulatory properties. Indeed, the subunit compositions of AMPA and NMDA receptors differ in different types of neurons. The subunit composition of these receptors within the single neuronal type also undergoes developmental changes and can be regulated by synaptic activity.

Cell-surface expression and physiological properties of AMPA receptors are regulated by **transmembrane AMPA receptor regulatory proteins (TARPs)**, which are intimately associated with the pore-forming subunits of the AMPA receptor channel and are thus called auxiliary subunits of the AMPA receptor (**Figure 3-26**A). The first TARP was discovered by investigating the physiological

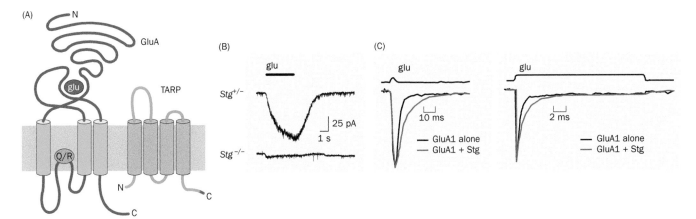

Figure 3-26 Transmembrane AMPA receptor regulator proteins (TARPs). (A) Schematic of TARP next to a pore-forming subunit of the AMPA receptor (GluA). Each TARP subunit consists of four transmembrane segments. The C-terminal tip contains a PDZ binding motif (red; see Section 3.17 for PDZ proteins). Recent cryo-EM-based structural studies indicate that each tetrameric AMPA receptor can associate with 1–4 TARP subunits. The Q/R site represents the amino acid change from glutamine (Q) to arginine (R) due to RNA editing. **(B)** Whole-cell recordings of cerebellar granule cells in slice; membrane potential was held at –80 mV. Glutamate (glu) application results in a large EPSC in the *Stargazer* heterozygous (Stg$^{+/-}$) neuron. EPSC is largely absent in the homozygous mutant (Stg$^{-/-}$) neuron. **(C)** Excised patch recordings (see Box 14-3 for details of this method) from *Xenopus* oocyte expressing GluA1 alone or co-expressing GluA1 and Stargazin (Stg). Stargazin slows down deactivation (how fast the channel closes after a short pulse of glutamate; left) and desensitization (how fast the channel closes in response to prolonged glutamate application; right) of the AMPA receptor channel. (A, adapted from Jackson AC & Nicoll AN [2011] *Neuron* 70:178–199. With permission from Elsevier. B, adapted from Chen L, Chetkovich DM, Petralia RS, et al. [2000] *Nature* 408:936–943. With permission from Springer Nature. C, from Tomita S, Adesnik H, & Sekiguchi M, et al. [2005] *Nature* 435:1052–1058.)

deficits in the *Stargazer* mutant mouse, so named because mutant mice exhibit unusual, repeated head elevations. AMPA receptor-mediated excitatory postsynaptic currents were found to be virtually absent in *Stargazer* mutant cerebellar granule cells (Figure 3-26B). Further investigation indicated that Stargazin, the protein encoded by the *Stargazer* gene, has a multitude of functions: (1) it regulates the trafficking of the AMPA receptor from ER to the cell surface, (2) it helps anchor the AMPA receptor to the postsynaptic density scaffold (see the next section), and (3) it regulates a variety of physiological properties of the AMPA receptor, such as slowing the deactivation and desensitization time course (Figure 3-26C), allowing more inward current through the channel in response to the same amount of glutamate release. Most neurons express multiple TARPs. Cerebellar granule cells are an exception in that Stargazin is the only TARP expressed there; thus mutation of this single gene led to dramatic effects. TARPs have also been found in *C. elegans*. *C. elegans* TARPs can substitute for Stargazin in modifying receptor function, suggesting evolutionary conservation of TARPs. AMPA receptor auxiliary subunits belonging to gene families other than TARPs have since been identified, as have auxiliary proteins for the NMDA and kainate receptors.

3.17 The postsynaptic density is organized by scaffold proteins

Just as the presynaptic terminal is highly organized by active-zone scaffold proteins, the postsynaptic density is highly organized by postsynaptic proteins. At glutamatergic synapses, for example, the postsynaptic density consists of not only glutamate receptors but also other associated proteins (**Figure 3-27**). These include (1) trans-synaptic adhesion proteins that align active zones with postsynaptic densities (see also Section 3.7), (2) proteins that participate in signal transduction cascades, and (3) a diverse array of scaffold proteins that connect glutamate receptors and trans-synaptic adhesion molecules to signaling molecules and cytoskeletal elements. The resulting protein network controls glutamate receptor localization, density, trafficking, and signaling, all of which affect synaptic transmission and synaptic plasticity. Synaptic scaffolds are also present in

Figure 3-27 The organization of the postsynaptic density at glutamatergic synapses. At the cell surface, the postsynaptic density of a mature glutamatergic synapse is enriched in AMPA and NMDA receptors (AMPAR and NMDAR) as well as trans-synaptic cell adhesion molecules like cadherins and neuroligins (which, respectively, bind presynaptic cadherins and neurexins; Figure 3-10). The scaffold proteins of the PSD-95 family, named for their localization to the postsynaptic density and molecular weight, bind many proteins, including the GluN2 subunit of the NMDAR, AMPAR-associated TARPs, the neuroligin synaptic adhesion molecules, the signal-transducing enzyme CaMKII, and other scaffold proteins that bind metabotropic glutamate receptors (mGluRs) and other postsynaptic density proteins (not shown). The diagram depicts only a small subset of known components and interactions in the postsynaptic density. (Adapted from Sheng M & Kim E [2011] *Cold Spring Harb Perspect Biol* 3:a005678.)

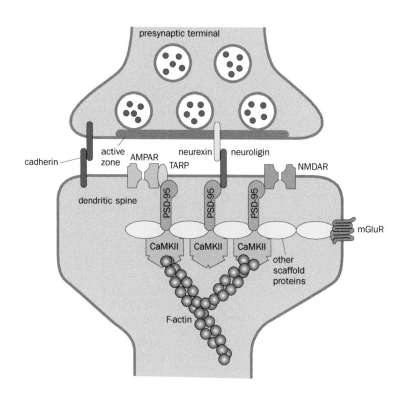

GABAergic postsynaptic terminals; GABAergic and glutamatergic postsynaptic densities utilize common as well as unique scaffold proteins. We will learn more about the postsynaptic density protein network in the contexts of development and synaptic plasticity in Chapters 7 and 11, respectively, and how its dysfunction contributes to brain disorders in Chapter 12.

We use one of the most abundant scaffold proteins at the glutamatergic synapse, **PSD-95** (postsynaptic density protein-95 kDa), to illustrate the organizational role of scaffold proteins in dendritic spines, where glutamatergic synapses are usually located (Figure 3-27). PSD-95 contains multiple protein–protein interaction domains, including three **PDZ domains**, which bind C-terminal peptides with a sequence motif that occurs in many transmembrane receptors. (PDZ is an acronym for three proteins that share this domain: PSD-95, identified from biochemical analysis of the postsynaptic density; Discs-large, which regulates cell proliferation in *Drosophila* and is also associated with the postsynaptic density; and ZO-1, an epithelial tight junction protein.) These protein–protein interaction domains enable PSD-95 to bind directly to the GluN2 subunit of the NMDA receptor, AMPR receptor auxiliary subunits (TARPs; Figure 3-26), the trans-synaptic adhesion molecule neuroligin, and Ca^{2+}/calmodulin-dependent protein kinase II (CaMKII, an enzyme highly enriched in postsynaptic densities, whose role in signal transduction will be introduced in Section 3.20). PSD-95 also binds other PDZ-domain-containing scaffold proteins that in turn associate with other postsynaptic components such as metabotropic glutamate receptors and the actin cytoskeleton. Thus, the scaffold protein network stabilizes neurotransmitter receptors at the synaptic cleft by placing them close to the trans-synaptic adhesion complex apposing the active zone (Figure 3-10). It also brings enzymes (for example, CaMKII) close to their upstream activators (for example, Ca^{2+} entry through NMDA receptors) and downstream substrates, and organizes the structure of the dendritic spine by bridging the trans-synaptic adhesion complex and the underlying actin cytoskeleton. Recent studies suggest that PSD scaffold proteins exhibit a property called "liquid–liquid phase separation," where multivalent interactions among PSD proteins cause them to self-organize into a separate phase in solution. This property likely contributes to the PSD organization *in vivo*.

3.18 Ionotropic GABA and glycine receptors are Cl⁻ channels that mediate inhibition

The role of inhibition in nervous system function was first established in the study of spinal cord reflexes over a century ago (Section 1.9). In the 1950s, when intracellular recording techniques were applied to the study of spinal motor neurons, it was found that stimulating certain input resulted in a rapid hyperpolarization (termed an **inhibitory postsynaptic potential** or **IPSP**) due to outward current flow across the motor neuron membrane (called an **inhibitory postsynaptic current** or **IPSC**). In a revealing experiment (**Figure 3-28**), researchers set the membrane potential of the motor neuron at different initial values by injecting different constant currents through an electrode, and the membrane potential was measured by a second electrode in response to stimulation of an inhibitory input. When the initial membrane potential was equal to or more depolarized than the resting potential of about –70 mV, stimulation of the inhibitory input caused hyperpolarization, but when the initial membrane potential was more hyperpolarized than –80 mV, stimulation of the inhibitory input produced depolarization. The reversal potential of around –80 mV is close to the equilibrium potential for Cl⁻ (E_{Cl}), suggesting that the IPSC is carried by Cl⁻ flow. Indeed, by increasing intracellular Cl⁻ concentration, the reversal potential became less negative following the change in E_{Cl}, as predicted by the Nernst equation. This experiment suggested that inhibition of the spinal motor neuron is mediated by an increase in Cl⁻ conductance across the motor neuron membrane.

Subsequent studies have shown that the fast IPSC/IPSP is mediated by the neurotransmitters glycine (used by a subset of inhibitory neurons in the spinal cord and brainstem) and GABA (used by most inhibitory neurons), which act on ionotropic **glycine receptors** and **GABA$_A$ receptors**, respectively. The structure of GABA$_A$ receptors is similar to that of nicotinic AChRs (Figure 3-20), comprising a pentamer with two α subunits, two β subunits, and one γ subunit. Each subunit has multiple isoforms encoded by different genes (Table 3-3), and other subunits such as δ and ε can be used in lieu of γ. Many pharmaceutical drugs act on GABA$_A$ receptors to modulate inhibition in the brain. As we will learn in Chapter 12, the most widely used anti-epilepsy, anti-anxiety, and sleep-promoting drugs all bind to and enhance the functions of GABA$_A$ receptors. Glycine receptors also make up a pentamer. Both GABA$_A$ and glycine receptors are ligand-gated ion channels selective for anions, primarily Cl⁻.

How does an increase in Cl⁻ conductance resulting from the opening of GABA$_A$ or glycine receptor channels on postsynaptic neurons cause inhibition? In most neurons, E_{Cl} is slightly more hyperpolarized than the resting potential, as in the spinal motor neurons we just studied. Thus, an increase in Cl⁻ conductance causes Cl⁻ influx (which is equivalent to an outward current because Cl⁻ carries a negative charge), resulting in a small hyperpolarization (**Figure 3-29**A, left panel). Importantly, if the neuron also simultaneously receives an excitatory input that produces an EPSP (for example, via opening of glutamate receptor channels), the more depolarized membrane potential enhances the driving force for Cl⁻ influx. This increases the outward current triggered by GABA, which counters the EPSP-producing inward current, making it more difficult for the cell's membrane potential to reach the threshold for firing action potentials (Figure 3-29A, middle and right panels).

The interaction of excitatory and inhibitory input can also be modeled by an electrical circuit wherein each input is represented by a branch consisting of a switch (representing neurotransmitter release), a conductance (g_e or g_i, representing EPSC or IPSC conductance), and a battery (representing the reversal potential for the excitatory glutamate receptors, E_{e-rev}, or the GABA$_A$ receptor, E_{Cl}). When only the inhibitory input is switched on, because E_{Cl} is more hyperpolarized than the resting potential (E_r), a small outward current will be produced by the g_i branch, resulting in a small hyperpolarizing IPSP (Figure 3-29B, left). When only the excitatory input is switched on, a large inward current will be produced by the g_e branch, as E_{e-rev} is much more depolarized than E_r, resulting in a large depolarizing EPSP (Figure 3-29B, middle). When both the excitatory and inhibitory

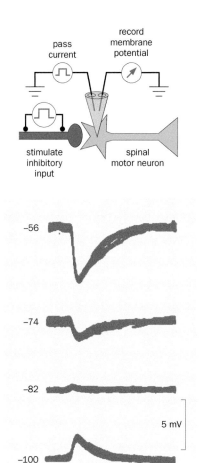

Figure 3-28 Inhibitory postsynaptic potentials (IPSPs). Top, experimental setup. Two electrodes were inserted into a spinal motor neuron, one for passing current to change the holding membrane potential and the other to measure the membrane potential in response to electrical stimulation of an inhibitory input. (The schematic is simplified; the sensory afferent from an antagonist muscle was stimulated, which inhibits the recorded motor neuron through intermediate inhibitory interneurons; see Figure 1-19.) Bottom, IPSPs recorded at four different holding membrane potentials. Each record represents the superposition of about 40 traces. At membrane potentials of –74 mV or above, stimulation of inhibitory input resulted in hyperpolarizing IPSPs, with increasing amplitudes as holding membrane potentials became less negative. At membrane potentials of –82 mV or below, stimulation of inhibitory input resulted in depolarizing IPSPs, with increasing amplitudes as holding membrane potentials became more negative. (Graphs adapted from Coombs JS, Eccles JC, & Fatt P [1955] *J Physiol* 130:326–373.)

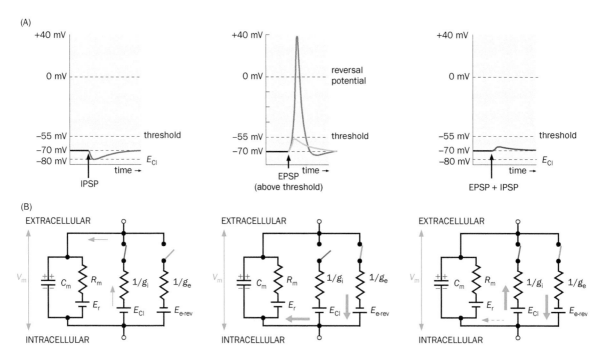

Figure 3-29 The inhibitory effect of Cl⁻ conductance mediated by GABA_A receptors. (A) In this neuron, the Cl⁻ equilibrium potential, E_{Cl}, is slightly more hyperpolarized than the resting potential. Left, an IPSP from GABA_A receptors causes hyperpolarization of this postsynaptic neuron toward E_{Cl}. Middle, an EPSP from glutamate receptors causes depolarization of the postsynaptic neuron, as the reversal potential (~0 mV) is far above the resting potential. If the EPSP amplitude exceeds the threshold, it produces an action potential. Right, an IPSP can cancel the effect of an EPSP when excitatory and inhibitory inputs are present at the same time, thus preventing postsynaptic firing. **(B)** Circuit models for the three scenarios in Panel A. Two branches, representing inhibitory and excitatory neurotransmitter receptors with conductances of g_i and g_e when neurotransmitter binding opens the receptor channels, are added to the resting neuronal model represented by the membrane capacitance (C_m), resistance (R_m), and resting potential (E_r). Left, when only the inhibitory branch is switched

on (GABA release activating GABA_A receptors), a small outward current results (upward arrow) because E_{Cl} is more hyperpolarized than E_r. This causes more charge to build up at C_m, thus hyperpolarizing the membrane potential (V_m). Middle, when only the excitatory branch is switched on (glutamate release activating glutamate receptors), a large inward current results (downward arrow) because the reversal potential for the excitatory ionotropic glutamate receptors (E_{e-rev}) is far more depolarized than E_r. This causes discharge of C_m, thus depolarizing the membrane potential. Right, when both the inhibitory and excitatory branches are switched on (GABA and glutamate released at the same time), a large fraction of the inward current in the excitatory branch is diverted by the outward current in the inhibitory branch. As a result, the current discharging C_m (dashed arrow) is smaller. (Note that the more depolarized V_m is, the larger the outward current is, because of the larger driving force for Cl⁻.)

inputs are switched on, part of the inward current in the g_e branch will flow outward through the g_i branch (Figure 3-29B, right), leading to a smaller depolarization than when the g_e branch is active alone. Indeed, as can be seen from the circuit model, even when E_{Cl} equals E_r, meaning that no net Cl⁻ influx or efflux occurs at rest, GABA_A receptor opening creates an extra path (an extra conductance) that tends to hold the membrane potential near E_{Cl}, counteracting the inward current created by excitatory inputs, and thereby diminishing the voltage change across the membrane. This so-called shunting contributes to GABA's potent inhibitory effect.

A noteworthy exception to GABA's inhibitory effects can occur in developing neurons. The intracellular Cl⁻ concentration is high in many developing neurons because their Cl⁻ transporters (Figure 2-12B) are not yet expressed at a high level as in mature neurons. When the intracellular Cl⁻ concentration is high, E_{Cl} is substantially more depolarized than the resting potential, such that an increase in Cl⁻ conductance results in Cl⁻ efflux, causing depolarization that can exceed the threshold for action potential generation. Under these circumstances, GABA may thus act as an excitatory neurotransmitter.

As we will learn soon, another inhibitory action of GABA is mediated by metabotropic **GABA_B receptors**, which act through intracellular signaling pathways to cause the opening of K⁺ channels. Because E_K is always more negative than the resting potential, opening of K⁺ channels always causes hyperpolarization, making the neurons less likely to reach the threshold for an action potential in response to excitatory input.

3.19 Metabotropic neurotransmitter receptors trigger G protein cascades

We now turn to metabotropic receptors, which act through intracellular signaling pathways rather than mediating ion conduction directly (Figure 3-22B). These receptors, which belong to the **G-protein-coupled receptor (GPCR)** superfamily, participate in signaling cascades involving a heterotrimeric guanine nucleotide-binding protein (**trimeric GTP-binding protein**, or simply **G protein**). ACh, glutamate, and GABA all bind to their own metabotropic receptors: muscarinic AChRs, metabotropic GluRs (mGluRs), and GABA$_B$ receptors, respectively, each with several variants. Other GPCRs include the receptors for dopamine, norepinephrine, serotonin (most subtypes), ATP (P2Y subtypes), adenosine, and neuropeptides (Table 3-3) as well as sensory receptors for vision and olfaction and a subset of taste receptors, which we will study in Chapters 4 and 6. Indeed, GPCRs, encoding receptors with diverse functions, constitute the largest gene family in mammals (**Figure 3-30**). GPCRs are crucial for neuronal communication, responses to external stimuli, and many other physiological processes. As a result, they are targets of many pharmaceutical drugs.

All GPCRs possess seven transmembrane helices (**Figure 3-31**A), and almost all are activated by binding of specific extracellular ligands. (A notable exception is rhodopsin in photoreceptors, which is activated by light absorption, as will be discussed in greater detail in Chapter 4.) Ligand binding triggers conformational changes in the transmembrane helices and allows the cytoplasmic domain to associate with a trimeric G protein complex consisting of three different subunits: **Gα**, **Gβ**, and **Gγ** (Figure 3-31B).

Before GPCR activation, the G protein heterotrimer preassembles and binds GDP via the Gα nucleotide-binding site (Figure 3-31C, Resting state). Because Gα and Gγ are both lipid-modified, this ternary complex associates with the plasma membrane. Ligand activation of the GPCR triggers the binding of its cytoplasmic domain to Gα. This stabilizes a nucleotide-free conformation of Gα and thereby catalyzes the replacement of GDP with GTP (Figure 3-31C, Steps 1 and 2). Next, GTP binding causes Gα to dissociate from Gβγ. Depending on the cellular context, Gα-GTP, Gβγ, or both can trigger downstream signaling cascades (Figure 3-31C, Step 3). Gα not only binds to GDP and GTP but also is a **GTPase**, hydrolyzing GTP to GDP. This GTPase activity provides a built-in termination mechanism for G protein signaling (Figure 3-31C, Step 4) and is often facilitated by additional proteins. GDP-bound Gα has a strong affinity for Gβγ, which promotes reassembly of the ternary complex, thereby returning to the resting state (Figure 3-31C, Step 5) and readying the trimeric G protein for the next round of GPCR activation (**Movie 3-3**).

Figure 3-30 G-protein-coupled receptors (GPCRs) in the human genome. The human genome contains about 800 GPCRs separated into five major branches according to the sequence similarities in their transmembrane domains. The dot at the center represents the root of the branches. Numbers in parentheses indicate the number of genes within a specific branch. Names of some representative GPCRs discussed in this book are given. The glutamate branch includes mGluRs and GABA$_B$ receptors as well as sweet and umami taste receptors. The frizzled/TAS2 branch includes bitter taste receptors. The secretin branch includes neuropeptide corticotropin-releasing factor (CRF) receptors involved in stress response. The adhesion branch includes receptors that signal across the synaptic cleft. The largest branch, rhodopsin, is further divided into four clusters. These include many GPCRs important in neurobiology: opsins and melanopsins for vision and receptors for serotonin, dopamine, acetylcholine (muscarinic), and epinephrine/norepinephrine (α cluster); many neuropeptide receptors (β and γ clusters); receptors for thyroid-stimulating hormone (TSH), follicle-stimulating hormones (FSH), and a large number of rapidly evolving odorant receptors (δ cluster). (Based on Fredriksson R, Lagerstrom MC, Lundin LG, et al. [2003] *Mol Pharmacol* 63:1256–1272; Alexander SPS, Christopoulos A, Davenport AP, et al. [2017] *Br J Pharmacol* 174 Suppl 1:S17.)

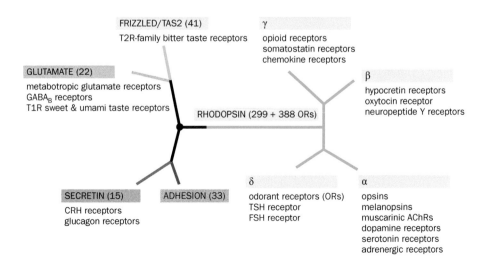

FRIZZLED/TAS2 (41)
T2R-family bitter taste receptors

γ
opioid receptors
somatostatin receptors
chemokine receptors

GLUTAMATE (22)
metabotropic glutamate receptors
GABA$_B$ receptors
T1R sweet & umami taste receptors

β
hypocretin receptors
oxytocin receptor
neuropeptide Y receptors

RHODOPSIN (299 + 388 ORs)

SECRETIN (15)
CRH receptors
glucagon receptors

ADHESION (33)

δ
odorant receptors (ORs)
TSH receptor
FSH receptor

α
opsins
melanopsins
muscarinic AChRs
dopamine receptors
serotonin receptors
adrenergic receptors

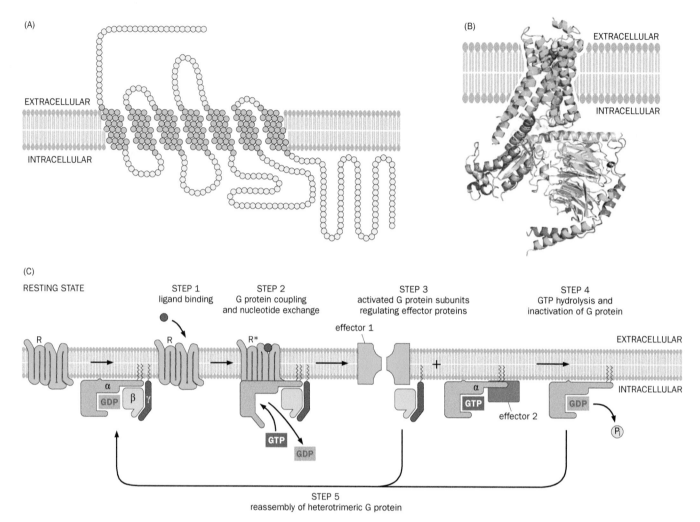

(A)

EXTRACELLULAR

INTRACELLULAR

(B)

EXTRACELLULAR

INTRACELLULAR

(C)

RESTING STATE

STEP 1
ligand binding

STEP 2
G protein coupling
and nucleotide exchange

STEP 3
activated G protein subunits
regulating effector proteins

STEP 4
GTP hydrolysis and
inactivation of G protein

effector 1

EXTRACELLULAR

INTRACELLULAR

STEP 5
reassembly of heterotrimeric G protein

Figure 3-31 Structure and signaling cascade of GPCRs. (A) Primary structure of the β2-adrenergic receptor, the first cloned ligand-gated GPCR. Each circle represents an amino acid. All GPCRs span the lipid membrane seven times, with N-termini on the extracellular side and C-termini on the intracellular side. **(B)** Crystal structure of the β2-adrenergic receptor in complex with trimeric G proteins. Green, β2-adrenergic receptor, with seven transmembrane helices spanning the lipid bilayer; yellow, agonist in its binding pocket; orange, Gα; cyan, Gβ; blue, Gγ. The Gα part in the foreground contains binding sites for GDP/GTP, Gβ, and the β2-adrenergic receptor. The Gα part in the background can swing relative to the part in the foreground, allowing exchange of GDP and GTP. **(C)** Schematic of the GPCR signaling cascade. Resting state, the preassembled trimeric G protein complex in the GDP-bound state associates with the plasma

membrane because Gα and Gγ are covalently attached to lipids (zigzag lines). Step 1, ligand binding. Step 2, a conformational change of the GPCR induced by ligand binding (R → R*) creates a binding pocket for Gα. R* (activated GPCR) catalyzes the exchange of GDP for GTP on Gα. Step 3, GTP-bound Gα dissociates from R*, releasing Gα-GTP and Gβγ to trigger their respective effector proteins that transduce and amplify signals; here, effector 1 is an ion channel that binds Gβγ and effector 2 is an enzyme that binds Gα-GTP. Step 4, the intrinsic GTPase activity of Gα converts Gα-GTP to Gα-GDP. Step 5, Gα-GDP reassociates with Gβγ, returning to the resting state. (A, adapted from Dohlman HG, Caron MG, & Lefkowitz RJ [1987] *Biochemistry* 26:2657–2668. B & C, adapted from Rasmussen SG, DeVree BT, Zou Y, et al. [2011] *Nature* 477:549–555. With permission from Springer Nature.)

Such cycling between GTP- and GDP-bound forms is a general signaling mechanism employed by a superfamily of G proteins (**Box 3-3**).

3.20 A GPCR signaling paradigm: β-adrenergic receptors activate cAMP as a second messenger

β-adrenergic receptors (Figure 3-31A, B) have been the most extensively studied ligand-activated GPCRs. They are activated by epinephrine and norepinephrine (also known as adrenaline and noradrenaline, from which the name of the receptors originates). Whereas norepinephrine is produced by neurons and acts as a neurotransmitter in both the CNS and the autonomic nervous system (Section 3.11), **epinephrine** is produced primarily by adrenal chromaffin cells; it circulates through the blood and acts as a hormone, mediating systemic responses to

Box 3-3: G proteins are molecular switches

The G protein cycle outlined in Figure 3-31C is a universal signaling mechanism of GPCRs, which are used in many biological contexts. Indeed, the switch between a GDP-bound form and a GTP-bound form defines the G protein superfamily, which includes not only trimeric G proteins but also small monomeric GTPase families such as the **Rab**, **Ras**, and **Rho** families. These small GTPases resemble part of the Gα subunit of the trimeric G protein. Rab GTPases regulate different steps of intracellular vesicular trafficking (Section 2.1); we encountered a family member, Rab3, in the context of bridging the synaptic vesicle with the presynaptic active zone scaffold proteins (Figure 3-10). The Ras family of GTPases contains key signaling molecules involved in cell growth and differentiation. As will be discussed in Box 3-4, Ras GTPases play crucial roles in transducing signals from the cell surface to the nucleus. Rho GTPases are pivotal regulators of the cytoskeleton; we will study them in the context of growth cone signaling and neuronal wiring in Chapter 5.

All members of the G protein superfamily are molecular switches. For the trimeric G proteins as well as the Ras and Rho families of GTPases, the GDP-bound form is inactive and the GTP-bound form is active in downstream signaling. The transitions between the GTP-bound and GDP-bound forms are usually facilitated by two types of proteins: the **guanine nucleotide exchange factors** (**GEFs**), which switch GTPases *on* by catalyzing the exchange of GDP for GTP, and **GTPase activating proteins** (**GAPs**), which switch GTPases *off* by speeding up the endogenous GTPase activity, converting GTP to GDP (Figure 3-32; Movie 3-4). As will be discussed in Section 3.23, proper signal termination is an important aspect of signaling.

In the context of trimeric G protein signaling discussed in Section 3.19, ligand-activated GPCRs act as GEFs for the trimeric G proteins. By stabilizing the transition state of the nucleotide-free conformation of Gα (Figure 3-31B), GPCRs catalyze the exchange of GDP for GTP on Gα (Step 2 of Figure 3-31C). The reaction is driven in the direction of Gα-GTP production by the dissociation of Gα-GTP from the GPCR and from Gβγ. We will learn more about GAPs in GPCR signaling in the context of visual transduction in Chapter 4.

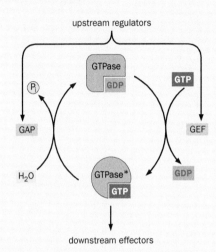

Figure 3-32 The GTPase cycle. GTPases cycle between a GDP-bound form and a GTP-bound form. For signaling GTPases such as trimeric G proteins as well as Ras and Rho subfamilies of small GTPases, the GTP-bound form usually binds effectors and activates downstream signaling. The guanine nucleotide exchange factor (GEF) catalyzes the exchange of GDP for GTP, thus activating the GTPases. The GTPase activating protein (GAP) accelerates the G protein's endogenous GTPase hydrolysis of the bound GTP, thus inactivating the GTPases. GEFs and GAPs are regulated by upstream signals.

extreme conditions, the so-called fright, fight, and flight responses. (A small number of CNS neurons also use epinephrine as a modulatory neurotransmitter.) Classic biochemical studies demonstrated that epinephrine activates β-adrenergic receptors to produce an intracellular second messenger called **cyclic AMP** (**cAMP**). cAMP is synthesized from ATP by a membrane-associated enzyme called **adenylate cyclase** (Figure 3-33). In fact, studies of mechanisms by which β-adrenergic receptors activate adenylate cyclase, together with parallel investigations of the signal transduction pathways downstream from rhodopsin activation (to be discussed in Section 4.4), led to the discovery that trimeric G proteins are essential intermediates in GPCR signaling.

Originally identified as a second messenger in the context of epinephrine action, cAMP is a downstream signal for many GPCRs. cAMP can directly gate ion channels, as will be discussed below as well as in Chapters 4 and 6. However, the most widely used cAMP effector is the **cAMP-dependent protein kinase** (also called A-kinase, **protein kinase A**, or **PKA**). PKA is a **serine/threonine kinase**, which means that it adds phosphates onto specific serine or threonine residues of target proteins, thereby changing their properties. PKA comprises two regulatory and two catalytic subunits; in the absence of cAMP, these subunits form an inactive tetramer usually associated with various **AKAPs** (for A-kinase anchoring proteins) located in specific parts of the cell. cAMP binding to the regulatory subunits triggers the dissociation of the catalytic subunits from the regulatory subunits; the catalytic subunits become free to phosphorylate their substrates

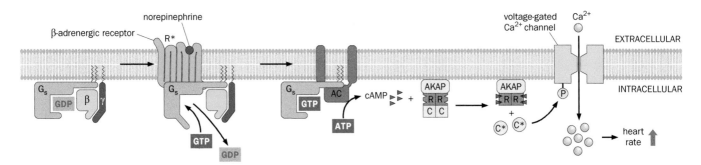

Figure 3-33 Norepinephrine speeds up the heart rate: GPCR signaling through cyclic AMP (cAMP) and protein kinase A (PKA). From left, norepinephrine binding to a β-adrenergic receptor activates G_s, a Gα variant, in cardiac pacemaker cells. G_s-GTP associates with and activates the membrane-bound adenylate cyclase (AC). AC catalyzes the production of cAMP from ATP. cAMP activates PKA. Each PKA consists of two regulatory (R) and two catalytic (C) subunits. Each regulatory subunit contains two cAMP binding sites and is associated with the A-kinase anchoring protein (AKAP). When all four cAMP binding sites on the regulatory subunits are occupied, the catalytic subunits of PKA are released from the complex, become active (C*), and phosphorylate their substrates, such as voltage-gated Ca^{2+} channels. PKA phosphorylation of Ca^{2+} channels increases their open probabilities, facilitates Ca^{2+} influx and depolarization of the pacemaker cells, and thus speeds up heart rate. Note that the central portion of this pathway, from G_s to PKA activation, is widely used in other cellular contexts.

(Figure 3-33; Movie 3-5). PKA phosphorylates many substrates with short- or long-lasting effects on neuronal excitability.

As a specific example, we discuss the mechanism by which norepinephrine released from the axon terminals of neurons in the **sympathetic nervous system** (an arm of the autonomic nervous system) speeds up heart rate. In cardiac pacemaker cells, a special type of cardiomyocytes located in the sinoatrial node, norepinephrine binds to and activates a β-adrenergic receptor, which associates with a Gα variant called **G_s** (for **stimulatory G protein**). G_s-GTP triggers cAMP production by binding to and activating an adenylate cyclase. Elevated cAMP levels lead to PKA activation. PKA phosphorylates a variety of substrates, including voltage-gated Ca^{2+} channels on the plasma membrane of pacemaker cells, which increases their open probability. Ca^{2+} entry depolarizes pacemaker cells, shortens the duration between action potentials they produce, and thus speeds up heart rate (Figure 3-33). In parallel to PKA activation, cAMPs also bind directly to and open hyperpolarization-activated cyclic nucleotide-gated channels (HCN channels; Box 2-4) in pacemaker cells. Since HCN channels are nonselective cation channels, their opening leads to depolarization of pacemaker cells and increased heart rate. In summary, this signaling cascade provides a paradigm relating how a neurotransmitter (norepinephrine) elicits a physiological response (an increase in heart rate) by binding a metabotropic receptor (β-adrenergic receptor), leading to the activation of a second messenger (cAMP) and its downstream effectors (PKA, voltage-gated Ca^{2+} channels, and HCN channels).

3.21 α and βγ G protein subunits trigger diverse signaling pathways that alter membrane conductance

The human genome encodes 16 Gα, 5 Gβ, and 13 Gγ variants. Their different combinations give rise to myriad trimeric G proteins coupled to different GPCRs that can trigger diverse signaling pathways. For example, in addition to the G_s we just discussed, a variant of Gα called **G_i** (for **inhibitory G protein**) also binds adenylate cyclase, but inhibits its activity, causing a decline in intracellular cAMP concentration. Different Gα variants are associated with different receptors and regulate distinct downstream signaling pathways. In postsynaptic neuronal compartments, the ultimate effectors of GPCRs are usually ion channels that regulate membrane potential or neurotransmitter release, most notably K^+ and Ca^{2+} channels (Box 2-4). In the previous section, we discussed a classic example of how norepinephrine activates a Ca^{2+} channel via cAMP and PKA to increase heart rate. In the next two sections, we examine more examples to highlight the diverse outcomes of GPCR signaling.

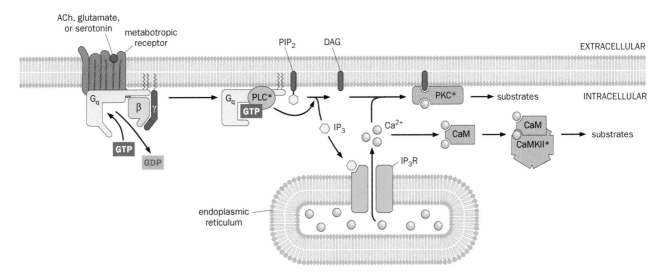

Figure 3-34 GPCR signaling through phospholipase C (PLC) and Ca²⁺. From left, activation of a variety of metabotropic receptors can activate G_q, a variant of Gα. G_q-GTP in turn activates PLC, which catalyzes the conversion of PIP_2 to DAG and IP_3. IP_3 activates the IP_3 receptor (IP_3R, an IP_3-gated Ca²⁺ channel) on the ER membrane, allowing Ca²⁺ release from ER to the cytosol. DAG and Ca²⁺ co-activate PKC. Ca²⁺ also binds to calmodulin (CaM), and the resulting complex activates CaMKII and other CaM kinases. Asterisks represent activated components.

An important G protein effector of many metabotropic receptors (for example, receptors for acetylcholine, glutamate, and serotonin) is a membrane-associated enzyme called **phospholipase C** (**PLC**) (Figure 3-34; Movie 3-6). PLC is activated by G_q, a Gα variant. Activated PLC cleaves a membrane-bound phospholipid called PIP_2 (phosphatidyl 4,5-bisphosphate) to produce two important second messengers: **diacylglycerol** (**DAG**) and **inositol 1,4,5-triphosphate** (**IP₃**). DAG binds to and activates **protein kinase C** (**PKC**), a serine/threonine kinase. PKC activation also requires a rise in intracellular Ca²⁺ concentration. This is achieved via IP_3, which binds to an IP_3-gated Ca²⁺ channel (the **IP₃ receptor**) on the membrane of the endoplasmic reticulum (ER), triggering release of ER-stored Ca²⁺ into the cytosol. In addition to activating PKC, Ca²⁺ interacts with many additional effectors. A key effector is a protein called **calmodulin**. The Ca²⁺/calmodulin complex can regulate diverse signaling pathways, including the activation of Ca²⁺/calmodulin-dependent protein kinases (CaM kinases), another important class of serine/threonine kinases. Like PKA, both PKC and CaM kinases phosphorylate many downstream target proteins, including ion channels and receptors, to modulate their activity. A specific subtype of CaM kinases, **CaM kinase II** (**CaMKII**), is one of the most abundant proteins in the postsynaptic density (Section 3.17). Ca²⁺ can also directly increase the open probability of Ca²⁺-dependent K⁺ channels (Box 2-4). Thus, activation of PLC activates PKC and at the same time causes a rise in intracellular Ca²⁺ concentration, both of which can alter neuronal excitability (Figure 3-34).

Historically, Gα was identified as the first signaling intermediate between the GPCR and any effectors. Subsequently, it was found that Gβγ can also mediate signaling, as in the case of acetylcholine regulation of heart rate. In fact, the concept of a chemical neurotransmitter was first established in this context in a classic experiment conducted in 1921 by Otto Loewi. It was known that stimulating the **vagus nerve**, a cranial nerve connecting the medulla to internal organs, slows heart rate. Loewi collected fluid from a frog heart whose vagus nerve had been stimulated, added it to an unstimulated heart, and found that beating of the second heart also slowed. This experiment showed that vagus nerve stimulation released a chemical transmitter, later identified to be acetylcholine, to slow heart rate. (According to Loewi, the initial idea for this experiment came from a dream in the middle of the night. He wrote it down on a piece of paper and went back to sleep. The next morning he remembered dreaming about something important but could not remember what it was or decipher what he had written. The

Figure 3-35 **Acetylcholine slows down the heart rate: direct action of Gβγ on a K⁺ channel.** From left, ACh activation of a muscarinic ACh receptor (mAChR) on a cardiac pacemaker cell causes the dissociated Gβγ to bind directly to and activate a G-protein-coupled inward-rectifier K⁺ (GIRK) channel, leading to K⁺ efflux and hyperpolarization of the pacemaker cell, which slows down heart rate. Based on Clapham DE & Neer EJ [1997] *Annu Rev Pharmacol Toxicol* 37:167–203.)

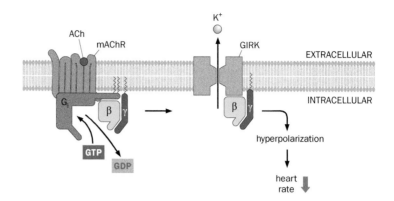

next night, the dream returned, and this time he got up immediately and went to the lab to perform the experiment.)

Subsequent work has shown that ACh binds to a specific muscarinic AChR and triggers the dissociation of the trimeric G protein complex. βγ subunits then bind to and activate a class of K⁺ channels called GIRKs (<u>G</u>-protein-coupled <u>i</u>nward-<u>r</u>ectifier <u>K</u>⁺) channels, resulting in K⁺ efflux, hyperpolarization of cardiac pacemaker cells, and a slowing of heart rate (**Figure 3-35**). In addition, the Gα activated by the muscarinic AChR is a G_i variant, which inhibits adenylate cyclase and counteracts the effect of β-adrenergic receptors discussed in Section 3.20.

We have seen that two different neurotransmitters, norepinephrine and ACh, act on different GPCRs, G proteins, and effectors on the same cell to speed up or slow down heart rate, respectively (compare Figures 3-33 and 3-35). These neurotransmitters are used in the two opposing arms of the autonomic nervous system. The sympathetic arm, which uses norepinephrine as a neurotransmitter, and the **parasympathetic** arm, which uses ACh as a neurotransmitter, often have antagonistic functions (see Section 9.1 for more details).

3.22 Metabotropic receptors can act on the presynaptic terminal to modulate neurotransmitter release

In addition to acting on dendrites and cell bodies, metabotropic receptors can also act directly on presynaptic terminals to modulate neurotransmitter release. In the simplest case, neurons can use metabotropic receptors to modulate their own neurotransmitter release, as in the case of sympathetic neurons that release norepinephrine (**Figure 3-36**). The presynaptic terminals of these neurons express α-adrenergic receptors that bind norepinephrine released into the synaptic cleft. Activation of these presynaptic α-adrenergic receptors rapidly inhibits voltage-gated Ca^{2+} channels at the active zone by direct binding of Gβγ to the Ca^{2+} channel, which reduces the depolarization-induced Ca^{2+} entry essential for triggering neurotransmitter release. This negative feedback loop results in diminishing levels of neurotransmitter release, leading to presynaptic depression. This is one of multiple mechanisms of short-term plasticity that alters the probability of neurotransmitter release, as noted in Section 3.10.

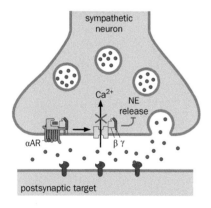

Figure 3-36 **Action of norepinephrine on a presynaptic Ca²⁺ channel.** From left, released norepinephrine (NE, blue dots) binds and activates a presynaptic α-adrenergic receptor (αAR). Activated Gβγ binds and inhibits the presynaptic voltage-gated Ca²⁺ channel, reducing Ca²⁺ influx in response to depolarization and thereby inhibiting neurotransmitter release.

A presynaptic terminal of a given neuron can also contain metabotropic receptors for neurotransmitters produced by other neurons. In this case, the presynaptic terminal of a neuron acts as the postsynaptic site for these other neurons (**Figure 3-37**). Depending on the nature of the neurotransmitter, the type of receptor, the signaling pathway, and the final effector, the net effect can either be facilitation or inhibition of neurotransmitter release. Accordingly, these effects are called **presynaptic facilitation** or **presynaptic inhibition**. Presynaptic facilitation can be achieved by closing K⁺ channels, which depolarizes the presynaptic membrane potential and facilitates activation of voltage-gated Ca^{2+} channels so that Ca^{2+} entry can trigger neurotransmitter release; we will see an example of this in Chapter 11 where serotonin mediates presynaptic facilitation in the sea slug *Aplysia* to enhance the magnitude of a reflex to a noxious stimulus. Presynaptic

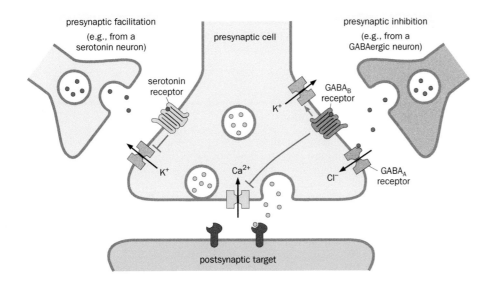

Figure 3-37 Presynaptic facilitation and inhibition. Left, an example of presynaptic facilitation by activation of a metabotropic receptor, such as a serotonin receptor; facilitation can be achieved by decreasing K^+ conductance (red inhibitory sign). Right, examples of presynaptic inhibition by GABA. Activation of the $GABA_A$ receptor increases Cl^- conductance and thereby counters depolarization. Activation of the $GABA_B$ receptor can act by increasing K^+ conductance (green arrow) or decreasing Ca^{2+} conductance (red inhibitory sign).

inhibition can be achieved by opening K^+ channels or closing voltage-gated Ca^{2+} channels, both of which inhibit neurotransmitter release. For example, in Chapter 6, we will learn that *Drosophila* olfactory receptor neurons (ORNs) activate GABAergic local interneurons, which synapse back onto ORN axon terminals to provide negative feedback control of ORN neurotransmitter release. (Presynaptic inhibition can also be achieved through GABA acting on ionotropic $GABA_A$ receptors present on the presynaptic terminals of some neurons.) Presynaptic facilitation and inhibition are also widely used in vertebrate nervous systems.

3.23 GPCR signaling features multiple mechanisms of signal amplification and termination

As we have seen in previous sections, metabotropic neurotransmitter receptors have diverse functions that depend on their locations and their coupling to different G proteins, signaling pathways, and effectors. Their effects unfold more slowly than the rapid effects of ionotropic receptors. However, second messenger systems feature an important property: signal amplification. For example, activation of a single adrenergic receptor can trigger multiple rounds of G protein activation, each activated adenylate cyclase can produce many cAMP molecules, and each activated PKA can phosphorylate many substrate molecules.

Signals must be properly terminated in order for cells to respond to future stimuli. Indeed, all signaling events we have discussed so far are associated with built-in termination mechanisms. The GPCR is deactivated when its ligand dissociates; Gα-GTP is deactivated by its intrinsic GTPase activity, often facilitated by GAPs; Gβγ is deactivated by reassociation with Gα-GDP; adenylate cyclase is deactivated in the absence of Gα-GTP; cAMP produced by adenylate cyclase is metabolized into AMP by an enzyme called **phosphodiesterase**; the catalytic subunits of PKA reassociate with regulatory subunits and become inactive when cAMP concentration declines; and **protein phosphatases** remove phosphates from phosphorylated proteins, thus counteracting the actions of kinases. While some of these termination mechanisms are constitutive, others are regulated by signals.

Another important mechanism of terminating G protein signaling is via the binding of **arrestin**, first discovered in the context of rhodopsin signaling but subsequently found to apply widely to GPCR signaling. Specifically, activated GPCRs are phosphorylated by a GPCR kinase. Phosphorylated GPCRs allow arrestin binding, which competes with GPCR binding to G proteins. In addition, arrestin binding facilitates endocytosis of GPCRs, thus reducing the number of GPCRs on the cell surface. Interestingly, arrestin binding to GPCRs, while terminating GPCR signaling through the G protein, can activate separate downstream signaling pathways.

Signal amplification and termination apply generally to signal transduction pathways (Box 3-4). In Chapter 4, we will see a salient example of signal amplification and termination when we study how photons are converted to electrical signals in vision.

Box 3-4: Signal transduction and receptor tyrosine kinase signaling

In response to extracellular signals, cells utilize many pathways to relay such signals to varied effectors to produce specific biological effects; this process is generally referred to as **signal transduction**. In the context of synaptic transmission, we have focused on the actions of ionotropic and metabotropic receptors that change the membrane potential of the postsynaptic cell. In this box, we expand our scope by placing neurotransmitter receptor signaling in the general framework of signal transduction and by discussing receptor tyrosine kinase signaling pathways, which are crucial for nervous system development and function.

In a typical signal transduction pathway (**Figure 3-38**A), an extracellular signal (a **ligand**) is detected by a **cell-surface receptor** in the recipient cell. (We will learn of an exception in Chapter 10: steroid hormones diffuse across the cell membrane to bind receptors *within* the cell.) The extracellular signal is then relayed through one or a series of intracellular signaling proteins to reach the effector(s), producing

cellular responses to the extracellular signal. The final effectors are diverse, but usually fall into one of the following categories: (1) enzymes that change cellular metabolism; (2) regulators of gene expression that alter chromatin structure, gene transcription, mRNA metabolism, or protein translation and degradation; (3) cytoskeletal proteins that regulate cell shape, cell movement, and intracellular transport; or (4) ion channels that alter the cell's membrane potential and excitability. Indeed, we can map what we have learned about metabotropic and ionotropic receptor signaling onto this general framework of signal transduction (Figure 3-38B).

The extracellular signal can come from different sources. If the signal is produced by the recipient cell itself (as is the case of presynaptic norepinephrine receptor signaling; Figure 3-36), it is called an **autocrine** signal. If the signal comes from nearby cells, it is called a **paracrine** signal; neurotransmitters can be considered specialized paracrine signals

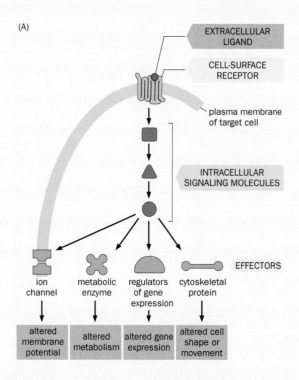

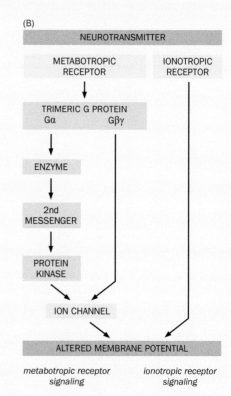

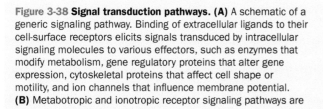

Figure 3-38 Signal transduction pathways. (A) A schematic of a generic signaling pathway. Binding of extracellular ligands to their cell-surface receptors elicits signals transduced by intracellular signaling molecules to various effectors, such as enzymes that modify metabolism, gene regulatory proteins that alter gene expression, cytoskeletal proteins that affect cell shape or motility, and ion channels that influence membrane potential. **(B)** Metabotropic and ionotropic receptor signaling pathways are

mapped onto the generic signaling pathway in Panel A, with colors indicating the components of the signaling pathway. Note that ionotropic receptors are simultaneously receptors and effectors that change the membrane potential, thus representing the shortest and fastest (within milliseconds) signaling pathway. (A, adapted from Alberts B, Johnson A, Lewis J, et al. [2015] Molecular Biology of the Cell, 6th ed. Garland Science.)

Box 3-4: continued

with postsynaptic neurons corresponding to target cells. If the signal comes from a remote cell through circulating blood, it is called an **endocrine** signal or a **hormone** (as is the case of epinephrine). When the signal comes from a neighboring cell, it can either be a diffusible molecule such as a neurotransmitter or a secreted protein or be a membrane-bound protein that requires cell–cell contact for signal transduction. Secreted and membrane-bound protein ligands are widely used in cell–cell communication during development, which will be discussed in detail in Chapters 5 and 7.

In addition to the ionotropic and metabotropic receptors, many cell-surface receptors used in the nervous system feature intracellular domains with enzymatic activity. As an example, we discuss here a widely used family of enzyme-coupled receptors called **receptor tyrosine kinases (RTKs)**—transmembrane proteins with an N-terminal extracellular ligand-binding portion and a C-terminal intracellular portion possessing a **tyrosine kinase** domain as well as tyrosine phosphorylation sites (Figure 3-39A). About 60 genes in the mammalian genome encode RTKs. We focus here on RTK signaling involving the neurotrophin receptors, but the principles are generally applicable to other RTK signaling pathways.

Neurotrophins are a family of secreted proteins that regulate the survival, morphology, and physiology of target neurons (we will discuss the biological effects of these proteins in Section 7.15). They bind to and activate the **Trk receptor**

family of RTKs. How does neurotrophin binding to Trk activate signaling? Neurotrophins naturally form dimers. When each neurotrophin binds a Trk receptor, the neurotrophin dimer brings two Trk receptors into close proximity, such that the kinase domain of one Trk can phosphorylate tyrosine residues on the other Trk. Phosphorylation of key tyrosine residues creates binding sites for specific adaptor proteins. These adaptor proteins contain either an SH2 (src homology 2) domain or a PTB (phosphotyrosine binding) domain, which enables the adaptors to bind phosphorylated tyrosines in the context of specific amino acid sequences and thereby initiate downstream signaling. In the Trk receptors, for instance, two key tyrosine residues recruit the binding of several specific adaptor proteins, eliciting separate transduction pathways that can also cross-talk with each other (Figure 3-39A).

One such signaling pathway is initiated by binding of the adaptor Shc (Figure 3-39B), which binds tyrosine-phosphorylated Trk via its PTB domain and becomes tyrosine phosphorylated by Trk. This further recruits the binding of Grb2, an SH2-domain-containing adaptor protein. Grb2 is associated with Sos, a guanine nucleotide exchange factor for the small GTPase Ras (Box 3-3). Ras normally associates with the membrane due to lipid modification like that of Gα. Thus, Trk activation recruits Sos to the plasma membrane to catalyze the exchange of GDP for GTP on Ras. Ras-GTP binds a downstream effector called Raf, a serine/threonine protein kinase. Raf phosphorylates and activates another

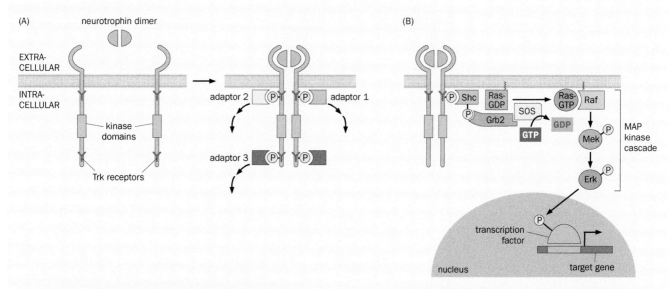

Figure 3-39 Neurotrophin receptor signaling as an example of receptor tyrosine kinase signaling. (A) In the absence of neurotrophin, Trk receptors are present as monomers with unphosphorylated tyrosine residues (Y). Binding of a neurotrophin dimer brings two Trk receptors into close proximity, allowing the kinase domain of each Trk to phosphorylate specific tyrosine residues on the other Trk. Tyrosine phosphorylation recruits binding of specific adaptor proteins, each eliciting a downstream signaling event. Different adaptors can bind the same phosphorylated tyrosine (as in the case of adaptor 1 and adaptor 2). **(B)** Details

of one adaptor pathway. Shc binds to a membrane-proximal phosphorylated tyrosine on Trk, leading to tyrosine phosphorylation of Shc. This helps recruit the binding of the Grb2-Sos complex. Sos acts as a guanine nucleotide exchange factor that catalyzes the conversion of Ras-GDP to Ras-GTP (red zigzag lines indicate lipid modification of Ras). Ras-GTP binds to and activates the MAP kinase cascade, including Raf, Mek, and Erk. Activated Erk phosphorylates a number of transcription factors, which activate or repress transcription of target genes.

(Continued)

Box 3-4: continued

serine/threonine protein kinase Mek, which in turn phosphorylates and activates a third serine/threonine kinase Erk. Activated Erk phosphorylates and activates a number of **transcription factors** (DNA-binding proteins that activate or repress transcription of target genes), which leads to transcription of specific genes that promote neuronal survival and differentiation, two major biological effects of neurotrophin signaling during development.

Erk is also called MAP kinase (<u>m</u>itogen-<u>a</u>ctivated <u>p</u>rotein kinase), and therefore Mek is a MAP kinase kinase (since it phosphorylates MAP kinase), and Raf is a MAP kinase kinase kinase. The Raf-Mek-Erk kinase cascade is often referred to as the **MAP kinase cascade**, which acts downstream of Ras and other signaling molecules, such as arrestin (Section 3.23). The Ras-MAP kinase cascade is a widely used signaling pathway that serves many functions, including cell survival and differentiation, as discussed earlier; cell fate determination (Section 5.17); and cell proliferation. It is also used in activity-dependent transcription (Section 3.24).

3.24 Postsynaptic depolarization can induce new gene expression

In addition to changing the membrane potentials and excitability of postsynaptic neurons at the time scales of milliseconds (through ionotropic receptors) or tens of milliseconds to seconds (through metabotropic receptors), neurotransmitters can also trigger long-term (hours to days) changes in the physiological states of postsynaptic neurons by inducing expression of new genes. For example, transcription of *Fos* was induced by ionotropic AChR activation within 5 minutes of nicotine application to cultured cells (**Figure 3-40**). *Fos* encodes a transcription factor, and its transient activation can change the expression of many downstream target genes.

Fos is the prototype of a class of genes called **immediate early genes** (**IEGs**), whose transcription is rapidly induced by external stimuli in the presence of protein synthesis inhibitors; this means that no new protein synthesis is required to turn on IEGs. In neurons, IEGs can be rapidly induced by depolarization of postsynaptic neurons in response to presynaptic neurotransmitter release. Some IEGs, such as *Fos* or *Egr1* (<u>e</u>arly <u>g</u>rowth <u>r</u>esponse-1), encode transcription factors that regulate the expression of other genes. Other IEGs encode regulators of neuronal communication that act more directly. Among them, **brain-derived neurotrophic factor** (**BDNF**) is a secreted neurotrophin that regulates the morphology and physiology of target neurons (Box 3-4). **Arc** (<u>a</u>ctivity-<u>r</u>egulated <u>c</u>ytoskeleton-associated protein) is a cytoskeletal protein present at the postsynaptic density that regulates trafficking of glutamate receptors, thus contributing to synaptic plasticity. As will be discussed in later chapters, **activity-dependent transcription** (regulation of gene expression by neuronal activity) plays a prominent role in the maturation of synapses and neural circuits during development and in their modulation by experience in adulthood. Because of the rapid induction of IEGs by neuronal activity, their expression has also been used to identify and manipulate neurons activated by specific experiences, behavioral episodes, and internal states (see Section 14.11 for details).

Many signaling pathways linking neurotransmitter receptors to transcription have been identified. A rise in intracellular Ca^{2+} concentration ($[Ca^{2+}]_i$) is often a key step. $[Ca^{2+}]_i$ increases can be accomplished by several means: via the NMDA receptor at the postsynaptic density in dendritic spines (Figure 3-24), via voltage-gated Ca^{2+} channels enriched on the dendritic trunk and cell body, and via IP_3 receptors (Figure 3-34) or the related **ryanodine receptors** on the ER membrane. (Instead of being activated by IP_3, ryanodine receptors are activated by a rise in $[Ca^{2+}]_i$ and thus amplify the Ca^{2+} signal; ryanodine is a plant-derived agonist of this ER-resident Ca^{2+} channel.) Although free Ca^{2+} ions usually do not diffuse far from the source of entry into the cytosol, they can associate with various Ca^{2+}-binding proteins, most notably calmodulin (CaM) (Figure 3-34), and initiate signals that can be transduced to the nucleus (**Figure 3-41**). For example, Ca^{2+}/CaM activates CaM kinases, such as CaMKII, which is enriched in postsynaptic densities, and

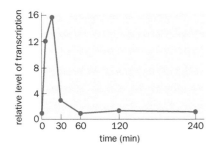

Figure 3-40 Nicotinic AChR activation induces transcription of *Fos*, an immediate early gene. Nicotine application to a cultured neuronal cell line at time 0 induces rapid and transient transcription of *Fos*, indicated by relative levels of newly synthesized *Fos* RNA. (From Greenberg ME, Ziff EB, & Greene LA [1986] *Science* 234:80–83. With permission from AAAS.)

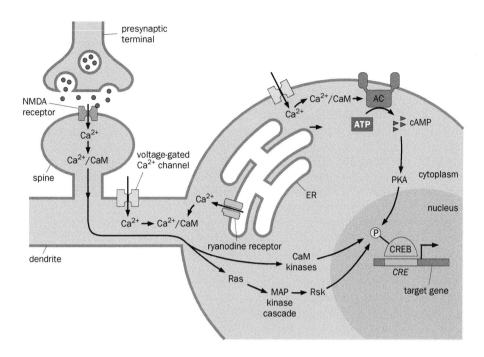

Figure 3-41 Signaling pathways from the synapse to the nucleus. Shown are pathways from postsynaptic terminals and the somatodendritic plasma membrane to the nucleus that involve Ca²⁺ and lead to the phosphorylation and activation of a transcription factor, CREB. An increase in $[Ca^{2+}]_i$ can result from an influx of extracellular Ca²⁺ through NMDA receptors concentrated in dendritic spines or voltage-gated Ca²⁺ channels enriched on the somatodendritic plasma membrane or can be mediated by the release of Ca²⁺ from internal stores in the ER through ryanodine receptors. Ca²⁺ bound to calmodulin (CaM) activates CaM kinases, Rsk (via the Ras-MAP kinase cascade), and PKA (via Ca²⁺-activated adenylate cyclase and cAMP production). CaM kinases, Rsk, and PKA can all phosphorylate CREB, promoting its activity to induce transcription of target genes with cAMP-response elements (*CRE*s) in their promoters. (Based on Cohen S & Greenberg ME [2008] *Annu Rev Cell Dev Biol* 24:183–209 and Deisseroth K, Mermelstein PG, Xia H, et al. [2003] *Curr Opin Neurobiol* 13:354–365.)

CaMKIV, which is enriched in nuclei. A specific isoform, γCaMKII, can transport Ca²⁺/CaM from the plasma membrane near voltage-gated Ca²⁺ channels to the nucleus so that Ca²⁺/CaM can activate nuclear effectors such as CaMKIV. In addition, Ca²⁺/CaM can activate several adenylate cyclase subtypes, leading to the production of cAMP and activation of PKA. The Ras-MAP kinase cascade (Box 3-4) is yet another signaling pathway that can be activated by Ca²⁺/CaM.

As a specific example, we discuss how these pathways lead to activation of a transcription factor called **CREB**. CREB was originally identified because it binds to a DNA element (*CRE*) in the promoter of the gene that produces the neuropeptide somatostatin, rendering somatostatin's transcription responsive to cAMP regulation. (*CRE*, for <u>c</u>AMP <u>r</u>esponse <u>e</u>lement; CREB, for <u>CRE</u> <u>b</u>inding protein.) *CRE* was subsequently found in the promoter of many IEGs including *Fos*. Biochemical studies indicate that phosphorylation of a particular serine residue is crucial for the activity of CREB as a transcriptional activator. This serine can be phosphorylated by several kinases, including PKA, CaMKIV, and a protein kinase called Rsk (<u>r</u>ibosomal protein <u>S6</u> <u>k</u>inase), a substrate of MAP kinase. Although each of these kinases can be activated by Ca²⁺ (Figure 3-41), each pathway has unique properties. For example, the CaM kinase–mediated pathway is more rapid, resulting in CREB phosphorylation that peaks within minutes after a transient neuronal depolarization, whereas the MAP kinase pathway mediates a gradual increase in CREB phosphorylation over an hour following a transient neuronal depolarization.

In addition to CREB, other Ca²⁺-responsive transcription factors bind different IEG promoters. Thus, neuronal activity can reach nuclei and change the transcriptional programs of postsynaptic cells via many routes. Furthermore, neuronal activity and Ca²⁺ can also affect chromatin structures through enzymes that control methylation of DNA and posttranslational modifications of histones (for example, methylation, demethylation, acetylation, and deacetylation), the protein component of chromatin. These **epigenetic modifications** (so named because they do not alter DNA sequence) also alter gene expression patterns through regulation of chromatin structures and accessibility of promoters to specific transcription factors. Another form of epigenetic modification is methylation of mRNAs, which can alter their stability and translation efficiency—for example, methylation promotes translation from mRNAs induced by neuronal activity. As will be discussed in Chapter 12, mutations in many components of synapse-to-nucleus signaling pathways have been found to cause human brain disorders, highlighting the important role of activity-dependent transcription in human brain functions.

3.25 Dendrites are sophisticated integrative devices

Aside from regulating gene expression, the primary function of synaptic transmission is to influence the firing patterns of postsynaptic neurons. This is the means by which information is propagated from one neuron to the next within a neural circuit. As a way of integrating what we've learned about neuronal communication in Chapter 2 and this chapter, in the final two sections we discuss how a postsynaptic neuron integrates synaptic inputs to produce its firing pattern, thus completing a full round of neuronal communication (Figure 1-18). We start our discussion with excitatory inputs.

Most excitatory inputs to a neuron increase its membrane conductance (for example, by opening ionotropic glutamate receptor channels), producing EPSCs and thus EPSPs (Figure 3-23). To influence the postsynaptic cell's firing pattern, these electrical signals need to travel to the axon initial segment, where action potentials are usually initiated. As we learned in Section 2.8, electrical signals evolve over time and decay over distance, specified by the passive (cable) properties of neuronal fibers such as the time (τ) and length constants (λ). Theoreticians have used model neurons to predict amplitudes of somatic EPSPs produced by synaptic input at different locations in dendrites. In the model neuron shown in **Figure 3-42**A, for example, the complex dendritic tree is simplified to 10 compartments with varying distances from the soma in order to predict the amplitudes of somatic EPSPs in response to dendritic inputs. A fixed transient increase in synaptic conductance, equivalent to a transient opening of ionotropic glutamate receptors, produces somatic EPSPs with different shapes and amplitudes when applied to different locations in the dendrites (Figure 3-42B). More distant synapses produce smaller, slower, and broader EPSPs. This is because EPSPs produced at more distant synapses decay more substantially, as they need to travel longer distances to reach the soma. In this model neuron, synaptic inputs given at compartments 4 and 8 produce peak somatic EPSP amplitudes only 29% and 10% of the somatic EPSP amplitude when the same input is given at the soma (Figure 3-42B).

A mammalian CNS neuron receives on average thousands of excitatory synaptic inputs along its dendritic tree. A single EPSP at one synapse is usually insufficient to depolarize the postsynaptic neuron above the action potential firing threshold, due to the small size of an individual EPSP when it arrives at the axon initial segment. Indeed, at any given time, the postsynaptic neuron integrates many excitatory inputs in order to reach the firing threshold. Such integration takes two forms. In **spatial integration**, nearly simultaneously activated synapses at different spatial locations sum their excitatory postsynaptic currents when they converge along the path to the soma, producing a larger EPSP (**Figure 3-43**A). In

Figure 3-42 Somatic EPSPs from dendritic inputs in a model neuron.
(A) The soma and dendritic tree of this neuron are simplified to 10 compartments for the purpose of mathematical modeling. Compartment 1 represents the soma, and compartments 2–10 represent dendritic segments with increasing distance from the soma, with the length constant (λ) as the unit. Dashed lines illustrate divisions between every two compartments.
(B) When a transient excitatory input of the same size and shape (dashed curve, with y axis to the right) is provided at compartments 1, 4, or 8, the shapes of EPSPs at the soma show distinct profiles. The somatic EPSP produced by the somatic input has the largest amplitude and fastest rising and decay time, the somatic EPSP produced by the input given at compartment 8 has the smallest amplitude and slowest rising and decay times, and the somatic EPSP produced by the input given at compartment 4 has the intermediate amplitude and temporal spread. Time is represented in the unit of the time constant τ. (Adapted from Rall W [1967] *J Neurophysiol* 30:1138–1168.)

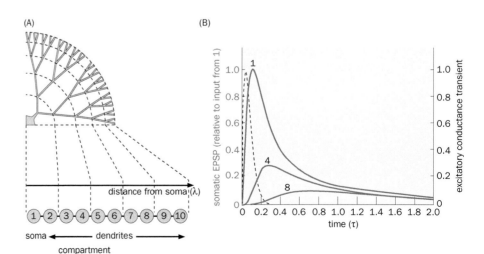

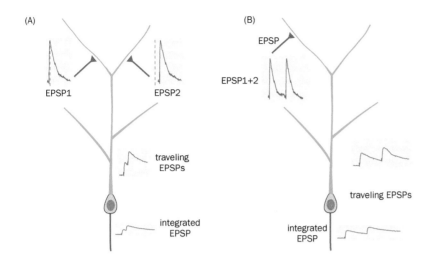

Figure 3-43 Spatial and temporal integration of synaptic inputs. (A) Spatial integration. Two excitatory inputs from two branches of the dendritic tree that arrive shortly after one another (vertical dotted lines in the EPSP traces indicate the same time) summate their signals when they converge, producing depolarization of the membrane potential at the axon initial segment that exceeds the amplitude produced by each alone (compare the heights of the second and first peaks). **(B)** Temporal integration. Two discrete EPSPs produced at the same synapse (top left) become gradually integrated as they travel from distal dendrites toward the soma due to the temporal spread of electrical signals (Figure 3-42B). At the axon initial segment, the integrated EPSP produces a peak potential (the second peak) greater than that produced by a single EPSP (the first peak).

temporal integration, synapses activated within a specific window (including successive activation at the same synapse) sum their postsynaptic currents, producing a larger EPSP (Figure 3-43B).

As we see from the model neuron in Figure 3-42, inputs from proximal synapses contribute more to the firing of the neuron because they are less attenuated. In some mammalian neurons, distal synapses are stronger in order to compensate for such distance-dependent attenuation. Importantly, inputs from distal synapses also have a longer window during which to contribute to temporal integration (Figure 3-43B). Without considering inhibitory inputs, we can already see that individual dendrites are sophisticated integrative devices. At any given moment, a spiking neuron converts analog signals from the many inputs it receives into digital signals (to spike or not to spike).

While the passive properties of neuronal membranes discussed thus far provide a foundation for understanding how synaptic inputs regulate firing of postsynaptic neurons, dendritic integration is more complex and nuanced. Voltage-gated Na^+, Ca^{2+}, and K^+ channels are present on the dendrites of many mammalian CNS neurons. EPSPs can open dendritic voltage-gated Na^+ or Ca^{2+} channels, causing further depolarization and thus signal amplification. Co-activation of nearby excitatory synapses in dendritic branches can produce dendritic spikes that actively propagate across dendritic segments. Although these dendritic spikes are not fully regenerative as are axonal action potentials and may not propagate all the way to the soma (because of the lower density of voltage-gated Na^+ channels in dendrites compared to axons), they nevertheless amplify synaptic input and propagate membrane potential changes across greater distances with smaller attenuation than passive spread. Finally, action potentials generated at the axon initial segment can back-propagate into dendrites via the participation of dendritic voltage-gated Na^+ channels, and these back-propagated action potentials can interact with EPSPs in interesting ways.

As a specific example, we study an experiment in which a cortical pyramidal neuron in an *in vitro* brain slice was subjected to dual patch clamp recording at the soma and at the apical dendrites (**Figure 3-44**A). Electrical stimulation of presynaptic axons produced a subthreshold EPSP at the soma (Figure 3-44B). However, if an action potential was induced in the recorded neuron 5 ms before presynaptic stimulation (by injecting a pulse of depolarizing current through the somatic patch pipette), the back-propagated action potential synergized with the dendritic synaptic potential to reach the threshold of a dendritic spike, which greatly amplified the synaptic potential. This allowed the neuron to produce two additional somatic action potentials (Figure 3-44C). Assuming that under physiological conditions, the pyramidal neuron fires action potentials in response to proximal dendritic inputs, this integration mechanism can enable amplification of near-synchronous input at the proximal and distal dendrites by producing a

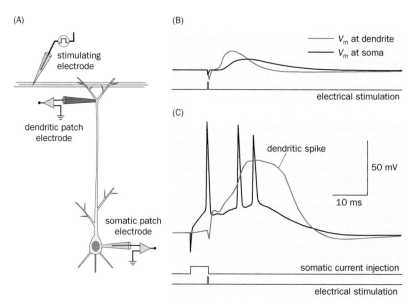

Figure 3-44 **Interactions between synaptic input and a back-propagating action potential.** **(A)** Experimental setup. A cortical pyramidal neuron in a brain slice is being recorded by patch electrodes at the apical dendrite (red) and soma (gray), both in whole-cell recording mode. A stimulating electrode delivers electrical stimulation to presynaptic axons. **(B)** Electrical stimulation (bottom trace) produces a dendritic EPSP recorded by the dendritic patch electrode (red), and an attenuated somatic EPSP recorded by the somatic patch electrode. The somatic EPSP is below the threshold for firing an action potential. **(C)** A 5-ms depolarizing current pulse injected into the soma before electrical stimulation produces an action potential (the first black spike), which propagates back to the dendrites and integrates with the dendritic EPSP, reaching the threshold for producing a dendritic spike (red trace). The propagation of the dendritic spike to the soma produces two additional somatic action potentials (the second and third black spikes). Thus, the back-propagating action potential synergizes with the dendritic EPSP to produce additional output spikes. (Adapted from Larkum ME, Zhu JJ, & Sakmann B [1999] *Nature* 398:338–341. With permission from Springer Nature.)

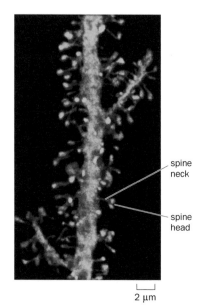

Figure 3-45 **Dendritic spines.** A dendritic segment of a human cortical pyramidal neuron ~100 μm from the cell body, showing dendritic spines with long necks. Imaged after intracellular injection of a fluorescent dye. (From Yuste R. Dendritic Spines, Cover Image, © 2010 Massachusetts Institute of Technology, by permission of The MIT Press.)

burst of action potentials that could not be generated by either the proximal or distal inputs alone.

Active properties of mammalian CNS neurons can vary in different neuronal types or even in different compartments of the same neuron because of particular distributions and densities of voltage-gated ion channels. We are far from a complete understanding of how synaptic potentials are integrated in light of these active properties; this is an active area of research fundamental to our understanding of neuronal communication and neural circuit function.

3.26 Synapses are strategically placed at specific locations in postsynaptic neurons

In addition to excitatory inputs, each neuron also receives inhibitory and modulatory inputs. How these inputs shape the output of a postsynaptic neuron depends on which subcellular compartments of the postsynaptic neuron these inputs contact.

In general, most excitatory synapses are located on dendritic spines distributed throughout the dendritic tree (**Figure 3-45**). The various presynaptic terminals targeting a given postsynaptic neuron may originate from many different presynaptic partner neurons, but each dendritic spine typically receives synaptic input from a single excitatory presynaptic terminal. The thin spine neck chemically and electrically compartmentalizes each synapse such that it can be modulated semi-independently from neighboring synapses. These semi-independent compartments enable neurons to encode information in the strengths of individual synapses with different input neurons. The strength of each synapse can be

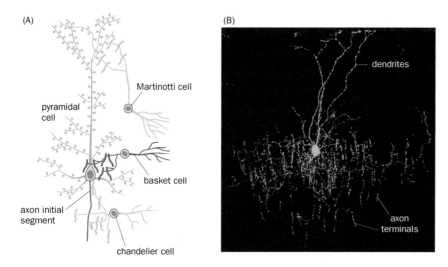

Figure 3-46 Inhibitory inputs onto a cortical pyramidal neuron. (A) Schematic of presynaptic terminals (shown as short strings of beads) from three different types of GABAergic inhibitory neurons onto a cortical pyramidal neuron, whose dendrites are in cyan and axons are in red. The Martinotti (yellow), basket (blue), and chandelier (green) cells form synapses respectively onto the distal dendrites, the cell body and proximal dendrites, and the axon initial segment of the pyramidal neuron (and other pyramidal neurons not shown). **(B)** A chandelier cell in the mouse cerebral cortex. Each group of presynaptic terminals, which looks like a candle on an old-fashioned chandelier, wraps around the initial segment of a pyramidal neuron. Each chandelier cell thus controls the firing of many pyramidal neurons. (B, courtesy of Z. Josh Huang. See also Taniguchi H, Lu J, & Huang ZJ [2013] *Science* 339:70–74.)

modified by prior synaptic activity of that particular synapse, a property crucial for memory as we will discuss in Chapter 11.

In contrast to excitatory synapses, which are most enriched on spines, inhibitory synapses form on dendritic spines, dendritic shafts, the cell body, and the axon initial segment. These distributions allow inhibitory synapses to oppose the action of EPSPs as they pass by (Figure 3-29). Let's use a typical pyramidal neuron in the cerebral cortex to describe inhibitory inputs it receives from three types of GABAergic neurons (**Figure 3-46**A). **Martinotti cells** target the distal dendrites of the pyramidal neuron. Thus, activation of Martinotti cells can inhibit the production or propagation of dendritic spikes. By contrast, **basket cells** target the cell body of the pyramidal neuron and thereby influence the overall integration of synaptic input from all dendritic branches. Last, **chandelier cells** target the axon initial segment of the pyramidal neuron (Figure 3-46B), meaning that chandelier cells have the most direct impact on the production of action potentials. In summary, inhibitory neurons have specialized functions partly because they make synapses onto specialized locations on postsynaptic cells.

A postsynaptic neuron can also receive synaptic input at its own axon terminals, as discussed in Section 3.21. Here, inputs do not control action potential firing but rather the efficacy with which action potentials lead to neurotransmitter release. The presynaptic partners in these cases can be modulatory neurons using transmitters such as acetylcholine and the monoamines. GABAergic and glutamatergic neurons can also affect their target neurons via both ionotropic and metabotropic receptors on their axon terminals (Figure 3-37).

In summary, individual neurons are complex, highly organized integrators. Each neuron receives inputs from numerous presynaptic partners at different parts of its complex dendritic tree, its cell body, its axon initial segment, and its axon terminals (**Figure 3-47**). The interactions of excitatory, inhibitory, and modulatory inputs together shape the neuron's output patterns, which are communicated to its postsynaptic target neurons by the frequency and timing of action potentials and the probability of neurotransmitter release induced by each action potential. Some neurons also receive input (and send output) through electrical synapses

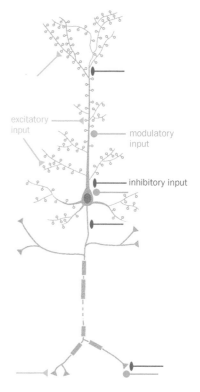

Figure 3-47 Subcellular distribution of synaptic input. In a typical mammalian neuron, excitatory inputs are received mostly at dendritic spines (and along the dendrites for neurons lacking dendritic spines) and presynaptic terminals. Inhibitory inputs are received at the dendritic spines and shaft, cell body, axon initial segment, and presynaptic terminals. Modulatory inputs are received at dendrites, cell bodies, and presynaptic terminals.

(Box 3-5). At a higher level, individual neurons are parts of complex neural circuits that perform diverse information-processing functions underlying processes ranging from sensory perception to behavioral control. Having studied the basic concepts and principles of neuronal communication, we are now ready to apply them to fascinating neurobiological processes in the following chapters.

Box 3-5: Electrical synapses

Although chemical synapses are the predominant form of interneuronal communication, electrical synapses are also prevalent in both vertebrate and invertebrate nervous systems. The morphological correlate of the electrical synapse is the **gap junction**, which usually contains hundreds of closely clustered channels that bring the plasma membranes of two neighboring cells together (Figure 1-14B) and allow passage of ions and small molecules between the two cells. In mammalian neurons, electrical synapses usually occur at the somatodendritic compartments of two partner neurons.

In vertebrates, gap junctions are made predominantly by a family of **connexin** proteins, encoded by about 20 genes in the mammalian genome. Each gap junction channel has 12 connexin subunits, with 6 subunits on each apposing plasma membrane forming a hemichannel. Each connexin subunit has four transmembrane domains with an additional N-terminal domain embedded in the membrane. As revealed by the crystal structure of connexin-26 (**Figure 3-48**), extensive interactions between the extracellular loops of the hemichannels bring the two apposing membranes from neighboring cells within 4 nm of each other, and align the two hemichannels, forming a pore with an innermost diameter of 1.4 nm. Invertebrate gap junctions are made by a different family of proteins called **innexins** (invertebrate connexin). A third family of proteins called pannexins may contribute to gap junctions in both vertebrates and invertebrates.

Electrical synapses differ from chemical synapses in several important ways. First, whereas chemical synapses transmit signals with a delay on the order of 1 ms between depolarization in the presynaptic terminal and synaptic potential generation in the postsynaptic cell, electrical synapses transmit electrical signals with virtually no delay. Second, whereas chemical synapses are activated only by presynaptic depolarization (and, in spiking neurons, only suprathreshold signals that produce action potentials), electrical synapses transmit both depolarization and hyperpolarization. Third, whereas chemical synapses are asymmetrical—membrane potential changes in the presynaptic neuron produce membrane potential changes in the postsynaptic neuron, but not vice versa—electrical signals can flow in either direction across electrical synapses. Exceptions exist to this rule, however; some electrical synapses prefer one direction over the opposite direction and are thus called rectifying electrical synapses. Finally, many electrical synapses allow small molecules such as peptides and second messengers to pass through; indeed, the diffusion of small-molecule dye from one cell to another, called **dye coupling**, is often used as a criterion to identify the presence of gap junctions between two cells. The conductance of electrical synapses can be mod-

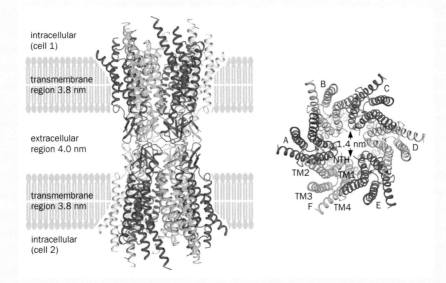

Figure 3-48 Structure of a gap junction channel. Summary of crystal structures of connexin-26. Left, view from the side. Each hemichannel consists of six subunits, which are differentially colored. Each subunit has four transmembrane helices and an N-terminal helix (NTH) embedded in the membrane. Right, surface view, with the transmembrane helices and NTH labeled for subunit F. The central passage allows molecules with a linear dimension smaller than 1.4 nm to pass freely between two cells. (Adapted from Maeda S, Nakagawa S, Suga M, et al. [2009] *Nature* 458:597–602. With permission from Springer Nature.)

Box 3-5: continued

ulated by several factors, such as the membrane potential, the transjunctional voltage (the difference between membrane potentials across the electrical synapse), and other factors such as phosphorylation state, pH, and Ca^{2+} concentration.

The special properties of electrical synapses discussed here are utilized in many circuits in invertebrates and vertebrates. For instance, electrical synapses are found in circuits where rapid transmission is essential, such as in vertebrate retinal circuits that processes motion signals (where analog signals are transmitted between nonspiking neurons), and in predator avoidance escape circuits. Indeed, electrical synapses were first characterized between the giant axon and motor neuron in the crayfish escape circuit in the 1950s. Another use of electrical synapses is to facilitate synchronized firing between electrically coupled neurons (another term for neurons that form electrical synapses with each other). As a specific example, we study electrical synapses in the mammalian cerebral cortex.

Using whole-cell recording techniques (see Box 14-3 and Section 14.21 for details) in a cortical slice preparation, researchers found that when two fast-spiking (FS) inhibitory neurons (corresponding mostly to basket cells in Figure 3-46A) were recorded simultaneously with patch electrodes, current injection into one cell caused nearly synchronous mem-

brane potential changes in both cells; both depolarization and hyperpolarization could result, depending on the sign of the injected current (**Figure 3-49**A), indicating that these two cells formed electrical synapses with each other. Paired recordings of many cell types indicated that electrical synapses form with substantial cell-type specificity. For example, low-threshold-spiking (LTS) inhibitory neurons (corresponding mostly to Martinotti cells in Figure 3-46A) also form electrical synapses with each other at a high probability, but rarely with FS neurons. Excitatory pyramidal neurons, which could be identified as regular-spiking (RS) cells by their electrophysiological properties, did not form any electrical synapses with other pyramidal neurons or with FS or LTS neurons (Figure 3-49B). Furthermore, while injecting subthreshold depolarizing currents into one of the two electrically coupled FS cells did not elicit action potentials, injecting the same subthreshold depolarizing currents into both FS cells elicited synchronous action potentials (Figure 3-49C). This suggested that a network of FS cells can act as detectors for synchronous inputs and their synchronous firing can further strengthen network synchrony. Subsequent work indicated that in addition to FS and LTS cells, other specific types of inhibitory neurons also form type-specific electrical synapse networks, thus providing a rich substrate for coordinating electrical activity in the cerebral cortex.

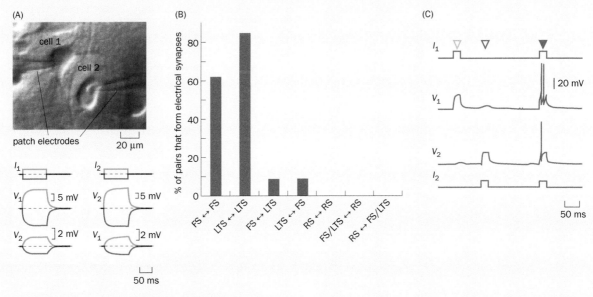

Figure 3-49 Electrical synapses between inhibitory neurons in the rat cerebral cortex. (A) Top, image of a rat cortical slice with two cells and two patch electrodes, taken with differential interference contrast microscopy. Bottom left, when positive (blue) or negative (red) current (I_1) was injected into cell 1, it depolarized or hyperpolarized, respectively, the membrane potential of cell 1 (top or bottom traces of V_1). In addition, cell 2 was correspondingly depolarized or hyperpolarized at the same time (top or bottom traces of V_2). Note the reduced amplitude and slower rising time of V_2 compared to V_1 due to attenuation across the gap junction and the time taken to charge the membrane capacitance of cell 2. Bottom right, positive or negative current injected into cell 2 (I_2) also caused depolarization or hyperpolarization, respectively, of both cells. Thus, these two cells form electrical synapses.

(B) Quantification of electrical synapses between specific types of cells based on paired recordings in Panel A. Cells were classified based on their firing patterns into fast spiking (FS), low-threshold spiking (LTS), or regular spiking (RS), corresponding roughly to basket, Martinotti, and pyramidal cells, respectively. Arrows indicate the directionality of electrical synapses tested in paired recordings. **(C)** The top and bottom traces show the injection of small depolarizing currents into cell 1 (top) or cell 2 (bottom). Injection into a single cell (open arrowheads) did not cause firing of either cell. Injecting the same current into both cells simultaneously (filled arrowhead) caused both cells to fire action potentials. (Adapted from Galarreta M & Hestrin S [1999] *Nature* 402:72–75. With permission from Springer Nature. See also Gibson JR, Beierlein M, & Connors BW [1999] *Nature* 402:75–79.)

SUMMARY

Neurons communicate with each other through electrical and chemical synapses. Electrical synapses allow rapid and bidirectional transmission of electrical signals between neurons via gap junctions. Although less prevalent than chemical synapses, electrical synapses are widely used in both invertebrates and vertebrates in neural circuits that require rapid information propagation or synchronization. Chemical synapses are unidirectional: electrical signal in the presynaptic neuron is transmitted to the postsynaptic neuron or muscle via the release of a chemical intermediate, the neurotransmitter.

At the presynaptic terminal, neurotransmitter release is mediated by fusion of the synaptic vesicle with the presynaptic plasma membrane. Action potential arrival depolarizes the presynaptic terminal, opening voltage-gated Ca^{2+} channels at the active zone. Ca^{2+} influx, acting through the synaptic vesicle-associated Ca^{2+} sensor synaptotagmin, releases a break on partially assembled SNARE complexes. The full assembly of the SNARE complex provides the force that drives membrane fusion and transmitter release from within the synaptic vesicle to the synaptic cleft. Excess neurotransmitters are rapidly degraded or recycled through reuptake mechanisms. Synaptic vesicles are rapidly recycled and refilled with neurotransmitter, enabling continual synaptic transmission in response to future action potentials.

Nervous systems across the animal kingdom utilize a common set of neurotransmitters. In the vertebrate CNS, glutamate is the main excitatory neurotransmitter, while GABA and glycine are the main inhibitory neurotransmitters. Acetylcholine is the excitatory neurotransmitter at the vertebrate neuromuscular junction, but can also act as a modulatory neurotransmitter in the CNS. Other neuromodulators include monoamines and neuropeptides. The specific actions of neurotransmitters are determined by the properties of their receptors on the postsynaptic neurons.

Neurotransmitter receptors are either ionotropic or metabotropic. Ionotropic receptors are ion channels gated by neurotransmitter binding and act rapidly to produce synaptic potentials within a few milliseconds of presynaptic action potential arrival. Ionotropic acetylcholine and glutamate receptors are nonselective cation channels; upon neurotransmitter binding, these receptors produce depolarization in the form of excitatory postsynaptic potentials. NMDA receptors act as coincidence detectors, because channel opening depends on both presynaptic glutamate release and a depolarized state of the postsynaptic neuron. Ionotropic GABA and glycine receptors are Cl^- channels. Their opening usually produces Cl^- influx, which impedes postsynaptic neurons from reaching the threshold at which they fire action potentials.

All metabotropic receptors are G-protein-coupled receptors. Neurotransmitter binding activates trimeric G proteins associated with the receptors. $G\alpha$-GTP and $G\beta\gamma$ each can activate different effectors, depending on G protein variants and cellular contexts. $G\beta\gamma$ can act on K^+ and Ca^{2+} channels directly, whereas $G\alpha$ usually acts via second messengers such as cAMP and Ca^{2+} to activate protein kinases that phosphorylate ion channels to change the membrane potential and excitability of postsynaptic neurons. Metabotropic receptor activation causes membrane potential changes within tens of milliseconds to seconds. Longer-term changes of postsynaptic neurons in response to neurotransmitter release and neuronal activity involve synapse-to-nucleus signaling and alterations of gene expression.

Chemical synapses are highly organized. At the presynaptic terminal, the active zone protein complexes bring synaptic vesicles to the immediate vicinity of voltage-gated Ca^{2+} channels such that Ca^{2+} influx rapidly triggers neurotransmitter release. Trans-synaptic cell adhesion proteins align presynaptic active zones with postsynaptic high-density neurotransmitter receptor clusters. Postsynaptic density scaffold proteins further link neurotransmitter receptors to their regulators and effectors for efficient synaptic transmission and for regulating synaptic plasticity.

Integration of excitatory, inhibitory, and modulatory inputs at the dendrites, cell bodies, and axon initial segments of postsynaptic neurons collectively determine their own action potential firing patterns. Synaptic inputs to axon terminals further modulate the efficacy with which postsynaptic action potentials lead to neurotransmitter release. These mechanisms are used extensively by the nervous system in all the processes, from sensation to action, that we will study in the following chapters.

OPEN QUESTIONS

- How is neuropeptide release regulated?

- When a neuron produces multiple neurotransmitters, can it separately regulate the release of each? How?

- Are there more neurotransmitters and receptors to be discovered? How would you discover them?

- How can we build a model of a synapse depicting all of its molecular components, their three-dimensional structures, and their precise localizations in the pre- and postsynaptic compartments and the synaptic cleft?

- How do different GPCR ligands differentially activate different downstream signaling cascades?

- What are the generalizable rules that govern how complex dendrites of mammalian CNS neurons integrate thousands of synaptic inputs?

FURTHER READING

Books and reviews

Cohen S & Greenberg ME (2008). Communication between the synapse and the nucleus in neuronal development, plasticity, and disease. *Annu Rev Cell Dev Biol* 24:183–209.

Hille B (2001). Ion Channels of Excitable Membranes, 3rd ed. Sinauer.

Huang EJ & Reichardt LF (2003). Trk receptors: roles in neuronal signal transduction. *Annu Rev Biochem* 72:609–642.

Katz B (1966). Nerve, Muscle, and Synapse. McGraw-Hill.

Sheng M, Sabatini BL, & Südhof TC (2012). The Synapse. Cold Spring Harbor Laboratory Press.

Südhof TC & Rothman JE (2009). Membrane fusion: grappling with SNARE and SM proteins. *Science* 323:474–477.

Unwin N (2013). Nicotinic acetylcholine receptor and the structural basis of neuromuscular transmission: insights from *Torpedo* postsynaptic membranes. *Q Rev Biophys* 46:283–322.

Presynaptic mechanisms

Augustine GJ, Charlton MP, & Smith SJ (1985). Calcium entry and transmitter release at voltage-clamped nerve terminals of squid. *J Physiol* 367:163–181.

Bennett MK, Calakos N, & Scheller RH (1992). Syntaxin: a synaptic protein implicated in docking of synaptic vesicles at presynaptic active zones. *Science* 257:255–259.

Del Castillo J & Katz B (1954). Quantal components of the end-plate potential. *J Physiol* 124:560–573.

Fernandez-Chacon R, Konigstorfer A, Gerber SH, Garcia J, Matos MF, Stevens CF, Brose N, Rizo J, Rosenmund C, & Südhof TC (2001). Synaptotagmin I functions as a calcium regulator of release probability. *Nature* 410:41–49.

Heuser JE & Reese TS (1981). Structural changes after transmitter release at the frog neuromuscular junction. *J Cell Biol* 88:564–580.

Ichtchenko K, Hata Y, Nguyen T, Ullrich B, Missler M, Moomaw C, & Südhof TC (1995). Neuroligin 1: a splice site-specific ligand for beta-neurexins. *Cell* 81:435–443.

Katz B & Miledi R (1967). The timing of calcium action during neuromuscular transmission. *J Physiol* 189:535–544.

Koenig JH & Ikeda K (1989). Disappearance and reformation of synaptic vesicle membrane upon transmitter release observed under reversible blockage of membrane retrieval. *J Neurosci* 9:3844–3860.

Kuffler SW & Yoshikami D (1975). The number of transmitter molecules in a quantum: an estimate from iontophoretic application of acetylcholine at the neuromuscular synapse. *J Physiol* 251:465–482.

Llinas R, Sugimori M & Silver RB (1992). Microdomains of high calcium concentration in a presynaptic terminal. *Science* 256:677–679.

Sabatini BL & Regehr WG (1996). Timing of neurotransmission at fast synapses in the mammalian brain. *Nature* 384:170–172.

Schiavo G, Benfenati F, Poulain B, Rossetto O, Polverino de Laureto P, DasGupta BR, & Montecucco C (1992). Tetanus and botulinum-B neurotoxins block neurotransmitter release by proteolytic cleavage of synaptobrevin. *Nature* 359:832–835.

Schneggenburger R & Neher E (2000). Intracellular calcium dependence of transmitter release rates at a fast central synapse. *Nature* 406:889–893.

Söllner T, Whiteheart SW, Brunner M, Erdjument-Bromage H, Geromanos S, Tempst P, & Rothman JE (1993). SNAP receptors implicated in vesicle targeting and fusion. *Nature* 362:318–324.

Sutton RB, Fasshauer D, Jahn R, & Brunger AT (1998). Crystal structure of a SNARE complex involved in synaptic exocytosis at 2.4 A resolution. *Nature* 395:347–353.

Takamori S, Holt M, Stenius K, Lemke EA, Gronborg M, Riedel D, Urlaub H, Schenck S, Brugger B, Ringler P, et al. (2006). Molecular anatomy of a trafficking organelle. *Cell* 127:831–846.

Tang AH, Chen H, Li TP, Metzbower SR, MacGillavry HD, & Blanpied TA (2016). A trans-synaptic nanocolumn aligns neurotransmitter release to receptors. *Nature* 536:210–214.

Watanabe S, Mamer LE, Raychaudhuri S, Luvsanjav D, Eisen J, Trimbuch T, Sohl-Kielczynski B, Fenske P, Milosevic I, Rosenmund C, et al. (2018). Synaptojanin and endophilin mediate neck formation during ultrafast endocytosis. *Neuron* 98:1184-1197.

Postsynaptic mechanisms

Chen L, Chetkovich DM, Petralia RS, Sweeney NT, Kawasaki Y, Wenthold RJ, Bredt DS, & Nicoll RA (2000). Stargazin regulates synaptic targeting of AMPA receptors by two distinct mechanisms. *Nature* 408:936–943.

Galarreta M & Hestrin S (1999). A network of fast-spiking cells in the neocortex connected by electrical synapses. *Nature* 402:72–75.

Larkum ME, Zhu JJ, & Sakmann B (1999). A new cellular mechanism for coupling inputs arriving at different cortical layers. *Nature* 398:338–341.

Lipscombe D, Kongsamut S, & Tsien RW (1989). Alpha-adrenergic inhibition of sympathetic neurotransmitter release mediated by modulation of N-type calcium-channel gating. *Nature* 340:639–642.

Lu W, Du J, Goehring A, & Gouaux E (2017). Cryo-EM structures of the triheteromeric NMDA receptor and its allosteric modulation. *Science* 355:eaal3729.

Ma H, Groth RD, Cohen SM, Emery JF, Li B, Hoedt E, Zhang G, Neubert TA, & Tsien RW (2014). GammaCaMKII shuttles Ca2+/CaM to the nucleus to trigger CREB phosphorylation and gene expression. *Cell* 159:281–294.

Magee JC & Cook EP (2000). Somatic EPSP amplitude is independent of synapse location in hippocampal pyramidal neurons. *Nat Neurosci* 3:895–903.

Magleby KL & Stevens CF (1972). A quantitative description of end-plate currents. *J Physiol* 223:173–197.

Montminy MR, Sevarino KA, Wagner JA, Mandel G, & Goodman RH (1986). Identification of a cyclic-AMP-responsive element within the rat somatostatin gene. *Proc Natl Acad Sci U S A* 83:6682-6686.

Nowak L, Bregestovski P, Ascher P, Herbet A, & Prochiantz A (1984). Magnesium gates glutamate-activated channels in mouse central neurones. *Nature* 307:462–465.

Rall W (1967). Distinguishing theoretical synaptic potentials computed for different soma-dendritic distributions of synaptic input. *J Neurophysiol* 30:1138–1168.

Rasmussen SG, DeVree BT, Zou Y, Kruse AC, Chung KY, Kobilka TS, Thian FS, Chae PS, Pardon E, Calinski D, et al. (2011). Crystal structure of the beta2 adrenergic receptor-Gs protein complex. *Nature* 477:549–555.

Takeuchi A & Takeuchi N (1960). On the permeability of end-plate membrane during the action of transmitter. *J Physiol* 154:52–67.

Taniguchi H, Lu J & Huang ZJ (2013). The spatial and temporal origin of chandelier cells in mouse neocortex. *Science* 339:70–74.

Twomey EC, Yelshanskaya MV, Grassucci RA, Frank J, & Sobolevsky AI (2017). Channel opening and gating mechanism in AMPA-subtype glutamate receptors. *Nature* 549:60–65.

Walsh RM Jr., Roh SH, Gharpure A, Morales-Perez CL, Teng J, & Hibbs RE (2018). Structural principles of distinct assemblies of the human alpha4beta2 nicotinic receptor. *Nature* 557:261–265.

Wang R, Walker CS, Brockie PJ, Francis MM, Mellem JE, Madsen DM, & Maricq AV (2008). Evolutionary conserved role for TARPs in the gating of glutamate receptors and tuning of synaptic function. *Neuron* 59:997–1008.

Zeng M, Chen X, Guan D, Xu J, Wu H, Tong P, & Zhang M (2018). Reconstituted postsynaptic density as a molecular platform for understanding synapse formation and plasticity. *Cell* 174:1172–1187.

CHAPTER 4
Vision

Seeing is believing.

Anonymous, ancient proverb

The primary purpose of the nervous system is for animals to sense the world and respond accordingly. Thus, sensory systems are of prime importance for animal survival, as individuals and as species. Studies of sensory systems have also revealed many general principles of nervous system operation. We therefore devote this chapter and Chapter 6 to studying how sensory systems work. We begin with vision because many important concepts were first established by studies of the visual system and then extended to other sensory systems.

Vision is the sense on which humans are most reliant, as suggested by the chapter's epigraph. Most multicellular organisms, from jellyfish to mammals, utilize vision; many single-cell organisms also respond to light (see Chapter 13 on the evolution of visual systems). Vision helps animals identify food sources, seek mates, and avoid dangers such as predators. All visual systems share a common source of input—namely, photons hitting photoreceptors in spatial and temporal patterns. Their job is to extract useful features from these light signals in forms that help guide animal behavior. Among the tasks the visual system performs are differentiating objects from background, locating objects of interest, detecting motion, and navigating in the environment.

This chapter's organization follows the sequence of events that make vision possible. We trace the visual pathway from photoreceptors to retinal circuits to the visual cortex. Given the complexity of the topic and the knowledge we have obtained, the details can sometimes seem overwhelming. At each level of analysis it is useful to keep in mind the ultimate purpose of vision: to navigate the world and optimize survival.

HOW DO RODS AND CONES DETECT LIGHT SIGNALS?

Human and most vertebrates have only one organ—the eyes (**Figure 4-1**)—that senses light. Light enters the eye through the pupil—the opening or aperture at the center of the iris—and is focused through the lens. The pupil adjusts size

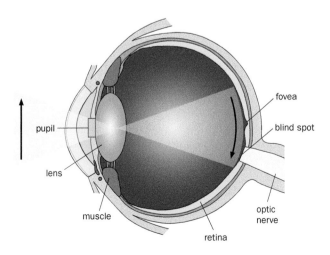

Figure 4-1 Cross section of a human eye. Light travels past the pupil and is focused by the lens onto the retina. An external object (arrow) is projected as an inverted image on the retina. Visual signals are transmitted to the brain via the optic nerve. A blind spot is formed at the head of the optic nerve. The fovea, at the center of the retina, has the densest population of cones, the photoreceptors responsible for high-acuity vision and color vision (see Section 4.8 for more details).

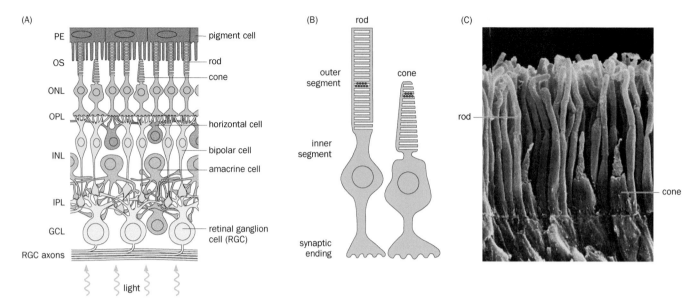

Figure 4-2 Organization of the retina and photoreceptor cells.
(A) The layered structure of the vertebrate retina. In this schematic, light enters the retina from the bottom. PE, pigment epithelium; OS, outer segments of photoreceptors; ONL, outer nuclear layer, containing rod and cone cell bodies; OPL, outer plexiform layer where rods and cones synapse with bipolar and horizontal cells; INL, inner nuclear layer, containing cell bodies of bipolar, horizontal, and amacrine cells; IPL, inner plexiform layer where bipolar, amacrine, and retinal ganglion cells (RGCs) form synapses; GCL, ganglion cell layer, containing cell bodies of RGCs and some amacrine cells. Cell classes are listed on the right. Whereas photoreceptors (rods and cones), bipolar cells, and RGCs are all glutamatergic excitatory neurons, horizontal cells and amacrine cells are mostly inhibitory neurons. (*Note that in this book,* *we use green tones for excitatory neurons and red tones for inhibitory neurons.*) We will encounter many cell types within each class later in the chapter. **(B)** Schematic of a rod and a cone. Within the outer segment, photosensitive molecules (dots) are highly concentrated in an orderly stack of membranes. In rods these membranes are intracellular discs, whereas in cones they are continuous with the plasma membrane. (Dots reflect the distribution of only a small fraction of photosensitive molecules to highlight the type of membrane on which photoreception occurs.) **(C)** A scanning electron micrograph showing three cones in the foreground among many rods. (B, adapted from Baylor DA [1987] *Inv Ophthal Vis Sci* 28:34–49. C, courtesy of William Miller.)

automatically according to the light level, and the curvature of the lens also automatically adjusts so that the eye can focus on a range of distances. The outside world is projected as images onto a thin layer of neural tissue at the back of the eye, the **retina**, which is the most important part of the eye to neurobiologists. The retina is where visual input is received as spatiotemporal patterns of photons, converted to electrical signals, processed by exquisitely precise retinal circuits, and then sent to the brain through the **optic nerve**.

The vertebrate retina is a layered structure made of five classes of neurons (**Figure 4-2**A). The input layer, which is actually at the back of the retina, consists of **photoreceptors** that detect photons and convert them to electrical signals, a universal form of information understood by the rest of the nervous system, as we learned in Chapter 2. The output layer comprises **retinal ganglion cells** (**RGCs**), which transmit information from the eye to the brain through their axons; RGC axons collectively make up the optic nerve. In between are **bipolar cells**, which transmit information from the photoreceptors to the RGCs, and **horizontal cells** and **amacrine cells**, whose actions influence the signals transmitted from photoreceptors to bipolar cells and then to RGCs. **Pigment cells** at the back of the eye absorb extra photons and prevent light scattering.

We start our journey through the visual system with the photoreceptors. Vertebrates have two types of photoreceptors, **rods** and **cones**, named after their shapes (Figure 4-2B, C). As will be discussed in detail later, cones are responsible for high acuity, daylight, and color vision; in primates, cones are concentrated in the **fovea**, the central part of the retina (Figure 4-1). Rods are more numerous, more sensitive to photons, and specialized for night vision. We begin our story with rods.

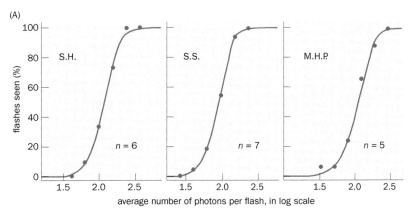 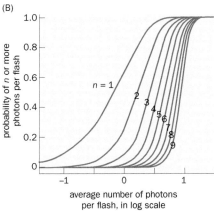

Figure 4-3 Determining how many photons can be seen. (A) In these three psychometric plots, the percentage of light flashes seen by the observers (*y* axis) is plotted against the average number of photons per flash (*x* axis, in log scale— *log scales are all 10-based in this book unless otherwise noted*). Each dot represents the average of 35–50 independent measurements. **(B)** A series of Poisson distributions plotting the probability of *n* or more photons per flash against the average number of photons per flash. The best fits (curves) for each of the investigators (designated by their initials) were *n* = 6, 7, and 5, respectively, representing the number of quanta needed to elicit a flash sensation. (From Hecht S, Shlaer S, & Pirenne MH [1942] *J Gen Physiol* 25:819–840. With permission from Rockefeller University Press.)

4.1 Psychophysical studies revealed that human rods can detect single photons

How sensitive are our rods? Given the quantum nature of light, we can ask a more precise version of this question: how many photons must be absorbed by a rod in order to elicit a biological effect? A decisive answer was obtained by an experiment of human **psychophysics**, which investigates the relationship between physical stimuli and the sensations or behaviors they elicit.

Taking into consideration the knowledge available from both the physics and biology of vision, Selig Hecht and co-workers utilized experimental conditions maximizing the sensitivity of the human retina—that is, the conditions allowing the detection of the lowest amount of light input. These included having the subjects (the investigators themselves) adapt to the dark by sitting in a completely dark room for at least 30 minutes before the onset of the experiments, shining lights on a small area of the peripheral retina where rods were known to be the densest, and using short (1 ms) exposures of light with a wavelength (510 nm) to which human rods were known to be most sensitive. They then varied the number of photons in these flashes and scored whether the subjects said they saw the light flash or not.

The researchers found that the frequency with which the flash was seen was proportional to the probability with which it yielded *n* or more quanta on the retina. They measured this frequency at each of several intensities to produce frequency-of-seeing plots (**Figure 4-3**A)—**psychometric functions** quantifying the frequency of light flashes seen against the relative light intensity. At the low light intensities used in this experiment, the process of absorbing photons could be modeled as a series of random, independent events following a Poisson distribution (Box 3-1). So the frequency-of-seeing plots could be compared to the plots of the probability that *n* or more photons were absorbed by each flash as predicted by a Poisson distribution (Figure 4-3B). The best fit of *n* was found to be between 5 and 7 for the three investigators. Thus, to reliably perceive a light flash, an average of five to seven independent events (photon absorptions) must occur in a retinal field of about 500 rods. Since the probability of two photons being absorbed by the same rod under these circumstances is very small, each rod must be able to report the absorption of a single photon.

4.2 Electrophysiological studies identified the single-photon response of rods: light hyperpolarizes vertebrate photoreceptors

Rods have cytoplasmic extensions at both ends (Figure 4-2B). As is true for most neurons, a rod's output end is its presynaptic terminal, where information is transmitted to the rod's postsynaptic partners—the bipolar and horizontal cells. At the input end is a highly specialized photon detection apparatus, the **outer segment**,

Box 4-1: Vision research uses diverse animal models

Researchers use different animal models to study vision because each model offers unique experimental advantages or special properties (see Sections 14.1–14.5 for a general discussion). For example, retinal explants of amphibians (Figure 4-4) and reptiles (Figure 4-12) are relatively easy to maintain *in vitro*, enabling researchers to characterize electrophysiological properties of retinal neurons with exquisite experimental control. Bovine retinas provide abundant outer segments for protein purification, facilitating structural studies of rhodopsin (Figure 4-6B) and biochemical analysis of phototransduction (Figure 4-8). Cats were chosen as a mammalian model of vision research (Figure 4-20; Figure 4-41) because they were easily available. Mice are increasingly being used for vision research because they offer precise deletion of individual genes, labeling of specific neuronal populations and their axonal projections (Figure 4-36), and the ability to manipulate the activity of specific cell types (Figure 4-50). Macaque monkeys have been used because, among all model organisms, their visual system

most resembles the human visual system (they are trichromatic and possess a fovea) and because they can be trained to perform sophisticated behavioral tasks allowing researchers to probe neural mechanisms of visual perception (Figure 4-55). Finally, humans can be subjects of vision research because they can directly report what is seen in psychophysical studies (Figure 4-3) and because human genetic variants can provide information about how the visual system functions (Figure 4-19). In Box 4-3 and in Chapter 5, we discuss the use of invertebrate visual systems to study their function and development.

In addition to the experimental advantages each model organism offers, comparative studies of diverse visual systems also provide insight into how visual systems came about and how diverse, as well as convergent, solutions may be used to tackle a common problem. We will expand on this evolutionary perspective in Chapter 13.

which is made of tightly stacked membrane disks enriched in photosensitive molecules.

Given the prediction from human psychophysical studies that individual rods can detect single photons, experiments were designed to measure the light sensitivity of single rods. The outer segment of a single rod from dissected frog retina (see **Box 4-1** for the use of different animal models for studying vision) was sucked by negative pressure into an electrode (a suction electrode) that formed a tight seal with the rod at the bottom. In this configuration, electric current passing through ion channels in the plasma membrane of the outer segment of a single rod could be measured in response to a light beam directed at a small fraction of the outer segment (**Figure 4-4**A). By systematically reducing the light intensity, a condition was reached in which most light flashes did not produce any response. The occasional responses that did occur had a rather uniform size (Figure 4-4B). The probability distribution of the response amplitude fit nicely with a Poisson distribution (Box 3-1), indicating that the responses were caused mostly by absorption of single photons and occasionally by absorption of two photons (Figure 4-4C).

These single-rod measurements confirmed that photon absorption results in *hyperpolarization* of rods—that is, current flows out of the rod in response to light (Figure 4-4B). This is an unusual strategy for reporting sensory stimulation; one

Figure 4-4 Single-photon responses of a rod. (A) The outer segment of a single rod from a piece of frog retina was collected in a suction electrode to measure its electrical response to stimulation by a narrow beam of green light. **(B)** Responses of a single rod (top) to light flashes (bottom, marked as vertical lines), measured as currents (in picoamperes, pA). The currents constitute outward flow, as indicated by the positive peaks. Each row represents responses to 10 light flashes. A total of 40 responses are shown. At this light intensity, most flashes produced no response. The remaining stimuli produced unitary responses. The time scale of light flashes is given at the bottom. **(C)** A histogram of light-induced current was fitted to a Poisson distribution (the smooth curve), revealing the magnitude of single-photon responses. (A, courtesy of Denis Baylor. B & C, from Baylor DA, Lamb TD, & Yau KW [1979] *J Physiol* 288:613–634. With permission from the John Wiley & Sons, Inc.)

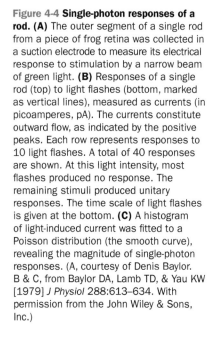

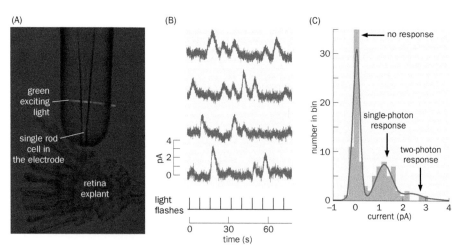

would expect that sensory stimuli should depolarize and activate the neurons, rather than hyperpolarize and inhibit them. This is indeed the case for invertebrate photoreceptors and for most other sensory systems (see Chapter 13), but vertebrate rods and cones use hyperpolarization to report light stimulation.

The single-photon responses not only offered a satisfying physiological explanation for the human psychophysical observations but also provided a quantitative measurement of exactly what a single photon does to a rod photoreceptor. Each photon absorption results in ~1 picoampere (pA) of net outward current across the membrane of the rod (Figure 4-4C), which is equivalent to a blockade of ~10^7 positive ions (mostly Na^+ and some Ca^{2+}) that would otherwise flow into the cell (**Figure 4-5**). How is this feat achieved? To answer this question, we turn to biochemistry of rods.

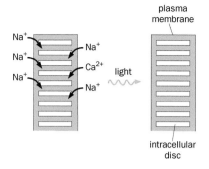

Figure 4-5 Light hyperpolarizes rods. Schematic of a portion of the rod outer segment. Absorption of each photon blocks the inward flow of about 10^7 cations.

4.3 Light activates rhodopsin, a prototypical G-protein-coupled receptor

Rhodopsins are the photosensitive molecules highly enriched in the stacked disk membranes of the rod outer segment (Figure 4-2B). Each rhodopsin consists of an **opsin** (from *opsis,* meaning "vision" in Greek) protein and a small molecule called **retinal**, which is closely related to and derived from vitamin A. Retinal is covalently attached to a lysine residue of the opsin.

Retinal is the **chromophore** (light-absorbing portion of a molecule) that gives rhodopsin its rosy color (in Greek, *rhodo* means "rose"). Retinal exists in two isomers in photoreceptors: all-*trans* and 11-*cis*. Photon absorption by rhodopsin causes a switch of 11-*cis* retinal to all-*trans* retinal (**Figure 4-6**A). This large structural change then triggers a corresponding conformational change in its protein partner, the opsin.

Opsin is a prototypical G-protein-coupled receptor (GPCR). It has seven transmembrane helices, a design commonly used in receptors for sensory stimuli, hormones, and neurotransmitters (Section 3.19). Rhodopsin was the first GPCR to be purified to homogeneity and to have its primary structure determined. It was also the first GPCR whose structure was solved with atomic resolution by X-ray

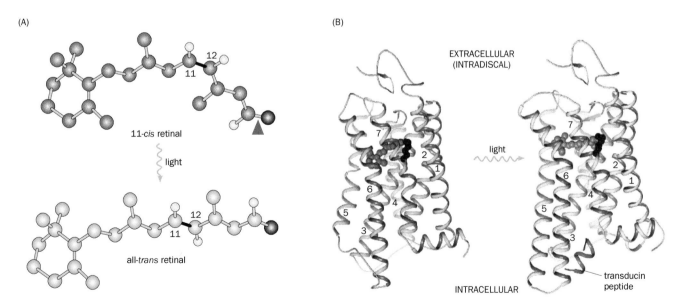

Figure 4-6 The structure of rhodopsin. (A) Light causes the chromophore retinal to switch from the 11-*cis* to an all-*trans* configuration. Note the difference to the right of the dark bond between carbons 11 and 12. The red arrowhead indicates the bond between retinal and opsin (the black circle is from a lysine residue of the opsin). **(B)** Crystal structures of inactive rhodopsin with 11-*cis* retinal (red, left panel) and active rhodopsin with all-*trans* retinal (blue, right panel) in complex with a fragment of the G protein transducin at the intracellular face. The lysine residue of the opsin (black) to which the retinal attaches is rendered in a space-filling format. Light-induced isomerization of retinal creates a binding site for transducin by causing conformational changes in the transmembrane helices, which are numbered 1–7. (A, structures based on Wald G [1968] *Science* 162:230–239. B, adapted from Choe HW, Kim YJ, Park JH, et al. [2011] *Nature* 471:651–656. With permission from Springer Nature.)

Figure 4-7 Light activates phosphodiesterase, which hydrolyzes cGMP. Guanylate cyclase (GC) catalyzes the synthesis of cGMP from GTP. Phosphodiesterase (PDE) catalyzes cGMP hydrolysis. Both enzymes are membrane bound. The font sizes of cGMP and GMP represent their concentrations. **(A)** In the dark, PDE is inactive (dashed arrow) and cGMP is abundant. **(B)** Light triggers activation of PDE (* designates active), which hydrolyzes cGMP.

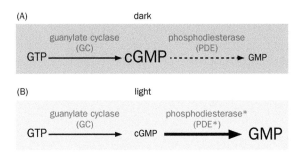

crystallography. Retinal isomerization from 11-*cis* to all-*trans* induces a structural change in the transmembrane helices (Figure 4-6B), which creates a binding site on the cytosolic side for the α subunit of a heterotrimeric G-protein called transducin, which we will discuss in detail in the next section.

4.4 Photon-induced signals are greatly amplified by a transduction cascade

Studies of rhodopsin have been greatly facilitated by the abundance of rhodopsin proteins in the rod outer segments and the ease of obtaining large amounts of rod outer segments from bovine retinas. In fact, isolated rod outer segments can respond to light activation *in vitro,* which enabled detailed studies of the **phototransduction** process: the biochemical reactions triggered by photon absorption.

Biochemical investigations first revealed that light induces a decline in the level of **cyclic GMP** (**cGMP**), a cyclic nucleotide similar to the second messenger cAMP we studied in the context of adrenergic receptor signaling (Section 3.20). cGMP is produced from GTP by the enzyme **guanylate cyclase** and is hydrolyzed by a second enzyme, **phosphodiesterase** (**PDE**) (Figure 4-7A). Importantly, direct injection of cGMP into rod outer segments caused rod depolarization, mimicking darkness. So a decline in the cGMP level appeared to be responsible for light-triggered rod hyperpolarization, suggesting that light activates PDE (Figure 4-7B). But how is this achieved?

Biochemical experiments revealed that light-activated rhodopsin in rod outer segments catalyzes the exchange of bound GDP for GTP. A GTP-binding protein complex named **transducin** was identified as the intermediate linking light-induced rhodopsin activation to PDE activation and the subsequent decrease in cGMP levels. Specifically, purified rhodopsin and transducin (composed of Tα, Tβ, and Tγ subunits) could together reconstitute light-induced GDP–GTP exchange without the involvement of PDE (Figure 4-8, Steps 1 and 2). Furthermore, purified Tα, when GTP bound, could trigger PDE activation in the absence

Figure 4-8 Schematic summary of the phototransduction mechanism. Step 1, light triggers activation of rhodopsin (R to R*). Step 2, activated rhodopsin (R*) catalyzes the exchange of the transducin (T)-bound nucleotide, replacing GDP with GTP; Tα-GTP subsequently dissociates from Tβγ and R*. Step 3, Tα-GTP binds to and activates phosphodiesterase (PDE*) by sequestering the inhibitory PDE γ subunits. (Each Tα-GTP molecule binds to one PDE γ subunit; for simplicity, only one Tα-GTP/PDE γ subunit complex is drawn here.) This leads to hydrolysis of cGMP to GMP. Step 4, the intrinsic GTPase activity of Tα hydrolyzes Tα-GTP to Tα-GDP, which reassociates with Tβγ and is ready for another round of phototransduction. The deactivation of R* and regeneration of R is discussed in Section 4.6. (Adapted from Stryer L [1983] *Cold Spring Harbor Symp Quant Biol* 48:841–852.)

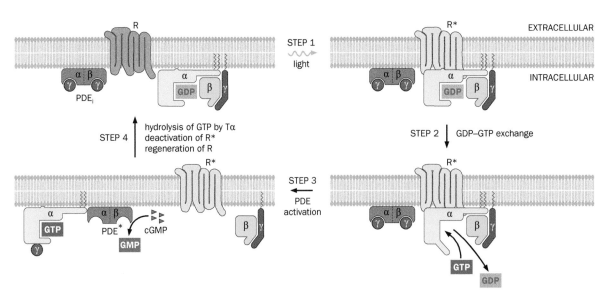

of light stimulation (Figure 4-8, Step 3). The intrinsic GTPase activity of Tα causes its return to the GDP-bound state and its re-association with Tβγ, re-forming the trimeric complex needed for the next round of activation (Figure 4-8, Step 4). This mechanism bears a striking resemblance to the activation of adenylate cyclase by the β-adrenergic receptor via trimeric G proteins (Figure 3-31; Figure 3-33; Movie 3-3). Indeed, these parallel studies established a model for how GPCRs, the largest family of receptors, transduce signals.

In the phototransduction cascade (Figure 4-8), a single light-activated rhodopsin (R*) catalyzes the activation of >20 transducin molecules, each of which activates a PDE, together resulting in hydrolysis of tens of thousands of cGMP molecules per second. Signal amplification in this biochemical cascade begins to explain how a single photon triggers a large decrease in the cGMP level.

4.5 The light-triggered decline in cyclic-GMP levels leads directly to the closure of cation channels

How does cGMP hydrolysis cause a change in current flow across the rod membrane? It had been assumed that cGMP, like its famous cousin cAMP, acts as a second messenger by activating protein kinases (Figure 3-33). However, in the case of phototransduction, cGMP binds directly to the cytoplasmic side of a cation channel, increasing its conductance. This mechanism of action was discovered using the patch clamp technique (**Figure 4-9**A; see Box 14-3 for details). When a piece of rod outer segment plasma membrane containing a channel was exposed from the cytoplasmic side to solutions containing increasing concentrations of cGMP, channel conductance increased correspondingly (Figure 4-9B). This effect did not depend on the presence of ATP, ruling out the involvement of protein kinases. The effect of cGMP was specific: the channel could not be opened by cAMP at concentrations orders of magnitude higher than that of cGMP. This direct activation of ion channels by cGMP allows changes in cGMP concentration to rapidly affect the membrane potential of rods in response to light.

Subsequent cloning of the gene encoding the rod cGMP-gated channel, a member of the <u>c</u>yclic <u>n</u>ucleotide-<u>g</u>ated channel (**CNG channel**) family, revealed that its primary structure resembles that of the voltage-gated K^+ channels we discussed in Chapter 2, with six transmembrane domains and a pore loop between S5 and S6 (Figure 2-34). The C-terminal cytoplasmic domain of CNG channels contains an additional cyclic nucleotide-binding domain. Like K^+ channels, functional CNG channels comprise four subunits with four cyclic nucleotide-binding sites. Channel opening is gated by direct binding of cyclic nucleotides, rather than by membrane potential changes. Like the nicotinic acetylcholine receptor (nAChR) discussed in Section 3.13, the rod CNG channel is a nonselective cation channel: its opening in the dark causes influx of both Na^+ and Ca^{2+} and efflux of K^+. Because the rod CNG channel has a reversal potential near 0 mV (also like the nAChR), the resulting Na^+ influx far exceeds K^+ efflux. Thus, the net effect of channel opening is depolarization. As will be discussed shortly, Ca^{2+} entry in the dark is crucial for recovery and adaptation.

The identification of the cGMP-gated cation channel provides the final piece in the puzzle of how light is converted into electrical signals in the rod photoreceptor (**Figure 4-10**). The light-triggered conformational change of rhodopsin activates transducin, which then activates PDE, resulting in cGMP breakdown. A

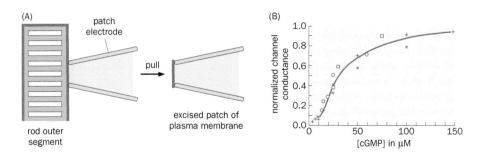

Figure 4-9 cGMP directly activates a cation channel. (A) A small piece of rod outer segment plasma membrane can be excised by a patch clamp electrode to study its properties *in vitro*. **(B)** The excised membrane increases its conductance in response to increasing concentrations of cGMP in the solution (equivalent to the intracellular side of the membrane patch). Different symbols represent measurements performed at different Ca^{2+} concentrations, which did not affect the conductance. (B, adapted from Fesenko EE, Kolesnikov SS, & Lyubarsky AL [1985] *Nature* 313:310–313. With permission from Springer Nature.)

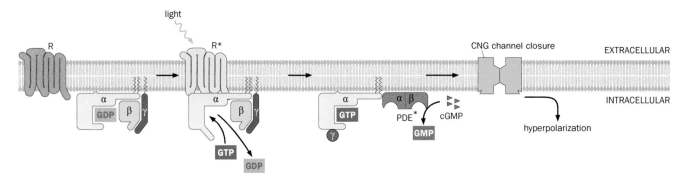

Figure 4-10 A summary of the phototransduction cascade. A * indicates an activated component. Rightward arrows indicate the successive steps of signal transduction.

decline in cGMP concentration causes the closure of cGMP-gated cation channels, decreasing the steady cation influx. This leads to hyperpolarization of the rod, causing declines in glutamate release at rod → bipolar cell and rod → horizontal cell synapses.

4.6 Recovery enables the visual system to continually respond to light

So far, we have emphasized the sensitivity of rods, but sensory systems optimize many other features, such as speed, dynamic range, selectivity, and reliability. A combination of electrophysiological and biochemical studies in rods has enabled precise measurement and mechanistic understanding of many of these features. In this section and the next, we study two related processes: recovery and adaptation.

Recovery refers to the process by which light-activated photoreceptor cells return to the dark state (see Section 3.23 for a general discussion of signal termination). Recovery ensures that each visual stimulus produces only a transient signal, such that the same rod can encode the next visual stimulus in a matter of seconds. All rod components activated in the phototransduction cascade must rapidly return to the dark state, readying the rod for light reception and transduction. This requires the action of several parallel molecular pathways.

The cGMP level must increase during recovery to reopen the CNG channels that depolarize the rod. This is achieved by activation of guanylate cyclase (GC), the enzyme that catalyzes synthesis of cGMP from GTP (Figure 4-7). GC activity is highly sensitive to intracellular Ca^{2+} concentration ($[Ca^{2+}]_i$). It remains largely inactive when $[Ca^{2+}]_i$ is greater than 100 nM, but becomes highly active when $[Ca^{2+}]_i$ drops below 60 nM (**Figure 4-11**A). In the dark, $[Ca^{2+}]_i$ is ~400 nM, balanced by influx via CNG channels and efflux via Na^+/Ca^{2+} exchange proteins on the plasma membrane. When light causes cGMP levels to decline by closing CNG channels, $[Ca^{2+}]_i$ drops below 50 nM, as the exchange proteins remain active. This decrease in $[Ca^{2+}]_i$ activates GC through an intermediate protein called GCAP (guanylate cyclase activating protein), a Ca^{2+}-binding protein that in its Ca^{2+}-free form binds to and activates GC. Thus, closure of CNG channels by light triggers a negative-feedback loop leading to elevated cGMP, which opens CNG channels and restores the dark state (Figure 4-11B, pathway 1).

Tα-GTP must be deactivated. Tα-GTP has intrinsic GTPase activity and can convert itself to a GDP-bound state. Tα-GDP then dissociates from PDE, inactivating PDE. The intrinsic GTPase of Tα-GTP is greatly facilitated by RGS9 (**RGS** stands for **regulator of G protein signaling**) (Figure 4-11B, pathway 2). RGS9 acts as a GTPase-activating protein (GAP) for Tα-GTP (see Box 3-3 for a general discussion of GAPs).

Rhodopsin must be deactivated to prevent further activation of the transduction cascade. To shut off rhodopsin, rhodopsin kinase specifically binds to and phosphorylates the cytoplasmic tail of R* but not R; this phosphorylation recruits a protein called **arrestin** to R*; binding of arrestin to phosphorylated R* competes with Tα binding and therefore prevents R* from activating more transducin molecules (Figure 4-11B, pathway 3; Movie 3-3). First discovered in the visual system,

(A)

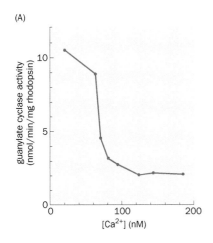

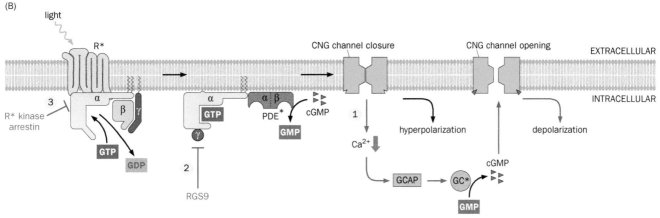

Figure 4-11 **Mechanisms of recovery. (A)** Guanylate cyclase activity is highly sensitive to intracellular Ca^{2+} concentration, with GC activity increasing as $[Ca^{2+}]_i$ decreases. **(B)** Summary of three major mechanisms of recovery (red) superimposed on the phototransduction pathway (black) reproduced from Figure 4-10. (1) The closure of CNG channels reduces $[Ca^{2+}]_i$, which causes GCAP to activate guanylate cyclase (GC*), leading to increased cGMP production. cGMP binding to CNG channels leads to channel opening and membrane depolarization. (2) RGS9 facilitates hydrolysis of Tα-bound GTP, thereby deactivating PDE. (3) Rhodopsin kinase (R* kinase) specifically phosphorylates R*. Phosphorylated R* recruits binding of arrestin. R* phosphorylation and arrestin binding deactivate R*. (A, adapted from Koch KW & Stryer L [1988] *Nature* 334:64–66. With permission from Springer Nature.)

arrestin has since been widely found to compete with G proteins for binding to activated GPCRs (Section 3.23).

Finally, for rhodopsin to return to the dark state, all-*trans* retinal must be converted back to 11-*cis* retinal. This process is slow (on the scale of minutes) because it takes place in pigment cells in the pigment epithelium (Figure 4-2) and involves the exchange of retinal molecules between rods and pigment cells. (The conversion of all-*trans* retinal to 11-*cis* retinal for cones occurs in a specialized cell type in the retina called the **Müller glia**.) However, given that only a small fraction of a rod's rhodopsin molecules are typically activated by low-light stimulation, the activities of rhodopsin kinase and arrestin ensure that within seconds the rod is again ready for visual excitation using the remaining rhodopsin molecules in the same rod.

4.7 Adaptation enables the visual system to detect contrast over a wide range of light levels

Object detection requires separating the object from its background. **Luminance contrast**, the difference in *light intensity* between adjacent spaces, is a primary mechanism for object detection. (In later sections, we will discuss the other primary mechanism, **color contrast**, the difference in *light wavelengths* between adjacent spaces.) Our visual system has a very large **dynamic range**, the ratio of the largest to the smallest value of light intensity, enabling us to detect single photons in near darkness and discriminate objects from background over a 10^{11}-fold range of ambient light level. Rods contribute to the lower part of this dynamic range, discerning luminance contrast over a 10^4-fold range of ambient light levels. How can rods detect single photons, yet simultaneously not be saturated by light intensity four orders of magnitude greater? **Adaptation**, the adjustment of a photoreceptor's sensitivity according to the background light level, allows this feat.

Figure 4-12 Light adaptation of gecko rods. Each dot corresponds to a peak hyperpolarization amplitude measured by intracellular recordings (*y* axis) in response to a 0.1 s light flash of a given intensity (*x* axis). Experiments were grouped in a series of curves according to background light intensities given at the bottom of the curves. DA, dark adapted. Four groups on the left (red) and two groups on the right (brown) were recorded from two separate rods. As background light increases, producing the same response magnitude (for example, a 1 mV hyperpolarization indicated by the dashed line) requires increasing light stimulation intensities. A rightward shift on a log axis means that the same fractional intensity changes produced the same response in the cell after adaptation, consistent with Weber's law. (From Kleinschmidt J & Dowling JE [1975] *J Gen Physiol* 66:617–648. With permission from Rockefeller University Press.)

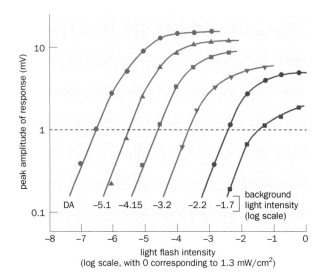

Adaptation is a universal phenomenon in sensory biology, first described by Ernst Weber in the 1820s in an observation now called **Weber's law**: the *just-noticeable difference* between two sensory stimuli is proportional to the magnitude of the stimulus. For example, human subjects can distinguish between weights of 100 g and 105 g (a just-noticeable difference of 5 g, or 5% of the total weight) but cannot distinguish 1000 g from 1005 g. At the 1000-g range, it typically takes a difference of 50 g for the distinction to be clear (a just-noticeable difference of 50 g, again 5% of the total weight).

In the visual system, light adaptation means that photoreceptors become less sensitive to the same intensity of stimulation when the background illumination is higher. In other words, to achieve the same amount of hyperpolarization at higher background illumination, a stronger stimulus is required. This is quantitatively illustrated by the experiment shown in **Figure 4-12**. To reach a 1 mV hyperpolarization (indicated in the figure by the dashed line), a light flash with a relative intensity of ~$10^{-6.5}$ was adequate for dark-adapted rods. As the background light intensity increased, the flash intensity required to reach the same 1 mV hyperpolarization also increased. Note also that at low background light levels, the light responses saturated at low light intensity. That is, photocurrents did not increase in response to further increases in light intensity, as seen in the flattening of the response curve. However, as background light increased, saturation shifted rightward on the curve. Thus, adaptation expands the dynamic range of light intensity within which rods can distinguish the intensity differences between background and objects.

What is the basis of visual adaptation? Interestingly, Ca^{2+} plays a key role in adaptation, as it does in recovery (Figure 4-11B). High levels of background illumination cause some cGMP-gated channels to close, resulting in a decline in $[Ca^{2+}]_i$. A decline in $[Ca^{2+}]_i$ leads to the following biochemical changes: (1) The basal level of GC activity increases because of the action of GCAP (Figure 4-11A) and (2) R* phosphorylation increases because it is normally inhibited by high $[Ca^{2+}]_i$, which leads to more arrestin binding that competes with R* binding with Tα. These changes make phototransduction less efficient, and thus stronger activation of PDE by more light is required to promote effective hyperpolarization. Note that these events occur in the outer segment of the photoreceptor, so $[Ca^{2+}]_i$ changes don't complicate its role in regulating synaptic transmission, which occurs in axon terminals at the opposite end of the photoreceptor.

In summary, elaborate biochemical mechanisms provide rods with exquisite sensitivity not only for converting light stimuli into electrical signals but also for negatively regulating the phototransduction cascade once it is triggered. These mechanisms ensure that light-induced signals are transient and have a wide dynamic range.

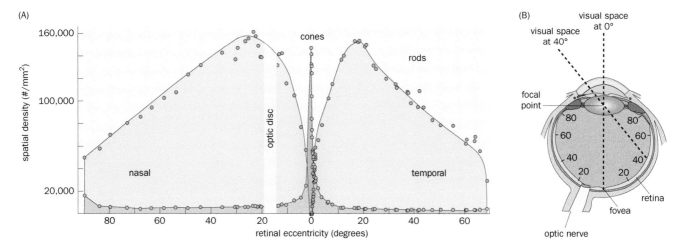

Figure 4-13 **Distribution of rods and cones in the human retina.**
(A) Cones (orange) are highly concentrated in the fovea at the center of the primate retina. Rods (blue) are most enriched in areas peripheral to the fovea. No photoreceptors are present at the optic disc, creating a visual blind spot at the site where the optic nerve leaves the eye. **(B)** Illustration of retinal eccentricity. Visual space at 0 degrees corresponds to the center of gaze, and greater angles correspond to objects farther from the center of gaze. (A, adapted from Rodieck RW [1998] The First Steps in Seeing. Sinauer. From Osterberg G [1935] *Acta Othalmol* 6 Suppl.)

4.8   Cones are concentrated in the fovea for high-acuity vision

We now turn to the second type of photoreceptors, the cones (Figure 4-2B). The cone signaling pathway, from light to electrical signals, is similar to that of rods. However, variations allow cones to serve complementary functions: vision in bright light, vision with high spatial acuity, a wider range of motion vision, and color vision. Indeed, cones are much more important than rods in the daily lives of modern humans.

Here is an experiment you can do to test your high-acuity and color vision. Look straight ahead at a point on the wall, extend your arm on your side to eye level, place in your palm small objects of different colors and shapes, and move your extended arm slowly to the center of gaze. You will find that you cannot tell the exact color and shape of the objects until they are quite close to the center of gaze. This is because cones, which are responsible for high-acuity and color vision, are highly concentrated in the fovea at the center of your retina (**Figure 4-13**A). The retinal eccentricity refers to the location of the retina corresponding to a particular space in the **visual field** (the entire external world that can be seen at a given time). Cones are highly concentrated within a few degrees of retinal eccentricity. By contrast, rods are densest in areas of the retina adjacent to the fovea (Figure 4-13B). That is why a dim star in the night sky can be seen more easily if you look slightly to the side of the star.

High-acuity vision in primates is supported by the high density of cones packed in the fovea. Neighboring cones at the fovea can thus resolve much closer light spots than neighboring cones in the peripheral retina, where cone density is lower. The primate retina and central visual system devotes much of its processing power to foveal signals: although the fovea makes up only 1% of the total retinal area, about half of all retinal ganglion cell axons transmit information from the fovea to the brain. In addition, unlike in the rest of the retina, where light must pass through layers of cells before reaching photoreceptors (Figure 4-2A), at the center of the fovea, other cell and synaptic layers are displaced peripherally (**Figure 4-14**). The fovea is also devoid of blood vessels. These properties enable light to reach cones with minimal obstruction, further enhancing high-acuity vision. In patients with **macular degeneration**, a disease that causes photoreceptors in the fovea to die, high-acuity vision is preferentially impaired, with devastating consequences.

4.9   Cones are less sensitive but much faster than rods

Comparing responses to light flashes in single rods or cones (**Figure 4-15**) revealed important differences in their sensitivity and speed. Cones are much less

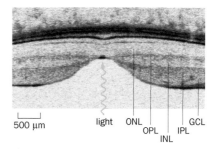

Figure 4-14 **Inner retinal layers are displaced at the human fovea.** Retina near the fovea of a live human eye imaged by optical coherence tomography. At the foveal pit, the inner retinal cell layers as well as the plexiform layers are displaced, such that light can access cones directly. ONL, outer nuclear layer; OPL, outer plexiform layer; INL, inner nuclear layer; IPL, inner plexiform layer; GCL, ganglion cell layer. (Adapted from Cucu RG, Podoleanu AG, Rogers JA, et al. [2006] *Optics Lett* 31:1684–1686.)

Figure 4-15 Flash responses of single rod and cone in macaque monkeys. Currents from a rod and a cone in response to increasing magnitudes of light flashes are plotted in a series of curves. Each curve represents a response to 2 times stronger than the intensity of flash of the curve immediately below it. The bottom curve for the cone starts at a light intensity level about 65 times higher than the bottom curve for the rod, indicating that cones are less sensitive than rods. Cones recover faster than rods. Cones also produce an "undershoot" during the recovery phase. (From Baylor DA [1987] *Inv Ophthal Vis Sci* 28:34–49. With permission from the Association for Research in Vision and Ophthalmology.)

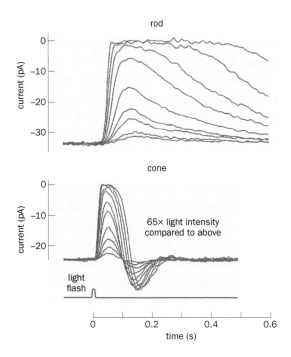

sensitive than rods. Individual cones require flash intensities about two orders of magnitude greater to produce currents similar to those produced by individual rods. However, cones respond much faster than rods. Cones reach their peak current 50 ms after a light flash; rods require about 150 ms. Moreover, cones return to baseline much more rapidly than rods, so cones can react to repeated light stimuli faster than rods and therefore have better temporal resolution, which is crucial for motion detection. These differences reflect variations in phototransduction cascades and recovery mechanisms (Figure 4-11B) as well as morphological differences between rods and cones. For example, the opsin kinase expressed by cones has a much higher specific activity toward cone opsins than rhodopsin kinase has toward rhodopsin. RGS9, which facilitates the hydrolysis of Tα-bound GTP and is a rate-limiting factor for recovery in rods, is expressed at a much higher level in cones. Finally, the larger plasma membrane surface-to-volume ratio of cones (Figure 4-2B) makes the $[Ca^{2+}]_i$ decline and return more rapidly in response to light stimulation. All of these factors contribute to a lower sensitivity but speedier recovery in cones compared to rods.

Cones can also adapt to a broader range of light intensities than rods. As we discussed in Section 4.7, rods can discern luminance contrast over a 10^4-fold range of ambient light levels. Cones extend this range by an additional factor of 10^7, enabling us to see over a 10^{11}-fold range of ambient light levels. Such differences in rods and cones contribute to this disparity. In addition, bright light causes retinal isomerization in a significant fraction of cone opsins (we will discuss cone opsins in more detail in Section 4.11), a process called *pigment bleaching*. Pigment bleaching reduces the effective concentration of cone opsins that can still respond to light, which also contributes to adaptation, as more light is needed to produce the same photocurrent in a cone population.

4.10 Photoreceptors with different spectral sensitivities are needed to sense color

In principle, the cone functions we have discussed thus far could be accomplished by a new type of photoreceptor that differs from rods in density, distribution, morphology of the outer segment, and phototransduction pathway parameters. However, if this hypothetical photoreceptor responded to light as a function of wavelength the same way rods do—that is, if it shared the rods' **spectral sensitivity**—it would fail to serve one important cone function: color vision.

Our sense of color depends on the detection and *comparison* of light of different wavelengths. Visible light corresponds to wavelengths ranging from ~370 nm (violet) to ~700 nm (red). When white light from the sun, which contains a mixture of different wavelengths, shines on an object, some wavelengths (that is, a subset of spectra) are absorbed more than others; the unabsorbed light that arrives at the eye gives rise to a sensation of color. Color vision enables the distinction between objects based on the wavelengths of light they reflect, refract, transmit, or fluoresce.

The spectral sensitivity of rhodopsin covers much of the visible light range, peaking around 500 nm (gray curve in **Figure 4-16**A); for this reason, Hecht and colleagues used 510-nm light for their psychophysics experiment (Section 4.1). Using rhodopsin alone, however, it is not possible to differentiate between a certain amount of light at a wavelength of 500 nm, or 10 times that amount of light at a wavelength of 570 nm, because both conditions would excite rods to the same extent. The presence of multiple photoreceptors with distinct spectral sensitivities makes it possible for an animal to extract color information by comparing the *relative* excitation of two different kinds of photoreceptors. We will discuss the retinal circuits that perform this function in Section 4.17.

In addition to rods, most mammals have two types of cones with different spectral sensitivities: a short-wavelength S-cone and a longer-wavelength cone. Catarrhines, a group of primates comprising humans, apes, and Old World monkeys, have three cones—the S-cone, M-cone, and L-cone—and are therefore called **trichromats**. Trichromacy enables primates to more easily identify objects of similar luminance but different colors, such as red fruits among green leaves. The idea of trichromacy was first suggested in 1802 by noted polymath Thomas Young, who deduced that "as it is almost impossible to conceive of each sensitive point of the retina to contain an infinite number of particles, each capable of vibrating in perfect unison with every undulation, it becomes necessary to suppose the number limited . . . each sensitive filament of the nerve may consist of three portions, one for each principal color."

Studies since the 1960s have confirmed Young's hypothesis and identified three types of human cones with different spectral sensitivities. We summarize these findings by returning to a familiar preparation: electrophysiological recording of single photoreceptors using the suction electrode (Figure 4-4). Stimulation by monochromatic light flashes with systematically varied wavelengths revealed that individual cones from macaque monkeys (an Old World monkey) belonged to one of three types with distinct spectral sensitivities: S, M, or L (Figure 4-16A). The variance of spectral sensitivity was so small among cones of the same type that

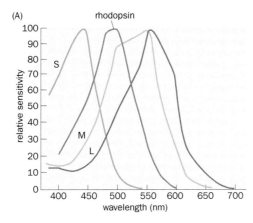

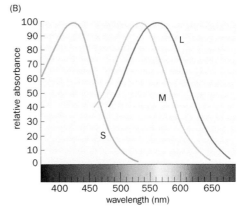

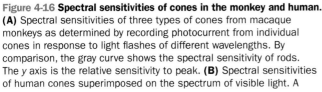

Figure 4-16 Spectral sensitivities of cones in the monkey and human.
(A) Spectral sensitivities of three types of cones from macaque monkeys as determined by recording photocurrent from individual cones in response to light flashes of different wavelengths. By comparison, the gray curve shows the spectral sensitivity of rods. The y axis is the relative sensitivity to peak. **(B)** Spectral sensitivities of human cones superimposed on the spectrum of visible light. A microspectrophotometer was used to measure the absorption of different wavelengths by individual cones obtained during eye surgery. (A, from Baylor DA, Nunn BJ, & Schnapf JL [1987] *J Physiol* 390:145–160; Baylor DA, Nunn BJ, & Schnapf JF [1984] *J Physiol* 357:575–607. With permission from John Wiley & Sons, Inc. B, adapted from Nathans J [1989] *Sci Am* 260[2]:42–49. With permission from Springer Nature.)

Figure 4-17 Spatial distribution of three types of cones in the human retina. Pseudocolor images of trichromatic cone mosaics in the fovea of two individuals determined by comparing retinal images before and after treatment with a saturating amount of light with a specific short, medium, or long wavelength. Blue, green, and red dots represent S-, M-, and L-cones, respectively. (From Roorda A & Williams DR [1999] *Nature* 397:520–522. With permission from Springer Nature.)

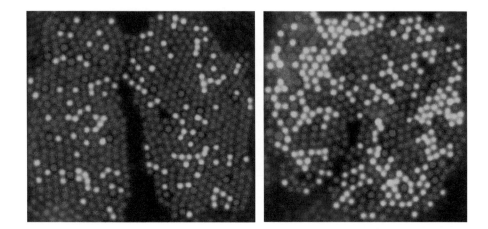

each cone could be unambiguously assigned to one of these three types. Experiments using retina from humans with normal color vision revealed spectral sensitivities similar to that of the monkey (Figure 4-16B).

How are the three different cones distributed in the retina? This was determined in living human eyes using a technique called retinal densitometry. By comparing reflections from the retina before and after providing a saturating amount of light at a specific wavelength, images of "bleached" retina revealed distributions of each of the pigments at single-cone resolution (**Figure 4-17**). S-cones are much sparser than M- and L-cones, making up less than 5% of the cones at the fovea. This explains why we lack high spatial resolution for blue objects. Indeed, because lights of different wavelengths refract differentially through the single lens we have at the front of our eye, they cannot be focused with equal sharpness on the retina; this phenomenon is called chromatic aberration. We have evolved sharper focus for long-wavelength light, so we would not gain better resolution of blue light even if the S-cone density was higher. The ratio of M- and L-cones differs among individuals, and the spatial distributions of M- and L-cones appear random. In Chapter 13, we will study how evolution has shaped these properties.

4.11 Cloning of the cone opsin genes revealed the molecular basis of color detection and human color blindness

By the early 1980s, experimental evidence had provided strong support for the existence of three types of cones in some primate retinas, each with distinct spectral sensitivities (Figure 4-16). But what is the molecular basis for this? A simple hypothesis was that there exist three separate cone opsin genes, with each cone cell expressing one and only one opsin with a specific spectral sensitivity. A further hypothesis was that the genes encoding the cone opsins are similar to the rod opsin gene, as they likely descend from a common ancestor. The isolation of cone opsin genes validated these hypotheses.

The abundance of rhodopsin in the outer segments of rods (Section 4.3) enabled the purification and determination of partial amino acid sequences for bovine rod opsin protein, which led to identification of the corresponding gene. The DNA encoding the bovine rod opsin was then used as a probe to identify human opsin genes using a strategy called *low-stringency hybridization*, which enables the isolation of DNA based on similar but not identical sequences. Four separate genes were discovered that encode human opsins: one for the rod opsin and three for the S-, M-, and L-cone opsins (**Figure 4-18**), as predicted by the original hypothesis.

The three cone opsins each share about ~40% amino acid sequence identity with the rod opsin. The S-opsin also shares ~40% sequence identity with the M- and L-opsins. The M- and L-opsins, however, are 96% identical (Figure 4-18). Among the sequence variations are charged amino acids located in the transmembrane domains that interact with retinal; these variations account for the different spectral sensitivities of different opsins when coupled to retinal.

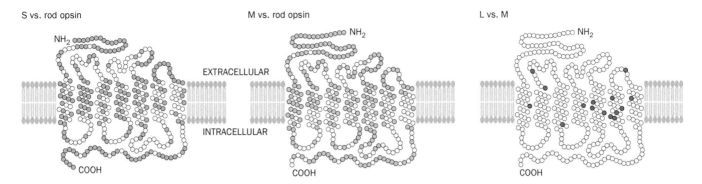

S vs. rod opsin   M vs. rod opsin   L vs. M

EXTRACELLULAR

INTRACELLULAR

Figure 4-18 Pair-wise comparisons of human opsin proteins. Amino acid sequences of rod opsin (the protein component of rhodopsin) and three cone opsins (S, M, L) were optimally aligned in a generic G-protein-coupled receptor configuration with seven transmembrane helices (Figure 4-6B). The N-termini face the disc lumen or extracellular environment, and the C-termini face the cytoplasm. In these pair-wise comparisons, open circles indicate identical amino acids, while filled circles indicate different amino acids. (From Nathans J, Thomas D, & Hogness DS [1986] *Science* 232:193–202. With permission from AAAS.)

The M- and L-opsin genes map next to each other on the X chromosomes. This property, together with their very high sequence identity, suggest that the M- and L-opsins evolved from a common ancestor fairly recently. This is in keeping with the notion that trichromacy, which emerged about 35 million years ago during primate evolution, likely results from a gene duplication event that occurred in a common ancestor of Old World monkeys, apes, and humans, after Old World monkeys diverged from New World monkeys (see Section 13.16 for details).

It is well known that red–green color blindness (that is, individuals cannot distinguish red from green) occurs more commonly than blue color blindness and much more frequently in men than women (~7% of men and 0.4% women in the United States are red–green color-blind). Before opsin gene cloning, genetic mapping had already pinpointed green–red color blindness to the X chromosome. This explains the predominance of color blindness in males, who have only one copy of the X chromosome compared to two copies in females. After the M- and L-opsin genes were cloned, molecular alterations in these genes were found in a majority of color-blind males tested (**Figure 4-19**A). Given the high degree of sequence similarity and the short distance between the M- and L-opsin genes and their flanking DNA, unequal crossing over could occur frequently during meiotic recombination, resulting in the loss or alteration of the M- or L-opsin genes (Figure 4-19B).

Thus, the identification and analysis of human cone opsin genes provided a satisfying explanation for the molecular-genetic basis of color detection as well as variant color vision, including color blindness, in humans.

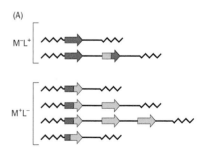

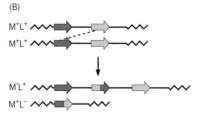

Figure 4-19 M- and L-opsin gene rearrangements responsible for the most common forms of variant human color vision. (A) Top, individuals who have normal L-cone sensitivity and are missing M-cone sensitivity (M$^-$L$^+$) have an intact L-opsin gene (red) in the head-to-tail array and an altered or missing M-opsin gene (green). Bottom, by contrast, individuals who have normal M-cone sensitivity and are missing L-cone sensitivity (M$^+$L$^-$) have an intact M-opsin gene and an altered or missing L-opsin gene. (Note that amino acids that determine spectral sensitivity are clustered at the C-terminal half of the opsin protein, such that a hybrid made of N-terminal L-opsin and C-terminal M-opsin confers M-cone sensitivity.) **(B)** Recombination between L- and M-opsin genes at meiosis leads to gene arrangements that produce variant color vision. High sequence similarity between the L- and M-opsin genes and their flanking DNA leads to occasional unequal crossing over (dashed line) between two copies of the X chromosome in maternal germ-line cells. This results in gene duplication in one gamete and deletion in the other gamete. In some cases, the recombination event occurs within the genes and generates L/M and M/L hybrid genes, which encode cone opsins with sensitivities differing from normal M- and L-cone opsins (M' refers to an altered M-cone sensitivity). Since males have only one X chromosome, a male child who inherits a variant X chromosome from his mother would exhibit variant color vision that depends on the number and nature of remaining opsin genes. (Adapted from Nathans J, Piantanida TP, Eddy RL, et al. [1986] *Science* 232:203–210.)

HOW ARE SIGNALS FROM RODS AND CONES ANALYZED IN THE RETINA?

During image detection, photoreceptors report light intensity as two-dimensional arrays that change over time. But the visual system does not simply provide a faithful representation of the external world. Its purpose is to extract from light signals *useful features*, such as motion, depth, color, and size of visual objects, to aid animals' survival and reproduction. Furthermore, animals devote finite neural resources to vision due to various biological constraints. For example, humans have ~100 million photoreceptors but only ~1 million retinal ganglion cells (RGCs) that transmit visual information from the retina to the brain. So the visual system discards as much irrelevant information as it can and encodes behaviorally relevant information as efficiently as possible.

The rest of the chapter deals with a central question: how do the retina and the brain extract behaviorally relevant information from visual stimuli? Specifically, how do the two-dimensional arrays of electrical signals from photoreceptors collectively inform the animal as to what is seen? Where and how does feature extraction occur?

4.12 Retinal ganglion cells use center–surround receptive fields to analyze contrast

As we highlighted in Chapter 1, asking the right question is often an important step toward making an important discovery. While the broad questions we just raised are key to understanding vision, they do not necessarily help us get started. Instead, we need questions that can be clearly answered with decisive experiments. One key question that has facilitated our understanding of visual information processing is: what visual signals best excite a given cell at a specific stage in the visual processing pathway?

Before answering this question, we must introduce an important concept in sensory physiology, the **receptive field**. This concept originated from studies of somatosensation, referring to the area of the body where stimuli could influence the firing of a neuron in the somatosensory pathway. The receptive field of a neuron in the visual system refers to the area of visual field (or corresponding area of the retina; Figure 4-13B) from which activity of a neuron can be influenced by visual stimuli. To measure the receptive field of a spiking neuron, one can place an electrode in the extracellular space next to the neuron and record its firing, a technique called **single-unit extracellular recording** (see Section 14.20 for details). The experimenter can then systematically alter the visual stimuli the retina receives to identify the kinds of stimuli that can influence the activity of the neuron being recorded. Thus, identifying the receptive field of a neuron means to identify the visual stimuli that alter its firing. As we move along the neural pathways that process visual signals, we will find that the receptive field changes in interesting and illuminating ways. We start with RGCs, the output cells of the retina (Figure 4-2A).

In the 1950s, shortly after electrodes for single-unit extracellular recording were invented, Stephen Kuffler performed a classic experiment to identify the RGC receptive field in the mammalian retina. He inserted an electrode from the side of the eye into the retina of an anesthetized cat and placed it next to an RGC. Then a spot of light was moved across the retina to identify the area corresponding to the RGC, as indicated by changes in its firing rate. The size and pattern of the spot were then systematically varied to determine how such variations changed the firing pattern of the RGC.

Figure 4-20 shows two kinds of cells Kuffler identified. For each of these cells, the receptive field can be described in terms of two concentric circles: a small circular *center* and a broader ring located around the center, called the *surround*. Both cells had basal firing in the absence of visual stimuli (top traces). Cell 1 was best excited by a small spot of bright light at the center, as measured by the number of action potentials fired during the stimulation period (second trace). As the

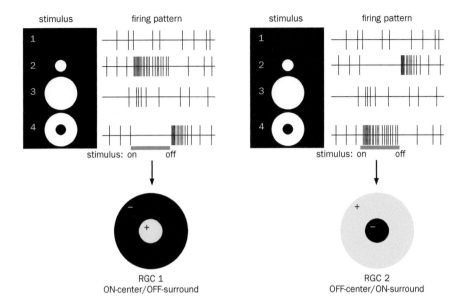

Figure 4-20 **Receptive fields of two kinds of retinal ganglion cells.** A series of stimuli (1–4) were applied to the retina. The firing patterns for two different kinds of RGCs were recorded as action potentials (vertical bars) in response to these stimuli. The stimulus duration is indicated by the horizontal bars below the fourth action potential trace. The receptive fields of the two cells are characterized by two concentric circles: the center and the surround. +, excited by light; –, inhibited by light. (Adapted from Hubel DH [1995] Eye, Brain, and Vision [Scientific American Library, No 22], W.H. Freeman. See also Kuffler SW [1953] *J Neurophysiol* 16:37–68.)

diameter of the spot increased, the firing rate *decreased* (third trace); after the spot reached a certain size, further increase of the diameter no longer changed the firing rate. This is because light stimuli that fall in the small circle at the center of Cell 1's receptive field activate the cell, whereas stimuli that fall in the surround suppress the firing of the cell. The suppressive effect was best illustrated by the fourth stimulus: a central dark spot surrounded by light failed to elicit action potentials and instead decreased the RGC's basal firing rate. In addition to responses to light onset, suppressive responses to the fourth light stimulus were followed by increased firing shortly after light offset, before settling into the baseline firing pattern. The responses of Cell 2 were precisely the opposite of the responses of Cell 1.

Kuffler found that cells he recorded from either behaved like Cell 1 or Cell 2 in Figure 4-20, suggesting that they belong to two distinct subclasses of RGCs. Using their responses to light onset as the criterion, the Cell 1 subclass is called ON-center/OFF-surround, while the Cell 2 subclass is called OFF-center/ON-surround. We now know that there are many types of RGCs within each of these subclasses (see Figure 4-25). Indeed, modern technology has enabled the mapping of RGC receptive fields at single-cone resolution (**Figure 4-21**).

The fact that RGCs have **center–surround** receptive fields has important implications for visual information processing. RGCs do not simply respond to light but start to analyze the spatial pattern, contrasting light and dark over a small area of the retina. Similar principles—namely that retinal neurons respond to changes in spatial patterns of luminance—were also found in earlier studies of lower vertebrates and invertebrates. But what causes the division between ON and OFF visual signals? And how do RGCs acquire their center–surround receptive fields? The following two sections address these two questions.

4.13 Bipolar cells are either depolarized or hyperpolarized by light based on the glutamate receptors they express

To understand how RGCs acquire their receptive fields, we need a better understanding of information flow between cells in the retina (Figure 4-2). We use pathways downstream of cones as examples, as most of the retina is devoted to analyzing information delivered from cones.

Before getting started, one important property of information processing in the retina is worth emphasizing. As discussed in Section 1.8, neurons use two forms of electrical signaling: all-or-none action potentials and graded local potentials. Most electrical signaling within the vertebrate retina utilizes graded potentials. Only RGCs and a subset of amacrine cells fire action potentials. Two disadvantages of

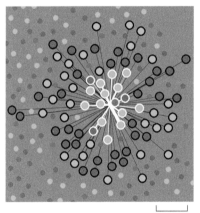

25 μm

Figure 4-21 **Connectivity map between cones and an RGC in the peripheral monkey retina.** Connectivity was reconstructed by extracellular recording of an OFF midget RGC in a retinal explant in response to light stimuli with a 5-μm × 5-μm pixel size, sufficient to resolve individual cones (blue, green, and red circles, representing S-, M-, and L-cones determined by spectral sensitivities). White and black outlines represent cones in the center and surround of the receptive field, respectively. The thickness of the white line represents connection strength; the thickness of the black lines is 5× magnified compared to that of white lines for visibility. (Adapted from Field GD, Gauthier JL, Sher A, et al. [2010] *Nature* 467:673–677. With permission from Springer Nature.)

graded potentials are (1) reduced speed of signal propagation due to temporal spread of the *R-C* circuit and (2) attenuation of signal magnitude during conduction (Section 2.8). However, with the exception of RGCs, retinal neurons synapse onto partners located within the retina, so these graded potentials travel only short distances, and the losses associated with them are insubstantial. On the other hand, graded signals can transmit more information than action potentials. This is because the unit of information transfer for graded potentials is individual synaptic vesicles, as opposed to individual spikes for action potentials. Graded potentials can transmit more vesicles per second than spiking neurons can transmit action potentials per second. So neurons within the retina use graded potentials to transmit information before the most behaviorally relevant information is selected for transmission to the brain by RGC axons.

Another useful concept to introduce as we trace information flow through neural circuits is the **sign** of signals. We can give any given neuron a sign (for example, positive for becoming depolarized in response to a sensory stimulus). If a second neuron is activated by the given neuron, it maintains the sign of that neuron (also becoming depolarized by the sensory stimulus); if a second neuron is inhibited by the given neuron, it inverts the sign of the given neuron (becoming hyperpolarized by the sensory stimulus). With these notes, let's now examine what happens to bipolar cells when light stimulates cones.

Cones release glutamate that directly acts on two classes of postsynaptic target cells: bipolar cells and horizontal cells (Figure 4-2A). Bipolar cells are themselves glutamatergic excitatory neurons that synapse directly onto RGCs. There are two major subclasses of bipolar cells, **OFF bipolar** and **ON bipolar**, which project their axons to different laminae (sublayers) of the inner plexiform layer (Figure 4-22). OFF bipolars express ionotropic glutamate receptors (Section 3.15) and are depolarized by glutamate released from the photoreceptors. Therefore, OFF bipolars keep the sign of photoreceptors and are hyperpolarized by light. In ON bipolars, glutamate activates a metabotropic glutamate receptor and a G protein cascade (Section 3.19), which results in closure of cation channels. Hence a sign inversion occurs between the photoreceptors and ON bipolars. ON bipolars are *inhibited* by glutamate release from photoreceptors and are therefore depolarized by light.

The ON and OFF bipolar cell types transmit excitatory output respectively onto the ON and OFF RGCs we discussed in the previous section. These ON and OFF pathways enable the visual system to analyze dark-to-light and light-to-dark

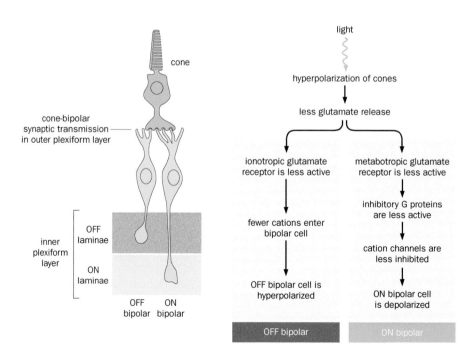

Figure 4-22 OFF and ON bipolar cells respond to light in opposite manners. Left, schematic of the pathway from cones to OFF and ON bipolar cells, which project their axons to OFF and ON laminae of the inner plexiform layer, respectively. Right, flow charts of responses to light stimulation from cones to OFF and ON bipolar cells. OFF bipolar cells maintain the sign of photoreceptors by using an ionotropic glutamate receptor. ON bipolar cells invert the sign of photoreceptors by using metabotropic glutamate receptors coupled to inhibitory G proteins, activation of which leads to closure of cation channels.

transitions in parallel. As introduced in Section 1.12, parallel information processing is a common strategy in the nervous system. We will encounter many more examples later in this textbook.

4.14 Lateral inhibition from horizontal cells constructs the center–surround receptive fields

We now address the origin of the center–surround receptive field. As we saw in Section 4.12, RGCs respond not just to light intensity of the stimulus, but rather to a *light intensity difference* within a small area of the retina. Horizontal cells contribute critically to this property.

Horizontal cells are inhibitory neurons that extend their processes laterally to connect to many photoreceptors (Figure 4-2A). Individual horizontal cell dendrites both receive excitatory input from photoreceptors and send inhibitory output back to the presynaptic terminals of photoreceptors to reduce their glutamate release. (Note here an exception to the general rule discussed in Section 1.7 regarding information being received by dendrites and sent by axons; in horizontal cells as well as the amacrine cells we will soon discuss, *dendrites can both receive and send information*.) Furthermore, the inhibitory output also spreads to presynaptic terminals of neighboring photoreceptors (**Figure 4-23**). This inhibitory

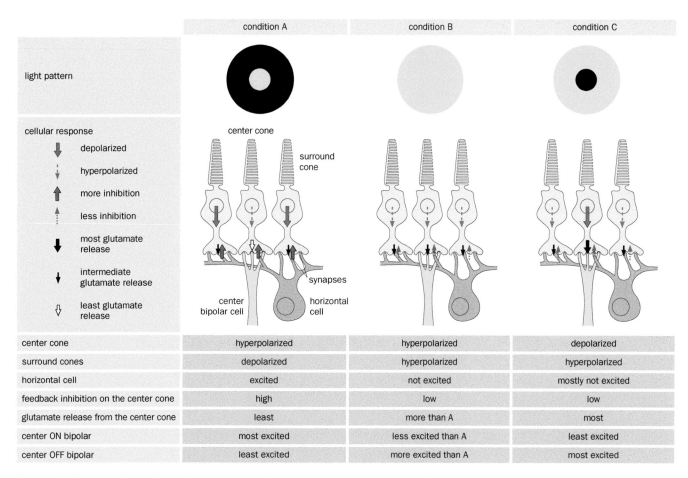

	condition A	condition B	condition C
center cone	hyperpolarized	hyperpolarized	depolarized
surround cones	depolarized	hyperpolarized	hyperpolarized
horizontal cell	excited	not excited	mostly not excited
feedback inhibition on the center cone	high	low	low
glutamate release from the center cone	least	more than A	most
center ON bipolar	most excited	less excited than A	least excited
center OFF bipolar	least excited	more excited than A	most excited

Figure 4-23 Bipolar cells acquire their center–surround receptive field through lateral inhibition of horizontal cells. Condition A, a small spot of light is directed on the center cone. The center cone is hyperpolarized whereas many surround cones (only two are drawn here for simplicity) are depolarized. The horizontal cell, which receives excitatory input from and sends inhibitory feedback output to many cones, is maximally excited. The center cone releases minimal glutamate due to its hyperpolarized state and inhibition from the horizontal cell. Condition B, a larger spot of light covers both center and surround cones. All cones become hyperpolarized, causing the horizontal cell to hyperpolarize and thereby reducing its inhibitory output. The center cone releases more glutamate than in condition A. Condition C, light activates the surround cones but not the center cone. The center cone releases maximal glutamate because of its depolarized state and minimal inhibition from the horizontal cell.

(A)

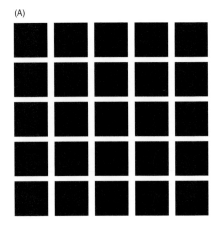

(B)

Figure 4-24 Visual illusions due to lateral inhibition. (A) The Hermann grid consists of black squares separated by vertical and horizontal white bands. Ghost gray blobs appear at the intersections of the white bands. **(B)** The Mach bands consist of a series of juxtaposed grayscale bands. At each border, the edge on the dark side appears darker and the edge on the light side appears lighter, thus exaggerating the contrast.

spread, or **lateral inhibition** (Figure 1-20I), enhances the difference between the light-stimulated photoreceptor and the surrounding unstimulated photoreceptors, providing a mechanism by which their downstream bipolar cells can acquire a center–surround receptive field. Indeed, lateral inhibition is used throughout the nervous system to sharpen the differences between parallel inputs as information is propagated through neural circuits.

Let's study how lateral inhibition produces a center–surround receptive field by examining three conditions of light input. First, suppose that a small spot of light is directed on a center cone that synapses onto a downstream bipolar cell (Figure 4-23, condition A). Light hyperpolarizes the center cone, causing it to release less glutamate. At the same time, the surround cones do not receive light input, are depolarized, and release lots of glutamate. Because each horizontal cell receives input from many cones, it is maximally excited and consequently inhibits all cones it connects to, including the center cone. Thus, due to intrinsic hyperpolarization by light and inhibition by horizontal cells, the center cone releases minimal glutamate. Second, when the spot of light becomes larger (Figure 4-23, condition B), nearby cones also become hyperpolarized. The horizontal cell is less excited, and its inhibitory action is diminished. This relief of inhibition, or **disinhibition**, of the center cone causes it to release more glutamate than in condition A. Third, when the surrounding cones but not the center cone receive light input (Figure 4-23, condition C), the horizontal cell remains mostly hyperpolarized because it receives many more inputs from the surround cones than from the center cone. The center cone remains mostly disinhibited, while it is intrinsically depolarized due to not receiving light input. Therefore the center cone releases maximal glutamate. If the bipolar cell postsynaptic to the center cone were an ON bipolar, it would have a receptive field similar to the ON-center/OFF-surround RGC that Kuffler described. If the center bipolar cell were an OFF bipolar, it would have a receptive field similar to the OFF-center/ON-surround RGC (Figure 4-23 bottom).

In summary, the inhibitory actions of horizontal cells and their lateral connectivity, in combination with the existence of the two types of bipolar cells, result in bipolar cells with ON-center/OFF-surround and OFF-center/ON-surround receptive fields (Movie 4-1). These bipolar cells then impart similar receptive structures to their postsynaptic RGCs. Amacrine cells contribute to additional surround mechanisms for RGCs, as will be discussed in later sections.

The center–surround receptive fields mediated by lateral inhibition underlie some striking visual illusions. For instance, in the Hermann grid (Figure 4-24A), gray blobs appear at the intersections of white bands in between black squares. In the Mach bands (Figure 4-24B), the borders between any two neighboring gray bands appear to have stronger differences in their gray levels, creating the perception of a sharp edge. As an exercise, try to compare receptive fields consisting of concentric circles at different locations in the Hermann grid or the Mach bands and see whether you can come up with satisfactory explanations for these visual illusions. The Mach bands also illustrate a useful function of lateral inhibition—to enhance edge detection.

4.15 Diverse retinal cell types and their precise connections enable parallel information processing

The retina is arguably the best-studied piece of vertebrate neural tissue in terms of understanding the composition of cell types, their connections, and their functions. Indeed, studies of the retina have made important contributions to our understanding of what constitutes a neuronal cell type (Box 4-2). The action of horizontal cells in constructing the center–surround receptive fields for ON and OFF bipolar cells is but one of many feats that retinal circuits accomplish. There are more than 10 types of bipolar cells (Figure 4-25). Of these, one—an ON bipolar—receives information from rods, and the rest receive information from cones. Among the different ON and OFF cone bipolars, some respond transiently and others in a sustained fashion to changes in synaptic transmission from cones, such that they represent information across different temporal domains. Bipolar

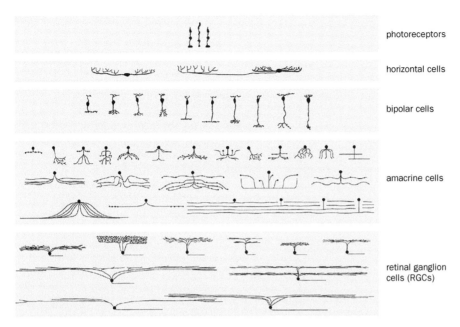

photoreceptors

horizontal cells

bipolar cells

amacrine cells

retinal ganglion cells (RGCs)

Figure 4-25 Cell types in the mammalian retina. Retinal cell types have been defined by a combination of morphological and physiological properties. Photoreceptors include rods and two or three types of cones with different spectral sensitivities. Horizontal cells provide lateral inhibition to construct the center–surround receptive fields of bipolar and ganglion cells. One type of horizontal cell (right) has distinct axonal and dendritic processes at different distances from the soma and form reciprocal synapses with rods and cones, respectively. Different types of bipolar cells have receptive fields of different sizes and connect to distinct types of RGCs by terminating their axons in specific laminae of the inner plexiform layer. Amacrine cells and ganglion cells have the most types. Many amacrine cells and ganglion cells of the same type cover the retina once and only once, with varying laminar specificities and receptive field sizes. Recent studies have identified more cell types than are depicted here (Box 4-2). (From Masland RH [2001] *Nat Neurosci* 4:877–886. With permission from Springer Nature.)

cells also vary in the size of their receptive fields and therefore their spatial coverage of photoreceptors.

The parallel streams of information carried by different bipolar cells are delivered to distinct RGC types through precise connections in the retinal circuits. For example, the ON and OFF bipolar cells form synapses with, respectively, ON and OFF RGCs at distinct laminae in the inner plexiform layer (Figure 4-22, left). Some RGCs (bistratified or ON–OFF RGCs) elaborate their dendrites in two separate laminae and therefore receive input from both ON and OFF bipolars (e.g., the right RGC in the penultimate row in Figure 4-25). The bipolar → RGC synapses are further modulated by the actions of ~40 different types of amacrine cells, most of which are GABAergic or glycinergic inhibitory neurons that elaborate their dendrites within the inner plexiform layer. (Like the horizontal cells introduced earlier, dendrites of amacrine cells both receive input and send output, and many amacrine cells do not have axons; in fact, the word *amacrine* means "without a long fiber"). As a consequence, different types of RGCs carry distinct visual information regarding contrast, size, motion, and color to the brain.

Interestingly, dendrites of many types of RGCs and amacrine cells cover the entire retina once and only once, in the same way that tiles cover a kitchen floor (**Figure 4-26**). This property is aptly named **dendritic tiling**; it allows a particular type of retinal neuron to sample the entire visual world with minimal redundancy. With the introduction of new tools in molecular genetics in recent years (see Sections 14.11 and 14.13 for details), an expanding number of RGC and amacrine types have been identified and their functions in retinal information processing described. As specific examples, we discuss in the following sections how retinal circuits extract motion and color information via the actions of specific retinal cells and their connection patterns.

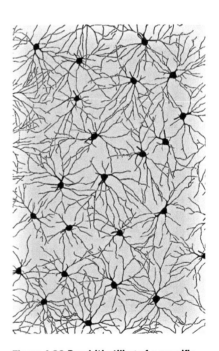

Figure 4-26 Dendritic tiling of a specific RGC subtype in the cat. RGC cell bodies are regularly spaced. Dendrites from neighboring cells overlap minimally and together cover the retina uniformly. (From Wässle H [2004] *Nat Rev Neurosci* 5:1–11. With permission from Springer Nature.)

Box 4-2: How to define a neuronal cell type?

Generally speaking, biologists classify cells into types based on their functions: cells of the same type perform the same biological function. However, we still do not know the exact functions that many neurons perform, making it difficult to use this criterion to define cell type. Neurons that perform the same function usually have similar somal positions, dendritic morphologies, axonal projections, physiological properties, and gene expression patterns. Thus, these parameters (often in combination) have traditionally been used to classify cell types. In recent years, great strides have been made in defining cell types using new techniques, notably single-cell RNA-sequencing (**single-cell RNA-seq**; see Section 14.13 for more details), in which mRNAs expressed by individual cells can be determined and quantified at the level of the entire **transcriptome** (collection of all expressed mRNAs). We discuss one study that exemplifies these advances.

In the mouse retina, about a dozen bipolar cell types had been defined using a combination of morphology, dendritic connectivity with specific photoreceptors, axonal projection to specific laminae of the inner plexiform layer, temporal response to light stimulation, and expression of specific marker genes (**Figure 4-27**A). In a recent single-cell RNA-seq study, researchers took advantage of transgenic mice expressing green fluorescent protein (GFP) in all bipolar cells, and collected GFP+ cells from dissociated retina to enrich for bipolar cells using fluorescence-activated cell sorting. Individually sequenced bipolar cells were separated into 15 discrete transcriptomic clusters based on their gene expression similarity (Figure 4-27B). Researchers were able to match a subset of transcriptomic clusters with known bipolar cell types based on prior knowledge about marker expression. For the rest of the bipolar types without known molecular markers, researchers simultaneously used sparse labeling methods (Section 14.16) to visualize the morphology of individual neurons and *in situ* hybridization (Section 14.12) to label neurons with cluster-specific genes. They found nearly one-to-one matching between previously identified cell types and transcriptomic clusters (compare Figure 4-27A–C); a few previously known cell types contained more than one transcriptomic cluster and could be further divided. Thus, for the retinal bipolar cells, which make up a well-characterized neuronal class, there is an

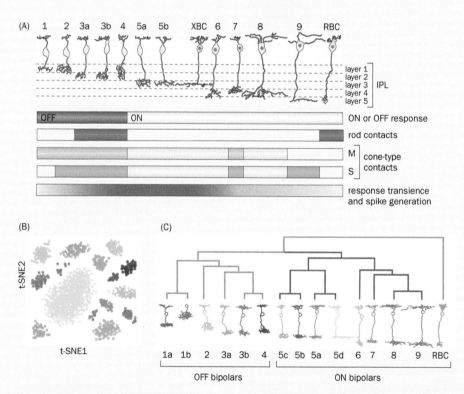

Figure 4-27 Retinal bipolar cell types in in the mouse. (A) Bipolar cells had been classified into 13 types based on their dendritic connectivity with rods and two types of cones, their axonal targeting to specific laminae of inner plexiform layer (IPL), and their physiological responses to light. Another property not illustrated is marker gene expression. **(B)** In this plot resulted from a single-cell RNA-seq experiment, each dot represents a single cell. Cells from the same cluster (same color) share more similarities in their transcriptomes than cells from different clusters. The high-dimensional gene expression data (each gene defines one dimension, and the level of expression defines the value on that axis) are visualized by a two-dimensional representation using a statistical method called t-distributed stochastic neighbor embedding (t-SNE; see Section 14.13 for more details). Mouse retinal bipolar cells can be separated into 15 transcriptomic clusters. **(C)** Hierarchical clustering of bipolar cells based on their transcriptomes. The transcriptomic clusters in Panel B and morphological types in Panel C are color matched based on matching morphology and cluster-specific gene expression. (A, from Euler T, Haverkamp D, Schubert T, et al. [2014] *Nat Rev Neurosci* 15:507–519. With permission from Springer Nature. B & C, from Shekhar K, Lapan SW, Whitney IE, et al. [2016] *Cell* 166:1308–1323. With permission from Elsevier Inc.)

Box 4-2: continued

excellent match between transcriptomes determined by single-cell RNA-seq and morphological and physiological properties.

Similar single-cell RNA-seq experiments are being performed to characterize the amacrine and retinal ganglion cell classes

in the retina, estimated to be at around 40 and 30 types, respectively. We have far more knowledge about cell types in the retina than in most brain regions; still, single-cell RNA-seq has helped define cell types in other brain regions (see examples of cortical neurons in Box 4-4 and somatosensory neurons in Section 6.27).

4.16 Direction-selectivity of RGCs arises from asymmetric inhibition by amacrine cells

Direction-selective retinal ganglion cells (**DSGCs**) were first discovered in the 1960s in rabbits via single-unit recordings of RGCs. The firing rate of a DSGC depends on the direction in which a spot of light travels through its receptive field: the highest rate is elicited by the movement of light in the **preferred direction** and the lowest firing rate in the **null direction** (Figure 4-28). The most abundant and well-studied DSGCs are the ON–OFF DSGCs, which send dendritic tufts to both the ON and OFF laminae of the inner plexus layer, where they receive input from ON and OFF bipolar cells, respectively (Figure 4-29A).

A crucial cell type that shapes the direction-selective responses of DSGCs is the **starburst amacrine cell** (**SAC**), a type of GABAergic inhibitory neuron that also releases acetylcholine. Two subtypes of SACs send their dendrites to the same ON or OFF lamina as the ON–OFF DSGCs, where they form extensive synaptic connections (Figure 4-29B). In the mouse, SACs receive excitatory input from bipolar cells preferentially in the proximal dendrites and send output from the distal third of their dendritic trees to DSGCs. Ablating SACs eliminates the direction-selective response of DSGCs. Interestingly, a SAC's output from a dendrite is strongest in response to light stimulation in a centrifugal direction (excitation spreading from soma to the tip of dendrites). So how do SACs regulate the direction selectivity of DSGCs?

We use an experiment that employed paired **whole-cell patch recording** of SACs and DSGCs to illustrate their relationship. Whole-cell recordings can measure not only DSGC output in the form of action potentials but also the excitatory and inhibitory synaptic inputs, and thereby they can be used to investigate synaptic mechanisms that produce neuronal output (see Section 14.21 for details). Stimulating SACs revealed that they provide inhibitory input to DSGCs that depends on the position of the cells in the retina. Strong inhibition occurred only if the SAC was located on the null side of the DSGC (Figure 4-29C).

Given the symmetrical shape of the dendritic fields of both SACs and DSGCs (Figure 4-29B), how is this asymmetric inhibition achieved? This question was investigated using **laser-scanning two-photon imaging** of Ca^{2+} signals followed by **serial electron microscopic (EM) reconstruction** (see Sections 14.22 and 14.19 for details of these techniques). Two-photon imaging of Ca^{2+} signals enabled

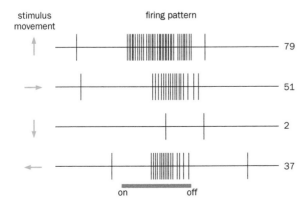

Figure 4-28 **Direction-selective ganglion cells (DSGCs).** Single-unit recording of a ganglion cell in the rabbit reveals that it is most active when a light spot moves upward across its receptive field and least active when the light spot moves downward. Arrows on the left indicate the direction in which the light moves. Numbers correspond to the spikes fired during the period of motion stimulus (bar at the bottom). (Adapted from Barlow HB, Hill RM, & Levick WR [1964] *J Physiol* 173:377–407.)

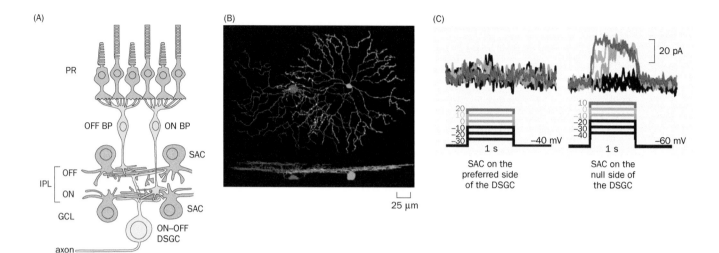

(A)

PR

OFF BP   ON BP

SAC

IPL   [OFF
      [ON

GCL

SAC

ON–OFF
DSGC

axon

(B)

25 μm

(C)

20 pA

1 s   –40 mV

SAC on the
preferred side
of the DSGC

1 s   –60 mV

SAC on the
null side of
the DSGC

Figure 4-29 Asymmetric inhibition of ON–OFF direction-selective ganglion cells (DSGCs) by starburst amacrine cells (SACs).
(A) Schematic of input to the DSGC. PR, photoreceptor; BP, bipolar cell; IPL, inner plexiform layer; GCL, ganglion cell layer. **(B)** Visualization of a DSGC (red) and a neighboring SAC (green) after fluorescent dye fill. The top view is taken from the retinal surface, while the bottom view shows a retinal cross section as in Panel A. White dots indicate potential synaptic contacts between the two cells. **(C)** In a patch

clamp experiment on an SAC–DSGC pair, inhibitory inputs were measured in the DSGC (top traces) in response to depolarizing steps applied to the SAC (below). Strong inhibition occurred when the SAC was located on the null side of the DSGC (right), but not when the SAC was located on the preferred side of the DSGC (left). (B, from Wei W & Feller MB [2011] *Trends Neurosci* 34:638–645. With permission from Elsevier Inc. C, from Fried SI, Münch TA, & Werblin FS [2002] *Nature* 420:411–414. With permission from Springer Nature.)

the determination of direction selectivity of many individual DSGCs in a retinal explant. Subsequent EM reconstruction of their synaptic connections revealed that SACs preferentially formed synaptic connections on the null side of the DSGC dendrites, despite being in physical proximity to DSGC dendrites in other directions (**Figure 4-30**A). This asymmetric connectivity provides a potential anatomical

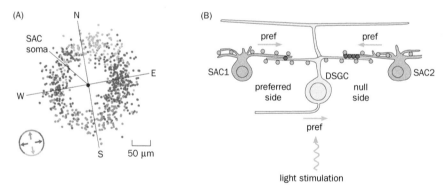

(A)

N

SAC
soma

W   E

S   50 μm

(B)

pref   pref

SAC1   DSGC   SAC2

preferred   null
side   side

pref

light stimulation

Figure 4-30 Asymmetric synaptic connections between SACs and DSGCs and a model of direction selectivity. (A) The cumulative distribution of output synapses of 24 SACs is displayed relative to their respective somata (center). Synapses are color coded according to the direction selectivity of DSGCs (bottom left). SAC output synapses are preferentially found at the distal branches and align antiparallel to the preferred direction of DSGCs. For instance, purple synapses (onto east-preferred DSGCs) are mostly to the west of the SAC soma. Thus, from the perspective of a DSGC, SAC input synapses form preferentially on the DSGC's branches in the null direction. **(B)** A model accounting for direction selectivity of a DSGC whose directional preference (pref) is indicated by the green arrow. Although each DSGC receives input from many SACs, only two are shown for simplicity—one synapses on the dendrites on the preferred side of the DSGC (SAC1) and the other synapses on the null side (SAC2). Only the ON response is modeled. Both the DSGC and SACs receive excitatory synaptic input from ON bipolars (green dots). The DSGC receives preferential inhibitory input from the SAC (red dots) on the null side. The centrifugal preference of

SACs' synaptic output is also indicated by the yellow arrows. When the light stimulus moves from left to right, SAC1 is excited first, followed by the DSGC, and then SAC2. Although the dendrites of SAC1 are maximally excited, its inhibitory input to the DSGC is minimal because of the weak SAC1 → DSGC connection. The inhibitory input from SAC2 to the DSGC comes after DSGC's own excitation and is also diminished because the direction is opposite to the preferred direction of SAC2's dendritic output. Thus, the DSGC is maximally activated. When the light stimulus comes from right to left, SAC2 is activated first, and its dendrites release much GABA because its preferred direction matches that of the light stimulus. Because of the strong SAC2 → DSGC connection, the DSGC receives strong inhibitory input from SAC2 right before, and at the same time as, it receives excitatory bipolar input. Inhibition cancels excitation (Figure 3-29) and thus the DSGC is minimally activated. (A, from Briggman KL, Helmstaedter M, & Denk W [2011] *Nature* 471:183–188. With permission from Springer Nature. B, adapted from Wei W & Feller MB [2011] *Trends Neurosci* 34:638–645. With permission from Elsevier Inc.)

basis for the asymmetric inhibition of DSGCs by SACs. The mechanism by which this asymmetric connectivity is established remains to be explored.

With these different lines of information, we can now piece together a model accounting for the direction selectivity of DSGCs. Let's use the ON response of a DSGC as an example (Figure 4-30B). When light moves in the null direction, SACs located on the null side of the DSGC are excited first by ON bipolar cells. The centrifugal excitation of SACs' dendrites cause them to release much GABA, and their preferential connections with the DSGC provide strong inhibitory input right before, and at the same time, as the DSGC receiving excitatory input from bipolar cells. This effectively cancels out light-induced excitation. When light moves in the preferred direction, the DSGC receives excitatory input from bipolar cells before it receives GABAergic input from SACs because SACs located on the preferred side do not deliver strong inhibitory input. This sequence causes maximal excitation of the DSGC.

Motion detection is a fundamental function of the visual system of all animals. Do different animals use similar or different strategies? Box 4-3 discusses motion vision in the fruit fly, a model organism with many experimental advantages.

Box 4-3: Motion vision in the fly

Insects have superb motion vision: consider the difficulty of swatting an annoying housefly! The fruit fly *Drosophila melanogaster* is a particularly facile model organism in which to study visual information processing because of the abundant circuit analysis tools. Moreover, a large fraction of its synaptic connections have been reconstructed by serial electron microscopy. In this box, we summarize insights learned from these studies with a focus on motion vision.

The compound eye of *Drosophila* comprises 800 repeating units called **ommatidia** (see Figure 5-34A). Within each ommatidium, six out of the eight photoreceptors, R1–R6, are essential for motion vision. They project their axons to the **lamina**, the first **neuropil** layer underneath the retina, where they synapse onto lamina neurons L1 and L2 (**Figure 4-31A**). (In invertebrates, neuropils are regions of neural tissue densely packed with neuronal and glial processes but with no cell bodies.) L1 and L2 neurons target their axons to specific layers of the **medulla** neuropil, where they synapse onto specific types of Mi (medulla interneuron) or Tm (trans-medulla) neurons. The retinotopic organization of the eye is maintained in the lamina and medulla, each of which contains 800 columns, one for each ommatidium. Information related to motion vision is relayed to T4 and T5 neurons, which receive inputs from specific types of Mi and Tm neurons across several columns and project axons to the **lobula plate**, a neuropil-containing neuron with dendritic trees that cover a wide visual field (Figure 4-31A, B). Direction selectivity first emerges in T4 and T5 neurons, which respond selectively to moving light edges and dark edges, respectively. There are four subtypes of T4 and T5 neurons; each subtype is preferentially activated by motion in one of the four cardinal directions and projects axons to one specific layer of the lobula plate neuropil (Figure 4-31C).

How do T4 and T5 neurons acquire their response properties? We first discuss their selective light versus dark responses, which resemble the ON and OFF pathways in the vertebrate retina (Section 4.13). Genetic silencing of L1 and L2 neurons indicate that the L1 pathway is responsible for detecting moving light edges, whereas the L2 pathway is responsible for detecting moving dark edges. Serial EM reconstructions revealed that Mi1 and Tm3 neurons, which are postsynaptic to L1, provide extensive input to T4, while Tm1 and Tm2, which are postsynaptic to L2, provide extensive input to T5 (Figure 4-31A). Recording the activity of these neurons in response to light flashes using a genetically encoded voltage indicator (Section 13.22) revealed that L1 and L2 respond similarly to light flashes, with a hyperpolarization phase followed by a depolarization phase. (*Drosophila* photoreceptors depolarize to light flashes and use histamine as their neurotransmitter, which opens a histamine-gated Cl^- channel and thus hyperpolarizes postsynaptic lamina neurons.) However, whereas the L2 → Tm1/Tm2 synapses are sign preserving, the L1 → Mi1/Tm3 synapses are sign inverting. Thus, Mi1 and Tm3 depolarize to light, whereas Tm1 and Tm2 depolarize to dark (Figure 4-31D, E). Furthermore, Ca^{2+} imaging of Mi1/Tm3 and Tm1/Tm2 revealed a nonlinear transformation between the biphasic voltage signal and Ca^{2+} signal: only voltage increments are converted to Ca^{2+} signals for synaptic transmission to downstream targets (Figure 4-31F). (Note that, as in the vertebrate retina, all neurons we have discussed so far use graded potentials rather than action potentials to transmit information.) This upstream processing ensures that postsynaptic T4 and T5 neurons respond preferentially to light and dark, respectively.

How does direction selectivity arise in T4 and T5 neurons? All models of motion detection involve integration of inputs from at least two photoreceptors sequentially activated by a moving stimulus. Let's use path I and path II to refer to the pathways downstream of the photoreceptor activated first and second, respectively, by a moving stimulus in the preferred direction. In one conceptual model (**Figure 4-32A**), both paths provide excitatory signals, and there is a temporal delay in path I, such that signals from both paths arrives at the integration site simultaneously if the stimulus moves in the preferred direction. This results in an amplified signal

(Continued)

Box 4-3: continued

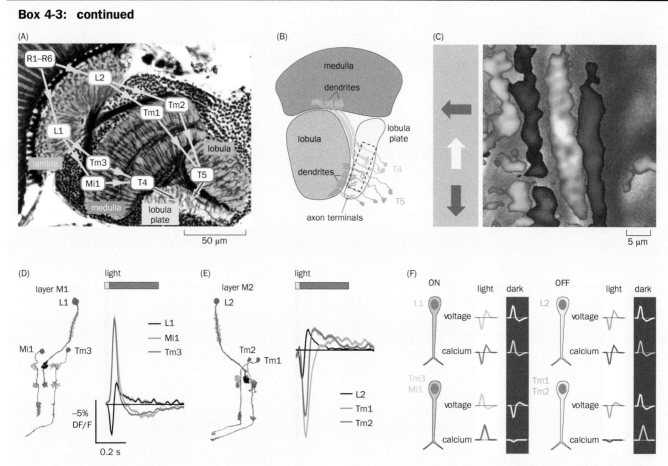

Figure 4-31 *Drosophila* motion vision pathways. (A) Schematic of the motion vision circuit in *Drosophila* superimposed on a histological section of the fly visual system. Neuropils (highlighted by colors) are surrounded by cell bodies (individual nuclei are stained in black). The connections between the various neuronal types are schematized by yellow arrows and were revealed by serial EM reconstruction. Only a subset is shown for clarity. **(B)** Schematic displaying the morphology of T4 and T5 cells. Both have cell bodies next to the lobula plate neuropil. T4 extends its process first to the medulla, where its dendritic branches receive inputs from Mi1 and Tm3; the axon then extends back to the lobula plate neuropil, where it terminates in one of the four layers. T5 has a similar organization except that its dendrites receive input from Tm1 and Tm2 in the lobula. Dashed box represents imaging area in Panel C. **(C)** Two-photon Ca²⁺ imaging of T4 axon terminals in the lobula plate. Stimuli moving in four cardinal directions activate T4 subtypes, which project axons to four color-matched layers. **(D)** Left, schematic of two-photon voltage imaging of L1 axons (black), Mi1 dendrites (cyan), or Tm3 dendrites (red) at the M1 layer of the medulla. Right, voltage traces in response to a 25-ms light flash. A genetically encoded voltage indicator,

ASAP2f, was expressed in L1, Mi1, or Tm3 neurons in separate flies. DF/F, change of fluorescence intensity over basal level fluorescence intensity. (Because ASAP2f *decreases* its fluorescence in response to depolarization, the *y* axis unit is –5%.) Their voltage traces are color coded and superimposed onto the same graph for comparison. Note that Mi and Tm3 invert their signs compared to L1 and are depolarized by light. **(E)** Same as in Panel D, except imaging was performed on L2 axons or dendrites of Tm1 or Tm2 at the M2 layer of the medulla. Note that Tm1 and Tm2 preserve the L1 sign and are hyperpolarized by light. **(F)** Schematic summary of voltage and Ca²⁺ imaging in the L1 and L2 pathways. In L1 and L2 neurons, Ca²⁺ signals maintain the biphasic nature of the voltage signals. However, in their postsynaptic targets, Ca²⁺ signals maintain only voltage increments, thus selectively transmitting light signals in the L1 pathway and dark signals in the L2 pathway. (A, from Takemura S, Nern A, Chklovskii DB, et al. [2017] *eLife* 6:e24394. B & C, adapted from Maisek MS, Haag J, Ammer G, et al. [2013] *Nature* 500:212–216. With permission from Springer Nature. D–F, adapted from Yang HH, St-Pierre F, Sun X, et al. [2016] *Cell* 166:245–257. With permission from Elsevier Inc.)

that crosses a threshold for transmission to downstream neurons. A signal from either path alone is insufficient to cross the threshold; as a result, stationary stimuli and stimuli moving in the null direction do not produce responses. According to this model, direction selectivity results from preferred direction enhancement. In a second conceptual model (Figure 4-32B), the delay occurs in path II, which provides an inhibitory signal at the integration site. A stimulus moving in the preferred direction allows path I's excitatory signal to avoid the inhibition from path II, whereas path I's

excitatory signal is canceled by the inhibition by path II if the stimulus moves in the null direction. According to this model, direction selectivity results from null direction suppression. Finally, other conceptual models use more complex temporal filtering strategies of both paths and combine signals to achieve both preferred direction enhancement and null direction suppression.

Recent studies suggest that both preferred direction enhancement and null direction suppression contribute to direction

Box 4-3: continued

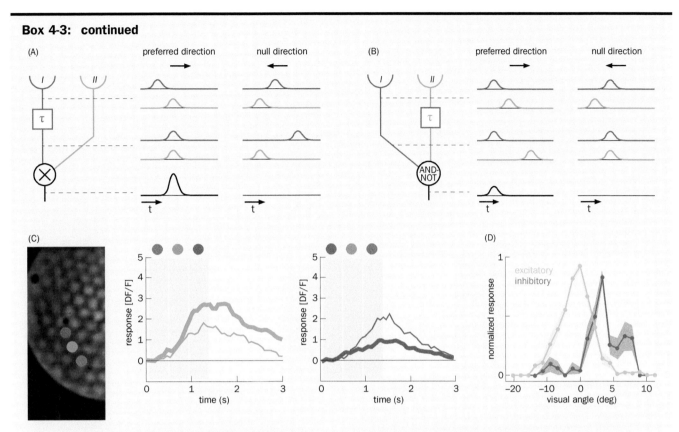

Figure 4-32 Models and mechanisms of direction selectivity.
(A & B) Two models of direction selectivity through signal integration from two pathways. In both cases, path I (red) is activated by a moving stimulus in the preferred direction before path II (blue). In model A, signal from path I experiences a delay (τ) before integrating with signal from path II. When the stimulus moves in the preferred direction, signals in path I and path II reach the integration site simultaneously, producing an amplified signal that passes the threshold for transmission to downstream neurons. In model B, signal from path II experiences a delay and is inhibitory at the integration site. When the stimulus moves in the null direction, signal produced in path I coincides with the inhibitory signal in path II and is thus selectively suppressed. **(C)** A genetically encoded Ca^{2+} indicator was specifically expressed in T4 neurons that respond to upward motion, and three adjacent ommatidia (left, colored spots) were stimulated. Middle, when the three ommatidia were activated in the pink–green–brown sequence, mimicking a stimulus moving in the preferred direction, the resulting Ca^{2+} signal (thick curve) is greater than the sum of signals from stimulation of individual ommatidia (thin curve). Right, when the three ommatidia were activated in the brown–green–pink sequence, mimicking a stimulus moving in the null direction, the resulting Ca^{2+} signal (thick curve) is less than the sum of signals from stimulation of individual ommatidia (thin curve). DF/F, change of fluorescence intensity over basal level fluorescence intensity. **(D)** Whole-cell recordings of T4 neurons in response to light flashes in different visual areas indicated that the peaks of excitatory and inhibitory input are spatially displaced, with the inhibitory input peak shifted toward the preferred direction of the T4 neuron (positive angles). (A & B, from Yang HH & Clandinin TR [2018] *Ann Rev Vis Sci* 4:143–163. With permission from Annual Reviews. See also Hassenstein B & Reichardt W [1956] *Z Naturforsch B* 11:513–524; Barlow HB & Levick WR [1965] *J Physiol* 178:477–504. C, adapted from Haag J, Arenz A, Serbe E, et al. [2016] *eLife* 5:e17421. D, adapted from Gruntman E, Romani S, & Reiser MB [2018] *Nat Neurosci* 21:250–257. With permission from Springer Nature.)

selectivity. For example, using light stimuli restricted to a single ommatidium, researchers measured Ca^{2+} signals in T4 cells in response to activation of three adjacent ommatidia sequentially, mimicking moving stimuli in the preferred or null direction, and compared those to simply adding Ca^{2+} signals by activating three ommatidia individually (a linear summation). They found that stimuli moving in the preferred direction produced a signal greater than the linear summation, consistent with preferred direction enhancement. Stimuli moving in the null direction produced a signal smaller than the linear summation, consistent with null direction suppression (Figure 4-32C). Measuring excitatory and inhibitory inputs of T4 and T5 neurons revealed that they are spatially displaced along the preferred–null direc-

tion axis (Figure 4-32D); this property could contribute to null direction suppression, as in mammalian DSGCs.

While we still do not know what cells correspond to specific aspects of either of the conceptual models, what gives rise to the temporal delay, or whether these models are sufficient to account for all stimulus conditions, recent progress suggests that answers to these questions are within reach. Comparing motion vision in flies and mammals will also be instructive. We have already observed parallel ON and OFF pathways in both flies and mammals for processing visual information as well as the use of null direction suppression for direction selectivity, despite substantial differences in the anatomical organization of the circuits.

Figure 4-33 Color perception is influenced by the background. The two crosses in this image are the same color (see their junction at the top). However, the left cross appears gray on a yellow background, and the right cross appears yellow on a gray background. (Adapted from Albers J [1975] Interaction of Color. Yale University Press.)

4.17 Color is sensed by comparing signals from cones with different spectral sensitivities

As discussed in Section 4.10, humans and some other primates have three types of cones: S, M, and L. Light of different wavelengths differentially activates each type. How does this differential activation give rise to color perception? According to the color-opponency theory proposed by Ewald Hering in the 1870s, human color vision is achieved by three sets of antagonistic color receptors interacting with each other: black–white, blue–yellow, and green–red. The color-opponency theory successfully explains many phenomena, such as the influence that visual background has on our color perception (**Figure 4-33**). What is the neural basis of color opponency?

Mechanisms similar to those underlying the center–surround receptive field for detecting luminance contrast are used to compare differential excitation of cones of distinct spectral sensitivities, the output of which is delivered to the brain by **color-opponent RGCs**. As predicted by Hering's theory, two major types of color-opponent RGCs have been identified, the blue–yellow and green–red. The blue–yellow opponent system is used by all mammals to compare short- and long-wavelength light, whereas the green–red opponent system is used by trichromatic primates to further distinguish the two long-wavelength colors. (The luminance contrast–detecting RGCs we discussed in Section 4.12 have mixed color input to both center and surround and hence underlie the black–white opponency system in Hering's theory.) We first illustrate how the blue–yellow opponent system is established.

Glutamate release from each S-cone is already "color opponent." This results from both its intrinsic phototransduction tuning and its inhibition by horizontal cells via mechanisms analogous to the center–surround receptive field construction discussed earlier (Figure 4-23). Specifically, when a retinal field containing a mosaic of S-, M-, and L-cones (**Figure 4-34**A) receives pure short-wavelength (blue) light input, S-cones hyperpolarize, which reduces their glutamate release. Furthermore, horizontal cells are activated because M- and L-cones, which constitute the vast majority of cones (Section 4.10), are depolarized when they receive minimal light input; activation of horizontal cells further inhibits glutamate release from S-cones. As long-wavelength light (green, red, or a mixture, which produces yellow) is added to the short-wavelength light input, horizontal cells become less excited; S-cones are disinhibited and release more glutamate, even though their phototransduction status has not changed. When the retinal field receives only long-wavelength light, S-cones release much glutamate because of their depolarized status and the disinhibition from horizontal cells. Thus, the output of S-cones is influenced by short-wavelength (blue) and long-wavelength (yellow) light in an antagonistic manner, or is blue–yellow color opponent (Figure 4-34A).

The S-cone output is received by a specialized type of **blue-ON bipolar cell**, which selectively connects with S-cones (Figure 4-34B). These blue-ON bipolars inherit the color opponency of the S-cones, except that they are excited by blue light and inhibited by yellow light because of the sign inversion at the cone → ON bipolar synapse (Figure 4-22). A major target of the blue–yellow color opponent system is the blue-ON/yellow-OFF **small bistratified RGCs**, which have ON-layer dendrites that receive input from blue-ON bipolars, and OFF-layer dendrites that receive input from OFF bipolars. Thus, the small bistratified RGCs are excited by blue light and inhibited by yellow light based on their input from blue-ON bipolars alone. This color opponency is further enhanced by input from OFF bipolars, which carry mostly additional yellow signals from M-/L-cones (Figure 4-34B).

Physiological studies have also identified blue-OFF/yellow-ON RGCs. Recent studies have suggested that they too derive their color opponency from the blue-ON bipolars, except that the connection between bipolars and RGCs involves an intermediate glycinergic inhibitory amacrine cell, which inverts the sign of the signal.

The green–red color-opponent signal in trichromatic primate retina is read out by **midget ganglion cells**. In the fovea, which contains the highest cone density, each midget RGC is excited by a single midget bipolar cell, which in turn receives

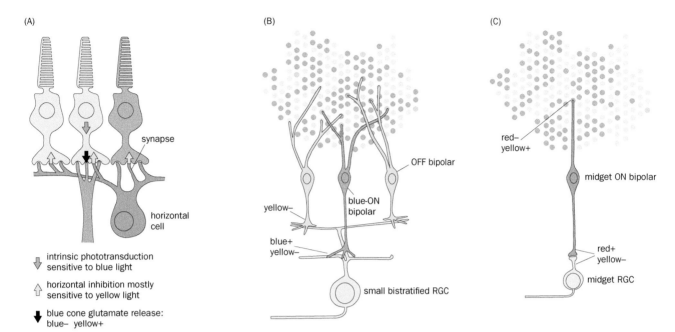

Figure 4-34 The blue–yellow and green–red color-opponent systems.
(A) S-cone output is color opponent due to a combination of its intrinsic phototransduction and horizontal cell-mediated lateral inhibition. It intrinsically responds to short-wavelength (blue) light. Because the surround consists mostly of M- and L-cones, the inhibitory signal from horizontal cells carries a long-wavelength (yellow) signal, and hence S-cone output is designated as blue– yellow+ with regard to whether light increases (+) or decreases (–) its glutamate release. **(B)** The retinal circuit transmits information from S-cones to blue-ON/yellow-OFF small bistratified RGC. Blue-ON bipolars collect information only from S-cones. They send excitatory input to small bistratified retinal ganglion cells, and their output is blue+ yellow–

as they invert the sign of S-cones. The small bistratified RGC also receives excitatory input from OFF bipolars, which carry additional yellow– signal. **(C)** An example of the green–red color-opponent system in the trichromat fovea. The L-cone output is red– yellow+. The midget ON bipolar inverts the sign, such that the midget RGC is red+ yellow–; thus the long-wavelength (red) light elicits a net excitation and medium-wavelength (green) light elicits an inhibition onto this midget RGC. (Adapted from Mollon JD [1999] *Proc Natl Acad Sci U S A* 96:4743–4745, based on data from Dacey DM & Lee BB [1994] *Nature* 367:731–735; Verweij J, Hornstein EP, & Schnapf JL [2003] *J Neurosci* 23:10249–10257; Packer OS, Verweij J, Li PH, et al. [2010] *J Neurosci* 30:568–572.)

input from a single M- or L-cone (Figure 4-34C). Just like the S-cones (Figure 4-34A), the M- and L-cones already produce color-opponent output as a consequence of their phototransduction cascades, which preferentially report green or red light stimuli, and inhibition from horizontal cells encoding a mixture of mostly green and red stimuli. Thus, the green–red color opponent system, at least in the fovea, does not require further specialized retinal circuitry or cell types: bipolars and RGCs simply inherit the color-opponency of the cone output made possible by horizontal cell-mediated lateral inhibition.

4.18 The same retinal cells and circuits can be used for different purposes

The last two sections may have created the impression that specific retinal neurons and circuits are dedicated to specialized functions, such as analyzing motion or color signals. However, ample evidence indicates that the same retinal cells and circuits can serve distinct functions under different stimulus conditions. For example, none of the color-opponent RGCs we just discussed are dedicated exclusively to color vision. The green–red opponent system is also used for high-acuity vision in the center of the retina: the midget RGC in the fovea receives information from a single cone, endowing these RGCs exquisite spatial resolution (Figure 4-34C). Trichromatic primates created another use—green–red opponency—for this existing system after the introduction of a new type of cone during their evolution.

Even the more ancient blue-ON/yellow-OFF opponent RGC system has other functions. In the dark, blue-ON/yellow-OFF RGCs receive signals from rods. Indeed, even though rods are more numerous than cones, they do not have dedicated RGCs. They send signals mostly through rod bipolar cells, which are ON

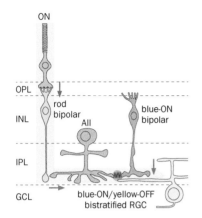

Figure 4-35 Rods piggyback on the cone circuit. Rod signals are transmitted by a single type of rod ON bipolar cell. When light is directed onto the rod, the rod becomes hyperpolarized. The sign-inverting rod → rod bipolar synapses (red arrow) depolarize the rod bipolar, which excites the AII amacrine cell through excitatory synapses (horizontal green arrow). The AII amacrine cell depolarization spreads to the blue-ON bipolar via gap junctions (symbolized by the zigzagged line). Depolarization of the blue-ON bipolar excites ON-center RGCs, including the bistratified RGC, through excitatory synapses (vertical green arrow). Thus, light signals through the rod to excite the blue-ON/yellow-OFF bistratified RGC. (Adapted from Kolb H & Famiglietti EV [1974] *Science* 186:47–49. See also Field GD, Greshner M, Gauthier JL, et al [2009] *Nat Neurosci* 12:1159–1166.)

bipolars. Rod bipolars excite **AII amacrine cells**, whose depolarization spreads to blue-ON bipolar cells via gap junctions. Light signals from rods thus excite the blue-ON/yellow-OFF RGCs we just studied (**Figure 4-35**). This pathway explains a psychophysical observation: in dim light, visual scenes have a bluish hue.

In summary, retinal circuits and their output RGCs separate visual information into parallel streams. At the same time, their multifarious pathways deliver different forms of information to the brain under different light conditions, highlighting the multifunctionality and efficiency of retinal cells and circuits. We discuss another fascinating example of the multifunctionality of retinal cells in the final section on the retina.

4.19 Intrinsically photosensitive retinal ganglion cells have diverse functions

Intrinsically photosensitive retinal ganglion cells (**ipRGCs**) are an extreme example of the multifunctionality of retinal cells and circuits. Like other RGCs, ipRGCs extend dendrites into the inner plexiform layer and receive input from rods and cones via bipolar cells. They also send axons to common targets of many RGC types, such as the lateral geniculate nucleus of the thalamus, which relays information to the visual cortex (see Section 4.20). Thus, ipRGCs participate in aspects of the retina's image analysis function just like other RGCs.

What distinguishes ipRGCs from all other RGCs is that they are directly activated by light. In other words, ipRGCs are also *photoreceptors*. This remarkable discovery was made in 2002 from several convergent lines of evidence. In addition to forming images, light also plays an important role in entraining the circadian clock via the RGC axon projection to the **suprachiasmatic nucleus** (**SCN**) of the hypothalamus, a subject that will be discussed in more detail in Chapter 9. While the eye is essential for this function, blind mice in which rods and cones are degenerated can still sense light and entrain their clock. It was discovered first in frogs and then in mice that some RGCs express a special G-protein-coupled receptor called **melanopsin** (**Figure 4-36A**), which resembles opsins used in invertebrate vision (see Section 13.14 for details). Using mice genetically engineered to express an axon label specifically in melanopsin-expressing RGCs, researchers found that their axon terminals were highly enriched in the SCN (Figure 4-36B). Finally, both melanopsin-expressing RGCs and RGCs that project axons to the SCN were

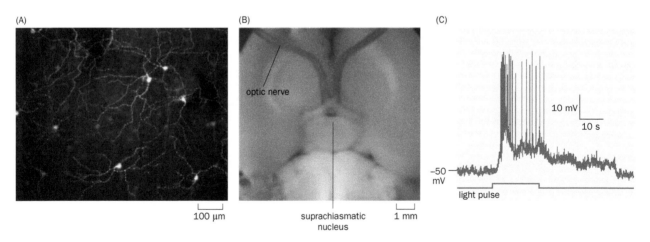

Figure 4-36 Intrinsically photosensitive retinal ganglion cells (ipRGCs). **(A)** Mouse retina stained with an antibody against melanopsin, revealing its distribution throughout the dendritic trees of ipRGCs. **(B)** The optic nerve is visualized by the activity of β-galactosidase (blue), which labels the axonal projections of ipRGCs and is expressed from a transgene knocked into the melanopsin locus (see Section 14.7 for the knock-in method). Melanopsin-expressing ipRGCs project their axons into the suprachiasmatic nucleus of the hypothalamus (SCN). **(C)** Whole-cell recording from an ipRGC retrogradely labeled by injecting fluorescent beads into the SCN. A light flash (bottom trace) depolarizes the ipRGC, causing it to fire action potentials (top trace). This response persisted after synaptic transmission was blocked (not shown), indicating that ipRGCs are directly depolarized by light. (A, from Provencio I, Rollag MD, & Castrucci A [2002] *Nature* 415: 493–494. With permission from Springer Nature. B, from Hattar S, Kumar M, Park A, et al. [2002] *Science* 295:1065–1070. With permission from AAAS. C, adapted from Berson DM, Dunn FA, & Takao M [2002] *Science* 295:1070–1073. With permission from AAAS.)

strongly depolarized by light (Figure 4-36C), even when transmission from rods and cones was blocked completely, indicating that melanopsin-expressing RGCs are intrinsically photosensitive. Subsequent analyses of melanopsin knockout mice confirmed the function of ipRGCs in circadian clock entrainment.

Further studies revealed that ipRGCs are less sensitive than rods and cones to light because melanopsin density is much lower than rhodopsin and cone opsin density, but ipRGCs have a slower phototransduction cascade and therefore integrate light signals over a longer time. There are multiple subtypes of ipRGCs that express melanopsin at different levels and serve diverse functions, ranging from a pupillary light reflex that enables automatic adjustment of pupil size in response to different light intensities to entrainment of circadian rhythms and mood regulation. Researchers are still discovering new ways in which ipRGCs impact the visual system.

HOW IS VISUAL INFORMATION PROCESSED BEYOND THE RETINA?

We now travel along the visual pathway from the eye to the brain. Axons of RGCs form the optic nerves, which carry visual information from the retina into the brain (**Figure 4-37**A). At the **optic chiasm**, where the optic nerves from the left and right eyes converge, axons of RGCs on the **nasal** side (close to the nose) project **contralaterally**, which refers to an axonal projection that crosses the midline and therefore terminates on the opposite side as the soma. Axons on the **temporal** side (close to the temple) project **ipsilaterally**, which refers to an axonal projection that stays on the same side as the soma. (As will be discussed in Section 5.6, the fractions of RGC axons that cross the midline differ between species.) After passing the optic chiasm, RGC axons continue to travel along the **optic tract**, while sending off subpopulations that terminate at several distinct central targets that serve diverse functions. These include the suprachiasmatic nucleus, which regulates circadian rhythms; the **pretectum** in the brainstem, which regulates pupil and lens reflexes and reflexive eye movements; and the **superior colliculus**, which regulates head orientation and eye movements. We will focus our discussion on one important central target, the **lateral geniculate nucleus** (**LGN**) of the thalamus, which acts as a processing station for information transmitted to the **visual cortex**. (Note that the LGN has a dorsal and a ventral division. Only the dorsal LGN relays information to the visual cortex, and that is what we refer to throughout the chapter.)

4.20 Retinal information is topographically represented in the lateral geniculate nucleus and visual cortex

RGC axons form synapses with LGN neurons, which send their own axons into the visual cortex (Figure 4-37). These axonal projections are organized topographically,

Figure 4-37 Visual pathways from the eye to the brain. (A) The human visual system as seen from the bottom side of the brain. Axons of retinal ganglion cells (red line) collectively form the optic nerve, which travels past the optic chiasm and continues along the optic tract to distinct targets in the brain. These include the suprachiasmatic nucleus (SCN) of the hypothalamus, the pretectum, the superior colliculus, and the lateral geniculate nucleus (LGN) of the thalamus, which relays information to the visual cortex. For clarity, only RGC axons from the eye on the left side are depicted; a subset of those that stay on the left side of the brain after the optic chiasm are incompletely depicted. Gray lines indicate the axons of LGN neurons forming the optic radiation and projecting to the primary visual cortex (V1). **(B)** Illustration of how visual information from two eyes is organized. RGCs from the left and right eyes looking at the same point in space project their axons to similar locations but different layers of the LGN (illustrated are two layers separated by a dashed line). Visual information from the two eyes converges in cortical neurons, either through directly convergent LGN axonal input from both eyes to the same cortical neuron (purple), or convergent input from cortical neurons (red and blue) that receive input from only one eye.

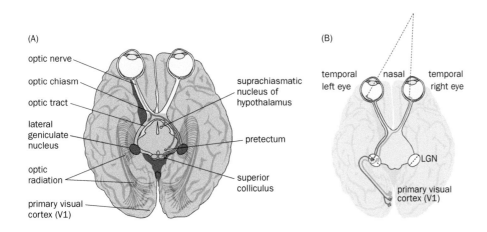

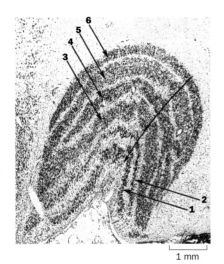

1 mm

Figure 4-38 Cell layers in the monkey lateral geniculate nucleus. Nissl staining (see Section 14.15 for details), which labels cell bodies, shows that the monkey LGN is organized into six layers. Layers 1 and 2 have larger cell bodies and are called the magnocellular layers; layers 3–6 have smaller cell bodies and are called the parvocellular layers. Layers 1, 4, and 6 receive contralateral input, while layers 2, 3, and 5 receive ipsilateral input. The six maps of the visual field are in precise register; for example, cells in different layers along the unnumbered arrow correspond to the same space in the visual field. (From Hubel DH & Wiesel TN [1977] *Proc R Soc Lond B* 198:1–59. With permission from the Royal Society.)

such that nearby RGCs in the retina project to nearby LGN neurons, which in turn connect with nearby cells in the **primary visual cortex** (also called **V1**, or striate cortex). **Retinotopy**—the topographical arrangement of cells in the visual pathway according to the position of the RGCs transmitting signals to them—enables spatial information from the retina to be faithfully represented in the brain and is a fundamental organizational principle of the visual system. We will study how this is achieved in great detail in Chapter 5.

Since nasal RGC axons project contralaterally and temporal RGC axons project ipsilaterally, the left LGN receives input from the temporal side of the left retina and nasal side of the right retina, both of which correspond to the right half of the visual field (Figure 4-37B). Thus, the LGN and visual cortex in the left half of the brain receive input from the right visual field, whereas the LGN and visual cortex in the right half of the brain receive input from the left visual field.

In addition to retinotopy, the primate LGN is organized into six layers, which can be visualized by density of cell staining (Figure 4-38). Layers 1 and 2 have larger cell bodies (termed *magnocellular layers*) and receive input from RGCs with large receptive fields. Layers 3–6 have smaller cell bodies (termed *parvocellular layers*) and receive input from RGCs with small receptive fields. In addition, each layer receives input primarily from one specific eye: layers 1, 4, and 6 receive contralateral input, whereas layers 2, 3, and 5 receive ipsilateral input. Thus, LGN layers help segregate inputs of RGC axons according to their cell types and eye of origin. In Section 4.26 we will discuss the functional differences between magnocellular and parvocellular layers. In other mammals with fewer LGN layers (such as cats and ferrets) or no anatomically apparent layers (such as rodents), inputs from RGCs representing different cell types and eye of origin are nevertheless segregated in LGNs, suggesting that this is a general organizational principle. Generally speaking, individual LGN neurons represent information received from one eye; an exception to this rule has recently been found in mouse LGN neurons. Information from the left and right eyes is integrated in the primary visual cortex (Figure 4-37B).

In primates, the representation of the fovea is highly expanded in V1; as the eccentricity increases (that is, receptive fields move toward the periphery; Figure 4-14), the corresponding area in V1 for the unit area of the retina decreases (Figure 4-39). Thus, as in the sensory and motor homunculi discussed in Section

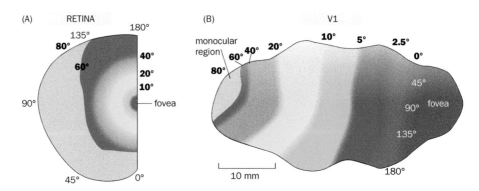

Figure 4-39 Topographic representation of the retina in the primary visual cortex (V1) of the macaque monkey. (A) The retina is diagrammed in a polar coordinate system, with the radius representing eccentricity (in bold) and circumference representing the dorsoventral position (0° being most ventral). **(B)** V1 is diagrammed in a Cartesian coordinate system, with the *x* axis representing eccentricity (bold) and the *y* axis representing the dorsoventral position. Colors in the V1 map correspond to those in the retina map. For instance, representation of the fovea (purple), which occupies a small area of the retina, is greatly enlarged in V1. As eccentricity increases, the corresponding area in V1 per unit area of the retina decreases. Gray areas represent monocular regions, where only the left eye (shown) receives visual stimuli. The retinotopy of V1 was determined by extracellular single-unit recording of V1 neurons in anesthetized monkeys in response to stimuli targeted to different parts of the retina. The recording sites were determined by histology, and V1 is represented as a two-dimensional map that maintains the relative positioning of the three-dimensional cortex. (Adapted from Van Essen DC, Newsome WT, & Maunsell JHR [1984] *Vis Res* 24:429–448. With permission from Elsevier Inc.)

1.11, the visual cortex does not represent the retina proportionally. More cortical resources are devoted to the central part of the visual field.

4.21 Receptive fields of LGN neurons are similar to those of RGCs

What visual stimuli best excite LGN neurons? David Hubel and Torsten Wiesel investigated this question using the same single-unit recording technique to monitor neurons from the LGN that their mentor Stephen Kuffler had applied to the study of RGCs in the retina. To precisely control visual stimuli, they anesthetized their subject (a cat) to minimize eye movement, so they could systematically map the receptive field by altering the locations of visual stimuli on a screen facing the anesthetized cat. LGN neurons were found to have similar response properties as RGCs. They have center–surround concentric receptive fields, with ON-center/OFF-surround receptive fields, or the reverse (Figure 4-21). Thus, at a gross level, LGN neurons appear to have similar receptive field organization as RGCs.

LGN neurons likely function much more than as a simple relay station from the eye to the visual cortex. For example, using virus-mediated retrograde **trans-synaptic tracing** (see Section 14.19 for details), researchers have recently determined how individual LGN neurons integrate information from their presynaptic partner RGCs. While some LGN neurons receive direct input from just one RGC type, consistent with them serving as a relay, other LGN neurons receive direct input from multiple RGC types (**Figure 4-40**). Some LGNs even receive binocular RGC inputs. Furthermore, electron microscopy studies revealed that only a small fraction of synapses onto LGN neurons come from RGCs; the remaining synapses are from cortical or brainstem neurons, or inhibitory neurons in the thalamus. The activity of LGN neurons is modulated by cortical feedback and brain state, enabling LGN control over information flow into the cortex. The thalamus also plays a role in multisensory integration, as all sensory systems, except olfaction, access the cortex via the thalamus. Much work is still needed to understand these functions of the thalamus.

4.22 Primary visual cortical neurons respond to lines and edges

Although the receptive fields of LGN neurons resemble those of RGCs, Hubel and Wiesel encountered something very different when they applied single-unit recordings to the primary visual cortex (V1), where LGN axons terminate. **Figure 4-41** and **Movie 4-2** illustrate the procedure for mapping the receptive field of a V1 neuron. After the microelectrode was inserted into the cortex, visual stimuli were projected onto a screen representing the entire visual field of the cat to identify an area that could excite a cortical neuron next to the electrode tip. Then, the shape, size, and position of the stimulus were systematically varied with the aim of finding the stimuli that most effectively changed the firing pattern of the neuron being recorded. Circular spots of light were no longer the best stimuli no matter their size or location (Figure 4-41A). The best stimulus was a bar of light with a specific orientation—for example, the neuron being tested in Figure 4-41 responded best to a vertical bar (Figure 4-41B). Once the bar reached a certain length, further lengthening had no additional effects, but widening the bar in either direction reduced the firing rate. Thus, the receptive field of this cell can be drawn as shown in Figure 4-41C, with an ON-center region, flanked by two OFF-surround regions. Cells with such a receptive field in V1 are called **simple cells**.

The OFF regions on either side of the ON region are not necessarily symmetric; in fact, some cells have only one OFF-surround region. Despite these variations, the receptive fields of simple cells share certain properties: (1) they are spatially restricted to a narrow area of the visual field, (2) they are best stimulated by bars of light, (3) they are highly sensitive to orientation, and (4) they have ON–OFF antagonism. Diffuse light covering both the ON and OFF regions does not excite these cells. Simple cells can therefore be considered line or edge detectors in specific regions of the visual field.

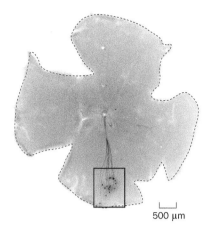

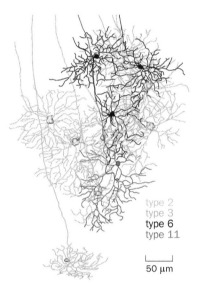

500 μm

type 2
type 3
type 6
type 11

50 μm

Figure 4-40 Thalamic integration of RGC inputs. Top, a mouse retina that has been cut and flat mounted, revealing a group of RGCs that provide direct input to a single LGN neuron, revealed by rabies virus–mediated trans-synaptic tracing (Figure 14-33). The boxed area is magnified in the bottom panel, where dendritic morphology is depicted. RGCs are color coded by type, as determined by dendritic morphology and laminar targeting within the inner plexiform layer. (Adapted from Rompani SB, Mullner FE, Wanner A, et al. [2017] *Neuron* 93:767–776. With permission from Elsevier Inc.)

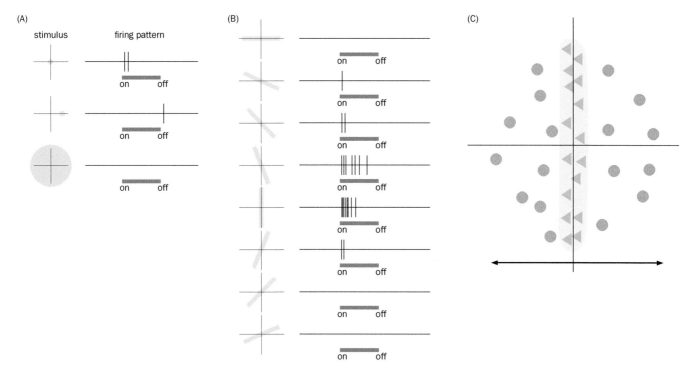

Figure 4-41 Receptive field of a simple cell in the cat primary visual cortex (V1). (A) In response to the visual stimuli on the left, the vertical bars in the plots on the right represent action potentials recorded by extracellular electrodes; stimulation durations (1 s) are represented as horizontal bars below the action potential plots. Circular spots of various sizes elicit minimal responses from this V1 simple cell. **(B)** Simple cells are most responsive to a bar of light with a specific orientation. In this example, a bar of light oriented vertically elicits the most intense burst of action potential firing during the stimulus period. Response plots are formatted as in Panel A. **(C)** The receptive field of the V1 simple cell in (B). Triangles represent excitatory (ON) responses, and circles represent inhibitory (OFF) responses. (Adapted from Hubel DH & Wiesel TN [1959] *J Physiol* 148:574–591.)

A different class of cells, also encountered in V1, is called the **complex cell**. These cells have more complex and more variable receptive fields than simple cells (see **Figure 4-42** for an example). In general, complex cells do not have mutually antagonistic ON and OFF regions as do simple cells. But like simple cells, complex cells are also highly orientation selective, responding to light bars on a dark background or dark bars on an illuminated background. Complex cells are excited by bars with a specific orientation falling on any part of their relatively large receptive fields (compared to single cells). Thus, complex cells can be considered more abstract line/edge detectors than simple cells, as they generalize over space, responding to the line/edge ON or OFF anywhere in their receptive field.

A fraction of both simple and complex cells is also highly sensitive to movement of a bar with a specific orientation. Some cells are direction selective: they are activated more by the bar moving in one direction than in the opposite direction. Recent studies in mice suggest that while some V1 cells inherit their direction selectivity from direction-selective RGCs (Section 4.16), the direction selectivity of most V1 cells is produced *de novo*.

4.23 How do visual cortical neurons acquire their receptive fields?

Hubel and Wiesel proposed that the receptive field of a simple cell can form by combining the receptive fields of a series of LGN neurons arranged in a line as a consequence of their specific connection patterns (**Figure 4-43**A). Likewise, the receptive field of a complex cell can be constructed by combining a series of simple cells located in adjacent areas with the same orientation (Figure 4-43B). These proposals have been called a **feedforward** model, as the receptive field of a neuron is shaped primarily by neurons at an earlier visual processing stage. Accord-

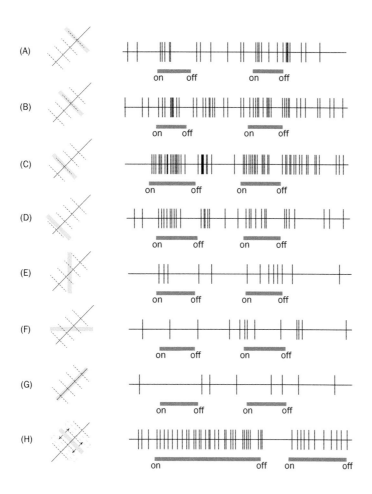

Figure 4-42 **Receptive field of a complex cell.** This complex cell is best excited by a bar of light oriented at a 135° angle. It is not sensitive to the exact position of the bar within a relatively broad area **(A–D)**, is highly orientation selective **(E–G)**, and responds to motion of the 135°-bar within the receptive field **(H)**. Horizontal bars below each of the firing pattern plots indicate the periods of stimulation. (Adapted from Hubel DH & Wiesel TN [1962] *J Physiol* 160:106–154.)

ingly, receptive fields of neurons along the visual pathway are organized in a hierarchy, from light intensity detection (photoreceptors), to contrast detection (RGC, LGN), to line and edge detection (simple and complex cells in V1). Through each transformation, light information comes closer to the ultimate purpose of vision we emphasized at the beginning of the chapter: extraction of features used to guide animal behaviors essential for survival. Individual V1 neurons, which encode lines and edges, can thus send their information to multiple higher visual cortical areas for specialized functions such as object recognition and motion detection, as will be discussed later.

More than half a century since these classic studies were conducted, the neural circuit basis of these feedforward models in the visual cortex is still not fully elucidated, although experimental data are consistent with feedforward projections

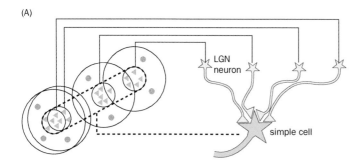

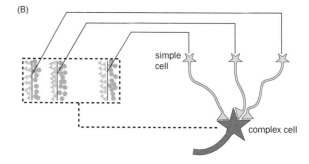

Figure 4-43 **Models of receptive field construction for simple and complex cells. (A)** The receptive field of a simple cell can arise from the convergent input of a series of LGN neurons in which several circular receptive fields in combination form a line. **(B)** The receptive field of a complex cell can arise from convergent input of a series of spatially aligned simple cells with the same orientation selectivity. (From Hubel DH & Wiesel TN [1962] *J Physiol* 160:106–154. With permission from John Wiley & Sons Inc.)

Figure 4-44 Evidence supporting the simple cell receptive field model.
(A) Summary of receptive fields of 23 functionally connected pairs of LGN neurons and V1 simple cells in the cat. (Functional connections were determined by cross correlations of their firing patterns in response to visual stimuli.) The receptive field for each simple cell was transformed into an idealized receptive field, with center (red) flanked by two surrounds (blue). The receptive field center of each LGN neuron (circle) was shown in relation to the idealized simple cell, with diameter representing the receptive field size, thickness representing the strength of correlation with the partner simple cell, and color representing sign (excitatory or inhibitory) relative to the V1 center (red, same sign; blue, opposite sign). **(B)** Top, schematic of the experiment. V1 neurons of anesthetized mice were recorded via whole-cell patch clamp to measure excitatory input in response to a moving grating. Cortical input was silenced using optogenetics (blue light beam; see Box 4-4 and Section 14-25 for details), allowing researchers to specifically measure the thalamic input. Bottom, thalamic excitatory postsynaptic current (EPSC_thal) in response to a grating cycle is orientation sensitive (maximal response at 150°). Blue, cycle average traces; green, best fitting sinusoids. (A, adapted from Reid RC & Alonso JM [1995] *Nature* 378:281–284. With permission from Springer Nature. B, adapted from Lien AD & Scanziani M [2013] *Nat Neurosci* 9:1315–1323.)

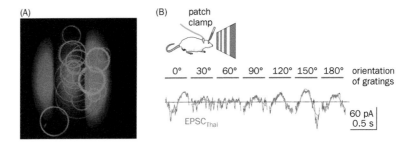

of LGN neurons contributing to receptive field properties of simple cells. For example, paired recordings of LGN neurons and V1 simple cells in the cat revealed that receptive fields of LGN neurons map onto synaptically connected simple cells, as predicted by the Hubel–Wiesel model (**Figure 4-44**A). Moreover, patch clamp recordings of mouse V1 neurons while silencing cortical inputs revealed that thalamic input to individual V1 neurons was already orientation selective (Figure 4-44B). Because responses of individual thalamic neurons are not orientation selective, this finding suggests that V1 neurons receive input from LGN neurons whose receptive fields are aligned along the direction of the tuned orientation, again supporting the Hubel–Wiesel model (Figure 4-43A). The thalamic input is likely amplified by intracortical excitatory networks. Local inhibitory networks and feedback from higher visual cortical areas also help sculpt the receptive fields of V1 neurons. Further understanding of the neural circuit basis of cortical receptive fields will require a combination of physiology and circuit tracing, exemplified by the study of direction-selective RGCs discussed in Section 4.16. Regardless of the final answers, it is clear that the responses of cortical neurons must reflect the physiological characteristics of individual neurons and their precise wiring patterns.

4.24 Cells with similar properties are organized in vertical columns in visual cortices of large mammals

V1 is highly organized based on two basic principles. The first is retinotopy: specific parts of the visual cortex receive information from specific parts of the retina and therefore represent specific parts of the visual field (Figure 4-39). The second is so-called **functional architecture**: within a given retinotopic area, neurons that share similar properties are further organized into distinct modules along the cortical surface.

When Hubel and Wiesel performed their single-unit recordings, they recorded not only the visual responses of individual cells but also their positions in the cortex. In a typical experiment, they inserted an electrode from the cortical surface until they found a responding cell, measured its receptive field and orientation selectivity, then advanced the electrode farther until they found and recorded another cell, and so on. At the end of the recording, they passed current to create a lesion marking the end of the electrode and then examined the brain in histological sections (**Figure 4-45**A). Plotting the orientation selectivity of individual cells along the electrode path revealed a remarkable organization. If the electrode was inserted perpendicular to the cortical surface, then cells exhibited similar orientation selectivities at different depths of the cortex. However, if the electrode was inserted at an oblique angle to the cortical surface, then the orientation selectivity of recorded cells changed gradually (Figure 4-45A). Together with earlier findings from the somatosensory cortex, these studies suggest that the cerebral cortex is organized into vertical columns from the surface to the white matter underneath, and cells within the same vertical column share similar properties.

Modern imaging experiments have greatly improved the resolution at which orientation columns can be observed. Whereas single-unit recordings can target only one cell a time, optical imaging can record activities of many neurons simultaneously (see Section 14.22 for details). For example, **intrinsic signal imaging** utilizes metabolic activities, including blood flow and oxygenation levels, as indi-

(A)

brain surface

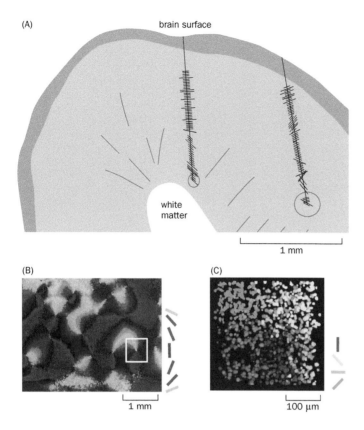

white matter

1 mm

(B)

1 mm

(C)

100 μm

Figure 4-45 Orientation columns in the cat visual cortex revealed by three different methods. (A) Two examples of single-unit recordings followed by histological tracing of the electrode paths. Lines intersecting the electrode paths represent the preferred orientation, with longer lines representing individual cortical cells and shorter lines representing regions with unresolved background activity. Circles at the end represent lesion sites. The left electrode penetration path is more perpendicular to the cortical surface, and the orientation preferences change little between different cells along the path. The right electrode penetration path is more oblique, and orientation preferences change markedly along its path. **(B)** Surface view of orientation columns visualized by intrinsic signal imaging. Visual stimuli with different orientations (color-coded at right) resulted in activation of different cell patches superimposed onto the same image. The white square highlights a pinwheel, with the center of the square representing the center of the pinwheel. **(C)** Surface view of orientation columns near the pinwheel center visualized by two-photon Ca^{2+} imaging. Visual stimuli using bars with different orientations (right) increased fluorescence of a Ca^{2+} indicator, which enabled individual cells with specific orientation selectivity to be mapped. This image is an overlay of cells recorded 130–290 μm from the surface, corresponding to layers 2 and 3 of V1; the significance of these layers is discussed further in Section 4.25. (A, adapted from Hubel DH & Wiese TN [1962] *J Physiol* 160:106–154. With permission from John Wiley & Sons Inc. B, from Bonhoeffer T & Grinvald A [1991] *Nature* 353:429–431. With permission from Springer Nature. C, from Ohki K, Chung S, Kara P et al. [2006] *Nature* 442:925–928. With permission from Springer Nature.)

cators of neuronal activity. When applied to the visual cortex in response to stimuli of different orientations, intrinsic signal imaging revealed that orientation-selective columns are organized in a pinwheel-like structure in cats and monkeys. The transitions between different orientations are continuous, except at the centers of pinwheels, where cells with different orientations converge (Figure 4-45B). Two-photon fluorescence imaging of intracellular Ca^{2+} concentration changes, which permits neuronal activity to be recorded at single-cell resolution, further demonstrated that the orientation columns and pinwheel centers are highly ordered at the level of individual neurons (Figure 4-45C). The significance of this pinwheel organization and how such a precise organization is constructed are not yet understood.

Columnar organization was also found when cortical neuron responses to the left and right eyes were examined in the primary visual cortex. Whereas most LGN neurons respond to stimuli from one eye, individual visual cortical neurons can respond to input from both eyes that corresponds to the same spatial position (Figure 4-35B), creating **binocular vision**, which is important for depth perception. However, many cortical neurons in the input layer 4 (see next section) still have a strong preference for input from either the left or right eye, a property called **ocular dominance**. As will be discussed in greater detail in Chapter 5, cells that respond preferentially to each eye are organized in a columnar fashion (Figure 5-18).

Not all mammals exhibit anatomically recognizable orientation or ocular dominance columns within a retinotopic field. In rodents, many individual cells are orientation selective, but no columnar organization is discernible: neurons with the same orientation selectivity do not cluster, and neighboring neurons often have distinct orientation selectivity (compare **Figure 4-46**A with Figure 4-45C). However, when researchers combined *in vivo* Ca^{2+} imaging of mouse V1 neurons with subsequent paired patch-clamp recording in slices of the same neurons to measure their connectivity, they found that neurons exhibiting similar orientation selectivity were more likely to be connected to each other (Figure 4-46B). Furthermore, when presented with natural stimuli, V1 neurons that had higher

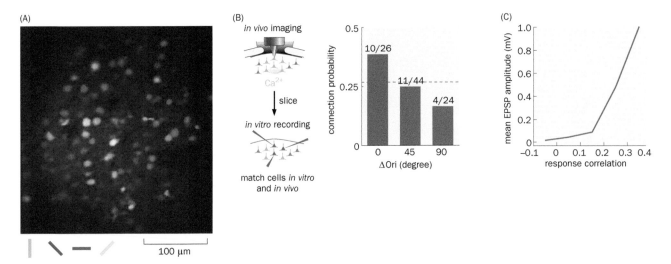

Figure 4-46 Organization of orientation-selective neurons in rodent V1. (A) Two-photon Ca²⁺ imaging of V1 neurons in a rat in response to orientation-specific stimuli. Cells are color coded according to their orientation selectivities (below). In the rat, cells with the same orientation do not form organized columns but instead appear scattered. **(B)** Left, schematic of experimental procedures. Mouse V1 neurons were first recorded *in vivo* using two-photon Ca²⁺ imaging to determine their orientation selectivities or response properties to natural scenes. Then acute slices containing the same recorded neurons were prepared; cells recorded *in vivo* were identified in the slice, and simultaneous whole-cell recordings were performed to measure synaptic connectivity and strength (see Section 14.26 for details). Right, connection frequency (e.g., 10 out of 26 paired recordings for the first column) is plotted against difference in orientation selectivity (ΔOri). Neurons with similar orientation selectivities are more likely to be connected. **(C)** Mean excitatory postsynaptic potential (EPSP) amplitude of paired recordings (including unconnected pairs) in the slice is plotted against response correlations to a set of natural stimuli measured by *in vivo* Ca²⁺ imaging. EPSP amplitude was measured by evoking spikes in the presynaptic partner and measuring peak depolarization value in the postsynaptic partner. (A, from Ohki K, Bhung S, Ch'ng YH, et al. [2005] *Nature* 433:597–603. With permission from Springer Nature. B & C, from Ko H, Hofer SB, Pichler B, et al. [2011] *Nature* 473:87–91; and Cossell L, Iscaruso MF, Muir DR, et al. [2015] *Nature* 518:399–403. With permission from Springer Nature.)

correlations in their responses *in vivo* were also more strongly connected (Figure 4-46C).

A parsimonious explanation for the difference between V1 organization in the cat and monkey versus that in rodents is that neurons sharing similar response properties are preferentially connected with each other in all species, but only when the visual cortex is large does it become necessary to group together neurons of similar properties in anatomically distinct columns, so as to limit the length of the axons needed to connect them. We now discuss how signals flow between cortical neurons located at different depths from the surface.

4.25 Information generally flows from layer 4 to layers 2/3 and then to layers 5/6 in the neocortex

The mammalian **neocortex** (the largest part of the cerebral cortex, including the visual cortex) is typically divided into six layers based on cell density differences that can be visualized using histological stains (Figure 4-47A; see Section 14.15 for more details). Studies of the visual cortex using *in vivo* intracellular recordings (Section 14.21) have provided insight into how information flows between different layers of the neocortex. Intracellular recordings can not only map the receptive field as single-unit extracellular recordings do but can also fill the recorded neurons with histochemical tracers such that their dendritic and axonal projection patterns can be reconstructed. These experiments revealed that neurons located in different layers of the primary visual cortex have distinct dendritic and axonal projection patterns (Figure 4-47B), which can be used to construct a model of information flow (Figure 4-47C).

Let's start with the input to V1 from the LGN, whose axons predominantly terminate in layer 4 of V1. In primates, axons of different LGN neuron types terminate in different sublayers of layer 4 (axons of cells *a* and *b* in Figure 4-47B). Layer 4 cortical neurons (cells *c* and *d*) restrict their dendrites mostly to within layer 4 so as to receive the LGN input and send their axons predominately within their own

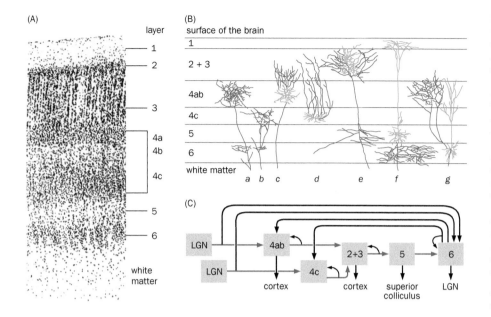

(A) layer 1 2 3 4a 4b 4c 5 6 white matter

(B) surface of the brain 1, 2 + 3, 4ab, 4c, 5, 6, white matter — a b c d e f g

(C) LGN → 4ab → 2+3 → 5 → 6 ; LGN → 4c ; cortex, cortex, superior colliculus, LGN

Figure 4-47 **Information flow in the primary visual cortex. (A)** Nissl staining of human V1 reveals that cortical neurons are organized in distinct layers based on cell size, shape, and density (see also Figure 14-22). **(B)** Representative axons of cat LGN neurons that provide input to V1 (*a, b*); and axonal (red) and dendritic (blue) morphologies of V1 neurons in layers 4ab (*c*), 4c (*d*), 2/3 (*e*), 5 (*f*), and 6 (*g*), respectively, revealed by tracer injected during intracellular recordings. Axons and dendrites were distinguished from each other according to their morphological characteristics. **(C)** Schematic representation of information flow, constructed using the assumption that overlapping axons and dendrites in the same cortical layer form synaptic connections. The cells depicted in Panel B correspond to the cells diagrammed immediately beneath them in Panel C. The feedforward information flow within V1 is highlighted in red. (A, adapted from Brodmann K [1909] *Vergleichende Localisationslehre der Grosshirnrinde in ihren Prinzipien dargestellt auf Grund des Zellenbaues.* Barth: Leipzig. B & C, from Gilbert CD [1983] *Ann Rev Neurosci* 6:217–247. With permission from Annual Reviews.)

layer and vertically to layers 2/3. Layer 2/3 neurons (cell *e*) send their axons within layers 2/3, to layer 5, and through the white matter to other visual cortical areas. Layer 5 neurons (cell *f*) project their axons to layer 6, to other cortical and subcortical areas such as the basal ganglia, and to the superior colliculus for cortical control of eye movement. Layer 6 neurons (cell *g*) send axons back to layer 4 and to the LGN, providing feedback control of LGN input. Thus, information flows generally from layer 4 to layer 2/3 and then to layer 5/6 within the cortex. Physiological data support this model of general information flow. For instance, simple cells are enriched in layer 4, consistent with them being direct recipients of LGN input. Complex cells are not usually found in layer 4, but are present in layers 2/3 and layers 5/6, consistent with them receiving input from simple cells.

It is useful to keep in mind that the circuit is, in reality, far more complex than the general LGN → L4 → L2/3 → L5/6 scheme, given the extensive interconnections between neurons in the same layer (recurrent connections; Box 1-2), the complex, multilayer-spanning dendritic trees of some neurons, and the participation of diverse inhibitory neurons. Nevertheless, information flow within an area of the neocortex, first outlined in studies of V1, appears to apply to other sensory cortical areas as well (Box 4-4). Identifying the principles of information processing in the cerebral cortex remains one of the most important challenges in modern neuroscience.

Box 4-4: Cracking neocortical microcircuits

We've learned that in the six-layered V1, signals follow a general pathway, from thalamus → L4 → L2/3 → L5/6. Layer 1 contains mostly apical dendrites, axons from other cortical areas, and some local GABAergic neurons. Since the six-layered organization applies to much of the mammalian neocortex, is this pattern of information flow a general property of the neocortex? Studies of other cortical areas have lent support to this proposal. For example, in the part of the mouse somatosensory cortex that represents whiskers (see Box 5-3 for more details), excitatory connections between neurons in different layers have been studied by paired whole-cell patch recording in brain slices. As summarized in Figure 4-48A, the signal flow, as determined by the direc-

tion and strength of synaptic connections between pair of neurons, also follows the general pathway of L4 → L2/3 → L5/6. A comparison of V1 and somatosensory cortex (Figure 4-48 and 4-47C) supports the notion of a "canonical microcircuit" that applies to information flow within a neocortical area. Depending on the specific input to and output from a given cortical area, there could be significant variations on this canonical microcircuit in different parts of the neocortex. For example, primary sensory cortices tend to have a well-elaborated layer 4, reflecting its role in receiving and processing sensory input from the thalamus. The primary motor cortex, on the other hand, has a less prominent layer 4 but a more prominent layer 5, reflecting the importance of
(Continued)

Box 4-4: continued

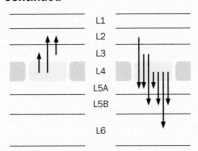

Figure 4-48 Connectivity between excitatory neurons of mouse primary somatosensory cortex. Arrows indicate the origin and destination of direct connections as determined by paired whole-cell patch recordings (see Section 14.26 for details). Rectangles in layer 4 represent anatomically distinct areas within the somatosensory cortex that represent individual whiskers. Layer 5 has been subdivided into L5A and L5B. (Adapted from Lefort S, Tomm C, Floyd Sarria JC, et al. [2009] *Neuron* 61:301–316. With permission from Elsevier Inc.)

its subcortical projections for motor control, which we will discuss in detail in Chapter 8.

In addition to excitatory glutamatergic neurons, which constitute about 80% of cortical neurons and have been the focus of our discussion so far, GABAergic neurons, which make up the remaining 20%, also play many essential functions in shaping cortical activity. Most cortical GABAergic neurons act locally: they receive local and long-range excitatory input and local inhibitory input from other GABAergic neurons and send inhibitory output to nearby glutamatergic and GABAergic neurons. Cortical GABAergic neurons can be subdivided into many types based on their morphologies, electrophysiological properties, layer and synapse distributions, and expression of molecular markers (Section 3.25). Recent advances in single-cell RNA-sequencing technology have begun to allow researchers to classify cortical neurons based on their transcriptomes (see Section 14.13 for details). In a study aimed at comparing cortical neurons in V1 and premotor cortex in the mouse, for example, researchers

identified ~50 GABAergic transcriptomic clusters common to V1 and premotor cortex; by contrast, the 20–30 glutamatergic clusters identified in each area differed in the finest branches of the classification, even though they could be similarly grouped based on layer and projection specificity (**Figure 4-49**). How well these transcriptomic clusters correspond to cell types defined by anatomy, physiology, and function is still a subject of investigation.

We use a specific example of investigating visual information processing in mouse V1 (**Figure 4-50**A) to integrate several concepts we have learned and to showcase modern techniques for mapping circuit function. A powerful approach for investigating the function of specific neuronal populations in circuit function is to selectively activate or silence them and observe the consequences. Researchers have established many transgenic mouse lines that express the Cre recombinase in specific neuronal populations. These Cre lines can be used in combination with virally mediated expression of Cre-dependent effector genes (see Sections 14.10, 14.12, and 14.23–14.25). Some of the most powerful effectors are microbial light-activated ion channels and pumps, which enable researchers to control neuronal activity by shining light directly on neurons that express these effectors. For example, blue light activates neurons expressing the effector **channelrhodopsin-2 (ChR2)**, a light-activated cation channel from green algae. Orange light silences neurons expressing the effectors **archaerhodopsin** and **halorhodopsin**, proton and chloride pumps from archaea, respectively. Genetically targeted expression of these effectors allows researchers to control activity of specific neuronal populations by light stimulation, a procedure called **optogenetics** (see Section 14.25 for details).

When V1 layer 6 thalamus-projecting pyramidal neurons engineered to express ChR2 were artificially activated by blue light, visual responses of V1 neurons from other layers were attenuated during the blue light stimulation period (Figure 4-50B, top). When layer 6 neurons engineered to express archaerhodopsin and halorhodopsin were silenced

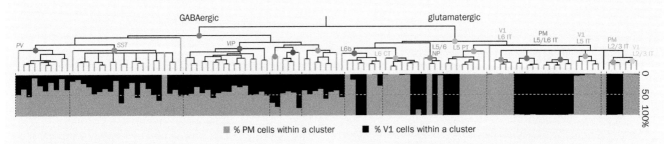

Figure 4-49 Classification of cortical cell types based on single-cell RNA-sequencing. Comparing single-cell RNA-seq data from primary visual cortex (V1) and premotor cortex (PM) of the mouse using hierarchical clustering reveals that GABAergic neuronal clusters in two areas are mostly indistinguishable. Each terminal branch (purple) has mixed contributions of PM and V1 (percentage quantified below). GABAergic neurons comprise several subclasses defined by molecular markers such as PV (pavalbumin+, which includes basket and chandelier cells; Figure 3-46), SST (somatostatin+, which includes Martinotti cells; Figure 3-46), and VIP (expressing

vasoactive intestinal peptide, whose main targets are other GABAergic neurons). Glutamatergic transcriptomic clusters from the two areas are distinct at the finest branch. Orange and blue terminal branches are made up of mostly PM and V1 cells, respectively. These glutamatergic neurons are nevertheless organized by layer (L) and projection types—IT, intracortical projecting; PT, subcortical projecting; CT, thalamus projecting; NP, nearby projecting. For example, L2/3 IT represents layer 2/3 intracortical projecting neurons. (Adapted from Tasic B, Yao Z, Graybuck LT, et al. [2018] *Nature* 563:72–78. With permission from Springer Nature.)

Box 4-4: continued

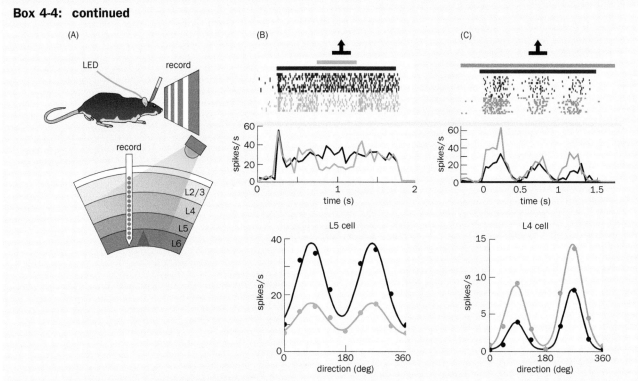

Figure 4-50 Effects of activating and silencing layer 6 pyramidal neurons on visual responses of V1 neurons. (A) Experimental setup. An anesthetized transgenic mouse expressing the Cre recombinase in a large subset of layer 6 pyramidal neurons that project to thalamus is given visual stimuli of drifting gratings on a screen. An extracellular multielectrode array is inserted into V1 to record visual responses of neurons in various layers (L2/3 to L6; each dot corresponds to a recording site). A light-emitting diode (LED) is used to photostimulate channelrhodopsin-2 (ChR2) with blue light, or, in separate experiments, a combination of archaerhodopsin and halorhodopsin with orange light. These optogenetic effectors are only expressed in Cre+ layer 6 pyramidal neurons (red triangle). **(B)** A layer 5 (L5) cell in mouse V1 is activated by an upward drifting grating (indicated by the arrow on the top). The black horizontal bar represents the visual stimulation period. Individual spikes are shown below the horizontal bar as vertical ticks; each row is a repetition. Rows in blue indicate trials in which a blue light stimulation was given (during the time indicated by the blue horizontal bar at the top);

blue light depolarizes L6 neurons expressing ChR2. The plot below quantifies the spike frequency. As is evident, blue light activation of L6 neurons decreases the spike frequency of this L5 cell. The bottom plot is a tuning curve of the L5 cell, obtained by stimulating the neuron with drifting gratings of different orientations (90° and 270° represent motion in the upward and downward directions, respectively). Thus, optogenetic activation of layer 6 neurons does not change the orientation selectivity of the L5 neurons but reduces its response magnitude. **(C)** Similar experiments as in Panel B, except that visual responses were recorded from a layer 4 (L4) cell and orange light was applied to suppress the activity, throughout the visual stimulation period, of L6 neurons expressing archaerhodopsin and halorhodopsin. Optogenetic silencing of L6 neurons does not change the orientation selectivity of the L4 neuron but enhances its response magnitude. (Adapted from Olsen SR, Bortone DS, Adesnik H, et al. [2012] *Nature* 483:47–54. With permission from Springer Nature.)

by orange light, visual responses of V1 neurons from other layers were enhanced (Figure 4-50C, top). Interestingly, the orientation selectivity of other cortical neurons was unaffected by the activity of layer 6 neurons, as seen in the similarities of their tuning curves despite the magnitude differences (Figure 4-50B, C, bottom panels). Thus, both loss-of-function (neuronal silencing) and gain-of-function (neuronal acti-

vation) experiments suggested that layer 6 modulates the rate at which the firing frequency of other cortical neurons increases in response to increasing excitatory input. This function is called **gain control** and is discussed further in Section 6.9. It remains to be determined whether layer 6 neurons in other cortical areas serve a similar function.

4.26 Visual information is processed in parallel streams

Vision consists of perception of different aspects of the visual world. As discussed in Section 4.15, visual information is processed in parallel in the retina. At the level of retinal output, an estimated 20 (for primates) or more than 30 (for mice) different RGC types, each having different morphology and connectivity within the retina, carry different kinds of processed information to the brain. Although some RGCs project their axons to the hypothalamus or superior colliculus for

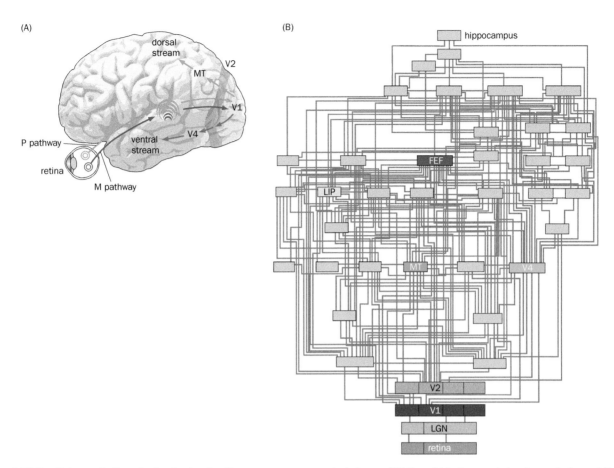

Figure 4-51 Parallel organization of primate visual pathways.
(A) Schematic of two major parallel pathways from the eye to the cortex. The P and M pathways start with retinal ganglion cells with small (P) and large (M) receptive fields and engage different layers of the LGN and different sublayers in layer 4 of V1. After V1, information diverges into the dorsal and ventral streams. The dorsal stream derives information mainly from the M pathway and is responsible for analyzing motion and depth. The ventral stream derives information from both the P and M pathways and analyzes shape and color. V2 (gray band), an area immediately adjacent to V1 that relays information to both ventral and dorsal streams; V4, a major analysis station in the ventral stream; MT, the middle temporal visual area that specializes in analyzing motion signals (see Section 4.28 for more details).
(B) Connection diagram of different visual cortical areas derived from anatomical tracing experiments in macaque monkeys. The >30 areas form >300 connections, mostly reciprocal, and establish highly intertwined cortical processing streams. The dorsal stream going to the parietal lobe is to the left, and the ventral stream going to the temporal lobe is to the right. FEF, frontal eye field; LIP, lateral intraparietal area (see Section 4.28 for more details). (B, from Felleman DJ & Van Essen DC [1991] *Cereb Cortex* 1:1–47. With permission from Oxford University Press.)

subconscious functions (Figure 4-37A), most RGC subtypes also send information via the LGN to the cortex, eventually giving rise to conscious perception of object color, shape, depth, and motion. How are these different streams of information organized?

Primate visual pathways from the retina to the cortex have traditionally been divided into two main components: the P and M pathways (**Figure 4-51**). The **P pathway** originates from RGCs that have small receptive fields (such as the midget cells), engages the parvocellular layers of the LGN, and carries information about high acuity and color vision. The **M pathway** originates from RGCs that have large receptive fields (such as the **parasol cells**), engages the magnocellular layers of the LGN, carries information about luminance, and has excellent contrast and temporal sensitivity; thus M signals are particularly appropriate for encoding motion. Recent discovery of new RGC types and their projections indicate that this two-pathway system is a gross simplification—each pathway likely contains many subpathways carrying distinct information. Nevertheless, the major division of labor between the M and P pathways illustrates the principle that visual information is processed along parallel paths from the retina to the cortex.

When these parallel information streams arrive at V1, they terminate in different sublayers of layer 4 (Figure 4-47). Information exiting V1 is again divided roughly into two major streams (Figure 4-51A). The **ventral stream** (often called the *what* stream) leads to the temporal cortex and is responsible for analyzing form and color, whereas the **dorsal stream** (the *where* stream) leads to the parietal cortex and is responsible for analyzing motion and depth. The M and P pathways each contribute to both dorsal and ventral streams, with the dorsal stream receiving input mostly from the M pathway and ventral stream receiving about equal amounts of input from the M and P pathways. Another view of the division of labor between the two streams emphasizes their functional outputs: the dorsal and ventral streams provide vision for action and vision for perception, respectively. These parallel streams are consistent with observations from human patients with cortical lesions: people with parietal lesions suffer from spatial neglect, whereas people with temporal lesions suffer from object agnosia.

The general direction of information flow among cortical layers within an area we discussed in Section 4.25 and Box 4-3 can be extended to infer information flow between different cortical areas. For example, when the connections of V1 and the adjacent V2 (secondary visual cortex) were examined by **anterograde** (from cell bodies to axon terminals) and **retrograde** (from axon terminals to cell bodies) tracing methods (see Section 14.18 for details), it was found that V1 → V2 axonal projections terminated mostly in layer 4 of V2, while V2 → V1 axonal projections terminated in superficial (including layer 1) and deep layers but avoided layer 4. Since visual information in the cortex originates mostly from V1 and spreads to higher cortical areas, these experiments suggest a general connection rule between intercortical areas: feedforward projections (e.g., V1 → V2) terminate in layer 4, whereas feedback projections (e.g., V2 → V1) avoid layer 4.

Analyses of extensive anatomical tracing data based on this rule have coarsely subdivided the visual cortex of the macaque monkey into more than 30 areas (Figure 4-51B). These cortical areas are extensively interconnected, often via reciprocal connections (such as V1 → V2 and V2 → V1). They nevertheless appear to follow a hierarchical organization, where visual information from V1 undergoes ~10 levels of processing through cortical areas with multiple distinct processing streams that are highly intertwined (Figure 4-51B).

The bewildering complexity of these connections raises important questions: at what level and in what form does visual perception arise? One can ask a more specific question about the parallel processing streams. Suppose that one sees a white cat chasing after a black mouse, which is itself moving toward a piece of yellow cheese. According to what we have learned, the motion, color, and form of these three objects are processed in separate streams going primarily to different cortical areas. How does one correctly associate *white* and *fast moving* with *cat* or *yellow* and *stationary* with *cheese*? This poses a binding problem: how does the visual system correctly link up (or bind) different features of the same objects?

At present we do not have a satisfying answer to this question. One property that can contribute to the solution to the binding problem is visual attention. **Attention** refers to the cognitive function in which a subset of behaviorally relevant sensory information is selected for further processing at the expense of irrelevant information. Electrophysiological recording experiments in awake, behaving monkeys have indicated that visual responses to an object being attended to are characterized by increased firing rates and stimulus selectivity and decreased trial-to-trial variability, compared to objects not being attended to. Another observation is that attention decreases correlations in trial-to-trial variability in simultaneously recorded neurons; this makes individual neurons more independent of each other in encoding the information and enhances the signal-to-noise ratio (see Section 14.31 and Box 14-5 for more detailed discussions). One area implicated in this attentional control is the **frontal eye field** (FEF in Figure 4-51B) of the **prefrontal cortex**, a frontal neocortical area anterior to the motor cortex implicated in high-level cognitive functions. The frontal eye field receives extensive *feedforward* connections from visual areas in both the dorsal and ventral streams, and sends *feedback* projections to many visual cortical areas. Evidence suggests that the FEF is causally related to attentional modulation of signal processing, such as firing rate and stimulus selectivity, in specific areas of the visual cortex.

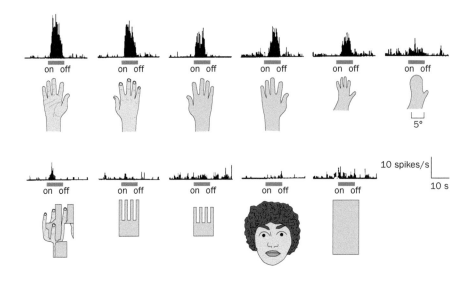

Figure 4-52 Object-selective neurons in the temporal cortex. Single-unit recordings in an anesthetized monkey revealed that the inferior temporal cortex contains neurons that selectively respond to specific objects. In this example, the neuron being recorded is activated (fires more spikes) in response to human hands (top row, first four images from left) and a monkey hand (top row, fifth image). A monkey hand with digits fused (top row, sixth image) and the images shown in the bottom row were less effective at stimulating the neuron. Horizontal bars underneath action potential plots indicate the stimulation period. Other neurons (not shown) responded selectively to faces. (Adapted from Desimone R, Albright TD, Gross CG, et al. [1984] *J Neurosci* 4:2051–2062. Copyright © 1984 The Society for Neuroscience.)

How the cerebral cortex processes information is still mostly unknown. This is a major unsolved question in science, and the study of vision has brought us closest to its frontier. In the final two sections of this chapter, we give two specific examples of research into the function of higher visual cortical areas.

4.27 Face-recognition cells form a specialized network in the primate temporal cortex

Starting in the 1970s, single-unit recordings in monkeys have shown that some neurons in the ventral visual processing stream of the temporal lobe recognize specific objects, including faces (**Figure 4-52**). Advances in **functional magnetic resonance imaging (fMRI)** (Section 1.10) led to the discovery that a specific area of human temporal cortex, the **fusiform face area**, is preferentially activated by images of human faces. Subsequently, face recognition areas were also discovered in macaque monkeys using fMRI. Because fMRI samples the average activity of many neurons, these findings implied that face recognition areas may contain concentrated populations of face-recognition cells. This hypothesis was tested by combining fMRI with single-unit recordings in macaque monkeys: fMRI permitted accurate targeting of the microelectrode to the face-recognition area. Indeed, a large majority of recorded neurons in certain areas of the monkey temporal cortex appeared to be preferentially tuned to faces (**Figure 4-53**).

Six patches of face-recognition cells—known as PL, ML, MF, AF, AL, and AM (named after their relative anatomical locations)—have been identified in the inferior temporal lobe of macaque monkeys (**Figure 4-54**A). Although the exact locations of these patches differ from individual to individual, their relative positions are stereotyped, such that each patch can be individually localized. This stereotyped positioning permitted further investigation of the relationship between these face-recognition patches using electrical **microstimulation**—the application of currents to activate neurons nearby the recording electrode (see Section 4.28 for further discussion about this method). Microstimulation of a given face-recognition patch was followed by fMRI to determine which other brain areas were activated. Interestingly, stimulation of one face-recognition patch preferentially activated other face-recognition patches (Figure 4-54A), whereas stimulating an area adjacent to a face-recognition patch activated patches of other, non-face-recognition areas (Figure 4-54B). These experiments suggested that the face-recognition patches are preferentially connected to each other.

Further studies suggested a hierarchical processing model for face recognition. Neurons in the middle patches (ML and MF) recognize faces from specific angles. Neurons in the anterior-lateral (AL) patch not only recognize faces from specific angles but also recognize the mirror images of those faces. Neurons in the anterior-medial (AM) patch respond to faces from all views. Thus, information

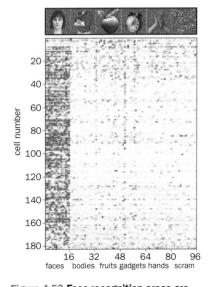

Figure 4-53 Face-recognition areas are highly enriched for face-recognition cells. Targeted single-unit recordings after functional magnetic resonance imaging revealed that face-recognition cells are highly concentrated in the middle face patch of the monkey temporal cortex. Firing patterns of 182 cells in response to 16 images each of faces, bodies, fruits, gadgets, hands, and scrambled images. Orange pixels represent action potentials. (From Tsao DY, Freiwald WA, Tootell RBH, et al. [2006] *Science* 311:670–674. With permission from AAAS.)

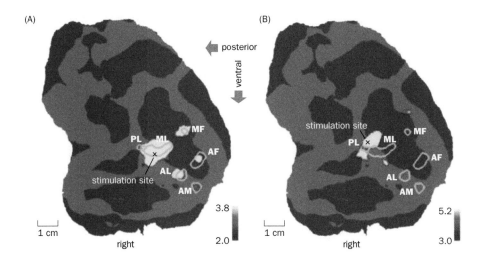

Figure 4-54 Face-recognition patches are preferentially interconnected.
Face-recognition patches were first identified by functional magnetic resonance imaging (fMRI) and shown as a flat map (by flattening the cortical surface into a two-dimensional representation, with light and dark gray representing gyri and sulci, respectively) outlined in green: PL, ML, MF, AF, AL, AM. **(A)** A microelectrode was inserted into the ML (marked by ×) and first used to confirm that the neuron being recorded was face selective. The electrode was then used to stimulate that area, and activation of other brain areas, as indicated by false color (see bottom right for a key in arbitrary units), was superimposed on the face-recognition patches. Note the coincidence between the face-recognition areas identified by fMRI and those activated by microstimulation. **(B)** As a control, a stimulation electrode was placed outside the face-recognition patches. Microstimulation also activated two patches outside the stimulation area, but these did not overlap with the face-recognition areas. (From Moeller S, Freiwald WA, & Tsao DY [2008] *Science* 320:1355–1359. With permission from AAAS.)

flow appears to follow a general pathway of ML/MF → AL → AM, achieving more abstract face recognition (face recognition independent of viewing angles) in the AM region. The latency (time delay) of neuronal activation in response to face stimuli supports this model, with ML/MF neurons activated before AL neurons and AL neurons activated before AM neurons.

Studies of face-recognition cells and patches have raised many interesting questions about visual perception for future investigation. How are the receptive fields of neurons along the ventral processing stream (Figure 4-51) sequentially transformed from line and edge detection in V1 to face recognition in the inferior temporal cortex? Why are face-recognition areas clustered into distinct patches? What are the functions of cortical areas located between these face-recognition patches? Do different face-recognition cells in one area recognize different faces? Recent studies have begun to shed light onto some of these questions. For example, one study suggested that cells in ML/MF and AM face patches encoded faces as linear combinations of features related to shape and appearance. Using recording data from 205 neurons in response to 2000 faces, researchers generated a linear model that could, with high accuracy, predict responses to new faces as well as reconstruct faces based on recording data (see Section 14.31 for a discussion of encoding and decoding). Inferior temporal cortical neurons outside face patches appear to represent non-facial objects following a similar organization to that of face recognition cells. The fact that primates devote cortical areas specifically to face recognition may reflect the abundance and importance of social interaction in these species. Whatever the evolutionary significance, face recognition offers an outstanding system for further exploring mechanisms of visual perception.

4.28 Linking perception to decision and action: microstimulation of MT neurons biased motion choice

The ultimate purpose of visual perception is to guide animal behavior. So where and how is sensation converted to action? How can we study this? Thus far we have discussed insights into visual information processing obtained by studying what stimuli can best excite visual pathway neurons. But how do we know that the activity of specific neurons contributes to the animal's perceptions? Unlike humans, experimental animals cannot directly tell us about their perceptions. One approach is to design behavioral assays that depend on the activity of specific visual pathway neurons and then use those behavioral assays to test the consequences of perturbing activity in these neurons.

Along the dorsal stream in the higher-order visual cortex of primates is an area called the **middle temporal visual area** (**MT**; Figure 4-51). MT neurons are especially sensitive to the direction of motion: individual neurons fire in response to motion in particular directions, and neurons tuned to the same direction of motion are clustered in vertical columns. To investigate the function of these

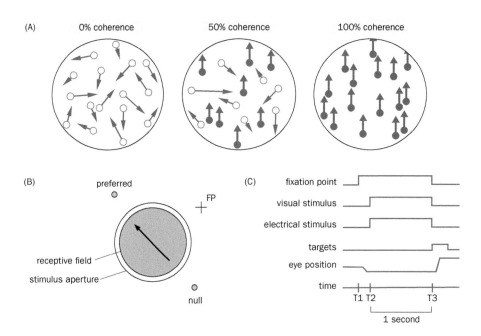

Figure 4-55 Experimental paradigm for testing monkeys' motion perception. **(A)** Schematic of the random dot display. The direction (indicated by arrows) and coherence of moving dots are adjusted by experimenters. Shown here are three examples in which 0%, 50%, or 100% of the dots exhibited coherent upward motion. **(B)** In an actual experiment, the monkey was trained to fix its eyes on the fixation point (FP). A neuron in MT was first recorded to determine its receptive field (orange circle) and directional motion preference (arrow). Then the moving dot stimuli were applied within an aperture (large circle) centered on the receptive field of the recorded neuron; the direction of coherent motion was either in the cell's preferred direction or in the opposite (null) direction. **(C)** The temporal sequence of the microstimulation experiment. The monkey first fixed its eyes on the fixation point (T1). Then the random dot display was applied to the receptive field, along with electrical stimulation for 1 second (T2 to T3). At the end of the visual and electrical stimulation, the monkey chose between one of two indicator lights, moving the eyes to one of the targets to signal its perception of the motion direction. Microstimulation was applied in a randomly selected half of the trials and withheld in the other half. (A, adapted from Salzman CD, Murasugi SM, Britten KH, et al. [1992] *J Neurosci* 12:2331–2355. Copyright © 1992 The Society for Neuroscience. B & C, adapted from Salzman CD, Britten KH, & Newsome WT [1990] *Nature* 346:174–177. With permission from Springer Nature.)

neurons in motion perception, researchers trained monkeys to report the direction of motion in a random dot display on a TV screen (**Figure 4-55**A). The monkey would first fix its eyes on a screen where the moving dots were displayed. After the display was over, it was given two choices represented by two indicator lights on the screen, one located in the direction of motion, the other in the opposite direction. The monkey would make a **saccade** (that is, move the fixation point of its eyes rapidly) toward what it judged was the direction of motion (Figure 4-55B, C) and would receive a juice reward for each correct choice. Dots with coherent motion (indicated by filled circles in Figure 4-55A) were mixed with dots that move in random directions (indicated by open circles in Figure 4-55A). The task is easy when 100% or 50% of moving dots are coherent, is difficult if only 10% of moving dots are coherent, and becomes impossible when 0% of moving dots are coherent. After extensive training, monkeys could make correct choices above chance even if only 10% of dots were coherent. Using this behavioral assay, researchers found that MT lesions markedly elevated coherence needed for monkeys to make correct choices, indicating that MT is required for optimal motion perception. Indeed, as detailed in Box 14-5, the firing pattern of individual MT neurons in single trials can be used to decode monkey's behavioral choice with high accuracy under ideal experimental conditions.

To probe the underlying neural mechanism, experimenters placed an electrode in MT and identified the receptive field and preferred direction of the neuron recorded by the electrode. The coherent motion of moving dots in the behavioral experiment was then applied either in the preferred direction of the recorded neuron or in the opposite (null) direction. A psychometric function could be established by plotting the percentage of choices (decisions) by the monkey favoring the preferred direction against the percentage of dots with coherent motion (**Figure 4-56**, purple lines). The same electrode was then used for microstimulation to activate nearby neurons while the monkey viewed the moving dots in a randomly selected half of the trials (Figure 4-55C). Microstimulation shifted the psychometric function leftward, indicating an overall increase in preferred direction decisions for all visual stimulus conditions (Figure 4-56, orange lines). In other words, microstimulation helped the monkey make correct choices when correlated motion was in the preferred direction of stimulated neurons but caused the monkey to make more errors when correlated motion was in the null direction. Both findings support the notion that the net effect of microstimulation is to bias the monkey's choice toward the preferred direction of the stimulated neurons.

Figure 4-56 Microstimulation of MT neurons biases motion perception. Psychometric curves depicting the proportion of decisions in which two different monkeys (top and bottom graphs) moved their eyes toward the recorded neuron's preferred direction against the percentage coherence of the moving dots, following the experimental setup in Figure 4-55. Positive values on the x axis represent coherent motion applied in the same direction as the preferred direction of the neuron; negative x axis values correspond to coherent motion in the null direction. The monkeys had different biases when no microstimulation was applied. The first monkey displayed a small bias for the preferred direction (~56% at 0% coherence), while the second monkey displayed a large bias for the null direction (~33% at 0% coherence); an unbiased result would be ~50% at 0% coherence. In both cases, microstimulation shifted the curve leftward. For example, after viewing 5%-coherent dots moving in the preferred direction (green dashed line), the first monkey chose the preferred direction ~70% of the time in the absence of microstimulation; these correct choices increased to ~80% with microstimulation. On the other hand, after viewing 5%-coherent dots moving in the null direction (red dashed line), the monkey's erroneous preferred-direction choices increased from 40% without microstimulation to 65% with microstimulation. Microstimulation therefore consistently increased the number of decisions favoring the neuron's preferred direction. (Adapted from Salzman CD, Britten KH, & Newsome WT [1990] *Nature* 346:174–177. With permission from Springer Nature.)

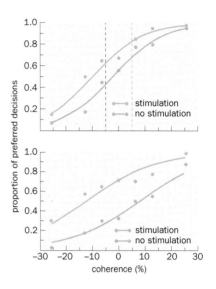

This is an important conceptual advance, as we have moved from correlation to causation. Not only does the activity of MT neurons report the direction of motion but the activation of MT neurons is also *sufficient* to bias the animal's behavior in a specific manner in accord with their response properties. Given that there are many neurons in the MT and elsewhere influenced by the direction of motion, it is remarkable that the effect of microstimulation of a small area could cause a significant behavioral bias. One explanation is that since MT neurons with similar properties (direction-of-motion selectivity) are arranged in columns, as we have seen for V1 neurons, microstimulation may have activated many nearby neurons encoding the same information and can thereby collectively affect the animal's perceptual choice.

Subsequent studies suggest that motion information flows from MT to a network of cortical and subcortical areas that are good candidates for converting the sensory evidence from MT into decisions to focus the eyes on one or the other target. Several of these brain areas—including the **lateral intraparietal area (LIP)** of the parietal cortex, frontal eye field in the prefrontal cortex, superior colliculus, and basal ganglia—have been studied via single-unit recordings while monkeys perform the motion discrimination task, producing insight into how decisions are formed. In the experiment shown in **Figure 4-57**, for example, the firing rates of LIP neurons predicted the direction the eyes would move, signaling the monkey's perception of the moving dots' direction. LIP neurons corresponding to eye movement toward the direction of the monkey's eventual choice (response direction) ramped up their firing rates as the monkey was viewing the moving dots and during the delay period before the saccade, as if LIP neurons were gradually accumulating evidence from MT about the noisy visual stimulus. The power of LIP neurons to predict the direction of the saccade improved with time and stimulus strength.

Thus, by focusing on a specific and quantitative behavioral task and identifying relevant neurons through physiological recording and activity manipulation, researchers have gained important insight into the neural basis of motion perception and decision making. Microstimulation during behavioral tasks has become a powerful method that complements physiological recordings of neuronal activity, as it tests the causal relationship between the activity of neurons and their functions. Importantly, these kinds of causal links between neuronal activity and animal behavior require the experiments to be performed in awake, behaving animals. Indeed, neurons in the **association cortex** (cortical areas that associate information from multiple sensory areas and link to motor output), such as LIP and prefrontal cortex, tend to be silent in anesthetized animals but become highly active in awake, behaving animals.

In recent years, exciting technological developments have made possible manipulations of neuronal activity that is more precise than microstimulation by recording electrodes. For example, as illustrated in Box 4-4, ChR2 can be targeted

(A)

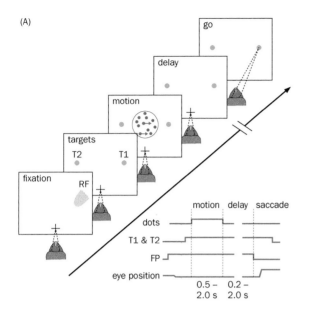

(B)

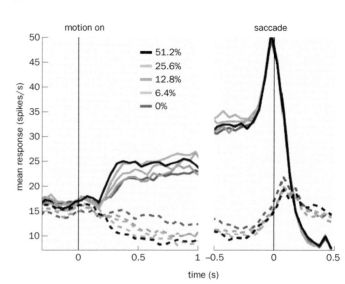

Figure 4-57 Neurons in the lateral intraparietal area (LIP) and perceptual decisions. (A) Behavioral task. Monkeys focus their eyes on the fixation point (+). Two targets appear, one of which is within the receptive field (RF) of a LIP neuron being recorded. After a period of moving dot stimuli and a variable delay, the fixation point disappears, and the monkey signals its perception of the direction of motion by making a saccade toward one of the two targets. **(B)** Average spike rates of 104 LIP neurons during the motion-viewing period (left panel), during the delay period before the eye movement (before saccade on the right panel), and afterward (after saccade on the right panel). When the monkey judged the direction as being toward the receptive field (solid lines), the firing rates of LIP neurons increased over time during the motion-viewing and delay periods. In the color-coded inset key, the percentages refer to the percent coherence of the moving dots. The strength of the stimulus (percentage of coherent moving dots) affected the speed and magnitude of LIP neurons' firing rate increases during the motion-viewing period. The firing rates of LIP neurons decreased when the monkey judged the direction as being away from the receptive field (dashed lines). Not shown here, LIP neurons in the opposite half of the brain (with receptive fields corresponding to target 2) demonstrated the opposite response—ramping up for target 2 choices and ramping down for target 1 choices. (Adapted from Shadlen MN & Newsome WT [2001] *J Neurophysiol* 86:1916–1936. With permission from the American Physiological Society.)

to and activate specific neuronal populations, enabling researchers to establish causal relationships between the activity of the targeted neurons and circuit or behavioral outputs. Combining these new tools with rigorous behavioral experiments discussed in this chapter will surely enable researchers to probe deeper into the neural mechanisms of perception and its transformation into action.

SUMMARY

Sensory systems transform environmental stimuli into electrical signals. These signals are transmitted to the brain to form an internal representation, or perception, of the sensory stimuli, with the ultimate goal of helping animals survive and reproduce. In vision, sensory stimuli are the spatiotemporal patterns of light projected onto a two-dimensional retina, from which animals extract information about the identity, location, and motion of objects in the external world that are of behavioral significance.

The first step in vision is to detect light stimuli and convert them into electrical signals. This is mostly achieved by rods and cones, which together cover a very wide range of light intensities. Photon absorption triggers a signaling cascade utilizing trimeric G proteins and phosphodiesterase and leading to the closure of cGMP-gated cation channels and hyperpolarization of rods and cones. cGMP-gated channel closure also causes a decrease in the intracellular Ca^{2+} concentration, which is central to recovery and adaptation. Variations of the phototransduction properties enable rods and cones to serve different functions. Rods are more sensitive to light due to greater amplification in the phototransduction cascade and are used mostly for night vision. Cones recover more rapidly, have a larger adaptation range, and serve daylight vision. The high density of cones in the primate fovea enables cones to serve high-acuity vision. Finally, different kinds of cones with opsins that confer different spectral sensitivities enable color vision.

Signals from rods and cones are analyzed by exquisitely precise retinal circuits before information is delivered by retinal ganglion cell (RGC) axons to the brain. Two subclasses of bipolar cells receive synaptic input from photoreceptors and deliver excitatory output to RGCs: OFF bipolars are hyperpolarized by light, whereas ON bipolars are depolarized by light. Horizontal cells are excited by photoreceptors and send inhibitory feedback laterally to many photoreceptors. This lateral inhibition causes the output of the photoreceptor to reflect not only their intrinsic phototransduction state but also the activity of surrounding photoreceptors. This center–surround antagonism enables bipolar cells and RGCs to detect luminance contrast. Lateral inhibition also contributes to color detection: a comparison of light signals from cones with different spectral sensitivities.

In addition to detecting luminance and color contrast, retinal circuits extract many other kinds of signals through parallel actions of different types of bipolar cells, amacrine cells, and RGCs. For example, asymmetric inhibition by starburst amacrine cells makes certain RGCs sensitive to the direction of motion. Intrinsically photosensitive RGCs not only receive light input from rods and cones but can also be directly depolarized by light through melanopsin. One of many functions of ipRGCs is to entrain the circadian clock.

A key target for RGC axons is the lateral geniculate nucleus (LGN) of the thalamus, which relays information to the visual cortex for further analysis of form, color, and motion. A principal organization in the LGN and visual cortical areas is retinotopy—nearby neurons represent nearby points in visual space in an orderly manner. The receptive fields of individual neurons, however, are transformed along the visual pathway. Simple and complex cells in V1 respond most vigorously to bars with specific orientations, suggesting that V1 cells detect lines and edges. Cells in MT are highly tuned to motion in specific directions, whereas cells in patches of the inferior temporal cortex are highly tuned to faces, suggesting further specialized functions in these higher cortical visual areas.

Studies of the visual cortex have informed mammalian neocortex function in general. Within a cortical area, information generally flows from LGN axons → layer 4 → layers 2/3 → layers 5/6 vertically. Between cortical areas, feedforward or feedback inputs usually terminate in or avoid layer 4 of the recipient areas, respectively. These rules have enabled the construction of a hierarchical model for visual information streams beyond V1. Our understanding of the general principles of information processing in neocortex, and specifically how visual response properties in one visual area are transformed into those of another area, are still rudimentary. Electrophysiological recordings, circuit tracing, activity manipulation, quantitative studies of behavior, and computational modeling must be combined to tackle the complexity of neocortical function.

OPEN QUESTIONS

- How do intrinsically photosensitive retinal ganglion cells balance their roles of sensing light directly and analyzing signals transmitted from rods and cones?

- How are motion detection algorithms implemented by the connectivity and physiological properties of neurons and circuits in the mammalian retina, cortex, and insect visual system? What are their similarities and differences?

- How do LGN neurons integrate feedforward information from the retina and feedback information from the visual cortex?

- How do the receptive fields of neurons within the ventral processing stream sequentially transform line and edge detection in V1 to object and face recognition in the inferior temporal cortex?

- What are the functions of feedback pathways from higher to lower visual cortical areas in the hierarchical model?

- What is the neural basis of visual perception—activity of neurons in specific brain regions or ensemble activity of neurons across many regions?

FURTHER READING

Books and reviews

Do MTH (2019). Melanopsin and the intrinsically photosensitive retinal ganglion cells: biophysics to behavior. *Neuron* 104:205–226.

Hubel D & Wiesel T (2004). Brain and Visual Perception: The Story of a 25-Year Collaboration. Oxford University Press.

Luo DG, Xue T, & Yau KW (2008). How vision begins: an odyssey. *Proc Natl Acad Sci USA* 105:9855–9862.

Mauss AS, Vlasits A, Borst A, & Feller M (2017). Visual circuits for direction selectivity. *Annu Rev Neurosci* 40:211–230.

Reynolds JH & Desimone R (1999). The role of neural mechanisms of attention in solving the binding problem. *Neuron* 24:19–29.

Sanes JR & Masland RH (2015). The types of retinal ganglion cells: current status and implications for neuronal classification. *Annu Rev Neurosci* 38:221–246.

Stryer L (1988). Molecular basis of visual excitation. *Cold Spring Harb Symp Quant Biol* 53 Pt 1:283–294.

Light detection in rods and cones

Baylor DA, Lamb TD, & Yau KW (1979). Responses of retinal rods to single photons. J Physiol 288:613–634.

Baylor DA, Nunn BJ, & Schnapf JL (1987). Spectral sensitivity of cones of the monkey *Macaca fascicularis*. *J Physiol* 390:145–160.

Fesenko EE, Kolesnikov SS, & Lyubarsky AL (1985). Induction by cyclic GMP of cationic conductance in plasma membrane of retinal rod outer segment. *Nature* 313:310–313.

Fung BK, Hurley JB, & Stryer L (1981). Flow of information in the light-triggered cyclic nucleotide cascade of vision. *Proc Natl Acad Sci U S A* 78:152–156.

Hecht S, Shlaer S, & Pirenne MH (1942). Energy, quanta, and vision. *J Gen Physiol* 25:819–840.

Nathans J, Thomas D, & Hogness DS (1986). Molecular genetics of human color vision: the genes encoding blue, green, and red pigments. *Science* 232:193–202.

Palczewski K, Kumasaka T, Hori T, Behnke CA, Motoshima H, Fox BA, Le Trong I, Teller DC, Okada T, Stenkamp RE, et al. (2000). Crystal structure of rhodopsin: a G protein-coupled receptor. *Science* 289:739–745.

Signal analysis in the retina

Barlow HB & Levick WR (1965). The mechanism of directionally selective units in rabbit's retina. *J Physiol* 178:477–504.

Berson DM, Dunn FA, & Takao M (2002). Phototransduction by retinal ganglion cells that set the circadian clock. *Science* 295:1070–1073.

Briggman KL, Helmstaedter M, & Denk W (2011). Wiring specificity in the direction-selectivity circuit of the retina. *Nature* 471:183–188.

Dacey DM & Lee BB (1994). The 'blue-on' opponent pathway in primate retina originates from a distinct bistratified ganglion cell type. *Nature* 367:731–735.

Euler T, Detwiler PB, & Denk W (2002). Directionally selective calcium signals in dendrites of starburst amacrine cells. *Nature* 418:845–852.

Field GD, Greschner M, Gauthier JL, Rangel C, Shlens J, Sher A, Marshak DW, Litke AM, & Chichilnisky EJ (2009). High-sensitivity rod photoreceptor input

to the blue-yellow color opponent pathway in macaque retina. *Nat Neurosci* 12:1159–1164.

Hattar S, Liao HW, Takao M, Berson DM, & Yau KW (2002). Melanopsin-containing retinal ganglion cells: architecture, projections, and intrinsic photosensitivity. *Science* 295:1065–1070.

Kuffler SW (1953). Discharge patterns and functional organization of mammalian retina. *J Neurophysiol* 16:37–68.

Nawy S & Jahr CE (1990). Suppression by glutamate of cGMP-activated conductance in retinal bipolar cells. *Nature* 346:269–271.

Packer OS, Verweij J, Li PH, Schnapf JL, & Dacey DM (2010). Blue-yellow opponency in primate S cone photoreceptors. *J Neurosci* 30:568–572.

Shekhar K, Lapan SW, Whitney IE, Tran NM, Macosko EZ, Kowalczyk M, Adiconis X, Levin JZ, Nemesh J, Goldman M, et al. (2016). Comprehensive classification of retinal bipolar neurons by single-cell transcriptomics. *Cell* 166:1308–1323.

Verweij J, Hornstein EP, & Schnapf JL (2003). Surround antagonism in macaque cone photoreceptors. *J Neurosci* 23:10249–10257.

Yang HH, St-Pierre F, Sun X, Ding X, Lin MZ, & Clandinin TR (2016). Subcellular imaging of voltage and calcium signals reveals neural processing *in vivo*. *Cell* 166:245–257.

Information processing in the visual cortex

Chang L & Tsao DY (2017). The code for facial identity in the primate brain. *Cell* 169:1013–1028.

Cossell L, Iacaruso MF, Muir DR, Houlton R, Sader EN, Ko H, Hofer SB, & Mrsic-Flogel TD (2015). Functional organization of excitatory synaptic strength in primary visual cortex. *Nature* 518:399–403.

Desimone R, Albright TD, Gross CG, & Bruce C (1984). Stimulus-selective properties of inferior temporal neurons in the macaque. *J Neurosci* 4:2051–2062.

Felleman DJ & Van Essen DC (1991). Distributed hierarchical processing in the primate cerebral cortex. *Cereb Cortex* 1:1–47.

Ferrera VP, Nealey TA, & Maunsell JH (1992). Mixed parvocellular and magnocellular geniculate signals in visual area V4. *Nature* 358:756–761.

Gilbert CD & Wiesel TN (1979). Morphology and intracortical projections of functionally characterised neurones in the cat visual cortex. *Nature* 280:120–125.

Hillier D, Fiscella M, Drinnenberg A, Trenholm S, Rompani SB, Raics Z, Katona G, Juettner J, Hierlemann A, Rozsa B, et al. (2017). Causal evidence for retina-dependent and -independent visual motion computations in mouse cortex. *Nat Neurosci* 20:960–968.

Hubel DH & Wiesel TN (1962). Receptive fields, binocular interaction and functional architecture in the cat's visual cortex. *J Physiol* 160:106–154.

Kanwisher N, McDermott J, & Chun MM (1997). The fusiform face area: a module in human extrastriate cortex specialized for face perception. *J Neurosci* 17:4302–4311.

Moeller S, Freiwald WA, & Tsao DY (2008). Patches with links: a unified system for processing faces in the macaque temporal lobe. *Science* 320:1355–1359.

Moore T & Armstrong KM (2003). Selective gating of visual signals by microstimulation of frontal cortex. *Nature* 421:370–373.

Newsome WT, Britten KH, & Movshon JA (1989). Neuronal correlates of a perceptual decision. *Nature* 341:52-54.

Ohki K, Chung S, Ch'ng YH, Kara P, & Reid RC (2005). Functional imaging with cellular resolution reveals precise micro-architecture in visual cortex. *Nature* 433:597-603.

Olsen SR, Bortone DS, Adesnik H, & Scanziani M (2012). Gain control by layer six in cortical circuits of vision. *Nature* 483:47-52.

Rockland KS & Pandya DN (1979). Laminar origins and terminations of cortical connections of the occipital lobe in the rhesus monkey. *Brain Res* 179:3-20.

Salzman CD, Britten KH, & Newsome WT (1990). Cortical microstimulation influences perceptual judgements of motion direction. *Nature* 346:174-177.

Shadlen MN & Newsome WT (2001). Neural basis of a perceptual decision in the parietal cortex (area LIP) of the rhesus monkey. *J Neurophysiol* 86:1916-1936.

Tasic B, Yao Z, Graybuck LT, Smith KA, Nguyen TN, Bertagnolli D, Goldy J, Garren E, Economo MN, Viswanathan S, et al. (2018). Shared and distinct transcriptomic cell types across neocortical areas. *Nature* 563:72-78.

CHAPTER 5
Wiring the Visual System

There are billions of neurons in our brains, but what are neurons? Just cells. The brain has no knowledge until connections are made between neurons. All that we know, all that we are, comes from the way our neurons are connected.

Tim Berners-Lee (2000),
Weaving the Web: The Original Design and Ultimate Destiny of the World Wide Web

The human brain contains about 10^{11} neurons, and each makes on average $>10^3$ synapses with other neurons. Thus, there are $>10^{14}$ synaptic connections in the brain that give rise to our ability to sense, think, remember, and act. How are such vast numbers of connections established during development?

The brain is often compared with another complex system that has impressive problem-solving power: the computer (Section 1.12). Their functions are both specified by wiring diagrams constructed from much simpler units: the neuron and the transistor. A key difference is the wiring process. Each connection on a computer chip is prespecified by engineers. By contrast, the brain self-assembles during development, starting from a single cell. The instructions for the entire developmental process, including brain wiring, are embedded in an organism's genome. In fact, the information contained in the human genome sequence is surprisingly small: 3×10^9 nucleotides of DNA with four possible bases for each nucleotide, or about 750 megabytes. How does this information direct the wiring of the brain? Is it sufficient?

An enduring debate among neuroscientists, psychologists, and even philosophers is whether brain wiring, which underlies brain function and behavior, is shaped more by nature (genes) or nurture (experience). From our discussion in Chapter 1, we know that both factors contribute, but how do genes and experience specify brain wiring? Do their relative contributions differ in different parts of the nervous system, at different stages of development, and in different animals?

In this chapter and Chapter 7, we will explore the mechanisms of nervous system development, with an emphasis on wiring. We devote this chapter to wiring the visual system as a primer, since many fundamental insights about brain wiring have come from research on the developing visual system. After we have learned more about other sensory systems in Chapter 6, we will expand our scope to study the construction of the nervous system and summarize the general principles in Chapter 7.

HOW DO RETINAL GANGLION CELL AXONS FIND THEIR TARGETS?

Imagine that you are one of about a million retinal ganglion cells (RGCs) near the center of a human retina, for example, a midget RGC with a red-ON center receptive field (Figure 4-34C) located in the fovea of the left eye (**Figure 5-1**). Your goal is to transmit information to the brain about the presence of long-wavelength light within your visual field. Proper wiring of your dendrites and axon is essential for achieving this goal. Your dendrites must connect with the correct types of bipolar and amacrine cells to construct your receptive field, and your long axon must exit the eye, travel along the left optic nerve, decide whether to stay in the left hemisphere or cross the midline at the optic chiasm, travel along the optic tract, and send branches to the lateral geniculate nucleus (LGN) and superior colliculus. As we learned in Chapter 4, visual information in the retina is topographically represented in the brain as retinotopic maps. To preserve the spatial relationships of the visual image on the retina, your axon must terminate at appropriate retinotopic positions within the superior colliculus and LGN according to your location

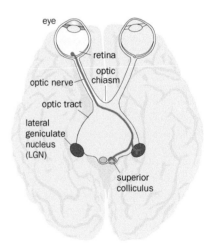

Figure 5-1 The journey of an axon of a retinal ganglion cell. The axon of an RGC must make a series of choices during development: to exit the eye and join the optic nerve, to cross (or not cross) the midline at the optic chiasm, to form terminal branches at specific brain targets (for example, the lateral geniculate nucleus and superior colliculus), to terminate at specific layers and positions within these targets, and to form synapses with appropriate partners in these targets, according to the RGC's subtype and retinal location.

Figure 5-2 The retinotopic map. The retina forms a two-dimensional visual image of an object in the external world. This image is reconstructed in a brain target, the tectum in this example, by point-to-point topographic projections of retinal ganglion cell axons. The tectum in nonmammalian vertebrates is equivalent to the superior colliculus in mammals.

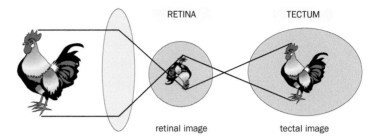

in the eye. Finally, you must choose the correct synaptic partners from specific layers within these targets according to your identity as a red-ON center midget cell.

As this example illustrates, a single neuron makes many decisions during the wiring process. It is astonishing to consider that in the developing nervous system, numerous neurons are making so many decisions simultaneously. In the first part of this chapter, we will focus largely on one specific aspect of visual system wiring: how do RGC axons terminate at appropriate spatial locations in their target areas to ensure the formation of accurate retinotopic maps (**Figure 5-2**)? Studies of this specific problem have yielded important insights into the general principles of brain wiring.

5.1 Optic nerve regeneration experiments suggested that RGC axons are predetermined for wiring

Historically, two broad mechanisms by which neurons become connected to their targets were proposed. In one mechanism, axons initially connect to many different targets; through trial and error, a subset of these connections are selected by their functions to establish the final connection pattern (**Figure 5-3**, left). In the other mechanism, axons are predetermined to choose their targets directly without functional selection (Figure 5-3, right). The dominant view in the early twentieth century was that functional selection plays a primary role in establishing neuronal connections. However, a decisive experiment on optic nerve regeneration performed by Roger Sperry in the 1940s strongly suggested that neurons are predetermined to choose their targets.

Amphibians have the amazing ability to **regenerate** their nerve connections after damage by reextending truncated axons back to their targets. The optic nerve, which consists of the RGC axons that connect the eye to the brain, can be transected and then allowed to regenerate; upon completion of regeneration, vision is

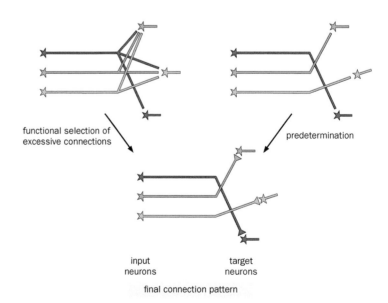

Figure 5-3 Two contrasting mechanisms for axon targeting. Left, axons initially overproduce branches and make excessive connections, and the correct connections are later selected by function and inappropriate connections are pruned. Right, axons are predetermined to find their targets.

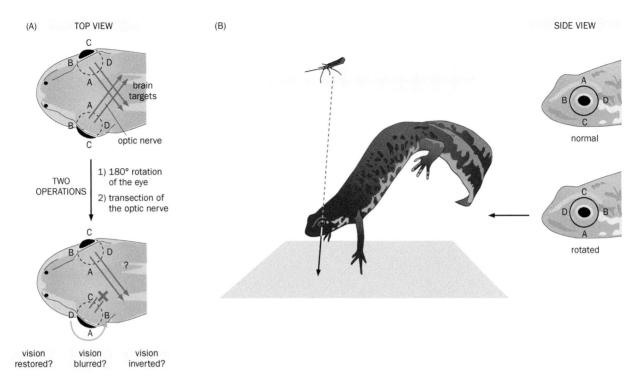

Figure 5-4 Sperry's optic nerve regeneration experiment.
(A) Schematic of experimental design. The left eye was rotated 180°, followed by axon transection. Letters surrounding the eyes are with respect to the *original* position. Red lines and arrows represent RGC axons and the direction of their brain target. The question mark (?) at the target of regenerating axons represents possible experimental outcomes listed at the bottom. Note that RGC axons project mostly to contralateral brain targets (Section 5.6). **(B)** Illustration of the newt's behavior based on its inverted vision from the manipulated eye. When food was presented above, the newt swam downward to fetch it. (Note that before the behavioral test, the other eye was removed so the newt could see only with the rotated eye.) Side views of the normal and rotated eyes are drawn on the right. (Based on Sperry RW [1943] *J Comp Neurol* 79:33–55.)

completely restored. Taking advantage of this ability, Sperry surgically rotated one of a newt's eyes 180° such that the visual world was upside down and front-side back; each RGC now looked at a point in space that was 180° different from the point it looked at before surgery. He then severed the optic nerve and allowed the RGC axons from the rotated eye to regrow into the brain to form connections with their targets (**Figure 5-4**A). After allowing time for regeneration, he used behavioral experiments to test what the newt saw with its manipulated eye.

Three possible outcomes could result from this manipulation. First, vision could be restored completely. This would provide strong evidence supporting the functional selection hypothesis. In this scenario the location of the object in visual space determines where the axons would terminate. Hence, a different RGC goes to the same target before and after rotation. Second, vision could be blurred. This would mean that regenerated axons only partially successfully find their targets in the brain following eye rotation (recall that vision could be restored completely if no eye rotation was performed). Third, vision could be restored, but in an inverted manner. In this scenario, the location of the RGC in the retina determines where the axon terminates. Hence, the same RGC axon connects to the same target before and after eye rotation. Sperry observed the third outcome during behavioral experiments after regeneration. When food was presented at the surface of the aquarium, above the newt's head, the newt would swim downward to fetch it, bumping itself against the bottom of the tank (Figure 5-4B).

This experiment strongly suggested that RGC axons carry specific information corresponding to their *original* positions in the eye, and that the brain contains information corresponding to those positions. Despite the rotation of the eye, which changes the spatial receptive fields of individual RGCs, such positional information still enabled these RGC axons to connect with their *original* target neurons in the brain, restoring vision, but in an inverted manner.

5.2 Point-to-point connections between retina and tectum arise via chemoaffinity

Over the next 20 years, Sperry and colleagues collected more information on how regenerating RGC axons grow into a structure called the **tectum**, the major target of RGCs in the brains of lower vertebrates, equivalent to the mammalian superior colliculus. For example, they transected the optic nerve, ablated half of all RGCs, and then examined the axon terminations of the remaining RGCs in the tectum. These experiments allowed them to determine which parts of the retina connected to which parts of the tectum. Their key findings are summarized in **Figure 5-5**. Ventral RGCs project to the medial half of the tectum, dorsal RGCs to the lateral half. Anterior RGCs (also called **nasal** RGCs because they are close to the nose) project to the posterior tectum, while posterior RGCs (also called **temporal** RGCs because they are close to the temple) project to the anterior tectum. Thus, the point-to-point retinotopic map between the retina and the tectum is enabled by the orderly projections of RGC axons. The axonal projection of nasal RGCs to the posterior tectum is particularly illuminating: here the axons passed through an empty field of tectum—because the temporal RGCs that normally terminate here were ablated—and still homed in on their original targets. These observations provided conclusive evidence that RGC axons are predetermined to connect with specific targets in the brain following regeneration.

This evidence led Sperry to propose the **chemoaffinity hypothesis** in 1963:

It seems a necessary conclusion from these results that cells and fibers of the brain and (spinal) cord must carry some kind of individual identification tags, presumably cytochemical in nature, by which they are distinguished one from another almost, in many regions, to the level of the single neuron; and further, that the growing fibers are extremely particular when it comes to establishing synaptic connections, each axon linking only with certain neurons to which it becomes selectively attached by specific chemical affinity.

Figure 5-5 Orderly projections of regenerating retinal axons in the tectum. Each of the four pairs of drawings illustrates targeting of RGC axons from a portion of a goldfish retina to the tectum, following the transsection of the retina's optic nerve, the ablation of half of all RGCs, and regeneration of axons from the remaining RGCs. Thus, RGCs from the ventral retina project their axons to the medial tectum (top pair); dorsal RGCs project their axons to the lateral tectum (second pair); posterior (temporal) RGCs project their axons to the anterior tectum (third pair); and anterior (nasal) RGCs project their axons to the posterior tectum (bottom pair). Gray areas represent ablated RGCs in the retina and empty target fields in the tectum. (Adapted from Sperry RW [1963] *Proc Natl Acad Sci U S A* 50:703–710.)

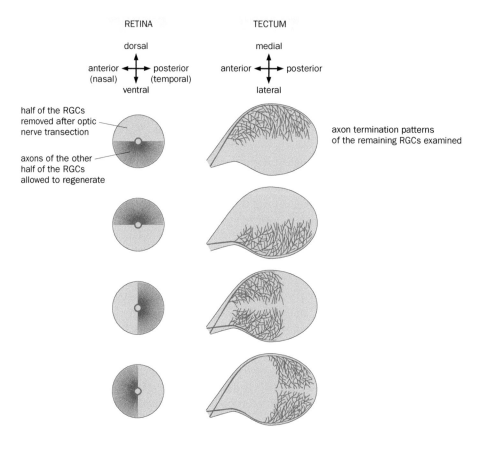

RETINA

dorsal

anterior (nasal) ←→ posterior (temporal)

ventral

TECTUM

medial

anterior ←→ posterior

lateral

half of the RGCs removed after optic nerve transection

axons of the other half of the RGCs allowed to regenerate

axon termination patterns of the remaining RGCs examined

A caveat of Sperry's hypothesis is that connection mechanisms during normal development could differ considerably from those used in regeneration. Indeed, regeneration can be considered a high bar for molecular recognition, as connection specificity during regeneration cannot result from a normal developmental milieu. In retrospect, mechanisms used during normal development turned out to be similar to those Sperry proposed based on regeneration experiments.

An apparent problem with the chemoaffinity hypothesis is that our genome does not have enough information to encode tags, or cell-surface recognition proteins, for every neuron and/or connection, if each must carry an individual identification tag to be distinguished from others. Sperry realized this problem and proposed, specifically for the retinotopic mapping, that protein gradients could be used to provide positional information in the retinal and tectal fields. This means that *different levels of the same protein* could be used to enforce precise targeting of many different neurons. We will see soon that these predictions were prescient and will discuss additional mechanisms that can be used to address this numeric problem in Chapter 7.

5.3   The posterior tectum repels temporal retinal axons

The chemoaffinity hypothesis inspired scientists to search for the cytochemical tags that guide growing axons toward their targets. The retinotectal mapping has been one of the leading model systems in this search. In this context, the chemoaffinity hypothesis predicts that there must be molecular differences between cells from different parts of the tectum in order for retinal axons to differentially select their targets. Likewise, there must be molecular differences between retinal axons originating from distinct parts of the retina so that they react differently to the molecular cues in the tectum.

Indeed, biochemical studies showed that membrane proteins extracted from the posterior tectum of chicks differed from those extracted from the anterior tectum. (The chick has a large tectum, making it an excellent source of starting material for biochemical analysis.) This was elegantly illustrated using an assay in which membranes from the anterior or posterior tectum, which presumably include cell-surface recognition proteins, were laid down in 50 μm alternating stripes to serve as substrates for retinal axons to grow onto *in vitro* (**Figure 5-6**). Nasal retinal axons grew without selectivity, whereas temporal retinal axons grew preferentially on

Figure 5-6 Temporal RGC axons are repelled by the posterior tectal membrane. **(A)** Experimental setup. Membranes from anterior (A) and posterior (P) tectum were laid down in alternating 50 μm stripes. Explants of temporal (T) or nasal (N) retina were placed above the stripes, allowing RGC axons to grow down onto the stripes. **(B)** Temporal axons (bottom image) grew onto stripes containing anterior tectal membrane proteins (labeled A in the top image) and avoided posterior tectal membranes (P). **(C)** Nasal axons grew indiscriminately. The top and bottom images in Panels B and C are from the same samples double-labeled by fluorescent markers that recognize posterior tectal membrane preparation (top) and retinal axons (bottom). (B & C, from Walter J, Kern-Veits B, Huf J, et al. [1987] *Development* 101:685–696. With permission from The Company of Biologists, Ltd.)

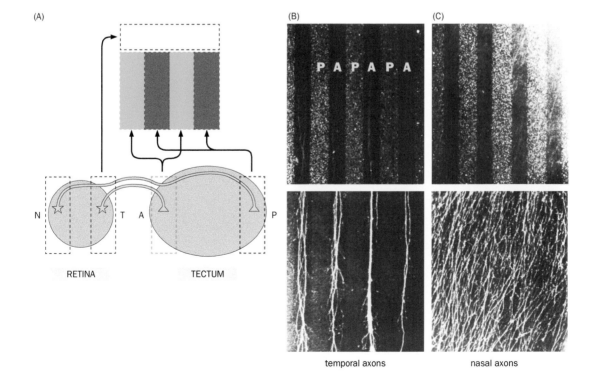

(A)

(B)

(C)

RETINA   TECTUM

temporal axons   nasal axons

membranes from the anterior tectum. Interestingly, in the *absence* of membranes from the anterior tectum, temporal retinal axons could grow on membranes from the posterior tectum; this demonstrated the importance of alternative choices to reveal selectivity: axon growth is really a preference for one substrate over the other.

The observation that temporal retinal axons prefer the anterior tectum to the posterior tectum could result from two alternative possibilities: either temporal retinal axons are *attracted* by the anterior tectum, where they normally terminate *in vivo*, or they are *repelled* by the posterior tectum. A simple experiment was carried out to address these possibilities. Since these putative attractants and repellents are most likely proteins, and protein activities are usually abolished by heating, anterior or posterior tectal membrane proteins were selectively heat inactivated before placement on the stripes. Temporal RGC axons lost selectivity when posterior (but not anterior) tectal membrane proteins were heat inactivated. Thus, it appeared that temporal axons respond to an activity from posterior tectal membranes: they normally target to the anterior tectum because they are repelled by protein component(s) present on the posterior tectal membrane.

5.4   Gradients of ephrins and Eph receptors instruct retinotectal mapping

The repellent activity of the posterior tectal membrane was next used to biochemically purify the specific protein(s) responsible for repulsion. This led to the identification of ephrin-A5, a member of a protein family called **ephrins**. Ephrin-A5 is an extracellular protein attached to the plasma membrane by a glycosylphosphatidylinositol (GPI) lipid anchor. Ephrins bind to the **Eph receptors (Ephs)**, transmembrane proteins with cytoplasmic tyrosine kinase domains (Box 3-4) that are highly expressed in the nervous system. Importantly, purified ephrin-A5 mimicked the posterior tectal membrane in repelling temporal retinal axons in the stripe assay.

Consistent with previous findings suggesting that a repellent is present in the posterior tectal membrane, mRNA of ephrin-A5 is expressed in the tectum in a posterior > anterior gradient—that is, ephrin-A5 mRNA is more abundant in the posterior and less abundant in the anterior tectum. Remarkably, EphA3, a receptor for ephrin-A5, is expressed in the retina in a temporal > nasal gradient (**Figure 5-7A**). Thus, a scenario emerged that can account for targeting selectivity along the anterior–posterior tectal axis: RGC axons originating in the most temporal retina express the highest amount of EphA3 and are therefore most sensitive to its repellent ligand ephrin-A5, distributed in a posterior > anterior gradient in the tectum.

Figure 5-7 Expression gradients for an ephrin in the tectum and an Eph in the retina. (A) Left, in the retina, *in situ* hybridization (see Section 14.12 for details) shows that EphA3 mRNA is distributed in a temporal > nasal gradient (top panel). Right, in the tectum, *in situ* hybridization shows that ephrin-A5 mRNA is distributed in a posterior > anterior gradient (P > A, top panel). The hybridization signal, schematically illustrated in the dashed boxes in the middle panels, is quantified in the bottom panels. The cup-shaped spherical retina was cut along several radial lines (like a clover leaf) to allow flat mounting for imaging and quantification. N, nasal; T, temporal; D, dorsal; V, ventral. **(B)** A model for retinotopic mapping. Temporal RGCs express high levels of EphA3 and can target to only the anterior tectum, which expresses low levels of the repellent ephrin-A5. Nasal axons express low levels of EphA3 and can target to the posterior tectum because they are less sensitive to repulsion by high levels of ephrin-A5. (A, adapted from Cheng H, Nakamoto M, Bergemann AD, et al. [1995] *Cell* 82:371–381. With permission from Elsevier Inc. See also Drescher U, Kremoser C, Handwerker C, et al. [1995] *Cell* 82:359–370. B, adapted from Tessier-Lavigne M [1995] *Cell* 82:345–348. With permission from Elsevier Inc.)

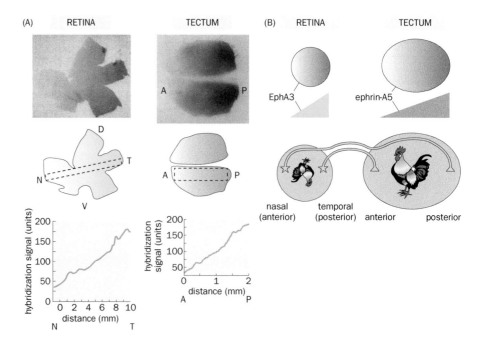

Hence, the temporal-most RGCs target the anterior-most region of the tectum. RGCs from progressively more nasal positions in the retina express progressively less EphA3 and are therefore less sensitive to repulsion by ephrin-A5. As a result, they target to progressively more posterior regions of the tectum (Figure 5-7B; Movie 5-1).

The function of ephrins in retinal axon targeting *in vivo* was confirmed by viral misexpression of ephrins in chicks and ephrin knockout experiments in mice. In the mouse, both ephrin-A5 and a related ligand ephrin-A2 are expressed in posterior > anterior gradients in the superior colliculus (the mammalian equivalent of the tectum). In *EphrinA5/A2* double-knockout mice, temporal axons no longer exhibited selective targeting to the posterior superior colliculus; instead, their axons were scattered along the entire anterior–posterior axis (Figure 5-8). (Curiously, nasal axons were also affected, and mistargeted axons formed clusters; we will return to this observation in later sections.) The double-knockout experiments demonstrated that ephrin-A2 and ephrin-A5 indeed play essential roles *in vivo* in instructing RGC axon targeting along the anterior–posterior axis of the superior colliculus. Targeting along the medial–lateral axis was intact, indicating that ephrin-A/EphA signaling is important for only one axis.

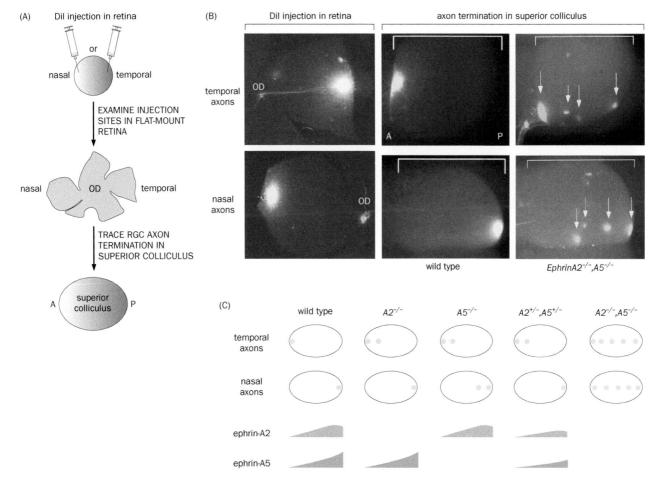

Figure 5-8 Genetic validation of ephrin-A function *in vivo*.
(A) Experimental procedure. Top, fluorescently labeled DiI, a lipophilic dye that diffuses along the lipid bilayer and serves as an axon tracer, was injected into either the temporal or nasal retina. Middle, the retina was flat-mounted to validate DiI injection sites. OD, optic disc where RGC axons exit the retina to project to the brain. Bottom, DiI-labeled axon termination patterns in the superior colliculus were examined. A, anterior; P, posterior. **(B)** Analysis of retinotopic projections in wild-type and *EphrinA2/A5* double-knockout mice. Left panels show DiI injection sites in the temporal and nasal retina. Right panels show RGC axon termination patterns in the superior colliculus. In wild-type mice, the temporal and nasal axons project to the anterior and posterior superior colliculus, respectively. In *EphrinA2/A5* double-knockout mice, both temporal and nasal axons are scattered along the anterior–posterior axis of the superior colliculus (white arrows). **(C)** Schematic summary of axon targeting for five different genotypes: wild type, single *EphrinA* knockouts (*A2−/−* and *A5−/−*), double heterozygotes (*A2+/−,A5+/−*), and double knockouts (*A2−/−,A5−/−*). Dots represent terminations of temporal and nasal axons. Bottom panels summarize the protein gradients of ephrin-A2 and ephrin-A5. As the number of disrupted *EphrinA* gene copies increases, the targeting defects become more severe. (Adapted from Feldheim DA, Kim Y, Bergemann AD, et al. [2000] *Neuron* 25:563–574. With permission from Elsevier Inc.)

5.5   A single gradient is insufficient to specify an axis

The tectal ephrin-A and retinal EphA gradients satisfactorily account for the targeting of temporal axons to the anterior tectum, because temporal axons express the highest amounts of EphA, the receptor for the ephrin-A repellent (Figure 5-7). But nasal axons also express EphA, albeit at low levels. Why aren't they repelled by ephrin-A, instead targeting to the posterior tectum highly enriched in the repellent activity?

One possibility is that competition between RGC axons results in the filling in of the target space. Although this did not appear to be the case in the regeneration experiments (Figure 5-5), the filling in of target space is important during development. Due to axon–axon competition, nasal axons are "pushed" more posteriorly because the anterior tectum is already occupied by temporal axons with the highest expression level of the receptor for a posterior repellent. Studies in genetically engineered mice overexpressing EphA in a subset of RGCs provided strong evidence to support this model. About 40% of RGCs across the retina express a transcription factor called Islet 2 (Isl2). Using a knock-in strategy (Section 14.7), a constant level of additional EphA receptor was expressed in RGCs under the control of the *Isl2* promoter. This created a gain-of-function condition: intercalated among RGCs expressing EphAs in a natural gradient were a population of about 40% RGCs expressing an additional, fixed level of EphA3 driven by the *Isl2* promoter (Figure 5-9A, top), thus generating two intercalated gradients with different levels of EphA expression.

What happened to the retinotopic map in these knock-in mice? Unlike wild-type mice, which form one-to-one retina-target connections, the homozygous *EphA3* knock-in (ki/ki) mice had a duplicated map in the superior colliculus: dye injections at a single small spot in the retina labeled two separate target areas along the anterior–posterior axis of the superior colliculus (Figure 5-9A, bottom). Both target areas continued to follow retinotopy (Figure 5-9B).

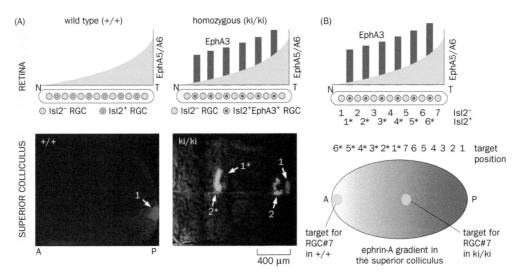

Figure 5-9 RGC axon targeting position is determined by relative Eph expression levels. (A) Top, schematic of Eph levels in RGCs from wild-type or homozygous knock-in (ki/ki) mice. Roughly 40% of RGCs (symbolized by cells with unfilled inner circles) express the transcription factor Islet2 (Isl2), and Isl2+ cells are intercalated among Isl2– RGCs across the retina. In ki/ki mice, these Isl2+ cells (symbolized by cells with filled inner circles) express additional EphA3 under the control of the *Isl2* promoter. This EphA3 overexpression (dark blue) is superimposed on the endogenous EphA gradient (light blue), mostly composed of EphA5 and EphA6. N, nasal; T, temporal. Bottom, targeting positions of RGC axons in the superior colliculus (A, anterior; P, posterior) in mice with corresponding genotypes after dye injection into the nasal retina at the positions indicated by numbers in Panel B. A red dye was injected into retinal area 1/1* in a wild-type mouse. In a

ki/ki mouse, a red dye was injected into retinal area 1/1*, and a green dye into retinal area 2/2*. Knock-in mice have a duplicated map in the superior colliculus corresponding to the respective terminations of Isl2– (1, 2) and Isl2+ (1*, 2*) RGC axons. **(B)** Schematic interpretation of the data from ki/ki mice. Isl2+ RGC axons (1* to 6*) push Isl2– RGC axons toward the posterior superior colliculus. For example, temporal-most RGC#7 (Isl2–) axons normally target to the anterior-most superior colliculus because RGC#7 expresses the most EphA. In ki/ki mice, RGC#7 is pushed toward the middle of the superior colliculus because its EphA level is exceeded by all Isl2+ RGCs. Although neither its EphA nor the target ephrin levels are altered, RGC#7 targeting is altered due to axon–axon competition. (A, from Brown A, Yates PA, Burrola P, et al. [2000] *Cell* 102:77–88. With permission from Elsevier Inc.)

The simplest interpretation of these data is that, in these ki/ki mice, Isl2-negative RGCs formed a map in the superior colliculus according to their endogenous EphA levels, while Isl2-positive RGCs formed a separate map according to EphA levels that were the sum of the graded endogenous EphAs and the constant additional EphA3 expressed from the *Isl2* promoter. This experiment illustrated an important property regarding EphA/ephrin-A interaction: the target positions of RGC axons in the superior colliculus are determined by the *relative* rather than *absolute* levels of EphA. For example, an Isl2-negative temporal RGC (number 7 in Figure 5-9B) axon normally targets to the anterior-most area of the superior colliculus due to its high level of EphA expression. However, in the ki/ki mice, it targeted to the middle of the superior colliculus, despite its EphA expression level not having changed. This mistargeting occurred because its EphA level was surpassed by those of Isl2-positive RGCs overexpressing EphA3. Thus, RGC axons must compare relative EphA levels with each other to make their target selection. Exactly how this is achieved is unclear.

Mechanisms other than axon–axon competition also contribute to RGC axon targeting along the anterior–posterior axis. One mechanism is the bidirectional signaling of ephrin-A/EphA. As well as expressing a temporal > nasal EphA gradient, RGC axons also express a nasal > temporal ephrin-A countergradient. Likewise, the tectum (superior colliculus) also expresses an anterior > posterior EphA countergradient in addition to the posterior > anterior ephrin-A gradient (**Figure 5-10**A). Besides ephrin-A → EphA signaling, which was discovered first and called *forward signaling*, EphA in the superior colliculus can also serve as a repulsive ligand for retinal axons expressing ephrin-A, a process called *reverse signaling* (Figure 5-10B). Reverse signaling explains why nasal axons prefer the posterior superior colliculus, as these axons express the highest ephrin-A levels, forcing them to choose target regions expressing the lowest EphA levels. When ephrin-As were globally knocked out, no receptors in the nasal RGCs were present to detect the EphA gradient in the superior colliculus, thus abolishing their targeting selectivity (Figure 5-8C). A recent study suggested that ephrin-A from RGC axons can repel EphA-expressing RGC axons, and this axon–axon repulsion contributes to retinocollicular mapping.

How are graded expression patterns of axon guidance molecules like ephrin-A and EphA established during development? Searches for other molecules expressed in a gradient revealed a transcription factor called Engrailed-2 (En2), which regulates the expression of ephrin-A. Graded En2 expression is in turn regulated by members of the **fibroblast growth factor (FGF)** family of secreted proteins. During early development, FGF mRNA is expressed at the midbrain–hindbrain junction,

Figure 5-10 Countergradients and bidirectional signaling of ephrin-A and EphA. (A) Schematic illustration showing both ephrin-A and EphA expressed as countergradients in the retina and the superior colliculus (SC). Forward signaling from ephrin-A expressed in the SC to EphA-expressing RGCs causes temporal axons to avoid the posterior SC, while reverse signaling from EphA expressed in the SC to ephrin-A-expressing RGCs causes nasal axons to avoid the anterior SC. The arrows indicate RGC axon targeting, not the direction of signaling. **(B)** Illustration of forward and reverse ephrin-A/EphA signaling. In forward signaling, ephrin-A acts as a ligand to send a signal to EphA-expressing growth cones, where EphA acts as a receptor tyrosine kinase (see Figure 3-39 for more details of receptor tyrosine kinase signaling). In reverse signaling, EphA acts as a ligand to send a signal to ephrin-A-expressing growth cones. Red zigzag line indicates that ephrin-A is anchored to the membrane by glycosylphosphatidylinositol. (A, adapted from Rashid T, Upton AL, Blentic A, et al. [2005] *Neuron* 47:57–69. With permission from Elsevier Inc. B, adapted from Egea J & Klein R [2007] *Trends Cell Biol* 17:230–238. With permission from Elsevier Inc.)

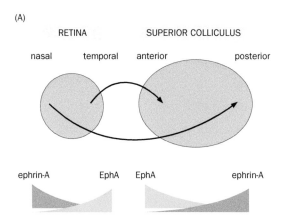

(A)

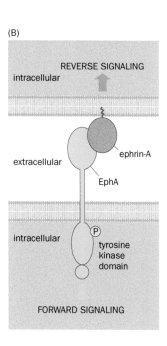

(B)

coinciding with the posterior edge of the tectum, such that secreted FGF proteins form a posterior > anterior gradient in the tectum. In tectum explant cultures, FGFs can upregulate En2 and ephrin-A expression and downregulate EphA expression. Thus, developmental patterning molecules such as FGFs help set up graded expression of axon guidance molecules via transcriptional regulation, a theme we will revisit in Chapter 7.

In summary, the identification of ephrins and Ephs as ligand–receptor pairs that are expressed in complementary gradients in the tectum/superior colliculus and retinal axons provided a satisfying proof of a key aspect of the chemoaffinity hypothesis proposed by Sperry 30 years earlier. More broadly, these and subsequent studies also suggested that similar mechanisms could operate in the wiring of the nervous system in general. These discoveries, along with the discoveries of several other axon guidance molecules around the same time, launched a new era in the molecular biology of axon guidance (Box 5-1).

Box 5-1: Molecular biology of axon guidance

In the 1990s, biochemical approaches in vertebrates, cellular studies in grasshopper embryos, and genetic analysis in the roundworm *C. elegans* and the fruit fly *Drosophila* led to the identification of a set of classic **axon guidance molecules**. Studies of these guidance molecules established a general framework regarding the mechanisms by which axons are guided to their targets (Figure 5-11A).

Axons are guided away from **repellents** (repulsive molecular cues) and toward **attractants** (attractive molecular cues). Each of these two categories includes both **long-range cues**—secreted proteins that act at a distance from their cells of origin, and **short-range cues**—usually cell-surface-bound proteins that require contact between the cells that produce them and the axons they guide to exert their effects. Short-range cues can also be secreted proteins bound to the extracellular matrix near their secretion sites. These guidance cues act as **ligands** to activate **receptors** expressed on the surface

of axonal growth cones (Figure 5-11B; see Box 5-2 for more discussion of growth cones). For example, the ephrin-As are contact-mediated repulsive ligands, since these proteins are bound to membranes by a GPI anchor and repel RGC axons that express the EphA receptors. Some families of axon guidance cues contain both secreted and membrane-bound proteins; this is the case for the **semaphorins**, which were independently identified by monoclonal antibodies against antigens expressed in specific subsets of axon fascicles in grasshopper embryos and by biochemical purification of a repellent activity in vertebrate neurons. Some guidance cues can act as an attractant in one context and a repellant in a different context. An important class of contact-mediated attractive molecules is the **cell adhesion molecules**, named for their ability to cause cells to adhere to other cells or to the extracellular matrix. These include immunoglobulin superfamily cell adhesion molecules (**Ig CAMs**) and Ca²⁺-dependent cell adhesion proteins (**cadherins**). Some Ig

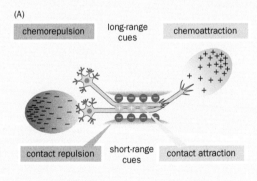

(A)

chemorepulsion — long-range cues — chemoattraction

contact repulsion — short-range cues — contact attraction

(B)

ligand	receptor
ephrin-A	EphA
ephrin-B	EphB
EphA	ephrin-A
EphB	ephrin-B
netrin/UNC-6	DCC/Unc40; Unc5
semaphorin	plexin; neuropilin; integrin
slit	robo
cadherin	cadherin
Ig CAM	Ig CAM
Hedgehog	Ptc; Ihog/Boc
Wnt	Frizzled; Derailed/Ryk

Figure 5-11 Molecular mechanisms of axon guidance. (A) Axons can be guided by repulsion or attraction. These repellents and attractants can be either secreted (from the red or green cell, respectively), and can thus act at a distance from the cells of origin, or be bound to the cell surface or the extracellular matrix, such that they require contact with axons (blue) to exert their effects. **(B)** A partial list of axon guidance cues (ligands) and their receptors. Note that some molecules can serve as both ligands and receptors. We will encounter these and other proteins in this chapter and Chapter 7. (A, adapted from Kolodkin AL & Tessier-Lavigne M [2011] *Cold Spring Harb Perspect Biol* 3:a001727.)

Box 5-1:   continued

CAMs and cadherins are **homophilic cell adhesion proteins**, which facilitate adhesion between cells via direct binding of the same proteins from apposing cells, whereas others are **heterophilic**, binding to different members of these protein families or proteins from other families. Cell adhesion proteins can promote axon guidance by providing a permissive environment for axons to extend or by stabilizing transient contact between axons and their targets. Remarkably, many of these axon guidance cues are evolutionarily conserved across worms, flies, chickens, and mammals.

We use **netrin/Unc6** as an example to illustrate evolutionary conserved mechanisms of axon guidance. In the vertebrate spinal cord, **commissural neurons** of the dorsal spinal cord send their axons ventrally toward the **floor plate**, a structure at the ventral midline of the spinal cord (**Figure 5-12**B). The floor plate is an intermediate target for commissural axons before they cross the midline and turn anteriorly to relay sensory information to the brain (see Section 7.6 for details). Using explant co-culture in a three-dimensional collagen matrix, researchers found that chick floor plate explants promoted axon outgrowth from commissural neurons in dorsal spinal cord explants and also caused these axons to turn toward floor plate explants. These findings suggested that

the floor plate contains chemoattractant(s) for commissural axons that are also outgrowth promoting. Biochemical purification of the growth-promoting activity in chick identified two related proteins called netrins (named after the Sanskrit word meaning "one who guides"). When expressed from heterologous cells, netrins could induce turning of commissural axons from spinal cord explants toward the netrin source (Figure 5-12A), indicating that netrin can indeed act as a chemoattractant.

In the mouse, only one netrin is present in the spinal cord; it is expressed in the floor plate and the ventricular zone (an area adjacent to the ventricle near the midline where neuroprogenitors reside; see Section 7.2 for details) of the ventral spinal cord (Figure 5-12B, left). In netrin knockout mice, commissural axons exhibit severe defects in their guidance toward the floor plate. Recent conditional knockout analysis revealed that ventricular zone-derived netrin, which is deposited along the path of commissural axons via the long processes of ventricular zone cells, plays a predominant role at a shorter range in guiding commissural axons on their trajectory toward the ventral midline. Subsequently, floor plate-derived netrin acts at a longer range, along with ventricular zone-derived netrin, to guide axons toward the

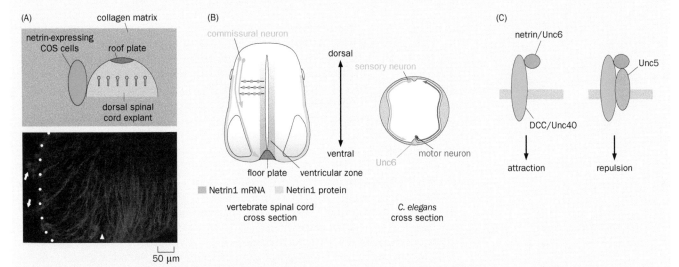

Figure 5-12 Netrin and evolutionary conserved axon guidance mechanisms from nematodes to mammals. (A) Top, schematic of experimental preparation for testing the activity of netrin on commissural axons. Bottom, a dorsal spinal cord explant containing the dorsal-most roof plate and commissural neurons (circles in the schematic) was cultured *in vitro* in a three-dimensional collagen matrix. Commissural axons normally grow ventrally, as illustrated in the schematic. When netrin-expressing COS cells were placed on the left of the dorsal spinal cord explant, the commissural axons (stained in red) located close to the lateral edge turned left toward the COS cells, and some exited the explant (arrows). Thus, netrin expressed from COS cells attracts commissural axons from a distance. The arrowhead marks the border where netrin causes commissural axons to turn, ~150 μm from the source. **(B)** Schematic summary of the evolutionarily conserved functions of netrin. Left, in mice, netrin is produced both in the floor plate and ventricular zone. Ventricular zone-derived netrin is transported along the processes of ventricular

zone cells (cyan dots along horizontal gray lines) and deposited near the periphery of the spinal cord, promoting ventral growth of commissural axons at early stages. Near the midline, floor plate-derived netrin attracts commissural axons toward the midline. Right, in *C. elegans*, Unc6 is produced by cells at the ventral midline and is required for ventral guidance of sensory axons and dorsal guidance of motor axons. **(C)** Netrin/Unc6 acts as an attractant through the DCC/Unc40 receptor and as a repellent when Unc5 is present as a co-receptor with DCC/Unc40. (A, adapted from Kennedy TE, Serafini T, de la Torre JR, et al. [1994] *Cell* 78:425–435. With permission from Elsevier Inc. B, based on Wu Z, Makihara S, Yam PT, et al. [2019] *Neuron* 101:635–647 and Hedgecock EM, Culotti JG, & Hall DH [1990] *Neuron* 2:61–85. See also Serafini T, Kennedy TE, Galko MJ, et al. [1994] *Cell* 78:409–424; Moreno-Bravo JA, Roig Puiggros S, Mehlen P, et al. [2019] *Neuron* 101:625–634; Ishii N, Wadsworth WG, Stern BD, et al. [1992] *Neuron* 9:873–881.)

(Continued)

Box 5-1: continued

ventral midline (Figure 5-12B, left). Thus, both gain- and loss-of-function experiments validated netrin's key role in the guidance of commissural axons toward the floor plate, via multiple cellular sources and mechanisms.

Netrin has strong sequence similarity to the *C. elegans* protein Unc6. The *Unc6* gene was originally identified by its mutant phenotype of <u>unc</u>oordinated movement (hence the name *Unc*). Like netrin, Unc6 expression is also enriched in the ventral midline. Unc6 plays an essential role in circumferential axon guidance along the dorsal–ventral axis. In wild-type worms, dorsally located sensory neurons target their axons ventrally, whereas ventrally located motor neurons target their axons dorsally (Figure 5-12B, right panel). In *Unc6* mutant worms, sensory axons exhibited defects in their ventral guidance whereas motor axons exhibited defects in their dorsal guidance. These observations suggested that Unc6 is a bidirectional axon guidance cue.

Genetic analysis in *C. elegans* revealed that **Unc40** and **Unc5** are also required in circumferential axon guidance. Interestingly, *Unc5* mutant worms exhibited defects specifically for dorsal guidance of motor axons but not ventral guidance of sensory axons. Further studies indicated that Unc40 and its mammalian homolog **DCC** (<u>d</u>eleted in <u>c</u>olon <u>c</u>ancer) are receptors for Unc6/netrin. When acting alone, Unc40/DCC mediates the attractive response. Together with Unc5 as a co-receptor, Unc40/DCC mediates the repulsive response of Unc6/netrin (Figure 5-12C). Remarkably, these guidance molecules and mechanisms are conserved in *C. elegans*, *Drosophila*, and mammals.

Since axon guidance cues were first identified in the 1990s, the list of cues has expanded considerably. Receptors for many axon guidance cues have also been identified (Figure 5-11B). Interestingly, some of these ligand–receptor pairs also function earlier in development to pattern cells, tissues, and body axes as well as to guide cell migration. In addition, many axon guidance molecules are used at later stages of development to regulate target selection and synapse development. We will return to many of these molecules later in this chapter and in Chapter 7.

5.6 To cross, or not to cross: that is the question

In all vertebrates, RGC axons crossing the midline form an **optic chiasm** (**Figure 5-13**). Midline crossing is likely an ancestral state because all RGC axons cross the chiasm in fish, tadpoles, and birds. (Indeed, midline crossing occurs in many axonal pathways in the CNS; Box 5-1.) In these animals, the eyes are typically lateral (located on the sides of the head) and two eyes sample mutually exclusive regions of visual space. Information is therefore sent from the left eye to the right brain and from the right eye to the left brain. However, in animals whose eyes face forward (as in humans), such that the left and right eyes view partially overlapping regions of the visual space, the left half of the visual space goes to the right half of the brain, and the right half of the visual space goes to the left half of the brain. Hence, information from each eye is sent to both sides of the brain, and converges in the primary visual cortex (Figure 4-37). This binocular visual processing helps us compute the depth of objects in our visual field.

The degree of binocularity in different animals results from different percentages of RGC axons crossing the midline at the optic chiasm (Figure 5-13) as well as different eye positions. In humans, axons originating in the nasal retina (~60% of total RGCs) cross the midline, constituting the **contralateral** projection, while axons from the temporal retina (~40% of total RGCs) remain on the same side, constituting the **ipsilateral** projection. This enables the projections of temporal RGCs from the ipsilateral eye and nasal RGCs from the contralateral eye that survey the same visual space to converge in the same brain hemisphere, allowing binocular vision (Figure 4-37). In carnivores such as cats and ferrets, 15–30% of axons are ipsilateral. In mice, only 3–5% of the RGCs located in the ventrotemporal retina project ipsilaterally, so mice have only a small degree of binocular vision.

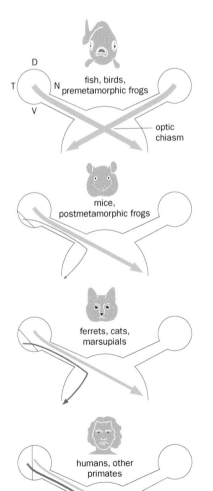

Figure 5-13 Midline crossing of RGC axons. In different animals, different fractions of RGCs project ipsilaterally (that is, to the hemisphere on the same side as the eye). As the fraction of ipsilateral projection increases (represented by the thickness of the red arrow), the animal's degree of binocular visual processing increases correspondingly. Thus, binocularity is absent in fish, birds, and tadpoles (top panel), but highly developed in primates (bottom panel). T, temporal; N, nasal; D, dorsal; V, ventral. (Adapted from Petros TJ, Rebsam A, & Mason CA [2008] *Annu Rev Neurosci* 31:295–315. With permission from Annual Reviews.)

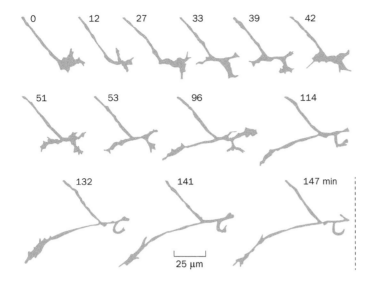

Figure 5-14 **Growth cone behavior at the optic chiasm.** Time-lapse images of a growth cone from an ipsilaterally projecting RGC in an embryonic mouse retina-chiasm explant, taken as the RGC commits to not crossing the midline (dotted line to the right of the last image). min, minutes. (From Godement P, Wang L, & Mason CA. [1994] *J Neurosci* 14:7024–7039. Copyright ©1994 Society for Neuroscience.)

What determines whether or not RGC axons cross the midline when they reach the optic chiasm?

Visualization of RGC axon growth cones as they travel near the optic chiasm in explants derived from mouse embryos provided clues (**Figure 5-14**). Growth cones from both ipsilateral and contralateral retinal axons slow down considerably at the chiasm. There, two actin-rich structures at the leading edge of the growth cones—thin projections (filopodia) surrounded by sheet-like webbing (lamellipodia)—undergo cycles of extension and retraction driven by the polymerization and depolymerization of actin and by the motor protein myosin (see more discussion of growth cone dynamics in **Box 5-2**). Contralateral axons then quickly move forward, whereas ipsilateral axons continue to extend and retract until ipsilaterally directed filopodia are consolidated, giving rise to new growth cones.

The molecular basis for the differing behaviors of ipsilaterally and contralaterally projecting axons has been identified in mice (**Figure 5-15**). During the time window in which RGC axons project to the brain, EphB1 (a member of the Eph receptor family we encountered earlier) is expressed specifically in the ventrotemporal retina, where ipsilateral RGC axons originate. Glial cells at the optic chiasm express ephrin-B2, which repels EphB1-expressing RGCs and prevents their crossing the midline. The restricted expression of EphB1 is enabled by specific expression of the transcription factor Zic2 in the ventrotemporal retina. Misexpression of Zic2 in other parts of the retina resulted in misexpression of EphB1, forcing otherwise contralateral axons to project ipsilaterally. Interestingly, Zic2 expression in different animal species precisely predicts the degree of ipsilateral projection and hence the degree of binocular vision. Thus, the evolution of the Zic2 expression pattern, combined with changes in eye position, was likely an important force behind the evolution of binocular vision.

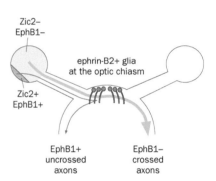

Figure 5-15 **Mechanism of midline crossing of RGC axons in mice.** Ventromedial RGCs specifically express the transcription factor Zic2, which activates EphB1 expression. At the chiasm, EphB1-expressing RGC axons choose the ipsilateral trajectory because they are repelled by glial cells expressing ephrin-B2. (Adapted from Petros TJ, Rebsam A, & Mason CA [2008] *Ann Rev Neurosci* 31:295–315.)

Box 5-2: Cell biology and signaling at the growth cone

How do guidance cues direct the trajectories of axons? The secret lies at the growing tip of the axon, the **growth cone**. Originally discovered by Santiago Ramón y Cajal via Golgi staining of developing nerve tissues and then subsequently studied by live imaging in tissue culture (Figure 1-13; **Movie 5-2**), the growth cone contains two prominent structures consisting primarily of filamentous actin (F-actin; Figure 2-5A). The thin, protruding **filopodia** are made of bundled F-actin, whereas the veil-like **lamellipodia** are a meshwork of branched F-actin (**Figure 5-16**A). Rapid polymerization and depolymerization of F-actin, in combination with myosin-induced flow of F-actin from the periphery to the center of the growth cone (**retrograde flow**), cause the growth cone to exhibit dynamic shape changes, including extension, retraction, and turning. The central part of the growth cone is rich in microtubules (Figure 2-5B), which

(Continued)

Box 5-2: continued

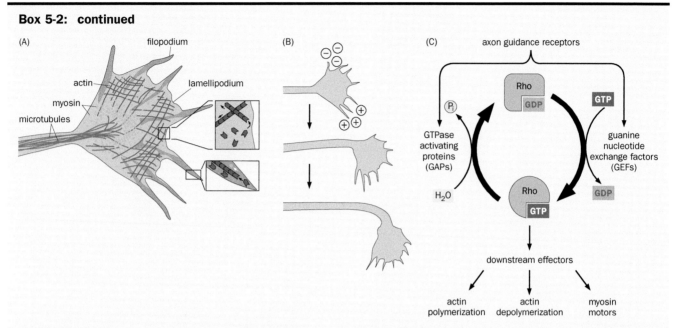

Figure 5-16 Cell biology and signaling at the growth cone. (A) A typical growth cone contains microtubules at its center and F-actin at its periphery. The F-actin fibers exist in two major forms: branched networks supporting lamellipodia and bundles enriched in filopodia. Insets highlight actin polymerization at the leading edge of both filopodia and lamellipodia, with plus ends (the concave or hollowed-out sides of each arrowhead-shaped monomer) facing the leading edge of the cell. Myosins (green dots) are distributed throughout the growth cone but are most concentrated at the interface between actin- and microtubule-rich domains. At the periphery, the myosin motors are responsible for retrograde flow of F-actin fibers (white arrow in the lower inset). Some microtubules also protrude to the leading edge, where they interact intimately with the actin cytoskeleton. **(B)** Attractive guidance cues (plus signs) stabilize filopodia, while repulsive guidance cues (minus signs) destabilize filopodia. The combination of these effects causes the growth cone to turn. **(C)** One of many growth cone signaling mechanisms involves the Rho family of small GTPases. Axon guidance receptors transmit signals via Rho GTPases to downstream effectors that modulate cytoskeletal activity. Guanine nucleotide exchange factors (GEFs) convert Rho to the active (GTP-bound) form, whereas GTPase activating proteins (GAPs) facilitate the intrinsic GTPase activity of Rho, converting it to the inactive (GDP-bound) form. Thus, by regulating GEFs and GAPs, axon guidance receptors control the activity levels of Rho GTPases. (Adapted from Luo L [2002] *Ann Rev Cell Dev Biol* 18:601–635. See also Dent EW, Gupton SL, & Gertler FB [2011] *Cold Spring Harb Perspect Biol* 3:a001800.)

provide structural support and a stabilizing force. Like F-actins, microtubules are also highly dynamic; a subset of microtubules protrudes to the leading edge of the growth cone and interacts intimately with the actin cytoskeleton.

The essence of growth cone guidance is to transform the detection of extracellular cues into changes in the underlying cytoskeleton. Suppose that the axon in Figure 5-16B turns downward in response to attractive cues below, repulsive cues above, or both. Turning can be achieved by preferentially stabilizing the filopodia that extend toward the bottom, destabilizing the filopodia that extend toward the top, or a combination of these events. This bias can then be reinforced by stabilizing the microtubules that extend to the stabilized filopodia and transporting membranous organelles and vesicles to fuel the growing filopodia via microtubule motors (Section 2.3). A growing axon extends long filopodia in different directions to better detect the differences in attractive and repulsive cues in its environment, allowing the growth cone to make the most informed comparison to direct its next move.

Axon guidance cues signal through their receptors to initiate downstream signaling pathways, which eventually modulate the cytoskeleton. Many signaling pathways link diverse axon guidance receptors to regulation of the actin cytoskel-

eton. For example, small GTPases of the Rho family, including Rho, Rac, and Cdc42, play crucial roles in regulating actin polymerization and depolymerization and actin–myosin interaction. As discussed in Box 3-3, these small GTPases are tightly regulated molecular switches activated by guanine nucleotide exchange factors (GEFs) and inactivated by GTPase activating proteins (GAPs). The activities and localizations of GEFs and GAPs are in turn regulated by axon guidance receptors (Figure 5-16C). For example, a guanine nucleotide exchange factor called ephexin binds EphA in the context of ephrin-A → EphA forward signaling. Upon EphA activation by extracellular ephrin-A binding, ephexin enhances its nucleotide exchange activity toward the small GTPase RhoA, whose activation leads to growth cone collapse, realizing the repulsive activity of ephrin-A. Some axon guidance receptors also regulate the cytoskeleton through other signaling pathways or more directly modulate the factors controlling actin polymerization and depolymerization as well as myosin-mediated retrograde flow of F-actin. Other axon guidance receptors signal to the regulators of microtubules and to the proteins linking the actin and microtubule cytoskeletons. How a growth cone integrates signals from different pathways to coordinate its next move is still very much an open question.

HOW DO EXPERIENCE AND NEURONAL ACTIVITY CONTRIBUTE TO WIRING?

So far, our studies of RGC axon guidance seem to suggest that the visual system is hard-wired, with molecular recognition able to explain nearly all aspects of axon guidance. However, around the time Sperry proposed the chemoaffinity hypothesis, David Hubel and Torsten Wiesel performed a series of experiments on visual cortical neurons that revealed a different force guiding the wiring of the visual system. These experiments have had just as much influence on generations of neuroscientists as did the chemoaffinity hypothesis.

5.7 Monocular deprivation markedly impairs visual cortex development

Recall from Chapter 4 that some neurons in the primary visual cortex manifest binocular responses by combining inputs from neurons of the lateral geniculate nucleus (LGN). Axons from different eyes segregate into eye-specific layers within the LGN, and thus each LGN neuron receives direct RGC input mostly from only one eye. Due to RGC axons crossing at the optic chiasm (Figure 5-13), some LGN neurons receive input from the temporal part of the ipsilateral eye, while other LGN neurons located in other layers receive input from the nasal part of the contralateral eye. LGN signals from the left and right eyes eventually converge onto individual cortical neurons to create binocular neurons, the basis for binocular vision (Figure 4-37B).

When Hubel and Wiesel mapped receptive fields of single neurons in the primary visual cortex (V1) of the cat, they found that individual cells responded preferentially to stimuli from the ipsilateral eye, the contralateral eye, or both eyes that see the same visual space. They plotted these responses according to a 1 to 7 **ocular dominance** scale (Figure 5-17A). Cells sharing the same ocular dominance clustered spatially in vertical columns called **ocular dominance columns**.

To probe whether visual experience has anything to do with the development of the receptive fields of cortical neurons, Hubel and Wiesel sutured one eyelid of the cat from birth, thus depriving that eye of form vision. (They conducted these experiments to create an animal model of human congenital cataract, where children born with serious clouding of the lens or cornea are blind for life unless surgery to correct vision is performed when they are very young.) They then opened the sutured eye when the cat was several months old and recorded the visual responses along the visual pathway. Little difference was found between responses recorded from the retinal neurons of the sutured and open eyes or from the neurons of the two LGNs, but a profound difference was found in V1 neurons. The

Figure 5-17 **Visual response of cortical neurons under normal and monocular deprivation conditions. (A)** Neurons in the cat primary visual cortex (V1) are divided into seven categories based on their response to visual stimuli from the ipsilateral and contralateral eyes. Those that respond exclusively to contralateral or ipsilateral stimuli are given a value of 1 or 7, respectively, while cells that respond equally well to stimulus from either eye receive the value 4. Values 2, 3, 5, and 6 represent cortical neurons that exhibit different degrees of contralateral or ipsilateral bias. The plot shows the distribution of scores given to 233 cells recorded from V1 of a normally reared cat. **(B)** When the contralateral eye was deprived by eyelid suture during the period indicated by the black bar underneath, virtually all neurons responded only to visual stimulation from the ipsilateral eye after the suture was reopened. Five cells (represented in the distribution by a dashed bar) could not be driven by either eye. (From Wiesel TN & Hubel DH [1963] *J Neurophysiol* 26:1003–1017. With permission from the American Physiological Society.)

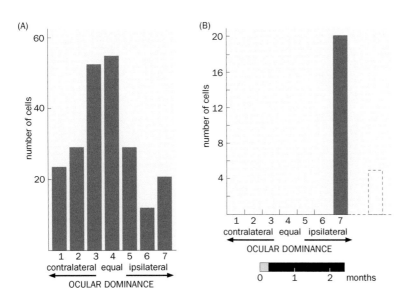

Figure 5-18 Ocular dominance columns visualized by autoradiography under normal and monocular deprivation conditions. Radioactively labeled amino acids were injected into one eye. These amino acids were incorporated into proteins, transported to RGC axons terminals, transferred to postsynaptic targets in the LGN, and ultimately transported to LGN axon terminals in layer 4 of V1. Results are visualized here by autoradiography of V1 histological sections. The white bands (autoradiographic labels) represent input to V1 from the injected eye. **(A)** In a normally reared monkey, ocular dominance columns receiving input from the injected and noninjected eyes were of equal width. **(B)** In a monkey in which one eye was surgically closed during the critical period, radioactively labeled amino acids were subsequently injected into the open eye. Note that the white bands are much wider than the black bands, indicating that V1 receives input mostly from the open eye. (From Hubel DH, Wiesel TN, & LeVay S [1977] *Phil Trans R Soc Lond B* 278:377–409. With permission from the Royal Society of London.)

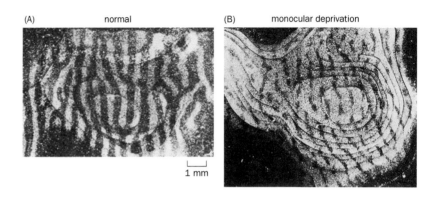

(A)   normal   (B)   monocular deprivation

⊢—⊣
1 mm

great majority of cortical cells were driven by the normal eye; very few cortical cells responded to stimulation from the deprived eye (Figure 5-17B). Nearly identical results were obtained in later experiments on monkeys. Behaviorally, these animals lost visual functions in their deprived eye, consistent with physiological recordings in the cortex.

Hubel and Wiesel further performed a series of time-course experiments to determine when during development monocular deprivation had the most significant effect. They found a window they called the **critical period**, which in cats begins abruptly around the fourth week after birth (about two weeks after kittens open their eyes) and ends gradually around 12 weeks of age. Monocular deprivation after the critical period had little effect compared to the large effect during the critical period. For example, closing one eye for just a few days during the fourth or fifth weeks had devastating effects both on vision and on cortical responses to the deprived eye, while much longer closure later in life had no significant consequences. For the most part, visual system defects resulting from monocular deprivation during the critical period were irreversible.

The drastic effect of visual deprivation was demonstrated when a transneuronal tracer was used to visualize ocular dominance columns in layer 4 of V1. In this technique, radioactively labeled amino acids were injected into one eye. They were taken up and incorporated into proteins in retinal neurons, transported by RGC axons to the LGN, and then further transferred to V1 terminals of LGN axons after crossing the RGC → LGN synapses. Eventually, LGN axons and cortical neurons connected to the injected eye were preferentially labeled by radioactivity. V1 sections were then exposed on a film to reveal the spatial distribution of radioactive labeling. This yielded a striking image of strongly labeled white stripes, representing input from the injected eye, alternating with largely black stripes, representing input from the uninjected eye. In normal animals, radioactively labeled and unlabeled stripes had similar widths, representing equal contributions of inputs from left and right eyes to layer 4 of V1 (**Figure 5-18**A). In monocularly deprived monkeys and cats, layer 4 of V1 was dominated by input from the non-deprived eye (Figure 5-18B).

These experiments have had profound implications on disciplines ranging from neuroscience and developmental psychology to ophthalmology. They revealed the powerful force of nurture, or experience, in sculpting the developing neural circuits. They also indicated that chemoaffinity is not the sole determinant of neural wiring. But how is neural wiring shaped by experience?

5.8 Competing inputs are sufficient to produce spatial segregation at the target

An important clue as to how experience influences brain wiring came from a control experiment Hubel and Wiesel did. When all vision was deprived by suturing the eyelids of *both* eyes during the critical period, cortical cells exhibited visual responses that were not as drastically affected as those resulting from monocular deprivation. Thus, the effect of deprivation was not simply atrophy due to disuse. This observation led to the idea that competition plays a key role in setting up

(A)  (B)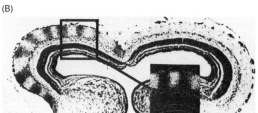

Figure 5-19 Competing retinal inputs from the third eye produce segregated RGC axons in the frog tectum.
(A) Transplantation of an eye primordium to a host embryo at an appropriate developmental stage can produce a frog with three eyes. **(B)** When axonal input from the third eye competes for tectal targets with axonal input from one of the normal eyes, axons segregate, as indicated by transneuronal tracing of radioactively labeled amino acids injected into a normal eye. Inset, enlargement showing axonal input segregation in a dark field, with the labeled axon termini appearing white. (From Constantine-Paton M & Law MI [1978] *Science* 202:639–641. With permission from AAAS.)

ocular dominance columns. Specifically, thalamocortical axons originating from eye-specific layers of the lateral geniculate nuclei (Figure 4-37B) may compete for space in the visual cortex. Hence, visual experience, in the form of correlated neuronal activity (see Sections 5.11 and 5.12), may provide a competitive advantage to the nondeprived eye. Under binocular deprivation, inputs from both eyes are weakened and the resulting defect is not as severe as that caused by monocular deprivation.

How can one experimentally test the idea that competition between the two eyes *causes* the formation of ocular dominance columns? As we noted in Section 1.14, two kinds of perturbation experiments, loss-of-function and gain-of-function, can be used to test causality between two phenomena. The experiments described so far employ loss-of-function perturbations, testing whether experience or activity in the eyes is necessary for the development of ocular dominance columns in V1. Can one design a gain-of-function sufficiency experiment to strengthen the causality between competing inputs and the generation of ocular dominance columns?

The key is to identify a normal condition in which there is minimal competition between inputs from two eyes. This condition can readily be found in nature. Recall that binocular vision exists to varying degrees in different vertebrates (Figure 5-13). The frog has few RGCs that project ipsilaterally, so its tecta receive input mostly from the contralateral eye, and there is little opportunity for inputs from both eyes to compete. Remarkably, researchers created novel circumstances for such competition by utilizing the superb embryological techniques available in frogs. A third eye was created by transplanting an eye primordium to a normal frog embryo at the appropriate location and developmental stage (**Figure 5-19**A). RGC axons from the third eye projected to one of the tecta, targeting appropriate retinotopic positions in response to the molecular cues discussed earlier in the chapter. RGC axons from the third eye then had to compete for space with RGC axons from a normal eye. When radioactively labeled amino acids were injected into the normal eye projecting to the same tectum as the third eye, inputs from two eyes projecting to the same tectum were found to segregate into bands of a characteristic width (Figure 5-19B), resembling the ocular dominance columns in the primary visual cortex of normal cats and monkeys.

This experiment elegantly demonstrated that competing inputs are sufficient to generate segregated bands at their postsynaptic targets. Follow-up experiments showed that blocking neuronal firing (by injecting TTX into the eye to block voltage-gated Na$^+$ channels and thus action potential propagation) results in desegregation of eye-specific inputs into the tecta of three-eyed frogs. These findings supported a role for neuronal activity in the generation and maintenance of segregated tectal inputs from each eye, similar to the role of neuronal activity in the development of ocular dominance columns in the visual cortex, to which we return.

5.9 Ocular dominance columns in V1 and eye-specific layers in LGN develop by gradual segregation of eye-specific inputs

The striking appearance of ocular dominance columns in layer 4 of V1 in cats and monkeys, and their perturbation during the critical period by monocular visual deprivation (Figure 5-17; Figure 5-18), led researchers to ask how the columns first form during development. The autoradiography method of visualizing ocular

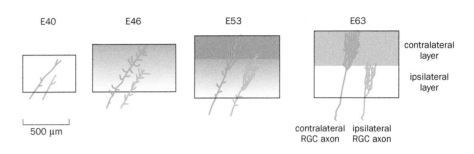

Figure 5-20 Formation of eye-specific layers of RGC axons in the LGN of the cat. Gradual appearance of eye-specific layers (segregation of gray and white backgrounds, as determined by labeling of radioactive amino acids injected into one eye) in the cat LGN occurs as individual axons prune their branches in the incorrect layer and elaborate their branches in the correct layer. Embryonic days after fertilization are noted above. (Adapted from Sretavan DW & Shatz CJ [1986] *J Neurosci* 6:234–251.)

dominance columns was used to address this question. During early development, monocular injection of radioactive amino acids into either eye labeled a continuous band in layer 4 of V1. This band gradually becomes intermittent across development, until it achieves the degree of segregation observed in adults. Likewise, physiological recordings revealed that during early development, most cortical neurons respond, albeit weakly, to inputs from both eyes. However, subsequent experiments revealed that radioactive amino acids show much greater lateral diffusion (spillover) in young brains, leading to an overestimate of initial exuberance. Experiments utilizing intrinsic signal imaging (Figure 4-45B) and anterograde tracing from small numbers of LGN neurons suggested that ocular dominance columns form earlier than revealed by the autoradiography method. The exact time ocular dominance columns form remains incompletely resolved due to technical limitations, but researchers agree that it occurs during early development (see the next section) and the boundaries of these columns become sharper as development proceeds.

Inputs from the eyes are conveyed to the cortex via neurons in the LGN (Figure 4-37), where eye-specific input segregation has been more precisely defined. The autoradiography method suggested that RGC axons from both eyes intermix in the LGN during early development. This input gradually becomes refined and eventually segregates into eye-specific LGN layers. These observations were confirmed by visualizing the terminal arborizations of individual RGC axons. For example, an RGC axon terminating in the LGN initially forms branches along its entire path across the LGN. As layer-specific segregation occurs, the RGC axon branches more densely within what will become the layer specific for its eye; at the same time, its branches (and synapses) within the layer destined to be occupied by the other eye are pruned (Figure 5-20).

5.10 Retinal neurons exhibit spontaneous activity waves before the onset of vision

In monkeys, ocular dominance columns form prenatally, when the eyes are still naturally occluded. Eye-specific segregation in the LGN occurs even earlier, well before the onset of vision. If competing inputs from the two eyes are responsible for eye-specific segregation of RGC axons in the LGN and for initial development of ocular dominance columns in V1, what is the nature of such competition?

Experiments utilizing TTX injection to block neuronal activity suggested the importance of **spontaneous activity** in ocular dominance column development. Binocular TTX injection blocked the formation of ocular dominance columns in V1 more completely than dark rearing or binocular eyelid suturing, suggesting that in addition to patterned visual stimulation, which is degraded by dark rearing or suturing, spontaneous retinal activity plays an important role. Likewise, blocking prenatal spontaneous activity in the retina prevented eye-specific segregation of RGC axons in the LGN. But how does spontaneous activity create competition between the two eyes?

It turns out that before the onset of vision, RGCs in the retina are spontaneously active, firing action potentials that are transmitted to the LGN. The RGC activity is highly correlated, with nearby RGCs firing together. Experiments conducted in ferrets were particularly informative. Ferrets have more advanced

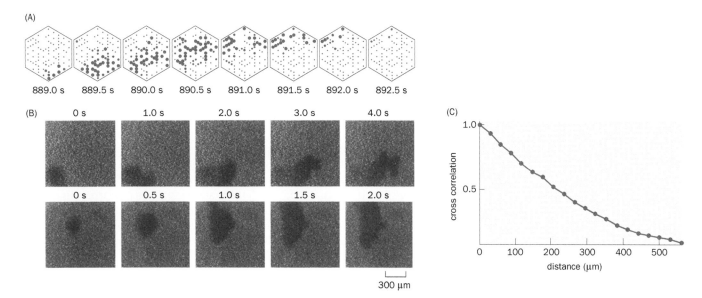

Figure 5-21 Retinal neurons exhibit waves of spontaneous action potentials. (A) A retinal wave is shown sweeping through the retina from bottom right to top left over 3.5 seconds (s). Each dot represents one electrode in a multielectrode array on which a postnatal day 5 ferret retinal explant was placed. Larger dots represent higher frequencies of RGC action potential firing. **(B)** Retinal waves can also be detected by Ca^{2+} imaging of retinal explants from young ferrets. Dark patches represent increased intracellular Ca^{2+} concentrations, which indicate higher neuronal activity. Two separate waves are shown (top and bottom panels), with different origins and directions of spread. **(C)** Cross correlation of retinal neuron activity as a function of distance from Ca^{2+} imaging experiments. When a neuron in a defined central position is active, a cross correlation measures how often another neuron at a given distance from the central neuron is also active; a cross correlation of 1 means that two neurons are always co-active. As is evident, nearby neurons are more likely to be co-active. (A, from Meister M, Wong ROL, Baylor DA, et al. [1991] *Science* 252:939–943. With permission from AAAS. B, from Feller MB, Wellis DP, Stellwagen D, et al. [1996] *Science* 272:1182–1187. With permission from AAAS. C, adapted from Feller MB, Butts DA, Aaron HL, et al. [1997] *Neuron* 19:293–306. With permission from Elsevier Inc.)

binocular vision than rodents (Figure 5-13). The ferret visual system is much less mature at birth than that of cats and monkeys, so experiments investigating early visual system development can be effectively carried out in postnatal ferrets. The correlated activities of RGCs could be measured by multielectrode recording of isolated retina from young ferrets: extracellular electrodes 50–500 μm apart detected sequential action potentials of nearby retinal neurons (**Figure 5-21**A). This spread of spontaneous excitation across the developing retina is called a **retinal wave**.

Retinal waves were more dramatically demonstrated by Ca^{2+} imaging experiments (Figure 5-21B; **Movie 5-3**), which enable researchers to visualize the activities of many neurons simultaneously by imaging fluorescence changes due to rising intracellular Ca^{2+} concentrations (Section 14.22). As waves of neuronal activation sweep across the retina, nearby RGCs in the same retina are activated at about the same time (Figure 5-21C). These waves are spontaneously initiated about once per minute. The likelihood of left and right retinas generating waves at the same time is very low, so activity patterns from RGCs in different retinas are uncorrelated. As we will learn in Section 5.12, correlated activity patterns from the same eye strengthen synaptic connections whereas uncorrelated activity patterns weaken synaptic connections. Thus, correlation of activity patterns in the same eye by retinal waves could account for the segregation of RGC axons into eye-specific layers in the LGN before the onset of vision.

Further studies of retinal waves identified several distinct phases of wave propagation during development before eye opening. The most prominent phase, which features cholinergic retinal waves involving connections between cholinergic amacrine cells and RGCs, coincides with RGC axons making their layer choices in the LGN and superior colliculus. During this period, rods and cones are immature, and bipolar cells have not yet connected to RGCs, consistent with the idea that these retinal waves are spontaneous in RGCs rather than evoked by visual experience.

5.11 Retinal waves and correlated activity drive segregation of eye-specific inputs

The roles of retinal waves in the wiring of the visual system was tested by pharmacological perturbations. Epibatidine, an alkaloid found on the skin of an endangered species of poison dart frog, is a long-lasting cholinergic agonist that blocks retinal waves by binding tightly to the nicotinic acetylcholine receptor (nAChR) expressed by RGCs and amacrine cells and causing receptor desensitization. Thus, epibatidine blocks synaptic transmission and cholinergic retinal waves. Injection of epibatidine into the eye therefore provided a loss-of-function manipulation for establishing the role of retinal waves in visual system wiring.

Eye-specific segregation of RGC axons occurs prenatally in cats (Figure 5-20), but during the first 10 postnatal days in ferrets, allowing these perturbation experiments. Eye-specific segregation was visualized by injecting the left and right eyes with different axonal tracers for visualizing their RGC axons (Figure 5-22A). At postnatal day 1, RGC axons from the left and right eyes were completely intermingled in the binocular region of the LGN (Figure 5-22B). Axons from the ipsilateral and contralateral eyes normally segregate into layers by postnatal day 10 (Figure 5-22C), but binocular epibatidine injection prevented segregation of eye-specific input (Figure 5-22D), suggesting that cholinergic retinal waves are essential for the eye-specific segregation of RGC axons in the LGN.

Interestingly, monocular injection of epibatidine caused RGC axons from the injected eye to lose territory to axons from the uninjected eye (Figure 5-22E). Conversely, a gain-of-function condition in which one eye was injected with a cAMP

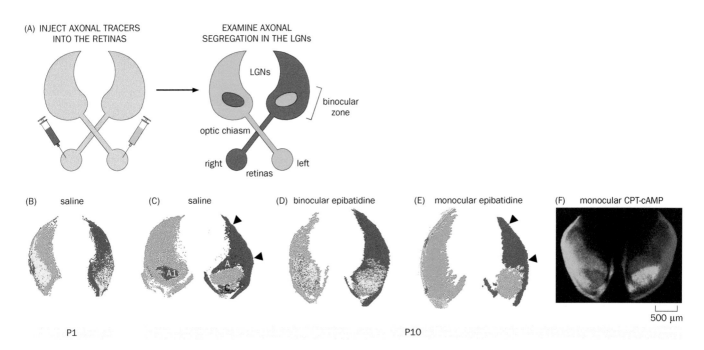

Figure 5-22 **Retinal waves regulate eye-specific segregation of RGC axons at the LGN. (A)** Schematic of experimental procedure. Two different axonal tracers (pseudocolored in red and green) were injected into the left and right retina of the ferret to label RGCs from each eye, and RGC axon termination patterns in LGNs were examined a day later. Shown in the schematic is a control experiment examined at postnatal day 10 (P10). The binocular zone, where eye-specific inputs segregate into distinct layers, is bracketed. Panels B–F show RGC axon terminals in LGNs under different experimental conditions. **(B)** In a control (saline injected) ferret at postnatal day 1 (P1), RGC axons have not segregated, as seen by the yellow patches in the future binocular zone. **(C)** By P10, the binocular zone is well segregated in a control ferret, with ipsilateral axons occupying layer A1, flanked by contralateral layers A and C. Between the arrowheads lies a monocular zone that represents only the contralateral eye. **(D)** When both eyes were injected with epibatidine to block cholinergic retinal waves, eye-specific segregation did not occur; the binocular zone at P10 is largely yellow, as at P1. **(E)** When epibatidine was injected only into the right eye, axons from the left eye (green) take over the binocular zone in both LGNs. **(F)** cAMP agonist CPT-cAMP augments retinal waves. Injecting CPT-cAMP into the left eye expands the areas occupied by RGC axon terminals from the left eye (green) in the binocular zones of both LGNs. (B–E, from Penn AA, Riquelme PA, Feller MB, et al. [1998] *Science* 279:2108–2112. With permission from AAAS. F, from Stellwagen D & Shatz CJ [2002] *Neuron* 33:357–367. With permission from Elsevier Inc.)

agonist known to increase the frequency of retinal waves caused RGCs from the injected eye to outcompete axons from the normal eye for territory in the LGN (Figure 5-22F). Together, these experiments demonstrated that competitive interactions between inputs from two eyes, requiring spontaneous activity generated by retinal waves, play a critical role in the proper segregation of eye-specific axons in the LGN.

As alluded to in Section 5.10, retinal waves likely act in the context of eye-specific segregation by providing correlated activity patterns within the same eye. In the mouse, a direct test of the role of correlated activity in eye-specific layer segregation of RGC axons in the superior colliculus has been performed using optogenetic stimulation. In the small binocular zone of the mouse superior colliculus, ipsilateral and contralateral RGCs normally segregate their axons in different layers between postnatal days 5 and 9, before the onset of natural vision. Combining tools in viral delivery and mouse genetics, light-activated channelrhodopsin (ChR2; Section 14.25) was expressed in mouse RGCs, and blue light that activates ChR2 was delivered to both eyes in either synchronous or asynchronous patterns from postnatal day 5 on. Synchronous activation of RGCs of both eyes disrupted eye-specific segregation of their inputs in the superior colliculus, while asynchronous activation facilitated segregation (**Figure 5-23**). The simplest interpretation of these results is that correlated activity patterns, whether provided naturally by retinal waves within the same eye or artificially by synchronous optogenetic stimulation of different eyes, cause axons to stabilize projections in the same layer. By contrast, uncorrelated activity patterns, provided naturally by uncorrelated retinal waves from different eyes or artificially by asynchronous optogenetic stimulation, facilitate axon segregation.

In vivo imaging in neonatal mice before eye opening revealed that waves of spontaneous activity occur in many regions of the visual system, including the superior colliculus, primary visual cortex (V1), and higher-order visual cortical areas. Waves of spontaneous activity in the superior colliculus and V1 are often coordinated not only in timing but also with respect to their retinotopy, suggesting that they both originate from the retina. Indeed, epibatidine injection into the contralateral eye (recall that >95% RGCs in the mouse project contralaterally)

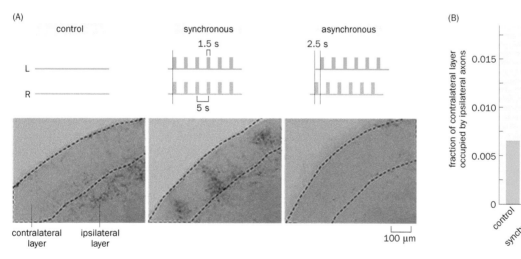

Figure 5-23 Synchrony of firing patterns and eye-specific axon segregation. (A) Top panels, schematic of the experimental procedure. ChR2 was expressed in RGCs from both eyes of transgenic mice. In the two experimental treatments, light was delivered to the two eyes either in synchrony (middle panel: vertical bars indicate 1.5 s synchronous optogenetic stimulation of both eyes every 5 s) or asynchrony (right panel: 1.5 s optogenetic stimulation every 5 s, with the onset of stimuli to each eye 2.5 s apart) from postnatal days 5 through 7. In the control treatment at left, no light was delivered. Bottom panels, ipsilateral axons (dark signal) in the contralateral layer (between the two dashed lines) were quantified; images were taken at P9. In control animals, the contralateral layer contained few residual ipsilateral axons. The amount of ipsilateral axons in the contralateral layer was enhanced by synchronous optogenetic activation of both eyes but diminished by asynchronous activation of the left and right eyes. **(B)** Quantification of these effects. (Adapted from Zhang J, Ackman JB, Xu H, et al. [2012] *Nat Neurosci* 15:298–309. With permission from Springer Nature.)

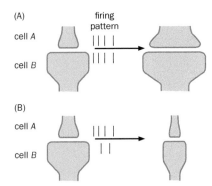

Figure 5-24 Schematic depiction of Hebb's rule. A presynaptic terminal of cell *A* and a postsynaptic site of cell *B* are schematized. **(A)** According to Hebb's rule, if cell *A* repeatedly or persistently takes part in exciting cell *B*, so that their action potential firing patterns are correlated, as shown, the synaptic connection between them strengthens. **(B)** Hebb's rule extends to the converse: if cell *A* repeatedly and persistently fails to excite cell *B*, so that their firing patterns are uncorrelated, then the synaptic connection between them weakens. Changes in connection strength are depicted here as changes in the size of the synapse, but connection strength changes in the developing nervous system can also be realized by turning transient contacts into new synapses or by eliminating existing synapses between two cells. (Based on Hebb DO [1949] The Organization of Behavior. LEA Inc.; Stent GS [1973] *Proc Natl Acad Sci U S A* 70:997–1001.)

severely diminished spontaneous waves in V1, supporting the notion that they are triggered by cholinergic retinal waves propagating along the ascending visual pathway. Spontaneous activity has also been observed in other parts of the developing nervous system, such as in cells along the developing auditory pathway.

5.12 Hebb's rule: correlated activity strengthens synapses

How do correlated and uncorrelated activity patterns affect neuronal wiring? In his 1949 book titled *The Organization of Behavior*, psychologist Donald Hebb made a "neurophysiological postulate" to explain how learning can be transformed into lasting memory: "When an axon of cell *A* is near enough to excite a cell *B* and repeatedly or persistently takes part in firing it, some growth process or metabolic change takes place in one or both cells such that *A*'s efficiency, as one of the cells firing *B*, is increased."

Hebb had no biological data to support his postulate—he simply thought that such a rule, if it existed, would provide a structural basis for memory. He could not have imagined that his postulate would have a long-lasting influence not only on the study of memory and learning (the topic of Chapter 11) but also on activity-dependent wiring of the brain. His postulate, now called **Hebb's rule** (**Figure 5-24**A), has repeatedly received experimental support. So has its converse (Figure 5-24B): when the presynaptic axon of cell *A* repeatedly and persistently *fails* to excite the postsynaptic cell *B* while cell *B* is firing under the influence of other presynaptic axons, metabolic change takes place in one or both cells such that cell *A*'s efficiency at firing cell *B* is *decreased*. A further extension of Hebb's rule that applies to activity-dependent wiring during development is that strengthening a weak synapse can convert a transient connection into a more stable one, while continual weakening of an existing synapse can eventually eliminate it altogether.

As a specific example, let's examine how Hebb's rule and its extensions explain the segregation of RGC axons into eye-specific layers in the binocular zone of the LGN (**Figure 5-25**). Early during development, RGC axons from both the left and the right retinas form connections with postsynaptic target neurons residing in areas that will eventually develop into left eye– or right eye–specific layers (Figure 5-20). Consider an LGN neuron in the left eye–specific layer. Although it receives synaptic input from RGCs from the left and right eyes, it receives more inputs from left eye RGCs (Figure 5-25, top), perhaps in response to molecular cues. The retinotopic mapping we studied in the superior colliculus also applies to the LGN (see Section 5.15), such that inputs originating from nearby RGCs are activated together by retinal waves (Figure 5-21C). These coordinated inputs are sufficient to depolarize the LGN target neuron, causing it to fire action potentials. According to Hebb's rule, the synapses between left eye RGCs and the LGN neuron will be strengthened as a result. RGCs originating from the right eye fire action potentials with different temporal patterns, since their retinal waves are not coordinated with the left eye's retinal waves. Given the relative weakness of their initial connections, the right eye RGCs are not only out of sync with those from the left eye but also often fail to depolarize the target neuron sufficiently to cause it to fire, further weakening these connections, according to the converse of Hebb's rule. Repeated rounds of this process, enabled by persistent retinal waves, ensure that the left eye RGCs win the battle for connecting to this neuron; the right eye RGCs withdraw their connections (Figure 5-25, bottom). In another part of the LGN, right eye RGCs have the initial advantage and out-compete left eye RGCs. This explains why competition between inputs from two eyes to drive the firing of the postsynaptic neurons is essential in the segregation of eye-specific layers. Colloquially, these phenomena are often expressed as "Fire together, wire together. Out of sync, lose your link."

The same principles apply to the formation of ocular dominance columns in layer 4 of V1, although the particular neurons involved differ, with the presynaptic LGN neurons targeting postsynaptic V1 layer 4 neurons. Note that during development, eye-specific segregation of RGC axons in the LGN precedes eye-specific

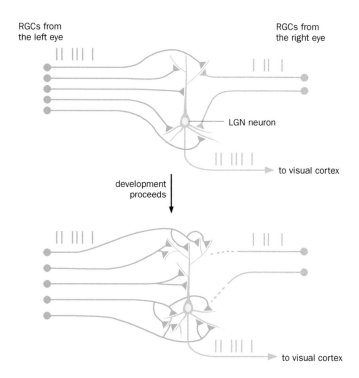

Figure 5-25 Application of Hebb's rule to eye-specific connections between RGCs and an LGN neuron. At an early developmental stage, the LGN neuron (green) receives RGC input from both eyes (top). More RGCs from the left eye (blue) are connected to the LGN neuron, making the LGN neuron more likely to fire according to left eye RGC firing patterns (as indicated by the similar temporal patterns of vertical bars in blue and green). Coincident firing of presynaptic left eye RGCs and the postsynaptic LGN neurons strengthens their synapses. Noncoincident firing of right eye RGCs (orange) and the LGN neuron weakens their synapses. As development proceeds, right eye RGC synapses are lost while left eye RGC axons and LGN dendrites form new branches and connections; eventually, the LGN neuron connects only to the left eye (bottom). Molecular cues likely create the initial connection bias between LGN neurons and left or right eye RGC axons, as eye-specific layers are stereotyped across animals.

segregation of LGN axons in V1. This ensures that LGN axons have already solidified their ocular preference by the time they are making their target choices in V1, and nearby LGN axons may already have acquired similar firing patterns. The LGN axons can then impose these patterns on individual V1 layer 4 neurons, just as RGC axons impose their patterns on LGN neurons. As detailed in Section 14.33, computational modeling that implement Hebb's rule could produce ocular dominance columns that resemble those found *in vivo*.

We can extend this logic to explain the sensitivity of these eye-specific connections during the critical period in V1. The ocular dominance columns take a long time to solidify, such that the vision-driven activity of the RGCs and LGNs can participate in this process after retinal waves recede and the retina is dominated by light- and photoreceptor-initiated activity. When deprived of this activity, application of Hebb's rule and its extensions would result in the strengthening of connections with the open eye and withering of connections with the closed eye.

5.13 A Hebbian molecule: the NMDA receptor acts as a coincidence detector

What mechanisms implement Hebb's rule? More specifically, how is correlated pre- and postsynaptic activity detected? How does correlated firing of pre- and postsynaptic neurons lead to growth? What leads to withering of synaptic connections when coincident firing does not occur? How widely applicable is Hebb's rule in the developing nervous system? We have only partial answers to some of these important questions.

Recall we learned in Chapter 3 that most excitatory transmission in the vertebrate CNS is carried out by the neurotransmitter glutamate. This includes synaptic transmission by RGCs and LGN neurons to their postsynaptic targets. One of the ionotropic glutamate receptors we studied in Section 3.15, the **NMDA receptor**, has an interesting property: its pore is normally blocked by Mg^{2+}, and depolarization in the postsynaptic neuron can relieve this block (Figure 3-24). Thus, the NMDA receptor serves as a molecular coincidence detector: it is activated only when (1) the presynaptic neuron releases glutamate and (2) the postsynaptic neuron is simultaneously depolarized to an extent that relieves the Mg^{2+} block. Importantly, the NMDA receptor is not the only glutamate receptor in postsynaptic neurons; glutamate can also activate the AMPA receptor, resulting in depolarization.

We use Figure 5-25 to explain how the NMDA receptor can implement Hebb's rule. Suppose that a series of action potentials from RGCs arrives at their axon terminals that synapse with LGN neurons, reflecting retinal waves or visual stimuli. Glutamate release induced by the first set of action potentials causes the LGN neuron to depolarize to the extent that the Mg^{2+} block of the NMDA receptor is relieved. The next set of action potentials can now open the NMDA receptor channel. Unlike most AMPA receptors, which depolarize the postsynaptic cell but do not conduct Ca^{2+}, the NMDA receptor allows Ca^{2+} entry, which can trigger a series of biochemical reactions in the postsynaptic cells. These include local changes at the synapse as well as altered gene expression in the nucleus (Section 3.23). These changes can lead to the stabilization and growth of synapses between the RGC axons and LGN dendrites.

Pharmacological experiments support a role of the NMDA receptor in activity-dependent wiring of the visual system, from eye-specific segregation in three-eyed frogs to shifts in ocular dominance columns induced by monocular deprivation. Recent live imaging experiments further suggest a role of the NMDA receptor in stabilizing RGC axonal branches in the developing frog tectum in response to synchronous visual stimuli. The molecular mechanisms by which NMDA receptor activation strengthens synapses and lack of correlated firing weakens synapses during development are still not well understood. We will revisit this topic in Chapter 11, where the role of the NMDA receptor in synaptic plasticity and learning will be discussed in more detail.

A powerful way to establish the necessity of a molecule in a biological process is a gene knockout experiment, but the role of the NMDA receptor in the development of eye-specific layers and ocular dominance columns has not been examined by such an experiment due to technical limitations. Hebb's rule applies to activity-dependent wiring in many other parts of the nervous system, and the role of the NMDA receptor in the development of the whisker-barrel system has been investigated in mice (Box 5-3).

Box 5-3: Activity-dependent wiring of the rodent whisker-barrel system requires the NMDA receptor

Rodents use their whiskers to perform many tasks, such as assessing the quality and distance of objects. Whiskers are represented as discrete units, termed **barrels**, in layer 4 of the mouse somatosensory cortex. (The part of the rodent somatosensory cortex that represents whiskers is called the **barrel cortex**.) Each barrel represents input primarily from a single whisker (Figure 5-26). Discrete units corresponding to individual whiskers are also found along the ascending pathways from whiskers to the barrel cortex. Sensory neurons that innervate whiskers project their axons to discrete units called *barrelettes* in the brainstem, with each barrelette composed of axon terminals originating from sensory neurons that mostly innervate a single whisker. Brainstem neurons then carry whisker-specific information to the thalamus, where their endings form discrete units called *barreloids*, again with each barreloid representing sensory information mostly from a single whisker. Last, thalamic neurons relay such information via their axonal projections to specific barrels in the somatosensory cortex (Figure 5-26).

Barrel formation in layer 4 of the somatosensory cortex occurs during the first postnatal week in mice. At birth

(postnatal day zero, or P0), when **thalamocortical axons** (TCAs—that is, axons of thalamic neurons that project to the cortex) first arrive at the somatosensory cortex, TCA terminals and their major postsynaptic targets, stellate cells, are evenly distributed across layer 4 (Figure 5-27, left). By postnatal day 7 (P7), the barrels are clearly delineated (Figure 5-27, right), and TCAs representing a single whisker are confined to a single barrel center. The cell bodies of layer 4 stellate cells cluster mostly at the periphery of the barrels, with each stellate cell sending dendrites into a single barrel. Thus, the segregation of TCAs into discrete barrels is somewhat analogous to segregation of eye-specific layers and formation of ocular dominance columns in the visual system.

How is the barrel cortex patterned within the first postnatal week? As in the visual system, sensory experience plays an essential role in the segregation of TCAs into discrete barrels. Newborn mice already have well-developed whiskers. When whisker follicles were ablated in newborn mice to deprive sensory input, the segregation of TCAs and stellate cells did not occur. Ablating a specific row of whisker follicles was especially revealing. TCAs corresponding to the ablated

Box 5-3: continued

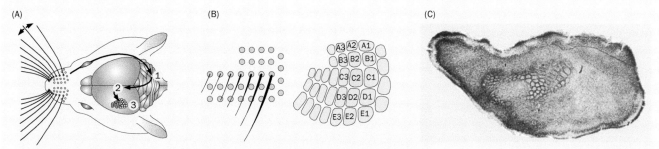

(A)　　　　　　　　　(B)　　　　　　　　　(C)

Figure 5-26 The whisker-barrel system in mice. (A) Information flow from the whiskers to the barrel cortex, from top view of a mouse. Sensory neurons that innervate individual whiskers project axons to the brainstem barrelettes (1). Brainstem neurons project axons to the contralateral somatosensory thalamus to form the barreloids (2). Thalamocortical axons then project to the primary somatosensory cortex to form the barrels (3). **(B)** Correspondence of individual whiskers and barrels. The C2 whisker and its corresponding barrel are highlighted in yellow. **(C)** The organization of barrels in somatosensory cortex can be visualized in a mouse brain section by Nissl staining (Section 14.15), which labels the cell bodies. (A & B, adapted from Petersen CCH [2007] *Neuron* 56:339–355. With permission from Elsevier Inc. C, adapted from Woolsey TA & van der Loos H [1970] *Brain Res* 17:205–242. With permission from Elsevier Inc.)

whiskers did not segregate into individual barrels. They also occupied much smaller cortical areas compared to those representing adjacent, intact whiskers (**Figure 5-28**A). In fact, it was this experiment that initially revealed which whisker corresponds to which barrel.

The NMDA receptor plays an essential role during multiple steps in the formation of the whisker-barrel system. When the obligatory GluN1 subunit of the NMDA receptor was knocked out in mice, the animals died at birth, so barrel cortex development could not be examined. The barrelettes in the brainstem normally develop prenatally but were absent in GluN1-knockout mice, indicating an essential role for the function of the NMDA receptor in barrelette formation. Using a conditional knockout strategy (Section 14.8), GluN1 was knocked out only in cortical neurons, allowing these mice to survive past the first week. Barrels in layer 4 somatosensory cortex were diffusely organized: TCAs no longer formed tight clusters, and stellate cells were no longer displaced to the sides of the barrels (Figure 5-28B). In these mice, barrelettes and barreloids in the brainstem and thalamus developed normally, since the NMDA receptor was intact outside of the cortex. Thus, this experiment indicated that the NMDA receptor in cortical neurons is essential for segregating TCA terminals and stellate cells.

Since almost every neuron in the cortex expresses the NMDA receptor, it was still difficult to determine whether the knockout effect was due to NMDA receptors acting in individual neurons or to a network effect. To address NMDA receptor function in individual cells, GluN2B, a subunit of the NMDA receptor that plays a prominent role during development, was knocked out in isolated single cells using a genetic mosaic strategy that also specifically labels these knockout cells (see Section 14.16 for details); this allows the assessment of the function of the NMDA receptor made by individual neurons, or the **cell-autonomous** function (we will expand this concept in Section 5.17) of the NMDA receptor. Since only a very small fraction of isolated neurons lost GluN2B, barrel cortex pattern formation was normal. However, whereas control stellate neurons confined their dendrites to a specific barrel, stellate neurons that lost GluN2B extended dendrites into multiple adjacent barrels (Figure 5-28C), indicating that each stellate neuron requires its own

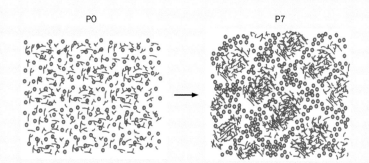

P0　　　　　　　　　　　　　P7

Figure 5-27 Patterning of the barrel cortex during the first postnatal week. At postnatal day 0 (P0, left), thalamocortical axons (TCAs, red) and stellate cells (blue) are evenly distributed in layer 4 of the somatosensory cortex. One week later (P7, right), TCAs have segregated into individual barrels according to the whiskers they represent, and stellate cells are displaced to the sides of the barrels. (From Molnár Z & Molnár E [2006] *Neuron* 49:639–642. With permission from Elsevier Inc.)

(Continued)

Box 5-3: continued

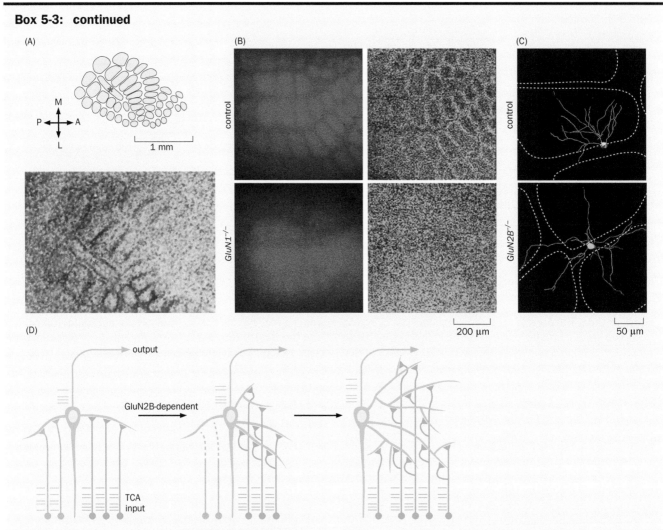

Figure 5-28 Sensory experience and the NMDA receptor regulate the patterning of both the barrel cortex and individual neurons' dendrites. (A) When a particular row of whisker follicles was ablated in a newborn mouse, the barrels representing those whiskers were fused. In both the schematic (top) and the Nissl-stained brain section (bottom), the star marks the fused barrels. A, anterior; P, posterior; M, medial; L, lateral. **(B)** Compared to control mice (top panels), conditional knockout mice whose cortical cells lack the NMDA receptor subunit GluN1 (bottom panels) no longer have properly segregated thalamocortical axons (TCAs) and stellate cells. In the left panels, TCAs are labeled by bulk injection of the axon tracer DiI into the thalamus; in the right panels, nuclear staining labels stellate cells. **(C)** A control stellate cell (top) extends its dendrites into a single barrel; dotted lines outline the barrels. By contrast, when the NMDA receptor subunit GluN2B is knocked out from an individual stellate cell (bottom), the cell distributes its dendrites into multiple barrels. **(D)** Interpretation of the results in

Panel C according to Hebb's rule. In early development, dendrites of a stellate cell (green) are contacted by TCAs representing multiple whiskers. If TCAs representing one whisker (blue) provide more input to the stellate cell than TCAs representing another (orange), the stellate cell is more likely to fire action potentials (represented as bars perpendicular to the axon) according to the blue TCA firing pattern. Over time, correlated firing leads to strengthening of the synapses and growth of dendritic branches representing the blue whisker; uncorrelated firing in other dendritic branches leads to destabilization of synapses and pruning of the dendrites representing other whiskers. (A, from van der Loos H & Woolsey TA [1973] *Science* 179:395–398. With permission from AAAS. B, from Datwani A, Iwasato T, Itohara S, et al. [2002] *Mol Cell Neurosci* 21:477–492. With permission from Elsevier Inc. C & D, adapted from Espinosa JS, Wheeler DG, Tsien RW, et al. [2009] *Neuron* 62:205–217. With permission from Elsevier Inc.)

NMDA receptor to pattern its dendrites and ensure that the TCA inputs it receives represent sensory information from one whisker. These results support the notion that NMDA receptor function is required for patterning dendritic arbor-izations of individual neurons according to the input they receive (Figure 5-28D), likely through a Hebbian mechanism analogous to the sorting of axons into eye-specific layers or ocular dominance columns (Figure 5-25).

HOW DO MOLECULAR DETERMINANTS AND NEURONAL ACTIVITY COOPERATE?

In the first part of this chapter, we used retinotopic projections of RGCs as an example to illustrate how molecular determinants enable developing axons to find their targets. In the second part, we used ocular dominance columns and eye-specific segregation of RGC axons to illustrate how experience and spontaneous neuronal activity sculpt the final connectivity. If activity-independent and activity-dependent mechanisms both make essential contributions to nervous system development, when and where is each mechanism used?

Generally speaking, activity-independent molecular determinants lay down the coarse framework of the nervous system, directing axons to specific regions; in some species and circuits they can also direct very precise connections, down to single cells and subcellular compartments. Neuronal activity and experience play an important role in refining this coarse framework and determining its quantitative features, such as the number and strengths of synapses between two connecting neurons. Activity-independent mechanisms usually precede activity-dependent mechanisms, though they can temporally overlap considerably. We know little about how these two mechanisms work together. The wiring of the visual system offers a case in which both mechanisms have been investigated in sufficient detail to offer a glimpse of how they might cooperate.

5.14 Ephrins and retinal waves act in parallel to establish the precise retinocollicular map

From what we've learned in the first part of the chapter, gradients of ephrin-As and EphAs seem to satisfactorily account for the precise targeting of RGC axons in the superior colliculus along the anterior–posterior axis (Figure 5-7; Figure 5-8). However, researchers found that knockout mice for **β2 nAChR**, a subunit of the nicotinic acetylcholine receptor essential for cholinergic retinal wave propagation, exhibited profound RGC axon wiring errors at the superior colliculus. Although coarse topography was maintained in β2 nAChR knockout mice, RGC axons originating in a small part of the retina produced a more diffuse termination zone in the superior colliculus compared to controls (**Figure 5-29**A).

A likely explanation for the β2 nAChR knockout phenotypes is that cholinergic retinal waves confer correlated activity on neighboring RGC neurons to strengthen their connections with neighboring target neurons, ensuring that nearby RGCs project to target neurons that are likewise nearby. This explanation dovetails with a curious finding from ephrin-A2/A5 double-knockout mice: axon terminals from

Figure 5-29 Retinal waves and ephrins cooperate to establish the retinocollicular map. (A) In a wild-type mouse (top panel), a focal injection of the fluorescent axon tracer DiI in the temporal retina resulted in a focal axon termination zone (TZ) in the anterior superior colliculus. A, anterior; P, posterior; L, lateral; M, medial. Focal injection of DiI into the temporal retina of a mouse lacking the β2 nicotinic acetylcholine receptor (β2 nAChR⁻/⁻) resulted in more diffuse labeling of RGC axon terminals in the superior colliculus beyond the normal termination zone (arrow in the bottom panel), indicating a defect in RGC axon refinement. **(B)** Summary of experiments examining the combined roles of ephrin/Eph and retinal waves in RGC axon targeting. Relative concentrations of ephrin-As (blue) and EphAs (orange) across tissues are represented by colored triangles indicating gradients. Groups of RGCs from four positions in the retina are shown as colored dots. In wild-type animals (left), these four groups of RGC axons project in an orderly fashion to four different areas of the superior colliculus. When ephrin-A2, ephrin-A5, and cholinergic retinal waves are disrupted in triple-knockout mice (right), RGC axons from a specific retinal position along the nasal–temporal (N–T) axis project diffusely along the entire anterior–posterior (A–P) axis of the superior colliculus. Remarkably, RGC axon projections along the dorsal–ventral (D–V) axis are unperturbed, indicating additional mechanisms at play. (A, from McLaughlin T, Torborg CL, Feller MB, et al. [2003] *Neuron* 40:1147–1160. With permission from Elsevier Inc. B, adapted from Pfeiffenberger C, Yamada J, & Feldheim DA [2006] *J Neurosci* 26:12873–12884.)

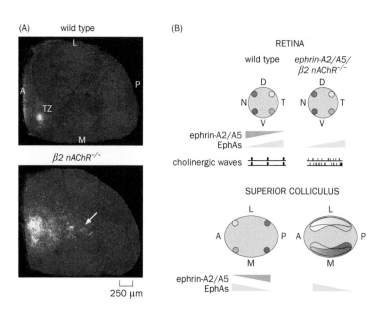

nearby RGCs still exhibited clustering at incorrect positions rather than completely diffuse projections along the anterior–posterior axis (Figure 5-8); this most likely occurred because intact retinal waves in these mutant mice still ensured that nearby RGC axons target to nearby targets.

To test the relationship between ephrins and retinal waves, triple-knockout mice lacking ephrin-A2, ephrin-A5, and β2 nAChR were generated. In these mice, a diffuse pattern of RGC axons along the anterior–posterior axis was observed, whether from RGCs originating in nasal or temporal retina (Figure 5-29B). This phenotype was much more severe than phenotypes exhibited in ephrin-A2/A5 double-knockout (Figure 5-8) or β2 nAChR knockout mice, indicating that ephrins and retinal waves cooperate: while ephrins direct global topographic RGC axon projections into the superior colliculus, retinal waves refine and preserve neighboring relationships of RGC axons.

5.15 Ephrins and retinal waves also cooperate to establish the retinotopic map in the visual cortex

As we learned in Chapter 4, the visual system is organized as a series of retinotopic maps such that spatial information in the retina is retained in the LGN, the primary visual cortex (V1), and higher areas of the visual cortex. We've also learned that spontaneous retinal waves propagate to V1. Do the mechanisms regulating retinotopic projections to the superior colliculus/tectum we have focused on thus far also apply to the construction of retinotopic maps in other areas? Indeed, gradients of ephrin-As and EphAs have been found in both the LGN and V1 (**Figure 5-30**). Through both forward and reverse signaling (Figure 5-12B), these gradients can in principle be used to establish topographic maps in multiple visual areas. They also help align different visual maps. For instance, information in the superior colliculus is used for reflexive control of eye movement, while information in the LGN is transmitted to V1 for detailed analysis of visual features. V1 also sends descending input to the superior colliculus. The topographic maps of descending input from V1 and ascending RGC projections to the superior colliculus align with each other (Figure 5-30).

The roles of ephrins and retinal waves in the construction of retinotopic maps in V1 have been examined using intrinsic signal imaging to map receptive fields of cortical neurons in response to grating bar stimuli in knockout mice (**Figure 5-31**). In wild-type animals, V1 has an orderly and continuous retinotopic map.

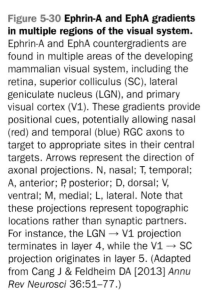

Figure 5-30 **Ephrin-A and EphA gradients in multiple regions of the visual system.** Ephrin-A and EphA countergradients are found in multiple areas of the developing mammalian visual system, including the retina, superior colliculus (SC), lateral geniculate nucleus (LGN), and primary visual cortex (V1). These gradients provide positional cues, potentially allowing nasal (red) and temporal (blue) RGC axons to target to appropriate sites in their central targets. Arrows represent the direction of axonal projections. N, nasal; T, temporal; A, anterior; P, posterior; D, dorsal; V, ventral; M, medial; L, lateral. Note that these projections represent topographic locations rather than synaptic partners. For instance, the LGN → V1 projection terminates in layer 4, while the V1 → SC projection originates in layer 5. (Adapted from Cang J & Feldheim DA [2013] *Annu Rev Neurosci* 36:51–77.)

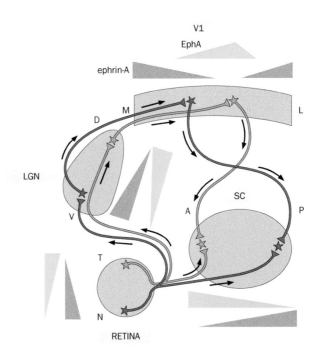

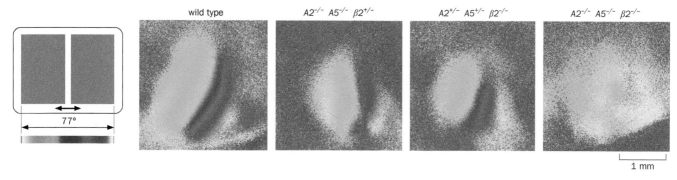

wild type   A2$^{-/-}$ A5$^{-/-}$ β2$^{+/-}$ A2$^{+/-}$ A5$^{+/-}$ β2$^{-/-}$ A2$^{-/-}$ A5$^{-/-}$ β2$^{-/-}$

77°

1 mm

Figure 5-31 Ephrins and retinal waves are required for the retinotopic map of V1. A grating bar that moves along the nasal–temporal axis was used as a visual stimulus (left). V1 responses were measured by intrinsic signal imaging (Figure 4-45B). Colors correspond to positions in the 77° of visual space shown on the left. Compared to the orderly map in wild-type mice, the maps in *ephrin-A2/A5* double-knockout or *β2 nAChR* knockout mice are more scattered; the map in triple-knockout mice is completely disrupted. (Adapted from Cang J, Niell CM, Liu X, et al. [2008] *Neuron* 57:511–523. With permission from Elsevier Inc.)

The map is less organized in ephrin-A2/A5 or β2 nAChR knockout mice and is not detectable in triple-knockout mice, suggesting that ephrin-As and retinal waves collaborate to establish the V1 map, just as they do for the superior colliculus map. The abnormal retinotopic map in V1 can result from a combination of an abnormal RGC → LGN map and additional defects in the LGN → V1 projections.

Together, these results support the notion that Ephrin-A/EphA interactions specify the rough topographic target location for RGCs, and retinal waves specify that neurons with correlated activity connect with nearby targets. Indeed, if molecular gradients such as those of Ephrin-A and EphA acted alone to specify precise positions, axons from neighboring RGCs would have to use a slight difference in their receptor levels to detect very small molecular differences in the target area. Such a mechanism might be challenging to establish. Acting together, molecular determinants and neuronal activity ensure that retinal positions along the anterior–posterior axis are precisely represented at multiple visual areas. Other molecules and mechanisms must work to establish the retinotopic map along the orthogonal dorsal–ventral axis; we know far less about these mechanisms than about retinotopic projections along the anterior–posterior axis.

5.16 Different aspects of visual system wiring rely differentially on molecular cues and neuronal activity

Are molecular cues and activity-dependent mechanisms similarly employed in wiring different parts of the nervous system? In this final section on the vertebrate visual system, we use a few examples to explore the relative contributions of these two wiring mechanisms.

As we learned in Chapter 4, specific types of RGCs receive synaptic input from specific types of bipolar and amacrine cells in the inner plexiform layer of the retina, which consists of distinct laminae. Dendrites from most RGCs and amacrine cells are limited to one or two specific laminae, as are axons from all bipolar cell types (**Figure 5-32**A). To determine whether neuronal activity and competition are involved in lamina-specific targeting, neurotransmission from ON bipolar cells was blocked in genetically modified mice throughout the period of bipolar cell synaptogenesis. This produced a competitive disadvantage for the ON pathway, a manipulation conceptually similar to suturing one eye shut in the context of competition between two eyes. Surprisingly, the connection specificity was not altered: ON bipolar cells still connected with ON RGCs at the correct lamina, albeit with reduced synapse numbers. Even ON–OFF bistratified RGCs, which normally elaborate two dendritic trees to two distinct laminae and synapse with both ON and OFF bipolar cells, still did so despite the selective silencing of ON bipolar cells. These findings suggest that lamina-specific targeting of dendrites and axons in the retina is more hard-wired than the formation of eye-specific layers in the LGN or ocular dominance columns in V1.

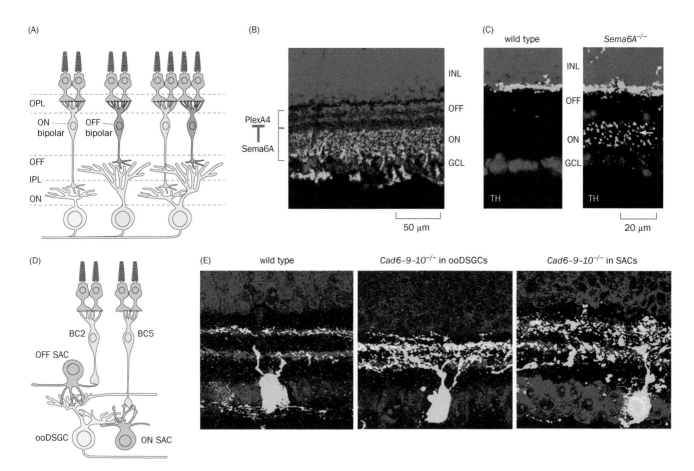

Figure 5-32 Lamina-specific targeting in the retina is specified by attractive and repulsive molecular interactions. (A) In wild-type mice, ON bipolar cells connect with ON ganglion cells (left), and a subset of the dendrites of ON–OFF ganglion cells (right) in the ON laminae of the inner plexiform layer (IPL). OFF bipolar cells connect with OFF ganglion cells (middle) and a subset of ON–OFF ganglion cell dendrites in the OFF laminae (far right). When neurotransmission from ON bipolar cells is blocked, disparate input to the ON and OFF laminae results, but connection specificity is not altered (not shown). **(B)** Sema6A (green) is expressed in mouse retinal neurons that target processes to the ON laminae, while its repulsive receptor PlexA4 (red) is expressed in retinal neurons that target processes to the OFF laminae. The repulsive interaction ensures that PlexA4-expressing neurons restrict their processes to the OFF lamina. INL, inner nuclear layer; GCL, ganglion cell layer. **(C)** A type of amacrine cell that expresses the enzyme tyrosine hydroxylase (TH) targets its processes to the outermost OFF lamina in wild type, as visualized by antibody staining against TH (green). In *Sema6A* mutants, TH-positive processes mistarget to ON laminae (arrow). **(D)** Schematic of a subset of cells that constitute a direction-selectivity circuit. BC2 and BC5 are OFF and ON bipolar cells,

respectively, that are presynaptic to two subtypes of ON–OFF direction-selective RGCs (ooDSGC) with dorsal or ventral direction selectivity. SAC, starburst amacrine cells. These cells express specific types of cadherins and Ig CAMs. **(E)** Compared to wild type (left), when *Cad6, Cad9,* and *Cad10* (which encode cadherins of the same subfamily and can serve redundant functions) are conditionally knocked out in ooDSGCs (middle) or SACs (right), ooDSGC dendrites (green) are no longer restricted to two specific laminae where SAC dendrites reside (red staining for vesicular acetylcholine transporter, a marker of SACs). Blue stains nuclei in Panels B, C, and E, which label the INL and GCL. (A, adapted from Kerschensteiner D, Morgan JL, Parker ED, et al. [2009] *Nature* 460:1016–1020. With permission from Springer Nature. B & C, from Matsuoka RL, Nguyen-Ba-Charvet KT, Parray A, et al. [2011] *Nature* 470:259–264. With permission from Springer Nature. D & E, from Duan X, Krishnaswamy A, Laboulaye MA, et al. [2018] *Neuron* 99:1145–1154. With permission from Elsevier Inc. See also Duan X, Krishnaswamy A, De la Huerta I, et al. [2014] *Cell* 158:793–807; Peng YR, Tran NM, Krishnaswamy A, et al. [2017] *Neuron* 95:869–883; Ray TA, Roy S, Kozlowski C, et al. [2018] *eLife* 7:e34241.)

Indeed, molecular determinants that specify the lamina-specific targeting of retinal neurons have been identified. For example, the semaphorins, evolutionarily conserved repulsive axon guidance molecules, and their receptors, the **plexins** (Box 5-1), play key roles in determining OFF- versus ON-layer targeting of retinal neurons in the inner plexiform layer. PlexA4 is expressed in retinal neurons that target processes to the OFF layer, while its repulsive ligand is expressed in retinal neurons that target processes to the ON layer (Figure 5-32B). Deleting either PlexA4 or Sema6A caused OFF-layer-targeting amacrine and RGC dendrites to mistarget to the ON layer (Figure 5-32C). Another semaphorin–plexin pair prevents inner plexiform layer–targeting retinal neurons from targeting their processes to the outer plexiform layer. In these examples, repulsive cues establish barriers that constrain layer selection of retinal neuron processes.

How do retinal neurons select specific laminae and, ultimately, synaptic partners? Attractive (adhesive) interactions mediated by two classes of proteins, the cadherins and the immunoglobulin superfamily cell-adhesion molecules (Ig CAMs) (Box 5-1), play essential roles. In the retinal circuit for direction selectivity (Figure 5-32D; Section 4.16), for example, each neuronal type expresses a specific subset of cadherins and Ig CAMs. Specifically, starburst amacrine cells (SACs) and ON–OFF directionally selective RGCs (ooDSGCs), which are synaptic partners, both express cadherin-6 (Cad6). Conditional knockout of Cad6 and closely related cadherins in ooDSGCs or in SACs resulted in similar mistargeting phenotypes for ooDSGC dendrites (Figure 5-32E). In these manipulations, SAC dendrite targeting was not affected, and other experiments indicated that SAC dendrites pattern the inner plexiform layer before ooDSGC dendrite extension. Thus, SAC dendrites provide a scaffold for ooDSGC dendrite targeting, likely through homophilic adhesion (that is, cadherins on SAC dendrites bind to cadherins on ooDSGC dendrites). SAC dendrites also serve as a scaffold for ON and OFF bipolar cells targeting their axons to specific laminae in the direction-selective circuit. In separate experiments, contactin-5, a homophilic Ig CAM, was found to be expressed in ooDSGCs and in ON SACs, but not OFF SACs. Contactin-5 acts in both ooDSGCs and ON SACs to promote stabilization of ooDSGC dendrites in the ON laminae.

RGCs of a given type not only extend their dendrites to specific laminae of the inner plexiform layer to receive information but also project their axons to specific laminae of their central targets, such as the superior colliculus and LGN, to send information. Tracing axonal projections of a specific RGC type labeled in a transgenic mouse line revealed both lamina-specific targeting as well as column-like lateral organization of the RGC axon terminations in the superior colliculus. Interestingly, blocking cholinergic retinal waves affected the refinement of the columnar organization but not the lamina-specific targeting of these axon terminations (Figure 5-33). Thus, neuronal activity can selectively affect a subset of wiring properties of one neuronal type. Parallel studies in zebrafish also indicated that lamina-specific targeting of RGC axons in the tectum is activity independent but relies on target-derived molecular cues, such as secreted proteins slit and reelin (Box 5-1) acting as a repellent and attractant, respectively.

Studies thus far have suggested that spontaneous activity and visual experience have large influence on wiring the visual cortex. For example, orientation selectivity, a salient feature in the primary visual cortex (Section 4.22), is influenced by spontaneous neuronal activity before eye opening and by visual experience afterward, although it is less malleable than ocular dominance. The direction selectivity that certain cortical neurons possess appears late during development after eye opening. Visual experience can influence the proper development of direction-selective columns. For example, exposing ferret pups selectively to motion in a specific direction can selectively strengthen the direction selectivity of cortical neurons and their organization into columns with respect to the direction of the training motion.

These examples suggest that the relative contributions of activity-independent molecular determinants, spontaneous activity, and visual experience differ in the wiring of different visual circuits. An important parameter that determines whether activity-independent or activity-dependent mechanisms are employed may be whether the connection specificity reflects qualitative differences between

Figure 5-33 Retinal waves are required for refinement of RGC axons into columns but not specific layers. Left, in control mice heterozygous for the β2 nAChR gene, a genetically defined subpopulation of RGCs expressing green fluorescent protein (GFP) targets its axons (labeled in green) to a specific layer of the superior colliculus (yellow bracket). These axons are additionally clustered into a columnar organization. All RGC axons are labeled red using an anterograde tracer. Right, when cholinergic retinal waves are blocked in β2 nAChR knockout mice, columnar organization is disrupted, but layer-specific targeting remains normal. D, dorsal, L, lateral. (From Huberman AD, Manu M, Koch SM, et al. [2008] *Neuron* 59:425–438. With permission from Elsevier Inc.)

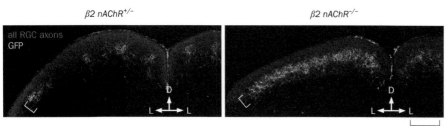

β2 nAChR$^{+/-}$ β2 nAChR$^{-/-}$

all RGC axons
GFP

D

L L

D

L L

250 μm

discrete cell types or quantitative differences within the same cell types. Connection specificities that differentiate discrete cell types, such as ON or OFF types of bipolar cells, amacrine cells, and RGCs in the retina, is more likely specified by molecular determinants. Conversely, connection specificity involving quantitative differences within similar cell types, such as the connection of a specific LGN neuron to the same types of RGCs from the left or right eye or a V1 neuron to the same type of projection neurons from the left or right LGN, may rely more on activity-dependent mechanisms. Because cell types are better defined in the retina than in the cortex at present, we see more examples of hard wiring in the retina. As more effort is devoted to identifying cortical cell types (Figure 4-49), it will be interesting to see whether cases of hard wiring are similarly prevalent in the cortex and how they interact with activity-dependent mechanisms.

Given the importance of cell type in our discussion of wiring specificity, a key question is how cell types are established during development. This question has been particularly well studied in the *Drosophila* visual system, to which we turn as the final topic of the chapter.

VISUAL SYSTEM DEVELOPMENT IN *DROSOPHILA*: LINKING CELL FATE TO WIRING SPECIFICITY

As introduced in Box 4-3, insects have exquisitely wired visual systems that grant them superb vision. Powerful molecular genetic approaches in *Drosophila* have made its visual system a model for addressing general questions regarding cell fate determination and wiring specificity.

The compound eye of *Drosophila* (**Figure 5-34**A) comprises ~800 **ommatidia**. Each ommatidium contains eight photoreceptors: R1–R6 at the periphery, and R7 and R8 at different depths in the center (Figure 5-34B). R1–R6 photoreceptors are responsible for motion detection; they send their axons to the **lamina**, where they synapse with lamina neurons (Figure 5-34C, D). R7 and R8 are responsible for color vision; they send their axons to the M6 and M3 layers, respectively, of the **medulla**. Lamina neurons also send their axons to specific medulla layers (Figure 5-35D). Different types of transmedulla neurons send dendrites to specific medulla layers to receive input from R7 and R8, and project their axons to higher visual centers in the lobula complex (Figure 4-31). Thus, the layer-specific organization of neuronal processes in the fly **optic lobe** (consisting of retina, lamina, medulla, and lobula complex) resembles that of the vertebrate retina. The *Drosophila* visual circuit is thought to be hardwired, as mutations disrupting phototransduction, action potential propagation, and synaptic transmission do not lead to obvious wiring defects. However, as in the vertebrate visual system, stimulus-independent spontaneous activity occurs in the developing fly optic lobe before the onset of vision. It will be interesting to determine whether such spontaneous activity is used to refine connection patterns and strength.

5.17 Cell–cell interaction determines *Drosophila* photoreceptor cell fates: R7 as an example

The nervous system contains many types of neurons and glia. For example, the vertebrate retina has ~90 types of neurons (Box 4-2), each with a unique morphology, connection pattern, neurotransmitter profile, and gene expression pattern. How a neuron or glial cell acquires its **cell fate** (the type of cell it is to become) is a fundamental question in developmental neurobiology, and studies of photoreceptor fate determination in *Drosophila* have provided rich mechanistic insight into this question.

Cell fates can be determined by two broad mechanisms. The first mechanism relies on **cell lineage**, or the relationship of cells by birth. A widely used means of lineage-based fate determination is **asymmetric cell division**: a cell is born with a fate different from its sibling because an intrinsic determinant segregates asymmetrically during cell division. The second mechanism is **induction**, or cell–cell interaction: a cell is born with the same potential as its sibling or cousins, and its

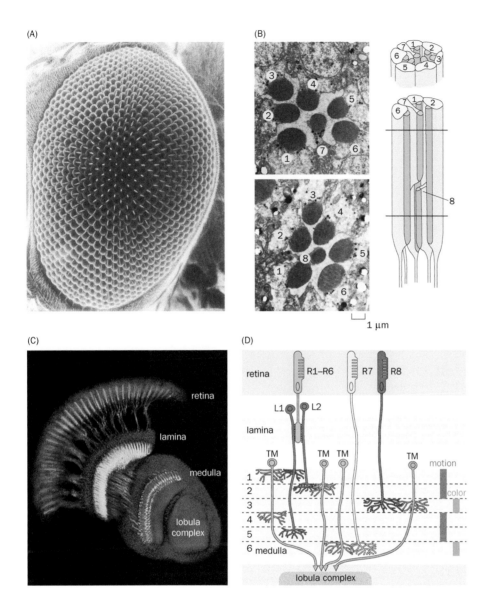

Figure 5-34 The *Drosophila* eye and visual circuit. (A) Scanning electron micrograph of a *Drosophila* compound eye, which consists of ~800 ommatidia. **(B)** Electron micrographs of ommatidial cross sections at superficial (top) and deep (bottom) levels, as indicated by the horizontal lines on the right. The electron-dense structures are the rhabdomeres, which are enriched in rhodopsins, analogous to the outer segments of vertebrate photoreceptors. Rhabdomeres from each photoreceptor (R1–R8) occupy stereotyped positions in the ommatidium; note that the central photoreceptor in the superficial cross section is R7, while R8 is central in the deep cross section. **(C)** Section of the *Drosophila* optic lobe showing its neuronal organization. Photoreceptors are stained in green. Their cell bodies occupy the outermost layer, the retina. Depending on the cell types, their axons project to two neuropils, the lamina or the medulla (stained in red with a synapse marker; because of the extensive overlap between photoreceptor axons and the synapse marker staining, the lamina appears yellow). The lobula complex receives input from transmedulla neurons. Blue, nuclear staining indicating the locations of cell bodies. **(D)** Diagram of the fly visual circuit. Photoreceptors R1–R6 target axons to the lamina, where they synapse with L1 and L2 neurons that send axons to specific layers of the medulla. R7 and R8 send axons directly to layers 6 and 3 of the medulla, respectively. This layer-specific targeting resembles that of amacrine cells, bipolar cells, and RGCs in the vertebrate retina (Figure 5-32A). TM, different types of transmedullary neurons. The medulla contains additional layers (7–10) that are not shown. Note that all neurons except the photoreceptors shown here are unipolar, with dendrites and axons originating from a single process (Section 1.7). (A, adapted from Ready DF, Hanson TE, & Benzer S [1976] *Dev Biol* 53:217–240. With permission from Elsevier Inc. B, adapted from Reinke R & Zipursky SL [1988] *Cell* 55:321–330. With permission from Elsevier Inc. C, from Williamson WR, Wang D, Haberman AS, et al. [2010] *J Cell Biol* 189:885–899. With permission from Rockefeller University Press. D, adapted from Sanes JR & Zipursky SL [2010] *Neuron* 66:15–36. With permission from Elsevier Inc.)

eventual fate is achieved via reception of external inductive signals. The stereotyped positions of R1–R8 within each ommatidium (Figure 5-34B), as well as a wealth of molecular markers, enabled researchers to determine when specific R cells become neurons and whether R cells are related by birth.

Using a pan-neuronal marker to stain developing ommatidia, researchers found that R cells undergo neuronal differentiation in a stereotyped sequence. R8 is the first to become a neuron, followed sequentially by the R2/R5 pair, the R3/R4 pair, and the R1/R6 pair. R7 is the last cell to adopt a neuronal fate (Figure 5-35A). To probe their lineage relationships, one can use a method called **clonal analysis**, which randomly labels an early progenitor in such a way that all its progeny are also labeled (see Section 14.16 for details). If specific R cells within an ommatidium are always co-labeled, one could deduce that those cells share an immediate common progenitor. Strikingly, no such relations were found in any R cell pair, suggesting that their fates are acquired by induction, rather than through cell lineage. Indeed, the inductive mechanisms for R cell fate determination are among the best-studied in developmental biology. Here we examine the determination of R7 cell fate as an example.

Analyses of mutations in two *Drosophila* genes, **Sevenless** and **Bride of Sevenless (Boss)**, have been particularly instructive. Flies lacking either gene have a very specific phenotype: all ommatidia lack the R7 photoreceptor. The presumptive R7 cell turns into a nonneuronal support cell in the absence of induction.

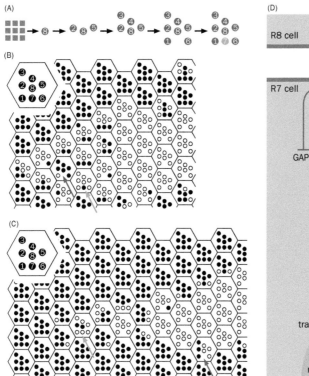

Figure 5-35 Determination of R7 cell fate. (A) From a field of undifferentiated cells (gray squares), R8 is the first cell to become a neuron (red) in an ommatidium. This is followed by R2/R5, R3/R4, R1/R6, and finally R7 (blue). **(B)** Mosaic analysis of *Sevenless*. The scheme represents a cross section through many ommatidia, each enclosed by a hexagon with R1–R8 arranged according to the key at top left. Open dots represent *Sevenless* mutant cells marked by their lack of a cell marker; these were derived from a single precursor cell (a clone). Filled dots represent wild-type cells. These mosaic clones can be used to determine which cell must be wild type in order for an ommatidium to have an R7 (the middle cell in the bottom row of each ommatidium). The red arrow points at an ommatidium in which R1–R5 and R8 are all wild type, yet R7 is missing. The blue arrow points to an ommatidium in which all cells except R7 are mutant, yet R7 is present. Collectively, these data indicate that R7 determination requires *Sevenless* only in the presumptive R7. **(C)** Mosaic analysis of *Boss* follows the same scheme introduced in Panel B, except that open dots represent *Boss* mutant cells. The blue arrow points to an ommatidium in which only R2 and R8 are wild type, and R7 is present. The red arrow points to an ommatidium in which R2 is wild type, R8 is mutant for *Boss,* and R7 is absent. Collectively, these data indicate that *Boss* is required in R8 for R7 determination. **(D)** A summary of signal transduction pathways for R7 determination. Boss binding to Sevenless causes tyrosine phosphorylation of Sevenless, which recruits Drk and Son of sevenless (Sos) to the membrane. Sos acts as a guanine nucleotide exchange factor for the small GTPase Ras. GTP-bound Ras activates Raf and the MAP kinase cascade, which in turn activates transcription factors to initiate R7-specific transcription. Activated Sevenless also inhibits GAP1, a GTPase-activating protein that inactivates Ras. (B, adapted from Tomlinson A & Ready DF [1987] *Dev Biol* 123:264–275. With permission from Elsevier Inc. C, adapted from Reinke R & Zipursky SL [1988] *Cell* 55:321–330. With permission from Elsevier Inc. D, based on Simon MA, Bowtell DD, Dodson GS, et al. [1991] *Cell* 67:701–716; Zipursky SL & Rubin GM [1994] *Ann Rev Neurosci* 17:373–397.)

Mosaic analysis was used to determine which cells require *Sevenless* or *Boss* in order for an ommatidium to develop an R7 cell. Genetic mosaic flies were created in which a specific population of marked cells carried mutations, while the rest of the cells were wild type (see Section 14.8 for details). Although *Sevenless* is normally expressed in several R cells, if the wild-type gene was present in the presumptive R7 cell, the ommatidium developed an R7 cell; the *Sevenless* gene could be absent in any other R cell without affecting R7 cell development in that ommatidium (Figure 5-35B). Thus, *Sevenless* acts **cell autonomously**: it is needed only in the presumptive R7 cell to make an R7 photoreceptor. By contrast, analysis of ommatidia that are mosaic for *Boss* indicated that as long as the R8 in an ommatidium has wild-type *Boss,* an R7 develops (Figure 5-35C). Thus, *Boss* acts **cell nonautonomously**: it is required in R8 for R7 formation. Indeed, subsequent molecular cloning and biochemical studies indicated that *Sevenless* encodes a receptor tyrosine kinase that acts in R7 to receive inductive signals, and *Boss* encodes a ligand expressed only in R8 cells; the Boss protein, which localizes to the R8 cell membrane, binds to and activates the Sevenless protein in the neighboring presumptive R7 cell (Figure 5-35D).

Genetic studies have also elucidated intracellular signaling pathways by which Sevenless transduces signals to regulate R7 fate (Figure 5-35D). As with many receptor tyrosine kinases, such as neurotrophin receptors (Figure 3-39), activated Sevenless proteins dimerize, and the homodimer subunits cross-phosphorylate each other at tyrosine residues. Phosphorylated tyrosine on Sevenless recruits the binding of an adaptor protein called Drk (homolog of Grb2; Figure 3-39), which brings to the membrane a Drk-binding partner called Son of sevenless (Sos), a guanine nucleotide exchange factor for the small GTPase Ras. This switches Ras into the GTP-bound, active form. At the same time, phosphorylated Sevenless inhibits the GTPase-activating protein GAP1 and thereby helps maintain Ras in its GTP-bound state. Ras activates Raf and the MAP kinase cascade, leading to activation of specific transcription factors that implement an R7-specific gene expression program. Remarkably, this signaling pathway, from receptor tyrosine kinase to Ras and MAP kinase cascade, is highly conserved in many cell fate determination processes across the animal kingdom (Box 3-4). It is also used in synaptic activity–induced gene expression (Figure 3-41). Last, it is a major cell proliferation pathway in other biological contexts; many components, such as Ras and Raf, when abnormally activated, lead to cancer.

After a cell becomes an R7, cell fate is not complete. As we learned in Chapter 4, color vision requires animals to have photoreceptors with different spectral sensitivities. *Drosophila* R7 and R8 express different rhodopsin (Rh) proteins that allow them to detect light in the UV and blue–green ranges, respectively. The onset of Rh expression occurs well after photoreceptors have acquired their R7 or R8 fates. In fact, both R7 and R8 have two subtypes. About 30% of R7 cells express Rh3; the remaining 70% express Rh4. The distribution of Rh3- and Rh4-expressing R7s appears stochastic (**Figure 5-36A**), analogous to the stochastic distribution of L- and M-cones in the human retina (Figure 4-17). The distribution of R7 rhodopsins results from stochastic expression in R7 of Spineless, a transcription factor that activates Rh4 expression and represses Rh3 expression (Figure 5-36B). There are also two populations of R8: 30% of R8 cells express Rh5, and 70% express Rh6. Remarkably, rhodopsin expression in R7 and R8 is coordinated within each ommatidium: an Rh3-expressing R7 always couples to an Rh5-expressing R8, while an Rh4-expressing R7 always couples to an Rh6-expressing R8. In *Sevenless* or *Boss* mutant flies where R7 is absent, all R8 cells express Rh6, indicating that signaling from Rh3-expressing R7 induces Rh5 expression in R8. In *Spineless* mutants, all R8 cells express Rh5. Thus, in addition to regulating Rh expression in R7, Spineless also inhibits an R7 → R8 signal that orchestrates the coupling of R7 and R8 subtypes (Figure 5-36B).

In summary, a series of cell–cell interaction events determines the cell fates of *Drosophila* photoreceptors. Some fate decisions are made soon after a cell is born, such as whether to become an R8 or an R7. Other decisions are made well after a cell is born, such as whether an R7 expresses Rh3 or Rh4. The presumptive R7 receives an inductive signal from R8 to acquire the R7 fate; an R7 subtype later

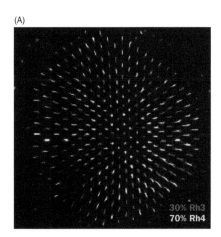

(A)

30% Rh3
70% Rh4

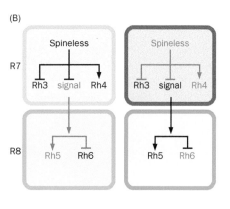

(B)

Figure 5-36 Determination of R7 and R8 subtypes expressing rhodopsins with different spectral sensitivities. (A) About 30% of R7 cells express Rh3, and the rest express Rh4. The image shows that the retinal distribution of Rh3- and Rh4-expressing R7s appears random. (Note that color labels here do not correspond to the spectral sensitivities of R7 photoreceptors.) **(B)** Stochastic expression of Spineless in about 70% of R7 cells activates Rh4 expression and inhibits Rh3 expression. Spineless also inhibits an R7 → R8 signal that activates Rh5 expression and inhibits Rh6 expression in R8. Thus, expression of Spineless in R7 results in Rh4 expression in R7 coupled to Rh6 expression in R8 (left). Lack of Spineless expression in R7 results in expression of Rh3 in R7 coupled to Rh5 expression in R8 (right). Black, active components; gray, inactive components. (From Johnston RJ & Desplan C [2010] *Ann Rev Neurosci* 26:689–719.)

sends an inductive signal back to R8 to instruct it to become a specific subtype. Similar cell–cell interaction events occur throughout the developing nervous system to determine numerous cell types and subtypes.

5.18 Multiple pathways direct layer-specific targeting of photoreceptor axons

Connecting with proper synaptic partners is one of the most important facets of a neuron's fate. Thus, a major outcome of a neuron's fate decision is the expression of a unique repertoire of guidance receptors, such that its growth cones react to environmental cues differently from the growth cones of a different neuronal type. As we have seen in both the vertebrate retina and fly optic lobe, targeting axons to specific layers is an essential step in connecting with proper synaptic partners. In the following we use layer-specific targeting of R7 and R8 to illustrate how various molecules and mechanisms cooperate to establish wiring specificity.

In wild-type flies, R7 axons target to the M6 layer of the medulla, while R8 axons target to the M3 layer (Figure 5-34D). Forward genetic screens (see Section 14.6 for details) identified several genes necessary for correct targeting of R7 or R8 axons. For example, N-cadherin, a Ca^{2+}-dependent homophilic cell adhesion protein enriched in the nervous system (in the same protein family as Cad6/9/10, which we discussed in the context of lamina-specific targeting in the mouse retina), is essential for R7 targeting to M6. Genetic mosaic analysis indicated that when N-cadherin was removed from isolated single R7 neurons, their axons mistargeted to the M3 layer (**Figure 5-37**), demonstrating that N-cadherin acts cell autonomously in R7 for its layer selection. Given that N-cadherin is a homophilic cell adhesion protein, it should also be expressed and required in the target neurons for R7 axon targeting. Indeed, N-cadherin is widely expressed and is required for many targeting events, in part through highly regulated timing of expression. In several cases, it has been shown that N-cadherin is in fact required in both axons and their target neurons for specific targeting.

Loss of N-cadherin does not affect the ability of R8 axons to target to M3, likely due to redundancy with other molecules. Indeed, several other molecules have been identified to regulate R8 targeting. Netrin, an attractive axon guidance cue (Box 5-1), is normally enriched in the M3 layer due to its expression in specific lamina neurons that target that layer. R8 cells express the *Drosophila* netrin receptor **Frazzled** (homolog of DCC/Unc40), and the attraction mediated by the netrin–Frazzled interaction contributes to the layer specificity of R8 axon targeting (**Figure 5-38**). In mosaic animals in which Frazzled was removed from singly labeled R8 cells, the mutant R8 axons terminated at more superficial layers. When netrin was misexpressed in target layers M1 and M2, R8 axons mistargeted to these layers instead of M3. These experiments indicate that Frazzled acts cell autonomously in R8 neurons to target axons to the netrin source. Interestingly, a membrane-bound form of netrin is sufficient for R8 targeting, consistent with live imaging studies, suggesting that the principal function of netrin–Frazzled signaling is to stabilize extended growth cones at the target layer, rather than to attract growth cones to the target layer. R8 but not R7 cells also specifically express **Capricious**, a transmembrane protein containing extracellular leucine-rich repeats. Removal of Capricious resulted in termination of R8 axons prior to the M3 layer,

Figure 5-37 N-cadherin cell-autonomously regulates layer-specific targeting of R7 axons. All photoreceptor axons are labeled red. Green signal (which appears yellow, as it overlaps with red) highlights isolated wild-type R7 axons (left) or R7 axons mutant for *N-cadherin* (right) in a genetic mosaic animal (see Section 14.16 for details of the labeling method). Wild-type R7 axons (arrowheads) extend their growth cones to the M6 layer (lower dashed line), whereas individual R7 axons mutant for *N-cadherin* (arrows) in an otherwise normal environment terminate their axons near the M3 layer (upper dashed line), where R8 axons normally terminate. Thus N-cadherin is cell autonomously required for layer-specific targeting of R7 axons. (From Clandinin TR, Lee C-H, Herman T, et al. [2001] *Neuron* 32:237–248. With permission from Elsevier Inc. See also Lee C-H, Herman T, Clandinin RT, et al. [2001] *Neuron* 30:437–450.)

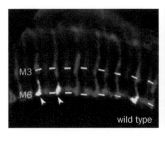
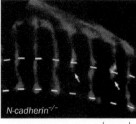

20 µm

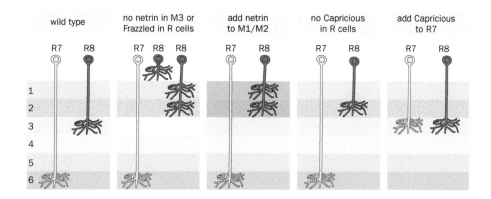

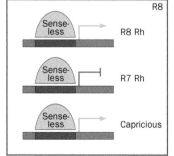

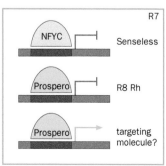

while misexpression of Capricious in R7 caused mistargeting of R7 axons to the M3 layer, where R8 axons normally target (Figure 5-38). These experiments suggest that netrin–Frazzled and Capricious play *instructive* roles in layer-specific targeting—that is, the specific expression of these guidance molecules instructs the axons to make specific targeting decisions. Since disruption of each of these molecules leads to targeting errors in a just subset of R8 axons, these studies also suggest that multiple ligand–receptor systems act in concert to ensure the fidelity of axon targeting in wild-type animals, a common theme in axon guidance and target selection.

How does Capricious acquire its R8 expression pattern, which appears to be a key determinant in targeting specificity? Researchers found that Capricious expression is activated by the transcription factor Senseless, which is specifically expressed in R8 and acts as a key R8 fate determinant. Senseless also promotes expression of R8-specific rhodopsins and inhibits expression of R7-specific rhodopsins. Co-regulation of axon targeting receptors and rhodopsins ensures that wiring specificity is linked to spectral sensitivity (**Figure 5-39**). R7 cells express their own specific transcription factor, Prospero, which represses R8-specific rhodopsin expression. Expression of Senseless is inhibited in R7 by the activity of a transcriptional repressor called NFYC (Figure 5-39). When NFYC was deleted from R7, misexpression of Senseless in mutant R7 turned on Capricious and led to mistargeting of R7 axons to the M3 layer.

Studies of layer-specific targeting of R7 and R8 axons exemplify a general principle: cell fate determination ensures that each cell type expresses a unique set of guidance molecules, which in turn instruct their axons to make specific wiring choices. Recent cell-type specific RNA-sequencing studies revealed that each retinal neuronal type expresses hundreds of cell surface molecules likely involved in establishing neuronal connectivity and that about 50 cell surface molecules are differentially expressed in R7 and R8 neurons. This molecular complexity likely ensures not only faithful targeting of R7 and R8 to the correct layers but also precise connections with specific postsynaptic targets. We remain far from understanding this molecular complexity. We will revisit these topics in Chapter 7, where we extend lessons learned from studies of the developing visual system to the developing nervous system in general.

Figure 5-38 Selected cell-surface proteins that instruct layer-specific targeting of R8 axons. In wild type, R8 axons target to the M3 layer. Deletion of netrin, which is normally expressed in lamina neurons that project to the M3 layer, or the netrin receptor Frazzled in photoreceptor cells in mosaic animals cause R8 axons to stall at the border of the medulla or mistarget to the M1 and M2 layers. Misexpressing netrin in the M1 and M2 layers causes R8 axons to mistarget to these layers. Loss of Capricious, a transmembrane protein with leucine-rich repeats, causes R8 axons to mistarget, mostly to the M2 layer. Note that R7 targeting in all these conditions is unaltered. Misexpressing Capricious in R7 cells causes R7 axons to mistarget to the M3 layer. These loss- and gain-of-function experiments indicate that both netrin and Capricious play instructive roles in R8 axon targeting. (Modified from Sanes JR & Zipursky SL [2010] *Neuron* 66:15–36. With permission from Elsevier Inc. See also Shinza-Kameda M, Takasu E, Sakurai K, et al. [2006] *Neuron* 49:205–213; Timofeev K, Joly W, Hadjieconomou D, et al. [2012] *Neuron* 75:80–93; Akin O & Zipursky SL [2016] *eLife* 5:e20762.)

Figure 5-39 Linking cell fate and targeting specificity. In R8 (left), the R8-specific transcription factor Senseless activates the expression of Capricious and R8 rhodopsins and inhibits the expression of R7 rhodopsins. In R7 (right), NFYC represses Senseless expression, and R7-specific transcription factor Prospero represses R8 rhodopsin expression. Links between Prospero and axon-targeting molecules remain to be established. (Based on Morey M, Yee SK, Herman T, et al. [2008] *Nature* 456:795–799.)

SUMMARY

Two principal mechanisms for wiring the nervous system have been discovered. The first involves recognition of extracellular molecular cues by receptors on axonal growth cones. These extracellular cues can be attractive or repulsive. They can be secreted and act at a distance or be cell surface bound and require contact between growth cones and cue-producing cells. Guidance receptors act by altering cytoskeleton in growth cones to mediate attraction and repulsion. Many axon guidance mechanisms, such as those involving netrin and semaphorin, are highly conserved in evolution.

The formation of the retinotopic map in the visual system offers an excellent experimental system for dissecting the mechanisms by which molecular cues instruct wiring specificity. Retinal ganglion cell (RGC) axons express a temporal > nasal gradient of EphA receptors. The tectum/superior colliculus, a major target of RGC axons, expresses a posterior > anterior gradient of the repellent ephrin-A. Ephrin-A/EphA interactions play a key role in defining the positions at which RGC axons terminate along the anterior–posterior axis in the target according to their cell body positions in the retina. Molecular cues play essential roles in many other aspects of visual system wiring, including determining whether RGC axons cross or do not cross the optic chiasm and defining the lamina(e) within the retina to which RGCs, amacrine cells, and bipolar cells target their dendritic and axonal processes.

Activity-dependent wiring constitutes the second principal mechanism for establishing neuronal connections. This has been investigated in ocular dominance columns in the primary visual cortex (V1) and eye-specific segregation of RGC axons in the lateral geniculate nucleus (LGN). Spontaneous activity in the form of retinal waves plays a crucial role in driving eye-specific segregation in the LGN and V1. Visual experience during the critical period consolidates ocular dominance columns. Hebb's rule of "Fire together, wire together" provides a cellular mechanism by which neuronal activity sculpts wiring specificity. The NMDA receptor acts as a molecular coincidence detector for correlated firing. Studies in the whisker-barrel system have provided strong support for the function of the NMDA receptor in executing Hebb's rule during activity-dependent neuronal wiring.

Molecular determinants alone can specify connectivity with great precision, as exemplified by the wiring of the fly visual system. However, in wiring much of the vertebrate visual system, molecular determinants and activity-dependent wiring act in concert, as exemplified by the collaboration of ephrins/Eph receptors and retinal waves in establishing global positions and refining local near-neighbor relations in the retinotopic maps. Different circuits rely differentially on the contributions of molecular cues and activity-dependent mechanisms. Defining when, to what extent, and why each mechanism is used in wiring different parts of the nervous system will be interesting topics for future research.

Neuronal wiring is a crucial output of a cell fate decision. Fate determinants, which often are transcription factors, control the expression of guidance receptors such that different types of neurons chart different paths through the same environment. We have seen examples in the layer-specific targeting of fly photoreceptors and in the chiasm crossing of vertebrate RGCs. We will expand on many of these principles in our study of neural development in Chapter 7.

OPEN QUESTIONS

- What mechanisms mediate axon–axon competition during RGC axon targeting?

- How do growth cones integrate signals from diverse attractive and repulsive cues?

- How do correlated and uncorrelated firings of pre- and postsynaptic partners lead to growth and pruning, respectively, of synapses during development?

- Are there non-Hebbian mechanisms that underlie the effects of neuronal activity on circuit wiring? If so, how do they work?

- How is spontaneous activity produced during development?

FURTHER READING

Books and reviews

Dent EW, Gupton SL, & Gertler FB (2011). The growth cone cytoskeleton in axon outgrowth and guidance. *Cold Spring Harb Perspect Biol* 3:a001800.

Hebb DO (1949). The Organization of Behavior: A Neuropsychological Theory John Wiley & Sons Inc.

Huberman AD, Feller MB, & Chapman B (2008). Mechanisms underlying development of visual maps and receptive fields. *Annu Rev Neurosci* 31:479-509.

Kolodkin AL & Tessier-Lavigne M (2011). Mechanisms and molecules of neuronal wiring: a primer. *Cold Spring Harb Perspect Biol* 3:a001727.

Sanes JR & Zipursky SL (2010). Design principles of insect and vertebrate visual systems. *Neuron* 66:15-36.

Shatz CJ (1992). The developing brain. *Sci Am* 267(3):60-67.

Sperry RW (1963). Chemoaffinity in the orderly growth of nerve fiber patterns and connections. *Proc Natl Acad Sci U S A* 50:703-710.

Zhang C, Kolodkin AL, Wong RO, & James RE (2017). Establishing wiring specificity in visual system circuits: from the retina to the brain. *Annu Rev Neurosci* 40:395-424.

Molecular determinants in vertebrate visual system development

Brown A, Yates PA, Burrola P, Ortuno D, Vaidya A, Jessell TM, Pfaff SL, O'Leary DD, & Lemke G (2000). Topographic mapping from the retina to the midbrain is controlled by relative but not absolute levels of EphA receptor signaling. *Cell* 102:77-88.

Chen Y, Mohammadi M, & Flanagan JG (2009). Graded levels of FGF protein span the midbrain and can instruct graded induction and repression of neural mapping labels. *Neuron* 62:773-780.

Di Donato V, De Santis F, Albadri S, Auer TO, Duroure K, Charpentier M, Concordet JP, Gebhardt C, & Del Bene F (2018). An attractive reelin gradient establishes synaptic lamination in the vertebrate visual system. *Neuron* 97:1049-1062.

Drescher U, Kremoser C, Handwerker C, Loschinger J, Noda M, & Bonhoeffer F (1995). *In vitro* guidance of retinal ganglion cell axons by RAGS, a 25 kDa tectal protein related to ligands for Eph receptor tyrosine kinases. *Cell* 82:359-370.

Duan X, Krishnaswamy A, Laboulaye MA, Liu J, Peng YR, Yamagata M, Toma K, & Sanes JR (2018). Cadherin combinations recruit dendrites of distinct retinal neurons to a shared interneuronal scaffold. *Neuron* 99:1145-1154.

Feldheim DA, Kim YI, Bergemann AD, Frisen J, Barbacid M, & Flanagan JG (2000). Genetic analysis of ephrin-A2 and ephrin-A5 shows their requirement in multiple aspects of retinocollicular mapping. *Neuron* 25:563-574.

Herrera E, Brown L, Aruga J, Rachel RA, Dolen G, Mikoshiba K, Brown S, & Mason CA (2003). Zic2 patterns binocular vision by specifying the uncrossed retinal projection. *Cell* 114:545-557.

Matsuoka RL, Nguyen-Ba-Charvet KT, Parray A, Badea TC, Chedotal A, & Kolodkin AL (2011). Transmembrane semaphorin signalling controls laminar stratification in the mammalian retina. *Nature* 470:259-263.

Peng YR, Tran NM, Krishnaswamy A, Kostadinov D, Martersteck EM, & Sanes JR (2017). Satb1 regulates contactin 5 to pattern dendrites of a mammalian retinal ganglion cell. *Neuron* 95:869-883.

Sperry RW (1943). Visuomotor coordination in the newt (*Triturus viridescens*) after regeneration of the optic nerve. *J Comp Neurol* 79:33-55.

Suetterlin P & Dreshcer U (2014). Target-independent EphrinA/EphA-mediated axon-axon repulsion as a novel element in retinocollicular mapping. *Neuron* 84:740-752.

Walter J, Henke-Fahle S, & Bonhoeffer F (1987). Avoidance of posterior tectal membranes by temporal retinal axons. *Development* 101:909-913.

Activity-dependent wiring in the visual system

Ackman JB, Burbridge TJ, & Crair MC (2012). Retinal waves coordinate patterned activity throughout the developing visual system. *Nature* 490:219-225.

Constantine-Paton M & Law MI (1978). Eye-specific termination bands in tecta of three-eyed frogs. *Science* 202:639-641.

Feller MB, Wellis DP, Stellwagen D, Werblin FS, & Shatz CJ (1996). Requirement for cholinergic synaptic transmission in the propagation of spontaneous retinal waves. *Science* 272:1182-1187.

Kerschensteiner D, Morgan JL, Parker ED, Lewis RM, & Wong RO (2009). Neurotransmission selectively regulates synapse formation in parallel circuits *in vivo*. *Nature* 460:1016-1020.

Li Y, Van Hooser SD, Mazurek M, White LE, & Fitzpatrick D (2008). Experience with moving visual stimuli drives the early development of cortical direction selectivity. *Nature* 456:952-956.

McLaughlin T, Torborg CL, Feller MB, & O'Leary DD (2003). Retinotopic map refinement requires spontaneous retinal waves during a brief critical period of development. *Neuron* 40:1147-1160.

Meister M, Wong RO, Baylor DA, & Shatz CJ (1991). Synchronous bursts of action potentials in ganglion cells of the developing mammalian retina. *Science* 252:939-943.

Munz M, Gobert D, Schohl A, Poquerusse J, Podgorski K, Spratt P, & Ruthazer ES (2014). Rapid Hebbian axonal remodeling mediated by visual stimulation. *Science* 344:904-909.

Sretavan DW & Shatz CJ (1986). Prenatal development of retinal ganglion cell axons: segregation into eye-specific layers within the cat's lateral geniculate nucleus. *J Neurosci* 6:234-251.

Stryker MP & Harris WA (1986). Binocular impulse blockade prevents the formation of ocular dominance columns in cat visual cortex. *J Neurosci* 6:2117-2133.

Wiesel TN & Hubel DH (1963). Single-cell responses in striate cortex of kittens deprived of vision in one eye. *J Neurophysiol* 26:1003-1017.

Drosophila visual system development

Akin O, Bajar BT, Keles MF, Frye MA, & Zipursky SL (2019). Cell-type-specific patterned stimulus-independent neuronal activity in the *Drosophila* visual system during synapse formation. *Neuron* 101:894-904.

Hiesinger PR, Zhai RG, Zhou Y, Koh TW, Mehta SQ, Schulze KL, Cao Y, Verstreken P, Clandinin TR, Fischbach KF, et al. (2006). Activity-independent prespecification of synaptic partners in the visual map of *Drosophila*. *Curr Biol* 16:1835-1843.

Ready DF, Hanson TE, & Benzer S (1976). Development of the *Drosophila* retina, a neurocrystalline lattice. *Dev Biol* 53:217-240.

Reinke R & Zipursky SL (1988). Cell-cell interaction in the *Drosophila* retina: the bride of sevenless gene is required in photoreceptor cell R8 for R7 cell development. *Cell* 55:321–330.

Simon MA, Bowtell DD, Dodson GS, Laverty TR, & Rubin GM (1991). Ras1 and a putative guanine nucleotide exchange factor perform crucial steps in signaling by the sevenless protein tyrosine kinase. *Cell* 67:701–716.

Timofeev K, Joly W, Hadjieconomou D, & Salecker I (2012). Localized netrins act as positional cues to control layer-specific targeting of photoreceptor axons in *Drosophila*. *Neuron* 75:80–93.

Axon guidance and activity-dependent wiring in other systems

Babola TA, Li S, Gribizis A, Lee BJ, Issa JB, Wang HC, Crair MC, & Bergles DE (2018). Homeostatic control of spontaneous activity in the developing auditory system. *Neuron* 99:511–524.

Espinosa JS, Wheeler DG, Tsien RW, & Luo L (2009). Uncoupling dendrite growth and patterning: single-cell knockout analysis of NMDA receptor 2B. *Neuron* 62:205–217.

Hedgecock EM, Culotti JG, & Hall DH (1990). The *unc-5, unc-6,* and *unc-40* genes guide circumferential migrations of pioneer axons and mesodermal cells on the epidermis in *C. elegans. Neuron* 4:61–85.

Iwasato T, Datwani A, Wolf AM, Nishiyama H, Taguchi Y, Tonegawa S, Knopfel T, Erzurumlu RS, & Itohara S (2000). Cortex-restricted disruption of NMDAR1 impairs neuronal patterns in the barrel cortex. *Nature* 406:726–731.

Ng J, Nardine T, Harms M, Tzu J, Goldstein A, Sun Y, Dietzl G, Dickson BJ, & Luo L (2002). Rac GTPases control axon growth, guidance and branching. *Nature* 416:442–447.

Serafini T, Kennedy TE, Galko MJ, Mirzayan C, Jessell TM, & Tessier-Lavigne M (1994). The netrins define a family of axon outgrowth-promoting proteins homologous to *C. elegans* UNC-6. *Cell* 78:409–424.

Shamah SM, Lin MZ, Goldberg JL, Estrach S, Sahin M, Hu L, Bazalakova M, Neve RL, Corfas G, Debant A, et al. (2001). EphA receptors regulate growth cone dynamics through the novel guanine nucleotide exchange factor ephexin. *Cell* 105:233–244.

Van der Loos H & Woolsey TA (1973). Somatosensory cortex: structural alterations following early injury to sense organs. *Science* 179:395–398.

CHAPTER 6
Olfaction, Taste, Audition, and Somatosensation

All sensory systems share common tasks, including the transformation of sensory stimuli into electrical signals; the optimization of detection sensitivity, selectivity, speed, and reliability; and the extraction of salient features with the ultimate purpose of helping animals survive and reproduce. At the same time, each sensory system has unique properties related to the physical nature of its sensory stimuli and how the sense serves the organism. Our detailed study of vision in Chapter 4 provides a framework that we will expand here as we examine the remaining major senses: olfaction, taste, audition, and somatosensation. We will study olfaction first and in greatest detail, as it differs from the other senses in some important ways, such as the large number of odorant receptors and the direct path by which olfactory signals are transmitted to the cortex. We begin our story with salmon homing.

HOW DO WE SENSE ODORS?

Salmon have a fascinating life cycle. They are born and live their first 1–2 years in freshwater streams. They then swim out to the ocean to feed on rich food sources and grow. When they are reproductively mature, they return from the ocean to the very stream in which they were born, overcoming obstacles and passing many similar streams along the way (**Figure 6-1**). Once in their native stream, they spawn, completing their cycle, and die shortly afterward.

The remarkable homing behavior of salmon relies primarily on their sense of smell. Local differences in soil and vegetation give each stream a unique chemical composition and a distinctive odor. The memory of these odors becomes imprinted on young salmon, particularly in the period when they leave their native streams and swim toward the ocean. Adult salmon then use these odor memories to find their native streams during the journey home.

In a field study, young Coho salmon raised in a Wisconsin state fish hatchery were exposed to one of two unnatural chemicals in trace amounts early in their life. They were then released into Lake Michigan at an age when salmon normally leave their native streams. Eighteen months later, when the salmon returned to spawn, one stream was scented with morpholine and another with phenethyl alcohol at concentrations similar to those the salmon were exposed to when young. More than 90% of captured salmon previously exposed to morpholine were caught in the stream scented with morpholine, and more than 90% of captured salmon previously exposed to phenethyl alcohol were caught in the stream scented with phenethyl alcohol (**Figure 6-2**). This striking specificity provided strong support for the odor-imprinting hypothesis.

Salmon homing behavior provides just one example of how the sense of smell shapes animal life. Because the olfactory system has somewhat regressed during primate evolution, we may be less aware of its importance. For many animals, however, the sense of smell is the primary means of finding food, avoiding predators,

Figure 6-1 Salmon homing. Adult salmon swimming back from the ocean toward their native streams to reproduce. (Courtesy of Marvina Munch/USFWS.)

Figure 6-2 **Salmon homing relies on olfaction. (A)** Map of a field study site where salmon homing was monitored. Hatchery-raised salmon that had been exposed to a trace amount of either morpholine (M) or phenethyl alcohol (PA) when young were monitored after they matured and swam upstream to spawn. The map shows the release site for the three groups of tagged young salmon (M-exposed, PA-exposed, and untreated controls). The two labeled streams were scented with M or PA, respectively, at the time when mature salmon returned to spawn. More than 20 rivers along the coast of Lake Michigan (beyond the map shown) were also monitored for salmon capture. **(B)** Quantification of where M-treated, P-treated, and control salmon were captured, as a percentage of total salmon captured from each treatment group. Most M-treated and PA-treated salmon were caught in the M-treated and PA-treated streams, respectively. By contrast, most control salmon were captured in other streams. (Adapted from Scholz AT, Horrall RM, Cooper JC, et al. [1976] *Science* 192:1247–1249.)

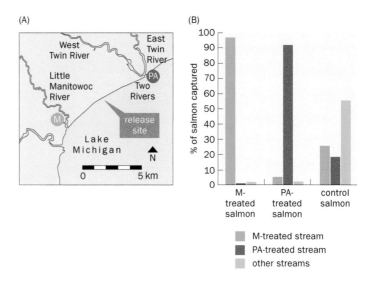

and seeking mates, all of which are essential for the survival of individuals and the propagation of species.

6.1 Odorant binding leads to opening of a cyclic nucleotide-gated channel in olfactory receptor neurons (ORNs)

Odorants, molecules that elicit olfactory perception, are usually airborne and volatile when detected by terrestrial animals. Diffusing through the air and into the nose, odorants pass through a layer of mucus and land on the surfaces of **olfactory cilia**. Each cilium is a dendritic branch of an **olfactory receptor neuron** (ORN, also called an olfactory sensory neuron or OSN) that resides within the **olfactory epithelium**. Olfaction begins with the binding of odorants to **odorant receptors** on olfactory cilia. This triggers depolarization of ORNs and produces action potentials propagated by ORN axons to the **olfactory bulb**, the first olfactory processing center in the brain. Within the olfactory bulb, ORN axons terminate in discrete, ball-like structures called **glomeruli**, where they form synapses with the dendrites of their postsynaptic target neurons (**Figure 6-3**).

A key question to investigate in any sensory system is how sensory stimuli are converted into electrical signals, the universal form of neuronal communication. For the olfactory system, the question becomes: how does the binding of an odorant to its receptor(s) initiate an electrical signal? Before their identification, evidence already suggested that odorant receptors were **G-protein-coupled receptors** (**GPCRs**), like rhodopsins in photoreceptor cells. A key difference between these two systems is the trigger: odorant receptors are activated by binding of odorants, whereas rhodopsins are activated by absorption of photons.

The signal transduction pathway through which odorant receptor activation generates electrical signals has been delineated in detail (**Figure 6-4**). Odorant

Figure 6-3 **Organization of the peripheral olfactory system. (A)** Odorants pass through the nostrils and bind to odorant receptors on olfactory cilia, the dendrites of olfactory receptor neurons (ORNs). ORN axons terminate in discrete glomeruli in the olfactory bulb, the first olfactory processing center in the brain. **(B)** Scanning electron micrograph of rat olfactory cilia. (B, from Menco BPM & Farbman AI [1985] *J Cell Sci* 78:283–310. With permission from The Company of Biologists, Ltd.)

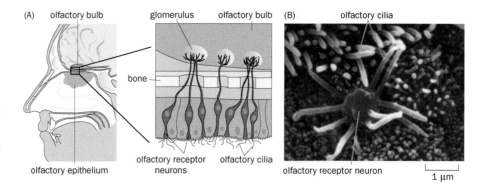

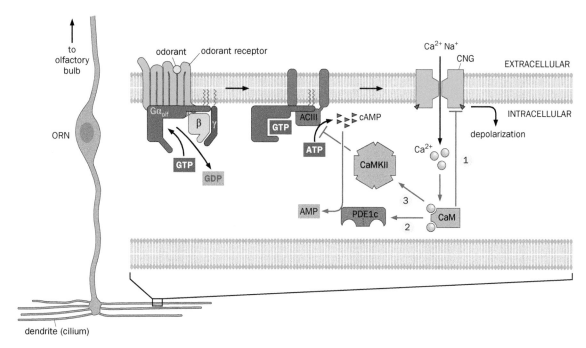

Figure 6-4 The olfactory transduction pathway and its regulation.
Left, an olfactory receptor neuron (ORN). Right, a magnified image of the cilium, a dendritic extension of the ORN, where odorants bind to odorant receptors. This binding activates a G protein cascade involving $G\alpha_{olf}$ and an adenylate cyclase ACIII that catalyzes production of cAMP. cAMP binds to and activates a cyclic nucleotide-gated (CNG) channel, leading to influx of Na^+ and Ca^{2+} and depolarization of the ORN. Negative regulatory pathways are colored red, including (1) inhibition of the CNG channel by the Ca^{2+}/calmodulin complex (CaM), (2) activation of a phosphodiesterase (PDE1c) by CaM, and (3) inhibition of ACIII by CaM kinase II (CaMKII). (Courtesy of Dr. Haiqing Zhao. See also Firestein S [2001] *Nature* 413:211–218.)

binding to odorant receptors triggers activation of $G\alpha_{olf}$, a specific kind of $G\alpha$. $G\alpha_{olf}$ then activates ACIII, a specific type of membrane-bound adenylate cyclase, which leads to an increase in the concentration of cyclic AMP (cAMP). cAMP binds to a specific cyclic nucleotide-gated (CNG) channel and causes it to open, thereby allowing influx of cations (Na^+, Ca^{2+}) down their electrochemical gradients into ORNs, causing depolarization. Amplification via opening of Ca^{2+}-activated Cl^- channels may also contribute to depolarization (Cl^- is more concentrated in the ORN than in the extracellular environment, so opening of Cl^- channels leads to Cl^- efflux). In addition, ORNs can be depolarized by mechanical stimuli; this mechanosensitivity also requires odorant receptors and their transduction cascade and can thus further amplify responses to odorants during sniffing. Hence ORNs are depolarized as a result of odorant binding and receptor activation.

As in the visual system, olfactory responses must terminate after the odorant is withdrawn so that ORNs can respond to future olfactory stimuli. This process is termed *olfactory recovery*. Olfactory responses are also modified by prior experience of the same odorant, a phenomenon called *olfactory adaptation*. Although olfactory recovery and adaptation refer to different biological phenomena, they share molecular pathways, as do visual recovery and adaptation (Sections 4.6–4.7). Indeed, Ca^{2+} entry via the CNG channel plays a central role in olfaction (Figure 6-4), as in vision (Figure 4-11), but acts on different effectors. In ORNs, intracellular Ca^{2+}, in a complex with the Ca^{2+}-binding protein calmodulin, has at least three functions in regulating olfactory recovery and adaptation. First, Ca^{2+}/calmodulin binds to and directly inhibits the CNG channel. Second, Ca^{2+}/calmodulin activates the phosphodiesterase PDE1c, facilitating cAMP hydrolysis. Third, Ca^{2+}/calmodulin activates Ca^{2+}/calmodulin-dependent protein kinase II, which phosphorylates ACIII and downregulates production of new cAMP. Thus, these negative feedback loops return cAMP to the baseline level, close CNG channels, and prepare ORNs for future stimuli (Figure 6-4).

The importance of the signal transduction pathway for olfaction has been demonstrated by analysis of mutant mice. For example, olfactory CNG channel

Figure 6-5 A cyclic nucleotide-gated (CNG) channel is essential for olfaction.
(A) Illustration of the electroolfactogram (EOG) in a side view of the mouse olfactory system. An extracellular electrode was inserted into the olfactory epithelium (pink) to measure the collective activity of many ORNs in response to odorants delivered to the nostrils. **(B)** Mice with a knockout (ko) of the gene encoding the olfactory CNG channel (CNGA2) show no EOG responses to any of the odorants (red traces). In comparison, EOG traces from wild-type (wt) animals (blue) show a reduction of the extracellular potential (a consequence of ORN depolarization), followed by a recovery phase. Odorant stimuli are shown in black. (B, adapted from Brunet LJ, Gold GH, & Nagai J [1996] *Neuron* 17:681–693. With permission from Elsevier Inc.)

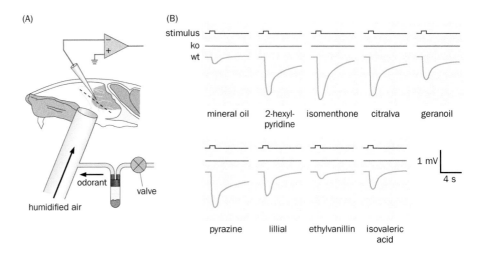

knockout mice are **anosmic**—that is, unable to perceive odors, because they cannot transform odorant binding into electrical signals (**Figure 6-5**). Mice lacking $G\alpha_{olf}$ or ACIII are similarly anosmic. The recovery and adaptation pathways have likewise been validated. For example, genetic disruption of calmodulin binding to the CNG channel and disruption of PDE1c in mice cause preferential deficits in recovery and adaptation, respectively.

6.2 Odorants are represented by combinatorial activation of ORNs

We now turn to a central question in olfaction: how are odorants recognized by the olfactory system? Odorants are usually mixes of volatile chemicals with diverse structures and properties (**Figure 6-6**), and the olfactory system can detect and discriminate between a vast number of odorants. How is this feat accomplished?

Physiological studies of responses of individual ORNs to odorants suggest that odorant recognition is a complex process. A comparison of the responses of different ORNs to a panel of chemically diverse odorants revealed that each odorant activates many ORNs, and each ORN is activated by multiple odorants (**Figure 6-7**). Therefore, the identity of an odorant is not encoded by a single ORN, but rather by the outputs of many different ORNs. In other words, odorants are represented *combinatorially* at the level of ORNs. This is reminiscent of color coding we learned in Chapter 4, wherein color information is extracted by comparing the activities of cones expressing opsins with different spectral sensitivities. Whereas color vision utilizes only two or three input channels, olfaction uses many hundreds of parallel input channels.

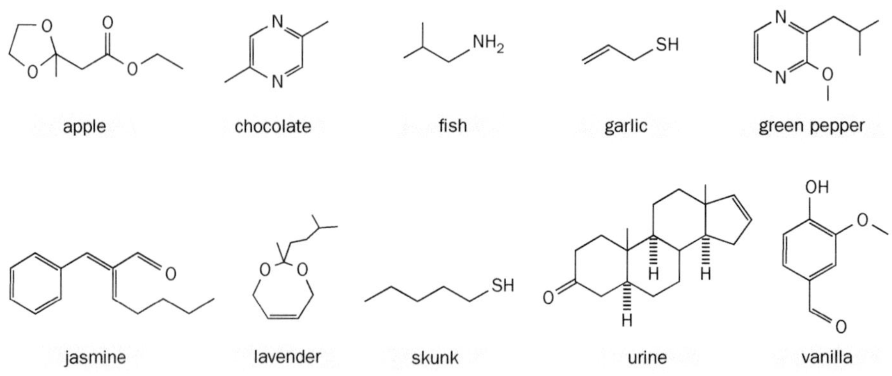

Figure 6-6 Structural formulas of molecules that produce selected smells. Odorants have diverse chemical structures. Shown here are characteristic compounds that contribute to specific smells. (Courtesy of Linda Buck.)

apple   chocolate   fish   garlic   green pepper

jasmine   lavender   skunk   urine   vanilla

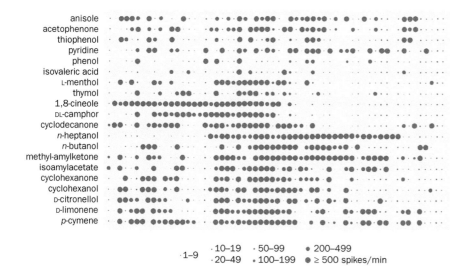

Figure 6-7 **Combinatorial coding of odorants by ORNs.** Responses of 60 individual frog ORNs (columns) to a panel of 20 odorants (rows). The different odorants, which exhibit diverse structures, were delivered at high concentrations. Spot sizes represent spike rate indicated at the bottom. Each odorant stimulates a different set of ORNs, and the magnitude of the responses depends on both the ORN and the odorant. (Adapted from Sicard G & Holley A [1984] *Brain Res* 292:283–296. With permission from Elsevier Inc.)

6.3 Odorant receptors are encoded by many hundreds of genes in mammals

Just as the identification of cone opsin genes was a key step in understanding the molecular basis of color detection, the identification of genes encoding odorant receptors provided insight into the molecular basis of odorant detection. It also led to the discovery of the organizational principles of the olfactory system (see Sections 6.5–6.6).

As noted in Section 4.11, amino acid sequences derived from purified bovine rhodopsin were used in the cloning of opsin genes. However, no purified odorant receptor was available to provide amino acid sequence information. To isolate odorant receptor cDNAs, an assumption was made that odorant receptors are G-protein-coupled receptors (**Figure 6-8**A), based on studies implicating G protein and adenylate cyclase in olfactory transduction (Section 6.1). In 1991, researchers took advantage of what was then a recent innovation in molecular biology, a highly sensitive DNA amplification technique called **polymerase chain reaction** (**PCR**). Oligonucleotide primer pairs were designed to correspond to highly conserved sequences within the transmembrane domains of known GPCRs. These primers were used to amplify cDNAs generated from mRNAs isolated from rat nasal epithelia. These cDNA clones were then used as probes to examine expression of these genes across tissue types. mRNAs of this new gene family were expressed exclusively in the olfactory epithelium (Figure 6-8B), supporting the idea that these cDNAs corresponded to odorant receptors. Analysis of the PCR amplification products suggested a very large repertoire (Figure 6-8C) of an estimated 1,000 odorant receptors encoded in the rat genome.

Now, in the post-genome era, we have confirmed the extraordinary diversity of mammalian odorant receptor genes (**Figure 6-9**). There are ~1,400 genes encoding odorant receptors in mice; of these, 1,063 are functional and 328 are **pseudogenes** rendered nonfunctional by stop codons in the coding sequences or other disrupting mutations. This means that functional odorant receptors account for >4% of the entire protein-coding genes in mice (~23,000 total). Indeed, odorant receptor genes constitute the largest gene family in mammalian genomes.

The human genome contains ~800 genes for odorant receptors, but only 388 are functional; more than half are pseudogenes. Having such a large fraction of pseudogenes in a gene family usually indicates that these genes are not under selection and are in the process of being eliminated. Accordingly, our sense of smell may not be as advanced as that of mice, rats, or pigs, whose genomes each encode >1,000 functional odorant receptors (Figure 6-9).

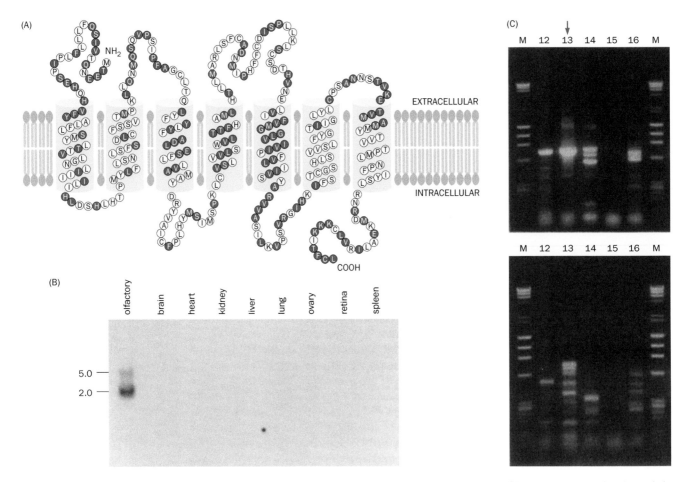

Figure 6-8 Identification of genes encoding odorant receptors.
(A) Primary structure of an odorant receptor. The seven cylinders represent seven transmembrane helices. Yellow circles represent conserved amino acids. Blue circles represent amino acids that are highly variable among odorant receptors, many of which are located in the transmembrane helices where odorants bind. **(B)** A northern blot (see Section 14.12 for details) showing that mRNAs of odorant receptors are expressed specifically in the olfactory epithelium. Left, molecular weight markers in kilobases. **(C)** PCR products (top) and their restriction enzyme digestion patterns (bottom). Lanes 12–16 each show the PCR products from a particular set of primers. In lane 13 (arrow), the sum of the molecular weights of the PCR digestion products greatly exceeds the molecular weight of the PCR product itself, suggesting that the PCR product is a mixture of different DNA species. M, molecular weight marker. (Adapted from Buck L & Axel R [1991] *Cell* 65:175–187. With permission from Elsevier Inc.)

6.4   Polymorphisms in odorant receptor genes contribute to individual differences in odor perception

Significant individual differences exist in people's olfactory perception. For instance, some people can detect a strong smell in their urine after eating asparagus while others cannot. As another example, androstenone, a mammalian social odorant commonly used in animal husbandry to induce female pigs to mate, smells pleasant (vanilla-like) or hardly at all to some people and sickening to others. What accounts for individual differences in smell? Twin studies suggest a strong genetic component, as identical twins have a much more similar threshold for androstenone detection than fraternal twins (**Figure 6-10**A). Differences in androstenone perception are now known to be associated with DNA sequence **polymorphisms** (variations among individuals) in the human odorant receptor Or7D4 (Figure 6-10B), which is strongly activated by androstenone. In the human population, 16% of alleles encode substitutions at amino acid positions 88 and 133 that change the more common arginine/threonine (RT) variant to the tryptophan/methionine (WM) variant. Biochemical studies indicated that the WM variant Or7D4 has a much lower affinity for androstenone than the RT variant. Psychophysical tests showed that humans who carry two RT alleles are more sensitive to androstenone and are more likely to consider the smell sickening, whereas humans carrying the WM allele are more likely to consider androstenone vanilla-like (Figure 6-10B). Thus, amino acid changes in a single odorant receptor contribute to different odor perceptions.

6.5 Each ORN expresses a single odorant receptor

So far, we have examined how odorant binding leads to the activation of individual ORNs, and how odorant molecules are detected by up to 1000 different receptors. But how does the brain recognize odors as specifically perceived objects (**percepts**)? How do animals distinguish between different odors so that they respond differently to the smells of food, mates, and predators? The organization of the olfactory system offers insights that help us begin to answer these questions.

There are about 5 million ORNs in the nasal epithelium of the mouse. How is the expression of 1000 different odorant receptors partitioned over the 5 million ORNs? Given that each ORN responds to many odorants (Figure 6-7), each ORN may express multiple odorant receptors; alternatively, each ORN expresses a single odorant receptor capable of binding multiple odorants.

The identification of odorant receptors and characterization of their expression patterns led to the unequivocal resolution of this issue. *In situ* **hybridization**, in which nucleic acid probes are used to visualize sites of gene expression in native tissues (see Section 14.12 for details), revealed that cells expressing mRNA of a particular odorant receptor are widely distributed in the nasal epithelia. On average, each odorant receptor is expressed in about 0.1% of ORNs, consistent with the hypothesis that each ORN expresses a single odorant receptor. (Recent

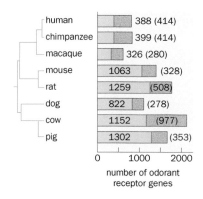

Figure 6-9 Odorant receptor genes are the largest gene family in the mammalian genome. Hundreds of odorant receptor genes (blue) and pseudogenes (red, numbers in parentheses) have been identified in the sequenced genomes of mammalian species. On the left is a phylogenetic tree for these mammalian species. (Adapted from Nei M, Niimura Y, & Nozawa M [2008] *Nat Rev Genet* 9:951–963. With permission from Springer Nature. The data for pig are from Groenen MA, Archibald AL, Uenishi H, et al. [2012] *Nature* 491:393–398.)

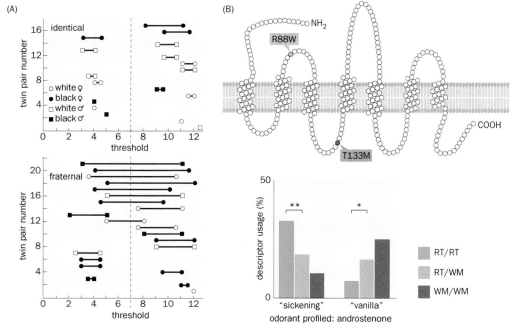

Figure 6-10 Polymorphisms in an odorant receptor contribute to individual differences in androstenone smell. (A) Twin studies revealed a strong genetic component in the detection threshold of androstenone. Results for the two members of each twin pair are linked by a horizontal line. A single entry for a twin pair indicates an identical threshold. The x axis shows the response thresholds to androstenone in a series of binary steps, with the most diluted (1) concentration being 1.79 µM and the most concentrated (12) being 3.67 mM. The concentration indicated by the dashed vertical line divides subjects into two groups: androstenone-sensitive (to the left of the line) and androstenone-insensitive (to the right of the line). The

threshold varies much less for identical twins (top) than for fraternal twins (bottom), regardless of race and sex. **(B)** Top, the structure of human Or7D4, with the locations of the polymorphic amino acids indicated. Bottom, humans with different Or7D4 alleles use different descriptors for androstenone: subjects with the WM allele have increased probabilities of considering androstenone vanilla-like. The differences are statistically significant (*, $p < .05$; **, $p < .01$; χ^2 test with Bonferroni correction). (A, adapted from Wysocki CJ & Beauchamp GK [1984] *Proc Natl Acad Sci U S A* 81:4899–4902. With permission from the authors. B, adapted from Keller A, Zhuang H, Chi Q, et al. [2007] *Nature* 449:468–472. With permission from Springer Nature.)

(A)

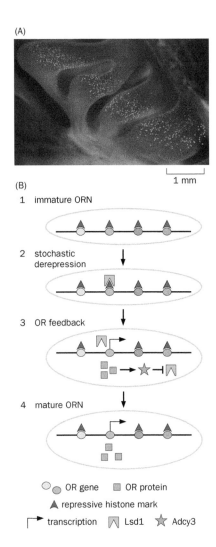

1 mm

(B)

1 immature ORN

2 stochastic
   derepression

3 OR feedback

4 mature ORN

○◉ OR gene   ■ OR protein

▲ repressive histone mark

⌐ transcription   ⫢ Lsd1   ☆ Adcy3

Figure 6-11 Each ORN expresses a single odorant receptor. (A) Three types of ORNs, each expressing a different odorant receptor, are shown here as blue, yellow, and green. All three types are broadly distributed along the long axis of the olfactory epithelium. Along the short axis, they are expressed in different quarters and therefore are mostly segregated. Summarized from *in situ* hybridization data. **(B)** Model of odorant receptor (OR) expression. (1) In immature ORNs, all OR genes (four are shown on one chromosome) are decorated with histone marks characteristic of repressed gene expression. (2) Transient expression of histone demethylase Lsd1 causes stochastic expression of an OR gene. (3) The functional OR protein activates a signaling pathway resulting in the expression of the adenylate cyclase Adcy3, which in turn inhibits expression of Lsd1, such that (4) all OR genes except one remain in the repressed state. (A, from Vassar R, Ngai J, & Axel R [1993] *Cell* 74:309–318. With permission from Elsevier Inc.; see also Ressler KJ, Sullivan SL, & Buck LB [1993] *Cell* 73:597–609; Miyamichi K, Serizawa S, Himura HM, et al. [2005] *J Neurosci* 25:3586–3592. B, based on Dalton RP & Lomvardas S [2015] *Annu Rev Neurosci* 38:331–349.)

single-cell RNA-sequencing experiments provided definitive evidence supporting this hypothesis.) The cell bodies of a given ORN type are broadly distributed along the long axis of the nasal epithelium without apparent spatial organization. Along the orthogonal axis, each ORN type localized within a zone about a quarter the length of the axis (Figure 6-11A). Thus, different ORN types expressing distinct receptors are completely intermingled along the long axis and are considerably intermingled even along the short axis. The broad distribution of a given ORN type enables the sampling of odorant molecules across a large surface area of nasal epithelium, thereby increasing the sensitivity of odorant detection.

Interestingly, within a given ORN, the functional odorant receptor mRNA is transcribed exclusively from one chromosome of a homologous pair; this property is called **allelic exclusion**. Recent studies indicate that genes encoding odorant receptors are in chromosomal regions subject to complex histone modifications, such that they are inaccessible to the transcriptional machinery. Transient expression of a histone demethylase, Lsd1, allows just one odorant receptor gene to be expressed. Expression of a functional odorant receptor then triggers a negative feedback pathway involving activation of an adenylate cyclase (Adcy3) which turns off Lsd1 so no other odorant receptor can be expressed in the same ORN (Figure 6-11B).

Given that each ORN expresses a single odorant receptor, mouse ORNs can be divided into ~1000 types based on the specific odorant receptor each one expresses. Since individual ORNs respond to many odorants, each odorant receptor must be activated by multiple odorants, and the identities of odorants are encoded by a combinatorial receptor code. The power of such combinatorial coding is enormous. For example, if each odorant activates two independent receptors, and each combinatorial activation of any two receptors generates a unique percept, then a repertoire of 1000 receptors (and receptor-expressing ORNs) can, in principle, generate half a million different percepts ($1000 \times 999 \div 2$—that is, the number of ways to select 2 out of 1000). This number can further expand if different percepts are generated by the activation of more than two receptors, by differential extents of activation of individual receptors or by different time courses of activation. This is how the olfactory system can in principle recognize and distinguish a vast number of odorant molecules.

6.6 ORNs expressing the same odorant receptor project their axons to the same glomerulus

How does the brain decode signals representing the mixture of odorants in the nose? ORN axons terminate in spherical structures called glomeruli in the olfactory bulb (Figure 6-3A). Within each glomerulus, ORN axons form synaptic connections with the dendrites of **mitral cells** and **tufted cells**, distinct types of output neurons of the olfactory bulb. Mitral cells and tufted cells have many similarities—both are glutamatergic excitatory neurons that send apical dendrites to a single glomerulus and project long axons to relay information to multiple regions of the **olfactory cortex**, where olfactory information is further analyzed. However, these cell types differ in their cell body locations, response properties,

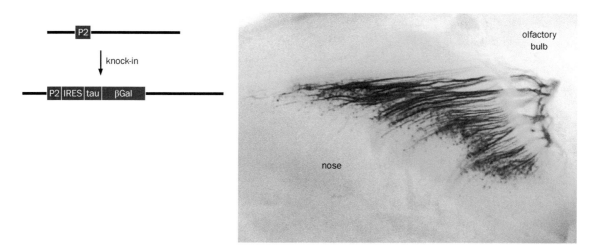

Figure 6-12 **Visualizing ORN axon convergence.** Left, the gene encoding the odorant receptor P2 (the box indicates its coding region) was engineered in embryonic stem cells by inserting a piece of DNA that enables the tracing of P2-expressing ORN axons. IRES, internal ribosome entry sequence that allows the translation of the axon tracer protein tau-βgal from the modified P2 mRNAs; tau, a microtubule binding protein for efficient labeling of axons; βgal, β-galactosidase enzyme from bacteria. Right, β-galactosidase activity, visualized by adding a colored substrate, reveals distributed ORN cell bodies in the nose (dots) and the convergence of their axonal projections onto one glomerulus in the olfactory bulb. (Adapted from Mombaerts P, Wang F, Dulac C, et al. [1996] *Cell* 87:675–686. With permission from Elsevier Inc.)

and axonal projection patterns (Figure 6-17). For simplicity, we use mitral cells as exemplary olfactory bulb output neurons in subsequent discussions.

What kind of organizational scheme could allow the nervous system to determine which odorants are present in an animal's surroundings? As seen in Figure 6-11A, ORNs expressing a given odorant receptor are broadly distributed in the nasal epithelia. However, subsequent studies revealed that ORNs expressing the same odorant receptor send their axons to the same glomeruli. The most direct evidence came from visualization of axons from ORNs expressing a specific odorant receptor. In these experiments, a specific odorant receptor gene was engineered via the knock-in procedure (see Section 14.7 for details) to co-express an axonal tracer, the β-galactosidase enzyme fused to the microtubule binding protein **tau**, which is highly enriched in axons. ORNs expressing the engineered odorant receptor gene could thus be traced from their cell bodies all the way to their axon terminals in the olfactory bulb by visualizing a colored substrate of the β-galactosidase enzyme (**Figure 6-12**).

Many ORN types have been tagged via this method. The resulting data indicate that the axons of ORNs that express the same odorant receptor converge on specific glomeruli, usually one each on the medial and lateral surfaces of the olfactory bulb. Moreover, in different animals, the target glomeruli for the same ORN type (expressing the same odorant receptor) are positioned with **stereotypy**— that is, little variation between individuals.

Thus, olfactory information is represented as a spatial map of glomerular activation in the olfactory bulb. The problem of identifying which odorants are present in the environment is equivalent to deciphering which glomeruli are active. Indeed, using neuronal activity monitoring techniques such as the intrinsic signal imaging (Figure 4-45B), specific odorants have been shown to activate specific ensembles of glomeruli (**Figure 6-13**).

These remarkable discoveries illustrate a general principle in the organization of the nervous system: the use of maps to represent information. In studying the visual system, it was intuitive to assume that the brain uses a spatial map to represent information gathered in the retina, as vision is inherently a representation of spatial properties of the world. By contrast, it was not apparent, a priori, how odorants would be represented in the brain. Despite also using neural maps to represent sensory information, however, the glomerular map of odorants is qualitatively different from the visual map: it does not represent space in the external world, but rather the nature of the chemicals present.

amyl acetate

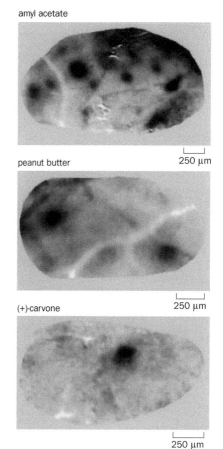

peanut butter

250 μm

(+)-carvone

250 μm

250 μm

Figure 6-13 **Intrinsic signal imaging of olfactory responses in the mouse olfactory bulb.** Each dark patch represents a glomerulus activated in response to specific odorants, as indicated. Each odorant elicits a specific spatial pattern of glomerular activation. (From Rubin BD & Katz LC [1999] *Neuron* 23:499–511. With permission from Elsevier Inc.)

6.7   Olfactory bulb circuits transform odor representations through lateral inhibition

We now have a good understanding of the first steps of olfaction. Convergent axonal projections of ORNs expressing the same odorant receptors allow olfactory information to be represented by patterns of glomerular activation in the olfactory bulb. How do glomerular activation patterns inform higher olfactory centers, eventually leading to odor perception and odor-driven behavior? To date, we have no satisfactory answers to these important questions. The next two sections address progress and challenges in this area of research, focusing on the olfactory bulb and the olfactory cortex, respectively.

The neural circuitry of the olfactory bulb share similarities with that of the retina (**Figure 6-14**). Within each glomerulus, ORN axons form glutamatergic synapses with the apical dendrites of mitral cells. Each mitral cell sends apical dendrites to a single glomerulus. ORN → mitral cell thus constitutes an excitatory pathway similar to the photoreceptor → bipolar cell → ganglion cell pathway in the retina. Because each glomerulus receives input from ORNs expressing the same odorant receptor and because each mitral cell sends apical dendrites to a single glomerulus, each mitral cell is linked to a single odorant receptor. Given this orderly anatomy, each glomerulus, with its associated ORNs and mitral cells, can be considered a discrete **olfactory processing channel**. On average, each glomerulus receives input from thousands of ORNs, and its output is conveyed by ~25 mitral cells. Each mitral cell receives pooled input from many hundreds of ORNs, and there is thus an information convergence that increases sensitivity and decreases noise for mitral cell representations of olfactory information (Section 1.12).

The olfactory bulb performs many more functions than simply pooling information. Like the retina, the olfactory bulb contains many local interneurons, whose projections are confined to the olfactory bulb. Altogether, olfactory bulb local interneurons outnumber mitral and tufted cells by ~100-fold in mammals, and consist of many types. Most are GABAergic and therefore inhibitory. **Periglomerular cells** make up a large group of interneurons that receives direct input from ORN axons within the glomeruli or from apical dendrites of mitral cells and spreads inhibition to nearby glomeruli (Figure 6-14A). **Granule cells** make up another large group that receives input from mitral cells rather than ORNs. In addition to having apical dendrites that innervate single glomeruli, mitral cells also send out **secondary dendrites** that cover a sizable portion of the olfactory bulb. Dendrites of granule cells and secondary dendrites of mitral cells form reciprocal **dendro-**

Figure 6-14 Schematic diagrams comparing olfactory bulb and retinal circuits. Simplified diagram of neurons and connections in the olfactory bulb (left) and retina (right, see Sections 4.14–4.18). In both diagrams, thin lines represent axons and thick lines represent dendrites. Axon and dendrite terminals are enlarged to highlight connections. Green represents excitatory neurons, and orange inhibitory neurons. In the olfactory bulb circuit, ORNs and mitral cells constitute the excitatory pathway delivering olfactory signals from the nose to the brain through discrete olfactory processing channels, two of which are shown here. Each glomerulus receives axonal input from ORNs expressing the same odorant receptor, and mitral cells send their apical dendrites to a single glomerulus. Granule cells and most periglomerular cells are GABAergic and spread inhibition laterally across different glomeruli. Green and red arrows indicate excitatory and inhibitory, respectively, synaptic connections between cells. Dashes on ORN axons indicate that the distances from cell bodies to glomeruli are much longer than schematized. Not shown are many additional types of olfactory bulb local interneurons, some of which extend processes over larger areas of the olfactory bulb than periglomerular and granule cells. Note that the dendrites of mitral, granule, periglomerular, horizontal, and amacrine cells not only receive but also send information, representing exceptions to the theory of dynamic polarization discussed in Section 1.7. Despite the apparent similarities, there are notable differences. The olfactory circuit utilizes one fewer layer of excitatory neurons than the retinal circuit. Horizontal cells mainly inhibit presynaptic terminals of photoreceptors (although they can also directly inhibit bipolar cells), whereas periglomerular cells primarily inhibit mitral cell dendrites. (Adapted from Shepherd GM [ed] [2004] The Synaptic Organization of the Brain, 6th ed. Oxford University Press.)

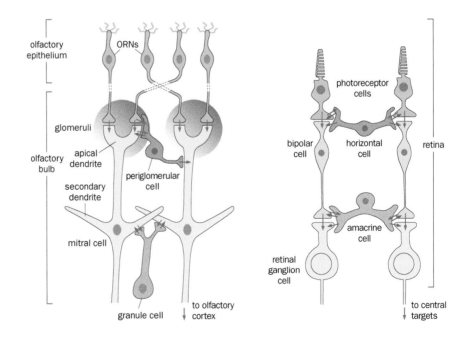

dendritic synapses: mitral cells excite granule cells, which in turn inhibit both the mitral cells that excited them and other mitral cells with secondary dendrites in the vicinity (Figure 6-14A; Figure 6-15). Thus, periglomerular and granule cells in the olfactory bulb circuit play similar roles as horizontal and amacrine cells in the retinal circuit, respectively (Figure 6-14).

In additional to GABAergic periglomerular and granule cells, some olfactory bulb local interneurons use glutamate or dopamine as neurotransmitters. Their functions are less well characterized. Another notable property of olfactory bulb interneurons is that a subset of these are products of **adult neurogenesis**. As we will learn in Chapter 7, the vast majority of neurons in the mammalian CNS are born during embryonic and early postnatal development. Olfactory bulb interneurons represent an exception in the rodent brain: they are continuously produced throughout adult life.

The similarity of olfactory bulb and retinal circuitries suggests that olfactory bulb GABAergic neurons play a role in lateral inhibition (Section 4.14). Specifically, each mitral cell collects at its apical dendrites excitatory input from the axons of presynaptic ORNs in its cognate glomerulus and receives at both its apical and secondary dendrites inhibitory input from nearby glomeruli representing other olfactory processing channels, in analogy to the center–surround receptive fields of retinal ganglion cells. Thus, for each olfactory processing channel, olfactory bulb interneurons transform the "receptive field" of input ORNs into that of the output mitral cells. (The olfactory receptive field is defined as the odorant repertoire that can alter the firing pattern of a cell in the olfactory system.) However, whereas the center–surround receptive field has a clear role in analyzing contrast, color, or motion by comparing activation levels of nearby photoreceptors, the exact role of lateral inhibition in the olfactory bulb is less clear.

One model proposes that lateral inhibition sharpens odorant representation against similar odorants. This model is unlikely to be generally applicable, in part because neighboring glomeruli do not necessarily represent chemically similar odorants and in part because secondary dendrites of mitral cells cover a large portion of the olfactory bulb. Furthermore, the *odor space* (that is, the entire repertoire of odorant molecules) is high dimensional: there are numerous chemically distinct odorants, each of which possesses multiple functional groups that activate multiple ORNs and glomeruli (Figure 6-7; Figure 6-13). This high dimensionality makes it difficult to project odorants onto the two-dimensional olfactory bulb surface in an orderly fashion.

An alternative view is that lateral inhibition executes **gain control** of mitral cells, even if surround inhibition is nonselective. (Gain control refers to the modulation of the relationship between the input to a system and its output, thereby keeping output within a limited dynamic range; see also Box 4-4). For example, when the animal encounters an odoriferous environment in which many ORNs are activated simultaneously, many GABAergic inhibitory neurons would also be activated, such that the gain of all olfactory processing channels is reduced, thereby preventing saturation in neuronal spike rates. Gain control could enable the olfactory processing channel(s) activated most strongly to stand out from the rest.

Yet another hypothesis is that lateral inhibition serves mainly to regulate the timing of mitral cell activity, which can carry information about odorant identity and concentration. For example, lateral inhibition could enable first-activated olfactory processing channels to suppress later-activated ones, which enables odor identity to be extracted by the timing of mitral cell activation. In summary, while we are far from a complete understanding, local olfactory bulb circuitry could play a variety of roles in information processing through lateral inhibition, all with the goals of sharpening the representation of ethologically meaningful olfactory information to be further processed by downstream olfactory cortex.

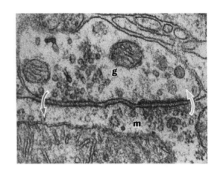

Figure 6-15 Dendrodendritic synapses between mitral cells and granule cells. Electron micrograph of adjacent reciprocal synapses between a granule cell process (g) and a mitral cell process (m). Arrows indicate the direction of information flow. (From Rall W, Shepherd GM, Reese TS, et al. [1966] *Exp Neurol* 14:44–56. With permission from Elsevier Inc.)

6.8 Olfactory inputs are differentially organized in different cortical regions

The olfactory pathway is the only sensory system that does not pass through the thalamus: mitral and tufted cells receive input from ORN axons and project their

Figure 6-16 Schematic of the olfactory system in mice. In this sagittal view of the mouse brain, the organization of the olfactory system from the olfactory epithelium to the olfactory bulb highlights the convergent axonal projections of two types of ORNs to two discrete glomeruli. In the olfactory bulb, each glomerulus is innervated by both mitral and tufted cells. Each mitral (blue, red) and tufted (green) cell sends apical dendrites into a single glomerulus and secondary dendrites laterally. Mitral cells project long-distance axons to multiple olfactory cortical regions, including the anterior olfactory nucleus, piriform cortex, olfactory tubercle, cortical amygdala, and entorhinal cortex. Although the two mitral cells originate in different parts of the olfactory bulb, their axonal projection patterns are not easily distinguishable in the piriform cortex. The dotted line in the piriform cortex divides this largest region of the olfactory cortex into anterior (A) and posterior (P) portions. D, dorsal; V, ventral. The cortical amygdala receives preferential input from the dorsal olfactory bulb. Tufted cell axons innervate anterior olfactory cortices and terminate in a special region of the olfactory tubercle. (Courtesy of Kazunari Miyamichi. The mitral and tufted cell drawings follow original data from Igarashi KM, Ieki N, An M, et al. [2012] *J Neurosci* 32:7970–7985.)

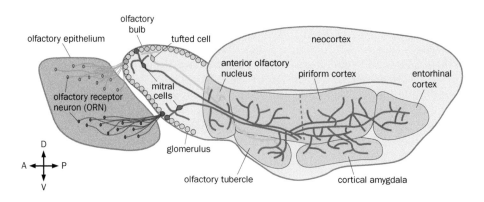

axons directly to various olfactory cortical regions (**Figure 6-16**). Thus, as olfactory cortical neurons are only two synapses away from the sensory world, the sense of smell has more direct access to the cortex than any other sense. Convergent axonal projections to the same glomerulus from ORNs of the same type enable the organization of olfactory information into a spatial map in the olfactory bulb. How is this glomerular map represented in the olfactory cortex?

Electrophysiological recording and optical imaging studies have suggested that the largest olfactory cortical region, the **piriform cortex**, displays a surprising lack of spatial organization. For example, when two-photon Ca^{2+} imaging (see Section 14.22 for details) was applied to the piriform cortex, researchers found that individual cortical neurons activated by specific odorants were distributed broadly across the piriform cortex with no discernible spatial patterns (**Figure 6-17**). These physiological studies are consistent with anatomical studies suggesting that the axons of individual mitral cells project broadly throughout the piriform cortex (Figure 6-16). This scheme sharply contrasts with the organization of the primary visual cortex, where cortical neurons are spatially organized according to retinotopy, and in some species further according to ocular dominance and orientation selectivity.

Why does the piriform cortex discard the spatial organization of the olfactory bulb? One possibility is that unordered connectivity between mitral cells and piriform cortical neurons creates a blank slate that allows piriform cortical neurons to collectively receive many combinations of different olfactory processing channels from the olfactory bulb. Indeed, in addition to receiving input from mitral cell axons, piriform cortex pyramidal neurons are extensively *recurrently connected*—pyramidal neurons send axons to synapse onto many other pyramidal neurons.

Figure 6-17 Piriform cortical neurons activated by specific odorant stimuli are broadly distributed with no apparent spatial order. (A) Window for *in vivo* imaging of the mouse piriform cortex. More than 1000 cells can be imaged simultaneously in the same anesthetized animal using a fluorescent indicator reporting changes in the intracellular Ca^{2+} concentration. **(B–D)** Piriform cortical cells responsive to three odorants are intermingled throughout the imaging window, indicating that the spatial map of discrete glomeruli in the olfactory bulb is not retained in the piriform cortex. (From Stettler DD & Axel R [2009] *Neuron* 63:854–864. With permission from Elsevier, Inc.)

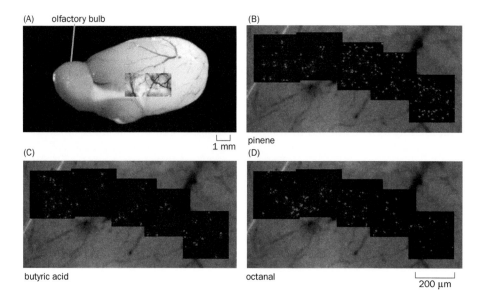

Figure 6-18 **Differentially distributed olfactory bulb input to the piriform cortex and cortical amygdala. (A)** Viruses enabling retrograde trans-synaptic tracing were injected into small volumes of different olfactory cortical regions. Transduction of cortical neurons led to the labeling of their presynaptic partner mitral cells and the corresponding glomeruli in the olfactory bulb. **(B)** Two separate injections into the piriform cortex (magenta, red) resulted in labeling of glomeruli distributed throughout the olfactory bulb; injections into the cortical amygdala (green) resulted in labeling of glomeruli located preferentially in the dorsal olfactory bulb. Labeled glomeruli from each experiment were reconstructed and superimposed onto a standard olfactory bulb model. A, anterior; P, posterior; D, dorsal; V, ventral. (Adapted from Miyamichi K, Amat F, Moussavi F, et al. [2011] *Nature* 472:191–196. With permission from Springer Nature. See also Sosulski DL, Bloom ML, Cutforth T, et al. [2011] *Nature* 472:213–216; Ghosh S, Larson DS, Hefzi H, et al. [2011] *Nature* 272:217–220.)

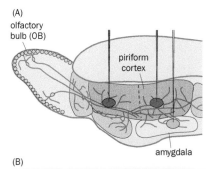

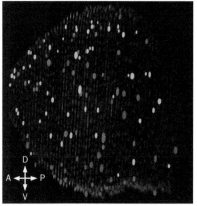

(These recurrent excitations are balanced by piriform cortex GABAergic neurons, which mediate lateral inhibition and gain control as olfactory bulb interneurons.) These connection matrices may serve as substrates on which olfactory experiences act by consolidating or eliminating connections or adjusting the strengths of connections to create odor representations reflecting an individual's experience. We will discuss these concepts in greater detail in Chapter 11.

Not all cortical regions represent odors as does the piriform cortex. Tracing mitral cell axons revealed projection patterns in the piriform cortex that were similar for mitral cells originating from different glomeruli, consistent with the physiological data discussed earlier. However, projections to the **cortical amygdala**, the part of the amygdala complex that receives direct mitral cell input, displayed more orderly patterns. These findings were reinforced by a study utilizing **retrograde trans-synaptic tracing**, which allows the labeling of direct presynaptic partners of a given neuron (see Section 14.19 for details). Individual cortical neurons were found to receive direct input from multiple mitral cells representing distinct glomeruli and thus olfactory processing channels. Whereas piriform cortex neurons receive direct input from glomeruli throughout the olfactory bulb with no obvious bias, neurons from the cortical amygdala receive biased input that strongly favors glomeruli in the dorsal olfactory bulb (**Figure 6-18**). Interestingly, in an experiment where ORNs that project to the dorsal olfactory bulb were selectively ablated, mice no longer avoided odorants that otherwise elicit innate avoidance behavior, such as chemicals from fox urine or spoiled food. Thus, some olfactory processing channels in the dorsal bulb may be particularly important in transforming detection of specialized odorants into innate avoidance behavior, and this information may be preferentially represented in cortical amygdala.

In summary, outputs of the olfactory bulb are differentially represented in different cortical regions, which could serve distinct purposes. An analogous organization appears in the fruit fly olfactory system, which will be discussed in the next part of this chapter. Finally, the accessory olfactory system in mammals, a system entirely parallel to the main olfactory system we have discussed thus far, is used to detect nonvolatile chemicals of special biological significance (**Box 6-1**).

Box 6-1: The mammalian accessory olfactory system specializes in detecting pheromones and predator cues

Most mammalian species possess an **accessory olfactory system** (also called the **vomeronasal system**), which is anatomically and biochemically distinct from the main olfactory system. Sensory neurons of the accessory olfactory system are housed in a special structure called the **vomeronasal organ (VNO)** located at the front of the nose (**Figure 6-19**A). Whereas ORNs in the main olfactory epithelia detect airborne odorants, VNO neurons detect nonvolatile stimuli through a narrow aqueous duct and project their axons to glomeruli in the **accessory olfactory bulb**, which is adjacent to the main olfactory bulb (Figure 6-19A). Mitral cells

of the accessory olfactory bulb deliver information to the olfactory cortex, but their axons innervate different regions from those innervated by mitral cells of the main olfactory bulb (see Figure 10-26 for details).

Each VNO neuron expresses a single vomeronasal receptor belonging to either the V1R or V2R families. These two GPCR families are distinct from each other and from the main olfactory system's odorant receptors. The signal transduction pathways downstream of V1Rs and V2Rs are also distinct from those used by the main olfactory system: whereas
(Continued)

Box 6-1: continued

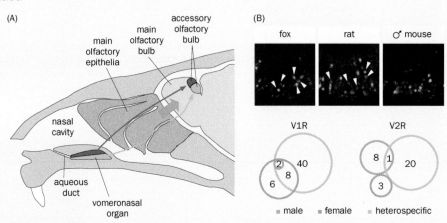

Figure 6-19 The accessory olfactory system and vomeronasal receptors. (A) The main and accessory olfactory systems in mice in a sagittal view. The vomeronasal organ (VNO) is located at the front of the nose and connected to the nasal cavity by an aqueous duct. Apically (red) and basally (green) located VNO neurons express V1R and V2R receptors and project axons (red and green arrows) to the anterior and posterior accessory olfactory bulb, respectively. The blue arrow represents ORN axon projections from the main olfactory epithelium (pink) to the main olfactory bulb. **(B)** Top, double *in situ* hybridization to identify receptors for specific chemical cues. VNO neuronal expression of immediate early gene *Egr1* in response to beddings of fox, rat, or male mice for a female recipient appears in green; expression of a specific V2R subfamily is shown in red. The yellow signals (arrowheads) represent overlap between the V2R probe and neurons activated by fox or rat bedding (note the lack of yellow cells with male mouse stimuli). Thus, these V2R receptors are activated by predator odors but not conspecific odors. Bottom, Venn diagrams for 56 V1R and 32 V2R receptors with respect to their activation by stimuli from male mice, female mice, and heterospecifics (predators and other species of mice). V2Rs appear more selectively tuned to one of the three types of cues. (A, adapted from Brennan PA & Zufall F [2006] *Nature* 444:308–315. With permission from Springer Nature. B, from Isogai Y, Si S, Point-Lezica L, et al. [2011] *Nature* 478:241–245. With permission from Springer Nature.)

activation of ORNs in the main olfactory system is mediated by opening of cyclic nucleotide-gated (CNG) channels (Figure 6-4), VNO neuron activation is mediated primarily by opening of a specific TRP channel (Box 2-4) called TRPC2. *Trpc2* knockout mice lose the function of nearly the entire accessory olfactory system.

The number of genes encoding intact V1Rs and V2Rs varies considerably between different mammalian species, with ~300 in mice but only a few in primates. Thus, the function of the accessory olfactory system varies considerably across species. Using immediate early gene expression to approximate active neurons (Section 3.23), stimuli that activate many mouse vomeronasal receptors have been identified (Figure 6-19B). Some of these vomeronasal receptors are activated by **pheromones** present in the urine, tear, and skin secretions of mice. (Pheromones are substances produced by an individual to elicit a specific reaction from other individuals of the same species. The word *pheromone* is derived from the Greek *pherein*, "to transfer," and *hormon*, "to excite.") Many other mouse vomeronasal receptors are devoted to detection of chemical cues from other species, such as mammalian and avian predators (Figure 6-19B).

We use a specific example to illustrate the functions of the accessory olfactory system. Urine is an important source of chemicals for social communication among many mammals, including mice. Male mice mark their territories by spreading urine. A resident male will exhibit aggression toward a sexually mature male intruder but not toward a castrated male. However, a castrated intruder swabbed with urine from a sexually mature male can elicit aggression from a resident male (**Figure 6-20**A). This response requires a functioning accessory olfactory system, as *Trpc2*-knockout resident mice do not exhibit aggression toward intruder males. The behavioral response of resident males toward intruders served as an assay for biochemical purification of the active components in mouse urine that elicit aggression. **Major urinary proteins (MUPs)**, which are highly stable proteins that can mark territory for a long time, were identified as one active component. Swabbing castrated male intruders with purified MUPs produced in *E. coli* elicited aggression from wild-type resident males (Figure 6-20A).

MUPs represent a means for social communication not only within a species but also between species. Mice are innately fearful of cats and rats, their predators. Rat urine elicited robust defensive responses in mice, including avoidance and the release of stress hormones. Biochemical purification of rat urine revealed the active component that elicits defensive responses in mice to be a specific rat MUP. Again, *Trpc2* mutant mice failed to respond to rat and cat MUPs (Figure 6-20B), indicating that these MUPs also activate the accessory olfactory system. Identification of the vomeronasal receptors mediating the effects of these MUPs and tracing of the corresponding neural circuits in the brain will advance our understanding of the neural basis by which MUPs from different species elicit aggressive or defensive behaviors.

As mentioned in Section 6.3, a large fraction of the genes encoding odorant receptors in the main olfactory system

Box 6-1: continued

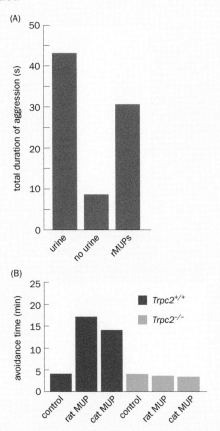

Figure 6-20 Major urinary proteins (MUPs) elicit aggressive or defensive responses. **(A)** Duration of aggression (in a 10-minute interval) exhibited toward a castrated male intruder by a resident male. Castrated male intruders that had been swabbed with urine from a sexually mature male elicited more aggression from the resident than those not swabbed with urine. Swabbing castrated male intruders with recombinant MUPs expressed in bacteria (rMUPs) also increased the aggression of resident males. **(B)** Mice avoid an area in a cage containing rat or cat MUPs. This behavior is abolished in mice lacking TRPC2 and thus requires functioning VNO neurons. (A, adapted from Chamero P, Marton TF, Logan DW, et al. [2007] *Nature* 450:899–902. With permission from Springer Nature. B, from Papes F, Logan DW, & Stowers L [2010] *Cell* 141:692–703. With permission from Elsevier Inc.)

have become pseudogenes in humans. The situation is more drastic in the accessory olfactory system: humans have five intact V1R genes and zero intact V2R genes; the remainder (>100) are all pseudogenes. The gene encoding the TRPC2 channel has also become a pseudogene in humans. The VNO itself is not present in adult humans, though it appears transiently during development. Thus, the accessory olfactory system appears to be extinct in humans.

Given the evolutionary loss of the accessory olfactory system, do humans use pheromones for social communication at all? Studies in mice indicate that, in addition to the accessory olfactory system, the main olfactory system can also detect pheromones (see Chapter 10 for more detail), which leaves open the possibility that humans can still use the main olfactory system for pheromone signaling. It has been observed that women living together tend to synchronize their menstrual cycles. Such synchrony can be induced by compounds from the armpits of donor women. Ovulation can be accelerated or delayed by donor extracts obtained at different times in the menstrual cycle. Identification of the chemicals responsible for these effects will consolidate our understanding of potential human pheromones and to explore their biological functions.

HOW DO WORMS AND FLIES SENSE ODORS?

An understanding of the mammalian brain can come from studies of simpler organisms. We saw an excellent example of this in Chapter 2: the use of the squid giant axon to determine the ionic basis of the action potential. The principles we learned from squid apply to almost all animals with a nervous system. Can an understanding of complex neurobiology problems such as olfactory perception also benefit from studies of simpler organisms?

The use of model organisms (Sections 14.1–14.5) in neurobiology research has considerable merit. Model organisms usually offer biological simplicity, technical ease, or both. We can borrow insights learned from simpler organisms to investigate analogous problems in mammals. If the mechanisms are similar, simpler systems can inform us about common solutions to neurobiology problems and speed up discoveries in mammals. And if the mechanisms are different, we nevertheless learn about alternative solutions to a common problem and, in the process, enrich our understanding of the diversity of life.

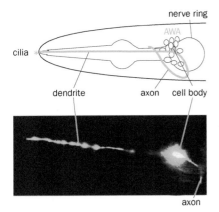

Figure 6-21 An olfactory neuron in *C. elegans*. Schematic (top) and fluorescence image (bottom) of one of the two AWA sensory neurons. The cell body of the AWA is in the nerve ring, which contains many neuronal cell bodies; its dendrite extends as a cilium, which ends in the anterior opening of the worm (*). (Adapted from Sengupta P, Chou JH, & Bargmann CI [1996] *Cell* 84:875–887. With permission from Elsevier Inc.)

6.9   *C. elegans* encodes olfactory behavioral choices at the sensory neuron level

The nematode *Caenorhabditis elegans* has emerged as an important model organism for the study of the nervous system (Section 14.2). Hermaphrodite *C. elegans* has 302 neurons, and the wiring diagram for its nervous system at the synaptic level (**connectome**)—the approximately 6393 chemical synapses between neurons, 1410 neuromuscular junctions, and 890 gap junctions—has been mapped by serial electron microscopic reconstructions. Neurons essential for *C. elegans* olfaction were identified by behavioral studies of odorant-induced attraction and repulsion coupled with laser ablation and genetic manipulation. Three pairs of sensory neurons—AWAs, AWBs, and AWCs—are responsible for detection of most volatile chemicals. These neurons send their dendrites to the anterior opening of the worm for odorant detection (**Figure 6-21**). Interestingly, there is a clear functional separation for these three pairs of neurons: AWAs and AWCs are required for odor-mediated attraction, while AWBs are required for odor-mediated repulsion. These neurons collectively express many of the ~100 G-protein-coupled receptors, which are the presumed odorant receptors. This simple organization poses an interesting question about odor perception and odor-mediated behavior: is it activation of a particular odorant receptor or a particular neuron that determines behavioral output?

Odr10 has been identified as a receptor responsible for detecting diacetyl, an attractant, and is expressed in AWAs. *Odr10* mutant worms are no longer attracted to diacetyl. Using a transgenic approach (see Section 14.9 for details), Odr10 was reintroduced into specific neuronal types in worms otherwise lacking the receptor. Transgenic reintroduction into AWAs of *Odr10* mutant worms rescued attraction to diacetyl, as expected. But if *Odr10* was reintroduced into AWBs of *Odr10* mutant worms, transgenic worms were *repelled* by diacetyl (**Figure 6-22**). Thus, it is neither the odorant nor its receptor, but rather the responding neuron, that determines the behavioral output.

This elegant experiment suggested a simple logic for *C. elegans* odor perception: receptors for attractants are expressed in AWAs or AWCs and receptors for repellents are expressed in AWBs. The response of the particular type of sensory neuron then determines the behavioral output, presumably through its specific connections with the motor programs mediating attraction or repulsion. This solution may not easily extend to the more complex mammalian olfactory system, which has many more sensory neurons, the capacity to distinguish between many different odorants, and a different organizational principle (that is, one receptor per neuron). It nevertheless illustrates how a simple organism with a small number of total neurons manages to solve the problem of olfactory perception to guide its behavior: by encoding the **hedonic values** (whether potentially beneficial or

Figure 6-22 *C. elegans* olfactory neurons determine its behavioral output.
(A) Odorant preference assay. Worms start at the center of a plate (origin) and are allowed to move freely for a defined period of time. A preference index is calculated according to the number of worms distributed in different sectors: (A + B – E – F) / total number. **(B)** Wild-type worms are attracted to diacetyl (left). Worms lacking the *Odr10* odorant receptor gene are insensitive to diacetyl (middle). When *Odr10* is reintroduced into AWB neurons of *Odr10* mutant worms, these worms are repelled by diacetyl (right). (Adapted from Troemel ER, Kimmel BE, & Bargmann CI [1997] *Cell* 91:161–169. With permission from Elsevier Inc.)

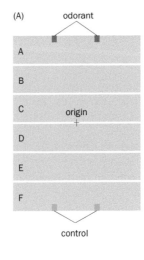

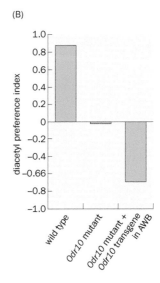

harmful to the animal) of odorants at the starting point of the sensory system—the sensory neurons. We will see later in this chapter that a similar logic applies to the mammalian taste system.

6.10 *C. elegans* sensory neurons are activated by odorant withdrawal and engage ON and OFF pathways

Given that we know the synaptic connections of all 302 *C. elegans* neurons, one might expect that the circuit mechanisms that transform the activity of AWA/AWC or AWB neurons into attractive or repulsive behaviors would be readily deciphered. However, the following studies illustrate that having a wiring diagram is just one step in understanding how a neural circuit operates.

The first surprise came when it was found that AWC is activated by odorant withdrawal rather than odorant application. *C. elegans* neurons are very small and notoriously difficult to access for electrophysiological recording. This discovery was made using Ca²⁺ imaging of restrained worms responding to odorants delivered via a microfluidic device (**Figure 6-23**A, B, top). A genetically encoded Ca²⁺ indicator, GCaMP (Section 14.22), was selectively expressed in AWC neurons, such that increases in fluorescence intensity reflected activation. Whereas odorant application caused a slight *decrease* in the Ca²⁺ signal (Figure 6-23A, blue trace), odorant withdrawal caused a large *increase* in the Ca²⁺ signal (Figure 6-23B, blue trace), indicating that AWC is inhibited by odorant application and activated by odorant withdrawal.

According to the wiring diagram, AWC synapses with a small number of interneurons, including AIB and AIY. Ca²⁺ imaging in AIB or AIY (via selective GCaMP expression in AIB or AIY, respectively) revealed that AIB was also activated by odorant withdrawal, with a greater latency than AWC (Figure 6-23B, red trace). By contrast, AIY was activated by odorant application (Figure 6-23A, green trace).

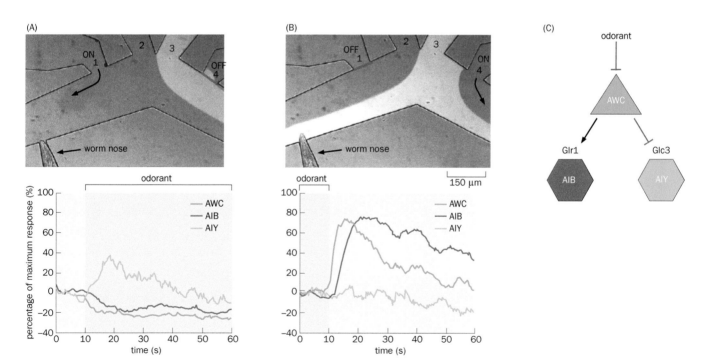

Figure 6-23 Olfactory responses of AWC and downstream interneurons. (A, B) Top, schematic of the microfluidic device used to control odorant delivery to restrained worms for Ca²⁺ imaging in specific neurons. When streams 1–3 are open (Panel A), odorant (gray) reaches the worm nose. When streams 2–4 are open (Panel B), buffer from stream 3 (white) reaches the worm nose. Bottom, AWC and AIB are inhibited by application of an attractive odorant (isoamyl alcohol) and activated by odorant withdrawal. AIY is activated by odorant application and unresponsive to odorant withdrawal. The *y* axis depicts the percentage of maximum response in the fluorescence intensity of a genetically encoded Ca²⁺ indicator expressed in the respective neurons. These curves represent averages from separate experiments. **(C)** Summary of the circuit diagram. AWC releases glutamate as a neurotransmitter and is inhibited in the presence of odorants. AWC activates AIB through Glr1, an AMPA-like ionotropic glutamate receptor, but inhibits AIY through Glc3, a glutamate-gated Cl⁻ channel. (Adapted from Chalasani SH, Chronis N, Tsunozaki M, et al. [2007] *Nature* 450:63–70. With permission from Springer Nature.)

Genetic perturbation experiments indicated that AWC uses glutamate as a neurotransmitter; when glutamate release by AWC was inhibited, odorant responses in AIB and AIY were abolished, indicating that AIB and AIY receive odorant information via AWC. How could AWC signal to two downstream neurons with opposite effects? Further analysis indicated that AIB expresses Glr1, an ionotropic glutamate receptor similar to the AMPA receptor, which causes depolarization of AIB in response to glutamate. By contrast, AIY expresses Glc3, a glutamate-gated Cl⁻ channel, which causes hyperpolarization of AIY in response to glutamate (Figure 6-23C).

Thus, the AWC circuit resembles the vertebrate rod and cone circuits: (1) sensory neurons are hyperpolarized by stimuli and (2) parallel ON and OFF pathways utilize the same neurotransmitter (glutamate) but different receptors in postsynaptic neurons (compare Figure 6-23C with Figure 4-22). The second property reflects a convergent strategy (discussed in detail in Chapter 13) for solving diverse sensory processing problems. The parallel ON and OFF pathways increase the contrast for odorant detection and the sensitivity of *C. elegans* to odorant onset and offset. Both are beneficial for navigation toward food sources and away from harmful substances.

6.11 The olfactory systems of insects and mammals share many similarities

Unlike *C. elegans,* which uses a small number of neurons to assess its olfactory environment, insects, with hundreds of thousands to many millions of neurons, possess olfactory systems remarkably similar to those of mammals (**Figure 6-24**). Insect **antennal lobes**, the sites of ORN axon termination, have a similar glomerular organization as mammalian olfactory bulbs. Recent studies, particularly in the fruit fly *Drosophila melanogaster,* have provided insight into the role of glomerular organization in the antennal lobe in olfactory information processing.

Figure 6-24 Comparing the mouse and fly olfactory systems. The olfactory systems of the mouse (top, a side view) and fly (bottom, a frontal view) share similar glomerular organizations, as seen when comparing the mouse olfactory bulb and fly antennal lobe, both of which are targeted by olfactory receptor neuron (ORN) axons: (1) each ORN expresses one specific odorant receptor, (2) ORNs expressing the same odorant receptor project their axons to the same glomeruli (on both antennal lobes in the fly), and (3) each second-order mitral cell or projection neuron (PN) sends dendrites to a single glomerulus and thus receives direct input from a single ORN type. (Exceptions to this in the fly: a small number of ORNs express two odorant receptors; some [atypical] PNs innervate multiple glomeruli.) Antennae (AT) and maxillary palps (MP) house ORNs. Local interneurons (LN) in the antennal lobe extend elaborate dendrites that cover many glomeruli (see Figure 14-26 for a singly labeled LN). PNs send axons mainly to two higher olfactory centers, the mushroom body (MB) and lateral horn (LH). Numbers in parentheses indicate the numbers of cells (ORN, PN, LN, mitral cells), types (ORN, PN, mitral cells), or structures (glomeruli). (Adapted from Komiyama T & Luo L [2006] *Curr Opin Neurobiol* 16:67–73. With permission from Elsevier Inc. The scanning electron microscopic image of the fly head is courtesy of John R. Carlson.)

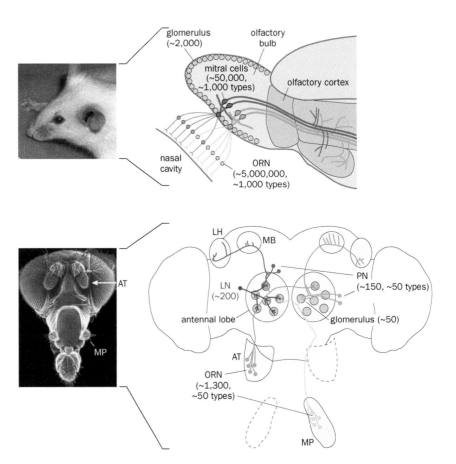

Two pairs of external sensory organs, the antennae and maxillary palps, are the "noses" of the fly and house ORN cell bodies (Figure 6-24 bottom left). Like mammals, most fruit fly ORNs express a single odorant receptor, and ORNs expressing the same odorant receptor project their axons to the same glomerulus in the antennal lobe (Figure 6-24 bottom right). Information is then relayed to higher olfactory centers by antennal lobe **projection neurons** (**PNs**). Like mammalian mitral/tufted cells, most PNs send their dendrites to a single glomerulus and thus receive direct input from a single ORN type. The antennal lobe also has many **local interneurons** (**LNs**), most of which are GABAergic. Thus, the organizations of the fly and mammalian olfactory systems share remarkable similarities.

As there are only ~50 ORN types and ~50 glomeruli in each fly antennal lobe, compared with ~1000 ORN types and ~2000 glomeruli in each mouse olfactory bulb, the fly olfactory system is numerically considerably smaller. All glomeruli are recognizable by their stereotyped sizes, shapes, and relative positions. The correspondence of ORN type and glomerular identity has been completely mapped. Furthermore, the odorant response properties of a large fraction of the ORN repertoire have been determined *in vivo* using an elegant strategy combining mutant and transgene expression in a similar way to the *C. elegans* Odr10 experiments. Specifically, in a mutant strain lacking the endogenous *Or22a* gene (which encodes a specific odorant receptor), the *Or22a* promoter is used to drive expression of a different odorant receptor gene, such that the electrophysiological responses of Or22a-expressing ORNs to various odorants then reflect the properties of the odorant receptor expressed via the transgene (**Figure 6-25**A). The firing properties of a majority of odorant receptors encoded in the *Drosophila* genome in response to a large panel of odorants have been determined in this way (Figure 6-25B).

As in vertebrates (Figure 6-7), each odorant activates multiple odorant receptors and each odorant receptor is activated by multiple odorants. Some odorant receptors are broadly tuned, meaning that they are activated by many odorants, while others are narrowly tuned, meaning that they are more selectively activated by a small number of odorants. The tuning curve is also shaped by odorant

Figure 6-25 Olfactory coding by the *Drosophila* odorant receptor repertoire. **(A)** Experimental strategy. Left, in wild-type flies, Or22a ORNs express the odorant receptor Or22a. Middle, in mutant flies lacking the *Or22a* gene, Or22a ORNs become "empty" ORNs that do not express a functional odorant receptor. Right, a transgene encoding a different odorant receptor (OrX) driven by the *Or22a* promoter can be introduced into the *Or22a* mutant, the resulting Or22a ORNs express OrX; *in vivo* extracellular recordings then measure the responses of OrX to various odorant stimuli. Bottom, transgene expression utilizes the GAL4/UAS binary system; from the first transgene, the *Or22a* promoter drives the expression of the yeast GAL4 transcription factor, which binds to UAS and activates OrX expression from a second transgene (see Section 14.9 for more details). **(B)** Olfactory responses of specific odorant receptors (names in the top row) as assayed by measuring ORN firing rates *in vivo*. The symbols ·, +, ++, and +++ represent, respectively, <50, 50–100, 100–150, and 150–200 spikes per second elicited by a 10^{-2} dilution of a specific odorant (rows) for Or22a ORNs lacking endogenous *Or22a* and expressing a specific odorant receptor (columns). The – represents a >50% decrease in the basal firing rate in response to odorant application. This figure depicts only a fraction of the results from this experiment, which assayed 110 odorants and 24 odorant receptors, or about half of the fruit fly receptor repertoire. Subsequent studies have characterized the response profiles of the remainder of the receptor repertoire. **(C)** Concentration-dependent response of ORNs expressing the Or35a odorant receptor. As the odorant concentration decreases, the tuning of Or35a ORNs becomes narrower. (Adapted from Hallem EA & Carlson JR [2006] *Cell* 125:143–160. With permission from Elsevier Inc. See also Silbering AF, Bell R, Galizia CG, et al. [2011] *J Neurosci* 31:13357–13375.)

(A)

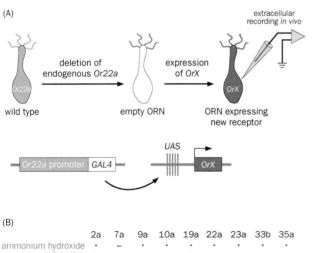

extracellular recording *in vivo*

deletion of endogenous *Or22a* → expression of OrX

wild type   empty ORN   ORN expressing new receptor

Or22a promoter | GAL4 — UAS — OrX

(B)

	2a	7a	9a	10a	19a	22a	23a	33b	35a
ammonium hydroxide	·	–	·	·	·	·	·	·	·
putrescine	·	–	·	–	·	·	·	·	·
cadaverine	·	·	·	·	·	·	–	·	·
γ-butyrolactone	·	·	+	·	·	+	·	·	+
γ-hexalactone	·	–	++	·	·	++	·	·	+++
γ-octalactone	·	–	·	·	·	·	·	·	·
γ-decalactone	·	–	·	·	–	·	·	·	·
δ-decalactone	·	–	·	·	–	–	·	·	–
methanoic acid	·	·	·	·	·	·	–	·	·
acetic acid	·	–	·	·	·	·	·	·	·
propionic acid	·	·	·	·	·	·	·	·	·
butyric acid	·	·	++	·	·	++	·	·	·
pentanoic acid	·	·	·	·	·	·	·	·	·
hexanoic acid	·	·	·	·	·	++	·	·	·
heptanoic acid	·	·	·	·	–	·	·	·	·
octanoic acid	·	·	·	·	·	·	·	–	·
nonanoic acid	·	–	·	·	·	·	·	·	·
linoleic acid	–	++	·	·	–	·	·	·	·

(C)

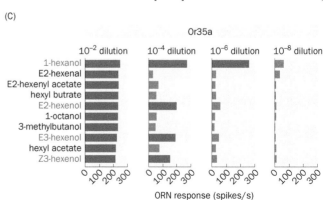

Or35a

| 10⁻² dilution | 10⁻⁴ dilution | 10⁻⁶ dilution | 10⁻⁸ dilution |

1-hexanol
E2-hexenal
E2-hexenyl acetate
hexyl butyrate
E2-hexenol
1-octanol
3-methylbutanol
E3-hexenol
hexyl acetate
Z3-hexenol

ORN response (spikes/s)

concentration—lower concentrations result in more selective tuning (Figure 6-25C). While most odorant receptors are activated by odorants, some odorant receptors are *inhibited* by specific odorants (Figure 6-25B)—the basal firing rates of ORNs expressing these odorant receptors decrease when the fly is exposed to specific odorants. (Due to the high sensitivity of *in vivo* recordings in flies, both activation and inhibition were detected, in contrast to the studies exemplified in Figure 6-7, in which only activation by odorants was reported.) These studies have made the fruit fly olfactory system the most completely characterized with regard to sensory stimuli and olfactory input channels (ORN types).

6.12 The antennal lobe transforms ORN input for more efficient representation by projection neurons

How is olfactory information processed in the antennal lobe? This question was answered by systematically comparing the firing patterns of ORNs and PNs that are direct synaptic partners to assess whether and how odorant coding is transformed between ORNs and PNs (**Figure 6-26**A). These experiments are very difficult in mammals due to technical limitations but can be readily performed in flies because of the identifiable glomeruli and tools allowing genetic access to many ORN and PN types. The following properties have been identified from such comparisons.

First, spike rates of individual PNs are less variable than those of their partner ORNs in response to the same stimulus (Figure 6-27B, C). This is because, on average, 60 ORNs (30 each from ipsi- and contralateral sides) synapse onto three PNs within each glomerulus, and each ORN axon forms strong synapses with the dendrites of all PNs within the glomerulus. In addition, PNs projecting to the same glomerulus form electrical synapses with each other, further synchronizing their activity. These properties allow individual PNs to pool information from many ORNs and thus encode sensory information as spike rates (Section 1.8) more reliably than their presynaptic partner ORNs.

Second, PNs preferentially report the rising phase of ORN activity. This can be seen from the raw data on spikes (Figure 6-26B) or in the **peristimulus time**

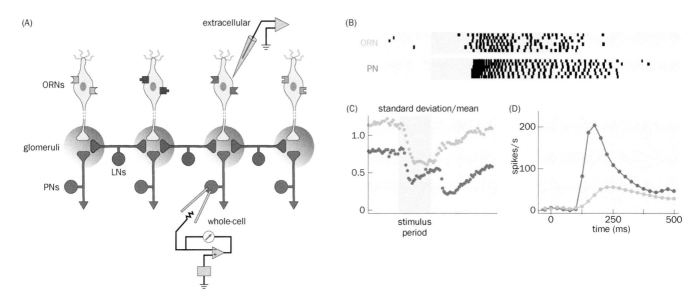

Figure 6-26 PN spike rates are less variable than ORN spike rates and selectively report the onset of ORN firing. (A) Schematic of experimental setup. Extracellular recording in the antenna or maxillary palp can measure spike rates of specific ORN types in response to olfactory stimuli. Whole-cell patch recording can measure spike rates of partner PNs genetically engineered to express a fluorescent marker. In this basic circuit schematic, ORNs and partner PNs form type-specific excitatory connections in specific glomeruli. Local interneurons (LNs), mostly GABAergic, laterally link multiple glomeruli. **(B)** ORN and PN firing in response to 500 ms odorant stimulation (yellow shading).

Each row represents a separate trial with identical stimuli. Each vertical tick is a spike. **(C)** PN responses are less variable than ORN responses, as indicated by a lower coefficient of variation, the ratio of standard deviation to the mean of spike counts. **(D)** Peristimulus time histogram plotting spike rates of ORNs (light green) and PNs (dark green) as functions of time after odorant onset (at 0 ms). The PN spike rate peaks and decays faster than the ORN firing rate. (Adapted from Bhandawat V, Olsen SR, Gouwens NW, et al. [2007] *Nat Neurosci* 10:1474–1482. With permission from Springer Nature.)

histogram (**PSTH**), which plots spike rates of neurons as a function of time after stimulus onset (Figure 6-26D). At least two separate factors account for this property: (1) ORN → PN synapses exhibit short-term depression (Section 3.10) such that, during a train of action potentials, ORN spikes arriving later cause smaller postsynaptic potentials than earlier spikes and (2) ORNs also activate local interneurons (LNs; Figures 6-24), which are mostly GABAergic inhibitory neurons. LNs synapse back onto ORN axon terminals and reduce the likelihood of transforming ORN action potentials into neurotransmitter release, a process called **presynaptic inhibition** (Section 3.22). Thus, LN activation provides negative feedback to control the synaptic output of ORNs and hence of input to PNs. Together, these factors allow PNs to be highly selective to odorant onset rather than to the constant presence of the same odorant, thereby informing the organism about *changes* in its environment. This is olfactory adaptation at the circuit level, distinct from adaptation at the level of ORN signal transduction discussed in Section 6.1.

Third, by spreading their representations of different odorants across a larger range of spike rates than ORNs, PNs use the available coding space more efficiently than ORNs. In the **coding space** of a neuronal population, the activity state is represented as a point in multidimensional space; each axis typically represents the spike rate of one constituent neuron (**Figure 6-27**A). When the spike rates of individual ORN types are plotted against those of partner PNs for different odorants, most ORN–PN pairs produce nonlinear curves. Weak ORN signals are selectively amplified in PNs, whereas strong ORN signals produce saturating levels of PN activation (Figure 6-27B). Two factors already discussed, short-term depression and presynaptic inhibition by GABAergic LNs, contribute to these properties. In addition, at low odorant concentrations, a small number of excitatory LNs boost weak ORN signals. Together, these properties allow PNs to use the coding space more efficiently than ORNs, thus enhancing odor discrimination.

Let's use a specific example to illustrate these points. The responses to 18 diverse odorants were measured for seven pairs of ORNs and PNs. The odorant identity could thus be represented by the spike rates of seven ORN types in a seven-dimensional ORN space, or, alternatively, by the spike rates for seven PN types in a seven-dimensional PN space. Since it is difficult to visualize a seven-dimensional space, we often use a statistical method called **principal component analysis (PCA)** to analyze and represent such spaces. PCA transforms the representation of a complex dataset in high dimensional space using a set of orthogonal axes called principal components according to data spread; data are most spread along the axis of the first principal component, followed by an orthogonal axis of the second principal component, and so on. High-dimensional datasets can then be represented in lower-dimensional space along the first few principal components. When the 18 different odorants were plotted in the space of the first two principal components, their representations were more clustered in the ORN space than in the PN space (**Figure 6-28**), meaning that it is easier for a downstream decoder to discriminate between these odorants based on the PN population activity than the ORN population activity.

6.13 Odors with innate behavioral significance use dedicated olfactory processing channels

As we discussed earlier in the chapter, combinatorial coding, in which odor identity is distributed among several olfactory processing channels, expands coding capacity beyond the number of parallel olfactory processing channels defined by the number of odorant receptors. This strategy allows the 50 olfactory channels in flies to discriminate between many more than 50 odorants. An important challenge lies in having "decoder" neurons in higher olfactory centers capable of distinguishing the activation of different combinations of PN populations (see Section 14.30 for a general discussion of encoding and decoding). However, certain odorants are of such importance to an organism that dedicated olfactory processing channels may be devoted to representing those odorants. This strategy, called **labeled lines**, whereby dedicated sensory processing channels carry specific information from the periphery to the brain, has also been proposed in other

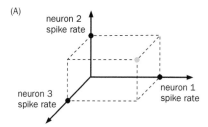

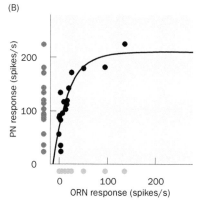

Figure 6-27 Coding space and the relationship between ORN and PN spike rates. (A) Illustration of coding space for a neuronal population consisting of just three neurons. The activity state of this neuronal population at any given time is represented by the green dot in three-dimensional space; its *x*, *y*, and *z* values (black dots) correspond to the spike rates of each component neuron. **(B)** Nonlinear transformation of ORN responses into PN responses. Each black dot represents one of the 18 odorants whose evoked responses were measured independently for a specific ORN–PN pair. In this nonlinear transformation, weak ORN inputs are selectively amplified, whereas strong ORN inputs produce saturated PN responses. Thus, the spike rate is more distributed in the PN space than in the ORN space for this ORN–PN pair. Most ORN–PN pairs across different glomerular channels exhibit similar curves. (B, adapted from Bhandawat V, Olsen SR, Gouwens NW, et al. [2007] *Nat Neurosci* 10:1474–1482. With permission from Springer Nature.)

Figure 6-28 Odorant coding is more widely distributed in populations of PNs than ORNs. Spike rates in response to 18 distinct odorants (different colors) are represented by the first two principal components from the seven-dimensional coding space of spike rates of ORNs (left) or PNs (right). Data were obtained from recording olfactory responses of seven ORN–PN pairs. Different odorants are more widely distributed in the PN coding space than in the ORN coding space, which means that they are more easily distinguishable by downstream neurons. (Adapted from Bhandawat V, Olsen SR, Gouwens NW, et al. [2007] *Nat Neurosci* 10:1474–1482. With permission from Springer Nature.)

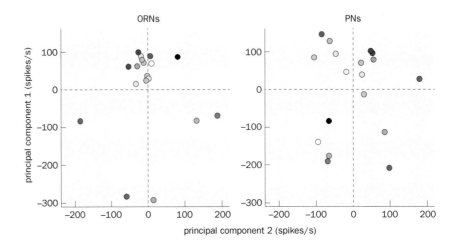

sensory systems (see Figure 6-33 below). Labeled lines simplify the process of decoding, as specific higher-order neurons receive information predominantly from a single processing channel. We provide two examples that suggest the use of this strategy by the fly olfactory system.

When stressed (such as being vigorously shaken in a test tube), fruit flies release odors that naive flies avoid. This avoidance can be measured in the laboratory by placing naive flies in a T-maze (**Figure 6-29**A) with one arm containing air released from shaken flies and the other arm containing fresh air. About 90% of naive flies choose the side containing fresh air (Figure 6-29B; **Movie 6-1**). A major component of this stress odor was identified to be carbon dioxide (CO_2), which flies avoid in a concentration-dependent manner (Figure 6-29B).

ORNs expressing the olfactory receptor Gr21a appear to be dedicated to CO_2 sensation. Two experiments have tested the function of Gr21a-expressing ORNs in mediating the CO_2 avoidance behavior. In the first experiment, Gr21a ORNs were acutely and selectively silenced by expressing a temperature-sensitive mutant protein, *Shibire^{ts}*, which blocks synaptic transmission at high temperatures (Sections 3.9 and 14.23). Flies no longer avoided CO_2 under this condition.

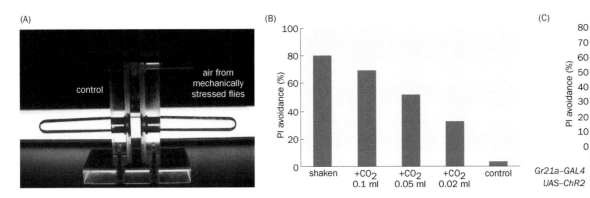

Figure 6-29 Activation of CO_2-sensitive ORNs produces avoidance behavior. (A) A T-maze can be used to test flies' preference for odorants. Flies were introduced to the center of the choice point from an "elevator" in the middle and allowed for one minute to choose between two arms (see Movie 6-1). Performance index (PI) = [(number of flies in the control arm – number of flies in the experimental arm) / total number] × 100 (%). Thus, if the distribution of 100 flies is 90 versus 10, PI = 80%. **(B)** Flies avoid the tube containing air from mechanically stressed flies (shaken). They also avoid the CO_2-containing tubes in a concentration-dependent manner (columns 2–4). In the control case, both tubes contain fresh air. **(C)** In this experiment, a transgene encoding ChR2 is expressed in Gr21a neurons using the GAL4/*UAS* system. Five experimental conditions are assessed; in the

first three conditions, blue light is delivered to one of the two tubes. Control flies with only the *GAL4* or *UAS* transgenes do not express ChR2 and do not respond to light. Flies that have both transgenes such that ChR2 is expressed in Gr21a ORNs avoid the tube exposed to blue light (column 3), just as they avoid the tube containing CO_2 (column 5); column 4 is a no-light control. Thus, activation of Gr21a ORNs causes a repulsive behavioral response, whether activated naturally by CO_2 or artificially by light-induced depolarization. (A, courtesy of David J. Anderson; B, adapted from Suh GSB, Wong AM, Hergarden AC, et al. [2004] *Nature* 431:854–859. With permission from Springer Nature. C, adapted from Suh GSB, Ben-Tabou de Leon S, Tanimoto H, et al. [2007] *Curr Biol* 17:905–908. With permission from Elsevier Inc.)

This loss-of-function experiment demonstrated the necessity of Gr21a ORNs for CO_2 avoidance. In the second experiment, the light-activated channelrhodopsin, ChR2, was expressed only in Gr21a ORNs, such that these neurons would be selectively activated by blue light (Section 14.25). When flies were placed in a T-maze in which one of the two arms was bathed in blue light for depolarizing the CO_2-sensitive ORNs, the flies avoided that arm, just as they avoided an arm containing CO_2 (Figure 6-29C). This gain-of-function experiment indicated that activation of Gr21a ORNs is sufficient for the avoidance behavior. Together, these experiments suggest that repulsion by CO_2 is already encoded in this special ORN type, much like repulsion is encoded in AWB neurons in *C. elegans* (Figure 6-22).

Flies use mating pheromones to communicate with each other about their sex and mating status (see Chapter 10 for more detail). A pheromone produced by male *Drosophila*, 11-*cis*-vaccenyl acetate (cVA), an ester of a long-chain alcohol, inhibits the courtship of males toward other males. cVA also inhibits courtship of males toward mated females, which contain cVA transferred from their previous male partner during mating (a cunning way for males to minimize sperm competition). ORNs expressing the Or67d odorant receptor play a major role in this behavior. Silencing the activity of Or67d ORNs reduced male–male courtship inhibition, while expression of a moth pheromone receptor in Or67d ORNs reduced courtship of male flies toward virgin females treated with the corresponding moth pheromone. Thus, Or67d ORNs specialize in sensing cVA, just as Gr21 ORNs specialize in sensing CO_2. Electrophysiological experiments indicated that the postsynaptic partner PNs of Or67d ORNs are also narrowly tuned to cVA and minimally activated by the other odorants tested.

Thus, while combinatorial coding allows fruit flies to distinguish between thousands of different odorants using only ~50 odorant receptors and processing channels, some processing channels may be devoted principally to detection of single, behaviorally important stimuli.

6.14 Odor representations in higher centers for innate and learned behaviors are stereotyped and stochastic, respectively

How is information carried by PNs represented in higher olfactory centers that discriminate among odors and induce appropriate behavioral responses? While we still do not have complete answers to this fundamental question, the numerical simplicity of the fly olfactory system and an abundance of genetic tools have yielded a more advanced understanding of higher olfactory centers in *Drosophila* than in the mammalian olfactory cortex.

The output neurons of the antennal lobe, PNs, send their axons to two major higher olfactory centers: the **mushroom body** and the **lateral horn** (Figure 6-24). The mushroom body is a center for olfactory learning and memory, whereas the lateral horn mediates innate olfactory behaviors. To investigate how olfactory input to these centers is organized, PNs that send dendrites to different glomeruli were individually labeled using genetic techniques (Section 14.16). Systematic analyses revealed striking stereotypy of axonal branching patterns and locations of terminal axonal arbors in the lateral horn for all observed PN types (Figure 6-30). Individual PN axons invade different parts of the lateral horn, and axons of different types of PNs project to overlapping portions of the lateral horn. These features could enable the integration of information conveyed by different PNs to the same postsynaptic third-order lateral horn neurons (convergence), while allowing individual PNs to deliver information to multiple lateral horn neurons (divergence). Recent physiological recordings of lateral horn neurons in response to olfactory stimuli and PN activation have validated these predictions.

Available data on odorant response profiles of most ORN types (Figure 6-25) and their glomerular targets enabled assignment of odorant specificity to most PN types. One can then model odorant response maps based on PN axon termination patterns. Results from such analyses indicate that representations of fruit odors and pheromones (such as cVA) are spatially segregated in the lateral horn; this segregation could enable higher-order neurons to more easily distinguish between

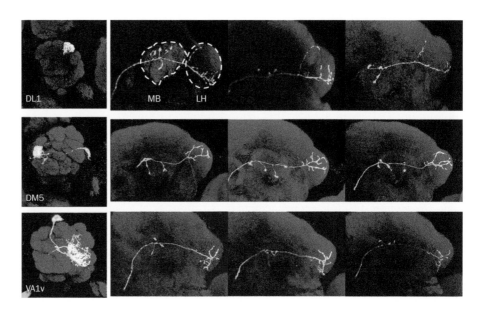

Figure 6-30 Stereotyped PN axon termination patterns in the lateral horn but not the mushroom body. Left panels, singly labeled PNs send dendrites to the DL1, DM5, and VA1v glomeruli of the antennal lobe, respectively. Right panels, axonal projection patterns in the mushroom body (MB) and lateral horn (LH) are shown for three DL1 PNs (top), three DM5 PNs (middle), and three VA1v PNs (bottom) from nine flies. Note that LH arborization patterns are similar among PNs of the same glomerular type and distinct between different PN types. By contrast, MB arborization patterns are much more variable between PNs of the same type. (From Marin EC, Jefferis GSXE, Komiyama T, et al. [2002] *Cell* 109:243–255. With permission from Elsevier Inc. See also Wong AM, Wang JW, & Axel R [2002] *Cell* 109:229–241; Jefferis GSXE, Potter CJ, Chan AM, et al. [2007] *Cell* 128:118–1203.)

these two types of olfactory stimuli and drive different behavioral outputs (for example, foraging versus mating).

Interestingly, PN axon arborization in the mushroom body is much less stereotyped (Figure 6-30). Anatomical mapping of the connectivity between PNs and **Kenyon cells**, the intrinsic cells of the mushroom body and postsynaptic targets of PNs, did not reveal any apparent structure: individual Kenyon cells appear to stochastically sample the glomerular repertoire (**Figure 6-31**). Physiological studies also suggest that odorant responses of Kenyon cells lack stereotypy: the odorant response of a genetically defined small subpopulation is as variable as the entire Kenyon cell population. Thus, each fly may generate a unique odor representation in the mushroom body, the meaning of which is likely acquired by experience. In Chapter 11, we will discuss in more detail how the mushroom body participates in olfactory learning and memory.

In summary, the two higher olfactory centers in the fly, the lateral horn and mushroom body, represent odors in distinct manners. The highly organized representation in the lateral horn suits its role in regulating innate behavior; this connectivity was selected in the course of evolution and develops according to a predetermined genetic plan. By contrast, stochastic representation in the mushroom body suits its role in associative learning; the meaning of the representation is likely acquired during an individual's life experience. A similar principle may apply to the vertebrate olfactory system. For example, the smell of the native stream may be imprinted in the salmon's equivalent of mushroom bodies in their juvenile period, enabling them to find the same stream years later. Indeed, the mammalian piriform cortex appears to resemble the fly mushroom body in its

Figure 6-31 Random convergence of glomerular input to mushroom body Kenyon cells. Each row represents a Kenyon cell. Each column represents a specific glomerulus. Each red or yellow bar represents respectively one or two PN inputs from a glomerulus to a Kenyon cell. While some glomeruli are represented more than others, each Kenyon cell receives input from an apparently random collection of glomeruli. (From Caron SJ, Ruta V, Abbott LF, et al. [2013] *Nature* 497:113–117. With permission from Springer Nature.)

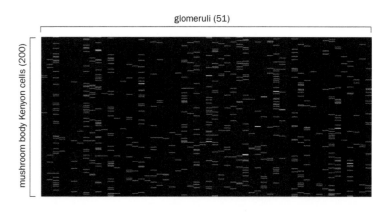

representation of odorants and mitral cell input (Section 6.8). We will return to these topics in Chapters 10 and 11 in the context of innate and learned behavior.

TASTE: TO EAT, OR NOT TO EAT?

Chemicals can also be detected by the taste and trigeminal chemosensory systems in vertebrates. The trigeminal chemosensory system (a part of the somatosensory system discussed later in this chapter) warns the animal of noxious stimuli and contributes to food preference. The taste system detects **tastants—** nonvolatile and hydrophilic molecules in saliva—using taste receptors in the tongue and oral cavity. Whereas olfaction can detect chemicals at a distance, taste does so only at closer range. In mammals, taste is dedicated to regulating feeding by revealing the content and safety of potential food. (In other species, including the fruit fly, taste also regulates mating behavior; see Chapter 10.) The identification of taste receptors has paved way for investigating how different tastes are sensed on the tongue.

6.15 Mammals have five basic taste modalities: bitter, sweet, umami, salty, and sour

Taste begins with the binding of chemicals to the tips of **taste receptor cells** (**TRCs**) on the surface of the tongue and oral cavity. Clusters of tens of TRCs form a **taste bud**, with the apical extensions of TRCs forming the **taste pores** at the surface of the tongue (**Figure 6-32**A). Groups of taste buds form several kinds of papilla distributed at different parts of the tongue. Tastant binding to taste receptors induces depolarization of TRCs. Unlike olfactory receptor neurons, TRCs are sensory epithelial cells that *lack axons*. They release neurotransmitter at their bases, activating terminal branches of gustatory ganglion neurons, whose axons form **gustatory nerves** innervating taste buds. Information is relayed to the **nucleus of the solitary tract** (**NTS**) in the brainstem, then to the thalamus, and eventually to a part of the **insular cortex** specialized for taste sensation (Figure 6-32B).

Taste has been divided into five basic modalities based largely on human perception: **bitter**, **sweet**, **umami**, **salty**, and **sour**. (Note the distinction between taste and flavor; **flavor** refers to a synthesis of taste and olfaction with additional contribution from the trigeminal somatosensory system for special chemicals,

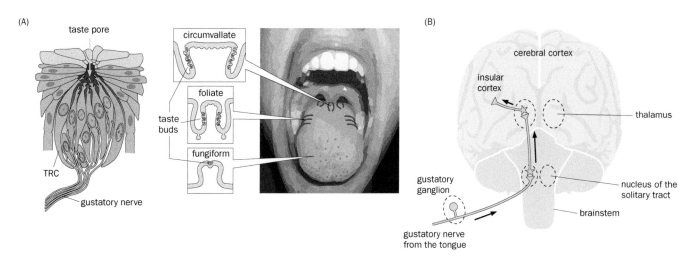

(A)

(B)

Figure 6-32 Organization of the mammalian taste system. (A) Left, each taste bud consists of up to ~100 taste receptor cells (TRCs) extending apical processes toward the taste pore, which opens to the oral cavity. Terminals of the gustatory nerve innervating the taste bud contact the basal side of the TRCs. Middle and right, taste buds are located in several kinds of papillae—circumvallate, foliate, and fungiform—which are located at different parts of the tongue. **(B)** Schematic of the central taste system. Gustatory nerves from

the front of the tongue, back of the tongue, and pharynx (not shown) originate from neurons in separate ganglia, all of which project their central axons to the nucleus of the solitary tract (NTS) in the brainstem. Taste information is relayed by NTS neurons to the thalamus and then to the insular cortex. Arrows indicate the direction of information flow. (A, adapted from Chandrashekar J, Hoon MA, Ryba NJP, et al. [2006] *Nature* 444:288–294. With permission from Springer Nature.)

Figure 6-33 Three models of taste system organization. (A) Each TRC expresses multiple receptors, and thus each ganglion neuron carries mixed information of multiple taste modalities. **(B)** Each TRC expresses a single receptor, but each ganglion neuron receives input from multiple types of TRCs and thus carries mixed information. **(C)** Each TRC expresses a single receptor, and each ganglion neuron receives input from only one kind of TRC and therefore carries information about one specific taste modality. This is the labeled line model. Experiments described in subsequent sections, with some notable exceptions, have largely supported the labeled line model. (Adapted from Chandrashekar J, Hoon MA, Ryba NJ, et al. [2006] *Nature* 444:288–294. With permission from Springer Nature.)

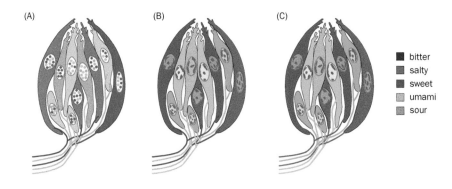

bitter
salty
sweet
umami
sour

temperature, and texture, and therefore has many more than five kinds.) These modalities are likely to be universal. Bitter and sour tastes usually warn animals of potentially toxic chemicals and spoiled food, respectively, and are generally aversive. Sweet and umami (savory, meaty) tastes provide information about the nutritional content of food (sugar and amino acids), and are generally appetitive. The salty taste allows animals to regulate their sodium levels and can be appetitive or aversive depending on the salt concentration and the animal's current physiological state.

Receptors for all five taste modalities are distributed throughout the tongue, and therefore there is no "tongue map" with distinct areas of the tongue for different tastants. In fact, each taste bud contains cells responsive to different modalities. This raises questions as to how the taste system is organized at the periphery (**Figure 6-33**). Is each TRC tuned to a specific modality, or is it more broadly tuned? What kind of taste information does each afferent axon carry to the brain?

6.16 Sweet and umami are sensed by heterodimers of the T1R family of G-protein-coupled receptors

The quest for taste receptors began in the late 1990s, following the spectacular success of the identification of odorant receptors and the insights they brought to the study of olfaction. Using molecular biology techniques allowing identification of genes specifically expressed in TRCs, two G-protein-coupled receptors with large N-terminal extracellular domains, **T1R1** and **T1R2**, were identified. The mRNAs of T1R1 and T1R2 were detected in taste buds, and T1R1 and T1R2 proteins were concentrated in the taste pore (**Figure 6-34**). Thus, these GPCRs were strong candidates for taste receptors. But what modalities do they detect?

In mice, the gene for a third member of the T1R family, **T1R3**, was subsequently identified. T1R3 corresponds to the *Sac* locus, the genomic location of a spontaneous mutation making mice insensitive to sweet tastes. Introducing a normal T1R3 transgene into the *Sac* mutant mouse restored sugar sensitivity, as assayed by preference in a two-bottle choice test (**Figure 6-35**A). The function of T1Rs was further tested by introducing these receptors along with a promiscuous G protein into heterologous cells that normally do not respond to sugars. In this gain-of-function assay, cells expressing T1R3 alone did not respond to sugar application, but cells expressing T1R2 and T1R3 together responded robustly (Figure 6-35B). These cells responded not only to natural sweets such as sucrose but also to artificial sweeteners such as saccharin. Thus, T1R2 and T1R3 together constitute the mammalian sweet taste receptors.

Similar heterologous expression studies identified T1R1 + T1R3 as the mammalian umami receptor. These results were confirmed by physiological and behavioral responses to specific tastants in mutant mice. Mice lacking T1R1 were insensitive to umami stimuli such as the meaty tastant monosodium glutamate (MSG), but maintained responses to sweet stimuli. Mice lacking T1R2 still responded to umami but not to sweet tastant. Mice lacking T1R3 responded to neither sweet nor umami, but had intact bitter, sour, and salty responses. Thus, T1R3 acts as a co-receptor for both sweet and umami tastants, conferring sweet taste when acting together with T1R2 and umami taste when acting together with T1R1.

taste pore   taste pore

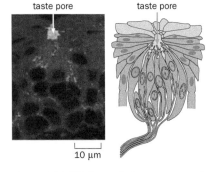

10 μm

Figure 6-34 T1R1 protein is concentrated at the taste pore. T1R1 protein (green in the image on the left), revealed by immunostaining using an antibody against T1R1, is concentrated at the taste pore; red staining is for filamentous actin, which labels the cytoplasm. The schematic on the right shows a taste bud from a similar view. (Adapted from Hoon MA, Adler E, Lindemeier J, et al. [1999] *Cell* 96:541–551. With permission from Elsevier Inc.)

(A)

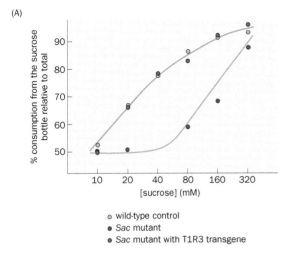

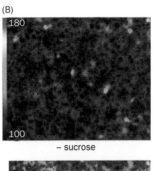

(B)

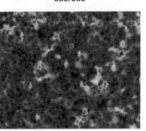
− sucrose

+ sucrose

○ wild-type control
● *Sac* mutant
● *Sac* mutant with T1R3 transgene

Figure 6-35 Identification of the sweet receptor. (A) Mice were given two bottles from which to drink, one containing water and the other containing sucrose solutions of specified concentration (*x* axis). Higher *y* axis scores indicate stronger preference for sucrose; a *y* axis score of 50 means no preference. Wild-type mice (orange circles) and *Sac* mutant mice expressing a T1R3 transgene (red circles) prefer sucrose solutions at much lower concentrations (red trace) than *Sac* mutant mice (purple trace). **(B)** HEK293 cells transfected with T1R2 and T1R3, as well as a promiscuous G protein, respond to sucrose application with a rise in intracellular Ca^{2+} concentration, as measured by a fluorescence indicator (scale to the left of the top panel, in nM). (Adapted from Nelson G, Hoon MA, Chandrashekar J, et al. [2001] *Cell* 106:381–390. With permission from Elsevier Inc.)

Not all mammals have intact sweet and umami tastes (Figure 6-36). Cats are indifferent to sweet food. The genes encoding T1R2 in domestic cats, as well as their wild feline relatives, were all found to carry the same mutations rendering them pseudogenes. These mutations occurred in a common ancestor of these feline species. Similarly, when the giant panda genome was sequenced, T1R1 was found to have become a pseudogene. This means that giant pandas do not have umami taste. These mutations may have driven a common ancestor of cats to be an obligate carnivore and led pandas to a vegetarian diet even though they are closely related to bears and dogs, which are omnivores. Alternatively, the special diets of cats and pandas may have evolved first and removed selection pressure on their sweet and umami receptors, respectively, such that the unutilized receptors mutated and became pseudogenes.

6.17 Bitter is sensed by a large family of T2R G-protein-coupled receptors

In 1931, a synthetic chemist at DuPont discovered that the dust of phenylthiocarbamide (PTC; Figure 6-37) he synthesized tasted bitter to the occupant of the next bench but was tasteless to himself even though he was much closer to the source of the compound. Subsequent genetic studies showed that PTC tasters and nontasters exist among diverse human populations. These two groups exhibit a difference of at least 16-fold in their sensitivity thresholds for PTC detection, and these traits are transmitted from parents to progeny in a Mendelian fashion. Genetic mapping of taste sensitivity to PTC and other bitter tastants provided clues leading to the identification of bitter taste receptors as a family of G-protein-coupled receptors called **T2Rs**. Mutations in specific human T2R genes account for different sensitivities to specific bitter tastants.

Although sweet, umami, and bitter tastes are all conferred by G-protein-coupled receptors, there are notable differences between bitter taste receptors and sweet/umami receptors. These differences are instructive about the behavioral and evolutionary relevance of these different taste modalities to animals. First, the T2Rs are encoded by a large family of proteins—about 30 in humans and 40 in mice—compared with the three T1R receptors that account for all sweet and umami tastes. The diversity of bitter receptors may reflect animals' needs to detect a diverse repertoire of potentially toxic compounds using the bitter taste system.

Second, most T2Rs have much higher affinities for their tastants than do T1Rs. For example, the T2R5 receptor from a normal mouse strain is activated by its ligand cycloheximide (a potent protein synthesis inhibitor produced by bacteria) in the hundreds of nanomolar range. By contrast, T1R2 + T1R3 is activated by

Figure 6-36 Guess which taste receptor is missing in these animals? Cats and tigers have lost T1R2 function and thus the ability to taste sweetness. Giant pandas do not have functional T1R1 and thus cannot taste umami. (Left, courtesy of Sumeet Moghe/Wikipedia. Right, courtesy of Chen Wu/Wikipedia. See Jiang P, Josue J, Li X, et al [2012] *Proc Natl Acad Sci U S A* 109:4956–4961.)

Figure 6-37 Structure of phenylthiocarbamide (PTC), a compound famous in taste research. PTC tastes extremely bitter to some people but is tasteless to others. Mutations in a single gene encoding a T2R receptor account for the difference between tasters and nontasters. (See Fox AL [1932] *Proc Natl Acad Sci U S A* 18:115–120; Blakeslee AF [1932] *Proc Natl Acad Sci U S A* 18:120–130; Kim U-k, Jorgenson E, Coon H, et al. [2003] *Science* 299:1221–1225.)

sucrose at the tens of millimolar range (Figure 6-35A), or at a >10⁵ higher concentration. This difference reflects the different functions of these taste modalities. Bitter compounds are potentially toxic and to be avoided, so the higher the affinity, the less the animal has to taste before rejecting it. For sweet and umami tastants that bring nutrients to animals, a low affinity helps ensure that the amount of nutrients is sufficiently high to be worth the effort of eating.

Third, different bitter taste receptors are co-expressed in the same TRCs (**Figure 6-38**A). This may make it difficult for animals to distinguish between different bitter tastants, but it may hardly matter. It is far more important to avoid toxic compounds than to distinguish between them. T2Rs are not expressed in the same TRCs as T1Rs (Figure 6-38B); indeed, T1R1 and T1R2, conferring umami and sweet tastes, respectively, are not expressed in the same cells. This segregation makes the distinction between umami, sweet, and bitter modalities unequivocal at the level of individual TRCs.

6.18 Sour and salty tastes involve specific ion channels

Whereas sweet, umami, and bitter tastes are conferred by GPCRs, the sour and salty (sodium) tastes involve specific ion channels (**Figure 6-39**). Sour taste requires TRCs expressing a TRP channel called PKD2L1. Genetic ablation of PKD2L1-expressing TRCs in mice abolished physiological responses to sour tastants but did not affect responses to sweet, umami, bitter, or low-salt tastants, indicating that these cells are tuned to sour taste. However, genetically knocking out the channel itself has no effect on sour taste reception, arguing against PKD2L1 in conferring sour taste. An evolutionary conserved proton channel, Otopetrin1 (Otop1), was recently found to be highly enriched in PKD2L1-expressing TRCs; mutations in Otop1 severely reduced proton currents in these cells. Furthermore, Otop1 knockout mice did not respond to sour taste, and expressing Otop1 in sweet TRCs caused them to also respond to sour stimuli. Thus, Otop1 is the sour taste receptor that senses proton concentration, the primary determinant of sourness.

The salty taste has two subsystems. The first responds only to Na⁺, a physiologically important ion, and elicits appetitive responses to relatively low Na⁺ concentrations (<100 mM NaCl). This low-salt or sodium taste is inhibited by amiloride, an inhibitor of the **epithelial Na⁺ channel** (**ENaC**) (Figure 6-39). Indeed, ENaC is essential for low-salt taste in mice. Salt-deprived ENaC knockout mice were no longer attracted to low-salt solutions, but were still repelled by high concentration salt solutions, revealing a second salty taste subsystem, which responds to high concentrations of NaCl (>300 mM) and other salts. This high-salt system is not inhibited by amiloride and usually elicits aversive responses. Recent studies suggest that the high-salt system activates both bitter and sour TRCs to elicit aversive responses. The mechanisms by which high concentrations of salt activate bitter and sour TRCs remain to be determined.

Figure 6-38 Bitter receptors are co-expressed in the same cells, which do not co-express the sweet or umami receptors. (A) Two different T2R receptors are co-expressed in the same taste cells. The top and bottom panels are the same tongue section showing *in situ* hybridization signals using probes against T2R3 and T2R7. **(B)** Bitter receptors (red, a mix of 20 T2R probes) and the T1R3 co-receptor for sweet and umami (green) are expressed in different cells. (A, from Adler E, Hoon MA, Mueller KL, et al. [2000] *Cell* 100:693–702. With permission from Elsevier Inc. B, from Nelson G, Hoon MA, Chandrashekar J, et al. [2001] *Cell* 106:381–390. With permission from Elsevier Inc.)

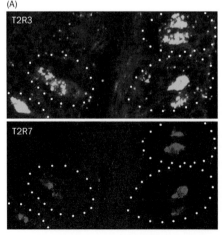

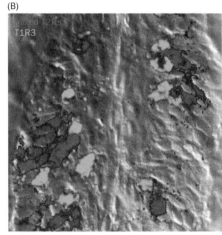

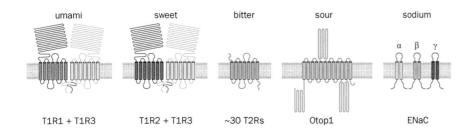

umami   sweet   bitter   sour   sodium

T1R1 + T1R3   T1R2 + T1R3   ~30 T2Rs   Otop1   ENaC

Figure 6-39 Summary of mammalian taste receptors. Umami and sweet taste are each sensed by two GPCRs, with T1R3 as a shared co-receptor. Bitter taste is sensed by the T2R family of GPCRs. Sour is sensed by the proton channel Otop1. Appetitive low-salt taste requires the epithelial Na+ channel ENaC, which consists of α, β, and γ subunits. (Adapted from Yarmolinsky DA, Zuker CS, & Ryba NJP [2009] *Cell* 139:234–244. With permission from Elsevier Inc. See Tu Y-H, Cooper AJ, Teng B, et al [2018] *Science* 359:1047–1050 for more detail on Otop1. Not shown here, aversive high-salt taste requires both bitter and sour taste receptor cells; see Oka Y, Butnaru M, von Buchholtz L, et al [2013] *Nature* 494:472–475.)

The human tongue does not appear to express ENaC channels, and salty taste in humans is not affected by amiloride. These observations suggest additional mechanisms at work in human salty taste. Indeed, even in the best characterized mouse taste system, a sizable fraction of TRCs do not express molecules characteristic of sweet, umami, bitter, sour, or salty tastes (Figure 6-39), raising the possibility that taste modalities beyond the basic five exist and remain to be discovered.

6.19 Sweet and bitter engage mostly segregated pathways from the tongue to the gustatory cortex

Having studied the receptors and cells for five basic taste modalities (Figure 6-39), we now turn to the question raised earlier regarding cellular organization for taste perception (Figure 6-33). We have already discussed the evidence that sweet, umami, and bitter receptors are expressed in distinct TRCs, and that killing sour-sensing cells does not affect the other modalities. These findings indicate that each modality, with the exception of high-salt taste, is represented by its specific TRCs. The following experiment provided a functional demonstration that the activation of specific TRCs confers specific taste perceptions.

Phenyl-β-D-glucopyranoside (PDG) tastes bitter to humans but does not elicit responses in wild-type mice. The human receptor for PDG, hT2R16, was identified and used to produce two kinds of transgenic mice. In the first kind, hT2R16 was driven by a promoter of a mouse T2R receptor, so it was expressed in TRCs that normally respond to bitter tastants. In the second kind, hT2R16 was driven by the promoter for T1R2, so it was expressed in TRCs that normally respond to sweet tastants. In a two-bottle choice test, transgenic mice expressing hT2R16 in bitter TRCs appeared to perceive PDG as bitter, as they avoided PDG-containing water. However, transgenic mice expressing hT2R16 in sweet TRCs were *attracted* to PDG-containing water, as if PDG were a sweet tastant (**Figure 6-40**). Thus, introducing an exogenous human bitter receptor into sweet TRCs reprogrammed the behavioral response to the bitter tastant. This experiment shows that sweet and bitter tastes are reflections of the selective activation of sweet or bitter TRCs, rather than properties of the receptors or tastants. This logic is remarkably similar to that of olfactory perception in *C. elegans* (Figure 6-22).

Given the separation of taste modalities at the level of TRCs, it seemed likely that individual taste afferents would represent specific modalities. Indeed, Ca²⁺ imaging experiments suggested that different parts of the insular cortex were enriched in cells preferentially activated by sweet, bitter, umami, and NaCl tastants, forming a gustatory map (**Figure 6-41**A). Furthermore, when paired with licking, optogenetic activation of cells in the bitter cortical field reduced drinking in thirsty mice, whereas artificial activation of cells in the sweet cortical field enhanced drinking in mildly water-satiated mice (Figure 6-41B). These experiments support a labeled line model (Figure 6-33C) in which sweet and bitter tastes not only are represented in separate TRCs but also use largely separate processing channels to reach different parts of the insular cortex to specify taste perception. However, individual cells that respond to multiple taste modalities along the ascending pathway to the insular cortex have also been reported, and at least the high-salt taste engages two different kinds of TRCs at the periphery. Thus, how information from different taste modalities is integrated remains an interesting question to explore.

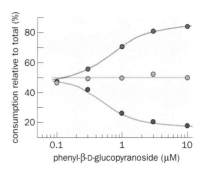

- hT2R16 expressed from the T1R2 promoter
- control (no hT2R16)
- hT2R16 expressed from a T2R promoter

Figure 6-40 Bitter and sweet tastes are determined by activation of specific taste receptor cells. In the two-bottle choice assay, one bottle contains water and the other contains a phenyl-β-D-glucopyranoside (PDG) solution at concentrations specified on the x axis. A value of 50 on the y axis indicates no preference, >50 indicates appetitive, and <50 indicates aversive. PDG is a natural ligand for the human bitter receptor T2R16 but does not elicit a response in control mice (blue trace). Transgenic mice expressing T2R16 under the control of a bitter receptor (T2R) promoter avoid PDG (purple trace). Transgenic mice expressing T2R16 under the control of a sweet receptor (T1R2) promoter prefer PDG (red trace). (Adapted from Mueller K, Hoon MA, Erlenbach I, et al. [2005] *Nature* 434:225–229. With permission from Springer Nature.)

Figure 6-41 A gustatory map in the insular cortex. (A) Top left, schematic representation of a gustatory map in the insular cortex, where sweet, umami, low-salt (NaCl), and bitter tastants activate spatially clustered cells in different regions of the insular cortex. R, rostral, V, ventral. Red and blue lines represent the middle cerebral artery and the rhinal vein, respectively, as landmarks. In the NaCl hot spot, many cells are activated by NaCl (bottom left) but few by sweet or bitter tastants (right), as measured by two-photon Ca^{2+} imaging. Gray spots are cells visualized by a fluorescent Ca^{2+} indicator injected into the insular cortex. Colored cells represent those activated by specific tastants above a certain threshold. **(B)** Channolrhodopsin (ChR2) was virally transduced in neurons in the bitter (left) and sweet (right) fields of the insular cortex in separate groups of mice. Head-restrained mice were trained to lick water in response to a sensory cue. In half of the trials, licking also triggers a laser that delivers blue light to activate ChR2-expressing neurons. Each circle represents average licking of a mouse during trials with and without photostimulation. Photostimulation of the bitter cortical field reduced water licking in thirsty mice (left), whereas photostimulation of the sweet cortical field enhanced water licking in mildly satiated mice (right). (A, from Chen X, Gabitto M, Peng Y, et al. [2011] *Science* 333:1262–1266. With permission from AAAS. B, from Peng Y, Gillis-Smith S, Jin H, et al. [2015] *Nature* 527:512–515. With permission from Springer Nature.)

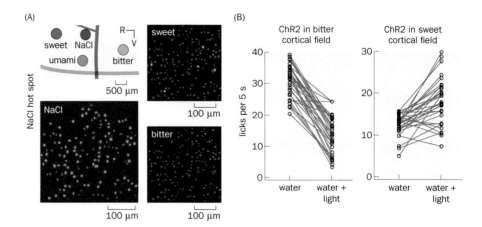

Figure 6-42 Multisensory integration in our choice of food. Ma-po tofu, a famous dish in Sichuan cuisine, engages multiple sensory modalities, including olfaction, taste, vision (its colorful display), and trigeminal somatosensation (for texture, temperature, hot chili pepper, and tingling Sichuan peppercorn).

Indeed, perceptions of different tastes, such as sweet, umami, and bitter, need to be integrated for animals to make the decision to eat or not to eat. Taste perception also interacts extensively with other sensory modalities, such as olfaction and the trigeminal somatosensory system, to shape our food choices and enjoyment of dining (**Figure 6-42**). Finally, past experiences and current physiological states also play important roles in food preferences. Thus, studies of taste and eating provide opportunities not only to delineate the neural mechanisms of taste perception but also to reveal how multiple senses and physiological states are integrated. We will study eating and its regulation in more depth in Chapter 9.

AUDITION: HOW DO WE HEAR AND LOCALIZE SOUNDS?

Audition is a particularly important sense in humans, as we communicate extensively using spoken languages and music. In the animal world, audition is also most widely used to communicate with conspecifics—to identify and locate mating partners, competitors, parents, and progeny. Other major functions of audition include alerting animals to predators or helping them locate prey. These functions are similar to those of vision and olfaction; indeed, these senses complement each other. Vision requires light, while audition and olfaction operate in both light and dark but deal with sensory stimuli that travel at different speeds. Odors, usually carried by the wind, travel at a speed of around 1 m/s, while sounds travel at a 340 m/s in the air, two orders of magnitude faster. Indeed, as we will see, a prominent feature of audition is the extraction and representation of information in the temporal domain. For example, music engages our perception of auditory stimuli at different time scales, ranging from >1 s for melodies, ~0.1 s for notes, to <0.01 s for pitch.

Sounds are transmitted as airborne pressure waves at specific frequencies (**Figure 6-43A**). In mammals, sounds are collected by the part of the ear external to the head, called the **auricle** or pinna. The auricle amplifies high-frequency sounds coming from in front of it, causing the ear to be directionally sensitive to those high frequencies. This increases the sensitivity of an animal's hearing and provides information about the location of the sound source. The auricle funnels sound into the ear canal, which focuses the sound on the tympanic membrane (**eardrum**) at the intersection of the outer ear (which includes the auricle and ear canal) and the middle ear. Sound-triggered vibrations of the eardrum are transmitted by three tiny bones in the air-filled middle ear. The bones tap on an elastic membrane (called the oval window) of the fluid-filled chambers of the **cochlea** in the inner ear (Figure 6-43B, C). A separate elastic membrane at the interface between the middle and inner ears, called the round window, moves in phasic opposition to the oval window and provides an outlet for the pressure changes due to the tapping on the oval window. Each cycle of a sound stimulus evokes a

(A)

(B)

(C)

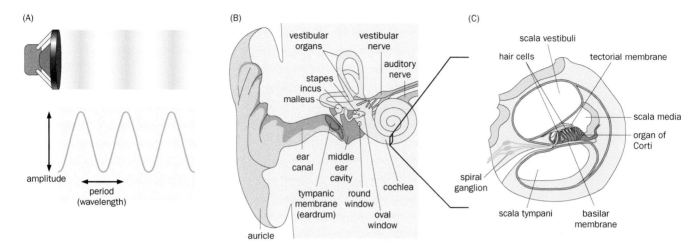

Figure 6-43 Sound and its reception by the mammalian ear.
(A) Sounds are transmitted as compressions and rarefactions of air particles. For 1000 Hz (1 kHz, or 1000 cycles per second) sounds, the period between two adjacent wave peaks is 1 ms. The wavelength is 0.34 m, given that the speed of sound in the air is 340 m/s. We perceive the amplitude of the wave as the loudness of the sound.
(B) After passing through the ear canal, pressure waves are transmitted by the tympanic membrane into vibrations that in turn cause vibrations of three bones in the air-filled cavity of the middle ear: the malleus, incus, and stapes. These vibrations are transmitted as displacement waves that travel down the fluid-filled chambers of the cochlea in the inner ear, from the oval window (attached to the scala vestibuli) to the round window (attached to the scala tympani). Above the cochlea are

the vestibular organs and their associated nerve, which is discussed in detail in Box 6-2. **(C)** Magnified cross section of the cochlea. In between three fluid-filled chambers (the scala vestibuli, scala media, and scala tympani) is the organ of Corti, which consists of hair cells (light green), support cells (brown), and the basilar membrane (yellow); see Figure 6-48 for a further magnified schematic of the organ of Corti. Displacement of the basilar membrane due to fluid movement produces relative movement of hair cells against the overlying tectorial membrane (blue). This changes the membrane potential of hair cells and resulting electrical signals are transmitted to the brain by the axons of spiral ganglion neurons, which make up the majority of the auditory nerve.

cycle of up-and-down movement of a small volume of fluid in each of the three chambers in the inner ear.

The cochlea (from the Greek *cochlos,* meaning "snail") is a coiled structure resembling a snail's shell anchored in the temporal bone. In a cross section of the cochlea (Figure 6-43C), **hair cells**, the primary sensory cells of audition, are embedded in a sheath of epithelial cells sitting on top of an elastic membrane, the **basilar membrane**. Hair cells, the surrounding support cells, and the basilar membrane constitute the **organ of Corti**. Cyclic fluid movement induced by sound stimuli displaces the basilar membrane in a cyclical fashion, producing shear force between hair cells and the overlying **tectorial membrane**. This mechanical stimulus is then converted into electrical signals via membrane potential changes in hair cells, as we discuss in the following (Movie 6-2).

6.20 Sounds are converted into electrical signals by mechanically gated ion channels in the stereocilia of hair cells

Humans can sense air pressure waves of up to 20,000 Hz (20 kHz), or 1 cycle every 50 μs; the sensitivity of some animals (for example, bats) extends to frequencies well above 100 kHz. How rapidly are sounds converted into electrical signals? In the late 1970s, it was demonstrated that frog hair cells depolarize in response to mechanical stimuli with a latency of 40 μs (Figure 6-44); subsequent studies indicated that in mammals the latency could be <10 μs. As second messenger systems operate on the time scale of tens to hundreds of milliseconds, these measurements ruled out the possibility that **mechanotransduction**, the process by which mechanical stimuli are converted into electrical signals, is mediated by a second messenger system, as is the case in vision and olfaction. Instead, they suggested that mechanical stimuli directly open ion channels to depolarize the membrane potentials of hair cells. How is this achieved?

Hair cells in all vertebrates share a similar structure and transduce mechanical stimuli in the same way. Each hair cell extends a bundle of hairs, **stereocilia**, from its apical surface. Each stereocilium is a rigid cylinder, rich in F-actin fibers,

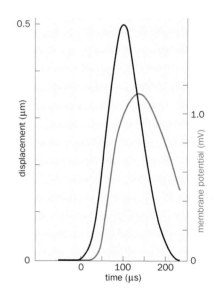

Figure 6-44 Rapid conversion of mechanical stimuli into electrical signals. Hair cells from a dissected bullfrog vestibular saccule (see Box 6-2 for detail) were stimulated with a glass probe attached to an electrically driven motor. Hair cell responses were measured by recording electrical potentials across the two chambers between which the dissected saccule was placed. The latency of the response is about 40 μs. (Adapted from Corey DP & Hudspeth AJ [1979] *Biophys J* 26:499–506. With permission from Elsevier Inc.)

with a narrow constriction at the base that enables it to pivot. The stereocilia comprising a single hair cell are arranged in rows of increasing heights, much like a staircase (**Figure 6-45**A). In the 1980s, electron microscopy studies revealed that at the apical end, each stereocilium is connected to its taller neighbor by a structure called the **tip link** (Figure 6-45B). The staircase-like arrangement of stereocilia and the tip links explain how mechanical force causes changes in the membrane potentials of hair cells.

According to a well-accepted model, sound-wave-induced relative movements between neighboring stereocilia causes the tip links to open **mechanosensitive channels** (Figure 6-45C). Stereocilia are exposed to a fluid with an unusually high K^+ concentration (~150 mM, unique for extracellular $[K^+]$ in the nervous system). Opening of the mechanosensitive channel causes K^+ influx and depolarization of the hair cell. This triggers Ca^{2+} entry through voltage-gated Ca^{2+} channels and glutamate release at the base of the hair cell (Figure 6-45D), leading to depolarization

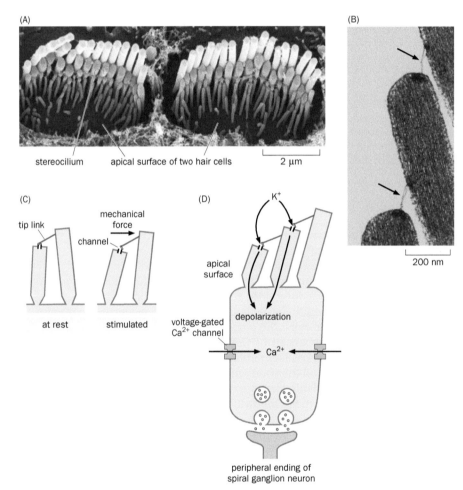

Figure 6-45 **Mechanisms by which hair cells convert mechanical stimuli into electrical signals.** **(A)** A scanning electron micrograph shows the stereocilia of two adjacent inner hair cells from the mouse cochlea. These stereocilia contain actin-based bundles of successive heights arranged like staircases. **(B)** Two tip links (arrows) between three adjacent stereocilia visualized by transmission electron microscopy. **(C)** A model of a mechanosensitive channel gated by the relative movement of two adjacent stereocilia. Mechanical force in the direction of the ascending staircase of stereocilia opens the channel. **(D)** Schematic of a vertebrate hair cell in the inner ear. The stereocilia protrude from the apical surface and serve as detectors of mechanical stimuli. A mechanical deflection in the direction of the ascending staircase causes the opening of mechanosensitive channels, entry of K^+ ions (which are highly enriched in the apical extracellular fluid), and depolarization of the hair cell. This triggers opening of voltage-gated Ca^{2+} channels, Ca^{2+} entry, and neurotransmitter (glutamate) release at the base of the hair cell. Glutamate depolarizes the peripheral endings of spiral ganglion neurons, which transmit signals as action potentials to the brain. (A, from Kazmierczak P & Müller U [2011] *Trends Neurosci* 36:220–229. With permission from Elsevier Inc. B, from Hudspeth AJ [2013] *Neuron* 80:536–537. With permission from Elsevier Inc.)

of the peripheral endings of **spiral ganglion neurons** (Figure 6-43C). Spiral ganglion neurons (residing in the spiral ganglion next to the cochlea) are bipolar: their peripheral axons collect auditory information from hair cells, and their central axons form the **auditory nerve** and transmit auditory information to the brain. Sound amplitude regulates the magnitude of hair cell depolarization and transmitter release and, in turn, the spike rate of spiral ganglion neurons.

The recent convergence of human genetic studies and physiological and cell biological studies in animal models has shed light on the molecular nature of the mechanotransduction apparatus. Deafness is the most common sensory deficit of genetic origin, affecting one in every 500 people. Mutations in about 100 genes have been associated with syndromic and nonsyndromic hearing loss. Many deafness genes affect the development, structure, and function of hair cells in the organ of Corti, including those encoding actin, actin-associated proteins, and myosin motors. Two deafness genes encode Ca^{2+}-dependent cell adhesion molecules called cadherin-23 (Cdh23) and protocadherin-15 (Pcdh15), which constitute the structural components of the tip link. Biochemical and immunoelectron microscopy (see Section 14.17 for details) studies determined that the tip link is formed by Cdh23 from the taller stereocilium binding to Pcdh15 from the shorter stereocilium (Figure 6-46A).

Although well characterized by electrophysiological and biophysical studies since the 1980s, the molecular identity of the mechanosensitive channel has remained mysterious long after the human genome was sequenced (in 2001). Recent studies suggest that transmembrane channel-like 1 and 2 proteins (Tmc1 and Tmc2), whose sequences do not resemble classic ion channels (Box 2-4), are essential components of the mechanosensitive channel. Multiple missense mutations in the human *Tmc1* gene cause recessive and dominant forms of deafness. Mice lacking both *Tmc1* and *Tmc2* lack mechanically induced depolarization of hair cells. Moreover, modification of amino acids in several transmembrane domains of Tmc1 results in changes in ion permeation properties of the mechanosensitive channel, suggesting that Tmc1 forms part of the pore of the channel. Mutations that alter ion selectivity can also cause human deafness.

In summary, auditory transduction is mediated by direct coupling of mechanical stimuli with opening of ion channels through the tip links between stereocilia of hair cells. Recent studies have identified the molecular constituents of the tip link and candidate mechanosensitive channel components. Further studies of how these and other associated proteins work together will paint a more complete picture of how mechanical forces are rapidly converted into electrical signals.

6.21 Sound frequencies are represented as a tonotopic map in the cochlea

The hearing range of humans is between 20 Hz and 20,000 Hz, with the highest sensitivity around 4,000 Hz. When played in sequence, most people can easily distinguish middle C (at 261.6 Hz) from the neighboring B or C-sharp keys on a piano, which are 15.7 Hz below and 15.5 Hz above middle C, respectively. Individuals with perfect pitch can identify the exact key upon hearing a single note. How are these feats accomplished?

Electrophysiological recordings revealed that individual auditory nerve fibers (each fiber corresponds to the axon of one spiral ganglion neuron) are most sensitive to sound of a **characteristic frequency**. This property is called **frequency tuning** and can be displayed on a frequency–amplitude plot as a V-shaped curve (Figure 6-47A). Different auditory fibers exhibit different characteristic frequencies. Hair cells are tuned to specific frequencies based on their locations within the cochlea (Figure 6-47B), leading to frequency tuning in postsynaptic spiral ganglion neurons. In mammals, the thickness and stiffness of the basilar membrane vary systematically along the curved structure of the cochlea. As a result of these mechanical variations, different frequencies of sound cause maximal vibrations of the basilar membrane at different locations along its length. The narrow, rigid base of the basilar membrane (near the oval window) is most sensitive to high-frequency sounds, whereas the wide, flexible apex (farthest from the oval

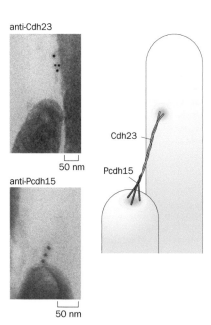

anti-Cdh23

50 nm

anti-Pcdh15

50 nm

Cdh23

Pcdh15

Figure 6-46 Molecular constituents of the tip link. Left, gold particles associated with antibodies against Cdh23 (top) and Pcdh15 (bottom). Right, a model. The tip link comprises cadherin Cdh23 from the taller stereocilium and protocadherin Pcdh15 from the shorter stereocilium. These two proteins bind to each other via the N-termini in their extracellular domains. (Adapted from Kazmierczak P, Sakaguchi H, Tokita J, et al. [2007] *Nature* 449:87–91. With permission from Springer Nature.)

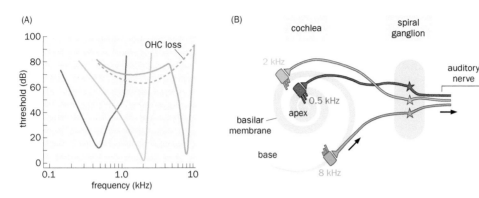

(A)

(B)

Figure 6-47 Sound frequencies are represented as tonotopic maps.
(A) Frequency tuning curves of three auditory nerve fibers. The x axis represents sound frequency. The y axis represents the sound level in dB (decibel, sound pressure level in a logarithmic unit). Zero dB approximates the lowest sound level audible by humans at 1 kHz, and every 10 dB increment represents a 10-fold increase in power. These curves are generated by presenting sound at different frequencies and identifying the minimal sound level that causes the auditory nerve to spike above baseline. Thus, the troughs represent the most sensitive frequencies for the auditory nerve fibers (0.5, 2, and 8 kHz for the three fibers, respectively). The dotted blue line represents the tuning curve of the 8 kHz auditory nerve fiber after loss of outer hair cells

(OHCs), discussed in detail in Section 6.22. **(B)** In the cochlea, hair cells are tuned to progressively lower frequencies from the base to the apex because of the mechanical properties of the basilar membrane. From the base to the apex, the basilar membrane (in blue) becomes progressively wider and less stiff, and thus resonates at progressively lower vibration frequencies. Three representative hair cells tuned to 8, 2, and 0.5 kHz are shown. Spiral ganglion neurons receive input from individual hair cells via orderly axonal projections and send output through the auditory nerve to the dorsal and ventral cochlear nuclei (separated by the dashed line) in the brainstem. (A, adapted from Fettiplace R & Hackney CM [2006] *Nat Rev Neurosci* 7:19–29. With permission from Springer Nature.)

window) is most sensitive to low-frequency sounds. Hair cells along the cochlea are thus tuned to different frequencies along the length of the basilar membrane, from its high-frequency base to its low-frequency apex, forming a **tonotopic map** (Figure 6-47B).

The mammalian organ of Corti contains two types of hair cells—one row of inner hair cells and three rows of outer hair cells (**Figure 6-48**). More than 95% of

Figure 6-48 The inner and outer hair cells of the mammalian organ of Corti.
(A) Scanning electron micrograph of the apical surface of a cochlear segment, showing stereocilia from three rows of outer hair cells and one row of inner hair cells. (A magnified surface view of two inner hair cells can be seen in Figure 6-45A). **(B)** Schematic of a cochlea cross-section (see Figure 6-43C for a zoomed-out view). Each inner hair cell receives ~10 afferent axons (three are shown in dark green), each originating from one neuron in the nearby spiral ganglion; these afferents deliver auditory information to the brainstem. A small fraction of afferent axons (not shown) innervate outer hair cells, but most of the axons projecting to outer hair cells are efferents (one is shown in red) originating from brainstem neurons. The basilar membrane moves up and down in response to sound stimulation, causing relative motion of hair bundles and the tectorial membrane. (A, from Hudspeth AJ [2013] *Neuron* 80:536–537. With permission from Elsevier Inc. B, adapted from Fettiplace R & Hackney CM [2006] *Nat Rev Neurosci* 7:19–29. With permission from Springer Nature.)

(A)

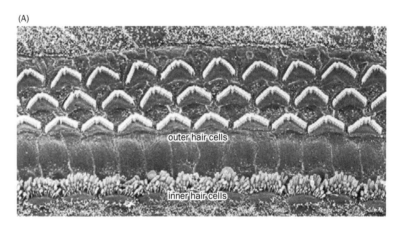

(B)

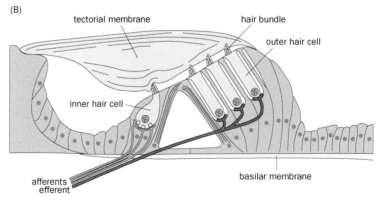

peripheral axons of spiral ganglion neurons contact inner hair cells. Each spiral ganglion neuron sends a peripheral axon (afferent fiber) that terminates on a single hair cell and is thereby tuned to the frequency of that one hair cell. The tonotopic map of the cochlea is relayed by the orderly central projections of spiral ganglion neurons to the **cochlear nuclei** in the brainstem (Figure 6-47B). Each inner hair cell is innervated by ~10 afferents of spiral ganglion neurons, and depending on the position of the termination on the hair cell, each afferent is differentially sensitive to the magnitude of hair cell depolarization. Sound amplitude is thus encoded by which of the ~10 spiral ganglion neurons innervating a particular inner hair cell are activated, in addition to the spike rates of the activated spiral ganglion neurons. In summary, inner hair cells and spiral ganglion neurons are respectively responsible for detecting auditory signals and for transmitting information about sound frequency and amplitude to the brain.

An important property of spiral ganglion neurons representing low-frequency sounds is that their firing pattern is periodic, reflecting the cyclic nature of sound stimuli. This can be seen in the periodic distribution of interspike intervals recorded from individual auditory fibers, with the period equaling the sound period (**Figure 6-49**A). This property is called **phase locking**, as spikes of spiral ganglion neurons occur at a specific phase of each cycle of the sound stimulus. Phase locking originates from the cyclic fluctuations of inner hair cell membrane potentials across each cycle of low-frequency sound (Figure 6-49B). (Phase locking in

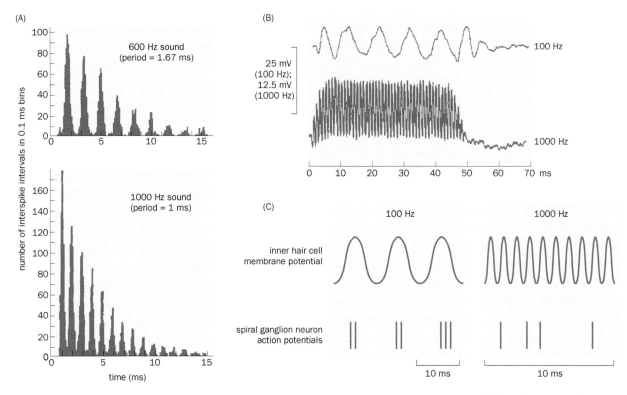

Figure 6-49 Phase-locking properties of spiral ganglion neurons and inner hair cells. (A) Histogram of interspike intervals recorded from a single auditory fiber of a squirrel monkey in response to 600 Hz (top) or 1000 Hz (bottom) sound stimuli. Note the periodicity of the distribution; the period of the histogram in each case matches the sound period noted on the graph. This means that firing of the auditory nerve occurs at a specific phase of the sound wave, even though not every cycle of sound produces a spike, which is why interspike intervals can be multiples of the sound period. **(B)** Membrane potential of a guinea pig inner hair cell in response to 100 Hz (top) or 1000 Hz (bottom) sound stimuli, measured by intracellular recording. Note that the period of the membrane potential fluctuation matches the period of sound stimuli. **(C)** Schematic illustration of membrane potential fluctuations of an inner hair cell (top) leading to periodic action potentials

(vertical bars) in a postsynaptic spiral ganglion neuron (bottom). In the case of the 100 Hz sound, several spikes are produced following each cycle of membrane potential elevation. In the case of the 1000 Hz sound, because of the refractory period of action potentials (Section 2.12), each cycle of membrane potential fluctuation produces at most one spike. Not every spiral ganglion neuron fires a spike at every cycle, but firing of a population of spiral ganglion neurons can represent all cycles. The timing of each action potential corresponds to a specific phase of hair cell membrane potential and is thus more precise (within tens of μs) than the duration of an action potential itself (~1 ms). This property enhances precision of sound localization. (A, adapted from Rose JE, Brugge JF, Anderson DJ, et al. [1967] *J Neurophysiol* 30:769–793. B, adapted from Palmer AR & Russell IJ [1986] *Hearing Res* 24:1–15. With permission from Elsevier Inc.)

mammals, constrained by the time constant of the hair cell membrane, does not occur at frequencies higher than 2–3 kHz.) Neurotransmitters are released by hair cells during the depolarized phase of each cycle. This produces action potentials in the postsynaptic spiral ganglion neurons in a cyclic manner (Figure 6-49C). Phase locking enables the *timing* of action potentials to provide higher temporal precision than the *duration* of the action potential. For instance, for a 1 kHz sound, with a period of 1 ms, the onset of action potentials occurs at a specific phase of the 1 ms cycle, which can be much less than 1 ms. Phase locking is also observed in central auditory neurons, endowing the temporal precision of neuronal firing in the auditory system essential for sound localization (Sections 6.24 and 6.25).

Outer hair cells are innervated by <5% of afferent fibers, each of which innervates multiple outer hair cells. The function of these afferent fibers is not well understood. Conversely, outer hair cells receive the bulk of efferent fibers from the brainstem (Figure 6-48). Thus, outer hair cells do not appear to play a prominent role in transmitting information to the brain. Instead, their major function is to amplify auditory signals through their interesting motor properties, as we discuss in the next section.

6.22 Motor properties of outer hair cells amplify auditory signals and sharpen frequency tuning

As discussed in Section 4.4, signals derived from photon absorption in photoreceptors are greatly amplified by a second messenger system. Given that the mechanosensitive channels responsible for changing the membrane potentials of hair cells are directly gated by sound stimuli, are auditory signals also amplified? If so, how?

Auditory signals are, indeed, greatly amplified in the cochlea. In mammals, this is due to the remarkable function of outer hair cells. Experiments carried out in the 1970s indicated that when outer hair cells representing certain sound frequencies were focally ablated, inner hair cells representing the same frequencies exhibited greatly increased detection thresholds and broadened frequency selectivities (Figure 6-47A). Two amplification mechanisms have been discovered since, both involving motility of outer hair cells triggered by sound stimuli (**Figure 6-50**A).

The first form of outer hair cell motility results from the hair bundle motor. When a positive deflection of the hair bundle (deflection along the direction of the ascending staircase) opens the mechanosensitive channel, the opening of the channel itself generates force in the direction of the stimulus, which leads to the opening of more mechanosensitive channels, thus creating a positive feedback loop (Figure 6-50B). Ca²⁺ entry into the stereocilia through mechanosensitive channels plays an important role in hair bundle motility, as do myosin motors at the tip of the stereocilia.

Figure 6-50 Motor functions of outer hair cells amplify auditory signals.
(A) Schematic of two motor functions of outer hair cells. Hair bundle motility refers to how channel opening in response to a sound stimulus produces force in the direction of the stimulus, causing hair bundles to move farther (dotted red outlines). Somatic motility (electromotility) refers to how the conformational changes of the highly enriched membrane protein prestin shorten the hair cell along its long axis when it depolarizes. **(B & C)** Schematic of positive feedback loops (on yellow background) involving hair bundle motility (Panel B) and electromotility (Panel C). Note that the positive feedback loop in Panel B is part of one step (red arrow) in the positive feedback loop in Panel C. The amplification of auditory signals in outer hair cells via these positive feedback loops is transmitted to nearby inner hair cells via enhanced basilar membrane displacement. (A, adapted from Fettiplace R & Hackney CM [2006] *Nat Rev Neurosci* 7:19–29. With permission from Springer Nature.)

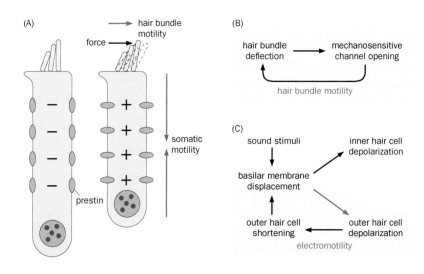

(A)

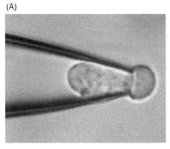

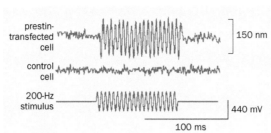

(B)

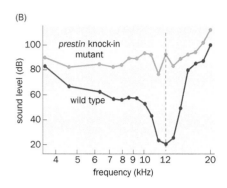

Figure 6-51 Prestin transduces the electromotility essential for auditory signal amplification. (A) Left, image of the experimental setup for testing prestin-mediated motility. Control or prestin-transfected cultured cells were partially drawn into a microchamber to mimic the elongation of outer hair cells. Voltage pulses were applied between the solutions in the microchamber and the surroundings, and movement of cell edges was recorded by a video camera. Right, 200 Hz voltage pulses produce 200 Hz motility of a prestin-transfected cell but not a control cell. **(B)** Average tuning curves of wild-type (brown) and *prestin* knock-in mutant mice in which two amino acids essential for electromotility were replaced (blue). The knock-in mutant exhibited a greatly increased threshold and degraded frequency selectivity. (A, adapted from Zheng J, Shen W, He DZZ, et al. [2000] *Nature* 405:149–155. With permission from Springer Nature. B, adapted from Dallos P, Wu X, Cheatham MA, et al. [2008] *Neuron* 58:333–339. With permission from Elsevier Inc.)

The second form of outer hair cell motility results from changes in the shapes of outer hair cells: hyperpolarization causes the outer hair cell to lengthen along its long axis, while depolarization causes it to shorten. This remarkable property, discovered in the 1980s, is called **electromotility**, and this movement enhances the displacement of the basilar membrane, creating another positive feedback loop that amplifies the sound signals transmitted to nearby inner hair cells (Figure 6-50C).

The electromotility of outer hair cells has not been seen in any other parts of the nervous system, including the neighboring inner hair cells. Indeed, this difference between outer and inner hair cells enabled a differential gene expression screen, leading to the discovery of the gene encoding **prestin** (after the music notation *presto*, for "fast tempo") that mediates electromotility. Ectopic expression of prestin in cultured cells derived from human kidney was sufficient to grant these cells high-frequency motility in response to membrane potential changes (**Figure 6-51**A). Prestin has a primary structure resembling that of Cl⁻ transporters and is highly enriched on the outer hair cell plasma membrane. Membrane potential changes are thought to alter the conformation of prestin (Figure 6-50A), which underlies its function in electromotility.

To examine the function of electromotility *in vivo*, knock-in mice were created in which two amino acids of prestin necessary for electromotility were replaced. While mechanotransduction was unaffected, these knock-in mice had drastically elevated threshold for sound detection and severely degraded frequency tuning (Figure 6-51B), as in the case of outer hair cell loss (Figure 6-47A). These experiments provided compelling evidence that prestin-mediated electromotility is essential for auditory signal amplification and sharpened frequency tuning.

6.23 Auditory signals are processed by multiple brainstem nuclei before reaching the thalamus

So far, we have learned how the cochlea detects and processes auditory signals. Hair cells convert mechanical stimuli into electrical signals. Positions of activated hair cells along the cochlea represent sounds of different frequencies in a tonotopic map. While outer hair cells amplify auditory signals and sharpen frequency tuning, inner hair cells produce electrical signals sent to the brain via ordered projections of spiral ganglion neurons. In the following sections, we discuss how the brain analyzes these auditory signals and extracts behaviorally relevant information.

Axons of spiral ganglion neurons terminate in the dorsal and ventral cochlear nuclei of the brainstem. Projection neurons in the dorsal cochlear nucleus send

Figure 6-52 Central auditory pathways.
Axons of spiral ganglion neurons terminate in the dorsal and ventral cochlear nuclei. Projection neurons of the dorsal cochlear nucleus send axons directly to the contralateral inferior colliculus, or indirectly via intermediate neurons in the lateral lemniscus. Projection neurons of the ventral cochlear nucleus terminate in ipsilateral and contralateral superior olivary nuclei. Neurons from superior olivary nuclei integrate ipsilateral and contralateral auditory signals and project to the inferior colliculus. Inferior colliculus projection neurons send axons to the superior colliculus (not shown) and medial geniculate nucleus of the thalamus, which in turn projects to the auditory cortex. For simplicity, only pathways conveying auditory signals from the left cochlea are depicted, and only excitatory projections are represented. See Figure 6-55 for a more detailed depiction of circuits of the superior olivary nuclei.

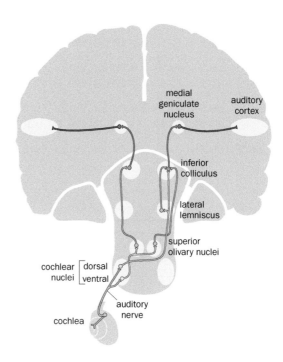

information directly to the contralateral **inferior colliculus** of the midbrain, or via intermediate neurons in the lateral lemniscus. Projection neurons in the ventral cochlear nucleus send information to the **superior olivary nuclei** on both the ipsi- and contralateral sides of the brainstem, where auditory signals from the left and right ears are first integrated. Projection neurons from the superior olivary nuclei also terminate in the inferior colliculus, an important center that integrates all auditory information. Auditory information is relayed from the inferior colliculus through the **medial geniculate nucleus** of the thalamus to the auditory cortex in the temporal lobe (**Figure 6-52**).

A prominent organizational feature along the ascending auditory processing pathway is the preservation of the tonotopic map generated in the cochlea. The orderly projections of spiral ganglion neuron axons impose tonotopic maps onto their postsynaptic targets in both the dorsal and ventral cochlear nuclei (Figure 6-47B). Orderly tonotopic maps also appear in the superior olivary nucleus, inferior colliculus, medial geniculate nucleus, and primary auditory cortex, enabled by serial topographic projections that probably result from mechanisms similar to those that give rise to the retinotopic projections (see Chapter 5). The maintenance of tonotopy indicates that sound frequency is an important feature of the auditory system, as spatial position is in the visual system.

In contrast to the visual system, where information leaving the retina reaches the primary visual cortex through just one intermediate synapse in the thalamus (Figure 4-37B), the auditory system employs several additional brainstem nuclei for information processing (Figure 6-52). These nuclei have many different functions. For instance, the superior olivary nuclei send axons back to outer hair cells in the cochlea (Figure 6-48) to regulate auditory amplification. The dorsal cochlear nucleus receives input from the somatosensory and vestibular systems in addition to the cochlea. Thus, output from the dorsal cochlear nucleus already carries integrated signals from multiple senses. In addition to the excitatory neurons outlined in Figure 6-52, inhibitory neurons at multiple stages of the auditory pathways play important roles in shaping auditory signals. For instance, inhibitory circuits in the inferior colliculus underlie the **precedence effect**, whereby a sound suppresses processing of sounds arriving immediately afterward. The precedence effect helps distinguish a sound coming from its original source (which arrives first) from its echoes (which arrive later). Indeed, the best-studied function of the brainstem auditory nuclei is the extraction of information about sound location, which we discuss in the next two sections.

6.24 In the owl, sound location is determined by comparing the timing and levels of sounds reaching two ears

Audition provides animals with information about sound location, in addition to its frequency and amplitude. Recall the superb ability of the barn owl to catch prey in complete darkness using auditory information alone (Figure 1-5). A number of different mechanisms for sound localization have been identified. We start with sound localization in the horizontal plane.

When a sound comes from the left, it reaches the left ear earlier than it reaches the right ear, generating an **interaural time difference (ITD)**. (For humans, the distance between the two ears is about 20 cm; given that the speed of sound is 340 m/s, this gives a maximal ITD of about 600 µs. The maximal ITD for barn owls is about 200 µs.) As the source of sound moves toward the midline of the head, the ITD decreases systematically, and when it is located directly ahead or behind the head, ITD is zero. This strict correspondence between ITD and the horizontal positions of sounds enables the auditory system to construct a map of space based on ITDs.

A theory was proposed as to how the brain can construct such a map. Suppose that some central neurons in the brain act as **coincidence detectors**, their activity peaking when signals from the left and right ears activate these neurons simultaneously. These coincidence detector neurons can be arranged such that different neurons receive input from left and right ears with different time delays. Peak activity of specific neurons would then report specific sound locations in the horizontal plane (**Figure 6-53**A; **Movie 6-3**).

This theory, known as the Jeffress model (named after its original proposer), was beautifully validated decades later in the barn owl, where a brainstem structure called the **nucleus laminaris (NL**; analogous to the mammalian medial superior olivary nucleus) receives input from bilateral cochlear nuclei. Ipsilateral input enters the NL from the dorsal side, and contralateral input from the ventral side. Both pathways traverse the NL, forming synapses onto the same neurons along each path (Figure 6-53B). *In vivo* extracellular recordings showed that sound signals reach the dorsal surface of the NL from the ipsilateral ear and the ventral surface of the NL from the contralateral ear at about the same time, ~3 ms after sound onset. (The contralateral fibers are larger, with faster action potential conduction speeds compensating for the longer distance traveled.) It then takes ~200 µs for signals to travel through thin axon fibers across the NL; this duration matches the barn owl's maximal ITD. These thin axon fibers form **delay lines**, because each axon fiber carries auditory signals to NL neurons located at different dorsoventral positions with different time delays. Specifically, NL neurons located closer to the dorsal surface receive ipsilateral input before contralateral input, while NL neurons located closer to the ventral surface receive contralateral input before ipsilateral input (Figure 6-53B). *In vivo* intracellular recordings confirmed that individual NL neurons were most strongly activated by sounds with specific ITDs (Figure 6-53C). Because the inputs to NL neurons from two ears are phase-locked to sound stimuli (Section 6.21; phase locking extends to ~9 kHz in barn owls), the temporal precision for coincidence detection can be tens of microseconds, far shorter than the duration of an action potential (Figure 6-49C).

Thus, the properties of NL neurons in the barn owl match very closely the Jeffress model, and the distribution of ITD signals along the dorsoventral axis of the NL forms the first spatial map of the auditory environment in the owl's brain. This map is relayed to the inferior colliculus. Because of the cyclic nature of sound waves, the coincidence detector neurons in the NL reach maximal response either when signals from the left and right ears arrive at the same time or when they arrive one or more tonal periods apart from each other (Figure 6-53C). This creates phase ambiguity for ITD-based sound localization, because individual NL neurons are maximally activated by more than one time difference. Interestingly, while individual NL neurons are frequency tuned (along an axis orthogonal to the

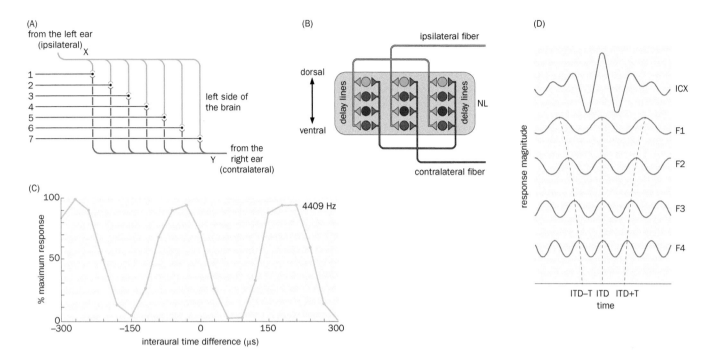

Figure 6-53 Interaural time difference can be used to localize sound. **(A)** The Jeffress model for the neural basis of sound localization. Suppose that sound signals from the left (ipsilateral) and right (contralateral) ears reach point X and Y simultaneously. Neurons 1–7 are postsynaptic targets of both ipsilateral and contralateral fibers. Because of its location, neuron 1 receives ipsilateral input before contralateral input; neuron 4 receives both inputs at the same time; neuron 7 receives contralateral input before ipsilateral input. Furthermore, suppose that neurons 1–7 are best excited when the two inputs are precisely coincident. Thus, neuron 1 is best excited when the sound is closer to the contralateral ear than the ipsilateral ear. Likewise, neuron 4 is best excited when the sound is along the midline, and neuron 7 is best excited when the sound is closer to the ipsilateral ear than to the contralateral ear. The output signals of neurons 1–7 thereby carry information about specific interaural time differences (ITDs) that reflect sound locations in the horizontal plane. **(B)** Schematic of barn owl nucleus laminaris (NL). Fibers from the ipsilateral and contralateral cochlear nuclei enter NL from the dorsal and ventral sides, respectively. Sound signals from the ipsilateral or contralateral ears reach the dorsal or ventral NL surfaces at the same time and then travel within the NL (delay lines). Both fibers form synapses with NL neurons along their paths. NL neurons located along the dorsoventral axis are therefore tuned to different ITDs. (NL neurons along the orthogonal axis, left-to-right in this schematic, are

tuned to different frequencies.) **(C)** An ITD curve of an NL neuron obtained via *in vivo* intracellular recording of action potentials (*y* axis, spike rate relative to maximum) in response to systematically varying ITDs (*x* axis). This neuron is best activated when the contralateral ear leads the ipsilateral ear by 30 μs (contralateral lead is represented by negative values). Note that the neuron also spikes maximally at −270 μs or +180/210 μs. This is because sound waves are periodic, and the peaks at −270 μs and +210 μs correspond to coincidence with the two neighboring wave peaks (the 240 μs time difference between peaks reflects the frequency of the sound stimuli of 4409 Hz). **(D)** Model for resolution of phase ambiguity by ICX neurons. An ICX neuron (top) integrates inputs from NL neurons with the same ITD but representing multiple frequencies (F1–F4). Dotted lines show how peaks at the ITD line up across frequency, while the peaks at ITD ± T (for tonal period) diverge with frequency. Therefore, the ICX neuron is maximally active only at the ITD in response to broadband stimuli. Experimental data (not shown) support this model. (A, adapted from Jeffress LA [1948] *J Comp Physiol Psychol* 41:35–39. B, adapted from Ashida G & Carr CE [2011] *Curr Opin Neurobiol* 21:745–751. With permission from Elsevier Inc. C, adapted from Carr CE & Konishi M [1990] *J Neurosci* 10: 3227–3245. Copyright ©1990 Society for Neuroscience. D, after Peña JL & Konishi M [2000] *Proc Natl Acad Sci U S A* 97:11787–11792. Copyright National Academy of Sciences, U.S.A.)

dorsoventral axis in Figure 6-53B), space-specific neurons in the external nucleus of the inferior colliculus (ICX) integrate ITD information across multiple frequencies. This integration resolves the phase ambiguity of individual frequencies: when the owl hears a broadband signal (a signal containing a wide range of frequencies), a space-specific neuron in the ICX is most excited by the true time difference between the two ears and not by other time differences one or more tonal periods apart. This is because only at the true ITD do the peak responses of all NL neurons with the same ITD tuning, but representing different frequencies, align (Figure 6-53D).

As aerial predators, barn owls must locate sounds in both horizontal and vertical planes in order to catch prey in complete darkness. Indeed, single-unit recordings indicated that individual ICX neurons in the owl are tuned to sounds originating from specific vertical and horizontal positions (**Figure 6-54**A). Neurons tuned to different horizontal and vertical positions collectively form an audi-

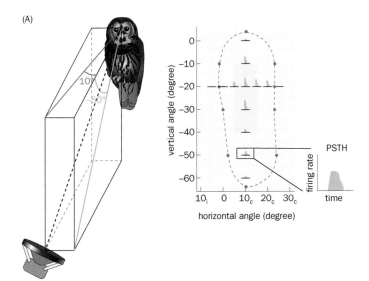

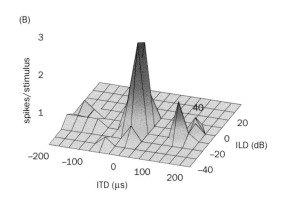

Figure 6-54 Auditory space map and integration of ITD and ILD signals in the inferior colliculus. (A) Left, schematic of two angles that define the direction of a sound in the owl's auditory space. Draw a line connecting the owl with the source of the sound (dark dashed line) and project this line onto the horizontal (magenta line) and vertical planes (blue line). The angles between these projected lines and a line straight forward define the sound direction on the horizontal (magenta angle) and vertical (blue angle) planes. Right, receptive field of an ICX neuron measured by single-unit recording in a lightly anesthetized owl in response to sounds from a speaker systematically moved to different locations. Dashed lines mark the borders of the neuron's receptive field projected from the actual measurement sites (red dots). The neuron is most excited by sounds from the dark yellow area (best area). Each small diagram is a peri-stimulus time histogram (PSTH, magnified in the inset on the right) for sound stimuli provided at that position. The best area for this particular neuron is around 10° contralateral (10_c) on the horizontal plane (x axis) and –20° on the vertical plane (y axis; negative values represent positions below the straightforward horizontal plane). **(B)** Quantification of *in vivo* intracellular recording of an ICX neuron from an anesthetized owl in response to auditory stimuli played from speakers placed on the owl's two ears, such that ITDs and ILDs could be precisely controlled and systematically varied. The neuron is maximally active at a specific combination of ITD and ILD levels. Minor peaks on the ITD axis represent ITD ± one tonal period. (A, adapted from Knudsen EI & Konishi M [1978] *Science* 200:795–797, B, adapted from Peña JL & Konishi M [2001] *Science* 292:249–252. With permission from AAAS.)

tory spatial map in the ICX. How do owls localize sounds in the vertical plane? The barn owl's left and right ears are vertically displaced from each other, such that they are differentially sensitive to sounds from above and below. The **interaural level differences (ILDs)** arising from the differences in the amount of attenuation and amplification of signals between the ears provide information about the vertical positions of sounds. ILD signals are analyzed along an auditory pathway in parallel with the ITD pathway we just discussed. A brainstem nucleus called posterior dorsal lateral lemniscus (LLDp) (analogous to the mammalian lateral lemniscus) receives excitatory input from the contralateral cochlear nucleus and inhibitory input from the ipsilateral cochlear nucleus. Different neurons along the dorsoventral axis of the LLDp are sensitive to different ILDs, creating a map for sounds from different vertical positions. LLDp neurons also project to the inferior colliculus, sending level difference signals to the same ICX neurons that receive time difference signals from the NL. This way, individual space-specific ICX neurons are tuned to both ITD and ILD (Figure 6-54B). ICX neurons project to the nearby tectum (equivalent to the mammalian superior colliculus), where the auditory and visual maps are aligned (Section 1.3).

6.25 Sound localization in mammals utilizes different circuit mechanisms from those in the owl

Although the general principles of sound location revealed in the barn owl apply to other animals, specific circuits and mechanisms differ considerably between different species. In mammals, including humans, information about sound location in the vertical plane is mostly derived from monaural cues (signals from a single ear) rather than binaural comparisons. Because of the complex shape of the auricle, sounds with different frequencies from above or below are differentially attenuated or amplified, changing the amplitude spectrum of the sound in a characteristic manner (**Figure 6-55**A). These changes in sound spectra provide cues for determining the vertical positions of sounds. Sound localization in the horizontal plane is achieved by a combination of ITDs and ILDs in mammals. ITDs are preferentially used to localize low-frequency sounds (Figure 6-55B), because the precise timing provided by phase locking of auditory neurons (Section 6.21) works well only for sounds with frequencies < 2 kHz. ILDs are preferentially used to localize high-frequency sounds (>2 kHz) because high-frequency (short-wavelength) sounds are more readily collected by the auricles and deflected by the head, creating larger level differences between the two ears (Figure 6-55C).

In the mammalian auditory system, ITDs of low frequency sounds are analyzed primarily in the medial superior olivary (MSO) nucleus of the brainstem (Figure 6-55D). The MSO receives excitatory inputs from both ipsilateral and contralateral ventral cochlear nuclei and plays a role analogous to that of the owl's NL

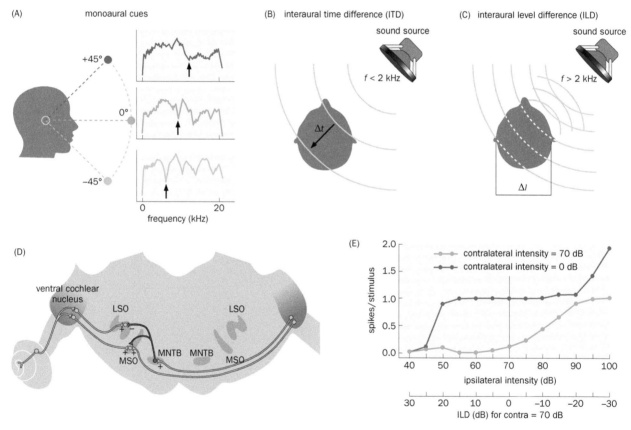

Figure 6-55 Sound localization systems in mammals. (A) Spectral analysis for sound localization in the vertical plane in mammals. Illustrated in color are interactions of a broadband sound with the vertically asymmetric auricle; the same trough (black arrow) in the effective spectrum of the sound shifts to higher frequencies (rightward on the x axis) when the sound source is shifted from below to above the horizon. **(B)** Interaural time difference (ITD) is preferentially used to localize sounds of low frequencies (f < 2 kHz), as auditory neurons phase lock only to low-frequency sounds, which also create larger time differences (Δt). **(C)** Interaural level difference (ILD) is preferentially used to localize sounds of high frequencies (f > 2 kHz), as high-frequency sounds are more effectively collected by the auricle and blocked by the head, creating larger interaural level differences (Δl). **(D)** Schematic brainstem circuits that analyze ITD and ILD in mammals. ITDs are analyzed in the medial superior olivary (MSO) nuclei, which receive excitatory input from both ipsilateral and contralateral cochlear nuclei and inhibitory input from the medial nucleus of the trapezoid

body (MNTB). ILDs are analyzed in the lateral superior olivary (LSO) nuclei, which receive excitatory input from the ipsilateral cochlear nucleus and inhibitory input from the ipsilateral MNTB. MNTB receives excitatory input from the contralateral cochlear nucleus. For simplicity, inputs only to the left LSO and MSO are shown. **(E)** LSO neurons encoding ILD. Spike numbers of an LSO neuron measured during sound (100 μs clicks) delivery from two speakers, one on each ear. In the experimental condition (cyan curve), the contralateral ear received sounds of constant amplitude of 70 dB, while the ipsilateral ear received sounds of increasing amplitude (top x axis). The LSO neuron fired when the sound level from the ipsilateral ear exceeded that from the contralateral ear, and the spike numbers reflected the ILD (bottom x axis). In the control condition (red curve), the contralateral ear received no sound. (A–C, adapted from Grothe B, Pecka M, & McAlpine D [2010] *Physiol Rev* 90:983–1012. With permission from the American Physiological Society. E, adapted from Irvine DRF, Park N, & McCormick L [2001] *J Neurophysiol* 86:2647–2666.)

in measuring ITDs, although the mechanism is different and less well understood (the binaural nuclei evolved independently in birds and mammals). ILDs are primarily analyzed in the lateral superior olivary (LSO) nucleus, where individual neurons receive glutamatergic *excitatory* input from the ipsilateral ventral cochlear nucleus and glycinergic *inhibitory* input from the ipsilateral medial nucleus of the trapezoid body (MNTB), which in turn receives excitatory input from the contralateral ventral cochlear nucleus (Figure 6-55D). Thus, LSO neurons are excited by ipsilateral auditory signals and inhibited by contralateral auditory signals. Figure 6-55E illustrates how ILDs can be decoded by LSO neurons on the left side of the brainstem. Sounds from the left side produce stronger excitatory input than inhibitory input and, accordingly, activate LSO neurons; sounds from the right side produce stronger inhibitory input than excitatory input and thus prevent LSO neurons from firing. Since LSO neurons on the right side of the brainstem do the exact opposite, LSO neurons collectively represent location in the horizontal plane.

The mammalian brainstem sound localization system utilizes large axons and synapses to achieve high speed and reliability. For instance, the synapse from the cochlear nerve onto ventral cochlear neurons that project to the MSO and LSO is called the endbulb of Held, and the synapse from cochlear nucleus neurons onto MNTB neurons is called the **calyx of Held**, both named after the nineteenth-century anatomist Hans Held, who described these synapses using the then newly invented Golgi stain. The presynaptic terminals at the calyx of Held can be >15 μm in diameter, wrapping around the postsynaptic cell bodies with many hundreds of release sites (**Figure 6-56**). These properties make the calyx of Held a model synapse for studying mechanisms of synaptic transmission in the mammalian CNS. For example, one can record directly from the presynaptic terminals with intracellular or patch clamp electrodes much like one could from the squid giant synapses (Figure 3-5). Each action potential arriving at the calyx of Held is rapidly converted into an action potential in postsynaptic MNTB neurons and rapidly transmitted to the LSO and MSO, as the diameters of MNTB axons are among the largest in the CNS. This rapidity allows LSO and MSO neurons to compare ipsilateral excitatory and contralateral inhibitory inputs (MNTB neurons are glycinergic) arriving at the same time.

6.26 The auditory cortex analyzes complex and biologically important sounds

By the time auditory signals reach the inferior colliculus, information regarding sound frequency, amplitude, and timing has been extracted and represented by specific neurons. Such information is channeled to the superior colliculus to select stimuli to attend to and to direct head and eye orientation and to the auditory cortex via the medial geniculate nucleus (Figure 6-52).

The auditory cortex analyzes the meanings of sounds and generates the perception of sound location in space. A prominent organization in the **primary auditory cortex** (**A1**) is a coarse tonotopy along the rostral–caudal axis. A1 neurons also represent many other auditory properties. Some are excited by signals from one ear while others are excited by signals from both ears. Some A1 neurons respond to auditory signals within a specific frequency bandwidth, a certain intensity range, or a specific temporal interval. Some A1 neurons respond to the rate and direction of frequency modulation (for example, sounds with increasing or decreasing frequencies). Syntheses of these properties allow neurons in higher-order auditory cortices to analyze complex sounds such as those used in human speech. Much work is required to understand how these properties of A1 neurons arise and how they are integrated by higher-order auditory cortical neurons for the analysis of complex sounds that enable us to comprehend speech and enjoy music.

The best understood examples of auditory cortical organization come from studies of **echolocation** in bats, which detect insects, their nightly meal, in complete darkness. Echolocating bats emit ultrasonic pulses (biosonar) and use echoes of these pulses to derive the size, distance, velocity, and activity of their targets.

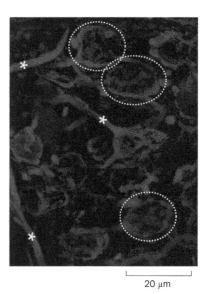

20 μm

Figure 6-56 The calyx of Held. Each calyx of Held (dashed outline) comprises a presynaptic terminal of a globular bushy cell from the ventral cochlear nucleus (here labeled by a red fluorescent protein in a transgenic mouse) wrapping around the cell body of an MNTB principal cell (not labeled) with 300–700 release sites. *, axons of globular bush cells. Scale, 20 μM. (From Joris PX & Trussell LO [2018] *Neuron* 100:534–549. With permission from Elsevier Inc. See also Borst JG & Soria van Hoeve J [2012] *Annu Rev Physiol* 74:199–224 for a summary of physiological properties of the calyx of Held.)

The auditory cortex of echolocating bats has specialized areas dedicated to analyzing different aspects of these echoes. Some bats emit constant-frequency (CF) pulses; others emit frequency-modulated (FM) pulses; yet others, such as the mustached bat, which hunts insects in vegetation, emit pulses with both components (**Figure 6-57**A). The mustached bat emits ultrasonic pulses at a **fundamental frequency** of ~30 kHz, and its second, third, and fourth harmonics (with frequencies 2, 3, and 4 times that of the fundamental frequency), with the second harmonic being the most dominant harmonic. When a bat is searching, it emits these pulses at low rates (~10/s) with long durations (~35 ms). As the bat approaches a target, the rate increases and the duration shortens. In the terminal phase, the rate increases to ~100/s and the duration shortens to just a few milliseconds (Figure 6-57A).

To determine the neural basis of echolocation, single-unit recordings in the auditory cortex of lightly anesthetized bats were performed while experimenters applied ultrasonic stimuli. In a cortical area called the FM-FM area, specialized

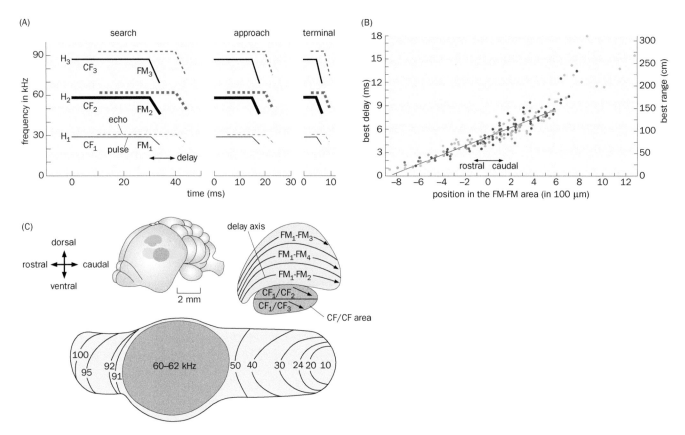

Figure 6-57 Organization of the mustached bat auditory cortex for analyzing ultrasonic echoes. (A) Ultrasonic pulses emitted by the mustached bat. In these time-frequency graphs, each pulse consists of a constant frequency (CF) component (horizontal lines), and a frequency-modulated (FM) component (diagonal lines at the end of each pulse). The pulses are emitted at a fundamental frequency of ~30 kHz (H_1), and second (H_2), third (H_3), and fourth (not shown) harmonics, with the second harmonic being the strongest (represented by the heaviest lines). Their respective echoes are shown in red, with a time delay. The durations of the pulses (and their echoes) progressively shorten from the search phase to the approach and terminal phases of hunting. **(B)** Relationship of neuronal position along the rostral–caudal axis in the FM-FM area (x axis) to best delay (left y axis) or best target range (right y axis). 0 corresponds to 5-ms best delay. The regression line is based on 152 neurons, with 0 to 10 ms best delay recorded from six different animals (neurons from each animal are designated by a specific color). Thus, neurons in the rostral part of the

FM-FM area represent shorter best delay, likely functioning in the terminal phase, while neurons in the caudal part represent longer best delay, likely functioning in the search phase. **(C)** Schematic of mustached bat auditory cortex (top left), and a magnified view (right and bottom). The FM-FM area (green) is organized along the rostral–caudal axis according to best delay (delay axis, with arrows pointing in the direction of increasing best delay), as illustrated in Panel B. The CF/CF area (orange) is organized along the rostral–caudal axis according to target velocity (with arrows pointing in the direction of increasing velocity). In both areas, bands along the dorsal–ventral axis correspond to echoes in the frequency range of different harmonics. Representation of 60–62 kHz, corresponding to the echoes of the second (and dominant) harmonic, is drastically overrepresented in the tonotopic map. (A, adapted from O'Neill WE & Suga N [1982] *J Neurosci* 2:17–31. B, adapted from Suga N & O'Neill WE [1979] *Science* 206:351–353. C, adapted from Suga N [1990] *Sci Am* 262(6):60–68. With permission from Springer Nature.)

in analyzing the FM part of the echo, researchers found that many neurons did not respond to a pulse or an echo presented alone but responded only when an echo followed a pulse with a specific time delay. The "best delay," the time interval between the pulse and the echo that elicits the most excitation, corresponds to the distance of the target (which equals time delay multiplied by the speed of sound divided by 2, to account for the round trip). Moreover, the best delays of individual neurons varied systematically across the cortex, creating a map of target distances along the rostral–caudal axis of the brain (Figure 6-57B). Each FM-FM neuron is tuned to analyze echoes of a specific frequency: either the second, third, or fourth harmonic. Neurons tuned to different echo harmonics are arranged in bands orthogonal to the axis of best delays (green in Figure 6-57C). The relative strengths of the different echo harmonics indicate the size of the object. Thus, the auditory cortex of the mustached bat is highly organized for analyzing object distance and size based on echoes of the sonar signals.

Additional studies indicated that the function of the CF component is to analyze the relative velocity of its targets based on frequency changes caused by the **Doppler effect** (a wave property describing how sound frequency detected by an observer increases if the sound-emitting object moves toward the observer and decreases if the sound-emitting object moves away.) The bat's auditory cortex also has a highly expanded representation of sound frequencies between 60 and 62 kHz (pink in Figure 6-57C), as this narrow frequency range encompasses the dominant second harmonic echoes from which bats collect detailed information, such as wing beats, much like the overrepresented foveal portion of the visual field in the primate visual cortex encompasses high-acuity and color signals (Figure 4-39). Collectively, auditory cortical neurons allow the mustached bat to extract information regarding size, distance, velocity, and other object details, so as to identify and capture flying insects in vegetated areas at night.

Lessons learned from studying auditory specialists such as the mustached bat can illuminate general principles of auditory cortex function. For instance, the auditory cortex of other mammals is similarly tuned to sounds of particular biological significance, such as the social calls of mates, parents, and progeny. Some mouse A1 neurons are tuned to ultrasonic vocalizations emitted by distressed pups, thus facilitating pup retrieval (see Section 10.16 for details). Likewise, human auditory cortex has specialized areas that allow us to analyze specific elements of our language. For example, multielectrode recordings in the human superior temporal gyrus, a high-order auditory area near Wernicke's area, revealed that intermingled individual neurons selectively represent specific aspects of speech, such as intonation (which syllable to stress), sentences (combinations of phonetic features), and speakers (with differing fundamental and formant frequencies) (**Figure 6-58**). As mentioned in Section 1.10, patients with damage to Wernicke's area exhibit selective language comprehension deficits. A future goal is to understand the organizational principles of the auditory cortex in other mammals, including humans, at the level of detail that we understand it in echolocating bats.

Whereas signals in the auditory system originate from the cochlea, its sister organs in the inner ear are part of the vestibular system, which serves entirely different functions (**Box 6-2**).

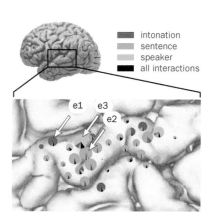

Figure 6-58 Representation of speech elements in human auditory cortex. Bottom figure depicts results from a multielectrode recording from the superior temporal gyrus (corresponding to the rectangle in the top image) of a patient hearing four different sentences, each with four variant intonations, spoken by three different speakers. Recording data were analyzed using regression to obtain the variance explained by specific features, including intonation, sentence content, speaker identity, and interactions between these features. Each circle represents a recording site. The diameter of each pie chart represents the magnitude of variance explained collectively, and slices of pie chart represent proportions of variance explained by individual features. For instance, activity recorded at electrode 1 (e1) largely encodes intonation, while activity at e2 and e3 encodes mostly sentence content and speaker identity, respectively. (From Tang C, Hamilton LS, & Chang EF. [2017] *Science* 357:797–801. With permission from AAAS.)

Box 6-2:   The vestibular system senses movement and orientation of the head

In addition to the cochlea, the inner ear contains sensory organs of the **vestibular system**, which in many ways parallels the auditory system. For example, both systems use stereocilia in hair cells to convert mechanical stimuli into electrical signals. In fact, the vestibular system is the more ancient of the two systems and is present in all chordates; the auditory end organ was derived from the vestibular sensory organs in bony fish, and became more elaborate in terrestrial animals, especially mammals. The vestibular system senses movement and orientation of the head and uses these signals to regulate balance and spatial orientation, coordinate head and eye movement, and perceive self-motion. Because most functions it performs are subconscious, the vestibular system is sometimes called a silent sense.

The vestibular system has five sensory organs: two **otolith organs** and three **semicircular canals** (Figure 6-59A). The otolith organs, called the utricle and the saccule, sense linear acceleration and stationary head tilts. The stereocilia of

hair cells in both the utricle and saccule are embedded in a membrane densely packed with calcium carbonate crystals known as otoliths ("ear stone" in Greek). The hair cells are anchored in the bony structure of the inner ear, and therefore move in sync with the head. The inertia of otoliths resulting from linear acceleration in a specific direction creates a relative movement between hair cells and the overlying otoliths, which bend stereocilia whose staircases ascend in the same direction as the direction of motion (Figure 6-59B). This opens the mechanosensitive channels in the hair bundles and causes depolarization of the hair cells. Stationary head tilts also activate hair cells with ascending stereocilia aligned with the direction of the tilt due to gravity of the otoliths. Because of the different orientations of hair cell stereocilia with respect to the three axes of space, hair cells in the utricle and saccule are collectively sensitive to linear accelerations and head tilts in all directions (Figure 6-59A, middle). Left–right accelerations and head tilts are sensed by utricular hair cells, up–down accelerations (for example, at the beginning and end of an elevator ride) by saccular hair

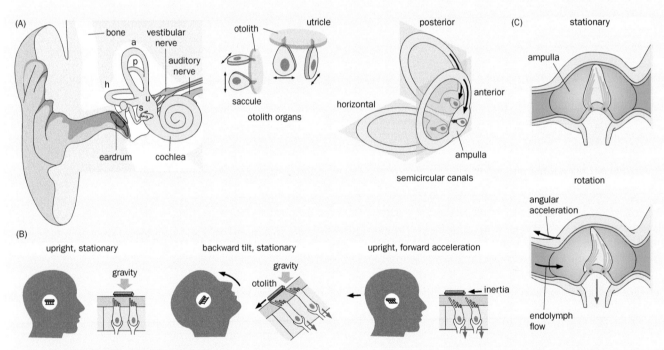

Figure 6-59 The vestibular sensory organs. (A) Left, schematic of the structure of the vestibular organs and the cochlea in the inner ear. u, utricle; s, saccule; h, p, a, horizontal, posterior, anterior semicircular canals. Middle, magnified utricle and saccule. The stereocilia are anchored in membranes rich in calcium carbonate crystals (otoliths). As a result of the specific orientations of hair cell stereocilia, the utricle senses linear accelerations or head tilts along the left–right and front–back axes while the saccule senses linear accelerations along the top–bottom and front–back axes. Right, magnification of semicircular canals. The horizontal semicircular canal senses angular accelerations (rotations) in the horizontal plane while the anterior and posterior semicircular canals sense angular accelerations in the vertical planes. Hair cells are located in the bulges, or ampulla, at the bases of semicircular canals.

(B) Compared to an upright stationary position, stationary backward tilt and upright forward acceleration both bend the stereocilia, via gravity or the inertia of otoliths, respectively; this opens mechanosensitive channels, depolarizes the hair cells, and increases synaptic transmission to vestibular ganglion neurons (red arrows). **(C)** Compared to a stationary position, angular acceleration toward the left causes the stereocilia to bend rightward because of the endolymph flow within the semicircular canal and the inertia of the stereocilia. This relative movement opens mechanosensitive channels, depolarizes the hair cells, and increases synaptic transmission to vestibular ganglion neurons (red arrow). (A, adapted from Day BL & Fitzpatrick RC [2005] *Curr Biol* 15:R583–R586. With permission from Elsevier Inc.)

Box 6-2: continued

cells, and front–back accelerations and head tilts by hair cells of both organs.

The three semicircular canals sense angular accelerations. The horizontal semicircular canal senses head rotation in the horizontal plane; the anterior and posterior semicircular canals, which are orthogonal to each other, collectively sense head rotations in various vertical planes (Figure 6-59A, right). The semicircular canals are filled with fluid contacting the stereocilia of hair cells anchored in the bulges at the bases of the canals (Figure 6-59C). Angular acceleration in the same plane as a semicircular canal creates relative movement of fluid that bends the stereocilia of hair cells. This causes opening or closing (depending on the direction) of mechanosensentive channels in hair cells, initiating electrical signaling. Signals from three semicircular canals collectively provide information about head rotations in all axes in three-dimensional space.

As in the auditory system (Figure 6-45), depolarization of hair cells in the vestibular system activates voltage-gated Ca^{2+} channels, resulting in glutamate release detected by afferent nerves of **vestibular ganglion neurons**. Electrical signals are delivered as action potentials along the **vestibular nerve**, which travels alongside the auditory nerve and terminates in the **vestibular nuclei** of the brainstem. Electrophysiological recordings of individual vestibular nerve fibers connected to hair cells in both the otolith organs and semicircular canals indicated that vestibular ganglion neurons spontaneously spike at a high rate (~40 Hz) in the absence of head movement and tilt. Thus, each neuron can report movement in opposite directions via an increase or decrease in spike rate. This bidirectional signaling of the vestibular system is unique—in other vertebrate sensory systems, sensory neurons are mostly either activated (as in olfaction, taste, audition, and somatosensation) or inhibited (as in vision) by sensory stimuli. Thus, this silent sense is in fact constantly at work even though we are unaware of it.

In addition to input from vestibular ganglion neurons, the vestibular nuclei also receive input from other sensory systems, notably the somatosensory system. Thus, vestibular nuclei integrate information about the position and movement of both the head and the body. Projection neurons in the vestibular nuclei send their axons to many parts of the CNS to carry out distinct functions (Figure 6-60). For example, axons descending to the spinal cord regulate the activity of motor neurons, which in turn control muscle contraction in the neck, trunk, and limbs to regulate balance and posture. Ascending axons deliver information to the ventroposterior thalamic nucleus and a number of cortical areas for multisensory integration. For example, vestibular signals reach specific areas of the parietal cortex, where they are integrated with signals from the visual system to help distinguish visual signals produced from motion in the external world from self-motion. Other important targets of projections from the vestibular nuclei include brainstem nuclei

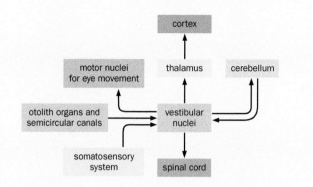

Figure 6-60 Flow of vestibular signals in the CNS. Axons of vestibular ganglion neurons deliver vestibular signals from sensory organs to the vestibular nuclei in the brainstem, which have several divisions that receive differential inputs from the otolith organs and semicircular canals as well as input from the somatosensory system. Descending vestibular signals to the spinal cord regulate posture, balance, and head position through a number of reflexes. Ascending signals to the thalamus and cortex contribute to the integration of vestibular signals with signals from other senses, such as visual signals. Vestibular signals are also sent to brainstem motor nuclei that control eye movement and to the cerebellum for movement modulation via feedback connections to the vestibular nuclei (see Figure 6-61 and Figure 8-28).

that control eye movement (Figure 6-60) and are essential for coordinating head and eye movements. We illustrate this function of the vestibular system with a specific example.

When the head moves, the eye sockets move with it. The **vestibulo-ocular reflex (VOR)** is a reflexive eye movement that stabilizes images on the retina by moving the eyes in the opposite direction of the head movement. The neural circuit basis for the VOR is well established. Let's use head rotation in the horizontal plane as an example (Figure 6-61). When the head rotates toward the right, hair cells in the horizontal semicircular canal on the right side depolarize, increasing the firing of their corresponding vestibular ganglion neurons. Hair cells in the horizontal semicircular canal on the left hyperpolarize, decreasing the firing of their corresponding vestibular ganglion neurons. The increase in firing from vestibular ganglion neurons of the right horizontal semicircular canal is transmitted by excitatory neurons in the medial vestibular nucleus to the contralateral (left) abducens nucleus, activating motor neurons that cause contraction of the lateral rectus muscle, leading to leftward movement of the left eye. The left abducens nucleus neurons also activate the motor neurons of the contralateral (right) oculomotor nucleus, causing contraction of the medial rectus muscles of the right eye, leading to its leftward movement. Simultaneously, the decrease in firing of vestibular ganglion neurons of the left horizontal semicircular canal is transmitted by a parallel pathway causing the relaxation of the medial rectus muscle of the left eye and the lateral rectus muscle of the right eye (Figure 6-61). Thus, the VOR enables image stabilization on the retina during head movement by

(Continued)

Box 6-2:   continued

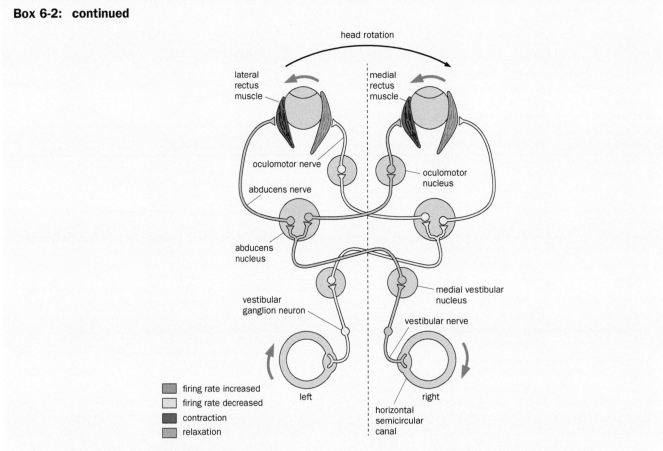

Figure 6-61 Circuit basis of the vestibulo-ocular reflex. When the head rotates to the right, it increases the spike rate of vestibular ganglion neurons associated with the right horizontal semicircular canal and decreases the spike rate of vestibular ganglion neurons associated with the left horizontal semicircular canal. The signals from the right vestibular nucleus increase the spike rate of neurons in the left abducens nucleus, which causes contraction of the lateral rectus muscle of the left eye and the medial rectus muscle of the right eye via motor neurons in the oculomotor nucleus. In parallel, a decrease in spike rate from the left vestibular nucleus causes relaxation of the medial rectus muscle of the left eye and the lateral rectus muscle of the right eye. These muscle actions cause both eyes to rotate toward the left, so that images in the retinas remain stable while the head rotates to the right.

producing compensatory eye movements. **VOR gain**, the degree to which the eyes move in response to head movement, can be adjusted precisely by experience; for instance, near-sighted eyeglasses make images look smaller, thus

eliciting a smaller VOR gain than when not wearing such glasses. As will be discussed in Section 8.11, the cerebellum, which also receives vestibular input and sends output back to the vestibular nuclei (Figure 6-60), adjusts VOR gain.

SOMATOSENSATION: HOW DO WE SENSE BODY MOVEMENT, TOUCH, TEMPERATURE, AND PAIN?

Of all the senses, the somatosensory system has the largest sensory organs—the entire skin and musculature. It also responds to the most diverse set of sensory stimuli, including mechanical, thermal, and chemical stimuli. From these stimuli, the somatosensory system derives information about position and movement of body parts (**proprioception**), temperature (**thermosensation**), touch (cutaneous mechanosensation), as well as pain (**nociception**) and itch (**pruriception**). (The sense of internal organ function, or **interoception**, will be discussed in Chapter 9 when we study the autonomic nervous system.) As introduced in Chapter 1, many fundamental discoveries in neurobiology, such as the use of spike rates to encode information (Figure 1-17), the circuit basis of spinal reflexes (Figure 1-19), and the topographic organization of the neocortex (Figure 1-25), were made by studying somatosensation. In the following sections, we first discuss the general organization of the somatosensory system. We then use the identification of sensory receptors as entry points to investigate touch and pain, from sensory neurons to the brain.

The cell bodies of all sensory neurons of the vertebrate somatosensory system are located in either the **dorsal root ganglia** (**DRG**; 31 pairs in humans) parallel to the spinal cord for sensation of the body, or a pair of **trigeminal ganglia** adjacent to the brainstem for sensation of the face (**Figure 6-62**). Each DRG contains on the order of 10,000 sensory neurons in the mouse; this number is much larger in humans to accommodate a larger body size. Each sensory neuron extends a single process, which bifurcates, giving rise to a peripheral axon that collects information from the skin or musculature and a central axon that sends information to the spinal cord and/or brainstem (see Figure 1-15D; Figure 6-63). The receptor potentials produced by mechanical, thermal, and chemical stimuli are converted into action potentials near a sensory neuron's peripheral endings; action potentials then transmit sensory information along the peripheral and central axons into the CNS. The spatiotemporal patterns of the action potentials of somatosensory neurons inform the brain about what kinds of stimuli are being sensed, where the stimuli come from, and how intense the stimuli are. As discussed in Sections 1.8 and 1.11, the stimulus intensity is encoded by the spike rate of sensory neurons, and the stimulus location is represented by somatotopic maps in the brain (see also Box 5-3 for a specific example: the whisker-barrel system in rodents). We focus here on how the somatosensory system distinguishes what *kinds of stimuli* are being sensed.

6.27 Different types of sensory neurons encode diverse somatosensory stimuli

The somatosensory system is one of the most complex and least well understood sensory systems. This complexity arises because there are many types of sensory neurons; each type can act alone or in combination with other types to encode information about a specific type of sensory stimulus. New types of sensory neurons are still being discovered. Historically, sensory neurons were classified by their electrophysiological properties, such as the speed of action potential conduction, which is dictated by the diameter of their axon fibers and degree of myelination (Box 2-3), and the adaptation they display in response to sustained stimuli. Different neuronal types also innervate distinct end organs, determining what kind of sensory stimuli they receive (**Table 6-1**).

The most numerous sensory neurons in the somatosensory system are **mechanosensory neurons**, which account for proprioception, touch, and a subset of pain sensations (Table 6-1). **Proprioceptive neurons** embed their peripheral endings in muscle spindles, tendons, and joints, allowing these neurons to sense

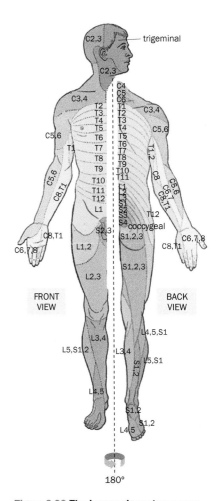

Figure 6-62 The human dermatome map. Each dermatome is an area of skin mainly innervated by a single nerve. The face dermatome is innervated by the trigeminal nerve originating from the trigeminal ganglion. Dermatomes in the rest of the body are numbered according to the dorsal root of the spinal nerves (axons of the sensory neurons in the dorsal root ganglia) that innervate them (C2–C8, cervical; T1–T12, thoracic; L1–L5, lumbar; S1–S5, sacral; coccygeal). The first cervical nerve (C1) mostly carries motor information and does not innervate skin.

Table 6-1: Somatosensory neuron classification

Sensation	Stimulus type	Axon fiber[a]	Peripheral endings (spike pattern)
Proprioception	Mechanical	Aα, Aβ	Muscle spindles, tendons, joints
Touch	Mechanical	Aβ	Merkel cells (slowly adapting)
		Aβ	Meissner corpuscles (rapidly adapting)
		Aβ	Pacinian corpuscles (rapidly adapting)
		Aβ, Aδ, C	Hair follicles
Temperature, pain, itch	Mechanical	Aδ, C[b]	Free endings in skin
	Heat, cold, chemical	Aδ, C[b]	Free endings in skin and internal organs

All categories listed in the table include multiple cell types.

[a] The diameter of the axon fiber correlates with action potential conduction speeds. The conduction speeds of various fibers—Aα (70–120 m/s), Aβ (30–70 m/s), Aδ (5–30 m/s), and C (0.2–2 m/s)—reflects their degrees of myelination (Figure 6-63). The conduction speeds listed above are from human sensory neurons.

[b] While most neurons in these categories have Aδ and C fibers, neurons with Aβ fibers have also been reported.

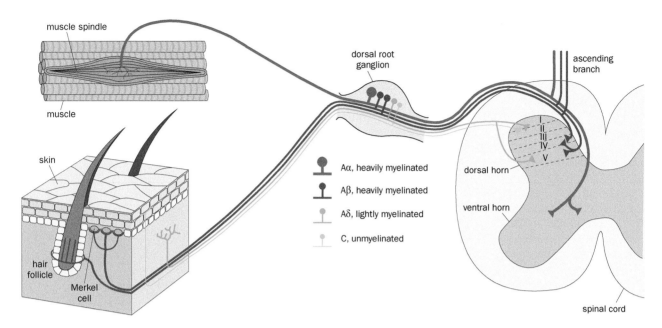

Figure 6-63 Parallel organization of the somatosensory system. All sensory neurons in the somatosensory system representing the body are located in the dorsal root ganglia (DRG). Each sensory neuron type exhibits unique termination patterns of its peripheral axon in the skin or musculature, and projects its central axon to specific lamina(e) of the spinal cord. Five representative sensory neurons are shown in this diagram. The red proprioceptive neuron has the thickest axon and heavy myelination (Aα fiber), innervates a muscle spindle, and terminates in the central and ventral spinal cord. Two purple low-threshold mechanoreceptors (LTMRs) that sense touch have thickly myelinated axons (Aβ fibers), innervate hair follicles and Merkel cells, respectively, and terminate in laminae III–V of the dorsal horn. The blue temperature-sensing neuron has a thinly myelinated axon (Aδ fiber) and free peripheral endings and terminates in laminae I and V of the dorsal horn. The green nociceptive neuron has an unmyelinated axon (C fiber) and free peripheral endings in the epidermis and terminates in dorsal horn laminae I–II. The proprioceptive neuron and LTMRs also form branches that ascend to the brainstem.

muscle stretch and tension (Figure 6-63, top left; see also Figure 1-19 for their function in the knee-jerk reflex). They provide information about the movements and relative positions of one's body parts—for example, they allow us to touch our own nose with eyes closed. Proprioceptive neurons employ the largest diameter and most heavily myelinated axons with the fastest conduction speed. This provides rapid feedback to the motor system regarding muscle strength and tension during movement, the importance of which will be discussed in detail in Chapter 8.

Touch sensory neurons innervate hair follicles, specialized epithelial cells, and encapsulated corpuscles in the skin. Different neuron types sense different kinds of innocuous touch, including vibration, indentation, pressure, and stretch of the skin as well as movement or deflection of hairs (Figure 6-63, bottom left). Compared with neurons that sense noxious mechanical stimuli (discussed later in the chapter), touch sensory neurons are more sensitive to low-force mechanical stimuli, and are therefore called **low-threshold mechanoreceptors (LTMRs)**. LTMRs that innervate glabrous (hairless smooth) skin, such as the palm of a hand, are categorized based on their peripheral ending and adaptation properties (Figure 6-64). For example, the SAI (slowly adapting type I) LTMRs innervate **Merkel cells** at the junction between the epidermis and dermis and fire persistently during a sustained stimulus (hence slowly adapting); SAI-LTMRs sense skin indentation and are responsible for fine discrimination of textures because each SAI-LTMR has a relatively small receptive field (the area of skin that influences the firing of a sensory neuron) compared to other types of LTMRs. The rapidly adapting type I (RAI) LTMRs terminate in a special structure called the Meissner corpuscle, fire only at the onset and offset of a static stimulus (hence rapidly adapting), and sense movement of the skin such as low-frequency vibration. Type II LTMRs terminate in encapsulated structures deep in the dermis and sense stretch (SAII-LTMRs) and high-frequency vibration (RAII-LTMRs; Figure 6-64). These LTMRs employ heavily myelinated Aβ fibers to transmit signals rapidly to the CNS. In addition to innervating glabrous skin, several distinct types of LTMRs innervate hair follicles and

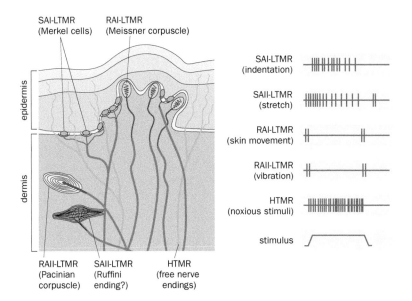

Figure 6-64 **Different types of mechanosensory neurons and their sensory endings in glabrous skin.** Left, schematic of nerve termination patterns for five types of neurons; right, names of the neuronal types, their primary stimuli, and their adaption properties. SAI-LTMRs (slow-adapting type I low-threshold mechanoreceptors) and SAII-LTMRs exhibit sustained firing (each vertical bar represents a spike) in response to a static stimulus (bottom right). SAI-LTMRs innervate Merkel cells. SAII-LTMRs are proposed to terminate in Ruffini endings, but this is controversial. RAI and RAII neurons innervate Meissner and Pacinian corpuscles, respectively, and their firing pattern is more transient in response to a static stimulus. Note that SAI and RAI fibers terminate more superficially than SAII and RAII fibers; all of these employ thickly myelinated Aβ fibers (myelin sheaths are represented in gray). Mechanosensory nociceptive neurons (high-threshold mechanoreceptors, or HTMRs) have free nerve endings (green). A separate set of LTMRs innervate hairy skin (not shown). (Adapted from Abraira VE & Ginty DD [2013] *Neuron* 79:618–639. With permission from Elsevier Inc.)

Merkel cells in hairy skin; while some types use Aβ fibers, others use lightly myelinated Aδ fibers or unmyelinated C fibers (Table 6-1).

Some mechanosensory neurons are **nociceptive neurons** (also called **nociceptors**), which specialize in detection of stimuli perceived as painful. Nociceptors that can detect mechanical stimuli usually require high-intensity mechanical stimuli to elicit action potentials and are thus also called **high-threshold mechanoreceptors (HTMRs)**. Most HTMRs have free nerve endings in the skin, employ Aδ or C fibers, and fire persistently during a static stimulus (Figure 6-64). Activation of HTMRs elicits a painful sensation.

In addition to mechanosensory neurons, many somatosensory neurons sense temperature and chemicals. A large fraction of **thermosensory neurons** are also nociceptors; their activation by noxious heat or cold produces a painful sensation to motivate animals to avoid extreme temperatures. Most chemosensory neurons in the somatosensory system are also nociceptive; they respond to environmental irritants and/or endogenously released chemicals following injury or inflammation. For example, the taste of spices such as hot chili pepper (Figure 6-42) is caused by activation of trigeminal chemosensory neurons that innervate the oral cavity. Notably, as we will discuss in more detail in Section 6.29, some neurons can sense stimuli of more than one sensory modality and are thus called **polymodal neurons**.

Thermosensory and nociceptive neurons have free nerve endings in the periphery and are heterogeneous in their axon fibers and stimulus specificities (Table 6-1). Nociceptive neurons can be broadly classified into two groups based on their axon fiber size: those that have myelinated Aδ fibers are activated by heat, noxious mechanical stimuli, or both and mediate acute and well-localized pain (also called first or fast pain); those that have smaller, unmyelinated C fibers are activated by hot and cold temperatures, as well as by endogenous chemicals released by injury and tissue inflammation, and mediate poorly localized slow pain and chronic inflammatory pain.

Although pain is predominantly mediated by C and Aδ fibers and touch by Aβ fibers, all three fibers contribute to both touch and pain (Table 6-1). The categorization of somatosensory neuron types is very much still a work in progress. Recent studies have classified DRG neurons in mice into about a dozen or more cell types using single-cell RNA-seq (Section 14.13), many of which matched cell types defined previously by morphology and physiology.

As in olfaction and taste, the identification of sensory receptors in recent years has shed light onto how somatosensory stimuli are converted into electrical signals and how the somatosensory system is organized. We discuss these advances in the next three sections.

6.28 Merkel cells and touch sensory neurons employ the Piezo2 channel for mechanotransduction

Since a large fraction of somatosensory neurons respond to mechanical stimuli, a key question is how mechanical forces are converted into electrical signals. Just as in audition (Section 6.20), studies had shown that somatosensory mechanotransduction requires ion channels directly gated by mechanical forces, but only recently was the first mammalian mechanosensitive channel identified (see **Box 6-3** for mechanosensitive channels in *C. elegans* and *Drosophila*).

None of the classic ion channels (Box 2-4), including homologs of invertebrate mechanosensitive channels, appears to mediate mechanotransduction in mammals. Researchers therefore hypothesized that mechanotransduction in mammals might use new types of ion channels, which should be transmembrane proteins. Using a neuronal cell line that produces inward currents in response to mechanical stimuli, an **RNA interference (RNAi)** screen (see Section 14.7 for details on RNAi) was carried out to knock down the expression of predicted transmembrane proteins in the mouse genome. Knocking down one protein named **Piezo1** (after *piesi* in Greek, meaning "pressure") led to a marked reduction of mechanically induced inward current (**Figure 6-65**A). Moreover, expression of Piezo1 or its close homolog Piezo2 in cells that normally do not respond to mechanical stimuli could confer mechanically induced inward current on these

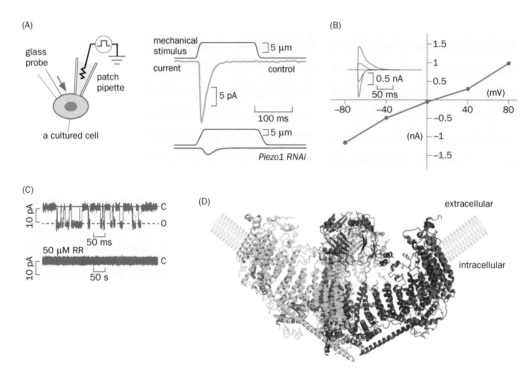

Figure 6-65 Piezos are mechanosensitive channels. (A) Left, experimental setup. A cell (from the mechanosensitive Neuro2A cell line) is stimulated with an electrically driven glass probe (stimulation indicated by a red arrow) while whole-cell patch recording is performed via a patch pipette. Top right, in a control cell, a mechanical stimulus (top trace) induced an inward current (bottom trace) when the cell was held at –80 mV. Bottom right, in a Piezo1 knockdown cell, the inward current is greatly diminished despite a larger mechanical force (note the different vertical scales). Mechanical stimulation is measured as the distance of the stimulating probe from the unperturbed cell membrane; positive values indicate indentation. **(B)** The current–voltage relationship measured from whole-cell recordings of mechanical stimulus-induced current flow (y axis) at different membrane potentials (x axis). The current traces in response to mechanical stimuli are shown in the top left quadrant; each trace corresponds to a current measurement at a different membrane potential. These experiments were conducted by transfecting DNA encoding Piezo2 into cells not otherwise mechanosensitive; thus the current–voltage relationship reflects the properties of Piezo2. **(C)** Single-channel current recorded from a patch of membrane consisting of artificial lipid bilayer and purified Piezo1 protein. Single-channel openings (O) and closings (C) are apparent (top trace). Channel openings were abolished following application of ruthenium red (RR), a Piezo channel blocker (bottom trace). **(D)** Ribbon diagram of an atomic model of the trimeric Piezo1 channel in a closed conformation based on Cryo-EM structure. Each color corresponds to one subunit. The Piezo1 trimer deforms the surrounding lipid bilayer (gray) into a dome. (A & B, adapted from Coste B, Mathur J, Scmidt M, et al. [2010] *Science* 330:55–60. C, adapted from Coste B, Xiao B, Santos JS, et al. [2012] *Nature* 483:176–181. With permission from Springer Nature. D, adapted from Guo YR & MacKinnon R [2017] *eLife* 6:e33660; see also Saotome K, Murthy SE, Kefauver JM, et al. [2018] *Nature* 554:481–486; Zhao Q, Zhou H, Chi S, et al. [2018] *Nature* 554:487–492)

cells. Mechanosensitive channels in Piezo-expressing cells were nonselectively permeable to cations, with a reversal potential around 0 mV (Figure 6-65B), much like nicotinic acetylcholine receptors in skeletal muscle (Figure 3-17) and AMPA-type glutamate receptors in neurons. Opening of these channels thus causes depolarization of cells with a negative resting potential.

Both Piezo1 and Piezo2 are predicted to have >30 transmembrane segments, though they do not exhibit sequence similarity to known ion channels. To test if the Piezos are ion channels themselves, rather than activators of mechanosensitive channels made from a different protein but somehow inactive in the absence of Piezos, researchers purified Piezo1 proteins and placed them in artificial lipid bilayers. Patch recordings from such reconstituted lipid bilayers validated that Piezo1 forms an ion channel by itself (Figure 6-65C). Recent structural studies using cryo-EM revealed that the mechanosensitive channel includes a Piezo1 trimer that deforms the membrane into a dome in its closed state (Figure 6-65D); membrane stretching is hypothesized to flatten the dome, leading to conformational changes that open the channel.

As discussed in Section 6.27, the peripheral endings of LTMRs are associated with specialized structures. For instance, endings of SAI fibers in both glabrous and hairy skin form close contacts with Merkel cells, which are specialized epidermal cells. Whether the Merkel cell or the SAI terminal is the site where mechanical stimuli are converted into electrical signals has long been a subject of debate. Recent studies demonstrated that isolated Merkel cells can respond to mechanical stimuli by conducting inward current, and this response was abolished in mice in which *Piezo2* was conditionally knocked out in Merkel cells (**Figure 6-66**A). Thus, Merkel cells are intrinsically mechanosensitive and depend on Piezo2 for mechanotransduction. However, in a skin-nerve explant preparation in which mechanical stimulation of Merkel cells induced robust spikes recorded from SAI fibers in wild-type mice, SAI fibers from mice in which *Piezo2* was knocked out in Merkel cells still exhibited an initial response to mechanical stimuli, but lost the more sustained response (Figure 6-66B). This experiment indicated that Merkel cells and SAI nerve terminals independently transduce mechanical signals: SAI nerve terminals are responsible for the rapid, dynamic response to mechanical stimuli, whereas Merkel cells are necessary for sustained firing of SAI axons. Merkel cells have recently been shown to use norepinephrine to signal to SAI

Figure 6-66 Mechanotransduction in Merkel cells and sensory neurons.
(A) Left, illustration of the experiment. A dissociated Merkel cell (marked with GFP) was subjected to whole-cell patch recording *in vitro* and a glass probe was used to apply mechanical stimuli. Right, mechanical stimuli induced robust inward current when the membrane potential was held at –80 mV (black trace), but this current was abolished when *Piezo2* was conditionally knocked out in Merkel cells (blue trace). Thus, Merkel cells require Piezo2 to transduce mechanical stimuli into electrical signals. **(B)** Left, illustration of the skin-nerve preparation. Fluorescently labeled Merkel cells (green) were stimulated with a glass probe while action potentials from the associated SAI fibers (peripheral axons of SAI-LTMRs) were recorded. Right, the mechanical stimulus is notated according to the distance of the stimulating probe from the skin (top). At 0 (dotted line), the probe touches the skin; above 0, the probe creates an indentation in the skin. An SAI fiber from a wild-type mouse responded to the stimulus with sustained firing, while in a mutant mouse in which *Piezo2* was conditionally knocked out (cKO) of Merkel cells, the initial response of an SAI fiber persisted but the sustained firing was disrupted. (A, adapted from Maksimovic S, Nakatani M, Baba Y, et al. [2014] *Nature* 509:617–621. With permission from Springer Nature. B, adapted from Woo SH, Ranade S, Weyer AD, et al. [2014] *Nature* 509:622–626. With permission from Springer Nature; see also Ranade SS, Woo SH, Dubin AE, et al. [2014] *Nature* 516:121–125.)

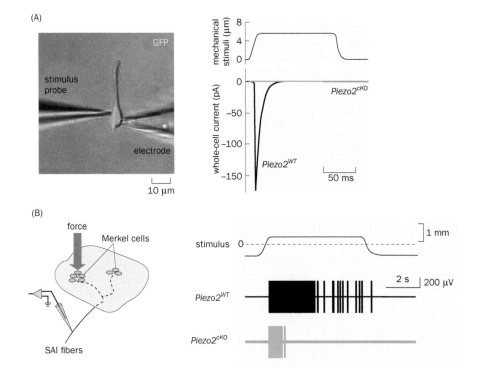

fibers through a G-protein-coupled β-adrenergic receptor, accounting for the slower but more sustained activation of SAI fibers by Merkel cells compared to the direct mechanotransduction by SAI terminals.

What molecule is responsible for mechanotransduction in SAI terminals? Expression studies indicated that mouse Piezo2 (but not Piezo1) is expressed in a subset of DRG neurons, including LTMRs innervating Merkel cells and Meissner corpuscles as well as Aβ-, Aδ-, and C-LTMRs innervating hair follicles. Mice in which the *Piezo2* gene was conditionally knocked out of DRG neurons exhibited profound defects in their behavioral responses to low-force mechanical stimuli, indicating that Piezo2 is required for mechanotransduction in neurons that sense gentle touch. Additional studies revealed that Piezo2 is also instrumental in proprioception. Human patients with *Piezo2* mutations exhibit profound defects in touch sensation and proprioception. For example, they have great difficulty touching their nose with a finger when they are blindfolded. As we will learn in Section 8.2, Piezo1 and Piezo2 are both involved in mechanosensation in the interoceptive systems. However, mice in which the *Piezo2* gene was conditionally knocked out in DRG neurons still exhibited normal responses to high-force mechanical stimuli. This suggests the presence of other, yet-to-be-identified mechanosensitive channels that work in parallel with Piezo2 to convert mechanical stimuli into electrical signals in sensory neurons.

Box 6-3: Mechanosensitive channels in worms and flies

Mechanotransduction is essential for many physiological processes in diverse organisms. Researchers have therefore carried out forward genetic screens (see Section 14.6 for details) in *C. elegans* and *Drosophila* for mutants defective in sensing mechanical stimuli, with the expectation that some of the corresponding genes would encode mechanosensitive channels. A large number of *Mec* genes have been identified in *C. elegans* based on mechanosensory phenotypes in mutant worms. A protein complex consisting of four membrane-associated proteins—Mec2, Mec4, Mec6, and Mec10—is essential for mechanotransduction. Mec2 and Mec6 are accessory proteins, while Mec4 and Mec10 both belong to the epithelial Na+ channel (ENaC) family and appear to be the pore-forming subunits of a mechanosensitive channel. Loss of Mec4 abolishes mechanical stimulus-induced inward current in touch sensory neurons *in vivo* (**Figure 6-67**A), while single amino acid changes in Mec4 alter conductance and ion selectivity.

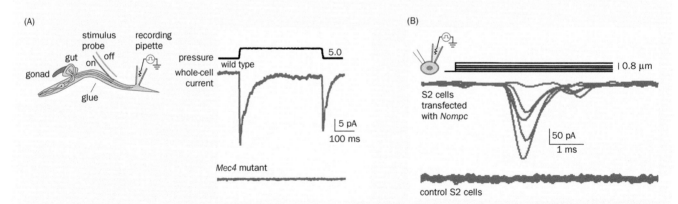

Figure 6-67 Mechanosensitive channels in worms and flies.
(A) Left, schematic of experimental preparation for measuring mechanotransduction in *C. elegans*. A larva was immobilized on glue. Internal hydrostatic pressure was released (some internal organs, such as the gut and gonad, were externalized) such that a touch sensory neuron (green) could be subjected to whole-cell patch recording in response to mechanical stimulation via a stimulus probe. Right, a touch sensory neuron from a wild-type larva responded to the onset and offset of pressure application (top trace, in nanoNewton/μm²) by passing inward current (middle trace), while a touch sensory neuron from a *Mec4* mutant larva did not respond to pressure (bottom trace). **(B)** Cultured *Drosophila* S2 cells were subjected to whole-cell patch recording while mechanical displacement of the membrane was applied via an electrically driven mechanical probe. In response to an increasing series of displacements, S2 cells transfected with *Nompc* cDNA exhibited inward currents with increasing magnitudes (superimposed traces), while control S2 cells did not respond to mechanical stimuli. (A, adapted from from O'Hagan R, Chalfie M, & Goodman MB [2005] *Nat Neurosci* 8:43–50. With permission from Springer Nature. B, adapted from Yan Z, Zhang W, He Y, et al. [2013] *Nature* 493: 221–225. With permission from Springer Nature; see also Walker RG, Willingham AT, & Zuker CS [2000] *Science* 287:2229–2234.)

Box 6-3: continued

In *Drosophila*, genetic screens for touch-insensitive mutant larvae identified *Nompc* (no mechanoreceptor potential C), which encodes a TRP channel (Box 2-4) and is necessary for mechanical stimulus–induced current in sensory neurons. Recent work demonstrated that NompC senses gentle touch in a specific type of larval body wall sensory neuron called class III dendritic arborization neurons. When ectopically expressed, NompC confers touch sensitivity to other sensory neuron types and mechanically induced inward currents to cultured cells (Figure 6-67B).

Mechanotransduction in *C. elegans* and *Drosophila* appears to utilize similar proteins, albeit in different contexts. For instance, TRP4, the *C. elegans* homolog of *Drosophila* NompC, acts as a stretch-activated mechanosensitive channel. The *Drosophila* homologs of an ENaC channel and Piezo work together in class IV dendritic arborization neurons of the larval body wall to sense nociceptive mechanical stimuli; thus gentle touch and noxious mechanical stimuli are sensed independently by two types of sensory neurons using different mechanosensitive channels. With the exception of Piezo, none of the mechanosensitive channels in worms and flies has been shown to transduce mechanical stimuli in mammals. (The ENaC channels are involved in salty taste in mammals; Figure 6-39). However, Piezos are clearly not the only mechanosensitive channels in mammals (Section 6.28). Thus, fundamental discoveries still lie ahead as we advance toward a more complete characterization of mechanosensitive channels, their mechanisms of action, and the cellular logic and evolutionary processes underlying their differential utilization in diverse organisms.

6.29 TRP channels are major contributors to temperature, chemical, and pain sensations

In addition to mechanical stimuli, the somatosensory system also detects chemical irritants and temperature in both the innocuous and noxious ranges. Remarkably, these diverse stimuli can be sensed by the same receptors.

To identify receptors that mediate nociception, researchers took advantage of the fact that capsaicin, the main pungent ingredient in hot chili peppers, activates nociceptive neurons. cDNAs made from dorsal root ganglia were transfected into cultured cells, and Ca^{2+} imaging was used to identify the presence of a single cDNA encoding an active capsaicin receptor (**Figure 6-68A**). This **expression cloning** strategy revealed the capsaicin receptor to be a TRP family channel (Box 2-4), **TRPV1**, permeable to Ca^{2+}, Na^+, and K^+. TRPV1 is activated not only by capsaicin but also by temperatures >43°C (Figure 6-68B, C), in the noxious range that produces heat pain. A similar expression cloning strategy led to the identification of another TRP channel, **TRPM8**, activated by menthol (known to evoke a cool sensation) and a wide range of cool to cold temperatures (<26°C). When expressed in *Xenopus* oocytes, TRPV1 and TRPM8 were activated by hot and cool temperatures, respectively (Figure 6-68C; **Movie 6-4**). These TRP channels are expressed in largely nonoverlapping populations of somatosensory neurons, suggesting that TRPV1- and TRPM8-expressing somatosensory neurons constitute independent pathways for delivering information regarding noxious heat and cool/cold temperature sensation to the CNS.

This hypothesis has been supported by physiological and behavioral defects in knockout mice. For example, similar fractions of trigeminal ganglion neurons dissociated from wild type and *Trpm8* mutant mice were activated at 45°C, a temperature that produces heat pain, but 45°C activated a much smaller fraction of neurons from *Trpv1* mutant mice (Figure 6-68D). Conversely, similar fractions of trigeminal neurons from wild type and *Trpv1* mutant mice were activated at 22°C, an innocuous cool temperature, but this temperature failed to activate any trigeminal neurons from *Trpm8* mutant mice. Nevertheless, a sizable fraction from *Trpv1* mutant mice were still activated by noxious heat, and a small population from *Trpm8* mutant mice were still activated at 12°C (a temperature that produces cold pain) (Figure 6-68D), indicating the existence of additional sensors for noxious hot and cold temperatures. A recent study suggested that TRPA1 and TRPM3, two other members of TRP ion channel family, also function in heat sensing. Only triple-mutant mice devoid of TRPV1, TRPA1, and TRPM3 exhibited a complete disruption of heat sensing. The partially redundant functions of these three TRP channels could serve as a fault-tolerant mechanism for avoiding burns.

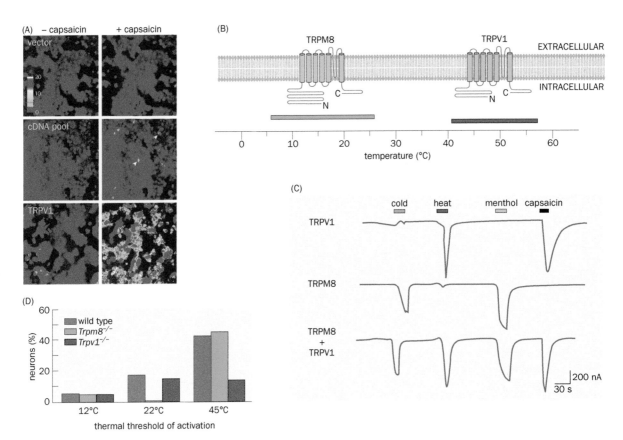

Figure 6-68 TRP channels sense heat, cold, and chemicals.
(A) Discovery of TRPV1 by expression cloning. Top row, control HEK293 cells did not respond to 3 μM capsaicin application; relative intracellular Ca²⁺ concentrations ($[Ca^{2+}]_i$) are indicated by the heat map on the left. Middle row, a small fraction of cells (arrowheads) transfected with a mixture of cDNAs (cDNA pool) derived from dorsal root ganglia responded to capsaicin with a rise in $[Ca^{2+}]_i$, as measured by the Fura-2 Ca²⁺ indicator (Section 14.22). Bottom row, after repeatedly dividing the cDNA pool into smaller and smaller subdivisions, a single cDNA from the pool that produced capsaicin-induced $[Ca^{2+}]_i$ increases and encoded TRPV1 was identified. **(B)** Schematic of TRPV1 and TRPM8 cation channels and the temperature ranges that activate them (blue for TRPM8, red for TRPV1). Other temperature ranges are covered by other TRP or Cl⁻ channels not shown. **(C)** Response of *Xenopus* oocytes injected with mRNA encoding TRPV1, TRPM8, or both to cold, heat, menthol, and capsaicin. As measured by the production of inward current in a voltage clamp setting, TRPV1 is activated by heat and capsaicin, whereas TRPM8 is activated by cold temperatures and menthol. **(D)** Activation of trigeminal ganglion neurons dissociated from wild-type, *Trpv1* mutant (*Trpv1⁻/⁻*), and *Trpm8* mutant (*Trpm8⁻/⁻*) mice was measured by Ca²⁺ imaging when exposed to different temperatures. *Trpv1* mutant mice exhibited a marked reduction in the fraction of neurons activated at 45°C, whereas neurons activated at 22°C were absent in *Trpm8* mutant mice. (A, from Caterina MJ, Schumacher MA, Tominaga M, et al. [1997] *Nature* 389:816–824. With permission from Springer Nature. B & C, adapted from McKemy DD, Neuhausser WN, & Julius D [2002] *Nature* 416:52–58. With permission from Springer Nature. D, adapted from Bautista DM Siemens J, Glazer JM et al. [2007] *Nature* 448:204–208. With permission from Springer Nature)

Additional TRP channels have been identified to sense chemicals. For instance, TRPA1 discussed above is also a receptor for pungent components of mustard oil and garlic and environmental irritants such as acrolein. TPRA1 (as well as TRPV1) can also act as an intracellular signal–activated ion channel that mediates depolarization in response to activation of other sensory receptors (analogous to the TRPC2 channel in the accessory olfactory system; Box 6-1). Expression of TRPA1 overlaps with expression of TRPV1 in DRG and trigeminal neurons. Thus, somatosensory neurons can be polymodal via at least two distinct mechanisms: first, they can express a receptor such as TRPV1 that senses stimuli in more than one modality; second, they can express multiple distinct receptors, each of which responds to sensory stimuli of one or more modalities.

In summary, studies of the TRP channels have brought together the sensations of temperature, pain, and chemicals. Hot and cool temperatures appear to be sensed mainly by two separate types of DRG and trigeminal neurons expressing TRPV1 and TRPM8, respectively. However, we still have incomplete knowledge of temperature sensation across all ranges and the mechanistic bases of the distinction between innocuous and noxious temperatures. For example, we have not yet identified sensory neurons dedicated to sensing innocuous warmth; one hypothesis is that warmth sensation is synthesized from the integrated activity of heat- and cool-sensing neurons in the central nervous system.

6.30 Sensation can result from central integration: the distinction of itch and pain as an example

Given the numerous types of distinct sensory stimuli, one way of organizing the somatosensory system is to assign a dedicated type of sensory neuron to each type of sensory stimulus; activation of a particular sensory neuron type informs the brain that a particular sensory stimulus is present. Our discussion so far has suggested that the somatosensory system employs this organization. Second-order neurons either can be highly dedicated to specific types of sensory neurons following the labeled line strategy, as in the mammalian taste system (Figure 6-33), or can integrate information from multiple types of sensory neurons, as in color perception (Sections 4.10 and 4.17). While we do not fully know the extent to which somatosensory information is transmitted to the brain via labeled lines or integration, the integrative strategy is clearly used and can indeed produce sensations not specifically represented by primary sensory neurons. We illustrate this using studies of itch.

Itch is defined as an unpleasant sensation eliciting a desire or reflex to scratch. It is usually induced by **pruritogens**, chemicals that elicit such a desire or reflex. These chemicals can be produced endogenously (such as histamine), from the external environment (such as spicules from the tropical legume *Mucuna pruiens*), or drugs (such as chloroquine, which is widely used to prevent and treat malaria). Itch has long been considered a sub-modality of pain, and, indeed, most neurons that express histamine receptors responsible for histamine-induced itch co-express TRPV1. However, the following experiments suggest that itch is distinct from pain.

Members of the Mrgpr (<u>M</u>as-<u>r</u>elated <u>G</u>-protein-coupled <u>r</u>eceptors) family of GPCRs are expressed in a subset of TRPV1-expressing DRG neurons. One member, **MrgprA3**, is a receptor for chloroquine and mediates chloroquine-induced itch. MrgprA3 is expressed in about 4% of DRG neurons, which also express histamine receptors and TRPV1. Indeed, *in vivo* recording indicated that MrgprA3-expressing neurons are highly polymodal—they are activated not only by chloroquine and histamine but also by capsaicin and noxious mechanical stimuli. However, ablation of MrgprA3 neurons selectively reduced scratching induced by chloroquine or histamine without affecting behavior elicited by noxious heat or mechanical stimuli. To investigate the underlying mechanisms, genetically engineered mice were produced in which *Trpv1* was deleted in the entire animal but added back in MrgprA3 neurons (**Figure 6-69**A). These mice were then given capsaicin through cheek injection and assayed for behavior. In this paradigm, mice elicit distinct, stereotypical responses to painful or itchy stimuli: they wipe their cheeks using their forepaws in response to pain but scratch their cheeks using their hind paws in response to itch. In wild-type mice, cheek injection of capsaicin elicited a strong wiping response but only a minimal scratching response, consistent with the notion that capsaicin activates TRPV1 in nociceptive neurons and causes a pain sensation. *Trpv1* mutant mice displayed minimal wiping and scratching responses. Interestingly, mice that expressed TRPV1 only in MrgprA3 neurons produced a robust scratching response and a minimal wiping response (Figure 6-69B), suggesting that activation of MrgprA3 neurons produces itch but not pain sensation.

Further insights into itch sensation came from studies of a group of spinal cord neurons that express the **gastrin-releasing peptide receptor (GRPR)**, a G-protein-coupled receptor activated by gastrin-releasing peptide (GRP). Pharmacological ablation of GRPR neurons selectively eliminated scratching responses to all tested pruritic agents, including histamine and chloroquine (Figure 6-69C), without affecting pain behavior. MrgprA3 neurons send signals to GRPR neurons in the spinal cord via an intermediate, GRP-expressing spinal cord interneurons. Thus, a model accounting for the experimental data discussed thus far is that pruriceptive

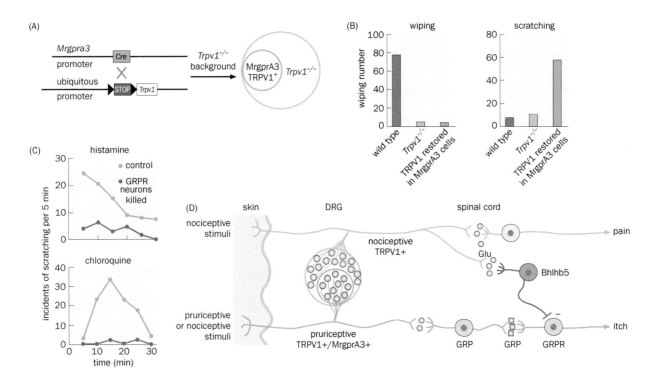

Figure 6-69 Itch-sensing cells and circuits. (A) Schematic depicting selective expression of TRPV1 in MrgprA3-expressing neurons. Left, transgenic mice expressing Cre recombinase under the control of the *Mrgpra3* promoter were crossed to mice carrying a second transgene in which TRPV1 expression requires Cre/loxP (triangles)-mediated excision of the stop sequence between a ubiquitous promoter and the *Trpv1* coding sequence (see Section 14.9 for details). Additionally, these mice lacked endogenous *Trpv1* gene. Right, the green circle represents neurons that normally express TRPV1 but lack functional TRPV1 due to the *Trpv1* mutant background. The blue circle represents neurons that normally express MrgprA3 (a subset of TRPV1-expressing neurons) expressing TRPV1 due to the genetic strategy employed. **(B)** Cheek injection of capsaicin elicited in wild-type controls a robust wiping response but minimal scratching, consistent with a pain response; *Trpv1* mutant mice displayed minimal responses; and mice in which TRPV1 was only expressed in MrgprA3 neurons displayed minimal wiping but robust scratching, suggesting that activation of MrgprA3 neurons produced the itch sensation. **(C)** Compared to controls, mice in which gastrin-releasing peptide receptor (GRPR) expressing neurons in the spinal cord were ablated did not respond to any tested pruritogens, including histamine (top) and chloroquine (bottom), suggesting that GRPR neurons are necessary for a pruriceptive response. **(D)** A model for the relationship between pain and itch sensation. Pruriceptive stimuli such as chloroquine activate MrgprA3-expressing neurons, which activate GRPR neurons through spinal cord gastrin-releasing peptide (GRP) releasing neurons, producing itch sensation. Nociceptive stimuli activate both MrgprA3-expressing neurons that co-express TRPV1 (blue) and other TRPV1-expressing neurons (green). These latter TRPV1 neurons activate spinal cord inhibitory neurons, such as those expressing Bhlhb5, to suppress the pruriceptive pathway. Glu, glutamate. (A & B, from Han L, Ma C, Liu Q, et al. [2013] *Nat Neurosci* 16:174–182. With permission from Springer Nature. C, from Sun Y, Zhao Z Meng X, et al. [2009] *Science* 325:1531–1534. D, from Dong X & Dong X [2018] *Neuron* 98:482–494.)

neurons, such as those expressing MrgprA3, are activated by both itchy and painful stimuli. Painful stimuli also activate MrgprA3-negative nociceptive neurons, which suppress the activity of GRPR neurons through inhibition within the spinal cord, thereby producing pain but not itch. Itchy stimuli activate only pruriceptive neurons, which in turn activate GRPR neurons dedicated to itch sensation (Figure 6-69D). In support of this model, selective loss of a subset of inhibitory interneurons expressing Bhlhb5, a transcription factor, exhibited highly elevated spontaneous scratching behavior, suggesting that these neurons inhibit itch. Mechanosensory neurons may also send inhibitory signals to the itch pathway, which might explain why scratching can temporarily relieve itch.

In summary, spinal circuitry causes some spinal cord neurons, such as GPRP neurons, to become more selective to itch than pruritogen-detecting primary sensory neurons themselves. The strategy employed in distinguishing between pain and itch is likely also used by other sensory modalities, such as thermosensation and touch. Indeed, as will be discussed in the next section, the dorsal spinal cord harbors a rich variety of cell types for processing somatosensory information. Integration of the activity of different types of sensory neurons can produce the many types of sensation we derive from our somatosensory system.

6.31 Touch and pain signals are transmitted to the brain in parallel

How does the central nervous system make sense of the activity in different types of somatosensory neurons? The answer comes partly from the fact that different types of DRG neurons terminate in specific locations within the spinal cord (Figure 6-63) and synapse with distinct types of postsynaptic neurons. The gray matter of the spinal cord is divided into 10 different laminae along the dorsal–ventral axis. Laminae I–V are located in the **dorsal horn** (the dorsal part of the spinal gray matter), which is devoted to processing somatosensory information. Laminae I and II are predominantly targeted by axons of nociceptive and thermosensory neurons, which use mostly unmyelinated C fibers and some lightly myelinated Aδ fibers. Some LTMRs with C and Aδ fibers also terminate in lamina II at sites ventral to the terminations of nociceptive fibers. Laminae III–V are targeted mostly by LTMRs with Aδ and Aβ fibers (**Figure 6-70A**). Proprioceptive sensory neurons send their axons further into the spinal cord, including the **ventral horn** (the ventral part of the spinal gray matter, where motor neurons reside); some proprioceptive neurons form synapses directly with motor neurons, forming stretch reflex circuits (Figure 1-19).

Within each lamina, further connection specificity must exist between input axons from different types of sensory neurons and their spinal target neurons. Recent single-cell RNA-seq studies have identified 15 types of both excitatory and inhibitory dorsal horn neurons, each with characteristic spatial distributions (Figure 6-70B). Only a small fraction of these are **dorsal horn projection neurons** (dorsal horn neurons that project axons to the brain); the vast majority are interneurons involved in local processing of information. Functional analysis of dorsal horn interneurons in sensory perception has already begun but there is a lot more to be learned. In addition to receiving sensory input, the dorsal horn is innervated by descending axons from the brain (Figure 6-70A, bottom left panel). A recent

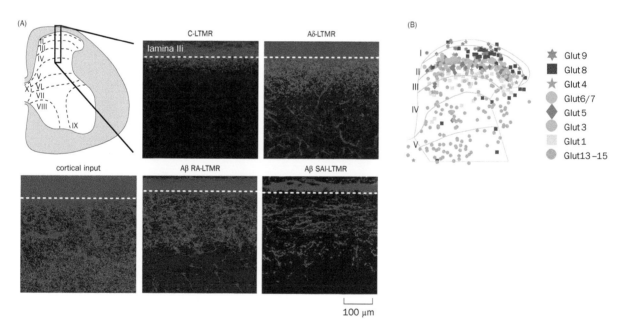

100 µm

Figure 6-70 Sensory axon termination in the spinal cord and spinal target neurons. (A) Spinal projections of LTMRs with C, Aδ, or two classes of Aβ fibers. Axons are labeled by a reporter expressed from genetically defined LTMR types utilizing Cre/*loxP*-mediated recombination (Section 14.9). Each image is a lumbar dorsal horn sagittal section at a position marked in the schematic on the top left. C-LTMR fibers terminate in lamina II ventral to a lamina IIi marker (the ventral border of the marker expression is depicted by the dashed line. Aδ-LTMR fibers extend further ventral, and the two types of Aβ-LTMR fibers innervate laminae III–V. The dorsal horn also receives extensive descending axon fibers from the cortex, as depicted in the bottom left panel. **(B)** Distribution of glutamatergic neurons in the lumbar region of the dorsal horn, based on *in situ* hybridization using marker genes derived from single-cell RNA-seq that define 15 types of glutamatergic excitatory neurons. Single-cell RNA-seq also identified 15 types of inhibitory neurons, each with a characteristic spatial distribution (not shown). (A, adapted from Abraira VE, Kuehn ED, Chirila AM, et al. [2017] *Cell* 168:295–310. With permission from Elsevier Inc. B, adapted from Häring M, Zeisel A, Hochgerner H, et al. [2018] *Nat Neurosci* 21:869–880. With permission from Springer Nature.)

study showed that descending input from the somatosensory cortex, acting on a specific type of local excitatory neuron, plays an important role in adjusting touch sensitivity.

Aβ-LTMRs and proprioceptive neurons, in addition to terminating in specific spinal cord laminae (Figure 6-63), also send ascending branches through the **dorsal column pathway** of the spinal cord to the **medulla** (the most caudal part of the brainstem); this constitutes the direct dorsal column pathway. In parallel, LTMR terminals in the spinal cord provide direct and indirect input to dorsal horn projection neurons that also send ascending axons along the dorsal column pathway to the medulla; this constitutes the indirect dorsal column pathway. Since C and Aδ fibers do not have ascending branches, the indirect pathway is the major route for ascending touch information originating from C- and Aδ-LTMRs. Axons of both direct and indirect dorsal column pathways synapse with medulla neurons that project to the contralateral thalamus. Thalamic neurons then relay these touch signals to the primary somatosensory cortex in a topographic manner to form the sensory homunculus (**Figure 6-71**, red pathway; see also Figure 1-21; Figure 1-25). Touch signals from hairy skin are relayed not only via the dorsal column pathway, which transmits signals from both glabrous and hairy skin, but also via distinct dorsal horn projection neurons with ascending axons that form the **spinocervical tract pathway** and terminate in the lateral cervical nucleus. Projection neurons from the lateral cervical nucleus then relay information to the contralateral thalamus.

Pain, itch, and temperature sensations are primarily conveyed through a distinct ascending path: lamina I dorsal horn projection neurons receive input from sensory axons, send their own axons across the ventral spinal cord, and ascend in the contralateral **anterolateral column pathway**. Some axons in the anterolateral column terminate in thalamic nuclei, through which information is relayed to the primary somatosensory cortex (Figure 6-71, blue pathway). Other axons in the anterolateral column pathway innervate brainstem target neurons. One target is the **parabrachial nucleus**, which relays pain information to the amygdala and hypothalamus to control autonomic responses to pain stimuli (see Chapter 9 for more detail). From the amygdala, pain information is also sent to the insular cortex (Figure 6-71, cyan). This pathway contributes to affective aspects of pain perception. Another major brainstem target of the anterolateral pathway is the **periaqueductal gray** in the midbrain, which initiates descending modulation of pain (Figure 6-71, purple) and will be discussed in the next section.

Figure 6-71 Central pathways for touch and pain. Schematics of major central pathways for touch (red) and pain (blue), where each neuron represents one or more types of neurons. Touch sensory neurons (Aβ-LTMRs) send branches that ascend to the brainstem within the dorsal column pathway (the direct pathway) and also synapse with dorsal horn neurons (orange), some of which also send ascending branches in the dorsal column pathway (the indirect pathway). Neurons in both the direct and indirect pathways synapse with medulla neurons that project their axons across the midline, terminating in the contralateral thalamus. (The spinocervical tract pathway, which transmits touch information to the contralateral thalamus in parallel with the dorsal column pathway, is not shown.) Nociceptive sensory neurons synapse with second-order dorsal horn spinal neurons, which send their axons across the midline of the spinal cord and ascend within the anterolateral column pathway to the contralateral thalamus. Distinct thalamic neurons relay touch and pain information to the primary somatosensory cortex. Some second-order pain pathway neurons terminate within different brainstem nuclei (blue branches). At the parabrachial nucleus, pain information is relayed by projection neurons to the amygdala and subsequently to the insular cortex for affective aspects of pain sensation (cyan pathway). At the periaqueductal gray, target neurons of the ascending anterolateral column pathway initiate descending modulation of pain through the autonomic and neuromodulatory systems (purple). (Based on Abraira VE & Ginty DD [2013] *Neuron* 79:618–639; Basbaum AI, Bautista DM, Scherrer G, et al. [2009] *Cell* 139:267–284.)

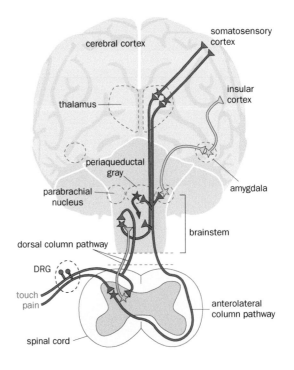

Thus, the organization of central pathways that transmit touch and pain signals to the brain is somewhat analogous to the M and P pathways in the visual system that respectively transmit motion and high-acuity/color signals from the retina to the visual cortex (Section 4.27). Just as the two-pathway system in vision is a gross simplification given the existence of many types of retinal ganglion cells, so too is this simple division of the touch and pain pathways a gross simplification of the somatosensory system. We've already seen that touch signals are transmitted by several parallel pathways. In fact, in addition to the major contribution by lamina I neurons, the anterolateral column pathway also contains the axons of projection neurons in deep laminae that might also transmit crude (nondiscriminative) touch information. Ongoing efforts to identify specific projection neuron types will enable precise manipulation of their activity patterns using modern circuit analysis tools (Sections 14.11 and 14.23–13.25), thereby facilitating the dissection of role of each ascending pathway in transmitting somatosensory information to the brain.

6.32 Pain is subjected to peripheral and central modulation

In our studies of sensory systems, we have emphasized information flow from sensory organs in the periphery to higher-order neurons in the brain. However, information can also flow from higher-order neurons back to lower-order neurons for feedback regulation of the sensory system. Sensory systems are also powerfully modulated by brain states (Figure 1-26B); for example, the sensitivity of our sensory systems drops markedly when we are drowsy or asleep, or when we are paying attention to something else. The study of pain has offered insight into such modulation, which also has significant clinical implications.

Pain modulation begins at the peripheral endings of sensory neurons. Inflammation is responsible for much of the pain following tissue damage or pathogen infection. Under inflammatory conditions, exposure of injured tissue to gentle touch or innocuous temperature can cause pain, a phenomenon called **allodynia**. In addition, injury or inflammation can trigger **hyperalgesia**, an enhanced response to noxious stimuli (Figure 6-72). Distinct pain states utilize mechanisms of peripheral and central sensitization to trigger allodynia and/or hyperalgesia, as discussed in the following.

The inflammatory response, while important for combating infections and repairing damaged tissues, also produces molecules such as **bradykinin** (a peptide) and **prostaglandin** (a lipid). Both bind GPCRs on the peripheral endings of nociceptive neurons and activate nociceptive neurons through second messenger systems that open TRP channels; in this case, TRP channels serve as effectors for GPCRs (Figure 6-73A). Proinflammatory molecules also sensitize nociceptive neurons by decreasing the threshold of TRP channel opening in response to sensory stimuli; in this case, TRP channels themselves serve as sensory receptors. In fact, active nociceptive neurons secrete neuropeptides such as **substance P** and calcitonin gene-related peptide (**CGRP**) from their peripheral endings. These neuropeptides induce local immune and epithelial cells to produce bradykinin and prostaglandin, which in turn activate or sensitize nociceptive neurons (Figure 6-73A). This phenomenon, called **neurogenic inflammation**, underscores the bidirectional signaling of nociceptive neurons: in addition to transmitting pain signals from the periphery to the CNS, their peripheral endings release neuropeptides in response to their own activity, thus also contributing to local immune responses. Some of the most widely used analgesic (pain-relieving) drugs, such as aspirin and ibuprofen, target Cox-1 and Cox-2, cyclooxygenase enzymes that synthesize prostaglandin, thereby reducing inflammation-induced activation and sensitization of nociceptive neurons at the periphery.

Our understanding of pain modulation in the CNS comes from studies of the mechanisms of another class of powerful analgesic drugs, the **opioids**. The analgesic effects of opiates, natural opioids harvested from the opium poppy, *Papaver somniferum,* have been known for thousands of years. Although the active ingredient, **morphine**, was isolated in 1804, the mechanisms of opiate action remained unknown until the 1970s, when radioactively labeled opiates were found to bind

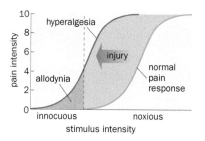

Figure 6-72 Modulation of pain sensitivity. Injury and inflammation shift the normal pain sensitivity curve (blue) leftward (red). This causes innocuous stimuli to induce pain (allodynia) and intensifies pain for noxious stimuli (hyperalgesia).

Figure 6-73 Peripheral and central modulation of pain. (A) Inflammation induces release of molecules such as bradykinin and prostaglandin, which activate G-protein-coupled receptors (blue) at the peripheral endings of nociceptive neurons. This triggers second messenger systems that activate TRPV1 and TRPA1 channels (green) or sensitize these channels to reduce their activation threshold, causing hypersensitivity to pain. Second messenger systems also regulate other local effectors and gene expression, resulting in long-term changes in pain sensitivity. Sensory nerve endings also release neuropeptides such as substance P and calcitonin gene-related peptide (CGRP), which act on nearby cells to trigger additional release of bradykinin and prostaglandin. **(B)** In the dorsal horn of the spinal cord, synaptic transmission between nociceptive sensory neurons and dorsal horn recipient neurons is regulated by local neurons that release endogenous opioid peptides, which act through opioid receptors (orange dots) to dampen both release of glutamate from sensory neuron terminals and depolarization of postsynaptic dorsal horn neurons in response to glutamate release. Opioid-releasing neurons in turn are modulated by descending axons from brainstem GABA, serotonin, and norepinephrine neurons. These neurons also innervate dorsal horn recipient neurons and presynaptic terminals of sensory neurons. +/− represent the idea that modulation can enhance or suppress excitability, depending on the types of descending and target neurons and on the pain modality. (See Corder G, Castro DC, Bruchas MR, et al. [2018] *Annu Rev Neurosci* 41:453–743 for opioid signaling in pain.)

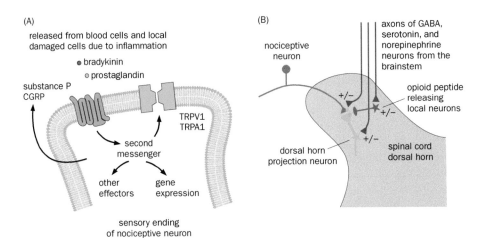

to specific receptors in the nervous system. We now know that **opioid receptors** constitute a subfamily of four GPCRs that are distributed widely across the nervous system and are highly concentrated in laminae I and II of the dorsal horn, both in presynaptic terminals of sensory neurons and in postsynaptic dorsal horn neurons that relay pain signals to the brain. Activation of opioid receptors reduces neurotransmitter release from sensory neurons, decreases the responsiveness of dorsal horn neurons, or does both (Figure 6-73B), resulting in opioids' analgesic effects. The discovery of opioid receptors also triggered the search for their endogenous ligands. Several endogenous opioid peptides, including enkephalin, endorphin, and dynorphin, have been identified that modulate pain pathways much like exogenous opioids do (Figure 6-73B).

Application of opioids in the brain also relieves pain—indeed, the regions most powerfully modulated by opioids are specific brainstem nuclei with high levels of opioid receptor expression. It is thought that activation of opioid receptors modulates the activity of descending GABA, serotonin, and norepinephrine projection neurons from the brainstem to the spinal cord. These descending projection neurons act on presynaptic terminals of pain sensory neurons, postsynaptic dorsal horn neurons, and opioid receptor-expressing dorsal horn neurons (Figure 6-73B). The sign of the modulation (whether it relieves or sensitizes pain) depends on the types of descending and recipient neurons, and the pain modality (mechanical versus thermal). Some of these descending feedback pathways may contribute to **placebo effects**, wherein the belief of taking pain medication without actual treatment effectively reduces the pain experienced by some patients. The detailed cellular and circuit mechanisms of pain modulation are still very much unknown. Further investigations should have large payoffs, both for understanding how sensory systems are modulated in general and for developing more effective and specific analgesic treatments.

6.33 Linking neuronal activity with touch perception: from sensory fiber to cortex

Sensory systems transform external stimuli into internal representations in the brain, giving rise to perception. We have assumed that the activity of sensory neurons and neurons in the ascending pathways contribute to perception, but how do we actually know this? Experiments in the somatosensory system have contributed important insights toward answering this question.

A powerful approach for establishing a link between the activity of specific neurons in the sensory system and perception is to compare the *detection threshold of neurons* and the *perceptual threshold of the organism* to the same specific stimuli. As discussed in Chapter 4, the perceptual threshold of visual stimuli can be determined by testing the psychometric functions of human subjects (Figure 4-3) or trained monkeys (Figure 4-56). In the somatosensory system, Vernon Mountcastle and co-workers applied sinusoidal mechanical stimuli (flutter) to

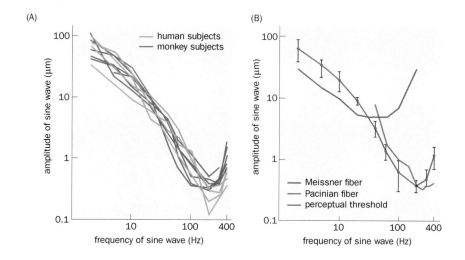

(A) human subjects / monkey subjects

(B) Meissner fiber / Pacinian fiber / perceptual threshold

Figure 6-74 Comparing perceptual thresholds with the thresholds of sensory neurons presented with the same stimuli. (A) Frequency–amplitude curves of perceptual threshold based on psychophysical experiments on six monkey (red) and five human (blue) subjects. Each point on the curve represents the lowest amplitude of sinusoidal mechanical stimuli of a given frequency applied to the fingertip that was perceived by human and monkey subjects. The human subjects reported if they felt the stimulus, while the monkeys had been trained to indicate the presence of a stimulus applied to the fingertip of a restrained hand by using the other hand to press a button. The remarkable similarities indicate that similar physiological mechanisms underlie flutter perception in both monkeys and humans. **(B)** Superimposed on the perceptual threshold curve (red, average from data in Panel A, error bars indicate standard error of the mean) are thresholds of two types of sensory neurons recorded from individual Aβ fibers innervating the fingers in response to the same sinusoidal mechanical stimuli as in the psychophysical experiments in Panel A. The blue and green curves represent the lowest bounds, or the most sensitive of each type of sensory neuron, at a given frequency. Comparing the curves for both neuron types and the perceptual threshold curve across a range of frequencies, we can see that at each frequency the perceptual threshold matches well the lower of the two neuronal thresholds, suggesting that the limits of psychophysical performance are set by the most sensitive individual neurons. (Adapted from Mountcastle VB, LaMotte RH, & Carli G [1972] *J Neurophysiol* 35: 122–136.)

the fingertips of human and monkey subjects and varied stimulation frequency to determine the threshold amplitude at which these stimuli could be detected. They observed that the thresholds at which specific frequencies of flutter produced perception—plotted as frequency–amplitude curves—were very similar for human and monkey subjects (**Figure 6-74**A). In parallel, electrophysiological recordings of single sensory axons (Figure 1-17) were carried out to determine the detection thresholds of individual sensory neurons in response to the same set of sensory stimuli applied to the fingertips of anesthetized monkeys. Two types of sensory neurons were activated, depending on the frequency of mechanical stimuli; low- and high-frequency stimuli activated sensory neurons terminating in Meissner and Pacinian corpuscles, respectively (Figure 6-64). Remarkably, the lower bound of the detection threshold of each type of sensory neuron matched very well the perceptual thresholds of the monkeys in the respective frequency ranges (Figure 6-74B). These experiments suggested that these two types of sensory neurons are responsible for the flutter perception across different frequency ranges and provided an excellent example of the **lower envelope principle**: the limits of psychophysical performance are set by the most sensitive individual neurons. Similar observations were made in subsequent **microneurography** experiments in which individual mechanosensory axons were recorded from awake human subjects while they attempted to detect brief indentation of the skin at the most sensitive point of the fiber's receptive field. The threshold of neuronal response matched the perceptual threshold again, suggesting that activity of sensory fibers carries the information used for perception of skin indentation.

Similar approaches comparing the perceptual threshold with the threshold of neuronal firing induced by the same stimulus have also been applied in the somatosensory cortex. As introduced in Chapter 1, the somatosensory cortex has a gross topographic map of the body surface (Figure 1-25). Within a particular area of the somatosensory cortex, neurons representing the same kinds of information—for instance, rapidly adapting neurons that detect vibratory stimuli at a fingertip—are organized in vertical columns perpendicular to the cortical surface, a discovery made by Mountcastle in the 1950s that inspired similar findings in the primary visual cortex (Section 4.24). To explore the relationship between neuronal activity and perception, thirsty monkeys were trained to perform a behavioral task in which they must indicate correctly the presence or absence of a stimulus in order to receive a drop of liquid as a reward (**Figure 6-75**A, B). Simultaneously, single-unit recordings were performed in specific areas of the somatosensory, premotor, and motor cortices (Figure 6-75C). It was found that rapidly adapting neurons in the primary somatosensory cortical areas (corresponding to area 1/3b of Brodmann's map; Figure 14-22) representing the fingertip indeed had a detection threshold matching the perceptual threshold, consistent with the notion that these neurons carry information used for perception, as do sensory fibers in the periphery.

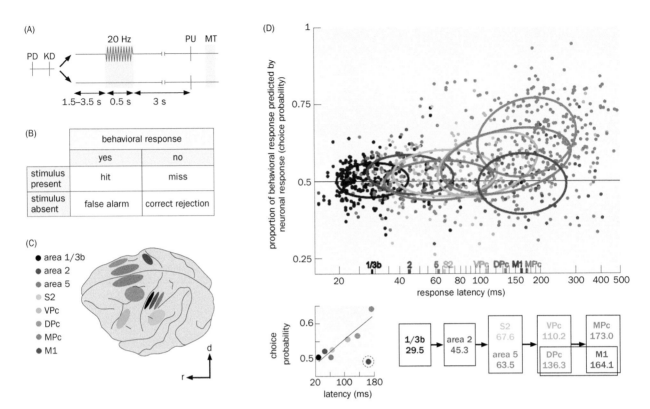

Figure 6-75 Neuronal activity and touch perception in the cortex.
(A) Behavioral task. A trial begins when a stimulator probe indents the skin of one fingertip of the restrained hand (probe down, PD). The monkey then places its other hand on a key (key down, KD). After a variable interval, on a randomly selected half of the trials, a 20 Hz vibratory stimulus near the detection threshold is applied. After a fixed delay, the probe is moved up (PU), signaling to the monkey that it can move its free hand to target (MT), which is one of the two push buttons indicating whether it felt the stimulus (yes) or not (no). **(B)** The trial outcome is classified into one of the four categories, depending on whether the stimulus is present or not and whether the behavioral response is yes or no. During training, the monkey receives rewards for trials resulting in a hit or correct rejection, but not for trials resulting in a miss or false alarm. **(C)** While the monkey was performing the behavioral task, single-unit recordings were performed in different areas of the primary somatosensory cortex (areas 1/3b, 2, 5), secondary somatosensory cortex (S2), ventral, dorsal, and medial premotor cortex (VPc, DPc, MPc), and primary motor cortex (M1). These areas are indicated by different colors in a schematic monkey

brain (r, rostral; d, dorsal). **(D)** Top, summary of response latency (x axis, with 0 representing stimulus onset) and perceptual signals (y axis) across cortical areas. Each dot represents a single unit recorded in a particular cortical area color matched to the schematic in (C). The y axis value represents a choice-probability index, an indicator of the accuracy with which neuronal responses predict behavioral response, with 0.5 being at chance level and 1 being perfect prediction. Neurons in each area were fitted with two-dimensional normal distributions (ovals represent the standard deviations). Color markings along the x axis represent the mean response latencies of neurons in each cortical area. Bottom left, response latency positively correlates with proportion of predicted behavioral response in all cortical areas except M1 (dotted circle). Bottom right, proposed information flow across cortical areas based on mean latencies (numbers underneath cortical areas, in milliseconds). Each rectangle groups the areas with latencies statistically indistinguishable from each other. (Adapted from de Lafuente V & Romo R [2006] *Proc Natl Acad Sci U S A* 39:14266–14271. Copyright National Academy of Sciences, U.S.A.).

By definition, neuronal responses to near-threshold stimuli are variable (on average, a neuron responds 50% of the time to a stimulus at its threshold). Likewise, perceptual responses are also variable (on average, the monkey detects a stimulus at its perceptual threshold 50% of the time). Thus, co-variability between different trials provides valuable information about the relationship between neuronal activity and perception: how well can a neuron's response predict the monkey's perceptual response? It was found that activity of primary somatosensory cortex neurons could predict behavioral outcome at only chance level (Figure 6-75D), suggesting that, while contributing essential information to perception, the primary somatosensory cortex is not the site where perceptual decisions are made. By contrast, activity of medial premotor cortex (MPc) neurons predicted behavioral response well above chance. Indeed, there is a gradual increase in the ability of neuronal activity along a pathway from somatosensory cortex in the parietal lobe to premotor cortex in the frontal lobe to predict behavioral response, correlated with increasing latency of firing to stimulus onset (Figure 6-75D). Additional evidence suggested that activity in premotor cortical neurons represented a perceptual decision, and not just motor preparation; indeed, neuronal activity in the primary motor cortex (M1), which functions in motor preparation and execution, did not predict behavioral outcome.

In summary, these experiments suggest that perception develops over time and across cortical areas, from representations of sensory stimuli in the primary somatosensory cortex to perceptual decision in the premotor cortex. A fascinating future direction is to investigate the mechanisms by which such representations are achieved through the activity and connectivity of cortical neurons. These investigations will shed light on a central question in neurobiology: how do sensory stimuli produce perceptions, which in turn influence actions? We will return to this subject in Chapter 8 when we study sensorimotor integration.

SUMMARY

All sensory systems share common goals. They transform environmental (and bodily) stimuli into electrical signals and transmit these signals to the brain to form internal representations of the sensory stimuli. However, the physical characteristics of sensory stimuli vary widely, from chemical, mechanical, thermal, to light. This necessitates the evolution of multiple sensory systems with distinct properties best suited for detecting and extracting biologically significant features from these different kinds of stimuli.

The olfactory system detects volatile chemicals to provide animals with information about food, mates, and danger from a distance. The mammalian olfactory system employs many hundreds of G-protein-coupled receptors (GPCRs) for odorant detection. Odorant binding to these receptors activates a cyclic nucleotide-gated cation channel, leading to depolarization of olfactory receptor neurons (ORNs) in the nose and propagation of action potentials by ORN axons to the olfactory bulb. A combinatorial coding strategy enables the olfactory systems to distinguish between many more odorants than the number of odorant receptors they employ.

From insects to mammals, most ORNs express a single odorant receptor. ORNs expressing the same receptor project their axons to the same glomeruli and synapse with the same type of second-order projection neurons, creating parallel and discrete information processing channels. Through convergent projections, synaptic properties, and lateral inhibition by local interneurons, ORN input is transformed for more reliable and efficient representation by projection neurons. In parallel to the use of combinatorial coding, odors with innate behavioral significance, such as danger signals or mating pheromones, utilize dedicated information processing channels in flies and a specialized accessory olfactory system in most mammals. Studies in both mice and flies suggest that input organization is more stereotyped in higher olfactory centers that regulate innate behavior and more variable in centers that regulate learned behavior. Due to its limited number of neurons, the *C. elegans* olfactory system uses a different coding strategy: many odorant receptors are expressed in the same sensory neurons, which encode the hedonic values of odorants to guide behavior.

Taste operates at a closer range than olfaction. Each of the five basic tastants in mammals—sweet, umami, bitter, sour, and salty—is mostly detected by a distinct kind of taste receptor cell. Sweet and umami tastants, which inform animals about nutrient levels, are each detected by a heterodimeric pair of GPCRs of the T1R family with low-binding affinities. Several dozen GPCRs of the T2R family detect bitter compounds with high binding affinities, enabling animals to avoid trace amounts of toxins. Separable areas in the gustatory cortex are differentially activated by sweet and bitter tastants. Taste, olfaction, and the trigeminal somatosensory system innervating the oral cavity collaborate to generate the perceived flavors of food.

Two parallel sensory systems in the inner ear convert mechanical stimuli into electrical signals. The vestibular system utilizes otolith organs and semicircular canals to sense head orientation and movement; these sensory signals are used, often along with signals from other sensory systems, to regulate balance and spatial orientation, coordinate head and eye movement, and perceive self-motion. The auditory system utilizes the cochlea to detect sounds for communication with conspecifics and revelation of predators and prey.

Temporal precision is a key feature of the auditory system. Sounds are rapidly converted into electrical signals by direct opening and closing of mechanosensitive channels at the tips of stereocilia in hair cells. Cyclic hair cell depolarization causes postsynaptic spiral ganglion neurons to fire in sync with specific phases of sound cycles. This temporal precision is crucial for sound localization, which relies on sub-millisecond interaural time differences extracted by coincidence-detecting neurons in the brainstem. Using interaural time and level differences, animals construct a spatial map of sound locations in the inferior colliculus. Sounds of different frequencies activate hair cells at different locations in the cochlea. Through ordered projections, this tonotopic map is maintained through the ascending auditory pathway. Besides representing sound frequency, level, and location, the auditory cortex analyzes biologically important sounds, as exemplified by the study of echolocating bats.

The somatosensory system employs many types of sensory neurons to detect diverse mechanical, thermal, and chemical stimuli, giving rise to proprioception, touch sensation, thermosensation, nociception, and pruriception. Innocuous touch is sensed by distinct types of low-threshold mechanoreceptors with different specialized terminal endings, adaptation properties, and degrees of myelination. Mechanosensitive channels identified to date include Piezos in mammals and invertebrates, as well as ENaC- and TRP-family channels in invertebrates. Temperature, chemicals, and noxious stimuli are sensed mainly by fibers with free sensory endings. TRPV1 and TRPM8 are the major sensors of hot and cool temperatures, respectively. TRPV1 is also expressed in many nociceptive and pruriceptive multimodal neurons.

Somatosensory signals are transmitted to the CNS via ordered projections of specific types of somatosensory neurons to specific laminae of the spinal cord. Signal integration in the spinal cord contributes to sensory processing, as exemplified by the distinction between pain and itch. Touch and temperature/pain signals are transmitted mostly through parallel pathways to the brainstem, thalamus, and somatosensory cortex. Pain perception is modulated at peripheral sensory endings by molecules released by both nearby nonneural cells and sensory neurons themselves during the inflammatory response, and at central synapses by endogenous opioids and descending feedback pathways. Parallel physiological and psychophysical studies of touch revealed that perceptual thresholds are set by the most sensitive peripheral sensory neurons, and perceptual decision making develops over time and across cortical areas, from the somatosensory cortex to the premotor cortex.

OPEN QUESTIONS

- How does olfactory imprinting enable salmon to find their native streams?

- How are the glomerular maps in insects and mammals utilized in higher olfactory centers for odor discrimination and odor-guided behavior?

- Are there taste modalities in addition to sweet, umami, bitter, sour, and salty?

- How do taste, olfaction, and trigeminal somatosensation cooperate to produce flavor?

- What are the primary functions of the auditory thalamus and cortex in mammals?

- What is special about the human auditory cortex that allows us to comprehend language and enjoy music?

- What mechanosensitive channels besides Piezo underlie mechanotransduction in the mammalian somatosensory system?

- How do spinal dorsal horn neurons and circuits integrate somatosensory, modulatory, and descending inputs?

FURTHER READING

Books and reviews

Abraira VE & Ginty DD (2013). The sensory neurons of touch. *Neuron* 79:618–639.

Axel R (1995). The molecular logic of smell. *Sci Am* 273:154–159.

Bargmann CI (2006). Comparative chemosensation from receptors to ecology. *Nature* 444:295–301.

Basbaum AI, Bautista DM, Scherrer G, & Julius D (2009). Cellular and molecular mechanisms of pain. *Cell* 139:267–284.

Konishi M (2003). Coding of auditory space. *Annu Rev Neurosci* 26:31–55.

Parker AJ & Newsome WT (1998). Sense and the single neuron: probing the physiology of perception. *Annu Rev Neurosci* 21:227–277.

Shepherd GM (2004). The Synaptic Organization of the Brain, 5th ed. Oxford University Press.

Wilson RI (2013). Early olfactory processing in *Drosophila*: mechanisms and principles. *Annu Rev Neurosci* 36:217–241.

Yarmolinsky DA, Zuker CS, & Ryba NJ (2009). Common sense about taste: from mammals to insects. *Cell* 139:234–244.

Vertebrate olfaction

Bolding KA & Franks KM (2018). Recurrent cortical circuits implement concentration-invariant odor coding. *Science* 361:eaat6904.

Brunet LJ, Gold GH, & Ngai J (1996). General anosmia caused by a targeted disruption of the mouse olfactory cyclic nucleotide-gated cation channel. *Neuron* 17:681–693.

Buck L & Axel R (1991). A novel multigene family may encode odorant receptors: a molecular basis for odor recognition. *Cell* 65:175–187.

Isogai Y, Wu Z, Love MI, Ahn MH, Bambah-Mukku D, Hua V, Farrell K, & Dulac C (2018). Multisensory logic of infant-directed aggression by males. *Cell* 175: 1827–1841 e1817.

Lyons DB, Allen WE, Goh T, Tsai L, Barnea G ,& Lomvardas S (2013). An epigenetic trap stabilizes singular olfactory receptor expression. *Cell* 154:325–336.

Malnic B, Hirono J, Sato T, & Buck LB (1999). Combinatorial receptor codes for odors. *Cell* 96:713–723.

Miyamichi K, Amat F, Moussavi F, Wang C, Wickersham I, Wall NR, Taniguchi H, Tasic B, Huang ZJ, He Z, et al. (2011). Cortical representations of olfactory input by trans-synaptic tracing. *Nature* 472:191–196.

Mombaerts P, Wang F, Dulac C, Chao SK, Nemes A, Mendelsohn M, Edmondson J, & Axel R (1996). Visualizing an olfactory sensory map. *Cell* 87:675–686.

Papes F, Logan DW, & Stowers L (2010). The vomeronasal organ mediates interspecies defensive behaviors through detection of protein pheromone homologs. *Cell* 141:692–703.

Rubin BD & Katz LC (1999). Optical imaging of odorant representations in the mammalian olfactory bulb. *Neuron* 23:499–511.

Scholz AT, Horrall RM, Cooper JC, & Hasler AD (1976). Imprinting to chemical cues: the basis for home stream selection in salmon. *Science* 192:1247–1249.

Wilson CD, Serrano GO, Koulakov AA, & Rinberg D (2017). A primacy code for odor identity. *Nat Commun* 8:1477.

Invertebrate olfaction

Bhandawat V, Olsen SR, Gouwens NW, Schlief ML, & Wilson RI (2007). Sensory processing in the *Drosophila* antennal lobe increases reliability and separability of ensemble odor representations. *Nat Neurosci* 10:1474–1482.

Caron SJ, Ruta V, Abbott LF, & Axel R (2013). Random convergence of olfactory inputs in the *Drosophila* mushroom body. *Nature* 497:113–117.

Chalasani SH, Chronis N, Tsunozaki M, Gray JM, Ramot D, Goodman MB, & Bargmann CI (2007). Dissecting a circuit for olfactory behaviour in *Caenorhabditis elegans*. *Nature* 450:63–70.

Frechter S, Bates AS, Tootoonian S, Dolan MJ, Manton J, Jamasb AR, Kohl J, Bock D, & Jefferis G (2019). Functional and anatomical specificity in a higher olfactory centre. *Elife* 8:e43079.

Hallem EA & Carlson JR (2006). Coding of odors by a receptor repertoire. *Cell* 125:143–160.

Jefferis GS, Potter CJ, Chan AM, Marin EC, Rohlfing T, Maurer CR Jr., & Luo L (2007). Comprehensive maps of *Drosophila* higher olfactory centers: spatially segregated fruit and pheromone representation. *Cell* 128:1187–1203.

Kurtovic A, Widmer A, & Dickson BJ (2007). A single class of olfactory neurons mediates behavioural responses to a *Drosophila* sex pheromone. *Nature* 446:542–546.

Suh GS, Ben-Tabou de Leon S, Tanimoto H, Fiala A, Benzer S, & Anderson DJ (2007). Light activation of an innate olfactory avoidance response in *Drosophila*. *Curr Biol* 17:905–908.

Troemel ER, Kimmel BE, & Bargmann CI (1997). Reprogramming chemotaxis responses: sensory neurons define olfactory preferences in *C. elegans*. *Cell* 91:161–169.

Taste

Blakeslee AF (1932). Genetics of sensory thresholds: taste for phenyl thio carbamide. *Proc Natl Acad Sci U S A* 18:120–130.

Chandrashekar J, Mueller KL, Hoon MA, Adler E, Feng L, Guo W, Zuker CS, & Ryba NJ (2000). T2Rs function as bitter taste receptors. *Cell* 100:703–711.

Mueller KL, Hoon MA, Erlenbach I, Chandrashekar J, Zuker CS, & Ryba NJ (2005). The receptors and coding logic for bitter taste. *Nature* 434:225–229.

Nelson G, Hoon MA, Chandrashekar J, Zhang Y, Ryba NJ, & Zuker CS (2001). Mammalian sweet taste receptors. *Cell* 106:381–390.

Peng Y, Gillis-Smith S, Jin H, Trankner D, Ryba NJ, & Zuker CS (2015). Sweet and bitter taste in the brain of awake behaving animals. *Nature* 527:512–515.

Tu YH, Cooper AJ, Teng B, Chang RB, Artiga DJ, Turner HN, Mulhall EM, Ye W, Smith AD, & Liman ER (2018). An evolutionarily conserved gene family encodes proton-selective ion channels. *Science* 359:1047–1050.

Zhang J, Jin H, Zhang W, Ding C, O'Keeffe S, Ye M, & Zuker CS (2019). Sour sensing from the tongue to the brain. *Cell* 179:392–402.

Audition

Carr CE & Konishi M (1990). A circuit for detection of interaural time differences in the brain stem of the barn owl. *J Neurosci* 10:3227–3246.

Chan DK & Hudspeth AJ (2005). Ca^{2+} current-driven nonlinear amplification by the mammalian cochlea *in vitro*. *Nat Neurosci* 8:149–155.

Corey DP & Hudspeth AJ (1979). Response latency of vertebrate hair cells. *Biophys J* 26:499–506.

Jeffress LA (1948). A place theory of sound localization. *J Comp Physiol Psychol* 41:35–39.

Kazmierczak P, Sakaguchi H, Tokita J, Wilson-Kubalek EM, Milligan RA, Müller U, & Kachar B (2007). Cadherin 23 and protocadherin 15 interact to form tip-link filaments in sensory hair cells. *Nature* 449:87–91.

O'Neill WE & Suga N (1979). Target range-sensitive neurons in the auditory cortex of the mustache bat. *Science* 203:69–73.

Pan B, Geleoc GS, Asai Y, Horwitz GC, Kurima K, Ishikawa K, Kawashima Y, Griffith AJ, & Holt JR (2013). TMC1 and TMC2 are components of the mechanotransduction channel in hair cells of the mammalian inner ear. *Neuron* 79:504–515.

Pena JL & Konishi M (2001). Auditory spatial receptive fields created by multiplication. *Science* 292:249–252.

Pickles JO, Comis SD, & Osborne MP (1984). Cross-links between stereocilia in the guinea pig organ of Corti, and their possible relation to sensory transduction. *Hear Res* 15:103–112.

Tang C, Hamilton LS, & Chang EF (2017). Intonational speech prosody encoding in the human auditory cortex. *Science* 357:797–801.

Zheng J, Shen W, He DZ, Long KB, Madison LD, & Dallos P (2000). Prestin is the motor protein of cochlear outer hair cells. *Nature* 405:149–155.

Somatosensation

Bautista DM, Siemens J, Glazer JM, Tsuruda PR, Basbaum AI, Stucky CL, Jordt SE, & Julius D (2007). The menthol receptor TRPM8 is the principal detector of environmental cold. *Nature* 448:204–208.

Caterina MJ, Schumacher MA, Tominaga M, Rosen TA, Levine JD, & Julius D (1997). The capsaicin receptor: a heat-activated ion channel in the pain pathway. *Nature* 389:816–824.

Coste B, Mathur J, Schmidt M, Earley TJ, Ranade S, Petrus MJ, Dubin AE, & Patapoutian A (2010). Piezo1 and Piezo2 are essential components of distinct mechanically activated cation channels. *Science* 330:55–60.

de Lafuente V & Romo R (2006). Neural correlate of subjective sensory experience gradually builds up across cortical areas. *Proc Natl Acad Sci U S A* 103: 14266–14271.

Han L, Ma C, Liu Q, Weng HJ, Cui Y, Tang Z, Kim Y, Nie H, Qu L, Patel KN, et al. (2013). A subpopulation of nociceptors specifically linked to itch. *Nat Neurosci* 16:174–182.

Häring M, Zeisel A, Hochgerner H, Rinwa P, Jakobsson JET, Lonnerberg P, La Manno G, Sharma N, Borgius L, Kiehn O, et al. (2018). Neuronal atlas of the dorsal horn defines its architecture and links sensory input to transcriptional cell types. *Nat Neurosci* 21:869–880.

Hoffman BU, Baba Y, Griffith TN, Mosharov EV, Woo SH, Roybal DD, Karsenty G, Patapoutian A, Sulzer D, & Lumpkin EA (2018). Merkel cells activate sensory neural pathways through adrenergic synapses. *Neuron* 100:1401–1413.

McKemy DD, Neuhausser WM, & Julius D (2002). Identification of a cold receptor reveals a general role for TRP channels in thermosensation. *Nature* 416:52–58.

Mountcastle VB, LaMotte RH, & Carli G (1972). Detection thresholds for stimuli in humans and monkeys: comparison with threshold events in mechanoreceptive afferent nerve fibers innervating the monkey hand. *J Neurophysiol* 35:122–136.

O'Hagan R, Chalfie M & Goodman MB (2005). The MEC-4 DEG/ENaC channel of *Caenorhabditis elegans* touch receptor neurons transduces mechanical signals. *Nat Neurosci* 8:43–50.

Ross SE, Mardinly AR, McCord AE, Zurawski J, Cohen S, Jung C, Hu L, Mok SI, Shah A, Savner EM, et al. (2010). Loss of inhibitory interneurons in the dorsal spinal cord and elevated itch in Bhlhb5 mutant mice. *Neuron* 65:886–898.

Sun YG, Zhao ZQ, Meng XL, Yin J, Liu XY, & Chen ZF (2009). Cellular basis of itch sensation. *Science* 325:1531–1534.

Vandewauw I, De Clercq K, Mulier M, Held K, Pinto S, Van Ranst N, Segal A, Voet T, Vennekens R, Zimmermann K, et al. (2018). A TRP channel trio mediates acute noxious heat sensing. *Nature* 555:662–666.

Yan Z, Zhang W, He Y, Gorczyca D, Xiang Y, Cheng LE, Meltzer S, Jan LY, & Jan YN (2013). *Drosophila* NOMPC is a mechanotransduction channel subunit for gentle-touch sensation. *Nature* 493:221–225.

CHAPTER 7

Constructing the Nervous System

千里之行，始于足下.

A journey of a thousand miles begins with a single step.

Laozi (~500 B.C.E.)

In this chapter, we return to the subject of nervous system wiring, a problem of central importance in neurobiology. We will apply what we learned in Chapter 5's discussion of visual system wiring to the rest of the nervous system and expand on the principles governing neural developmental processes. We will examine the daunting task of establishing trillions of synaptic connections among billions of neurons by addressing two deceptively simple questions: (1) How do individual neurons differentiate and connect with their partners? and (2) How do groups of neurons *coordinate* their wiring to form a functional circuit?

From the perspective of an individual neuron, neural development can be viewed as a sequential differentiation process (**Figure 7-1**). We first study the generic and conserved features of this differentiation process. Given the evolutionary conservation of many neurodevelopmental processes (see Box 5-1 for an example: axon guidance), we will study neuronal differentiation using particularly well-understood examples from different parts of the nervous systems of diverse animals. Next, we use the wiring of the olfactory system to explore how neural circuits are assembled, comparing and contrasting with the wiring of the visual system discussed in Chapter 5. We end the chapter by summarizing the strategies encountered in this chapter and Chapter 5, in the context of how animals use a limited number of genes to establish a much larger number of connections.

HOW DOES WIRING SPECIFICITY ARISE IN THE DEVELOPING NERVOUS SYSTEM?

7.1 The nervous system is highly patterned as a consequence of early developmental events

Development (**Figure 7-2A**) begins with **fertilization**, the fusion of sperm and egg to create a genetically new organism. The resulting fertilized egg, or **zygote**,

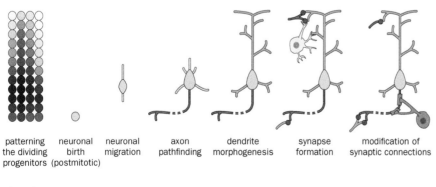

| patterning the dividing progenitors | neuronal birth (postmitotic) | neuronal migration | axon pathfinding | dendrite morphogenesis | synapse formation | modification of synaptic connections |

embryonic

early postnatal

throughout life

Figure 7-1 Schematic summary of neural development. Early developmental events pattern the nervous system, such that neural progenitor cells at different positions along the body axes produce different types of neurons. Neurons are born when they exit their final cell cycle (never to divide again). A postmitotic neuron undergoes a series of differentiation processes. These include migrating to its final destination, extending an axon toward its postsynaptic partners, establishing dendrite branching patterns, forming synaptic connections with its pre- and postsynaptic partners, and modifying those connections. Glia derive from the same neural progenitors as neurons and play important roles in regulating many of these steps. For example, astrocytes (light green) facilitate synapse formation while oligodendrocytes (brown) myelinate axons. Although the general sequence of events follows the left-to-right progression shown in the figure, some steps can overlap considerably. The bottom timeline shows that for most mammalian neurons, the first six steps are accomplished during embryonic and early postnatal development, and the last step stretches from late embryogenesis to the end of life. As an exception to what is depicted here, a small fraction of neurons in specific brain regions are born in the adult and subsequently undergo similar differentiation processes.

Figure 7-2 The early steps of development.
(A) Schematic of early development of the frog embryo illustrating key developmental stages. The egg and blastula (product of cleavage) are shown as external side views. The gastrula (product of gastrulation) is shown as a sagittal section to reveal the internal tissues that constitute the three germ layers: ectoderm, mesoderm, and endoderm. Also labeled is the future notochord, a mesoderm-derived structure that plays an important role in patterning the nervous system. The neurula (product of neurulation) is shown from a dorsal view. The red dotted line corresponds to the cross-sectional views shown in Panel B.
(B) Cell movement during neurulation. Cells of the neuroectoderm (cyan) overlying the notochord invaginate to form the neural fold and neural tube. Surrounding epidermal cells (red) migrate to the center, covering the neural tube. Arrows indicate the directions of cell movement. Neural crest cells (green), which are located at the dorsal neural tube during the final stage of this sequence, migrate out at a later stage (not shown) to produce neurons of the peripheral nervous system, among other cell types. Somites are mesoderm derived and give rise to skeletal muscles, cartilages, tendons, and connective tissues. (A, adapted from Wolpert L, Jessell TM, Lawrence P, et al. [2006] *Principles of Development*, 3rd ed. Oxford University Press. B, adapted from Schroeder TE [1970] *J Embryol Exp Morphol* 23:427–462.)

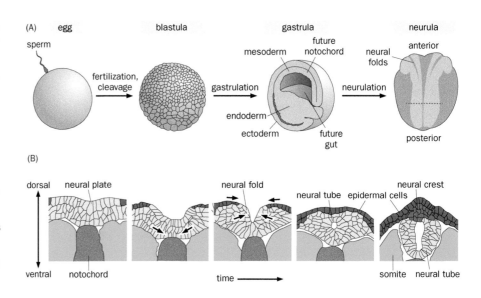

undergoes a process called **cleavage**, in which rapid cell division (typically without cell growth) generates a **blastula**, a hollow ball containing thousands of cells. Cleavage is followed by **gastrulation**, during which extensive cell rearrangement converts a blastula into a **gastrula** with three distinct germ layers: outer (**ectoderm**), middle (**mesoderm**), and inner (**endoderm**). In addition to creating three separate germ layers, gastrulation also establishes and consolidates the anterior-posterior and dorsal–ventral axes in vertebrates. Development follows the above steps in all vertebrates and most invertebrates, and often uses conserved molecular mechanisms (see Chapter 13 for a detailed discussion), although with large variations in details and timing. For example, a fruit fly embryo progresses from fertilization to completion of gastrulation in less than three hours while the same developmental events in a human embryo require nearly three weeks.

The construction of the nervous system in vertebrates begins during the next stage of development, **neurulation** (Figure 7-2A, B). On the dorsal side of the embryo, ectodermal cells overlying a mesoderm structure called the **notochord** thicken, forming the **neural plate**. The center of the neural plate then descends toward the interior of the embryo while the two edges move toward each other; this invagination creates the neural fold. The edges of the neural fold eventually fuse at the center, creating the **neural tube**, a hollow tube surrounded by layers of neuroectodermal cells. The neural tube is then covered by additional layers of ectoderm-derived epidermal cells (Figure 7-2B). The lumen of the hollow tube develops into **ventricles** housing cerebrospinal fluid. The neuroepithelial cells enclosing these ventricles are **neural progenitors**; they divide by the ventricles, producing intermediate progenitors and postmitotic neurons and glia. Neurons then migrate away from their birthplace near the ventricles to occupy different layers of neural tissues (Section 7.2). The **neural crest cells**, a special group of cells at the junction between the dorsal neural tube and the overlying epidermal cells (Figure 7-2B), migrate away from the neural tube to produce diverse cell types, including cells of the peripheral nervous system, such as sensory neurons of the dorsal root ganglia and postganglionic neurons of the autonomic nervous system.

The neural tube, which is located on the dorsal side of vertebrate embryos, is highly patterned along both its long anterior–posterior (rostral–caudal) axis and the orthogonal dorsal–ventral axis. Along the anterior–posterior axis, the neural tube is subdivided into the **forebrain**, **midbrain**, **hindbrain**, and spinal cord, each giving rise to different neural structures (**Figure 7-3**A). The cell types produced by the neural tube differ according to their position of origin along the dorsal–ventral and anterior–posterior axes. In the mammalian **telencephalon** (the anterior forebrain), for instance, the dorsal (or pallial) neural tube produces glutamatergic excitatory neurons of the neocortex, whereas the ventral (or subpallial) neural tube

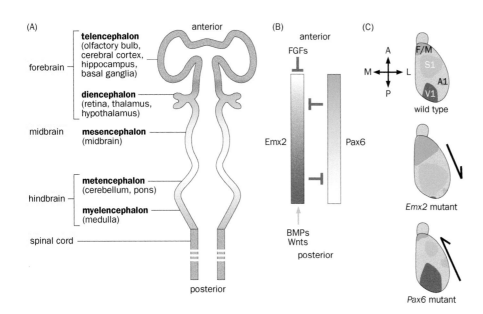

Figure 7-3 Patterning of the neural tube along the anterior–posterior axis. (A) The anterior neural tube is initially subdivided into the forebrain, midbrain, and hindbrain. It then develops into the five-division stage shown here. Parentheses include notable adult structures that each of these divisions gives rise to (see also Figure 1-8). **(B)** The telencephalon is patterned by secreted morphogens—fibroblast growth factors (FGFs) from the anterior neural tube and bone morphogenetic proteins (BMPs) and Wnt family proteins (Wnts) from the posterior—that specify expression gradients of the transcription factors Emx2 and Pax6 in neural progenitors along the anterior–posterior axis. Emx2 and Pax6 mutually inhibit each other. **(C)** Compared to wild-type mice, *Emx2* mutant mice have enlarged anterior cortical areas at the expense of posterior cortical areas, whereas *Pax6* mutant mice have diminished anterior cortical areas and enlarged posterior cortical areas relative to wild type. These data indicate that Pax6 and Emx2 are essential for the development of the anterior and posterior cortical areas, respectively. F/M, frontal/motor cortex (blue); S1, primary somatosensory cortex (green); A1, primary auditory cortex (orange); V1, primary visual cortex (red). A, anterior; P, posterior; M, medial; L, lateral. In the schematics for the knockout mutants, arrows indicate the direction of cortical area expansion due to loss of Emx2 or Pax6. (B & C, adapted from O'Leary DDM, Chou SJ, & Sahara S [2007] *Neuron* 56:252–269. With permission from Elsevier Inc.)

produces GABAergic neurons that migrate into the neocortex and basal ganglia (Section 7.2). In the spinal cord, the dorsal neural tube produces dorsal horn projection neurons that relay sensory information to the brain (Section 6.31), the ventral neural tube produces motor neurons that innervate muscles, and both the dorsal and ventral neural tubes produce many distinct types of interneurons (Section 7.4).

The neural tube is patterned by **morphogens**. These secreted proteins are synthesized at signaling centers (clusters of cells at specific locations, such as junctions between two brain compartments) and diffuse through tissues to regulate expression of target genes in a concentration-dependent manner. Key morphogen targets are often transcription factors in progenitors that control gene expression programs according to where the progenitors are located along the body axes. For example, the neural tube is patterned along the dorsal–ventral axis by the morphogen **Sonic Hedgehog**, which will be discussed in more detail in Section 7.4. Along the anterior–posterior axis, the developing telencephalon is patterned into different functional areas by **fibroblast growth factors (FGFs)** secreted from the anterior end of the neural tube and a combination of **bone morphogenetic proteins (BMPs)** and Wnt family proteins (**Wnts**) from the posterior neural tube. Through activation, repression, and mutual inhibition of target genes, these morphogens establish gradients of transcription factor expression in progenitors across the telencephalon. For example, the transcription factor Emx2 exhibits a posterior high–anterior low expression gradient in cortical progenitors, whereas the transcription factor **Pax6** has an anterior high–posterior low gradient (Figure 7-3B). These expression patterns suggest that Emx2 and Pax6 play important roles in specifying posterior and anterior cortical areas, respectively. Gene knockout experiments in mice supported this hypothesis: *Pax6* mutant mice exhibited an expansion of posterior cortical areas such as the primary visual cortex, with a concomitant shrinkage of the anterior frontal and motor cortices. *Emx2* mutant mice exhibited the opposite phenotype: an expansion of the frontal and motor cortices and primary somatosensory cortex at the expense of the visual cortex (Figure 7-3C). Thus, these transcription factors specify cortical areas via their expression patterns.

7.2 Orderly neurogenesis and migration produce many neuronal types that occupy specific positions

The adult nervous system has a highly organized structure. For example, neurons in different layers of the neocortex have different connection patterns and serve distinct functions (Box 4-4). How are different types of neurons generated from an initial sheet of progenitor cells lining the neural tube? How do they come to occupy specific positions? We use studies of the mammalian neocortex to explore these questions. As we will see, one rule of development is that cells are often born far away from their final destinations, as is the case for excitatory and inhibitory neurons in the neocortex.

Cortical excitatory neurons, which use glutamate as their neurotransmitter and are mostly pyramidal in cell body shape, originate in the dorsal telencephalon from progenitor cells that reside in the **ventricular zone**, a cellular layer next to the ventricles. The final layer in which a given cortical excitatory neuron resides is predicted by its birthday, the time when a neuron exits its last cell cycle and becomes postmitotic. This was originally discovered by injecting pregnant mice with radioactively labeled thymidine at specific developmental stages and using autoradiography to track the positions of strongly labeled cells in the postnatal cortex. (Radioactively labeled thymidine incorporates into newly synthesized DNA during cell division cycles; the cells most strongly labeled are those that became postmitotic shortly after injection and never divide again.) Cells born early reside in deep cortical layers; cells born later reside in progressively more superficial layers (Figure 7-4A). This inside-out pattern ruled out the possibility that late-born cells simply displace early-born cells away from the ventricular zone.

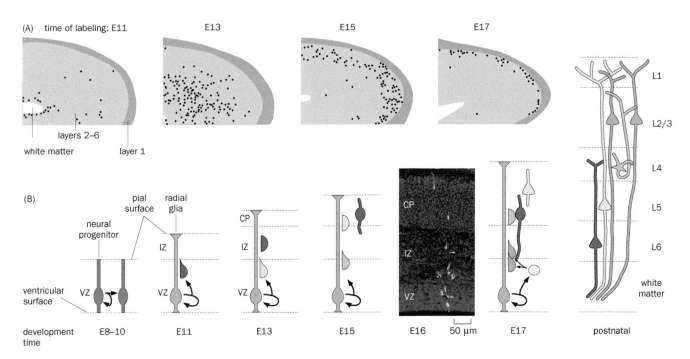

Figure 7-4 Neurogenesis and migration of cortical excitatory neurons.
(A) Autoradiography of ³H-thymidine-labeled neurons in the visual cortex of mice examined at postnatal day 10. Sections are labeled according to the embryonic day on which ³H-thymidine was injected into the pregnant mouse (E11, E13, E15, or E17). Each dot represents a heavily labeled neuron. Thus, neurons born on later days reside in progressively more superficial cortical layers. **(B)** Schematic summary of cortical neurogenesis and migration using the mouse as a model. At early embryonic stages (E8–10), neural progenitors in the ventricular zone (VZ) divide symmetrically, expanding the progenitor pool; at later stages they become radial glia and extend processes that span from the ventricular surface to the pial surface. Neurons produced by asymmetric division from radial glia at E11 (red) are the first to initiate and complete migration along the radial glia; as a result, these E11-born neurons settle in the deepest layer (layer 6, or L6). Neurons born at

E13 (yellow), E15 (green), and E17 (blue) also follow this radial migration pattern, migrating past the early-born neurons to settle in increasingly superficial layers (L5, L4, and L2/3, respectively). Note that asymmetric division can produce intermediate progenitors (for example, the light blue cell at E17), which divide further to produce postmitotic neurons. After E17, a subset of radial glia continue to divide, producing oligodendrocytes and astrocytes (not shown). IZ, intermediate zone (which disappears after neurogenesis is complete); CP, cortical plate (which develops into cortical layers). Inset, a section of developing mouse neocortex at E16. Blue staining labels all nuclei, outlining the three layers of cells (VZ, IZ, and CP). Green stains developing neurons originating from a single radial glia clone produced by a mosaic labeling method (Section 14.16). (A, adapted from Angevine JB & Sidman RL [1961] *Nature* 192:766–768. With permission from Springer Nature. B, image courtesy of Simon Hippenmeyer.)

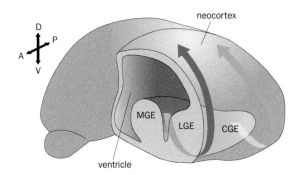

Figure 7-5 Origin and migration of cortical GABAergic neurons. Cortical GABAergic neurons are generated within the ventral telencephalon in the medial and caudal ganglionic eminences (MGE, CGE), and migrate dorsally and tangentially to the developing neocortex around the cerebral ventricle hollows, seen in the left hemisphere, which has been exposed by removing the anterior structure for illustration. D, dorsal; V, ventral; A, anterior; P, posterior. The lateral ganglionic eminence (LGE) produces olfactory bulb interneurons and striatal GABAergic projection neurons. (Courtesy of Edmund Au & Gord Fishell. See also Batista-Brito R & Fishell G [2009] *Curr Top Dev Biol* 87:81–118.)

Subsequent studies have provided a detailed understanding of orderly cortical neurogenesis and migration. At an early stage, neural progenitors undergo symmetric division, expanding the progenitor pool. As development proceeds, the division patterns become asymmetric (see Section 7.3 for more details on asymmetric cell division). At this stage, the progenitors are called **radial glia** because each extends two radial processes—one to the ventricle and the other to the pial surface of the developing cortex (Figure 7-4B). Each radial glia progenitor divides, producing another radial glia progenitor and a postmitotic neuron or an **intermediate progenitor** that resides in the **subventricular zone** (a cellular layer next to the ventricular zone) and divides further, producing postmitotic neurons (see Figure 13-37 for more details). Postmitotic neurons migrate radially along radial glia processes away from the ventricle and toward the pial surface. Neurons born later actively migrate past their early-born cousins and settle progressively farther from the ventricle (Figure 7-4B). As will be discussed in Chapter 13, variations in this basic plan account for the remarkable differences in brain size among mammalian species. After producing cortical excitatory neurons, a subset of radial glia then begins to produce astrocytes and oligodendrocytes.

Whereas cortical glutamatergic neurons originate in the ventricular zone of the dorsal telencephalon and migrate radially, cortical GABAergic neurons originate in the ventral telencephalon from structures called **ganglionic eminences** and migrate dorsally and tangentially across the developing cortex (Figure 7-5). Different types of GABAergic neurons are born at different times in different regions of the ganglionic eminences. They then migrate in an orderly fashion to their final destinations. For instance, neurons derived from the medial ganglionic eminence (MGE) account for the majority of GABAergic neurons in the neocortex, including fast-spiking basket and chandelier cells, which express the marker parvalbumin, as well as Martinotti cells, which express somatostatin (Section 3.25; Box 4-4). Neurons derived from the caudal ganglionic eminence (CGE) account for the remaining cortical GABAergic neurons, which express different markers and serve distinct functions from MGE-derived neurons. This migration process continues up to the second year of postnatal life in humans, especially in the prefrontal cortex. The MGE and CGE also collectively produce all GABAergic neurons in the basal ganglia and amygdala. The lateral ganglionic eminence (LGE) produces olfactory bulb interneurons and striatal GABAergic projection neurons (we will discuss the function of the latter population in Chapter 8).

Many interesting questions are currently being investigated: What controls the total number of cortical neurons born in a given species? How are their migrations directed and terminated? What ensures that each cortical area has the correct number and relative abundance of specific types of glutamatergic and GABAergic neurons? How do GABAergic neurons with a common origin acquire appropriate connectivity within a wide variety of target regions? While these questions are being addressed in animal models, recent technological advances also allow these topics to be explored in human brain organoids developed *in vitro* from pluripotent stem cells (Box 7-1).

Box 7-1: Mimicking neural development *in vitro*: from induced pluripotent stem cells to brain organoids

As we learned in Section 7.1, neurons are produced from progenitors located in defined parts of the neuroepithelia and are specified by signals that pattern the nervous system. The ectodermal progenitors giving rise to neural progenitors derive from **pluripotent stem cells**, which can produce cell types in all three germ layers of the embryo. These pluripotent stem cells, called **embryonic stem (ES) cells**, can be cultured *in vitro*. By overexpressing specific genes and culturing ES cells with specific growth factors or small molecules under conditions mimicking development *in vivo*, researchers can coax ES cells to differentiate in culture into specific types of neurons or glia (**Figure 7-6**, left). A major breakthrough came in 2006, when researchers reported the ability to reprogram somatic cells, such as skin fibroblasts, into **induced pluripotent stem (iPS) cells** by expressing combinations of transcription factors involved in maintaining the pluripotency of ES cells (Figure 7-6, middle top). These iPS cells resemble ES cells in many ways—they can be guided to differentiate into many different cell types *in vitro* and support germline transmission after transplantation into blastocysts (see Section 14.7 for details). Fibroblast-derived iPS cells can also be instructed to differentiate into neurons (Figure 7-6, middle bottom). Finally, fibroblasts

can be induced to trans-differentiate directly into neurons by overexpressing a set of transcription factors important for neuronal differentiation, bypassing the reprogramming to, and subsequent step-by-step differentiation from, iPS cells (Figure 7-6, right).

Remarkably, researchers have also developed protocols for coaxing ES or iPS cells to differentiate *in vitro* into specific tissues that resemble parts of intestine, retina, or other regions of the nervous system. When these differentiation methods are performed in three-dimensional (3D) cultures that enable a higher level of self-organization, the resulting structures are called **organoids**. For example, following a protocol outlined **Figure 7-7A**, researchers used a combination of specific media, growth factors, and Matrigel-based 3D culture to produce organoids resembling various regions of the nervous system, including the developing neocortex. At an early time point in culture, a section of a region resembling the dorsal forebrain revealed a layer adjacent to the hollow interior expressing a neural progenitor marker and a superficial layer expressing a postmitotic neuronal marker (Figure 7-7A, bottom left), mimicking embryonic neocortical development (Figure 7-4). At a late time point in culture, a marker enriched in late-born layer 2/3 neurons was expressed more superficially than a marker for early-born layer 5 neurons (Figure 7-7A, bottom right), in line with their inside-out migration patterns.

These experiments demonstrate the remarkable ability of pluripotent cells to self-organize into tissues *in vitro* given the appropriate environment. However, initial organoid approaches either used undirected methods of differentiation to generate a large diversity of brain cell types or directed differentiation into specific brain regions. To model the more complex interactions in the developing central nervous system, researchers have also developed approaches involving generating organoids of specific brain regions and then assembling them *in vitro*. For instance, when organoids resembling the pallium and subpallium were placed next to each other in 3D culture, substantial migration of cells from the subpallial region to the pallial domain, but not vice versa, was observed (Figure 7-7B), mimicking the migration of GABAergic neurons from the ganglionic eminences to the developing cortex *in vivo* (Figure 7-5). Similar strategies can be used to study interactions between other brain regions, including long-distance axonal projections.

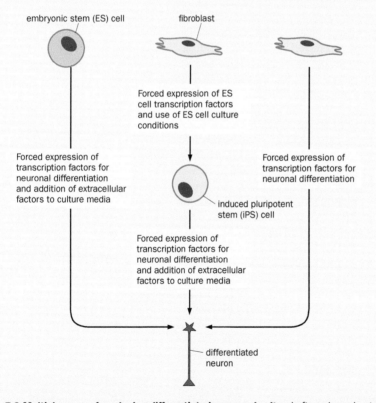

Figure 7-6 Multiple ways of producing differentiated neurons *in vitro*. Left, embryonic stem (ES) cells expressing specific transgenes can be guided to become particular types of neurons by providing specific extracellular factors and defined culture conditions (e.g., Kim et al. [2002] *Nature* 418:50). Middle, fibroblasts can be reprogrammed into iPS cells by forced expression of a cocktail of transcription factors normally expressed in ES cells (Takahashi & Yamanaka [2006] *Cell* 126:663). Resulting iPS cells can then be guided to differentiate into neurons following a protocol similar to the ES cell → neuron path shown on the left. Right, fibroblasts can be directly converted into neurons by expressing a cocktail of genes associated with neuronal differentiation (e.g., Vierbuchen et al. [2010] *Nature* 463:1035).

Box 7-1: continued

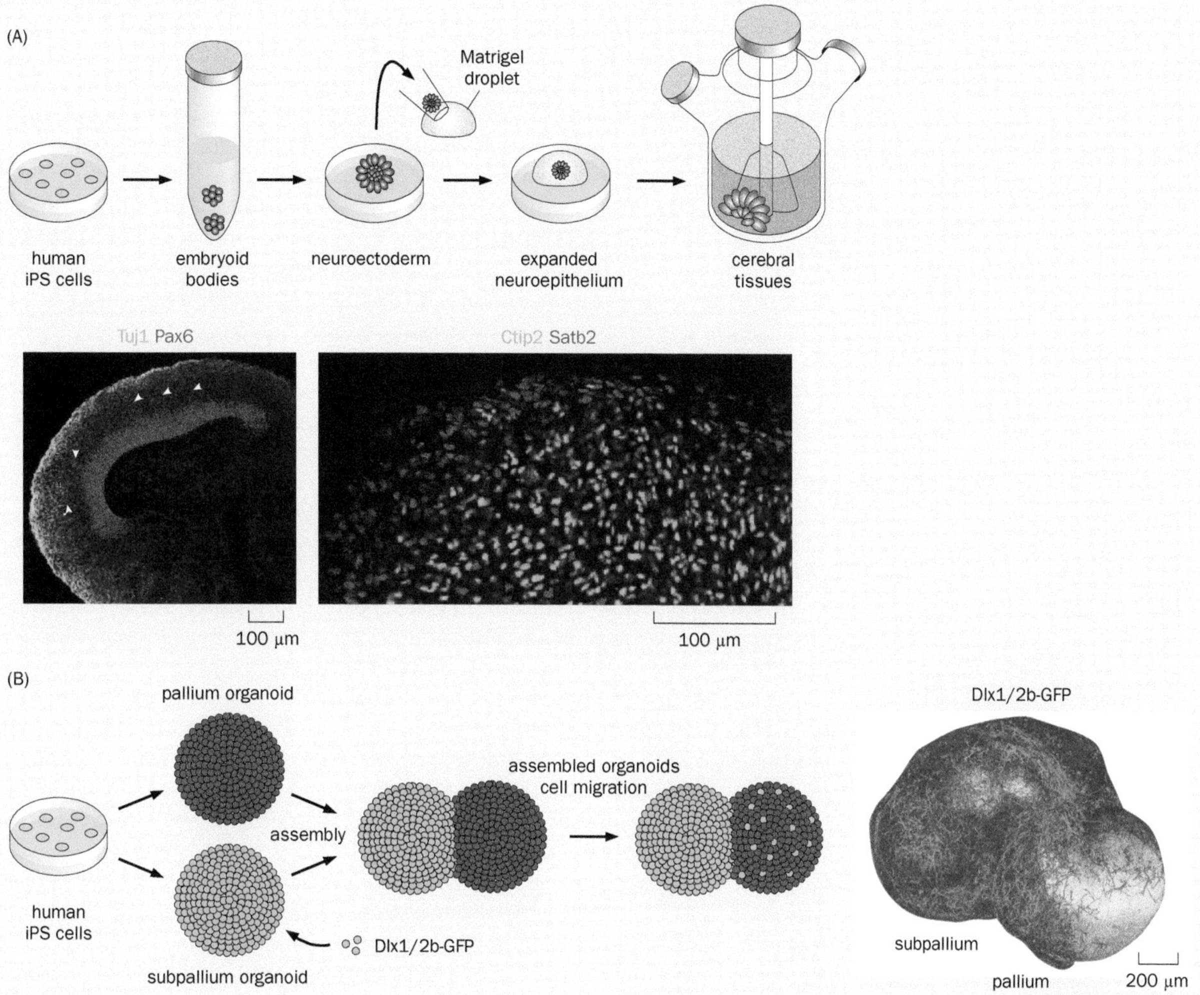

(A)

(B)

Figure 7-7 Mimicking human brain development in 3D culture.
(A) Top, schematic of a procedure for producing organoids enriched in brain tissues. Human-induced pluripotent stem (iPS) cells are plated in suspension culture, forming embryoid bodies. Upon differentiation into neuroectoderm, tissues are placed in a Matrigel droplet to maintain a three-dimensional architecture and then transferred to a spinning bioreactor (right) to maximize nutrient and oxygen access. At each stage, specific media are used to mimic *in vivo* differentiation conditions. Bottom left, a section showing a region resembling the dorsal forebrain. The neural progenitor marker Pax6 is enriched in a layer adjacent to the interior hollow of the organoid, whereas the differentiated neuron marker Tuj1 is enriched superficially, mimicking early cortical patterns. Bottom right, a section of dorsal cortical organoid showing that late-born Satb2+ cells are located more superficially than early-born Ctip2+ cells, mimicking the inside-out migration pattern. **(B)** Left, human iPS cells can be guided to differentiate into region-specific organoids enriched in cells of the dorsal telencephalon (pallium) or ventral telencephalon (subpallium) by different growth factors and small molecules. These organoids can then be placed next to each other to examine their interactions. Many cells from the subpallium-like region migrate to the pallium-like region, but not vice versa, in the assembled organoid, mimicking the migration of GABAergic neurons *in vivo*. Right, whole-mount image of a 30-day postfusion organoid using a tissue-clearing method (Section 14.15), revealing the migration of GABAergic interneurons (labeled with a Dlx1/2b-GFP+ reporter) from the subpallium-like to the pallium-like region. (A, from Lancaster MA, Renner M, Martin CA, et al. [2013] *Nature* 501:373–379. With permission from Springer Nature. B, from Birey F, Andersen J, Markinson CD, et al. [2017] *Nature* 545:54–59. With permission from Springer Nature.)

An important application of brain organoids is to investigate the pathogenesis of human brain disorders. Specifically, iPS cells derived from patients can be used to generate brain organoids that can be compared with those from controls. Identified abnormalities can be used as bioassays for drug testing. For example, researchers found that GABAergic cells derived from patients with a neurodevelopmental disorder caused by mutations in a Ca^{2+} channel migrate abnormally in assembled forebrain organoids, and this can be corrected by pharmacologically modulating the Ca^{2+} channel. For brain disorders caused by mutations in specific genes, such mutations can also be corrected at the iPS-cell stage by replacing the mutant gene with a wild-type copy through genetic and genome editing technologies (see Box 14-1 for details). Correction of a phenotype observed in organoids enables researchers to establish causality between the specific mutation and the phenotype. We will expand on these applications in Chapter 12.

7.3 Cell fates are diversified by asymmetric cell division and cell–cell interactions

By the time a neuron extends its axon and dendrites, it has usually acquired a **cell fate**, the outcome of the developmental decision as to what kind of cell it is. Some of the important (and related) fate choices include what neurotransmitter it will use and what pre- and postsynaptic partners it will select. As discussed in Section 5.17, cell fate can be determined via asymmetric segregation of fate determinants or cell–cell interaction. Investigation of these questions have greatly benefited from pioneering studies in *C. elegans* and *Drosophila* because of their relative simplicity, short life cycle, and the abundant genetic tools available in these invertebrate model organisms. The study of cell fate determination in the *Drosophila* sensory organ, for instance, provides an excellent example of the interplay between asymmetric segregation of fate determinants and cell–cell interaction (**Movie 7-1**).

Each external sensory organ in *Drosophila* consists of a socket cell, a hair cell, a sheath cell, and a sensory neuron. These cells derive from a common sensory organ precursor (SOP) through a series of asymmetric cell divisions (**Figure 7-8**A). In a *Drosophila* mutant called *Numb,* sensory neurons and sheath cells were missing (hence the name) and were replaced by extra socket and hair cells (Figure 7-8B, left); in extreme cases, all cells became sockets (Figure 7-8B, middle). Overexpression of a *Numb* transgene during a specific time window caused an opposite transformation of cell fates—extra neurons and sheath cells were produced at the expense of socket and hair cells (Figure 7-8B, right). These data suggested that the **Numb** protein product normally acts to diversify cell fate within the SOP lineage.

The mechanisms by which *Numb* regulates cell fate became clear when the distribution of Numb protein was examined. Shortly before the SOP divides, Numb becomes localized to one side of the SOP such that it is preferentially inherited by one of the two daughter cells (Figure 7-6C). Asymmetric segregation of newly produced Numb occurs in subsequent divisions as well, giving rise to cells with distinct fates within the SOP lineage.

How does asymmetric segregation of Numb confer distinct fates on different daughter cells? Genetic experiments suggest that one mechanism of action of Numb is to inhibit the function of **Notch**, a transmembrane receptor widely used for diversifying cell fates during development. Notch is activated by the transmembrane ligand **Delta** from neighboring cells. Binding to Delta induces intramembrane proteolytic cleavage of Notch and translocation of its intracellular domain to the nucleus, where it binds to its partners Mastermind and Suppressor of Hairless in *Drosophila* and activates target gene expression (**Figure 7-9**A). Among these target genes are *Notch* and *Delta* themselves—*Notch* transcription is

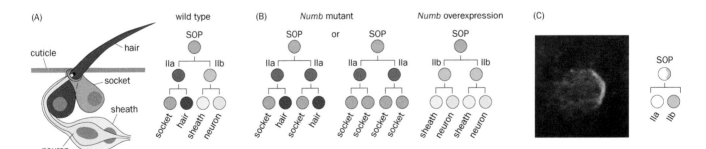

Figure 7-8 Asymmetric segregation of Numb during cell division diversifies sensory organ cell fates. (A) Schematic of an external sensory organ in *Drosophila* made up of four cells—socket cell, hair cell, sheath cell, and neuron—produced from a sensory organ precursor (SOP) via intermediate precursors IIa and IIb. More recent studies have identified an extra division after IIb that produces a glial cell (not depicted here) and an immediate precursor that produces the sheath cell and neuron. **(B)** In *Numb* mutants, neurons and sheath cells are replaced by socket and hair cells (left) or all four cells become socket cells (middle). When *Numb* is transiently overexpressed in the SOP, the opposite result ensues: socket and hair cells are replaced by sheath cells and neurons (right). **(C)** The Numb protein (green) is asymmetrically localized to one side of the SOP shortly before it divides. The schematic on the right shows that this asymmetric segregation results in two daughter cells with different levels of the Numb protein. (Adapted from Rhyu MS, Jan LY, & Jan YN [1994] *Cell* 76:477–491. With permission from Elsevier Inc. See also Uemura T, Shepherd S, Ackerman L, et al. [1989] *Cell* 58:349–360.)

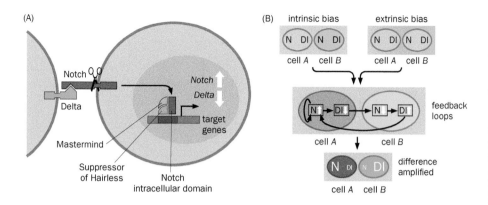

Figure 7-9 Notch/Delta mediated lateral inhibition diversifies cell fates. (A) In response to binding of its ligand Delta from a neighboring cell, Notch activation leads to the cleavage and translocation of its intracellular domain to the nucleus (red), where it interacts with its nuclear partners, Mastermind and Suppressor of Hairless, resulting in target gene expression. Among these gene expression changes, *Notch* is upregulated and *Delta* is downregulated. **(B)** A model for Notch/Delta-mediated lateral inhibition. Two cells initially have the same Notch (N) and Delta (Dl) levels. Due to an intrinsic (left) or extrinsic (right) biasing signal, Notch is slightly inhibited in cell B on the right; this leads to further decreased Notch expression and increased Delta expression. Neighboring cell A is thus exposed to more Notch ligand and has higher Notch activity, which further increases cell A's expression of Notch and decreases its expression of Delta. These feedback loops amplify initial differences in Notch and Delta levels. (Based on Artavanis-Tsakonas S, Rand MD, & Lake RJ [1999] *Science* 284:770–776. See also Struhl G & Adachi A [1998] *Cell* 93:649–660.)

upregulated and *Delta* transcription is downregulated by Notch signaling. These feedback loops can account for how Notch/Delta signaling diversifies cell fates through a process called **lateral inhibition**. (Note that this is distinct from the lateral inhibition in information processing by neural circuits that we discussed in Sections 4.14 and 6.7.)

Suppose that two neighboring cells, *A* and *B*, initially express the same amount of Notch and Delta. If Notch activity is inhibited in *B* by intrinsic factors (such as Numb) or extrinsic signals, inhibition of Delta expression is relieved. As Delta expression in *B* increases, Notch signaling in *A* increases, while Delta expression in *A* decreases, so that *B* receives less ligand for its own Notch (Figure 7-9B). Thus, Notch/Delta signaling amplifies small initial differences between neighboring cells, leading to full activation of Notch signaling in one cell and suppression of it in the other. This results in differential expression of Notch target genes and, eventually, different fates among neighboring cells.

While lateral inhibition utilizes feedback loops to amplify small initial differences, Numb biases Notch/Delta signaling to drive stereotyped decisions. Mechanisms similar to those used in sensory organ cell fate specification are widely used throughout development. Indeed, at an earlier developmental stage, the fate determination mechanism that differentiates the SOP from neighboring epidermal cells also utilizes Notch/Delta-mediated lateral inhibition. Vertebrate homologs of Notch, Delta, and Numb, as well as nuclear partners of Notch and some of the Notch target genes, are implicated in many fate determination events, including cortical neurogenesis discussed in Section 7.2.

7.4 Transcriptional regulation of guidance molecules links cell fate to wiring decisions

From a molecular perspective, the fate decision is usually realized by the expression of a unique set of fate determinants that distinguish a cell from other cells expressing different sets of fate determinants. Important among these determinants are transcription factors. Cell-type-specific transcription factors can dictate the expression and relative abundance of guidance molecules. As a result, axons from neurons with different fates may contain different populations of guidance molecules and respond differently to a common environment during their guidance and target selection.

We have already encountered examples of this in visual system wiring. In the mammalian retina, retinal ganglion cells (RGCs) that project ipsilaterally express transcription factor Zic2, which specifies the expression of EphB1, allowing these axons to respond to the repellent ephrin-B2 in the chiasm and adopt an ipsilateral course (Figure 5-15). In the chick tectum, graded expression of ephrin-As along the anterior–posterior axis is regulated by graded expression of the transcription factor Engrailed-2, which is in turn regulated by the FGF morphogens produced at the midbrain–hindbrain junction at the posterior edge of the developing tectum (Section 5.5).

The mechanisms by which neuronal fates are specified in vertebrates is best understood in the spinal cord, thanks to a combination of studies using chick

Figure 7-10 Morphogens and transcription factors specify spinal cord neuronal fates. (A) Distribution of Sonic hedgehog (Shh) protein. Shh (in white) is produced in the notochord and floor plate. It diffuses dorsally within the spinal cord, forming a concentration gradient. D, dorsal; V, ventral. **(B)** Shh acts by repressing (⊣) and activating (→) different transcription factors listed on the top (class I) and bottom (class II) in a concentration-dependent manner, thereby specifying the expression of these transcription factors in progenitor cells at different D–V positions. For example, a low level of Shh can repress Pax7 expression, whereas a high level is needed to repress Pax6 expression. Thereby Pax7 expression is restricted to the dorsal-most cells, whereas Pax6 expression extends more ventrally. **(C)** Two pairs of class I/class II transcription factors mutually repress each other's expression, thereby sharpening the borders of their expression domains along the D–V axis. **(D)** Five progenitor domains created by differential expression of these transcription factors give rise to five classes of spinal neurons, the V0, V1, V2, and V3 interneurons, and motor neurons (MN). **(E)** Motor neurons are produced in the D–V domain that expresses Pax6 and Nkx6.1, but not Irx3 and Nkx2.2. (Adapted from Jessell TM [2000] *Nat Rev Genet* 1:20–29. With permission from Springer Nature. See also Briscoe J, Pierani A, Jessell TM, et al. [2000] *Cell* 101:435–445.)

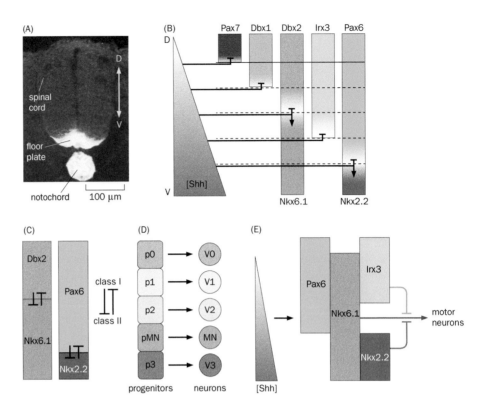

embryology and mouse genetics. Morphogens secreted from the dorsal roof plate or ventral floor plate reach different parts of the spinal cord at different concentrations and specify distinct cell fates along the dorsal–ventral axis in a concentration-dependent manner. A particularly well-studied morphogen, the floor-plate-derived **Sonic Hedgehog (Shh)**, determines fates of neural progenitors at different positions in the ventral spinal cord by regulating expression of specific transcription factors (**Figure 7-10**A, B). Shh represses class I transcription factors Pax7, Dbxl, Dbx2, Irx3, and Pax6, and activates class II transcription factors Nkx6.1 and Nkx2.2, at different concentrations (Figure 7-10B). (All of these transcription factors contain a DNA-binding domain called **homeodomain**, the origin of which will be discussed in Chapter 13.) Furthermore, class I and class II transcription factors mutually repress each other's expression (Figure 7-10C), sharpening the borders between different progenitor pools created initially by differential responses to the Shh morphogen gradient. Collectively, differential expression of these transcription factors specifies the fates of five progenitor types, four that produce ventral spinal cord interneurons and one that produces motor neurons (Figure 7-10D).

As a specific example of the transcription factor code, progenitors that express Pax6 and Nkx6.1 but do not express Irx3 or Nkx2.2 become motor neuron progenitors (Figure 7-10E). They activate motor neuron–specific transcriptional programs as their progeny exit the cell cycle. These include expression of a series of transcription factors, which in turn regulate expression of specific guidance receptors that direct their axons out of the spinal cord and into muscle fields. Motor neurons that innervate specific muscles are further diversified. For example, the fate of a motor neuron that projects into the dorsal or ventral limb is determined by its expression of one of two specific transcription factors containing the LIM domain: Lim1 and Isl1 (**Figure 7-11**A). Lim1-expressing motor neurons are located laterally and project axons to the dorsal limb, whereas Isl1-expressing motor neurons are located medially and project axons to the ventral limb. Lim1 and Isl1 mutually repress each other's expression, such that each motor neuron expresses only one of these two transcription factors (Figure 7-11B). At the target field, the dorsal limb expresses a third LIM transcription factor called Lmx1b. Genetic perturba-

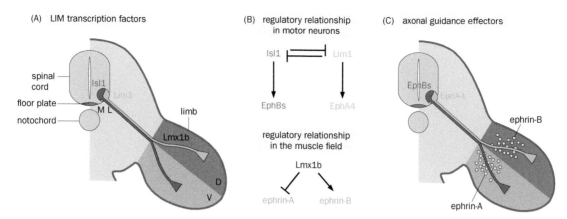

Figure 7-11 LIM transcription factors control the expression of ephrin/Eph to specify motor axon targeting. (A) Motor neurons located more medially (M) in the spinal cord express LIM transcription factor Isl1 (red) and target their axons selectively to the ventral limb (V), whereas motor neurons located laterally (L) express a different LIM transcription factor, Lim1 (green), and target their axons to the dorsal limb (D). A third LIM transcription factor, Lmx1b, is expressed in the dorsal limb. **(B)** Mutual repression of Isl1 and Lim1 ensures that they are expressed in different motor neurons. Furthermore, Isl1 promotes expression of EphBs, while Lim1 promotes expression of EphA4. In the limb, Lmx1b activates ephrin-B expression and represses ephrin-A expression. **(C)** EphB-expressing motor axons are repelled by dorsal limb ephrin-B and selectively target the ventral limb, whereas EphA-expressing motor axons are repelled by ventral limb ephrin-A and target the dorsal limb. (Adapted from Kania A & Jessell TM [2003] *Neuron* 38:581–596. With permission from Elsevier Inc. See also Luria V, Krawchuk D, Jessell TM, et al. [2008] *Neuron* 60:1039–1053.)

tions of Lim1 in motor neurons or Lmx1b in targets disrupt the trajectory choice of motor axons innervating muscles, indicating that these transcription factors play essential roles in the wiring decisions of these motor neurons.

How do these transcription factors regulate connectivity between motor axons and their targets? It turns out that they regulate expression of guidance molecules ephrins and Ephs, which we first encountered in our discussion of visual system wiring. In the dorsal limb, Lmx1b promotes expression of ephrin-B2 and represses expression of ephrin-As. The target field is thus patterned by binary expression of two ephrins: ephrin-As in the ventral limb and ephrin-B2 in the dorsal limb. In motor neurons, Lim1 promotes expression of the EphA4 receptor while Isl1 promotes expression of EphB receptors (Figure 7-11B). When axons arrive at the target field, Lim1/EphA4-expressing axons are repelled by ephrin-As in the ventral limb and thereby target dorsally, whereas Isl1/EphBs-expressing axons are repelled by ephrin-B2 in the dorsal limb and thereby target ventrally (Figure 7-11C). Thus, coordinated expression and function of these dual ephrin/Eph systems provides a robust mechanism ensuring that two subpopulations of motor neurons connect correctly to two distinct groups of muscle targets.

Transcription factors that specify neuronal fates and axon projection patterns in more complex tissues such as the mammalian neocortex have also been discovered. The projection patterns of cortical pyramidal neurons can be divided into three major classes (**Figure 7-12**A). The first class, including all of layer 2/3 and a fraction of layer 5 and 6 projection neurons, projects their axons to other cortical areas. Of these corticocortical projection neurons, a subset projects axons across the **corpus callosum** (which is composed of axon bundles linking the two cerebral hemispheres) to the contralateral cortex; these neurons are called callosal projection neurons (CPNs). The rest project to cortical areas within the same hemisphere. The second class of neurons project their axons to subcortical targets such as the pons, superior colliculus, and spinal cord; these are called subcerebral projection neurons (SCPNs). SCPNs constitute a large subset of layer 5 neurons. The third class of neurons project their axons to the thalamus and are therefore called corticothalamic projection neurons (CTPNs). CTPNs constitute a large subset of layer 6 neurons (Figure 7-12A).

As in the spinal cord, fates and projection patterns of cortical projection neurons are specified by a set of transcription factors that mutually inhibit each other. For example, the transcription factor **Satb2** is expressed in CPNs and is required

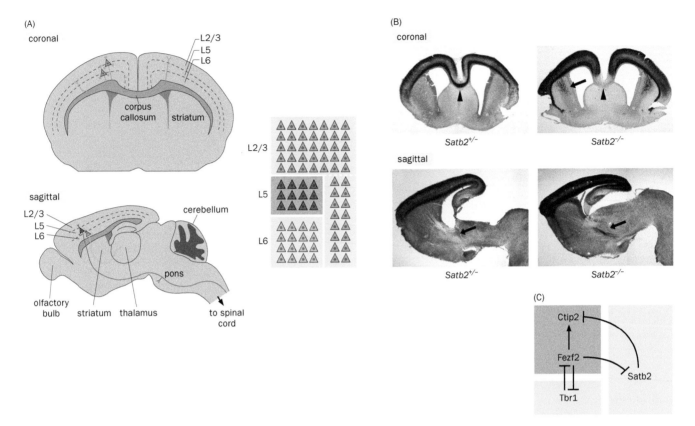

Figure 7-12 Transcription factors specify axonal projections of cortical neurons. (A) Left, schematic coronal (top) and sagittal (bottom) views of the adult mouse brain, showing axonal projection patterns of three major classes of cortical projection neurons. Corticocortical projection neurons (blue) can project axons across the corpus callosum to the contralateral hemisphere (CPNs) and/or to other cortical areas within the same hemisphere. Subcerebral projection neurons (SCPNs, red) project their axons to subcortical areas such as the pons and spinal cord. Corticothalamic projection neurons (CTPNs, green) project their axons to the thalamus. Right, schematic layer distribution of three projection neuron classes. Note that in reality, layer 5 (L5) and L6 SCPNs and CTPNs intermingle with corticocortical projection neurons. **(B)** Satb2 is required for callosal projections. In heterozygous controls, axons of Satb2-expressing neurons (visualized in blue, as a result of a marker inserted into the *Satb2* locus) cross the corpus callosum (arrowhead) but do not extend to subcerebral structures (arrow). In *Satb2* homozygous mutants, axons no longer cross the corpus callosum (arrowhead), but instead extend to subcerebral structures (arrows). These samples were examined at embryonic day 18.5, so the sections look different from the adult schematic in (A). **(C)** A simplified summary of transcription factors that regulate projection patterns of CPNs (blue), SCPNs (red), and CTPNs (green) in cortical layers 5/6, as well as their mutually repressive relationships. (A & C, adapted from Greig LC, Woodworth MB, Galazo MJ, et al. [2013] *Nat Rev Neurosci* 14:755–769. With permission from Springer Nature. B, from Alcamo EA, Chirivella L, Dautzenberg M, et al. [2008] *Neuron* 57:364–377. With permission from Elsevier Inc.)

for their callosal projection patterns. Satb2-expressing CPNs normally project axons across the corpus callosum but not to subcerebral structures. However, axons of these same neurons in *Satb2* mutant mice failed to project across the corpus callosum and instead projected to subcortical structures (Figure 7-12B). Likewise, the transcription factor **Fezf2** is expressed in SCPNs and specifies subcerebral projections, while the transcription factor **Tbr1** is expressed in CTPNs and regulates corticothalamic projections. Fezf2 and Tbr1 mutually repress each other; Fezf2 also represses Satb2, which in turn represses Ctip2, a downstream transcription factor that regulates SCPN identity and subcerebral projections (Figure 7-12C). Mutual repression of transcription factors ensures that each projection neuron class express a unique set of guidance molecules that specify their distinct projections. As in the case of spinal motor neurons, these transcription factors likely regulate expression of axon guidance molecules that determine cell-type-specific projections. For example, a Fezf2 target gene is *Ephb1*, which promotes axon projections along the corticospinal track but inhibits axon projections across the corpus callosum. Finally, mutations in some of these transcription factors in humans are strongly associated with neurodevelopmental defects, further highlighting their importance.

7.5 Crossing the midline: combinatorial actions of guidance receptors specify axon trajectory choice

As we learned in Chapter 5, axons are guided by attractive and repulsive cues. These cues act on guidance receptors at the surfaces of growth cones to regulate underlying cytoskeletal dynamics (Boxes 5-1 and 5-2). In the next two sections, we use investigations of midline crossing and pathway choices in insects and vertebrates to explore mechanisms of axon guidance that expand on these principles.

Axons often follow specific trajectories in a highly complex environment. This has been examined closely in the insect **ventral nerve cord**, which is analogous to the vertebrate spinal cord. The grasshopper embryo has been used as a model because the large size of its neurons allows dyes to be injected into single, **identified neurons** (meaning that the same neurons can be recognized from animal to animal because of their stereotyped locations, sizes, and shapes). It was found that axons from individually identified neurons follow specific pathways. Many of the identified neurons in the grasshopper ventral nerve cord have subsequently been found in *Drosophila*. Although *Drosophila* neurons are much smaller in size (and more difficult to inject dye into), *Drosophila* offers the power of forward genetic screens for identifying genes required for axon guidance from their loss-of-function phenotypes (see Section 14.6 for more details). This approach complements the biochemical approaches for identifying axon guidance molecules discussed in Section 5.4 and Box 5-1.

When stained with an antibody that recognizes all axons of the *Drosophila* embryonic ventral nerve cord, wild-type axons appear as segmentally repeated anterior and posterior commissures across the midline as well as longitudinal axonal tracts that run along the anterior–posterior axis (**Figure 7-13**A). This is because axons of most embryonic ventral nerve cord neurons cross the midline once, forming the commissures, and then turn either anteriorly or posteriorly, forming the longitudinal tracts. Mutants that exhibit interesting phenotypes in regard to midline crossing have been identified. For example, in *Slit* mutants, all axons collapse at the midline (Figure 7-13B). In *Roundabout* (*Robo*) mutants, individual axons cross the midline multiple times instead of turning anteriorly or posteriorly after midline crossing, thus forming thick commissures and thin longitudinal tracts (Figure 7-13C). In *Commissureless* (*Comm*) mutants, no axons ever cross the midline (Figure 7-13D).

Subsequent molecular genetics, biochemical, and cell biological analyses have identified the mechanisms by which the protein products of *Slit*, *Robo*, and *Comm* regulate midline crossing. **Slit** is a secreted protein produced by midline glia (stained in blue in Figure 7-13A) that acts as a repulsive axon guidance ligand. **Robo** is a receptor for Slit. **Comm** acts in the secretory pathway (Figure 2-2) to downregulate cell-surface expression of Robo. When axons initiate midline crossing, expression of Comm prevents Robo from being on the cell surface, thus enabling axons to cross the midline. After midline crossing, downregulation of Comm results in upregulation of Robo on the cell surface, thus preventing axons from recrossing the midline because of the repulsive Slit/Robo interaction. Hence, the mutant phenotypes can be explained as follows. In the absence of Comm,

Figure 7-13 Ventral nerve cord of wild-type and mutant *Drosophila* embryos. **(A)** The wild-type *Drosophila* embryonic ventral nerve cord is organized into repeated segments of anterior and posterior commissures (AC and PC, composed of axon bundles that cross the midline), longitudinal tracts (L), and segmental (S) and intersegmental (IS) nerves that connect the ventral nerve cord with peripheral tissues. Brown marks antibody staining that recognizes all axons. Blue stains midline glia. **(B)** In homozygous *Slit* mutant embryos, all axons collapse at the midline. **(C)** In homozygous *Roundabout (Robo)* mutant embryos, longitudinal tracts are thinner and commissures are thicker. **(D)** In homozygous *Commissureless (Comm)* mutant embryos, axons do not cross the midline and therefore commissures are missing. (From Seeger M. Tear G, Ferres-Marco D, et al. [1993] *Neuron* 10:409–426. With permission from Elsevier Inc.)

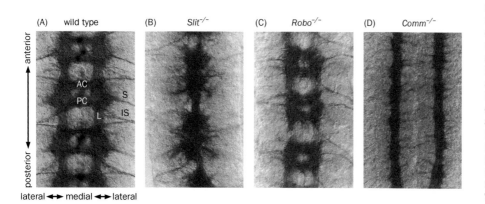

(A) wild type (B) *Slit*⁻/⁻ (C) *Robo*⁻/⁻ (D) *Comm*⁻/⁻

anterior ↑ posterior ↓

AC
PC
L
S
IS

lateral ◄► medial ◄► lateral

(A)

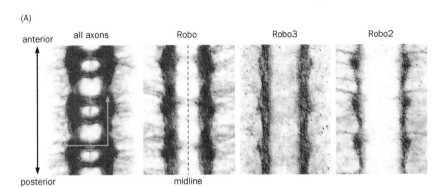

(B)

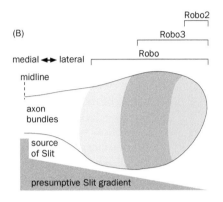

Figure 7-14 The combined activity of three Robo proteins specifies axon trajectories in *Drosophila* embryos. (A) The first micrograph shows the same antibody staining as in Figure 7-11A, which labels all axons. The axon path of a typical neuron is shown in cyan, with the circle indicating the cell body. The remaining micrographs show the three Robo proteins expressed in different subsets of longitudinal axon tracts. Note that Robo proteins are absent from the commissures, as they are downregulated by Comm. **(B)** Model for how expression of Robo family proteins specifies axon position along the medial–lateral axis, seen from a cross section of the fly ventral nerve cord. In response to a presumptive Slit gradient, axons are repelled to different distances from the midline according to the total amount of Robo proteins they express. (Adapted from Simpson JH, Bland KS, Fetter RD, et al. [2000] *Cell* 103:1019–1032. With permission from Elsevier Inc. See also Rajagopalan S, Vivancos V, Nicholas E, et al. [2000] *Cell* 103:1033–1045; Spitzweck B, Brankatschk M, & Dickson BJ [2010] *Cell* 140:409–420.)

Robo is always expressed on the cell surface and axons cannot cross the midline. (In fact, a small fraction of neurons whose axons naturally do not cross the midline do not express Comm.) In the absence of Robo, a large number of axons recross the midline after initial crossing because they have lost the receptor for a midline repellent. In the absence of Slit, no midline repellent is present, so all axons stay at the midline because of midline attractants (the fly homologs of *C. elegans* Unc6 and vertebrate netrins introduced in Box 5-1). The reason *Slit* and *Robo* mutants do not exhibit identical phenotypes is because additional Robo proteins act together with Robo, as discussed in the following.

After midline crossing, axons join the longitudinal tract projecting anteriorly or posteriorly at specific distances away from the midline. What determines the distance? This is achieved by a second function of the Slit/Robo system. Flies have three *Robo* genes that encode receptors for Slit: Robo, Robo2, and Robo3. Antibody staining indicated that these three proteins are differentially expressed in different axons at varying distances from the midline. Robo is expressed in all axons; Robo3 in the lateral two-thirds of axons, and Robo2 in the lateral-most one-third of axons (**Figure 7-14**A). Given that the repellent Slit is concentrated at the midline, it was proposed that the total amount of Robo proteins on an axon determines its lateral position in its trajectory: the more total Robo proteins an axon expresses, the more lateral the position that axon should take (Figure 7-14B). This model was validated by a series of loss-of-function and gain-of-function experiments: removing and adding Robo proteins caused medial and lateral shifts of axon positions, respectively. Thus, the total amount of Robo proteins expressed by an axon specifies the medial–lateral position of its trajectory.

7.6   Crossing the midline: axons switch their responses to guidance cues at intermediate targets

Many axons travel long distances to reach their final destinations. Intermediate targets are frequently used to divide these long journeys into segments. For example, commissural axons in the spinal cord first extend ventrally to the floor plate (Box 5-1); they then cross the floor plate and extend anteriorly along the spinal cord. The floor plate is thus an intermediate target for commissural axons. An interesting question arises: if commissural axons are so attracted to the floor plate, why do they ever leave after reaching it?

As we learned in Box 5-1, the evolutionarily conserved netrin pathway plays a key role in guiding commissural axons to the floor plate. As another striking example of evolutionary conservation of axon guidance molecules and mechanisms,

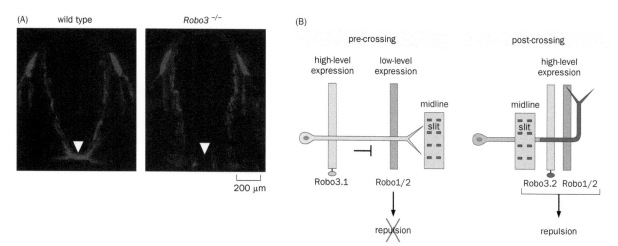

Figure 7-15 Commissural axons switch their responses to guidance cues at the floor plate, an intermediate target. (A) Embryonic day 11.5 mouse spinal cord stained with an axon marker, showing robust midline crossing of commissural axons at the floor plate (arrowheads) in wild type but lack of midline crossing in *Robo3* mutant. See Figure 7-16A for a schematic of the axonal path of a commissural neuron. **(B)** Schematic summary. Alternative splicing produces two Robo3 isoforms that differ in C-termini (green and red). Robo3.1 is highly enriched in pre-crossing axons and acts to inhibit Robo1/2 repulsion to midline Slit. Robo3.2 is highly enriched in post-crossing axons and acts together with Robo1/2 to facilitate repulsion to midline Slit. In addition to the change in Robo3 isoforms, post-crossing axons also express higher Robo1/2 levels than pre-crossing axons. (A, from Sabatier C, Plump AS, Ma L, et al. [2004] *Cell* 117:157–169. With permission from Elsevier Inc. B, adapted from Chen Z, Gore BB, Long H, et al. [2008] *Neuron* 58:325–332. With permission from Elsevier Inc.)

the floor plate also makes the vertebrate homolog of *Drosophila* **Slit**, which acts as a chemorepellent. Three Robo proteins have been characterized in the mouse. Whereas Robo1 and Robo2 act as repulsive guidance receptors, Robo3, a distant member of the Robo family, when knocked out in mice, revealed a striking phenotype: no commissural axons ever crossed midline (**Figure 7-15**A). Further analyses revealed that Robo3 has two alternatively spliced isoforms that differ in their cytoplasmic domain. Robo3.1 is highly enriched in pre-crossing axons, antagonizes the action of low-level Robo1/2 expression there, and thus prevents pre-crossing axons from responding prematurely to midline repellent Slit. Robo3.2 is highly enriched in post-crossing axons and acts together with elevated Robo1/2 expression in post-crossing axons to repel them away from the midline (Figure 7-15B). Thus, after axons cross the midline, a switch in Robo3 splicing isoform, along with upregulation of Robo1/2 expression, contributes to the different responses of commissural axons to the midline repellent Slit.

Additional factors and mechanisms regulate midline crossing. For example, activation of Robo by Slit can inactivate cytoplasmic signaling of DCC, the netrin receptor, and thus switch off attractive responses to netrin once axons reach the midline. Moreover, as discussed in Section 7.4, the floor plate produces the morphogen Shh to pattern the spinal cord for cell fate determination; later in neural development, Shh is used as a floor-plate-derived axon guidance cue that attracts commissural axons to the midline. Shh signaling can also induce repulsion by yet another floor-plate-produced guidance cue, the secreted semaphorin repellent **Sema3F** (Box 5-1). Binding of Shh initiates an intracellular signaling cascade that potentiates signaling by Sema3F in commissural neurons. Thus, as axons approach the midline, Sema3F-induced repulsion, potentiated by Shh signaling, contributes to the extension of midline-crossing axons away from the floor plate.

After crossing the midline, commissural axons turn anteriorly to project toward the brain (commissural neurons include dorsal horn projection neurons that transmit temperature and pain signals to the brain; Figure 6-71). What causes them to turn anteriorly rather than posteriorly? In the developing ventral spinal cord, Wnt4 (Section 7.1), a secreted morphogen, is distributed in an anterior > posterior gradient and acts as an attractant for commissural axons through the Frizzled3 receptor (Box 5-1). In an "open book" preparation wherein the behavior of commissural axons after midline crossing can be examined *in vitro,* wild-type axons all turned anteriorly, whereas axons from *Frizzled3* mutant mice projected

Figure 7-16 Wnt-Frizzled signaling directs commissural axons to turn anteriorly after midline crossing. (A) Schematic of the axonal path of commissural neurons, which first grows ventrally toward the midline and, after midline crossing, turns anteriorly. **(B)** Illustration of the open book preparation, wherein the spinal cord is bisected from the dorsal side and laid open. Turning of fluorescently labeled commissural axons can be examined *in vitro*. **(C)** Whereas all wild-type commissural axons turn anteriorly in the open book preparation (left), commissural axons of *Frizzled3* mutants (*Fz3−/−*) turn both anteriorly and posteriorly (middle), or stall. Percentage of explants that exhibit these specific phenotypes are quantified in the right panel. A, anterior; P, posterior; fp, floor plate. (From Lyuksyutova AI, Lu CC, Milanesio N, et al. [2003] *Science* 302: 1984–1988. With permission from AAAS.)

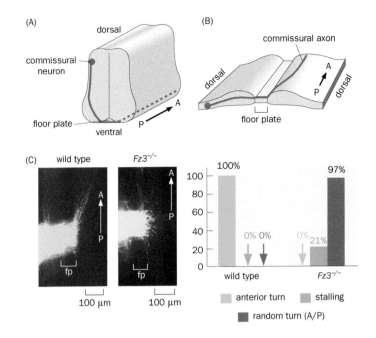

in all directions (**Figure 7-16**). This and other experiments indicate that the Wnt-Frizzled interaction plays an instructive role in directing commissural axons anteriorly after midline crossing.

Several lessons can be drawn from the examples discussed thus far. First, molecules and mechanisms are highly conserved from invertebrates to mammals, as exemplified by the netrin/DCC and Slit/Robo systems. Second, the same molecules can be used at multiple stages of development, as exemplified by the reuse of morphogens such as Shh and Wnts, which pattern early embryonic development and serve as axon guidance molecules, and by the dual function of Slit/Robo in midline crossing and lateral positioning of axonal trajectories. Third, attractants and repellents often act in parallel to ensure the accuracy of decisions made by the growth cone, as exemplified by the actions of netrin and Shh in attracting axons to the midline and of Slit and Sema3F in repelling axons from the midline. Fourth, precise regulation of guidance receptor expression, as exemplified by upregulation of Robo receptors after midline crossing and by the switch of splicing isoforms of mammalian Robo3, can alter axonal responses to the same cues before and after intermediate targets. Together, these mechanisms guide axons with remarkable precision along their complex journeys. In the case of vertebrate commissural neurons, for example, these mechanisms ensure that the axons first grow ventrally toward the midline, then cross the midline, and finally turn anteriorly toward their brain targets (Figure 7-16A).

7.7 The cell polarity pathway determines whether a neuronal process becomes an axon or a dendrite

Thus far we have focused on the role of axons in determining wiring specificity. In general, axons travel longer distances than dendrites, so they perform a greater share of the work involved in wiring the nervous system. Dendrites can also play an active role in determining wiring specificity, as will be discussed later in the context of olfactory circuit wiring. Axons and dendrites generally serve distinct functions: sending and receiving signals, respectively. In addition, axons and dendrites have different morphologies: a typical vertebrate neuron has only a single axon but multiple dendrites with elaborate branches near the cell body. Before we study mechanisms of dendrite morphogenesis, we first take a step back to an earlier stage of development and address the following question: how are axons and dendrites specified during development, such that they acquire different morphologies and functions?

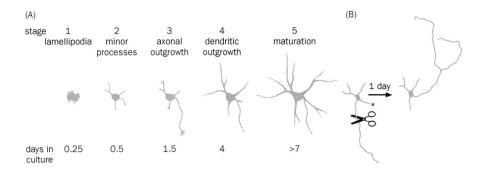

(A)

stage | 1 lamellipodia | 2 minor processes | 3 axonal outgrowth | 4 dendritic outgrowth | 5 maturation

(B)

1 day

days in culture | 0.25 | 0.5 | 1.5 | 4 | >7

Figure 7-17 **Differentiation of the axon and dendrites in dissociated hippocampal neurons. (A)** After embryonic hippocampal neurons are dissociated and plated in a culture dish, they undergo a characteristic sequence of developmental events schematized here in five stages. Time after plating is indicated below. By day 1.5, one process has become an axon. The rest of the processes develop into dendrites over subsequent days. **(B)** When an axon is severed, another process (*) develops into an axon; the drawing to the right shows the same neuron 24 h later. (A, adapted from Dotti CG, Sullivan CA, & Banker GA [1988] *J Neurosci* 8:1454–1468. Copyright ©1988 Society for Neuroscience. B, adapted from Dotti CG & Banker GA [1987] *Nature* 330:254–256. With permission from Springer Nature.)

Studies of hippocampal neurons in culture have helped address this question. When dissociated embryonic hippocampal neurons are plated on a culture dish under specific conditions, they extend numerous lamellipodia (Box 5-2) that consolidate into a few short processes. Then, one of these processes extends rapidly and becomes an axon. Subsequently, the remaining processes start to extend and branch; these become dendrites (**Figure 7-17**A). Hippocampal cultures can be maintained for weeks, during which time axons and dendrites develop pre- and postsynaptic properties and form synaptic connections.

The ease with which this culture system can be observed and manipulated has enabled researchers to probe the mechanisms that establish neuronal polarity: the distinction between axons and dendrites (Sections 2.2 and 2.3). For example, when a growing axon is severed near the cell body, a new axon develops from one (and only one) of the remaining short processes (Figure 7-17B). This experiment suggested that axon determination is plastic, and that once one process becomes the axon, others are inhibited from acquiring the same fate. Subsequent time-lapse imaging studies revealed that shortly before one of the processes extends rapidly and becomes an axon, its growth cone is particularly dynamic in extending and retracting filopodia and lamellipodia, accompanied by rapid turnover of F-actin. Indeed, researchers can turn any process of a stage-2 cultured hippocampal neuron (Figure 7-17A) into an axon by locally applying a low dose of cytochalasin D, which causes depolymerization of the actin cytoskeleton at its growth cone. This experiment suggested that destabilization of the actin cytoskeleton in the growth cone can induce axon formation. Likewise, locally stabilizing microtubules at the growth cone of a stage-2 process can induce axon formation.

Many molecular manipulations can perturb neuronal polarity in the hippocampal culture system. The functions of a small subset of these molecules has been validated *in vivo* via gene knockout in mice. A central pathway involves *Par* genes, which were originally identified in genetic screens in *C. elegans* for their role in asymmetric partitioning of cytoplasmic components in the early embryo. *Par* genes have subsequently been shown to regulate cell polarity in many different tissues and cell types. In cultured hippocampal neurons, the mammalian **LKB1** kinase, a homolog of the *C. elegans* protein Par4, is concentrated in the short process that is destined to become the axon, but not in the other short processes (**Figure 7-18**A). Conditional knockout of the mouse *Lkb1* gene drastically reduces the abundance of axons *in vivo* (Figure 7-18B). LKB1 is a component of a protein kinase cascade. Protein kinase A phosphorylates and activates LKB1, which in turn phosphorylates a set of SAD kinases related to Par1, another cell polarity regulator identified in *C. elegans* (Figure 7-18C). Many of these components are conserved from worm to mammal, suggesting that the establishment of neuronal polarity uses similar mechanisms across the animal kingdom.

A conceptual framework for establishing neuronal polarity has emerged from these and other studies. Asymmetry in signaling, due to asymmetric exposure to an external signal, asymmetric localization of an intrinsic factor, or both, singles out one process to become the axon. This initial determination is likely reinforced by a positive feedback loop to augment the growth of the axon, and by a negative (inhibitory) signal that spreads to neighboring processes to prevent them from becoming axons. The nature of these positive and negative signals remain to be

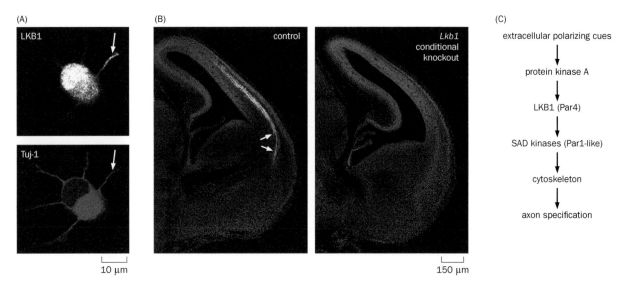

Figure 7-18 A cell polarity kinase cascade regulates axon specification. (A) Before overt axon differentiation in cultured hippocampal neurons, the LKB1 protein (top) is highly concentrated in only one process (arrows) out of several processes equally enriched in a different marker, Tuj-1 (bottom). **(B)** Conditional knockout of *Lkb1* in the embryonic neocortex leads to a severe reduction in axon abundance (green, indicated by arrows in control). **(C)** A working model for the signaling pathway from extracellular polarizing cues to the regulation of cytoskeleton for axon specification. (A, from Shelly M, Canceddia L, Heilshom S, et al. [2007] *Cell* 129:565–577. With permission from Elsevier Inc. B, from Barnes AP, Lilley BN, Pan YA, et al. [2007] *Cell* 129:549–563. With permission from Elsevier Inc.)

investigated. An important consequence of these signaling events is the modulation of cytoskeletal elements—F-actin and microtubules—that enable the rapid extension of axons. The polarity signal also leads to the differential organization of microtubule polarity in axons and dendrites (Section 2.3); this helps maintain neuronal polarity by facilitating transport of various cargos to specific compartments appropriate to their functions, as will be discussed in the next section.

7.8 Local secretory machinery promotes dendrite morphogenesis and microtubule organization

What makes dendrites morphologically different from axons (Figure 1-15)? As in the search for axon guidance molecules discussed in Section 7.5, researchers can address this question by conducting unbiased forward genetic screens (Section 14.6) to determine which genes, when mutated, preferentially affect the growth of dendrites but not of axons. One such screen in *Drosophila* embryonic sensory neurons revealed that mutations in components of a well-studied cell biological process, vesicle trafficking from the endoplasmic reticulum (ER) to the Golgi apparatus (Section 2.1), preferentially disrupted dendrite growth (**Figure 7-19**A). The Golgi apparatus was traditionally thought to be near the nucleus. However, fragments of the Golgi apparatus called **Golgi outposts** are distributed in the dendrites of cultured mammalian hippocampal neurons (Figure 7-19B) and *Drosophila* sensory neurons but are undetectable in axons. Disruption of Golgi outposts impaired dendrite growth and branch extension in mammalian as well as *Drosophila* neurons.

As discussed in Section 2.3, a key distinction between dendrites and axons is the orientation of their microtubules. In axons, the growing (plus) ends of the microtubules face the termini, such that microtubules are plus-end-out, whereas dendrites contain mixed populations of plus-end-out and minus-end-out microtubules. In nonneuronal cells, the minus ends are usually nucleated in the centrosome near the center of the cells. Interestingly, Golgi outposts can also serve as microtubule nucleation centers. Thus, Golgi outposts likely serve two separate functions: they fuel the growth of dendrites by facilitating delivery of lipid and membrane proteins from the ER to the plasma membrane, and they organize microtubules to facilitate bidirectional transport of cargos between cell bodies and dendritic terminals (Figure 7-19C).

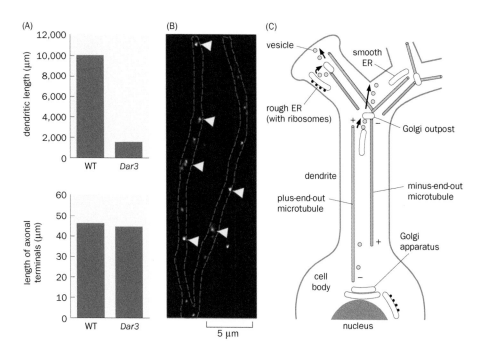

Figure 7-19 **Golgi outposts, dendrite elaboration, and microtubule nucleation.** (A) Mutations in *Dar3* (*dendritic arbor reduction*), which encodes a *Drosophila* GTPase essential for ER → Golgi vesicle trafficking, preferentially affect dendrite elaboration, with minimal effects on axon growth. WT, wild type. (B) Golgi outposts (arrowheads), visualized by a fluorescently tagged Golgi marker, are distributed along the dendrites (outlined) of a cultured rat hippocampal neuron. (C) Schematic summarizing the function of the secretory pathway in dendrites. Vesicle trafficking (black arrows) from ER to plasma membrane via Golgi outposts fuels growing dendrites with new membrane lipids and proteins. Some of these proteins may be synthesized locally on the rough ER in dendrites. Golgi outposts may also serve as nucleation centers for microtubules, producing both plus-end-out (orange) and minus-end-out (blue) microtubules, with their minus ends interacting with the Golgi outposts and accounting for their bidirectional organization in the dendrites. (A, from Ye B, Zhang Y, Song W, et al. [2007] *Cell* 130:717–729. B, from Horton AC & Ehlers MD [2003] *J Neurosci* 23:6188–6199. Copyright ©2003 Society for Neuroscience. See also Ori-McKenney KM, Jan LY, & Jan YN [2012] *Neuron* 76:921–930.)

ER → Golgi trafficking is an essential step in the secretory pathway for delivering transmembrane and secreted proteins to the plasma membrane. In principle, neuronal proteins destined for the secretory pathway can be delivered to axonal and dendritic compartments after being synthesized and processed by the ER and Golgi apparatus in the cell body. However, the presence of mRNA, ribosomes, ER, and Golgi outposts in the dendrites enables neurons to locally synthesize transmembrane and secreted proteins in dendrites far from their cell bodies (Section 2.2). Local protein synthesis confers flexibility in response to synapse-specific signaling and thus plays an important role in synaptic plasticity, as will be further discussed in Chapter 11.

7.9 Homophilic repulsion enables self-avoidance of axonal and dendritic branches

Dendrites, as their name (from *dendron* in Greek, meaning "tree") indicates, are often highly branched, enabling them to efficiently sample the input space. Axons also often branch along their long journey so that an individual neuron can send information to multiple postsynaptic targets at distinct locations. Dendritic and axonal branching utilizes one of two mechanisms: (1) growth cone splitting or (2) **interstitial branching**—that is, extending a collateral from the side of a process (**Figure 7-20**). In either case, the two branches must segregate in order for them to extend in different directions. Studies of *Drosophila* **Dscam** proteins suggest that, in many neurons, an active process is employed to prevent dendritic or axonal branches of the same neuron from sticking to each other.

Dscam proteins are evolutionarily conserved cell adhesion proteins with multiple extracellular immunoglobulin (Ig) domains and fibronectin repeats. (It is named after <u>D</u>own <u>s</u>yndrome <u>c</u>ell <u>a</u>dhesion <u>m</u>olecule because the human gene is located on chromosome 21, which is triplicated in Down syndrome; it is unknown whether an extra copy of *Dscam* contributes to symptoms of Down syndrome.) Remarkably, alternative splicing of a single *Dscam* gene can produce 38,016 different isoforms, with 19,008 extracellular domain variants (**Figure 7-21A**). Important insights into the function of *Dscam* diversity came from the convergence of analyses of *Dscam* mutant phenotypes, expression patterns of *Dscam* isoforms, and biochemical and structural studies of Dscam proteins. When *Dscam* was deleted from whole flies or from individual neurons within genetically mosaic flies, sister dendritic and axonal branches from the same neuron stuck together after branching, as exemplified by the dendrites of sensory neurons that innervate

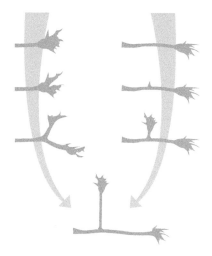

Figure 7-20 **Two distinct branching mechanisms.** A dendrite or an axon can generate two branches via growth cone splitting (left), or interstitial branching, wherein a branch is initiated from the trunk of an existing process (right).

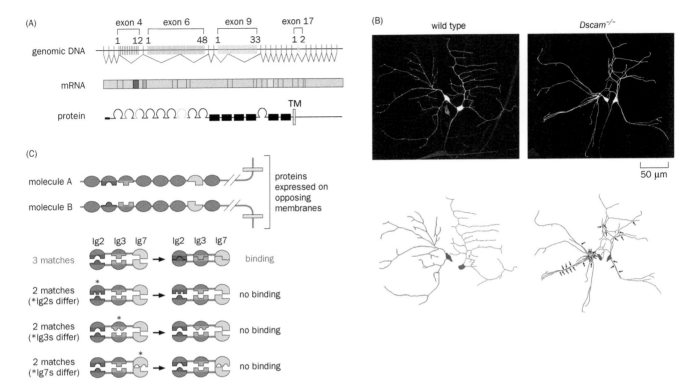

Figure 7-21 *Dscam* diversity and self-avoidance of axonal and dendritic branches. (A) Top, genomic structure of the *Drosophila Dscam* gene. Each vertical bar represents an exon. Middle, alternative mRNA splicing ensures that in the mature mRNA, exon 4 (red) is chosen from one of 12 variants, exon 6 (blue) from one of 48 variants, exon 9 (green) from one of 33 variants, and exon 17 (yellow) from one of 2 variants. Because these choices are made independently, the *Dscam* gene can produce 38,016 isoforms (= 12 × 48 × 33 × 2). Bottom, protein structure of Dscam. The extracellular portion of Dscam protein comprises 10 immunoglobulin (Ig) domains (arches) and 6 fibronectin type III domains (rectangles). The second, third, and seventh Ig domains include amino acids encoded by variable exons 4, 6, and 9, respectively. TM, transmembrane domain. **(B)** Dendritic branches of wild-type sensory neurons extend away from each other without overlap (left). In a *Dscam* mutant (right), dendritic branches from the same neuron stick together (arrows). Original images (top) and interpretive drawings (bottom) are shown for two neighboring neurons. **(C)** Summary of Dscam binding. Top, structural studies reveal that the three variable Ig domains contribute to the binding interface of two Dscam molecules located on opposing membranes. Bottom, biochemical studies show that Dscam mediates strong homophilic binding only when all three variable Ig domains are identical. (A, adapted from Schmucker D, Clemens JC, Shu H, et al. [2000] *Cell* 101:671–684. With permission from Elsevier Inc. B, from Matthews BJ, Kim ME, Flanagan JJ, et al. [2007] *Cell* 129:593–604. With permission from Elsevier Inc. C, adapted from Sawaya MR, Wojtowicz WM, Andre I, et al. [2008] *Cell* 134:1007–1018. With permission from Elsevier Inc. See also Wojtowitz WM, Wu W, Andre I, et al. [2007] *Cell* 130:1134–1145.)

the larval body wall (Figure 7-21B). Furthermore, the Ig domains encoded by variable exons enable Dscam proteins to bind one another *in trans* (from opposing membranes) in an isoform-specific manner, such that only identical *Dscam* isoforms bind strongly to each other (**homophilic binding**; Figure 7-21C). Interestingly, homophilic binding of Dscam extracellular domains leads to cytoplasmic signaling that results in *repulsion* between binding partners. Another clue to the puzzle is that individual neurons appear to express a stochastic set of alternatively spliced *Dscam* isoforms.

Taken together, these findings suggest a model by which molecular diversity is used to regulate dendritic and axonal branching: each neuron expresses multiple *Dscam* isoforms; when a nascent branch forms, the sister branches have the same Dscam proteins and hence bind to and repel each other. This homophilic repulsion ensures **self-avoidance**: different axonal branches from the same neuron go their own ways to innervate distinct targets, and different dendritic branches from the same neuron maximize the territory they cover and minimize overlap. Different neurons express distinct sets of *Dscam* isoforms, so no strong repulsion exists between their axons or dendrites. This enables different neurons to either fasciculate (bundle) their axons along a common pathway or have overlapping dendritic territories.

Do mammals use similar mechanisms? There are only two Dscam isoforms in mammals, not enough to distinguish dendrites of the same neuron from many

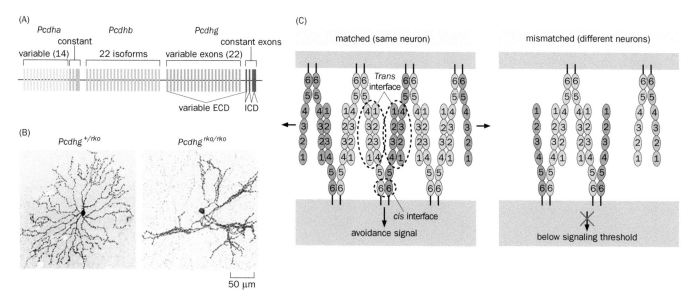

Figure 7-22 Clustered protocadherins in dendritic self-avoidance of mouse retinal neurons. (A) The mouse genome encodes three protocadherin clusters. In the *Pcdha* and *Pcdhg* clusters, an individual variable exon encodes the extracellular domain (ECD), the transmembrane domain, and a portion of the intracellular domain (yellow or cyan); this exon is joined to exons encoding the constant intracellular domain (ICD) via splicing. In the *Pcdhb* cluster, each exon (green) encodes the entire protein. **(B)** While the dendrites of a control starburst amacrine cell (heterozygous for a retinal knockout, or *rko*) spread broadly outward (left), homozygous retinal knockout of the entire *Pcdhg* cluster causes the dendrites of a starburst amacrine cell to clump together (right). **(C)** Model of Pcdh action based on biochemical and structural studies. Extracellular cadherin domain (EC) 6 mediates *cis* dimerization promiscuously for different isoforms (distinct colors represent different isoforms). EC1–EC4 mediate *trans* interaction in an isoform-specific manner. Left, when two cellular processes (gray) are from the same neuron and hence express the same Pcdh isoforms, Pcdhs can form large hetero-multimeric complexes sufficient to trigger avoidance signaling. Leftward and rightward arrows indicate that the complex extends in both directions. Right, when two processes are from different neurons, mismatched Pcdhs prevent the formation of large adhesion complexes and do not trigger avoidance signaling. (A & B, adapted from Lefebvre JL, Kostadinov D, Chen WV, et al. [2012] *Nature* 488:517–521. With permission from Springer Nature. C, from Rubinstein R, Thu CA, Goodman KM, et al. [2015] *Cell* 163:629–642.)

other neurons nearby. Nevertheless, in a remarkable example of convergent evolution, the **protocadherins**, analogous cell-surface molecules exhibiting molecular diversity in mammals, function similarly to *Drosophila* Dscam to regulate dendrite morphogenesis. Cadherins are ancient calcium-dependent cell adhesion proteins that play numerous important roles in cell–cell adhesion and morphogenesis during development (Box 5-1; Sections 5.16 and 5.18). Cadherin proteins form a sizable family that can be used for cell–cell recognition: a typical vertebrate genome contains about 20 different cadherin genes and additional genes encoding protocadherins that share sequence homology with cadherins. In particular, three genetic loci, *Pcdha, Pcdhb,* and *Pcdhg,* produce 14 α-, 22 β-, and 22 γ-protocadherin isoforms, respectively, for a total of 58 extracellular domain variants in mice (**Figure 7-22**A). Knockout of the *Pcdhg* cluster (which encodes 22 γ-protocadherins) in the retina caused the dendrites of individual amacrine cells to form clumps (Figure 7-22B), suggesting that clustered protocadherins mediate homophilic repulsion just as *Drosophila* Dscam does.

Each protocadherin contains six extracellular cadherin (EC) domains. Biochemical and structural studies indicate that EC6 mediates promiscuous *cis* dimerization while EC1–EC4 mediate isoform-specific *trans* interaction. Modeling suggests that if two processes interacting in *trans* are from the same neuron and express the same combination of clustered protocadherin isoforms, a large hetero-multimer is assembled, triggering avoidance. However, cells that share limited isoforms cannot assemble strong enough adhesion complexes for avoidance signaling (Figure 7-22C). Furthermore, combinatorial expression of multiple isoforms in the same cell can in principle produce many more than 58 distinct protein complexes for specific binding.

A phenomenon related to dendritic self-avoidance is dendritic tiling, introduced in Section 4.15 in the context of retinal neuronal types. Here, dendrites from different individual retinal neurons of the same type avoid each other, such that, collectively, any retinal neuronal type samples the entire visual world without redundancy (Figure 4-26). Dendritic tiling also occurs in sensory neurons that innervate the *Drosophila* larval body wall and belong to the same type (Figure 7-21B). Ablating one neuron causes dendrites of neighboring neurons of the same type to invade the empty space that was occupied by the ablated neuron, suggesting that mutual repulsion of dendrites from neighboring neurons of the same type results in dendritic tiling. Despite playing a key role in dendritic self-avoidance, neither Dscam in flies nor clustered protocadherins in the mammalian retina appear to regulate dendritic tiling. Thus, dendritic tiling might involve yet-to-be-discovered cell-surface recognition molecules and mechanisms. Interestingly, *Drosophila* Dscam2 (a paralog of Dscam1 with only two alternative spliced isoforms in the extracellular domain) and mouse clustered Pcdhα appear to regulate axonal tiling of specific neurons; in both cases, expression of a single, specific isoform in a given neuronal type ensures that axons of the same neuronal type avoid each other.

7.10 Selection of subcellular sites for synaptogenesis uses both attractive and repulsive mechanisms

As axons reach their targets and dendrites elaborate their branches, the next phase of neural development begins: the formation of synapses that allow neurons to communicate with each other or with their muscle targets (Figure 7-1). In this section, we discuss how sites for synapse formation are selected, and subsequent sections address how synaptogenesis occurs.

Many axons in the mammalian central nervous system (CNS) must find not only their postsynaptic partners but also target synapses to specific subcellular sites on postsynaptic neurons. For instance, three classes of neocortical GABAergic interneurons form synapses on different subcellular compartments of pyramidal neurons: basket cells, chandelier cells, and Martinotti cells form synapses at the somata, axon initial segments, and distal dendrites of target pyramidal neurons, respectively (Figure 3-46). How do GABAergic neurons selectively form synapses on specific subcellular compartments of their target neurons? Insights into this problem have come from studies of cerebellar basket cells, which synapse selectively onto the somata and axon initial segments of their postsynaptic target Purkinje cells. Neurofascin, an Ig superfamily cell adhesion molecule (Ig CAM), serves as a cue to target basket cell axons. It forms a subcellular gradient along the surface of the Purkinje cell, with the highest concentration at the axon initial segment. This is because the cytoplasmic domain of neurofascin is associated with

Figure 7-23 Subcellular distribution of neurofascin directs basket cell presynaptic terminal formation. (A) In a wild-type Purkinje cell labeled in green by Purkinje cell marker calbindin (Calb), neurofascin (NF, in red) is concentrated at the axon initial segment (AIS, between the two arrows). Neurofascin is also distributed along the Purkinje cell body (arrowhead) in a gradient, with the highest concentrations being close to the axon. **(B)** In an *Ankg* mutant, neurofascin is distributed evenly along the Purkinje cell body and is present in the axon at distances farther from the soma, as indicated by the distance between the two arrows. The bracket indicates the length of the AIS in wild-type Purkinje cells. **(C)** Schematic showing basket cell axon terminals (green) and neurofascin (red) distribution in wild-type and *Ankg* mutant Purkinje cells. In the mutant, the neurofascin gradient in the Purkinje cell body and its restriction to the AIS segment of the axon are disrupted, causing basket cell axon terminals to overshoot the AIS and thus to branch abnormally. (Adapted from Ango F, di Cristo G, Higashiyama H, et al. [2004] *Cell* 119:257–272. With permission from Elsevier Inc.)

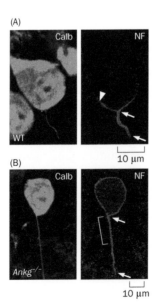

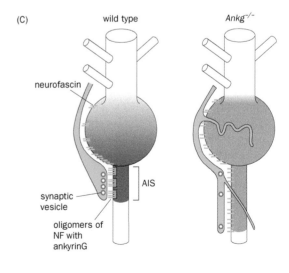

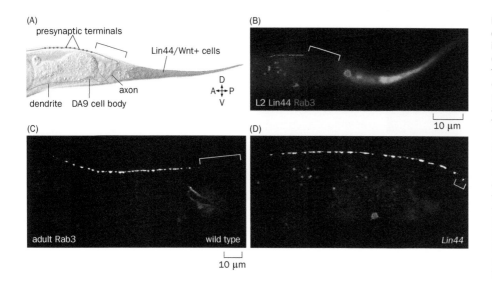

Figure 7-24 Presynaptic terminal distribution in the *C. elegans* DA9 neuron is regulated by an external Wnt gradient. (A) Schematic of the DA9 motor neuron in the posterior worm, depicting the location of its cell body, dendrite, axon, and presynaptic terminals. Bracket delineates an axonal region without presynaptic terminals in the wild-type worm. Lin44 is a *C. elegans* Wnt homolog produced by posterior cells. A, anterior; P, posterior; D, dorsal; V, ventral. (B) In wild-type second instar larvae (L2), posterior Lin44 expression (green) prevents presynaptic terminals (marked in red by the synaptic vesicle marker Rab3) from forming in the bracketed axon segment. (C & D) The bracketed synapse exclusion zone is longer in wild-type worms (Panel C) than in *Lin44* mutant worms (Panel D), as presynaptic terminal distribution extends more posteriorly. (From Klassen MP & Shen K [2007] *Cell* 130:704–716. With permission from Elsevier Inc.)

an intracellular scaffold protein called ankyrin G, which is highly concentrated at the axon initial segment. In *Ankg* mutant mice, which do not produce ankyrin G, neurofascin distribution became diffuse, as did the distribution of basket cell synapses (**Figure 7-23**). Thus, precise subcellular targeting of basket cell axons is directed by the selective subcellular distribution of an attractive cue in the target neuron. A recent study suggested that ankyrin G interaction with another Ig CAM, L1, regulates chandelier cell innervation of axon initial segments of neocortical pyramidal neurons.

A different mechanism for subcellular synaptic targeting has been uncovered in studies of the DA9 motor neuron of *C. elegans*. The DA9 neuron forms presynaptic terminals only along a specific segment of its axon's contact with its muscle targets (**Figure 7-24**). Synapse distribution in DA9 is directed by a posteriorly secreted Wnt protein. (In *C. elegans*, as in mammals, Wnt proteins are morphogens that specify global patterning in early developmental stages.) Wnt signaling mediated by a receptor expressed in the DA9 neuron inhibits synapse formation in the posterior segment of the axon. Synaptogenesis occurs only in an anterior zone, away from the influence of posterior Wnt. Thus, an extracellular protein that acts to pattern the global body plan also confines synapse formation to a specific segment along axons.

These examples illustrate that, as is the case for axon guidance (Figure 5-11A), the targeting of synapses to specific subcellular locations can be achieved through attractive or repulsive mechanisms and by using secreted or cell-surface-bound cues.

7.11 Bidirectional trans-synaptic communication directs the assembly of synapses

Once axons reach their final targets, they are ready to form synapses with their postsynaptic partners. What are the mechanisms that convert a dynamic growth cone into a stable presynaptic terminal? What initiates the differentiation of the postsynaptic specialization? Extensive communication between pre- and postsynaptic partners triggers the differentiation of both compartments, transforming an initial contact into what can be a lifelong union as synaptic partners.

The vertebrate neuromuscular junction (NMJ) has been used to study synapse development because of its experimental accessibility. A key step in postsynaptic development at the NMJ is the clustering of acetylcholine receptors (AChRs) on the muscle membrane apposing the presynaptic motor axon terminal, allowing efficient synaptic transmission (Section 3.1). Motor axon terminals produce a protein called **agrin**, named for its ability to cause AChR aggregation in cultured muscle cells (**Figure 7-25A**). Agrin acts through a receptor complex in muscle

Figure 7-25 Communication between motor axons and muscles in establishing the neuromuscular junction. (A) When chicken myotubes (control, top) are exposed to the agrin-containing fraction of *Torpedo* electric organ, AChR clusters form (bottom). AChRs are visualized by fluorescently tagged α-bungarotoxin, a snake-derived toxin that binds to and inhibits nicotinic AChR (Box 3-2). **(B)** Drawings depicting the staining results with α-bungarotoxin (red) in heterozygous control mice (top) and *Agrin* mutant mice (bottom). *Agrin* knockout results in severe disruption of AChR clustering (spread of red signals) and abnormal branching of motor axons (gray). *Musk* or *Lrp4* mutant mice exhibit similar phenotypes. **(C)** Schematic summary of communication between the motor axon terminal and the muscle cell. Muscle AChRs cluster spontaneously (1). The motor axon produces ACh (2) to disperse AChR clusters by endocytosis (3). A second product from the motor axon, agrin (4), acts via the MuSK/LRP4 receptor complex to inhibit AChR dispersion (5). Muscle LRP4 also serves as a ligand that signals back to the motor axon through an unknown receptor (yellow) to induce presynaptic differentiation (6). (A, from Godfrey EW. Nitkin RM, Wallace BG, et al. [1984] *J Cell Biol* 99:615–627. B, from Gautam M, Noakes PG, Moscoso L, et al. [1996] *Cell* 85:525–535. With permission from Elsevier Inc. C, adapted from Kummer TT, Misgeld T, & Sanes JR [2006] *Curr Opin Neurobiol* 16:74–82. With permission from Elsevier Inc. See also Zhang B, Luo S, Wang Q, et al. [2008] *Neuron* 60:285–297; Kim N, Stiegler AL, Cameron TO, et al. [2008] *Cell* 135:334–342; Yumoto N, Kim N, & Burden SJ. [2012] *Nature* 489: 438–442.)

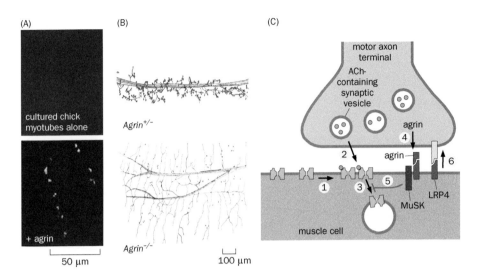

(A) cultured chick myotubes alone / + agrin / 50 μm

(B) *Agrin*$^{+/-}$ / *Agrin*$^{-/-}$ / 100 μm

(C) motor axon terminal / ACh-containing synaptic vesicle / agrin / MuSK / LRP4 / muscle cell

composed of **LRP4** (low-density lipoprotein receptor-related protein 4) and **MuSK** (muscle-specific receptor tyrosine kinase), which activates a signaling cascade that facilitates AChR clustering. Mice lacking *Agrin, Lrp4,* or *Musk* all exhibit severe defects in AChR clustering (Figure 7-25B) and die at the neonatal stage, providing *in vivo* support for the function of this ligand–receptor complex in neuromuscular junction development.

Interestingly, muscle are prepatterned in the mouse, as AChR clusters form at the right location in the absence of innervating motor axons, and motor axons appear to seek prepatterned AChR clusters to form synapses. In fact, the lack of AChR clustering in *Agrin* mutant mice can be suppressed by a complete lack of motor axon innervation. These experiments led to a model in which motor axon terminals send at least two signals to the muscle: one, likely the neurotransmitter ACh itself, to disperse AChR and the other, agrin, to counteract the dispersing signal and thereby reinforce AChR clustering near a functional (that is, ACh-producing) motor nerve terminal (Figure 7-25C). These findings highlight the extensive trans-synaptic communication necessary to establish just one aspect of synapse maturation, AChR clustering. Indeed, in addition to serving with MuSK as a co-receptor for agrin, muscle-derived LRP4 also induces presynaptic terminal differentiation independently of MuSK (Figure 7-25C). Additional motor axon- and muscle-derived signals likely act to stabilize and maintain the synaptic connections.

Of the trans-synaptic signaling molecules identified in the CNS, the transmembrane proteins **neurexin** and **neuroligin** have been studied most extensively. Neurexins are enriched on the presynaptic membrane, and their binding partner neuroligins are located on the postsynaptic membrane (**Figure 7-26**A). (See Figures 3-10 and 3-27 for the roles of neurexin and neuroligin in organizing presynaptic terminals and postsynaptic densities.) Like agrin at the neuromuscular junction, neurexins, even when expressed in cultured nonneuronal cells, can induce apposing neurons to develop postsynaptic specializations, including clustering of postsynaptic scaffold proteins and receptors for glutamate or GABA (Figure 7-26B). Conversely, neuroligins expressed in cultured nonneuronal cells can induce apposing neurons to develop presynaptic properties, including active zones and synaptic vesicle clusters (Figure 7-26C). These data suggest that neuroligin and neurexin act as trans-synaptic signals that induce pre- and postsynaptic assemblies, respectively.

Despite the robust synapse-inducing activity demonstrated by neuroligins *in vitro*, mice lacking all three neuroligin genes still have the normal number of synapses, albeit with severely reduced synaptic transmission. While these knockout experiments revealed an essential role of neuroligin in synaptic maturation and transmission, they suggest that neuroligins are not absolutely essential for synapse formation *in vivo*. A likely possibility is that neuroligins act in parallel with other trans-synaptic signaling pathways to induce synapse development;

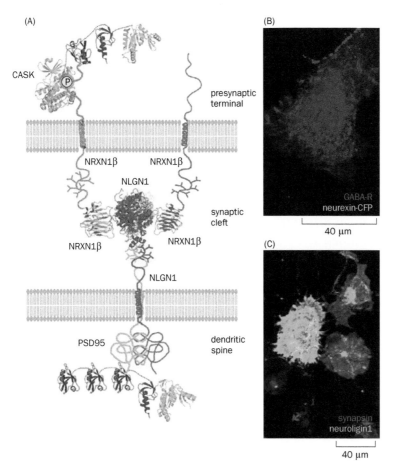

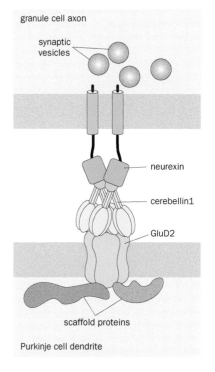

Figure 7-26 Neurexin/neuroligin-mediated trans-synaptic interaction can induce synapse assembly. (A) A structural model of trans-synaptic binding between a dimer of neuroligin1 (NLGN1) and two molecules of NRXN1β, a neurexin isoform. The cytoplasmic domain of NLGN1 binds one of the three PDZ domains (red) of PSD95, a postsynaptic scaffold protein (Figure 3-27); the cytoplasmic domain of NRXN1β binds the PDZ domain of CASK, a presynaptic scaffold protein. **(B)** Expression of neurexin 1β (tagged with cyan fluorescent protein, CFP) on the surface of a fibroblast (blue) induces GABA receptor clusters (red) in co-cultured neurons. **(C)** Expression of neuroligin1 (green) in HEK293 cells (the cell on the left express the highest amount) induces clustering of synapsin, a synaptic vesicle–associated protein (red), in passing axons of co-cultured neurons. Yellow indicates co-localization of synapsin and neuroligin. (A, from Südhof TC [2008] *Nature* 455:903–911. With permission from Springer Nature. B, from Graf E, Zhang X, Jin SX, et al. [2004] *Cell* 119:1013–1026. With permission from Elsevier Inc. C, from Scheiffele P, Fan J, Choih J, et al. [2000] *Cell* 101:657–669. With permission from Elsevier Inc. For a summary of other proteins that mediate trans-synaptic interactions, see Südhof TC [2018] *Neuron* 100:276–293.)

these other pathways may compensate for synapse development in the absence of neuroligins. Indeed, many other trans-synaptic interacting molecules have been identified. An interesting example involves a tripartite complex comprising neurexin, cerebellin1, and glutamate receptor δ2 (GluD2), best studied in the granule cell → Purkinje cell synapse in the cerebellum (**Figure 7-27**). Granule cell axons secrete cerebellin1, which binds simultaneously to presynaptic neurexin and postsynaptic GluD2, a member of the ionotropic glutamate receptor family that functions as a synaptic adhesion protein rather than a glutamate receptor. Mice deficient for cerebellin1 or GluD2 exhibit a severe reduction in the number of granule cell → Purkinje cell synapses, indicating that this tripartite complex is essential for the formation and/or maintenance of these synapses.

Trans-synaptic signaling molecules play diverse and important roles in the human nervous system, and mutations in many of them, including neurexins, neuroligins, and their associated scaffold proteins, have been linked to neurodevelopmental disorders, including autism and schizophrenia; we will discuss these links further in Chapter 12.

Figure 7-27 A tripartite trans-synaptic complex. Schematic of a tripartite complex comprising two neurexin molecules on the presynaptic (granule cell) membrane, two hexamers of cerebellin1, and four subunits of GluD2 on the postsynaptic (Purkinje cell) membrane. The soluble cerebellin1, secreted from presynaptic granule cells, binds both neurexin and GluD2 and thus serves as a bridge between pre- and postsynaptic membranes. (From Yuzaki M [2018] *Ann Rev Physiol* 80:243–262.)

7.12 Astrocytes stimulate synapse formation and maturation

Synapses are often wrapped by glia: astrocytes wrap CNS synapses and Schwann cells wrap neuromuscular junctions (Figure 3-3). In addition to regulating many aspects of synaptic function, such as neurotransmitter recycling (Figure 3-12), glia also regulate synapse development.

The roles of astrocytes in synapse development have been examined in cultured retinal ganglion cells. Purified RGCs cultured *in vitro* extend axons and dendrites, but exhibit little synaptic activity, as measured by spontaneous postsynaptic currents. (In this culture system, RGCs constitute both the presynaptic and postsynaptic cells even though RGCs do not form synapses with each other *in vivo*.) When RGCs were co-cultured with astrocytes, postsynaptic currents increased dramatically in both magnitude and frequency (Figure 7-28A). This was due both to increased synapse number and enhanced synaptic strength. Conditioned media from cultured astrocytes could mimic the effect of astrocytes; this finding provided a way to biochemically identify astrocyte-derived factors that could promote synapse number and strength. **Thrombospondins (TSPs)**, a class of astrocyte-derived factors, were found to enhance synapse number: adding TSP1 to the RGC culture mimicked the synaptogenic effects of astrocytes or astrocyte-conditioned media, as measured by RGC synapse numbers (Figure 7-28B). However, although TSP-induced synapses appeared structurally normal, they were functionally silent, as measured by spontaneous postsynaptic currents. Another class of astrocyte-derived factor—glypicans, glycosylated extracellular proteins anchored to the cell surface by GPI—was necessary to produce functionally mature synapses (Figure 7-28C). Further studies suggested that TSPs use a postsynaptic Ca^{2+} channel subunit called α2δ-1 as a receptor to promote synapse formation. (The role of α2δ-1 as a TSP receptor is unrelated to its Ca^{2+} channel function.) Glypican binds to a receptor protein tyrosine phosphatase (RPTPδ) on the presynaptic membrane, triggering the release of neuronal pentraxin 1 (NP1). NP1 enhances synaptic efficacy by binding to the AMPA receptor GluA1 and promoting its concentration on the postsynaptic surface. Importantly, mouse knockouts for TSPs, glypicans, α2δ-1, or RPTPδ all exhibit reduced synapse numbers or synaptic transmission efficacy *in vivo*, supporting the physiological function of these proteins in synapse formation and maturation.

In summary, synapse development requires extensive communication between pre- and postsynaptic partners and glia. Such communication can be achieved via secreted factors (e.g., agrin, cerebellin, thrombospondin) or transmembrane proteins that interact across the synaptic cleft (e.g., neurexin and neuroligin). These signaling events ensure that presynaptic terminals develop active zones for efficient neurotransmitter release and apposing postsynaptic densities are enriched

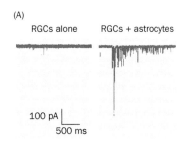

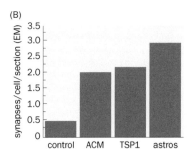

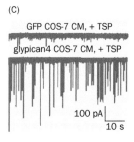

Figure 7-28 Secreted factors from astrocytes promote synapse formation and maturation. (A) Whole-cell recording of spontaneous postsynaptic currents from a retinal ganglion cell (RGC) cultured with other purified RGCs (left) or from an RGC co-cultured with purified RGCs and astrocytes (right). The presence of astrocytes increases the frequency and magnitude of spontaneous postsynaptic currents. **(B)** Co-culture with astrocytes (astros) causes a sixfold increase in RGC synapse number as quantified from electron micrographs. This effect is mimicked by astrocyte-conditioned media (ACM) or purified thrombospondin1 (TSP1). **(C)** TSP is insufficient to enhance synaptic strength, as measured by postsynaptic currents (top). Addition of media from COS-7 cells expressing glypican4 enhances synaptic strength (bottom). (A, from Ullian EM, Sapperstein SK, Christopherson KS, et al. [2001] *Science* 291:657–661. With permission from AAAS. B, adapted from Christopherson KS, Ullian EM, Stokes CC, et al. [2005] *Cell* 120:421–433. With permission from Elsevier Inc. C, adapted from Allen NJ, Bennett ML, Foo LC, et al. [2012] *Nature* 486:410–414. With permission from Springer Nature. See also Farhy-Tselnicker I, van Casteren ACM, Lee A, et al. [2018] *Neuron* 96:428–445.)

in neurotransmitter receptors. As discussed in Chapter 5, neuronal activity plays a major role in refining synaptic connections during development of the vertebrate nervous system. We next examine how neuronal activity influences synapse maturation.

7.13 Activity and competition sculpt synaptic connectivity

Our discussion so far has emphasized the *progressive* events in nervous system development, such as neurogenesis, axon extension, dendrite elaboration, and synapse formation. In the next three sections, we will discuss *regressive* events that are just as important for proper construction of the nervous system. These include elimination of supernumerary synapses, pruning of exuberant axons, and programmed death of excess neurons. We begin with **synapse elimination**, the removal of supernumerary synapses.

Not all synapses formed during development last. The establishment of mammalian neuromuscular connectivity provides a striking example of synapse elimination. In mature mice, each muscle fiber (an individual muscle cell) is innervated by a single motor axon, but in newborn mice, each muscle fiber is innervated by as many as 10 motor axons. During the first two weeks of postnatal life, poly-innervation is refined into mono-innervation through synapse elimination (**Figure 7-29**A). Time-lapse imaging studies have charted a detailed course. In newborn mice, multiple axons form synapses at the same neuromuscular junction, intermingling their terminals. Next, terminals originating from different motor axons gradually segregate: one terminal expands while others shrink. Eventually, all of the motor axons except one lose their territory and withdraw (Figure 7-29B). Blocking neuromuscular synaptic transmission slows down the process leading from poly-innervation to mono-innervation, suggesting that synapse elimination is an activity-dependent, competitive process. Indeed, if synaptic transmission is eliminated in a subset of axons by conditionally deleting choline acetyltransferase, an enzyme essential for ACh synthesis, mutant axons that cannot release ACh are usually unable to compete effectively against wild-type axons (Figure 7-29C).

As will be discussed in detail in Chapter 8, each muscle (composed of many muscle fibers) is innervated by multiple motor neurons belonging to the same **motor pool**. Classic embryological experiments in the chick revealed that motor neurons innervating different muscles already exhibit high specificity for their target muscles when they begin to invade the muscle field. This is likely mediated

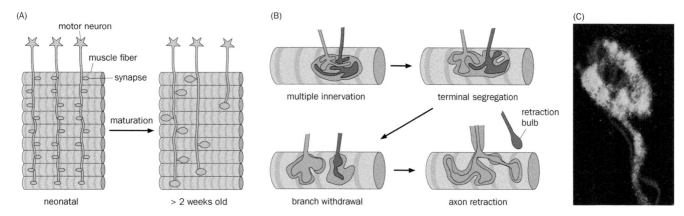

Figure 7-29 Competitive synapse elimination shapes motor neuron–muscle connectivity. (A) Schematic of neuromuscular junction maturation. In newborn mice, each muscle fiber is innervated by multiple motor neurons. During the first two postnatal weeks, synapses are eliminated through a local competitive process until each muscle fiber is innervated by a single motor neuron. **(B)** Summary of synapse elimination steps. Inputs from different motor neurons are initially intermingled. Their terminals then segregate. As one terminal expands, others shrink and withdraw. **(C)** Motor axons that lack choline acetyltransferase (ChAT), and thus cannot release ACh, do not compete effectively against co-innervating axons that express ChAT. In this micrograph, two axons innervating the same neuromuscular junction are stained in red. Only one axon contains ChAT protein (labeled cyan via an anti-ChAT antibody). The ChAT-containing axon is thicker, occupies more territory, and is predicted to win. (A, adapted from Tapia JC, Wylie JD, Kasthuri N, et al. [2012] *Neuron* 74:816–829. With permission from Elsevier Inc. B, from Sanes JR & Lichtman JW [1999] *Annu Rev Neurosci* 22:389–442. With permission from Annual Reviews. C, adapted from Buffelli M, Burgess RW, Feng G, et al. [2003] *Nature* 424:430–434. With permission from Springer Nature.)

Figure 7-30 A connectome consisting of a muscle pair and a complete set of innervating motor axons. Top, complete reconstruction of the connectivity of all motor neurons (15 on the left, 14 on the right) with a pair of small muscles (~200 muscle fibers per side) in the ear of a one-month-old mouse. Each motor axon is differentially colored, with branches ending on individual muscle fibers (not shown). Bottom, branching patterns of each motor neuron from the left (L1–L15) and right (R1–R14) are arranged according to the number of muscle fibers the neuron innervates (the size of motor unit, or MU), shown below the motor neuron number on the top right. The right-side neurons are flipped for ease of comparison with the left-side neurons. At the level of individual motor neurons, left- and right-side branching patterns do not show obvious stereotypy. (From Lu J, Tapia JC, White OL, et al. [2009] *PLoS Biol* 7:e1000032.)

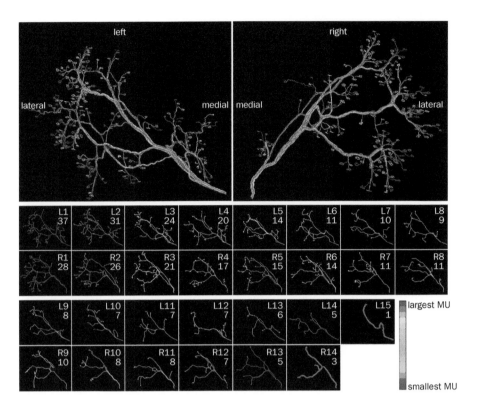

by molecular determinants that match different motor pools and muscles (see also Figure 7-9). However, within each motor pool–muscle pairing, an individual motor neuron innervates only a subset of muscle fibers; collectively, motor neurons of the same motor pool innervate all muscle fibers by making exuberant connections followed by synapse elimination (Figure 7-29A). What does the outcome look like after this process?

The **connectome** of a small ear muscle of a mouse—the pattern of synaptic connections formed between the entire set of motor neurons (about 15) and the entire set of muscle fibers they innervate (about 200)—has been reconstructed for each side (**Figure 7-30**). Although the motor neurons targeting the ear muscles on the left and right sides share identical genomes, the branching patterns of the left- and right-side motor axons differ considerably. These observations suggest that the detailed connection patterns *between individual motor neurons of a motor pool and individual muscle fibers of a muscle* are not genetically specified but are likely a result of synapse elimination based on local, activity-dependent competition. Patterns nevertheless emerge. For example, each muscle is innervated by motor neurons that synapse with a few to several dozen muscle fibers. The number of muscle fibers a motor neuron innervates correlates with its axon diameter. These properties provide the anatomical basis for the "size principle," which we will examine further in Chapter 8: motor neurons are usually progressively activated by increasing activity, with the smaller motor neurons (which innervate fewer muscle fibers) becoming activated before larger motor neurons (which innervate more muscle fibers). Future analysis of how this relatively simple connectome emerges during development will shed light on the rules by which neuronal activity sculpts synaptic connectivity.

Activity-dependent synapse elimination also occurs widely in the CNS. We have already seen salient examples in our study of the visual system. RGCs from both eyes initially connect to the same target neurons in the lateral geniculate nuclei (LGN); through activity-dependent mechanisms, the connections are refined such that each LGN neuron is innervated by RGCs from only a single eye in the adult (Figure 5-26). Recent studies have uncovered an important role for microglia, the resident immune cells of the nervous system (Figure 1-9), in the process of synapse elimination. Microglia engulf RGC presynaptic terminals in

Figure 7-31 Microglia engulfing synapses. (A) Microglia in the lateral geniculate nucleus are labeled in green. Fluorescently tagged anterograde tracers injected into the contralateral (red) and ipsilateral (cyan) eyes can be seen within the labeled microglia, presumably a result of engulfment. Grid size is 2 μm. Inset, high-magnification view of the microglia cell body. **(B)** Compared with wild-type control, the percentage of microglia with engulfed materials from RGC axons is markedly reduced in CR3 knockout (KO) mice. CR3 is the receptor for the complement component C3. (From Schafer DP, Lehrman EK, Kautzman AG, et al. [2012] *Neuron* 74:691–705. With permission from Elsevier Inc.; see also Paolicelli RC, Bolasco G, Pagani F, et al. [2011] *Science* 333:1456–1458; Lehrman EK, Wilton DK, Litvina EY, et al [2018] *Neuron* 100:120–134.)

(A)

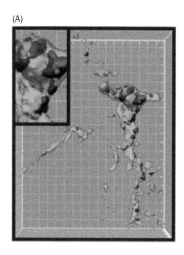

(B)

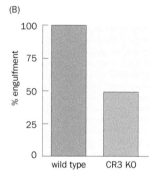

the LGN by recognizing components of the **complement cascade** on the surface of RGCs, which serve as "eat me" signals (**Figure 7-31**). (The complement cascade is part of an innate immune system that complements the ability of antibodies and phagocytic cells to clear pathogens and damaged cells from the organism.) RGCs also produce "don't eat me" signals, such as the cell surface antigen CD47, that inhibit microglia engulfment; CD47 preferentially localizes to more active pre-synaptic terminals, contributing to selective elimination of less active synapses. As we will see in Chapter 12, abnormalities in the complement cascade contribute to pathogenesis in both Alzheimer's disease and schizophrenia.

7.14 Developmental axon and dendrite pruning refines wiring specificity

The developing nervous system also employs **stereotyped axon pruning** to sculpt connectivity patterns. Stereotyped axon pruning differs from neuromuscular refine-ment in that the outcomes do not vary from animal to animal and the process usually involves elimination of long-distance axonal projections. One of the most striking examples of such pruning was discovered while studying mammalian neocortical layer 5 subcerebral projection neurons (SCPNs; Section 7.4) by inject-ing retrograde tracers into their target regions at different stages of development. SCPNs from both the motor and visual cortices first project axons to the spinal cord and then extend interstitial branches (Figure 7-20) to innervate the superior colliculus and several brainstem targets. Subsequently, SCPNs from the motor cortex prune their branches innervating the superior colliculus, while visual cor-tical SCPNs prune their branches innervating the spinal cord, consistent with their respective functions in controlling body and eye movement (**Figure 7-32**).

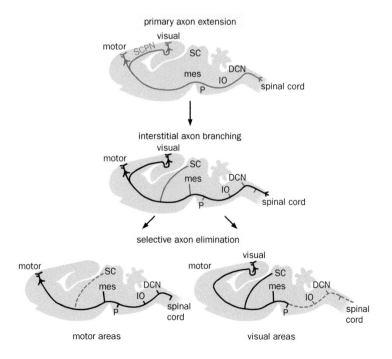

Figure 7-32 Stereotyped axon pruning refines connection specificity of cortical layer 5 subcerebral projection neurons. SCPNs in the motor and visual cortices initially develop similar projection patterns. Both extend axons to the spinal cord before sending interstitial branches to the superior colliculus (SC) and brainstem structures (mes, mesencephalon; p, pons; IO, inferior olive; DCN, dorsal column nuclei). Wiring specificity is shaped by subsequent axon pruning. Motor cortical SCPNs selectively prune their SC branches, while visual cortical SCPNs selectively prune their branches to the spinal cord and most brainstem targets. (Adapted from Luo L & O'Leary DM [2005] *Annu Rev Neurosci* 28:127–156. With permission from Annual Reviews.)

It remains unclear why certain neurons prune exuberant connections to achieve wiring specificity in their long-distance projections, while most neurons utilize specific guidance mechanisms from the outset (see Figure 5-3 for a comparison of these two mechanisms). One speculation for the stereotyped axon pruning strategy employed by SCPNs is evolutionary constraint: ancestral SCPNs might have innervated all subcortical targets, and after SCPNs diversified their functions in controlling, for example, body or eye movement, it was easier to evolve selective axon pruning than to alter initial guidance to achieve the final connection specificity of different SCPN subtypes.

Stereotyped axon pruning can be achieved by distinct cellular and molecular mechanisms. Axons can undergo distal-to-proximal retraction, such that materials from the retracting axons are recovered by the neuron. Repulsive axon guidance cues and receptors, such as semaphorins and plexins (Box 5-1) as well as Rho GTPase signaling (Box 5-2), play important roles in axon retraction. Axons can also undergo **developmental axon degeneration**, during which axons fragment into pieces subsequently engulfed by nearby glia. This mechanism is morphologically similar to **Wallerian degeneration** (first described by Augustus Waller in 1850), wherein, after axon transection, axons distal to the transection site fragment and then are eliminated. Although with different triggers, axon degeneration that occurs naturally during development and as a consequence of injury nevertheless shares mechanisms discussed in the following.

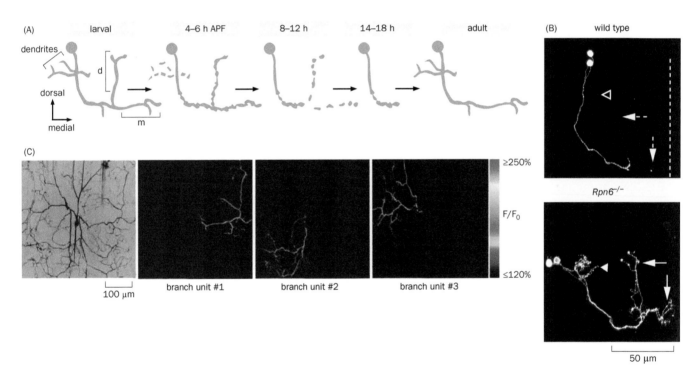

Figure 7-33 Axon and dendrite pruning during *Drosophila* metamorphosis via developmental degeneration. (A) Schematic of Kenyon cell (KC) dendrite and axon pruning during metamorphosis. In the larva (left panel), each KC extends a single process that gives rise to several dendritic branches near the cell body. The axon bifurcates to produce dorsal (d) and medial (m) branches. During the first 18 h after puparium formation (APF, the middle three panels), dendrites as well as the medial and dorsal axonal branches degenerate by breaking into pieces, retaining the axon before the original branching point. The axon then extends only medially to form the adult projection pattern (right). **(B)** Two wild-type (left) or *Rpn6* mutant (right) KCs from *Drosophila* at 18 h APF visualized by a mosaic labeling method (Section 14.16). Whereas larval-specific axonal branches (dashed arrows) and dendrites (open triangles) have been pruned in the wild type, they are inappropriately retained in *Rpn6* mutant neurons (arrows and arrowheads). Rpn6 is a key proteasome subunit conserved from yeast to human. Dotted lines indicate the midline. **(C)** Micrograph of dendrites of a live *Drosophila* body wall sensory neuron at an early pupal stage (left) and still images from a time-lapse movie (right) of three subsequent times showing Ca²⁺ transients localized to specific dendritic branch units. Degeneration of dendritic branch units follows the order of the appearance of Ca²⁺ transients. Ca²⁺ imaging was performed using a genetically encoded Ca²⁺ indicator. F/F_0, fluorescence intensity over basal fluorescence. (A, adapted from Luo L & O'Leary DM [2005] *Annu Rev Neurosci* 28:12–1567. With permission from Annual Reviews. B, from Watts RJ, Hoopfer ED, & Luo L [2003] *Neuron* 38:871–885. With permission from Elsevier Inc. C, from Kanamori T, Kanai MI, Dairyo Y, et al. [2013] *Science* 340:1475–1478. With permission from AAAS. See Zhai Q, Wang J, Kim A, et al. [2003] *Neuron* 39:217–225 and George EB, Glass JD, & Griffin JW, et al. [1995] *J Neurosci* 15:6445–6452 for the effect of proteasome and calpain inhibition on axon degeneration after injury.)

Axon and dendrite pruning occurs extensively in the insect nervous system during metamorphosis, when larval connectivity is transformed into adult connectivity to suit a new lifestyle (crawling versus flying). For example, at early pupal stages, larva-specific axonal branches of a specific type of Kenyon cell in the *Drosophila* mushroom body (Section 6.14) are eliminated via developmental axon degeneration, as are their dendrites; extension of adult-specific dendrites and axons follows (**Figure 7-33**A). Likewise, sensory neurons that innervate the larval body wall (Figure 7-21B) prune their dendrites in early pupa using a degenerative mechanism. Genetic analysis of these pruning processes identified molecular mechanisms shared between degeneration during development and after injury. For example, developmental degeneration was inhibited when the **ubiquitin-proteasome system**, a protein degradation system universally used by eukaryotes, was disrupted (Figure 7-33B). Proteasome inhibitors also slowed degeneration of distal axons following transection of the mouse optic nerve. Another shared mechanism is the activation of a Ca^{2+}-dependent protease called **calpain** during the degenerative process. For example, in the larval body wall sensory neuron, Ca^{2+} transients were observed in specific dendritic compartments shortly before degeneration (Figure 7-33C), and degeneration was delayed in calpain mutant flies. Likewise, blocking calpain activity interfered with degeneration of distal axons following transection.

These studies raised the possibility that an axon (and dendrite) self-destruction program may be employed during developmental degeneration and reactivated in adults in response to injury. This program may also be abnormally activated in certain neurodegenerative disorders, causing axon degeneration contributing to clinical symptoms. However, researchers have also found notable differences between naturally occurring and injury-induced degeneration. For example, forward genetic screen in *Drosophila* identified an evolutionarily conserved adaptor protein called Sarm, which promotes axon degeneration after injury, but the Sarm signaling pathway does not appear to be involved in developmental axon pruning. Further investigations of axon degeneration mechanisms will shed light on both neural development and disease.

7.15 Neurotrophins from target cells support survival of sensory, motor, and sympathetic neurons

Apoptosis (meaning "falling off" in Greek, as in leaves from a tree) is a form of programmed cell death and a widely occurring regressive event in development. Indeed, the phenomenon and core mechanisms of apoptosis are conserved from *C. elegans* to mammals and involve activation of a cascade of proteases called **caspases** that cleave many hundreds of protein substrates to bring about the demise of the cell. Apoptosis is employed in the development of most tissues and organs, including the nervous system. For example, about 40% of spinal motor neurons that innervate a chick's limbs die during the course of normal development.

Classic embryological manipulation experiments in chicks suggested that peripheral targets affect motor neuron numbers. Removal of a limb bud reduced the number of motor neurons innervating that limb, whereas transplantation of an extra limb bud increased the number of limb-innervating motor neurons (**Figure 7-34**A). The number of sensory neurons in the dorsal root ganglia is similarly affected by peripheral targets. In principle, changes of neuronal numbers in these experiments could result from peripheral targets influencing proliferation, differentiation, or survival of motor and sensory neurons. Systematic examination of neuronal numbers and morphology during the developmental time course suggested that regulation of neuronal survival plays a predominant role. Searching for the peripheral agent responsible for neuronal survival led to the discovery of a secreted protein named **nerve growth factor** (**NGF**), which dramatically stimulates axonal outgrowth of sensory and sympathetic ganglia cultured *in vitro* (Figure 7-34B). The effect of NGF on neuronal survival *in vivo* was validated by the observation that daily infusion of purified NGF into newly born mice greatly increased the size of sympathetic ganglia, whereas daily infusion of NGF antiserum, which inhibits the activity of NGF, drastically decreased their size, due to

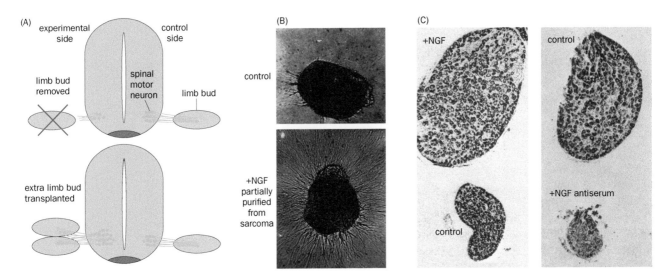

Figure 7-34 Neuron-target interaction and the effects of nerve growth factor. **(A)** Summary of interaction between spinal motor neurons and their limb targets in chick embryos. Compared to the control side, removal of a limb bud on the experimental side decreases the number of spinal motor neurons that innervate the limb bud, whereas transplantation of an extra limb bud increases the number. **(B)** Compared with an untreated control (top), a cultured sympathetic ganglion incubated for 18 h *in vitro* with nerve growth factor (NGF) exhibits abundant axonal outgrowth (bottom). **(C)** Left, injecting mice daily with purified NGF, starting at birth, causes a drastic increase in the size and neuronal number of sympathetic ganglia. Shown here are cross sections from thoracic ganglia of 19-day-old control and NGF-injected mice. Right, daily injection of NGF antiserum causes a drastic reduction in the size and neuronal number of sympathetic ganglia, shown here as cross sections from superior cervical ganglia of 9-day-old control and NGF antiserum-injected mice. (A, summarized from Hamburger V [1934] *J Exp Zool* 68:449–494; Hamburger V [1939] *Physiol Zool* 12:268–284. B, from Cohen S, Levi-Montalcini R, & Hamburger V [1954] *Proc Natl Acad Sci U S A* 40:1014–1018. C, from Levi-Montalcini R & Booker B [1960] *Proc Natl Acad Sci U S A* 46:373–384; 384–391.)

neuronal death (Figure 7-34C). Together, these experiments led to the influential **neurotrophic hypothesis** (from the Greek *trophos*, meaning "nourishment"): the survival of developing neurons requires neurotrophic factors produced by target cells. This ensures numerically matched and properly connected neurons and targets.

NGF is the founding member of the **neurotrophin** family, which also includes **brain-derived neurotrophic factor (BDNF)**, **neurotrophin-3 (NT3)**, and **neurotrophin-4 (NT4)**. Neurotrophins are produced via proteolytic cleavage of proneurotrophins. Proneurotrophins, as well as all mature neurotrophins, bind to the low-affinity neurotrophin receptor **p75NTR**. Each neurotrophin also binds preferentially to one of the three high-affinity **Trk** receptor tyrosine kinases (**Figure 7-35**A). Elegant *in vitro* studies demonstrated that NGF must be supplied to the axonal compartment, rather than the cell body, to maintain survival of cultured sympathetic neurons and stimulate axon growth (Figure 7-35B). As introduced in Box 3-4, neurotrophin dimers bring two Trk receptor molecules into close proximity so their cytoplasmic tyrosine kinases can cross-phosphorylate (Figure 3-39). This leads to activation of a number of downstream signaling pathways. One of these, involving the Sos, Ras, and MAP kinase cascade we encountered in Section 5.17, activates transcription of genes that promote neuronal survival and differentiation. Trk and Ras also activate phosphatidylinositol 3-kinase (PI3K) and a protein kinase named Akt (also known as protein kinase B), which inhibits apoptotic signaling and thus promotes survival. Interestingly, much of this signaling occurs while neurotrophins and their receptors are being transported from the axon terminal to the cell body (Figure 7-35C).

The role of neurotrophins in supporting the differentiation and survival of sensory, sympathetic, and motor neurons is widely supported by experimental evidence. Neurotrophins and Trk receptors are also widely expressed in the brain during development and in the adult. However, it remains unclear whether most CNS neurons require target-derived trophic support for their survival. Conversely, extensive evidence indicates that neurotrophins regulate many other important functions in the CNS, including dendrite morphogenesis, synapse development,

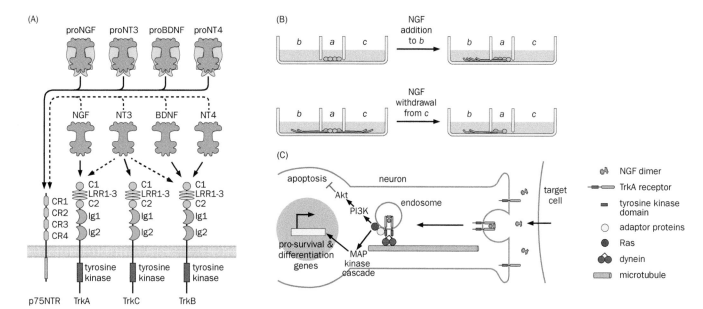

Figure 7-35 Neurotrophin signaling. (A) Neurotrophins and their receptors. NGF, BDNF, NT3, and NT4 are proteolytic products of secreted proneurotrophins (all shown as dimers), all of which bind to p75NTR. Mature neurotrophins bind to p75NTR with low affinity (dashed arrows) and to one of the three Trk receptors with high affinity. Ig, immunoglobulin domain; LRR, leucine-rich repeat; CR, C1, C2, cysteine-rich domains. **(B)** Testing where NGF is required for neuronal survival and axon growth. In a three-compartment culture dish, sympathetic neurons are placed in the center compartment *a*, and axons are allowed to grow into compartments *b* and *c*. However, media exchange between compartments is minimized by grease at the bottom of the compartment dividers. Top, neurons in chamber *a* extend axons only toward the chamber in which NGF is supplied. Bottom, at the start, sympathetic neurons have already extended their axons to NGF-supplied chambers *b* and *c*. If NGF is subsequently withdrawn from chamber *c*, the axons in chamber *c* degenerate, and neurons that sent axons only to compartment *c* die. These experiments indicate that NGF is required at the axon terminal to support neuronal survival and axon growth. **(C)** Schematic of NGF signaling pathways. Target-derived NGF binds TrkA at the axon terminal. The NGF/TrkA complex is endocytosed and retrogradely transported in endosomes from the axon terminal to the cell body. Activated TrkA receptors recruit adaptors and signaling molecules to the endosome. These signaling events alter the gene expression program and promote survival of the neuron. (A, from Reichardt LF [2006] *Phil Trans R Soc B* 361:1545–1564. With permission from the Royal Society. B, summarized from Campenot RB [1977] *Proc Natl Acad Sci U S A* 74:4516–4519. C, adapted from Zweifel LS, Kuruvilla R, & Ginty DD [2005] *Nat Rev Neurosci* 6:615–625. With permission from Springer Nature.)

and synaptic plasticity. Thus, this important class of proteins has evolved to play diverse roles in vertebrate neural development and function.

ASSEMBLY OF OLFACTORY CIRCUITS: HOW DO NEURAL MAPS FORM?

So far in the chapter, we have followed the developmental sequence of individual neurons, from patterning of progenitors and regulation of neurogenesis to establishment and refinement of synaptic connections. We have utilized examples from *C. elegans* motor neurons and *Drosophila* sensory neurons to mammalian spinal cord and cortical neurons to illustrate general principles of neural development. In the next part of the chapter, we study how olfactory maps are established to address the second question raised at the beginning of the chapter: How do groups of neurons coordinate their wiring to form a functional circuit? Our focus on the olfactory system will provide an informative comparison to the wiring of the visual system we studied in detail in Chapter 5.

7.16 Neural maps can be continuous, discrete, or a combination of the two

As introduced in Chapter 1, one of the most important organizational principles of the nervous system is that neurons are organized as maps to represent information. In the visual system, physical space in the world is first represented as a two-dimensional map in the retina. Through ordered retinotopic projections, this

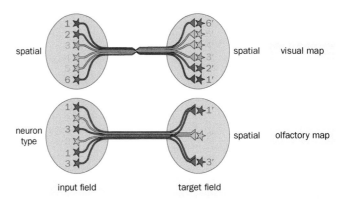

Figure 7-36 Continuous and discrete neural maps. (Top) Schematic of a continuous neural map, exemplified here by retinotopic projections in the visual system. Nearby neurons in the input field connect with nearby neurons in the target field, thereby preserving the spatial organization of the visual image. (Bottom) Schematic of a discrete neural map, exemplified here by the glomerular map of the olfactory system. Spatial organization in the target field does not reflect the position of input neurons, but is instead determined by input neuron type. Each of the three colors represents olfactory receptor neurons expressing the same odorant receptor and targeting their axonal projections to the same glomerulus. (Adapted from Luo L & Flanagan JG [2007] *Neuron* 56:284–300. With permission from Elsevier Inc.)

retinal map is relayed to and transformed in multiple higher-order visual centers (see Chapters 4 and 5). Similarly, the sensory and motor homunculi of the somatosensory and motor cortices represent maps of bodily sensation and motor output, respectively (Figure 1-25). The auditory system uses maps to represent the location and frequency of sounds (Sections 6.24 and 6.21). In the olfactory system, where the quality of stimuli rather than their spatial location is usually the primary concern, the brain utilizes a glomerular map to represent different odorants (Section 6.6). Neural maps are also used elsewhere in the brain. For example, as will be discussed in Box 11-2, place cells in the hippocampus and grid cells in the entorhinal cortex represent the spatial location of an animal in its environment. Thus, if we could understand how neural maps are established, we would have made considerable progress toward solving the problem of brain wiring.

Neural maps are of two kinds. **Continuous maps** are typified by the retinotopic projections that topographically represent spatial information from the input field, the retina; in the target fields, the tectum/superior colliculus (**Figure 7-36**, top); and throughout visual centers in the brain. The key property of continuous maps is that both the input and target fields use spatial relationships between neurons to represent information: nearby neurons in the input field connect with nearby neurons in the target field. Conversely, **discrete maps**, in which spatially dispersed neurons of a specific type project to a discrete target, are exemplified by the convergence of olfactory receptor neurons (ORNs) into glomeruli (Figure 7-36, bottom). No matter where they are spatially located, ORNs expressing the same odorant receptor project their axons to the same glomerulus.

Neural maps either resemble one of these two extremes or combine some aspects of both. The continuous tonotopic map representing sound frequency in the auditory system is similar to the retinotopic map. The taste system uses discrete information-processing channels for different taste modalities, like the olfactory map. The somatosensory cortex is, coarsely, a continuous representation of the bodily surface. However, representations of discrete entities can be superimposed on this continuous somatosensory map, as we saw in our discussion of rodent whiskers, which are represented by discrete barrels in the somatosensory cortex (Box 5-3). Even in the visual system, which is dominated by continuous retinotopic maps, discrete cell types at each level represent distinct information, such as ON–OFF responses, motion, and color. Distinct types of visual neurons usually target their processes to discrete neuropil layers (for example, within the inner plexiform, superior colliculus, and lateral geniculate nucleus) orthogonal to the retinotopic maps.

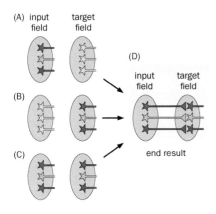

Figure 7-37 Developmental mechanisms for establishing wiring specificity. Three wiring strategies can give rise to an identical outcome **(D)**. **(A)** Input neurons are prespecified, and target neurons acquire their identities via connecting with specific input neurons. **(B)** Target neurons are prespecified, and input neurons acquire their identities from the target neurons with which they connect. **(C)** Input and target neurons are independently prespecified and find each other during the connection process. (Adapted from Jefferis GS, Marin EC, Stockers RF et al. [2001] *Nature* 414: 204–208. With permission from Springer Nature.)

We can envision three distinct developmental strategies for constructing neural maps (**Figure 7-37**). In the first strategy, input neurons are specified before connection, whereas target neurons are naive before connections form and acquire their identities via their connections with specific input neurons. The second strategy reverses this relationship, with naive input neurons acquiring their identities by connecting with specified target neurons. In the third strategy, both input and target neurons are specified independently before forming connections.

The retinotectal map we studied in Chapter 5 uses the third strategy, as neurons in both the input (the eye) and target fields (the tectum) of the retinal ganglion cells are prepatterned by expression of graded EphA and ephrin-A according to their spatial position (Figure 5-7). What about formation of discrete maps in the olfactory system? We now examine how these general concepts apply to the wiring of the mouse and *Drosophila* olfactory systems. Each system has provided insights into the establishment of wiring specificity in the assembly of neural circuits.

7.17 In mice, odorant receptors regulate guidance molecule expression to instruct ORN axon targeting

As we learned in Chapter 6, mouse ORNs expressing a given odorant receptor converge their axons onto 2 of the ~2000 glomeruli in each olfactory bulb. The positions of these glomerular targets are largely stereotyped from animal to animal. This feat is repeated for ~1000 ORN types. Thus, ORN axon targeting represents one of the most striking examples of wiring specificity in the developing nervous system. Along the long axis of the nasal epithelium, ORNs expressing the same odorant receptor are randomly distributed, yet their axons target to a precise location along the anterior–posterior (A–P) axis of the bulb. We focus on how ORN axons achieve this feat, highlighting differences with the position-based topographic targeting characterizing visual system development.

What molecular determinants could be responsible for the precise targeting of ~1000 distinct ORN types, each with an average population of ~5000 individual ORNs scattered randomly along the long axis of the nasal epithelium? One hypothesis is that the odorant receptors themselves are involved, since these distinctive molecules are shared by all ORNs projecting axons to a given glomerulus. This hypothesis was tested by changing the odorant receptor of a given ORN type to that of a different odorant receptor. The receptor swap was achieved by modifying the knock-in strategy that enabled tagging of all ORNs expressing a given receptor (Figure 7-38A). For example, the genetic locus normally expressing the odorant receptor P2 was engineered to express the coding sequence of another odorant receptor (M12 or M71) tagged with an axonal marker. Axons from ORNs in which M12 had replaced P2 (M12 → P2 ORNs) targeted to a new glomerulus not corresponding to the normal targets for either P2 or M12 ORNs; M71 → P2

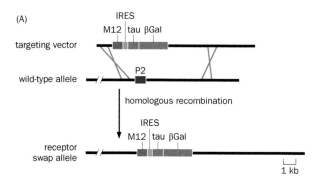

(A)

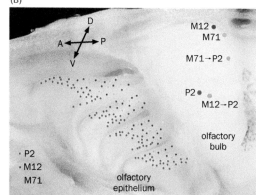

(B)

Figure 7-38 Alteration of ORN axon targeting after swapping of odorant receptors. (A) Genetic scheme of receptor swapping. The coding sequence of the P2 odorant receptor at the *P2* genomic locus was replaced by that of the M12 odorant receptor. In addition, an axonal marker, tau-βGal, following an internal ribosome entry sequence (IRES), was inserted to label ORN axons expressing the transgenic M12 odorant receptor (see Figure 6-12 legend for more details.) The pair of crosses symbolizes that homologous recombination between the targeting vector and the wild-type allele produced the receptor swap allele. **(B)** Replacing P2 with M12 or M71 coding sequences caused retargeting of ORNs to new sites designated as M12 → P2 and M71 → P2, respectively, which are distinct from the P2, M12, and M71 target glomeruli. ORN cell bodies in the olfactory epithelium and their target glomeruli in the olfactory bulb are color coded. A, anterior; P, posterior; D, dorsal; V, ventral. (Adapted from Wang F, Nemes A, Mendelsohn M, et al. [1998] *Cell* 93:47–60. With permission from Elsevier Inc.)

ORNs yielded similar results (Figure 7-38B). Since altering the odorant receptor changed the target choice of an ORN type, these experiments suggest that odorant receptors play an instructive role in ORN axon targeting. However, these instructions were insufficient to redirect the engineered ORNs to the target glomeruli corresponding to the donor receptors.

How do odorant receptors instruct ORN axon targeting? As discussed in Section 6.1, odorant receptors are G-protein-coupled receptors that regulate cAMP production in response to odorant binding. To test whether changes in G protein coupling and cAMP signaling affect ORN axon targeting, researchers generated a series of transgenic mice expressing a specific odorant receptor (I7) tagged with an axon marker. When a mutation blocking the ability to bind G proteins was introduced into the I7 receptor, axon convergence was disrupted. However, at the time of initial ORN targeting in embryos, mice have no sense of smell and do not yet express the olfactory-specific G_{olf} proteins necessary for olfactory transduction. How then do odorant receptors transduce signals? An alternative signaling pathway was suggested by the finding that the mutant I7 targeting defect could be corrected by co-expressing constitutively active forms of the G_s protein, the cAMP-dependent protein kinase A (PKA), or the CREB transcription factor (a substrate of PKA known to mediate cAMP-regulated gene expression; Section 3.23). Thus, the role of odorant receptors in regulating ORN axon targeting appears distinct from their role in odorant detection (**Figure 7-39**A). For ORN axon targeting, odorant receptors appear to regulate gene expression via G_s, cAMP, PKA, and CREB, all of which are present in immature ORNs (Figure 7-39B).

When the glomerular positions of these genetically engineered ORNs were examined, a correlation was found between the strength of G_s/cAMP/PKA signaling in the engineered ORNs and the location of their glomerular targets along the

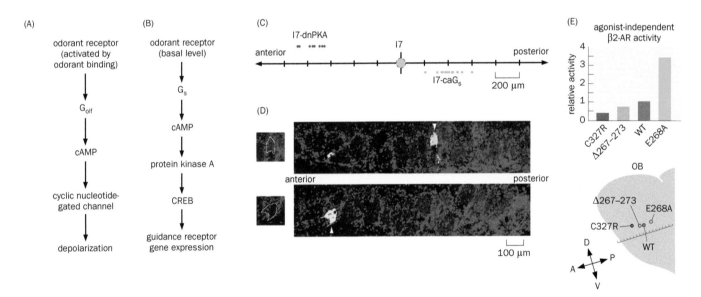

Figure 7-39 Signaling mechanisms by which odorant receptors regulate ORN targeting. (A) The olfactory transduction pathway (simplified from Figure 6-4). **(B)** A proposed signaling pathway by which odorant receptors regulate ORN axon targeting. **(C)** Summary of the effects of diminished or enhanced cAMP signaling on ORN axon targeting. The target glomerulus (green) of ORNs that express wild-type I7 serves as a landmark. Co-expressing I7 with a dominant-negative protein kinase A (dnPKA) causes ORNs to shift their glomerular targeting anteriorly (blue dots), whereas co-expressing I7 with a constitutively active G_s (caG$_s$) causes ORNs to shift their glomerular targeting posteriorly (yellow dots). Each dot represents data from an individual mouse. **(D)** These two panels are different olfactory bulb sections from the same transgenic mouse. Glomeruli (outlined by blue nuclear staining) in the posterior olfactory bulb have higher axonal neuropilin-1 expression (red), indicating that they are targeted by ORNs expressing higher levels of neuronpilin-1. The cyan-stained glomerulus

(top panel, arrowhead) is targeted by axons expressing I7, with cAMP signaling at a control level; these axons express neuropilin-1 (top left inset). The yellow-stained glomerulus (bottom panel, arrowhead) is targeted by cAMP-signaling-deficient axons co-expressing I7 and dnPKA; these axons express negligible levels of neuropilin-1 (bottom left inset), and their targeting is shifted to a more anterior glomerulus. **(E)** Target positions along the A–P axis of the olfactory bulb of ORNs that transgenically express one of four β2 adrenergic receptor (β2-AR) variants (bottom) correlate with the relative basal activity of these β2-AR variants *in vitro* (top). These variants are either single amino acid changes (C327R, E236A) or a deletion of a few amino acids (Δ267–273). (C & D, from Imai T, Suzuki M, & Sakano H [2006] *Science* 314: 657–661. With permission from AAAS. E, adapted from Nakashima A, Takeuchi H, Imai T, et al. [2013] *Cell* 154:1314–1325. With permission from Elsevier Inc.)

A–P axis of the olfactory bulb: the stronger the signaling, the more posterior the glomerular target (Figure 7-39C). Genes that are differentially expressed in ORNs whose axons target different positions along the A–P axis have been identified. One of these genes encodes the axon guidance receptor **neuropilin-1 (Nrp1)** (Box 5-1), which is expressed in a posterior > anterior gradient in ORN axon terminals (Figure 7-39D). Together, these data support a model in which odorant receptors regulate ORN axon targeting by controlling the expression of axon guidance molecules via the G protein/cAMP/CREB pathway.

How do different odorant receptors activate different levels of cAMP/PKA/CREB signaling? Because ORN axon targeting initiates prenatally, prior to odor exposure, ligand-independent basal activity of odorant receptors has been proposed to specify the strength of the signaling. This hypothesis was tested by substituting variants of the β2 adrenergic receptor (β2-AR), the best characterized GPCR (Sections 3.18 and 3.19), for an odorant receptor. Transgenic mice in which β2-AR replaced an endogenous odorant receptor resulted in ORNs targeting to a specific glomerulus in the middle of the olfactory bulb along the A–P axis. When different β2-AR mutants that exhibit distinct basal GPCR activity were used instead of the wild-type β2-AR, a strong correlation was found between basal activity level and targeting position along the olfactory bulb A–P axis (Figure 7-39E), supporting the hypothesis.

To summarize one current model (Figure 7-39B), each odorant receptor is presumed to have a specific basal GPCR activity determined by its sequence and structure, which can be coupled to G_s, a Gα protein present in developing ORNs. Therefore, ORNs expressing a given odorant receptor should have similar basal signaling levels, resulting in expression of similar levels of guidance molecules. No matter where the cell bodies of a given type of ORN are scattered within the olfactory epithelia, the same levels of guidance molecules they express can guide their axons to one glomerular target in the olfactory bulb. Other proposed models feature odorant receptors acting in ORN axon terminals to modulate the signaling of guidance receptors. These models are not mutually exclusive and could together account for how odorant receptors instruct the precise targeting of ORN axons.

7.18 ORN axons sort themselves by repulsive interactions before reaching their targets

Given the widespread cell body distribution of ORNs expressing the same odorant receptor, it still seems a daunting task for them to target their axons to a single glomerulus, even if they express similar levels of guidance receptors. Detailed studies of Nrp1 and its repulsive ligand semaphorin 3A (**Sema3A**; Box 5-1) offer an interesting mechanism: axons sort themselves out before they reach their targets.

Both Nrp1 and Sema3A are regulated by cAMP signaling in ORNs, but in opposite directions. Strong cAMP signaling leads to high Nrp1 and low Sema3A levels, whereas weak cAMP signaling promotes high Sema3A and low Nrp1 levels (**Figure 7-40**A). Thus, each ORN axon has specific levels of Nrp1 and Sema3A at the outset of its journey. ORN axon bundles leaving the olfactory epithelium were examined at different stages of their journey toward the olfactory bulb. At the early stages, axons with high levels of Sema3A were intermingled with high-Nrp1 axons, as expected, given the random distribution of their cell bodies in the olfactory epithelium. As axons progressed toward the olfactory bulb, high-Sema3A and high-Nrp1 axons gradually segregated. Before they entered the olfactory bulb, high-Nrp1 and high-Sema3A axons had already formed separate bundles and were poised to target to posterior and anterior parts of the olfactory bulb, respectively (Figure 7-40B). Repulsion between Sema3A- and Nrp1-expressing ORN axons appear to be essential for this sorting: in conditional knockout mice in which either Sema3A (Figure 7-40C) or Nrp1 were deleted from ORNs, both the order of axons traveling to the olfactory bulb and the final targeting positions of these axons were disrupted.

The use of signaling between developing axons to constrain their targeting choices has proved to be an important mechanism in brain wiring. We already introduced this idea in the context of retinotopic mapping (Section 5.5) and will

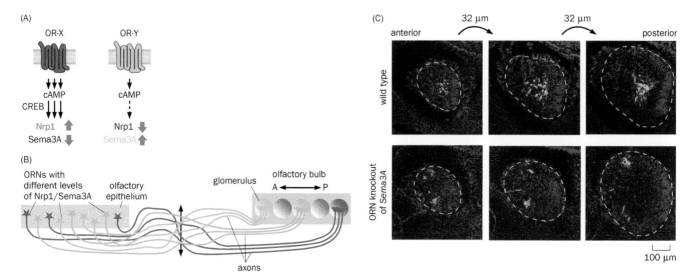

Figure 7-40 Sorting of ORN axons along their path via semaphorin/neuropilin mediated axon–axon repulsion. (A) Axon guidance cue Sema3A and its receptor Nrp1 are both regulated by cAMP signaling, but in opposite directions. ORNs with high basal cAMP signaling levels express low levels of Sema3A and high levels of Nrp1, while ORNs with low cAMP signaling levels are Nrp1-low and Sema3A-high. **(B)** At the outset the journey from the olfactory epithelium to the olfactory bulb, ORN axons with different Sema3A/Nrp1 levels are intermingled. They sort themselves out along the journey via Sema3A/Nrp1-mediated axon–axon repulsion. **(C)** Cross sections of ORN axon bundles taken at 32 μm intervals along the anterior (toward the nose) to posterior (toward the olfactory bulb) path. Nrp1-expressing (red) and Sema3A-expressing axons (green) gradually sort themselves out en route to the bulb in the wild type (top). However, when Sema3A is conditionally knocked out in ORNs (bottom), sorting is disrupted. (A & B, courtesy of Kazunari Miyamichi. C, from Imai T, Yamazaki T, Kobayakawa R, et al. [2009] *Science* 325:585–590. With permission from AAAS.)

see more examples in the wiring of the fly olfactory circuit (Section 7.22). The pre-target sorting of ORNs provides a striking example of the power of repulsive axon–axon interaction in establishing order among axons projecting from dispersed starting points.

7.19 Activity-dependent regulation of adhesion and repulsion refines glomerular targeting

Axon sorting provides patterned entry of ORN axons into the olfactory bulb along the A–P axis. Separate studies indicated that ORN axon targeting along the orthogonal dorsal–ventral (D–V) axis follows spatial topography that is genetically specified independently of odorant receptors. (This partially explains why swapping odorant receptors was insufficient to swap targeting specificity; Figure 7-38B). What is responsible for the eventual convergence of ORN axons expressing the same odorant receptors to *discrete* glomeruli located at various positions along the A–P and D–V axes? Is genetic hardwiring sufficient to achieve the precise wiring, or do experience and spontaneous neuronal activity play roles in refining these targeting events, as is the case in visual system wiring?

The role of odor-evoked activity in ORN axon targeting was examined by combining the cyclic nucleotide gated channel mutant that causes anosmia (Figure 6-5) with genetically labeled individual ORN types. In most cases, axon convergence of individual ORN types is grossly normal, suggesting that olfactory experience does not play a major role in ORN axon convergence. However, subtle defects were observed in a subset of ORN types. Indeed, even in normal mice, the convergence of axons to single glomeruli for some ORN types is incomplete at birth and requires refinement during postnatal development. This refinement was blocked by naris closure, a surgical procedure that prevents odorants from entering one of the two nostrils, suggesting that olfactory experience is required for uniglomerular convergence of some ORN types (**Figure 7-41**). Interfering with spontaneous activity of specific ORN types by expressing an inward rectifier K⁺ channel (Box 2-4)—which hyperpolarizes ORNs, rendering them less able to fire action potentials—

Figure 7-41 Refinement of glomerular targeting requires ORN activity. (A) ORNs normally refine their axon targeting to a single glomerulus (bulbᵒ). When one of the nostrils is closed so that ORNs projecting to the corresponding bulb (bulbˣ) are no longer exposed to odorants, ORN axons retain multiple glomerular targets (arrows). See Figure 6-12 for the strategy for labeling axons of an ORN type. **(B)** Quantification of axon refinement defects, as measured by the percentage of bulbs exhibiting multiple glomerular targets in normal (bulbᵒ) and experience-deprived (bulbˣ) conditions for two ORN types. (From Zou DJ, Feinstein P, Rivers AL, et al. [2004] *Science* 304:1976–1979. With permission from AAAS; see also Yu CR, Power J, Barnea G, et al. [2004] *Neuron* 42:553–566.)

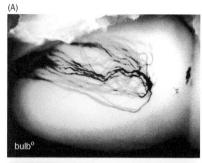
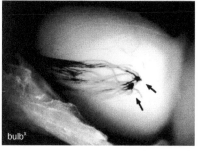
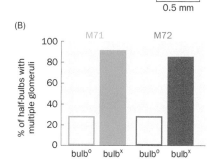

had a more drastic effect on ORN axon convergence, even though the general positioning of axon targeting was less perturbed.

These experiments indicate that spontaneous and odor-evoked ORN neuronal activities refine ORN axon targeting to individual glomeruli. What are the molecular bases? Searching for molecules whose expression was affected by naris closure revealed cell-surface proteins whose expression levels are regulated in an activity-dependent manner. These include ephrin-A and the EphA receptor as well as immunoglobulin superfamily cell adhesion molecules Kirrel2 and Kirrel3, which exhibit homophilic binding activity. Because they express the same odorant receptor, ORNs of the same type should have similar odor-evoked activity. They are therefore endowed with an "identity code" through the expression of specific levels of ephrin-A, EphA, and homophilic cell adhesion molecules. Adhesion among ORNs of the same types and repulsion between those of different types could help refine targeting of individual ORN types to discrete glomeruli (Figure 7-42). Thus, postnatal olfactory experience sharpens the expression of cell–cell interaction proteins in ORNs according to the odorant receptors they express.

So far, we have discussed ORN axon targeting in the olfactory bulb without considering any patterning that might exist in the bulb before ORN axons invade. Is the target field prepatterned (Figure 7-37C) or a blank slate (Figure 7-37A)? For the mouse olfactory bulb, we do not yet have a clear answer to this question. We

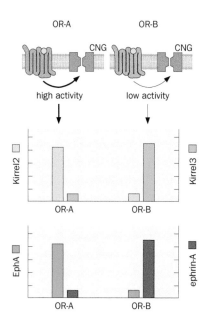
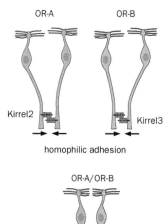
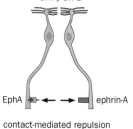

Figure 7-42 Activity-dependent expression of cell-surface proteins refines glomerular targeting. (Left) High levels of odorant-induced activity for a given olfactory receptor (for example, OR-A) induce expression of Kirrel2 and EphA, whereas low levels of odorant-induced activity (for example, OR-B) cause Kirrel3 and ephrin-A to be highly expressed. Thus, each ORN type has an identity code based on its expression levels of these surface proteins. (Right) Kirrel2 and Kirrel3 can cause axons of the same ORN type to adhere, while ephrin-A and EphA expressed on axons of different ORN types cause axon segregation. Both processes contribute to refinement of glomerular targeting. (Adapted from Serizawa S, Miyamichi K, Takeuchi H, et al. [2006] *Cell* 127:1057–1069. With permission from Elsevier Inc.)

now turn to the fly antennal lobe, the counterpart of the vertebrate olfactory bulb, where this topic has been investigated in detail.

7.20 Lineage and birth order specify the glomeruli to which the dendrites of *Drosophila* olfactory projection neurons target

As we learned in Chapter 6, the glomerular organization of the insect olfactory system resembles that of mammals (Figure 6-24). ORNs that express the same odorant receptor project axons to the same glomerulus in the antennal lobe. Information is then relayed to higher olfactory centers by olfactory **projection neurons (PNs)**, most of which send dendrites to a single glomerulus. The fly olfactory system offers several advantages for studying the wiring specificity of neural circuits: glomeruli are individually identifiable based on their stereotyped shapes, sizes, and relative positions, and genetic tools such as **MARCM** (mosaic analysis with a repressible cell marker; see Section 14.16 for details) allow investigators to visualize individual neurons, or neurons of the same lineage, and to delete or misexpress any gene in these labeled neurons. Given that each ORN type connects to one PN type at a specific glomerulus and vice versa, developmental mechanisms like those proposed in Figure 7-37 can be tested with single-cell resolution.

Neurons in the insect brain derive from progenitor cells called **neuroblasts**. MARCM-based lineage analysis revealed that most PNs derive from two neuroblasts located dorsally or laterally to the antennal lobe. PNs from the dorsal and lateral lineages project dendrites to stereotyped and mutually exclusive subsets of glomeruli (Figure 7-43). This observation suggested that a PN's lineage limits its glomerular choice. Furthermore, PNs within a lineage project dendrites to different glomeruli following a stereotyped birth order. Thus, lineage and birth order of a PN predict the glomerulus it sends dendrites to, the ORN type it forms synapses with, and the odorants it represents.

Further examination of olfactory circuit wiring during development revealed that dendrites of individual PNs extend to specific regions of the developing antennal lobe prior to the arrival of ORN axons. These developing dendritic projections occupy specific areas corresponding to their future target glomeruli. Thus, PNs have their own identities independently of ORNs and correspondingly pioneer a spatial map within the developing antennal lobe. Establishment of wiring speci-

(A)

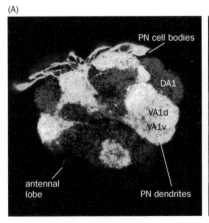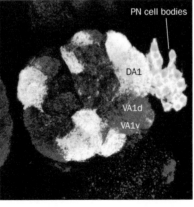

(B)

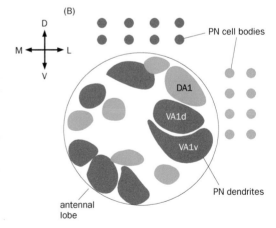

Figure 7-43 Lineage constrains glomerular choice of *Drosophila* projection neuron dendrites. (A) PNs derived from dorsal and lateral neuroblast lineages target different glomeruli. Left, dorsal lineage PNs and their dendrites express a membrane-targeted GFP (green), revealing the locations of their target glomeruli; magenta represents a synapse marker that labels all glomeruli. Right, lateral lineage PNs and their dendrites (green) highlight a complementary set of glomerular targets. These images are single optical sections of two antennal lobes at similar depths. Cell bodies of most dorsal PNs are in a different focal plane, and thus not visible. **(B)** Schematic summary of a subset of intercalated glomeruli targeted by PNs of the dorsal (red) and lateral (blue) lineages, respectively. D, dorsal; V ventral; M, medial; L, lateral. Three glomeruli, DA1, VA1d, VA1v (named after their positions, e.g., dorsal–anterior 1), are labeled in the schematic and the corresponding images in Panel A. (Adapted from Jefferis GS, Marin EC, Stockers RF, et al. [2001] *Nature* 414:204–208. With permission from Springer Nature.)

ficity in the olfactory circuit of the antennal lobe can therefore be divided into two phases: first, PN dendrites use ORN-independent cues to coarsely target their dendrites in a type-specific manner; second, ORN axons enter the prepatterned antennal lobe to find their partners. We will study both processes in the following sections.

7.21 Graded and discrete molecular determinants control the targeting of projection neuron dendrites

Semaphorins play important roles in PN dendrite targeting in flies, as they do in ORN axon targeting in mammals. Specifically, **Sema1A**, a transmembrane semaphorin, is distributed in a dorsolateral > ventromedial gradient in the developing antennal lobe, which, prior to ORN axon arrival, contains mostly PN dendrites (**Figure 7-44**A). Removing Sema1A from a single PN shifts its dendrite targeting ventromedially down the normal Sema1A gradient (Figure 7-44B), indicating that Sema1A plays a cell-autonomous role. Thus, unlike in most cases where semaphorins serve as guidance ligands, Sema1A acts as a *receptor* to regulate PN dendrite targeting. Furthermore, different PNs express different levels of Sema1A and are differentially repelled by two secreted semaphorins, **Sema2A** and **Sema2B**, which are distributed in ventromedial > dorsolateral countergradients. (Sema2A and Sema2B originate from several cellular sources, notably antennal lobes used for olfaction in larvae that undergo degeneration at the time dendrite targeting of adult antennal lobe occurs; see Section 7.14 regarding degenerative events during metamorphosis.) Thus, opposing gradients of different semaphorins cooperate to specify where PN dendrites target along the dorsolateral–ventromedial axis.

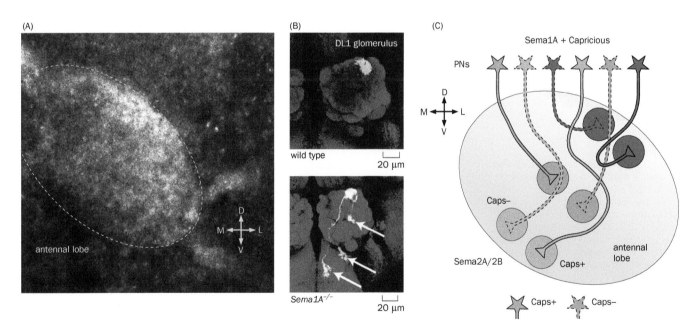

Figure 7-44 PN dendrite targeting is instructed by global gradients and local binary determinants. (A) Sema1A protein is distributed in a dorsolateral > ventromedial gradient in the developing antennal lobe (dashed line), which, prior to the arrival of ORN axons, consists mostly of PN dendrites. **(B)** Top, a single DL1 PN targets its dendrites exclusively to the DL1 glomerulus (located at the dorsolateral antennal lobe). Bottom, when Sema1A is removed from a single DL1 PN, some dendrites still target to the DL1 glomerulus, but the rest mistarget ventromedially (bottom, arrows). Thus, Sema1A cell autonomously directs the dendrites of DL1 PNs to the DL1 glomerulus. Both wild-type and mutant DL1 PNs are labeled green by MARCM; a synapse marker for all glomerular structures is stained magenta. **(C)** Summary of mechanisms controlling PN dendrite targeting. Different PNs express different levels of Sema1A, shown here as different intensities of red.

The blue to yellow background shading represents the ventromedial > dorsolateral gradients of Sema2A and Sema2B, two secreted semaphorins that repel Sema1A-expressing PN dendrites. Thus, PNs that express progressively higher levels of Sema1A target their dendrites to progressively more dorsolateral glomeruli, accounting for the Sema1A gradient in Panel A. In addition, some PNs express Capricious (Caps, solid outline), while others do not (dotted outline). Caps-positive and Caps-negative PNs locally segregate into discrete glomeruli. D, dorsal; V, ventral; M, medial; L, lateral. (A & B, adapted from Komiyama T, Sweeney LB, Schuldiner O, et al. [2007] *Cell* 128: 399–410. With permission from Elsevier Inc. C, summarized from Komiyama T, Sweeney LB, Schuldiner O, et al. [2007] *Cell* 128:399– 410; Sweeney LB, Chou YH, Wu, Z, et al. [2011] *Neuron* 72:734–747; Hong W, Zhu H, Potter CJ, et al. [2009] *Nat Neurosci* 12:1542–1550.)

What mechanisms ensure that each PN targets its dendrites to a single, discrete glomerulus? Graded determinants can set up a coarse map but cannot specify discrete outcomes with high precision, as neighboring glomeruli may exhibit only small differences in determinant level. A second kind of PN molecular label was found. About half of PNs express the leucine-rich-repeat containing cell-surface protein Capricious (Caps), which we already encountered in our discussion of visual system wiring in *Drosophila* (Figure 5-39). Glomerular targets of Caps-positive and Caps-negative PNs form a salt-and-pepper pattern. Loss of Caps in Caps-positive PNs caused them to mistarget selectively to nearby Caps-negative glomeruli. Conversely, misexpression of Caps in Caps-negative PNs caused them to mistarget selectively to nearby Caps-positive glomeruli. Further studies suggested that Caps mediates PN dendrite–dendrite interaction. Thus, Caps acts as a binary determinant to segregate PN dendrites into two distinct subgroups. This function is analogous to that of mouse Kirrel2/3- and ephrin-A/EphA-mediated axon–axon interaction in ORN axon refinement (Figure 7-42).

In summary, global graded determinants, such as semaphorins, and local binary determinants, such as Capricious, cooperate to drive dendritic prepatterning of the antennal lobe in a PN type-specific manner (Figure 7-44C).

7.22 Sequential interactions among ORN axons limit their target choice

Odorant receptors are not used for ORN axon targeting in *Drosophila,* in contrast to their use in mammals. The use of odorant receptors to instruct ORN targeting likely accompanied the large expansion of ORN types during vertebrate evolution. Indeed, most *Drosophila* ORNs express odorant receptors only after wiring specificity has been achieved. Thus, the wiring mechanisms for fly ORNs differ from odorant-receptor-dependent ORN axon targeting in mammals (Section 7.19).

However, like their mammalian counterparts, *Drosophila* ORNs do rely on extensive axon–axon interaction and pretarget sorting, even utilizing the same family of molecules, the semaphorins. When pioneering ORN axons from the antenna first arrive at the antennal lobe, individual axons circumnavigate the antennal lobe, taking either a dorsolateral or a ventromedial trajectory (**Figure 7-45**). The secreted Sema2B is enriched in axons that take the ventromedial trajectory but absent from axons that take the dorsolateral trajectory. In flies lacking Sema2B, ORN axons that normally take the ventromedial trajectory took the dorsolateral trajectory instead and eventually targeted ectopic glomeruli in the dorsolateral antennal lobe, far from their normal ventromedial targets. Genetic analyses suggested that Sema2B-expressing ORNs are first attracted to a ventromedial trajectory by the antennal lobe gradients of Sema2A and Sema2B that pattern PN dendrites at an earlier developmental stage (Figure 7-44C). Sema2B from these initial ventromedial ORN axons then attracts other Sema2B-expressing ORN axons to the same trajectory (Figure 7-45, left two panels). These axon–axon interactions, analogous to those observed in vertebrates, ensure correct trajectory choice, which is essential for eventual target selection.

Figure 7-45 Schematic summarizing that sequential axon–axon interactions limit ORN target choice. When antennal ORN axons first reach the antennal lobe, a ventromedial > dorsolateral gradient of antennal lobe Sema2A and Sema2B (red shading) attracts axons expressing Sema2B (red) to a ventromedial trajectory (arrow), while axons not expressing Sema2B (orange) take a dorsolateral trajectory. Sema2B from early-arriving ORN axons then attracts other Sema2B-expressing ORN axons to the same trajectory (curved arrow). Selection of a specific trajectory limits subsequent glomerular target choice. Later, when maxillary palp (MP) axons (cyan) arrive, the antennal lobe is already largely occupied by antennal ORNs. Sema1A from antennal ORN axons repels MP ORN axons (⊣), such that MP axons innervate glomeruli complementary to those innervated by antennal ORNs. D, dorsal; V, ventral; M, medial; L, lateral. (Adapted from Sweeney LB, Couto A, Chou YH, et al. [2007] *Neuron* 53:185–200; Joo WJ, Sweeney LB, Liang L, et al. [2013] *Neuron* 78:673–686. With permission from Elsevier Inc.)

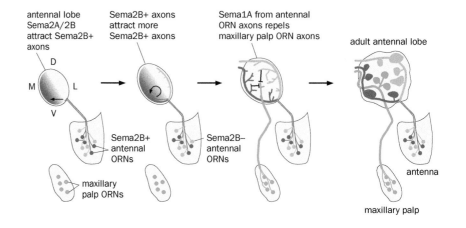

Axon–axon interaction also occurs at final target regions, as exemplified by targeting of ORN axons from the maxillary palp (MP), which reach the antennal lobe well after antennal ORN axons. (Flies have two spatially separated pairs of "noses" that house different ORN types: the antenna and the maxillary palp; Figure 6-24). When antennal ORNs were ablated during development, MP axons mistargeted to glomeruli normally occupied by antennal ORNs, suggesting that antennal ORN axons normally constrain target choice of the later-arriving MP ORN axons (Figure 7-45, right two panels). Deleting Sema1A from antennal, but not MP, ORN axons affected the targeting choice of MP ORN axons, revealing that Sema1A plays an essential role in ORNs by serving as a repulsive ligand, in contrast to its function in PNs as a receptor that instructs dendrite targeting. Thus, axon–axon interactions at multiple developmental stages limit the target choices of axons of a given ORN type.

7.23 Homophilic matching molecules instruct connection specificity between synaptic partners

What mechanisms establish the final, one-to-one connection between 50 types of ORNs and PNs? Given the extensive ORN axon–axon interactions we just discussed, it is conceivable that ORN axons may be patterned to create an ORN axon map that is superimposed onto the PN dendrite map established earlier. Alternatively, or in addition, ORN axons may recognize cues on their partner PN dendrites using "matching" molecules.

Genetic screens were carried out to identify potential synaptic partner matching molecules that, when misexpressed, would force nonpartners to match or prevent normal synaptic partners from matching. Such screens identified two evolutionarily conserved **Teneurins**. Each of the two fly Teneurins, Ten-m and Ten-a, is highly expressed in a distinct subset of matching ORN and PN types (Figure 7-46A, B) and exhibits homophilic adhesion properties. These data suggest that matching expression of Ten-m and Ten-a can be used to instruct synaptic partner matching (Figure 7-46B). This hypothesis has been supported by loss- and gain-of-function experiments. For example, loss of Ten-a in DA1 PNs, which normally express high Ten-a levels, caused their dendrites to mismatch with VA1v ORN axons, which express low Ten-a levels (Figure 7-46C, middle). Likewise, overexpression of Ten-m in DA1 PNs, which normally express low Ten-m levels, also caused their dendrites to mismatch with VA1v ORN axons, which express high Ten-m levels (Figure 7-46C, right). Further, simultaneous expression of Ten-m in nonpartner ORNs and PNs caused their axons and dendrites to form ectopic

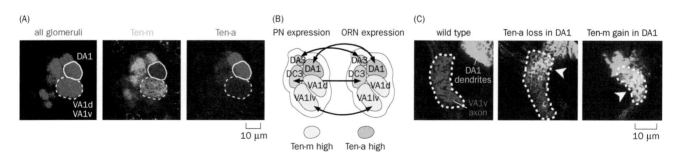

Figure 7-46 Teneurin-mediated homophilic attraction directs PN–ORN synaptic partner matching. (A) Expression patterns of Ten-m (green) and Ten-a (red) in the developing antennal lobe, with the magenta neuropil marker labeling all glomeruli evenly. The solid circles highlight the DA1 glomerulus, which has high Ten-a levels and low Ten-m levels. The dashed circles highlight the VA1d and VA1v glomeruli, which have high Ten-m levels and low Ten-a levels. **(B)** Schematic summary of matching Teneurin expression in five glomeruli. PNs and ORNs targeting the same glomerulus express matching levels of Ten-m (blue) and Ten-a (orange). Those targeting DA1 are Ten-a high and Ten-m low; those targeting VA1d and VA1v are Ten-m high and Ten-a low; those targeting DA3 are high in both; those targeting DC3 are low in both.

Double-ended arrows symbolize the hypothesis that matching Teneurin levels instruct synaptic partner matching between cognate ORNs and PNs. **(C)** Genetic evidence supporting the matching hypothesis. In wild type, DA1 dendrites (green) and VA1v axons (red) do not match (left). When Ten-a is knocked down in DA1 PNs, which normally express high Ten-a levels, DA1 dendrites mismatch with VA1v axons, which normally express low Ten-a levels (middle). When Ten-m is overexpressed in DA1 PNs, which normally express low Ten-m levels, their dendrites also mismatch with VA1v axons, which normally express high Ten-m levels (right). (Adapted from Hong W, Mosca TJ, & Luo L [2012] *Nature* 484:201–207. With permission from Springer Nature.)

connections. These synaptic-partner-matching molecules thus ensure final connection specificity between cognate ORNs and PNs.

In summary, genetic analyses of fly olfactory circuit assembly have identified a wealth of cellular and molecular mechanisms by which wiring specificity of a moderately complex neural circuit is achieved during development (Movie 7-2). The fly olfactory circuit clearly employs an independent specification mechanism (Figure 7-37C) that requires more complex molecular recognition strategies than the alternative mechanisms (Figure 7-37A, B). Why? As we learned in Chapter 6, PNs that project dendrites to specific glomeruli also exhibit stereotyped axon terminal arborizations in the lateral horn, a higher olfactory center (Figure 6-30). Indeed, many molecules involved in PN dendrite targeting also regulate PN axon terminal arborization. This hard-wiring strategy ensures that specific olfactory information received by specific ORN types is always delivered to specific regions of the lateral horn by independently specified PNs. This may be essential for innate odor-mediated behaviors such as feeding and courtship, which are robust and do not require learning. Likewise, mammalian mitral/tufted cells may also be genetically prespecified to some degree, such that information from specific odorant receptors can be faithfully transmitted to specific targets in higher olfactory centers where it can drive innate olfaction-mediated behaviors, such as avoidance of a predator odor. This hypothesis remains to be tested.

We have also seen that construction of the fly olfactory circuit employs wiring molecules, such as semaphorins and Capricious, that are used elsewhere. Likewise, synaptic partner matching molecules identified in the fly olfactory circuit, such as Teneurins, have also been shown to play analogous roles in the wiring of the mouse hippocampal system (Box 7-2). Reuse of molecules in different circuits is an important way by which a limited number of genes establishes many more precise connections, a topic we turn to in the final part of this chapter.

Box 7-2: Teneurin and wiring specificity in the hippocampal circuit

As we will learn in Chapter 11, the mammalian hippocampus plays key roles in spatial navigation and acquisition of new explicit memories. Information flow in the hippocampal circuit is well characterized (Figure 11-5), as are topographic connection patterns between different regions of the hippocampal circuit. For example, proximal CA1 neurons project axons to target neurons in the distal subiculum, while distal CA1 project to the proximal subiculum. Ten3, one of the four Teneurins in mouse, is highly expressed in proximal CA1 and distal subiculum (Figure 7-47A). Thus, as in the fly olfactory circuit, Ten3-high neurons connect with Ten3-high targets (Figure 7-47D, top). Furthermore, expression of Ten3 in nonadhesive cells can cause them to form aggregates *in vitro*, indicating that Ten3 promotes homophilic adhesion.

To test whether Ten3 promotes homophilic attraction between proximal CA1 axons and distal subicular target *in vivo*, conditional knockout *Ten3* mice that allow *Ten3* to be deleted in cells expressing Cre recombinase were generated (see Section 14.8 for details). In the first experiment, a Cre-expressing virus was injected into the CA1 region of newborn pups (before the CA1 → subiculum projection is established) and Cre+ proximal CA1 axons were traced in the adult. Control proximal CA1 axons targeted preferentially to distal subiculum, whereas *Ten3* knockout proximal CA1 axons spread more proximally in the subiculum (Figure 7-47B). Thus, Ten3 is required in CA1 for precise proximal

CA1 axon targeting. In the second experiment, the Cre-expressing virus was injected into a small region of the subiculum of newborn pups, followed by injection of anterograde axon tracers into proximal CA1 in the adult. Cre expression in the subiculum did not affect proximal CA1 axon targeting in wild-type control mice. In experimental *Ten3* conditional mice, however, proximal CA1 axons avoided Cre-expressing *Ten3* knockout regions of the distal subiculum (Figure 7-47C). The simplest interpretation is that Ten3+ proximal CA1 axons are no longer attracted to the *Ten3* knockout region in the distal subiculum but are instead attracted to Ten3+ distal subiculum next to the knockout region, creating a valley between nearby peaks (Figure 7-47C, D). Thus, the targeting phenotypes of both CA1- and subiculum-knockouts support a homophilic attraction model (Figure 7-47D).

In addition to the CA1 → subiculum projection, proximal CA1/distal subiculum and distal CA1/proximal subiculum also connect with specific subregions of several other nodes in the extended hippocampal network, including the entorhinal cortex and specific nuclei in the thalamus and hypothalamus. Remarkably, in all of these connections, Ten3+ neurons connect to Ten3+ targets. Thus, Ten3-mediated homophilic attraction may regulate wiring specificity across many nodes in the extended hippocampal network. Physiological studies have suggested Ten3+ and Ten3– subregions across the network constitute two parallel subnetworks that

Box 7-2: continued

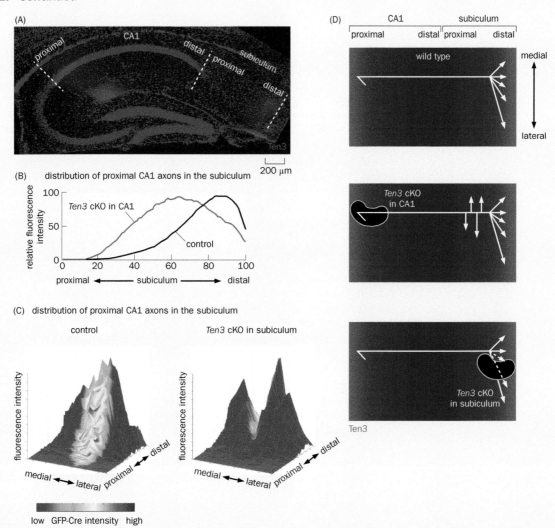

Figure 7-47 Ten3 controls proximal CA1 → distal subiculum targeting specificity. (A) Ten3 protein (red) is highly enriched in proximal CA1 and distal subiculum in the developing mouse hippocampus. Dashed lines denote the borders of CA1 and subiculum. Blue, nuclear staining showing cell body distribution. **(B)** Anterograde tracers were injected into proximal CA1, and their distribution in the subiculum along the proximal–distal axis was plotted. Control proximal CA1 axons project preferentially to distal subiculum (black curve). Proximal CA1 axons with Ten3 conditional knockout (cKO, red curve) spread more proximally in the subiculum. **(C)** In these "mountain" plots, the color represents the intensity of GFP marking Cre+ cells, and the height represents the fluorescence intensity of an anterograde tracer injected into proximal CA1. In the control (left), the color does not affect the height, as Cre-expressing target cells do not affect proximal CA1 axon targeting. In Ten3 cKO

in subiculum (right), proximal CA1 axons avoid the colored patch in the distal subiculum devoid of Ten3. **(D)** Schematic summarizing the conditional knockout phenotypes. Top, in wild-type control, Ten3+ (red) proximal CA1 axons (white line) project to Ten3+ distal subiculum, spreading along the orthogonal medial–lateral axis. Middle, when Ten3 is conditionally knocked out in a CA1 patch (black), Ten3– axons are no longer attracted to Ten3+ distal subiculum and therefore terminate more proximally. Bottom, when Ten3 is conditionally knocked out in a distal subicular patch, Ten3+ proximal CA1 axons lose their attraction to that patch and are instead attracted to Ten3+ distal subiculum next to the knockout patch (dashed arrows), thus creating a valley in the mountain plot. (A–C, from Berns DS, DeNardo LA, Pederick DT, et al. [2018] *Nature* 554:328–333. With permission from Springer Nature.)

preferentially process information regarding space and objects, respectively. In addition to the extended hippocampal network, Ten3 is expressed in the mouse visual system and is required for correct targeting of retinal ganglion cells to the lateral geniculate nucleus. Indeed, all four mouse Teneurins are expressed in diverse brain regions, suggesting that they may regulate wiring specificity across the developing brain.

HOW DO ~20,000 GENES SPECIFY ~10^{14} CONNECTIONS?

As we have seen, developing nervous systems employ many different strategies to precisely wire themselves up. Some strategies are highly conserved through evolution, employing the same molecules and mechanisms; others are more varied and adapted to the developmental programs of specific parts of the nervous system or specific organisms. As a summary of what we have learned in Chapter 5 and this chapter, we will now address the "wiring specificity problem" raised at the beginning of Chapter 5. The human brain contains ~10^{11} neurons that make, in total, >10^{14} synaptic connections. If each neuron carried a specific "identification tag" allowing it to be wired differently from any other neuron, as Sperry proposed in his chemoaffinity hypothesis, how is wiring specificity achieved, given that there are only ~20,000 protein-coding genes in the human genome? The following sections discuss various solutions to this problem (see Table 7-1 for a summary of the examples discussed in this chapter and Chapter 5).

7.24 Some genes can produce many protein variants

We encountered an extraordinary example of molecular diversity in *Drosophila Dscam*. This single gene can encode cell-surface proteins with 19,008 variants of the extracellular domains (Figure 7-21). Although vertebrate *Dscams* do not possess the extraordinary alternative splicing potential of insect *Dscams*, other classes of vertebrate proteins exhibit molecular diversity. For example, protocadherin α and γ clusters also utilize alternative extracellular domains to achieve molecular diversity and promiscuous *cis*-dimerization, but isoform-specific *trans*-interaction of protocadherins can generate many trans-cellular interaction complexes (Figure 7-22). Another example of extraordinary diversity was found in the trans-synaptic signaling molecule neurexin (Figure 7-26). Six major neurexin isoforms are encoded by three separate genes, each of which has two alternative promoters. Altogether, ~3000 neurexin variants can be produced through independent selection of several alternative exons encoding the extracellular domain.

So far, these genes with extraordinary protein variants have not been shown to play a role in matching pre- and postsynaptic partners. However, as we discussed in Section 7.9, *Drosophila* Dscam and mouse clustered protocadherins play key roles in dendritic and axonal self-avoidance, an important step in constructing the nervous system. Molecular diversity in neurexins allows different isoforms to specify distinct synaptic signaling properties.

7.25 Different levels of a single protein can specify multiple connections

The use of protein gradients reduces the number of molecular species required to wire the nervous system: different quantities of the same protein can specify different connections. This was first demonstrated in the ephrin/Eph receptor gradients that wire up the retinotopic map (Figure 5-7). The use of molecular gradients to specify continuous maps seems intuitive. Two neurons exhibiting a small difference in receptor levels might project to targets a short distance apart; as the difference in the receptor levels of the two neurons increases, the distance between their projection targets also increases.

Surprisingly, molecular gradients are also used in discrete maps, such as the olfactory map. In the case of mammalian ORN axon targeting, levels of Sema3A and its repulsive receptor neuropilin are used to sort different ORN axons along the path toward their final destination (Figure 7-40). In the fly antennal lobe, a projection neuron's level of Sema1A, a transmembrane semaphorin, controls the position of the PN's dendritic target along a particular axis according to extracellular gradients of two secreted semaphorins (Figure 7-44). However, in each case, these gradients regulate coarse targeting, which is further refined by cell-type-specific interactions to specify discrete targets.

Table 7-1:   Strategies for maximizing the number of wiring specification signals from a limited genome

Strategy	Example (Section)	Known function	Animal model
Many protein products from a single gene	Dscam (7.9)	Dendrite and axon self-avoidance	Fly
	Protocadherins (7.9)	Dendrite and axon self-avoidance	Mammal
	Neurexins (7.11, 7.24)	Synapse specification	Vertebrate
Many levels from a single protein	Ephrin-A/EphA (5.4)	Retinotopic map	Vertebrate
	Semaphorins (7.18, 7.21)	Olfactory maps	Mouse, fly
Multiple functions from a single protein	Ephrin-A and EphA as both ligands and receptors (5.5, 5.15)	Retinotopic map formation	Vertebrate
	Sema1A as a receptor or a ligand (7.21, 7.22)	Olfactory map formation	Fly
	Unc40, without or with Unc5 (Box 5-1)	Attractive or repulsive midline guidance	*C. elegans* to vertebrate
Same protein used in multiple places and times	Ephrin-A/EphA (5.4, 7.4, 7.19)	Retinotopic map; motor axon guidance; glomerular refinement	Vertebrate
	Sema1A, Sema2B (7.21, 7.22)	dendrite targeting; axon targeting in olfactory map formation	Fly
	Sema3 (7.6, 7.14, 7.18)	Midline repulsion; ORN axon sorting; axon pruning	Mouse
	Sonic hedgehog (7.4, 7.6)	Dorsoventral patterning in the spinal cord; midline guidance	Vertebrate
	Wnt (7.1, 7.6, 7.10)	Developmental patterning along the A–P axis; direction of turning along the A–P axis; subcellular presynaptic site selection	*C. elegans,* vertebrate
	Slit (7.5, 7.6)	Midline crossing Pathway choice	Fly to vertebrate Fly
	Netrin (Box 5-1, 5.18)	Midline guidance Photoreceptor axon targeting	*C. elegans* to vertebrate Fly
	Capricious (5.18, 7.21)	Visual system; olfactory system	Fly
	Teneurin (7.23, Box 7-2)	Olfactory system Visual system, hippocampal circuit	Fly Mouse
Combinatorial use of proteins	Semaphorins and cadherins (5.16)	Lamina-specific targeting	Mouse retina
	Semaphorins and Capricious (7.21)	Projection neuron dendrite targeting	Fly
	Semaphorins and ephrin-A/EphA (7.18, 7.19)	ORN axon targeting	Mouse
Use of experience and spontaneous neuronal activity	Ocular dominance column formation in V1 (5.7, 5.9)		Cat, ferret, monkey
	Eye-specific layer segregation in the LGN (5.9–5.11)		Mammal
	Neuromuscular synapse refinement (7.13)		Mouse
	Wiring of the whisker-barrel system (Box 5-3)		Mouse
	Refinement of glomerular targeting (7.19)		Mouse

7.26 The same molecules can serve multiple functions

The strategy of using the same molecule for multiple functions is exemplified by ephrin-A and EphA in the retinotopic map. Each molecule can serve as either a ligand or a receptor in forward or reverse signaling (Figure 5-12). One function of this bidirectional signaling is to sharpen the retinotopic map (Section 5.5). Bidirectional signaling also potentially enables both ephrin-A and EphA to serve as ligands to instruct the targeting of their presynaptic partners and as receptors to instruct their own axon targeting to match with their postsynaptic partners across multiple stages of the visual pathway (Figure 5-31). Likewise, Sema1A in *Drosophila* can serve as either a ligand or a receptor in different olfactory neurons (Figure 7-44; Figure 7-45). As a third example, expression of Unc40/DCC alone produces an attractive response to the midline guidance cue Unc6/netrin, whereas co-expression of Unc40/DCC with Unc5 produces a repulsive response to the same cue (Box 5-1).

Although we have not discussed specific examples, the activity and specificity of axon guidance receptors can also be modified extracellularly by glycosylation and proteolysis and intracellularly by changes in intracellular signaling molecules, such as Ca^{2+} or cAMP, or downstream effectors. These regulations allow axon guidance receptors to produce distinct responses to the same guidance cues.

7.27 The same molecules can be used at multiple times and places

We have seen many examples of this strategy. In vertebrates, for instance, ephrin/ Eph ligand–receptor pairs are used in the visual system for retinotopic mapping (Figure 5-7), in the olfactory system for refining ORN axon targeting in an odorant receptor- and activity-dependent manner (Figure 7-41), and in spinal cord motor neurons to specify limb innervation patterns (Figure 7-9). The spatial separation of these systems allows reuse of the same molecules without causing confusion. In other cases, temporal differences between targeting events make it possible for the same molecular signals to perform multiple functions in the same system. One such example is found in the *Drosophila* olfactory circuit, where the same semaphorins (Sema1A, Sema2B) drive wiring decisions of PN dendrites and ORN axons at different times in development (Figure 7-44; Figure 7-45). Furthermore, molecules active in early developmental patterning can find additional uses in neuronal wiring. For example, Sonic Hedgehog patterns the vertebrate spinal cord along the dorsal–ventral axis early in development (Figure 7-8) and later serves as a midline guidance cue for commissural neurons. In mice and *C. elegans*, Wnt globally patterns the anterior–posterior axis and later specifies the turning direction of commissural axons after midline crossing in mice (Figure 7-16) and the site selection of presynaptic terminals of *C. elegans* motor neurons (Figure 7-24). Finally, some axon guidance molecules, such as the semaphorins, are also used in later developmental steps, such as developmental axon pruning (Section 7.14) and synapse formation.

In addition to conserving molecules, this mechanism of multiple uses suggests that it may be easier to evolve new spatial and temporal expression patterns for existing molecules than to invent new molecules for the purpose of wiring different parts of the nervous system. This idea will be discussed further in Chapter 13.

7.28 Combinatorial use of wiring molecules can markedly reduce the number of molecules needed

Another strategy for maximizing the use of a finite number of signaling molecules takes advantage of the combinatorial actions of different wiring molecules. For the sake of simplicity, let's assume that input and target neurons use homophilic adhesion molecules to establish wiring specificity. If each input–target neuron pair uses only one homophilic adhesion molecule to specify its connection, then 25 molecules are needed to specify 25 connections. If, however, molecules 1 through 5 and A through E can be independently recognized by synaptic partners, and synapses form preferentially only when both match, then only 10 molecules can specify 25 connections (**Figure 7-48**).

By the same token, 10^{12} connections can be specified by 2,000,000 molecules if there is a mechanism for combinatorial recognition of two molecules; 30,000 via combinatorial recognition of three molecules; and only 600 via combinatorial recognition of six molecules. While this strategy can rapidly reduce the number of molecules needed to specify connections, it also creates new difficulties: cells must impose thresholds for connections so that only when multiple labels simultaneously match will connections be allowed or highly preferred; cells must also devise mechanisms to interpret multiple signals independently and unambiguously.

One could envision many cases in which combinatorial coding might be used. For instance, ephrin-A/EphA signaling specifies anterior–posterior positions in the retinotopic map, and molecules that specify dorsal–ventral positions likely work

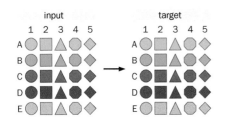

Figure 7-48 A combinatorial strategy. If all input and target neurons express two homophilic adhesion molecules (represented here by a number and a letter) and if both molecules must match for preferential formation of a connection, then only 10 molecules can specify 25 connections. As illustrated here, this means that five colors (letters) and five shapes (numbers) can generate 25 unique colored shapes (letter–number combinations). A blue triangle input neuron (that is, a neuron expressing molecules B and 3) would preferentially connect with a blue triangle (B3) target neuron, but would fail to connect with target neurons expressing only one of the two specificity molecules, for example, red triangles (which express C and 3), or blue squares (which express B and 2).

combinatorially with ephrin-A/EphA, such that each retinal axon has specific values on each of the two axes. Thereby, a two-dimensional retinal map can project in an orderly fashion to higher-order visual centers in the brain. Several other strategies, including temporal separation (Section 7.27) and stepwise assembly (Section 7.29), can be used to overcome the difficulties that combinatorial coding imposes on signal thresholding and interpretation.

7.29 Dividing wiring decisions into multiple steps conserves molecules and increases fidelity

We have seen how combining multiple molecular recognition events can decrease the number of molecules needed to specify a wiring decision. The same molecule-conserving result can also be achieved by combining sequential steps. Suppose that 25 input neurons must connect to 25 target neurons, as in Figure 7-48. If all connections are made simultaneously, 25 specificity molecules are needed. If, however, the decisions are divided into two steps, with each step having five possible outcomes, then the first step can use five molecules to divide 25 input neurons into five subgroups. In the second step, the same five molecules can be used to specify the connections of each subgroup (**Figure 7-49**). In addition to conserving molecules, such stepwise decisions could also increase robustness and reduce errors at each decision point, as there are fewer simultaneous choices.

These steps can be temporal or spatial: we have encountered examples of both. In dendrite targeting of the fly olfactory circuit, an initial coarse map is established by global gradients of Sema1A and Sema2A/2B. Next, local binary choices are specified by cell-type-specific expression of Capricious. Capricious-positive and -negative glomeruli are distributed widely the antennal lobe in a salt-and-pepper fashion, suggesting that the presence or absence of Capricious works with graded semaphorin levels to locally sort out dendrite targeting (Figure 7-44). Although a strict temporal ordering has not been proven, it is likely that semaphorin-mediated coarse targeting occurs before Capricious-mediated glomerular sorting. Likewise, in fly ORN axon targeting, Sema2B-mediated binary trajectory choice divides ORNs into two groups that preferentially innervate the dorsolateral or ventromedial halves of the antennal lobe. At subsequent steps, these two groups could conceivably utilize the same local cues without ambiguity.

The pretarget sorting of mammalian ORN axons exemplifies spatial steps. Instead of postponing their wiring decision until they reach their final destination, the growing axons sort themselves out along their journey from the olfactory epithelium to the olfactory bulb. As they travel toward their targets, axons expressing identical odorant receptors, and thus similar levels of neuropilin-1 and Sema3A, are mutually repelled by axons expressing different levels of these molecules, such that axons from ORNs of the same type preferentially associate with each other. This repulsive interaction can be thought of as a continuous series of steps. When the axons finally reach the olfactory bulb, each axon is constrained by its presorting and can thus choose from only a limited set of targets (Figure 7-40).

7.30 Many connections do not need to be specified at the level of individual synapses or neurons

The combined use of molecular and developmental strategies described above allows a limited set of molecules to wire many complex neural circuits with precision. Indeed, these strategies also enable considerable molecular redundancy in neural wiring. For example, recent RNA-sequencing studies revealed that closely related cell types, such as *Drosophila* R8 and R7 photoreceptors or olfactory projection neurons that send dendrites to neighboring glomeruli, differentially express dozens of cell-surface molecules when they are making wiring decisions. Mutating individual genes usually causes only partial mistargeting phenotypes. Thus, redundancy must make circuit wiring less sensitive to disruption of individual genes. After all, neuronal wiring is not a product of top-down engineering to minimize molecules involved; rather, different molecules are recruited

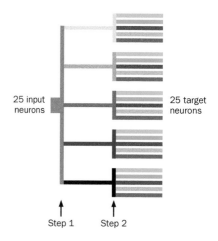

Figure 7-49 Sequential decisions conserve molecules and increase fidelity. In Step 1, input neurons are divided into five groups specified by five molecules symbolized by different shades of gray. In Step 2, five molecules symbolized by different colors further divide each group into five subgroups.

gradually in an *ad hoc* basis during evolution, a theme we will explore further in Chapter 13.

Neuronal wiring needs to be specified only as precisely as is useful. In the vertebrate neuromuscular system, the connections between specific motor pools and muscles are precisely specified. However, within a motor pool–muscle pair, details of connections between individual motor neurons and muscle fibers vary considerably, as exemplified by the connectome of a mouse ear muscle (Figure 7-30). In the olfactory system, great precision is required during ORN–PN matching—no ORN or PN types can be confused with any other types, as the different types represent distinct sensory information. However, the wiring of individual ORNs within one type does not need to be differentiated, as these ORNs project to the same glomerulus and represent the same sensory information. Thus, ~30 or ~5000 individual ORNs of the same type in the fly and the mouse, respectively, can use the same set of wiring instructions.

In certain neural circuits, it is not necessary or even beneficial for wiring to be specified precisely. The connection matrix between *Drosophila* olfactory PNs and Kenyon cells of the mushroom body (Figure 6-31) provides a good example. Here, the circuit's function can be served as long as individual PNs connect stochastically with several Kenyon cells. The meaning of these connections is acquired by individual experience, as we will learn in Section 11.18. It is possible that the role of stochastic wiring expands with increasing brain complexity; this interesting hypothesis remains to be tested.

7.31 Wiring can be instructed by neuronal activity and experience

Spontaneous and experience-driven neuronal activities play important roles in wiring of certain neural circuits. This allows the strength and pattern of synaptic transmission to directly shape neuronal wiring. We have studied this most extensively in the wiring of the visual system (Sections 5.7–5.11) but have also seen this in the wiring of the somatosensory system (Box 5-3) and the olfactory system (Section 7.19) and in synapse refinement at the vertebrate neuromuscular junction (Section 7.13). In complex circuits, molecular determinants guide the wiring of the basic circuit scaffold, with cellular and sometimes subcellular specificities. The number and strength of connections, which contribute to the optimization of circuit function, are then refined following Hebb's "Fire together, wire together" rule. Indeed, the brain changes its wiring as we learn throughout life, mostly by altering the strengths of synapses between connected neurons (Figure 7.1; see Chapter 11 for more discussion).

A remaining challenge is to determine the relative contributions of the different wiring strategies described in this chapter and in Chapter 5, in particular the relative contributions of activity-independent molecular specification and activity-dependent processes. What general rules determine their differential use? How do these two broad mechanisms cooperate? One hypothesis is that the proportion of activity-dependent wiring increases as the nervous system gets more complex, as is observed in comparing the wiring of worm or fly nervous systems with those of mammals. Another hypothesis is that neuronal circuits underlying innate behaviors, such as avoiding predator odors, are more likely to be hardwired by molecular determinants selected by evolution, whereas those underlying acquired abilities, such as human language, are sculpted more by life experiences and neuronal activity.

SUMMARY

The construction of the nervous system occurs in the context of a highly coordinated developmental program. The vertebrate nervous system derives from the neural tube. Extrinsic morphogens and intrinsic transcription factors pattern the anterior–posterior and dorsal–ventral axes of the neural tube, from which diverse types of neural progenitors and neurons arise. Cell fate diversification can be achieved by both asymmetric cell division and cell–cell interaction. Coordinated expression and mutual repression of transcription factors are frequently used to specify fates of distinct cell types. Postmitotic neurons often migrate long distances from their birthplaces to their final destinations. By the time individual neurons begin connecting to one another, a blueprint specifying the numbers and types of neurons in all regions of the nervous system has already been laid down.

The first step of a neuron's morphological differentiation is to send out processes, one of which becomes an axon. One critical output of the neuronal fate decision is the expression of a specific repertoire of guidance receptors. Guidance receptors in axons' growth cones respond to attractive and repulsive cues along their path and can change responses at intermediate targets. Multiple cue–receptor systems act simultaneously and sequentially to guide axons to their final destinations. Dendrites differ from axons in length, branching pattern, microtubule polarity, and dependence on local secretory machinery. Dendrites and axons must both branch out to effectively cover the receptive space and to send output to distinct targets, respectively. Homophilic repulsion mediated by molecules with extraordinary diversity enables self-avoidance of axonal and dendritic branches.

A critical step in brain wiring is synapse formation. Extensive bidirectional communication between pre- and postsynaptic partners utilizing trans-synaptic molecular interactions occurs at both neuron–muscle and neuron–neuron synapses. Signals from surrounding glia are also critical for synapse formation and maturation. Subcellular site selection of synapses employs molecules and mechanisms similar to those involved in axon guidance and developmental patterning. Final connectivity patterns are also sculpted by regressive events, including elimination of extra synapses, pruning of exuberant axons and dendrites, and death of excess neurons.

The precise one-to-one pairing of axons of a given type of olfactory receptor neuron (ORN) with dendrites of a given type of second-order projection neuron (PN) in the mouse and fly olfactory systems provides excellent models for studying how wiring specificity of neural circuits arises. In the mouse, the basal signaling levels of odorant receptors control ORN axon targeting by regulating the expression levels of guidance molecules. Repulsive axon–axon interaction sorts ORN axons along their path. Activity-dependent expression of adhesive and repulsive molecules further refines uniglomerular targeting of ORN axons. In the fly, ORNs and PNs are independently specified, and both cell types play active roles in regulating wiring specificity. PN dendrites coarsely prepattern the antennal lobe, utilizing a combination of global gradients and local binary determinants. ORN axons sort themselves through axon–axon interactions and recognize cues on partner PNs to establish the final one-to-one synaptic matches.

The cellular and molecular mechanisms we have studied in this chapter and in Chapter 5 begin to explain how a limited number of recognition molecules can specify an astronomical number of synaptic connections. Different levels of the same protein can specify different connections. The same molecule can have multiple functions and can be used in different parts of the nervous system and at different developmental stages. A combinatorial coding strategy can conserve molecules; this strategy is often implemented by dividing wiring decision into multiple steps. Many connections are instructed by neuronal activity and experience. Some connections do not need to be specified at the level of individual neurons or synapses, as neuronal connections need to be specified only as precisely as is useful.

For ease of experimental observation and manipulation, most studies of neuronal wiring have thus far been conducted in relatively simple model neurons and organisms. The lessons learned and approaches acquired can now be applied to more complex neural circuits in mammalian nervous systems. Determining how a neural circuit is organized in the adult often precedes deciphering how wiring specificity arises. Elucidating the organizational principles of neural circuits and the mechanisms of their assembly will in turn help us understand the function of these circuits in perception and behavior.

OPEN QUESTIONS (see related Open Questions in Chapter 5)

- How is the total number of neurons and glia in a given region regulated? How are the proportions of specific cell types, such as excitatory and inhibitory neurons, controlled?

- What signals ensure that each neuron has a single axon originating from the soma?

- How does trans-synaptic signaling drive the assembly of presynaptic terminals and postsynaptic densities during synaptogenesis?

- To what extent are second-order projection neurons in the mammalian olfactory system (mitral and tufted cells) specified before they connect with olfactory receptor neurons?

- How does differential expression of transcription factors give rise to differential expression of the cell-surface proteins to generate cell-type-specific wiring patterns?

- How redundant is the cell-surface code directing wiring specificity?

- To what extent is the wiring pattern shaped stochastically? What mechanisms control the balance between stereotypy and stochasticity?

FURTHER READING

Reviews

Hong W & Luo L (2014). Genetic control of wiring specificity in the fly olfactory system. *Genetics* 196:17–29.

Jan YN & Jan LY (2010). Branching out: mechanisms of dendritic arborization. *Nat Rev Neurosci* 11:316–328.

Levi-Montalcini R (1987). The nerve growth factor 35 years later. *Science* 237:1154–1162.

O'Leary DD, Chou SJ, & Sahara S (2007). Area patterning of the mammalian cortex. *Neuron* 56:252–269.

Pasca SP (2018). The rise of three-dimensional human brain cultures. *Nature* 553:437–445.

Sanes JR & Zipursky SL (2020) Synaptic specificity, recognition molecules, and assembly of neural circuits. *Cell* 181:536–556.

Südhof TC (2018). Towards an understanding of synapse formation. *Neuron* 100:276–293.

Neural patterning and fate determination

Anderson SA, Eisenstat DD, Shi L, & Rubenstein JL (1997). Interneuron migration from basal forebrain to neocortex: dependence on *Dlx* genes. *Science* 278:474–476.

Briscoe J, Pierani A, Jessell TM, & Ericson J (2000). A homeodomain protein code specifies progenitor cell identity and neuronal fate in the ventral neural tube. *Cell* 101:435–445.

Cohen S, Levi-Montalcini R, & Hamburger V (1954). A nerve growth-stimulating factor isolated from sarcom as 37 and 180. *Proc Natl Acad Sci U S A* 40:1014–1018.

Gao P, Postiglione MP, Krieger TG, Hernandez L, Wang C, Han Z, Streicher C, Papusheva E, Insolera R, Chugh K, et al. (2014). Deterministic progenitor behavior and unitary production of neurons in the neocortex. *Cell* 159:775–788.

Lancaster MA, Renner M, Martin CA, Wenzel D, Bicknell LS, Hurles ME, Homfray T, Penninger JM, Jackson AP, & Knoblich JA (2013). Cerebral organoids model human brain development and microcephaly. *Nature* 501:373–379.

Lodato S, Molyneaux BJ, Zuccaro E, Goff LA, Chen HH, Yuan W, Meleski A, Takahashi E, Mahony S, Rinn JL, et al. (2014). Gene co-regulation by Fezf2 selects neurotransmitter identity and connectivity of corticospinal neurons. *Nat Neurosci* 17:1046–1054.

Rhyu MS, Jan LY, & Jan YN (1994). Asymmetric distribution of numb protein during division of the sensory organ precursor cell confers distinct fates to daughter cells. *Cell* 76:477–491.

Axon and dendrite development

Barnes AP, Lilley BN, Pan YA, Plummer LJ, Powell AW, Raines AN, Sanes JR, & Polleux F (2007). LKB1 and SAD kinases define a pathway required for the polarization of cortical neurons. *Cell* 129:549–563.

Bradke F & Dotti CG (1999). The role of local actin instability in axon formation. *Science* 283:1931–1934.

Charron F, Stein E, Jeong J, McMahon AP, & Tessier-Lavigne M (2003). The morphogen sonic hedgehog is an axonal chemoattractant that collaborates with netrin-1 in midline axon guidance. *Cell* 113:11–23.

Chen WV, Nwakeze CL, Denny CA, O'Keeffe S, Rieger MA, Mountoufaris G, Kirner A, Dougherty JD, Hen R, Wu Q, et al. (2017). Pcdhαc2 is required for axonal tiling and assembly of serotonergic circuitries in mice. *Science* 356:406–411.

Chen Z, Gore BB, Long H, Ma L, & Tessier-Lavigne M (2008). Alternative splicing of the Robo3 axon guidance receptor governs the midline switch from attraction to repulsion. *Neuron* 58:325–332.

Joo W, Hippenmeyer S, & Luo L (2014). Dendrite morphogenesis depends on relative levels of NT-3/TrkC signaling. *Science* 346:626–629.

Lefebvre JL, Kostadinov D, Chen WV, Maniatis T, & Sanes JR (2012). Protocadherins mediate dendritic self-avoidance in the mammalian nervous system. *Nature* 488:517–521.

Luria V, Krawchuk D, Jessell TM, Laufer E, & Kania A (2008). Specification of motor axon trajectory by ephrin-B:EphB signaling: symmetrical control of axonal patterning in the developing limb. *Neuron* 60:1039–1053.

Lyuksyutova AI, Lu CC, Milanesio N, King LA, Guo N, Wang Y, Nathans J, Tessier-Lavigne M, & Zou Y (2003). Anterior-posterior guidance of commissural axons by Wnt-frizzled signaling. *Science* 302:1984–1988.

Ori-McKenney KM, Jan LY, & Jan YN (2012). Golgi outposts shape dendrite morphology by functioning as sites of acentrosomal microtubule nucleation in neurons. *Neuron* 76:921–930.

Osterloh JM, Yang J, Rooney TM, Fox AN, Adalbert R, Powell EH, Sheehan AE, Avery MA, Hackett R, Logan MA, et al. (2012). dSarm/Sarm1 is required for activation of an injury-induced axon death pathway. *Science* 337:481–484.

Rubinstein R, Thu CA, Goodman KM, Wolcott HN, Bahna F, Mannepalli S, Ahlsen G, Chevee M, Halim A, Clausen H, et al. (2015). Molecular logic of neuronal self-recognition through protocadherin domain interactions. *Cell* 163:629–642.

Scheiffele P, Fan J, Choih J, Fetter R, & Serafini T (2000). Neuroligin expressed in nonneuronal cells triggers presynaptic development in contacting axons. *Cell* 101:657–669.

Schmucker D, Clemens JC, Shu H, Worby CA, Xiao J, Muda M, Dixon JE, & Zipursky SL (2000). *Drosophila* Dscam is an axon guidance receptor exhibiting extraordinary molecular diversity. *Cell* 101:671–684.

Seeger M, Tear G, Ferres-Marco D, & Goodman CS (1993). Mutations affecting growth cone guidance in *Drosophila*: genes necessary for guidance toward or away from the midline. *Neuron* 10:409–426.

Spitzweck B, Brankatschk M, & Dickson BJ (2010). Distinct protein domains and expression patterns confer divergent axon guidance functions for *Drosophila* Robo receptors. *Cell* 140:409–420.

Tosney KW & Landmesser LT (1985). Specificity of early motoneuron growth cone outgrowth in the chick embryo. *J Neurosci* 5:2336–2344.

Watts RJ, Hoopfer ED, & Luo L (2003). Axon pruning during *Drosophila* metamorphosis: evidence for local degeneration and requirement of the ubiquitin-proteasome system. *Neuron* 38:871–885.

Wojtowicz WM, Wu W, Andre I, Qian B, Baker D, & Zipursky SL (2007). A vast repertoire of Dscam binding specificities arises from modular interactions of variable Ig domains. *Cell* 130:1134–1145.

Ye B, Zhang Y, Song W, Younger SH, Jan LY, & Jan YN (2007). Growing dendrites and axons differ in their reliance on the secretory pathway. *Cell* 130:717–729.

Synapse development

Aoto J, Martinelli DC, Malenka RC, Tabuchi K, & Südhof TC (2013). Presynaptic neurexin-3 alternative splicing trans-synaptically controls postsynaptic AMPA receptor trafficking. *Cell* 154:75–88.

Buffelli M, Burgess RW, Feng G, Lobe CG, Lichtman JW, & Sanes JR (2003). Genetic evidence that relative synaptic efficacy biases the outcome of synaptic competition. *Nature* 424:430–434.

Christopherson KS, Ullian EM, Stokes CC, Mullowney CE, Hell JW, Agah A, Lawler J, Mosher DF, Bornstein P, & Barres BA (2005). Thrombospondins are astrocyte-secreted proteins that promote CNS synaptogenesis. *Cell* 120:421–433.

Farhy-Tselnicker I, van Casteren ACM, Lee A, Chang VT, Aricescu AR, & Allen NJ (2017). Astrocyte-secreted glypican 4 regulates release of neuronal pentraxin 1 from axons to induce functional synapse formation. *Neuron* 96:428–445 e413.

Graf ER, Zhang X, Jin SX, Linhoff MW, & Craig AM (2004). Neurexins induce differentiation of GABA and glutamate postsynaptic specializations via neuroligins. *Cell* 119:1013–1026.

Klassen MP & Shen K (2007). Wnt signaling positions neuromuscular connectivity by inhibiting synapse formation in *C. elegans*. *Cell* 130:704–716.

Lehrman EK, Wilton DK, Litvina EY, Welsh CA, Chang ST, Frouin A, Walker AJ, Heller MD, Umemori H, Chen C, et al. (2018). CD47 protects synapses from excess microglia-mediated pruning during development. *Neuron* 100:120–134 e126.

Lu J, Tapia JC, White OL, & Lichtman JW (2009). The interscutularis muscle connectome. *PLoS Biol* 7:e32.

Tai Y, Gallo NB, Wang M, Yu JR, & Van Aelst L (2019). Axo-axonic innervation of neocortical pyramidal neurons by GABAergic chandelier cells requires ankyrinG-associated L1CAM. *Neuron* 102:358–372 e359.

Yumoto N, Kim N, & Burden SJ (2012). Lrp4 is a retrograde signal for presynaptic differentiation at neuromuscular synapses. *Nature* 489:438–442.

Assembly of the olfactory circuits

Hong W, Mosca TJ, & Luo L (2012). Teneurins instruct synaptic partner matching in an olfactory map. *Nature* 484:201–207.

Imai T, Yamazaki T, Kobayakawa R, Kobayakawa K, Abe T, Suzuki M, & Sakano H (2009). Pre-target axon sorting establishes the neural map topography. *Science* 325:585–590.

Jefferis GS, Marin EC, Stocker RF, & Luo L (2001). Target neuron prespecification in the olfactory map of *Drosophila*. *Nature* 414:204–208.

Komiyama T, Sweeney LB, Schuldiner O, Garcia KC, & Luo L (2007). Graded expression of semaphorin-1a cell-autonomously directs dendritic targeting of olfactory projection neurons. *Cell* 128:399–410.

Li H, Horns F, Wu B, Xie Q, Li J, Li T, Luginbuhl DJ, Quake SR, & Luo L (2017). Classifying *Drosophila* olfactory projection neuron subtypes by single-sell RNA sequencing. *Cell* 171:1206–1220.

Nakashima A, Takeuchi H, Imai T, Saito H, Kiyonari H, Abe T, Chen M, Weinstein LS, Yu CR, Storm DR, et al. (2013). Agonist-independent GPCR activity regulates anterior-posterior targeting of olfactory sensory neurons. *Cell* 154:1314–1325.

Wang F, Nemes A, Mendelsohn M, & Axel R (1998). Odorant receptors govern the formation of a precise topographic map. *Cell* 93:47–60.

CHAPTER 8
Motor Systems

To move things is all mankind can do, for such the sole executant is muscle, whether in whispering a syllable or in felling a forest.

Charles Sherrington, 1924

The sensory systems we studied in Chapters 4 and 6 enable animals to know about the world they live in. But sensation is only the first step: to adapt to the world, animals must *act* upon sensory knowledge. For example, animals have evolved the ability to sense the presence of food so they can eat it and obtain energy essential for life. Likewise, animals have evolved the ability to detect danger so that they can fight, hide, or run from it. In the next two chapters, we study the output of the nervous system that makes these active responses to sensation possible.

A key output is the **motor system**, which controls the contraction of skeletal muscles and thereby enables movement, such as reaching for and grasping an object, walking, talking, or maintaining the body's posture. **Figure 8-1** illustrates just one example of how the nervous system exquisitely controls movement, in this case enabling a horse to trot gracefully. We study the motor system in this chapter, with a focus on locomotion control in vertebrates. We will discuss the other two output systems—autonomic and neuroendocrine—at the beginning of Chapter 9.

Movement is produced by coordinated activation of motor neurons driving coordinated contraction of skeletal muscles. In vertebrates, motor neuron cell bodies are located in the spinal cord and brainstem; their axons exit the CNS and innervate specific muscles in the body and head, respectively. Motor neurons are themselves under elaborate control: they receive direct input from proprioceptive

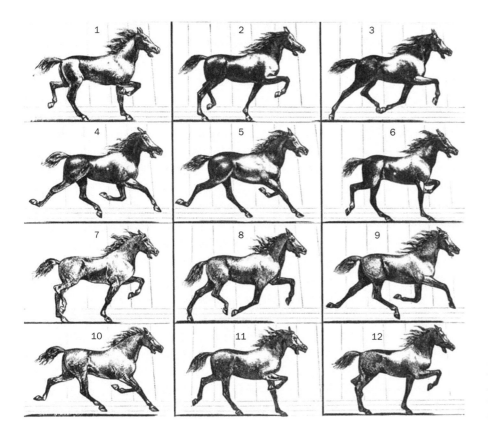

Figure 8-1 The movement of a trotting horse. A complete stride of a trotting horse in 12 frames, engraved after photos by Eadweard Muybridge. These photos (exposure time 1/500 s) showed clearly for the first time that the trotting horse is entirely in the air for part of the stride (frames 4, 5, 9, and 10). (From Muybridge E [1878] *Sci Am* 39[16]:241.)

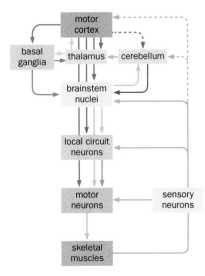

Figure 8-2 Hierarchical organization of movement control. Arrows indicate the direction of information flow. Solid arrows between two regions indicate that at least some connections are direct. Dashed arrows indicate connections through intermediate neurons. While the same neuron can send collaterals that innervate different targets (for example, motor cortex neurons that project to the brainstem and spinal cord often have collaterals in the basal ganglia), separate pathways are drawn for simplicity. The Y-shaped terminal of sensory neurons in the skeletal muscle symbolizes the peripheral endings of proprioceptive somatosensory neurons.

somatosensory neurons, spinal cord premotor neurons, brainstem nuclei specialized in initiating and modulating movement, and (in some species, especially humans) the motor cortex. These descending motor control centers are organized in a hierarchical manner, with the somatosensory system providing feedback signals at multiple levels within this hierarchy. In addition, there are important loops involving the basal ganglia, cerebellum, and thalamus, which add to the sophistication of motor control (**Figure 8-2**). In Chapters 4 and 6, we studied sensory systems by following the direction of information flow, starting with sensory neurons and moving into the brain. We will take the same peripheral-to-central approach in our studies of the motor system, but this time we will address topics in the direction opposite the information flow, starting with muscles and motor neurons and then discussing how layers of upstream neurons and circuits contribute to movement control.

8.1 Muscle contraction is mediated by sliding of actin and myosin filaments and regulated by intracellular Ca²⁺

Muscle contraction underlies all bodily movement, as reflected in the epigraph. The mechanism of muscle contraction is understood in molecular detail. The muscle cell is an elongated cell composed of thread-like **myofibrils** along its long axis (**Figure 8-3**A). Under electron microscopy, myofibrils are seen to consist of

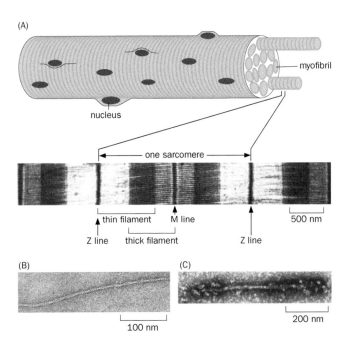

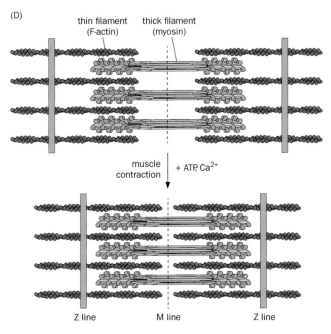

Figure 8-3 The molecular organization of a muscle cell and the sliding filament model of muscle contraction. (A) Top, schematic of a muscle cell, which has multiple nuclei (each muscle cell results from fusion of many myoblasts) and consists of parallel myofibrils. Bottom, as seen under electron microscopy, myofibrils are composed of repeating units called sarcomeres. Each sarcomere is formed by the intersection of thin filaments originating from the Z line with thick filaments originating from the M line. **(B)** A negatively stained electron micrograph shows a microfilament consisting of two F-actin polymer strands; in muscle, F-actin constitutes the major component of the thin filament. **(C)** A negatively stained electron micrograph shows an aggregation of

purified myosin, which forms a bare region in the middle and thick protrusions at each end, resembling the organization of thick filaments. **(D)** Illustration of the sliding filament model. The relative movement of thick and thin filaments, caused by myosin motors moving on the actin filaments, underlies muscle contraction. Ca²⁺ is required for myosin–actin interactions. ATP hydrolysis powers this movement. (A–C, micrographs from Huxley HE [1965] *Sci Am* 213(6):18–27. With permission from Springer Nature. See also Huxley AF & Niedergerke R [1954] *Nature* 173:971–973 and Huxley H & Hanson J [1954] *Nature* 173:973–976.)

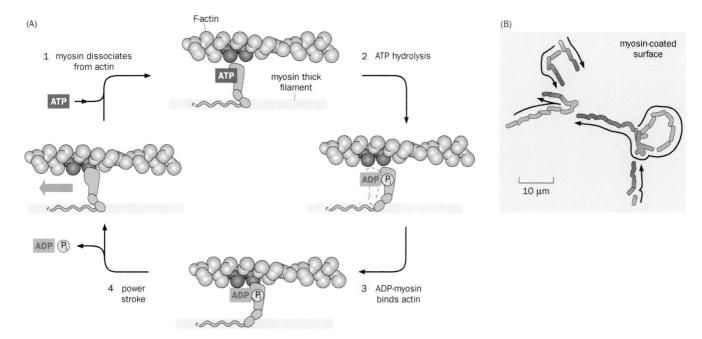

Figure 8-4 ATP hydrolysis powers the movement of myosin and actin filaments relative to each other. (A) The coupled cycles of actin–myosin interaction and ATP hydrolysis. (1) ATP binding to myosin triggers dissociation of myosin and actin. (2) ATP hydrolysis causes a conformational change of the myosin head such that it aligns with the binding surface of the next actin subunit (circles). (3) ADP-myosin binds actin again. (4) ADP and P_i release produces a power stroke—sliding of F-actin against myosin (pink arrow). Note that the myosin head's orientation relative to actin changes with the power stroke. The highlighted actin subunits are shifted one subunit to the right with respect to the actin filament before and after Step 1 to accommodate a new cycle. **(B)** Movement paths of fluorescently labeled actin filaments (rectangles) on a myosin-coated glass slide. Five actin filaments were tracked over the course of 38 s, and their positions are indicated at successive short intervals as they appeared on the video monitor. Arrows indicate the movement direction (darker actin filaments represent later in time). See also Movie 8-2. (A, based on Lymn RW & Taylor EW [1971] *Biochemistry* 10:4617–4624. With permission from the American Chemical Society. B, adapted from Kron SJ & Spudich JA [1986] *Proc Natl Acad Sci U S A* 83:6272–6276.)

repeating units called **sarcomeres**. Each sarcomere is made up of overlapping thick and thin filaments arranged in an orderly fashion. The thin filaments comprise **filamentous actin** (**F-actin**) fibers with several associated proteins (Figure 8-3B). The thick filaments comprise the **myosin** protein (Figure 8-3C). The sliding of thick and thin filaments over each other, mediated by physical interactions between actin and myosin, forms the basis of muscle contraction (Figure 8-3D).

As introduced in Section 2.3, myosin is an F-actin-binding **motor protein**. Myosin has a long tail embedded in the thick filament and a globular head that forms the cross bridge between the thick and thin filaments (Figure 8-3C, D). The head of myosin has an ATPase domain that hydrolyzes ATP. The hydrolysis cycle requires interaction of the myosin head with actin and converts chemical energy from ATP hydrolysis into mechanical force in the form of a **power stroke**, which underlies how myosin and actin move relative to each other (**Figure 8-4A; Movie 8-1**). This motility can be visualized in reduced preparations—for example, fluorescently labeled actin filaments were observed to move on a glass slide coated with pure myosin proteins (Figure 8-4B; **Movie 8-2**). Indeed, the movement produced by the interaction of myosin and actin, which was first investigated in the context of muscle contraction, underlies many aspects of cell motility, including cell migration and growth cone guidance (Box 5-2).

Actin/myosin-mediated contraction requires Ca^{2+} (Figure 8-3D), as the F-actin in thin filaments is coated by two proteins called tropomyosin and troponin, which prevent actin from binding to the myosin head under low intracellular Ca^{2+} concentrations. A rise in $[Ca^{2+}]_i$ causes a conformational change in the actin-tropomyosin complex, exposing the actin surface that interacts with myosin.

Ca^{2+} regulation of actin/myosin-mediated contraction forms the link between motor neuron activity and muscle contraction, a process called **excitation-contraction coupling** (**Figure 8-5**). As discussed in Sections 3.1 and 3.12, the arrival

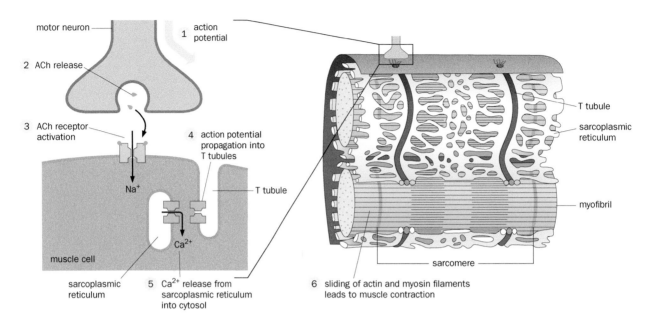

Figure 8-5 Sequence of events from motor neuron excitation to skeletal muscle contraction. The arrival of an action potential at the motor axon terminal (1) triggers acetylcholine (ACh) release (2). ACh binds to the nicotinic ACh receptor on the postsynaptic muscle surface, opening the nicotinic ACh receptor channel and triggering depolarization and action potential production in the muscle cell (3). Action potentials propagate within the muscle cell to T tubules (4) and trigger Ca^{2+} release from nearby sarcoplasmic reticulum into the cytosol (5). Elevated $[Ca^{2+}]_i$ causes muscle contraction (6). (The drawing on the right is adapted from Alberts et al. [2015] Molecular Biology of the Cell, 6th ed. Garland Science.)

of an action potential at the motor axon terminal of the vertebrate skeletal muscle junction causes release of the neurotransmitter acetylcholine (ACh). ACh binds to and opens the nicotinic ACh receptor channel at the neuromuscular junction, which causes depolarization of the muscle cell and production of action potentials within the muscle cell itself. Muscle cell depolarization triggers release of Ca^{2+} from the **sarcoplasmic reticulum**, a special endoplasmic reticulum derivative that extends throughout muscle cells. **Transverse tubules** (**T tubules**), invaginations of the plasma membrane that extend into the muscle cell interior, bring plasma membrane close to the sarcoplasmic reticulum, such that depolarization effectively triggers Ca^{2+} release throughout the large muscle cell. This causes nearly synchronous contraction of all sarcomeres within the same muscle cell, enabling muscles to respond rapidly to commands from motor neurons. Efficient reuptake of Ca^{2+} by the sarcoplasmic reticulum enables muscle cells to respond to repeated commands from motor neurons (**Movie 8-3**).

8.2 Motor units within a motor pool are recruited sequentially from small to large

Each vertebrate skeletal muscle consists of a few hundred to over a million individual muscle cells (also called **muscle fibers**). As discussed in Section 7.13, each muscle fiber is innervated by a single motor neuron in adults. However, each motor neuron innervates multiple muscle fibers, ranging from a few (in an eye muscle) to a few thousand (in a leg muscle). Individual muscle fibers innervated by a single motor neuron are dispersed within a given muscle, such that activation of that motor neuron produces force evenly across the muscle (**Figure 8-6**; Figure 7-30). A motor neuron and the set of muscle fibers it innervates are collectively called a **motor unit**. As neuromuscular junctions are powerful synapses that almost always convert presynaptic action potentials into neurotransmitter release and muscle contraction, the muscle fibers within a motor unit are nearly always activated together. Thus, the motor unit is the *elementary unit of force production* in the motor system.

Motor neurons innervating the same muscle cluster together in the ventral spinal cord or brainstem and constitute a **motor pool** (Figure 8-6); each motor

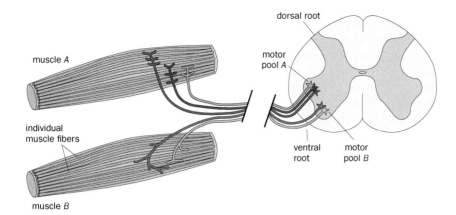

Figure 8-6 Motor pools and motor units.
A cluster of motor neurons in the ventral spinal cord constituting a motor pool innervating each muscle. The axons of these motor neurons exit the spinal cord via the ventral root. Two motor pools, *A* and *B*, are shown here. Within each motor pool, individual motor neurons innervate multiple muscle fibers, while each muscle fiber is innervated by a single motor neuron. A motor neuron and the set of muscle fibers it innervates together constitute a motor unit. Motor units vary in size, as is illustrated for the two motor units from motor pool *B*.

pool contains from dozens to thousands of neurons. The **motor unit size**, the number of muscle fibers a motor neuron innervates, varies considerably for neurons within the same motor pool (Figure 7-30). Collectively, motor units within a motor pool follow a **size principle**: neurons that have smaller motor unit sizes usually have smaller axon diameters and cell bodies and are recruited into action before neurons with larger motor unit sizes.

This size principle is illustrated by recordings of motor axon bundles' responses to sensory or electrical stimulation. For example, pressing on the Achilles tendon elicits a stretch reflex that activates the triceps surae muscle in the lower leg (**Figure 8-7**A). The magnitude of triceps surae activation increases as the pressure on the Achilles tendon increases. Researchers recorded from motor axon bundles innervating the triceps surae muscle in response to varying pressure applied to the Achilles tendon. They found that motor axons of smaller units were excited

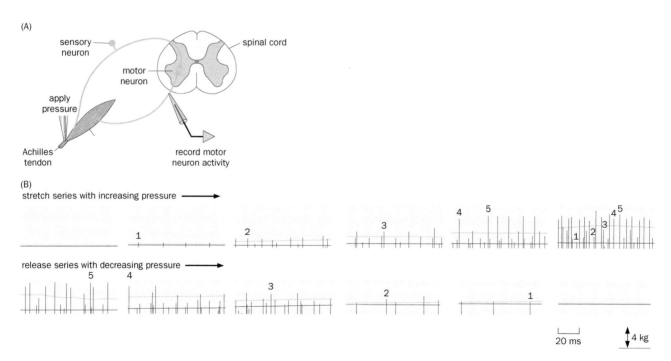

Figure 8-7 Motor units are recruited in order based on their size.
(A) Schematic of the experimental setup. Nerves innervating all hindlimb muscles except the triceps surae muscle were surgically severed. Pressure was applied to the Achilles tendon via a mechanoelectric transducer to elicit a stretch reflex. Responses of motor neurons innervating the triceps surae muscle were recorded at the ventral root of the spinal cord, where action potentials of different motor neurons can be recorded simultaneously via extracellular recording. The size of the signal (vertical bar in Panel B) correlates with axon diameter and hence with motor unit size. **(B)** Top, as the pressure increases (indicated by the distance between the blue and black horizontal lines, scale at bottom right), sensory neurons' firing rates increase, increasing motor neuron firing through monosynaptic transmission. The smallest motor neuron (1) fires first, followed by motor neurons of increasing sizes (2 through 5). Bottom, as the pressure decreases, the largest motor neuron (5) ceases firing first, followed by motor neurons of descending size (4 through 1). (B, adapted from HennemanE, Somjen G, & Carpernter DO [1965] *J Neurophysiol* 28:560–580.)

by a small amount of pressure and that motor axons corresponding to increasingly larger motor units were recruited sequentially as the pressure increased. In response to gradually reduced pressure on the Achilles, motor axons with the largest units ceased firing first, followed in an orderly manner by axons of decreasing motor unit size (Figure 8-7B). Importantly, this order of motor neuron recruitment does not usually vary—whether the muscles are stimulated by the natural stretch reflex or by direct electrical stimulation—suggesting that it is an intrinsic property of the motor pool.

The size principle of motor units enables incremental control of the magnitude of an individual muscle's contraction in response to excitatory and inhibitory inputs received by its motor pool. Thus, the size principle resembles sensory adaptation and Weber's law we discussed in the context of sensory perception (Section 4.7). When the magnitude of muscle contraction is small, adding or subtracting a small motor unit makes a notable difference; these differences are used to achieve fine motor control. As the magnitude of muscle contraction increases, the size of a motor unit required to make a notable difference in the total contraction strength also becomes larger. The size principle is also important from an energetics perspective. Most movements are small and use only small motor units that consume little energy, whereas large motor units, which exert greater force and consume more energy, are used more rarely.

8.3   Motor neurons receive diverse and complex inputs

We now turn to a key question in movement control: how is activation of different muscles coordinated? The principal goal of muscle contraction is to change the joint angle, as in the case of the knee-jerk reflex (Figure 1-19). The angle change is achieved by coordinated action of **extensor** muscles, whose contraction increases the angle (thus extending the joint), and **flexor** muscles, whose contraction decreases the angle (**Figure 8-8**). Extensors and flexors are **antagonistic muscles**, as they perform opposite actions, and they often fire in succession. For example, extension of a joint is initiated by contraction of the extensor and terminated by subsequent contraction of the flexor, so that the joint does not overextend. Complex movements such as the trotting of a horse involve coordinated contraction of many extensor–flexor pairs within each of the four legs, as well as coordination between the legs.

Our discussions so far have indicated that coordination of muscle contraction must be due to coordinated firing of specific motor pools. Therefore, we must learn more about how motor pools are organized and how motor neuron firing is controlled by their inputs. Motor pools that control trunk muscles are located in two bilaterally symmetric medial motor columns running the length of the rostral–caudal axis of the spinal cord, while motor pools that control muscles within each limb form lateral motor columns at the rostral–caudal positions in the spinal cord that innervate each limb (**Figure 8-9**). Within the lateral motor columns, the motor pools that innervate specific limb muscles are located at specific positions. These were originally determined by retrograde tracing (Section 14.18): dye injected into a specific muscle is taken up by axon terminals of motor neurons that innervate that muscle and is transported back to the neuronal cell bodies, resulting in labeling of the corresponding motor pool.

Motor neurons receive very complex sets of inputs. Each motor neuron elaborates a dendritic tree that covers a large area of the spinal cord (Figure 1-15C), enabling it to receive direct input from diverse sources (Figure 8-2). One major source of input comprises excitatory and inhibitory spinal interneurons known as spinal cord **premotor neurons**. The distribution of premotor neurons has been analyzed comprehensively via the retrograde trans-synaptic tracing method (see Section 14.19 for details). **Figure 8-10** shows the distribution of premotor neurons for the motor pool that innervates the quadriceps muscle in the dorsal thigh of the mouse hind limb. Premotor neurons can be ipsilateral or contralateral to the motor pools and are spread across many spinal segments. They consist of multiple transmitter types, including glutamate, GABA, glycine, and ACh. Motor pools

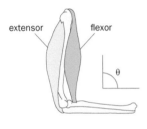

extensor   flexor

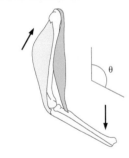

extensor contraction

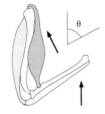

flexor contraction

Figure 8-8 Extensor–flexor muscle pairs control the joint angle. Contraction of the extensor muscle increases the joint angle (θ), leading to extension of the joint. Contraction of the flexor muscle decreases the joint angle. The extent and timing of extensor and flexor contraction are coordinated to precisely control the joint angle.

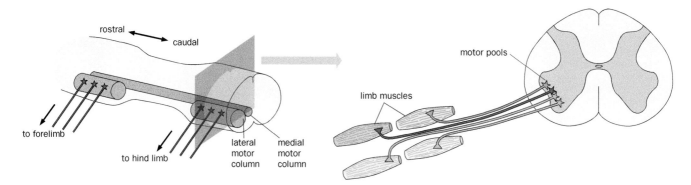

Figure 8-9 Organization of motor columns and motor pools in the spinal cord. Left, motor neurons are organized into motor columns in the ventral spinal cord along the rostral–caudal axis. The medial motor columns regulate trunk muscles, while the lateral motor columns are present at the levels of the spinal cord corresponding to the locations of the limbs and innervate limb muscles. Right, a cross section of the spinal cord at the level of the hind limb. Motor pools that innervate specific muscles (shown as four color-matched pairs) are located in stereotyped positions within the ventral spinal cord. This organization is bilaterally symmetrical (the arrangement is illustrated on only one side).

innervating different muscles receive input from distinct combinations of premotor neurons.

In addition to receiving input from spinal cord premotor neurons, motor neurons also receive monosynaptic input from a small subset of proprioceptive somatosensory neurons, which innervate muscle spindles and form simple reflex arcs (Figure 1-19). These sensory neurons are located in the dorsal root ganglia, and their central axons enter the spinal cord through the **dorsal root** (Figure 6-63), distinct from the **ventral root** where motor axons exit the spinal cord (Figure 8-6). Motor neurons also receive direct descending input from neurons in the brainstem and (in some species) motor cortex, whose axons travel down along distinct pathways in the spinal cord white matter. And, like the motor neurons they regulate, spinal cord premotor neurons themselves receive input from sensory neurons and neurons in the brainstem and motor cortex (Figure 8-2).

The bewildering complexity of spinal cord premotor neurons, along with descending input and sensory feedback, affords exquisite control of motor neuron firing and muscle contraction. It also poses great challenges for researchers trying to discover the principles underlying motor coordination. In the next three sections, we will study one important principle by which rhythmic activation of motor neurons and muscle contraction is achieved.

8.4 Central pattern generators coordinate rhythmic contraction of different muscles during locomotion

Locomotion requires coordinated and rhythmic contraction of many different muscles. For example, in a trotting horse, different leg muscles are activated in a specific sequence to enable each leg to step on the ground, leave the ground, extend forward, and step on the ground again. The four legs are highly coordinated. When the horse slows down to a walk or speeds up to a gallop, the cycle speed for each leg and the synchrony among the legs differ from when the horse is trotting (**Figure 8-11**). How are these motor programs controlled?

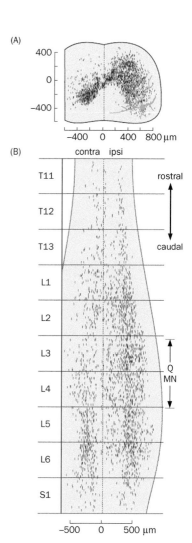

Figure 8-10 Distribution of spinal neurons presynaptic to the motor pool innervating the quadriceps muscle in the mouse right hind limb. Even though the cell bodies of quadriceps motor pools (Q MN) are restricted to the two spinal segments indicated, their presynaptic premotor neurons (dots) are widely distributed, as seen from **(A)** a transverse projection of the spinal cord and **(B)** a longitudinal projection covering thoracic (T), lumbar (L), and sacral (S) segments. A drawing of the dendritic arbor of a typical Q motor neuron (cyan) is superimposed in Panel A. The dotted red line represents the midline of the spinal cord. (Adapted from Stepien AE, Tripodi M, & Arber S [2010] *Neuron* 68:456–472. With permission from Elsevier Inc. Motor neuron drawing courtesy of Silvia Arber.)

Figure 8-11 Stepping patterns of a horse during walking, trotting, and galloping. From left to right, each horizontal bar represents for a single leg the time off (gray segments) and on (green segments) the ground. During trotting, the left hind leg (LH) and right foreleg (RF) move in sync, as do the right hind leg (RH) and left foreleg (LF). During galloping, the two forelegs are in sync, as are the two hind legs. (Adapted from Pearson K [1976] *Sci Am* 235[6]:72–86. With permission from Springer Nature.)

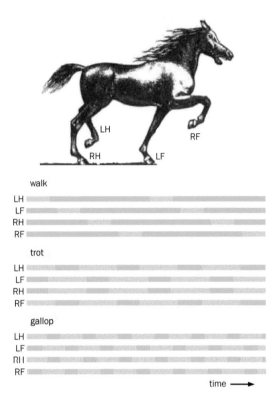

walk

trot

gallop

time ⟶

Recall the knee-jerk reflex discussed in Section 1.9: proprioceptive sensory neurons transduce mechanical stimulation in the spindle of the extensor muscle to activate extensor motor neurons through monosynaptic excitation and simultaneously inhibit flexor motor neurons through an intermediate spinal cord inhibitory neuron (Figure 1-19). An early hypothesis of "chained reflexes" proposed that rhythmic movement such as walking was due to sequential activation of spinal reflexes. Specifically, movement of a leg caused by contraction of a muscle would produce feedback from proprioceptive sensory neurons, which would activate a second motor pool and its corresponding muscle through the spinal cord reflex circuits. This would trigger activation of a second set of sensory neurons that in turn would activate a third motor pool and muscle, and so on, until the original muscle would be activated again, completing the chained reflex cycle.

The chained reflex hypothesis predicts that (1) the spinal cord with intact sensory feedback should be sufficient for rhythmic activation of muscles and (2) when sensory feedback is blocked, rhythmic activation of muscles should cease. The organization of the spinal cord made testing these predictions possible, since transecting dorsal roots would block sensory feedback to motor neurons without affecting connections between motor neurons and muscles, which leave the spinal cord through ventral roots (Figure 8-6). An experiment to test these hypotheses was performed a century ago. In this experiment, all hind limb muscles of an anesthetized cat were surgically inactivated except for one extensor–flexor pair in the lower leg, so that contraction of the extensor and flexor could be measured accurately. A spinal transection rostral to the segments controlling the hind limb was found to induce a transient rhythmic and alternate contraction of the remaining extensor and flexor that mimicked normal walking. (Spinal transection induces massive glutamate release, which mimics excitatory input from the brainstem, as we will discuss later.) Importantly, when dorsal roots were transected to eliminate sensory feedback, the pattern and frequency of alternate contraction of the extensor and flexor induced by spinal transection persisted (**Figure 8-12**A). These results supported prediction (1), as rhythmic contraction of the extensor and flexor could indeed be supported by an isolated spinal cord disconnected from descending control. (As we will learn later, the rhythmic activity must be *initiated* by the brain-

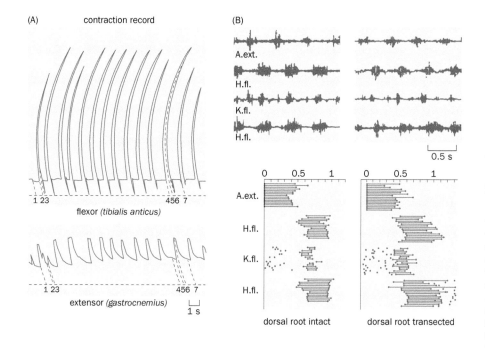

(A) contraction record

flexor (tibialis anticus)

extensor (gastrocnemius)

(B)

A.ext.

H.fl.

K.fl.

H.fl.

0.5 s

A.ext.

H.fl.

K.fl.

H.fl.

dorsal root intact   dorsal root transected

Figure 8-12 Rhythmic and coordinated muscle contraction in the absence of sensory feedback. (A) Contractions of a flexor muscle and an extensor muscle induced by spinal cord transection were recorded. Dorsal roots had been transected before the experiment to remove sensory feedback. The numbers and dashed lines below are used to synchronize the two records so the contraction of the two muscles can be compared in the same time frame. Within each cycle, the flexor contracted before the extensor, followed by a period when neither muscle contracted. Thus, rhythmic activation of these muscles persisted in the absence of sensory feedback. **(B)** A mesencephalic cat, in which the brainstem/spinal cord and the cerebral cortex/thalamus had been surgically disconnected, was induced to walk on a treadmill by brainstem stimulation (see Figure 8-17). Action potentials in four muscles of the same hind limb during the step cycle were recorded by electromyogram (EMG). The four muscles showed similar activity patterns when sensory feedback was intact (left) or removed by dorsal root transection (right), as seen in the EMG record (top) and in their timing of activation during the step cycles (bottom; 15 EMG recordings per muscle were compiled, with step cycle as the x axis unit). The four muscles, from top to bottom, are an ankle extensor, a hip flexor, a knee flexor, and a second hip flexor. (A, adapted from Graham Brown T [1911] *Proc R Soc London [B]* 84: 308–319. B, from Grillner S & Zangger P [1975] *Brain Res* 88:367–371. With permission from Elsevier Inc.)

stem; in this case, the spinal lesion initiated the rhythmic activity.) Contrary to prediction (2), however, this experiment revealed that rhythmic contractions persist in the absence of sensory feedback.

Further support for the idea that an autonomous spinal cord mechanism could produce rhythmic output came when technical advances in 1960–1970 made it possible to measure rhythmic contraction of many muscles during locomotion. For example, in a widely used experimental preparation, an incision is made at the level of the midbrain (mesencephalon; Figure 1-8) of a cat such that the cerebral cortex/thalamus and the brainstem/spinal cord are disconnected. Although the resulting "mesencephalic cat" can no longer voluntarily control its movement, it can still walk on a treadmill after brainstem stimulation. The contractions of many muscles during walking can be recorded simultaneously by their action potential patterns in electromyograms. The coordinated contractions of different leg muscles during stepping were found to be similar before and after dorsal root transection, as measured by the timing and duration of contraction for each muscle during the stepping cycle (Figure 8-12B).

Studies of rhythmic movements in invertebrate systems likewise found that rhythmicity originates from specific segments or ganglia in the central nervous system (see Section 8.5 below). Collectively, these experiments led to the concept of the **central pattern generator** (**CPG**), a central nervous system circuit capable of producing rhythmic output for coordinated contraction of different muscles without sensory feedback. The existence of CPGs does not mean that sensory feedback is unimportant. On the contrary, sensory feedback modulates and can override the CPG rhythm. For example, in the mesencephalic cat, sensory feedback produced by increasing the speed of the treadmill can modulate the speed of the stepping cycle and even trigger a transition of the motor patterns from walking to trotting or galloping. Nevertheless, experiments such as those described in Figure 8-12 indicate that rhythmic output can originate from neural circuits in the spinal cord upon initiation.

The concept of CPGs has extended beyond control of locomotion; CPGs have been proposed to control breathing (see Box 8-1), swallowing, and many other rhythmic movements. Indeed, the phenomenon of neural network oscillation goes beyond motor control. For example, different frequencies of rhythmic activity observed in the thalamus, cerebral cortex, and hippocampus have been proposed to play important roles in perception and cognition. How do neural circuits produce rhythmic output?

8.5 Intrinsic properties of neurons and their connection patterns produce rhythmic output in a model central pattern generator

Our best mechanistic understanding of rhythmic output production by CPGs has come from studies of invertebrate model circuits. These circuits usually consist of a small number of individually identifiable neurons that are large in size and easily accessible for intracellular recordings (Section 14.1). For example, the **stomatogastric ganglion (STG)** of crustaceans (lobster and crab) produces a pyloric rhythm to control the cyclic movement of a portion of the stomach. The pyloric rhythm can be seen in the triphasic firing patterns of four types of neurons—one interneuron (AB) and three types of motor neurons (PD, LP, and PY)—through simultaneous intracellular recordings (Figure 8-13A). Each neuron cycles between a hyperpolarized state and a depolarized state with bursts of action potentials. Importantly, the pattern seen in an intact lobster or crab can be faithfully reproduced when the stomatogastric nervous system is studied *in vitro* in the complete absence of sensory feedback, indicating that the rhythmic firing pattern is intrinsic to the STG.

The connection patterns between these four types of neurons (Figure 8-13B) have been determined by simultaneous electrophysiological recording and by cell ablation experiments. At the core of the pyloric rhythm is the AB interneuron, which exhibited rhythmic firing even when it was isolated from the rest of the circuit (Figure 8-13C). Thus, AB is a **pacemaker cell** which produces rhythmic output even in the absence of input. This pacemaker property results from the AB neuron's **intrinsic properties**, which are determined by the composition, concentration, and biophysical properties of the ion channels it expresses. Based on studies of other STG neurons, the transition from the depolarized state to the hyperpolarized state in AB is presumed to result from inactivation of voltage-gated cation channels and delayed opening of K^+ channels, much like the voltage-gated Na^+ and K^+ channels that underlie action potential production (Section 2.10). The rebound from the hyperpolarized state to the depolarized state likely results from opening of hyperpolarization-activated cation channels such as HCN channels (Figure 2-35), which causes depolarization, leading to activation of voltage-gated cation channels.

Figure 8-13 The pyloric circuit in the crustacean stomatogastric nervous system. (A) Simultaneous recordings of AB, PD, LP, and PY neurons in the stomatogastric ganglion (STG) show that each neuron cycles between depolarized and hyperpolarized states, with the AB/PD, LP, and PY neurons having offset activation phases. Action potentials (vertical spikes) are associated with the depolarized states. **(B)** The connection diagrams between 1 AB, 2 PD, 1 LP, and 5 PY neurons. AB and PD are electrically coupled through gap junctions (zigzag line). All chemical synapses are inhibitory. **(C)** In this experiment, the STG was first dissected out but remained connected with its central input nerve. The output spikes of PD and LP neurons were measured by extracellular recordings of their motor nerves, while the AB neuron was recorded by intracellular recording. All exhibited rhythmic output patterns (left). The conduction of the input nerve was then blocked, and the PD and LP neurons were ablated, as seen by the lack of spike output. The AB neuron continued to oscillate (albeit faster) in the absence of all functional connections (right). (A & B, adapted from Marder E & Bucher D [2007] *Ann Rev Physiol* 69:291–316. With permission from Annual Reviews. C, adapted from Miller JP & Selverston A [1982] *J Neurophysiol* 48:1378–1391.)

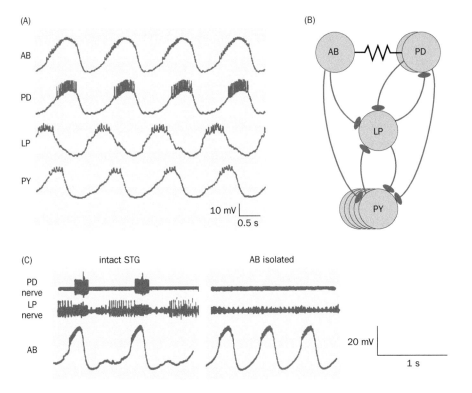

As diagrammed in Figure 8-13B, AB is electrically coupled to PD motor neurons through gap junctions; this causes PDs to fire in sync with AB. AB and PDs also form inhibitory chemical synapses onto LP and PYs, thereby inhibiting their firing. LP and PYs mutually inhibit each other; LP also inhibits PDs. Thus, when AB and PDs are in their depolarized state, their firing inhibits LP and PYs, forcing them to fire out of sync with AB/PDs. When AB/PDs stop firing, LP rebounds from inhibition before PY (this is because LP receives less inhibition from PDs and expresses fewer K$^+$ channels and more HCN channels than does PYs); at this point, LP inhibits firing of PDs and PYs. When the PYs eventually rebound from inhibition, they inhibit firing of LP, thereby disinhibiting PDs and facilitating the beginning of the next cycle. This chain of mutual inhibition among the three types of motor neurons produces the triphasic rhythm used to control coordinated contraction of the stomach muscles innervated by PD, LP, and PY neurons. In summary, the rhythmic firing of STG neurons is determined by the intrinsic properties of the constituent neurons as well as their connection patterns and strengths.

The simplicity of the crustacean STG system has enabled researchers to generate quantitative models based on the intrinsic properties of individual neurons and the connection strengths of the gap junction and inhibitory synapses. Such models have been used to simulate rhythmic output that matches experimental data. In one such study, the pyloric rhythm was simplified by lumping AB and PD together such that there are three neuronal types in the model: AB/PD, LP, and PY. From 20 million combinations of ion channel compositions (which determine membrane conductance) and connection strengths (which determine synaptic conductance) between the three types of neurons, nearly half a million distinct "solutions" that closely resembled the pyloric rhythm observed in animals were found. Two such solutions produced nearly identical triphasic rhythms (**Figure 8-14**, top), using notably different combinations of membrane and synaptic conductances (Figure 8-14, bottom). For example, the Na$^+$ conductance in the PY neuron is low in the case on the left and high in the case on the right, while the KCa conductance (reflecting the properties of Ca^{2+}-activated K$^+$ channels; Box 2-4) in the LP neuron is high on the left and low on the right. Variable parameters have

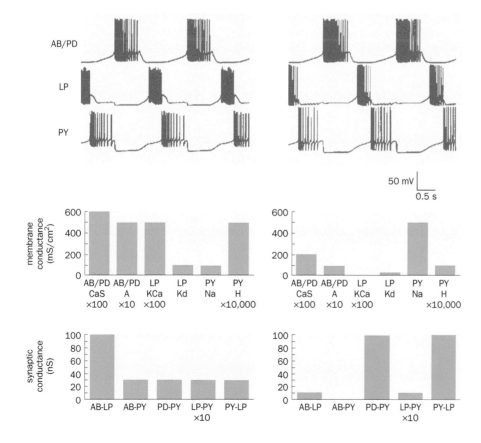

Figure 8-14 Similar network activity can be produced by distinct combinations of circuit parameters. Membrane potential traces from two model pyloric networks closely resemble each other (top), despite being produced by very different combinations of ion channels and synaptic strengths (bottom). Only a small subset of ion channel properties (represented as membrane conductance) and synaptic connection parameters (represented as synaptic conductance) are listed here. CaS, a voltage-gated slow and transient Ca^{2+} current; A, a voltage-gated transient K$^+$ current; KCa, a Ca^{2+}-dependent K$^+$ current; Kd, a delayed rectifier K$^+$ current; Na, a voltage-gated Na$^+$ current; H, a hyperpolarization-activated inward current. See Box 2-4 for more detail about the ion channels that produce some of these currents. (Adapted from Prinz AA, Bucher D, & Marder E [2004] *Nat Neurosci* 7: 1345–1352. With permission from Springer Nature.)

indeed been observed experimentally in natural populations of animals. Thus, distinct combinations of intrinsic neuronal properties and connection strengths can generate similar network properties, reflecting the flexibility and robustness of the network producing the pyloric rhythm. This finding suggests that the activity patterns of neurons and the network are *degenerate*, in that they can be achieved in multiple ways.

The crustacean STG system is also regulated by multiple neuromodulators, which act by changing membrane and synaptic conductances in specific neurons that express their receptors (Section 3.11; neuromodulation will be further discussed in Box 9-1). As many solutions produce similar rhythmic output patterns, each neuromodulator can in principle regulate a distinct set of ion channels or synaptic transmission components in different cells to help produce the same rhythmic output pattern.

8.6 The spinal cord uses multiple central pattern generators to control locomotion

The mechanisms by which CPGs in the vertebrate spinal cord control locomotion are more complex and less well understood than those of the crustacean STG. Nevertheless, insights obtained from less complex invertebrate circuits can be applied to test specific hypotheses. As discussed earlier, an elementary operation in motor coordination is the alternate contraction of the extensor and flexor controlling the same joint (Figure 8-8). In one model allowing this, the extensor and flexor motor neurons are each activated through mutually inhibitory premotor circuits (**Figure 8-15**A). At a higher level, different extensor–flexor pairs that control different joints of the same limb might be coordinated by analogous interactions among excitatory and inhibitory neurons to bring about coordinated movement of a limb; these constitute a CPG network for an individual limb. At an even higher level, CPG networks in the four limbs might be further coordinated to control different kinds of locomotion. For instance, during walking of most mammals, left and right limbs are out of sync, as are forelimbs and hind limbs on the same side. This may involve mutual inhibition of CPGs controlling these different limbs. Such models propose that mutual inhibition of left and right CPGs switches to mutual excitation during hopping or galloping so that left and right limbs are in sync (Figure 8-15B; Figure 8-11).

Physiological recording and perturbation experiments in preparations such as the mesencephalic cat have provided support for this conceptual framework of CPG organization. However, a deeper understanding of locomotion control requires identification of the constituent neurons and elucidation of the connection patterns of the CPG network, as has been achieved with the crustacean STG circuit. Since these CPG elements are composed of many individual neurons of multiple cell types, the methods that allowed researchers to crack the simpler STG circuit—such as simultaneous recording of multiple neurons and systematic ablation of identified neurons—are inadequate. The circuit functions performed by a single neuron in the STG are likely carried out by a specific type of neuronal populations in the vertebrate spinal cord. One promising approach is to use findings from gene expression and developmental studies (Figure 7-10) to gain genetic access to individual spinal neuron types. Genetic access could then allow connec-

Figure 8-15 Conceptual framework for mammalian central pattern generators that control locomotion. (A) In this model (left), flexor and extensor motor pools (represented by single motor neurons) are excited by corresponding excitatory premotor neurons (green). These excitatory premotor neurons inhibit each other and their antagonistic motor neurons through inhibitory interneurons (red), thus creating alternating patterns of excitation (right). **(B)** At a higher level, CPG networks for different limbs are proposed to be connected via inhibitory (red) and excitatory (green) interactions. The excitatory and inhibitory interactions between limbs may be switched on or off, depending on modes of locomotion. All CPG networks also receive descending excitatory input from the brainstem and motor cortex. While serving as good working hypotheses, these models are hypothetical, as specific constituent neuronal elements have not yet been identified. (A, adapted from Pearson K [1976] *Sci Am* 235[6]:72–86. With permission from Springer Nature. B, adapted from Grillner S [2006] *Neuron* 52:751–766. With permission from Elsevier Inc.)

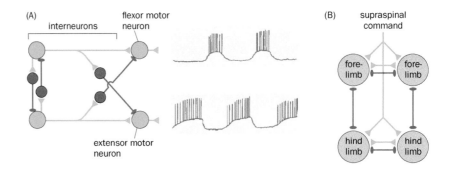

(A) interneurons — flexor motor neuron — extensor motor neuron

(B) supraspinal command — forelimb — forelimb — hindlimb — hindlimb

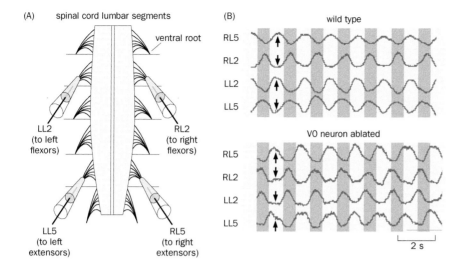

Figure 8-16 Ablation of V0 spinal cord interneurons disrupts left–right alternation. (A) Schematic of experimental setup. Simultaneous recordings of four ventral roots, which connect to extensors and flexors of hind limb muscles, as indicated, were performed in an *in vitro* explant. The locomotor-like pattern in this preparation was triggered by application of NMDA and serotonin, mimicking brainstem activation (Figure 8-17). **(B)** Top, in wild-type mice, the activities of left and right ventral roots of lumbar segment 2 (LL2 and RL2) alternated with each other, as did the activities of LL5 and RL5; these alternating activities account for the alternating movement of the left and right limbs during walking. Bottom, when V0 neurons were ablated, the activities of LL2 and RL2 synchronized, as did those of LL5 and RL5; this produced a hopping movement. Note that the activities of LL2 and LL5, and those of RL2 and RL5, also alternated because their flexor–extensor relationships remained unchanged when V0 was ablated. Dark and light alternating stripes represent the alternating phases of the locomotor cycles. (Adapted from Talpalar AE, Bouvier J, Borgius L, et al. [2013] *Nature* 500:85–88. With permission from Springer Nature.)

tivity tracing and activity manipulation of entire populations of specific neuronal cell types using modern neural circuit analysis tools (see Chapter 14).

We use one specific example to illustrate this approach. As discussed in Section 7.4, the transcription factor Dbx1 is expressed in developing spinal cord progenitors that eventually become V0 interneurons. Transgenic mice were produced in which Dbx1-expressing cells conditionally express diphtheria toxin A following Cre-induced recombination. These mice were crossed to a second transgenic mouse line in which Cre recombinase is expressed in the spinal cord, including Dbx1-expressing progenitors. The resulting double transgenic mice, in which spinal cord V0 neurons were specifically ablated by diphtheria toxin A expression, exhibited a striking phenotype: wild-type mice alternate their left and right limbs during walking, whereas transgenic mice synchronized their limbs such that they hopped rather than walked at all speeds tested (Movie 8-4). Recording the hind limb flexor and extensor activities in an *in vitro* spinal cord explant preparation indicated that the left–right alternating firing pattern was switched to a synchronous firing pattern (Figure 8-16). Thus, V0 interneurons play an essential role in alternating the activities of left and right limbs during normal walking. The flexor–extensor cycles within each limb remained intact despite the disruption of left–right alternation in the absence of V0 interneurons, suggesting that they are regulated independently.

V0 interneurons comprise both excitatory and inhibitory subpopulations. Interestingly, ablating inhibitory and excitatory V0 interneurons preferentially affected left–right alternation at low and high locomotion speeds, respectively. These observations suggest that different subpopulations are recruited to the same regulatory network for different locomotion speeds; in other words, the composition of CPGs regulating the same motor pattern is dynamic. Similar observations have been made in studies of zebrafish swimming on different molecularly identified neuronal subsets: as swimming speed increases, new interneuron populations become active, while those active during low speeds become inactive.

Recent studies have suggested that ventral spinal interneurons are highly heterogeneous, having diverse developmental origins, positions along the anterior–posterior axis, and gene expression patterns. For instance, spinal cord neurons derived from the V1 progenitor alone (Figure 7-10) may contain dozens of distinct subtypes. Functional interrogation of specific spinal interneuron subtypes, like that of the V0 interneuron discussed here, should provide new insight into how spinal cord circuits control movement.

8.7 The brainstem contains specific motor command nuclei

Although isolated spinal cord preparations can generate rhythmic output for locomotion, as discussed in Section 8.4, the initiation of rhythmic output requires excitatory stimuli, such as application of glutamate. Where does endogenous

(A)

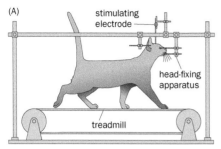

(B)

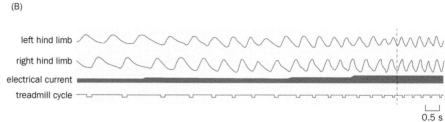

Figure 8-17 Electrical stimulation in the mesencephalic locomotor region (MLR) of the brainstem initiates stepping. (A) Experimental setup. An incision is made in the midbrain such that the cerebral cortex/thalamus is disconnected from the brainstem/spinal cord. The resulting mesencephalic cat cannot voluntarily control its locomotion. When placed on a treadmill with speed driven by the step cycle of the cat, electrical stimulation is applied to specific parts of the brain to initiate locomotion, which can be measured by limb movement and treadmill speed. **(B)** Electrically stimulating the MLR initiated movement, as seen in the cycles of the hind limbs. Furthermore, increasing the electrical stimulation current (thick trace below the hind limb cycles) sped up the step cycle, as seen in the increasing frequencies of the step cycles and treadmill cycle (bottom). Note that toward the end (after the vertical dashed line), trotting became galloping, with both hind limbs moving in sync. (Adapted from Shik ML, Severin FV, & Orlovski GN [1966] *Biophysics* 11:756–765. With permission from Springer Science & Business Media.)

excitation come from? Electrical stimulation of an area called the **mesencephalic locomotor region** (**MLR**) in the brainstem was found to initiate locomotion in mesencephalic cats. Indeed, increasing the intensity of MLR stimulation could speed up the walking and trigger transitions to trotting or galloping (**Figure 8-17**). Thus, the MLR appears to be a brainstem center that controls locomotion. MLR stimulation activates other brainstem neurons, which in turn send descending axons through the reticulospinal tract to innervate spinal interneurons and motor neurons.

Classic electrical stimulation, lesion, and anatomical tracing studies are limited by a lack of cellular resolution, as they are not restricted to specific cell types. As in the case of spinal cord CPG research, recent advances in neural circuit dissection have begun to change this picture. After locating the MLR in the midbrain of a head-fixed mouse, where electrical stimulation could facilitate locomotion, researchers resorted to optogenetic activation of neurons in that region with specific transmitter types by expressing channelrhodopsin (ChR2) only in cholinergic, GABAergic, or glutamatergic neurons. They found that photo-activating glutamatergic neurons (but not the other classes) could initiate or speed up locomotion (**Figure 8-18**A). Conversely, optogenetic inhibition of MLR gluta-

(A)

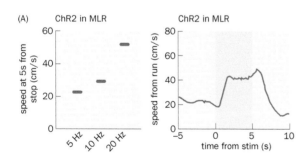

(B)

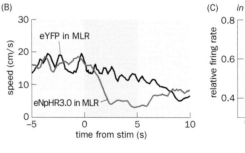

(C)

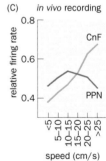

Figure 8-18 Dissecting the mouse mesencephalic locomotor region. (A) Left, optogenetic stimulation of channelrhodopsin- (ChR2-) expressing glutamatergic neurons in the MLR initiates locomotion from rest, the speed of which positively correlates with stimulation frequency. Right, optogenetic activation of these neurons also increases the speed of moving mice. **(B)** Optogenetic stimulation of halorhodopsin- (eNpHR3.0-) expressing MLR glutamatergic neurons, which silences their activity (Section 14.25), reduces the speed of locomotion. As a control, optogenetic stimulation of eYFP-expressing MLR glutamatergic neurons has no effect. **(C)** *In vivo* extracellular recordings of glutamatergic neurons (identified using a phototagging strategy; see Section 14.20 for details). Glutamatergic neurons in the cuneiform nucleus (CnF) have higher firing rates (relative to baseline) than those in the pedunculopontine nucleus (PPN) during high-speed locomotion, and the converse is true during low-speed locomotion. (A & B, from Roseberry TK, Lee AM, Laline AL, et al. [2016] *Cell* 164:526–537. With permission from Elsevier Inc. C, from Caggiano V, Leiras R, Goñi-Erro H, et al. [2018] *Nature* 553:455–460. With permission from Springer Nature. See also Josset N, Roussel M, Lemieux M, et al. [2018] *Curr Biol* 28:884–901; Capelli P, Pivetta C, Esposito MS, et al. [2017] *Nature* 551:373–377.)

matergic neurons slowed locomotion (Figure 8-18B). Further dissection of specific subregions within the MLR identified glutamatergic neurons from a dorsal cuneiform nucleus (CnF) as a major driver of high-speed locomotion, whereas glutamatergic neurons from a ventral pedunculopontine nucleus (PPN) appeared to drive slow gaits. *In vivo* extracellular recordings revealed that CnF and PPN glutamatergic neurons are preferentially active during locomotion at high and low speeds, respectively (Figure 8-18C). Taken together, these experiments suggest that glutamatergic neurons in the MLR (particularly the CnF) promote locomotion. Further experiments identified glutamatergic neurons in the lateral paragigantocellular nucleus, a small nucleus in the caudal brainstem, as an intermediate between the MLR and spinal cord that controls high-speed locomotion.

In addition to controlling locomotion, the brainstem also contains myriad neuronal groups that control many other bodily movements. For example, recent studies using modern circuit analysis tools have identified glutamatergic neurons in a brainstem nucleus called MdV (standing for medullary reticular formation, ventral part) that preferentially provide direct input to motor neurons controlling forelimb but not hind limb muscles, and thereby regulate skilled forelimb-based motor tasks. Additional brainstem nuclei regulate eye movement, licking, chewing, swallowing, whisking (a prominent behavior for rodents), and breathing. We discuss breathing in Box 8-1 as another example of a brainstem motor control circuit.

Box 8-1: How is breathing controlled?

From birth to death, animals must breathe continually, inhaling oxygen and exhaling carbon dioxide in order to maintain metabolic activity. While we do it without thinking much of the time, breathing can be under voluntary control (think of divers and opera singers). In mammals, a typical breath at rest consists of an inspiratory phase involving active contraction of muscles that control the diaphragm, followed by an expiratory phase. As metabolism increases (as when exercising), expiration muscles are recruited to produce active expiration to empty the lung so that the next inspiration can take in more air more quickly. Furthermore, there can be a third phase termed *post-inspiration*, an expiratory phase that slows the release of air through the active contraction of upper airway muscles (useful for singing long syllables, for example). The three phases are controlled by three interacting oscillator circuits located in nearby regions of the medulla (Figure 8-19A, B). We focus our discussion here on the region that controls the inspiratory phase, the **pre-Bötzinger complex** (**preBötC**), which is the best studied of the three.

The preBötC was discovered as a center that generates breathing rhythms in the early 1990s by systematic anatomical mapping. In a reduced neonatal rat preparation that includes brainstem and spinal cord, rhythmic output reflecting breathing cycles can be recorded from several nerves, including the hypoglossal nerve (cranial nerve XII, which innervates tongue muscles) and phrenic nerve (a cervical nerve that innervates the diaphragm). Researchers then sectioned away brain regions from the reduced preparation and found that removing sections rostral or dorsal to preBötC did not affect rhythmic output, but if the region containing the preBötC was sectioned away, rhythmic output was abolished. Importantly, a ~500 μm coronal slice containing the preBötC was sufficient to produce rhythmic

activity in the hypoglossal nerve, and whole-cell recordings confirmed that neurons within the preBötC produced rhythmic burst firing (Figure 8-19C). Further studies revealed that preBötC axons control inspiration by activating premotor neurons in the rostral ventral respiratory group (rVRG; Figure 8-19B), which activates inspiratory motor neurons that control diaphragm contraction.

The neural mechanisms underlying rhythm generation have not yet been resolved despite intense effort, in part due to the complexity and heterogeneity of preBötC neurons. Essential for rhythm generation are a few thousand excitatory neurons derived from progenitors that express the transcription factor Dbx1 (the same gene expressed by progenitors of V0 interneurons in the spinal cord; Figure 8-16; Figure 7-10). In *Dbx1* mutant mice, these preBötC neurons fail to develop, and mice die at birth because they cannot breathe. Indeed, rhythmic activation of preBötC neurons could be detected several days before pups were born in control but not *Dbx1* mutant slices using electrophysical recordings and Ca^{2+} imaging (Figure 8-20A). Furthermore, optogenetic activation of channelrhodopsin-expressing preBötC neurons derived from Dbx1+ progenitors could shift the phase or increase the magnitude of inspiration, depending on the timing of photostimulation within the breathing cycle (Figure 8-20B). Together, these experiments reveal a central role played by Dbx1+ progenitor–derived preBötC neurons in the generation of inspiration. *In vivo* single-unit recordings revealed further heterogeneity of these neurons, some of which fired just before inspiration, while others fired during or after inspiration. A current model posits that among Dbx1+ progenitor–derived preBötC neurons, some are involved in rhythmogenesis, whereas others send output to premotor neurons to execute the breathing pattern.

(Continued)

Box 8-1: continued

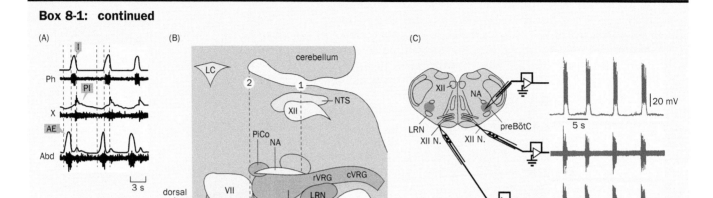

Figure 8-19 Brainstem nuclei that control breathing. (A) Breathing cycles consisted of three phases, inspiration (I), post-inspiration (PI), and active expiration (AE), shown here in a reduced rat preparation. The three phases were measured via electrical recordings from the phrenic (Ph) nerve, cranial nerve X, and lumbar abdominal (Abd) nerve, respectively; raw spike patterns (bottom) and spike rates (top) are shown. Vertical dashed lines separate the three phases for two cycles. Breathing cycles can also consist of the following combinations of active phases: I, I + PI, or I + AE. **(B)** A sagittal schematic of the rat brainstem. Nuclei that control breathing phases are highlighted in red: the pre-Bötzinger (preBötC) for inspiration, post-inspiratory complex (PiCo) for post-inspiration, and parafacial respiratory group (pFRG) for active expiration. The rostral ventral respiratory group (rVRG) contains inspiratory premotor neurons that receive input from the preBötC. The caudal ventral respiratory group (cVRG) contains expiratory premotor neurons that receive input from the pFRG. Other labeled anatomical landmarks are the locus coeruleus (LC), which contains norepinephrine modulatory neurons; cranial nuclei VII and XII, which contain motor neurons innervating the face and tongue, respectively; nucleus of solitary tract (NTS), which relays taste and interoceptive (for example, lung mechanosensory) information to higher brain centers; nucleus ambiguous (NA), which contains motor neurons that control the larynx, pharynx, and soft palate for swallowing and speech; and lateral reticular nucleus (LRN), which relays input to the cerebellum. **(C)** Schematic (left) and recordings (right) from a coronal slice containing the preBötC. Rhythmic burst firing of cranial nerve XII (XII N.; bottom two traces) and of a neuron within the preBötC are detected via extracellular recording and whole-cell recording, respectively. IO, inferior olive, which sends climbing fiber projections to the cerebellum (Section 8.10). (A & B, from Del Negro CA, Funk GD, & Feldman JL [2018] *Nat Rev Neurosci* 19:351–367. With permission from Springer Nature. C, from Smith JC, Ellenberger HH, Ballanyi K, et al. [1991] *Science* 254:726–729. With permission from AAAS.)

Recent genetic dissection of preBötC neurons has also revealed subsets that serve specialized functions. For example, a subset of Dbx1⁺ progenitor–derived preBötC neurons sends ascending axons to the locus coeruleus (Figure 8-19B), where they synapse onto norepinephrine neurons to regulate brain states such as arousal and placidity (see Box 9-1 for more details on norepinephrine neurons). Another group of preBötC neurons receives neuropeptide input and con-

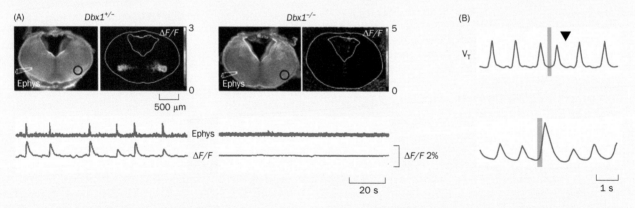

Figure 8-20 Dbx1⁺ progenitor–derived preBötC neurons control inspiration. (A) Top, micrographs of coronal slices from heterozygous control (left) and homozygous *Dbx1* mutant (right) embryonic day 15.5 pups. Within each pair, the bright-field images (left) show the extracellular electrophysiological recording site (Ephys) within the preBötC (circles on the right hemispheres). The fluorescence images (right) show elevated bulk Ca²⁺ signal, indicating active preBötC neurons. Bottom, traces of simultaneous electrophysiological recording and Ca²⁺ imaging show rhythmic activity of the preBötC. ΔF/F, change in fluorescence signal divided by baseline fluorescence. **(B)** Tidal volume (V_T) of airflow measured in a transgenic mouse expressing channelrhodopsin in neurons derived from Dbx1⁺ progenitors. Top, when photostimulation was applied in the middle of the expiration phase, the next breathing cycle was shifted forward (downward arrowhead indicates where the next cycle would have been without the photostimulus). Bottom, when photostimulation was applied at the beginning of an inspiration, the magnitude of the inspiration increased. (A, from Bouvier J, Thoby-Brisson M, Renier N, et al. [2010] *Nat Neurosci* 13:1066–1074. With permission from Springer Nature. B, from Cui Y, Kam K, Sherman D, et al. [2016] *Neuron* 91:602–624. With permission from Elsevier.)

Box 8-1: continued

trols sigh, a double-sized breath induced by physiological needs, such as reinflating collapsed alveoli in the lung, or by emotions, such as sadness or relief. Breathing rhythms also interact intimately with other orofacial activities, such as swallowing, licking, whisking (for rodents), and speech (for humans). Some rhythmic movements, such as whisking and licking, are coordinated with breathing, and it has been proposed that the preBötC-based inspiratory rhythm may function as a master clock to regulate other rhythmic orofacial movements in order to coordinate distinct motor outputs. For example, inspiration and swallowing do not occur simultaneously to ensure that food does not accidentally enter the airway. Thus, investigation into the neural mechanisms of breathing can help understand many aspects of motor control as well as their interactions with higher brain centers involved in arousal and emotional regulation.

8.8 The basal ganglia consist of two parallel pathways that are oppositely regulated by dopamine

So far, we have followed the motor system along a direction opposite to the information flow for motor execution: from muscles to motor neurons, premotor neurons in the spinal cord, and brainstem motor command centers. However, as seen from Figure 8-2, the motor system is far from a linear feedforward path from central to peripheral. There is sensory feedback at each stage. Moreover, within each stage there are local recurrent loops (e.g., Figure 8-15). In the next few sections, we will study higher motor control regions, including the basal ganglia, thalamus, cerebellum, and neocortex. Long-range recurrent loops between these regions are essential features for motor control (Figure 8-2). We discuss these regions one by one, but need to bear in mind their close interactions. Importantly, each of these regions employs generic circuit architecture for functions beyond motor control; nevertheless, our understanding of how basal ganglia and cerebellar circuits operate has come mostly from studies of motor control. We begin with the basal ganglia.

The **basal ganglia** are a collection of nuclei interior to the cerebral cortex (Figure 1-8). Two neurological disorders, Parkinson's disease and Huntington's disease, which primarily affect the basal ganglia, highlight the importance of these structures in motor control. Patients with Parkinson's disease have difficulty initiating movement, while patients with Huntington's disease cannot stop excessive movement. (We will discuss these diseases in more detail in Chapter 12.) The basal ganglia use a generic circuit design (**Figure 8-21**) for a wide range of functions that vary according to the specific sources of input and output. Before discussing its function in motor control, we first outline the basal ganglia circuit.

The input nucleus of the basal ganglia is the **striatum**, also called the caudate-putamen because in primates the striatum consists of two separable structures, the caudate and the putamen. The striatum receives convergent excitatory input from the neocortex and the thalamus. The dorsolateral striatum preferentially receives input from sensory and motor cortices and thus is most directly involved in motor control. The dorsomedial striatum preferentially receives input from association cortices and is involved in cognitive processes. The ventral striatum (also called the **nucleus accumbens**) preferentially receives input from the prefrontal cortex, hippocampus, and amygdala, and regulates motivated behavior.

The great majority of neurons in the striatum are two types of intermingled GABAergic **spiny projection neurons** (SPNs; also called medium spiny neurons) distinguished by their expression of different G-protein-coupled dopamine receptors. Those expressing the D_1 receptor constitute the **direct pathway**, which projects predominantly to the output nuclei of the basal ganglia—the **globus pallidus internal segment** (**GPi**) and the **substantia nigra pars reticulata** (**SNr**). Those expressing the D_2 receptor constitute the **indirect pathway**, which projects predominantly to the **globus pallidus external segment** (**GPe**), which in turn sends GABAergic input to the GPi either directly or through the **subthalamic nucleus** (**STN**) (Figure 8-21A). The two pathways can be visualized in transgenic mice expressing green fluorescence protein (GFP) under the control of regulatory elements from the D_1 or D_2 dopamine receptor genes (Figure 8-21B). GPi and SNr

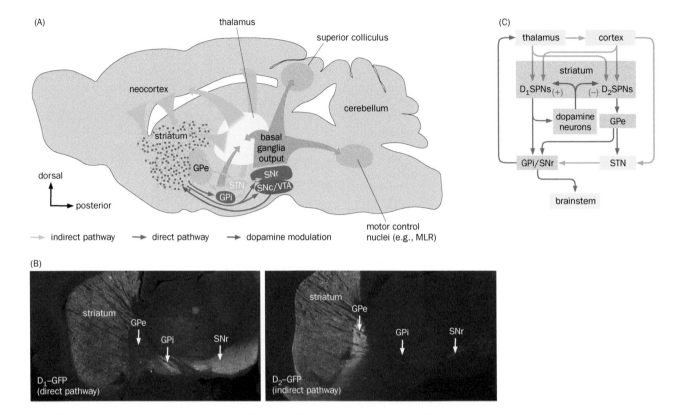

Figure 8-21 Organization of basal ganglia circuits. (A) A simplified model of basal ganglia circuits from a sagittal perspective of a mouse brain. The striatum receives excitatory inputs from the neocortex and thalamus and sends output via two types of GABAergic spiny projection neurons (SPNs). SPNs expressing the D$_1$ dopamine receptor (D$_1$SPN, red) constitute the direct pathway and project mainly to the globus pallidus internal segment (GPi) and substantia nigra pars reticulata (SNr). SPNs expressing the D$_2$ dopamine receptor (D$_2$SPN, orange) constitute mainly the indirect pathway and project to the globus pallidus external segment (GPe). The GPe sends GABAergic projections to the GPi and the subthalamic nucleus (STN, which also receives direct cortical input), which in turn projects to the GPi and SNr. The GPi and SNr send basal ganglia output to the thalamus, superior colliculus, and brainstem motor control nuclei, such as the mesencephalic locomotor region (MLR). Dopamine neurons in the substantia nigra pars compacta (SNc) and ventral tegmental area (VTA) also receive input from D$_1$SPNs and many other brain regions (not shown here) and send modulatory output back to the striatum. **(B)** The projection patterns of D$_1$SPNs (left) and D$_2$SPNs (right) are visualized in sagittal sections of transgenic mice in which green fluorescent protein (GFP) expression is driven by the regulatory elements of the D$_1$ and D$_2$ receptors, respectively. Cell bodies of both neuronal types are within the striatum; D$_1$SPNs project mainly to the GPi and SNr, whereas D$_2$SPNs project to the GPe, as seen by GFP fluorescence intensity. **(C)** A simplified basal ganglia circuit diagram. Green arrows represent excitatory projections. Red arrows represent inhibitory projections. Blue arrows represent dopaminergic projections, which promote D$_1$SPN (+) firing and inhibit D$_2$SPN (–) firing. (A & B, adapted from Gerfen CR & Surmeier DJ [2011] *Ann Rev Neurosci* 34:441–466. With permission from Annual Reviews.)

GABAergic projections target the thalamus, which itself projects to the neocortex and directly back to the striatum, forming two feedback loops (Figure 8-21A, C). The SNr also sends output to brainstem motor control nuclei and the superior colliculus. The actual connection patterns are more complex than the simplified model outlined here, but the model captures the major features of the basal ganglia circuit.

Dopamine modulates the excitatory synaptic connections through which the neocortex and thalamus provide input to striatal spiny projection neurons (see Box 9-1 for a general discussion of neuromodulatory systems). The dopamine neurons responsible for this modulation are located in the **substantia nigra pars compacta (SNc)** and the adjacent **ventral tegmental area (VTA)**, which project preferentially to the dorsal and ventral striatum, respectively. Dopamine release has opposite effects on the direct and indirect pathways. Activation of the D$_1$ receptor, which is coupled to a stimulatory G protein, depolarizes D$_1$ receptor-expressing spiny projection neurons (D$_1$SPNs), activating the direct pathway. Activation of the D$_2$ receptor, which is coupled to an inhibitory G protein, hyperpolarizes D$_2$ receptor-expressing spiny projection neurons (D$_2$SPNs), inhibiting the indirect pathway (Figure 8-21C). The activity of SNc/VTA dopamine neurons is in turn regulated by inputs from many brain regions, including direct input from D₁SPNs (Figure 8-21A, C).

8.9 The direct and indirect pathways act in concert to facilitate the selection and initiation of motor programs

Given the circuit properties discussed in Section 8.8, how do the basal ganglia regulate movement? *In vivo* recording studies indicate that striatal SPNs are mostly silent at rest. By contrast, output neurons in the GPi and SNr are active at rest, sending **tonic** inhibitory output to their targets. (Tonic refers to regularly timed and repetitive firing patterns; **phasic** refers to rapid and transient firing patterns.) Immediately before the onset of voluntary movement, cortical and thalamic excitatory input activates spiny projection neurons (e.g., Figure 8-22). This inhibits firing of GPi and SNr output neurons through the direct pathway, causing disinhibition of motor control centers in the superior colliculus and brainstem (Figure 8-21C), thus facilitating movement initiation. Likewise, tonic inhibitory output to the thalamus is also relieved, further promoting movement initiation. Early recording studies did not distinguish between D₁SPNs and D₂SPNs, and the interpretation of their findings is based on the projection patterns of D₁SPNs. Given their opposing roles in the circuit and regulation by dopamine, how do these two types of SPNs contribute to motor control?

Optogenetics (Section 14.25) has been used to test the causal relationship between movement and activation of the direct or indirect pathways. Channelrhodopsin was expressed in D₁SPNs or D₂SPNs in separate mice such that they could be activated by light. As predicted from the circuit model, activation of the direct and indirect pathways decreased or increased, respectively, the firing of SNr output neurons (Figure 8-23A). Behaviorally, direct pathway activation enhanced locomotion, whereas indirect pathway activation suppressed locomotion (Figure 8-23B). These experiments suggest that *global* activation of the basal ganglia direct or indirect pathway facilitates or inhibits movement, respectively.

Recent cell-type-specific *in vivo* recording revealed that both direct and indirect pathways are activated when animals initiate a specific motor behavior such as locomotion, suggesting a more nuanced picture than the simple idea that D₁SPNs promote movement while D₂SPNs inhibit movement. In one such experiment (Figure 8-24A), mice moved freely in an arena while Ca²⁺ imaging was performed on D₁SPNs or D₂SPNs via a head-mounted miniaturized microscope (see Section 14.22 for details). Body acceleration correlated positively with Ca²⁺ transients in both D₁ and D₂SPNs (Figure 8-24B). Furthermore, activity of individual D₁SPNs or D₂SPNs was selective for specific actions, such as acceleration, deceleration, and turning (Figure 8-24C). In a complementary experiment, bulk Ca²⁺ signals from D₁SPN or D₂SPN populations were recorded simultaneously using a technique called **fiber photometry** (Section 14.22) in freely moving mice; the behavior of mice was segmented into specific "syllables" based on their three-dimensional movements (Figure 8-24D). While both D₁SPNs and D₂SPNs were generally activated by syllable onset, onset of certain syllables exhibited *temporal differences* between these two neuronal populations (Figure 8-24E).

Together, these experiments suggest that striatal SPNs in the direct and indirect pathways act in concert to regulate many aspects of motor behavior. That activity changes of SPNs correspond to onset of behavioral changes supports their function in the selection and initiation of specific motor programs. Antagonism between the direct and indirect pathways, as predicted by basal ganglia circuitry (Figure 8-21C), could manifest at different levels: global movement (for example, go versus no-go), selection of specific motor programs and simultaneous inhibition of other competing programs, and fine temporal regulation within a specific motor program. Future physiological recordings and temporally precise manipulations of direct and indirect pathway neurons specific for particular motor programs, in conjunction with precise behavioral monitoring, will shed more light on how the basal ganglia regulate movement.

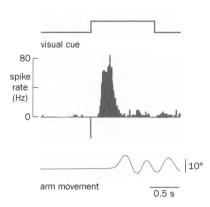

Figure 8-22 Firing of some striatal neurons anticipates movement onset. In this experiment, a monkey was trained to move its arm three times (bottom trace) in response to onset of a visual cue (top trace). The firing rate of a neuron in the arm control region of the striatum was recorded by an extracellular electrode. This striatal neuron did not track movement per se, but instead fired most vigorously *prior to* movement onset. Other striatal neurons in the same study did track movement (not shown). (Adapted from Kimura M [1990] *J Neurophysiol* 63:1277–1296.)

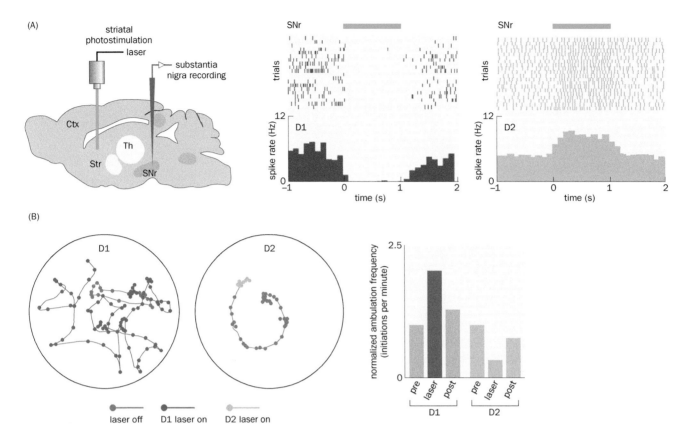

Figure 8-23 Function of the direct and indirect pathways investigated by optogenetic stimulation. (A) Left, schematic of the experiment depicted in a sagittal perspective of the mouse brain. Channelrhodopsin (ChR2) was specifically expressed in D_1SPNs or D_2SPNs in the striatum (Str). Middle and right, the top graphs are spike trains in which each row represents a trial; the bottom graphs illustrate the spike rate of the SNr neurons before, during, and after photostimulation. Photostimulation (blue bar) of D_1SPNs suppressed SNr neuron firing (middle panels), while stimulation of D_2SPNs increased SNr neuron firing (right panels). Ctx, cortex; Th, thalamus. **(B)** Photostimulation of D_1SPNs or D_2SPNs promotes or suppresses movement, respectively. The location of a mouse in a circular arena is measured at 300-ms intervals. The gray paths represent 20 s of activity before photostimulation; the colored paths (red or orange) represent 20 s of activity during photostimulation. When D_1SPNs were activated, the mouse moved more, as indicated by longer distances between the red dots. When D_2SPNs were activated, the mouse moved less, as indicated by the clustered orange dots. These effects are quantified in the right panel. (Adapted from Kravitz AV, Freeze BS, Parker PR, et al. [2010] *Nature* 466:622–626. With permission from Springer Nature.)

SNc dopamine neurons, which project to the dorsal striatum, have traditionally been thought to exhibit a tonic firing pattern for modulating striatal neuronal responses to excitatory input from cortex and thalamus. Recent *in vivo* recording and imaging experiments in behaving mice indicate that they too exhibit phasic firing preceding movement onset. Optogenetic manipulations suggest that SNc dopamine neurons promote initiation and vigor (for example, speed and amplitude) of subsequent movement. In addition to modulating movement, dopamine also regulates connection strengths of synapses between thalamic/cortical input and SPNs, which likely plays important roles in motor skill learning and habit formation. Dopaminergic projections from the VTA (which neighbors the SNc) to the ventral striatum are particularly important in carrying signals related to reward. We will return to the function of the basal ganglia when studying reward-based learning in Chapter 11 and movement disorders and addiction in Chapter 12.

8.10 The cerebellum contains more than half of all neurons in the brain and has a crystalline organization

The cerebellum (Latin for "little brain") is evolutionarily ancient in vertebrates and occupies a sizable chunk of the mammalian brain (Figure 1-8). Like the basal ganglia, the cerebellum has a generic circuit design that serves diverse functions. However, its best-characterized functions are fine control of movement and motor learning. Cerebellar defects in human patients and experimental animals cause

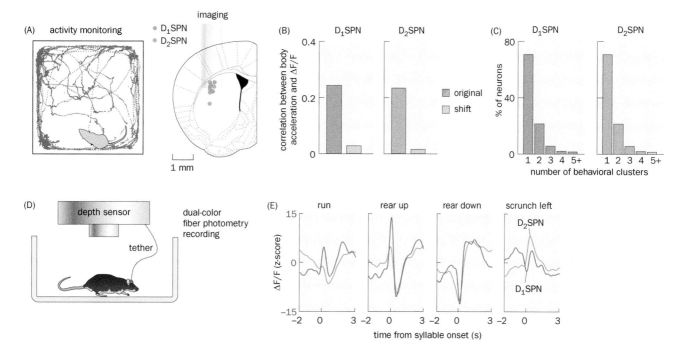

Figure 8-24 Measuring the activity of D₁SPNs or D₂SPNs during motor behavior. **(A)** Left, schematic of the experimental setup. Mice wearing a head-mounted microscope moved freely in an arena. Activity of D₁SPNs or D₂SPNs in dorsolateral striatum was imaged through a graded index (GRIN) lens via transgenic expression of the genetically encoded Ca²⁺ indicator GCaMP6 (Figure 14.41). **(B)** Analysis of time-dependent traces of body movement and Ca²⁺ signals in individual cells (as fluorescence change over baseline fluorescence, or ΔF/F) indicated that the activity of both D₁SPNs and D₂SPNs are positively correlated with body acceleration (red and blue bars). As a negative control, when the two traces were temporally shifted, the correlations disappear (gray bars). **(C)** Activity of most individual D₁SPNs and D₂SPNs was selective for one or two specific actions, represented by "behavioral clusters" generated by quantitative analysis of behavior from continuous videos. **(D)** Schematic of dual-color fiber photometry recording of bulk Ca²⁺ activity in dorsolateral striatum of freely moving

mice monitored by a sensor that records mouse three-dimensional movement. GCaMP6 and RCaMP1 (Ca²⁺ indicators with green and red fluorescence, respectively) were expressed in D₁SPNs and D₂SPNs, respectively. **(E)** Mouse behavior was segmented into syllables using an unsupervised machine-learning procedure (see Figure 14-52 for more detail; the descriptors on top reflect human observers' attempts to describe these syllables). D₁SPNs and D₂SPNs are generally activated at the onset of specific syllables. However, temporal differences between the activity of D₁SPNs and D₂SPNs were observed. For instance, the onset of "run" coincides with an increase in D₁SPN activity but a decrease in D₂SPN activity followed by an increase. Z-score is the difference from the mean divided by the standard deviation. (A–C, from Klaus A, Martins GJ, Paizao VB, et al. [2017] *Neuron* 95:1171–1180. With permission from Elsevier Inc. D & E, from Markowitz JE, Gillis WF, Beron CC, et al. [2018] *Cell* 174:44–58. With permission from Elsevier Inc.)

various kinds of motor system problems, such as **ataxia**, an abnormality in coordinated muscle contraction and movement. For example, transgenic mice with disrupted cerebellar connectivity cannot walk in a straight line; instead, they wobble from side to side (**Figure 8-25**). How does the cerebellum control movement? Before answering this question, we first need to examine its circuitry, which is in fact one of the best understood in the mammalian brain due to the small number of constituent cell types (**Figure 8-26**A).

The most morphologically complex neuron in the cerebellum is the **Purkinje cell** (Figure 1-11). Each Purkinje cell extends an elaborate planar dendritic tree that receives excitatory synapses from *10⁴–10⁵* **parallel fibers** that intersect with the Purkinje cell dendrites at right angles (Figure 8-26A). Parallel fibers are axons of **granule cells**, whose small cell bodies are densely packed in the granular layer and receive excitatory input from **mossy fibers** originating from heterogeneous populations of neurons in the pons, medulla, and spinal cord. (Remarkably, cerebellar granule cells are so numerous that they account for more than half of all

Figure 8-25 Ataxia caused by a cerebellar defect. The hind feet of mice were dipped in paint and their footprints recorded on paper. Compared with normal mice, which walk straight, mice with cerebellar defects—in this case due to connection abnormalities of Purkinje cells—typically wobble from side to side, with inconsistent step sizes and more widely spaced hind feet. (From Luo L, Hensch TK, Ackerman L, et al. [1996] *Nature* 379:837–840. With permission from Springer Nature.)

normal   defective Purkinje cells

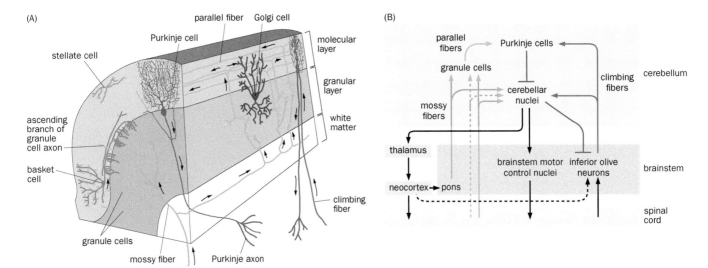

(A)

Figure 8-26 Organization of cerebellar circuitry. (A) Organization of the cerebellar cortex. Purkinje cell bodies form a single layer between the molecular and granular layers and send output to the cerebellar nuclei via the white matter. Their planar dendrites extend across the entire depth of the molecular layer. Granule cells are located in the granular layer. Their axons first ascend into the molecular layer, then bifurcate to form parallel fibers that span up to 2 mm (green dots in the cross section), each intersecting with hundreds of Purkinje cell dendrites. Two external inputs, mossy fibers and climbing fibers, synapse onto granule cells and Purkinje cells, respectively. Also drawn are the basket, stellate, and Golgi cells, three major types of local GABAergic neurons. Arrows indicate the flow of information. **(B)** Schematic summary of major connections within the cerebellum and between the cerebellum and other brain regions involved in motor control. Cerebellar granule cells receive input from mossy fibers originating from the pons, which receives descending input from the neocortex, and, directly (solid arrow) or indirectly (dotted arrow), from the spinal cord. Inferior olive neurons send climbing fibers to Purkinje cells. The cerebellar nuclei, which receive input from Purkinje cells and collaterals of mossy fibers and climbing fibers, send excitatory output to brainstem nuclei involved in motor control and to the neocortex via the thalamus. The cerebellar nuclei also send inhibitory output to the inferior olive neurons, which also receive input from the spinal cord and (indirectly) from neocortex. Note that this diagram focuses mostly on spinal cord–based motor control; input and output may vary in other systems; see Figure 8-28 for an example in vestibular control of eye movement.

neurons in the mammalian brain.) Each Purkinje cell is also innervated by just *one* **climbing fiber**, an axon originating from a neuron in the **inferior olive** nucleus in the caudal brainstem (Figure 8-19C) that "climbs" the major branches of the Purkinje cell's dendritic tree, forming numerous excitatory synapses along the way. Purkinje cells are GABAergic and send inhibitory signals to neurons in the **cerebellar nuclei**, the output nuclei of the cerebellum. Major projection targets of the cerebellar nuclei include the brainstem motor nuclei for descending motor control and the thalamus for communication with the neocortex. Both mossy fibers and climbing fibers also send collateral branches directly to the cerebellar nuclei (Figure 8-26B).

In addition to these projection neurons, the cerebellar cortex also contains three major types of local interneurons—the **basket cell** (Figure 1-15B), **stellate cell**, and **Golgi cell** (Figure 8-26A). All three cell types receive input from parallel fibers in the molecular layer. Basket and stellate cells send inhibitory output to Purkinje cells at their somata and distal dendrites, respectively, thus executing feedforward inhibition. Golgi cells send inhibitory output back onto granule cells, thus executing feedback inhibition (Box 1-2).

This circuit organization is repeated across the entire cerebellum, making the cerebellum a crystalline-like structure. The cerebellar cortex has a coarse somatotopic map. For instance, the medial and lateral parts of the cerebellum preferentially regulate trunk and limb movement, respectively. They also preferentially receive mossy fiber inputs from the spinal cord (in some cases via intermediate brainstem nuclei) and from the cerebral cortex (via the intermediate **pontine nuclei** located in the basal pons), respectively. However, such divisions are by no means absolute (see Figure 8-27A). Finally, Purkinje cells from the most posterior part of the cerebellum, the flocculus, directly innervate the vestibular nuclei instead of the cerebellar nuclei; the flocculus receives mossy fiber input preferentially from the vestibular nuclei in return (see Figure 8-28).

8.11 The cerebellum refines motor execution through feedback and feedforward regulations

How does the cerebellum regulate motor function? Shortly after the circuit architecture of the cerebellar cortex was delineated, David Marr and James Albus proposed a theory of cerebellar function around 1970 that is still influential today. The first element of the theory relates to the possible computational function of the numerous granule cells. As discussed in the previous section, 10^4–10^5 granule cells innervate each Purkinje cell. On the other hand, each granule cell receives just four mossy fiber inputs, one at each of its four dendritic claws (Figure 8-27A). It was proposed that the large number of granule cells enables random recombination of mossy fiber inputs, representing disparate information (related to, for example, motor commands, motor execution, or sensory feedback in the context of motor control), and hence create a high-dimensional representation of inputs. In other words, information encoded by mossy fibers is re-encoded by granule cells in a high-dimensional space, where each axis is defined by the firing rate of a single granule cell (see Figure 6-27 for an illustration, and Section 14.31 for a general discussion of encoding). This high-dimensional re-encoding can be used to separate similar patterns that are not linearly separable in low-dimensional representations (Figure 8-27B), a concept termed **pattern separation** in theoretical neuroscience. Recent anatomical and physiological studies support the notion that individual granule cells can receive input from disparate mossy fibers (Figure 8-27A).

The second key element of the theory posits that coincident firing of the granule cells and climbing fiber innervating the same Purkinje cell would result in changes in synaptic connection strength between parallel fibers arising from those specific granule cells and the Purkinje cell. This property would allow the climbing fiber to serve as a teaching signal that would regulate the transmission of information from a granule cell population representing a specific motor program to the Purkinje cell, thereby modulating cerebellar output during that motor program. Subsequent physiological studies have indeed shown that coincident firing of the climbing fiber and parallel fibers causes a *decrease* in synaptic efficacy between the active parallel fibers and the Purkinje cell, a process called **long-term depression**, which will be discussed in more detail in Chapter 11.

Let's use an example to illustrate how the cerebellar theory applies to a specific biological process: the vestibulo-ocular reflex (VOR) we introduced in Box 6-2.

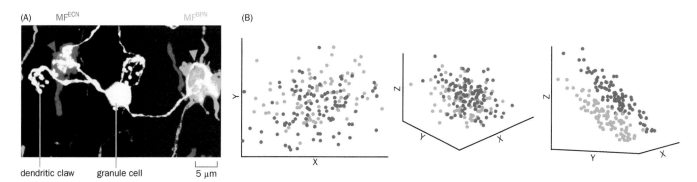

Figure 8-27 Input convergence onto granule cells and pattern separation by dimensionality expansion. (A) In this experiment, a sparse subset of granule cells (white) are labeled in a transgenic mouse, revealing dendritic claws that belong to the same granule cell. Adeno-associated viruses (AAVs) expressing green or red fluorescent proteins were injected into the external cuneate nucleus (ECN, which relays upper body proprioceptive input to the cerebellum) or basal pontine nucleus (BPN, which relays neocortical input to the cerebellum), resulting in axons (mossy fibers) labeled in green or red. The micrograph shows a single granule cell receiving convergent mossy fiber (MF) inputs from ECN and BPN onto two of its claws. **(B)** Graphs illustrating how high dimensional representations are used for pattern separation. The red and blue data points are intermingled in a two-dimensional space (left panel) and cannot be linearly separated (that is, we cannot draw a line with all the red dots on one side of it and all the blue dots on the other). However, after adding a third dimension (middle panel), the red and blue dots can be separated by a plane in three-dimensional space, as seen when rotating the axes (right). (A, from Huang CC, Sugino K, Shima Y, et al. [2013] *eLife* 2:e004000. B, courtesy of Mark Wagner. For the cerebellar theory, see Marr D [1969] *J Physiol* 202:437–470; Albus JS [1971] *Math Biosci* 10:25–61.)

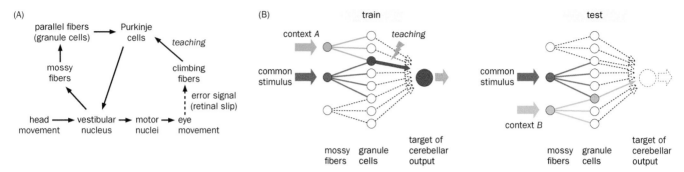

Figure 8-28 Cerebellar function in adjusting vestibulo-ocular reflex (VOR) gain. (A) In the VOR, the head movement signal is relayed to the vestibular nuclei to control eye movement (bottom pathway; see Figure 6-61 for more details), but the signal strength can be adjusted by the cerebellar cortex loop detailed here. The head movement signal reaches the cerebellar cortex via mossy fibers from the vestibular nucleus, and an error signal caused by image movement on the retina (retinal slip) reaches the cerebellar cortex via climbing fibers (dashed arrow represents indirect connection). Pairing of these signals modifies parallel fiber → Purkinje cell synaptic strengths, which alters the signals sent to the vestibular nuclei, thereby adjusting VOR gain. Thus, error signals "teach" circuits through modification of synaptic strengths. **(B)** Schematic illustrating the specificity of cerebellum-based learning. Filled circles and solid lines indicate active neurons and connections. Re-encoding of mossy fiber input signals by granule cells allows learning-related circuit changes to be confined to granule cells receiving input both from a common stimulus *and* a signal representing a specific training context A. The thick purple arrow symbolizes strengthened cerebellar output due to learning. (While the granule cell → Purkinje cell connection is weakened by long-term depression, cerebellar output is strengthened, as Purkinje cells are GABAergic inhibitory neurons.) After training, application of the common stimulus in context A increases the cerebellar output, triggering target activation. When the common stimulus is presented with a distinct context B, however, the altered granule cell → Purkinje cell synapses are not engaged and thus the cerebellar output signal (yellow) is below the threshold to trigger target activation. (Modified after Boyden ES, Katoh A, & Raymond JL [2004] *Ann Rev Neurosci* 27:581–609; see also Ito M [1982] *Ann Rev Neurosci* 5:275–296.)

Recall that the VOR causes a compensatory eye movement in the direction opposite to a head turn, thus stabilizing visual images on the retina during head turns. However, when a change in circumstances (for example, wearing a pair of glasses that shrink the size of visual images) causes a mismatch between the magnitude of eye rotation and head turn, the VOR gain (the ratio between eye and head velocities) is adjusted by cerebellum-based motor learning. Specifically, head movement produces vestibular signals in the semicircular canals sent via mossy fibers from the vestibular nucleus to the cerebellum. An error signal resulting from retinal slip (imperfect VOR that fails to stabilize images during a head turn) is sent to the cerebellum via climbing fibers. Repeated pairings of the error signal with the vestibular signal modify the synaptic connection strength between parallel fibers and Purkinje cells, which alters the signals sent back to the vestibular nuclei, thereby adjusting the strength of VOR signals sent to motor nuclei controlling the eye movement (**Figure 8-28**A).

Experimental data also indicate that VOR gain adjustment exhibits context specificity. For instance, VOR gain learned when the head is tilted at a specific angle does not apply to situations in which the head is tilted at a different angle, and VOR gain learned during low-speed head rotation does not apply to high-speed head rotation. How does such specificity arise? Here is where the large number of granule cells and high dimensionality come into play. Suppose that one mossy fiber input represents a common left-turning-head stimulus (which causes eyes to turn right via the VOR), and other mossy fibers represent specific contexts, such as angles of head tilt or speeds of head rotation. The large number of granule cells enables the common stimulus to be recombined with specific contexts in different granule cells (Figure 8-28B). Suppose further that granule cells are activated only by simultaneous activation of more than one mossy fiber. Then the teaching signal from the climbing fiber would adjust the synaptic strength only between the granule cell carrying the common stimulus *and* the signal representing specific context during training, and thus VOR gain is altered only in that specific context. Thus, the combination of two properties—re-encoding of mossy fiber input by granule cell populations and modification of connection strengths between specific granule cells and Purkinje cells—can in principle account not only for adjustment of VOR gain but also for its context specificity (Figure 8-28B).

These theoretical predictions have largely been supported by experimental data; however, evidence suggests that VOR gain adjustment is additionally influenced by changes in input–output connectivity within the vestibular nucleus instructed by signals from the cerebellum. In addition, recent studies indicate that granule cell re-encoding of mossy fibers may not generally apply; for instance, simultaneous imaging of motor cortical output and granule cells revealed that motor learning causes granule cells to faithfully transmit cortical inputs rather than extensively re-encode such inputs.

In addition to receiving *feedback* signals from sensory systems (such as retinal slip in the VOR example), the cerebellum also receives abundant *feedforward* signals from motor systems. In the context of spinal cord–mediated movement, for example, mossy fibers carry at least three types of signals: (1) signals related to motor performance (feedback from proprioceptive sensory neurons), (2) signals related to motor intent (output signals from premotor and motor neurons), and (3) signals related to motor commands (from the motor cortex relayed through the pons). The latter two are feedforward **efference copies** of motor signals, as they are sent to the cerebellum prior to motor execution. In motor control, sensory feedback is critical for motor planning, but the actual sensory feedback is out of date by the time it reaches the brain. Efference copy signals arising from motor output pathways are thought to allow the cerebellum to construct a **forward model** of *expected* sensory feedback to provide online (real-time) tuning of motor output. If the *actual* sensory feedback (reaching the cerebellum through different routes) differs from the expected sensory feedback, the cerebellum can modify the forward model using mechanisms analogous to the VOR gain adjustment we just discussed so as to provide more accurate online tuning of motor output.

Recent evidence has indicated that the cerebellum is involved in processing cognitive and reward signals, in addition to its classic role in motor control. Indeed, across the entire neocortex, layer 5 subcerebral projecting neurons innervate the basal pons (Figure 7-32) in a topographic manner and thereby communicate with the cerebellum via just one intermediate station. In return, cerebellar nuclei project axons to much of the neocortex via the thalamus (Figure 8-26B). Thus, the cerebellum may participate in all functions carried out by the neocortex through these reciprocal loops, utilizing a common circuit motif for feedforward and feedback controls. Much remains to be learned about the functions and mechanisms of such reciprocal communication between the neocortex and cerebellum, two structures that altogether contain 99% of all neurons in the human brain.

8.12 Voluntary movement is controlled by the population activity of motor cortex neurons

In mammals, the motor cortex, which includes the **primary motor cortex** (also termed **M1**) and the more anterior **premotor cortices**, provides the ultimate command to initiate voluntary movement and control complex movements. The motor cortex integrates information from multiple sensory systems (see Section 8.13) and sends descending axons to motor control regions of the brainstem, spinal cord interneurons, and (in certain primates, including humans) motor neurons themselves (Figure 8-2). The relative contributions of these pathways to motor control are not entirely clear. Evidence suggests that M1 tends to control fine movement of distal limbs more directly and trunk muscles more indirectly via the brainstem and spinal cord interneurons. The motor cortex also communicates extensively with the basal ganglia and cerebellum to refine motor output. How these circuits act in concert to orchestrate motor control is largely unknown.

How is the motor cortex functionally organized? As discussed in Section 1.11, M1 contains a somatotopic map, the motor homunculus (Figure 1-25); this was originally discovered in nonhuman primates and in human patients undergoing neurosurgery: application of brief electrical stimulation to specific areas of the motor cortex caused specific muscles on the contralateral side of the body to twitch. (The descending motor pathway crosses the midline once such that the left motor cortex controls the right side of the body.) Depiction of the motor homunculus

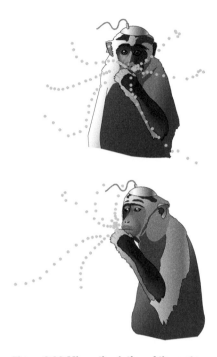

Figure 8-29 Microstimulation of the motor cortex can elicit complex movements. In this example, a 500-ms microstimulation at a specific site in the arm area of M1 caused the contralateral arm to move toward the mouth regardless of its initial position. At the same time, the hand was changed to a grip posture toward the mouth, and the mouth opened, mimicking a feeding behavior. The dotted lines represent 11 different trajectories traced from video recordings at 30 frames per second. (Adapted from Graziano M, Taylor C, & Moore T [2002] *Neuron* 34:841–851. With permission from Elsevier Inc.)

gives the impression that there is point-to-point representation of muscles in M1. While regions controlling legs, trunk, arms, and face are largely segregated in the motor cortex, precise topography does not exist at a finer scale. Indeed, while brief electrical stimulation causes muscle twitching, longer stimulation can produce complex and coordinated movements, such as moving the arm toward the mouth with the hand in a grip position and opening the mouth (**Figure 8-29**). As discussed in Section 4.28, a caveat of microstimulation experiments is that the number and types of neuronal cell bodies and axons being stimulated are poorly defined. Still, the finding that specific and ethologically relevant movements can be elicited by stimulating a single site suggests that motor cortical neurons and circuits can encode motor programs for specific behaviors.

In vivo electrophysiological recordings in behaving monkeys have provided important insights into the organization of the motor cortex at the single-neuron level. For example, individual cortical neurons in the finger representation region of M1 were recorded during an experiment in which monkeys were trained to move one finger at a time. Most individual neurons were broadly tuned to the movement of multiple fingers (**Figure 8-30**A). At the same time, neurons maximally tuned to the movement of a specific finger were distributed at multiple locations, intermingled with neurons maximally tuned to the movement of other fingers (Figure 8-30B).

Given the broad tuning of individual neurons, how does the motor cortex control specific movements? We use the control of arm reaching to illustrate. In a revealing experiment, monkeys were trained to move an arm from a starting position in the center of a cube to one of the eight corners of the cube, and thus in eight directions with equal distances in three-dimensional space. The spiking activities of hundreds of individual cortical neurons in the arm area of the motor cortex were recorded one neuron at a time during each of the eight kinds of movement. Activity of most individual neurons was modulated during more than one movement direction (**Figure 8-31**A), just as neurons in the finger area were active during movement of any of several different fingers. The "preferred direction" for each neuron, as a vector in a three-dimensional space, could nevertheless be determined based on its firing rates during the eight different movements. The direction of an arm movement could not be reliably predicted from the activity of single neurons, as each neuron was broadly tuned and exhibited variable spike rates. However, a **population vector**, constructed by summing the preferred direction vectors of several hundred neurons weighted by the magnitudes of their spike rates during movements, provided an excellent estimate of the actual arm movement (Figure 8-31B). In other words, it was possible to predict the arm movement direction based on the *population activity* of motor cortex neurons. Remark-

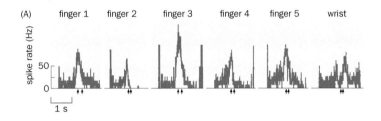

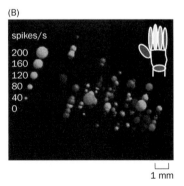

Figure 8-30 Representation of finger movement in the primary motor cortex of the monkey. (A) Activity of an individual motor cortex neuron during instructed movement of five fingers and the wrist, as determined by *in vivo* extracellular recordings. The first and second arrows beneath the plot signify the beginning and end, respectively, of each movement. The cell being recorded is maximally tuned to movement of finger 3, but is active during other movements as well. **(B)** Spatial distribution of active neurons in the hand area of M1. Each neuron is color-coded according to which of the fingers (or the wrist) elicited the maximal response, following the color scheme at top right. The size of the sphere represents the spike rate notated in the key on the left. Recorded neurons form parallel lines in the same direction as that of electrode paths, from top right to bottom left. Neurons tuned to movement of the same finger(s) do not cluster at this scale. (Adapted from Schieber MH & Hibbard LS [1993] *Science* 261:489–492.)

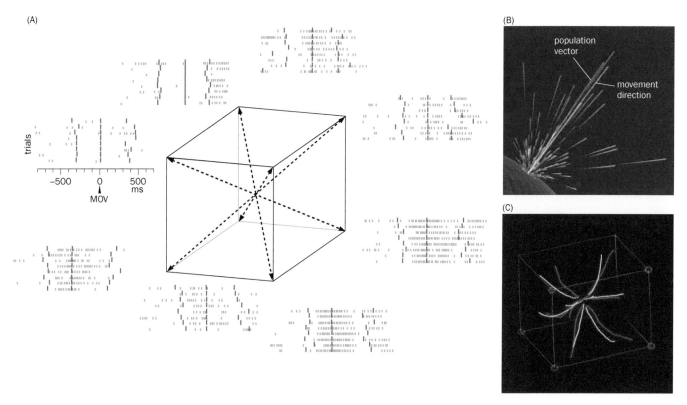

Figure 8-31 The population activity of motor cortex neurons determines movement directions. (A) A monkey was trained to move its arm from the starting position at the center of the cube to one of its eight corners upon the onset of a visual cue. The eight plots shown here illustrate spikes fired by a single neuron during each of the eight directions of arm movement. Within a plot, each row presents this neuron's spiking pattern during one trial. Each blue vertical bar represents a spike. The red vertical bars in the middle, which are used to align the plots, indicate the movement onset (MOV) and are labeled as $t = 0$ ms (milliseconds) in the time scale beneath the upper left plot. The red vertical bars to the left and right represent the onset of a visual cue (instructing the monkey to initiate movement) and the end of arm movement, respectively. The firing rate of this neuron changed most before and during arm movements toward the two corners at bottom right, but also when the arm moved in other directions. **(B)** The spike rates of individual neurons are represented as the lengths of individual vectors (blue) pointed in each neuron's preferred direction. The direction of the population vector of 224 individual neurons (orange, a weighted sum of individual vectors representing individual neurons) approximates the direction of arm movement (green). **(C)** The neural representations of trajectories (yellow; constructed based on each population vector over time) resemble the actual trajectories (red). (A & B, from Georgopoulos AP, Schwartz AB, & Kettner RE [1986] *Science* 233: 1416–1419. With permission from AAAS. C, from Georgopoulos AP, Kettner RE, & Schwartz AB [1988] *J Neurosci* 8:2928–2937. With permission from The Society for Neuroscience.)

ably, even the trajectory of arm movement from the reception of the start cue to the end of the movement could be approximated from time-varying spike rates of the motor cortex neuronal population (Figure 8-31C). These experiments provided strong evidence that movement direction is determined by the population activity of motor cortex neurons.

What drives activity in the motor cortex? Two major sources of cortical input to M1 are the parietal lobe, which relays integrated sensory information, and the frontal lobe, including the premotor cortex, which receives convergent input from much of the rest of the cortex and directs motor planning and volitional control of motor action. We discuss these two topics in the next two sections.

8.13 The posterior parietal cortex regulates sensorimotor transformations

The motor system often responds to sensory stimuli in a process called **sensorimotor transformation**. Returning to locomotion, if a cat sees an obstacle in its path, it must modify its stepping cycle to bypass the obstacle. Likewise, "simple" acts in our daily lives, such as catching a ball or reaching for a glass of water, require integration of the motor system with the visual system (to see the ball or the glass) and proprioceptive somatosensory system (to know where the hand is with respect to the ball or the glass). Interactions between motor and sensory systems can occur at

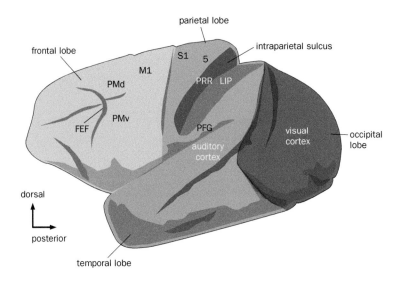

Figure 8-32 Lateral view of the monkey neocortex. Colors represent the four cortical lobes. Regions discussed in the next two sections are highlighted. The intraparietal sulcus (blue) is opened to reveal the cortical areas within the fold. FEF, frontal eye field; PMd, dorsal premotor area; PMv, ventral premotor area; M1, primary motor cortex; S1, primary somatosensory cortex; PRR, parietal reach region; LIP, lateral intraparietal area; PFG, a parietal area between PF and PG. Note that the visual cortex includes areas beyond the occipital lobe, whereas the auditory cortex occupies just a small fraction of the temporal lobe.

multiple levels. For example, proprioceptive neurons can communicate directly with motor neurons or via spinal interneurons in reflex circuits (Figure 1-19), or through many other intermediates (Figure 8-2). In this section, we study sensorimotor transformation at the level of the neocortex in the monkey (**Figure 8-32**), continuing a discussion we left off in sensory systems (Sections 4.28 and 6.35), with a focus on arm reaching.

The posterior parietal cortex, situated between the somatosensory, visual, and auditory cortices (Figure 8-32), receives input from these sensory systems and mediates multisensory integration; individual neurons often respond to stimuli of multiple sensory modalities. Interestingly, activity of posterior parietal neurons can also predict motor actions. For example, as we learned in Section 4.28, neurons in the **lateral intraparietal area** (**LIP**) can predict the direction of a saccade after integrating input from motion direction–sensitive neurons in the middle temporal (MT) visual area of the dorsal visual stream. A distinct region of posterior parietal cortex called the **parietal reach region** (**PRR**; Figure 8-32) is preferentially associated with monkeys' arm reaching: PRR neurons are activated by visual as well as proprioceptive stimuli, and in turn, their activity can predict reach movements.

In a revealing experiment, monkeys were trained to move their eyes or their hand (but not both) to a briefly presented peripheral target. Whether to move the eyes or the hand was signaled by the color of the peripheral target. A *delay* was enforced between the brief target presentation and the "go" cue; if the monkey initiated eye or hand movement before the go cue, the trial aborted and the monkey could not obtain a reward (**Figure 8-33**A). Single-unit recordings in the posterior parietal cortex identified cells that were preferentially active during the delay period before saccade (Figure 8-33B, cell 1) or before reach (cell 2). This delay period activity is exceptionally interesting to neuroscientists: it is initiated by onset of the peripheral visual target, but continues until the time of the movement, long after the target disappears. Because of its long duration, delay period activity is very different from motor neuron activity, which consists of a brief burst of action potentials immediately before movement onset. Delay period activity is thought to be related to cognitive processes such as attention, working memory, and movement preparation, as we will see later.

Analyses of recording sites indicated that the cells active before saccades were enriched in the LIP, whereas cells active before reaches were mostly found in the PRR. To address a causal role for the PRR in reach, researchers injected **muscimol**, an ionotropic GABA receptor agonist (Box 3-2), into the PRR to transiently silence neuronal activity. When PRR neurons were inactivated while monkeys performed the task, the magnitude of reach was reduced and the magnitude of saccade was unaffected (Figure 8-33C). Together, these experiments suggest that PRR neurons encode the intent to reach and are necessary for reach precision.

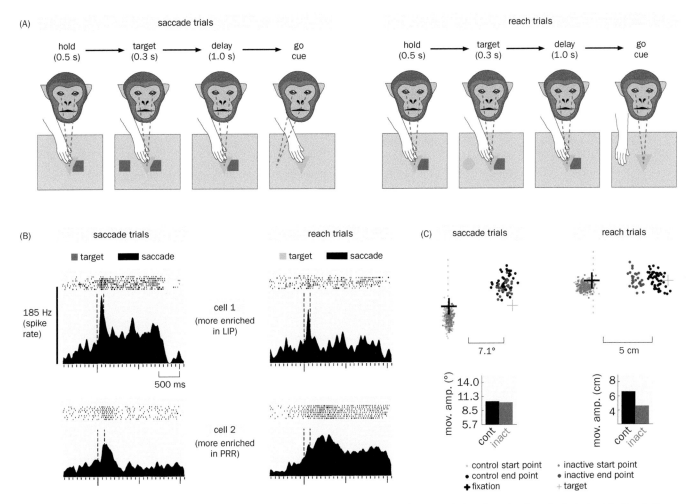

Figure 8-33 A posterior parietal cortex region involved in hand reaching. (A) Task structure. The monkey is trained to fix its eyes and place its hand at the center of the screen at the beginning of a trial. The red square and green triangles at the center are targets for eye and hand fixation, respectively. A new target transiently appears on the side, with a red square indicating a saccade trial (left) and a green circle indicating a reach trial (right). After the go cue (the disappearance of the central target for eye or hand fixation), the monkey either makes a saccade or a hand reach toward the area of the transient target. **(B)** Single-unit recordings of two representative cells in the posterior parietal cortex during saccade (left) or reach (right) trials. Each row at the top shows every third spike recorded during each of the eight trials, and the bottom shows the corresponding peri-event time histogram.

Both cells increase their spike rates transiently to target stimuli. The firing rate of cell 1 is elevated above baseline during the delay period leading to saccades, whereas the firing rate of cell 2 increases during the delay period leading to reaches. **(C)** Behavioral effect of silencing PRR neurons by local infusion of muscimol. Top, data from individual trials (each dot represents one trial). Bottom, average data for eye or hand movement. Compared to control, muscimol infusion does not affect saccades (left) but reduces the magnitude of reaches (right). (A & C from Hwang EJ, Hauschild M, Wilke M, et al. [2012] *Neuron* 76:1021–1029. With permission from Elsevier Inc. B, adapted from Snyder LH, Batista AP & Andersen RA [1997] *Nature* 386:167–170. With permission from Springer Nature.)

Sensorimotor transformation is more than simply relaying information between the sensory and motor systems. In visually guided reaching, for example, sensory information arrives in eye-centered coordinates, whereas motor output is produced in hand-centered coordinates (**Figure 8-34**). Single-unit recordings indicated that PRR neurons encode target location in eye-centered coordinates, whereas neurons in Area 5 of the posterior parietal cortex (Figure 8-32) encode target location in both eye- and hand-centered coordinates. In principle, hand-centered coordinates can be derived by subtracting hand location from target location in eye-centered coordinates (Figure 8-34), and neurons in Area 5 may represent an intermediate step in this transformation.

The posterior parietal cortex is also involved in online monitoring and correction of movement trajectories. As discussed in Section 8.11, effective online control requires a *forward model* using efference copy signals to estimate the upcoming states of the movement rather than relying solely on visual or somatosensory feedback, which have delays of 30–90 ms. Analyses of movement angle

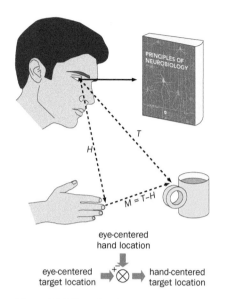

Figure 8-34 **Transforming eye-centered coordinates into hand-centered coordinates in visually guided reaching.** Top, schematic illustrating the target (coffee mug) and hand in eye-centered coordinates (vector T and H, respectively) and target in hand-centered coordinates (vector M). Bottom, illustration of the transformation from eye-centered target location to hand-centered target location. (From Buneo CA, Jarvis MR, Batista AP, et al. [2002] *Nature* 416:632–636. With permission from Springer Nature.)

encoding during reaching revealed that while some posterior parietal cortical neurons had lag times of 30–90 ms, consistent with their receiving sensory feedback, the majority of neurons had shorter or no lag times, supporting their participation in forward models. Taken together, these experiments suggest a key role for the posterior parietal cortex in sensorimotor transformation during visually guided reaching. Indeed, human patients with strokes affecting the posterior parietal cortex suffer from **optic ataxia**, an inability to guide the hand toward an object using visual information, even though other aspects of their movement and vision are less affected.

8.14 The frontal cortex regulates movement planning: a dynamical systems perspective

In addition to the parietal cortex, the frontal lobe anterior to the primary motor cortex houses many neurons whose activity, referred to as **preparatory activity**, precedes the onset of motor commands. For example, neurons in the dorsal premotor cortex (PMd; Figure 8-32) are well known to exhibit preparatory activity prior to arm-reach movement. Preparatory activity is most common in tasks involving an enforced delay before a go cue for movement initiation (Figure 8-33), but similar preparatory activity can also be found following sensory cues but before movement even without an enforced delay. Preparatory activity in premotor cortex is specific to the upcoming movement, can reduce reaction time and increase accuracy compared to movement without a preparatory period, and when disrupted, can delay movement onset. Thus, preparatory activity in premotor cortex plays an important role in movement planning.

Because many premotor cortex neurons are also active during the actual movement, and indeed some also send descending axons to the spinal cord, a question arises as to why preparatory activity does not lead to movement. Before addressing this question, we introduce a new perspective on how motor action is represented by activity of neuronal populations. As discussed in Section 8.12, *in vivo* recordings indicate that activity of most motor cortex neurons correlates with contraction of specific muscles as well as specific movement parameters such as direction, magnitude, and velocity. This phenomenon, known as multiple selectivity (individual neurons can encode multiple signals in their activity), raises the intriguing question of how downstream circuits decode specific signals in order to generate motor behavior. Instead of asking this question from the perspective of individual neurons, an alternative perspective on motor control treats the motor cortex as a **dynamical system**, composed of different **dynamical states** that evolve over time according to specific rules. Recall that in our discussion of olfactory coding, we introduced the concept that the activity state of a neuronal population can be represented as a point in a multidimensional space, with each axis representing the firing rate of one constituent neuron (Figure 6-28A). Likewise, the dynamical state of the motor cortex at any given time can be described as the firing rates of a population of motor cortex neurons at that time. The activity of motor cortex neurons over time can be represented as an evolving dynamical state that follows a specific trajectory in the activity state space (**Figure 8-35**A).

We use a specific example to illustrate how the dynamical systems perspective provides insight into cortical control of movement. In a reaching task, a monkey was trained to move its arm toward a target on the screen, but only after a go cue appeared (Figure 8-35B, top). Trajectories of neural states throughout each of many individual trials were computed from activity of hundreds of *simultaneously* recorded dorsal premotor cortex neurons (Figure 8-35B, bottom) using a **multi-electrode array** (see Figure 14-36 for details). Target appearance caused the starting positions of different trials, which were initially more variable (blue ellipse), to converge onto a smaller region in the state space during movement preparation (green ellipse). The onset of the go cue caused the neural states to move into yet another small region at the time of movement onset (gray ellipse), following similar trajectories across different trials. These observations suggest that stimulus onset reduces the variability of population activity of the premotor cortex.

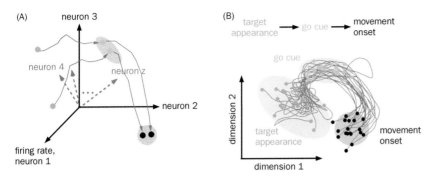

Figure 8-35 A dynamical systems perspective on motor control. (A) The dynamical state of the motor cortex at a given time is a vector in a high-dimensional activity state space, with values on each axis representing the firing rates of individual neurons. The first three dimensions are shown here as solid arrows, and other dimensions are symbolized by dashed gray arrows in the background. Time is implicit in this diagram; the passage of time is represented by a trajectory in the state space. Two example trajectories representing two trials of the behavioral task in Panel B are schematized. **(B)** Top, behavioral task. The monkey is trained to move its arm to reach a target shown on a screen only after a subsequent go cue appears. Bottom, motor cortex activity for 18 different trials, represented by 18 trajectories in the state space. Blue, green, and black circles represent neural states (computed from activity of hundreds of simultaneously recorded cortical neurons) just before target appearance, at the appearance of the go cue, and at movement onset, respectively. Trajectories between blue and green circles represent movement planning, while trajectories between green and black circles represent movement initiation. The high-dimensional state space in Panel A is simplified by projecting it onto two dimensions while preserving the relative size of the ellipsoids representing variance across trials. (Adapted from Shenoy KV, Sahani M, & Churchland MM [2013] *Ann Rev Neurosci* 36:337–359. See also Churchland MM, Yu BM, Cunningham JP, et al. [2010] *Nat Neurosci* 13:369–378.)

Having introduced the dynamical systems perspective, we now return to the question of why preparatory activity in the premotor cortex does not lead to movement. Let's begin with a simple system: a muscle receiving input from two neurons and producing a response according to the sum of their firing rates (**Figure 8-36A**). Each neuron can vary its firing rate, but if the sum of its firing rates remains the same (gray trajectories along the blue line in Figure 8-36B), no change in muscle contraction results. The blue line represents the "output-null" dimension because firing rate changes along this dimension do not cause movement. However, when additional inputs to the system cause the gray trajectories to move away from the blue line, the changes in the net output of the two-neuron system cause movement to occur by contracting or relaxing the muscle. Thus, the dimension orthogonal to the blue line (the green line) is the "output-potent" dimension in this simple two-dimensional state space.

Figure 8-36 features only two neurons, but it is easy to imagine systems of 10 or 1000 neurons in the motor cortex, where the same principles hold. As long as the sum of neural firing rates remains constant, no net motor output is transmitted to the spinal cord. Within the constraint of net equality, the activity of individual

Figure 8-36 Preparatory activity resides in an output-null dimension. (A) In this two-neuron system, the muscle responds to the sum of the firing rates of neurons 1 and 2. **(B)** When changes in the activity state occur along the blue line of the two-neuron system, the sum of the firing rates for neurons 1 and 2 remains constant, and no movement results. When changes in the activity state deviate from the blue line, movement ensues. Thus, the blue line represents the output-null dimension, while the orthogonal green line represents the output-potent dimension. Drawn in two different shades of gray are two trajectories representing movement planning (when the trajectories coincide with the blue line) and movement execution (after the go cue represented by circles at the upper left and lower right), obtained from simultaneous multielectrode recordings in a task similar to that described in Figure 8-35. The two trajectories represent neural activity state changes for reach left and reach right, respectively. (Adapted from Kaufman MT, Churchland MM, Ryu SI, et al. [2014] *Nat Neurosci* 17:440–448. With permission from Springer Nature. See also Elsayed GF, Lara AH, Kaufman MT, et al. [2016] *Nat Comm* 7:13239.)

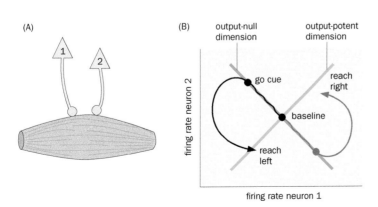

neurons in the system can move up and down, implementing computations that prepare the system for efficient movement to the correct target when the go signal is given (Figure 8-36B, upper left and lower right gray circles). This separation of output-null and output-potent dimensions in the neural activity space provides an elegant mathematical explanation for the separation of movement planning and movement execution in the same neuronal population, despite the fact that many neurons encode multiple signals in their activity (as discussed earlier). Recent multielectrode array recordings indicate that this principle is indeed implemented within premotor circuitry. The mechanisms by which population activity is restricted to the output-null dimension during movement planning and is converted to the output-potent dimension just before movement onset are unknown.

We have discussed the parietal cortex's involvement in sensorimotor transformation in Section 8.13 and the premotor frontal cortex's involvement in movement planning in this section, but these two processes are highly intertwined. Indeed, neurons in the parietal and frontal cortices communicate extensively with each other during sensorimotor transformation and movement planning and also communicate extensively with M1 during movement execution. For instance, the PRR and PMd have strong reciprocal connections that control arm reaching, while the LIP reciprocally connects with the **frontal eye field**, a premotor cortical region that controls eye movement (Figure 8-32; see also Figure 4-51). Premotor and primary motor cortical inputs to the parietal cortex provide efference copies of motor plans that help construct the forward models discussed earlier. Thus, the frontal, parietal, and primary motor cortices can be conceptualized as a giant dynamical system that underlies sensorimotor transformation, movement planning, and movement execution. Recent advances in neural circuit dissection tools in rodents have enabled more precise perturbations of neuronal dynamics, allowing examination of the causal role of preparatory activity in linking sensation to action (Box 8-2).

Interestingly, certain neurons in the rostral part of the ventral premotor cortex and the reciprocally connected PFG area of the posterior parietal cortex (Figure 8-32) are active not only when a monkey performs an action, such as reaching or grasping an object, but also when another monkey or a human performs the same action. These neurons are aptly named **mirror neurons**. The properties of mirror neurons have been proposed to enable monkeys to understand intention and imitate the actions of others using the framework of the monkey's own action planning. fMRI imaging studies suggest that a similar mirror neuron system is also present in the analogous regions of the human frontal and parietal cortices. The mirror neuron system provides a fascinating window into the interface between perception and action as well as the neural basis of learning by imitation.

Box 8-2   Probing the causal function of preparatory activity

Cortical control of movement planning and execution has historically been studied in nonhuman primates. Analogous investigations have recently been carried out in rodents, taking advantage of modern circuit dissection tools that are widely used in mice and rats (see Chapter 14). Here we discuss two examples of investigations into the role of preparatory activity in linking sensation to action.

In our first example, thirsty rats use the frequency of auditory stimuli to decide in which direction to orient their head to obtain a water reward (Figure 8-37A). A delay was enforced such that rats needed to wait until the go cue before initiating the movement. Rats can perform the task fairly well when the click frequency deviates substantially from a threshold (Figure 8-37B, solid line). When muscimol

was infused into a frontal lobe premotor area called the frontal orienting field (FOF) to silence neuronal activity, rats exhibited a marked deficit in orienting toward the side contralateral to the site of muscimol infusion (Figure 8-37B, dashed line). Single-unit recordings revealed that a large fraction of FOF neurons exhibited preparatory activity during the delay period that predicted subsequent orienting direction (Figure 8-37C). Furthermore, optogenetic silencing of halorhodopsin-expressing FOF neurons (Section 14.25) during the delay period disrupted contralateral orienting, like muscimol infusion. The temporal precision of optogenetic silencing allowed researchers to conclude that preparatory activity in one side of the FOF during the delay period plays a causal role in promoting orienting to the contralateral side.

Box 8-2: continued

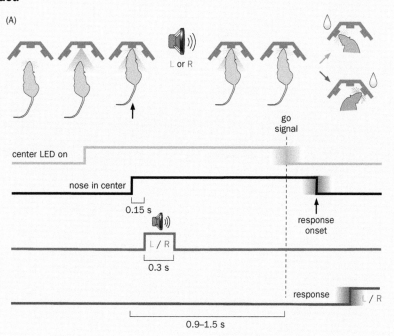

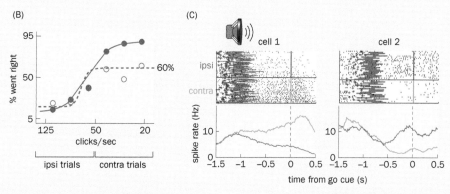

Figure 8-37 **Preparatory activity in the rat frontal orienting field (FOF) prior to orienting. (A)** Task structure. Water-restricted rats were trained to poke their nose into the center port upon LED light onset. This triggers a click tone with a specific frequency; a water reward is available at the left or right port if the frequency is >50 Hz or <50 Hz, respectively. A variable delay was introduced by requiring the rat to wait until after the LED turns off (the go cue) to orient its head toward a water port to earn its reward. **(B)** Psychometric curves for normal rats (solid curve) and rats with the left FOF infused with muscimol (dashed curve). Compared with normal rats, muscimol-infused rats perform poorly in orienting to the contralateral port.

(C) Single-unit recordings from two representative FOF cells. Top, each row shows spike patterns during one trial. Trials are aligned to the go cue (t = 0), and ipsi- and contralateral orienting trials are grouped separately. Tone periods are highlighted in brown. +, center poke onset. Bottom, peri-event time histograms of ipsi- and contralateral trials. Cell 1 exhibits more preparatory activity during contralateral trials, and cell 2 during ipsilateral trials. (From Erlich JC, Bialek M, & Brody CD [2011] *Neuron* 72:330–343. With permission from Elsevier Inc. See also Kopec CD, Erlich JC, Brunton BW, et al. [2015] *Neuron* 88:367–377.)

In our second example, head-fixed mice were trained to lick water from a left or right port depending on the position of the pole sensed by their whiskers, again after an enforced delay (Figure 8-38A). Researchers could silence the output of a given cortical area with high temporal precision by photostimulating transgenic mice expressing channelrhodopsin (ChR2; Section 14.25) in all GABAergic inhibitory neurons (Figure 8-38B; this is because the vast majority of GABAergic inhibitory neurons project locally, and therefore their activation silences nearby pyramidal neurons that send output from the photostimulated region to distant tar-

gets). This enabled researchers to examine the behavioral consequence of silencing different cortical areas during specific periods of the trial. For example, silencing of the barrel cortex (primary somatosensory cortical region representing whiskers; Box 5-3) during *the sample period* markedly reduced performance, validating an essential sensory function for the barrel cortex in the task. Researchers then screened for cortical areas that, when silenced during *the delay period*, would disrupt task performance. They found that a premotor area called the anterior lateral motor cortex (ALM) had the largest effect: mice displayed a marked

(*Continued*)

Box 8-2: continued

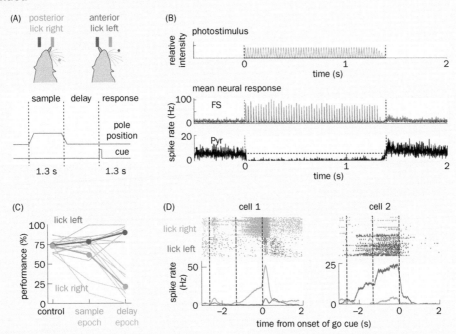

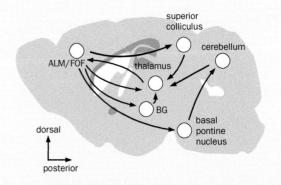

Figure 8-38 Preparatory activity in the mouse anterior lateral motor cortex (ALM) prior to licking. (A) Task structure. Water-restricted head-fixed mice were trained to respond to presentation of two vertical poles (blue and red circles) detected by their whiskers by licking the left or right water ports, as indicated. The poles were raised to the whisker level during the sample period. A delay was enforced between the end of the sample period and an auditory go cue. **(B)** Effect of silencing output of a cortical area by photostimulating inhibitory neurons using transgenic mice expressing ChR2 in all GABAergic cells. Top, photostimulation applied at 40 Hz. Middle, a neuron exhibiting fast spiking (FS), a signature of cortical GABAergic basket cells, fires action potentials time locked to photostimulation.

Bottom, photostimulation of inhibitory cells silences the firing of a pyramidal (Pyr) neuron. **(C)** Effects on lick performance when the left ALM was silenced during the sample or delay epochs. Silencing the ALM during the delay epoch drastically reduced contralateral licking and slightly increased ipsilateral licking. Thin lines represent the performances of individual mice, and thick lines their averages. **(D)** Single-unit recordings from two cells, as in Panel C. Cell 1 exhibits stronger preparatory activity for licking right, and cell 2 for licking left. (A, C, & D from Li N, Chen TW, Guo ZV, et al. [2015] *Nature* 519:51–56. With permission from Springer Nature. B, from Guo ZV, Li N, Huber D, et al. [2014] *Neuron* 81:179–194. With permission from Elsevier Inc.)

reduction in subsequent licking to the contralateral port (Figure 8-38C). Single-unit recordings identified preparatory activity in many ALM neurons, some of which also exhibited trial-specific activity patterns during the sample period (Figure 8-38D). Together, these experiments identified a causal role of ALM preparatory activity in transforming whisker-based somatosensation into lick movements.

The temporal delay between sensory stimuli and movement onset in these perceptual discrimination tasks requires the animals to remember the sensory stimuli at the time of movement execution. Such short-term (on the order of seconds) memory during a task is referred to as **working memory**. Persistent activity of frontal cortex neurons has been hypothesized to be the physiological substrate of working memory. Thus, preparatory activity for movement planning can also be viewed as representing working memory that links sensation to action. What is the neural basis of persistent activity? Persistent activity has been hypothesized to be due to either intrinsic biophysical properties of the neurons or mutually excitatory local connections. A surprising finding from the rodent models discussed here is that persistent activity in the ALM and FOF is likely contributed by a brain-wide network.

For example, the ALM forms strong reciprocal connections with medial dorsal (MD) thalamic nucleus (**Figure 8-39**), which also exhibit preparatory activity. Optogenetic perturbation experiments indicated that the preparatory activity

Figure 8-39 Simplified schematic of interacting brain regions likely involved in maintaining persistent activity between sensation and action in a rodent brain. Arrows indicate direct projections. ALM, anterior lateral motor cortex; BG, basal ganglia; FOF, frontal orienting field. (From Svoboda K & Li N [2018] *Curr Opin Neurobiol* 49:33–41. With permission from Elsevier Inc.)

Box 8-2: continued

of the MD thalamus and the ALM mutually reinforce each other and that both are required for task performance. The ALM (and the neocortex in general) also forms bidirectional connections with the cerebellum via the basal pontine nuclei and thalamic nuclei (Figure 8-26B). Preparatory activity is also found in cerebellar neurons, and disrupting cerebellar output also interferes with preparatory activity in ALM neurons. In the orienting task described earlier, preparatory activity has also been found in the superior colliculus, a projection target of the FOF that also sends projections back to the cortex via the thalamus (Figure 8-39). Simulta-

neous silencing of the FOF and superior colliculus synergistically disrupted task performance (that is, more so than simply adding the effects of the individual perturbations), suggesting that the activities of the FOF and superior colliculus enhance each other through their reciprocal connections. Altogether, these studies suggest that multiple brain regions interact to sustain the preparatory activity linking sensation to action. These findings also highlight the importance of simultaneously studying multiple brain regions in order to decipher the principles of information processing during sensorimotor transformations.

8.15 Population activity of motor cortex neurons can be used to control neural prosthetic devices

Investigating how the neocortex controls movement not only helps us understand the neural basis of action but also has important clinical applications. Specifically, the ability to predict movement directions using the firing patterns of cortical neurons (Figure 8-31) has inspired intense research into brain–machine and brain–computer interfaces, or **neural prosthetic devices**. A major goal of these devices is to help patients suffering from paralysis due to spinal cord injury or motor neuron disease to regain voluntary motor control. These patients' motor cortices are still active and can presumably send commands to control body movement. Unfortunately, either the axons that deliver the commands to the spinal cord are damaged or motor neurons have degenerated, such that the brain and muscles are disconnected. A general strategy for creating neural prosthetic devices starts with extracting the activity of motor cortex neurons. The most effective approach thus far is an implanted multielectrode array that can directly record spikes of hundreds of individual neurons simultaneously. The activity of these neurons is then fed into a computer to extract movement intent, which is used to control an output device, such as a robotic arm (**Figure 8-40**A) or a computer cursor. Visual feedback to the patients can help them learn to change the firing patterns of the recorded neurons for better performance. Remarkable progress has been made in recent years in the accuracy, complexity, and speed of such neural prosthetic devices. We give two examples to illustrate these advances.

In the first example, a multielectrode array implanted in a monkey's motor cortex recorded spikes from 116 neurons (Figure 8-40B, top). A computer decoded these patterns in real time by extracting movement intent using a method similar to the discussed in Section 8.12. The movement intent was then executed by a robotic arm with five degrees of freedom: three at the shoulder, one at the elbow, and one at the hand for gripping. After training by fetching marshmallows as a reward, the monkey was able to control the extension of the robotic arm by "thinking" about the movement (Figure 8-40B, bottom). Such thinking caused the monkey to extend the robotic arm toward a marshmallow placed at different locations in three-dimensional space, grab the marshmallow, and bring it to its mouth (Figure 8-40C; **Movie 8-5**). Remarkably, these actions took only a few seconds and had a 60% success rate. This and other such examples suggest that this approach can help human patients with motor deficits regain movement control.

The second example is the first human clinical trial using neural prosthetics to enable a patient with tetraplegia to move a computer cursor via cortical control. The patient suffered a spinal cord injury three years before having a 100-electrode array surgically implanted into his motor cortex (**Figure 8-41**A). During the training period following surgery, the patient was asked to imagine moving his hand along the trajectory of a cursor on a computer screen controlled by a technician (Figure 8-41B). Population activity of motor cortex neurons recorded during the imagined movement was then used to construct an algorithm best matching the

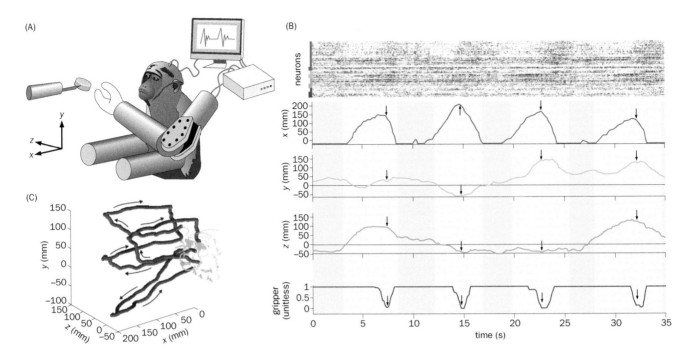

Figure 8-40 Cortical control of a prosthetic arm for self-feeding. **(A)** Schematic of a brain–machine interface for prosthetic control. A multielectrode array was implanted in the motor cortex of a monkey. The spiking patterns of many motor cortex neurons were used for real-time control of a robotic arm using a population vector algorithm (Figure 8-31). The monkey was first trained to control the robotic arm by moving a joystick with its own arm, before graduating to cortical control in this experiment, during which its own arms were restrained. **(B)** Top, spiking activity of 116 neurons used to control the robotic arm during four self-feeding trials. Each row shows a different neuron's spiking over time, with each of the four trials occurring consecutively (lighter yellow regions below). The neurons are grouped along the *y* axis based on which of the four dimensions of robot movement they responded most strongly to (red, neurons preferring motion along the *x* dimension; green, *y* dimension; blue, *z* dimension; purple, gripper movement) and also by whether they preferred movement in the negative (thin bars) or positive (thick bars) directions along that dimension. Bottom, four traces corresponding to robotic arm movement in the *x*, *y*, and *z* directions and the position of the gripper (1 = open, 0 = closed) during the four trials shown in Panel C. Arrows indicate the gripper closing on the target. **(C)** Spatial trajectories of the same four trials as in Panel B, with varied marshmallow positions. Red and blue parts of each trajectory represent open and closed grippers, respectively. Note that the monkey opened the gripper before the food reached its mouth because it learned that marshmallows tend to stick to the gripper so that it does not need to close the gripper for the full duration of the return trip. (Adapted from Velliste M, Perel S, Spalding MC, et al. [2008] *Nature* 453:1098–1101. With permission from Springer Nature.)

Figure 8-41 Movement control for a human tetraplegia patient via a brain-computer interface. (A) A magnetic resonance image of the patient's brain before surgical implantation of a 4 mm × 4 mm multielectrode array in the arm control region of the right motor cortex. The square indicated by the red arrow denotes the target position of the implant. **(B)** The computer cursor was controlled by the patient's motor cortex activity, as recorded by the multielectrode array. **(C)** Example of a cursor control task performed by the patient involving acquisition of targets (green circles) and avoidance of obstacles (red squares). Blue lines denote the cursor trajectory in four separate trials. (Adaped from Hochberg LR, Serruya MD, Friehs GM, et al. [2006] *Nature* 442:164–171. With permission from Springer Nature. See Pandarinath C, Nuyujukian P, Blabe CH, et al. [2017] *eLife* 6:e18554 for a recent example of typing with a brain–computer interface.)

motor intent. During the next phase, the patient imagined moving the cursor, and the algorithm transformed his motor cortex activity, recorded from the multielectrode array, into real-time control of the computer cursor. After extensive training, the patient could perform a variety of computerized tasks, such as moving a cursor to designated targets while avoiding obstacles (Figure 8-41C), opening files in an email inbox, and drawing circles (Movie 8-5), all via imagined motion.

Despite these remarkable achievements, human clinical trials have also revealed important limitations: the speed, accuracy, and level of control are considerably less than control of a computer cursor via a standard mouse by hand. The long-

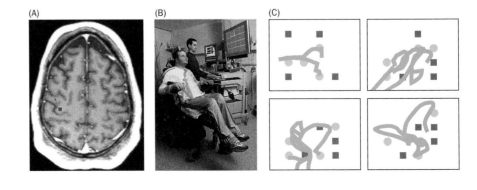

term stability of electrode arrays also limits such trials. Researchers are making remarkable progress in addressing these limitations. In a recent study, for example, brain–computer interfaces enabled some patients to type with a computer keyboard at a speed of ~30 correct characters per minute (about a third to half the speed of able-bodied subjects typing on a smartphone), thus greatly facilitating their communication capabilities.

In addition to helping patients, neural prosthesis research has also provided important insights into how the motor cortex controls voluntary movements. Each motor task appears to be controlled by an ensemble of cortical neurons— this is supported by the successful extraction of spiking activity from populations of neurons to control prosthetic devices and the dependence of this success on engaging a critical number of neurons (ranging from dozens to hundreds). At the same time, each individual neuron contributes to multiple tasks—for instance, movement in all three dimensions of space and grip, as shown in the example in Figure 8-40. These properties may explain why information from several hundred neurons—a very small fraction of all motor cortex neurons—can allow remarkable control of prosthetic devices. From a dynamical systems perspective, neural states extracted from the activity of a few hundred neurons can be useful approximations of neural states of an entire subregion in the relevant motor homunculus for executing well-trained tasks. The improvement in performance by learning through visual feedback suggests that cortical neuronal ensembles that control movement are plastic—indeed, this was observed when monitoring changes in the tuning properties of individual neurons and neuronal ensembles as animals learned to improve their performance via training. We will return to the subject of learning and plasticity in Chapter 11.

SUMMARY

Motor systems are organized hierarchically. The powerful neuromuscular junction converts nearly every action potential from motor neurons into muscle contractions via rises in intracellular Ca^{2+}, which triggers sliding of actin and myosin fibers. Thus, the end goal of motor control is to specify motor neuron firing patterns. Motor neurons integrate information from multiple sources, including input from spinal premotor neurons, descending commands from brainstem motor control nuclei and the motor cortex, and sensory feedback from proprioceptive neurons. In rhythmic motor programs such as locomotion, the rhythmic output pattern is produced by central pattern generators in the spinal cord in the absence of sensory feedback and activated by brainstem motor control nuclei. The mechanisms by which central pattern generators operate are best understood in invertebrate systems, where the biophysical properties of constituent neurons and their connection patterns and strengths determine the rhythmic output patterns. Modern genetic and circuit analysis tools have begun to enable dissection of complex circuits in the spinal cord and brainstem that control various motor behaviors, from locomotion to breathing.

Voluntary movement is controlled by the motor cortex via extensive collaboration with the basal ganglia and cerebellum, both employing generic circuit designs. In the striatum of the basal ganglia, two types of spiny projection neurons both receive cortical and thalamic inputs and separately constitute the direct and indirect pathways, which control basal ganglia output. The direct and indirect pathways are bidirectionally modulated by midbrain dopamine neurons and act in concert to regulate the selection and initiation of motor programs. The cerebellum utilizes a vast number of granule cells to integrate inputs concerning motor commands from the motor cortex and brainstem, motor performance from premotor neurons in the spinal cord, and feedback from sensory systems. The cerebellum constructs a forward model that predicts the sensory consequences of movement in order to adjust motor output. The motor cortex is grossly organized in somatotopy, but this somatotopy breaks down at fine scales. While each motor cortex neuron is broadly tuned to multiple motor tasks, population activity of motor cortex neurons can be predictive of movement parameters, such as the direction and trajectory of arm reaching. The population activity of motor cortex neurons has been used to control neural prosthetic devices with remarkable success.

The posterior parietal and premotor cortices form extensive connections with each other and with the primary motor cortex. Together, these form an extended dynamical system that plays crucial functions in sensorimotor transformation and motor planning. Recent studies have implicated additional brain regions, including the thalamus and cerebellum, in maintaining preparatory activities in the transition between sensation and action. A major future challenge is to understand how neuronal populations in different brain regions cooperate to control movement planning and execution.

OPEN QUESTIONS

- How do individual spinal motor neurons integrate distinct sources of inputs to control their firing pattern?

- What are the neural substrates for central pattern generators in the vertebrate spinal cord and brainstem?

- How do neurons in different brain regions, such as the motor cortex, cerebellum, and basal ganglia, cooperate to control movement planning and execution?

- How do local and long-range recurrent connections affect information flow in motor circuits?

FURTHER READING

Reviews

Ferreira-Pinto MJ, Ruder L, Capelli P, & Arber S (2018). Connecting circuits for supraspinal control of locomotion. *Neuron* 100:361–374.

Gerfen CR & Surmeier DJ (2011). Modulation of striatal projection systems by dopamine. *Annu Rev Neurosci* 34:441–466.

Ito M (2006). Cerebellar circuitry as a neuronal machine. *Prog Neurobiol* 78:272–303.

Kiehn O (2016). Decoding the organization of spinal circuits that control locomotion. *Nat Rev Neurosci* 17:224–238.

Marder E & Bucher D (2007). Understanding circuit dynamics using the stomatogastric nervous system of lobsters and crabs. *Annu Rev Physiol* 69:291–316.

Shenoy KV, Sahani M, & Churchland MM (2013). Cortical control of arm movements: a dynamical systems perspective. *Annu Rev Neurosci* 36:337–359.

Spinal cord, brainstem, and invertebrates

Bikoff JB, Gabitto MI, Rivard AF, Drobac E, Machado TA, Miri A, Brenner-Morton S, Famojure E, Diaz C, Alvarez FJ, et al. (2016). Spinal inhibitory interneuron diversity delineates variant motor microcircuits. *Cell* 165:207–219.

Esposito MS, Capelli P, & Arber S (2014). Brainstem nucleus MdV mediates skilled forelimb motor tasks. *Nature* 508:351–356.

Graham Brown, T (1911). The intrinsic factors in the act of progression in the mammal. *Proc R Soc Lond B* 84:308–319.

Henneman E, Somjen G, & Carpenter DO (1965). Functional significance of cell size in spinal motoneurons. *J Neurophysiol* 28:560–580.

McLean DL & Fetcho JR (2009). Spinal interneurons differentiate sequentially from those driving the fastest swimming movements in larval zebrafish to those driving the slowest ones. *J Neurosci* 29:13566–13577.

Prinz AA, Bucher D, & Marder E (2004). Similar network activity from disparate circuit parameters. *Nat Neurosci* 7:1345–1352.

Roseberry TK, Lee AM, Lalive AL, Wilbrecht L, Bonci A, & Kreitzer AC (2016). Cell-type-specific control of brainstem locomotor circuits by basal ganglia. *Cell* 164:526–537.

Smith JC, Ellenberger HH, Ballanyi K, Richter DW, & Feldman JL (1991). Pre-Bötzinger complex: a brainstem region that may generate respiratory rhythm in mammals. *Science* 254:726–729.

Stepien AE, Tripodi M, & Arber S (2010). Monosynaptic rabies virus reveals premotor network organization and synaptic specificity of cholinergic partition cells. *Neuron* 68:456–472.

Talpalar AE, Bouvier J, Borgius L, Fortin G, Pierani A, & Kiehn O (2013). Dual-mode operation of neuronal networks involved in left–right alternation. *Nature* 500:85–88.

Yackle K, Schwarz LA, Kam K, Sorokin JM, Huguenard JR, Feldman JL, Luo L, & Krasnow MA (2017). Breathing control center neurons that promote arousal in mice. *Science* 355:1411–1415.

Cortex, cerebellum, and basal ganglia

da Silva JA, Tecuapetla F, Paixao V, & Costa RM (2018). Dopamine neuron activity before action initiation gates and invigorates future movements. *Nature* 554:244–248.

Gallese V, Fadiga L, Fogassi L, & Rizzolatti G (1996). Action recognition in the premotor cortex. *Brain* 119 (Pt 2):593–609.

Gao Z, Davis C, Thomas AM, Economo MN, Abrego AM, Svoboda K, De Zeeuw CI, & Li N (2018). A cortico-cerebellar loop for motor planning. *Nature* 563:113–116.

Georgopoulos AP, Kettner RE, & Schwartz AB (1988). Primate motor cortex and free arm movements to visual targets in three-dimensional space. II. Coding of the direction of movement by a neuronal population. *J Neurosci* 8:2928–2937.

Graziano MS, Taylor CS, & Moore T (2002). Complex movements evoked by microstimulation of precentral cortex. *Neuron* 34:841–851.

Hochberg LR, Bacher D, Jarosiewicz B, Masse NY, Simeral JD, Vogel J, Haddadin S, Liu J, Cash SS, van der Smagt P, et al. (2012). Reach and grasp by people with tetraplegia using a neurally controlled robotic arm. *Nature* 485:372–375.

Hwang EJ, Hauschild M, Wilke M, & Andersen RA (2012). Inactivation of the parietal reach region causes optic ataxia, impairing reaches but not saccades. *Neuron* 76:1021–1029.

Kaufman MT, Churchland MM, Ryu SI, & Shenoy KV (2014). Cortical activity in the null space: permitting preparation without movement. *Nat Neurosci* 17:440–448.

Klaus A, Martins GJ, Paixao VB, Zhou P, Paninski L, & Costa RM (2017). The spatiotemporal organization of the striatum encodes action space. *Neuron* 95:1171–1180.

Markowitz JE, Gillis WF, Beron CC, Neufeld SQ, Robertson K, Bhagat ND, Peterson RE, Peterson E, Hyun M, Linderman SW, et al. (2018). The striatum organizes 3D behavior via moment-to-moment action selection. *Cell* 174:44–58.

Pandarinath C, Nuyujukian P, Blabe CH, Sorice BL, Saab J, Willett FR, Hochberg LR, Shenoy KV, & Henderson JM (2017). High performance communication by people with paralysis using an intracortical brain-computer interface. *Elife* 6:e18554.

Panigrahi B, Martin KA, Li Y, Graves AR, Vollmer A, Olson L, Mensh BD, Karpova AY, & Dudman JT (2015). Dopamine is required for the neural representation and control of movement vigor. *Cell* 162:1418–1430.

Velliste M, Perel S, Spalding MC, Whitford AS, & Schwartz AB (2008). Cortical control of a prosthetic arm for self-feeding. *Nature* 453:1098–1101.

Wagner MJ, Kim TH, Kadmon J, Nguyen ND, Ganguli S, Schnitzer MJ, & Luo L (2019). Shared cortex-cerebellum dynamics in the execution and learning of a motor task. *Cell* 177:669–682.

CHAPTER 9
Regulatory Systems

The nervous system has two output systems in addition to the motor system we discussed in Chapter 8: the autonomic nervous system and the neuroendocrine system. Whereas the motor system (also called the somatic motor system) controls contraction of skeletal muscles and mediates body movement, the **autonomic nervous system** (also called the visceral motor system) controls contraction of smooth and cardiac muscle and regulates internal organs. The **neuroendocrine system** secretes hormones in response to sensory stimuli and brain states and regulates animal physiology and behavior, such as food and water intake and responses to the daily light–dark cycle. Both of these output systems receive feedback regulation from the **interoceptive system**, which senses the state of internal organs. Together, the autonomic, neuroendocrine, and interoceptive systems maintain vital bodily functions. They operate mostly subconsciously, thus freeing our conscious brains to pursue other interests, as reflected in the epigraph.

We start this chapter with an overview of these regulatory systems. We then use the examples of eating, drinking, circadian rhythms, and sleep to illustrate how these most fundamental functions are regulated.

HOW DOES THE BRAIN REGULATE INTERNAL ORGANS?

The autonomic nervous system controls the internal organs that perform an animal's basic physiological functions, such as digestion of food and circulation of blood. The term *autonomic* reflects the involuntary nature of many of these functions. In the following sections, we review the two major divisions of the autonomic nervous system: the **sympathetic system** and **parasympathetic system**. (The third division, the **enteric nervous system**, is associated with the gastrointestinal tract and acts rather independently to regulate digestion, with some input from the sympathetic and parasympathetic systems.) We then discuss the regulation of the autonomic nervous system via feedback from the interoceptive system. Lastly, we study the hypothalamus, a command center for the autonomic and neuroendocrine systems.

9.1 The sympathetic and parasympathetic systems play complementary roles in regulating body physiology

The final effectors of the autonomic nervous system are (1) the **smooth muscle** that controls movement within the digestive, respiratory, vascular, excretory, and reproductive systems; (2) the **cardiac muscle** that controls heartbeat; and (3) various glands that excrete fluids locally through specific ducts (the **exocrine system**) or release hormones into the bloodstream to circulate throughout the body (the **endocrine system**). Whereas the effector of the motor system, skeletal muscle, is controlled by somatic motor neurons, the effectors of the autonomic nervous system are controlled by the sympathetic and parasympathetic systems in parallel, each with two layers of output neurons (**Figure 9-1**).

Imagine a mouse that, while searching for food, suddenly smells cat urine: its heart rate, breathing rate, and blood pressure all rise, while most of its housekeeping

Figure 9-1 Comparison of motor and autonomic output systems. The motor system consists of somatic motor neurons (green) projecting to their effectors, skeletal muscle. The sympathetic and parasympathetic divisions of the autonomic nervous system both use two layers of output neurons: (1) preganglionic neurons in the CNS (darker red and blue) that project to the ganglia, and (2) postganglionic neurons in the ganglia (lighter red and blue) that project to their effectors. The parasympathetic ganglia are usually closer to the final effector organs than are the sympathetic ganglia. While most postganglionic sympathetic neurons use norepinephrine (NE) as a neurotransmitter, almost all other types of autonomic neurons use acetylcholine (ACh). A notable exception is that some parasympathetic postganglionic neurons use nitric oxide instead of ACh to relax stomach muscles.

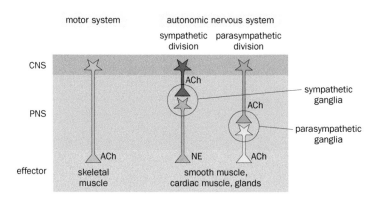

functions, such as salivation and digestion, are inhibited. The mouse is preparing for an emergency response, such as running away. These responses result from the activation of the sympathetic system. When the danger is over, the mouse's heart rate, breathing rate, and blood pressure return to baseline, and it resumes its normal activities. These recovery processes are enabled by the parasympathetic system.

The parasympathetic and sympathetic divisions can be considered the yin and yang of the autonomic nervous system to a first approximation, as their actions on many effector organs are complementary and often in opposition (**Figure 9-2**). For example, the parasympathetic system stimulates salivation and digestion, while the sympathetic system inhibits these processes; the parasympathetic system slows down the heartbeat, while the sympathetic system speeds it up; the parasympathetic system constricts airways in the lungs, while the sympathetic system causes airways to relax. In general, the sympathetic system facilitates energy expenditure to support enhanced activity in the face of extreme situations, such as the "fight or flight" responses of the mouse we just discussed, by inhibiting physiological functions not immediately required for survival and by activating emergency responses. Indeed, the sympathetic system drives production of **epinephrine (adrenaline)** and norepinephrine through innervation of the adrenal medulla; epinephrine is a hormone that circulates in the bloodstream to increase blood flow and glucose production to ready the body for rapid action. Conversely, the parasympathetic system returns the body's physiology to its non-emergency state and facilitates energy replenishment (rest and digest). Recent studies, however, have suggested that this historic view of the division of labor of the sympathetic and parasympathetic systems is simplistic—for example, the parasympathetic system itself contains different types of neurons that have opposing effects on physiology.

In both sympathetic and parasympathetic systems, the neurons innervating effector organs are called **postganglionic neurons**. Their cell bodies are located in the sympathetic or parasympathetic ganglia in the PNS. Postganglionic neurons are controlled by **preganglionic neurons**, whose cell bodies are located in the CNS and which project axons that synapse onto postganglionic neurons (Figure 9-1; Figure 9-2). (Pre- and postganglionic neurons are also collectively called **visceral motor neurons**, as opposed to somatic motor neurons, which control skeletal muscle contraction.) In the sympathetic system, preganglionic neurons are located in the thoracic and lumbar segments of the lateral spinal cord gray matter, and postganglionic neurons reside in nearby sympathetic trunk or prevertebral ganglia. By contrast, most parasympathetic preganglionic neurons are located in the brainstem; their axons, such as those constituting the **vagus nerve**, travel long distances to innervate parasympathetic ganglia, which are usually closely associated with their effector organs (Figure 9-2).

As with somatic motor neurons of the motor system, preganglionic neurons of both the sympathetic and parasympathetic systems use ACh as their neurotransmitter. Most postganglionic parasympathetic neurons also use ACh, whereas postganglionic sympathetic neurons mostly use norepinephrine (Figure 9-1; see Figure 3-33 and Figure 3-35 for the opposing actions of norepinephrine and ACh

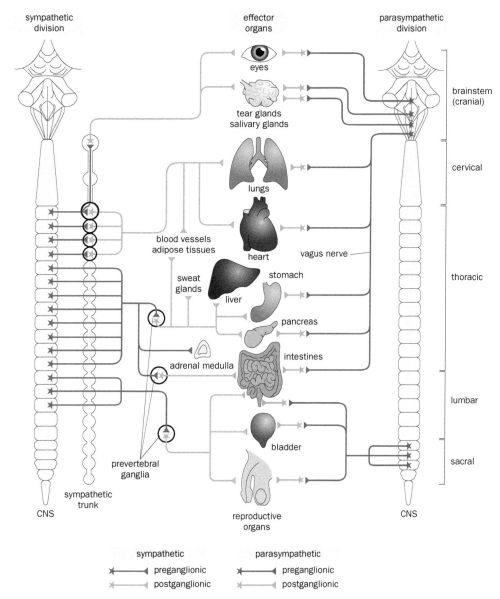

Figure 9-2 Organization of the sympathetic and parasympathetic systems. In most cases, the sympathetic system (red) and parasympathetic system (blue) regulate the actions of the same effector organs, often with opposing effects. In the sympathetic system, preganglionic neurons (dark red) are located in the thoracic and lumbar segments of the lateral spinal cord; postganglionic neurons (light red) are in the sympathetic trunk or prevertebral ganglia. In the parasympathetic system, preganglionic neurons are located in the brainstem and sacral segments of the spinal cord, while postganglionic neurons are located near the effector organs they control. Many of the parasympathetic preganglionic axons, such as those that exit the brainstem and constitute part of the vagus nerve, project long distances to their target neurons. Some effectors, such as blood vessels, brown and white adipose tissues, and sweat glands, receive only sympathetic input. Sympathetic preganglionic neurons directly innervate the adrenal medulla and activate endocrine cells (the equivalent of "postganglionic neurons") to release epinephrine and norepinephrine into the blood. Only a subset of the effectors of the autonomic system is depicted here. (Based on Jänig W [2006] The Integrative Action of the Autonomic Nervous System.)

on heart rate). Neuropeptides are also widely used to modulate the actions of the sympathetic and parasympathetic systems.

Like the motor system (Figure 8-2), the autonomic nervous system is also regulated at multiple levels by sensory feedback and central control (**Figure 9-3**). In the following sections, we will discuss this regulation in more detail.

9.2 The interoceptive system provides feedback about states of internal organs

The autonomic system receives feedback regulation from two distinct groups of **visceral sensory neurons** (spinal and brainstem) in the interoceptive system.

Figure 9-3 Multilayered control of the autonomic nervous system. Arrows indicate the direction of information flow through direct synaptic connections. Additional multisynaptic connections are not shown.

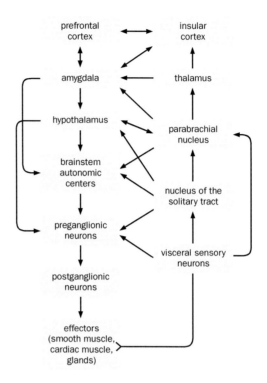

Both groups send peripheral axons (**visceral afferents**) to internal organs, where they are activated by mechanical and chemical stimuli. Spinal visceral sensory neurons are located in the dorsal root ganglia along with somatosensory neurons and projects central axons to the spinal cord (Figure 6-63). Their central axons can synapse onto sympathetic preganglionic neurons directly or via spinal interneurons for local feedback control of autonomic output. They also transmit information via spinal projection neurons to the **parabrachial nucleus** in the brainstem directly or via intermediate neurons in the **nucleus of the solitary tract** (**NTS**) of the caudal brainstem (Figure 9-3). Brainstem visceral sensory neurons are located in the petrosal, nodose, and jugular ganglia associated with cranial nerves (these ganglia can be considered analogous to dorsal root ganglia), and their central axons terminate mostly in the NTS (**Figure 9-4**A). Determining the division of labor between spinal and brainstem visceral sensory neurons is still a work in progress. In the following, we study regulation of blood pressure, breathing, and food intake to illustrate some of the functions they serve.

Increasing blood pressure triggers the *baroreflex*, which decreases heart rate, cardiac output, and peripheral resistance via vasodilation (widening of blood vessels). Blood pressure is sensed by peripheral terminals of mechanosensitive petrosal and nodose ganglion neurons called baroreceptors in the carotid sinus and aortic arch, respectively (Figure 9-4B, left). As we learned in Section 6.28, the mechanosensitive channel Piezo2 play essential roles in mechanotransduction in the somatosensory system. Conditional knockout experiments indicated that Piezo1 and Piezo2 act together in the nodose ganglion neurons innervating the aortic arch and the petrosal ganglion neurons innervating the carotid sinus to sense blood pressure. In transgenic mice expressing channelrhodopsin (ChR2) in Piezo2+ neurons, optogenetic activation of the glossopharyngeal nerve branch innervating the carotid sinus and the vagus nerve branch innervating the aortic arch reduced blood pressure and heart rate (Figure 9-4B, right), presumably through activation of NTS neurons, which in turn regulate autonomic output (Figure 9-3).

Breathing receives feedback control through both mechanical and chemical signals. When the lung is inflated, activation of stretch receptors triggers the *Hering–Breuer reflex*, which inhibits inspiration and reduces respiration rate (see Box 8-1 for a discussion of neural control of breathing). The lung is innervated

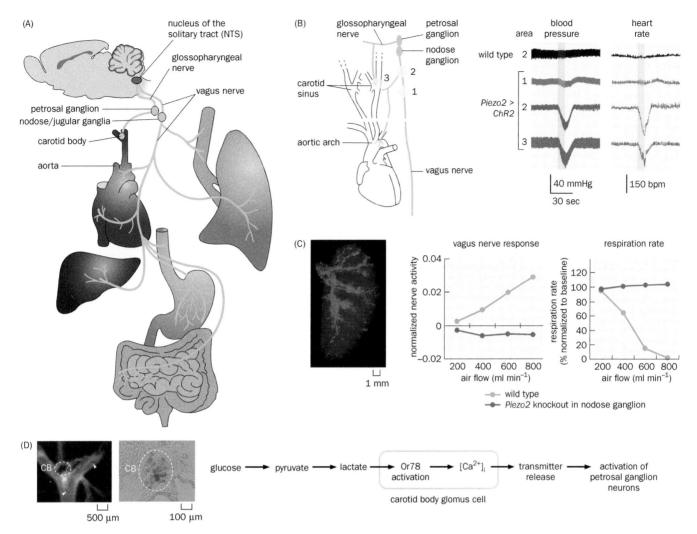

Figure 9-4 Organization of the brainstem visceral sensory system and its function in regulating blood pressure and breathing. **(A)** Schematic of the brainstem visceral sensory system. Visceral sensory neurons in the petrosal and jugular/nodose ganglia are associated with the glossopharyngeal and vagus nerves, respectively. Their peripheral axons receive signals from internal organs, and their central axons send these signals to the NTS. **(B)** Blood pressure is sensed by Piezo1/2-expressing neurons in the petrosal and nodose ganglia innervating the carotid sinus and aortic arch, respectively. In transgenic mice expressing ChR2 in Piezo2+ neurons (*Piezo2>ChR2*), optogenetic activation of the glossopharyngeal nerve branch that collects input from the carotid sinus (site 3) or the vagus nerve branch that collects input from the aortic arch (site 2), but not the vagus nerve before the joining of the aortic arch branch (site 1), causes reductions in blood pressure and heart rate. Stimulation at site 2 in wild-type mice not expressing ChR2 has no effect. mmHg, milliliter of mercury, a unit of pressure; bpm, beats per minute. **(C)** Left, distribution of visceral afferents in the lung

from a genetically defined subset of nodose ganglion neurons. Middle and Right, lung inflation (increasing air flow shown on the x axis) increases vagus nerve firing and triggers a reduction in respiration rate in wild-type mice but not in mice lacking *Piezo2* in nodose ganglion neurons. **(D)** Left, a reporter co-expressed with Or78 highlights glomus cells in the carotid body (CB) in low and high magnification images. Right, a summary of hypoxia-sensing mechanisms. Hypoxia induces a switch in glucose metabolism from the mitochondrial pathway to the glycolysis pathway using lactate as an intermediate. Lactate activates Or78 in carotid body glomus cells, leading to activation of petrosal ganglion neurons. (B, from Zeng WZ, Marshall KL, Min S, et al. [2018] *Science* 362:464–467. With permission from AAAS. C, from Chang RB, Strochlic DE, Williams EK, et al. [2015] *Cell* 161:622–633 and Nonomura K, Woo SH, Chang RB, et al. [2017] *Nature* 541:176–181. With permission from Elsevier Inc. and Springer Nature. D, from Chang AJ, Ortega FE, Riegler J, et al. [2015] *Nature* 527:240–244. With permission from Springer Nature.)

by visceral afferents of a subset of nodose ganglion neurons (Figure 9-4C, left). When *Piezo2* was knocked out from nodose ganglion neurons, lung inflation no longer activated the vagus nerve or inhibited inspiration (Figure 9-4C, right panels). These experiments indicated that nodose ganglion neurons sense air flow in the lung via the mechanosensitive channel Piezo2.

A reduction in blood oxygen level stimulates breathing within seconds. Blood oxygen level is sensed by glomus cells in the carotid body adjacent to but distinct from the carotid sinus discussed earlier (Figure 9-4A). One mechanism for hypoxia-induced increases in respiration rate involves Or78, a member of the olfactory receptor family whose known function was to detect odorants (Section 6.3). Or78 is expressed in glomus cells (Figure 9-4D, left). Hypoxia decreases activity in mitochondrial metabolism and increases production of the metabolite lactate, which binds Or78, depolarizing glomus cells. This causes Ca^{2+} entry into glomus cells, triggering neurotransmitter release, which activates visceral afferents of petrosal ganglion neurons, transmitting information to the brainstem (Figure 9-4D, right).

The gastrointestinal system provides information about food and nutrients essential for regulation of eating (to be discussed in detail later in the chapter). In general, the stomach provides mechanosensory input indicating emptiness or fullness, while the intestine provides mechanosensory input and chemosensory information about nutrients, irritants, and resident microorganisms. Mechanical stimuli are sensed by mechanosensitive cells residing along the gastrointestinal tract or directly by visceral afferents. Chemical stimuli are sensed mostly by **enteroendocrine cells** along the gastrointestinal tract, so named for their ability to release hormones. Notable hormones include **cholecystokinin (CCK)** and serotonin. CCK is an appetite-reducing neuropeptide which we will study further in the context of eating regulation. Serotonin is released by the enterochromaffin cell, a type of enteroendocrine cell, and regulates digestive functions in part through the enteric nervous system. Both CCK and serotonin also act as neurotransmitters on spinal and brainstem visceral afferents, which transmit information to the brain.

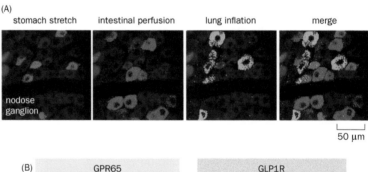

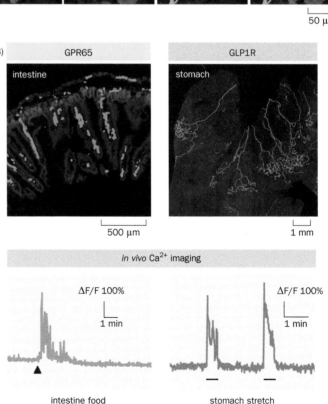

Figure 9-5 Feedback signals from the gastrointestinal tract. (A) *In vivo* Ca^{2+} imaging in the nodose ganglion revealed that different stimuli (labeled on the top) activate distinct subsets of visceral sensory neurons. Cells with fluorescent signals above a threshold are colored according to the sensory stimuli that activate them. In this experiment, a genetically encoded Ca^{2+} indicator (Section 14.22) was expressed in all excitatory neurons. **(B)** Nodose ganglion neurons expressing GPR65 and GLP1R, two G-protein-coupled receptors, send peripheral axons to intestinal villi (green) and stomach muscle (magenta), respectively (top row), and are selectively activated by intestinal food and stomach stretching, respectively, as assayed by *in vivo* Ca^{2+} imaging (bottom row; arrowhead, stimulus onset; horizontal lines, stimulus duration). In these experiments, *GPR65-Cre* and *GLP1R-Cre* drove expression of a Cre-dependent axonal tracer or a genetically encoded Ca^{2+} indicator. (From Williams EK, Change RB, Strochlic DE, et al. [2016] *Cell* 166:209–221. With permission from Elsevier Inc.)

How is visceral information organized at the level of sensory neurons? Exciting recent advances enable genetic dissection of visceral sensory neuron heterogeneity. *In vivo* Ca^{2+} imaging revealed that nodose ganglion neurons activated by distinct mechanical and chemical stimuli are spatially intermingled (**Figure 9-5**A). By generating knock-in mice expressing the Cre recombinase in specific neuronal subpopulations based on expression patterns of endogenous genes (see Sections 14.9–14.11 for the general strategy), researchers can trace axonal projection patterns, record physiological responses, and manipulate activity to examine the function of each subpopulation. For example, nodose ganglion neurons expressing GPR65 and GLP1R, two G-protein-coupled receptors, send their peripheral axons, respectively, to the intestinal lumen and stomach muscle (Figure 9-5B, top) and their central axons to distinct regions of the NTS. While GPR65$^+$ neurons are activated by intestinal food, GLP1R$^+$ neurons are activated by stomach stretching (Figure 9-5B, bottom). These and other recent data suggest a "labeled line" organization (see Chapter 6), wherein each neuronal subpopulation sends a specific kind of sensory information to the brain.

In summary, the interoceptive system employs a variety of receptors, cells, and parallel pathways to sense mechanical and chemical signals in internal organs, mirroring the organization of the somatosensory system (see Chapter 6). In some cases, peripheral fibers of visceral sensory neurons directly detect sensory stimuli, as in the case of baroreceptors sensing blood pressure and lung inflation. In other cases, peripheral fibers are postsynaptic to specialized chemosensory cells, such as glomus cells and enteroendocrine cells, which sense blood oxygen level and gut nutrients, respectively. These interoceptive signals converge in the NTS and parabrachial nucleus of the brainstem, which in turn provide feedback to autonomic output at multiple levels and send signals to higher brain centers including the hypothalamus, thalamus, and neocortex (Figure 9-3). In later sections of this chapter, we will encounter interoceptive cells in the hypothalamus that sense temperature and blood osmolarity. Altogether, the interoceptive system senses body states and provides feedback signals to stabilize vital functions.

9.3 The hypothalamus regulates diverse homeostatic processes

Ascending interoceptive information eventually reaches the **insular cortex**, which is located ventral and internal to the tongue-representing somatosensory cortex in humans (Figure 1-25). As we learned in Chapter 6, the insular cortex also receives input from the taste (Figure 6-33) and pain (Figure 6-71) systems. It interacts extensively with the prefrontal cortex, and together they constitute the cerebral control center of the autonomic nervous system (Figure 9-3).

Central control of the autonomic nervous system also involves two other important structures, the **amygdala** and the **hypothalamus**. The amygdala, an almond-shaped structure located in the ventral forebrain, receives extensive input from the thalamus, insular cortex, and prefrontal cortex, and provides feedback to the cortex and descending input to brainstem autonomic control centers and the hypothalamus. One of the best studied functions of the amygdala is its regulation of physiological responses to fear and emotional states. We will study the function and circuit organization of the amygdala in Chapter 11 in the context of emotion-related signal processing and memory. For now, we focus on the hypothalamus, a key regulator of the sympathetic and parasympathetic systems, and a key coordinator of hormone secretion.

Named for its location beneath the thalamus, the hypothalamus is made up of many discrete nuclei, each with specialized functions (**Figure 9-6**A). These functions include control of (1) energy balance through regulation of eating, digestion, and metabolic rates; (2) blood pressure and electrolyte composition through regulation of drinking and salt intake; (3) reproduction through hormonal regulation of sexual maturation, mating, pregnancy, lactation and other parental behaviors; (4) body temperature through various thermoregulatory processes; (5) emergency responses through stress hormone secretion and sympathetic activation; and (6) circadian rhythms and sleep.

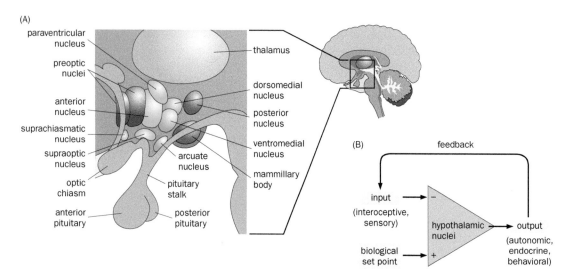

Figure 9-6 Organization and homeostatic functions of the hypothalamus.
(A) Schematic of the hypothalamus and pituitary from a sagittal perspective, depicting only the subset of nuclei close to the midsagittal plane. The functions of some of these nuclei will be discussed in the remainder of this chapter and in Chapter 10. For example, the arcuate nucleus, paraventricular nucleus, and lateral hypothalamus (not shown) regulate food intake and energy balance.

The suprachiasmatic nucleus regulates circadian rhythms, and the preoptic nuclei and ventromedial nucleus regulate sexual behavior.
(B) In this homeostatic model, the magnitude of hypothalamic output is proportional to the difference between the input and biological set point (symbolized by the amplifier). The net effect of the output is to reduce this difference, as indicated by the feedback arrow.

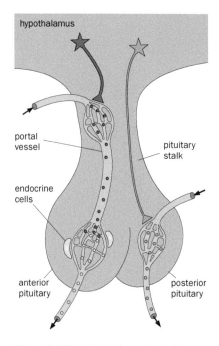

Figure 9-7 Hypothalamic control of hormone release through the pituitary.
Oxytocin- and vasopressin-producing neurons (blue) release hormones (blue dots) directly to the bloodstream in the posterior pituitary. Other hypothalamic neurons (red) release prehormones (red dots) that circulate through the portal vessel to the anterior pituitary, where they activate endocrine cells (yellow) to release hormones (yellow dots) that circulate throughout the body. Arrows indicate the direction of blood flow.

The hypothalamus is a collection of parallel circuits that regulate each of these basic bodily functions. Some hypothalamic nuclei receive input from sensory systems that provide information about the environment, such as light from the visual system to control circadian rhythms and odors and pheromones from the main and accessory olfactory systems to indicate the presence of potential mates or predators. A large fraction of the input received by hypothalamic nuclei is interoceptive, reflecting the state of the body. Some hypothalamic cells are interoceptive themselves. For example, hypothalamic cells that directly sense body temperature have been reported, and we will discuss hypothalamic cells that sense blood osmolarity in more detail in Section 9.9 in the context of regulation of drinking.

A key mechanism utilized by hypothalamic circuits is **homeostasis**, which refers to coordinated physiological processes that maintain the steady state of an organism. A biological set point, to borrow a term from control system engineering, refers to the target state of physiological parameters such as blood pressure, body temperature, and nutritional levels. Input to hypothalamic nuclei provides quantitative indicators of these physiological states. Differences between these values and biological set points trigger negative feedback in the form of activation of hypothalamic nuclei that control autonomic, behavioral, and endocrine output, thereby reducing these differences (Figure 9-6B). Such homeostatic regulation maintains physiological states within narrow ranges around their biological set points. For example, if our body temperature rises above 37°C due to intense exercise, the hypothalamus sends signals to the sympathetic system to enhance sweating, which cools the body via evaporation, and vasodilation, which increases heat loss via enhanced blood flow to the skin. The hypothalamus also drives thermoregulatory behaviors, causing us to feel hot so that we seek a cool place or remove clothing, thus facilitating a return to normal body temperature.

9.4 The hypothalamus and pituitary regulate hormone secretion

In parallel to its output to downstream neural circuits (Figure 9-3), the hypothalamus also controls hormone secretion from the nearby **pituitary**, the endocrine center of the brain. The hypothalamus employs two distinct mechanisms (**Figure 9-7**). Some hypothalamic neurons produce the peptide hormones **oxytocin** and

Table 9-1:   Hormones released from hypothalamic neurons and pituitary endocrine cells

Hormone		Function
Direct hormone release by hypothalamic neurons at the posterior pituitary		
Oxytocin		Regulates maternal and social behavior
Vasopressin		Regulates water balance; regulates social behavior
Stimulatory action of hypothalamic neurons on anterior pituitary endocrine cells		
Hypothalamus	*Anterior pituitary*	
Corticotropin-releasing hormone (CRH)	Adrenocorticotropin (ACTH)	Stimulates glucocorticoid release from the adrenal cortex and regulates responses to stress
Gonadotropin-releasing hormone (GnRH)	Luteinizing hormone (LH); follicle-stimulating hormone (FSH)	stimulates production of sex hormones, sexual maturation, and sexual behavior
Growth hormone-releasing hormone (GHRH)	Growth hormone (GH)	Stimulates growth
Thyrotropin-releasing hormone (TRH)	Thyroid-stimulating hormone (TSH, also called thyrotropin); prolactin	Stimulates metabolism; stimulates milk production
Inhibitory action of hypothalamic neurons on anterior pituitary endocrine cells		
Hypothalamus	*Anterior pituitary*	
Somatostatin	Growth hormone	
Dopamine	Prolactin	

vasopressin and release these hormones from their axon terminals in the **posterior pituitary** directly into the bloodstream for systemic circulation. (Oxytocin and vasopressin are also released from synaptic terminals elsewhere as peptide neurotransmitters that signal to postsynaptic neurons; Box 10-3.) Other hypothalamic neurons release prehormones into local blood circulation. These prehormones, carried by specialized portal vessels through the pituitary stalk, travel to the **anterior pituitary**, where they activate endocrine cells, inducing them to release their hormones, which then circulate throughout the body. Some hypothalamic neurons also release signaling molecules, such as somatostatin and dopamine, that inhibit hormone release by endocrine cells in the anterior pituitary. Hormones from the anterior and posterior pituitary regulate a variety of physiological functions, including growth, metabolism, reproduction, and responses to stress (Table 9-1).

The issue of how specific hypothalamic circuits maintain homeostasis can be divided into several questions: How are inputs sensed by hypothalamic nuclei? How are biological set points determined? How are differences between set points and physiological states compared and translated into output? In the rest of this chapter, we will study how these regulatory mechanisms control eating, drinking, sleep, and circadian rhythms; we will revisit these questions in Chapter 10 in the context of sexual behavior.

HOW ARE EATING AND DRINKING REGULATED?

Food and water are such essential ingredients of life that animals have evolved strong and innate behaviors to search for and consume food or water when they are hungry or thirsty. At the same time, the body also has negative feedback systems to stop eating or drinking when sufficient food or water has been consumed. In the following sections, we will study the neurobiological basis of these homeostatic regulations. We start with neural mechanisms that control eating, as much work has been done on this subject over many decades. We then study drinking as a comparison. We end with a discussion of how hunger and thirst, the two basic homeostatic need states, can powerfully drive motivated behavior.

9.5 Hypothalamic lesion and parabiosis experiments suggested that feedback signals from the body inhibit eating

We all experience this daily: when we are hungry, we crave food; after eating, we lose interest in the same food we craved while hungry. One possible explanation for this phenomena is that feedback signals from the body to the brain regulate our desire to eat. Experiments in a wide variety of animal species have shown that while starvation causes body weight to decrease and forced overfeeding causes body weight to increase, if food is provided *ad libitum* afterward (that is, if animals have free access to food and can eat as much or as little as they like), their body weights return to stable, pre-experimental values. When rats were fed food diluted with non-nutritious substances, the amount they ate corresponded not to the volume of the food but to its caloric content, such that their body weight was maintained. These striking examples of homeostasis suggested the existence of feedback signals relating to nutritional status that control eating. Unlike normal rats, however, rats with lesions in the ventromedial hypothalamus ate more when food was provided *ad libitum*, leading them to become obese. These experiments suggested that the ventromedial hypothalamus regulates eating. (Note that the critical site in these experiments was the *arcuate nucleus*, which we will study in detail later, not the *ventromedial hypothalamic nucleus*, which does not regulate food intake.) But how is eating regulated such that normal rats maintain their body weights?

Initial insight into this question came from a **parabiosis** experiment in which two rats were surgically joined at the abdomen when young, such that they could exchange signaling molecules that circulated in their blood during the rest of their lives together. After successful surgeries, parabiotically linked rats grew well and were healthy (**Figure 9-8**A). The ventromedial hypothalamus was then lesioned in one rat of the pair. The lesioned rat started to eat more food and became obese. Remarkably, its parabiotic partner lost interest in eating even when the food was placed under its nose; as a consequence, the non-lesioned partner lost weight and became emaciated over time. Postmortem analysis of weight and fat content revealed that the partners with hypothalamic lesions had dramatically increased body weight and fat content, as did single rats with similar lesions. By contrast, the non-lesioned partners had lost weight and had dramatically reduced fat content (Figure 9-8B). The simplest interpretation of these results is that the lesioned rat lost the ability to respond to negative feedback signals from the body and therefore kept eating, causing obesity. Excessive food intake led to continual production of negative feedback signals that, circulating through the blood, instructed the non-lesioned partner with a normal hypothalamus to stop eating, leading to its starvation.

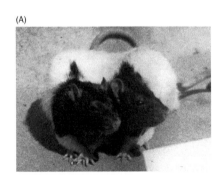

(A)

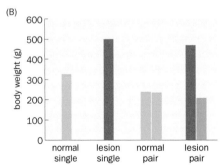

(B)

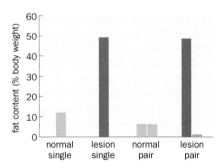

Figure 9-8 Parabiosis experiments suggest that feedback signals from the body inhibit eating. (A) A pair of parabiotic rats. The abdomens of these rats were surgically joined when they were young so that they could exchange signaling molecules via their bloodstreams. **(B)** Postmortem analysis of average body weight and fat content for four experimental conditions: normal single rats (green), single rats with lesions in the ventromedial hypothalamus (red), parabiotic pairs in which neither partner is lesioned (green/green), and parabiotic pairs in which one of the partners has a lesion in the ventromedial hypothalamus (red) and the other does not (blue). Both single and partnered rats with lesions overeat, leading to dramatic gains in weight and fat content. In parabiotic pairs, a hypothalamic lesion in one partner causes the non-lesioned partner to refuse food and starve, as indicated by the drastic loss of fat content. (From Hervey GR [1959] *J Physiol* 145:336–352. With permission from John Wiley & Sons Inc.)

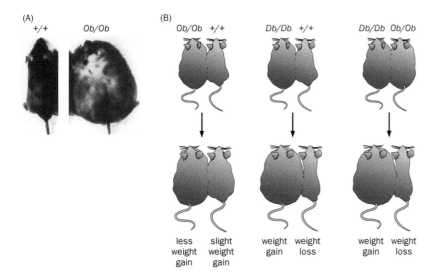

Figure 9-9 Parabiosis experiments with mutant mice that exhibit obesity. (A) At age 10 months, a normal mouse (left) weighed 29 grams. A mouse homozygous for the *Ob* mutation (right), which arose spontaneously in the Jackson Laboratory, a mouse repository center, weighed 90 grams. **(B)** Summary of three parabiosis experiments. The consequences of parabiotic pairing for each individual are shown below. Collectively, these experiments suggest that *Ob* might produce a feedback signal, and *Db* might be necessary for reception of the signal. (A, from Ingalls AM, Dickie MM, & Snell GD [1950] *J Hered* 41:317–318. With permission from Oxford University Press. B, based on Coleman DL & Hummel KP [1969] *Am J Physiol* 217: 1298–1304 and Coleman DL [1973] *Diabetologia* 9:294–298.)

9.6   Studies of mutant mice led to the discovery of the leptin feedback signal from adipose tissues

Clues to the nature of feedback signals came from studies of two spontaneously occurring mutant strains of mice, *Ob* (for *Obese*) and *Db* (for *Diabetic*). Mice homozygous for each of these mutations displayed almost identical obesity phenotypes (**Figure 9-9**A), but their corresponding genes were mapped to different genetic loci. Important insight regarding the nature of these mutations came again from a series of parabiosis experiments (Figure 9-9B). When *Ob/Ob* (homozygous for the *Ob* mutation) and wild-type (+/+) mice were joined parabiotically, *Ob/Ob* mice ate less and became less obese than non-parabiotic *Ob/Ob* mice, and +/+ mice gained more weight than non-parabiotic +/+ mice. When *Db/Db* and +/+ mice were joined parabiotically, *Db/Db* mice gained weight similarly to that of non-parabiotic *Db/Db* mice, whereas +/+ mice stopped eating and became emaciated, like the partners of the ventromedial hypothalamus–lesioned rats. Finally, when parabiosis was carried out between *Db/Db* and *Ob/Ob* mice, *Db/Db* mice kept gaining weight, whereas *Ob/Ob* mice stopped eating and lost weight (Figure 9-9B).

These remarkable results led researchers to propose that the normal *Ob* gene encodes a circulating negative feedback signal to the brain to inhibit eating, whereas the normal *Db* gene encodes a receptor in the hypothalamus that responds to this signal. Thus, when *Ob/Ob* mice were parabiotically linked with +/+ mice, the signal supplied by the +/+ partner helped the *Ob/Ob* partner control its eating and body weight; both *Ob/Ob* and +/+ mice gained weight because both partners shared the signal made by just the +/+ partner. When *Db/Db* mice were parabiotically linked to +/+ or *Ob/Ob* mice, *Db/Db* partners could not respond to the negative feedback signal and therefore kept eating. Continued production of the negative feedback signal inhibited +/+ and *Ob/Ob* partners from eating, leading to starvation.

This hypothesis was tested when a molecular-genetic method called **positional cloning** (Section 14.6) was used to identify the gene corresponding to the *Ob* mutation. The *Ob* gene was found to encode a secreted protein, named **leptin** (after the Greek *leptos,* meaning "thin"), which is produced specifically by fat tissue (**Figure 9-10**A) and which circulates throughout the body. When *Ob/Ob* mice were injected with leptin produced *in vitro* from the cloned cDNA, their food intake and body mass were markedly reduced (Figure 9-10B). Injection of leptin into normal mice also caused them to eat less and become thin. Shortly after the cloning of the *Ob* gene, its receptor was identified and found to correspond to the *Db* gene. The leptin receptor signals mainly through regulation of gene expression and is highly expressed in the ventromedial hypothalamus, in particular the **arcuate nucleus** (Figure 9-4A), which lesion studies had implicated in controlling

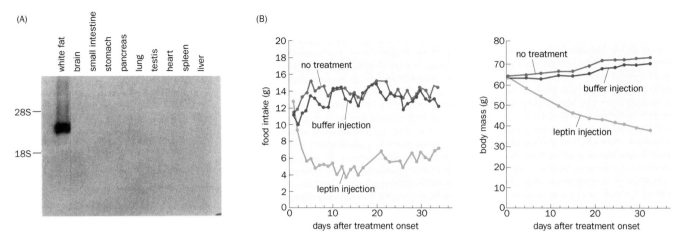

Figure 9-10 Leptin as a feedback signal that controls food intake. **(A)** A northern blot (Section 14.12) shows that *Leptin* mRNA is specifically produced in white fat but not in other tissues tested. 28S and 18S ribosomal RNAs are molecular weight markers. **(B)** Compared to no treatment (red traces) or buffer injection (purple traces), daily injection of 5 μg leptin per gram of body weight (blue traces) into *Ob/Ob* mice results in drastic reductions in food intake (left) and body mass (right). (A, from Zhang Y, Proenca R, Maffei M, et al. [1994] *Nature* 372:425–432. With permission from Springer Nature. B, from Halaas JL, Gajiwala KS, Maffei M, et al. [1995] *Science* 269:543–546.)

eating. The discoveries of leptin and its receptor provided satisfying evidence that adipose tissue produces a negative feedback signal for hypothalamic control of food intake and body weight. They also facilitated further studies of the neural mechanisms by which this is achieved (see below).

After the cloning of mouse *Leptin,* disruption of the human *Leptin* gene was found to cause a rare population of early-onset obesity cases. Daily injection of recombinant leptin markedly reduced weight gain (**Figure 9-11**) and improved the quality of life for these patients. However, most obese patients do not react to leptin treatment; in fact, serum leptin levels usually correlate with the degree of obesity. Understanding the cause of obesity requires a deeper understanding of how leptin and other feedback signals act in the nervous system to control eating behavior and energy balance, which we turn to next.

9.7 POMC neurons and AgRP neurons in the arcuate nucleus are key regulators of eating

The expression pattern of the leptin receptor encoded by the *Db* gene provided the first clues as to where leptin acts in the brain. The best-studied leptin target cells, which indeed regulate food intake, are two groups of intermingled neurons in the arcuate nucleus of the ventromedial hypothalamus. Leptin activates the first group of neurons, called **POMC neurons** because they express pro-opiomelano-cortin. POMC is a precursor protein that is cleaved to produce multiple neuropeptides, including the anorexigenic (appetite-reducing) α-melanocyte-stimulating hormone (**α-MSH**). α-MSH reduces food intake by activating the melanocortin-4 receptor (**MC4R**), a G-protein-coupled receptor expressed by target neurons. Disruption of the genes encoding POMC or MC4R leads to obesity in both mice and humans. Indeed, losing one copy of the MC4R gene is sufficient to cause obesity in humans, and MC4R disruption is the largest single genetic contribution to human obesity, accounting for up to 5% of severe obesity cases.

It is instructive to study how researchers established a link between leptin and POMC neurons in the arcuate nucleus in the context of body weight control. First, electrophysiological recordings of POMC neurons in hypothalamic slices *in vitro* showed increased firing of POMC neurons within minutes of leptin application (**Figure 9-12**A), indicating that leptin promotes POMC neuron activation. Second, the leptin receptor is highly expressed in POMC neurons, suggesting that leptin can act directly on POMC neurons. Third, while leptin application reduces the body weight of *Ob/Ob* mice (Figure 9-10B), it fails to do so in arcuate-lesioned *Ob/Ob*

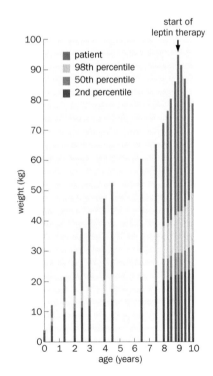

Figure 9-11 An example of leptin therapy in humans. A patient homozygous for a frameshift mutation in the *Leptin* gene developed early-onset obesity. She constantly felt hungry, demanded food, and was disruptive when denied food. At age nine, her body weight was about twice the 98th percentile level. Daily injection of recombinant leptin steadily reduced her body weight. (Based on Farooqi IS, Jebb SA, Langmack G, et al. [1999] *New Engl J Med* 341:879–884.)

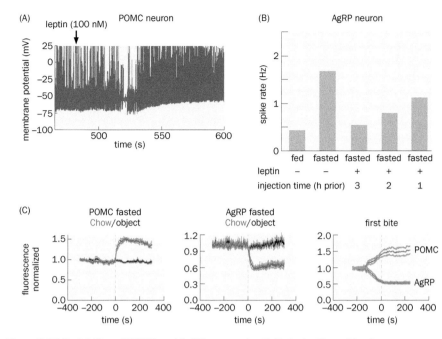

Figure 9-12 Regulation of POMC and AgRP neuronal activity by leptin and food sensory cues.
(A) Patch clamp recording of a POMC neuron in a hypothalamic slice from a transgenic mouse in which POMC neurons express GFP for their identification. *In vitro* leptin application increases action potential frequency. **(B)** Loose patch recording of fluorescently labeled AgRP neurons in hypothalamic slices prepared from mice that were fed, fasted (for 24 h), or fasted with leptin injection 3, 2, or 1 h before sacrifice. Fasting enhances the firing rate of AgRP neurons; leptin application suppresses this enhancement. **(C)** *In vivo* fiber photometry recording (Section 14.22) of arcuate POMC (left) or AgRP (middle) neurons expressing genetically encoded Ca^{2+} indicator GCaMP6 in fasted mice in response to presentation of food (mouse chow) or a non-food object. Activity of POMC neurons and AgRP neurons rapidly increases and decreases, respectively, when mice sense food. Recordings were aligned to presentation of food/object in the left and middle panels and to the first bite in the right panel. Black and red traces are the means, and gray represents mean ± standard error of the means. (A, from Cowley MA, Smart JL, Rubinsein M, et al. [2001] *Nature* 411:480–484. With permission from Springer Nature. B, from Takahashi KA & Cone RD [2005] *Endocrinology* 146:1043–1047. With permission from Oxford University Press. C, from Chen Y, Kin YC, Kuo TW, et al. [2015] *Cell* 160:829–841. With permission from Elsevier Inc.; see also Betley JN, Xu S, Cao ZFH, et al. [2015] *Nature* 521:180–185; Mandelblat-Cerf Y, Ramesh RN, Burgess CR, et al. [2015] *eLife* 4:e07122.)

mice, demonstrating that leptin reception in the arcuate nucleus is essential for body weight control. Fourth, conditional knockout of the leptin receptor in POMC neurons causes weight gain, revealing a causal relationship between leptin action on POMC neurons and body weight control. However, the body weight gain produced by deletion of the leptin receptor in POMC neurons is much milder than that produced by leptin receptor deletion in the entire animal. Together, these data indicate that, in addition to acting directly on POMC neurons, leptin must also act on other neurons in the arcuate nucleus (inferred from the third line of evidence) to regulate body weight.

The second group of arcuate nucleus neurons expressing the leptin receptor is the **AgRP neurons**, which release two orexigenic (appetite-stimulating) neuropeptides, the <u>ag</u>outi-<u>r</u>elated <u>p</u>rotein (AgRP) and <u>n</u>euro<u>p</u>eptide <u>Y</u> (NPY). In addition, AgRP neurons also release GABA. AgRP neurons are more active when animals are food-deprived, and their activity is inhibited by leptin application (Figure 9-12B). Loss- and gain-of-function experiments have further demonstrated the importance of AgRP neurons in regulating eating. In a loss-of-function approach, transgenic mice were created in which AgRP neurons selectively express a human receptor highly sensitive to bacterial diphtheria toxin. Application of diphtheria toxin in adult mice selectively killed AgRP neurons within days; this caused the mice to stop eating and rapidly lose body weight (**Figure 9-13**A). Thus, AgRP neurons are necessary for food intake. In a gain-of-function experiment, optogenetic

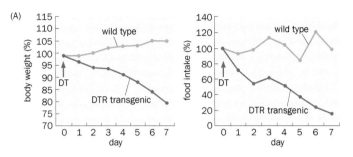

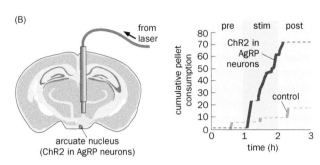

Figure 9-13 AgRP neurons promote eating. (A) Application of diphtheria toxin (DT) to wild-type mice (blue) does not affect body weight (left) or food intake (right). However, in transgenic mice (red) expressing the human diphtheria toxin receptor (DTR) in AgRP neurons, DT application selectively kills these neurons, resulting in a drastic reduction in body weight and food intake. **(B)** Left, adenoassociated virus encoding Cre-dependent channelrhodopsin (ChR2) was injected into the arcuate nucleus of transgenic mice expressing Cre in AgRP neurons, allowing optogenetic activation of AgRP neurons by blue laser light. Right, photostimulation (stim) leads to robust food pellet consumption during the stimulation period, compared to before (pre) and after (post) stimulation (dashed lines). The light trace shows a control mouse without ChR2 expression. (A, adapted from Luquet S, Perex FA, Hnasko TS, et al. [2005] *Science* 310:683–685. B, adapted from Aponte Y, Atasoy D, & Sternson SM [2011] *Nat Neurosci* 14: 351–355. With permission from Springer Nature.)

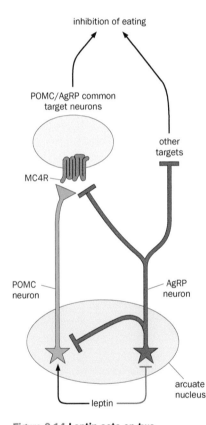

Figure 9-14 Leptin acts on two antagonistic populations of arcuate neurons to suppress eating. POMC neurons, which inhibit eating by activating the melanocortin-4 receptor (MC4R) in target neurons, are activated by leptin. Leptin also inhibits AgRP neurons, which promote eating via three different mechanisms: (1) inhibiting POMC neurons directly, (2) antagonizing POMC signaling on common target neurons through AgRP's competition with α-MSH produced by POMC neurons for binding the MC4R receptor, and (3) inhibiting additional target neurons that inhibit eating (see Figure 9-15 for details).

activation of channelrhodopsin-expressing AgRP neurons induced voracious eating within minutes (Figure 9-13B), regardless of whether the mice had just eaten or not. Thus, acute activation of AgRP neurons is sufficient to induce eating and can override inhibitory feedback signals from the body.

Given that their activities are modulated by leptin and other feedback signals from the body (see the next section), arcuate AgRP neurons and POMC neurons have long been considered interoceptive neurons that detect the nutritional state of the body to regulate eating and energy expenditure. A surprising finding from recent *in vivo* recording and imaging studies is that the activity of both neuron types is rapidly modulated by the sight or smell of food *before* onset of eating (Figure 9-12C). Thus, AgRP and POMC neuronal activity *anticipates* upcoming changes in the body's nutritional state, allowing adjustment of physiology and behavior more rapidly than feedback signals from the body after food consumption. This mode of regulation is conceptually similar to the forward model we introduced in the motor system (Section 8.11).

In summary, leptin acts on two intermingled neuronal populations within the arcuate nucleus: it activates POMC neurons, which inhibit eating, and inhibits AgRP neurons, which promote eating. Both contribute to leptin's appetite-reducing effect (**Figure 9-14**). (In addition to their opposing roles in regulating eating, POMC neurons and AgRP neurons promote and inhibit, respectively, energy expenditure; these effects also contribute to leptin's regulation of body weight.) AgRP and POMC neurons interact extensively: AgRP neurons directly inhibit POMC neurons through GABA and NPY release. Additionally, the AgRP peptide competes with α-MSH for MC4R binding, such that AgRP neurons antagonize the action of POMC neurons at their common targets. However, the function of AgRP neurons cannot be fully accounted for by their role in antagonizing POMC neurons. For example, neither the anorexigenic effect of ablating AgRP neurons nor the orexigenic effect of stimulating AgRP neurons was affected by blocking MC4R signaling. In addition, the action of AgRP neurons usually manifests at a time scale more rapid than that of POMC neurons. These observations suggest that AgRP neurons act on additional targets to exert their orexigenic effect through GABA and NPY (Figure 9-14). We discuss these targets in the next section.

9.8 Multiple neural pathways and feedback signals act in concert to regulate eating

Recent advances in circuit mapping as well as recording and manipulating activity of specific neuronal types (Sections 14.11, 14.18–26) have led to the identification of many brain regions that regulate eating (**Figure 9-15**). We begin our discussion with the arcuate AgRP neurons, which are GABAergic neurons that

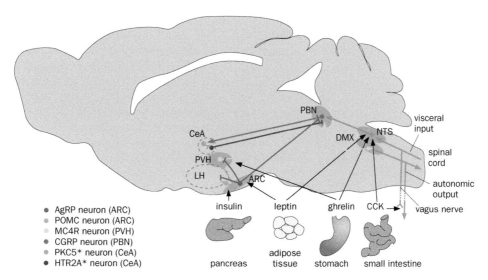

Figure 9-15 A simplified schematic of neural pathways involved in eating and their regulation by feedback signals from the body. Neurons are color coded (key in the bottom left), with red and green hues representing GABAergic and glutamatergic neurons, respectively. ARC, arcuate nucleus; PVH, paraventricular hypothalamic nucleus; PBN, parabrachial nucleus; DMX, dorsal motor nucleus of the vagus nerve; CeA, central amygdala; LH, lateral hypothalamus; NTS, nucleus of the solitary tract. LH and CeA are outlined in dashes because they are more lateral than the other brain regions represented in this midsagittal schematic of the mouse brain. Activation of ARC AgRP neurons and CeA HTR2A$^+$ neurons promotes eating, while activation of all other neurons shown here inhibits eating. Bottom, the body sends multiple feedback signals to the brain to control eating. Leptin and insulin reflect nutritional status (relating to fat and sugar, respectively), whereas ghrelin and cholecystokinin (CCK) reflect gastrointestinal status (signaling hunger and satiety, respectively). These signals act at multiple sites in the brain (arrows). (See Sternson SM & Eiselt AK [2016] *Annu Rev Physiol* 79:401–423 and Andermann ML & Lowell BB [2017] *Neuron* 95:757–778 for reviews.)

promote eating (Section 9.7). AgRP neurons project axons to multiple regions within and beyond the hypothalamus, inhibiting their target neurons. A critical target of AgRP neurons is the **paraventricular hypothalamic nucleus (PVH)**. Silencing PVH target neurons mimicked the effect of activating AgRP neurons in promoting voracious eating, while activating PVH target neurons mitigated the effects of AgRP neuron activation. A key group of AgRP target PVH neurons express MC4R and are also target of arcuate POMC neurons. These MC4R$^+$ PVH neurons promote satiety and project to brainstem targets, notably caudal medulla nuclei, which send parasympathetic output to regulate digestion.

A second target of AgRP neurons is the parabrachial nucleus (PBN) in the pons, which receives diverse somatic and visceral sensory input (Figure 9-3) as well as input from MC4R$^+$ PVH neurons discussed earlier (Figure 9-15). A specific population of glutamatergic PBN neurons that express the neuropeptide **calcitonin gene-related peptide (CGRP)** is activated by consumption of large meals. Silencing CGRP$^+$ PBN neurons increases meal size, suggesting that their activity promotes meal termination. (Thus, inactivating these neurons by AgRP directly or through MC4R$^+$ PVH neurons may contribute to the appetite-promoting effect of AgRP neurons.) A major target of CGRP$^+$ PBN neurons is the central amygdala (CeA), where CGRP$^+$ PBN neurons likely activate a population of GABAergic neurons that express protein kinase C δ isoform (PKCδ). Activation of PKCδ$^+$ CeA neurons inhibits eating. This effect is likely mediated by their inhibition of another group of CeA GABAergic neurons marked by expression of the serotonin receptor HTR2A, activation of which promotes eating. HTR2A$^+$ CeA neurons project back to the PBN (Figure 9-15).

While our journey so far has covered just a subset of the neuronal populations and connections known to regulate eating, several insights have emerged: (1) eating is regulated by neuronal populations distributed in multiple brain regions, (2) neurons that promote or inhibit eating interact extensively, and (3) a specific brain region often contains discrete populations with opposing functions (e.g., AgRP and POMC neurons in the arcuate nucleus; PKCδ$^+$ and HTR2A$^+$ neurons in the CeA). Another prominent brain region that has long been implicated in eating regulation is the **lateral hypothalamus**, which also receives direct input from arcuate AgRP and POMC neurons (Figure 9-15). Recent studies suggest that lateral hypothalamus GABA and glutamate neurons promote and inhibit eating, respectively. How neurons in these different brain regions cooperate to positively and negatively regulate different aspects of eating is an intense area of research.

We introduced eating circuits from a leptin-centric view in Section 9.7, but the body also produces a multitude of feedback signals that travel to the brain to regulate eating (Section 9.2). Some of these signals are hormones that circulate in the blood and act on some of the same neurons and circuits we have studied so far (Figure 9-15, bottom). For example, the pancreas produces **insulin** in response to rising blood glucose levels after meals. Leptin and insulin levels reflect the nutritional status of the body, and their actions in controlling food intake involve relatively long time scales. Short-term signals from the gastrointestinal system stimulate or inhibit food intake before, during, or after each meal; of these, the two best-studied are the neuropeptides **ghrelin** and **cholecystokinin** (**CCK**) (Figure 9-15). Ghrelin is mainly produced by stomach-associated glands in response to reduced glucose levels and acts as a hunger signal to stimulate eating at least in part by directly stimulating AgRP neurons. CCK is produced by enteroendocrine cells in the intestine in response to rises in fatty acid, amino acid, and sugar concentrations, and acts as a satiety signal to inhibit eating. A notable target for these short-term signals is the nucleus of the solitary tract (NTS) in the caudal medulla, which is a key source of excitatory input to the PBN. In addition to acting as a hormone, CCK also acts as a neurotransmitter on visceral afferents, transmitting signals through the interoceptive system to the NTS. Finally, mechanical signals from the gastrointestinal system (e.g., relaying stomach stretch) act via the vagus nerve to contribute to short-term feedback regulation of eating. The effects of these short-term signals can be influenced by long-term signals. For instance, leptin can act directly through its receptor on NTS neurons, or indirectly through its action on hypothalamic neurons, to increase the sensitivity of the NTS to the CCK satiety signal, such that when the nutritional level is high, animals are more likely to terminate their meals in response to satiety signals.

In summary, we have seen that the seemingly simple acts that we perform effortlessly multiple times a day—the initiation and conclusion of eating—are controlled by diverse signals and interacting neural pathways. Some of these pathways can compensate for one another. For instance, despite the importance of arcuate AgRP neurons, ablating them during early development has little effect, compared to the severe effects of ablating them in adult mice (Figure 9-13A). Even in the case of starvation induced by AgRP neuronal ablation in adults, transient rescue by GABA agonist application in the PBN for several days can lead to permanent rescue after GABA agonist withdrawal, likely by providing enough time to activate alternative pathways. Likewise, knocking out the leptin receptor in arcuate AgRP neurons from four-week-old mice results in substantially increased body weight gain afterward, but similar knockout early in development has much milder effects. These studies highlight the flexibility of the nervous system: alternative pathways can be employed to compensate for lost neurons or genes. They underscore the importance of eating in the life of an animal—multiple and partially redundant pathways have evolved to ensure its proper control.

Investigating the molecular and neural mechanisms of eating control has shed light on obesity. Obesity is a pressing health problem in modern societies, as it is a predisposing factor for many diseases and has become increasingly prevalent in recent decades. As the example of human mutations in leptin illustrates, genetic makeup contributes significantly to obesity. After all, eating is a basic drive essential for life. For our hunter-gatherer ancestors in the not-so-distant past, genotypes that generated less effective feedback to suppress eating might have promoted greater nutrient storage during times of food scarcity and have thus been favored by natural selection. In an age of food abundance and sedentary lifestyle, genetic compositions that had been advantageous may now predispose individuals to obesity. As the story of leptin shows, understanding the basic mechanisms of eating regulation is key to bringing obesity under control.

9.9   A hypothalamic thirst circuit controls drinking by integrating signals of homeostatic need and rapid sensory feedback

Just like hunger triggers eating to maintain energy homeostasis, thirst triggers drinking to maintain electrolyte homeostasis. In parallel with the discovery of brain centers that regulate food intake, decades of work using lesion, extracellular recording, and microstimulation methods have identified three brain regions in the anterior hypothalamus, collectively known as the **lamina terminalis**, as key regulators of water intake (**Figure 9-16**). Recent studies have begun to uncover specific neuronal populations and circuit mechanisms that regulate drinking.

Dehydration is sensed by two nuclei in the lamina terminalis: the **subfornical organ** (**SFO**) and the organum vasculosum of the lamina terminalis (**OVLT**). Unlike most neurons in the brain, which do not have direct access to substances circulating in the blood because of the **blood–brain barrier** (**BBB**; see Box 12-1 for more detail), SFO and OVLT neurons are outside the BBB. Thus, they can directly sense blood osmolarity as well as angiotensin II, a hormone circulating in the blood. The blood angiotensin II level and osmolarity are both elevated by dehydration, which activate specific populations of interoceptive SFO and OVLT neurons. Optogenetic activation of excitatory neurons within the SFO caused robust drinking even when mice were water satiated (**Figure 9-17**A), indicating that activation of these dehydration-sensing neurons is sufficient to promote drinking.

Key downstream targets of SFO and OVLT dehydration-sensing neurons are the excitatory neurons in the **median preoptic nucleus** (**MnPO**), the third nucleus of the lamina terminalis. Lesion experiments indicated that, of all nuclei, ablation of the MnPO resulted in the most severe symptoms: rats stopped drinking water even though they were dehydrated (Figure 9-17B). MnPO excitatory neurons act downstream of SFO excitatory neurons, as their ablation blocked the effects of SFO optogenetic stimulation-induced drinking. Optogenetic activation of excitatory MnPO neurons also caused robust drinking in water-satiated mice, as did optogenetic activation of their axon terminals in several hypothalamic and thalamic targets. Thus, MnPO excitatory neurons integrate input from the SFO and OVLT and broadcast the thirst state to multiple target regions (Figure 9-16).

Two key targets of MnPO neurons, which also receive direct input from the SFO and OVLT, are the supraoptic nucleus and the paraventricular hypothalamic nucleus (Figure 9-16). Both nuclei contain neurons that produce the neuropeptide **vasopressin**, which is released into the blood from their axon terminals in the posterior pituitary (Figure 9-16; Section 9.4). Vasopressin increases water retention from the kidney and thus helps water preservation.

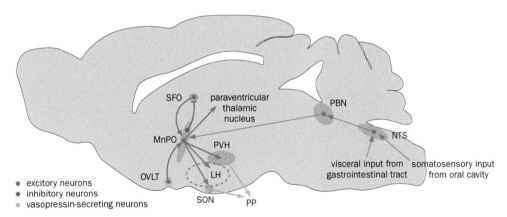

Figure 9-16 Neural circuits that regulate drinking. SFO, subfornical organ; OVLT, organum vasculosum of the lamina terminalis; MnPO, median preoptic nucleus; SON, supraoptic nucleus; PVH, paraventricular hypothalamic nucleus; LH, lateral hypothalamus, PP, posterior pituitary; PBN, parabrachial nucleus; NTS, nucleus of the solitary tract. Thick arrows represent pathways that are activated by dehydration and/or promote drinking; thin arrows represent pathways that inhibit drinking. Only a subset of known connections are depicted. (See McKinley MJ, Yao ST, Uschakov A, et al. [2015] *Acta Physiol* 214:8–32 and Zimmerman CA, Leib DA, Knight ZA [2017] *Nat Rev Neurosci* 18: 459–469 for reviews.)

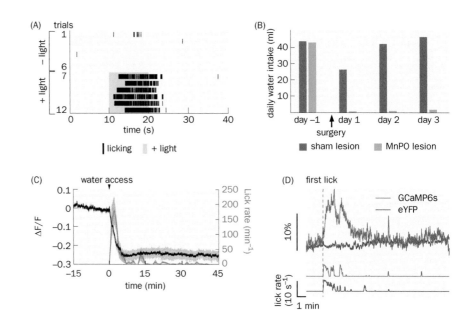

Figure 9-17 Properties of selected neurons in the drinking circuit.
(A) Optogenetic activation (blue shaded period between 10 and 20 s in trials 7–12) of channelrhodopsin- (ChR2-) expressing SFO excitatory neurons triggered intense drinking in a water-satiated mouse. **(B)** Rats with MnPO lesions stopped drinking after the surgery; their food intake was comparable to rats undergoing sham lesions without water access (not shown), suggesting that MnPO lesions do not affect food intake. **(C)** When thirsty mice were given water access, water intake (red trace) is accompanied by a rapid decline in SFO excitatory neuron activity (black trace), as seen by the fluorescence intensity change over baseline ($\Delta F/F$) of a genetically encoded Ca^{2+} indicator GCaMP6 recorded using fiber photometry (Section 14.22). **(D)** Similar recording as in Panel C but performed on GABAergic MnPO neurons, indicating that they are rapidly activated by water intake. Gray traces show recordings from control animals in which GABAergic MnPO neurons express GFP instead of GCaMP6. (A, from Oka Y, Mingyu Y, & Zuker C [2015] *Nature* 520:349–352. With permission from Springer Nature. B, from Johnson AK & Buggy J [1978] *Am J Physiol* 234:R122–R129. With permission from the American Physiology Society. C, from Zimmerman CA, Lin YC, Leib DE, et al. [2016] *Nature* 537:680–684. With permission from Springer Nature. D, from Augustine V, Gokce SK, Lee S, et al. [2018] *Nature* 555:204–209. With permission from Springer Nature.)

How are the activities of dehydration-sensing neuron regulated? Physiological recordings of excitatory neurons in the SFO and MnPO, as well as vasopressin⁺ neurons in the supraoptic nucleus, revealed that their activity *decreased* rapidly as soon as thirsty mice began to drink water (Figure 9-17C), well before changes in blood osmolarity or angiotensin II levels. Thus, the act of drinking provides rapid feedback signals to the hypothalamic thirst circuit to anticipate a reduced need state. Animals use this anticipation to adjust water intake, instead of relying purely on slower feedback signals from the blood. Drinking activates water sensors in the oral cavity and gastrointestinal tract to send signals to the brain (Figure 9-16). Indeed, a specific population of excitatory PBN neurons that inhibits water intake has been identified. Among their projection sites is the MnPO, where a population of GABAergic neurons inhibits SFO interoceptive neurons. Activity of both the PBN and GABAergic MnPO neurons is rapidly *elevated* by drinking (Figure 9-17D), consistent with these neurons constituting a rapid feedback pathway to regulate drinking (Figure 9-16).

We can see interesting parallels by comparing the neural circuits that regulate eating (Figure 9-15) and drinking (Figure 9-16). Food or water deficiency activates neurons that detect and integrate hunger or thirst signals, notably hypothalamic GABAergic AgRP or glutamatergic MnPO neurons. Activation of these neurons leads to robust eating or drinking. Food or water intake sends rapid feedback signals to the hypothalamus through ascending pathways to adjust eating or drinking behavior. Thus, behavioral changes occur well before actual changes in the body's nutritional or hydrational need states are reflected in slower homeostatic feedback signals (e.g., leptin level or blood osmolarity), to prevent excessive eating or drinking. The thirst and hunger systems also interact extensively. For example, thirsty animals usually eat less food. Indeed, SFO dehydration-sensing neurons are activated by intake of dry food, explaining the prevalence of drinking during meals. In the next section, we continue comparing hunger and thirst in the context of motivated behavior.

9.10 How do hunger and thirst drive motivated behavior?

Because of the utmost importance of food and water, animals exhibit strong innate behaviors to search for and consume food or water when they are hungry or thirsty. These basic, innate **drives,** or motives for action to maintain homeostasis, are also powerful means for acquiring learned skills and behaviors. Virtually all tasks recorded in this textbook involving awake, behaving animals were trained and implemented by giving water or food to water- or food-restricted animals as rewards. Psychologists in the 1930s and 1940s proposed the **drive reduction theory** for motivated behavior, which posits that deviations from homeostasis (such as dehydration) create aversive drives (thirst), which motivate animals to perform actions (searching for and consuming water) to reduce the aversive drives. However, there has been little experimental support in subsequent decades for this elegant theory. In fact, electrical self-stimulation experiments provided evidence opposing drive reduction theory. (In these experiments, the animal presses a lever to receive a dose of stimulation from an electrode implanted in a specific brain region; if stimulation is rewarding, the animal will keep pressing the lever. See Section 11.24 for more details.) Specifically, brain regions that promote eating or drinking when stimulated also tend to promote self-stimulation—that is, such stimulation appears rewarding to animals. Over the years, drive reduction theory has given way to the **incentive salience theory**, which posits that food and water are inherently rewarding to animals, and hunger and thirst amplify their reward value. Recent advances in our understanding of thirst and hunger circuits have allowed researchers to test these theories with more precision.

As discussed in Section 9.9, excitatory MnPO neurons are activated by dehydration, and their activation promotes drinking even when mice are water satiated. Interestingly, after mice were trained to press a lever to obtain water to drink (**Figure 9-18**A; this procedure is called **operant conditioning**, to be discussed further in Section 11.14), optogenetic activation of dehydration-activated MnPO neurons also promoted lever pressing even when the mice were water satiated (Figure 9-18B). Thus, MnPO activation does not simply turn on a drinking program, but creates a drive to motivate animals to perform an arbitrary task so as to obtain water to drink. Furthermore, once trained, mice would press lever to *turn off* optogenetic stimulation of dehydration-activated MnPO neurons (Figure 9-18C), which means that activation of these neurons is aversive. Finally, tracking the activity of MnPO neurons revealed that not only was their activity reduced by drinking (Section 9.9), but the amount of the reduction scaled with the amount of water intake. Thus, activity of dehydration-activated MnPO neurons encodes an aversive thirst drive that motivates behaviors to reduce this activity, in strong support of the drive reduction theory. Previous electrical stimulation failed to identify such neurons because the MnPO contains neuronal types with different (and often opposing) behavioral functions; electrical stimulation does not discriminate between these different neuronal types or axon fibers in passage.

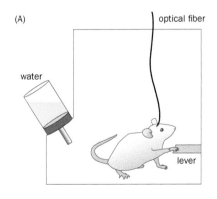

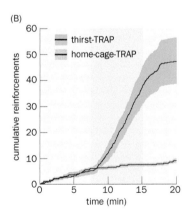

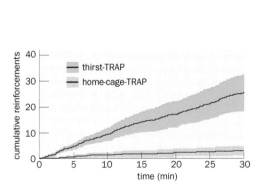

Figure 9-18 Activity of dehydration-activated MnPO neurons encodes an aversive thirst drive. (A) Experimental setup. ChR2 was selectively expressed in MnPO neurons active during a water deprivation condition (thirst-TRAP) or during a home-cage condition, when mice could drink water *ad libitum* (home-cage-TRAP; see Section 14.11 for details of the TRAP method). Mice were trained to press a lever to obtain a unit of water for drinking (Panel B), or to turn *off* optogenetic stimulation of MnPO neurons for 20 s (Panel C). **(B)** Optogenetic activation (during the period highlighted in darker yellow) of dehydration-activated (thirst-TRAP) MnPO neurons causes intense lever pressing in water-satiated mice. Each reinforcement represents lever presses that produce one unit of water for drinking. **(C)** In this experiment, optogenetic stimulation is constantly *on* unless the mouse presses the lever, which results in a 20 s break. Each reinforcement represents lever presses that produce a 20 s break in optogenetic stimulation. After training, mice voluntarily press the lever to turn *off* optogenetic activation of thirst-TRAP MnPO neurons during the entire 30 min session, indicating that activation of dehydration-activated MnPO neurons is aversive. Home-cage-TRAP serves as a negative control in both experiments. Values in Panels B and C represent mean ± standard errors of the mean. (From Allen WE, DeNardo LA, Chen MZ, et al. [2017] *Science* 357:1149–1155. With permission from AAAS. See also Leib DE, Zimmerman CA, Poormoghaddam A, et al. [2017] *Neuron* 96:1272–1281.)

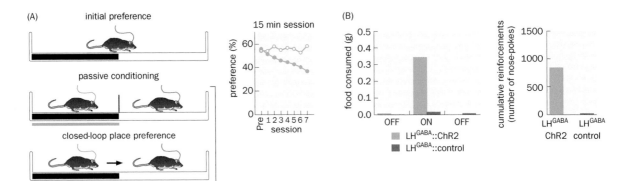

Figure 9-19 Probing the relationship between hunger and motivation. (A) Left, experimental setup. The mouse was initially tested for its preference of the two chambers by determining the time spent in each. It was then placed in one of the two chambers for optogenetic activation of AgRP neurons without food (passive conditioning), followed by closed-loop place preference, wherein AgRP neurons were optogenetically activated whenever the mouse entered the chamber previously associated with AgRP neuron activation. Right, mice preferentially avoid the chamber associated with AgRP neuron activation, suggesting that AgRP neuron activation is aversive. **(B)** Left, photostimulation of ChR2-expressing lateral hypothalamus (LH) GABAergic neurons induces robust food consumption in fed mice during but not before or after photostimulation. Right, photostimulation also appears to be rewarding; once trained, the same ChR2-expressing mice voluntarily nose-poke to receive more stimulation. Control mice express a fluorescent protein instead of ChR2 in LH GABAergic neurons. (A, from Betley JN, Xu S, Huang Cho ZF, et al. [2015] *Nature* 521:180–185. With permission from Springer Nature. B, from Jennings JH, Ung RL, Resendez SL, et al. [2015] *Cell* 160:516–527. With permission from Elsevier Inc.).

Analogous experiments in the eating circuit revealed more complexity. Like activation of dehydration-activated MnPO neurons, activation of arcuate AgRP neurons in food-restricted mice was aversive in a place preference assay, where mice avoided the chamber previously associated with photostimulation of AgRP neurons (Figure 9-19A). However, the aversion of AgRP neuron stimulation was not strong enough for operant conditioning, such as pressing a lever to turn off stimulation. Furthermore, optogenetic activation of AgRP neurons in the presence of food appeared to be rewarding when animals were allowed to eat. Conversely, optogenetic activation of GABAergic neurons in the lateral hypothalamus both was rewarding and promoted eating (Figure 9-19B), mimicking previous electrical self-stimulation experiments in the same area. Thus, hunger may drive motivated behavior through multiple, parallel mechanisms, some consistent with drive reduction theory and others with incentive salience theory.

The differences between thirst and hunger in driving motivated behavior may reflect our incomplete knowledge of both systems. It could also be that eating requires more elaborate, multilayered regulation than does drinking. It does not feel particular rewarding to drink water when we are not thirsty, but we still experience pleasure eating delicious food even when we are not hungry. Whatever answers future research will reveal, studying thirst and hunger circuits has enabled researchers to connect innate drives to behaviors acquired through learning, a topic we will expand on in Chapter 11.

HOW ARE CIRCADIAN RHYTHMS AND SLEEP REGULATED?

We've just studied eating, drinking, and the molecular and neural bases of hypothalamus-mediated homeostatic regulation of body weight and hydration. We now study circadian rhythms and sleep to further explore how regulatory systems operate. While we eat and drink to maintain energy and water homeostasis, the physiological functions of sleep remain largely enigmatic despite the fact that we spend about a third of our lives doing it. Sleep is timed by **circadian rhythms**, self-sustained oscillations in an organism's behavior, physiology, and biochemistry, with a period close to 24 hours. In the following sections, we first study the mechanisms that regulate circadian rhythms, which are well understood thanks to inspiring research over the past decades. We then explore sleep and discuss its possible functions.

9.11 Circadian rhythms are driven by auto-inhibitory transcriptional feedback loops conserved from flies to mammals

Circadian rhythms are found in many branches of life from bacteria to humans. These rhythms reflect the interaction of organisms living on Earth with a salient environmental signal—the daily cycle of light and dark caused by Earth's rotation around its own axis once every 24 hours. For example, most fruit flies eclose (emerge from their pupal case) in the morning, so that young flies have a whole day of light to search for their first meal. Mice are active at night, when they are safer from their predators. However, daily rhythms are not just passive reactions to environmental stimuli: circadian rhythms are sustained even in the absence of light. For instance, mice kept in constant darkness continue to be active only during their "subjective nights." The period of the circadian rhythm, as best seen by the onset of running wheel activity, is slightly shorter than 24 hours, but with minimal variation (**Figure 9-20**). (During the light–dark cycle, the period returns to 24 hours due to a process called light entrainment, which will be discussed later.) These phenomena indicate that an internal clock is running in the absence of environmental cues. How can we find out what this biological clock is?

The first insight into the molecular nature of this biological clock came from genetic studies in fruit flies around 1970. Hypothesizing that circadian rhythms are regulated by specific genes and can be disrupted by mutations in those genes, Seymour Benzer and colleagues conducted a forward genetic screen (Section 14.6) in fruit flies, utilizing their eclosion as an assay to identify which genes, when mutated, would disrupt circadian rhythms. Three mutations mapping to the same gene (named *Period*) were found to speed up, slow down, or abolish circadian rhythms (**Figure 9-21**A). This finding immediately suggested not only the existence of genes that regulate circadian rhythms, but also that *Period* must be related to the central control mechanism, since different disruptions of this gene could adjust the period of the clock in opposite directions. A similar genetic screen later carried out in mice identified a mutation named *Clock* (for <u>c</u>ircadian <u>l</u>ocomotor

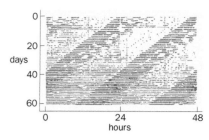

Figure 9-20 Circadian rhythm of running wheel utilization by a mouse. Each horizontal line is a 2-day record of running wheel utilization (vertical ticks) by a mouse in *constant darkness*. These records are double plotted for easier visualization: the record for day *n + 1* is both to the right of day *n* on the same horizontal line and immediately below day *n* on the next horizontal line. Despite being kept in constant darkness for 60 days, the mouse's wheel utilization exhibits a remarkably consistent circadian rhythm with a period slightly shorter than 24 h, as seen by the leftward shift of the daily onset of running wheel utilization. (From Pittendrigh CS [1993] *Ann Rev Physiol* 55:16–54. With permission from Annual Reviews.)

Figure 9-21 Circadian rhythm mutants in *Drosophila* and mice. (A) Eclosion rhythms of populations of normal and mutant flies. Adult flies emerge from their pupal cases in the morning. After being exposed to a 12 h light–12 h dark cycle during development and then kept in constant darkness, eclosion of normal adult flies occurred during the subjective morning, with a ~24 h rhythm. Arrhythmic mutant flies lost the circadian clock altogether, while short- and long-period mutant flies had rhythms of ~19 h and ~28 h, respectively. All three mutants disrupt the same gene, *Period*. **(B)** Rhythms of running wheel utilization by mice, as described in Figure 9-20. During the first 20 days, mice were kept on a 14 h light–10 h dark cycle (LD, indicated by the alternating light and dark bars above). They were then switched to constant darkness (DD). During LD, the rhythms were entrained by light, and the mice utilized the running wheel during the night. Upon shifting to DD, the period for the normal mouse was 23.7 h (the onset of running wheel activity shifted leftward). By contrast, the *Clock/+* heterozygous mutant had a period of 24.8 h (activity onset shifted rightward), and the *Clock/ Clock* homozygous mutant had a period of 27.1 h during the first 10 days in DD, and lost the circadian rhythm afterwards (A, from Konopka RJ & Benzer S [1971] *Proc Natl Acad Sci U S A* 68:2112–2116. B, adapted from Vitaterna MH, King DP, Chang A, et al. [1994] *Science* 264: 719–725.)

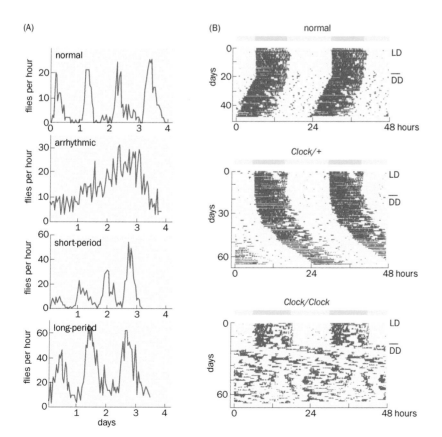

(A)

normal

arrhythmic

short-period

long-period

(B)

normal

Clock/+

Clock/Clock

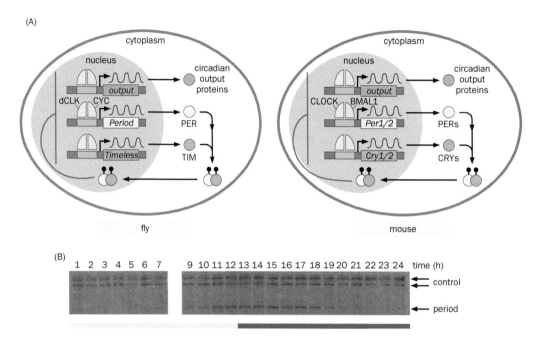

Figure 9-22 Auto-inhibitory transcriptional feedback in the central clocks of *Drosophila* and mice. (A) Central clock mechanisms are strikingly similar in the fly and mouse. A pair of transcription activators—dCLK *(Drosophila* CLOCK) and CYC in flies, CLOCK and BMAL1 in mice—activates transcription of *Period* and *Timeless* in flies or two *Per* and two *Cry* genes in mice. The resulting protein products form a complex, become phosphorylated (black dots attached to the proteins), and enter into the nucleus to negatively regulate their own transcription. Many other genes are similarly regulated in a circadian fashion, accounting for the circadian output. **(B)** *Per* mRNA levels cycle in a circadian fashion, with a peak at dusk and a trough at dawn. The levels of two control mRNA species remain constant. Populations of flies were maintained in a 12 h light (1–12), 12 h dark (13–24) cycle as indicated at the bottom. Fly heads were harvested hourly for mRNA extraction. (A, adapted from Mohawk JA, Green CB, & Takahashi JS [2012] *Ann Rev Neurosci* 35:445–462. B, after Hardin PE, Hall JC, & Rosbash M [1990] *Nature* 343:536–540.)

output cycles kaput), which slowed down the circadian rhythm in heterozygous mice, and disrupted the rhythm in homozygotes (Figure 9-21B).

Subsequent studies have shown that the protein products of the *Period* and *Clock* genes are key components of an auto-inhibitory transcriptional network conserved from flies to mammals. CLOCK is a DNA-binding transcription factor that, together with its partner (CYC in fly and BMAL1 in mouse, which are orthologs), binds to the promoters of *Period* and other rhythmically expressed genes to activate their transcription. Among the resulting products are the fly protein PER (encoded by the *Period* gene) and its binding partner TIM (encoded by the ***Timeless*** gene), and the two mouse PER proteins (encoded by two genes, *Per1* and *Per2*) and their binding partners, the CRY proteins (for **cryptochrome**; encoded by two *Cry* genes). Heterodimeric complexes of PER/TIM (fly) and PER/CRY (mouse) enter the nucleus and negatively regulate the activity of CLOCK, thereby forming negative feedback loops that downregulate their own production (**Figure 9-22A**). An initial insight leading to this model came from the finding that *Period* mRNA levels cycled in a circadian fashion (Figure 9-22B), suggesting that PER may negatively regulate its own transcription. Specifically, when the level of PER protein is high, PER inhibits its own transcription, so that the *Period* mRNA level declines. This leads to less production of PER protein, which then relieves the negative regulation of *Period* transcription. This in turn causes more production of *Period* mRNA, leading again to a high level of PER protein, and so on (**Movie 9-1**).

Research in recent decades has revealed details of the molecular basis of circadian clock regulation. In addition to the negative transcriptional feedback loop for regulating production of PER and its partners (TIM or CRY proteins), other regulatory mechanisms control the phosphorylation, nuclear entry, and degradation of these circadian rhythm regulators. Another transcriptional feedback loop produces a circadian rhythm in the expression of the CLOCK partner BMAL1 in mice, adding robustness to the network. One of the most striking findings is that circadian regulation of gene expression is not restricted to neurons in the brain but is instead found in nearly all cells in peripheral tissues, with a large fraction of all transcripts in flies and mammals being regulated in a circadian fashion. For example, under certain conditions, mammalian fibroblasts in culture maintain circadian gene expression utilizing the same molecules and mechanisms found in neurons that control the animal's circadian behavior. Indeed, circadian cycles of gene expression and auto-inhibitory transcriptional networks are also essential for circadian regulation in plants and unicellular fungi. Although these organisms do not utilize PER and CLOCK, the auto-inhibitory transcriptional feedback loop appears to be a universal strategy for regulation of circadian rhythms in diverse eukaryotic organisms.

9.12 Entrainment in flies is accomplished by light-induced degradation of circadian rhythm regulators

A universal feature of circadian rhythms is that their phase can be reset by light, a process called **entrainment**. (Phase refers to the specific times during a 24 hour period that a given physiological, behavioral, or biochemical activity reaches a particular stage of the cycle, such as the peak or trough.) This is why we recover from jetlag after being in a new time zone for a few days. For a typical 12 hour day/12 hour night cycle, light exposure during the first half of the night shifts the phase of the circadian rhythm to an earlier time (phase delay), whereas light exposure during the second half of the night shifts the phase to a later time (phase advance). How does light interact with the circadian clock to reset the phase? And given that many cells throughout the body have their own clocks, how are different clocks synchronized?

Unlike the conserved central clockwork mechanisms, light entrainment mechanisms differ considerably in flies and mice. Because of the thin insect cuticle, light can penetrate directly into most cells, including dedicated clock neurons in the brain. The photoreceptor protein is cryptochrome (CRY), an ancient blue-light sensor. (The mammalian homologs of fly CRY do not function as photoreceptors but curiously function as an integral part of the core circadian negative feedback loop; Figure 9-22A.) In response to light, fly CRY forms a complex with and induces rapid degradation of the PER binding partner, TIM. Thus, light degrades the negative regulatory PER/TIM complex (**Figure 9-23**).

Specifically, in the first half of the night, before a critical amount of the complex has entered the nucleus to downregulate its own transcription, degradation of the complex means that more time is needed to build up PER/TIM proteins

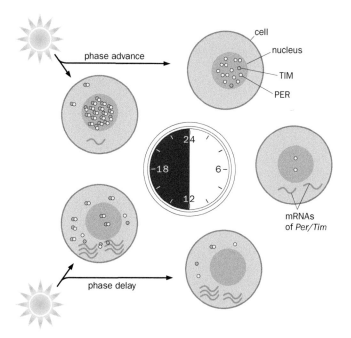

Figure 9-23 A model of light entrainment in flies. Light triggers TIM and PER degradation and thereby relieves inhibition of *Per/Tim* mRNA production. Thus, following a full day of accumulation, levels of *Per* and *Tim* mRNA peak in the early evening. In the middle of the night, PER/TIM proteins enter the nucleus and inhibit *Per/Tim* transcription. A light pulse given before PER/TIM nuclear entry degrades PER/TIM proteins while *Per/Tim* mRNA levels remain high, mimicking the state at dusk. A light pulse given after PER/TIM nuclear entry degrades the proteins while mRNA levels are already low, mimicking the state at dawn. (From Hunter-Ensor M, Ousley A, & Sehgal A [1996] *Cell* 84: 677–685. With permission from Elsevier Inc.)

from existing mRNAs. This resets the clock back to the beginning of the night, causing a phase delay. By the second half of the night, the PER/TIM complex has accumulated in the nucleus and downregulated transcription of *Period* and *Timeless,* such that little mRNA is left. A light pulse given at this time also degrades PER/TIM proteins, but in this case mimics natural light-mediated degradation of the PER/TIM complex that would normally occur at the beginning of the day, which initiates a new transcription and circadian cycle. Thus, a light pulse given during the second half of the night resets the clock forward to dawn, causing a phase advance (Figure 9-23; Movie 9-1).

9.13 Pacemaker neurons in the mammalian suprachiasmatic nucleus integrate input and coordinate output

Though many mammalian cells have their own clocks, most are inaccessible to light. The **suprachiasmatic nucleus (SCN)** of the hypothalamus (Figure 9-6A) is a master regulator of both circadian rhythms and light entrainment. SCN lesions disrupt behavioral rhythmicity, which can be restored by transplanted fetal SCN tissues. The following experiment established the SCN's position at the top of the circadian hierarchy in mammals. A spontaneous mutation (named *Tau*) was found in golden hamsters that shortened the circadian period to ~22 hours when heterozygous and ~20 hours when homozygous. When fetal SCN tissues from *Tau*-mutant mice were transplanted into SCN-lesioned wild-type hosts, the restored rhythm also had a shortened period. Similarly, a normal circadian period was produced in mutant hosts that had wild-type fetal SCN tissues transplanted after lesion (**Figure 9-24**). Thus, the *genotype* of the SCN determines the circadian *phenotype* of the mosaic animals, highlighting the central role of the SCN in controlling rhythmic behavior.

The SCN receives input from intrinsically photosensitive retinal ganglion cells (ipRGCs) in the retina (Section 4.19). Light directly activates ipRGCs, which release glutamate at their terminals on SCN neurons. Glutamate receptor activation triggers Ca^{2+} entry, kinase activation, and phosphorylation of the transcription factor CREB (Section 3.24) in SCN neurons. Phosphorylated CREB binds to the promoters of *Per1* and *Per2* and activates their transcription. Thus, transcriptional regulation of clock genes appears to play a major role in mammalian light-induced phase resetting, whereas light-mediated regulation of clock proteins in the fly is predominantly post-transcriptional.

While a direct connection to the retina affords the SCN a privileged position for interaction with the environment for phase resetting, the following properties contribute to the SCN being the master regulator of circadian rhythms. (1) The electrical activity of SCN neurons oscillates in a circadian fashion even in isolation, qualifying them as **circadian pacemaker neurons** (analogous to the pacemaker neurons in the central pattern generators discussed in Section 8.5). This is in part due to circadian regulation of K^+ channel activity, which determines the resting membrane potential of SCN neurons. (2) SCN neurons form a close network with other SCN neurons through gap junctions and neuropeptide signaling, such that SCN neurons are highly synchronized in their circadian electrical activity and gene expression. This network property has been observed in electrophysiological recordings and by imaging expression of a bioluminescent circadian reporter generated in transgenic mice. Whereas individually cultured SCN neurons exhibited circadian cycles with different phases and periods, neurons in SCN slices with preserved local connections displayed robust synchronization of circadian period and phase between cells (**Figure 9-25**A). A recent study found that SCN astrocytes also have a cell-autonomous clock. Remarkably, restoring the astrocyte clock in mutant mice with disrupted circadian rhythms could reinstate circadian rhythms in SCN neurons and behavioral rhythms in the mice. Thus, astrocytes also play a key role in producing rhythmic output via cellular interactions within the SCN. (3) SCN neurons secrete signaling molecules rhythmically and make extensive synaptic connections with many hypothalamic nuclei, such that they can impose their coherent rhythmic activity onto these targets, which

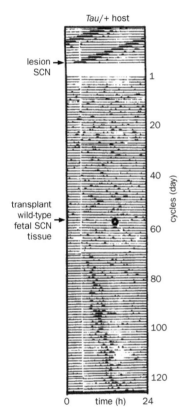

Figure 9-24 The genotype of the suprachiasmatic nucleus (SCN) determines the behavioral phenotype of mosaic animals. Here, a golden hamster heterozygous for a *Tau* mutation served as a host for SCN transplantation after lesion. All experiments were recorded in constant darkness after the animal had established a circadian rhythm assayed by running wheel utilization. Before the SCN lesion, it had a period of 21.7 h (top rows). After the lesion, the behavioral rhythm was abolished. After transplantation of fetal SCN tissue from a wild-type golden hamster (circle indicates time of transplantation), not only was rhythm restored, it also had a wild-type period of ~24 h. (From Ralph MR, Foster RG, Davis FC, et al. [1990] *Science* 247:975–978. With permission from AAAS.)

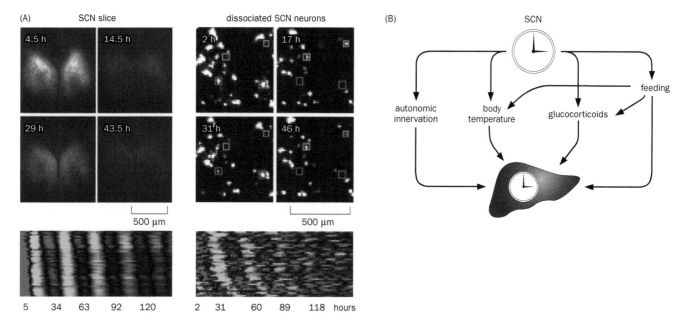

Figure 9-25 Local connections and global actions of suprachiasmatic nucleus (SCN) neurons. (A) Top, bioluminescence images of *in vitro* cultured SCN slices or dissociated SCN neurons from mice in which the luciferase gene was knocked-in at the *Per2* locus; the production of light by luciferase in the presence of its substrates reports the cyclic level of endogenous *Per2* expression. Images were taken at different times after culture preparation, as indicated in the upper-left corner of each panel. SCN neurons from a slice (left) exhibit synchronized activity: most cells have high *Per2* transcriptional activity at 4.5 h and 29 h after the onset of slice culture and low activity at 14.5 h and 43.5 h. In cultures of dissociated SCN neurons (right), local connections between cells are disrupted. Individual cells in these cultures still exhibit cyclic expression but do not oscillate in the same phase (for example, compare the cells in the red and green squares). Bottom, raster plots tracking the bioluminescence of 40 SCN cells over 5 days (each cell is represented by one horizontal line, with values above or below the average in red and green, respectively), showing synchronized rhythms in the SCN slice but much less so among the dissociated neurons. **(B)** Summary of pathways from the SCN to peripheral clocks (with the liver clock as an example). Through secreted signals and neuronal projections, SCN neurons send rhythmic output to hypothalamic nuclei that control autonomic innervation, body temperature, feeding, and hormone secretion, all of which influence peripheral clocks. (A, from Liu AC, Welsh DK, Ko CH, et al. [2007] *Cell* 129:605–616. With permission from Elsevier Inc. B, adapted from Mohawk JA, Green CB & Takahashi JS [2012] *Annu Rev Neurosci* 35:445–462.)

in turn regulate both clocks in peripheral tissues and the physiology and behavior of the organism (Figure 9-25B).

For example, SCN neurons send output to the paraventricular hypothalamic nucleus, a central regulator of autonomic output (Figure 9-15). SCN neurons also send output to the temperature control nucleus of the hypothalamus and regulate organismal physiology through temperature cycles, a powerful entrainment agent for peripheral clocks. (Our body temperature fluctuates about 0.5°C across the day, with its lowest temperature at around 4:30 A.M. and its highest at around 7 P.M.) SCN neurons also regulate the activity of hypothalamic CRH neurons, which control glucocorticoid production (Table 9-1) in a circadian fashion. Peripheral clocks, in addition to receiving signals from the SCN, can also be entrained by other cues. For example, the liver clock is entrained by food intake; in SCN-lesioned animals, food can serve as a powerful entrainment signal that regulates many aspects of physiology.

In summary, circadian regulation of physiology, biochemistry, and behavior of multicellular organisms utilizes a combination of mechanisms including cell-autonomous auto-inhibitory transcriptional regulation and extensive communication linking SCN pacemaker neurons, other hypothalamic nuclei, and peripheral tissues. A major contribution to our current understanding followed from identification of genes that control circadian rhythms in fruit flies and mice and follow-up studies of their mechanisms of action. Indeed, mutations in the human homologs of the fly *Period* gene and its regulators are responsible for **familial advanced sleep phase**—an extreme variant of the human circadian system characterized by very early morning waking and an early evening sleep onset (**Figure 9-26**). These findings highlight how studying basic neurobiological problems in

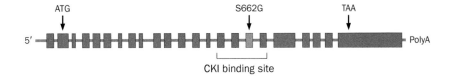

Figure 9-26 Familial advanced sleep phase in humans due to mutations in circadian regulators. Structure of the human *Per2* gene, with exons in blocks and introns in lines, and translational start (ATG) and stop (TAA) codons indicated. A point mutation at amino acid 662, which changes a serine (S), a potential phosphorylation site of casein kinase I (CKI), to a glycine (G), was identified to underlie familial advanced sleep phase in a family with an autosomal dominant inheritance pattern. Remarkably, mutations of casein kinase I in *Drosophila*, golden hamster (encoded by *Tau*; Figure 9-24), and human (in a different family with advanced sleep phase) also alter circadian rhythms. (From Toh KL, Jones CR, He Y, et al. [2001] *Science* 291:1040–1043. With permission from AAAS. For circadian mutations in casein kinases, see Kloss B, Price JL, Saez L, et al. [1998] *Cell* 94:97–107; Lowrey PL, Shimomura K, Antoch MP, et al. [2000] *Science* 288:483–490; Xu Y, Padiath QS, Shapiro RE, et al. [2005] *Nature* 434: 640–644.)

invertebrate model organisms can have profound relevance to human health, and how human genetics can contribute unique insights.

9.14 Sleep is widespread in animals and exhibits characteristic electroencephalogram patterns in mammals

We all know what sleep is in humans and are familiar with sleep in other mammals. But sleeplike phenomena are widespread in the animal kingdom. Researchers have agreed upon a set of behavioral criteria that define the sleep state: sleep is rapidly reversible (as opposed to coma); it is associated with decreased responsiveness to sensory stimuli; it is subject to homeostatic regulation—sleep deprivation results in more sleep afterward; it is also timed by the circadian system, which governs a more rigid pattern of daily alertness and sleepiness, including nocturnal and diurnal patterns. Using these behavioral criteria, scientists have extended the concept of sleep to organisms such as zebrafish and fruit flies. For example, fruit flies exhibit bouts of inactivity and characteristic postures that occur mostly at night. During this state, flies are less easily aroused by mechanical stimulation. This state is also homeostatically regulated; application of strong stimuli that prevents flies from sleeping at night induces a robust sleep rebound (an increase in sleep after deprivation) the next day (**Figure 9-27**).

In mammals, the traditional subjects of sleep research, sleep and sleep stages can be defined using **electroencephalography** (**EEG**). EEG was first discovered in the 1920s. It reports the collective activity of many neurons recorded by electrodes placed on the surface of the scalp (see Figure 12-43 for an illustration of the method; the resulting record is called an **electroencephalogram**, also abbreviated as EEG). During the waking period, EEG patterns are characterized by high-frequency, low-amplitude oscillations. During descent into deeper stages of sleep, low-frequency oscillations increase in amplitude, until we reach the deepest stage—slow-wave sleep, when low-frequency, large-amplitude oscillations are most prevalent (**Figure 9-28A**). In the 1950s, it was discovered that one stage of sleep, named **REM sleep**, is characterized by rapid eye movements and complete muscle relaxation. The rest of the sleep is accordingly termed non-REM sleep, or **NREM sleep**. When we fall asleep, we first enter light NREM sleep and descend into deep NREM sleep. We then ascend to lighter stages of NREM sleep before transitioning to REM sleep. We subsequently descend back into deep NREM sleep, and repeat this cycle several times before waking up. On average, each cycle in humans lasts for about 90 minutes (Figure 9-28B). REM sleep is also associated with vivid dreaming. Adult humans typically spend 25% of sleep time in REM sleep, but REM sleep occupies up to 80% of infants' total sleep time. Recently, it has been found that certain neural signatures of sleep previously thought to be characteristic of mammals, such as REM sleep, are also present in lizard and zebrafish.

Figure 9-27 Sleep in *Drosophila* is timed by circadian rhythms and regulated by homeostasis. During a typical circadian cycle, normal fruit flies sleep mostly at night (blue trace). Minutes of sleep (*y* axis) are plotted for each assayed hour (*x* axis; 1–12 h = lights on, 13–24 h = lights off). When sleep was disrupted during the previous night (not shown), marked sleep rebound occurred the next day (red trace). (Adapted from Shaw PJ, Cirelli C, Greenspan RJ, et al. [2000] *Science* 287:1834–1837; see also Hendricks JC, Finn SM, Panckeri KA, et al. [2000] *Neuron* 25:129–138.)

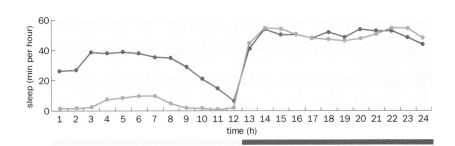

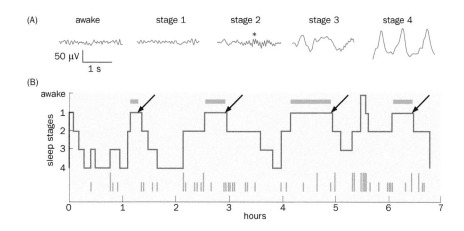

Figure 9-28 EEG patterns and sleep cycles in humans. (A) Representative EEG patterns for different sleep phases. From wakefulness to stage 4 sleep, there is a progressive increase in the amplitude of low-frequency oscillations in EEG patterns. * indicates a sleep spindle, which is characterized by 11–15 Hz oscillations that last more than 0.5 s. **(B)** Recording of a human subject during a 7 h period of normal sleep. Sleep stages are shown on the left. Blue bars represent periods of REM sleep. Arrows point to the end of a sleep cycle (the end of REM sleep) and the beginning of a new sleep cycle. Vertical bars below represent body movements, with tall bars representing major body movements. Note that sleep researchers have since combined stages 3 and 4 into a new stage 3 (slow-wave sleep). (From Dement W & Kleitman N [1957] *Electroenceph Clin Neurophysiol* 9:673–690. With permission from Elsevier Inc.)

The synchronized oscillatory EEG patterns during NREM sleep are contributed by activity produced by both local recurrent connections between cortical neurons and long-range thalamic input. Indeed, thalamic nuclei contain the best-characterized pacemaker cells in mammals, which can produce rhythmic firing patterns in the absence of input due to their intrinsic ion channel properties (Figure 2-35). During the waking period and REM sleep, inputs from sensory and neuromodulatory systems change the firing patterns of these thalamic pacemaker neurons from synchronous oscillatory patterns to high-frequency asynchronous patterns. In the next section, we discuss how neuromodulatory systems regulate the sleep–wake cycle.

9.15 The mammalian sleep–wake cycle is regulated by multiple neurotransmitter and neuropeptide systems

What neural mechanisms regulate sleep? This question has been investigated primarily in mammals, in which both behavior and EEG patterns can be used to describe the sleep state. Lesion and electrical stimulation studies have identified an **ascending arousal system** (also called the reticular activating system) from the brainstem to the forebrain essential for maintaining wakefulness (Figure 9-29A). The ascending arousal system consists of several parallel streams, each using different neurotransmitters. One stream is composed of cholinergic neurons in the tegmental nuclei that project to the thalamus and basal forebrain cholinergic neurons. Basal forebrain cholinergic neurons in turn project across many regions of the forebrain. These tegmental and basal forebrain cholinergic neurons are most active during wakefulness and REM sleep, and the activity of these neurons alters the firing of thalamic and cortical neurons. Additional streams in the ascending arousal system originate from groups of neurons that use monoamines as neurotransmitters and project directly to the cerebral cortex, hippocampus, and basal forebrain (Box 9-1). These streams include norepinephrine neurons in the **locus coeruleus**, serotonin neurons in the **raphe nuclei**, dopamine neurons in the ventral **periaqueductal gray** near the dorsal raphe nucleus, and histamine neurons in the **tuberomammillary nucleus** of the hypothalamus. These neurons are all active while animals are awake and decrease their activity during both NREM and REM sleep. Yet another stream involves neurons that express the neuropeptide hypocretin, which will be discussed in greater detail in the following.

A second system important for regulating the sleep–wake cycle consists of sleep-active neurons, which are, by definition, most active while animals are asleep. The best studied of these are GABAergic neurons located in the preoptic area (POA) of the hypothalamus (Figure 9-29; Figure 9-6A). Lesions of POA neurons cause substantial reductions in NREM and REM sleep. More precise optogenetic experiments revealed that activation of POA GABAergic neurons that project to the tuberomammillary nucleus increased both NREM and REM sleep (Figure 9-29B). Hence, GABAergic POA projection neurons are sleep-promoting, in addition to sleep-active, neurons.

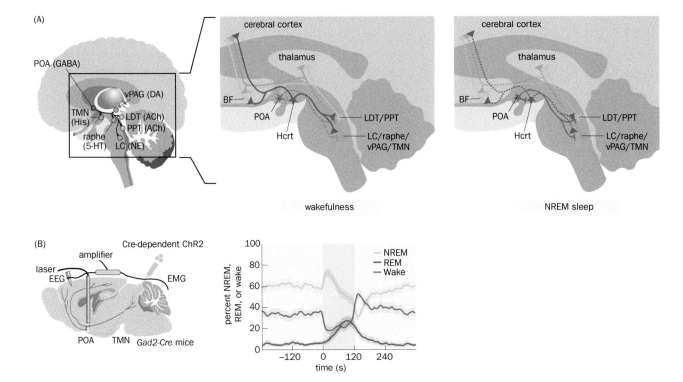

Figure 9-29 Neural systems that maintain states of wakefulness and sleep. (A) Schematic of neurons active during wakefulness (middle) and sleep (right) in midsagittal views of the human brain, magnified from the box on the left. During wakefulness, cholinergic (ACh) neurons in the laterodorsal and pedunculopontine tegmental nuclei (LDT and PPT), norepinephrine (NE) neurons in the locus coeruleus (LC), dopamine (DA) neurons in the ventral periaqueductal gray (vPAG), serotonin (5-HT) neurons in the raphe nuclei, and histamine (His) neurons in the tuberomammillary nucleus (TMN) are all active. These neurons activate the cerebral cortex and thalamus, either directly or indirectly via basal forebrain (BF) cholinergic neurons; they also inhibit GABAergic neurons in the hypothalamic preoptic area (POA). Hypothalamic hypocretin (Hcrt) neurons are likewise active and send excitatory output to many neurons, including the wake-promoting cholinergic and monoamine neurons. During NREM sleep, POA GABAergic neurons are active and inhibit tegmental cholinergic neurons, LC/DR/TMN monoamine neurons, and Hcrt neurons. In REM sleep (not shown), activity states are similar to NREM except that cholinergic neurons are active. In these schematics, active neurons are represented by solid projections and inactive neurons by dashed projections. These modulatory neurons can excite or inhibit their target neurons depending on the receptors expressed in target neurons; the symbol ⊣ denotes known

inhibition. For simplicity, TMN neurons are grouped with brainstem neurons even though they are located in the hypothalamus (see the left panel). **(B)** Left, experimental preparation. Only GABAergic neurons projecting to the TMN express channelrhodopsin (ChR2), as an axon terminal transducing virus expressing ChR2 in a Cre-dependent manner was injected into the TMN of transgenic mice (*Gad2-Cre*) in which Cre is expressed only in GABAeregic neurons. An optical fiber was targeted to the POA to specifically activate POA GABAergic neurons projecting to the TMN. EEG and EMG (electromyographic) recordings were performed to determine sleep–wake status. Right, optogenetic activation of TMN-projecting POA GABAergic neurons (blue period) reduces wakefulness and increases NREM and REM sleep. Additional experiments (not shown) revealed that optogenetic activation of all POA GABAergic neurons or TMN-projecting POA glutamatergic neurons promote wakefulness, highlighting the importance of targeting neuronal subpopulations based on cell type *and* projection patterns. (A, adapted from Saper CB, Fuller PM, Pederson NP, et al. [2010] *Neuron* 68: 1023–1042. With permission from Elsevier Inc. See also Weber F & Dan Y [2016] *Nature* 538:51–59. B, from Chung S, Weber F, Zhong P, et al. [2017] *Nature* 545:477–481. With permission from Springer Nature.)

Interestingly, GABAergic POA neurons also project axons to tegmental cholinergic neurons and all classes of the monoamine neurons in the ascending arousal system and inhibit their activity by releasing GABA. At the same time, POA neurons receive input from many neurons in the arousal system and are inhibited by acetylcholine, norepinephrine, dopamine, and serotonin. Thus, the arousal- and sleep-promoting systems form a mutually inhibitory circuit to maintain the stability of each state and facilitate rapid and complete transitions between states. This mutually inhibitory circuit motif resembles the central pattern generators controlling rhythmic movement (Figure 8-13; Figure 8-15). While the time scales differ, the logic of these systems is similar: animals activate flexor and extensor muscles in alternation but usually not simultaneously; likewise, animals are either awake or asleep but not both at once.

An important insight into sleep regulation came from studying a sleep disorder called **narcolepsy**. Narcoleptic patients have trouble staying awake, especially when excited. They can also switch from being awake directly into REM sleep

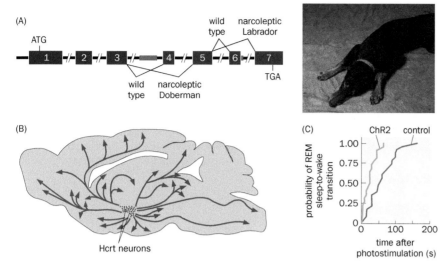

Figure 9-30 Hypocretin and hypocretin neurons promote wakefulness. (A) Recessive mutations identified in narcoleptic dogs (pictured on the right) map to the hypocretin receptor gene. The narcoleptic mutation in the Doberman strain is caused by an insertion in the third intron (long red bar) resulting in the skipping of exon 4 during splicing. The narcoleptic mutation in the Labrador strain occurs in intron 6 near the 5′ splice site (short red bar), causing exon 6 to be skipped. Both mutations produce nonfunctional receptor proteins. **(B)** Hypocretin-expressing (Hcrt) neurons (purple dots) cluster in a small area of the lateral hypothalamus, but their axons (arrows) project to many parts of the brain. **(C)** Channelrhodopsin (ChR2, blue trace) or a control fluorescent protein (mCherry, red trace) were expressed in hypocretin-expressing neurons in the lateral hypothalamus via viral delivery of a Cre-dependent expression vector in *Hcrt-Cre* transgenic mice. Photostimulation of ChR2-expressing-hypocretin neurons increases the probability of transition to wakefulness from REM sleep (as seen by the leftward shift of the probability distribution curve compared to control) and from non-REM sleep (not shown). (A, from Lin L, Faraco J, Li R, et al. [1999] *Cell* 98:365–376. With permission from Elsevier Inc. B, adapted from Peyron C, Tighe DK, van den Pol AN, et al. [1998] *J Neurosci* 18:9996–10015. C, from Adamantidis AR, Zhang F, Aravanis AM, et al. [2007] *Nature* 450:420–424. With permission from Springer Nature.)

without going through NREM sleep. A breakthrough in understanding narcolepsy came from identification of mutations that give rise to narcolepsy-like symptoms in certain breeds of dogs. Positional cloning, which was used to identify the *Obese* and *Clock* genes (Section 14.6), identified in narcoleptic dogs mutations disrupting the splicing of a gene (**Figure 9-30**A) encoding a G-protein-coupled receptor (GPCR) for the neuropeptide **hypocretin**. Hypocretin is also called **orexin**; it was independently identified via biochemical purification of hypothalamic neuropeptides that activate orphan GPCRs (GPCRs without known ligands) and found to stimulate rats' food intake when administered into the cerebral ventricles. Indeed, about the same time narcoleptic mutations in dogs were mapped to the hypocretin receptor, orexin-knockout mice were found to exhibit phenotypes resembling narcolepsy. Thus, hypocretin/orexin has dual roles in regulating sleep and food intake. Loss of hypocretin neurons and, in rare cases, mutations in the gene encoding hypocretin, have since been found to be causes of human narcolepsy.

Hypocretin-producing neurons are located exclusively in the lateral hypothalamus, but their axons project widely in the brain, including to acetylcholine-, norepinephrine-, serotonin-, and histamine-producing neurons in the ascending arousal pathway (Figure 9-30B). *In vivo* recordings indicated that hypocretin neurons are most active when animals explore the environment during the waking period and stop firing during sleep. Furthermore, optogenetic activation of hypocretin-producing neurons in mice increased the probability that animals wake up from NREM or REM sleep (Figure 9-30C). Thus, the hypocretin system promotes wakefulness, likely through activation of the ascending arousal system (Figure 9-29A) and direct action on target neurons throughout the brain (Figure 9-30B).

Box 9-1: Neuromodulatory systems

Neurotransmitters such as glutamate and GABA can elicit rapid excitation or inhibition of postsynaptic neurons by activating ionotropic receptors. Modulatory neurotransmitters, also called **neuromodulators**, act on slower time scales. Canonical neuromodulators include monoamines (dopamine, serotonin, norepinephrine, and histamine), acetylcholine (when it acts on muscarinic receptors), and neuropeptides (Section 3.11). In Chapter 8 and this chapter, we have seen many examples of neuromodulators: regulation of striatal output by dopamine; regulation of eating by α-MSH, NPY, AgRP, and CCK; and regulation of sleep by hypocretin, acetylcholine, and monoamines. These modulators usually act on G-protein-coupled receptors (Table 3-3) expressed by their target neurons and therefore have slower and longer-lasting effects than the more rapid signaling of ionotropic glutamate and GABA receptors. Neuromodulators can act on receptors in different subcellular compartments of their target neurons and can thus exert diverse effects. As an example, **Figure 9-31** illustrates how dopamine can bidirectionally modulate the efficacy of synaptic transmission and excitability of target neurons.

A striking feature of neuromodulatory systems is their broad reach. We highlight here four neuromodulatory systems that regulate forebrain function: the norepinephrine, serotonin, dopamine, and acetylcholine systems. Although the cell bodies of these neurons are clustered in discrete nuclei in the brainstem and basal forebrain, their axons project extensively in the forebrain (**Figure 9-32**). For example, norepinephrine neurons in the locus coeruleus give rise to axons innervating the neocortex, olfactory bulb, olfactory cortex, hippocampus, amygdala, thalamus (Figure 9-32A). Likewise, serotonin neurons in the raphe nuclei project diffusely throughout the forebrain (Figure 9-32B). (In addition to forebrain targets, norepinephrine and serotonin neurons also send extensive projections that broadly innervate the brainstem, cerebellum, and spinal cord.) Midbrain dopamine neurons have more focused projections than the norepinephrine and serotonin systems, but still project widely across the striatum, olfactory cortex, and prefrontal cortex (Figure 9-32C; see Figure 14-24B for the axonal arborization of a single dopamine neuron). Each monoamine transmitter acts on multiple receptors (Table 3-3) that are differentially distributed in different cell types and subcellular compartments, allowing them to modulate many aspects of signaling in their target neurons. Acetylcholine-producing neurons in the basal forebrain also project broadly across the neocortex, hippocampus, amygdala, and olfactory bulb (Figure 9-32D). While acetylcholine also signals through nicotinic receptors, which are ionotropic, some of its actions are mediated by muscarinic receptors, which are metabotropic and therefore modulatory.

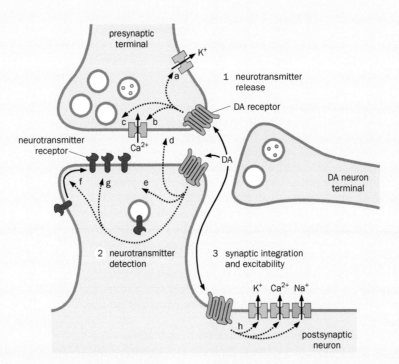

Figure 9-31 Diverse effects of neuromodulators on their target neurons. Here we use dopamine (DA) released from a DA neuron terminal (blue) to illustrate its effects on a pair of synaptically connected neurons (green). Dopamine can affect (1) neurotransmitter release by modulating presynaptic K⁺ channels (a), Ca²⁺ channels (b), or release machinery (c), or regulating production of retrograde messengers (d; will be discussed in Chapter 11); (2) postsynaptic neurotransmitter detection by regulating insertion (e), recruitment (f), or conductance (g) of neurotransmitter receptors; and (3) synaptic integration and excitability of postsynaptic neurons by modulating voltage-gated K⁺, Na⁺, and Ca²⁺ channels (h). All of these effects can be bidirectional, depending on the specific dopamine receptors these target neurons express—for example, D_1 receptors are coupled to G_s and their activation therefore increases cAMP concentration, whereas D_2 receptors are coupled to G_i and their activation therefore decreases cAMP concentration. (Adapted from Tritsch NX & Sabatini BL [2012] *Neuron* 76:33–50. With permission from Elsevier Inc.)

Box 9-1: continued

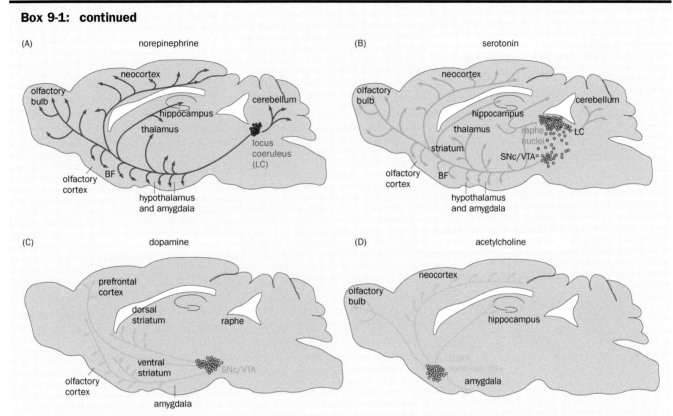

Figure 9-32 Forebrain projection patterns of monoaminergic and cholinergic neurons. (A) Norepinephrine neurons in the locus coeruleus (LC) project their axons diffusely across most regions of the forebrain and the cerebellum. **(B)** Serotonin neurons in the raphe nuclei likewise project extensively to almost the entire forebrain and cerebellum. **(C)** Dopamine neurons from the ventral tegmental area (VTA) and substantia nigra pars compacta (SNc) project widely across the striatum, olfactory cortex, and prefrontal cortex. **(D)** Acetylcholine-producing neurons in the basal forebrain (BF) project broadly to the neocortex, hippocampus, amygdala, and olfactory bulb. Note that only the major target sites of the forebrain and cerebellum projections are depicted. Note also that neurons that utilize these transmitters are also present in smaller clusters in parts of the brain not depicted. (Courtesy of Lindsay Schwarz, Brandon Weissbourd, and Kevin Beier. The drawings are sagittal views of the mouse brain based on axon projection data from the Allen Brain Atlas [www.brain-map.org].)

Trans-synaptic tracing experiments (Figure 14-33) revealed that these neurons also receive direct synaptic input from many brain regions. These findings raised an interesting question regarding the architecture of these neuromodulatory systems. Do individual neurons within each system act equivalently to integrate diverse input and broadcast a common output across all target regions or does each system contain subsystems with distinct properties? Recent experiments suggested that answers to this question differ for different monoamine systems. The locus coeruleus norepinephrine system appears to adopt the first strategy: populations of norepinephrine neurons that project to one target site also send collateral branches that reach all other target sites examined. Conversely, midbrain dopamine neu-

rons and dorsal raphe serotonin neurons consist of subsystems with distinct input distributions, projection patterns, physiological properties, and behavioral functions.

We are only beginning to understand the diverse functions of these neuromodulatory systems and their mechanisms of action. We have discussed the function of dopamine in regulating movement (Figure 8-21), and monoamine and acetylcholine neurons in regulating the sleep–wake cycle (Figure 9-29), but these are just the tip of the iceberg. The importance of these modulatory systems in human health is highlighted by the fact that most drugs currently used to treat psychiatric disorders interfere with their function, as will be discussed in more detail in Chapter 12.

9.16 Altering neuronal excitability in sleep regulatory centers affects the amount of sleep

What is the molecular basis of sleep and its homeostatic regulation? While we have far less understanding of sleep than we do of circadian rhythms, we use a few examples to highlight ongoing research that addresses this question.

Inspired by the success of forward genetic screens at uncovering evolutionarily conserved mechanisms of circadian rhythms, researchers have used fruit flies to identify mutations that disrupt sleep. One of the first sleep mutants in *Drosophila* mapped to the gene encoding the voltage-gated K⁺ channel Shaker (Section 2.15); specifically, a reduction in Shaker function reduced sleep amount in *Drosophila* (**Figure 9-33**A). Subsequently, depletion of an essential accessory subunit (encoded by the gene *Hyperkinetic*) of the Shaker K⁺ channel, as well as of a protein that promotes Shaker expression, were found to cause similar phenotypes.

Given that K⁺ channels generally reduce neuronal excitability, a simple explanation is that reduced K⁺ channel activity promotes neuronal excitability and thus wakefulness. However, detailed analyses of where and how these channels act provided a more nuanced picture. In parallel with genetic screens for mutants that affect sleep, researchers also utilized circuit analysis tools to pinpoint which fly neurons regulate sleep. A small group of neurons that project to the dorsal fan-shaped body (dFB), a central brain neuropil in insects, was found to induce sleep when artificially activated. Remarkably, depletion of Shaker or Hyperkinetic only in dFB neurons reduced sleep, and reexpression of Hyperkinetic only in dFB neurons could restore sleep in otherwise *Hyperkinetic* mutant flies (Figure 9-33B). dFB neurons are more excitable when flies are sleep deprived; Shaker and Hyperkinetic *enhance* excitability and firing of dFB neurons and, in so doing, promote sleep. Hyperkinetic contains an oxidoreductase domain, and its activity is promoted by oxidative byproducts that accumulate during the waking period, which may account for the increased excitability of dFB neurons resulting from increasing sleep debt. Reexpression of a *Hyperkinetic* transgene with a point mutation in its oxidoreductase domain failed to restore sleep in *Hyperkinetic* mutant flies (Figure 9-33B). Together, these experiments suggest that regulation of excitability of dFB neurons in the fly brain contributes to homeostatic regulation of sleep.

Studies in mammals also suggest links between neuronal excitability and sleep–wake regulation. A forward genetic screen in mice using EEG as a readout identified a dominant mutant, *Dreamless,* with reduced REM sleep (Figure 9-33C, left). Molecular-genetic and electrophysiological studies revealed that *Dreamless* was a gain-of-function mutation in a voltage-independent, nonselective cation channel that increased its conductance. *Dreamless* is highly expressed in deep mesencephalic nuclei neurons whose activity is downregulated during REM sleep (REM-off neurons), and the *Dreamless* mutation increased excitability of these REM-off neurons (Figure 9-33C, right).

The association of excitability and sleep–wake is also suggested by a genetic study of humans who require fewer hours of sleep than most but are otherwise

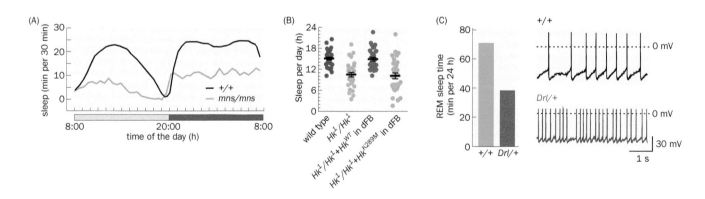

Figure 9-33 Sleep and neuronal excitability. (A) A forward genetic screen identified *minisleep* (*mns*), a mutation that disrupts the voltage-gated K⁺ channel Shaker and results in severely reduced sleep amount. The light–dark periods are indicated at the bottom. **(B)** Flies homozygous for a *Hyperkinetic* mutation (*Hk¹/Hk¹*) also exhibit reduced sleep amount. This phenotype is rescued by expressing a wild-type *Hk* transgene in dFB neurons, but not an *Hk* transgene with a point mutation (K289M) in the oxidoreductase domain. Each dot represents measurement of one fly; middle: mean ± standard error of the mean. **(C)** Left, mice heterozygous for *Dreamless* (*Drl/+*) exhibit reduced REM sleep compared to wild-type (+/+) control. Right, compared to control neurons, *Drl/+* neurons in a REM-off region of the deep mesencephalic nuclei exhibit more frequent spontaneous firing. (A, from Cirelli C, Bushey D, Hill S, et al. [2005] *Nature* 434:1087–1092. With permission from Springer Nature. B, from Kempf A, Song SM, Talbot CB, et al. [2019] *Nature* 568:230–234. With permission from Springer Nature. C, from Funato H, Miyoshi C, Fujiyama T, et al. [2016] *Nature* 539:378–383. With permission from Springer Nature.)

healthy. An autosomal dominant mutation that changes a single amino acid in the β_1 adrenergic receptor causes carriers to sleep on average only 5.7 hours per night, compared to 7.9 hours in noncarriers in the same family. When knocked into mice, the same mutation also reduces sleep by 1 hour. Further experiments suggested that the mutation enhances excitability of wake-promoting dorsal pons neurons that express the β_1 adrenergic receptor. Thus, in both the *Dreamless* and β_1 adrenergic receptor examples, sleep–wake behavior has been associated with excitability of sleep–wake-regulating neurons, although precise causality has yet to be established.

Finally, the neuromodulator adenosine has been proposed to contribute to homeostatic regulation of sleep based on the findings that: (1) the extracellular concentration of adenosine in the brain increases during wakefulness and declines during recovery sleep, and (2) caffeine, a potent wake-promoting stimulant, acts as an antagonist of G-protein-coupled adenosine receptors (Table 3-3). Disrupting the A_2A adenosine receptor in mice eliminates caffeine's wake-promoting effect, supporting a causal role of antagonistic adenosine signaling in promoting wakefulness. Questions remain regarding the extent to which endogenous adenosine contributes to regulation of sleep and what target neurons adenosine and caffeine act upon to regulate sleep–wake states.

We have only a coarse outline of the neural mechanisms that regulate sleep, and many questions remain to be answered by future investigations: How do different sleep regulatory neurons act? How are REM and non-REM sleep regulated? How is the activity of sleep regulatory neurons modulated by circadian rhythms? Are there evolutionarily conserved mechanisms that regulate sleep, as is the case for circadian rhythms?

9.17 Why do we sleep?

Perhaps the most enigmatic question regarding sleep is: Why do we need it? As introduced earlier, sleep is ubiquitous in mammals, birds, and reptiles, including prey animals that risk their lives to sleep. Indeed, some birds and marine mammals have developed uni-hemispheric sleep, in which half of the brain undergoes slow-wave sleep while the other half remains awake for the purpose of evading predators or maintaining flying or swimming during migration. Some invertebrates also sleep. "If sleep does not serve an absolutely vital function, then it is the biggest mistake the evolutionary process has ever made," remarked Allan Rechtschaffen, a sleep researcher whose following deprivation study demonstrated the importance of sleep.

To separate the effects of sleep disruption from the effects of the physical perturbations required to disrupt sleep, a pair of rats were housed on a shared disk over two water pans, and their EEGs were recorded continuously. Whenever the experimental rat was about to fall asleep, the change in its EEG pattern would trigger the disk to spin such that both rats had to move in order to stay on the disk. (Rats do not sleep in water.) In this experimental paradigm, both rats received identical physical stimuli, but the control rat could take naps when the experimental rat's EEG pattern indicated wakefulness (**Figure 9-34**). When totally deprived of sleep, experimental (but not control) rats died within a few weeks. This experiment thus demonstrated that sleep is essential for life; but exactly why this is so was not clear, since sleep deprivation induced a host of physiological changes, all of which could contribute to lethality. These included increased stress levels, increased energy expenditure, weight loss, compromised thermal regulation, a weakened immune system, and loss of gut integrity. It is difficult to pinpoint the primary cause of death and to separate causes and effects when so many parameters change at the same time.

Many interesting ideas have been proposed to explain why we sleep. One proposal is that animals may have evolved to perform at high levels during limited periods within the daily cycle, for instance, at times when food is most readily available; rest during other times might function to conserve energy. Indeed, many examples in nature can be found in which animals adapt their sleep patterns according to their ecological niche, as previously noted for fruit flies and mice.

Figure 9-34 The disk-over-water apparatus for studying the effects of sleep deprivation. Experimental and control rats are housed in separate cages sharing a ground disk. The disk spins whenever the EEG of the experimental rat indicates that it is about to fall asleep. When the disk spins, the rats cannot sleep because they must move to avoid falling into the water pans beneath the disk. However, the control rat can sleep when the experimental rat is alert. (Adapted from Rechtschaffen A, Gilliland MA, Bergmann BM, et al. [1983] *Science* 221:182–184.)

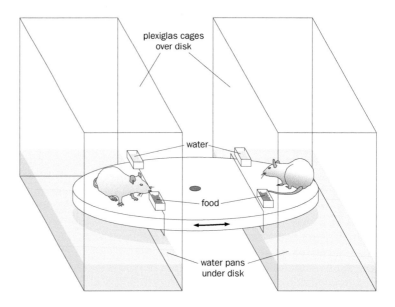

Another proposal is that sleep is necessary for restoration of key cellular components—such as machinery for protein synthesis, protein folding, and synthesis of lipids and membrane—that are consumed during the waking period. Indeed, changes in gene expression consistent with this hypothesis have been found: homologous genes from different species often exhibit similar shifts in expression pattern during sleep–wake cycles. However, neither of these proposals explains why quiet rest without sleep is insufficient. Another suggested function of sleep is to clear metabolic waste products that accumulate in the awake brain. Live imaging of fluorescent tracer-injected mouse brains revealed that the interstitial space (extracellular space surrounding neurons and glia) increases by 60% during sleep compared with the awake period; this facilitates the exchange of interstitial fluid with cerebral spinal fluid and helps clear waste products.

Perhaps the most intriguing proposal is that sleep facilitates memory, learning, and synaptic plasticity, topics that will be studied in detail in Chapter 11. Animal and human studies have shown that both declarative memory (required to recall names and events) and procedural memory (required to perform specific motor tasks) are enhanced by sleep, even by short naps. Physiological recordings in rodents have identified replay of hippocampal ensemble firing patterns resembling those observed during learning procedures that took place during the previous waking period, as if animals were rehearsing during sleep what they had just learned. As potential contributing factors to this phenomenon, levels and phosphorylation states of many synaptic proteins in flies and mice undergo specific changes as the waking period lengthens naturally or is prolonged experimentally and are restored to basal states after sleep.

It is possible that sleep originally evolved for one specific function and that more and more functions subsequently evolved to take advantage of the existence of the sleep state. Some of these later-evolved functions may have become just as vital as—or even more vital than—such an original function. With an expanding knowledge of how sleep is controlled by specific neural circuits and genes, we will have more ways of perturbing specific aspects of sleep in order to dissect its myriad functions.

SUMMARY

The nervous system senses and reacts not only to external stimuli but also to the internal milieu. The sympathetic and parasympathetic systems—the outputs of the autonomic nervous system—play complementary roles in regulating physiology. To a first approximation, the sympathetic system promotes energy expenditure and mobilizes the body for rapid action, whereas the parasympathetic system returns the body to a nonemergency state. The interoceptive system uses mechano- and chemosensitive receptors to sense parameters of body physiology and provides essential feedback to the autonomic nervous system. Hypothalamic nuclei integrate information from the sensory and interoceptive systems and, in turn, regulate autonomic output, hormone secretion, and organismal behavior to maintain homeostasis.

Eating is regulated by a number of neuronal types distributed across the brain, notably POMC and AgRP neurons in the arcuate nucleus of the hypothalamus. POMC neurons inhibit eating by secreting α-MSH, which activates the melanocortin-4 receptor on target neurons. AgRP neurons promote eating by antagonizing the actions of POMC neurons and inhibiting additional target neurons in the hypothalamus and brainstem. POMC and AgRP neurons are activated and inhibited, respectively, by leptin, an adipose tissue-derived feedback signal that inhibits eating. Whereas leptin and the pancreas-derived hormone insulin signal nutrient levels and act over long time scales to maintain energy balance, stomach-derived ghrelin and intestine-derived cholecystokinin (CCK) signal hunger and satiety, respectively, and act on short time scales to promote and inhibit eating, respectively.

Drinking is regulated by excitatory neurons in the median preoptic nuclei (MnPO). MnPO excitatory neurons are activated by interoceptive neurons in the nearby lamina terminalis that sense dehydration signals in the blood and broadcast the thirst signal across the brain. Water intake sends a rapid feedback signal to inhibitory neurons in the MnPO to reduce drinking well before dehydration signals in the blood decline. Hunger and thirst are powerful drives that motivate animals to seek and consume food and water. Activation of excitatory MnPO neurons is aversive and motivates animals to seek and consume water to reduce this activation, supporting the drive reduction theory of motivation. Hunger motivation may involve both drive reduction and incentive salience.

Circadian rhythms are self-sustained oscillations of an organism's biochemistry, physiology, and behavior that have a near 24 hour period and can be entrained by light. The central clockwork is highly conserved from insects to mammals and utilizes auto-inhibitory transcriptional feedback loops. Light entrainment is achieved by light-mediated degradation of circadian regulators in flies but is mediated by visual input that regulates transcription of circadian regulators in the suprachiasmatic nucleus (SCN) of the hypothalamus in mammals. While individually exhibiting pacemaker properties, SCN neurons also form an interacting network to control peripheral clocks via their interactions with other hypothalamic nuclei that control autonomic and neuroendocrine systems.

Like eating and drinking, sleep is homeostatically regulated; it is additionally timed by the circadian system. Sleep is universal in mammals, and sleeplike states are found in all vertebrates and some invertebrates. The sleep–wake cycle is controlled by mutually inhibitory groups of neurons in the hypothalamus and brainstem, with the arousal system utilizing several parallel pathways and signaling via monoamines, acetylcholine, and hypocretin. Sleep is vital to animals, with diverse functions that might have evolved to take advantage of this unique brain state.

OPEN QUESTIONS

- How do the interoceptive (input) and autonomic (output) systems interact with peripheral tissues?

- A biological set point is a useful concept for explaining homeostatic regulation of physiological functions, but are there neural substrates for such set points? What are they?

- How do the neural circuits that regulate eating and drinking integrate homeostatic needs and sensory feedback?

- What are the neural mechanisms by which hunger and thirst regulate motivated behavior?
- What are the respective functions of REM and NREM sleep?

FURTHER READING

Books and reviews

Andermann ML & Lowell BB (2017). Toward a wiring diagram understanding of appetite control. *Neuron* 95:757–778.

Cannon WB (1932). The Wisdom of the Body. W. W. Norton.

Mignot E (2008). Why we sleep: the temporal organization of recovery. *PLoS Biol* 6:e106.

Mohawk JA, Green CB, & Takahashi JS (2012). Central and peripheral circadian clocks in mammals. *Annu Rev Neurosci* 35:445–462.

Saper CB (2002). The central autonomic nervous system: conscious visceral perception and autonomic pattern generation. *Annu Rev Neurosci* 25:433–469.

Weber F & Dan Y (2016). Circuit-based interrogation of sleep control. *Nature* 538:51–59.

Interoceptive and neuromodulatory systems

Bai L, Mesgarzadeh S, Ramesh KS, Huey EL, Liu Y, Gray LA, Aitken TJ, Chen Y, Beutler LR, Ahn JS, et al. (2019). Genetic identification of vagal sensory neurons that control feeding. *Cell* 179:1129–1143.

Chang AJ, Ortega FE, Riegler J, Madison DV, & Krasnow MA (2015). Oxygen regulation of breathing through an olfactory receptor activated by lactate. *Nature* 527:240–244.

Ren J, Friedmann D, Xiong J, Liu CD, Ferguson BR, Weerakkody T, DeLoach KE, Ran C, Pun A, Sun Y, et al. (2018). Anatomically defined and functionally distinct dorsal raphe serotonin sub-systems. *Cell* 175:472–487.

Williams EK, Chang RB, Strochlic DE, Umans BD, Lowell BB, & Liberles SD (2016). Sensory neurons that detect stretch and nutrients in the digestive system. *Cell* 166:209–221.

Zeng WZ, Marshall KL, Min S, Daou I, Chapleau MW, Abboud FM, Liberles SD, & Patapoutian A (2018). PIEZOs mediate neuronal sensing of blood pressure and the baroreceptor reflex. *Science* 362:464–467.

Eating and drinking

Allen WE, DeNardo LA, Chen MZ, Liu CD, Loh KM, Fenno LE, Ramakrishnan C, Deisseroth K, & Luo L (2017). Thirst-associated preoptic neurons encode an aversive motivational drive. *Science* 357:1149–1155.

Atasoy D, Betley JN, Su HH, & Sternson SM (2012). Deconstruction of a neural circuit for hunger. *Nature* 488:172–177.

Balthasar N, Coppari R, McMinn J, Liu SM, Lee CE, Tang V, Kenny CD, McGovern RA, Chua SC Jr., Elmquist JK, et al. (2004). Leptin receptor signaling in POMC neurons is required for normal body weight homeostasis. *Neuron* 42:983–991.

Betley JN, Xu S, Cao ZFH, Gong R, Magnus CJ, Yu Y, & Sternson SM (2015). Neurons for hunger and thirst transmit a negative-valence teaching signal. *Nature* 521:180–185.

Campos CA, Bowen AJ, Schwartz MW, & Palmiter RD (2016). Parabrachial CGRP neurons control meal termination. *Cell Metab* 23:811–820.

Coleman DL (1973). Effects of parabiosis of obese with diabetes and normal mice. *Diabetologia* 9:294–298.

Hervey GR (1959). The effects of lesions in the hypothalamus in parabiotic rats. *J Physiol* 145:336–352.

Luquet S, Perez FA, Hnasko TS, & Palmiter RD (2005). NPY/AgRP neurons are essential for feeding in adult mice but can be ablated in neonates. *Science* 310:683–685.

Oka Y, Ye M, & Zuker CS (2015). Thirst driving and suppressing signals encoded by distinct neural populations in the brain. *Nature* 520:349–352.

Ryan PJ, Ross SI, Campos CA, Derkach VA, & Palmiter RD (2017). Oxytocin-receptor-expressing neurons in the parabrachial nucleus regulate fluid intake. *Nat Neurosci* 20:1722–1733.

Zhang Y, Proenca R, Maffei M, Barone M, Leopold L, & Friedman JM (1994). Positional cloning of the mouse obese gene and its human homologue. *Nature* 372:425–432.

Zimmerman CA, Huey EL, Ahn JS, Beutler LR, Tan CL, Kosar S, Bai L, Chen Y, Corpuz TV, Madisen L, et al. (2019). A gut-to-brain signal of fluid osmolarity controls thirst satiation. *Nature* 568:98–102.

Circadian rhythms and sleep

Adamantidis AR, Zhang F, Aravanis AM, Deisseroth K, & de Lecea L (2007). Neural substrates of awakening probed with optogenetic control of hypocretin neurons. *Nature* 450:420–424.

Brancaccio M, Edwards MD, Patton AP, Smyllie NJ, Chesham JE, Maywood ES, & Hastings MH (2019). Cell-autonomous clock of astrocytes drives circadian behavior in mammals. *Science* 363:187–192.

Chung S, Weber F, Zhong P, Tan CL, Nguyen TN, Beier KT, Hormann N, Chang WC, Zhang Z, Do JP, et al. (2017). Identification of preoptic sleep neurons using retrograde labelling and gene profiling. *Nature* 545:477–481.

Donlea JM, Pimentel D, & Miesenböck G (2014). Neuronal machinery of sleep homeostasis in *Drosophila*. *Neuron* 81:860–872.

Hardin PE, Hall JC, & Rosbash M (1990). Feedback of the *Drosophila* period gene product on circadian cycling of its messenger RNA levels. *Nature* 343:536–540.

Konopka RJ, & Benzer S (1971). Clock mutants of *Drosophila melanogaster*. *Proc Natl Acad Sci U S A* 68:2112–2116.

Lee AK & Wilson MA (2002). Memory of sequential experience in the hippocampus during slow wave sleep. *Neuron* 36:1183–1194.

Lin L, Faraco J, Li R, Kadotani H, Rogers W, Lin X, Qiu X, de Jong PJ, Nishino S, & Mignot E (1999). The sleep disorder canine narcolepsy is caused by a mutation in the hypocretin (orexin) receptor 2 gene. *Cell* 98:365–376.

Ralph MR, Foster RG, Davis FC, & Menaker M (1990). Transplanted suprachiasmatic nucleus determines circadian period. *Science* 247:975–978.

Rechtschaffen A, Gilliland MA, Bergmann BM, & Winter JB (1983). Physiological correlates of prolonged sleep deprivation in rats. *Science* 221:182–184.

Sehgal A, Price JL, Man B, & Young MW (1994). Loss of circadian behavioral rhythms and *per* RNA oscillations in the *Drosophila* mutant *timeless*. *Science* 263:1603–1606

Shi G, Xing L, Wu D, Bhattacharyya BJ, Jones CR, McMahon T, Chong SYC, Chen JA, Coppola G, Geschwind D, et al. (2019). A rare mutation of beta1-adrenergic receptor affects sleep/wake behaviors. *Neuron* 103:1044–1055.

Vitaterna MH, King DP, Chang AM, Kornhauser JM, Lowrey PL, McDonald JD, Dove WF, Pinto LH, Turek FW, & Takahashi JS (1994). Mutagenesis and mapping of a mouse gene, *Clock*, essential for circadian behavior. *Science* 264:719–725.

Xie L, Kang H, Xu Q, Chen MJ, Liao Y, Thiyagarajan M, O'Donnell J, Christensen DJ, Nicholson C, Iliff JJ, et al. (2013). Sleep drives metabolite clearance from the adult brain. *Science* 342:373–377.

CHAPTER 10

Sexual Behavior

When, as by a miracle, the lovely butterfly bursts from the chrysalis full-winged and perfect . . . it has, for the most part, nothing to learn, because its little life flows from its organization like a melody from a music box.

Douglas A. Spalding (1873)

Sex is nearly universal in the biological world. Even the bacterium *E. coli,* a unicellular prokaryote that reproduces asexually by rapid cell division, periodically engages in conjugation, exchanging genetic material between individual cells to produce recombinant progeny. In plants and animals, sexual reproduction becomes increasingly prevalent as organisms become more complex, and diverse strategies are employed to ensure mating success. As an interesting example, bee orchids have evolved flowers that resemble female bees and scents that mimic virgin bee pheromone, such that they attract male bees to "mate" with their flowers (**Figure 10-1**) and carry the orchid pollen from flower to flower. Thus, the bee orchid capitalizes on the bee's sexual behaviors to facilitate its own sexual reproduction.

In the preceding chapters, we have studied how animals perceive the world via their sensory systems, how they act via their motor systems, and how the nervous system is assembled during development. In the context of eating, drinking, and sleeping, we also started to address a central question in neurobiology: how does the nervous system generate behavior? In this chapter, we study sexual behavior to illustrate how sensory and motor systems and developmental processes are integrated to generate behaviors that are fundamental for the propagation of species.

Sexual behavior offers important experimental advantages to researchers investigating the neural bases of behavior. First, sexual behavior is robust and often **stereotyped** (varying little from one individual to another); both of these characteristics facilitate quantitative behavioral analyses. Second, sexual behavior has a strong innate component, as echoed in the epigraph from Spalding, is to a large extent specified by genetic programs, and can therefore be subjected to genetic analysis. Third, reproductive behaviors are **sexually dimorphic**, that is, they differ in females and males. In many species, sexually dimorphic behaviors result from differences in the **sex chromosomes**. These genetic differences between females and males have facilitated identification of elements critical for female- and male-typical behaviors.

In this chapter, we focus primarily on the sexual behavior of fruit flies and rodents, where the neural basis of sexual behavior has been best understood by utilizing the experimental advantages discussed earlier and by modern circuit dissection tools (see Chapter 14). By studying analogous behavior in animals that exhibit different levels of complexity, we can appreciate both the common principles and the diverse strategies that ensure animals' reproductive success.

HOW DO GENES SPECIFY SEXUAL BEHAVIOR IN THE FLY?

We have already encountered many examples in which the fruit fly *Drosophila melanogaster* has been used to address fundamental questions in neurobiology. The advanced genetic manipulations available in *Drosophila* make it an attractive model organism (Section 14.2). Studies of sexual behavior in flies have been further facilitated by the finding that individual genes can have large influences on sexual behavior.

Figure 10-1 A sexual decoy. The flower of the bee orchid *Ophrus apifera* resembles a female bee. The bee orchid evolved this deceptive flower pattern to attract male bees as pollinators. (Courtesy of Perennou Nuridsany/Science Source.)

10.1 *Drosophila* courtship behavior follows an instinctive, stereotyped ritual

Fruit flies congregate at food sources, where they find mating partners, carry out courtship, and mate. A male fruit fly uses mostly visual and chemosensory cues to find his appropriate mate: a virgin female of the same species. Male courtship consists of stereotyped behavioral elements (Figure 10-2; Movie 10-1): orienting toward the female, tapping her abdomen from behind, extending a single wing to produce a courtship song, licking her genitalia, and bending his abdomen to attempt copulation. If the female is receptive, she reacts by slowing down, letting the male tap, listening to his song, and allowing him to lick. For *Drosophila melanogaster*, the entire ritual from orienting to copulation usually lasts a few minutes.

The courtship ritual depends on multiple sensory modalities for communication between males and females. Orienting requires visual and chemosensory (pheromone) cues. Tapping transmits olfactory, taste, and somatosensory information. (In flies, taste receptor neurons are present not only in the proboscis—the insect equivalent of a mouth—but also on the forelegs.) Singing engages audition in both the female (the target audience for the courtship song) and the male (who is incited to court by hearing other males sing). Licking involves taste for males and somatosensation for females. Somehow, the brains of female and male fruit flies integrate these sensory cues and orchestrate a complex, extended interaction with their mating partners—but how?

Before we delve into the neural mechanisms, it is important to note that the courtship ritual is innate and genetically programmed. One can rear a male fly in isolation so that he has never seen or smelled a female. Within minutes of being introduced to a virgin female, the sight and smell of an appropriate partner will trigger mating behavior such that he can perform the entire ritual perfectly. On the other hand, not all aspects of the courtship behavior are fixed and immutable. For example, males adjust the amplitude of their courtship songs based on the distance of females they are courting (Figure 10-3). In blind males, this correlation breaks down when the distance is beyond 5 mm (about a body length of a fly). Thus, male flies use vision to estimate the distance to females and modulate their

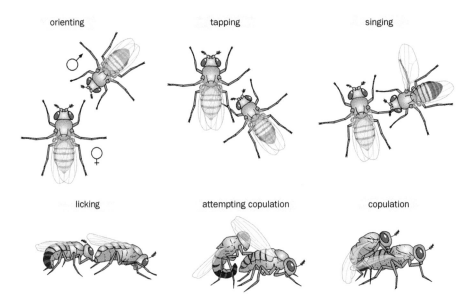

Figure 10-2 The courtship ritual of *Drosophila melanogaster*. Fruit fly courtship consists of stereotyped steps (not necessarily in a strict sequence) depicted here. The male finds the female and orients himself in front of her using visual and chemosensory cues. He moves to her side and taps her with his foreleg, which contains taste receptor neurons. He sings courtship songs by vibrating one of his wings. He licks her genitalia and attempts copulation. Successful courtship concludes with copulation. (From Greenspan RJ [1995] *Sci Am* 272[4]:72–78. With permission from Springer Nature.)

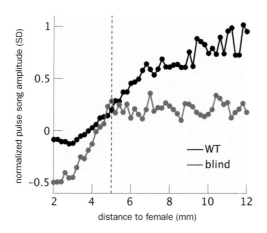

Figure 10-3 Males modulate the amplitude of their song according to female distance. Normalized amplitude of courtship songs (measured as standard deviation [SD] from the mean; see Figure 10-11 for a song record) produced by wild-type (WT) males is positively correlated with distance to females. In blind males, this correlation breaks down for distances larger than 5 mm (to the right of the dashed vertical line). (From Coen P. Xie M, Clemens J, et al. [2016] *Neuron* 89:629–644. With permission from Elsevier Inc.)

courtship song amplitude in real time, indicating remarkable control of acoustic communication during behavior.

10.2 *Fruitless (Fru)* is essential for many aspects of sexual behavior

A first step in the genetic dissection of a complex biological process is to isolate single-gene mutations that affect specific aspects of the process of interest. Identifying such mutations makes it possible to establish causal relationships between the genes and the biological processes they control, as we have seen in the case of *Period* and its control of circadian rhythms (Section 9.11).

A gene called ***Fruitless (Fru)*** has been found to regulate all aspects of the male courtship ritual in *Drosophila*. Males with mutations that eliminate a male-specific isoform of *Fru* (see the following section) usually do not exhibit courtship; under certain conditions, or with less severe mutations, specific aspects of the courtship ritual are affected, such as sex recognition, song production, and sperm transfer (Figure 10-4; Movie 10-1). By contrast, females are not morphologically or behaviorally affected by these mutations. What explains these the sexually dimorphic phenotypes?

In the fruit fly, sex is determined by the ratio of the copy numbers of X chromosomes and **autosomes** (non-sex chromosomes, A). Females (XX) have two copies of the X chromosome and two copies of each autosome and thus an X/A ratio of 1. Males (XY) have only one X chromosome and therefore an X/A ratio of ½. (The Y chromosome makes no contribution to sex determination in flies.) These different ratios cause sex-specific alternative splicing and gene expression through a regulatory hierarchy involving two splicing factors (Figure 10-5A). The transcript from the first promoter of *Fru* exhibits a sex-specific splicing pattern, producing a nonfunctional protein in females and a functional **Fru^M** protein in males (Figure 10-5B). Fru^M is a DNA-binding protein that controls the expression of other genes, consistent with its regulatory role.

The sex determination hierarchy also regulates sex-specific alternative splicing of a second gene called ***Doublesex (Dsx)***, producing Dsx^F in females and Dsx^M in males (Figure 10-5A). Dsx^F and Dsx^M are both transcription factors that regulate expression of many downstream target genes to specify the differentiation of female and male bodies and, to some extent, female and male brains, as we will see later.

10.3 Expression of male-specific Fru^M in females is sufficient to generate most aspects of male courtship behavior

As we have seen from its loss-of-function phenotypes, Fru^M is necessary for multiple aspects of the male courtship ritual. Is Fru^M also sufficient to generate male sexual behavior? This was tested by deleting the genomic DNA fragment that enables female-specific splicing of the endogenous *Fru* gene (we refer to the resulting variant as *Fru^Δ*; see Figure 10-5B legend). As a result, both males and

Figure 10-4 Male *Fruitless (Fru)* mutants exhibit altered sexual behavior. Whereas wild-type males court only females, some *Fru* mutant males court males and females indiscriminately. These *Fru* mutant males sometimes form a courtship chain: each male courts the male in front of him while being courted by the male behind him. (From Hall JC [1994] *Science* 264: 1702–1714. With permission from AAAS.)

Figure 10-5 The sex determination hierarchy and its regulation of *Fruitless*. **(A)** In females, an X/A ratio of 1 leads to expression of functional Sex lethal (Sxl), a splicing factor required to produce functional Transformer (Tra). Together with Tra2, Tra controls the alternative splicing of the *Dsx* and *Fru* genes, resulting in expression of Dsx^F protein. The male X/A ratio of ½ does not yield functional Sxl and consequently generates no Tra. This leads to male-specific splicing that produces Dsx^M and Fru^M. Dsx^F and Dsx^M specify sexual dimorphic female and male bodies, respectively. Fru^M along with Dsx specifies sexual dimorphism of brain and behavior. **(B)** Top, the *Fruitless* transcript from the first promoter; exons are depicted as blocks joined by introns (thin lines). Note the sex-specific splicing of the second exon. In males, the default splicing joins the blue portion of the second exon with the 3′ exons (simplified here as a single gray exon); the resulting translation product is a functional, male-specific Fru^M protein (bottom). Female-specific splicing caused by binding of Tra/Tra2 (indicated by the star) yields a transcript including the second exon's white segment, which has multiple stop codons and does not produce a functional protein. Deleting this white segment produces *Fru^Δ* (see Figure 10-6). (Based on Baker BS, Taylor BJ, & Hall JC [2001] *Cell* 105:13–24.)

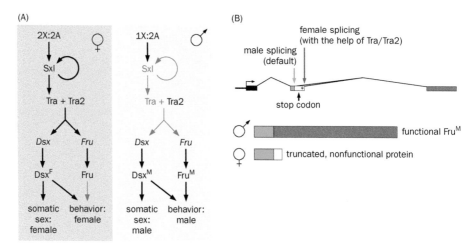

females utilize the male form of splicing and thus produce functional Fru^M proteins (**Figure 10-6**A).

As expected, *Fru^Δ* males acted like normal males, since their splicing pattern was unaffected by the genetic modification. What happened to *Fru^Δ* females, which were forced to express Fru^M protein? Although they looked like females, they acted like males. *Fru^Δ* females vigorously courted other females and performed the male courtship ritual like normal males (Figure 10-6B, C; Movie 10-1), except that they did not copulate because their external anatomy was still female. This remarkable result demonstrated that changing the splicing pattern of a single gene is sufficient to confer many aspects of male-typical courtship behavior to females.

10.4 Activity of Fru+ neurons promotes male courtship behavior

How does Fru^M expression in females confer male-typical behavior? To answer this question, we must step back and examine which cells in males normally express Fru^M. Antibodies specific for the Fru^M isoform were used to label groups of neuronal nuclei in the male brain (Figure 10-6A) and ventral nerve cord (the

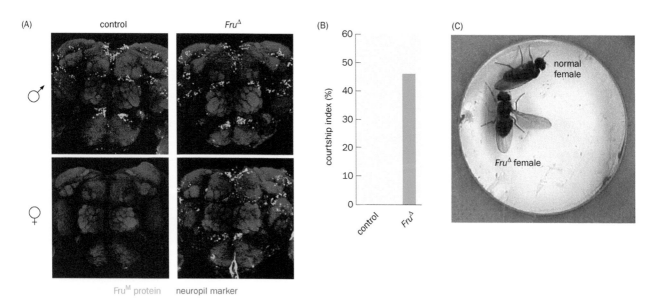

Figure 10-6 Expression of Fru^M in females confers male-typical sexual behavior to females. **(A)** In wild-type control flies, the Fru^M protein (green, visualized by antibody staining) is produced in males but not females. By contrast, both male and female *Fru^Δ* flies produce Fru^M. Magenta, staining of a synaptic marker that highlights the neuropil organization of the *Drosophila* central brain. **(B)** Whereas control females do not court other females, *Fru^Δ* females robustly court wild-type females, as indicated by the high courtship index, the percentage of time spent on courtship when two flies are placed in a chamber. **(C)** A *Fru^Δ* female courting a wild-type female by singing a courtship song. (From Demir E & Dickson BJ [2005] *Cell* 121:785–794. With permission from Elsevier Inc.)

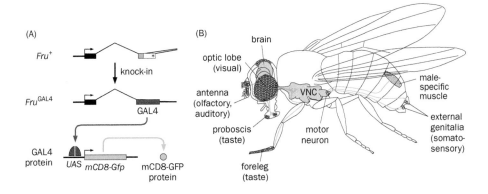

Figure 10-7 **Strategies to investigate Fru^M expression and function.** (A) *Fru^GAL4*, in which the yeast transcription factor GAL4 is knocked into the *Fru* locus after the first promoter, mimics the expression pattern of Fru^M. Bottom, in flies carrying both *Fru^GAL4* and *UAS-mCD8-Gfp* transgenes, GAL4 proteins bind to *UAS* to activate expression of a membrane-tethered green fluorescence protein, allowing visualization of axonal projections of Fru^M-expressing neurons. Other *UAS* transgenes have been used to silence or activate these neurons or to impair endogenous Fru^M expression. (B) Using *Fru^GAL4* to drive expression of mCD8-GFP, it was found that *Fru^GAL4* (and, by inference, Fru^M) is expressed in subsets of olfactory receptor neurons and auditory sensory neurons in the antenna, taste receptor neurons in the proboscis and foreleg, visual neurons in the optic lobe, mechanosensory neurons in the external genitalia, and motor neurons in the ventral nerve cord (VNC) that innervate a male-specific muscle. Green lines highlight the axonal projections of these neurons. (B, adapted from Billeter JC, Rideout EJ, Dornan AJ, et al. [2006] *Curr Biol* 16: R766–R776. With permission from Elsevier Inc.)

insect equivalent of the vertebrate spinal cord). In total, Fru^M was found to be expressed in the nuclei of about 2000 neurons, constituting 1% of the ~200,000 total neurons in *Drosophila* CNS.

Fru^M distribution was also studied via a binary transgene expression method (see Section 14.9 for details) featuring knock-in of the yeast transcription factor GAL4 under the control of the first *Fru* promoter. This transgene, denoted as *Fru^GAL4*, can drive the expression of any transgene containing a GAL4-binding upstream activation sequence (*UAS*), resulting in an expression pattern mimicking Fru^M expression (**Figure 10-7**A). Using *Fru^GAL4* to drive expression of a membrane-tethered green fluorescent protein that labels not only cell bodies but also axons and dendrites, researchers discovered that Fru^M is also expressed in a subset of olfactory receptor neurons and most auditory neurons in the antenna, a subset of taste receptor neurons in the foreleg and proboscis, a subset of visual neurons in the optic lobe, somatosensory neurons in the external genitalia, and the motor neurons innervating a male-specific muscle (Figure 10-7B). Fru^M expression is thus consistent with the likely functions of these neurons in different aspects of courtship (Section 10.1).

To assess the function of Fru^M-expressing neurons (designated as Fru+ neurons hereafter) in courtship behavior, *Fru^GAL4* was used to express an effector transgene that silences neuronal activity (a loss-of-function manipulation). A widely used neuronal silencer in flies is a temperature-sensitive mutant protein called Shibire^ts (Shi^ts). At a low temperature (19°C), Shi^ts expression is innocuous. At a high temperature (29°C), Shi^ts expression blocks synaptic vesicle recycling (Section 3.9) and therefore shuts off synaptic transmission. Thus, Shi^ts can be used to silence neurons expressing it by moving Shi^ts-expressing flies from low to high temperature. When all Fru+ neurons were transiently silenced in male flies, courtship behavior was severely impaired (**Figure 10-8**A), indicating that the activity of Fru+ neurons is acutely required in adult males for execution of courtship

Figure 10-8 **Testing the necessity and sufficiency of Fru+ neurons in male courtship.** (A) Male flies carrying both *Fru^GAL4* and *UAS-Shi^ts* transgenes (red, *GAL4 + UAS* transgenes) exhibit reduced courtship toward virgin females at the restrictive temperature (29°C) but not the permissive temperature (19°C). Control males (*GAL4* or *UAS* transgene only) exhibit courtship at both temperatures. This indicates that blocking synaptic transmission from Fru+ neurons impairs courtship. (B) As temperature rises, males expressing dTRPA1 in Fru+ neurons increasingly elicit courtship-like behavior in the absence of a female partner, including unilateral wing extension, and motor actions resembling tapping, licking, and attempted copulation. (A, from Stockinger P, Kvitsiani D, Rotkopf S, et al. [2005] *Cell* 121:795–807. With permission from Elsevier Inc. B, from Kohatsu S, Koganezawa M, & Yamamoto D [2011] *Neuron* 69:498–508. With permission from Elsevier Inc.)

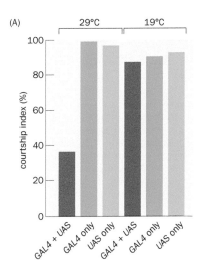

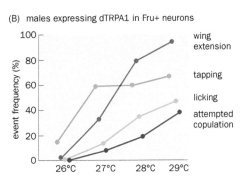

(B) males expressing dTRPA1 in Fru+ neurons

behavior. However, many other aspects of fly behavior, including locomotion, flight, phototaxis, and olfaction and taste in general, were not affected. Thus, Fru+ neurons appear to be largely dedicated to male courtship performance.

Remarkably, artificial activation of Fru+ neurons in isolated males can promote multiple steps of the courtship ritual in the *absence* of a mating partner. This gain-of-function manipulation was achieved by expressing a heat-activated cation channel, dTRPA1, in Fru+ neurons. (Some fly TRP channels are used to sense temperatures, as in mammals; Section 6.29.) At higher temperatures, Fru+ neurons became increasingly depolarized and activated, and isolated males increasingly exhibited motor actions resembling tapping, singing, licking, and even attempted copulation (Figure 10-8B), although these were not assembled into coordinated, sequenced displays in the absence of a partner. Thus, both loss- and gain-of-function experiments indicate that Fru+ neurons play a central role in promoting male courtship behavior.

10.5 Fru+ neurons orchestrate multiple aspects of male courtship behavior

How do Fru+ neurons control courtship behavior in males? Given that Fru^M is expressed in specific subsets of sensory, motor, and central neurons (Figure 10-6A; Figure 10-7B), one hypothesis is that these neurons mediate sensory perception, motor performance, and central coordination in the context of sexual behavior. This hypothesis has received strong experimental support, and these experiments have outlined the neural circuitry that drives different aspects of male courtship behavior (**Figure 10-9**A).

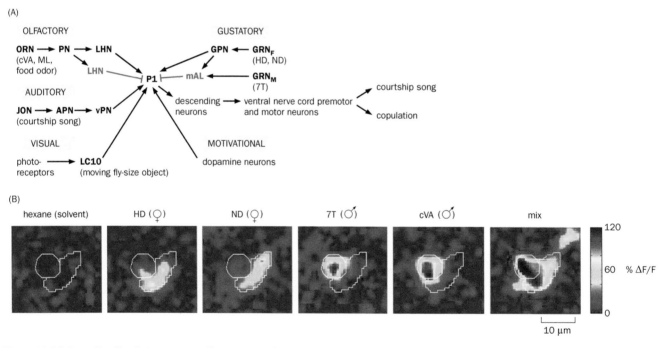

Figure 10-9 Information flow between some Fru+ neurons that regulate aspects of male courtship behavior. (A) Schematic summarizing several groups of Fru+ neurons (in bold font) that participate in different aspects of courtship. Arrows indicate information flow via both direct and indirect connections. Red, inhibitory neurons and their output. Representative sensory stimuli are in parentheses below the neurons (see text for details). ORN, olfactory receptor neuron; PN, projection neuron; LHN, lateral horn neuron; GRN$_F$ and GRN$_M$, gustatory receptor neuron that responds to female and male cues, respectively; GPN, gustatory projection neuron; mAL, a cluster of Fru+ neurons in the gustatory processing center of the brain; JON, Johnston's organ neuron; APN, auditory projection neuron; vPN, ventromedial protocerebrum-projecting neuron; LC10, Fru+ neurons in the optic lobe. **(B)** Changes in the fluorescence intensity over baseline (ΔF/F) of a genetically encoded Ca^{2+} indicator (Section 14.22) in response to different taste stimuli reveal the activity of two Fru+ gustatory receptor neurons (outlined) in a male foreleg. One neuron is activated by female hydrocarbons HD and ND, and the other is activated by male hydrocarbons 7T and cVA. Hexane is the solvent for the hydrocarbons and serves as a negative control. (A, based on Auer TO & Benton R [2016] *Curr Opin Neurobiol* 38:18–26. B, from Thistle R, Cameron P, Ghorayshi A, et al. [2012] *Cell* 149: 1140–1151. With permission from Elsevier Inc.)

Fru^M is expressed in 3 of the 50 types of olfactory receptor neurons (ORNs; Section 6.11). Silencing all three types of Fru+ ORNs impairs courtship in the dark, indicating that the olfactory cues processed by these ORN types play important roles in courtship. One type of Fru+ ORN senses 11-*cis*-vaccenyl acetate (cVA), a pheromone produced by males that is transferred from male to female during mating (Section 6.13) and inhibits male–male and male–mated female courtship. The other two types of Fru+ ORNs both promote male–virgin female courtship: one senses the fly odor methyl laurate (ML) and the other senses aromatic odors enriched in food sources, which act as aphrodisiacs to promote mating, as fruit flies mate near their food sources. The downstream circuits for the cVA processing channel have been delineated, through olfactory projection neurons and third-order excitatory and inhibitory lateral horn neurons to the integrative P1 neurons (Figure 10-9A; we will discuss P1 neurons shortly).

The taste system in flies resembles the mammalian taste system we studied in Chapter 6, with different gustatory receptor neurons activated by sweet and bitter tastants. An important difference is that a subset of fly gustatory receptor neurons (GRNs) appears to be dedicated to sensing hydrocarbon molecules on the bodies of other flies for discriminating sex and species, for the purpose of selecting appropriate mating partners. Some Fru+ GRNs on the foreleg are activated by male-enriched hydrocarbons 7T (7-tricosene) and cVA while others are activated by female-enriched hydrocarbons HD (7,11-heptacosadiene) and ND (7,11-nonacosadiene) (Figure 10-9B). Ascending pathways that transmit this taste information to P1 neurons have been identified, with female cues activating both excitatory and inhibitory Fru+ interneurons and male cues activating Fru+ inhibitory interneurons (Figure 10-9A).

The auditory system is highly enriched in Fru+ neurons, from mechanosensory neurons in the Johnston organ in the second antennal segment to second-order auditory project neurons. Additionally, a group of high-order Fru+ ventrolateral protocerebrum projection neuron (vPNs) may relay auditory information to P1 neurons. Finally, a cluster of Fru+ neurons in the optic lobe (Box 4-3) named LC10 has been found to be tuned to fly-size moving objects and is required for orienting toward and maintaining proximity to the female (Figure 10-9A).

We have seen that information from multiple sensory systems converges onto **P1 neurons**, a cluster of ~20 Fru+ neurons in the central brain. What are these neurons? The discovery of P1 neurons as a central regulator of male courtship behavior showcased the advanced genetic tools available in *Drosophila*. P1 neurons were originally identified in a mosaic genetic screen (see Sections 14.8 and 14.16 for details) for small groups of neurons that, when masculinized in an otherwise female fly, confer male-typical courtship behavior. Subsequently, researchers found that artificial activation of P1 neurons alone was sufficient to elicit tapping, unilateral wing extension, and licking behavior with high probability (**Figure 10-10**A, B). As we already learned, P1 neurons receive sensory input from olfactory, gustatory, auditory, and likely visual systems, thus serving as a multisensory integration center. Accordingly, physiological studies revealed that P1 neurons were activated by gustatory cues from the female body, and this activation was inhibited by cVA signals processed by the olfactory system (Figure 10-10C). This may account for why mated females are less attractive to males. Optogenetic studies indicated that transient P1 neuron activation had a long-lasting effect of promoting male courtship (Figure 10-10D), suggesting that activity of P1 neurons represents a brain state conducive to courtship behavior. (Some P1 neurons also participate in male aggression toward other males.) A specific population of dopamine neurons innervate P1, and its activity reflects the motivational drive for mating. When these neurons are very active, dopamine release activates dopamine receptors in P1 neurons, tuning their sensitivity to GABA down and their sensitivity to acetylcholine (the main excitatory neurotransmitter in the fly CNS) up, thus promoting courtship initiation and continuation.

P1 neurons connect with neurons that project axons to the ventral nerve cord (Figure 10-10A), and their activation can thus promote mating-related motor programs. A hallmark of male *Drosophila* courtship is the courtship song. Given that

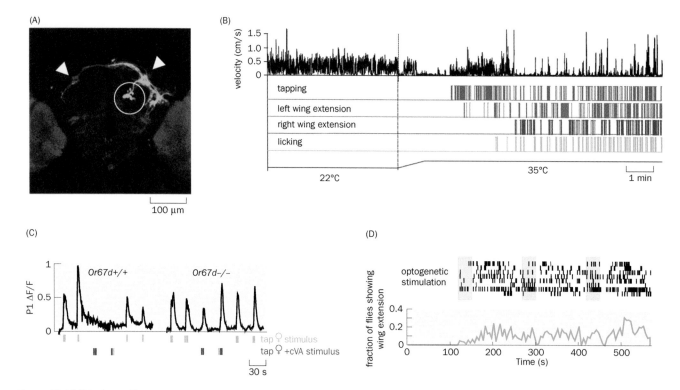

Figure 10-10 P1 cluster Fru+ neurons integrate multisensory input and control courtship behavior. (A) A neuroblast clone containing the P1 cluster of Fru+ neurons shows the location of their cell bodies (circle) and elaborate projections both ipsilaterally (right hemisphere) and contralaterally (left hemisphere), visualized by the MARCM method (see Section 14.16 for details). Arrowheads indicate the areas where P1 processes overlap with another group of Fru^M neurons (not shown) that send descending projections to the ventral nerve cord. **(B)** An ethogram (quantitative plot of behavior over time) of a male fly with a P1 neuroblast clone expressing dTRPA1, which causes neuronal depolarization in response to heat. After raising the temperature (and hence activating P1 neurons), the fly slows down (top) and begins to exhibit tapping, unilateral wing extension, and licking behaviors in isolation. **(C)** Ca^{2+} imaging of P1 neurons reveals their activation by tapping females (orange), but this activation is suppressed by simultaneous presentation of cVA (purple) in wild type (left). cVA-mediated suppression is absent in flies lacking the cVA receptor Or67d (right). **(D)** Optogenetic activation of P1 neurons (green bars) expressing a channelrhodopsin variant shows that P1 activation elicits unilateral wing extension, and this behavior manifests beyond the optogenetic stimulation windows. Top, raster plot with each row representing one fly and each tick one wing extension. Bottom, quantification. (A & B, from Kohatsu S, Koganezawa M, & Yamamoto D [2011] *Neuron* 69:498–508. With permission from Elsevier Inc. C, from Clowney EJ, Iguchi S, Bussell JJ, et al. [2015] *Neuron* 87: 1036–1049, With permission from Elsevier Inc. D, from Inagaki HK, Jung Y, Hoopfer ED, et al. [2014] *Nat Methods* 11:325–332. With permission from Springer Nature.)

Fru^M is expressed in a subset of neurons in the ventral nerve cord, it was hypothesized that these neurons control the unilateral wing beat that produces the courtship song. This hypothesis was supported by an experiment utilizing light-induced uncaging of ATP to activate an ATP-gated cation channel expressed in Fru+ neurons. Activation of Fru+ neurons in the ventral nerve cord in headless flies generated a courtship song resembling the song of a normal courting fly (**Figure 10-11**). The effectiveness of these songs was tested in a behavioral assay. When a male's wings are removed, his copulation success is drastically reduced because he cannot attract his partner's attention via a courtship song. Playing back a recorded courtship song complements the mating defect of a wingless male, and the recorded song resulting from artificial activation of Fru+ neurons was as effective as the song of a wild-type courting fly.

In summary, studies of Fru+ neurons have supported a model in which these neurons play important roles in many aspects of male courtship behavior: Fru+ sensory neurons process mating-related sensory cues; Fru+ neurons in the central brain integrate these sensory cues and transmit them to the ventral nerve cord; Fru+ ventral nerve cord neurons execute mating-related motor programs. Thus, studies of Fru+ neurons have provided an excellent example of the dissection of neural circuits underlying a complex suite of behaviors.

While we have focused on the key roles of Fru+ neurons in regulating male courtship behavior, male-specific Doublesex (Dsx^M), a component parallel to Fru in the sex determination hierarchy (Figure 10-5), is also expressed in the brain and contributes to male courtship behavior. The role of Dsx^F-expressing neurons in female courtship behavior has also been extensively studied, which we now turn to.

10.6 Fru+ and Dsx+ neurons promote female receptivity to courtship

How is female sexual behavior controlled? As noted previously, females do not express Fru^M protein. Are the groups of neurons expressing Fru^M in males entirely absent in females, or are they present in females but just unable to produce Fru^M protein as a result of sex-specific alternative splicing (Figure 10-5)? Fru^{GAL4} was used to distinguish between these possibilities, as it mimics transcription from the first *Fruitless* promoter but is not subjected to regulation by sex-specific alternative splicing (Figure 10-6A). Researchers found that females did indeed express Fru^{GAL4} in specific populations of neurons that grossly resemble the Fru+ neurons of males, which we designate as Fru+ neurons in females. However, as we will learn in the next section, substantial dimorphisms exist in both the number and wiring patterns of Fru+ neurons in males and females. In addition, when Dsx^{GAL4} (GAL4 knocked in at the *Doublesex* locus) was used to visualize *Dsx*-expressing cells in the brain, it was found that *Dsx* is expressed in neurons of both the male and the female CNS (we designate these as Dsx+ neurons, which express Dsx^M in males and Dsx^F in females). Indeed, there is substantial (though incomplete) overlap between Dsx+ and Fru+ neurons in both males and females. Investigating the functions of Dsx+ and Fru+ neurons in females provided entry points into studying the neural circuits that regulate female receptivity (**Figure 10-12**A).

Systematic examination of Dsx+ neurons in females identified a cluster of Dsx+/Fru– neurons in the central brain named **pC1 neurons**, which are located in an analogous position to and have similar projection patterns as P1 neurons in males (Figure 10-12A; we will discuss the relationship between P1 and pC1 neurons in the next section). Activation of pC1 neurons enhanced female receptivity to copulation (Figure 10-12B), indicating that pC1 neurons promote female receptivity. Like male P1 neurons, female pC1 neurons receive input from multiple sensory modalities, including the male pheromone cVA and courtship song, both of which promote receptivity of virgin females. Among targets of pC1 neurons is a pair of descending neurons called vpoDN (Figure 10-12A). Optogenetic activation of vpoDNs triggers vaginal plate opening (hence the name of the neuron) required for copulation, and genetic ablation of vpoDNs causes a complete blockade of copulation.

Whereas virgin females are receptive to male courtship stimuli, mated females actively reject suitors. Interestingly, when Fru+ neurons in virgin females were reversibly silenced at a high temperature via *UAS-Shi^{ts}* expression, their receptivity to courtship was greatly diminished, becoming comparable to that of mated females (Figure 10-12C). Thus, Fru+ neurons promote courtship receptivity in virgin females. A key group of Fru+ neurons that contribute to this effect are six to eight sensory neurons that innervate the female reproductive tract. These neurons express a G-protein-coupled receptor for the **sex peptide**, which is produced in the male and transferred in seminal fluid to females during mating in order to reduce female receptivity. (This strategy, along with transferring cVA to mated females to inhibit attraction of mated females to potential suitors, has obvious evolutionary significance: successfully mated males can minimize sperm competition and maximize the likelihood of propagating their own genes.) In mated females, RNAi knockdown (see Section 14.7 for details on RNAi) of sex peptide receptor expression in Fru+ neurons enhanced receptivity; however, suppressing the knockdown, and therefore restoring expression of the sex peptide receptor only in these sex peptide sensory neurons (SPSNs), was sufficient to reduce receptivity in mated females (Figure 10-12D). In virgin females, SPSNs are active; binding of the sex peptide to its receptor on SPSNs reduced SPSN activity. Thus, silencing SPSNs recapitulates sex peptide binding, accounting for the effect of silencing Fru+ neurons discussed earlier.

Using a combination of circuit dissection tools featuring genetic access to specific cell types; anatomical tracing, physiological recording, and manipulation

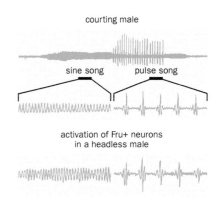

Figure 10-11 Fru+ neurons in the ventral nerve cord control the production of courtship song. Top, the courtship song of a normal courting fly (top) consists of alternating sequences of sine song and pulse song. Bracketed segments are magnified below. Bottom, ventral nerve cord Fru+ neurons in headless male flies expressing an ATP-gated cation channel and injected with caged ATP can be activated by light, which induces ATP uncaging (see Figure 14-46 for details). This produces a song closely resembling the song of a courting male fly. (Adapted from Clyne JD & Miesenböck G [2008] *Cell* 133:354–363. With permission from Elsevier Inc.)

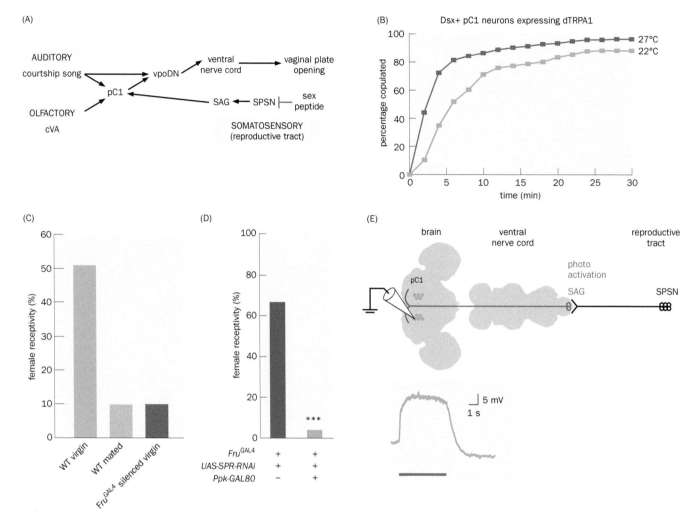

Figure 10-12 Regulation of female courtship receptivity. (A) Schematic of the female nervous system depicting a subset of neurons that regulate courtship receptivity. In virgin females, pC1 neurons integrate auditory, olfactory, and somatosensory inputs and in turn activate vpoDNs (vaginal plate opening descending neurons, which also receive auditory input more directly). vpoDNs project axons to the ventral nerve cord to promote vaginal plate opening, which is essential for copulation. pC1 neurons receive tonic excitation from SAG neurons of the ventral nerve cord, which are activated by sex peptide sensory neurons (SPSNs) in the reproductive tract. After mating, the sex peptide inhibits SPSNs, thereby reducing the excitatory drive onto SAG neurons and pC1 neurons. **(B)** dTRPA1 is expressed in Dsx+ pC1 neurons in female virgins using an intersectional approach (Section 14.11). At the higher temperature, when dTRPA1-expressing neurons are more depolarized, mating occurs more quickly. **(C)** Reversible silencing of Fru+ neurons at a high temperature using Shi^ts (following the same procedure as in Figure 10-8A) causes experimental virgin females (column 3) to behave like wild-type (WT) mated females (column 2), with markedly reduced receptivity compared to WT virgins (column 1).

(D) GAL4/UAS-mediated RNAi knockdown of sex peptide receptor (SPR) expression in Fru+ neurons in mated females causes them to exhibit high courtship receptivity (column 1). Inhibiting this knockdown in SPSNs (by expressing GAL80, an inhibitor of GAL4; GAL80 is driven by the Ppk promoter expressed in SPSNs) reduces receptivity to the level of normal mated females (column 2). **(E)** Top, schematic of the experiment. Whole-cell recording was performed on a pC1 neuron responding to optogenetic stimulation of SAG neurons, whose cell bodies are located in the abdominal ganglion of the ventral nerve cord and whose axons project to the brain. Bottom, optogenetic activation of SAG neurons depolarizes the pC1 neuron. (A, courtesy of Barry Dickson. B, from Zhou C, Pan Y, Robinett CC, et al. [2014] *Neuron* 83:149–163. C, from Kvitsiani D & Dickson BJ [2006] *Curr Biol* 16:R355–R356. With permission from Elsevier Inc.; D, from Yang CH, Rumpf S, Xiang Y et al. [2009] *Neuron* 61:519–526. With permission from Elsevier Inc. See also Häsemeyer M, Yapici N, Heberlein U, et al. [2009] *Neuron* 61:511–518 and Yapici N, Kim YJ. Ribeiro C, et al. [2008] *Nature* 451:33–38. E, courtesy of Barry Dickson. See also Wang F, Wang K, Forknall N, et al. [2020] *Nature* 579:101–105.)

of these cell types; and serial EM reconstruction (see Chapter 14), researchers have made remarkable progress in understanding the neural basis of the change in female receptivity. SPSNs project their axons to the ventral nerve cord and synapse onto a group of neurons called SAG neurons, which in turn project their axons to the brain, where they synapse directly onto pC1 neurons. SPSNs and SAG neurons are both excitatory, and their activation leads to activation of pC1 neurons (Figure 10-12E). Taken together, these studies provide a satisfactory account for receptivity differences in virgin and mated females. Active SPSNs in virgin females, through SAG neurons, provide an excitatory drive for pC1 neurons and thus promote receptivity of virgin females to courtship. Silencing of SPSNs by sex peptide binding inhibits this excitatory drive, thus decreasing the receptivity of mated females (Figure 10-12A). Interestingly, a recent study revealed that the physical act of copulation itself, independent of sex peptide, causes a short-term reduction of female receptivity. This effect is mediated by mechanosensory neurons in the reproductive tract activating a specific population of ventral nerve cord excitatory neurons that project to the brain. Thus, multiple pathways act in parallel to modify female receptivity based on their mating status.

10.7 Fru^M and Dsx regulate sexually dimorphic neuronal numbers and wiring

In previous sections, we learned that Fru+ and Dsx+ neurons promote male courtship behavior, whereas Fru+ and Dsx+ neurons in virgin females promote receptivity. How do males and females exhibit these dimorphic sexual behaviors?

Sexual dimorphism in *Drosophila* originates in the sex-specific splicing of *Fru* and *Dsx* (Figure 10-5). Given the principles of neural development we learned in Chapter 7, *Fru* and *Dsx* could differentiate the male and female nervous systems by generating different (1) numbers of neurons, (2) wiring patterns, or (3) functional properties of neurons. These possibilities are not mutually exclusive; while the third mechanism remains to be tested, the first two mechanisms have received ample experimental support.

Overall, there are about 900 Dsx+ neurons in males but only about 700 in females. Thus, there is global sexual dimorphism in neuronal numbers. However, in some areas of the nervous system there are more Dsx+ neurons in females than males; one such example is the posterior ventral nerve cord, which is implicated in regulating female reproductive behaviors such as egg laying. Likewise, when discrete subsets of Fru+ neurons were examined in males and females, sexual dimorphism was also found to be prevalent. For example, P1 neurons, which play a central role in integrating sensory information and regulating behavioral output (Figure 10-9A), are present only in males (Figure 10-13A, B). In fact, P1 neurons, which are Fru+/Dsx+, derive from the same neuroblast (neural progenitor) as pC1 neurons, which are Dsx+/Fru– and play a key role in regulating female courtship receptivity (Figure 10-12A). However, the same neuroblast lineage contains far fewer neurons in females than in males.

As we learned in Chapter 7, neuronal numbers in the developing nervous system can be regulated by controlling division of progenitor cells or programmed cell death after neuronal birth. The latter was found to be the predominant mechanism for producing sexual dimorphism. When genes essential for programmed cell death were eliminated in the neuroblast that produces P1 neurons, these neurons persisted in the female brain (Figure 10-13C). Genetic analysis indicates that Dsx^F expression in developing female P1 neurons activates a cell death program that eliminates these neurons after they are born. Indeed, differences in the number of Dsx+ CNS neurons in males and females were largely eliminated when *Dsx^{GAL4}* was used to drive a transgene that inhibits programmed cell death.

(A) (B) (C)

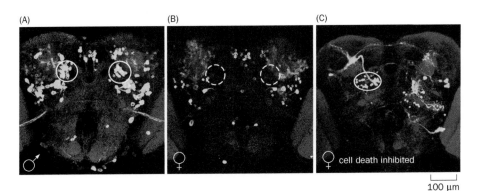

100 µm

Figure 10-13 **Programmed cell death accounts for the sexual dimorphism of P1 neurons.** The P1 cluster of Fru+ neurons (circled) is present in males **(A)** but absent in females **(B)**. When programmed cell death is inhibited in the neuroblast lineage that produces P1 in the left hemisphere, these neurons are retained in females **(C)**. (From Kimura K, Hachiya T, Koganezawa M, et al. [2008] *Neuron* 59:759–769. With permission from Elsevier Inc.)

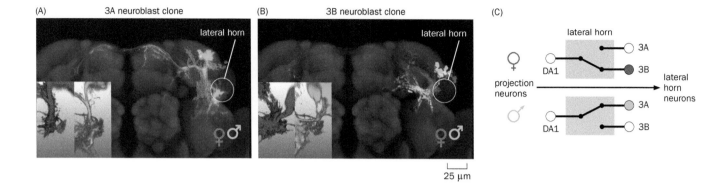

25 μm

Figure 10-14 A sexually dimorphic circuit switch in a pheromone-processing pathway. (A) Neuroblast clones containing 3A neurons in a male (green) and a female (magenta) are registered to a standard brain based on neuropil staining (gray). Male and female 3A neurons exhibit substantial differences in their dendrite projections in the lateral horn. The inset shows that male but not female 3A dendrites overlap substantially with axons of DA1 projection neurons (yellow). **(B)** 3B neurons also exhibit sexual dimorphism, with female but not male dendrites overlapping substantially with axons of DA1 projection neurons (inset). **(C)** A circuit switch model. Along the olfactory pathway from projection neurons to lateral horn neurons, DA1 projection neurons connect with 3B lateral horn neurons in females (top) but 3A lateral horn neurons in males (bottom). (Adapted from Kohl J, Ostrovsky AD, Frechter S, et al. [2013] *Cell* 155:1610–1623. With permission from Elsevier Inc.)

In addition to differences in neuronal numbers, some Fru+ and Dsx+ neurons also exhibit sexually dimorphic wiring. For example, while eliminating programmed cell death permitted P1 neurons to survive in females, the projection patterns of these neurons differed from those of male P1 neurons. Additional expression of Fru^M in these surviving neurons instructed male-typical projection patterns. Thus, Dsx and Fru^M act together to regulate the differentiation of P1 neurons in males: the lack of Dsx^F is responsible for their survival, and the presence of Fru^M instructs their wiring patterns.

Processing of the male-produced pheromone, cVA, also requires sexually dimorphic neuronal wiring. As discussed in Section 10.5, cVA inhibits male courtship of other males or mated females. cVA also promotes male–male aggression, an evolutionary conserved behavior related to mating; males exhibit aggression to defend territory and compete for female partners. Interestingly, in females, cVA activates pC1 neurons (Figure 10-12B) and promotes female courtship, and virgin females with mutant cVA receptors exhibit reduced mating behavior. How does the same pheromone produce sexually dimorphic behaviors? cVA activates olfactory receptor neurons (ORNs) and projection neurons (PNs) that project to the DA1 glomerulus. While DA1 ORNs and PNs in males and females produce indistinguishable physiological responses to cVA, sexual dimorphisms between connections of DA1 PNs and third-order neurons in the lateral horn have been identified. A group of lateral horn neurons called 3A exhibit sexually dimorphic dendrite projections such that only male 3A dendrites have substantial overlap with axons of DA1 PNs conveying cVA signals (**Figure 10-14**A). Another group of lateral horn neurons, 3B, also exhibit sexually dimorphic dendrite projections, with only female 3B dendrites overlapping substantially with DA1 PN axons (Figure 10-14B). Electrophysiological recordings indicated that cVA preferentially activates 3A neurons in males and 3B neurons in females. Both 3A and 3B neurons express Fru^M in males and require Fru^M to generate male-typical dendritic projections. Thus, Fru^M controls a circuit switch, promoting DA1 PN → 3A connections and inhibiting DA1 PN → 3B connections in males, resulting in sexually dimorphic wiring patterns (Figure 10-14C); this circuit switch may contribute to sexually dimorphic behavior elicited by the same pheromone.

In summary, sex-specific Fru and Dsx isoforms regulate cell death, resulting in the generation of sexually dimorphic neuronal populations in the male and female nervous systems. In addition, Fru and Dsx also control sexually dimorphic neuronal wiring, such that the same sensory input can be channeled into different neural pathways to generate sexually dimorphic behavior.

10.8 Even innate behavior can be modified by experience

The epigraph by Spalding is only partially correct in its description of the life of a butterfly, as even innate behaviors such as courtship are modifiable by experience. We discuss two examples in this final section on fly sexual behavior.

Males lacking Fru^M (*Fru^M* null) do not exhibit mating behavior when they first encounter females. However, co-housing with females allowed expression of some courtship behaviors, and the amount of courtship they exhibited increased with

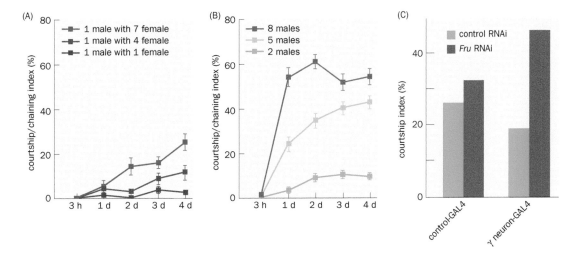

Figure 10-15 Innate courtship behavior can be modified by experience. (A & B) In these experiments, Fru^M null males were housed individually right after eclosion. When first encountering virgin females or males, they did not exhibit courtship behavior. However, co-housing with females (Panel A) or males (Panel B) increased their courtship index. The x axes represent the duration of co-housing, and the numbers of co-housed flies are indicated. **(C)** After being repeatedly rejected by mated females, three control groups of males with RNA interference (RNAi) against *Gfp* or *Fru* in control neurons (columns 1 and 2) and RNAi against *Gfp* in a subset of Kenyon cells (the γ neurons; column 3) have reduced courtship indices. Flies in which Fru^M is knocked down in γ neurons (column 4) still exhibit a high courtship index, indicating a deficiency in courtship conditioning. (A & B, adapted from Pan Y & Baker BS [2014] *Cell* 156:236–248. With permission from Elsevier Inc. C, adapted from Manoli DS, Foss M, Villella A, et al. [2005] *Nature* 436:395–400. With permission from Springer Nature.)

increased duration of co-housing and increased number of cohabited females (Figure 10-15A). Likewise, group housing with other males increased their chaining behavior (Figure 10-15B; Figure 10-3). Thus, social interactions with other flies can modify innate courtship behavior. Interestingly, this experience-dependent behavior requires Dsx^M, as Fru^M null males also lacking functional Dsx^M failed to display chaining behavior even when group housed with other males.

As another example of the plasticity of innate behavior, when a male fly tries to court a mated female, he is rejected repeatedly because of the change in the female's receptivity (Section 10.6). Wild-type males that have been repeated rejected reduce their courtship attempts in a process called **courtship conditioning**. The Kenyon cells of the mushroom body, which receive olfactory input from projection neurons (Figure 6-25; 6-31), are essential for this experience-dependent modification of behavior. Remarkably, courtship conditioning is also regulated by Fru^M, which is expressed in a subset of Kenyon cells. Courtship conditioning was blocked by RNAi knockdown of Fru^M in this subset of Kenyon cells: males did not learn from repeated failures and did not reduce courtship after being repeatedly rejected (Figure 10-15C).

Both examples illustrate the intricate interplay between nature and nurture: innate behaviors can be modified by experience, and the genetic program that specifies innate behavior also regulates its experience-dependent modification. The production of courtship song by songbirds provides another excellent example of the interplay between nature and nurture (Box 10-1).

In summary, studies of two key genes, *Fruitless* and *Doublesex,* have revealed how regulatory genes specify complex behaviors by building into the nervous system circuit elements enabling sensory perception, multisensory integration, motor action, and experience-dependent modification. Molecular-genetic approaches have been used to assemble an impressive list of circuit elements. Applying modern circuit dissection tools (see Chapter 14), researchers have begun to investigate how these circuit elements operate and how information flow in these circuits produces exquisitely coordinated behavior. What can sexual behavior in flies teach us about sexual behavior in mammals, including our own species? Are there common principles in sexual behavior and its neural control that are shared across diverse organisms? Let's continue our journey to find out.

Box 10-1: Bird song: nature, nurture, and sexual dimorphism

Thousands of bird species use songs to communicate; often only the males sing. A male songbird sings to females to signal his species, individual identity, location, and readiness to mate. His song also conveys to other males information regarding territory and neighbor-versus-stranger status. Bird songs can be represented as notes and syllables in time-frequency sound spectrograms (Figure 10-16A). Each species has its characteristic song(s). In the past decades, bird song research has made contributions to many areas of neurobiology.

The song a mature bird sings is a product of extensive interplay between nature and nurture. During an early sensory stage, a young male bird hears and memorizes the song of a tutor, usually his father. Then, during a sensorimotor stage, the young bird starts producing his own immature song. He uses auditory feedback to compare his own song to the tutor song template he has memorized. Through trial-and-error, his song comes to more closely resemble the tutor song, until it acquires its mature, crystallized form. In some species, such as the white-crowned sparrow, the sensory and sensorimotor stages are completely separated in time, whereas in other species, such as the extensively studied zebra finches, these stages overlap (Figure 10-16B).

While many features of bird song are learned, some aspects appears to be innate. If a bird is raised in acoustic isolation during the sensory stage, he can sing only a rudimentary **innate song**, which is nevertheless species specific, reflecting the nature aspect of bird song acquisition. If a bird is exposed only to the song of a different species during the sensory period, his song can take the other species' form, reflecting the influence of nurture. If a bird is exposed to songs of his own species and others, he preferentially learns

from and sings the song of his species; thus, birds are predisposed to learn their species' song. Finally, auditory input is required not only during the sensory period, when the bird listens to the tutor song, but also during the sensorimotor stage, when the bird must compare what he sings to what he has memorized. If a bird is deafened after the sensory stage but before the sensorimotor stage, his song remains immature. Indeed, if a bird raised in acoustic isolation is additionally deafened during the sensorimotor stage, he sings a song that is different from the innate song of acoustically isolated birds that have not been deafened. Thus, trial-and-error learning during the sensorimotor stage is essential even for the innate song.

Lesion, anatomical, and physiological studies have identified neural circuits in the brain involved in song production, auditory feedback, and song learning (Figure 10-16C). Two pallial (dorsal forebrain) nuclei analogous to motor cortical areas in mammals, the **HVC** (high vocal center) and **RA** (robust nucleus of the arcopallium), are essential for song production. HVC neurons project to the RA, and RA neurons project to brainstem motor nuclei that control muscle contraction in the vocal organ and breathing center for song production. While not essential for song production, forebrain nuclei **LMAN** (lateral magnocellular nucleus of the anterior nidopallium) and **area X** are instrumental for song learning. Area X is located in an anterior forebrain structure analogous to the basal ganglia in mammals. Area X receives input from the HVC and sends output to the RA via the DLM (a thalamic nucleus) and LMAN to regulate transmission of information between the HVC and RA. Like its counterpart in the mammalian basal ganglia (Figure 8-21), area X receives modulatory input from midbrain dopamine neurons; this input enables trial-and-error-based learning. Auditory infor-

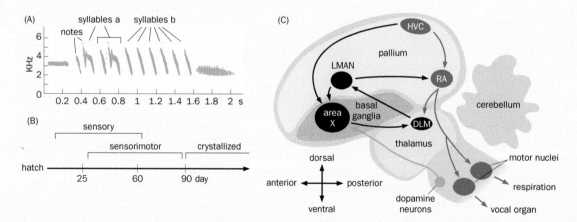

Figure 10-16 Bird song: composition, ontogeny, and neural circuits. **(A)** Time-frequency sound spectrogram of a white-crowned sparrow's song. A note is a continuous marking on the spectrogram; two or more notes may together form a syllable. **(B)** Developmental stages of zebra finches in days after hatching. During the sensory stage, a young bird listens to and memorizes a tutor song. During the sensorimotor stage, he sings an immature song and compares it to the tutor song he memorizes. The song takes its mature form at the crystallized stage. **(C)** A side view of a songbird brain highlighting neural circuits for song production and learning. The song motor

pathway (red), consisting of HVC, RA, and the brainstem motor nuclei that regulate muscle contraction in the vocal and respiratory systems, is responsible for song production. The anterior forebrain pathway (black), consisting of LMAN, area X, and DLM, is essential for song learning. Dopamine neurons project to and modulate neurons in area X. (A, from Konishi M [1985] *Ann Rev Neurosci* 8:125–170. With permission from Annual Reviews Inc. B, from Brainard MS & Doupe AJ [2002] *Nature* 417:351–358. With permission from Springer Nature. C, adapted from Brainard MS & Doupe AJ [2013] *Ann Rev Neurosci* 36:489–517.)

Box 10-1: continued

mation is processed in forebrain areas analogous to the mammalian auditory cortex, which sends input to the song system through the HVC.

Scientists have used songbird models to investigate a wide range of fascinating topics, such as the neural mechanisms of song production and song learning. Indeed, birdsong is one of the best models for human language. We end this box on a question related to the theme of this chapter: why is singing a male-typical trait in many bird species? In the 1970s, it was discovered that brain nuclei associated with song production and learning exhibit robust sexual dimorphism in canaries, where males sing much more than females, and in zebra finches, where only males sing (Figure 10-17). In zebra finches, for example, the HVC, RA, and area X are rudimentary or unrecognizable in females. The mechanisms underlying these sexual dimorphisms have been more extensively investigated in canaries. Interestingly, new neurons in these regions are produced in the adult brain every year

and recruited into functional circuits; for instance, adult-born neurons in the HVC respond to auditory input and project axons to the RA. (This exception to the rule that neurons are born during early developmental stages is called **adult neurogenesis**, which also occurs in limited regions of the mammalian brain.) These new neurons likely contribute to the ability of canaries to modify their songs every year. Further studies indicate that both the genotype of nuclei (whether they are made up of cells that are genetically male or female) and the male sex hormone testosterone (Section 10.9) play important roles in producing these sexual dimorphisms. Testosterone promotes the survival of adult-born neurons in the HVC by upregulating expression of brain-derived neurotrophic factor (BDNF; Section 7.15). Providing females with testosterone or infusing female brains with BDNF both promote the survival of new HVC neurons in females. We will study the mechanisms of sex hormone action in mammalian sexual dimorphism and sexual behavior in detail in the second part of this chapter.

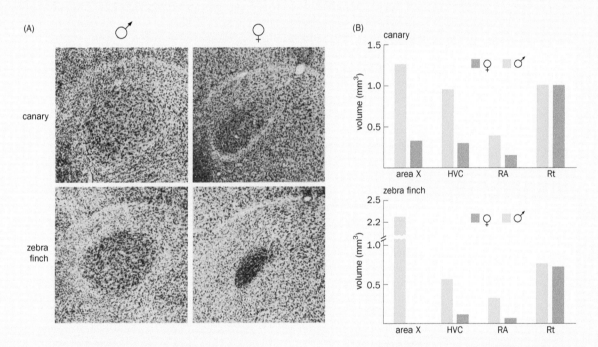

Figure 10-17 Sexual dimorphism in the vocal control areas of songbirds. (A) The RA nucleus in the canary (top) and zebra finch (bottom) is larger in males (left) than in females (right). These sections are stained with cresyl violet, a basic dye that stains cell bodies. The RA nuclei are more darkly stained in the center of each section, surrounded by lightly stained bands. **(B)** Volumes of nuclei making up the song circuits (area X, HVC, and RA) are larger in males than in females in both canaries (top) and zebra finches (bottom). Area X is unrecognizable in zebra finch females. The Rt nucleus (nucleus rotundus) is unrelated to vocal control and has a similar volume in males and females. (From Nottebohm F & Arnold AP [1976] *Science* 194:211–213. With permission from AAAS.)

HOW ARE MAMMALIAN SEXUAL BEHAVIORS REGULATED?

In the second part of this chapter, we study the mechanisms that regulate mammalian sexual behavior. Just like fruit flies, the mammals studied to date—mostly rodents—engage in sexual and other reproductive behaviors that have a large innate component. Male rodents **mount** females, while female rodents exhibit

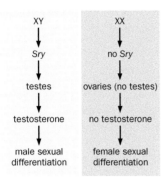

Figure 10-18 Sex determination and sexual dimorphism in mammals. In most mammals, the *Sry* gene on the Y chromosome specifies testis development. Testosterone produced by the testes specifies male differentiation. Female differentiation occurs as a default in the absence of the Y chromosome, the *Sry* gene, testes, and testosterone.

lordosis behavior when sexually aroused, assuming a posture that facilitates sexual intercourse. In addition, males behave aggressively toward intruders (especially sexually mature males) to defend their territory. Females exhibit maternal behavior, including nest building, pup retrieval, and nursing. How does the nervous system control these behaviors? What is the origin of sexually dimorphic mammalian behaviors? As in the case of the fly, the genetic origins of sexual dimorphism in mammals are the sex chromosomes, which provide an experimental entry point into investigation of the mechanisms of sexual behavior.

10.9 The *Sry* gene on the Y chromosome specifies male differentiation via testosterone production

Whereas sex in fruit flies is determined by the ratio of X chromosomes to autosomes (Figure 10-4), in most mammals (including mice and humans), sex is determined by the presence or absence of the Y chromosome (**Figure 10-18**). People with a single X chromosome but no Y chromosome develop into females (Turner syndrome, occurring in 1 out of 2500 girls), while people with two X chromosomes and a Y chromosome develop into males (Klinefelter's syndrome, occurring in 1 out of 1000 boys).

A key event in mammalian sex differentiation occurs during mid-embryogenesis when a structure called the genital ridge differentiates into testes in males and ovaries in females. It was discovered in the 1950s that when the genital ridge was removed *in utero* before overt sexual differentiation, all embryos differentiated as females. This experiment suggested that female differentiation constitutes the default pathway and that the presence of testes instructs male differentiation. Thus, the Y chromosome must contain genetic factor(s) that implement its sex determination effect by promoting testis development.

A single gene, **Sry** (*Sex determining region Y*), was identified in the 1990s as the testis-determining factor (Figure 10-18). *Sry* is located on the Y chromosome in all mammals whose sex is determined by the presence or absence of the Y chromosome and is expressed in the genital ridge during sexual differentiation. Human XY subjects with loss-of-function *Sry* mutations develop into females despite having the rest of the Y chromosome intact, indicating that *Sry* is *necessary* for male differentiation. Moreover, XX mice carrying an *Sry* transgene on an autosome develop into males, with characteristic male reproductive systems and male sexual behaviors, indicating that *Sry* is *sufficient* for mice to develop as males. *Sry* encodes a transcription factor that regulates gene expression. As we will see, however, the signals downstream of Sry-mediated control of sex determination—the sex hormones testosterone, estrogen, and progesterone—have more complex and interactive roles in their control of sexual dimorphisms and behaviors in mammals than do the cell-autonomous mechanisms that mediate sexual differentiation in flies.

A major function of the testes during embryonic development is to produce **testosterone**, a steroid hormone that promotes the development of the male reproductive system (masculinization) and inhibits the development of the female reproductive system (de-feminization) (Figure 10-18). In fact, much of the testes' effect on the development of the male reproductive system can be mimicked by applying exogenous testosterone at certain stages of embryonic development. So, the key questions are: How does testosterone act to specify sexual differentiation? Does it also regulate sexual behavior? What is the relationship between sexual differentiation during development and male- and female-typical sexual behaviors in adults? We will investigate these questions in the following sections.

10.10 Testosterone and estradiol are the major sex hormones

Testosterone (**Figure 10-19**A) was originally purified from adult male testes as an **androgen** (that is, a male sex hormone). Sexually inexperienced, **castrated males** (males with their testes removed) do not exhibit male-typical sexual behaviors, such as mounting of females and aggression toward intruder males. However,

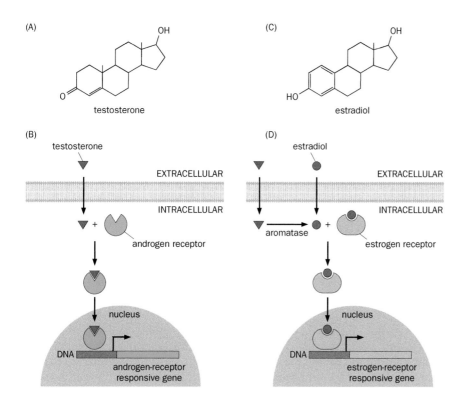

Figure 10-19 Sex hormones and their mechanisms of action. (A) Structure of testosterone, a male sex hormone (androgen). **(B)** Testosterone diffuses across the plasma membrane, binds to the androgen receptor, and causes nuclear translocation of the complex. The complex binds to promoters of target genes to regulate their transcription. **(C)** Structure of 17β-estradiol, a female sex hormone (estrogen). **(D)** Estradiol acts similarly to testosterone and regulates expression of estrogen-receptor responsive genes. Intracellular estradiol can derive either from circulating estradiol or from circulating testosterone converted to estradiol by the enzyme aromatase. The androgen and estrogen receptors both act as homodimers; for simplicity, they are represented here as single molecules. Not shown here, the steroid hormone progesterone has a similar mechanism of action.

injection of testosterone into adult castrated males can restore these behaviors, indicating an *activational* effect of testosterone in male sexual behavior.

Testosterone is synthesized from cholesterol in the testes. Due to its hydrophobicity, testosterone can freely diffuse across the plasma membrane to enter target cells, where it binds to the **androgen receptor**. This binding triggers nuclear translocation of the testosterone–androgen receptor complex, which acts as a sequence-specific DNA-binding transcription factor to activate or repress target gene expression (Figure 10-19B). A testosterone metabolite, dihydrotestosterone (DHT), is a more potent activator of the androgen receptor and is responsible for external genital masculinization during development.

Estradiol (abbreviation for 17β-estradiol; Figure 10-19C), the major **estrogen** (female sex hormone), is a steroid hormone made by the ovaries of sexually mature females. While estradiol is often regarded as the female counterpart of testosterone, we will see that it also plays important roles in *male development* in rodents. As with the effects of testosterone on castrated males, provision of estradiol to **ovariectomized females** (females from which the ovaries have been removed) can restore female-typical sexual behaviors such as lordosis. (Co-provision of **progesterone**, another steroid hormone that regulates female sexual behavior, augments the effect of estradiol.) Estradiol's mode of entry into cells—diffusion across the plasma membrane—likewise reflects its biochemical similarity to testosterone. Once inside a cell, estradiol binds its own receptors, the **estrogen receptors** 1 and 2, which are encoded by two different genes. The estrogen receptors are also DNA-binding transcription factors, but their sequence specificity differs from that of the androgen receptor, so they regulate expression of a different set of target genes (Figure 10-19D). Estradiol can also cause more rapid effects on target tissues, such as changes in intracellular Ca^{2+} concentration and protein phosphorylation. A third, G-protein-coupled estrogen receptor, which localizes to the endoplasmic reticulum membrane, mediates some of these rapid effects.

Importantly, cells that express an enzyme called **aromatase** can convert testosterone (but not DHT) to estradiol (Figure 10-19D). In these cells, therefore, testosterone can act by two means: (1) by binding to an androgen receptor directly, or (2) by binding to an estrogen receptor after being converted to estradiol. This may explain why estradiol, just like testosterone, stimulates males to fight with intruder males in rodents. Given these equivalent responses, the mechanism by which testosterone promotes aggression toward intruders likely involves its conversion into estradiol and subsequent action through estrogen receptors. However, provision of estradiol to females stimulates lordosis toward intruding males. The sexually dimorphic behavioral effects elicited by the same hormone suggest inherent differences between male and female brains. What are these differences, and what is their origin?

10.11 Early exposure to testosterone causes females to exhibit male-typical sexual behaviors

In 1959, scientists tested the effects of early testosterone exposure on the sexual behavior of adult guinea pigs. Testosterone was injected into pregnant female guinea pigs. The genitalia of some female progeny were partially or fully transformed to resemble those of males, in accord with the role of testosterone in regulating development of the male reproductive system. To determine the long-lasting effects of prenatal exposure to testosterone, gonads were removed from adult females that were prenatally exposed to testosterone, as well as from control females that were not exposed to testosterone. Sex hormones were then administered to these females to elicit sexual behavior. When given a combination of estradiol and progesterone to induce sexual arousal, control gonadectomized females exhibited robust lordosis; females that were exposed to testosterone prenatally had drastic impairment of lordosis, rather displaying mounting behavior. When given exogenous testosterone during testing, females that were exposed to testosterone prenatally, but not control gonadectomized females, displayed mounting behavior similar to that of castrated males receiving the same amount of exogenous testosterone. These changes in sexual behavior appeared to be permanent and did not occur if testosterone was given to females after early development.

This experiment suggested that prenatal testosterone has *organizational* activity that biases the nervous system toward a male configuration, like its effect on external genitalia. Females exposed to testosterone during embryonic development had their brains configured in a male-typical manner and therefore exhibited male sexual behavior when supplied as adults with testosterone or estrogen to "activate" such behavior. This **organization–activation model** for the dual roles of sex hormones (**Figure 10-20**) has received much experimental support in the past 60 years and has become a central principle of endocrinology.

The organizational effect of testosterone has a sensitive window of action. The 1959 study indicated that this window is prenatal in guinea pigs. Analogous experiments on rats and mice revealed that their sensitive period starts shortly before birth and spans the first 10 days after birth (Figure 10-20). This sensitive period coincides with peak testosterone production in male rodents. Due to the ease with which sex hormones can be provided in conjunction with gonadectomy in postnatal animals, rats and mice have become the model organisms of choice for studying hormonal regulation of sexual behavior.

During embryonic development in rodents, the testosterone produced by the testes of male fetuses circulates through the blood to reach the brain. By contrast, estradiol, which originates in the placenta, is bound to and sequestered by circulating α-fetoprotein, a protein secreted by fetal liver cells into the bloodstream, such that estradiol cannot reach the brain via circulation. However, estradiol can be produced locally within the brain cells of males by the action of aromatase on testosterone (Figure 10-19D). During the early postnatal period, while the testes continue to produce testosterone, estradiol production by female ovaries is negligible. Thus, only male brains are affected by estradiol during early development.

In principle, testosterone could exert its organizational effects in normal male brains (and in female brains with early exposure to testosterone) through activation of the androgen receptor, or aromatization to estradiol and subsequent activation of estrogen receptors. Several lines of evidence suggest that its organizational effect is mainly through the latter mechanism in rodents. First, female

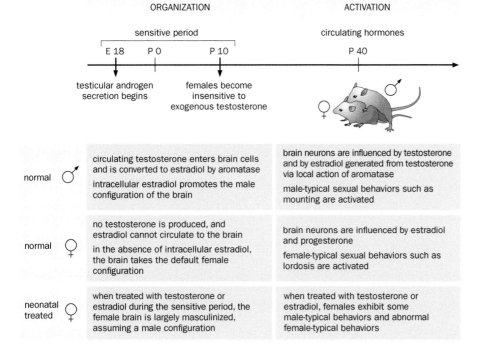

ORGANIZATION   ACTIVATION

Figure 10-20 **The organization–activation model for sex hormone action.** Originally proposed by Phoenix et al. (*Endocrinology* 65:369, 1959) to explain the behavior of guinea pigs treated prenatally with testosterone, this model has been validated and expanded by many subsequent experiments. The timing of events shown here corresponds to the timelines of rats and mice. E, embryonic; P, postnatal.

mice treated with estradiol during the first 10 days after birth exhibit male-typical behaviors such as territorial marking (**Figure 10-21**) and aggression toward intruders. Early estradiol treatment also reduces female receptivity to mating. Second, aromatase knockout male mice exhibit profound defects in male-typical behaviors. Although this experiment does not by itself reveal whether the defects are caused by an organizational role of estradiol during development or an activational role of estradiol in adults, in conjunction with the first experiment, it suggests an organizational role. Third, although the androgen receptor is expressed in the adult brain, it has very low expression in the developing brain during the sensitive period of organization. Indeed, male mice with brain-specific knockout of the androgen receptor still exhibit qualitatively similar male-typical mating behavior, albeit at a reduced level.

Importantly, hormonal control of sexual differentiation has substantial interspecies differences. For example, while the testosterone surge occurs neonatally in rats and mice, in humans testosterone is elevated from gestational week 8 to week 24, well before birth. This is likely because rats and mice are born at earlier developmental stages than primates. In addition, whereas testosterone's organizational effects in mice are largely mediated by estradiol, as discussed earlier, evidence suggests that the direct action of testosterone on the androgen receptor is more important for its organizational functions in humans.

10.12 Dialogue between the brain and the gonads initiates sexual maturation at puberty and maintains sexual activity in adults

In mammals, including rodents and humans, sexual maturation occurs at puberty. A specialized group of neurons in the preoptic area of the hypothalamus starts to release, in a pulsating fashion, high levels of **gonadotropin-releasing hormone** (**GnRH**, also called LHRH for luteinizing hormone-releasing hormone). GnRH travels, via portal vessels in the pituitary stalk, to the anterior pituitary, where it stimulates release of the **gonadotropins—luteinizing hormone (LH)** and **follicle-stimulating hormone (FSH)**—by pituitary endocrine cells. LH and FSH then circulate via the blood to reach the gonads, stimulating maturation of the male testes or female ovaries (**Figure 10-22**; Section 9.4). In response, the gonads release sex hormones: testosterone in males and estradiol in females. Testosterone and estradiol

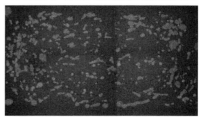

Figure 10-21 **Territorial marking behavior in male, female, and neonatal estrogen-treated female mice.** Male mice urinate throughout their cages (top panel), a typical territory-marking behavior. Female mice urinate in a corner (arrow in the middle panel). Female mice treated neonatally with estrogen (NE) partially adopt the male urination pattern (bottom). (From Wu MV, Manoli DS, Fraser EJ, et al. [2009] *Cell* 139:61–72. With permission from Elsevier Inc.)

Figure 10-22 Dialogue between the brain and the gonads during puberty and adulthood. Puberty is marked by activation of gonadotropin-releasing hormone (GnRH) neurons in the hypothalamus. Kisspeptins released from Kiss1 neurons in the arcuate nucleus (ARC) and anteroventral periventricular nucleus (AVPV) stimulate GnRH release from GnRH neurons, which express the Kiss1R. GnRH circulates to the anterior pituitary, where it stimulates endocrine cells to release luteinizing hormone (LH) and follicle-stimulating hormone (FSH). The bloodstream transports LH and FSH to the gonads (testes or ovaries), which respond by producing steroid sex hormones that regulate the differentiation of secondary sex characteristics during puberty. Sex hormones also circulate to the brain to affect further sexual differentiation during puberty and activate sexual behavior in adults. Specifically, negative feedback to Kiss1 neurons in ARC provides homeostatic regulation of GnRH release, while positive feedback to Kiss1 neurons in AVPV in females facilitates a preovulatory surge of GnRH release during the estrous cycle. (From Sisk CL & Foster DL [2004] *Nat Neurosci* 7:1040–1047. With permission from Springer Nature. See also Pinilla L, Aguilar E, Dieguez C, et al. [2012] *Physiol Rev* 92:1235–1316.)

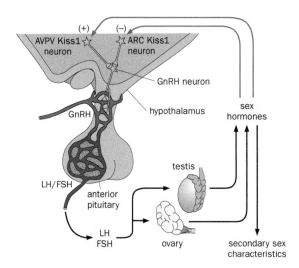

act in the periphery to trigger development of secondary sex characteristics. They also circulate to the brain to influence sexual maturation during puberty. Indeed, recent evidence has suggested that sex hormones can still have some organizational effects on the brain during puberty. This dialogue between the brain and gonads continues into adulthood, when sex hormones have an *activational* role in promoting sexual behavior in adults.

What triggers GnRH neurons to release GnRH during puberty? Insights came from studies of a human condition called **hypogonadotropic hypogonadism** (HH), in which patients display delayed, reduced, or absent puberty due to reduced gonadotropin levels. HH has diverse causes, including mutations in the genes that produce GnRH or the GnRH receptor, or failure of GnRH neurons to migrate during development. Remarkably, GnRH neurons are born in the epithelia of the vomeronasal organ in the accessory olfactory system (Box 6-1) and migrate to preoptic areas of the hypothalamus during embryonic development. Other causes of HH include mutations in the *Kiss1* gene, which encodes neuropeptides called **kisspeptins**, and in its G-protein-coupled receptor Kiss1R (also called GPR54). *Kiss1* and *Kiss1r* knockout mice also display HH phenotypes like those seen in human patients. Importantly, a lack of LH release due to a *Kiss1r* mutation in humans can be suppressed by supplying exogenous GnRH, indicating that kisspeptins/Kiss1R act upstream of GnRH release. Thus, the kisspeptins/Kiss1R system serves as an important upstream activator of GnRH neurons during puberty (Figure 10-22). Additional factors that trigger the activation of GnRH and Kiss1 neurons during puberty include nutritional factors such as leptin (see Chapter 9).

How does feedback from the gonads control GnRH release in adults? While GnRH neurons do not express sex hormone receptors themselves and thus cannot be directly regulated by sex hormones, Kiss1 neurons express sex hormone receptors and provide a key step in the feedback regulation of GnRH release by gonads in adults. Testosterone and estradiol both downregulate *Kiss1* expression in the arcuate nucleus, thereby providing a negative regulatory loop for homeostatic regulation (Figure 9-6B) of GnRH release and sex hormone production. At the same time, *Kiss1* in the **anteroventral periventricular nucleus** (AVPV) of the hypothalamus is positively regulated by estradiol in rodents; this positive feedback loop is proposed to produce the preovulatory surge of gonadotropin release during the estrous cycle (Figure 10-22).

In summary, extensive dialogue between the brain and the gonads occurs during both development and adulthood. During early development, testosterone exerts a major organizational effect to configure the brain into a male-typical form. In females, estrogen is excluded from the brain during early development and the brain is therefore configured into a female-typical form in the absence of sex hormone activity. This organizational period may extend into puberty, when a surge of sex hormones in both males and females due to GnRH release continues to configure the brain into a sex-typical manner. In adults, testosterone and estradiol act on male and female brains, respectively, to activate male- and female-typical reproductive behaviors.

10.13 Sex hormones specify sexually dimorphic neuronal numbers and connections

Having introduced the organizational roles of sex hormones during development, we now explore the differences between male and female brains that might account for the differential responses of adult males and females to sex hormones. Like fruit flies, mammals exhibit sexual dimorphism in neuronal numbers and projection patterns.

Sexual dimorphism in neuronal number was first discovered in a small nucleus within the **medial preoptic area** (**MPOA**) in the anterior hypothalamus; MPOA is implicated by lesion studies as essential for male courtship behaviors, including mounting, intromission, and ejaculation. Nuclear staining indicated that the sexually dimorphic nucleus of the MPOA (SDN-POA) was several-fold larger in male rats than female rats (**Figure 10-23**A). Castration immediately after birth caused a reduction in the size of the male SDN-POA, whereas treating females with testosterone at birth caused an increase in the size of the SDN-POA (Figure 10-23B). Further studies indicated that this difference results from programmed cell death that occurs only in females. In the male brain, sex hormones act during the sensitive period around birth to inhibit cell death of SDN-POA neurons. Similar sexual dimorphism in the anterior hypothalamus has also been observed in humans.

Since the discovery of sexual dimorphism in the MPOA, additional sexual dimorphisms have been found in other brain areas. For example, the **medial amygdala** and **bed nucleus of the stria terminalis** (**BNST**) have more cells in males than females. Both of these basal forebrain nuclei have been implicated in regulating male courtship behavior. On the other hand, the anteroventral periventricular nucleus (AVPV) in the hypothalamus is larger in females than males, consistent with its major role in regulating the ovulatory cycle. Sexual dimorphism results from the action of neonatal hormones either preventing (medial amygdala, BNST) or promoting (AVPV) programmed cell death in the male brain.

In addition to neuronal number, sexually dimorphic projections have been documented. Recent molecular-genetic advances in mice allowed visualization of sexual dimorphism with higher sensitivity than classic histology methods. Using knock-in strategies in the mouse (Section 14.7), markers can be expressed in the same pattern as key genes implicated in sexual differentiation like the gene encoding aromatase (**Figure 10-24**A). More aromatase-expressing neurons were found in a subregion of the medial amygdala in males than females (Figure 10-24B). Moreover, aromatase-expressing neurons have denser projections to several parts

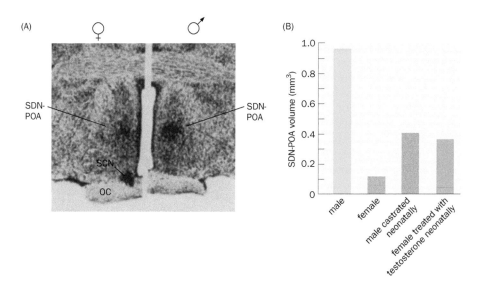

Figure 10-23 **Sexual dimorphism in the medial preoptic area of the hypothalamus.** **(A)** Nuclear staining of female (left) and male (right) rat brains revealed a striking size difference in a densely stained nucleus, the SDN-POA, of the anterior hypothalamus. The absence of the suprachiasmatic nucleus (SCN) in the male brain is an artifact of the section plane. OC, optic chiasm. **(B)** The SDN-POA is eight times larger in males than females. The SDN-POA size is markedly reduced in males that were castrated as neonates, whereas females treated neonatally with testosterone have increased SDN-POA size. (Adapted from Gorski RA, Gordon JH, Shryne JE et al. [1978] *Brain Res* 148: 333–346. With permission from Elsevier Inc.)

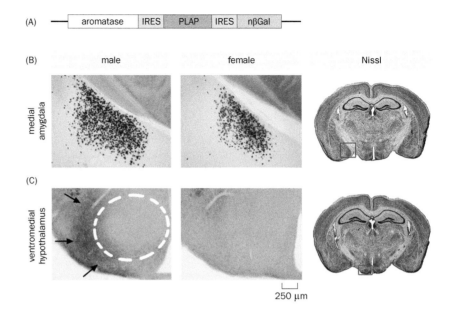

Figure 10-24 Visualizing aromatase-expressing neurons and their projections.
(A) Schematic of an engineered aromatase reporter gene introduced into the mouse. Aromatase-positive neurons express two additional marker genes from an IRES (internal ribosome entry site; Figure 6-13). Placenta alkaline phosphatase (PLAP; an enzyme that labels plasma membrane) is used to visualize neuronal projections, and nuclear-targeted β-galactosidase (nβGal) labels nuclei. **(B)** In the posterodorsal medial amygdala, males have more aromatase-expressing neurons than females, as revealed by βGal activity. **(C)** Near the ventromedial hypothalamic nucleus (circled), males have denser projections than females, as revealed by PLAP staining (arrows). The red boxes in the right column of Panel B and Panel C indicate where the images in the left panels are from. (Adapted from Wu MV, Manoli DS, Fraser EJ, et al. [2009] *Cell* 139:61–72. With permission from Elsevier Inc.)

of the brain in males than females, including the **ventromedial hypothalamic nucleus** (**VMH**; Figure 10-24C), a key center for regulating mating behavior (see Section 10.15). Females that undergo neonatal estrogen treatment displayed male-typical neuronal numbers and projection patterns, indicating that these dimorphisms are largely produced by the action of neonatal estradiol.

In addition to their organizational role during development, sex hormones also regulate neuronal connections in adulthood. As discussed in Chapter 3, excitatory synapses that provide input to cortical and hippocampal pyramidal neurons are formed mainly onto dendritic spines. Thus, spine density can be used as a proxy for the density of excitatory synapses onto a pyramidal neuron. Interestingly, the density of dendritic spines in hippocampal CA1 pyramidal neurons of female rats fluctuates across the estrous cycle (**Figure 10-25**): spine density is 30% greater during proestrus, when rats have higher estrogen levels, than during estrus, when rats have lower estrogen levels. Ovariectomy reduces spine density, which can be partially restored by estradiol provision. Thus, the natural fluctuation of spine density appears to be regulated by estradiol. The function of this drastic change in spine density remains unknown. Recently, it has been shown that axon terminal density of the progesterone receptor–expressing (PR+) VMH neurons at the AVPV changes threefold across the 5-day mouse estrous cycle. This axon terminal density change occurs in adult females but not in males and is regulated by estrogen signaling in PR+ VMH neurons (which co-express estrogen receptor 1), the activity of which promotes female receptivity.

In summary, as in the fly brain, sexual dimorphisms occur extensively in the mammalian brain in both neuronal numbers and connections, and programmed cell death is a key mechanism used by both systems to regulate sexual dimorphisms in neuronal numbers. However, unlike the cell-autonomous specification of sexual dimorphisms by transcription factors during development in flies, sexual dimorphisms in mammals are regulated non-cell-autonomously by circulating sex hormones, which can act at different stages of development and in adults. While the causal roles of sexual dimorphisms in the mammalian brain in regulating sexual behavior remain to be demonstrated, the brain regions where sexual dimorphisms occur and their connection patterns provide clues as to how they might contribute to sexually dimorphic behavior.

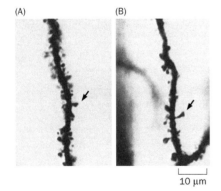

Figure 10-25 Density of dendritic spines in hippocampal CA1 pyramidal neurons fluctuates across the estrous cycle. Golgi staining shows that in CA1 hippocampal neurons the density of dendritic spines (arrows) is about 30% greater during proestrus **(A)** than during estrus **(B)**. (From Woolley CS, Gould E, Frankfurt M, et al. [1990] *J Neurosci* 10:4035–4039. Copyright ©1990 Society for Neuroscience.)

10.14 Sexually dimorphic nuclei define neural pathways from olfactory systems to the hypothalamus

What are the connection patterns of the sexually dimorphic nuclei? Gross brain connections can be studied by classic neuroanatomical methods, such as inject-

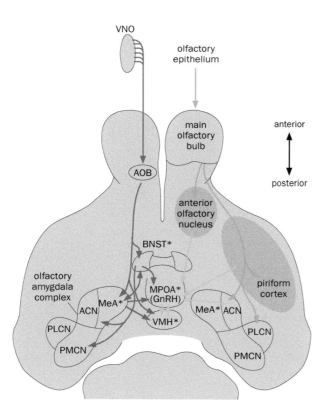

Figure 10-26 **Neural pathways linking the main and accessory olfactory systems to the hypothalamus in a rodent brain.** The accessory olfactory system (left) innervates sexually dimorphic nuclei (indicated by *) such as the MeA (medial amygdala) and BNST (bed nucleus of the stria terminalis), which both project axons to the hypothalamic MPOA (medial preoptic area) and VMH (ventromedial hypothalamic nucleus). Axons from the main olfactory bulb (right) project to distinct targets that also send input to the hypothalamus, including MPOA GnRH (gonadotropin-releasing hormone) neurons. GnRH neurons form reciprocal connections with VMH neurons. These pathways were delineated by a combination of classic anterograde/retrograde tracing and retrograde transneuronal tracing from GnRH neurons. The olfactory amygdala complex is divided into the MeA and the anterior, posterolateral, and posteromedial cortical amygdala (ACN, PLCN, PMCN). AOB, accessory olfactory bulb; VNO, vomeronasal organ. (Adapted from Yoon H, Enquist LW, & Dulac C [2005] *Cell* 123:669–682. With permission from Elsevier Inc. See also Boehm U, Zou Z, & Buck LB [2005] *Cell* 123:683–695.)

ing into specific brain regions dyes that travel from neuronal cell bodies to axons (**anterograde tracers**) or from axons and terminals back to cell bodies (**retrograde tracers**) (Section 14.18). Transneuronal tracing methods (Section 14.19) have also been used to study the inputs to specific neuronal populations. These studies have identified strong, bidirectional connections between the medial amygdala and BNST (**Figure 10-26**), both of which also project extensively to the hypothalamus, including the MPOA. The medial amygdala and BNST are both direct targets of mitral cells in the accessory olfactory bulb (Box 6-1), suggesting that sexually dimorphic nuclei play a role in pheromone perception. In addition, hypothalamic nuclei involved in regulating sexual behavior, such as the GnRH neurons in the MPOA, also receive direct input from the main olfactory system (Figure 10-26). Together, these studies suggest that the main and accessory olfactory systems provide major input to sexually dimorphic nuclei.

Olfactory and pheromonal information has long been implicated in regulating mating behavior. As discussed in Chapter 6, a cyclic nucleotide-gated channel expressed in the olfactory epithelia (CNGA2) is essential for olfactory transduction in the main olfactory system. Mice lacking CNGA2 are anosmic (Figure 6-5). The accessory olfactory system utilizes a different transduction mechanism, culminating in the opening of the TRPC2 channel to depolarize sensory neurons in the vomeronasal organ (Box 6-1). Thus, by studying *Cnga2* or *Trpc2* mutant mice, one can separately test the contributions of the main and accessory olfactory systems to sexual behavior.

Cnga2 mutant male mice exhibited virtually no mating behaviors, including sniffing to show interest, mounting, or intromission (**Figure 10-27**A). They also did not exhibit aggression toward intruder males, like normal male mice do. These experiments indicate that olfactory cues detected by the main olfactory system are essential to mating and territorial defense responses. Indeed, physiological studies have identified mitral cells in the main olfactory bulb that are specifically tuned to components of female urine that are attractive to males. *Cnga2* mutant female mice are deficient in retrieving pups, a parental behavior we will discuss in Section 10.16.

Interestingly, *Trpc2* mutant male mice exhibited normal mating behaviors toward female partners. However, they also exhibited equally robust mating

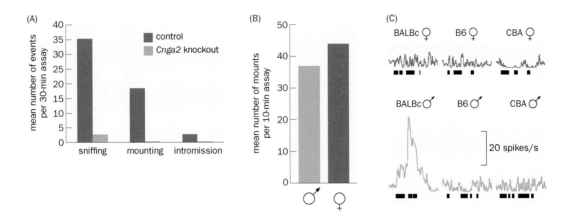

Figure 10-27 The main olfactory system is essential for mating, whereas the accessory olfactory system detects conspecific sex.
(A) Compared to control mice, *Cnga2* knockout mice show little interest in sexually mature females (expressed as sniffing) and do not exhibit mounting or intromission. **(B)** *Trpc2* knockout males mount females and intruder males with equal frequency given a simultaneous choice. The intruder males, which were castrated to reduce aggressive and sexual behavior, were swabbed with urine from intact males, which normally would elicit aggressive rather than mating behavior from the resident male (Figure 6-21). **(C)** *In vivo* single-unit recording of an AOB mitral cell in a behaving CBA male mouse. The firing rate of the cell (*y* axis) was selectively elevated when the mouse investigated and made physical contact (black horizontal bars on the *x* axis for time) with a BALBc male, but not a BALBc female or B6 and CBA mice of either sex (all partner mice were anesthetized). BALBc, B6 (C57Bl6), and CBA are names of specific mouse strains. (A, from Mandiyan VS, Coats JK, & Shah NM [2005] *Nat Neurosci* 12:1660–1662. With permission from Springer Nature. B, adapted from Stowers L, Holy TE, Meister M, et al. [2002] *Science* 295:1493–1500. C, from Luo M, Fee MS, & Katz LC [2003] *Science* 299:1196–1201.)

behaviors toward intruder males (Figure 10-27B). *Trpc2* mutant male mice also produce ultrasonic vocalization toward other males, a behavior usually reserved for courtship and directed toward females. Thus, male mice without accessory olfactory system function appear unable to distinguish the sex of their partners. This behavioral deficit suggests that for a male mouse, mating is a default behavior upon encountering another mouse. His accessory olfactory system enables him to recognize the partner mouse's sex and switch his behavior from mating to fighting if the other mouse is a male. In support of this model, sensory neurons and mitral cells of the accessory olfactory system have been identified that are specifically tuned to a combination of sex and strain signals from other mice (Figure 10-27C; Box 6-1).

What does the accessory olfactory system do in females? Analysis of *Trpc2* mutant female mice revealed a surprising phenotype: *Trpc2* mutant females produce ultrasonic vocalizations resembling those of males, and they mount other females and males (Movie 10-1). They are also impaired in female-typical behaviors, such as nest building and caring for pups. Thus, *Trpc2* mutant females act as if they were partially sex-transformed. Importantly, surgical ablation of vomeronasal organs in adult females produces similar phenotypes as *Trpc2* knockout, indicating that these phenotypes are not due to the loss of *Trpc2* during development. These observations imply that females have retained neural circuits for male-typical sexual behaviors, and signals from the accessory olfactory system normally inhibit these male-typical circuits. This notion is supported by the observation that, when given high doses of testosterone, wild-type females exhibit as much mounting behavior toward sexually receptive females as castrated males treated with the same testosterone doses. Indeed, sex reversals in mating patterns, such as mounting by untreated females, have been observed in nature from reptiles (Box 10-2) to mammals.

How do we reconcile these results with the organizational roles of sex hormones during development? A parsimonious model is that in both male and female brains, there exist primordial circuits that control male- and female-typical sexual behaviors. Sexual dimorphisms in the form of neuronal numbers and connection patterns, specified by the organizational function of steroid hormones during development, bias these circuits toward eliciting male- or female-typical sexual behaviors in response to hormone activation in adults. The function of the accessory olfactory system is to gate these circuits based on the sex identities of

Box 10-2: Courtship in unisexual lizards

The presence of primordial circuits underlying sex-typical behaviors provides animals flexibility in their behavioral repertoire to adapt to the changing environment. A fascinating case is illustrated in some whiptail lizard species, such as *Cnemidophorus uniparens,* which descend from bisexual ancestors but are unisexual, in that all individuals are females. They reproduce by **parthenogenesis**, in which embryos develop from unfertilized eggs without exchange of genetic material. Interestingly, these unisexual lizards still display courtship behavior, with one lizard acting "male-like" by approaching and mounting a "female" partner (**Figure 10-28**A). This courtship behavior has been shown to stimulate ovulation and thereby promote reproduction, despite no resulting exchange of genetic material.

Further studies have revealed interesting similarities and differences between unisexual lizards and most bisexual

vertebrate species. There is no detectable testosterone in unisexual lizards. However, their ovulation cycles persist, with fluctuation of estrogen and progesterone levels. Indeed, the female-like and male-like behaviors coincide with stages during the ovulation cycle with high- or low-estrogen levels (Figure 10-28B). Evidence suggests that progesterone, which peaks after ovulation, activates male-like courtship behavior, as testosterone does in most bisexual vertebrates. Hormone implants in different brain regions suggest that progesterone acts in an MPOA-equivalent area in the lizard brain to activate male-like behavior, whereas estrogen acts in a VMH-equivalent area to activate female-like courtship behavior. These studies provide interesting insights into the adaptation of courtship behaviors and their hormonal control to changes in reproductive strategies.

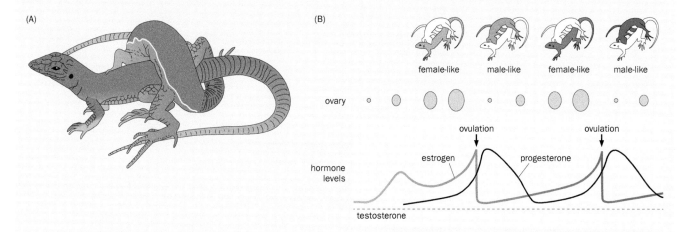

Figure 10-28 Courtship in unisexual lizard. (A) The pseudo-copulation position exhibited by a pair of unisexual whiptail lizards, with the lizard on the top displaying a male-like mounting behavior and the lizard on the bottom assuming a female-like mating position; the ritual closely resembles copulation of bisexual whiptail lizards. **(B)** Individual lizards alternate in their display of female-like and

male-like behavior coinciding with their ovulation cycles. The bottom panels show the size of ovary and hormone levels corresponding to the red lizard in the top panel. It displays female-like behavior when the estrogen level is high and male-like behavior when the estrogen level is low but the progesterone level is high. (From Crews D [1987] *Sci Am* 257[6]:116–121. With permission from Springer Nature.)

the partner and self. Thus, the accessory olfactory system activates the circuit appropriate for the animal's own sex and inhibits the circuit for the opposite sex. The fact that most salient sexually dimorphic nuclei are targets of the accessory olfactory system (Figure 10-26) supports this model.

Among targets of the accessory olfactory system, the BNST is a key brain region that discriminates the sex of the partner in sexually naive males. For example, aromatase+ BNST neurons in males exhibited higher activity when encountering novel females than males, and silencing this activity reduced mounting, intromission, and ejaculation of resident males toward female partners. This activity difference was eliminated in *Trpc2* mutant residents (**Figure 10-29**A). Interestingly, whereas a resident male typically attacks an intruder male, transient activation of aromatase+ BNST neurons in the resident male caused it to mount, instead of attack, an intruder male (Figure 10-29B). Different levels of activation of aromatase+ BNST neurons by male and female partners may transmit information to hypothalamic targets for eliciting mating or fighting behaviors.

Tracing the information flow of specific pheromone signals has also highlighted the pathway from the olfactory system to the hypothalamus. For example,

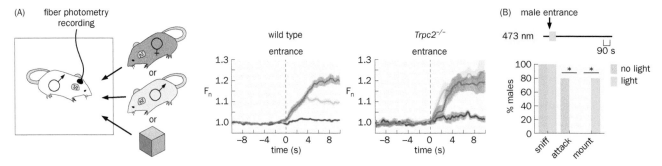

Figure 10-29 Aromatase-expressing BNST neurons in males discriminate sex partners. (A) Left, schematic of the experiment. The resident male expresses the genetically encoded Ca²⁺ indicator GCaMP6 in its aromatase+ BNST neurons and is subjected to fiber photometry recording (Section 14.22) during the entrances of an intruder male, a female, and an inanimate object. Aromatase+ BNST neurons in wild-type residents are activated more by females (red trace) than males (green) and are not activated by inanimate objects (gray; middle). The difference in activity elicited by male and female partners is absent in *Trpc2* mutant residents (right). F$_n$, change of fluorescence from baseline. **(B)** In this experiment, aromatase+ BNST neurons express channelrhodopsin so that they are activated by photostimulation. Transient optogenetic activation (cyan) causes the resident male to mount an intruder male; in the absence of optogenetic stimulation (gray), the resident male typically attacks the intruder male. (From Bayless DW, Yang T, Mason MM, et al. [2019] *Cell* 176:1190–1205.)

the pheromone ESP-1 (exocrine gland-secreted peptide 1) from male tears can enhance female lordosis. Transneuronal tracing from vomeronasal sensory neurons expressing the ESP-1 receptor identified neuronal populations in the medial amygdala, BNST, MPOA, and VMH as candidate neurons that mediate EPS-1's effects. Functional analysis further suggests that ESP-1 activated neurons in the VMH enhances lordosis.

10.15 The ventromedial hypothalamic nucleus regulates both female and male sexual behavior

The ventromedial hypothalamic nucleus (VMH; Figure 10-26) has long been implicated in regulating both female and male sexual behavior. In this section, we study the VMH to illustrate how researchers have used progressively more refined techniques to investigate its function in rodent sexual behavior.

VMH has been linked to controlling lordosis in female rodents, a posture for copulation triggered by tactile stimulation of the flank and perineal areas of females by mounting males. In rats, lordosis occurs only when females have a high level of circulating estradiol. By the 1970s, the following lines of evidence supported a role for the VMH in regulating lordosis: (1) female rats with VMH lesions exhibit impaired copulation, (2) radioactively labeled estradiol is taken up strongly by VMH neurons, and (3) an estrogen implant at the VMH can restore lordosis in ovariectomized rats. To further investigate a causal role of the VMH in regulating lordosis, researchers used microelectrodes to stimulate VMH neurons in ovariectomized rats treated with estradiol. They found that electrical stimulation enhanced lordosis triggered by appropriate tactile stimuli (**Figure 10-30**A) or in response to mounting by male rats, indicating that VMH neurons facilitate lor-

Figure 10-30 The ventromedial hypothalamic nucleus (VMH) facilitates lordosis in female rats. (A) Each data point on the graph represents lordosis behavior of an estradiol-treated, ovariectomized rat in response to five manual stimuli applied to the flank/perineal regions, on a scale from 0 (no lordosis) to 3 (maximal lordosis). Electrical stimulation of VMH neurons (10 Hz, 0.2 ms pulse at 50 μA; durations of the three experiments are shown at the bottom) increased the lordosis score within 15 min and had a long-lasting effect after the stimulation was terminated. **(B)** Bilateral lesion of the VMH by passing large currents (1 mA, 15 s) through stimulating electrodes caused gradual reduction of the lordosis score within the next 48 h. A slow, incomplete recovery followed, possibly due to activation of compensatory pathway(s). Data were collected from three rats (color coded). Estradiol was injected daily to enable lordosis. (From Pfaff DW & Sakuma Y [1979] *J Physiol* 288:189–202; 203–210.)

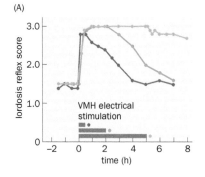

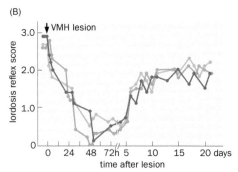

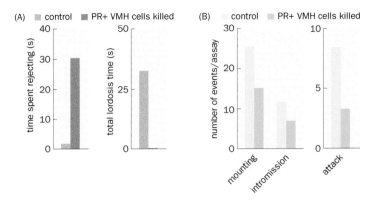

Figure 10-31 Progesterone receptor–expressing (PR+) VMH neurons regulate female and male mating behaviors as well as male aggression. (A) Compared with control females (wild-type mice injected with an AAV vector encoding Cre-dependent Caspase 3), females in which PR+ VMH neurons were killed (*PR-Cre* mice injected with the same AAV) rejected courting males and did not exhibit lordosis. **(B)** Compared with control males, males with ablated PR+ VMH neurons exhibited reduced mounting and intromission toward females, and reduced aggression toward intruder males. (Adapted from Yang CF, Chiang MC, Gray DC, et al. [2013] *Cell* 153:896–909. With permission from Elsevier Inc.)

dosis. Interestingly, increased lordosis behavior occurred about 15 minutes after electrical stimulation and returned to baseline *gradually* after terminating stimulation. Complementing this gain-of-function experiment, VMH lesions resulted in impaired lordosis, again with a time delay (Figure 10-30B). Together, these experiments suggest that rather than directly controlling the execution of lordosis, VMH neuronal activity promotes a brain state conducive to lordosis, possibly by activating a downstream execution circuit that integrates VMH signals over a relatively long time period. This is analogous to P1 neurons in the context of fly male sexual behavior (Figure 10-10).

Subsequent experiments revealed that the VMH is heterogeneous in neuron type composition, and activation of different subpopulations can have opposing effects on lordosis. In addition, electrical and optogenetic stimulation (Movie 14-5), along with *in vivo* activity recording, indicate that the VMH also regulates sexual and aggressive behavior in males. Are these behaviors controlled by the same or different subpopulations of VMH neurons? Using transgenic mice in which Cre recombinase was knocked into the *PR* locus (encoding the progesterone receptor, which augments the effect of estradiol on lordosis), researchers could inject Cre-dependent viruses into specific brain regions to mark and genetically manipulate only PR-expressing (PR+) neurons. One such manipulation was to kill these cells through activation of caspase-3, a key enzyme that promotes programmed cell death. When PR+ VMH neurons were ablated, females actively rejected courting males and did not engage in lordosis behavior (**Figure 10-31**A), indicating that PR+ VMH neurons promote lordosis. Some VMH neurons in males also express PR. In males, ablation of PR+ VMH neurons impaired male-typical behaviors, including mounting and intromission during courtship with females, and aggression toward intruder males (Figure 10-31B). Thus, PR+ VMH neurons regulate both female- and male-typical sexual behaviors.

PR+ VMH neurons may still be heterogeneous, such that lordosis-promoting neurons in females and aggression- and mating-promoting neurons in males may be distinct. Future manipulations of more specific cell types defined by single-cell transcriptomics (Section 14.13) may help address these possibilities. Within the male, the relationship between aggression- and mating-promoting neurons could be (1) distinct, responding selectively to male or female cues, respectively, or (2) the same, but differentially activated by male and female cues and promoting mating or fighting, depending on activity levels. Besides perturbation of more refined cell types, these possibilities can be addressed by recording the activity of *individual* neurons during different behaviors. These recording experiments have revealed some surprises. In one study, individual VMH neurons expressing the estrogen receptor 1 (Esr1) in male mice (which largely overlaps with PR+ VMH neurons) were imaged during behavior using a head-mounted miniature endoscope (Figure 14-43C). In sexually inexperienced males, a large fraction of Esr1+ VMH neurons was activated by both male and female intruders. As mice gained social experience, more neurons were activated by males only or females only, while the fraction of neurons activated by both males and females diminished (**Figure 10-32**). Consequently, the Esr1+ VMH neuron ensemble was better at

active Esr1+ VMH cells (%)

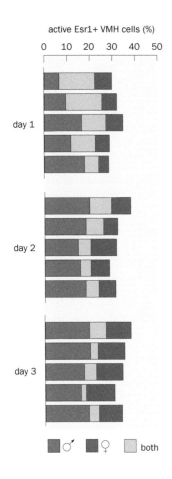

day 1

day 2

day 3

■♂ ■♀ ☐ both

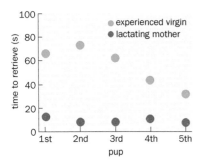

Figure 10-33 Pup retrieval in females.
Five pups were sequentially placed in a
cage with an adult female mouse, and the
average times it took for the female to
retrieve the pup are plotted on the y axis.
Lactating mothers efficiently retrieve pups.
Naive virgins usually ignore pups (not
shown; Movie 10-2), whereas experienced
virgins that have been co-housed with
lactating mothers and pups retrieve pups
at a reduced speed. (From Cohen L,
Rothschild G, & Mizrahi A [2011] *Neuron*
72:357–369. With permission from
Elsevier Inc.)

Figure 10-32 Sexual experience alters encoding by estrogen receptor–expressing (Esr1+) VMH neurons. Ca²⁺ imaging was carried out in Esr1+ VMH neurons using microendoscopes in resident males. Active cells had Ca²⁺ transients two standard deviations above the mean when exposed to a male intruder, a female intruder, or both. Each row represents an individual male resident. Before Day 1, these males were sexually naive. (From Remedios R, Kennedy A, Zelikowsky M, et al. [2017] *Nature* 550:388–392. See also Li Y, Mathis A, Grewe BF, et al. [2017] *Cell* 171:1176–1190 for similar observations in the neurons of the medial amygdala.)

discriminating between male and female intruders. Further analysis revealed that mating experience was crucial in promoting separation of male- and female-representing Esr1+ VMH neuronal ensembles. It remains to be determined whether these male- and female-specific ensembles are causally involved in aggression and mating behavior.

Similar findings—that neuronal encoding is influenced by sexual experience—were also made in the medial amygdala, another nucleus implicated in regulating sexual behavior (Figure 10-26). Together, these findings underscore that neural circuits regulating innate behavior in mammals are still modifiable by experience, echoing similar findings in flies (Section 10.8). Experience-based modification is further exemplified in parental behavior, to which we now turn in the final section of this chapter.

10.16 Parental behavior is activated by mating and regulated by specific populations of hypothalamic neurons

In addition to mating, mammals exhibit other sexually dimorphic behaviors. Typically, for example, males defend their territory by exhibiting aggression toward intruders, while females care for their pups. Most mammals are born immature and require extensive parental care to survive and prosper. In rodents such as mice and rats, mothers usually do most of the parental care, feeding and providing a warm, safe nest. Motherhood is associated with significant physiological and behavioral changes. For example, mothers retrieve displaced pups back into their nest, whereas naive virgin females usually do not. After being co-housed with lactating mothers and pups, however, virgin females learn to do so, albeit less quickly (**Figure 10-33**). Significant changes have been reported in the auditory and olfactory systems of new mothers that might make them more sharply tuned to alert calls and smells of displaced pups (**Movie 10-2**). A recent study identified a role for oxytocin, a neuropeptide involved in maternal behavior among many other functions (**Box 10-3**), in enhancing the detection of pup calls by auditory cortex neurons.

Male mice exhibit a qualitatively different behavior toward pups: some male mice (the fraction of which depends on the strain) attack and kill pups. Infanticide is hypothesized to have an evolutionary advantage in facilitating survival of infanticidal males' own progeny. Experiments have shown that infanticide speeds up mating of infanticidal males with mothers of the pups, as ovulation is inhibited in lactating mothers but is restored more quickly when the pups are gone. Interestingly, a behavioral switch occurs in males after mating such that they preferentially exhibit parental rather than infanticide behavior, facilitating survival of the male's own progeny.

As with mating behaviors, the accessory olfactory system plays a key role in regulating these sexually dimorphic parental-related behaviors. Vomeronasal organ ablated or *Trpc2* mutant male mice exhibit drastically reduced infanticide and enhanced parental behavior. Researchers have begun to dissect downstream neural circuits that regulate different aspects of parental behavior. For example, GABAergic neurons in the medial amygdala promote pup grooming in lactating mothers and virgin females. Further downstream in the MPOA (Figure 10-26), researchers have identified neuropeptide galanin-expressing neurons (MPOA^Gal neurons) to be preferentially activated after mice engaged in parental behavior. Remarkably, ablating MPOA^Gal neurons severely impaired parental behavior of fathers, while optogenetic stimulation of these neurons in virgin males inhibited infanticidal behavior (**Figure 10-34**A, B). Furthermore, ablating MPOA^Gal neurons

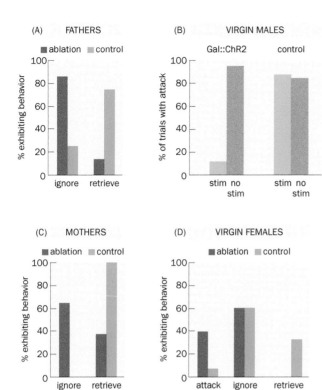

Figure 10-34 Regulation of parental behavior by medial preoptic area (MPOA) galanin-expressing neurons. (A) Ablation of MPOA[Gal] neurons decreased pup retrieval behavior of fathers. **(B)** Photostimulation of MPOA[Gal] neurons expressing channelrhodopsin (ChR2) reduced pup-attacking in virgin males. **(C)** Ablation of MPOA[Gal] neurons reduced pup-retrieval behavior in mothers. **(D)** Ablation of MPOA[Gal] neurons caused virgin females to attack pups. In all the experiments, control mice expressed GFP in MPOA[Gal] neurons. Experimental animals expressed in MPOA[Gal] neurons either ChR2 for photostimulation (Panel B) or diphtheria toxin (Panels A, C, and D) for killing. **(E)** Simplified scheme of input–output organization of three subpopulations of MPOA[Gal] neurons. Each receives different proportions of inputs from >20 brain regions (only 3 are shown here), and project axons to three distinct target regions to promote the motor action of pup grooming, motivate pup retrieval across a barrier, or inhibit social interaction with adult conspecifics. PAG, periaqueductal gray; VTA, ventral tegmental area; MeA, medial amygdala. MPOA[Gal] neurons also project to additional target regions (not shown); the behavioral functions of these projections remain to be determined. (A–D, from Wu Z, Autry AE, Bergan JF, et al. [2014] *Nature* 509:325–330. With permission from Springer Nature. E, from Kohl J, Babayan BM, Rubinstein ND, et al. [2018] *Nature* 556:326–331. With permission from Springer Nature.)

in mothers impaired their maternal behavior, and ablating these neurons in virgin females caused them to attack pups (Figure 10-34C, D). Together, these experiments suggest that MPOA[Gal] neurons promote parental behavior in both males and females and inhibit pup-attacking behavior in virgin males and females. Further circuit dissection revealed that MPOA[Gal] neurons comprise several distinct projection types to distinct targets, each promoting a specific aspect of parental behavior, such as the motor action of pup grooming, motivation to retrieve pups across barriers, and inhibition of social interaction with adult conspecifics (Figure 10-34E).

In summary, neural pathways involved in parental behavior in rodents appear to be organized in a similar fashion as those involved in mating, from the olfactory systems to the hypothalamus, involving some of the same sexually dimorphic nuclei and perhaps overlapping neuronal populations (Figure 10-26). These findings raise many interesting questions to be answered by future investigations: How does a neuronal population in a specific brain region participate in multiple behaviors? To what extent do individual neurons participate in specialized behaviors? How do sexually dimorphic circuits give rise to sexually dimorphic behaviors? Finally, some of the circuits and molecules (e.g., oxytocin; Box 10-3) that regulate mating and parental behavior are also implicated in social interactions in general. Thus, investigating the neural mechanisms of sexual and parental behaviors may contribute to our general understanding of the neural basis of social interactions.

Box 10-3: Oxytocin and vasopressin: neuropeptides with diverse functions and ancestral roles in sexual behavior

Oxytocin and **vasopressin** are two neuropeptides with diverse functions. Both peptides are synthesized by specialized groups of hypothalamic neurons. They are released from the posterior pituitary into the bloodstream and act as hormones (Section 9.4) to regulate a number of physiological processes such as labor and lactation (for oxytocin) or water retention (for vasopressin) through their G-protein-coupled receptors in peripheral tissues. Oxytocin- and vasopressin-expressing neurons also project to multiple brain regions and release these peptides as neurotransmitters to modulate the activity of target neurons expressing their receptors. Among their best-studied functions as peptide neurotransmitters, oxytocin promotes mother–infant bonding in multiple mammalian species and males' preferences of sex partners for mating, while vasopressin is implicated in aggression toward territorial intruders. Oxytocin is also implicated in nonsexual/reproductive social behaviors. For instance, oxytocin knockout mice exhibit selective social memory deficit (failure to distinguish new and familiar mice) with apparently normal nonsocial memory.

Both oxytocin and vasopressin also play important roles in pair bonding, a fascinating behavior that occurs in 3–5% of mammalian species, including humans, in which each member of a mating pair displays selective (though not necessarily exclusive) affiliation and copulation with his or her partner. Pair bonding is usually associated with biparental care of offspring. Although mice and rats, the most widely used mammalian models, do not exhibit pair bonding behavior, another rodent, the prairie vole, is a paragon of monogamy. After losing a partner, a prairie vole in the wild might never mate again. Such pair-bonding behavior can be reproduced in the laboratory: when offered a choice, prairie voles that have mated spend significantly more time huddling with their partners than with other voles.

The central roles of oxytocin and vasopressin in regulating pair bonding were demonstrated by perturbation experiments. Mating induces males to release vasopressin and females to release oxytocin. When oxytocin was infused into the brain of an ovariectomized (and therefore not sexually receptive) female prairie vole in the presence of a male, she would develop a preference toward this male in subsequent partner choice tests, even without having mated. Likewise, when vasopressin was infused into the brain of a male prairie vole in the presence of an ovariectomized female, he would develop partner preference toward this female, even without having mated (**Figure 10-35**). Conversely, when a male prairie vole was treated with a vasopressin antagonist immediately before exposure to a sexually mature female, he exhibited no preference toward this female even after mating with her. Likewise, administrating an oxytocin antagonist to a female prairie vole disrupted her partner preference after mating. Thus, in establishing a pair bond, application of oxytocin or vasopressin can replace mating itself. The mechanisms by which oxytocin and vasopressin achieve these feats are not entirely clear.

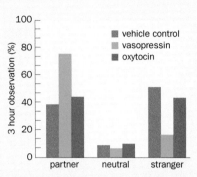

Figure 10-35 Infusion of vasopressin into the male prairie vole brain is sufficient to establish pair bonding. Male prairie voles were infused with vehicle control (artificial cerebrospinal fluid), vasopressin, or oxytocin while placed together with ovariectomized female partners without mating. Subsequently, each male was then given a choice between its partner, a novel female (stranger), and an empty cage (neutral). Vasopressin infusion induced partner preference in this choice assay. (From Winslow JT, Hastings N, Carter CS, et al. [1993] *Nature* 365:545–548. With permission from Springer Nature.)

Oxytocin and vasopressin are ancient molecules. In early vertebrate evolution, a single gene encoding an oxytocin/vasopressin-like neuropeptide underwent a duplication event. Over time, the duplicated genes for oxytocin and vasopressin evolved separately to regulate, respectively, female and male sexual and reproductive behavior. Oxytocin/vasopressin-like neuropeptides have also been identified in invertebrates; one example is **nematocin**, present in the nematode *C. elegans,* which has two G-protein-coupled receptors homologous to mammalian oxytocin and vasopressin receptors. Interestingly, nematocin and its receptors also play important roles in *C. elegans* mating.

C. elegans has two sexes: males and hermaphrodites. Hermaphrodites produce both sperm and eggs and can self-fertilize in addition to mating with males. As in fruit flies, *C. elegans* mating follows specific steps (**Figure 10-36**A). Mutant males without a functional *Nematocin* (*Ntc1*) gene exhibited abnormalities in all steps of mating (Figure 10-36B): they took longer to mate, encountered more hermaphrodites before initiating mating, and made more turns around hermaphrodites during mating. Whereas nearly all normal males leave a food-rich environment lacking a hermaphrodite within 24 hours in search of a mate, only half of all *Ntc1* mutant males did so. Males lacking the nematocin receptors exhibited essentially the same phenotypes. Nematocin and its receptors are expressed in neurons responsible for different aspects of the mating process, and deletion of *Ntc1* in specific neurons yielded specific behavioral defects. Thus, while not absolutely essential for mating, nematocin is a neuromodulator required for optimal execution of all mating steps.

Box 10-3: continued

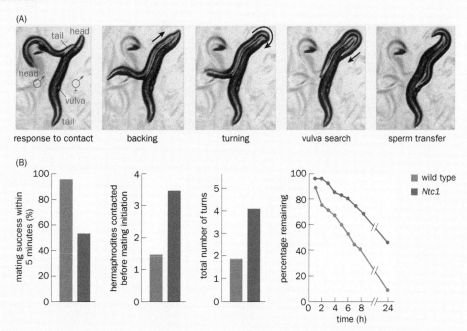

Figure 10-36 Nematocin is required for optimal execution of multiple steps of male mating. (A) *C. elegans*'s mating ritual. When the tail of a male encounters a hermaphrodite, he backs his body around the potential mate, initiates turning, and searches until his tail locates the vulva for sperm transfer. **(B)** Compared with wild-type controls (gray bar/trace), *Ntc1*-deficient males (red bar/trace) were less likely to mate successfully within 5 minutes of encountering a hermaphrodite (first panel). Mutant males also contacted more hermaphrodites before initiating mating (second panel) and made more turns during mating (third panel), and a larger fraction failed to leave hermaphrodite-free food lawns to search for a mating partner (fourth panel). (From Garrison JL, Macosko EZ, Bernstein S, et al. [2012] *Science* 338:540–543. With permission from AAAS.)

Oxytocin/vasopressin-like neuropeptides can also induce mating-like behavior in medicinal leeches, which are non-self-fertilizing simultaneous hermaphrodites (that is, mating occurs between two hermaphrodites, each of which acts as a male and a female simultaneously). Thus, oxytocin/vasopressin-like neuropeptides appear to have ancient roles in regulating mating behavior in animals with diverse reproductive strategies.

SUMMARY

In fruit flies, sexual behavior is specified by sex-specific splicing of two transcription factors, Fruitless (Fru) and Doublesex (Dsx). Expression of the male-specific Fru^M in the female brain is sufficient to generate most male-typical courtship behaviors in females. Fru^M is expressed in subsets of sensory neurons that detect mating-related cues, neurons in the central brain that integrate these cues, and neurons in the ventral nerve cord that produce mating-related motor commands. Fru+ and Dsx+ neurons in females promote female receptivity to courtship. During development, Fru^M and Dsx regulate sexually dimorphic execution of programmed cell death and neuronal wiring, presumably underlying sexually dimorphic behaviors.

In rodents, sexual behavior is controlled primarily by the dual actions of sex hormones, which play an organizational role during development and an activational role in adults. At neonatal stages, testosterone configures the brain into a male-typical manner by regulating sexually dimorphic programmed cell death and neuronal wiring. Testosterone and estradiol respectively activate male and female sexual behavior in adults. Sexually dimorphic nuclei, including pheromone-processing centers from the accessory olfactory system to the hypothalamus, regulate mating and other reproductive functions.

Flies and rodents thus share substantial similarities in how sexual behaviors are regulated. In both cases, sexually dimorphic behaviors originate from differences in genetic sex, which result in sexually dimorphic expression of key transcription factors (Fru^M and Dsx in flies and Sry in mammals) or sex hormones that activate transcription factors (androgen and estrogen receptors). These transcription factors in turn generate sexual dimorphism in the numbers of specific neurons and/or their projection patterns. A notable difference is the employment of sex hormones and their two-stage action in rodents. This may reflect the larger size and more sophisticated organ systems of mammals and the prolonged period between early development and sexual maturity. Future research on the neural basis of sexual behavior in these model organisms will further reveal both common principles and divergent strategies that ensure their reproductive success. Studies of other species and other aspects of sexual behaviors, including song production in songbirds and pair bonding in prairie voles, further enrich our understanding of the diversity of life and uncover themes and variations in how sexual behavior is regulated.

Finally, as we transition to the next chapter, which examines memory and learning, it is important to note that while sexual behavior is mostly innate, animals nevertheless learn from experience and modify even their sexual behavior. Indeed, song production in songbirds, which is required for reproductive success, has been a paradigm for studying learning. While learning offers flexibility that allows animals to adjust to changing environments, genetically programmed innate behaviors, selected by evolution, offer robustness and allow animals to adapt to their niche. Thus, genetically programmed innate behaviors and postnatally acquired learned behaviors interact dynamically throughout an animal's life. Future studies will further reveal how genes and neural circuits regulate, and are shaped by, these interactions.

OPEN QUESTIONS

- How do Fruitless and Doublesex in flies and estrogen and androgen receptors in mammals regulate sexually dimorphic neuronal wiring?

- How do sexually dimorphic neuronal numbers and connections give rise to sexually dimorphic behavior?

- How do social and mating experience modify sexual behavior and neuronal representation of social cues?

- How have the expression patterns of key regulators of sexual behavior, such as *Fruitless*, evolved?

FURTHER READING

Reviews

Auer TO & Benton R (2016). Sexual circuitry in *Drosophila*. *Curr Opin Neurobiol* 38:18-26.

Baker BS, Taylor BJ & Hall JC (2001). Are complex behaviors specified by dedicated regulatory genes? Reasoning from Drosophila. *Cell* 105:13-24.

Dickson BJ (2008). Wired for sex: the neurobiology of *Drosophila* mating decisions. *Science* 322:904-909.

Kohl J & Dulac C (2018). Neural control of parental behaviors. *Curr Opin Neurobiol* 49:116-122.

Morris JA, Jordan CL, & Breedlove SM (2004). Sexual differentiation of the vertebrate nervous system. *Nat Neurosci* 7:1034-1039.

Yang CF & Shah NM (2014). Representing sex in the brain, one module at a time. *Neuron* 82:261-78.

Sexual behavior in the fly

Clowney EJ, Iguchi S, Bussell JJ, Scheer E, & Ruta V (2015). Multimodal chemosensory circuits controlling male courtship in *Drosophila*. *Neuron* 87:1036-1049.

Clyne JD & Miesenböck G (2008). Sex-specific control and tuning of the pattern generator for courtship song in *Drosophila*. *Cell* 133:354-363.

Coen P, Xie M, Clemens J, & Murthy M (2016). Sensorimotor transformations underlying variability in song intensity during *Drosophila* courtship. *Neuron* 89:629-644.

Demir E & Dickson BJ (2005). fruitless splicing specifies male courtship behavior in *Drosophila*. *Cell* 121:785-794.

Dweck HK, Ebrahim SA, Thoma M, Mohamed AA, Keesey IW, Trona F, Lavista-Llanos S, Svatos A, Sachse S, Knaden M, et al. (2015). Pheromones mediating copulation and attraction in *Drosophila*. *Proc Natl Acad Sci U S A* 112:E2829-2835.

Feng K, Palfreyman MT, Hasemeyer M, Talsma A, & Dickson BJ (2014). Ascending SAG neurons control sexual receptivity of *Drosophila* females. *Neuron* 83:135–148.

Hoopfer ED, Jung Y, Inagaki HK, Rubin GM, & Anderson DJ (2015). P1 interneurons promote a persistent internal state that enhances inter-male aggression in *Drosophila*. *eLife* 4: e11346.

Kimura K, Hachiya T, Koganezawa M, Tazawa T, & Yamamoto D (2008). Fruitless and Doublesex coordinate to generate male-specific neurons that can initiate courtship. *Neuron* 59:759–769.

Kohatsu S, Koganezawa M, & Yamamoto D (2011). Female contact activates male-specific interneurons that trigger stereotypic courtship behavior in Drosophila. *Neuron* 69:498–508.

Kohl J, Ostrovsky AD, Frechter S, & Jefferis GS (2013). A bidirectional circuit switch reroutes pheromone signals in male and female brains. *Cell* 155:1610–1623.

Manoli DS, Foss M, Villella A, Taylor BJ, Hall JC, & Baker BS (2005). Male-specific fruitless specifies the neural substrates of *Drosophila* courtship behaviour. *Nature* 436:395–400.

Rideout EJ, Dornan AJ, Neville MC, Eadie S, & Goodwin SF (2010). Control of sexual differentiation and behavior by the *Doublesex* gene in *Drosophila melanogaster*. *Nat Neurosci* 13:458–466.

Shao L, Chung P, Wong A, Siwanowicz I, Kent CF, Long X, & Heberlein U (2019). A neural circuit encoding the experience of copulation in female *Drosophila*. *Neuron* 102:1025–1036.

Stockinger P, Kvitsiani D, Rotkopf S, Tirian L, & Dickson BJ (2005). Neural circuitry that governs *Drosophila* male courtship behavior. *Cell* 121:795–807.

Thistle R, Cameron P, Ghorayshi A, Dennison L, & Scott K (2012). Contact chemoreceptors mediate male-male repulsion and male-female attraction during *Drosophila* courtship. *Cell* 149:1140–1151.

Zhang SX, Miner LE, Boutros CL, Rogulja D, & Crickmore MA (2018). Motivation, perception, and chance converge to make a binary decision. *Neuron* 99:376–388.

Sexual behavior in mammals

Cohen L, Rothschild G, & Mizrahi A (2011). Multisensory integration of natural odors and sounds in the auditory cortex. *Neuron* 72:357–369.

Inoue S, Yang R, Tantry A, Davis CH, Yang T, Knoedler JR, Wei Y, Adams EL, Thombare S, Golf SR, et al. (2019). Periodic remodeling in a neural circuit governs timing of female sexual behavior. *Cell* 179:1393–1408.

Ishii KK, Osakada T, Mori H, Miyasaka N, Yoshihara Y, Miyamichi K, & Touhara K (2017). A labeled-line neural circuit for pheromone-mediated sexual behaviors in mice. *Neuron* 95:123–137.

Kimchi T, Xu J, & Dulac C (2007). A functional circuit underlying male sexual behaviour in the female mouse brain. *Nature* 448:1009–1014.

Kohl J, Babayan BM, Rubinstein ND, Autry AE, Marin-Rodriguez B, Kapoor V, Miyamishi K, Zweifel LS, Luo L, Uchida N, et al. (2018). Functional circuit architecture underlying parental behaviour. *Nature* 556:326–331.

Lin D, Boyle MP, Dollar P, Lee H, Lein ES, Perona P, & Anderson DJ (2011). Functional identification of an aggression locus in the mouse hypothalamus. *Nature* 470:221–226.

Luo M, Fee MS, & Katz LC (2003). Encoding pheromonal signals in the accessory olfactory bulb of behaving mice. *Science* 299:1196–1201.

Marlin BJ, Mitre M, D'Amour JA, Chao MV, & Froemke RC (2015). Oxytocin enables maternal behaviour by balancing cortical inhibition. *Nature* 520:499–504.

Pfaff DW & Sakuma Y (1979). Facilitation of the lordosis reflex of female rats from the ventromedial nucleus of the hypothalamus. *J Physiol* 288:189–202.

Phoenix CH, Goy RW, Gerall AA, & Young WC (1959). Organizing action of prenatally administered testosterone propionate on the tissues mediating mating behavior in the female guinea pig. *Endocrinology* 65:369–382.

Remedios R, Kennedy A, Zelikowsky M, Grewe BF, Schnitzer MJ, & Anderson DJ (2017). Social behaviour shapes hypothalamic neural ensemble representations of conspecific sex. *Nature* 550:388–392.

Seminara SB, Messager S, Chatzidaki EE, Thresher RR, Acierno JS Jr., Shagoury JK, Bo-Abbas Y, Kuohung W, Schwinof KM, Hendrick AG, et al. (2003). The GPR54 gene as a regulator of puberty. *N Engl J Med* 349:1614–1627.

Stowers L, Holy TE, Meister M, Dulac C, & Koentges G (2002). Loss of sex discrimination and male-male aggression in mice deficient for TRP2. *Science* 295:1493–1500.

Winslow JT, Hastings N, Carter CS, Harbaugh CR, & Insel TR (1993). A role for central vasopressin in pair bonding in monogamous prairie voles. *Nature* 365:545–548.

Wu MV, Manoli DS, Fraser EJ, Coats JK, Tollkuhn J, Honda S, Harada N, & Shah NM (2009). Estrogen masculinizes neural pathways and sex-specific behaviors. *Cell* 139:61–72.

Nematodes, lizards, and songbirds

Brainard MS & Doupe AJ (2013). Translating birdsong: songbirds as a model for basic and applied medical research. *Annu Rev Neurosci* 36:489–517.

Crews D (1987). Courtship in unisexual lizards: a model for brain evolution. *Sci Am* 257(6):116–121.

Garrison JL, Macosko EZ, Bernstein S, Pokala N, Albrecht DR, & Bargmann CI (2012). Oxytocin/vasopressin-related peptides have an ancient role in reproductive behavior. *Science* 338:540–543.

Rasika S, Alvarez-Buylla A, & Nottebohm F (1999). BDNF mediates the effects of testosterone on the survival of new neurons in an adult brain. *Neuron* 22:53–62.

CHAPTER 11

Memory, Learning, and Synaptic Plasticity

学而时习之，不亦说乎？

Is it not a pleasure, to have learned something, and to practice it at regular intervals?

Confucius (~500 B.C.E.)

A hallmark of the nervous system is its ability to change in response to experiences through **learning**. In the preceding chapters, we discussed how the nervous system processes sensory information and organizes motor output. However, the nervous system is much more than a giant sensorimotor circuit. In addition to acquiring sensory information from the environment and generating appropriate responses, animals are constantly learning from their sensory experiences and the consequences of their actions. Such learning processes can cause lasting changes in the nervous system, which retains the learned information we call **memory**. Learning enables animals to adapt to a changing environment much more rapidly than the evolutionary process, and its importance to animals and humans cannot be overstated. Memory endows us with much of our individuality, as we are profoundly shaped by what we remember of our past experiences.

Memory and learning have long fascinated human beings. The chapter epigraph, taken from the opening statement of the *Confucius Analects,* reveals that the importance of practicing what has been learned was already recognized 2500 years ago. The French philosopher René Descartes described memory as an imprint made on the brain by external experience (**Figure 11-1**). Over a century ago, psychologists had already established important concepts relating to memory, such as the distinct steps of acquisition, storage, and retrieval. But our understanding of the neurobiological basis of memory and learning comes mostly from research conducted during the past few decades, fostered by our increasing knowledge about the workings of the brain at the molecular, cellular, and systems levels.

PRELUDE: WHAT IS MEMORY AND HOW IS IT ACQUIRED BY LEARNING?

That different parts of the brain perform distinct functions seems obvious today, but historically it took a long time for this concept to take root (Section 1.10). Before the 1950s, the prevailing view was that memories for specific events and skills are distributed across the cerebral cortex. For example, in the 1920s, systematic lesions of the cerebral cortex were carried out in rats that had learned maze navigation to search for brain regions that, when removed, would affect the performance of the learned task. No specific region was found to be necessary; instead, task performance deteriorated progressively as increasingly larger regions were lesioned. From the 1950s onward, this concept of distributed memory changed, at

Figure 11-1 Memory as an imprint. According to René Descartes, memory can be analogized to the imprints left on a linen cloth after needles have passed through it; some of the needle holes stay open (as near points *a* and *b*), and for holes that are closed (as near points *c* and *d*), some traces remain that make it easier to reopen afterward. (From Descartes [1664] Treatise of Man.)

least in regard to acquisition of a specific form of memory, as a result of studies in human patients.

11.1 Memory can be explicit or implicit and short-term or long-term: Insights from amnesic patients

Henry Molaison, widely known as H.M. to protect his privacy until his death at the age of 82 in 2008, suffered from intractable seizures as a young man. In 1953, he underwent bilateral surgical removal of his medial temporal lobes to treat his seizure. While his seizures improved significantly, he emerged from the surgery with irreparable damage: he appeared to have lost the ability to form new memories. He did not recognize doctors who saw him frequently. Within half an hour of eating lunch, he could not remember a single item he had eaten; in fact, he could not remember having eaten lunch at all.

Extensive studies were performed on H.M. His personality and general intelligence—perception, abstract thinking, and reasoning abilities—were not affected by the surgery. In fact, his IQ improved slightly, from 104 presurgery to 112 postsurgery, likely because he was less afflicted by seizures after the surgery. However, he could not form memories in tasks such as remembering a three-digit number following repeated rehearsal; as soon as his attention shifted to a new task, he did not recall the old task or having ever been exposed to it. However, H.M. still had memories of his childhood until about three years before his surgery. For example, he remembered the address of his old house (but not the address of the house he moved to after the surgery).

Interestingly, not all forms of memory were impaired in H.M. In a mirror drawing task, subjects are asked to trace a line between the two borders of a double-outlined star (**Figure 11-2**A) while looking at their hands only in a mirror. People improve at this task with practice, so that the number of errors they make decreases in later trials. H.M. learned this task with a decreasing error rate within and across days (Figure 11-2B), although each day he could not recall ever having performed the task before.

Studies of amnesic patients like H.M. have provided important insights into memory. First, memory can be divided into two broad categories: explicit and implicit (**Figure 11-3**). **Explicit memory** (also called declarative memory) refers to memory requiring conscious recall, such as memories of names, facts, and events. When we say *memory* in daily life, we are usually referring to explicit memory. **Implicit memory** (also called nondeclarative or procedural memory) refers to memory in which previous experience aids in performance of a task without conscious recall. The skill that H.M. acquired in the mirror drawing task and the

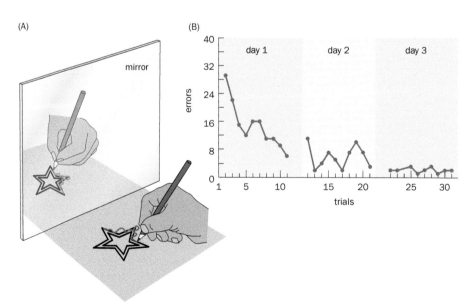

Figure 11-2 Memory of motor-skill learning displayed by H.M. (A) In this task, subjects are asked to view a double-outlined star in a mirror and draw a line in the space between its two borders. Subjects can see their hands only in the mirror. **(B)** With practice (number of trials, *x* axis), H.M. improved his performance in the task within and across days, as seen by the decreasing number of errors (occasions on which the traced line crosses a border, *y* axis). (B, adapted from Milner B, Squire LR, & Kandel ER [1998] *Neuron* 20:445–468. With permission from Elsevier Inc.)

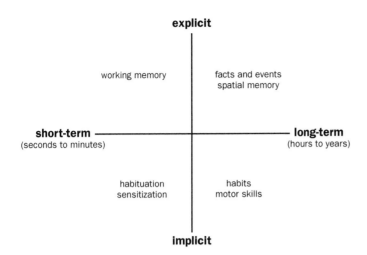

explicit

working memory

facts and events
spatial memory

short-term ─────────────┼───────────── **long-term**
(seconds to minutes)   (hours to years)

habituation
sensitization

habits
motor skills

implicit

Figure 11-3 Different types of memory.
One major division of memory is explicit
(facts and events that require conscious
recall) versus implicit (habits and motor
skills that do not require conscious recall).
Another distinction between different types
of memory is their duration: short-term
memory lasts for seconds to minutes,
while long-term memory can remain intact
throughout one's entire life.

ability to ride a bicycle involve implicit memory; so do habituation and sensitization, memory types that will be introduced later in this chapter. H.M. was selectively incapable of forming *new explicit memories* after his surgery.

Second, memory has different temporal phases, usually divided into **short-term memory** and **long-term memory** (Figure 11-3). The exact temporal window can vary for different types of memories and in different organisms, but typically the memories we define as short-term are retained for seconds to minutes, while long-term memories can last for hours to years. Both explicit and implicit memories have short- and long-term components. As we will learn later in the chapter, there are mechanistic differences between short-term and long-term memory. One example of explicit short-term memory is **working memory**, which refers to the updating, maintenance, and manipulation of information for short periods of time (such as doing multistep mental arithmetic). H.M. had intact working memory, which enabled him to engage in normal conversations with others, but he could not convert short-term memories of facts and events into long-term memories.

Third, distinct steps of the memory process and different types of memory engage different brain regions. Psychologists had divided memory into distinct steps. **Acquisition** is the initial formation of a memory as a consequence of experience and learning. **Retrieval** is the recall of a memory. **Storage** is the step in between acquisition and retrieval, during which memory is held somewhere in the nervous system. **Consolidation** occurs between acquisition and storage, during which newly acquired memory is stabilized for long-term use. Note that consolidation is not a one-time event following initial memory formation, but instead is repeated when the same memory is reactivated for retrieval, a process termed **reconsolidation**. Systematic comparisons of the lesions of H.M. and other amnesic patients revealed that an important region of the medial temporal lobe essential for acquisition of new explicit memories is the **hippocampus**, located underneath the cortical surface of the temporal lobes (Figure 1-8).

Importantly, following surgery, H.M. retained explicit memory for facts and events he had encountered years before surgery. This suggests that the hippocampus is required for acquisition of new explicit memories, but not for long-term storage or retrieval of remote explicit memories. This also implies that memories formed using the hippocampus are later stored elsewhere in the brain, such that they can be recalled even when hippocampal function is disrupted. The fact that H.M. appeared to have intact working memory, which enabled him to hold conversations, and implicit memory, which enabled him to perform the mirror drawing task, implies that working memory and implicit memory also do not require hippocampal function. It is generally accepted that the prefrontal cortex plays a central role in working memory (however, see Box 8-2), while the cerebellum and basal ganglia are instrumental for motor learning and implicit memory (Sections 8.8–8.11).

11.2 Hypothesis I: Memory is stored as strengths of synaptic connections in neural circuits

A key question that connects memory to the neurobiology we have studied in the preceding chapters is: What is the cellular basis of memory storage? A satisfactory answer to this question would allow researchers to then study the mechanisms by which memory is acquired and retrieved. A leading hypothesis, which is strongly supported by experimental evidence, is that memory is stored as strengths of synaptic connections in neural circuits.

Let's first discuss this hypothesis from a theoretical perspective. Suppose that we have a synaptic connection matrix between five input neurons and five output neurons, which have the potential to form 25 synaptic connections. To simplify our discussion, we use a binary code for the connection matrix, where 1 designates a connection (purple dots in **Figure 11-4**, left) and 0 indicates no connection. Suppose further that the firing threshold of each output neuron obeys the following integration rule: if two or more presynaptic input neurons connected to it fire simultaneously, then it will fire its own spike. The input–output function of this circuit, determined by the synaptic connection matrix, can then be used for event-triggered memory recall, as each input pattern can be considered an event and each output pattern can be considered a memory recall. Each input pattern is represented by a specific combination of firing patterns of the five input neurons at a given time. Three input patterns (X_1, X_2, and X_3; Figure 11-4, right) are shown, where 1 signifies a presynaptic neuron firing a spike, and 0 signifies a presynaptic neuron not firing a spike. After passing through the connection matrix, each input pattern produces a corresponding output pattern (Y_1, Y_2, and Y_3), representing the firing patterns of output neurons, as determined by the integration rule. Through this synaptic connection matrix, each input pattern produces a defined output pattern; in other words, each event (X_1, X_2, X_3, and so forth), by interacting with this synaptic matrix, triggers a specific memory recall (Y_1, Y_2, Y_3, and so forth) (**Movie 11-1**).

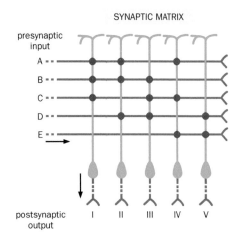

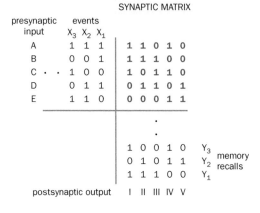

Figure 11-4 The synaptic weight matrix as a memory device. Left, a highly simplified model illustrating how a synaptic matrix can store memory. In this synaptic matrix, axons of five presynaptic input neurons (A–E, red) form specific connections with dendrites of five postsynaptic output neurons (I–V, blue), represented by a binary code: each purple dot signifies a synaptic connection (value = 1); the absence of a purple dot indicates no synaptic connection (value = 0); these values are indicated in purple in the synaptic matrix scheme shown on the right. The cell bodies of the postsynaptic output neurons (blue) are shown below the matrix. Arrows indicate the direction of information flow. Right, this synaptic matrix can transform specific events, represented by the firing patterns of five input axons at any given time, into memory recall, represented by the firing patterns of output neurons. For example, three specific input patterns, X_1, X_2, and X_3, are transformed into three corresponding output patterns, Y_1, Y_2, and Y_3. In these input and output patterns, 1 and 0 represent a spike and no spike, respectively. The integration rule of each postsynaptic neuron is set such that it fires when two or more of its presynaptic partners fire a spike at a given time. For example, for event X_1, presynaptic neurons A, B, and D fire; neurons C and E do not. Event X_1 will activate synapses between neurons A and B on postsynaptic neuron I, and thus activate neuron I as shown in output Y_1. Likewise, event X_1 will activate neurons II and III because three (A–II, B–II, D–II) and two (B–III, D–III) synapses are activated, respectively. By contrast, event X_1 will not activate neurons IV or V because only one of their synapses (A–IV or D–V, respectively) is activated in each case. The resulting Y_1 is that neurons I, II, and III fire a spike, while neurons IV and V do not. This 5 × 5 matrix has 2^{25} or ~30 million binary codes that can be used as a memory device to mediate event (X_N)-triggered recall (Y_N). As will be discussed in Section 11.3, learning alters the weights of synaptic matrices such that the same X_N might produce different Y_Ns before and after learning. In Section 11.18, we will study a synaptic weight matrix in *Drosophila* olfactory conditioning (see Movie 11-1).

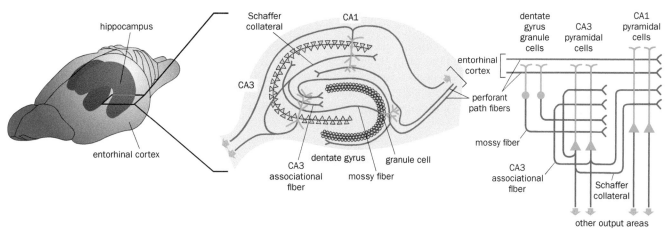

Figure 11-5 The hippocampal circuit. Left, location of the hippocampus and entorhinal cortex in the rat brain. A magnified section of the hippocampus (middle) and a circuit diagram (right) highlight the principal neurons (circles, granule cells; triangles, pyramidal neurons) and their major connections. Blue, dendrites and cell bodies; red, axons. Synapses can form where blue and red lines intersect. Perforant path axons from the superficial layers of the entorhinal cortex reach hippocampal CA1 pyramidal neurons directly via a monosynaptic pathway and indirectly via a trisynaptic pathway in which dentate gyrus granule cells and CA3 pyramidal neurons act as intermediates. CA3 pyramidal neurons also form extensive recurrent connections through associational fibers. Both CA3 and CA1 axons project to subcortical regions (middle panel, bottom left; right panel, bottom). In addition, CA1 axons project directly and via intermediate neurons in the adjacent subiculum to the deep layers of the entorhinal cortex (middle panel, top right). For simplicity, additional regions (e.g., CA2, subiculum) and known connections are not depicted.

Instead of only five input and five output neurons, as in the previous example, neural circuits in the mammalian brain usually comprise many more neurons. As the number of neurons increases, the number of possible synaptic connections goes up astronomically. While the 5×5 matrix in Figure 11-4 has $2^{(5 \times 5)}$ or ~30 million possible binary codes, a 100×100 matrix has $2^{(100 \times 100)}$ or ~10^{3000} possible binary codes, more than there are atoms in the universe. At the same time, suppose that input patterns are represented by the simultaneous firing of 10 out of 100 input neurons; choosing 10 active input fibers out of 100 provides ~10^{13} distinct events. Even if the input fibers encode a different event each millisecond, the system can run for more than 300 years without repeating an event. Furthermore, we have modeled the synaptic connection matrix as consisting of 0/1 binary codes, but in reality the strengths or weights of synaptic connections can have any value at between 0 (no connection) and 1 (maximal connection strength). This greatly expands coding capacity. In summary, these **synaptic weight matrices** can in principle store enormous amounts of information that can be used to transform specific input patterns (events) into specific output patterns (recalled memories). In Section 11.18, we will study a specific example of how memory of odors is encoded in a synaptic weight matrix in the fly brain and read out to instruct approach or avoidance behavior.

As an example of synaptic weight matrices in the mammalian brain, let's examine the circuit organization of the hippocampus (**Figure 11-5**). The hippocampus receives input from the neocortex via the adjacent **entorhinal cortex**. Axons projecting from neurons in the superficial layers of the entorhinal cortex, constituting the **perforant path**, synapse onto the dendrites of **granule cells** in the **dentate gyrus**, the input layer of the hippocampus. The axons of dentate gyrus granule cells, called **mossy fibers** because of their elaborate axon terminals, form synapses onto the dendrites of CA3 pyramidal neurons. The axons of CA3 pyramidal neurons in turn form branches called **Schaffer collaterals**, which synapse onto the dendrites of CA1 pyramidal neurons. In addition to receiving tri-synaptic input (perforant path → granule cells → CA3 → CA1), CA1 dendrites also receive direct input from the entorhinal cortex via the perforant path (Figure 11-5).

Thus, the hippocampus contains not just one but multiple synaptic matrices. These include the perforant path → granule cell synapses, the granule cell mossy fiber → CA3 synapses, the recurrent network among CA3 neurons, the CA3 Schaffer collateral → CA1 synapses, and the direct perforant path → CA1 synapses. In the rat hippocampus, there are hundreds of thousands of CA1 and CA3 pyramidal neurons and over a million dentate gyrus granule cells. Each neuron is connected with thousands to tens of thousands of other neurons. Altogether, these connections can provide a huge capacity for memory acquisition and storage.

11.3 Hypothesis II: Learning modifies the strengths of synaptic connections

If memory is stored as weights of synaptic matrices, then the essence of learning is to alter such weights based on experience. In Chapter 5, we already studied two such mechanisms—Hebb's rule and its extensions: When the firing of a presynaptic neuron repeatedly participates in causing a postsynaptic neuron to fire, their synaptic connection is strengthened; conversely, when the firing of a presynaptic neuron repeatedly fails to elicit firing of a postsynaptic neuron, their synaptic connection is weakened (Figure 5-24). In principle, these rules can be used to modify the weights of synaptic connection matrices, including formation of new synapses and dismantling of existing ones. In a synaptic weight matrix, a change in synaptic weight at specific synapses means that the same input can generate different outputs before and after learning. We use the term **synaptic plasticity** to describe changes in the strengths of synaptic connections in response to experience and neuronal activity.

From an algorithmic perspective, learning can be divided into supervised, unsupervised, and reinforcement based (see Section 14.33 for details). In supervised learning, the output of a circuit is compared with a desired target output, and synaptic weights are modified to minimize discrepancies. Cerebellar gain adjustment of the vestibular-ocular reflex (Figure 8-28) is a good example of supervised learning. In unsupervised learning, synaptic weight changes result from experience without specific instructions. Hebb's rule is particularly useful for unsupervised learning, as it allows co-active neurons to form associations and is fundamental to hippocampus-based learning, as we shall soon see. We will also study two examples of reinforcement learning: olfactory learning in fruit flies (Section 11.18) and reward-based learning in mammals (Section 11.24).

In summary, synaptic connections can be modified (formed, dismantled, strengthened, or weakened), and neurobiologists hypothesize that these modifiable synaptic connections represent a major form of plasticity underlying memory and learning. In the rest of this chapter, we will examine how well experimental evidence supports this conceptual framework. In addition to synaptic plasticity, other plastic changes also contribute to memory and learning. Learning can manifest in changes of excitability of neurons by modulating the expression level and subcellular distribution of ion channels (Section 8.5); one specific example is modulating the concentration of voltage-gated Na^+ channels at the axon initial segment, which determines the efficacy by which input (collective synaptic potentials) is transformed into output (action potentials) (Sections 3.25–3.26). Learning can also manifest in the degree of axon myelination, which determines how efficiently action potentials are propagated to its presynaptic terminals (Box 11-5).

Memory and learning have been studied at a variety of levels of analysis, including genes and proteins, individual neurons and their synapses, the circuits associated with those neurons, and the animal behaviors affected by the activity of those circuits. Researchers have studied memory and learning by taking two complementary approaches: a *top-down* approach that deconstructs complex phenomena to reveal the underlying mechanisms; and a *bottom-up* approach that starts with more basic, smaller-scale phenomena and explores how they relate to higher-level events (**Figure 11-6**). A full understanding of memory and learning requires investigation at all of these levels of analysis. We begin at the level of neurons and synapses, focusing on the mechanisms underlying synaptic plasticity.

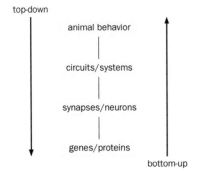

Figure 11-6 Memory and learning can be studied at multiple levels. When researchers start by observing a complex, high-level phenomenon and work to discover its underlying mechanisms, the approach is described as top-down or reductionist. By contrast, when researchers start by examining a low-level phenomenon and try to elucidate its relationship with more complex, high-level events, the approach is termed bottom-up or integrative.

HOW IS SYNAPTIC PLASTICITY ACHIEVED?

The ability of synapses to change their strengths according to experience is one of the most remarkable properties of the nervous system. Every synapse that has been examined shows some form of plasticity. Historically, most mechanistic studies of synaptic plasticity in mammals have centered on the hippocampus. This focus has been prompted by human (Section 11.1) and animal studies indicating that the hippocampus plays an essential role in memory acquisition, by the highly organized architecture of the synaptic input and output of hippocampal principal neurons (its excitatory projection neurons; Figure 11-5), by the opportunity to investigate many synaptic connections *in vitro* using brain slices, and by the discovery of the plasticity phenomena to which we now turn.

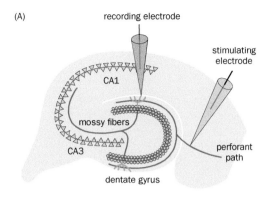

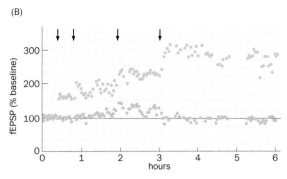

Figure 11-7 Long-term potentiation (LTP) induced by high-frequency stimulation. (A) Experimental setup. The stimulating electrode was placed at the perforant path, which consists of axons innervating dentate gyrus granule cells. A second electrode was placed in the dentate gyrus to record field excitatory postsynaptic potentials (fEPSPs), which represents the collective EPSPs of the population of granule cells near the recording electrode. Mossy fibers, granule cell axons.

(B) High-frequency stimulation (downward arrows, each representing 10 s of 15 Hz stimulation) caused increases in the amplitude of fEPSPs produced by subsequent single stimuli (green dots) compared to controls (yellow dots, no high-frequency stimulation). (Adapted from Bliss TVP & Lømo T [1973] *J Physiol* 232:331–356. With permission from John Wiley & Sons.)

11.4 Long-term potentiation (LTP) of synaptic efficacy can be induced by high-frequency stimulation

In the early 1970s, it was discovered that the connection strengths of hippocampal neurons could be altered by high-frequency stimulation (Figure 11-7). In these experiments, an extracellular recording electrode was implanted in the dentate gyrus of anesthetized rabbits to record the activity of granule cell populations near the electrode. A stimulating electrode was placed in the perforant path to provide synaptic input to granule cells. A single stimulus applied to the stimulating electrode would depolarize granule cells via the perforant path → granule cell synapses. This was recorded as a **field excitatory postsynaptic potential** (fEPSP; see Section 3.15 for EPSP and Section 14.20 for field potential), whose amplitude (or in later experiments, initial slope) is a measure of the strength of synaptic transmission between the stimulated axons of the perforant path and the granule cell population near the recording electrode. After brief trains of high-frequency stimulation were delivered by the stimulating electrode, each single stimulus thereafter produced an fEPSP with a two- to threefold greater magnitude than the baseline. This indicated that the strength of synaptic transmission (**synaptic efficacy** in short) between the perforant path axons and granule cells was enhanced as a result of the high-frequency stimulation. Importantly, this enhancement could last for many hours to several days (Figure 11-7). This phenomenon was thus called **long-term potentiation** (**LTP**).

LTP in response to high-frequency stimulation has since been observed at all excitatory synapses in the hippocampus, including the mossy fiber → CA3 synapse, the CA3 → CA3 recurrent synapse, the CA3 Schaffer collateral → CA1 synapse (hereafter referred to as the **CA3 → CA1 synapse**), and the perforant path → CA1 synapse (Figure 11-5). LTP also exists in many other regions of the nervous system, including the neocortex, striatum, amygdala, thalamus, cerebellum, and spinal cord. Importantly, LTP can be reproduced *in vitro* in **brain slices**, which largely preserve the local three-dimensional architecture of brain tissue *in vivo* while allowing easier experimental access for mechanistic studies. These studies revealed that LTP at different synapses can exhibit different properties through distinct mechanisms. Next, we focus on LTP at the CA3 → CA1 synapse, which is one of the most studied synapses in the mammalian brain.

11.5 LTP at the hippocampal CA3 → CA1 synapse exhibits input specificity, cooperativity, and associativity

Reproduction of LTP in hippocampal slices has enabled many studies to probe its properties. In one experiment, two separate electrodes were placed on the

Figure 11-8 Input specificity, cooperativity, and associativity of long-term potentiation. In each experiment, two sets of presynaptic axons from CA3, *a1* and *a2*, form synapses with the same postsynaptic CA1 neuron *b*. * indicates potentiated synapses. **(A)** LTP exhibits input specificity. In the schematic shown here, only the *a1* → *b* synapses that have undergone high-frequency stimulation (represented by repeated vertical bars) exhibit LTP. **(B)** LTP exhibits cooperativity (see text for the original definition). Depolarization (blue) of postsynaptic cell *b* by current injection enables a weak stimulus (single vertical bar) at axon *a1* to induce LTP. **(C)** LTP exhibits associativity. A weak *a2* stimulus normally would not induce LTP at *a2* → *b* synapses. However, when the timing of a weak *a2* stimulus coincides with high-frequency stimulation of *a1*, *a2* → *b* synapses also become potentiated, as depolarization at *a1* → *b* synapses spreads to *a2* → *b* synapses (blue represents the spread of depolarization). (Based on Nicoll RA, Kauer JA, & Malenka RC [1988] *Neuron* 1:97–103; Bliss TVP & Collingridge GL [1993] *Nature* 361:31–39.)

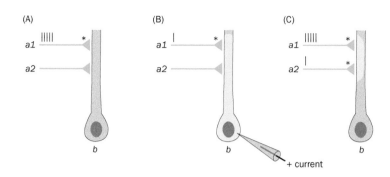

Schaffer collaterals to stimulate two sets of presynaptic axons (from two groups of CA3 neurons), *a1* and *a2,* which synapsed onto the dendrites of cell *b*, a CA1 postsynaptic neuron that was being recorded. LTP was induced by high-frequency stimulation of *a1* (**Figure 11-8**A). When synaptic efficacy was measured afterward, only the *a1* → *b* connection was potentiated, while the *a2* → *b* connection remained unchanged. Thus, LTP exhibits **input specificity**: it occurs at synapses that have experienced high-frequency stimulation, but not at inactive synapses onto the same postsynaptic neuron.

Cooperativity, a second property of LTP, refers to the observation that high-frequency stimulation of one or few axons is insufficient to induce LTP; rather, "cooperation" of many active axons is needed. The following experiment revealed the cellular mechanism underlying cooperativity: when a weak axonal stimulation insufficient to induce LTP was paired with coincident injection of depolarizing current into the postsynaptic cell from the recording electrode, LTP was induced (Figure 11-8B). Thus, LTP is induced at a synapse when two events coincide: (1) the presynaptic cell fires, releasing neurotransmitter, and (2) the postsynaptic cell is in a depolarized state.

The postsynaptic mechanism of LTP cooperativity explains why high-frequency stimulation can induce LTP in the first place. Early in the train, action potentials from *a1* depolarize cell *b* at *a1* → *b* synapses, such that the arrival of action potentials late in the train coincides with a depolarized state of the postsynaptic cell, hence potentiating *a1* → *b* synapses. This mechanism can also explain a third property of LTP illustrated in the following experiment. While high-frequency stimulation was applied to *a1* to induce LTP at *a1* → *b* synapses, *a2* was also stimulated at a level that, by itself, would not reach the threshold of inducing LTP. The coincident stimulation, however, was found to potentiate *a2* → *b* synapses as well (Figure 11-8C). This is because high-frequency stimulation of *a1* causes depolarization in a region of cell *b* that includes the sites of *a2* → *b* synapses. If *a2* receives a weak stimulus at the same time that *b* is depolarized at *a2* → *b* synapses, then those synapses will potentiate. This potentiation of synapses that experience a weak stimulus by a coincident strong stimulus is called **associativity** of LTP.

These properties of LTP make it suitable for adjusting the synaptic weight matrix hypothesized to underlie memory. In Figure 11-4, for example, input specificity allows the strengths of different synapses of an output neuron with *distinct input neurons* to be altered independently, while cooperativity allows a given input neuron to alter the strengths of synapses with a specific subset of co-active *output neurons*. Together, these properties allow experience to adjust synaptic weights in the matrix on a synapse-by-synapse basis. Associativity makes it possible for coincident inputs to influence each other's synaptic strengths and is particularly well suited for associative learning, as we will discuss later in the chapter.

11.6 The NMDA receptor is a coincidence detector for LTP induction

The requirement of postsynaptic depolarization for LTP is consistent with Hebb's rule (Figure 5-24). Indeed, this property made the CA3 → CA1 synapse the first known example of what is now called a **Hebbian synapse**, a synapse whose strength can be enhanced by co-activating pre- and postsynaptic partners. Recall

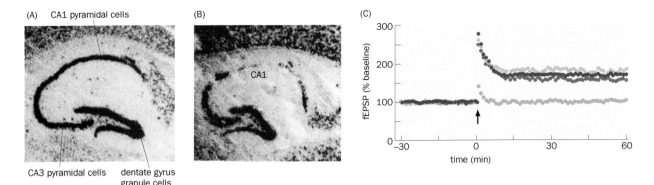

Figure 11-9 The NMDA receptor in the postsynaptic neuron is essential for LTP induction at the CA3 → CA1 synapse. (A) *In situ* hybridization shows that mRNA for the GluN1 subunit of the NMDA receptor is highly expressed in CA3 and CA1 pyramidal neurons and dentate gyrus granule cells. GluN1 is also expressed in the cerebral cortex above CA1. **(B)** Conditional knockout (see Section 14.8 for details) of GluN1 using a transgene expressing Cre recombinase in CA1 neurons selectively disrupts GluN1 mRNA expression in CA1 pyramidal neurons. **(C)** In *CA1-Cre*-mediated GluN1 conditional knockout mice, CA3 → CA1 LTP is blocked (blue trace) compared to normal LTP exhibited by control mice that are wild type (yellow trace), that have the GluN1 conditional allele alone (red trace), or that have the *CA1-Cre* transgene alone (brown trace). The upward arrow indicates high-frequency stimulation for inducing LTP at *t* = 0. (Adapted from Tsien JZ, Huerta PT, & Tonegawa S [1996] *Cell* 87:1327–1338. With permission from Elsevier Inc.)

that we have already studied a molecule capable of implementing Hebb's rule: the **NMDA receptor**. The opening of the NMDA receptor channel requires simultaneous glutamate release from the presynaptic terminal and depolarization of the postsynaptic neuron to remove the Mg^{2+} block (Figure 3-24). This property accounts for the cooperativity and associativity of LTP. Indeed, ample evidence supports a key role for the NMDA receptor in the establishment of LTP (called *LTP induction*) at the CA3 → CA1 synapse.

First, the NMDA receptor is highly expressed in hippocampal neurons (**Figure 11-9A**). Second, pharmacological inhibition of the NMDA receptor by a specific NMDA receptor antagonist, **2-amino-5-phosphonovaleric acid** (**AP5**), blocks LTP induction in hippocampal slices without affecting baseline synaptic transmission. Third, when the gene encoding GluN1, an essential subunit of the NMDA receptor channel, was selectively knocked out in hippocampal CA1 neurons of mice (Figure 11-9B), LTP at the CA3 → CA1 synapse was abolished (Figure 11-9C), but basal synaptic transmission was unaffected. Because GluN1 was knocked out only in the postsynaptic CA1 neurons and remained functional in the presynaptic CA3 neurons, this experiment also demonstrated the postsynaptic requirement for the NMDA receptor in LTP induction at the CA3 → CA1 synapse.

11.7 Recruitment of AMPA receptors to the postsynaptic surface is a major mechanism of LTP expression

At most CNS synapses, LTP induction occurs through postsynaptic activation of the NMDA receptor. (A notable exception is the mossy fiber → CA3 synapse, where LTP induction does not require the NMDA receptor and instead involves a largely presynaptic mechanism in which cAMP and protein kinase A act to regulate neurotransmitter release probability.) The mechanism by which NMDA receptor activation leads to long-lasting increases in synaptic efficacy, called *LTP expression*, has been the subject of intense debate. Two major types of mechanisms have been proposed: a presynaptic mechanism involving an increase in the neurotransmitter release probability (Section 3.10), and a postsynaptic mechanism involving an increase in the sensitivity of the postsynaptic cell to the release of the same amount of neurotransmitter. These two mechanisms are not mutually exclusive.

At the CA3 → CA1 synapse, accumulating evidence suggests that the predominant mechanism of LTP expression is an increase in the number of AMPA-type glutamate receptors at the postsynaptic surface. As discussed in Chapter 3 (Figure 3-24), the AMPA receptor is essential for basal synaptic transmission under conditions in which postsynaptic cells are insufficiently depolarized to activate the

NMDA receptor. Within minutes following activation of the NMDA receptor during LTP induction, more AMPA receptors are inserted into the postsynaptic membrane. Subsequent glutamate release can thus trigger the opening of more AMPA receptors and hence stronger depolarization.

In fact, some glutamatergic CNS synapses, including a large fraction of the CA3 → CA1 synapses, initially contain only NMDA receptors on the postsynaptic surfaces. These synapses cannot be activated by presynaptic glutamate release alone and are thus called **silent synapses**. However, coincident postsynaptic depolarization (presumably through AMPA receptors at other synapses in the same postsynaptic neuron) and presynaptic glutamate release activate the NMDA receptors at silent synapses and thereby cause insertion of AMPA receptors into the postsynaptic membrane, transforming silent synapses into synapses that can be activated by presynaptic activity alone (**Figure 11-10**A–C). LTP expression involves both the activation of silent synapses (Figure 11-10D) and insertions of AMPA receptors into synapses that already have some AMPA receptors.

In LTP and other forms of synaptic plasticity (discussed later in the chapter), AMPA receptor trafficking is subjected to many forms of regulation as a consequence of NMDA receptor activation. These include increasing exocytosis of AMPA-receptor-containing vesicles, leading to an increase in the number of cell-surface AMPA receptors; enhancing the binding of AMPA receptors to postsynaptic density scaffold proteins to increase their residence time at the postsynaptic surface; facilitating lateral diffusion of AMPA receptors toward the synaptic surface; and altering the subunit composition and phosphorylation status of AMPA receptors

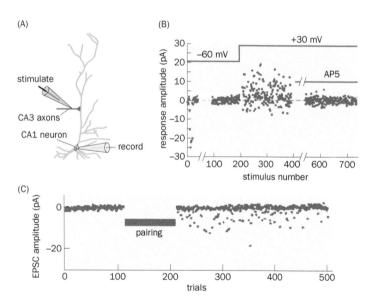

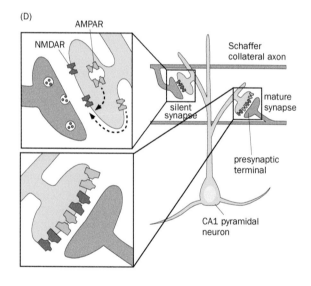

Figure 11-10 Silent synapses and their activation by LTP.
(A) Experimental schematic for Panel B and Panel C. In a hippocampal slice, a CA1 neuron's responses to stimulation of a set of CA3 axons were measured by whole-cell patch clamp recording (see Box 14-3 for details). **(B)** Demonstration of silent synapses. At the beginning of this experiment, the CA1 cell was held at –60 mV, and after obtaining small excitatory postsynaptic currents (EPSCs) by stimulating CA3 axons, the stimulation strength was reduced, resulting in stimulation of fewer axons, so that stimuli 100–200 did not produce any EPSCs. This means that no AMPA receptors were activated by the weak stimulus. However, when the cell was held at +30 mV, the same weak stimulus now evoked EPSCs that were blocked by AP5, indicating that the stimulated synapses contained NMDA receptors. In other words, the weak stimulus activated synapses containing NMDA but not AMPA receptors. **(C)** Activation of silent synapses. In this experiment, for the first 100 trials, CA1 neurons were held at –65 mV so that only AMPA currents could be induced by CA3 axon stimulation. Prior to pairing, EPSCs were not elicited, indicating that either the stimulated CA3 axons did not connect with the recorded CA1 neurons or that they were connected via silent synapses. After repeated pairing of CA3

axon stimulation with depolarization of the postsynaptic CA1 neurons, a condition that induces LTP (Figure 11-8B), a subset of CA3 stimulations elicited EPSCs, indicating that this subset was previously connected via silent synapses, which were activated (unsilenced) by pairing of presynaptic stimulation and postsynaptic depolarization. Note that EPSCs were outward (positive) when the cell was clamped at +30 mV (Panel B) and inward (negative) at –60 mV (Panel B) and –65 mV (Panel C), as the reversal potentials of AMPA and NMDA receptors are near 0 mV (Section 3.15). **(D)** Summary. Top left, silent synapses have only NMDA receptors (NMDARs) at their postsynaptic surfaces. LTP causes net insertion of AMPA receptors (AMPARs) into the postsynaptic surface via exocytosis of AMPAR-containing vesicles, recruitment of AMPARs from extra-synaptic areas, or both (dashed arrows). Bottom left, mature synapses contain both AMPA and NMDA receptors. (B, from Isaac JTR, Nicoll RA, & Malenka RC [1995] *Neuron* 15:427–434. With permission from Elsevier Inc. C, from Liao D, Hessler NA, & Mallnow R [1995] *Nature* 375:400–404. With permission from Springer Nature. D, from Kerchner GA & Nicoll RA [2008] *Nat Rev Neurosci* 9:813–825. With permission from Springer Nature.)

to increase their conductance. Exactly how each of these regulations are triggered by NMDA receptor activation is a subject of intense research; we turn now to one mechanism involving activation of a specific protein kinase.

11.8 CaMKII auto-phosphorylation creates a molecular memory that links LTP induction and expression

As we learned in Chapter 3, a key property of the NMDA receptor is its high Ca^{2+} conductance. NMDA receptor activation causes an increase in $[Ca^{2+}]_i$, which activates a number of signaling pathways; for example, Ca^{2+}-activated adenylate cyclases increase production of cAMP and activation of protein kinase A (Figure 3-41). Another key signaling molecule is Ca^{2+}/calmodulin-dependent protein kinase II (**CaMKII**), which is activated by Ca^{2+}/calmodulin binding and is highly enriched in the postsynaptic density (Figure 3-27; Figure 3-34). The holoenzyme of CaMKII has 12 subunits. Each subunit contains a catalytic domain and an auto-inhibitory domain that binds to the catalytic domain and inhibits its function. Binding of Ca^{2+}/calmodulin to CaMKII transiently displaces the auto-inhibitory domain and thus activates the kinase. When $[Ca^{2+}]_i$ decreases, Ca^{2+}/calmodulin dissociates, deactivating CaMKII (**Figure 11-11**A, top).

The combination of a multisubunit structure and auto-inhibitory domains regulated by phosphorylation endows CaMKII with an interesting property: active CaMKII can phosphorylate a threonine residue at amino acid 286 (T286) in the auto-inhibitory domain of a neighboring CaMKII subunit. T286 phosphorylation impairs auto-inhibition, so that the activity of the phosphorylated subunits persists even after Ca^{2+}/calmodulin dissociates. Thus, if the initial Ca^{2+} signal is strong enough to cause T286 phosphorylation at multiple subunits, subsequent CaMKII cross-phosphorylation can lead to sustained activity that outlasts Ca^{2+}/calmodulin binding. This process creates a "memory" in the CaMKII molecule—a historical record of Ca^{2+} signaling—until phosphatases erase the memory through T286 dephosphorylation (Figure 11-11A, bottom). This molecular memory contributes to sustained changes in synaptic efficacy after transient NMDA receptor activation. In support of this proposal, mice in which auto-phosphorylation of CaMKII at T286 was prevented by mutating T286 to an alanine exhibited profound LTP deficits (Figure 11-11B).

Activation of CaMKII also appears to be sufficient for LTP induction. When a truncated, constitutively active form of CaMKII lacking the auto-inhibitory domain was injected directly into CA1 pyramidal neurons, CA3 → CA1 synaptic transmission was potentiated. Furthermore, synapses potentiated by constitutively active

Figure 11-11 Auto-phosphorylation of CaMKII and its requirement in LTP. (A) The CaMKII holoenzyme has 12 subunits; only 6 are shown here for simplicity. Top, binding of Ca^{2+}/calmodulin to a particular subunit transiently activates that subunit (rightward direction; * denotes an active subunit). Ca^{2+}/calmodulin dissociates after $[Ca^{2+}]_i$ drops, and the subunit becomes inactive (leftward direction). Bottom, if a sufficient number of CaMKII subunits become activated in response to prolonged $[Ca^{2+}]_i$ elevation, active subunits phosphorylate specific threonine residues (T286) of neighboring subunits. This cross-subunit phosphorylation maintains CaMKII in an activated state after $[Ca^{2+}]_i$ drops and Ca^{2+}/CaM complexes dissociate, until phosphatase activity overrides the auto-activation. **(B)** LTP at CA3 → CA1 synapses can be induced by 10 Hz or 100 Hz high-frequency stimulation or two theta bursts (2TB), each consisting of four stimuli at 100 Hz with 200 ms separating the onset of each burst, which mimics endogenous firing of hippocampal neurons. In mutant mice in which T286 of CaMKII was replaced with an alanine (T286A), both forms of LTP were disrupted. (A, from Lisman J, Schulman H, & Cline H [2002] *Nat Rev Neurosci* 3:175–190. With permission from Springer Nature. B, adapted from Giese KP, Federov NB, Filipkowski RK, et al. [1998] *Science* 279:870–873.)

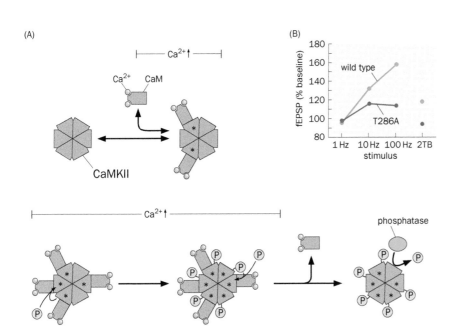

(A)

Ca^{2+}↑

Ca^{2+} CaM

CaMKII

Ca^{2+}↑

phosphatase

(B)

wild type

T286A

fEPSP (% baseline)

1 Hz   10 Hz   100 Hz   2TB
stimulus

Figure 11-12 LTP induction occludes CaMKII-induced synaptic potentiation. Top, experimental design; bottom, experimental data. The arrow linking the two schematics denotes that Panel B was a continuation of Panel A in the same preparation. **(A)** High-frequency stimulation was applied via the S1-stimulating electrode at the time indicated in the graph by the upward arrow. Only S1 → CA1 synapses were potentiated (brown trace), while the efficacy of S2 → CA1 synapses remained unchanged (yellow trace), showing input specificity. An extracellular recording electrode was used to measure field excitatory postsynaptic potential (fEPSP) in response to S1 or S2 stimulation. **(B)** Following potentiation and extracellular recording in Panel A, a postsynaptic CA1 neuron was patched for whole-cell recording, and constitutively active CaMKII enzyme was injected into the CA1 neuron through the patch electrode (at $t = 0$). Only the previously unpotentiated S2 synapses were substantially potentiated, as indicated by the gradually increased excitatory postsynaptic current (EPSC) in response to stimulation of S2 but not S1. Thus, CaMKII potentiation of S1 synapses was occluded by prior LTP. (Adapted from Lledo P, Hjelmstad GO, Mukherji S, et al. [1995] *Proc Natl Acad Sci U S A* 92:11175–11179.)

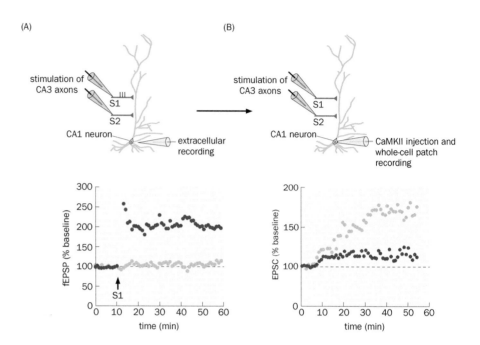

CaMKII could no longer be induced to exhibit LTP by high-frequency stimulation, while synapses at which LTP had been induced by high-frequency stimulation could no longer be potentiated by constitutively active CaMKII (**Figure 11-12**). Thus, the two mechanisms of synaptic potentiation—high-frequency stimulation and CaMKII activation—occluded each other. These occlusion experiments support the notion that CaMKII activation is an integral component linking LTP induction and expression (maintenance).

CaMKII activity contributes to regulation of synaptic transmission strength through multiple mechanisms. For example, CaMKII-catalyzed phosphorylation of AMPA receptors increases their ion conductance and influences their trafficking (see Section 11.9). CaMKII also phosphorylates postsynaptic scaffold proteins (Section 3.16), which creates docking sites for AMPA receptors in the postsynaptic membrane. Another key output mediated by CaMKII and other signaling molecules essential for long-lasting changes in synaptic efficacy involves activating transcription factors and expressing new genes (Figure 3-41). One process these genes likely regulate is the structural modification of synapses (see Section 11.13).

11.9 Long-term depression weakens synaptic efficacy

So far we have focused on LTP and its mechanisms of induction and expression. However, if synaptic connections could only be made stronger, the entire synaptic weight matrix would eventually become saturated, with no room to encode new memories. Indeed, many additional plasticity mechanisms co-exist with LTP so that synaptic weight can be adjusted bidirectionally, as discussed here and in the next sections.

One counterbalancing mechanism is **long-term depression**, or **LTD**. Like LTP, LTD has been found at many CNS synapses. It was first discovered in the cerebellum, where coincident activation of parallel fibers and climbing fibers causes depression of parallel fiber → Purkinje cell synapses. This property contributes to cerebellar cortex–mediated motor learning (Section 8.11). At hippocampal CA3 → CA1 synapses, LTD can be induced by low-frequency stimulation of presynaptic axons; note that the same synapses exhibit LTP in response to high-frequency stimulation (**Figure 11-13**A). Like LTP induction, LTD induction requires the NMDA receptor and Ca^{2+} influx. The increase in $[Ca^{2+}]_i$ resulting from low-frequency stimulation is lower than that resulting from high-frequency stimulation. This lower increase in $[Ca^{2+}]_i$ is thought to preferentially activate Ca^{2+}-dependent phosphatases, which do the opposite of what LTP-activated kinases do: the phosphatases reduce the number of AMPA receptors at the postsynaptic plasma membrane

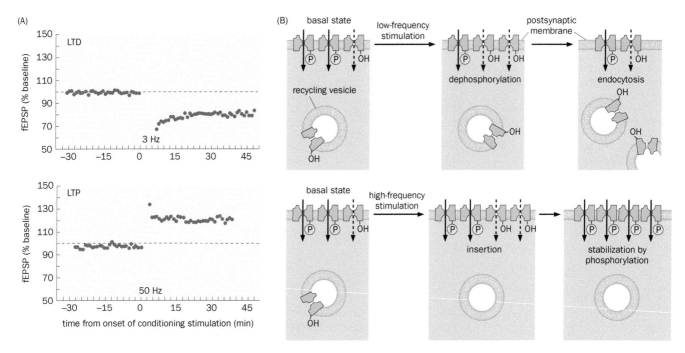

Figure 11-13 Long-term depression at the CA3 → CA1 synapse.
(A) Whereas high-frequency (50 Hz) stimulation induces LTP (bottom), low-frequency (3 Hz) stimulation of CA3 axons innervating a CA1 neuron causes long-term depression (LTD) of synaptic efficacy (top). **(B)** In this model, AMPA receptors are in a dynamic equilibrium between the cell surface and intracellular recycling vesicles. Low-frequency stimulation induces GluA1 dephosphorylation, which promotes endocytosis of AMPA receptors (top). High-frequency stimulation induces GluA1 phosphorylation, which stabilizes AMPA receptors at the postsynaptic membrane (bottom). In addition, AMPA receptors with phosphorylated GluA1 subunits have higher channel conductance (solid arrow for larger ion flow) than those with non-phosphorylated GluA1 subunits (dashed arrow). (A, adapted from Dudek SM & Bear MF [1992] *Proc Natl Acad Sci U S A* 89:4363–4367. B, from Lee HK, Takamiya K, Han JS, et al. [2003] *Cell* 112:631–643. With permission from Elsevier Inc.)

so that subsequent glutamate release from the presynaptic terminal induces less depolarization.

LTD and LTP can affect the same synapse sequentially. Low-frequency stimulation can depress a synapse that has previously been potentiated by LTP; high-frequency stimulation can potentiate a synapse that has previously been depressed by LTD. Regulation of phosphorylation of the AMPA receptor GluA1 subunit at specific amino acid residues by CaMKII, protein kinase A (PKA), and protein kinase C (PKC) likely plays a role in LTP and LTD expression. One model proposes that GluA1 phosphorylation in the context of LTP not only increases the channel conductance of AMPA receptors but also stabilizes AMPA receptors newly added to the postsynaptic membrane, whereas GluA1 dephosphorylation triggers endocytosis of AMPA receptors from the postsynaptic membrane, leading to LTD (Figure 11-13B). Indeed, knock-in mice in which phosphorylation sites of two serines on GluA1 were replaced with alanines (so that neither could be phosphorylated) had significantly reduced LTP and LTD expression. These and other experiments support the notion that, at a given synapse, LTD and LTP represent a continuum of modifications of synaptic strength. The ability to control synaptic weights bidirectionally via LTP and LTD greatly increases the flexibility and storage capacity of synaptic memory matrices.

Just as LTP can be produced by diverse mechanisms, LTD can also be produced by mechanisms other than a reduction in postsynaptic AMPA receptor signaling. In Section 11.12, we will encounter a presynaptic mechanism of LTD expression.

11.10 Spike-timing-dependent plasticity can adjust synaptic efficacy bidirectionally

Although high- and low-frequency stimulations are commonly used to experimentally induce synaptic plasticity, under physiological conditions, neurons are

Figure 11-14 Spike-timing-dependent plasticity (STDP). If the presynaptic neuron repeatedly fires before the postsynaptic neuron, the synapse is potentiated (left). If the presynaptic neuron repeatedly fires after the postsynaptic neuron, the synapse is depressed (right). Data here were taken from retinotectal synapses in developing *Xenopus in vivo*, where the presynaptic neuron was a retinal ganglion cell and the postsynaptic neuron was a tectal neuron. (From Zhang Li, Tao HW, Holt CE, et al. [1998] *Nature* 395:37–44. With permission from Springer Nature.)

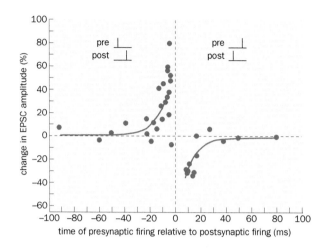

not usually activated at those precise frequencies. In reality, interconnected neurons fire action potentials at many frequencies. Another plasticity mechanism that can influence synaptic strength is **spike-timing-dependent plasticity (STDP)**. Originally discovered in the 1990s by researchers using patch clamp methods to study pairs of pyramidal neurons in rat cortical slices and cultures of dissociated hippocampal neurons, STDP has since been found in many different systems. In STDP, the precise timing of pre- and postsynaptic firing is critical in determining the sign of the synaptic strength change. For a typical synapse between two excitatory neurons, if the presynaptic neuron fires *before* the postsynaptic neuron within a narrow time window (usually tens of milliseconds) and if these pairings are repeated, then synaptic efficacy increases; if repeated firing of the presynaptic neuron takes place within tens of milliseconds *after* firing of the postsynaptic neuron, then synaptic efficacy decreases (**Figure 11-14**). Thus, STDP has features of both LTP and LTD. Indeed, it has many similarities to LTP and LTD, such as dependence on NMDA receptor activation.

STDP is well suited for implementing Hebb's rule. If the presynaptic cell fires repeatedly before the postsynaptic cell, then it is likely that firing of the presynaptic cell contributes to the stimuli that cause the postsynaptic cell to fire; the synapses between the two cells should be strengthened. If the presynaptic cell fires repeatedly after the postsynaptic cell, then it is unlikely that the presynaptic cell contributed to causing the postsynaptic cell to fire; synapses between the two cells should be weakened. In addition to serving a role in balancing potentiation and depression of synaptic strength in synaptic weight matrices, the timing dependence of STDP can be used for other purposes, including activity-dependent wiring of neural circuits (see Chapters 5 and 7).

11.11 Homeostatic synaptic plasticity adjusts overall synaptic strengths according to overall activity levels

LTP, LTD, and STDP are all synapse-specific phenomena that are manifestations of Hebbian plasticity. Hebbian plasticity is essentially a *positive-feedback* process, whereby strong synapses become stronger and weak synapses become weaker. If this were the only form of synaptic plasticity, then cells with mostly weak input synapses would develop progressively lower firing rates, and their output synapses would thus be weakened until the cells gradually become silent and drop out of the circuit altogether. Other cells with mostly strong input synapses would develop progressively higher firing rates that could saturate their synapses and cause epilepsy. Additionally, by driving synapses to their maximum or minimum values, such positive-feedback processes erode the synapse-specific differences in synaptic weights postulated to encode memories. Fortunately, there also exist *negative-feedback* processes for adjusting synaptic strengths to overcome these problems. One such process is **homeostatic synaptic scaling**.

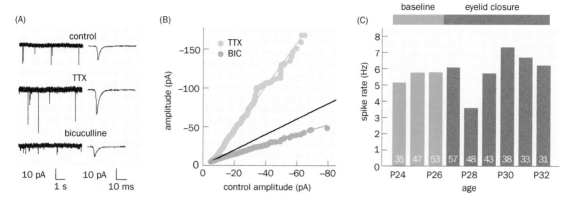

Figure 11-15 Homeostatic synaptic plasticity of mammalian neurons.
(A) Spontaneous miniature excitatory postsynaptic current (mEPSC) measured from cultured cortical pyramidal neurons in control and 48 h after TTX or bicuculline treatment. Left, raw traces; right, average mEPSC waveforms. **(B)** The x axis represents rank-ordered mEPSC amplitudes (each neuron is a dot) under the control condition, and the y axis represents rank-ordered mEPSC amplitudes after TTX (green) or bicuculline (orange) treatment. Compared with control plotted against control (black line), TTX treatment potentiates mEPSCs, while bicuculline depresses mEPSCs. These plots support a uniform multiplicative model of mEPSC adjustment. **(C)** Rats were visually deprived by eyelid closure at postnatal day 27 (P27). The average firing rate of V1 neurons, as measured by chronic single-unit recording in freely moving rats (number of units indicated in each column), undergoes a transient decrease on the second day after eyelid closure caused by sensory deprivation but returns to predeprivation levels afterward. (A, from Turrigiano GG, Leslie KR, Desai NS, et al. [1998] *Nature* 391:892–896. With permission from Springer Nature. B, from Abbott LF & Nelson SB [2000] *Nat Neurosci* 3:1178–1183. With permission from Springer Nature. C, from Hengen KB, Lambo ME, Van Hooser SD, et al. [2013] *Neuron* 80:335–342. With permission from Elsevier Inc.; see also Keck T, Keller GB, Jacobsen RI, et al. [2013] *Neuron* 80:327–334; Hengen KB, Torrado Pacheco A, McGregor JN, et al. [2016] *Cell* 165:180–191.)

Homeostatic synaptic scaling was originally described in cortical neurons in dissociated culture. Applying tetrodotoxin (TTX) to suppress action potential firing for 48 hours caused an increase in average unitary synaptic strength, as measured by a miniature excitatory postsynaptic current (mEPSC; Section 3.2). Conversely, application of bicuculine (a GABA$_A$ receptor antagonist) to reduce inhibition and thereby increase firing caused a decrease in mEPSC amplitude 48 hours later (**Figure 11-15**A). Thus, changes in network activity cause compensatory changes to the excitability of neurons so as to maintain firing rates within a certain range. An interesting property of synaptic scaling as studied in cortical neurons in culture is that all synapses appear to be adjusted by a multiplicative factor—up or down in proportion to the original strength of the synapse (Figure 11-15B). This property preserves the information stored as relative differences in strength among individual synapses that have been modified by LTP, LTD, or STDP. Importantly, homeostatic adjustment of the firing rate has also been observed *in vivo*. As one specific example, chronic single-unit recording was performed on pyramidal neurons in the primary visual cortex (V1) of freely moving rats following eyelid closure. After a transient decline in response to visual deprivation, the firing rate of V1 neurons returned to the predeprivation set point (Figure 11-15C).

Besides homeostatic synaptic scaling, there are other forms of homeostatic synaptic plasticity with different properties and likely distinct underlying mechanisms. Evidence suggests that the global adjustment of synaptic strengths in response to changes in network activity discussed earlier occurs over hours to days and is likely achieved through cell-autonomous mechanisms, such as insertion or removal of AMPA receptors in the postsynaptic density. Other forms of homeostatic synaptic plasticity can be more local, occur on a shorter time scale, and involve cell–cell communication. For example, at the *Drosophila* neuromuscular junction (Figure 3-11), partial blockade of the glutamate receptor on the muscle membrane, which caused a decrease in the amplitude of miniature postsynaptic potentials (mEPSPs), induced an increase in quantal content (Section 3.3) within 10 minutes (**Figure 11-16**A), likely due to increased neurotransmitter (glutamate) release probability in the motor axon terminals. Remarkably, this change in quantal content brings EPSP amplitude (a product of quantal content and mEPSP

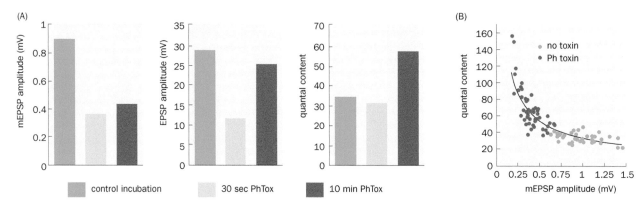

Figure 11-16 Rapid synaptic homeostasis at the *Drosophila* neuromuscular junction. (A) Intracellular recording of a *Drosophila* third instar larval body wall muscle was used to measure spontaneous mEPSPs and motor axon stimulation-evoked EPSPs. A 30 s application of philanthotoxin (PhTox), which partially blocks the muscle glutamate receptor, causes a reduction in both average mEPSP and EPSP amplitudes. However, incubating in PhTox for 10 min caused an increase in the average quantal content, such that the average EPSP magnitude was restored to the control level. **(B)** Scatter plot of quantal content versus mEPSP amplitude of individual experiments (each experiment is a dot). The line represents the function corresponding to perfect compensation for maintaining a constant EPSP amplitude (quantal content × mEPSP amplitude). (From Frank CA, Kennedy MJ, Goold CP, et al. [2006] *Neuron* 52:663–677. With permission from Elsevier Inc.)

amplitude) back to the value prior to receptor blockade (Figure 11-16A, B). The fact that a postsynaptic manipulation (receptor blockade) induces a change in neurotransmitter release implies retrograde signaling from postsynaptic muscle to presynaptic motor neurons, a concept we will expand upon in the next section.

11.12 Postsynaptic cells can produce retrograde messengers to regulate neurotransmitter release by their presynaptic partners

Our discussions thus far have largely focused on postsynaptic mechanisms for modifying synaptic efficacy, but synaptic plasticity can also engage presynaptic mechanisms. As discussed in Section 3.10, synapses can be facilitated or depressed due to increases or decreases in neurotransmitter release probability in response to a train of action potentials. Longer-term changes in synaptic efficacy, such as LTP at the hippocampal mossy fiber → CA3 synapse, can also be induced by presynaptic mechanisms, resulting in enhancement of neurotransmitter release probability. In other cases, however, modulation of presynaptic release probability is triggered by an initial change in the postsynaptic neuron, as exemplified by synaptic homeostasis at the *Drosophila* neuromuscular junction that we just discussed (Figure 11-16). This implies that the postsynaptic neuron must send a retrograde messenger back to its presynaptic partner, against the canonical direction of information flow across the chemical synapse.

Endocannabinoids (endogenous cannabinoids) are among the best-studied retrograde messengers: they are produced by postsynaptic neurons to regulate presynaptic neurotransmitter release probability. For example, hippocampal CA1 pyramidal neurons rapidly produce endocannabinoids upon depolarization. These lipophilic molecules, which include anandamide and 2-arachidonylglycerol, are ligands for a G-protein-coupled receptor, **CB1**, which is widely expressed in the brain and was first identified as the receptor for cannabinoids from the marijuana plant (genus *Cannabis*). In the 1990s, while some researchers discovered endocannabinoids and investigated their properties, others identified an interesting plasticity phenomenon called **depolarization-induced suppression of inhibition** (**DSI**) in hippocampal CA1 pyramidal neurons. CA1 pyramidal neurons receive inhibitory input from GABAergic neurons in addition to excitatory input from CA3 neurons and entorhinal cortex. During intracellular recording of CA1 neurons in hippocampal slices, it was found that depolarization elicited by intracellular current injection or high-frequency stimulation of CA3 axons caused transient suppression of inhibitory input to CA1 neurons (Figure 11-17A). Further experi-

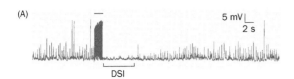

(A)

5 mV
2 s

DSI

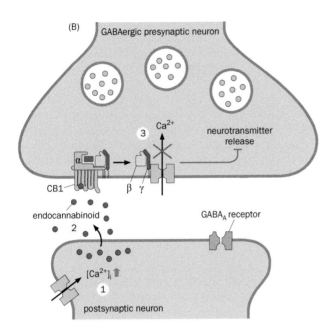

(B) GABAergic presynaptic neuron

Ca²⁺
3
neurotransmitter
release

α
CB1
β γ

endocannabinoid
2

GABA_A receptor

[Ca²⁺]ᵢ
1

postsynaptic neuron

Figure 11-17 Depolarization-induced suppression of inhibition (DSI) and endocannabinoid signaling. (A) Following stimulation by a train of action potentials (indicated by the horizontal red bar), a hippocampal CA1 neuron exhibited DSI, as evidenced in a transient reduction of the frequency of spontaneous inhibitory postsynaptic potentials (IPSPs). Because the intracellular recording electrode was filled with KCl, diffusion of Cl⁻ from the electrode into the cell reversed the Cl⁻ gradient and caused IPSPs to be positive. The slice was bathed in medium containing drugs that block excitatory synaptic transmission. **(B)** Schematic summary of endocannabinoid signaling in DSI. (1) CA1 neurons produce endocannabinoids in response to a rise in [Ca²⁺]ᵢ through voltage-gated Ca²⁺ channels or NMDA receptors (not shown) due to postsynaptic depolarization. (2) Endocannabinoids diffuse across the postsynaptic membrane and synaptic cleft, where they bind to the G-protein-coupled CB1 receptor enriched in presynaptic terminals of GABAergic neurons. (3) Activation of CB1 releases Gβγ, which binds to and causes closure of presynaptic voltage-gated Ca²⁺ channels, resulting in inhibition of GABA release. (A, from Pitler TA & Alger BE [1992] *J Neurosci* 12:4122–4132. Copyright ©1992 Society for Neuroscience. B, adapted from Wilson RI & Nicoll RA [2002] *Science* 296:678–682.)

ments indicated that DSI required Ca²⁺ influx into the postsynaptic CA1 neuron but did not affect the sensitivity of the CA1 neuron to exogenous GABA application. These data suggest that DSI is most likely mediated by a reduction in GABA release from its presynaptic partners.

Indeed, in the early 2000s, it was found that cannabinoid agonists could induce DSI in the absence of postsynaptic depolarization; accordingly, cannabinoid antagonists blocked DSI. Moreover, cannabinoid agonists and high-frequency stimulation of CA3 input occluded each other in causing DSI, and DSI was abolished in CB1 receptor knockout mice. These and other lines of evidence led to the model illustrated in Figure 11-17B. Depolarization of postsynaptic cells causes Ca²⁺ influx through voltage-gated Ca²⁺ channels (1), which triggers synthesis of endocannabinoids from their precursors. These lipid-soluble endocannabinoids diffuse across the postsynaptic membrane and the synaptic cleft (2) to activate CB1 receptors on the presynaptic membrane. CB1 activation triggers the release of G protein βγ subunits (3), which bind to and cause closure of voltage-gated Ca²⁺ channels in the presynaptic terminal, thereby inhibiting neurotransmitter release. DSI facilitates LTP at excitatory synapses. Specifically, depolarization of CA1 neurons due to excitatory input from CA3 induces DSI, which reduces inhibitory input onto CA1 neurons, in turn facilitating depolarization and thus LTP induction.

In addition to CA1 pyramidal neurons, cerebellar Purkinje cells also exhibit DSI, as well as an analogous phenomenon called DSE (depolarization-induced suppression of excitation), depending on whether inhibitory or excitatory inputs are examined. Endocannabinoid signaling is also responsible for cerebellar DSI and DSE. Besides DSI and DSE, which are transient changes in synaptic efficacy occurring in seconds (Figure 11-17A), endocannabinoids can also elicit long-term depression of presynaptic neurotransmitter release (termed eCB-LTD). Induction of eCB-LTD utilizes mechanisms similar to DSI, can occur at terminals of both excitatory and inhibitory presynaptic neurons, and requires presynaptic CB1 receptors. It remains to be determined how CB1 receptor activation causes long-term reduction in neurotransmitter release probability. Given the wide range of brain tissues in which the CB1 receptor is expressed, it is likely that many synapses use this retrograde system to adjust presynaptic input based on the activity of postsynaptic neurons. Indeed, eCB-LTD has been found in many brain regions, including the striatum, amygdala, hippocampus, neocortex, and cerebellum.

11.13 Long-lasting changes of connection strengths involve formation of new synapses

In addition to changing the probability of presynaptic neurotransmitter release and postsynaptic sensitivity to neurotransmitters—two major mechanisms that account for synaptic plasticity—long-lasting changes in synaptic efficacy can be accomplished via structural changes to synapses. These include altering the size of existing synapses, forming new synapses, and dismantling old ones. These long-lasting changes typically require new gene expression (**Box 11-1**). Structural changes in response to stimuli have been extensively documented in dendritic spines, where most excitatory synapses in the mammalian CNS are located, because of the relative ease of using fluorescence microscopy to image these structures in slice preparations and *in vivo*. For example, LTP was found to be accompanied by growth of existing dendritic spines and formation of new spines on CA1 pyramidal neurons in cultured hippocampal slices (**Figure 11-18**); these phenomena required NMDA receptor function, suggesting that these structural changes are also mediated by signaling events initiated by Ca^{2+} entry.

LTP-associated structural changes have also been studied by serial electron microscopy (see Section 14.19 for details). LTP-inducing high-frequency stimulation was found to cause a selective increase in incidences wherein a single axon terminal contacts multiple dendritic spines from the same dendrites at late (60 minute) but not early (5 minute) time points following stimulation (**Figure 11-19A**). Thus, while early stages of LTP involve modulations of AMPA receptors at existing synapses, late-stage LTP can manifest in structural modifications of synapses, namely duplication of spines contacted by the same axons, possibly followed by a split in presynaptic axon terminals resulting in synaptic duplication (Figure 11-19B). Because these structural changes occur specifically between pre- and postsynaptic partners that have undergone LTP, this mechanism enhances the dynamic range of synaptic connections between a pair of neurons while simultaneously maintaining input specificity. This mechanism may be particularly important during development, when synapse formation and dendritic growth are influenced by experience, conveyed through patterned activity in sensory pathways (for example, see Box 5-3).

In summary, a wealth of mechanisms underlying synaptic plasticity, including changes in presynaptic neurotransmitter release probability, postsynaptic sensitivity to neurotransmitter release, structure and number of synapses, and homeostatic synaptic plasticity, can be used to adjust connection strengths between two neurons. These mechanisms allow experience and activity to adjust connection strengths during both development and adulthood. Although we have focused largely on examples of mammalian hippocampal neurons and synapses, similar mechanisms occur throughout the nervous systems of both vertebrates and invertebrates. We next explore whether and how these plasticity mechanisms are linked to learning and memory.

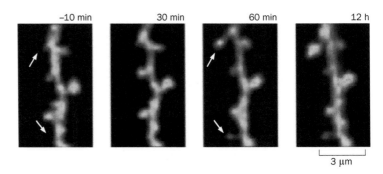

Figure 11-18 Growth of dendritic spines correlates with LTP. LTP is accompanied by formation of new spines (arrows) in CA1 pyramidal neurons from a cultured hippocampal slice imaged using two-photon microscopy. Time-lapse images were taken at −10, +30, +60 min, and +12 h relative to the onset of LTP induction (not shown). (From Engert F & Bonhoeffer T [1999] *Nature* 399:66–70. With permission from Springer Nature.)

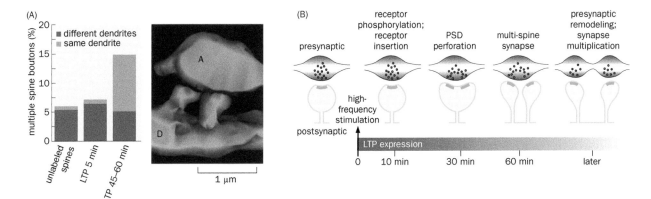

Figure 11-19 LTP correlates with formation of multiple-spine boutons.
(A) Left, quantification of the fraction of axon terminals that contact
multiple dendritic spines. Dendritic spines activated by LTP were
labeled by a staining procedure that produces precipitates in EM
micrographs of recently active spines to distinguish them from dendritic
spines unperturbed by LTP. A selective increase in the fraction of axon
terminals contacting two dendritic spines from the same dendrite can
be seen 45–60 min after LTP induction. Right, an example of serial EM
reconstruction, showing two dendritic spines from the same dendrite,
D, contacting the same presynaptic axon terminal, A. **(B)** A model of
the temporal sequence of LTP expression. The initial enhancement of
synaptic efficacy is caused by phosphorylation of AMPA receptors and
their insertion into the postsynaptic membrane. This is followed by
a split of the postsynaptic density (PSD), resulting in formation of a
multi-spine synapse. A further hypothetical split of the presynaptic
terminal results in duplication of synapses between the two neurons.
(A, adapted from Toni N, Buchs PA, Nikonenko I, et al. [1999] *Nature*
402:421–425. With permission from Springer Nature. B, from Lüscher C,
Nicoll RA, Malenka RC, et al. [2000] *Nat Neurosci* 3:545–550. With
permission from Springer Nature.)

Box 11-1: Synaptic tagging: maintaining input specificity in light of new gene expression

As discussed in Section 11.4, high-frequency stimulation
(HFS) can induce long-term potentiation (LTP) in the hippocampus *in vivo* that lasts for hours to days. Repeated HFS of
Schaffer collaterals can also induce LTP that lasts 8 hours or
more at CA3 → CA1 synapses in hippocampal slices *in vitro*.
Further studies suggest that LTP in this *in vitro* model can
be separated into two phases: an early-phase that decays
within 3 hours and is protein synthesis–independent, followed by a late-phase (called **late LTP**) that requires new
protein synthesis and gene expression. This property is
related to a lesson we will learn in Section 11.16: short-term
memory does not require new protein synthesis, whereas
long-term memory does.

How does LTP maintain its input specificity (Figure 10-8A)
in light of new protein synthesis and gene expression? For
new protein synthesis, one solution could be the use of local
protein synthesis from *preexisting mRNA* targeted to dendrites close to postsynaptic compartments (Section 2.2);
indeed, activity-dependent local protein synthesis is well
documented. However, for new gene expression, activity-
induced signals must go to the nucleus to trigger new
transcription, and the newly synthesized macromolecules
(mRNAs and/or their protein products) must determine
which synapses initiated the signal. To overcome this difficulty, a **synaptic tagging** hypothesis was proposed, which
states that in parallel with enhancing synaptic efficacy,
repeated HFS also produces local synaptic tags that selectively capture newly synthesized macromolecules, thereby
conferring input specificity. The following experiments (**Figure 11-20**) provided strong support for the synaptic tagging
hypothesis.

Two stimulating electrodes were placed at different depths
of the CA1 dendritic field in a hippocampal slice, ensuring
that they would stimulate different populations of CA3 →
CA1 synapses (S1 and S2) onto the same group of CA1 neurons, whose activity was monitored by a recording electrode.
In the first experiment (Figure 11-20A), only S1 received HFS
and only S1 synapses were potentiated, confirming input
specificity. In the second experiment (Figure 11-20B), 35
minutes after HFS at S1, protein synthesis inhibitors were
applied to the slice (prior experiments had shown that this
time lag would not inhibit late LTP formation at S1 synapses).
Then 25 minutes later, HFS was applied at S2 in the presence
of protein synthesis inhibitors, which would normally block
late LTP. However, S2 synapses exhibited normal late LTP
under this circumstance, thanks to the prior HFS at S1. The
simplest explanation is that HFS at S2 produced a synaptic
tag even in the presence of protein synthesis inhibitor, and
the tag captured newly synthesized macromolecules due to
HFS at S1. In the third experiment, researchers tested how
long the synaptic tag could last by first applying HFS at S1 in
the presence of a protein synthesis inhibitor, thus preventing it from inducing new gene expression but not inhibiting
its ability to produce a synaptic tag. Then the protein synthesis inhibitor was washed away, and HFS was applied to S2. If
the two HFSs were separated by 3 hours, then S1 no longer
exhibited late LTP (Figure 11-20C), suggesting that the synaptic tag lasts no more than 3 hours.

Although the molecular nature of the synaptic tag and the
newly synthesized macromolecules they interact with are not
fully understood (they may involve multiple molecular pathways in parallel), the concept of synaptic tagging is widely
(Continued)

Box 11-1: continued

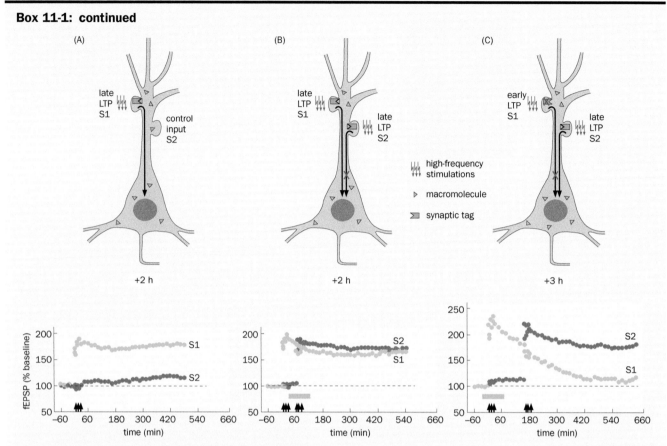

Figure 11-20 Experimental evidence for the synaptic tagging hypothesis. The top schematics illustrate the experimental conditions and summarize the results at 2 or 3 h after the first high-frequency stimulation (HFS) according to the synaptic tagging hypothesis. The bottom panels show the field EPSPs over time. **(A)** HFS was applied only to S1, and only S1 exhibited late LTP. This is because HFS at S1, while inducing new gene expression (downward arrow), also produced local synaptic tags, which captured newly synthesized macromolecules necessary for late LTP. **(B)** At 35 min after HFS was applied to S1, protein synthesis inhibitors were added to the slice (represented by the horizontal orange bar in the bottom panel), during which time HFS was applied to S2. Both S2 and S1 exhibited

late LTP. This is because HFS at S2, while incapable of inducing new gene expression (indicated by the red × on the downward arrow), nevertheless produced synaptic tags, which captured newly synthesized macromolecules due to HFS at S1. **(C)** HFS was applied at S1 in the presence of protein synthesis inhibitors, which affect late LTP but not early LTP. Then HFS was applied at S2 after protein synthesis inhibitors were washed away. When the two HFSs were 3 h apart, late LTP at S1 was disrupted, presumably because the synaptic tag at S1 decayed (as indicated by the red × on the synaptic tag) by the time newly synthesized macromolecules due to HFS at S2 arrived. (From Frey U & Morris RGM [1997] *Nature* 385:533–536. With permission from Springer Nature.)

accepted. Similar phenomena have also been observed in the *Aplysia* model for learning and memory that we will discuss in later sections. While the Hebbian mechanisms of synaptic plasticity relies on the precise timing between activation of pre- and postsynaptic neurons (within tens of milliseconds of each other), the synaptic tagging model suggests

that plasticity at one synapse may affect the plasticity of other synapses in the same neuron over a wider temporal window (for an hour or two). This may provide a cellular mechanism that explains why inconsequential events are remembered longer if they occur within a short window near consequential, well-remembered events.

WHAT IS THE RELATIONSHIP BETWEEN LEARNING AND SYNAPTIC PLASTICITY?

In this part of the chapter, we take a top-down approach to learning and memory, starting with animal behavior and seeking to link behavior to the function of circuits, neurons, synapses, and molecules (Figure 11-6). We first introduce different forms of learning and then study their underlying mechanisms in selected model organisms. We end with a discussion of spatial learning and memory in mammals, linking these processes to the mechanisms underlying hippocampal synaptic plasticity discussed in previous sections.

11.14 Animals exhibit many forms of learning

All animals face changing environments. Those that adapt well have greater chances of surviving and producing progeny. Consequently, many types of learning have evolved, each with specific properties, resulting in formation of explicit and implicit memories (Figure 11-3). Psychologists and behavioral biologists have used these properties to categorize learning into different forms.

The simplest form of learning is **habituation**, which refers to a decrease in the magnitude of responses to stimuli that have been presented repeatedly. For instance, we may be startled when we hear a noise for the first time, but we respond less strongly to subsequent instances of the same noise—we get used to it. Another simple form of learning is **sensitization**, which refers to an increase in the magnitude of response to a stimulus after a different kind of stimulus, often noxious, has been co-applied. Sensitization is more complex than habituation, as the behavioral response reflects an interaction between two different kinds of stimuli. We will study habituation and sensitization in the following two sections.

A more advanced form of learning is **classical conditioning** (also called Pavlovian conditioning), which enables animals to predict what will happen based on past experiences, facilitating adaptive anticipatory reactions. In classical conditioning, an association is learned between a neutral stimulus that was previously ignored (the **conditioned stimulus**, or **CS**) and an intrinsically rewarding or aversive stimulus (the **unconditioned stimulus**, or **US**) through reliable temporal pairing. A famous example is the experiment on salivation of dogs conducted by Ivan Pavlov, who discovered classical conditioning in the early twentieth century (**Figure 11-21**). Dogs always salivate in response to food in the mouth; this innate salivation is the **unconditioned response**. After repeated pairing of food with a sound, which did not produce salivation before pairing, the sound alone produced salivation. In this example, food is the US, sound is the CS, and the process of pairing food and sound is called conditioning; the eventual salivation response to sound alone is called the **conditioned response**.

Whereas sensitization merely changes the magnitude of the response to a stimulus due to the presentation of a second kind of stimulus, classical conditioning establishes a novel and qualitatively different stimulus–response (e.g., sound–salivation) relationship: classical conditioning requires forming an *association*

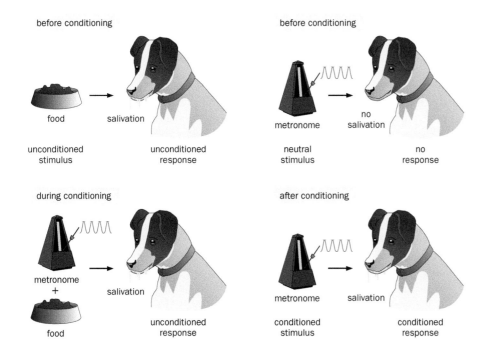

Figure 11-21 Classical (Pavlovian) conditioning. Before conditioning, the dog salivates in response to food in its mouth (top left), but not in response to a sound from a metronome (top right). During conditioning consisting of repeated pairing of the sound and food (bottom left), the dog learns to associate the sound with food, such that after conditioning, the dog salivates in response to the sound alone (bottom right). (Based on Pavlov IP [1926] Conditioned Reflexes. Dover Publications Inc.)

Figure 11-22 Operant conditioning. A hungry or thirsty rat placed in this chamber can learn through trial and error that pressing the lever results in dispensing of food or water (according to the particular experimental design); this reward reinforces lever pressing. (Based on Skinner [1938] The Behavior of Organisms. B.F. Skinner Foundation.)

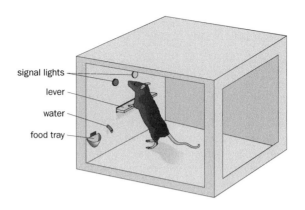

signal lights
lever
water
food tray

between the CS and US. For conditioning to be effective, proper timing of the CS and US is critical; the CS usually precedes the US. Classical conditioning is therefore a form of **associative learning**. It is observed across the animal kingdom, including in humans.

Another major form of associative learning, distinct from classical conditioning, is **operant conditioning** (also called **instrumental conditioning**). In operant conditioning, a reinforcer is given only when the animal performs a certain behavior, thus teaching the animal to learn to exert control over its environment. For instance, a hungry rat in a cage can be trained to press a lever to obtain a food pellet. Initially the rat may not "know" the association between the lever pressing and the food pellet; as the reinforcer (food pellet) is given each time the rat presses the lever, the rat gradually associates the lever pressing (its own action) with the food reward (**Figure 11-22**). After operant conditioning, the rat selects one action over many other possible actions in order to receive food pellets. A **law of effect** was proposed in the early twentieth century to explain the association process: behaviors followed by rewards will be repeated, whereas behaviors followed by punishments will diminish.

Operant conditioning is a prevalent learning mechanism in the animal kingdom and is widely used in the laboratory to train animals to perform tasks. Indeed, operant conditioning was used in many of the experiments discussed in this book, from motion perception to arm reaching (Figure 4-55; Figure 8-31), and in probing how thirst drives motivated behavior (Figure 9-18). Timing is crucial in operant conditioning—as in classical conditioning—and the effect is greatest when the reinforcer is presented shortly after the behavior. Another property shared by classical and operant conditioning is **extinction**: in classical conditioning, when the CS is repeatedly *not* followed by the US, the conditioned response will diminish; in operant conditioning, when the behavior is repeatedly *not* followed by the reinforcer, the behavior will diminish. Depending on the nature and timing of these stimuli, extinction can be temporary or long-lasting.

In our discussion so far, learning has been viewed as a modification of behavior after experience, with the outcome of learning measured by changes in behavior. There is a complementary view of learning. Psychologists refer to **cognitive learning** as acquisition of new knowledge rather than just modification of behavior. From this perspective, for instance, classical conditioning can be viewed as the animals having acquired the knowledge that the US follows the CS; the conditioned response is in fact a response to the predicted upcoming US rather than to the CS *per se*. While cognitive capabilities are usually thought to be limited to mammals with large cerebral cortices such as primates and particularly humans, the following example illustrates that even insects can master abstract concepts typically considered as cognitive learning.

Honeybees can be trained to perform a task called delayed match-to-sample, which is thought to utilize working memory. They first encounter a cue, such as a blue sign, after entering a Y-maze. After flying within the maze for a certain distance, they encounter a choice point, where the entrances into each arm of the

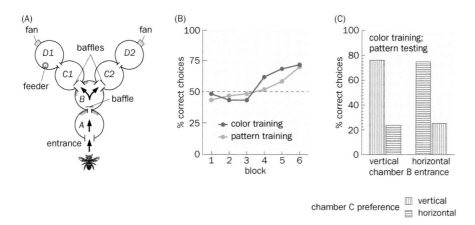

Figure 11-23 **Cognitive learning in honeybees. (A)** Experimental setup. At the entrance to chamber *B*, bees first encounter a stimulus (e.g., a blue sign). They then face a choice of two different *C* chambers, with the entrances marked with two different stimuli (e.g., blue and yellow signs). With the sugar solution in one of the *D* chambers as reward, bees can be trained after repeated trials to choose either the *C* chamber marked with a stimulus that matches the entrance to the *B* chamber (delayed match-to-sample, as shown here), or the *C* chamber marked with a stimulus that differs from the *B* chamber (delayed non-match-to-sample, which would apply if the feeder were placed in *D2*). **(B)** Learning curves for bees that performed color- or pattern-match tasks. Each block consisted of 10 consecutive training sessions. After six blocks, the percentage of correct choices for either task exceeded 70%, significantly above random chance (50%, dashed line). **(C)** After being trained for delayed-match-to-sample in color, bees were tested using a pattern-match task. Whether the entrance to *B* was marked with a vertical (left) or horizontal (right) grid pattern, bees preferentially chose the *C* chamber whose entrance was marked with the same pattern as the pattern at the entrance to *B*. (From Giurfa M, Zhang S, Jenett A, et al. [2001] *Nature* 410:930–933. With permission from Springer Nature.)

Y-maze are marked by blue or yellow signs. If they enter the arm marked by the same color as the color they encountered at the entrance of the maze, they will get a food reward (**Figure 11-23**A). After repeated training, bees could not only perform this task with a success rate well above chance (Figure 11-23B) but also applied this skill to a completely new set of cues. For example, when the maze was outfitted with grid patterns that trained bees had not encountered previously, they could perform a pattern-match task nearly as well as the original color-match task (Figure 11-23C). Moreover, bees could apply the learned skill across different sensory modalities; for instance, training with a pair of odors improved the test results for matching colors. Last, bees can be trained to obtain a reward by entering the maze arm marked by a cue that differs from the one at the entrance—a task called delayed non-match-to-sample—and can transfer the non-match skill from colors to patterns. Thus, honeybees appear to be able to apply abstract concepts of *sameness* and *difference* to guide their behavior.

What neurobiological bases underlie these different forms of learning? Do they share common mechanisms? How are they related to the synaptic weight matrix hypothesis we introduced early in this chapter? We now explore these questions, starting with simple forms of learning observed in the sea slug *Aplysia*.

11.15 Habituation and sensitization in *Aplysia* are mediated by changes in synaptic strength

Aplysia has been used as a model for studying the cellular and molecular basis of learning and memory since the 1960s. *Aplysia* has only 20,000 neurons, compared to about 10^8 neurons in the mouse. Many *Aplysia* neurons are large and individually identifiable, such that electrophysiological recordings can easily be performed on multiple neurons in the same animal and with reproducible results across animals (as in the crustacean stomatogastric ganglion discussed in Section 8.5). Importantly, *Aplysia* exhibits simple forms of learning and long-lasting memory that resemble those found in more complex organisms.

The **gill-withdrawal reflex** has been used as a model behavior (**Figure 11-24**A). When a tactile stimulus contacts the siphon, *Aplysia* reflexively withdraw their gill (and siphon) into the mantle shelf as a protective measure. This behavior exhibits habituation, as repeated siphon stimuli result in progressively smaller gill withdrawal magnitudes (Figure 11-24B, left). However, if an animal receives a noxious electric shock at the tail, gill withdrawal magnitude in response to the siphon stimulus applied shortly after the shock is greatly enhanced, indicating sensitization of the gill-withdrawal reflex by the tail shock (Figure 11-24B, right).

The neural circuits underlying the gill-withdrawal reflex have been mapped (Figure 11-24C), thanks to the ease of recording and manipulating neural activity using electrophysiology. Siphon stimulation activates 24 sensory neurons; activating these neurons artificially was found to mimic siphon stimulation and induce gill-withdrawal. Six motor neurons control the muscle contraction that causes

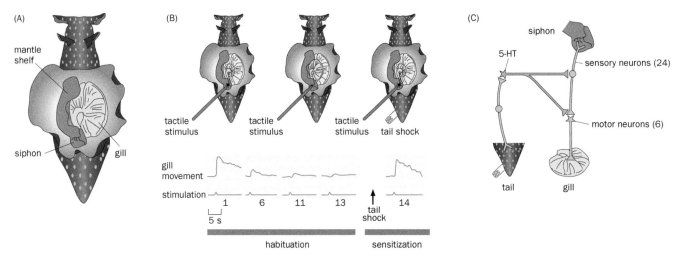

Figure 11-24 The *Aplysia* gill-withdrawal reflex and the underlying neural circuits. (A) Drawing of *Aplysia* highlighting the structures that mediate its gill-withdrawal reflex. **(B)** Top, drawing of the gill-withdrawal reflex and its habituation (middle) and sensitization (right). Bottom, recording of the gill movement (top traces) shows progressive decrement in response to repeated siphon stimulation (bottom traces). Numbers indicate repetitions. Shortly before the 14th stimulus, a tail shock was applied, which caused an increased response to stimulus 14. **(C)** Circuit diagram of the neurons mediating the gill-withdrawal reflex. The 24 sensory neurons that innervate the siphon connect directly with the six motor neurons that innervate the gill muscle. Sensory neurons activated by tail shock connect with serotonin (5-HT) neurons, which in turn innervate the siphon sensory neurons and their presynaptic terminals onto the gill motor neurons. (Adapted from Kandel ER [2001] *Science* 294:1030–1038.)

gill withdrawal. Activity of these motor neurons correlates with gill withdrawal, and direct electrical stimulation of these motor neurons is sufficient to cause gill withdrawal. These sensory and motor neurons form monosynaptic connections like those in the sensorimotor circuit controlling our knee-jerk reflex (Figure 1-19). A different group of sensory neurons transmits the tail-shock signal to a set of serotonin neurons, which in turn innervate the cell bodies of the sensory neurons that sense siphon stimuli and their presynaptic terminals on the motor neurons (Figure 11-24C). These connections allow tail shock to modulate the activity of sensory neurons or neurotransmitter release from sensory neurons onto their motor neuron targets (Figure 3-37).

Having mapped the neurons underlying the reflex circuit, researchers then asked what circuit changes underlie behavioral habituation—the reduction in the magnitude of gill withdrawal after repeated siphon stimulation. In principle, this could be caused by: (1) sensory neurons progressively reduce their response magnitude after repetitive stimuli, as in sensory adaptation (Section 4.7); (2) synaptic efficacy between sensory and motor neurons is depressed; (3) synaptic efficacy at the neuromuscular junction is depressed; or (4) the muscles become fatigued. A series of experiments using physiological recordings in conjunction with sensory stimulation and quantitative behavioral measurements were carried out to systematically examine these possibilities (Figure 11-25).

In experiment 1, gill-withdrawal responses to direct motor neuron stimulation were measured before and after behavioral habituation and were found to be the same. This ruled out the possibility that changes downstream of the motor neurons in the circuit, including depression of synaptic efficacy at the neuromuscular junction and muscle fatigue, were responsible for habituation. To test for sensory adaptation at peripheral sensory endings, a set of sensory stimuli were applied while motor neuron responses were recorded (experiment 2). Responses became smaller as more stimuli were applied, correlating with behavioral habituation. During stimuli 10 through 18, a segment of the siphon nerve that connects the sensory nerve endings to the sensory neurons was bathed in a sodium-free solution to block action potential propagation. After the nerve block was relieved, motor neuron response became larger, instead of becoming smaller, as would be predicted if sensory adaptation at the periphery was responsible for habituation.

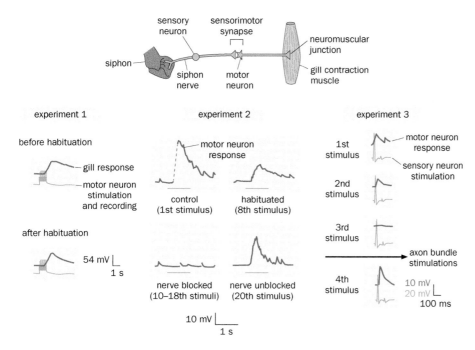

Figure 11-25 Neural mechanisms of habituation and sensitization of the *Aplysia* gill-withdrawal reflex. Top, diagram of information flow from siphon stimulation to gill withdrawal. Bottom, three experiments investigating the neural mechanisms of behavioral habituation. Experiment 1, gill responses (red traces) to direct motor neuron stimulation (spikes of motor neurons shown in blue) before and after habituation remained unchanged, arguing against the possibility that habituation affects processes downstream of the motor neuron. Experiment 2, intracellular recording of a motor neuron (red traces) in response to a series of 20 siphon stimuli (blue line represents the duration of one stimulus). The first 9 stimuli were applied under normal conditions, and the resulting motor neuron responses were depressed (compare the top right and top left traces), correlating with habituated behavioral responses. During stimuli 10–18, action potentials from the siphon nerve were blocked so that the motor neuron did not respond (bottom left). After the siphon nerve was unblocked, the 20th stimulus generated a larger response than the 8th stimulus (compare the top right and bottom right traces), thus arguing against the possibility that habituation affects processes upstream of the siphon nerve. Experiment 3, in a reduced preparation consisting of an isolated ganglion containing the sensory and motor neurons, motor neuron responses (red traces) were induced by sensory neuron stimulation that produced a single spike (blue traces). The top three pairs show three consecutive sensory neuron stimulations, which caused progressively reduced responses, mimicking behavioral habituation. In the bottom pair, the motor neuron response was facilitated due to stimulation of the axon bundle that includes axons of the serotonin neurons, mimicking behavioral sensitization. (From Kupfermann I, Catellucci V, Pinsker H, et al. [1970] *Science* 167:1743–1745; Castellucci V, Pinsker H, Kupfermann I, et al. [1970] *Science* 167:1745–1748. With permission from AAAS.)

This ruled out the possibility that habituation was due to an effect upstream of the sensory nerve. Collectively, these experiments suggested that changes at the sensorimotor synapses underlie behavioral habituation. Indeed, in studies carried out in an isolated ganglion, which facilitated stimulation and recording compared with intact *Aplysia,* motor neuron responses elicited by direct sensory neuron stimulation were found to undergo progressive depression after repeated trials (experiment 3), suggesting that depression of sensorimotor synaptic efficacy is the primary cause of behavioral habituation.

Analogous experiments were conducted to test the mechanism underlying sensitization by tail shock. Remarkably, the same sensorimotor synapses that were depressed during habituation were potentiated during sensitization (Figure 11-25, experiment 3). Together, these findings suggest that behavioral modifications, as measured by the magnitude of the gill-withdrawal reflex, are caused primarily by changes in synaptic efficacy between sensory neurons and motor neurons—habituation is caused by depression, while sensitization is caused by facilitation. These results support the hypothesis proposed in Section 11.3—namely, that changes in synaptic strengths underlie these simple forms of learning.

11.16 Both short-term and long-term memory in *Aplysia* engage cAMP signaling

Studies of the *Aplysia* gill-withdrawal reflex have also provided important insights into the mechanisms of short-term and long-term memory. Human behavioral studies suggest that repeated training can strengthen memories, causing them to become long lasting. Sensitization of the *Aplysia* gill-withdrawal reflex also exhibits these properties. While one tail shock caused a transient increase in gill-withdrawal magnitude that returned to baseline within an hour, four shocks produced a memory that lasted at least a day. The memory produced by four trains of four shocks within a day was retained even after four days. Four trains of four shocks every day for four days produced a drastic increase in response magnitude that persisted for more than a week (**Figure 11-26A**).

To facilitate mechanistic studies, researchers established an *in vitro* co-culture system consisting of a siphon sensory neuron and a gill motor neuron (named L7, which can be identified in each animal based on its stereotyped size, shape, and location) that form synaptic connections in a dish. In this system, repeated stimulation of the sensory neuron caused progressive decreases in the magnitude of the postsynaptic potential (PSP) recorded from the motor neuron, mimicking behavioral habituation and consistent with findings from studies in intact ganglia (for example, Figure 11-25, experiment 3). Sensitization could also be recapitulated in the co-culture system by applying serotonin to the culture (Figure 11-26B). While one serotonin pulse produced PSP facilitation that lasted for minutes, five serotonin pulses separated by 15 minute intervals produced PSP facilitation that lasted for 24 hours (Figure 11-26C), as in the outcome of repeated tail shock (Figure 11-26A). These short-term and long-term facilitations of synaptic efficacy have been used as cellular models of short-term and long-term memory.

Are there mechanistic differences between short-term and long-term memory? Studies in different animals from mice to goldfish indicated that long-term (but not short-term) memory formation is blocked by applying drugs that inhibit protein synthesis at the time of training, suggesting that long-term memory formation selectively requires new protein synthesis. Likewise, in the co-culture system in *Aplysia,* applying protein synthesis inhibitors during serotonin application blocked long-term (Figure 11-26C) but not short-term (Figure 11-26B) facilitation of synaptic transmission caused by serotonin application. Furthermore, applying protein synthesis inhibitors *before* or *after* serotonin application did not affect long-term facilitation. These studies support the notion that protein synthesis is required *during* acquisition of long-term facilitation.

We now have a good understanding of the molecular mechanisms that mediate short- and long-term facilitation in this system. During short-term facilitation, serotonin acts on a G-protein-coupled receptor in the presynaptic terminal of the sensory neuron to elevate the intracellular cAMP concentration through activation of an adenylate cyclase (Figure 3-33). Indeed, intracellular injection of cAMP into the sensory neuron enhanced synaptic transmission between sensory and

Figure 11-26 Long-term sensitization can be induced by repeated training or serotonin (5-HT) application and requires protein synthesis. (A) Duration of gill withdrawal above baseline in response to tail shock. Increased training produced longer-lasting sensitization of the gill-withdrawal reflex. **(B)** Behavioral habituation and sensitization can be recapitulated as changes in synaptic strength between a sensory neuron and a motor neuron co-cultured *in vitro*. Here the relative magnitude of the L7 motor neuron's postsynaptic potential (PSP) in response to sensory neuron stimulation is plotted against the stimulus number. A progressive decline in magnitude accompanied application of successive stimuli. Application of 5-HT, which mimics tail shock, increased PSP magnitude. Application of the protein synthesis inhibitor anisomycin (blue trace) had no effect on short-term depression or facilitation compared to the control (brown trace). **(C)** A single 5-HT application (1 × 5-HT) did not produce long-term facilitation, measured 24 h later, whereas a sequence of five 5-HT applications did (5 × 5-HT). Application of anisomycin during 5-HT application (5 × 5-HT + A) blocked long-term facilitation. (A, from Frost WN, Castelucci VF, Hawkins RD, et al. [1985] *Proc Natl Acad Sci U S A* 82:8266–8269. With permission from the authors. B & C, adapted from Montarolo PG, Goelet P, Castellucci VF, et al. [1986] *Science* 234:1249–1254.)

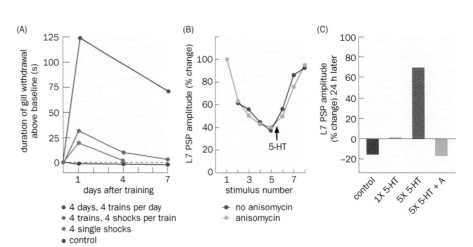

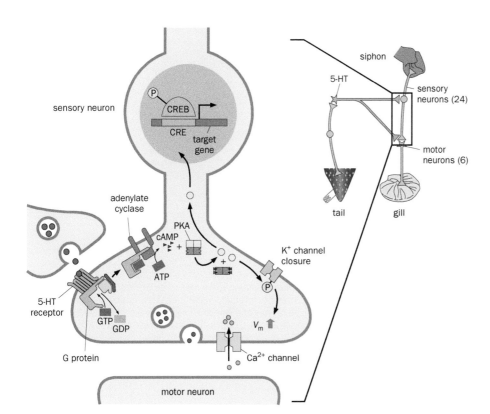

Figure 11-27 Both short-term and long-term facilitation in *Aplysia* involve cAMP and PKA. During short-term facilitation, tail shock induces serotonin (5-HT) release at the presynaptic terminal of the sensory neuron, which activates a G-protein-coupled 5-HT receptor. One of the downstream mechanisms is activation of adenylate cyclase, leading to cAMP production and PKA activation. PKA phosphorylates a specific type of presynaptic K⁺ channel, causing its closure, which elevates the resting potential and facilitates action potential–triggered neurotransmitter release. In conditions that produce long-term facilitation, the catalytic subunit of PKA enters the nucleus and phosphorylates nuclear substrates such as the transcription factor CREB, inducing new gene expression. The circuit diagram of the gill-withdrawal reflex and sensitization is shown on the right; the box indicates what the schematic on the left depicts. For simplicity, the 5-HT axon terminal at the cell body is omitted and the axon is shortened in the magnified schematic on the left. (Adapted from Kandel ER [2001] *Science* 294: 1030–1038.)

motor neurons. cAMP is a second messenger that activates protein kinase A (PKA). PKA phosphorylates a specific type of K⁺ channel that is active during the resting state, resulting in its closure. This raises the resting membrane potential, making it easier for action potentials arriving from the cell body to open voltage-gated Ca²⁺ channels at the presynaptic terminal of the sensory neuron, thus facilitating neurotransmitter release (**Figure 11-27**, bottom). Serotonin also activates other intracellular signaling pathways, notably protein kinase C (Figure 3-34), which can phosphorylate other substrates such as voltage-gated K⁺ channels, leading to spike broadening and increased neurotransmitter release per action potential. Thus, short-term facilitation alters synaptic strength by post-translational modification of ion channels, consistent with actions that take place on a timescale of seconds to minutes and do not require new protein synthesis.

Remarkably, cAMP and PKA also play key roles in long-term facilitation (Figure 11-27, top). Here, a widely used signaling pathway involving the transcription factor CREB is engaged (Figure 3-41): PKA phosphorylation activates CREB, which binds to the CRE (cAMP response element) sequences near the promoters of target genes to activate their transcription. How does PKA activation affect events both locally at the synapse and remotely in the nucleus? Whereas transient serotonin application causes transient PKA activation locally at the synapse, imaging experiments indicated that repeated or prolonged serotonin application induces translocation of the catalytic subunit of PKA to the nucleus, where it can phosphorylate nuclear substrates such as CREB. Just as LTP in the mammalian hippocampus is accompanied by structural changes (Section 11.13), long-term facilitation in *Aplysia* is also accompanied by growth of synaptic contacts between the sensory and motor neurons. Therefore, some transcriptional targets of CREB are likely responsible for regulating synaptic growth.

11.17 Olfactory conditioning in *Drosophila* requires cAMP signaling

While *Aplysia* offers large cells for physiological studies of learning and memory, the fruit fly *Drosophila* provides an unbiased way to identify genes required for learning and memory via genetic screening (Section 14.6). In this procedure, flies

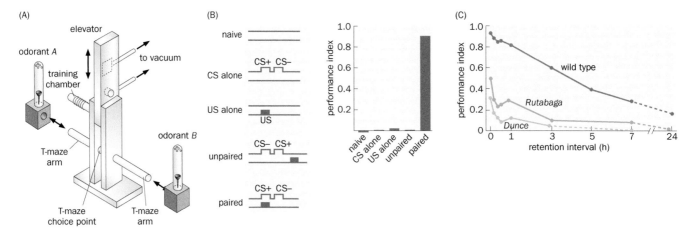

Figure 11-28 Olfactory conditioning in *Drosophila* and its disruption by mutations affecting cAMP metabolism. (A) Schematic of the olfactory conditioning procedure. About 100 flies in the training chamber were exposed to odorant *A* (CS⁺) paired with electric shock (US) and to odorant *B* (CS⁻) without electric shock. These flies were then transferred via a sliding elevator to the bottom T-maze arms, where flies freely choose tubes containing odorant *A* or odorant *B*. Performance index = [(number of flies in tube *B* − number of flies in tube *A*) / total number of flies] × 100. **(B)** Performance indices of flies under different training conditions. Flies learn the association only when the US is paired with the CS⁺. The CS and US each last 60 s. **(C)** Performance indices of wild-type and mutant flies. Performance indices represent learning when measured immediately after training (*t* = 0) and memory retention when measured at specific times thereafter. *Rutabaga* and *Dunce* mutant flies are defective in both learning and memory. (From Tully T & Quinn WG [1985] *J Comp Physiol* 157:263–277. With permission from Springer. See also Dudai Y, Jan Y, Byers D, et al. [1976] *Proc Natl Acad Sci U S A* 73:1684–1688 for the identification of the first learning mutant, *Dunce*.)

with mutations in random genes (produced by treating flies with a chemical mutagen, for example) can be screened using a behavioral assay that tests learning and memory. Mutant flies that perform poorly can be isolated, and the corresponding gene can be mapped using molecular-genetic procedures.

Flies can be trained to associate odors with electrical shocks. In a widely used classical conditioning paradigm, flies are exposed to odorant *A* while being shocked. They are also exposed to odorant *B* without shock. In this case, odorant *A* is designated as the CS⁺ as it is a conditioned stimulus that is associated with the unconditioned stimulus (US), electric shocks, while odorant *B* is designated as the CS⁻. To test their odorant preferences, flies are placed in a T-maze (**Figure 11-28**A; Movie 6-1), where they choose to enter one arm (exposed to odorant *A*) or the other (exposed to odorant *B*). Prior to the odorant–shock pairing, flies are as likely to choose odorant *A* as they are *B*. However, after the odorant–shock pairing, 95% of wild-type flies avoid the odorant associated with shock (Figure 11-28B). Timing of the CS–US paring is crucial (Figure 11-28B), as would be predicted from a classical conditioning paradigm. In addition to learning, which is measured as the behavioral performance immediately after training, flies can also be tested for memory at specific times after training. One odorant–shock pairing (for 1 minute) produces a memory that lasts for several hours (Figure 11-28C). Repeated pairings with proper intervals, called spaced training, can produce long-term memory that lasts for a week, similar to the *Aplysia* gill-withdrawal reflex following sensitization by tail shock (Figure 11-26A).

Two of the first mutations identified through genetic screening, named *Dunce* and *Rutabaga*, affected both learning and memory. Performance of flies carrying either of these two mutations was drastically reduced compared with normal flies immediately after training, indicating a learning defect. In addition, they quickly forgot what they learned (Figure 11-28C). Importantly, separate tests showed that these mutants could detect odorants and avoid shocks normally, indicating a specific defect in forming the odor–shock association. Molecular-genetic studies revealed that the *Rutabaga* gene encodes an adenylate cyclase, an enzyme that catalyzes cAMP synthesis, whereas *Dunce* encodes a phosphodiesterase, an enzyme that hydrolyzes cAMP (Figure 4-7). Thus, proper regulation of cAMP metabolism is essential for learning and memory in a classical conditioning paradigm in *Drosophila*. Subsequent experiments found that perturbation of CREB, the transcription factor regulated by cAMP, affects long-term but not short-term memory of olfactory conditioning. Together, these studies highlight remarkable similarity for cAMP signaling in fly olfactory conditioning and *Aplysia* gill-withdrawal reflex sensitization (Figure 11-27).

11.18 Modifications of the mushroom body synaptic weight matrix underlie *Drosophila* olfactory conditioning

The identification of molecules required for *Drosophila* olfactory learning and memory also provided an entry point for cellular and circuit studies. Both *Dunce* and *Rutabaga* are highly expressed in **Kenyon cells**, the intrinsic cells of the mushroom body and postsynaptic targets of olfactory projection neurons (Figure 6-24). Indeed, expression of a wild-type *Rutabaga* transgene in adult Kenyon cells was sufficient to rescue the memory deficits of *Rutabaga* mutant flies, demonstrating that cAMP regulation in Kenyon cells plays a crucial role in olfactory learning and memory.

A circuit model of olfactory learning has been proposed based on these studies and on the position of the mushroom body in the olfactory processing pathways (Section 6.14). According to this model (**Figure 11-29**), odorants (the CS) are represented by ensembles of Kenyon cells, whose connections with specific mushroom body output neurons are modified when the CS is paired with stimuli (the US) that are aversive (such as electric shocks) or appetitive (such as food when flies are hungry). The US has been hypothesized to be represented by modulatory (such as dopamine) neurons that contact Kenyon cell axons. This plasticity is cAMP-dependent. The US causes activation of a G-protein-coupled receptor in

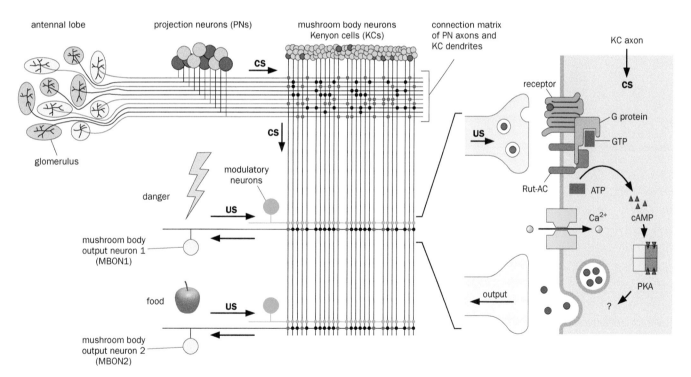

Figure 11-29 Circuit and molecular models of *Drosophila* olfactory conditioning. Left, a circuit model for olfactory conditioning. Odorants activate ensembles of glomeruli (ovals) in the antennal lobe (active glomeruli in yellow; Figure 6-24). Projection neurons (PNs) innervating active glomeruli become activated (red) and, in turn, activate a specific subset of mushroom body Kenyon cells (KCs). In this simplified connection matrix, each KC connects with three PN axons (dots); only when all three connected PNs are active would the KC become active (red cell on top, with three red dots in the matrix). The CS (odorant) information is represented by ensembles of active PNs and subsequently by ensembles of active KCs. Synapses between KC axons and dendrites of mushroom body output neurons (MBONs) are modified by nearby input from modulatory neurons that encode an aversive (electric shock) and appetitive (food) US (bottom). Each dot represents a synapse; red and blue dots represent active synapses. Co-activation of neurons representing the US and the CS modifies the efficacy of synapses between KC axons and MBON dendrites. Right, an enlarged diagram of the output synapses of KC neurons. Axons representing the US release modulatory transmitters that activate G-protein-coupled receptors, resulting in activation of the Rutabaga adenylate cyclase (Rut-AC). This causes increased cAMP production and PKA activation, leading to changes in efficacy of KC → MBON synapses through mechanisms yet to be determined. (Based on Heisenberg M [2003] *Nat Rev Neurosci* 4:266–275.)

Kenyon cell axons, which in turn activates the Rutabaga adenylate cyclase, leading to cAMP production and PKA activation. Co-activation of the CS and the US lead to altered transmitter release at Kenyon cell → mushroom body output neuron synapses, such that after learning, the CS alone would activate mushroom body output neurons in a pattern reflecting the modified synaptic strengths induced by the US.

Recent advances have provided remarkably detailed circuit mechanisms in line with this model. First, researchers have shown that optogenetic stimulation of specific dopamine neurons can mimic electrical shock or food presentation for olfactory conditioning, supporting the notion that different types of dopamine neurons provide different kinds of US. Second, comprehensive anatomical mapping identified 21 types of mushroom body output neurons, most of which connect with one of 15 axonal compartments of Kenyon cells. Each axonal compartment is also innervated by one or more of the 20 types of dopamine neurons, some of which carry punishment signals, whereas others carry reward signals (**Figure 11-30**A). Optogenetic activation of specific types of mushroom body output neurons led to approach or avoidance behavior (Figure 11-30A), suggesting that mushroom body output neuron ensembles collectively represent valence that can be used to bias memory-based action selection. Third, *in vivo* electrophysiological studies revealed that pairing olfactory stimuli and dopamine neuron activation caused a drastic reduction in the olfactory response in mushroom body output neurons innervating the same compartment as the dopamine neuron (Figure 11-30B), suggesting that depression of Kenyon cell → mushroom body output neuron synapses by dopamine signaling is the physiological basis of olfactory classical conditioning in flies.

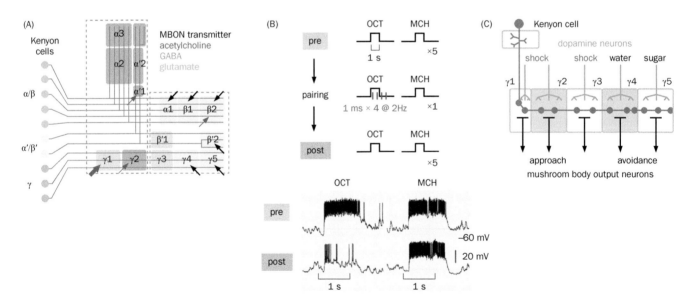

Figure 11-30 Circuit mechanisms of *Drosophila* olfactory conditioning. **(A)** Systematic circuit mapping using cell-type-specific drivers (Sections 14.9 and 14.11) revealed that Kenyon cell (KC) axons form 15 compartments that connect with distinct MBONs (shaded according to the neurotransmitters used by different MBONs). Each compartment also receives modulatory input from specific dopamine neurons (originating from two clusters outlined by dashed boxes). Color-coded arrows represent known reward (black) or punishment (red) signals carried by dopamine neurons, with the size of the arrow representing the magnitude of memory induced by their activation. Note that KCs belong to one of three major classes: γ, α′/β′, and α/β, based on their axonal branching patterns in adults (gray lines). Each α′/β′ and α/β KC forms one vertical and one horizontal branch, whereas γ KCs have only horizontal branches. **(B)** Physiological correlates of olfactory conditioning. In this experiment, *in vivo* whole-cell patch recording was performed on the γ1 MBON of immobilized flies responding to presentation of two odorants, 2-octanol (OCT) and 4-methylcyclohexanol (MCH). Top, experimental procedure. Olfactory conditioning consisted of pairing OCT with four pulses of optogenetic activation of dopamine neurons that innervate the γ1 compartment to mimic electrical shock. Bottom, compared to pretraining, the posttraining firing rate elicited by OCT was depressed while the posttraining firing rate elicited by MCH was unchanged. **(C)** Schematic showing different mushroom body compartments processing in parallel distinct reward and punishment signals to bias approach and avoidance behavior. For example, water and sugar signals for thirsty and hungry flies are carried by distinct dopamine neurons to depress KC → MBON synapses to distinct MBONs. (A, from Aso Y, Sitaraman D, Ichinose T, et al. [2014] *eLife* 3:e04580. See also Aso Y, Hattori D, Yu Y, et al. [2014] *eLife* 3:e04577; B, from Hige T, Aso Y, Modi MN, et al. [2015] *Neuron* 88:985–998. With permission from Elsevier Inc. See also Cohn R, Morantte I, & Ruta V [2015] *Cell* 163:1742–1755. C, based on Waddell S [2016] *Curr Biol* 26:R109–R112.)

If the output of the mushroom body simply facilitates approach and avoidance behavior in response to reward and punishment, why are there 15 Kenyon cell axonal compartments with distinct output neurons and dopamine modulatory neurons? It turns out that dopamine neurons that project to different compartments carry different information. For example, water and sugar as rewards to thirsty and hungry flies, respectively, are represented by dopamine neurons innervating distinct compartments (Figure 11-30C). Moreover, physiological and behavioral experiments indicated that different compartments follow different plasticity and learning rules, regarding, for example, how many repeated training sessions are needed to modify synaptic efficacy and behavior, and how long such modifications last. For example, using an experimental paradigm pairing optogenetic activation of dopamine neurons and odorant exposure to mimic olfactory conditioning, researchers found that activation of dopamine neurons innervating one compartment required only one training session to achieve nearly full performance immediately after training—but the resulting memory decayed rapidly. Activation of dopamine neurons innervating another compartment required 10 repeated training sessions to reach a similar performance level—but the resulting memory lasted much longer. The parallel circuits implemented by different compartments with distinct input specificity and plasticity rules thus provide rich neural substrates supporting different kinds of learning and memories with distinct properties.

In summary, in the fly olfactory conditioning paradigm, odorant information, representing the CS, enters the mushroom body Kenyon cell dendrites through excitatory input from olfactory projection neurons. Input from dopamine neurons, representing the US, modifies synapses linking Kenyon cells to downstream mushroom body output neurons. Thus, the connections between Kenyon cells and mushroom body output neurons represent an example of the synaptic matrix discussed in Figure 11-4. Here, input patterns represent specific odorants and, through the synaptic matrix, produce two major types (and 15 subtypes) of distinct output patterns, leading to approach or avoidance behavior. Before training, neutral odorants activate output neurons in a balanced fashion, such that no approach or avoidance behavior is elicited. During learning, coincident presentation of the CS and US modifies the connection strengths between a specific Kenyon cell ensemble representing the CS and specific mushroom body output neurons to shift the balance of output to promote approach or avoidance, such that after training, activation of same Kenyon cell ensemble alone would lead to either approach or avoidance (Movie 11-1).

Studies of olfactory conditioning in flies have also produced a molecular model (Figure 11-29, right) with remarkable similarities to that of sensitization of the *Aplysia* gill-withdrawal reflex (Figure 11-27). Mammals also use cAMP and PKA in learning and memory (see Section 11.20), including in hippocampus-dependent learning we now turn to. Many learning paradigms and memory tasks feature an important function of the hippocampus: spatial representation (Box 11-2).

Box 11-2: Place cells, grid cells, and representations of space

Navigation is essential for animals to find food and return home safely. Animals from insects to mammals use two types of navigation strategies: a **landmark-based strategy**, whereby animals use external cues to determine their location, and a **path-integration strategy**, whereby animals use information regarding the speed, duration, and direction of their own movement to calculate their current position with respect to their starting position. Both strategies require animals to have an internal representation of space.

In mammals, the hippocampus and entorhinal cortex are crucial for spatial representation. A seminal discovery was made in the 1970s when researchers performed single-unit recordings of hippocampal neurons in freely moving rats navigating an arena or a maze. Individual cells were found to fire robustly when the rat was at a particular location in the maze, regardless of what behavior the animal was performing (for example, passing through from various directions, exploring, or just resting); different cells fired at different locations (Figure 11-31A). These cells are called **place cells**, and the physical location that elicits place-cell firing is known as the cell's **place field**.

We now know that most hippocampal CA1 and CA3 pyramidal neurons can be place cells. Their place fields are influenced by external landmarks. For example, after the place

(Continued)

Box 11-2: continued

(A)

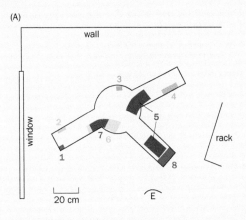

(B)

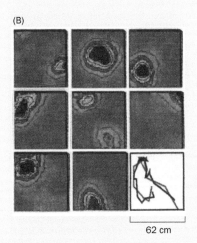

Figure 11-31 Hippocampal place cells. (A) Map of a maze showing the place fields (numbered and illustrated in different colors) of eight place cells in the hippocampus of a freely moving rat. Each place field represents the regions within the maze in which a given place cell exhibited increased firing. E, location of the experimenter. **(B)** The activity of place cells can be used to construct a map of a rat's trajectory. A multielectrode array was used to simultaneously record 80 hippocampal cells. The place fields of eight representative place cells are shown as eight heat maps: for each place field, the colors indicate the firing rate of the cell when the rat occupied a corresponding position in a 62 cm × 62 cm square arena (red, maximal firing rate; dark blue, no firing). Note that the place fields of different cells vary in size and are situated at different locations in the arena. Bottom right, the population vector (Figure 8-31) of firing rates of 80 hippocampal cells during a 30 s period was used to reconstruct the spatial trajectory of the rat. The calculated trajectory (red) closely matches the actual trajectory (black). (A, adapted from O'Keefe J [1976] *Exp Neurol* 51:78–109. B, from Wilson MA & McNaughton BL [1993] *Science* 261:1055–1058.)

field is established in a circular arena, if external landmarks are rotated, the place fields also rotate, preserving their *relative positions* to the external landmarks. However, once place fields form, place cells fire at the same locations in the dark, and place fields in the same environment can be stable for over a month (Movie 14-3). Since different place cells fire when the rat occupies different locations in the same arena, it is possible to reconstruct the path of a moving rat from simultaneous recordings of dozens of place cells using a multielectrode array (Figure 11-31B); in other words, a few dozen place cells contain sufficient information to reconstruct the rat's path. At the same time, a single place cell can be active in different environments, with differing place fields in each. Thus, each environment is represented by a **cell assembly**, a unique population of active place cells, and each cell participates in multiple cell assemblies representing different environments. These remarkable properties led to the proposal that hippocampal place cells collectively form cognitive maps that animals use to determine where they are in their environment and aid their navigation using landmark-based and path-integration strategies. Unlike the topographic map in the visual system (see Chapters 4 and 5), however, there is no obvious relationship between the positions of place cells in the hippocampus and the physical locations of their place fields.

The hippocampus is reciprocally connected with the nearby entorhinal cortex (Figure 11-5). Remarkably, studies in the 2000s revealed that many medial entorhinal layer 2/3 neurons also have space-modulated firing patterns. The locations where these cells fire most are distributed across the environment in a periodic manner, forming grids that tile the entire space; each cell's peak firing rate occurs at the api-

ces of the hexagonal grid unit (**Figure 11-32A**; **Movie 11-2**). These cells are aptly named **grid cells**. Each grid cell has a characteristic grid size that remains constant in arenas of differing sizes and shapes (Figure 11-32A). Neighboring grid cells share similar grid sizes but differ in the exact locations of their grid centers.

Grid cells and place cells have many similarities. The activity of simultaneously recorded grid cell populations, like those of place cells, can be used to reconstruct movement trajectories (Movie 11-2). As with place fields, grid patterns are influenced by external landmarks; when external landmarks are rotated in a circular arena, grid patterns rotate correspondingly. Grid patterns, like place fields, do not merely mirror sensory cues, since they are maintained when the animal moves in the dark. However, the properties of grid cells and place cells also differ in important ways. Grid cells tile space more efficiently: a few grid cells can cover a space requiring dozens or more of place cells. After animals are introduced into a novel environment, grid cells retain their grid size and populations of grid cells maintain the positions of their grid centers relative to each other across different arenas, while place cells remap more randomly. These observations suggest that grid cells provide a more fundamental metric of space and anchor the place fields of place cells.

In addition to grid cells, the entorhinal cortex contains **border cells**, which fire when the animal is at a specific edge of an arena (Figure 11-32B). Border cells provide information about the perimeters of the local environment, which can anchor grid patterns and place fields to geometric confines. **Head direction cells**, another intriguing cell type, fire when the animal's head faces a specific direction, independent

Box 11-2: continued

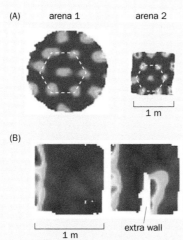

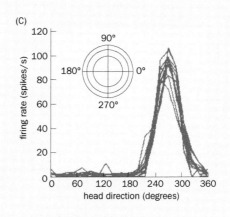

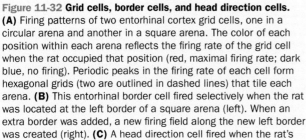

Figure 11-32 Grid cells, border cells, and head direction cells.
(A) Firing patterns of two entorhinal cortex grid cells, one in a circular arena and another in a square arena. The color of each position within each arena reflects the firing rate of the grid cell when the rat occupied that position (red, maximal firing rate; dark blue, no firing). Periodic peaks in the firing rate of each cell form hexagonal grids (two are outlined in dashed lines) that tile each arena. **(B)** This entorhinal border cell fired selectively when the rat was located at the left border of a square arena (left). When an extra border was added, a new firing field along the new left border was created (right). **(C)** A head direction cell fired when the rat's head was facing a specific direction (peaking at ~270°, or when the rat's head was facing south) regardless of where the rat was located in the arena. Each of the 12 traces represents the firing rate of the cell when the rat was in one of the 12 divisions of the circular arena (inset at top left). (A, from Hafting F, Fyhn M, Molden S, et al. [2005] *Nature* 436:801–806. With permission from Springer Nature. B, from Solstad T, Boccara CN, Kropff E, et al. [2008] *Science* 322:1865–1868. With permission from AAAS. C, from Taube JS, Muleer RU, & Ranck JB [1990] *J Neurosci* 10:420–435. Copyright ©1990 Society for Neuroscience.)

of the animal's location in the arena (Figure 11-32C). While grid cells and border cells have been found mostly in the entorhinal cortex, head direction cells are also present in brain regions that send input to the entorhinal cortex. Indeed, the entorhinal cortex receives diverse inputs representing visual, olfactory, and vestibular signals. In turn, intermingled populations of grid, border, and head direction cells in the entorhinal cortex all send direct projections via the perforant path (Figure 11-5) to the hippocampus, which integrates these diverse information streams to form place fields and send feedback signals to the entorhinal cortex. Exactly how information is integrated and how the place code is read in order to guide navigation remains unknown.

The remarkable properties of place and grid cells in the hippocampal–entorhinal network, far removed from the sensory world, have provided a glimpse into how abstract information such as space is represented in the brain. What is the relationship between spatial representation and memory, another important function of the hippocampus? One hypothesis is that the hippocampal–entorhinal network is used in parallel for navigation and memory. Explicit memory often involves the binding of disparate details into a coherent event; this could be conceptually analogous to the process by which hippocampal place cells extract spatial information from the activity of grid, border, and head direction cells. An alternative hypothesis is that the location of an experience is so essential to its explicit memory that the formation of a memory is intimately tied to the representation of space. Indeed, use of "memory palaces," organization of events into imaginary spaces, is an ancient and effective mnemonic technique, and space-based tasks have been among the most effective ways to assay memory in mammals. As researchers learn more about the functions of the hippocampal–entorhinal network in memory and spatial representation, the connections between these two systems will become clearer.

11.19 In rodents, spatial learning and memory require the hippocampus

Does synaptic plasticity underlie learning and memory in mammals as it does in *Aplysia* and *Drosophila*? In other words, do activity-dependent changes induced at given synapses during formation of a specific memory serve as a basis for information storage underlying that specific memory? In the following sections we will explore these questions using the mammalian hippocampus as a model, given the rich synaptic plasticity mechanisms discovered there (Sections 11.4–11.13) and its importance for forming explicit memories in humans (Section 11.1).

An essential step in linking memory and hippocampal synaptic plasticity is to establish hippocampus-dependent behavioral tasks that test memory in rodents,

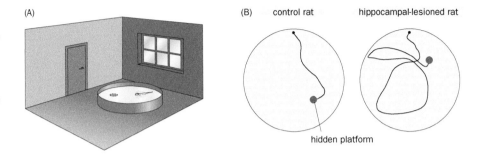

Figure 11-33 Spatial memory tested in the Morris water maze requires the hippocampus. (A) Schematic of the Morris water maze. After training, a rodent can use distant spatial cues to find a hidden platform (dashed circle) in a large pool of milky water. **(B)** After training, a control rat swam directly to the hidden platform, while a rat with a hippocampal lesion found the platform after taking a circuitous route. (Adapted from Morris RGM, Garrud P, Rawlins JNP, et al. [1982] *Nature* 297: 681–683. With permission from Springer Nature.)

the animal model in which synaptic plasticity has been most intensely investigated. Given that the mammalian hippocampus contains spatial maps of the external world (Box 11-2), several hippocampus-dependent behavioral assays that require spatial recognition have been established. One widely used assay is the **Morris water maze** (Figure 11-33A), a navigation task in which rodents learn to locate a hidden platform in a pool of milky water to avoid having to swim—despite being able to swim, rodents prefer not to. The rodent cannot see, hear, smell, or touch the platform until it finds the platform. Nevertheless, it can use distant cues in the room to learn the platform's spatial location, such that after training it can be placed at any position in the pool and will swim directly to the hidden platform (Figure 11-33B, left). Performance of this task requires the hippocampus: rats with hippocampal lesions no longer remembered the location of the hidden platform, even after extensive training (Figure 11-33B, right). When the platform was visible, control and hippocampal-lesioned rats found it with equal ease. Of the forms of memory tested in rodents, this spatial memory most closely resembles explicit memory in humans.

11.20 Many manipulations that alter hippocampal LTP also alter spatial memory

Development of spatial memory tasks such as the Morris water maze enabled researchers to determine whether manipulations that affect synaptic plasticity in the hippocampus also affect spatial memory. One of the first such manipulations was to block the function of the NMDA receptor with a specific antagonist, AP5. Infusion of AP5 into the hippocampus during the training session, at a concentration that blocked LTP *in vivo* (Section 11.6), disrupted subsequent recall of platform position in the Morris water maze. When the hidden platform was removed after training, control rats focused their search preferentially in the quadrant where the hidden platform had been, while AP5-treated rats swam randomly (Figure 11-34A). Conditionally knocking out the essential NMDA receptor subunit GluN1 in CA1 pyramidal neurons, which blocks LTP at the CA3 → CA1 synapse (Figure 11-9), also impaired performance in the water maze assay (Figure 11-34B). These experiments indicated an essential function for the NMDA receptor in the hippocampus, and specifically in CA1 pyramidal neurons, in spatial learning and memory.

Genetic manipulations in mice that disrupt hippocampal synaptic plasticity, notably LTP at CA3 → CA1 synapses, also interfere with hippocampus-dependent spatial memory tasks. For example, mice that lack CaMKII or have a point mutation in the CaMKII auto-phosphorylation site have impaired LTP at CA3 → CA1 synapses (Figure 11-11B) and perform poorly in the Morris water maze (Figure 11-34C). In concert with findings from *Aplysia* and *Drosophila*, the cAMP/PKA pathway is also essential for both hippocampal LTP and hippocampus-dependent memory. In mice carrying mutations in two Ca^{2+}-activated adenylate cyclases, which link Ca^{2+} entry to cAMP production (Figure 3-41), CA3 → CA1 LTP was impaired (Figure 11-35A), as was performance in a hippocampus-dependent memory task called passive avoidance. In this task, mice are placed in a chamber with two compartments, one of which is illuminated. Mice naturally prefer the dark compartment. During training, entry into the dark chamber is paired with electric shock. After training, mice are placed back in the illuminated compartment and the time

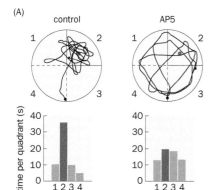
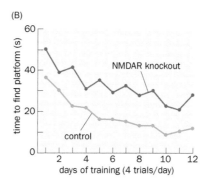
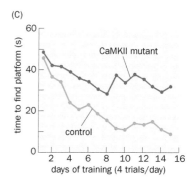

Figure 11-34 Manipulations that disrupt hippocampal LTP also interfere with performance in the Morris water maze. (A) Infusion of AP5, an antagonist of the NMDA receptor, disrupts spatial memory formation in rats. Rats were trained with the hidden platform (circle) in quadrant 2. During the test, the platform was removed and the rat's trajectory was recorded. Top, while the control rat focused its search near the phantom platform, the rat infused with AP5 *during training* swam randomly. Bottom, quantification of time spent in the four quadrants. **(B)** Throughout the training regime, mice with CA1-specific knockout of the GluN1 subunit of the NMDA receptor (red trace) were slower to find the hidden platform than *CA1-Cre*-only control mice (blue trace). **(C)** Mice in which the CaMKII auto-phosphorylation site was mutated (red trace) also took longer to find the hidden platform than controls (blue trace). See Figures 11-9 and 11-11 for LTP defects in the same mice as in Panel B and Panel C. (A, from Morris RGM, Anderson E, Lynch GS, et al. [1986] *Nature* 319:774–776. With permission from Springer Nature. B, from Tsien JZ, Huerta PT, & Tongegawa S [1996] *Cell* 87:1327–1338. With permission from Elsevier Inc. C, adapted from Giese KP, Federov NB, Filipkowski RK, et al. [1998] *Science* 279:870–873.)

it takes for a mouse to enter the dark compartment is used as a measure of their memory of the shock. Adenylate cyclase double knockout mice performed poorly 30 minutes after training compared with control mice (Figure 11-35B).

Since the 1990s, a number of genetic manipulations in mice have also been reported to enhance memory performance compared to controls in a variety of memory tasks such as the Morris water maze, passive avoidance, or fear conditioning (which we will discuss in more detail later). For example, transgenic mice overexpressing the GluN2B subunit, which is preferentially expressed in developing neurons (Box 5-3) and produces higher Ca²⁺ conductance than other GluN2 isoforms, exhibited superior performance in the Morris water maze and other memory tasks. Interestingly, GluN2B-overexpressing mice and other genetically engineered mice with enhanced memory performance also exhibited enhanced hippocampal LTP. The frequent co-occurrence of LTP and memory alterations due to various genetic and pharmacological manipulations suggests a strong link between memory and hippocampal synaptic plasticity.

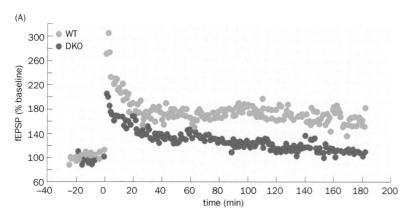
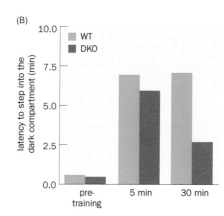

Figure 11-35 Interfering with the cAMP/PKA pathway affects hippocampal LTP and learning. (A) Compared with wild-type (WT) controls, mice in which two Ca²⁺-dependent adenylate cyclases were knocked out (DKO) exhibit a reduced LTP magnitude at CA3 → CA1 synapses. **(B)** DKO mice also exhibit impaired memory in a passive avoidance task. In this assay, mice are placed in the illuminated compartment of a two-compartment chamber. Before training, mice quickly move into the dark compartment naturally to avoid predators. After training (pairing electric shock with entrance into the dark compartment), DKO mice avoid the dark compartment similarly to controls 5 min after training but enter the dark compartment more quickly 30 min after training, suggesting impaired memory. (From Wong ST, Athos J, Figueroa XA, et al. [1999] *Neuron* 23:787–798. With permission from Elsevier Inc.)

11.21 From correlation to causation: the synaptic weight matrix hypothesis revisited

While establishing strong correlations between hippocampal LTP and spatial memory in rodents, none of the genetic and pharmacological manipulations discussed in the previous section demonstrated that synaptic changes *cause* memory formation. These manipulations could all affect synaptic plasticity and memory in parallel. To establish a causal link between modification of synaptic strength and learning, one would ideally specifically alter one and test the effect on the other.

One approach is to examine directly whether learning can induce hippocampal LTP. The key to this type of experiment is to identify which synapses in the hippocampus are related to a specific learning event. This difficult task was attempted in an experiment that combined passive avoidance training with use of a multielectrode recording array in rats. These rats were implanted with a multielectrode recording array at the CA1 dendritic fields and a stimulating electrode at the Schaffer collaterals (**Figure 11-36**A), such that synaptic transmission from Schaffer collaterals onto different populations of CA1 pyramidal neurons could be recorded before and after training. While none of the electrodes from control rats without training detected potentiation, a small fraction of electrodes from trained rats detected potentiation after behavioral training (Figure 11-36B). Moreover, synapses that were potentiated by behavioral training became less likely to be potentiated further by subsequent high-frequency stimulation of the Schaffer collateral, a process that induces LTP (Section 11.5). These results suggest that learning can produce synaptic potentiation that partially occludes subsequent LTP at the same synapses.

Another approach to investigate the relationship between LTP and learning is to test whether saturation of LTP prevents further learning. In one experiment, rats with unilateral hippocampal lesions (such that spatial memory must depend on the unlesioned hippocampus) were implanted with a multielectrode stimulating array at the perforant path in the unlesioned hippocampus. Repeated stimulation through this array could maximally induce and potentially saturate LTP at recording sites. Rats with nearly saturated LTP, measured *post hoc* by physiological recordings, were more impaired in the Morris water maze than rats that still exhibited residual LTP. In a different paradigm, chronic two-photon imaging of motor cortical neurons revealed dendritic spine dynamics associated with motor skill learning. In a recent experiment, molecular genetic tools in the mouse were used to target a photoactivatable Rac GTPase preferentially to dendritic spines potentiated after motor learning. Photoactivation of Rac GTPase, which causes selective shrinkage of these potentiated spines due to its activity on the actin cytoskeleton (Box 5.2), disrupted motor learning. Together, these experiments strengthen the link between learning and changes in the strength of specific synapses.

Let's revisit the hypotheses raised in Sections 11.2 and 11.3: memory is stored in the form of synaptic weight matrices in neural circuits, and learning is equivalent to altering the synaptic weight matrices as a result of experience. Strong cases for these hypotheses can be made in invertebrate models of learning such as the *Aplysia* gill-withdrawal reflex and fly olfactory conditioning. Modifications of the strength of the sensory neuron → motor neuron synapse underlie behavioral habituation and sensitization in *Aplysia*, while modification of the strength of the Kenyon cell → mushroom body output neuron synapse underlie olfactory memory. In the mammalian brain, the strongest evidence has come from studies of the hippocampus. One way to further strengthen the causal relationship between learning and alteration of synaptic weights would be to achieve the following: (1) identify the neurons and synapses in a circuit whose plasticity correlates with a learning experience, (2) determine the specific states of the synaptic weight matrix before the learning experience (state *A*) and after it (state *B*), (3) artificially change the synaptic weight matrix from state *A* to state *B* in a naive animal without learning, and (4) test whether the animal behaves as if the learning experience had occurred. This is a challenging task; the *in vivo* mimicry experiment has not been performed even in the simplest *Aplysia* gill-withdrawal paradigm. The complexity of the mammalian brain and its large number of neurons and synapses make this task

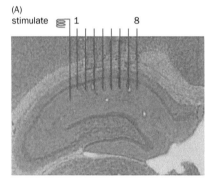

(A)
stimulate

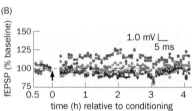

(B)

Figure 11-36 Learning can induce LTP.
(A) A multielectrode array (electrodes 1–8) was placed in the CA1 area of a rat hippocampus to record the responses of CA1 neurons to stimulation of Schaffer collaterals. **(B)** After passive avoidance training, recordings from the two electrodes indicated by red and orange dots showed an enhancement of field excitatory postsynaptic potentials (fEPSPs); recordings from the remaining six electrodes (represented by dots of other colors) did not demonstrate fEPSP enhancement. Additional experiments (not shown) revealed that synapses potentiated by passive avoidance training were less responsive to subsequent potentiation by high-frequency stimulation using the stimulating electrode. (From Whitlock JR, Heynen AJ, Shuler MG, et al. [2006] *Science* 313:1093–1097. With permission from AAAS.)

much more challenging. Nevertheless, researchers have employed modern circuit analysis tools to search for the potential physical substrates for memory (called **memory traces** or **engrams**). Box 11-3 provides an example of how this can be conducted at the cellular level.

Box 11-3: How to find a cellular engram

The search for engrams has had a long history. As mentioned at the beginning of this chapter, lesion experiments led to the conclusion that the engram for maze running was widely distributed in the rat cerebral cortex. Studies of human patients such as H.M. have led to the identification of the hippocampus as a site essential for forming new explicit memories. Tools in modern neuroscience have the potential to reveal engrams at the level of neurons (*cellular engram*) and synapses (*synaptic engram*, changes in the synaptic weight matrix as in Figure 11-4). Neural substrates representing a memory engram should become active during training, and their reactivation should mimic recall of the memory. We use a specific example to illustrate how researchers have utilized a combination of transgenic mice,

viral transduction, and optogenetic manipulation to search for a cellular engram (Figure 11-37).

To identify active neuronal populations, researchers used a *Fos-tTA* transgenic mouse in which expression of a **tetracycline-repressible transcriptional activator** (**tTA**) is controlled by the promoter of the immediate early gene *Fos*, such that tTA can be rapidly induced by neuronal activity (see Section 3.24 for this property of immediate early genes). tTA is a transcription factor that binds to DNA sequences called **tetracycline response elements** (**TREs**) to regulate gene expression; tTA activity is inhibited by tetracycline analogs like doxycycline (see Section 14.9 for more details of the tTA/*TRE* expression system). An adeno-associated virus

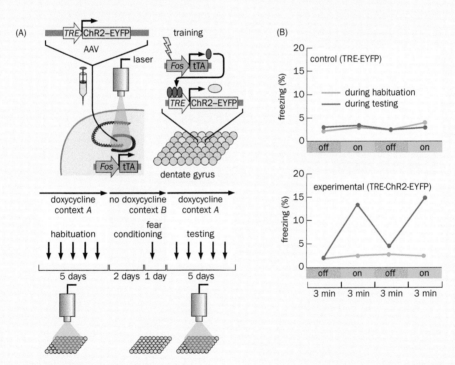

Figure 11-37 A cellular engram of fear memory. (A) Experimental design. Adeno-associated virus (AAV) encodes expression of channelrhodopsin (ChR2) fused to an enhanced yellow fluorescent protein (EYFP) under the control of a tetracycline response element (*TRE*). AAV was stereotactically injected into the dentate gyrus (indicated by the needle) of a transgenic mouse expressing the tetracycline-repressible transcription activator (tTA, red ovals) from the *Fos* promoter so that its expression is induced by neuronal activity. The matrix of circles represents dentate gyrus granule cells. Mice were first habituated to context *A* and photostimulation in the presence of doxycycline; the dentate gyrus granule cells activated in context *A* did not express ChR2, because tTA activity was inhibited by doxycycline. Fear conditioning was induced in context *B* in the absence of doxycycline such that dentate gyrus granule cells

activated by this experience expressed ChR2 (yellow circles); ChR2 expression persisted for several days, even after the mice were treated with doxycycline again to prevent new tTA-induced gene expression. Mice were then reintroduced into context *A* to test whether optogenetic stimulation of ChR2-expressing cells could induce fear memory recall. **(B)** In control mice, in which tTA induced expression of EYFP, photostimulation (light-on period in green) did not induce fear memory, as assayed by the percentage of time spent freezing (top). In experimental mice, photostimulation induced freezing in a light-dependent manner (bottom) during the testing period (red trace) but not during the earlier habituation period (blue trace). (Adapted from Liu X, Ramirez S, Pang PT, et al. [2012] *Nature* 484:381–385. With permission from Springer Nature.)

(Continued)

Box 11-3: continued

(AAV) encoding expression of channelrhodopsin (ChR2) under the control of *TRE* was used to transduce dentate gyrus granule cells, which provide input to CA3 pyramidal neurons (Figure 11-5). These mice were tested for hippocampus-dependent memory established by **contextual fear conditioning**. [In this paradigm, mice experience electric shock during training in a specific environment (context *A*). Mice subsequently placed in the same environment exhibit a freezing response: they remain immobile, an adaptive response of rodents to avoid being seen by predators. Mice placed in a different environment (context *B*, which differs from *A* in ceiling shape, flooring, and lighting) do not exhibit a freezing response.] Mice were first habituated to context *A* in the presence of doxycycline to prevent tTA/*TRE*-induced ChR2 expression. After doxycycline removal, mice were exposed to context *B* while they received electric shock to induce contextual fear conditioning. This resulted in tTA and ChR2 expression in a subset of dentate gyrus granule cells that were activated during fear conditioning in context *B*.

To test the effect of reactivation of neurons that were active during fear conditioning in context *B*, mice were given food containing doxycycline to prevent new tTA/*TRE*-induced ChR2 expression and were introduced to context *A* with or without optogenetic stimulation (Figure 11-37A). Control mice did not freeze in context *A*. However, ChR2-expressing mice froze in context *A* in response to optogenetic stimulation, as if they were in context *B* (Figure 11-37B). Thus, activation of a population of cells that was active during contextual fear conditioning induced fear recall in a different context, suggesting that this population of dentate gyrus granule cells contributes to the memory of context *B*. Therefore, this granule cell population constitutes a cellular engram for the experience of electrical shock in context *B*.

This experiment did not show which synapses were modified (synaptic engram) and what additional circuit properties changed to make mice fearful of context *B*. In principle, plasticity could occur anywhere in the neural pathway downstream of the granule cell population that leads to freezing. In light of the hippocampal plasticity findings discussed in this chapter, it is likely that plasticity occurs in the downstream circuits within the hippocampus, such as at the dentate gyrus → CA3 synapse, the CA3 → CA3 recurrent synapse, the CA3 → CA1 synapse, or several of the these. Plasticity can also occur in the amygdala, whose function in fear conditioning will be discussed in Section 11.23. Identifying synaptic engrams in these brain regions remains an open challenge.

WHERE DOES LEARNING OCCUR, AND WHERE IS MEMORY STORED IN THE NERVOUS SYSTEM?

So far, we have used *Aplysia, Drosophila,* and mammalian hippocampus as models for studying mechanisms of synaptic plasticity and memory. However, synaptic plasticity occurs throughout the nervous system. Likewise, different types of memory engage neural circuits throughout the nervous system. In the last part of this chapter, we study selected examples to broaden our discussion of memory systems in vertebrates.

11.22 The neocortex contributes to long-term storage of explicit memory

Although the medial temporal lobe, including the hippocampus, is essential for formation of new explicit memories, it does not appear to be required for the storage and retrieval of long-term, remote memories, as suggested by the ability of H.M. to recall memories of his childhood (Section 11.1). Then where is long-term explicit memory stored?

A widely accepted view is that long-term explicit memory is stored in the neocortex, and that specific types of memory engage specific cortical regions. This idea, first proposed in the late nineteenth century, states that remembering involves reactivating the neocortical sensory and motor components of the original event that led to formation of the memory. Two types of human studies are consistent with this view. First, lesions of specific parts of the neocortex lead to loss of specific types of memory. For example, patients with damage to the color- or face-processing areas of the visual cortex not only lose their ability to perceive colors or recognize faces but also exhibit retrograde memory deficits in those domains. Patients with adult-onset prosopagnosia (inability to distinguish faces) not only exhibit deficits in face perception but also cannot remember faces that were familiar to them before the onset of the disorder.

Second, functional imaging studies of healthy human subjects engaged in memory tasks revealed reactivation of cortical areas relevant to specific memory

Figure 11-38 Reactivation of specific sensory cortices during long-term memory recall. fMRI images of two subjects, each performing the task of vividly remembering an object when presented with a word that had been extensively paired with either a picture or a sound during prior training. For image-based recall, the high-order visual cortex was activated (arrows in the top panels). For sound-based recall, the high-order auditory cortex was activated (arrows in the bottom panels). The fMRI images in the top and bottom rows were taken at different horizontal planes. Both subjects had a bias for using the left cortex. (From Wheeler ME, Petersen SE, & Buckner RL [2000] *Proc Natl Acad Sci U S A* 97:11125–11129. Copyright National Academy of Sciences.)

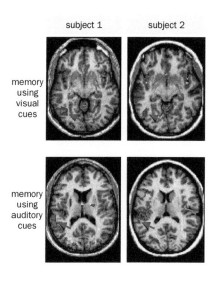

tasks. In one study, for example, subjects were first extensively trained to associate words (for example, *dog*) with either pictures (an image of a dog) or sounds (the bark of a dog). During subsequent testing, they were asked to vividly recall the items when given only the word as a cue, while their brains were being scanned via functional magnetic resonance imaging (fMRI). After the recall task/fMRI scan, subjects then indicated whether they vividly remembered an image or a sound; their answers to this question usually matched with their training. After pairing the word with an image during training, high-order visual cortical areas were selectively activated during recall (Figure 11-38, top), whereas after sound-based training, recall elicited selective activation of high-order auditory cortical areas (Figure 11-38, bottom). These data suggest that the act of remembering reactivates sensory-specific cortices.

Animal studies have strengthened the notion that the neocortex plays a role in remote memory and shed light on the interactions between the hippocampus and neocortex. As examples, we discuss two experiments on contextual fear memory (Box 11-3), a hippocampus-dependent form of memory that resembles human explicit memory. In the first experiment, rats received electric shock when placed in a specific environment; subsequently, their hippocampi were bilaterally lesioned 1, 7, 14, or 28 days after training. Seven days after surgery, rats were returned to the training environment to measure their fear memory. While control rats exhibited fear memory under all conditions, rats with hippocampal lesions lacked fear memory if lesioning was performed 1 day after training, but had progressively less severe deficits in memory recall as the duration between training and lesioning lengthened (Figure 11-39A). This result suggests that fear

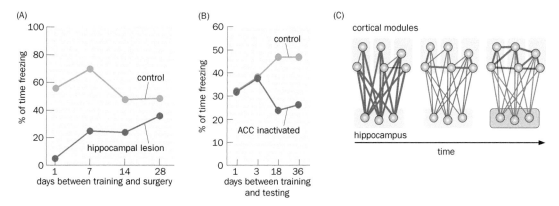

Figure 11-39 Interactions between the hippocampus and neocortex during long-term memory consolidation. (A) In this contextual fear-conditioning experiment, the x axis shows days elapsed between training and bilateral hippocampal lesion in experimental rats. All rats were tested for fear conditioning (y axis) 7 days after lesioning. Compared with unlesioned controls that went through the same surgical procedure (blue trace), lesioned rats (red trace) did not exhibit fear memory (quantified by the percentage of time spent freezing) when lesioning was performed 1 day after training. The effect of lesioning became less pronounced as the period between training and lesioning lengthened. **(B)** Injection of lidocaine, an anesthetic that blocks action potentials, into the anterior cingulate cortex (ACC; red trace), reduced contextual fear memory compared with controls (blue trace) when drug administration and testing were performed 18 or 36 days, but not 1 or 3 days, after training. This finding suggests that the ACC is required

for recall of remote memory but not recent memory. **(C)** A model of long-term memory consolidation. During initial memory formation (left), signals that pass through different cortical modules to the hippocampus establish links between the hippocampus and those cortical modules. Ongoing interactions between the hippocampus and the cortical modules after initial memory formation gradually establish links among the different cortical modules (middle), until these intracortical links are sufficient to represent the remote memory and the memory can be recalled independently of the hippocampus (right). Thick and thin lines represent strong and weak links, respectively. Links shown in gray at a given stage are not required for memory recall at that time. (A, adapted from Kim JJ & Fanselow MS [1992] *Science* 256:675–677. B, adapted from Frankland PW, Bontempi B, Talton LE, et al. [2004] *Science* 304: 881–883. C, adapted from Frankland PW & Bontempi B [2005] *Nat Rev Neurosci* 6:119–130. With permission from Springer Nature.)

memory becomes increasingly less dependent on the hippocampus as time passes after initial training.

Where is long-term fear memory stored? In the second experiment, researchers used immediate early gene expression to identify brain regions activated during retrieval of remote fear memory. Several frontal cortical areas were shown to have elevated expression of the immediate early genes *Fos* and *Egr1* (Section 3.24). Inactivation of these cortical areas by focal injection of lidocaine (an anesthetic that blocks action potential propagation by inhibiting voltage-gated Na$^+$ channels) identified the **anterior cingulate cortex** (**ACC**), which is located near the midline of the frontal lobe, as a neocortical site involved in remote fear memory retrieval. Inactivation of the ACC during testing caused significant loss of fear memory when testing occurred 18 or 36 days, but not 1 or 3 days, after initial training (Figure 10-39B). Accordingly, human fMRI studies consistently find activation of the frontal cortex, including the ACC, during different kinds of memory recall.

These experiments led researchers to propose the systems consolidation hypothesis (Figure 11-39C). Signals that lead to the original hippocampus-dependent formation of explicit memory also activate primary and association cortical areas. The hippocampus integrates distributed signals from multiple "cortical modules" during initial memory formation. (Different modules can be considered as distinct ensembles of active neurons with a cortical area or across different areas.) Over the course of long-term memory consolidation, the hippocampus trains the establishment of new connections among cortical modules, such that memories over the long term are no longer hippocampus dependent. Recent experiments in rodents using circuit analysis tools, such as the ability to manipulate cells activated by a specific experience or behavioral episode (Box 11-3), have supported and extended the systems consolidation model. For example, in fear conditioning paradigms in rodents, it has been shown that neurons in the medial prefrontal cortex (PFC) are activated during fear learning, and these activated neurons receive additional signals from the hippocampal network during memory consideration. Evidence also suggests that the PFC neuronal populations that participate in remote memory undergo dynamic changes during the first two weeks after learning, such that neurons activated during later retrieval are more effective at promoting remote memory recall than those activated during learning and early retrieval.

A candidate physiological mechanism that mediates communication between the hippocampus and the frontal cortex is **sharp-wave ripples** (SWRs)—large amplitude rapid oscillations in local field potential of the hippocampus during sleep or wakeful resting (**Figure 11-40A**). Evidence suggests that disruption of SWRs

Figure 11-40 Hippocampal sharp-wave ripples. (A) Top, local field potential recordings of CA1 dendritic fields showing that sharp-wave ripples (SWRs; arrows) occur after the rat transitions from walking to still. Bottom, SWR on an expanded times scale, showing large hyperpolarization when high-frequency signals are filtered out (brown) and rapid oscillations when high-frequency signals are included (blue, recorded in CA1 cell body region). **(B)** Simultaneous recording of hippocampal CA1 (green) and prefrontal cortex (PFC, black) from a rat that is stationary in a maze. From top to bottom: CA1 local field potential (LFP) filtered at 150–250 Hz to highlight SWRs (shaded); spiking raster of 18 CA1 single units (one row per unit); LFP in PFC; spiking raster of 7 PFC single units. Note that many CA1 units and one PFC unit (arrows) increase spiking during a subset of SWRs. (A, adapted from Buzsaki G [2015] *Hippocampus* 25:1073–1188 and Buzsaki G, Horvath Z, Urioste R, et al. [1992] *Science* 256:1025–1027. B, from Jadhav SP, Rothschild G, Roumis DK, et al. [2016] *Neuron* 90:113–127. With permission from Elsevier Inc.)

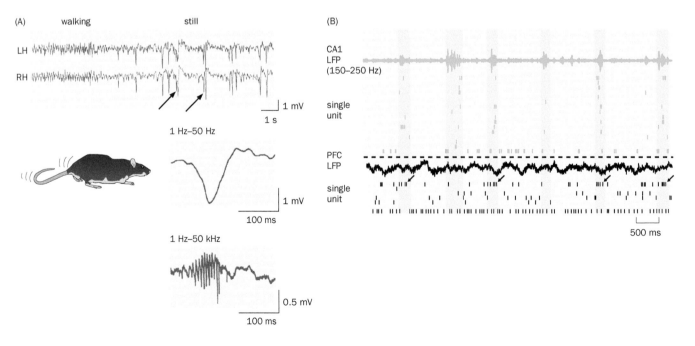

impair spatial memory in rats. Simultaneous recording of PFC and hippocampal CA1 neurons revealed that spiking of a subset of PFC neurons was modulated during hippocampal SWRs (Figure 11-40B), supporting the idea that hippocampal information is relayed effectively to PFC during SWRs. Because SWRs are prominent during sleep, an attractive hypothesis is that SWRs during sleep promote interaction between the hippocampus and the neocortex for consolidation of long-term memory (Section 9.17).

11.23 The amygdala plays a central role in fear conditioning

In Section 11.22 and Box 11-3, we studied hippocampus-dependent contextual fear conditioning. In another form of fear conditioning, called **cued fear conditioning**, an electric shock is applied at the end of a cue presented during training. The most commonly used cue is a sound, in which case cued fear conditioning is called **auditory fear conditioning** (Figure 11-41A). Auditory fear conditioning is a form of classical conditioning (Section 11.14); the tone and shock serve as the CS and US, respectively. When compared side by side in control animals (Figure 11-41B, left), auditory conditioning was more rapidly acquired during training than contextual conditioning and was more resistant to extinction during testing. Lesions indicate that while contextual fear conditioning requires the hippocampus, auditory fear conditioning does not. However, both forms of conditioning require the amygdala (Figure 11-41B). These studies suggest that tone–shock and context–shock associations are formed in the amygdala, while the hippocampus synthesizes contextual information required to create place memories.

Extensive anatomical, physiological, and perturbation studies have delineated the amygdala circuits that contribute to fear conditioning (Figure 11-42). The amygdala complex consists of several major divisions: the lateral amygdala, basal amygdala (collectively referred to as the **basolateral amygdala**), and **central amygdala**. (Note that this complex is adjacent to but distinct from the olfactory amygdala; Figure 10-26.) The basolateral amygdala is cortical-like, consisting of mostly excitatory pyramidal neurons with a small population of GABAergic interneurons. The central amygdala is striatal-like, composed of GABAergic neurons that make both local and long-range connections. In auditory fear conditioning, information about the tone (CS) reaches the lateral amygdala via a direct pathway

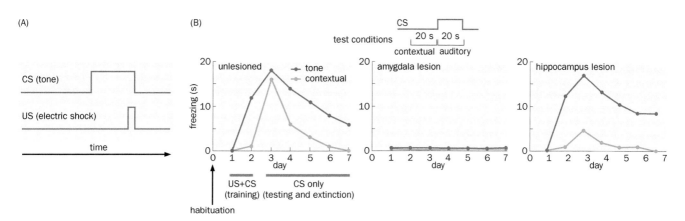

Figure 11-41 Both auditory and contextual fear conditioning require the amygdala. (A) Schematic of auditory fear conditioning, a form of classical conditioning in which the tone (~20 s) serves as a conditioned stimulus (CS) and the electric shock (~0.5–1 s) serves as an unconditioned stimulus (US) that co-terminates with the CS. **(B)** Left, learning curve as measured by average time spent freezing within a 20 s period (*y* axis) for control rats. On day 0, rats were introduced into the conditioning chamber for habituation without shock. On days 1 and 2, a 20 s tone was paired with a 0.5 s electric shock at the end of the tone, with two pairings per day. On days 3–7, only the tone was presented in the same conditioning chamber. As shown in the schematic above, contextual conditioning was measured as the

freezing time during the 20 s period immediately before the CS (tone) onset, whereas tone conditioning was measured as the freezing time during the 20 s tone period. Middle, bilateral amygdala lesions before training disrupted both contextual and tone conditioning. Right, bilateral hippocampal lesions disrupted contextual conditioning but not tone conditioning. Note that while the tone–shock association was tested in the conditioning chamber in this experiment, animals conditioned in one context to associate a tone with an electric shock also exhibited a robust fear response to the tone in a different context. (B, from Phillips RG & LeDoux JE [1992] *Behav Neurosci* 106:274–285. With permission from the American Psychological Association Inc.)

Figure 11-42 A simplified circuit diagram for fear conditioning. The tone signal (CS, blue) can reach the lateral amygdala via the auditory thalamus or the auditory cortex. The shock signal (US, red) can reach the lateral amygdala via the somatosensory thalamus or more directly the central amygdala through the pain pathway via the parabrachial nucleus (PBN). Contextual information from the hippocampus (green) enters through the basal amygdala. Within the amygdala complex, information flows from the lateral nucleus to the central nucleus either directly or via the basal amygdala. The central amygdala provides output to brainstem and hypothalamic targets to regulate behavioral, autonomic, endocrine, and neuromodulatory systems. (Based on LeDoux JE [2000] *Annu Rev Neurosci* 23:155–184 and Pape HC & Pare D [2010] *Physiol Rev* 90:419–463.)

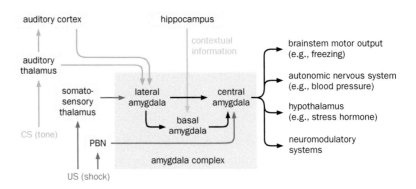

from the auditory thalamic nuclei and an indirect pathway from the high-order auditory cortex. The foot shock (US) signal also reaches the amygdala by multiple pathways, including projections from the somatosensory thalamic nuclei to the lateral amygdala and from the pain pathway via the parabrachial nucleus (PBN) to the central amygdala (Figure 6-71). Contextual input from the hippocampus enters the amygdala complex via the basal amygdala (Figure 11-42).

Historically, the basolateral and central nuclei of the amygdala were thought to function in series, with sensory information about the CS and US converging in the lateral amygdala, which served as the critical site of learning-related plasticity. The lateral amygdala in turn relayed CS–US associations onto the central amygdala directly or indirectly via the basal amygdala. The central amygdala coordinated the output of the amygdala complex for fear and defensive responses, coordinating behavioral output (e.g., freezing), autonomic responses (e.g., increased blood pressure), and neuroendocrine responses (e.g., stress hormone production) via distinct projections to the hypothalamus and brainstem (Figure 11-42). It has been recognized in recent years that, in addition to such serial function, the basolateral amygdala and central amygdala also function in parallel. For example, the central amygdala also receives direct sensory input, undergoes learning-related synaptic plasticity required for storage of fear memories, and contributes to behavioral processes independently of the basolateral amygdala.

What is the neural basis of behavioral conditioning? The classic framework suggests that the simultaneous presence of the CS and US during training strengthens, through Hebbian mechanisms, the connections between neurons representing the CS and neurons that generate defensive responses to the US; after conditioning, the CS alone can elicit defensive responses. The best evidence for this model has come from studies of the synapses connecting auditory thalamic input neurons and excitatory projection neurons of the lateral amygdala. Strong correlations have been established between auditory fear conditioning and LTP at these synapses: (1) auditory conditioning can enhance the response of lateral amygdala neurons to the shock-associated tone; (2) LTP can be induced by pairing presynaptic stimulation of thalamic axons with postsynaptic depolarization of lateral amygdala neurons; (3) fear conditioning and amygdalar LTP share molecular mechanisms with spatial learning and hippocampal LTP, including dependence on the postsynaptic NMDA receptor, CaMKII auto-phosphorylation, and AMPA receptor trafficking; and (4) when using optogenetic activation of input to lateral amygdala neurons as a substitute for a CS, after training, LTP or LTD of the synapses corresponding to the optogenetic input can activate or inactivate, respectively, the fear memory.

Recent *in vivo* Ca^{2+} imaging of the same basolateral amygdala neurons shed light on how learning is represented at the level of neuronal ensembles. Individual neurons that responded to the electric shock (US) and two different tones (CS^+, which was paired with shock during fear conditioning; and CS^-, which was not paired with shock) were identified and their activity monitored during auditory fear learning. Fear conditioning was found to preferentially affect CS^+ neurons (compared to CS^- neurons): similar numbers of neurons *potentiated* and *depressed* their CS⁺ responses during fear learning. While the potentiated responses are in line with the Hebbian model, the depressed responses suggest that the Hebbian model alone is insufficient to account for all plasticity in fear learning. Using multidimensional population vectors (one dimension per cell; Figure 6-27A) to plot each response to the CS or US, it was found that the population vector distance between the CS⁺ and the US was reduced as a consequence of fear learning, while the population vector distance between the CS⁻ and the US remained unchanged (Figure 11-43). Thus, at a population level, fear conditioning is consistent with a supervised learning framework (see Section 14.33 for details) in which the US encoding acts as a supervisor to change the CS encoding, causing neuronal representation of the CS⁺ to resemble that of the US. This would facilitate the ability of the CS⁺ to elicit a response (freezing) by activating downstream targets normally activated by the US.

Research on fear conditioning in rodent models has revealed the amygdala to be a key brain region for emotional memory and processing emotion-related signals (see Box 11-4 for further discussion). This has been substantiated in human studies. As in fear conditioning in rodents, the amygdala of human subjects is activated by presentation of an image (e.g., a blue square) previously paired with a mild electric shock to the wrist but not by presentation of a comparable image (e.g., a yellow circle) not paired with shock. In this fear conditioning paradigm, patients with amygdala lesions do not exhibit normal physiological responses, such as sweating due to activation of the sympathetic system as part of the fear response. Interestingly, the amygdala-lesioned patients remain aware of the explicit association of the CS (blue square) and the US (mild electric shock), suggesting that amygdala-dependent fear conditioning utilizes a form of implicit memory distinct from the explicit memory that remains intact in these patients. Furthermore, behavioral evidence indicates that rodents and humans rely on common mechanisms for fear memory regulation. When previously learned fear memories are reactivated during retrieval, they become labile and require **reconsolidation** (Section 11.1) to return to a stable state. Like the initial consolidation period following memory acquisition, reconsolidation requires CREB-mediated gene expression. During the relatively brief (tens of minutes) reconsolidation window, memories are susceptible to disruption via protein synthesis inhibitors and, remarkably, new learning, in both rodents and humans. Thus, reconsolidation represents an opportunity to rewrite emotional memories.

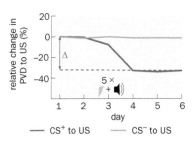

Figure 11-43 Fear learning causes population dynamics of the CS⁺ to resemble those of the US. Ca²⁺ activity of the same basolateral amygdala neurons expressing the genetically encoded Ca²⁺ indicator GCaMP6 was imaged using a head-mounted microendoscope (Figure 14-43C) before, during, and after fear learning. Mice were exposed to the conditioning chamber for 2 days with the CS⁺ and CS⁻ tones without electric shocks. On day 3, electric shocks were paired with the CS⁺ tone five times. On days 4–6, mice were exposed to the conditioning chamber with the CS⁺ and CS⁻ tones again. The y axis shows the relative change in the population vector distance (PVD; see Section 8.12 for the definition of a population vector) between the CS⁺ and US, and the CS⁻ and US. Δ is the difference in PVD values before and after training, representing recordings from 3655 cells. (From Grewe BF, Grimdemann J, Kitch LJ, et al. [2017] *Nature* 543:670–675. With permission from Springer Nature.)

Box 11-4: Neural substrates of emotion and motivation

By focusing on the amygdala in fear conditioning in Section 11.23 and midbrain dopamine neurons in reward-based learning in Section 11.24, we have had a glimpse into how information related to fear and reward is processed. However, amygdala and midbrain dopamine neurons do not process only information related to fear or reward, respectively; at the same time, fear- or reward-related signals are not processed exclusively by amygdala or midbrain dopamine neurons. In this box, we broaden our discussion of the neural circuits that process emotion and motivation in the rodent brain.

Some consider emotions such as happiness, fear, and disgust to be subjective, human-specific, and accessible only by self-reports. However, scientists since Charles Darwin have articulated observations regarding evolutionarily conserved emotional states in animals; for example, an external threat can produce similar physiological responses and even similar facial expressions in different animal species. One way to unify these views is to conceive of emotions as reflecting specific *brain states* in response to external or internal stimuli. These brain states in turn promote distinct adaptive behaviors. For example, electrical shock induces a fearful state that promotes freezing to reduce the chance of being captured by a predator. These brain states might be conserved from animals to humans even though only humans can self-report. This view enables studying the neurobiological mechanisms of emotion (e.g., fear) using animal models via behavioral (e.g., freezing) and physiological (e.g., increased blood pressure) readouts.

Psychology offers another way of thinking about how the brain instantiates emotional states. Contemporary theories suggest a conceptual framework wherein commonsense emotional categories such as "fear" and "happiness" are not considered stand-alone psychological or neural modules but are instead created from common, more basic parts. Chief among these is core affect, which is the simplest raw feeling

(Continued)

Box 11-4: continued

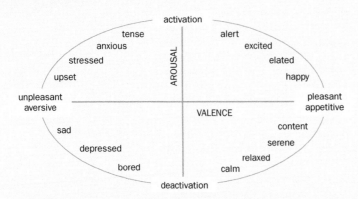

Figure 11-44 A two-dimensional representation of core affect. (Courtesy of Elizabeth Steinberg. See also Posner J, Russell JA, & Peterson BS [2005] *Dev Psychopathol* 17:715–734.)

that is consciously accessible in moods and emotions. Core affect can be described by the orthogonal dimensions of pleasure–displeasure (valence) and activation–deactivation (arousal) (**Figure 11-44**) and is realized by integrating external sensory information with interoceptive information to produce a unified mental state that is used to navigate the world. Many psychologists argue that the two-dimensional model of core affect explains variation in human emotion better than commonsense emotion categories. Critically, since the valence and intensity of stimuli can be systematically varied by an experimenter and the resulting neural, physiological, and behavioral responses can be objectively quantified, this conceptual framework can be productively applied to studying emotions in other species.

Emotion is also related to drive and motivation, concepts we introduced in the context of studying thirst and hunger (Section 9.10). Specifically, deviations from homeostasis create brain states that motivate animals to take actions to reduce such deviations. Indeed, thirst and hunger have been considered by some to be "homeostatic emotions." Reflecting the intricate links between emotion and motivation, studies of their underlying neural circuits have suggested that they both engage widely distributed and overlapping neuronal populations (**Figure 11-45**). The projection diagram only scratches the surface of the complex neural circuits involved in emotion and motivation. All depicted brain regions are heterogeneous and contain known and yet-to-be-identified cellular populations, which may differ in their input/output connection specificities, physiological response properties, and behavioral functions.

Let's use fear learning and the amygdala to illustrate the challenge of arriving at a satisfyingly mechanistic understanding of emotion and motivation. In Section 11.23, we learned that the amygdala is critical for fear conditioning and is strongly activated by fear-inducing stimuli in rodents and humans. However, it is inaccurate to describe the amygdala as being "for fear," as abundant evidence indicates that it also participates in positive emotional states and other cognitive functions. For example, fMRI studies show that the human amygdala is activated by positive images (e.g., pho-

tographs of a loving partner) and that the degree of activation is correlated with self-reported intensity of the emotional response. In rodents, electrophysiological studies indicate that individual amygdala neurons can encode valence (differential responses to positive and negative stimuli) and arousal/salience (similar responses to positive and negative stimuli). Lesioning the rodent amygdala, in addition to disrupting fear conditioning, also impairs reward-seeking behavior, and people with amygdala damage after a stroke have difficulty recognizing both positive and negative emotional facial expressions. Relevant to motivational drives and homeostasis, the central amygdala is also involved in appetite regulation, as we learned in Section 9.15.

Conversely, the amygdala is far from being the only brain region that participates in fear conditioning. Fear conditioning engages neurons in the prefrontal cortex, which can bidirectionally regulate conditioned fear behavior. Prefrontal cortical neurons are reciprocally connected with the basolateral amygdala and directly project to the periaqueductal gray and hypothalamus, two output regions of the central amygdala that mediate motor, autonomic, and neuroendocrine outputs (Figure 11-45). The paraventricular nucleus of the thalamus is strongly activated by physiological and psychological stressors and sends a projection to the central amygdala required for fear conditioning and a projection to the nucleus accumbens, which integrates information from the prefrontal cortex, basolateral amygdala, and midbrain dopamine neurons to regulate reward seeking and punishment avoidance. This incomplete, low-resolution list of brain regions that participate in fear conditioning highlights the complexity of understanding distributed and interconnected circuits that underlie even seemingly simple behaviors.

How is behavioral specificity achieved at any given moment in active, multifunctional brain regions? Highly structured local connectivity may be part of the answer. For example, the central amygdala contains several subnuclei, each containing heterogeneous populations of GABAergic neurons that often have antagonistic functions and mutually inhibit each other. These "microcircuits" are hypothesized to integrate diverse inputs to produce specific outputs of the desired magnitude.

In summary, neurons involved in emotion and motivation are distributed across the brain and are highly interconnected. Indeed, large-scale recording experiments revealed that simple behavioral tasks, such as lick for water in response to an odor cue in thirsty mice, engage dynamic activity changes across many of the brain regions depicted in Figure 11-45, and neurons carrying information about sensory

Box 11-4: continued

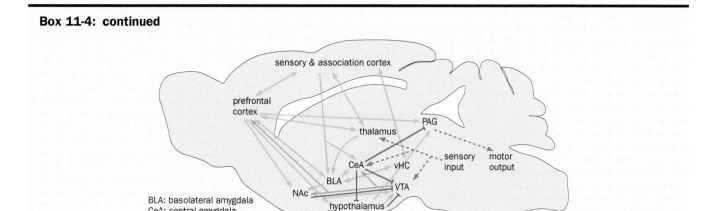

Figure 11-45 Simplified schematic of neural circuits involved in emotion and motivation. Brain regions and their projections involved in emotion and motivation depicted in a sagittal schematic of the mouse brain. Green, excitatory projections; red, inhibitory projections; blue, modulatory (dopamine) projections. Double arrows indicate bidirectional connections. See text for details. (Based on Tovote P, Fadok JP, Lüthi A [2015] *Nat Rev Neurosci* 16:317–331; Janak PH & Tye KM [2015] *Nature* 517:284–292; Adolphs R & Anderson DJ [2018] Neuroscience of Emotion, Princeton University Press.)

cues, motor actions, and brain states are widely distributed. A future challenge is to decipher how individual neurons acquire their physiological response properties—to what extent this is determined by their genetically defined cell types and connection patterns and to what extent by postnatal experience—and how neurons across many brain regions interact to collectively give rise to emergent, high-level phenomena such as emotion and motivation.

11.24 Dopamine plays a key role in reward-based learning

In Section 11.14 we introduced the law of effect in the context of operant conditioning: behaviors followed by rewards will be repeated, whereas behaviors followed by punishments will diminish. Since rewards can reinforce the actions that led to their receipt, this type of learning is also called **positive reinforcement**. What is the neural basis of positive reinforcement? Seminal experiments conducted in the 1950s offered important insights. Researchers implanted electrodes into different locations in the rat brain and placed the animal in an operant chamber equipped with a metal lever (**Figure 11-46**A). Whenever the rat pressed the lever, current passed through the electrode and excited neurons or axonal projections located near the electrode tip. In some experiments, the rat kept pressing the lever so as to receive more electrical stimulation, a behavior called intracranial self-stimulation. Intracranial self-stimulation was found to depend highly on the placement of the stimulating electrode: some brain regions evoked vigorous responses while others were ineffectual or even led to avoidance of the lever. In some intracranial self-stimulation experiments, rats kept pressing the lever at the expense of eating, drinking, and engaging in sex; they also withstood substantial foot shock to obtain more electrical stimulation (Figure 11-46B). Where are these presumed reward centers, the stimulation of which can override an animal's basic drives?

Systematic mapping revealed that the most effective self-stimulation sites coincide with midbrain dopamine neurons in the **ventral tegmental area** (**VTA**) and **substantia nigra pars compacta** (**SNc**) and their projections to the striatum (Figure 8-21), in particular to the ventral striatum, also called the **nucleus accumbens**. The role of dopamine neurons in self-stimulation was further supported by additional experiments. Lesions of dopamine neurons or their forebrain projections abolished self-stimulation, as did application of drugs that block dopamine synthesis. Addition of dopamine agonists into the nucleus accumbens, which bypassed blockade of dopamine synthesis, restored reward behavior. Finally,

Figure 11-46 Electrical and optogenetic self-stimulation. (A) Experimental design. When the rat presses the lever, the electrode implanted in its brain connects to a source of current so that the neurons and axon bundles near the electrode tip receive stimulation. If the activated neurons and axons signal reward, the rat will keep pressing the lever. **(B)** In this experiment, after the rat receives electric stimulation on one side of the arena, it must run across the center grid to the other side to receive more stimulation. After the rat had learned the task, electrical shock was applied in the central portion of the arena. If the electrode was implanted in certain reward centers, rats withstood more foot shock when crossing the grid to receive electrical stimulation than they did when crossing the grid for food after 24 h of food deprivation. **(C)** In this optogenetic self-stimulation experiment, a rat can poke its nose into two ports in a chamber. Nose poke at the active port results in photostimulation of the VTA, while nose poke at the inactive port has no consequences. Cre-dependent channelrhodopsin (ChR2) was virally delivered to either transgenic rats expressing Cre in dopamine neurons (Cre+) or control (Cre–) rats. Training results in robust nose poking in transgenic (but not control) rats at the active (but not inactive) port, suggesting that activation of VTA dopamine neurons reinforces nose-poking behavior. (A & B, adapted from Olds J [1958] *Science* 127:315–324. C, from Witten IB, Steinberg EE, Lee SY, et al. [2011] *Neuron* 72:721–733.)

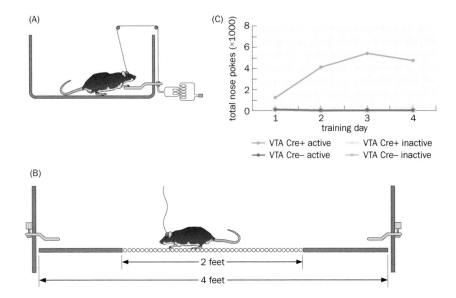

optogenetic stimulation of VTA dopamine neurons can drive positive reinforcement (Figure 11-46C). As we will learn in Chapter 12, most drugs of abuse act by enhancing the activity of midbrain dopamine neurons.

How does dopamine regulate reward and modify behavior? An important insight came from *in vivo* recording of dopamine neurons in alert monkeys performing behavioral tasks. Dopamine neurons normally fire in two different modes: in the **tonic** mode, dopamine neurons maintain a low, relatively constant basal firing rate; in the **phasic** mode, they fire in bursts in response to specific stimuli. In a specific example, a monkey was trained to associate the onset of light with a juice reward. Before and during the initial phase of training, dopamine neurons exhibited phasic firing in response to the juice reward. However, after extensive training, phasic firing was triggered by the light onset that predicted the reward, not the actual reward itself. Indeed, in trials in which the reward was omitted, the tonic firing rate was *depressed* at the time when reward was expected (**Figure 11-47**A). These data suggest that rather than signaling reward *per se*, phasic firing of dopamine neurons signals a **reward prediction error**: the difference between the actual reward and the predicted reward. Before training, reward came unexpectedly, resulting in a positive reward prediction error that triggered phasic firing. After training, reward was predicted by the sensory cue, such that the sensory cue became the unexpected reward signal; when reward was actually delivered, it was fully predicted, and hence there was no reward prediction error and little phasic firing; when reward was omitted, a negative prediction error resulted in depression of the tonic firing.

The close correspondence between dopamine reward prediction errors and the teaching signal required by computational models of reinforcement learning such as the temporal difference algorithm was quickly recognized, sparking a fruitful interplay between experimental and theoretical neuroscience (see Section 14.33 for more details). Importantly, the conclusions drawn from compelling correlations between dopamine neuron activity, prediction error signaling, and behavioral evidence of learning were supported by causal studies using optogenetic manipulations. Mimicking positive or negative prediction errors via optogenetic excitation or inhibition of VTA dopamine neurons was sufficient to drive associative learning, supporting the idea that brief changes in dopamine neuron firing can serve as bidirectional prediction error signals to change behavior when expectation and reality differ.

How are dopamine prediction error signals implemented at the level of synaptic weights? In one model of reward-based learning (Figure 11-47B), the connection between a signal neuron and a response neuron has an adjustable strength (ω). Through a negative feedback loop, the response magnitude (ωS) is compared

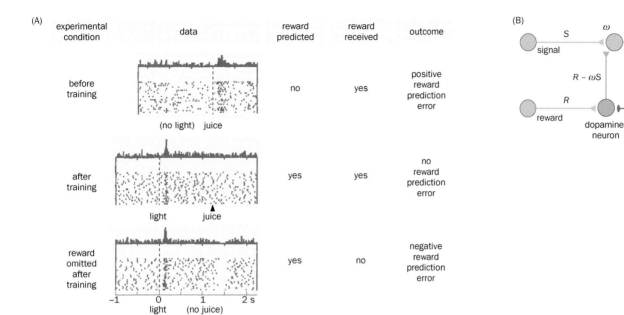

Figure 11-47 Dopamine neurons, reward prediction error, and reinforcement learning. (A) *In vivo* single-unit recordings of a midbrain dopamine neuron of a monkey trained to associate light onset with a reward (a drop of juice). Within each of the three blocks, each row is a separate trial and each vertical bar is a spike. Above the individual trials is a peri-event time histogram showing the firing rates from all trials. Top, before training, phasic firing was triggered by reward (juice) delivery (vertical dashed line). Middle, after training, phasic firing was triggered by light onset (vertical dashed line) but not reward delivery (arrowhead). Bottom, in trials when reward was omitted after training, tonic firing was depressed around the time reward was expected. Individual trials are aligned with reward delivery (top) or light onset (middle and bottom). Phasic firing of this dopamine neuron can be interpreted to signal the difference between the actual and expected reward summarized on the right. **(B)** A model for reward-based learning.

A signal neuron produces a signal with magnitude S and connects to a response neuron through a synapse whose strength ω can be adjusted, resulting in a response whose magnitude is the product of ω and S. In addition to sending information to downstream circuits (arrow), the response is also transmitted to an inhibitory feedback neuron (red), which in turn sends output to a dopamine neuron (blue). The dopamine neuron also receives an excitatory input that delivers a reward signal with magnitude R. Thus, the dopamine neuron positively adjusts ω with an output magnitude of R – ωS (for simplicity, we assume that synaptic transmission is faithful and integration is linear). (A, from Schultz W, Dayan P, & Montague PR [1997] *Science* 275:1593–1599. With permission from AAAS. B, adapted from Schultz W & Dickinson A [2000] *Ann Rev Neurosci* 23:473–500. See also Rescorla RA & Wagner AR [1972] Classic Conditioning II: Current Research and Theory. Appleton-Century-Crofts.)

to a reward signal (R). This difference, the reward prediction error carried by a dopamine neuron (blue in Figure 11-47B), is used to modify ω. Before training, ω is small, such that the reward prediction error ($R - \omega S$) is large. The dopamine neuron fires phasically, sending a large signal to increase ω. As learning proceeds, ω increases until $R - \omega S = 0$; at that point, dopamine neuron-mediated learning is accomplished, and the dopamine neuron no longer exhibits phasic firing in response to the reward.

How do dopamine reward prediction errors modify downstream circuits to modify behavior? As discussed in Section 8.8, the striatum is a major projection target of midbrain dopamine neurons, with VTA dopamine neurons projecting to the nucleus accumbens and SNc dopamine neurons projecting to the dorsal striatum. There, dopamine release regulates the strength of connections between cortical and thalamic excitatory inputs and spiny projection neurons (SPNs). In this case, the signal neuron in Figure 11-47B corresponds to cortical/thalamic projection neurons and the response neuron corresponds to SPNs. Since SPNs are GABAergic, they can signal directly to dopamine neurons (e.g., through D$_1$SPNs; Figure 8-21) to deliver ωS to those dopamine neurons. In support of this model, dopamine has been shown to regulate plasticity at cortical/thalamic → SPN synapses in acute slices. Importantly, the behavioral effects of dopamine neuron activity and cortico-/thalamo-striatal plasticity differ, depending on the anatomical location of the circuit in question. While dopaminergic input to the nucleus accumbens is associated with incentive value learning, dopamine-mediated synaptic plasticity in the dorsal striatum facilitates procedural learning and habit formation.

While reward prediction errors are a prominent feature of midbrain dopamine neuron activity, they are not the only signal present in this neuronal population. Recent studies have revealed substantial heterogeneity among midbrain dopamine neurons, with some signaling aversive stimuli and others salience of motivational stimuli; dopamine neurons in this latter group are activated by both strong appetitive and strong aversive signals and respond poorly to weak appetitive/aversive signals. Heterogeneity in dopamine neuron activity patterns and function is correlated with anatomical location within the VTA and SNc, projection targeting to striatal subregions, and sources of synaptic input (see also Box 9-1). For example, lateral SNc dopamine neurons that project to posterior lateral striatum signal threat, and their optogenetic activation led to avoidance in a choice assay. Such organization of different types of dopamine neurons into parallel, functionally distinct circuits resembles the mushroom body circuit that mediates olfactory learning in flies (Figure 11-30). Just as reward-based learning can increase the frequency of actions leading to reward, aversion-based learning can reduce the frequency of actions leading to punishment. Indeed, behavior leading to removal of a punishment can be reinforcing (see Figure 9-18 for an example); this is called **negative reinforcement**. Evidence suggests that both positive and negative reinforcements utilize dopamine signaling.

11.25 Earlier experience can leave behind long-lasting memory traces to facilitate later learning

We have seen that learning can occur and memory can be stored in neural circuits across the nervous system. In addition to synaptic plasticity, learning can also induce changes in axon myelination patterns that affect neuronal communications (**Box 11-5**). In the final section of this chapter, we further broaden the scope of learning and memory to developmental and structural plasticity by returning to the story of the barn owl introduced in Chapter 1, integrating what we have learned about the organization and wiring of the brain in the intervening chapters.

Recall that the owl's auditory map can adapt to match a visual map altered by wearing prisms and that this ability declines with age (Section 1.3). Recall further that if an owl had an earlier experience of auditory map adjustment, its auditory map readapted to an altered visual map more easily in adulthood (Figure 1-7). What is the neural basis for these phenomena? As we learned in Chapter 6, neurons in the nucleus laminaris of the owl's brainstem form a map that identifies sound locations on the horizontal plane based on interaural time differences (ITDs) (Figure 6-53). This ITD map projects topographically to the central nucleus of the inferior colliculus (ICC). ICC axons project further to the external nucleus of the inferior colliculus (ICX). ICX neurons then project to the optic tectum, where integration of auditory and visual information occurs topographically (**Figure 11-48**A, top). Tracing studies indicated that in juvenile prism-reared owls, ICC axonal projections to the ICX expand in the direction that matches the altered visual map in the optic tectum (Figure 11-48A, bottom). The expanded axons bear synaptic terminals and likely make functional connections with postsynaptic neurons in their new topographic location, thus realigning the auditory map with the prism-altered visual map. Although the mechanisms underlying this axonal expansion have not been examined in detail, it is likely that the connections made by the expanded axons are stabilized by synchronous firing with postsynaptic neurons that process altered visual information, as in the Hebb's rule-based synaptic strengthening in visual system wiring discussed in Chapter 5.

When the prisms were removed from the juvenile prism-reared owls, the auditory map was restored to normal so that it was realigned with the normal visual map (Section 1.3). Indeed, ICC neurons still maintain their normal axonal projections in the ICX during the prism-rearing period (Figure 11-48A bottom); these normal projections, which become topographically mismatched during the prism-wearing period, receive preferential GABAergic inhibition such that they are preferentially silenced. The persistence of these normal connections during prism rearing may account for the rapid restoration of the normal auditory map after the prisms are removed. The ICC axons that expanded into the topographi-

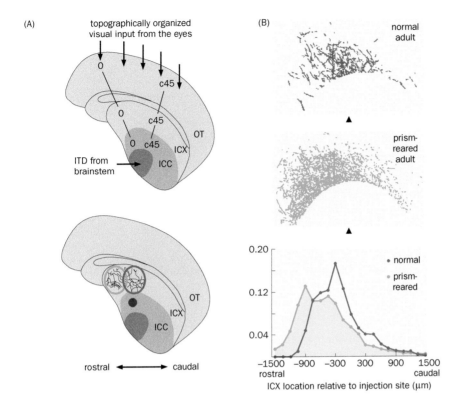

Figure 11-48 Adaptive axonal expansion in the inferior colliculus and auditory map adjustment in juvenile and adult owls. **(A)** Top, representation of auditory and visual information in the owl brain. Brainstem inputs to the central (ICC) and external (ICX) nuclei of the inferior colliculus are topographically organized according to interaural time differences (ITDs): 0 represents ITD = 0, c45 represents the contralateral side leading by 45 μs. ICX neurons project to the optic tectum (OT), where they align with topographically organized visual input. Bottom, axons of ICC neurons from the area indicated by the dark dot normally project to a topographically appropriate area of the ICX (red circle) but expand rostrally (blue circle) to align to a visual map altered by prism rearing in the juvenile owl. **(B)** Top and middle, anterograde tracing was used to examine ICC → ICX axonal projections in normal (top) and prism-reared (middle) adult owls. Arrowheads indicate the anterograde tracer injection site in the ICC. Bottom, normalized distribution of axonal projections in normal and prism-reared owls. The significant rostral shift in adult owls due to juvenile experience wearing prisms (prism-reared) likely accounts for the rapid auditory map adjustment during a second prism experience in adulthood (Figure 1-7B). (From Linkenhoker BA, von der Ohe CG, & Knudsen EI [2005] *Nat Neurosci* 8:93–98. With permission from Springer Nature. See also DeBello WM, Feldman DE, & Knudsen EI [2001] *J Neurosci* 21: 3161–3174.)

cally abnormal area of the ICX due to juvenile prism rearing are also maintained into adulthood (Figure 11-48B), well after prism removal and complete restoration of the normal auditory map as assayed by behavior. (It is unknown how the activity of these expanded axons is silenced after prism removal such that they do not interfere with proper sensorimotor behavior.) Thus, adaptive expansion of ICC axons as a consequence of juvenile prism rearing leaves behind *an anatomical trace* that likely facilitates readjustment of the auditory map in response to a similar visual displacement event later in adulthood.

A conceptually similar experiment examined structural changes in mammalian visual cortical neurons in response to monocular deprivation. As we learned in Chapter 5, monocular deprivation within a critical developmental period has a profound effect on wiring of the visual cortex. In mice, transient monocular deprivation for a few days during the critical period can modify the binocular area of the visual cortex, significantly shifting the relative representation of visual input from the two eyes in favor of the open eye. If the deprivation period is short, the normal balance of representation of the two eyes is restored after binocular vision is restored. Monocular deprivation in adult mice can also shift ocular dominance in response to a longer duration of monocular deprivation. Interestingly, ocular dominance shifts in response to monocular deprivation are more rapid in adult mice that previously experienced monocular deprivation than in those experiencing monocular deprivation for the first time, analogous to the finding in the owl.

To examine a structural basis for this monocular-deprivation-induced plasticity, repeated two-photon microscopic imaging was carried out through a window implanted in the binocular area of the mouse visual cortex (**Figure 11-49**A). Dendritic spines of pyramidal neurons were observed and quantified to determine spine gains and losses over time (Figure 11-49B). The first monocular deprivation resulted in substantial addition of new spines, a proxy for new synapse formation (Section 11.13); these spines were subsequently retained. The second deprivation, which caused a more rapid ocular dominance shift, did not change the spine density (Figure 11-49C). A simple interpretation of these data is that the anatomical traces left behind by the first monocular deprivation—the new spines—were reused for the ocular dominance shift during the second monocular deprivation, thus facilitating the neuron's more rapid adaptation. As with the owl studies, it

Figure 11-49 Spine dynamics in the adult mouse visual cortex in response to monocular deprivation. (A) Experimental protocol. After window implantation, intrinsic signal imaging (Figure 4-44B) was carried out to identify the binocular region. Repeated two-photon imaging of dendrites from a transgenic mouse expressing GFP in a sparse neuronal population in the binocular region was then carried out every 4 days, covering two 8-day rounds of monocular deprivation (MD1 and MD2). **(B)** Representative images of the same apical dendritic segment of a layer 5 pyramidal neuron. Blue and red arrowheads indicate spine loss and gain, respectively; these changes were inferred by comparing each image to the image acquired 4 days earlier. **(C)** A significant increase in spine gain was detected only during the MD1 (top) but not the MD2 (bottom), suggesting that new spines gained during MD1 are used to adjust ocular dominance during MD2. (From Hofer SB. Mrsic-Flogel TD, Bonhoeffer T, et al. [2009] *Nature* 457: 313–317. With permission from Springer Nature.)

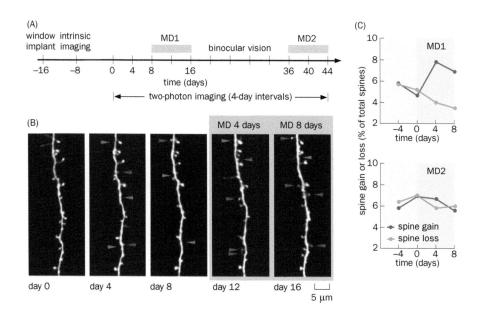

remains unknown whether and how the activity of the new spines is silenced during the period between the first and second rounds of monocular deprivation, such that the synaptic connections enabled by the new spines do not interfere with binocular vision during the intervening period.

In summary, these experiments suggest that structural changes in neural circuits in response to specific experiences—whether axonal arborization in the inferior colliculus or dendritic spines in the visual cortex—can provide long-lasting memory traces to facilitate future learning. These structural changes may underlie a widely occurring phenomenon called **savings**: less effort is required for an animal to relearn something it has previously learned. Altogether, modern research discussed in this chapter has provided rich neurobiological bases for Descartes's needle-through-the-cloth analogy of memory (Figure 11-1).

Box 11-5: White matter plasticity

In this chapter, we have focused mostly on synaptic plasticity as the cellular correlate of learning and memory. Experience and neural activity also cause other changes in the nervous system that affect neuronal communication. As we learned in Chapter 2, myelination speeds up action potential propagation. Thus, altering the degree of myelination can affect the speed and timing of information transmission in neural circuits. It had been assumed that the degree of myelination is an innate property of the nervous system, but recent studies indicated that ~20% of oligodendrocytes, which are responsible for myelinating axons in the brain and spinal cord (Section 1.4), are produced in the adult from oligodendrocyte precursor cells (OPCs). Studies in both human brain imaging and experimental model organisms suggest that experiences such as motor skill learning and social interaction can enhance myelination. We use two specific examples to illustrate.

In the first example, researchers found that brief optogenetic stimulation of the secondary motor cortex of transgenic mice expressing channelrhodopsin (ChR2) in layer 5 projection neurons caused a rapid increase in OPC proliferation within 3 hours (**Figure 11-50**A, B). Four weeks after an extended optogenetic stimulation paradigm, many of these OPCs developed into mature oligodendrocytes (Figure 11-50C), and myelin thickness within the deep layers of motor cortex and subcortical white matter increased. Furthermore, mice that had undergone optogenetic stimulation exhibited enhanced motor behavior performance. Thus, enhanced neuronal activity could lead to increased axonal myelination. In a complementary example, researchers found that training mice running on a complex wheel for three weeks resulted in a substantial increase of newly born oligodendrocytes in the corpus callosum, the white matter tract connecting the two cerebral hemispheres (Figure 11-50D). Conditional deletion in adult OPCs of a transcription factor, Mrf (myelin regulatory factor), required for new oligodendrocyte production impaired motor skill learning (Figure 11-50E), indicating that new oligodendrocyte production contributes to improved motor performance. Adult deletion of Mrf in OPCs has since been shown to impair water maze and contextual fear memories as well.

New oligodendrocyte generation is just one form of white matter plasticity. Other forms include changes in myelin

Box 11-5: continued

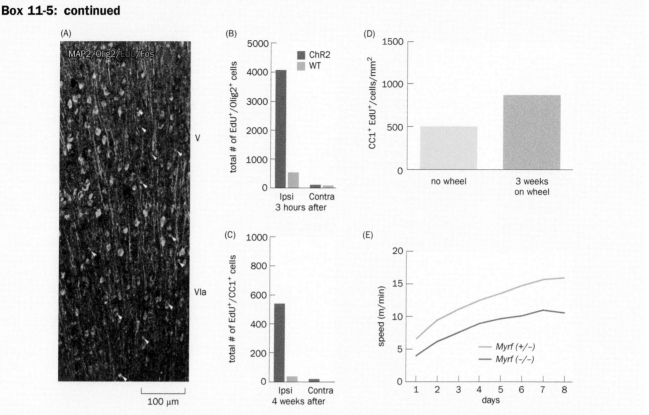

Figure 11-50 Oligodendrocyte production in response to activity and experience. (A) Optogenetic stimulation of secondary motor cortex in transgenic mice expressing ChR2 in layer 5 projection neurons results in rapid division of Olig2+ oligodendrocyte precursor cells (OPCs), as indicated by the incorporation of fluorescently labeled EdU (a nucleotide analog), a marker for newly born cells. Arrowheads indicate cells double positive for Olig2 and EdU. MAP2 labels apical dendrites, and Fos labels nuclei of neurons activated by optogenetic stimulation. V and VIa, layer 5 and 6a. **(B)** Quantification of total number of dividing Olig2+ cells in transgenic and wild-type (WT) control mice 3 h after photostimulation. Ipsi, experimental side; contra, contralateral secondary motor cortex. **(C)** Quantification of total number of mature oligodendrocytes (expressing the marker CC1) 4 weeks after daily optogenetic stimulation for 7 days. **(D)** Quantification of newly born oligodendrocytes during the 3 week period when mice were trained on a complex wheel, compared to control mice without wheel training. **(E)** Compared with heterozygous control mice, mice in which the gene *Mrf*, required for oligodendrocyte production, was conditionally deleted in adult OPCs have reduced running speed when trained on the complex wheel, indicating reduced motor skill learning. (A–C, from Gibson EM, Purger D, Mount CW, et al. [2014] *Science* 344:1252304.; With permission from AAAS. D & E, adapted from McKenzie IA, Phayon D, Li H, et al. [2014] *Science* 346:318–322.)

thickness, in the length between nodes of Ranvier, and in the properties of nodes of Ranvier. All these changes could be substrates for experience-dependent modification of the speed and timing of neuronal communication. With increas-ing knowledge about regulation of axon myelination and tools for manipulating neural circuits, researchers will surely discover more facets of white matter plasticity and their underlying mechanisms.

SUMMARY

In this chapter, we studied memory and learning at multiple levels: molecules, synapses, neurons, circuits, systems, animal behaviors, and theories. Diverse experimental models ranging from invertebrates to mammals have yielded data supporting two central theses: (1) memory is primarily stored as strengths of syn-aptic connections in neural circuits and (2) learning modifies synaptic weight matrices through a diverse set of plasticity mechanisms.

Studies of the *Aplysia* gill-withdrawal reflex suggested that depression and potentiation of the synaptic strength between the sensory and motor neurons mediate behavioral habituation and sensitization, respectively. Both short-term and long-term sensitization engage cAMP/PKA signaling, causing enhanced neurotransmitter release in presynaptic terminals and new gene expression that promote synapse growth, respectively. Genetic analysis of *Drosophila* olfactory conditioning also identified a central role for cAMP signaling in mushroom body Kenyon cells. Electric shock and food, as unconditioned stimuli, modulate the strengths of synaptic connections between ensembles of Kenyon cells representing conditioned stimuli (odors) and mushroom body output neurons through the release of dopamine. The compartmental organization of Kenyon cell axons, along with specifically associated dopamine neurons and mushroom body output neurons, provide substrates for different kinds of learning and memory through compartment-specific plasticity rules. Formation of new explicit memory in humans and spatial memory in rodents requires the hippocampus, which, along with the nearby entorhinal cortex, also plays a central role in spatial representation in mammals. Strong correlations have been established between spatial learning/memory and hippocampal synaptic plasticity.

Synapses onto hippocampal CA1 pyramidal neurons in rodents have been used as a model to investigate general mechanisms of synaptic plasticity. Long-term potentiation (LTP) of CA3 → CA1 synapses follows Hebb's rule: LTP is induced when presynaptic glutamate release coincides with postsynaptic depolarization. The NMDA receptor serves as a coincidence detector. Ca^{2+} entry through the NMDA receptor activates protein kinases. Auto-phosphorylation of the multisubunit CaMKII can convert a transient Ca^{2+} signal into persistent kinase activity. A key mechanism for LTP expression is an increase in AMPA receptor numbers at the postsynaptic membrane, which enhances response magnitude to presynaptic glutamate release. Long-term depression (LTD) preferentially activates phosphatases to counteract kinase activity. LTD, LTP, and spike-timing-dependent plasticity allow bidirectional adjustment of synaptic weights, while homeostatic synaptic plasticity provides negative feedback to ensure network stability. Long-term changes in the strengths of connections between pre- and postsynaptic neurons involve formation of new synapses as a result of long-lasting LTP.

Hippocampal synaptic plasticity mechanisms likely apply, with variations according to specific neuronal and circuit properties, to synapses throughout the nervous system where experience-dependent changes underlie different forms of learning and memory. For example, long-term storage of explicit memory engages specific neocortical areas that communicate with the hippocampus during memory acquisition. The amygdala processes emotion-related memory, such as auditory fear conditioning, a form of implicit memory. Many midbrain dopamine neurons signal reward prediction errors; they exhibit phasic firing when the actual reward exceeds the predicted reward. This property can reinforce the action that leads to dopamine release by adjusting the synaptic strengths of dopamine-modulated circuits, such as the synapses between cortical/thalamic input neurons and striatal spiny projection neurons. Reinforcement learning plays an important role in incentive value learning, procedural learning, and habit formation.

Learning has different forms, including habituation and sensitization, classical and operant conditioning, reward-based reinforcement learning, cognitive learning, and structural plasticity in developing and adult nervous systems. Most forms of learning involve changes in the synaptic weight matrices of relevant neural circuits. Additional forms of learning involve changes in neuronal excitability and axon myelination. These changes alter neural circuit function in information processing and ultimately modify behavior to enable animals to better adapt to changing environments.

OPEN QUESTIONS

- To what extent do learning and memory in adulthood and activity- and experience-dependent wiring during development share common mechanisms?

- What predicts whether a particular synaptic plasticity rule is used in a given neuron or synapse type?

- Are long-term memory and synaptic plasticity mediated by changing the expression of a set of genes common across different circuits and organisms? What are these genes?

- If learning involves synaptic plasticity in multiple brain regions, how are these changes coordinated to produce a learned state?

- How can we mimic learning by reconfiguring synaptic weights in specific neuronal circuits?

- Does the strategy of using compartment-specific learning rules in the *Drosophila* mushroom body also apply to the mammalian brain?

FURTHER READING

Reviews

Dan Y & Poo MM (2006). Spike timing-dependent plasticity: from synapse to perception. *Physiol Rev* 86:1033-1048.

Heisenberg M (2003). Mushroom body memoir: from maps to models. *Nat Rev Neurosci* 4:266-275.

Janak PH & Tye KM (2015). From circuits to behaviour in the amygdala. *Nature* 517:284-292.

Josselyn SA & Tonegawa S (2020). Memory engrams: recalling the past and imagining the future. *Science* 367:eaaw4325.

Martin SJ, Grimwood PD, & Morris RG (2000). Synaptic plasticity and memory: an evaluation of the hypothesis. *Annu Rev Neurosci* 23:649-711.

Milner B, Squire LR, & Kandel ER (1998). Cognitive neuroscience and the study of memory. *Neuron* 20:445-468.

Nicoll RA (2017). A brief history of long-term potentiation. *Neuron* 93:281-290.

Watabe-Uchida M & Uchida N (2018). Multiple dopamine systems: weal and woe of dopamine. *Cold Spring Harb Symp Quant Biol* 83:83-95.

Synaptic plasticity, spatial memory, and hippocampus

Bliss TV & Lømo T (1973). Long-lasting potentiation of synaptic transmission in the dentate area of the anaesthetized rabbit following stimulation of the perforant path. *J Physiol* 232:331-356.

Giese KP, Fedorov NB, Filipkowski RK, & Silva AJ (1998). Autophosphorylation at Thr286 of the alpha calcium-calmodulin kinase II in LTP and learning. *Science* 279:870-873.

Hafting T, Fyhn M, Molden S, Moser MB, & Moser EI (2005). Microstructure of a spatial map in the entorhinal cortex. *Nature* 436:801-806.

Hayashi-Takagi A, Yagishita S, Nakamura M, Shirai F, Wu YI, Loshbaugh AL, Kuhlman B, Hahn KM, & Kasai H (2015). Labelling and optical erasure of synaptic memory traces in the motor cortex. *Nature* 525:333-338.

Hengen KB, Torrado Pacheco A, McGregor JN, Van Hooser SD, & Turrigiano GG (2016). Neuronal firing rate homeostasis is inhibited by sleep and promoted by wake. *Cell* 165:180-191.

Isaac JT, Nicoll RA & Malenka RC (1995). Evidence for silent synapses: implications for the expression of LTP. *Neuron* 15:427-434.

Liao D, Hessler NA, & Malinow R (1995). Activation of postsynaptically silent synapses during pairing-induced LTP in CA1 region of hippocampal slice. *Nature* 375:400-404.

Liu X, Ramirez S, Pang PT, Puryear CB, Govindarajan A, Deisseroth K, & Tonegawa S (2012). Optogenetic stimulation of a hippocampal engram activates fear memory recall. *Nature* 484:381-385.

Marr D (1971). Simple memory: a theory for archicortex. *Philos Trans R Soc Lond B Biol Sci* 262:23-81.

Morris RG, Garrud P, Rawlins JN, & O'Keefe J (1982). Place navigation impaired in rats with hippocampal lesions. *Nature* 297:681-683.

Nabavi S, Fox R, Proulx CD, Lin JY, Tsien RY, & Malinow R (2014). Engineering a memory with LTD and LTP. *Nature* 511:348-352.

O'Keefe J (1976). Place units in the hippocampus of the freely moving rat. *Exp Neurol* 51:78-109.

Tsien JZ, Huerta PT. & Tonegawa S (1996). The essential role of hippocampal CA1 NMDA receptor-dependent synaptic plasticity in spatial memory. *Cell* 87:1327-1338.

Wilson MA & McNaughton BL (1993). Dynamics of the hippocampal ensemble code for space. *Science* 261:1055-1058.

Wilson RI & Nicoll RA (2001). Endogenous cannabinoids mediate retrograde signalling at hippocampal synapses. *Nature* 410:588-592.

Learning and memory in diverse invertebrate and vertebrate systems

Aso Y & Rubin GM (2016). Dopaminergic neurons write and update memories with cell-type-specific rules. *Elife* 5:e16135.

Bacskai BJ, Hochner B, Mahaut-Smith M, Adams SR, Kaang BK, Kandel ER, & Tsien RY (1993). Spatially resolved dynamics of cAMP and protein kinase A subunits in Aplysia sensory neurons. *Science* 260:222-226.

Dudai Y, Jan YN, Byers D, Quinn WG, & Benzer S (1976). dunce, a mutant of *Drosophila* deficient in learning. *Proc Natl Acad Sci U S A* 73:1684-1688.

Frankland PW, Bontempi B, Talton LE, Kaczmarek L, & Silva AJ (2004). The involvement of the anterior cingulate cortex in remote contextual fear memory. *Science* 304:881-883.

Grewe BF, Grundemann J, Kitch LJ, Lecoq JA, Parker JG, Marshall JD, Larkin MC, Jercog PE, Grenier F, Li JZ, et al. (2017). Neural ensemble dynamics underlying a long-term associative memory. *Nature* 543:670-675.

Hofer SB, Mrsic-Flogel TD, Bonhoeffer T, & Hubener M (2009). Experience leaves a lasting structural trace in cortical circuits. *Nature* 457:313-317.

Kupfermann I, Castellucci V, Pinsker H, & Kandel E (1970). Neuronal correlates of habituation and dishabituation of the gill-withdrawal reflex in *Aplysia*. *Science* 167:1743-1745.

Lin S, Owald D, Chandra V, Talbot C, Huetteroth W, & Waddell S (2014). Neural correlates of water reward in thirsty *Drosophila*. *Nat Neurosci* 17:1536-1542.

Linkenhoker BA, von der Ohe CG, & Knudsen EI (2005). Anatomical traces of juvenile learning in the auditory system of adult barn owls. *Nat Neurosci* 8:93-98.

McGuire SE, Le PT, Osborn AJ, Matsumoto K, & Davis RL (2003). Spatiotemporal rescue of memory dysfunction in *Drosophila*. *Science* 302:1765-1768.

Menegas W, Akiti K, Amo R, Uchida N, & Watabe-Uchida M (2018). Dopamine neurons projecting to the posterior striatum reinforce avoidance of threatening stimuli. *Nat Neurosci* 21:1421-1430.

Olds J (1958). Self-stimulation of the brain; its use to study local effects of hunger, sex, and drugs. *Science* 127:315–324.

Phillips RG & LeDoux JE (1992). Differential contribution of amygdala and hippocampus to cued and contextual fear conditioning. *Behav Neurosci* 106:274–285.

Schiller D, Monfils MH, Raio CM, Johnson DC, Ledoux JE, & Phelps EA (2010). Preventing the return of fear in humans using reconsolidation update mechanisms. *Nature* 463:49–53.

Schultz W, Dayan P, & Montague PR (1997). A neural substrate of prediction and reward. *Science* 275:1593–1599.

Steadman PE, Xia F, Ahmed M, Mocle AJ, Penning ARA, Geraghty AC, Steenland HW, Monje M, Josselyn SA, & Frankland PW (2020). Disruption of oligodendrogenesis impairs memory consolidation in adult mice. *Neuron* 105:150–164.

Steinberg EE, Keiflin R, Boivin JR, Witten IB, Deisseroth K, & Janak PH (2013). A causal link between prediction errors, dopamine neurons and learning. *Nat Neurosci* 16:966–973.

Tully T & Quinn WG (1985). Classical conditioning and retention in normal and mutant *Drosophila melanogaster*. *J Comp Physiol A* 157:263–277.

CHAPTER 12
Brain Disorders

In this chapter, we examine how nervous system dysfunction causes brain disorders. Brain disorders are a greater cause of disability than any other class of diseases in modern societies. An important and obvious goal in studying brain disorders is to identify therapeutic strategies that will decrease disability and alleviate human suffering. In addition, research that focuses on specific diseases offers unique perspectives on normal brain development and function in the same way that studying genetic mutants can reveal the normal functions of genes and the biological processes they control. Conversely, important progress toward understanding brain disorders has come from basic research seemingly unrelated to disease, as seen by numerous examples discussed in previous chapters, from myelination diseases (Box 2-3) to brain–machine interfaces (Section 8.15). Thus, basic and disease-focused investigations mutually enhance each other to help us understand the function and dysfunction of the nervous system. While focusing on specific brain disorders, this chapter also seeks to integrate and extend the knowledge and principles presented in all previous chapters.

Rather than comprehensively addressing the vast array of brain disorders, we will focus primarily on selected disorders, the principles of which can be applied broadly to those not discussed here. Some disorders are selected because they have a large impact on human society, others because their pathogenic mechanisms are better understood. We categorize these disorders as neurodegenerative, psychiatric, or neurodevelopmental. Although generally useful, these categories also reflect our lack of understanding of the underlying disease mechanisms. For instance, as we will see, some of the classic psychiatric disorders were thought to arise in adulthood, but in fact have a developmental origin. We start with Alzheimer's disease, the most common neurodegenerative disorder.

> *And men ought to know that from nothing else but the brain come joys, delights, laughter and sports, and sorrows, griefs, despondency, and lamentations. And by this, in an especial manner, we acquire wisdom and knowledge, and see and hear, and know what are foul and what are fair, what are bad and what are good, what are sweet, and what unsavory. . . . And by the same organ we become mad and delirious, and fears and terrors assail us. . . . All these things we endure from the brain, when it is not healthy.*
>
> Hippocrates (~400 B.C.E.)

NEURODEGENERATIVE DISORDERS

Alzheimer's disease (AD) is well known because of its prevalence: in the United States, it affects 1 in 10 people by age 65 and about half of the population above age 85. Thus, AD is a disease of aging. With the substantial global increase in lifespan over the past century, AD has become one of the leading causes of death in the aged population and an escalating burden on society. As in all **neurodegenerative disorders**, an increasing number of neurons become dysfunctional as AD progresses: synaptic connections are lost, dendrites and axons wither, and neurons eventually die. As a result, the brain undergoes significant atrophy (**Figure 12-1**). Memory loss, a symptom characteristic of AD, is followed by loss of other cognitive and intellectual capabilities, such as reasoning and language. Patients also exhibit personality changes and behavioral problems as the disease progresses. Gradually, patients lose their ability to cope with the activities of daily living, including eating, grooming, and sleeping, and often require around-the-clock care for years before they succumb to death.

Although AD may be one of the best-understood brain disorders in terms of its neuropathological underpinnings, thanks to biochemical and molecular-genetic studies since the 1980s, many key questions remain unanswered. It has been particularly difficult to translate laboratory findings into effective treatments

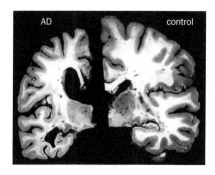

Figure 12-1 Brain atrophy in Alzheimer's disease (AD). Postmortem brain sections from an AD patient (left) and an age-matched, cognitively normal subject (right) reveal severe brain atrophy in the AD patient. (Courtesy of Nigel Cairns, Washington University, Department of Neurology.)

for AD. Nevertheless, AD research also offers valuable lessons that can be applied when studying other brain disorders.

12.1 Alzheimer's disease is defined by brain deposition of numerous amyloid plaques and neurofibrillary tangles

In 1906, a German psychiatrist and neuropathologist, Alois Alzheimer, reported a case of a patient who suffered from numerous psychiatric symptoms and severe memory loss, and who passed away 4.5 years after being admitted to an insane asylum. Using a newly invented silver-staining procedure to visualize postmortem brain specimens, Alzheimer described two major pathological features in the patient's cerebral cortex: numerous abnormal intracellular fibrils (now called **neurofibrillary tangles**) in a quarter to a third of all neurons, and extracellular plaques (now called **amyloid plaques**) distributed throughout the cerebral cortex. Although current clinical diagnosis is based on the patient's history and symptomatology, the presence in brain sections taken at autopsy of abundant neurofibrillary tangles and amyloid plaques remains the standard for a definitive diagnosis of the disease bearing Alzheimer's name (**Figure 12-2**A).

What is the molecular nature of neurofibrillary tangles and amyloid plaques? Could these pathological features offer clues about this devastating disease? With these questions in mind, researchers biochemically characterized these structures and identified their molecular nature in the mid-1980s. The neurofibrillary tangles consist of abnormal aggregates of hyperphosphorylated microtubule-binding protein **tau** (Figure 12-2B). Amyloid plaques consist mostly of a 39- to 43-amino-acid peptide called **amyloid β protein (Aβ)** for its strong tendency to form aggregates of β-pleated sheets (Figure 12-2C). While neurofibrillary tangles have also been found in several other neurodegenerative diseases, collectively called **tauopathies**, amyloid plaques are most characteristic of AD. Research focused on Aβ and tau has provided important insight in our investigation of AD. We begin our journey with Aβ.

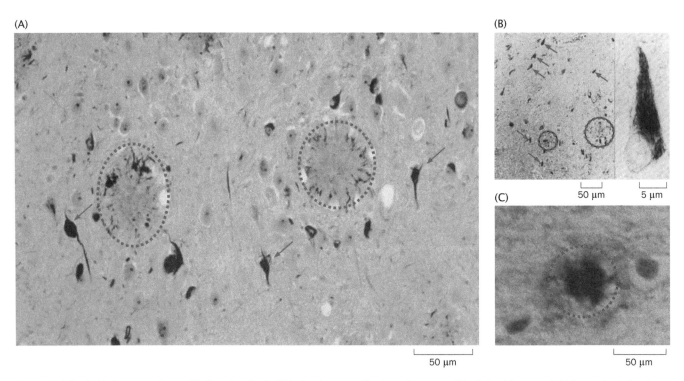

(A)

(B)

(C)

50 μm   5 μm

50 μm

50 μm

Figure 12-2 Amyloid plaques and neurofibrillary tangles in Alzheimer's disease. (A) Silver staining of a postmortem cortical section of an AD patient. Amyloid plaques (circles) and neurofibrillary tangles (arrows) are prevalent. **(B)** An antibody against the microtubule-associated protein tau strongly stains neurofibrillary tangles (arrows; a magnified view on the right) but not the cores of amyloid plaques (circles). **(C)** An antibody against a peptide derived from amyloid β protein stains the core of an amyloid plaque intensely (circle). (A, from Selkoe DJ [1999] *Nature* 399:A23–A31. With permission from Springer Nature. B, from Grundke-Iqbal I, Iqbal K, Quinlan M, et al. [1986] *J Biol Chem* 261:6084–6089. With permission from ASBMB. C, from Wong CW, Quaranta V, & Glenner GG [1985] *Proc Natl Acad Sci U S A* 82:8729–8732.)

12.2 Amyloid plaques consist mainly of aggregates of proteolytic fragments of the amyloid precursor protein (APP)

The peptide sequence of Aβ was used to isolate its gene from cDNA libraries, leading to the discovery that Aβ is part of a transmembrane protein called **amyloid precursor protein**, or **APP**. The predicted protein sequence indicated that APP has a large extracellular domain, a single transmembrane domain, and a small cytoplasmic domain (Figure 12-3). APP homologs have been found in the fly and worm, indicating that they are evolutionarily conserved, although their exact cellular function is still a subject of investigation. The sequence of the Aβ peptide itself is not conserved in the fly or worm, or in two other APP paralogs in humans that share similar overall structure and sequence. These data suggest that the normal function of APP does not require Aβ; rather, Aβ's amino acid sequence renders it particularly prone to aggregation after it is excised from APP.

The location of Aβ within APP is peculiar: two-thirds of the peptide is at the C-terminal end of the extracellular domain and one-third is part of the predicted transmembrane domain (Figure 12-3). This implies that to produce Aβ, APP must be cleaved by two different proteases, one of which must cut APP in the middle of the transmembrane domain. The existence of proteases capable of cleaving within the membrane was unknown when APP was first discovered. Indeed, the study of APP processing has enriched our understanding of the cell biology of regulated proteolysis.

The first protease identified as being able to process APP, named **α-secretase**, cuts APP in the middle of the Aβ peptide and therefore prevents the production of the pathology-associated Aβ (Figure 12-3, left). This proteolysis is likely related to the physiological function of APP, as the fly homolog of APP also produces a secreted form of APP by proteolysis near the end of the extracellular domain. Subsequently, the proteases that cut APP at the N- and C-termini of Aβ, producing the intact Aβ peptide, were identified and named **β-secretase** and **γ-secretase**, respectively (Figure 12-3, right). The cleavage site of γ-secretase is not fixed: it can produce Aβ of different lengths, ranging from 39 to 43 amino acids. The predominant forms have 40 or 42 amino acids and are called $A\beta_{40}$ and $A\beta_{42}$, respectively. γ-Secretase was subsequently found to be a general intramembrane protease complex important for many signaling events. Studies of APP and other γ-secretase substrates indicate that a major trigger for γ-secretase activation is extracellular cleavage of the substrate, yielding transmembrane proteins with short extracellular stubs (such as those produced by α- or β-secretase). APP is normally expressed in many cell types and is proteolytically processed by α- or β-secretases, followed by γ-secretase. So what goes wrong in Alzheimer's disease?

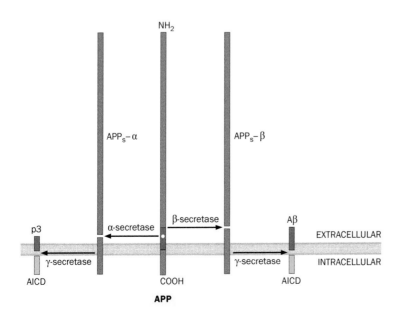

APP

Figure 12-3 Aβ is a proteolytic processing product of the amyloid precursor protein (APP). APP is synthesized as a transmembrane protein (middle), with an N-terminal large extracellular domain, a single transmembrane domain, and a short C-terminal cytoplasmic domain. Aβ (red) spans the junction of the extracellular and transmembrane domains. APP is usually cleaved by α-secretase (left; dot indicates the α-secretase cleavage site) or β-secretase (right) to produce a secreted form (APP$_s$-α or APP$_s$-β, respectively). The remaining portion is further processed by γ-secretase to yield an intracellular fragment (AICD). Whereas the combined actions of α- and γ-secretases produce a protein of 3 kilodaltons (p3), the combined actions of β- and γ-secretases produce intact Aβ.

12.3 Mutations in human APP and γ-secretase cause early-onset familial Alzheimer's disease

At this point you may ask: are APP and Aβ related to the *cause* of AD, or is Aβ plaque formation the *consequence* of a disease whose causes lie elsewhere? Genetic studies of early-onset **familial Alzheimer's disease** (**FAD**) have addressed this important question. Most AD cases are "late onset" (age 65 or older) and **sporadic** (from the Latin word for scattered), as most patients do not have an identifiable family history with the disease; nevertheless, as with many sporadic illnesses, genetic risk factors play an important role in AD, as we discuss in Section 12.5. However, patients with early-onset FAD usually develop AD symptoms in their 40s or 50s, following a Mendelian autosomal dominant inheritance pattern: an AD patient is heterozygous for a disease allele and imparts the disease allele to 50% of his or her progeny (see Box 12-3 for more details on the genetics of human disease). Those inheriting the disease allele invariably develop AD if they live long enough (**Figure 12-4**A). In these cases, genetic mapping has helped pinpoint mutations in specific genes that *cause* AD.

The first FAD mutation was mapped onto the *App* gene itself (Figure 12-4A): a missense mutation (I → V amino acid change) in the middle of the transmembrane domain near the C-terminus of Aβ$_{42}$ (Figure 12-4B). Subsequently, about 20 FAD mutations have been mapped onto the *App* gene. Interestingly, most mutations cluster near the γ- or β-secretase cleavage sites, with some in the middle of the Aβ peptide (Figure 12-4B). Biochemical studies indicate that mutations near the γ-secretase site increase the ratio of Aβ$_{42}$ over Aβ$_{40}$. Although Aβ$_{40}$ is the dominant form produced by γ-secretase, Aβ$_{42}$ has a higher tendency to form aggregates. FAD mutations in the middle of Aβ may also increase aggregation. An FAD mutation near the β cleavage site (KM → NL amino acid changes; also called the Swedish mutation, or APP$_{SWE}$) leads to an increase in Aβ production (**Figure 12-5**A, B). Most of these mutations cause early-onset AD. Notably, an A → T amino acid change near the β-secretase cleavage site found in an Icelandic population (green arrow in Figure 12-2B), which reduces β-secretase cleavage and Aβ production, confers *protection* against late-onset AD and age-related cognitive decline. Together, these genetic data suggest that increased Aβ production and aggregation are causally linked to at least some forms of AD.

Another piece of evidence supporting a causal link between increased Aβ production and AD came from **Down syndrome**, which is caused by having an extra copy of chromosome 21. Down syndrome patients invariably have high levels of

(A)

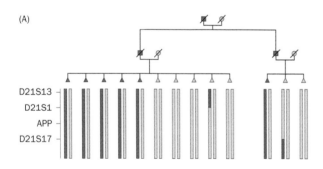

(B)

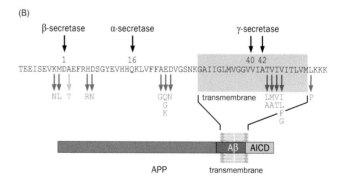

Figure 12-4 Mutations in the *App* gene cause familial Alzheimer's disease (FAD). (A) Pedigree of an early-onset FAD family (average onset 57 ± 5 years). Square, male; circle, female; triangle, either sex to preserve anonymity. Oblique lines indicate deceased individuals. Beneath the pedigree are maps of chromosome 21 in which segments of the chromosomes were mapped according to the markers on the left. The linkage data suggest that chromosome segments in red were inherited from the disease-causing chromosome of the affected fathers (red squares). Inheriting from the father the red chromosome segment that includes mutant *APP* correlates perfectly with having AD (red triangles). **(B)** Summary of FAD mutations in the APP protein, most of

which are located near Aβ cleavage sites or within the Aβ peptide. The cleavage sites for the three secretases are indicated. Numbers indicate amino acid residues starting from the beginning of the Aβ peptide. The green arrow points to an AD-protective A → T (alanine to threonine) mutation that reduces Aβ production. (A, from Goate A, Chartier-Harlin MC, Mullan M, et al. [1991] *Nature* 349:704–706. With permission from Springer Nature. B, adapted from Holtzman DM, John CM, & Goate A [2011] *Sci Transl Med* 3:77sr1. See Jonsson T, Atwal JK, Steinberg S, et al. [2012] *Nature* 488:96–99 for the AD-protective mutation.)

(A)

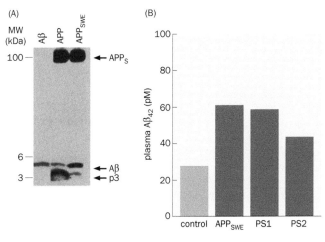

(B)

(C)

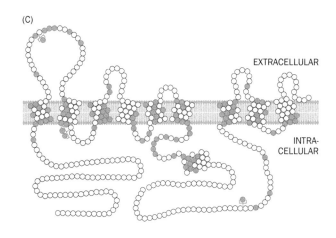

Figure 12-5 Mutations in APP and presenilins increase Aβ production.
(A) In these experiments, culture media containing radioactively labeled proteins from transfected cells was immunoprecipitated with an antibody against Aβ; immunoprecipitated protein was run on a gel. Left lane, synthetic Aβ as a control. Middle lane, transfecting cDNA expressing wild-type APP into cultured cells produced a major 3 kilodalton (kDa) band (p3) corresponding to the cleavage product of α- and γ-secretases, and a minor 4 kDa band corresponding to Aβ. Right lane, cells transfected with cDNA expressing APP$_{SWE}$ produced more Aβ than p3. MW, molecular weight in kilodaltons. **(B)** Compared with controls, the plasma Aβ$_{42}$ concentration (in picomoles per liter) of AD patients with the APP$_{SWE}$ mutation or pathogenic mutations in presenilin-1 (PS1) or presenilin-2 (PS2) was increased. **(C)** Structure of presenilin-1 and locations of FAD mutations. PS1 spans the membrane nine times; PS2, not shown here, has a similar structure. Mutations in PS1 resulting in FAD were first reported by Sherrington et al. in 1995 (*Nature* 375:754); by 2010, more than 170 mutations in PS1 causing FAD had been identified (red amino acid residues; red brackets denote insertions). (A, from Citron M, Oltersdorf T, Haass C, et al. [1992] *Nature* 360:672–674. With permission from Springer Nature. B, from Scheuner D, Eckman C, Jensen M, et al. [1996] *Nat Med* 2:864–870. With permission from Springer Nature. C, adapted from De Strooper B & Annaert W [2010] *Ann Rev Cell Dev Biol* 26:235–260.)

amyloid deposits in their 30s and 40s and exhibit Alzheimer's-like dementia in their 50s. The *App* gene is located on chromosome 21, so Down syndrome patients have an extra copy of *App*. Indeed, people with segment duplications of chromosome 21 that cover the *App* gene also have early-onset AD symptoms, suggesting that increasing the *App* gene dose is sufficient to cause AD.

Genetic mapping studies have also identified two other loci on human chromosomes 1 and 14 that harbor autosomal dominant FAD mutations. The causal genes encode similar transmembrane proteins named **presenilin-1** (Figure 12-5C) and **presenilin-2**. Subsequent biochemical and genetic studies showed that the presenilins, together with three associated proteins, constitute the γ-secretase protein complex responsible for cleaving APP near the C-terminus of Aβ within the transmembrane domain. AD patients with presenilin mutations have increased Aβ$_{42}$ levels compared to control subjects (Figure 12-5B). Thus, mutations in APP and its processing enzymes point to increases in Aβ production or aggregation as a common cause underlying AD pathogenesis, at least in early-onset FAD cases. This **amyloid β (Aβ) hypothesis** does not exclude the possibility that other mechanisms contribute to AD; for example, disruption of presenilin function could interfere with cellular processes unrelated to Aβ production that also contribute to AD pathogenesis.

How might excessive Aβ production and aggregation cause AD? While studies suggest that amyloid deposits are toxic to neurons, the severity of AD symptoms does not always correlate with the density of amyloid plaques. This indicates that amyloid plaques are not the only pathological factor. Indeed, as we will see in the next section, amyloid plaques and tau-enriched neurofibrillary tangles synergize. Furthermore, diffuse oligomeric forms of Aβ have been found to be potently toxic to neurons. For example, when applied to cultured neurons or brain slices, oligomeric Aβ has been shown to induce synaptic depression, spine loss, abnormal synaptic plasticity, and neuronal death, suggesting that nonaggregated Aβ oligomers may be potent AD-promoting agents. Future studies must further elucidate the relative contributions made by Aβ aggregates and different forms of oligomers to disease phenotypes *in vivo*.

Figure 12-6 Transgenic mice overexpressing human APP with pathogenic mutations exhibit amyloid plaques and cognitive deficits. (A) An amyloid plaque stained with an antibody against Aβ found in a 1-year-old transgenic mouse overexpressing APP with the Swedish mutation (APP$_{SWE}$). **(B)** As they age, mice transgenic for APP$_{SWE}$ (Tg⁺) exhibit learning defects in the Morris water maze (Figure 11-33). Mice were trained to locate a hidden platform using spatial cues. When the hidden platform was removed, control mice (Tg⁻) aged 9–10 months spent more time near the platform's previous location (indicating a spatial memory) than did APP$_{SWE}$ transgenic mice of the same age. (Adapted from Hsiao K, Chapman P, Nilsen S, et al. [1996] *Science* 274:99–103.)

(A)

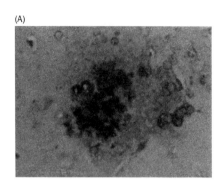

(B)

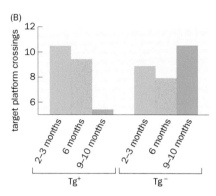

tauP301L

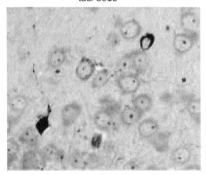

tauP301L + APP$_{SWE}$

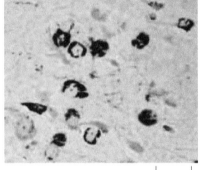

10 μm

Figure 12-7 Enhanced neurofibrillary tangle pathology in transgenic mice expressing mutant APP and tau. In humans, a proline → leucine mutation in tau (tauP301L) causes frontotemporal dementia with parkinsonism. Neurofibrillary tangles, as seen by intense silver staining (dark blue), are markedly increased in mice doubly transgenic for the Swedish allele of APP (APP$_{SWE}$) and tauP301L (bottom) when compared to transgenic mice expressing tauP301L alone (top). (From Lewis J, Dickson DW, Lin WL, et al. [2001] *Science* 293:1487–1491. With permission from AAAS.)

12.4 Animal models are crucial for investigating pathogenic mechanisms

Animal models are instrumental in human disease research. Appropriate animal models can validate causality between suspected pathogenic processes and disease outcomes. They can be used to trace disease progression, investigate disease mechanisms, test the effects of therapeutic agents, and determine the dose of a medicine required to engage a therapeutic target. The technical feasibility of performing precise genetic manipulations in mice (Sections 14.7–14.9) has made this species the dominant mammalian model for AD and many other brain disorders. However, the many differences between the mouse and human brains mean that even very useful mouse models do not tend to produce identical pathologies or treatment responses as human cases.

Mice do not naturally exhibit AD pathologies like amyloid plaques and neurofibrillary tangles. This is likely because the mouse's lifespan of about 2 years is not long enough for abnormal protein aggregates to cause sufficient insult to the nervous system, a process that usually takes decades in humans. However, these protein pathologies can be accelerated by overexpressing wild-type or FAD-mutation alleles of the human *App* gene in transgenic mice (**Figure 12-6**A). Moreover, some transgenic mice also develop age-dependent cognitive decline, such as deficits in spatial memory in the Morris water maze (Figure 12-6B). Thus, overproduction of human APP with FAD mutations is sufficient to produce AD-like pathology and cognitive defects, consistent with the Aβ hypothesis.

Studies utilizing transgenic mice have also revealed the relationship between amyloid plaques and neurofibrillary tangles, the two major pathological features of AD. As noted in Section 12.1, the major component of neurofibrillary tangles is the microtubule-associated protein tau. Interestingly, several neurodegenerative diseases collectively referred to as tauopathies feature abnormal accumulation of tau. Although mutations in tau have not been found in AD patients, they have been found in patients with a tauopathy called frontotemporal dementia with parkinsonism (FTDP). FTDP has a symptom complex quite different from AD, but like AD exhibits neurofibrillary tangle pathology. Transgenic mice expressing human tau with a dominant FTDP mutation at high levels can recapitulate tau aggregation similar to the neurofibrillary tangles observed in human AD and FTDP patients, without amyloid plaques.

Interestingly, mice overexpressing tau and either mutant APP (double transgenic) or both mutant APP and mutant presenilin (triple transgenic) developed both plaques and tangles. Moreover, neurofibrillary tangles were much more prevalent and widespread in double- and triple-transgenic mice than in transgenic mice expressing mutant tau alone (**Figure 12-7**). Thus, while plaques and tangles can form by independent mechanisms, increased Aβ production facilitates neurofibrillary tangle formation. Conversely, removing one or both copies of the endogenous *Tau* gene alleviated behavioral and synaptic defects caused by APP overexpression. Taken together, these experiments suggest that some symptoms caused by overproduction of APP or Aβ may be mediated by dysregulation of tau.

Despite the utility of mouse AD models in creating plaques and tangles and in investigating their relationships, these mouse models require artificial overexpression of mutant genes, often several in the same mouse. And importantly, they do not recapitulate the prominent neuronal death and brain atrophy of human AD (Figure 12-1). Furthermore, mouse models have so far not been useful in predicting efficacy of therapeutic agents in humans. This may result from the limited lifespan of mice and/or physiological differences between rodent and primate brains. Thus, development of primate models will aid further exploration of AD pathogenesis and testing of therapeutic strategies.

12.5 An apolipoprotein E (ApoE) variant is a major risk factor for Alzheimer's disease

Mutations in genes encoding APP and presenilins, while very helpful in establishing a causal relationship between Aβ production and FAD, account for only 1–2% of AD cases. These cases are usually early onset and dominantly inherited. Almost all other AD cases are late onset and sporadic. Although environmental factors are not negligible, AD has a strong genetic component. Twin studies of large AD populations indicate a high degree of heritability (Section 1.1)—about 70%. What additional genes contribute to AD?

Thus far, genetic analyses have not revealed additional FAD genes with Mendelian inheritance patterns beyond those encoding APP and presenilins. This suggests that most late-onset AD cases are caused by combinations of multiple genetic factors or by genetic factors interacting with environmental factors. By far the most important and common genetic risk factor identified to date is an individual's allele composition for a gene called *Apoe*, which encodes **apolipoprotein-E (ApoE)**. A component of high-density lipoproteins in the brain, ApoE is involved in lipid transport and metabolism. The most common *Apoe* allele is *ε3*. *ε4*, a less common allele, differs from *ε3* by a single amino acid. In the early 1990s, ApoE was found to bind Aβ and to be present in amyloid plaques. Genetic studies examining the relationship between *Apoe* allele composition and AD revealed that the *ε4* allele frequency, which is only approximately 15% in the general population, was 40% in AD patients. Compared to the most common *ε3/ε3* allele combination, individuals with a single *ε4* allele have a more than 3-fold greater chance of developing AD, while individuals with two *ε4* alleles have a 12-fold greater chance of getting AD. Furthermore, among AD patients, as the copy number of *ε4* alleles increases, the age of disease onset decreases (**Figure 12-8**).

Thus, *Apoe* is a **genetic susceptibility locus** for AD. Unlike the FAD mutations in APP or presenilins, which invariably cause AD if the carrier lives long enough, ApoE *ε4* does not definitively cause AD but does increase the likelihood of developing AD. However, given the relatively high frequency of the *ε4* allele in the human population, ApoE *ε4* contributes far more to the incidence of AD than do mutations in APP and presenilins. Interestingly, a less common allele, *Apoe ε2*, appears to be protective against AD, as the *ε2* allele frequency in the general population is 8%, but drops to 4% in the AD population.

One proposed function of ApoE is the regulation of Aβ metabolism and clearance. Further work is required to clarify how the *ε4* and *ε3* alleles differentially affect Aβ clearance. ApoE could also affect Aβ-independent processes. A recent study suggested that ApoE regulates tauopathy in an allele-dependent manner: neuronal death in transgenic mice overexpressing tau harboring a pathogenic mutation was more severe in the presence of the *ε4* allele than in the presence of the *ε3* or *ε2* allele of human *Apoe*. As we will see in the next section, ApoE is also implicated in neuroinflammation. Regardless of the detailed mechanisms of action, the discovery of ApoE's association with AD has provided a paradigm for studying complex genetic disorders, in which mutations in individual genes increase susceptibility to a disease rather than cause the disease outright. We will see many examples of such susceptibility variants later in this chapter.

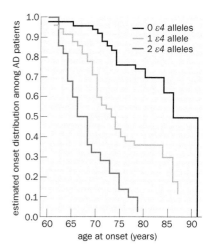

Figure 12-8 **The ε4 allele of the *Apoe* gene is a major risk factor for AD.** As the copy number of *Apoe ε4* increases, the age of AD onset decreases. For instance, about 50% of AD patients lacking the ε4 allele had been diagnosed with AD by age 86. The median age of AD onset drops to 73 years for patients with one ε4 allele and 66 years for those with two ε4 alleles. (From Corder EH, Saunders AM, Strittmatter WJ, et al. [1993] *Science* 261: 921–923. With permission from AAAS.)

12.6 Microglial dysfunction contributes to late-onset Alzheimer's disease

Recent genome-wide association studies and whole-genome sequence analyses (see Box 12-3 for details) have identified additional genes that increase the risk of developing late-onset AD. While most of these genetic variants have less impact on AD risk levels than does ApoE ε4, subjects with one copy of the R47H variant (arginine → histidine change at amino acid 47) in the gene encoding **TREM2** (triggering receptor expressed on myeloid cells 2) have an AD risk similar to those with one copy of *Apoe ε4*. However, far fewer patients are affected because the TREM2 variant frequency in the general population is far lower (<1%) than that of ApoE ε4. TREM2 is normally expressed in immune cells, including brain microglia that play an important role in clearing damaged cells and debris (Figure 1-9). A number of other AD risk alleles are also associated with microglia and immune function. These include genes encoding ABCA7 (ATP binding cassette subfamily A member 7) and CD33 (classification determinant 33), both cell surface receptors involved in phagocytosis, as well as CR1 (complement receptor type 1), which regulates the complement cascade. We use studies of TREM2 to illustrate how microglia can contribute to AD pathogenesis.

TREM2 is known to trigger microglial activation and stimulate phagocytosis, and evidence suggests that AD-associated variants reduce TREM2 function. When researchers combined loss of one or both copies of the endogenous *Trem2* gene with a transgenic AD mouse model, they found that loss of *Trem2* increased Aβ

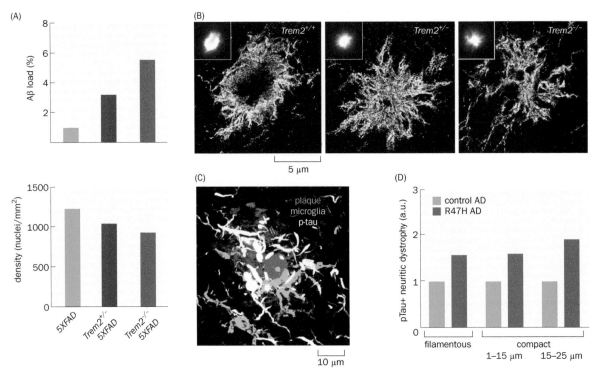

Figure 12-9 Contributions of microglial TREM2 to AD. (A) Compared to the *5XFAD* AD mouse model (which combines 3 FAD mutations in APP, 1 FAD mutation in presenilin-1, and 1 FAD mutation in presenilin-2 in the same mouse) alone (left column), losing one (middle column) or both (right column) copies of *Trem2* causes increased Aβ deposits (top) and reduced layer 5 cortical neuronal density (bottom), suggesting that TREM2 normally functions to reduce Aβ deposits and improve neuronal survival in this AD model. **(B)** STORM-based super-resolution imaging (Section 14.17) of immunolabeled Aβ plaque fibrils reveals that loss of *Trem2* causes more diffuse plaques. Insets show zoomed-out examples of plaques. **(C)** Postmortem brain section from a patient with a R47H mutation in *Trem2* showing microglia (green) in close contact with an amyloid plaque (magenta) but anti-colocalized with hyperphosphorylated tau (white), suggesting that microglia normally inhibit tau hyperphosphorylation. **(D)** Compared to control (sporadic) AD brains, R47H AD brains have increased neuritic dystrophy with hyperphosphorylated tau surrounding filamentous and compact amyloid plaques. (A, from Wang Y, Cella M, Mallinson K, et al. [2015] *Cell* 160:1061–1071. With permission from Elsevier Inc. B–D, from Yuan P, Condello C, Keene CD, et al. [2016] *Neuron* 90:724–739. With permission from Elsevier Inc.)

deposits and reduced neuronal density in a dose-dependent manner (**Figure 12-9**A). Further studies revealed a deficit in microglia recruitment by Aβ deposits in *Trem2* mutants; consequently, Aβ deposits were less compact than in controls (Figure 12-9B) and were associated with more hyperphosphorylated tau and axonal dystrophy (Figure 12-9C, D). These studies suggest that TREM2 promotes phagocytosis and clearance of Aβ aggregates and thus limits their neurotoxic effects.

Recent data suggest that among the identified extracellular ligands of TREM2 are lipoproteins, including ApoE and clusterin, another apolipoprotein that is an AD risk factor. Further, AD-associated TREM2 variants exhibit reduced binding to lipid-associated ApoE and clusterin. By binding both Aβ aggregates and TREM2 from microglia, these lipoproteins may facilitate clearance of Aβ aggregates. A current model is that Aβ aggregation triggers activation of microglia, an inflammatory response that functions to clear damaged cells and debris in part through elevated phagocytosis. High, persistent Aβ loads may overwhelm the normal homeostatic microglial functions, causing persistent activation and resulting in collateral damage, such as synaptic engulfment and release of neurotoxic substances. (Recall that synaptic pruning is also a physiological function of microglia during development; Figure 7-31). Indeed, recent single-cell RNA-sequencing studies identified a population of disease-associated microglia with marked transcriptomic alterations, including highly elevated expression of *Apoe* and *Trem2*. Similar microglia transcriptome changes have also been associated with other neurodegenerative disorders that we will discuss in the following sections. Clarifying which of these changes deter or accelerate disease progression will aid the design of effective therapeutic strategies.

12.7 How can we treat Alzheimer's disease?

The ultimate goal of studying human diseases is to find effective ways to cure or prevent them. Past decades of AD research have outlined potential pathways that lead to AD pathogenesis (**Figure 12-10**) and provided valuable clues about potential treatments. For example, multiple lines of evidence suggest that key pathogenic phenomena include increased Aβ levels and enhanced Aβ aggregation. Thus, this pathway has been the primary focus of intervention. Strategies for reducing Aβ toxicity include developing drugs to inhibit β- or γ-secretase activity, increase Aβ clearance, or neutralize Aβ activity. However, to date there are no disease-modifying therapies for AD.

Many factors must be considered when developing drugs for brain disorders (**Box 12-1**). While a drug should have its intended effects on its therapeutic target, it should have minimal side effects and toxicity at the therapeutic dose. For example, several γ-secretase inhibitors have been developed that effectively interfere with Aβ production but failed in clinical testing due to severe side effects, possibly because γ-secretase has many substrates other than APP. (One well-known γ-secretase substrate is the developmental signaling molecule Notch, which plays a key role in regulating cell fate discussed in Section 7.3; Notch also has many important functions in adults.) It is necessary to identify γ-secretase inhibitors that specifically interfere with its APP cleavage activity, or to modify γ-secretase activity to bias the product toward shorter, less toxic forms of Aβ. Other strategies include small molecules that target β-secretase activity or antibodies that target

Figure 12-10 Summary of known pathogenic pathways of Alzheimer's disease. The top pathway summarizes the Aβ hypothesis. The bottom pathway depicts production of neurofibrillary tangles. The arrow between the top and bottom pathways represents evidence that tau can mediate at least some of the Aβ toxicity in animal models. The middle pathway summarizes the involvement of microglia activation, which can help clear Aβ oligomer, trim plaques, and inhibit tau hyperphosphorylation to limit damage (red inhibitory pathways). However, abnormal microglia activation can cause collateral damage that contributes to neurodegeneration.

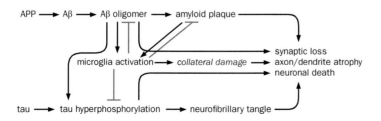

aggregated Aβ. Recent clinical trials for β-secretase inhibitors have been halted, as patients did not benefit, even though the drugs were effective at reducing Aβ production. Indeed, there was a trend toward cognitive worsening with β-secretase inhibitors, which may have resulted from inhibiting cleavage of other substrates. Clinical trial results for antibodies that target aggregated Aβ have had mixed results. Several anti-Aβ antibody therapies have been effective at reducing Aβ plaques but have shown little or no delaying of cognitive decline. Nevertheless, ongoing studies and analyses from large clinical studies are still testing this approach. Taken together, however, these failed clinical trials have led some researchers to question the Aβ hypothesis. They have also led scientists and drug developers to pursue other strategies for treating AD, such as targeting the tau pathway or modifying microglial function.

One reason clinical trials such as these may have failed could be that they were conducted on AD patients in which Aβ had already done irreversible damage— Aβ may already have induced substantial tauopathy, abnormal microglial activation, or neuronal death (Figure 12-10) in patients before they were treated. Indeed, by the time AD is typically diagnosed, neurodegeneration may be too extensive for the disease to be cured (**Figure 12-11**A). The most effective way to reduce the impact of AD may be early diagnosis and intervention. Delaying AD onset by even a few years can have a significant impact on the quality of life for patients and the burden on their families and society.

Even if effective drugs are developed, early diagnosis will be crucial in halting AD at the earliest possible stages. An important step is to identify **biomarkers**, measurable characteristic indicators of normal biological processes, pathogenic processes, and responses to therapeutic interventions. Important AD biomarkers include radioactive compounds that bind fibrillary Aβ (Figure 12-11B), allowing detection of amyloid deposits via **positron emission tomography** (**PET**) before onset of cognitive and behavioral symptoms. (PET is a noninvasive three-dimensional imaging technique that traces the distribution of positron-emitting probes introduced into the body.) Identifying biomarkers that signal AD progression, such as Aβ$_{42}$ and other metabolites in cerebrospinal fluid and plasma, can also contribute to early diagnosis. The goal is to have increasingly reliable diagnostic methods for treating preclinical AD before irreversible damage occurs (Figure 12-11A).

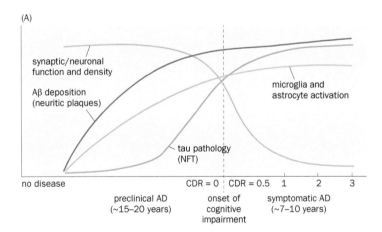

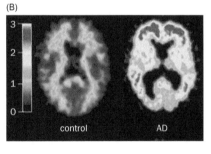

Figure 12-11 Temporal progression of Alzheimer's disease (AD) and positron emission tomography (PET) imaging as a biomarker.
(A) Timing of major AD pathophysiological events in relation to clinical course. CDR, Clinical Dementia Rating, where a score of 0 indicates normal cognition and scores of 0.5, 1, 2, and 3 indicate questionable, mild, moderate, and severe dementia, respectively. **(B)** Radioactively labeled benzothiazole, also known as Pittsburgh Compound-B (PIB), is used for PET imaging of AD brains, as it enters the brain rapidly, binds selectively to aggregated Aβ deposits, and is cleared rapidly. Compared with a 67-year-old normal subject (left), a 79-year-old AD patient (right) exhibits elevated standardized values (color-coded on the left) of PIB uptake. (A, from Long JM & Holtzman DM [2019] *Cell* 179:312–339. with permission from Elsevier Inc. B, from Klunk WE, Engler H, Nordberg A, et al. [2004] *Ann Neurol* 55:306–319. With permission from John Wiley & Sons.)

Box 12-1: Rational drug development to treat brain disorders

Many drugs currently used to treat brain disorders, including most drugs for psychiatric disorders, were discovered serendipitously, typically when clinicians noticed unintended but desirable effects during treatment of another condition (see Sections 12.15–12.17). However, basic research in neuroscience and advances in the pharmaceutical and biotechnology industries during the past decades are changing this picture. More effective therapeutics with limited side effects can be rationally developed. Identifying the mechanisms of disease pathogenesis is a prerequisite for rational therapeutic intervention. In this box, we focus on common steps in drug development (Figure 12-12), assuming that a key pathogenic process has already been identified.

The first step is to choose a molecular target for intervention. Some biological processes are more amenable than others to pharmacological intervention. For instance, cell-surface proteins are often preferred molecular targets because water-soluble chemicals and therapeutic antibodies can modulate their functions extracellularly. Once a target has been chosen, robust biological assays relevant to the disease process must be established to screen for candidate drugs, which usually belong to two large categories: small-molecule chemicals and large-molecule *biologics* (biologics are drugs produced from living organisms or contain components of living organisms), such as nucleic acids and antibodies. For small-molecule drugs, the first step is usually to establish *in vitro* assays that enable high-throughput screens of chemical libraries (which usually contain 10^5–10^6 synthetic, semi-synthetic, or naturally occurring compounds). Once a promising compound is identified, many variants can be synthesized to increase biological activity and target accessibility *in vivo* and to reduce potential side effects. Large-molecule biologics, while typically not selected

via high-throughput screens, nevertheless undergo a similar optimization process for affinity, activity, safety, and physiological processing.

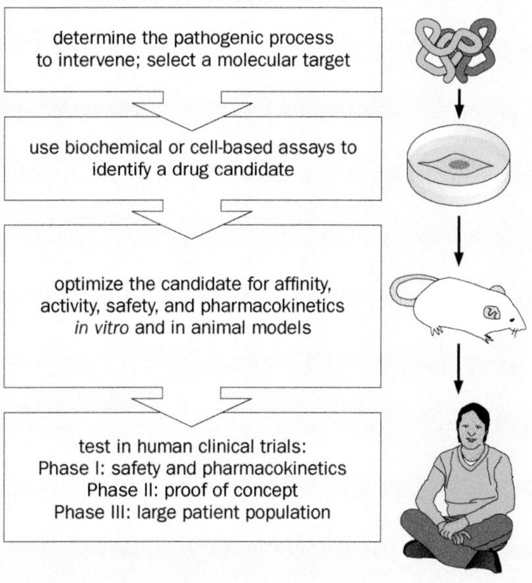

While some of optimization steps can be conducted *in vitro*, many steps must be performed in animal models *in vivo*. Two commonly used terms for body–drug interactions are **pharmacokinetics** (drug concentration), which characterizes what the body does with the drug, including its absorption, distribution, metabolism, and excretion, and **pharmacodynamics** (drug activity), which characterizes what the drug does to the body, including the intended effects on target molecules and processes as well as unintended side effects. Disease proxies, such as animal models, are very helpful in assaying a drug's therapeutic effects and establishing proof-of-concept. Animal models are also essential in evaluating potential drug toxicity. Early-stage animal models often utilize rodents, but toxicity is usually also evaluated in additional models with more human-like physiology, such as nonhuman primates.

If a target resides within the nervous system, an important step is to ensure that drugs, which usually access the body through circulation in the blood, can pass through the **blood–brain barrier** (**BBB**). Made up of endothelial cell tight junctions in the blood vessels of the brain, the BBB prevents the exchange of many substances between the blood and brain tissue. Small molecules can be chemically engineered to pass through the BBB or to enter cells directly, thereby reaching target molecules in any cellular compartments. Antibodies, on the other hand, usually access only cell-surface proteins and cross the BBB inefficiently. Intense research is being carried out to discover and develop ways of facilitating BBB crossing of antibodies and larger therapeutic molecules, including viruses for gene therapy.

If a drug passes preclinical tests in animals, including extensive safety studies, human clinical trials are next. Clinical trials are usually conducted in distinct phases, although some phases can be combined. Phase I studies are usually conducted on a small number of healthy volunteers and focus on drug metabolism. While Phase II studies continue to test the safety of a drug, they also gather preliminary data on the drug's effectiveness in a relatively small number of patients, comparing the drug with a placebo control. Phase III studies collect more information about safety and effectiveness from a large patient population. In the United States, the Food and Drug Administration (FDA) oversees clinical trials and approves drugs for the market. Drug development is a long process, averaging 10 years from initial target selection to approval for use in patients. This time window can be considerably longer for diseases that progress slowly, such as neurodegenerative diseases.

The process outlined here has facilitated the development of many drugs that treat diseases ranging from cancer to immune disorders and infectious diseases. Drugs rationally designed to treat neurodegenerative diseases are also in this pipeline (see Box 12-2 for an example of a successful therapy for motor neuron degeneration).

determine the pathogenic process to intervene; select a molecular target

use biochemical or cell-based assays to identify a drug candidate

optimize the candidate for affinity, activity, safety, and pharmacokinetics *in vitro* and in animal models

test in human clinical trials:
Phase I: safety and pharmacokinetics
Phase II: proof of concept
Phase III: large patient population

Figure 12-12 Flow chart depicting a typical rational drug development process.

12.8 Prion diseases results from propagation of protein-induced protein conformational changes

Just as abnormal Aβ aggregation is characteristic of Alzheimer's disease, **protein-opathy**—altered protein conformations, interactions, and homeostasis—appears to be a common feature of many neurodegenerative diseases. We discuss several examples in the next two sections. Among the most enigmatic causes of neurodegenerative disease are **prions** (pronounced *PREE-ons,* which stands for proteinaceous infectious particles). Three seemingly disparate diseases share prions as the causative agent. The first is **scrapie**, which infects sheep and goats after prolonged incubation following exposure to tissue from diseased animals; a variant of scrapie that affects cattle is colloquially known as "mad cow disease." The second is an infectious human disease called **kuru**, which occurred in Papua New Guinean tribes that practiced ritual cannibalism. The third is a rare inherited human disease called **Creutzfeldt–Jakob disease** (**CJD**). All three diseases are associated with massive neuronal death that causes sponge-like holes in the brain. For this reason, they are called spongiform encephalopathies (**Figure 12-13**A).

The nature of the infectious scrapie agent was heavily debated for several decades. A breakthrough came in the 1980s when reliable animal models were developed as bioassays for biochemical purification of the scrapie agent. Animals inoculated with infected brain tissues, or fractions of infected tissues following biochemical purification, exhibited spongiform encephalopathy. Moreover, brain extracts from such recently infected animals were highly infectious when inoculated into new animals. The infectivity of these extracts was not disrupted by treatments that destroy nucleic acids, suggesting that, unlike known infectious agents such as viruses and bacteria, the scrapie agent does not have a nucleic acid genome. This led to the **prion hypothesis**: the infectious scrapie agent is proteinaceous. This was considered heretical: how could a protein be infectious without an associated genome for replication?

The infectious agent was subsequently identified to be PrP^{Sc}, a conformational variant of the cell-surface protein PrP. (The ^{Sc} superscript stands for scrapie.) It was later shown that a noninfectious conformation of PrP (termed PrP^{C} for cellular PrP) is normally produced by most cell types. The presence of PrP^{Sc} can cause PrP^{C} to adopt the PrP^{Sc} conformation (Figure 12-13B, top), which is a highly stable β-pleated sheet (recall that Aβ also adopts a β-pleated sheet conformation in AD). PrP^{Sc} was proposed to propagate from cell to cell and even through the digestive system when diseased tissues are ingested by healthy animals, catalyzing conversion of PrP^{C} into PrP^{Sc} along the way. The inherited CJD does not require infec-

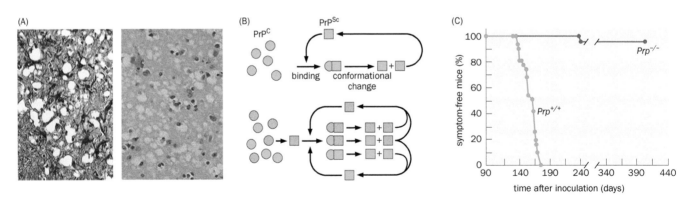

Figure 12-13 Prion diseases are caused by protein-induced protein conformational changes. (A) Pathologies in postmortem brain sections of a scrapie-affected sheep (left) and a human Creutzfeldt–Jakob disease (CJD) patient (right); note the numerous sponge-like holes. **(B)** The prion hypothesis. Top, infectious PrP^{Sc} (squares) can induce PrP^{C} (circles) to adopt the PrP^{Sc} conformation, thereby propagating PrP^{Sc}. Bottom, in genetic prion diseases such as CJD, mutant PrP^{C} proteins (orange circles) occasionally spontaneously adopt the PrP^{Sc} conformation (orange square), subsequently converting both mutant and wild-type (green) PrP^{C} into PrP^{Sc}. **(C)** Control mice (Prp$^{+/+}$) that receive intracerebral inoculation of mouse-adapted prions invariably die within 6 months. However, their Prp$^{-/-}$ littermates resist prion infection; following intracerebral inoculation, most live beyond a year with no prion pathology. Thus, endogenous PrP is essential for the effects of prion infection. (A, courtesy of Robert Higgins [left] and the CDC [right]. B, adapted from Prusiner SB [1991] *Science* 252:1515–1522. C, from Büeler H, Aguzzi A, Sailer A, et al. [1993] *Cell* 73:1339–1347. With permission from Elsevier Inc.)

tious protein agents, instead resulting from mutations in the *Prp* gene that make PrP^C more prone to spontaneously adopt the PrP^{Sc} conformation (Figure 12-13B, bottom). Thus, the prion hypothesis unified the causes of scrapie, kuru, and CJD, which are collectively called **prion diseases** (Movie 12-1).

Strong support for the prion hypothesis came from *Prp* knockout mice, which were found to resist PrP^{Sc} infection, indicating that PrP^{Sc} requires endogenous PrP^C to cause disease (Figure 12-13C). Indeed, the concept of protein-induced protein conformational change and its propagation has subsequently extended to other neurodegenerative diseases and to widespread phenomena in the normal physiology of organisms ranging from yeast to humans. The discovery of prion diseases has also resulted in alteration of certain medical practices to prevent iatrogenic transmission of prions (transmission via inadvertent medical exposure, such as via surgical instruments).

12.9 Aggregation of misfolded proteins is associated with many neurodegenerative diseases

Huntington's disease (**HD**) is a dominantly inherited disease that usually strikes patients during midlife; it is named after George Huntington, who first described the inheritance pattern in 1872. Its earliest symptoms are often depression and mood swings, followed by abnormal movements, since the striatum is the most vulnerable brain region in HD. Patients later develop cognitive deficits. Death occurs 10–20 years after symptom onset. Genetically speaking, HD is one of the simplest neurological diseases, as it is caused by alterations in a single gene encoding a widely expressed protein named **huntingtin**. The cause of HD is an expansion of a CAG (cytidine-adenosine-guanosine) trinucleotide repeat in the gene's coding sequence, resulting in an expanded polyglutamine (polyQ) repeat near the N-terminus of the huntingtin protein. (The CAG nucleotide triplet codes for the amino acid glutamine, abbreviated Q.) Healthy individuals have 6 to 34 polyQ repeats, whereas HD patients have 36 to 121. Greater numbers of polyQ repeats correlate with earlier onset of HD symptoms.

Since the discovery of polyQ repeats in HD, expanded polyQ repeats in eight other proteins have been shown to cause neurodegenerative diseases, all of which are dominantly inherited, including six forms of **spinocerebellar ataxia** that affect motor functions (Table 12-1). In each polyQ disease, there is a specific threshold for the number of polyQ repeats, over which disease occurs. Because of this common feature, it was hypothesized that expanded polyQ repeats could themselves cause disease. Indeed, *in vivo* transgenic overexpression of expanded polyQ repeats alone can cause degeneration of mouse and even *Drosophila* neurons. Proteins with long polyQ repeats form aggregates that accumulate in **inclusion**

Table 12-1:   Diseases caused by polyglutamine repeats

Disease	Gene product	Normal repeat length	Expanded repeat length
HD	Huntingtin	6–34	36–121
SCA1	Ataxin1	6–44	39–82
SCA2	Ataxin2	15–24	32–200
SCA3	Ataxin3	13–36	61–84
SCA6	Transcription factor α1ACT	4–19	10–33
SCA7	Ataxin7	4–35	37–306
SCA17	TATA-binding protein	25–42	47–63
SBMA	Androgen receptor	9–36	38–62
DRPLA	Atrophin	7–34	49–88

HD, Huntington's disease; SCA, spinocerebellar ataxia; SBMA, spinobulbar muscular atrophy; DRPLA, dentatorubral-pallidoluysian atrophy.

After Orr HT & Zoghbi HY (2007) *Annu Rev Neurosci* 30:575–621; Du X, Wang J, Zhu H, et al. [2013] *Cell* 154:118–133.

bodies, which may be present in the nucleus, cytoplasm, or axons, depending on the specific protein affected. The host proteins in which these polyQ repeats reside also play essential roles in pathogenesis *in vivo*. Inclusion bodies of abnormally aggregated host protein isoforms that contain expanded polyQ repeats can recruit additional proteins that normally interact with the host proteins. As a result, the normal functions of these interacting proteins are likely disrupted, accounting for the dominant nature of polyQ disorders. Participation of specific host proteins in the pathogenesis of polyQ disorders also explains why different polyQ diseases have different symptoms: Each disrupts a different set of interacting proteins. In HD, disruption of transcription, axonal transport, and mitochondrial function may all contribute to the eventual dysfunction and degeneration of neurons in the striatum, giving rise to the characteristic uncontrolled movements (Huntington's chorea).

Amyotrophic lateral sclerosis (ALS; also known as Lou Gehrig's disease) is a rapidly progressing motor neuron disease that usually kills patients within a few years after symptoms emerge, also usually at midlife. ALS patients exhibit selective degeneration of motor neurons in the brainstem and spinal cord that control skeletal muscle contraction, as well as motor cortical neurons innervating these motor neurons. As in AD, only a small fraction (~10%) of ALS cases are caused by inherited mutations; 90% are sporadic. Recent advances in molecular genetics uncovered more than 50 genes that potentially cause familial ALS. These mutations affect a range of cellular functions, including protein homeostasis, RNA metabolism, and cytoskeletal dynamics. The first identified familial ALS gene, *Sod1*, encodes the enzyme superoxide dismutase. However, ALS mutations in *Sod1* do not correlate with whether or not they disrupt enzymatic activity; rather, mutant proteins tend to form aggregates, suggesting that abnormal protein aggregations may be pathogenic. Another surprising finding is that although ALS is caused by motor neuron death, pathogenic Sod1 expressed in motor neurons alone is insufficient to cause disease. Expression of mutant Sod1 in glia is essential for disease progression, demonstrating that Sod1 has a non-cell-autonomous effect in pathogenesis.

Notably, mutations in several RNA-binding proteins, including TDP-43 (TAR DNA-binding protein of 43 kilodaltons, a DNA/RNA-binding protein), FUS (fused in sarcoma, another nucleic-acid-binding protein), and hnRNPA1 (heterogeneous nuclear ribonucleoprotein A1, involved in RNA processing) underlie certain forms of familial ALS. They are all members of the hnRNP family of proteins that bind many hundreds of RNAs and regulate their metabolism. These proteins, while normally distributed in the nucleus, are found in cytoplasmic inclusion bodies when pathogenic mutations are present. This disrupts the normal nuclear functions of these proteins and recruits RNAs and proteins that interact with them into inclusion bodies, which may account for why mutations in these RNA-binding proteins are dominant. Indeed, TDP-43-containing cytoplasmic inclusions have been found in motor neurons not only in familial ALS with TDP-43 mutations but also in most other familial ALS types, as well as the majority of sporadic ALS cases (**Figure 12-14**). Thus, TDP-43 aggregation may represent a common pathological event in ALS with diverse causes. Notably, a subset of AD and PD patients also have TDP-43 inclusions, illustrating the relatedness of neurodegenerative pathologies and diseases.

The most common form of familial ALS is caused by an expansion of hexanucleotide GGGGCC repeats in the intron of a previously unstudied gene, *C9orf72* (which stands for chromosome 9, open reading frame 72). These ALS patients also exhibit symptoms similar to frontotemporal dementia (FTD), which we introduced in the context of tauopathy (Section 12.4). Hexanucleotide repeat expansion in *C9orf72* is also found in 10% of sporadic ALS patients. Since the discovery of this hexanucleotide expansion in 2011, three pathogenesis models have been proposed to explain its effects (**Figure 12-15**): (1) reduced transcription of *C9orf72* and thus a decreased level of its normal protein product (loss of function from haploinsufficiency—not enough product from the remaining wild-type allele); (2) formation of RNA foci due to bidirectional transcription of repeat RNAs, which

(A)

(B)

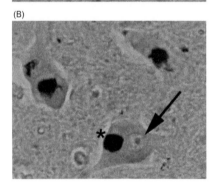

Figure 12-14 Abnormal TDP-43 cytoplasmic inclusions in amyotrophic lateral sclerosis (ALS). Spinal cord sections from an unaffected individual **(A)** and an ALS patient **(B)** immunostained for TDP-43 (brown). TDP-43 normally localizes to the nuclei of motor neurons (arrows in Panel A), but in some motor neurons of ALS patients, TDP-43 is absent from the nucleus (arrow in Panel B), having accumulated in cytoplasmic inclusion bodies (* in Panel B). (From Figley MD & Gitler AD [2013] *Rare Dis* 1:e24420. With permission from Landes Bioscience. See also Neumann M, Sampathu DM, Kwong LK, et al. [2006] *Science* 314:130–133.)

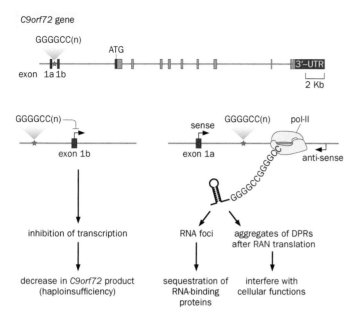

Figure 12-15 Three potential pathogenic mechanisms for expanded hexanucleotide repeats in C9orf72. Top, organization of the *C9orf72* gene. Vertical bars are exons and thin horizontal lines are introns. The GGGGCC hexanucleotide at intron 1 has <23 repeats in normal subjects but expands to hundreds to thousands of repeats in ALS patients. Bottom, three potential pathogenic mechanisms. Left, the expanded repeat inhibits transcription and thus reduces *C9orf72* protein product. Middle and right, bidirectional transcription of repeat RNA could produce either toxic RNA foci that sequester RNA-binding proteins (middle) or toxic dipeptide repeats (DPRs) through repeat-activated non-AUG (RAN) translation (right). (Based on Taylor JP, Brown RH, & Cleveland DW [2016] *Nature* 539:197–206; Gitler AD & Tsuiji H [2016] *Brain Res* 1647:19–29.)

sequester RNA-binding proteins (gain of toxic function due to RNA repeats); and (3) aggregates of dipeptide repeats translated from repeat RNA (gain of toxic function due to dipeptide repeats). Of these three models, the last one is the most unconventional; it requires a new type of translation called repeat-associated non-AUG (RAN) translation, which does not start from the AUG start codon. Evidence supporting all three models has accumulated, and it will be important to determine which one (or combination thereof) drives ALS pathogenesis.

12.10 Parkinson's disease results from death of substantia nigra dopamine neurons

Parkinson's disease (PD), first described by James Parkinson in 1817, is, after AD, the second most common neurodegenerative disease. PD primarily affects movement control: characteristic PD symptoms include shaking, rigidity, slowness, and difficulty with walking. The primary cause for these symptoms is the death of dopamine neurons in a midbrain structure called the **substantia nigra** (Latin for "black substance"), so named due to the high levels of melanin present in these dopamine neurons in healthy human subjects (**Figure 12-16**A). PD studies have benefited from, as well as contributed to, our understanding of dopamine regulation of the striatal circuits that control movement.

As we learned in Section 8.8, striatal GABAergic spiny projection neurons (SPNs) control movement through two parallel pathways: a direct pathway that inhibits the globus palladus internal segment (GPi) and the substantia nigra pas reticulata (SNr), and an indirect pathway that inhibits the globus pallidus external segment (GPe) and thereby relieves inhibition of the GPi/SNr (Figure 8-21). Dopamine neurons in the substantia nigra pars compacta (SNc) project to the striatum and regulate the direct and indirect pathways in opposite directions: they facilitate the direct pathway through the dopamine D_1 receptor and inhibit the indirect pathway through the dopamine D_2 receptor. Thus, loss of SNc dopamine neurons

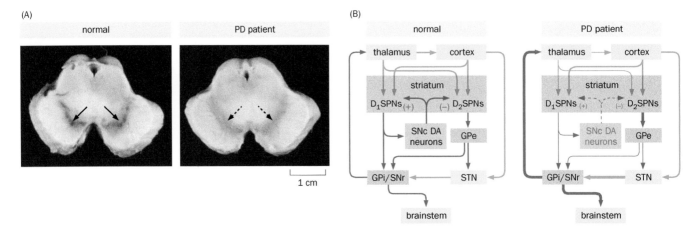

Figure 12-16 Parkinson's disease results from loss of substantia nigra dopamine neurons, leading to dysregulation of the basal ganglia. (A) Coronal sections of postmortem midbrains from a normal subject (left) and a PD patient (right). Arrows point to the substantia nigra, which is enriched in pigmented dopamine neurons that appear dark in normal brains and are selectively lost in PD brains (dashed arrows). **(B)** Basal ganglia circuit models in normal and PD brains. Left, dopamine neurons in the substantia nigra pars compacta (SNc DA neurons) promote synaptic transmission between cortical/thalamic input and D_1SPNs, which project directly to the globus pallidus internal segment and the substantia nigra pars reticulata (GPi/SNr). SNc DA neurons also suppress synaptic transmission between cortical/thalamic input and D_2SPNs, which project indirectly to GPi/SNr via the globus pallidus external segment (GPe) and the subthalamic nucleus (STN). Right, loss of SNc DA neurons in PD patients reduces the activity of D_1SPNs and enhances the activity of D_2SPNs, both of which contribute to hyperactivation of GABAergic output neurons in the GPi/SNr and excessive inhibition of their target neurons. Red, inhibitory projection; green, excitatory projection; blue, DA projection. Thicker and thinner arrows in the PD model indicate, respectively, increases and decreases in pathway strengths in PD compared to normal. (A, courtesy of the Duke University School of Medicine. B, based on Bergman H, Wichmann T, & DeLong MR [1990] *Science* 249:1436–1438 and Limousin P, Krack P, Pollak P, et al. [1998] *N Engl J Med* 339:1105–1111.)

in PD causes hypoactivation of the direct pathway and hyperactivation of the indirect pathway. Both lead to excessive activation of basal ganglia output neurons in the GPi/SNr, which in turn causes excessive inhibition of their target neurons in the thalamus and brainstem and thereby inhibits movement initiation (Figure 12-16B). This framework has inspired effective treatments such as deep brain stimulation, which will be discussed later.

12.11 α-Synuclein aggregation and spread are prominent features of Parkinson's pathology

What causes dopamine neuron death in PD? In many ways PD parallels what we have learned about AD and ALS. Most PD cases are late-adult onset and are sporadic, but a small fraction of PD cases are familial. In 1997, the first identified familial PD gene, inherited in an autosomal dominant fashion, was found to encode a presynaptic protein called **α-synuclein**. Soon afterward, α-synuclein was found to be the major component of **Lewy bodies**, intracellular inclusions that have been a defining feature of PD pathology since F. H. Lewy first described them in 1912 (although not all forms of PD exhibit Lewy body pathology). Familial PD mutations in α-synuclein (single amino acid changes) promote aggregation of α-synuclein *in vitro*. Indeed, increasing the copy number of the human α-synuclein gene is sufficient to cause PD with Lewy body pathology. Thus, PD can be caused by overproduction of normal α-synuclein or production of mutant α-synuclein prone to aggregation.

Postmortem analyses of brains from patients who died at different stages of PD suggest that the distribution of Lewy body pathology follows a stereotyped spatiotemporal sequence: neurons with α-synuclein pathology are found mostly in the brainstem of early-stage PD patients, but are more widespread in the forebrains of patients with more advanced PD. In fact, SNc dopamine neurons are not the first to exhibit α-synuclein pathology, though they might be the most vulnerable. This suggests that α-synuclein aggregates may spread from neuron to neuron. This hypothesis was spurred by findings from PD patients whose brains received

Neurodegenerative disorders

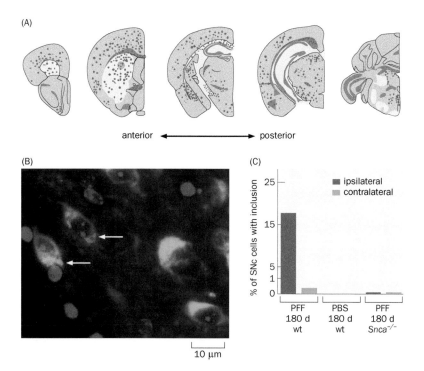

(A)

anterior ◄────────► posterior

(B)

10 μm

(C)

Figure 12-17 **Cell-to-cell spread of α-synuclein pathology. (A)** Serial coronal brain schematics of a wild-type mouse 180 days after injection of preformed α-synuclein fibrils (PFF) into the dorsal striatum reveal the development of α-synuclein pathology across the brain. Cells with α-synuclein aggregates are indicated in red; the injection site is marked by a light red circle in the second schematic from left. **(B)** A high-magnification image of the SNc, showing two dopamine neurons with α-synuclein aggregates in Lewy body–like inclusions (arrows). Dopamine neurons are visualized in green by immunostaining against tyrosine hydroxylase, a marker for dopamine neurons. α-synuclein is visualized in red using an antibody that preferentially binds to aggregated α-synuclein. Blue, DNA stain for visualizing cell nuclei. **(C)** SNc cells that exhibited α-synuclein pathology were located primarily ipsilaterally to the injection. Injecting phosphate-buffered saline (PBS) into wild-type (wt) mice or PFF into α-synuclein knockout (Snca⁻/⁻) mice did not cause α-synuclein pathology. (From Luk KC, Kehm V, Carroll J, et al. [2012] *Science* 338:949–953. With permission from AAAS.)

transplants of fetal dopamine neurons, a treatment strategy we will discuss in Section 12.13. When some of these patients died up to 14 years after surgery, the transplanted fetal neurons contained α-synuclein-enriched Lewy body-like structures. Additional support for this hypothesis came from studies in mice: focal injection into wild-type mice of α-synuclein fibrils preformed *in vitro* caused spread of α-synuclein pathology across the brain, including to SNc dopamine neurons (**Figure 12-17**). Injecting α-synuclein fibrils into α-synuclein knockout mice did not cause spread of α-synuclein pathology, suggesting that recruitment of endogenous α-synuclein is essential for the pathology, as in prion diseases (Figure 12-13C). These studies, which together suggest that PD may involve cell-to-cell spread of pathogenic proteins, raise important and as-yet-unresolved questions regarding the mechanisms by which such spread might occur for cytoplasmic proteins such as α-synuclein.

12.12 Mitochondrial and lysosomal dysfunctions contribute to Parkinson's disease pathogenesis

Before molecular-genetic studies of familial PD in the 1990s, Parkinson's disease was thought to be environmentally induced. Striking evidence came in the early 1980s, when MPTP (1-methyl-4-phenyl-1,2,3,6-tetrahydropyridine), a contaminant from chemical synthesis of the opioid-like drug MPPP (1-methyl-4-phenylpropionoxypiperidine), caused PD-like symptoms in drug users. Subsequent studies confirmed MPTP toxicity and established the mechanisms: MPTP freely crosses the BBB and is converted to MPP⁺ (1-methyl-4-phenylpyridinium) by **monoamine oxidase,** which normally oxidizes monoamine neurotransmitters, leading to their degradation. MPP⁺ is selectively accumulated via the plasma membrane dopamine transporter in dopamine neurons, where it potently inhibits its mitochondrial complex I function, thus selectively killing dopamine neurons (**Figure 12-18**). Biochemical assays of postmortem tissue from sporadic PD patients also indicated complex I deficiency, thereby implicating mitochondrial defects in dopamine neurons as a cause of PD.

While most PD cases are sporadic, human genetic studies since the 1990s have identified mutations in over 20 genes associated with familial PD, together accounting for 10–15% of PD cases. In addition to the dominant α-synuclein

MPTP   H₃C—N⟨⟩—⟨⟩

↓ monoamine oxidase

MPP⁺   H₃C—N⁺⟨⟩—⟨⟩

↓

inhibit mitochondrial function
dopamine neuron death

Figure 12-18 **PD-like symptoms caused by a chemical toxin.** After passing through the blood–brain barrier, MPTP (1-methyl-4-phenyl-1,2,3,6-tetrahydropyridine) is converted to MPP⁺ (1-methyl-4-phenylpyridinium) by monoamine oxidase. MPP⁺ selectively accumulates in dopamine neurons through the action of the plasma membrane dopamine transporter, inhibits mitochondrial function, and kills dopamine neurons.

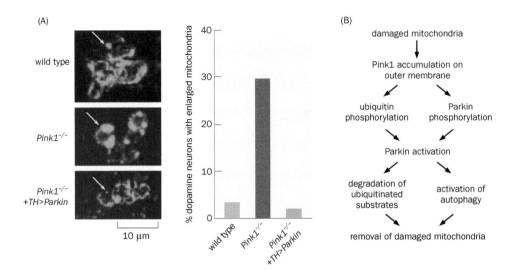

Figure 12-19 Pink1 and Parkin act in a common pathway in mitochondrial function. (A) Clusters of *Drosophila* dopamine neurons visualized by expressing a mitochondrion-targeted GFP transgene under the control of the tyrosine hydroxylase (TH) promoter. In each panel, an arrow points to an individual mitochondrion. In the *Pink1* mutant *(Pink1^{-/-})*, mitochondria were abnormally enlarged compared with wild type. This mitochondrial defect in *Pink1* mutant dopamine neurons was rescued by overexpression of Parkin, driven by the TH promoter *(TH>Parkin)*. Results are quantified at right. These data suggest that Pink1 acts upstream of Parkin in the same pathway. **(B)** Summary of the Pink1/Parkin pathway in the removal of damaged mitochondria. (A, from Park J, Lee SB, Lee S, et al. [2006] *Nature* 441:1157–1161. With permission from Springer Nature. See also Clark IE, Dodson MW, Jiang C, et al. [2006] *Nature* 441:1162–1166 and Yang Y, Gehrke S, Imai Y, et al. [2006] *Proc Natl Acad Sci U S A* 103:10793–10798. B, based on Pickrell AM & Youle RJ [2015] *Neuron* 85:257–273.)

mutations discussed earlier, familial PD can also be caused by autosomal recessive mutations. PD-linked recessive mutations have been identified in two evolutionarily conserved genes: *Pink1*, which encodes a mitochondrion-associated kinase, and *Parkin*, which encodes an enzyme in the ubiquitin-proteasome system for protein degradation. The relationship between *Pink1* and *Parkin* was first revealed by studying their homologs in *Drosophila*. *Pink1* mutant flies could not fly, died young, and exhibited degeneration of muscles and dopamine neurons with abnormal mitochondrial morphology and function (**Figure 12-19**A). *Parkin* mutant flies exhibited similar defects. Notably, defects in *Pink1* mutants were rescued by overexpression of Parkin, but defects in *Parkin* mutants were unaffected by overexpression of Pink1. These data suggested that Pink1 and Parkin act in a common pathway to regulate mitochondrial function, with Pink1 acting upstream of Parkin. This relationship has subsequently been confirmed and extended to mammalian systems. A primary function of the Pink1/Parkin pathway is to remove damaged mitochondria. Pink1 accumulates on the surface of damaged mitochondria, where it phosphorylates ubiquitin and Parkin. This leads to activation of Parkin on damaged mitochondria, which adds ubiquitin to substrates leading to their degradation. Parkin also activates **autophagy**, a cellular process that removes cytoplasm and damaged organelles—in this case damaged mitochondria—by the cell's own lysosome (Figure 12-19B). Studies of Pink1 and Parkin reinforced a prominent role for mitochondrial dysfunction in PD pathogenesis.

Autosomal dominant mutations in the gene encoding LRRK2 (leucine-rich repeat kinase 2) are the most common genetic cause of PD. Pathogenic mutations lead to hyperactivation of the kinase. LRRK2 has been implicated in the regulation of intracellular trafficking to lysosome through phosphorylation of small GTPases Rab. Indeed, several other familial PD genes and PD risk factors have been associated with lysosome function. For example, homozygous mutations in the gene *Gba*, encoding a lysosomal hydrolase called glucocerebrosidase essential for glycolipid metabolism, cause Gaucher's disease, a lysosomal storage disorder with neurological symptoms including parkinsonism. Heterozygous carriers of mutant *Gba* have a several-fold increased risk of developing PD. It is possible that lysosomal dysfunction disrupts the clearance of α-synuclein aggregates and/or damaged mitochondria, contributing to the death of dopamine neurons.

12.13 Treating Parkinson's disease: L-dopa, deep brain stimulation, and beyond

Despite the molecular complexity of Parkinson's disease, the selective loss of substantia nigra dopamine neurons as a common outcome suggested a possible treatment strategy: boosting dopamine levels. Starting in the 1960s, the successful use of **L-dopa** to treat PD symptoms has inspired generations of researchers and clinicians to find better treatments for PD and other devastating brain disorders.

Dopamine is synthesized from the amino acid L-tyrosine in two enzymatic steps (**Figure 12-20**). The first step, catalyzed by **tyrosine hydroxylase**, converts L-tyrosine to L-dopa, which is then acted upon by aromatic L-amino acid decarboxylase to yield dopamine. Tyrosine hydroxylase is the rate-limiting enzyme in the biosynthesis of **catecholamines**, a class of molecules that includes the neurotransmitters dopamine, norepinephrine, and epinephrine (Figure 12-20; epinephrine also acts as a hormone, see Section 3.19). First known as an intermediate precursor for norepinephrine and epinephrine, dopamine was recognized as a neurotransmitter itself in the late 1950s. Shortly thereafter, it was discovered that dopamine levels were markedly reduced in the striatum of postmortem PD brains, suggesting a selective deficit of the dopamine system in PD.

In PD, loss of dopamine neurons causes a decline in dopamine release. Thus, one strategy to increase dopamine release by the remaining dopamine neurons is to bypass the rate-limiting step of dopamine synthesis. While dopamine cannot effectively cross the blood–brain barrier, its immediate precursor L-dopa can. Through trial and error to optimize the therapeutic dose and reduce side effects, a treatment protocol was developed in the 1960s that is still widely used today: L-dopa administration drastically improves movement control in most early-stage PD patients.

Unfortunately, L-dopa is effective at ameliorating PD symptoms for only a few years. The eventual decline in L-dopa efficacy is likely because it must be converted to dopamine by the remaining dopamine neurons, whose progressive death is not halted by the treatment. An alternative treatment strategy is **deep brain stimulation** (**DBS**), which is designed to compensate for the alteration of circuit dynamics in PD due to the loss of dopamine modulation and excessive output of GPi/SNr (Figure 12-16B). For this treatment, electrodes are surgically implanted to stimulate neurons and axons in specific nuclei. The exact mechanisms by which deep brain stimulation affects the basal ganglia circuitry in the PD brain are likely to be complex. Excitatory and inhibitory neurons as well as axons-in-passage are all stimulated simultaneously, and firing of target neurons can be affected in opposite directions depending on stimulation frequency and distance to the electrode. Clinically, deep brain stimulation of neurons and axon fibers in the STN or GPi can alleviate movement-related symptoms in late-stage PD patients when L-dopa treatment becomes less effective.

A radically different strategy for PD treatment is to replace dying dopamine neurons with new dopamine neurons. This **cell-replacement therapy** has several requirements. First, a reliable source of dopamine neurons must be identified. Additionally, transplanted dopamine neurons must survive in the host, have access to their targets, and release appropriate levels of dopamine. Studies in animal PD models have shown that fetal tissues derived from midbrain regions that contain dopamine neurons can survive after being grafted into the host striatum. (To bypass the requirement for correct axonal projections to targets, dopamine neurons are usually transplanted directly to the striatum.) These grafted cells can release dopamine and improve motor control. Subsequently, small clinical trials in humans have reported improvement of clinical symptoms and long-term increase in dopamine release after transplantation (**Figure 12-21**). However, contamination by additional cell types is a major side effect of fetal tissue transplantation, impeding further clinical development. Recent progress in reprogramming

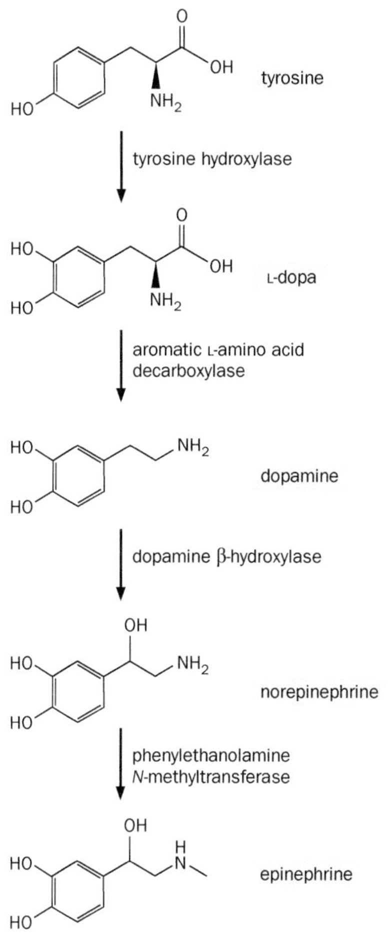

Figure 12-20 Biosynthetic pathway of catecholamines. Tyrosine hydroxylase is the rate-limiting enzyme in the pathway. Whereas dopamine does not cross the blood–brain barrier, L-dopa does. Injection of L-dopa has been used to boost the synthesis of dopamine by the remaining dopamine neurons as a therapy for dopamine deficiency in Parkinson's disease.

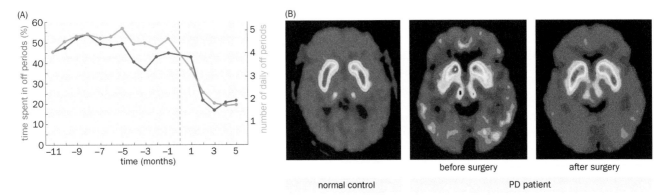

Figure 12-21 Embryonic dopamine neuron transplantation as a treatment of Parkinson's disease. (A) PD patients fluctuate between "off" periods when motor function is severely impaired, and "on" periods when motor function is relatively normal. This figure shows the self-report of a PD patient describing the fraction of awake time spent in off periods (red trace; scale on left) and the number of off periods per day (blue trace; scale on right) in the months before and after receiving transplanted dopamine neurons derived from fetal tissue. (Time of transplantation is indicated by the dashed line.) The duration and frequency of off periods are markedly reduced after transplantation. **(B)** Positron emission tomography scan of radioactive fluorodopa uptake into dopamine neuron axon terminals in the striatum of a normal subject (left) and a PD patient before (middle) and 12 months after (right) transplantation of fetal dopamine neurons. Before surgery, radioactivity (shown in red) in the brain of the PD patient was restricted to the caudate (medial striatum); after the bilateral transplantation, radioactivity was extended to the putamen (lateral striatum), more similar to the distribution in the normal control. (A, from Lindvall O, Brundin P, Widner H, et al. [1990] *Science* 247:574–577. B, from Freed CR, Greene PE, Breeze RE, et al. [2001] *N Engl J Med* 344: 710–719. With permission from the Massachusetts Medical Society.)

of dopamine neurons from induced pluripotent stem cells (Box 7-1) has renewed enthusiasm of the cell-replacement therapy.

12.14 The various neurodegenerative diseases exhibit both common themes and unique properties

Despite being associated with distinct proteins and disease symptoms, a broad suite of neurodegenerative diseases—including AD, prion diseases, polyQ diseases, ALS, and most forms of PD—have in common the abnormal aggregation of misfolded proteins or cleaved fragments. Familial mutations tend to facilitate such aggregation. These protein aggregates, or their intermediates, are either toxic by themselves or alter the localization or function of their normal interacting partners, thus disrupting protein homeostasis and causing toxic gain-of-function phenotypes. In some cases, loss-of-function of the misfolded protein may further exacerbate gain-of-function effects. Much remains to be learned about how misfolding occurs, what structure(s) define the toxic species, and which downstream effects are specific to each disease.

In the case of prions, PrP in the pathogenic conformation (PrP^{Sc}) serves as a seed to convert normal PrP^C into additional pathogenic PrP^{Sc}, causing the disease to spread and become infectious. Although no other neurodegenerative disorders are known to be infectious, the concept of seeding-induced conformational changes that result in misfolded protein aggregates and cell-to-cell spread of misfolded proteins applies to other diseases: evidence suggests that α-synuclein in PD (Section 12.11), tau in AD, and huntingtin in HD can spread from cell to cell, and this spreading may contribute to disease progression in each case. Much remains to be investigated about the mechanisms of cell-to-cell spread and the factors that promote or inhibit such spread.

One feature that distinguishes different diseases is the neuronal types that degenerate: AD causes widespread degeneration of many types of neurons in the cerebral cortex, hippocampus, and amygdala; HD primarily causes striatal neuron degeneration; ALS preferentially affects motor neurons; and PD results from degeneration of substantia nigra dopamine neurons, at least initially. Many of the causal genes mutated in familial forms of these diseases—such as APP and presenilins in AD, huntingtin in HD, TDP-43 in ALS, and α-synuclein in PD—are ubiquitously expressed. It remains largely a mystery how mutations in these widely expressed genes primarily damage specific neuronal cell types, thus causing specific diseases. One contributing factor could be that each disease is caused by aggregated proteins interacting with and interrupting the functions of a unique set of partners that may be differentially required in different neuronal types.

When the first edition of this textbook was published in 2015, the development of effective therapies for neurodegenerative diseases was limited to treating the symptoms of early-stage PD. Intense ongoing research discussed in previous sections may ultimately yield successful treatments for a broader range of neurodegenerative diseases. Indeed, recent success in treating spinal muscular atrophy (Box 12-2) has raised hope that these research and drug development efforts will pay off in treating other disorders of the nervous system.

Box 12-2: Spinal muscular atrophy, antisense oligonucleotides, and gene therapy

Spinal muscular atrophy (SMA) is the leading genetic cause of infant mortality and is caused by progressive motor neuron degeneration. In the most severe and common type (SMA type 1), disease onset occurs before 6 months of age. Patients cannot sit without support, and usually die within the first 2 years of life. Types 2, 3, and 4 have progressively later onsets and less severe symptoms. In 1995, it was discovered that SMA is caused by homozygous disruption of the *Smn1* (*survival motor neuron 1*) gene, which is conserved from yeast to human and encodes a widely expressed protein that regulates assembly of the ribonucleoprotein complex for RNA processing. Subsequently, it was found that the human genome also encodes one or more copies of *Smn2*, which differs at only 11 nucleotides, and no amino acids, from *Smn1*. However, one nucleotide difference, a C → T (U in RNA) change in exon 7, causes exon 7 to be skipped in ~90% of the splicing products, and exon 7-skipped transcripts produce a protein that is rapidly degraded (**Figure 12-22A**). Thus, the amount of full-length protein produced by *Smn2* cannot compensate for the loss of *Smn1*, resulting in motor neuron degeneration. The severity of SMA symptoms is determined largely by the copy number of *Smn2*; the vast majority of patients with two *Smn2* copies develop type 1 SMA, whereas patients with more *Smn2* copies develop less severe types of SMA.

The presence of an *Smn2* gene that can potentially make a functional protein suggested a therapeutic strategy: increasing the splicing efficiency of *Smn2* pre-mRNA such that exon 7 is included. Researchers identified a sequence in intron 7 that inhibits its inclusion via binding to the heterogeneous nuclear ribonucleoprotein (hnRNP) complex. This led to design of **antisense oligonucleotides (ASOs)** with sequences complementary to the intron 7 sequence, such that ASO binding would prevent hnRNP recruitment and could thus promote exon 7 inclusion (**Figure 12-22B**). After optimizing for stability and cellular uptake (ASOs are generally taken up by endocytosis and then escape from the

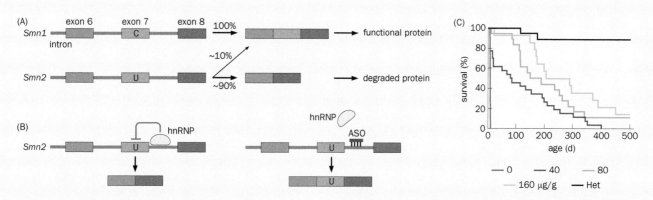

Figure 12-22 Treating spinal muscular atrophy (SMA) with antisense oligonucleotides. (A) Diagram of paralogous portions of pre-mRNA encoded by *Smn1* and *Smn2*, with exons color coded. SMA is caused by homozygous disruption of *Smn1*. Due to a C → U change in exon 7 of *Smn2*, exon 7 is skipped in most mRNAs after splicing, leading to rapid protein degradation. **(B)** Left, binding of a sequence in intron 7 by the heterogeneous nuclear ribonucleoprotein (hnRNP) inhibits inclusion of exon 7. Right, an antisense oligonucleotide (ASO) that binds the same sequence in intron 7 can displace hnRNP, allowing inclusion of exon 7 after splicing. **(C)** In a mouse model of severe SMA, all control mice die within 1–2 weeks (0 μg/g, or saline injection). When injected with ASO-10-27 (which binds the intron 7 sequence shown in Panel B) at 40, 80, and 160 μg/g of body weight, at postnatal days 0 and 3, there is a dose-dependent increase in survival. Het, heterozygous deletion of endogenous *Smn*, which does not affect survival. (B, adapted from Rigo F, Hua Y, Krainer AR, et al. [2012] *J Cell Biol* 199:21–25. C, from Hua Y, Sahashi K, Rigo F, et al. [2012] *Nature* 478:123–126. With permission from Springer Nature. See Finkel et al. [2017] *New Engl J Med* 377:1723 and Mercuri et al. [2018] *New Engl J Med* 378:625 for effect of ASO on SMA patients, and Mendell et al. [2017] *New Engl J Med* 377:1713 for effect of gene therapy on SMA patients.)

(Continued)

Box 12-2: continued

endosome into the nucleus), the therapeutic effects of ASOs were tested in animal models. In a mouse model of severe SMA (homozygous deletion of the single endogenous *Smn* gene and transgenic expression of human *Smn2*), subcutaneous injections of an ASO at the neonatal stage led to drastic improvement in survival (Figure 12-22C), accompanied by rescue of motor neuron survival and motor function. As ASOs do not efficiently cross the blood–brain barrie, treatment after BBB development is complete requires intrathecal injection (injection into the spinal canal so cerebrospinal fluid can carry the drug to cells in the spinal cord and brain). Clinical trials using intrathecal injection of intron-7-targeting ASOs produced substantial improvement in all forms of SMA, including extending survival and supporting motor skill development for patients with type 1 SMA. ASO treatment before symptom onset provides the greatest benefits. Because of these successes, the FDA approved the ASO strategy as an SMA therapy in 2016. These successes are also motivating similar approaches of developing ASO treatments for other neurodegenerative diseases, including ALS (against *Sod1* and *C9orf72*), Huntington's disease (against *huntingtin*), and Alzheimer's disease and other tauopathies (against *Tau*).

Another nucleic acid-based strategy for disease treatment is **gene therapy**, which refers to the general strategy of introducing DNA into patients' cells to correct a genetic defect. A simple way to treat genetic disorders due to loss of function of an endogenous gene is to transgenically reexpress the normal version. In addition, both loss- and gain-of-function mutations can in principle be corrected by transgene-mediated genome editing (see Box 14.1 for details). To be effective as a therapy for human use, however, gene therapy must overcome several obstacles: transgenes must be delivered to the right cell types at the right developmental time points and expressed at appropriate levels, and the procedure must not cause deleterious effects. Following decades of development, viral vectors such as adeno-associated virus (AAV, see Section 14.10 for details) have fulfilled safety requirements. One of the first gene therapy protocols approved by the FDA in late 2017 was to treat blindness caused by recessive mutations in *Rpe65*, which encodes an enzyme expressed in retinal pigment cells that catalyzes conversion of all-*trans* retinal to 11-*cis* retinal and is essential for recovery of phototransduction (Section 4.6). This was achieved by subretinal injection of an AAV vector encoding wild-type *Rpe65*.

A limiting factor of viral-based gene delivery for treating CNS disorders is BBB penetration. Intrathecal and intracerebroventricular delivery can bypass the BBB and are being used in ongoing clinical trials for treating CNS disorders. AAV variants that can penetrate the BBB more efficiently have also been developed. A recent gene therapy clinical trial using intravenous delivery of an AAV variant expressing wild-type *SMN1* yielded promising results for treating type 1 SMA in young children (with incomplete BBB development), and was approved by the FDA in 2019 as an SMA therapy for young children.

PSYCHIATRIC DISORDERS

Disorders of the nervous system have historically been divided into neurological and psychiatric. Neurological disorders are usually associated with structural, biochemical, or physiological symptoms, as in the neurodegenerative diseases we studied so far in this chapter. By contrast, psychiatric disorders have historically included those that affect the mind—how we perceive, feel, think, and act—without established physical bases. As the brain and the mind are inseparable, and as we gain more understanding of both, the distinctions between neurology and psychiatry become increasingly blurred and somewhat arbitrary. However, traditionally defined psychiatric and neurological disorders have historically been studied using different approaches. For example, studies of neurodegenerative diseases, traditionally considered neurological disorders, start with pathology. Scientists try to understand the mechanisms underlying pathological changes, in the hope of designing treatments to interfere with pathogenic processes. By contrast, most therapeutic drugs for psychiatric disorders have been discovered through serendipity. By studying how such drugs act, researchers attempt to uncover the mechanisms underlying these disorders. In the following, we illustrate this path of discovery in four classes of psychiatric disorders: schizophrenia, mood disorders, anxiety disorders, and addiction.

12.15 Schizophrenia can be partially alleviated by drugs that interfere with dopamine function

Schizophrenia has a lifetime prevalence of 1% in the general population. Its onset is usually adolescence or early adulthood, and patients are typically affected for the rest of their lives. Its most typical symptoms include hallucinations (sensory perception in the absence of external stimuli) and delusions (fixed false beliefs), including paranoia. These psychotic disturbances (**psychosis**) are the symptoms most commonly associated with the disorder and are termed *positive symptoms*, referring to their presence in patients but not in healthy people. Schizophrenia is also associated with a set of *negative symptoms* including social withdrawal and lack of motivation as well as cognitive impairment in memory, attention, and executive functions. The negative symptoms and cognitive impairments are usually more disabling to patients' quality of life because there is currently no efficacious medication for treating them.

Before the 1950s, there was no treatment for schizophrenia other than confining patients to mental asylums, often for decades. Then came the fortuitous discovery of the first antipsychotic drugs: chlorpromazine and reserpine. **Chlorpromazine** was chemically synthesized as a potential anesthetic, and **reserpine** is the active ingredient purified from the snakeroot plant used to treat hypertension. Both drugs were found to alleviate positive symptoms of schizophrenia patients, albeit with similar side effects: motor control deficits similar to those of Parkinson's disease (PD).

Subsequent studies showed that reserpine acts by interfering with the metabolism of all three monoamine neurotransmitters: dopamine, norepinephrine, and serotonin (**Figure 12-23**A; see also Section 3.11 and Box 9-1). After release into the synaptic cleft, these monoamine neurotransmitters are taken back into the presynaptic terminal by specific **plasma membrane monoamine transporters** (**PMATs**). A **vesicular monoamine transporter** (**VMAT**) shared by all three monoamines then transports these neurotransmitters from the cytosol into synaptic vesicles for future rounds of synaptic transmission. Reserpine inhibits VMAT, thus blocking monoamine neurotransmitter recycling. Monoamine transmitters retained in the cytosol after VMAT blockade are inactivated by **monoamine oxidase**, which removes the amine group and thereby allows its products to be further metabolized (Figure 12-23A). By inhibiting VMAT, reserpine effectively lowers the levels of dopamine, norepinephrine, and serotonin (Figure 12-23B). Indeed, lowering norepinephrine levels in the sympathetic neurons, which promote cardiac muscle contraction (Section 9.1), underlies reserpine's effect of relieving hypertension. Its ability to lower dopamine levels also explains why reserpine causes PD-like symptoms.

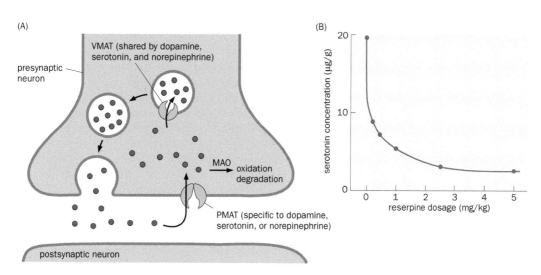

Figure 12-23 Monoamine neurotransmitter metabolism in the presynaptic terminal and the effect of reserpine. (A) Schematic of monoamine neurotransmitter metabolism in the presynaptic terminal. After release into a synaptic cleft, each monoamine neurotransmitter is taken up by its specific plasma membrane monoamine transporter (PMAT) into the presynaptic cytosol. There, neurotransmitter molecules are either taken up by the vesicular monoamine transporter (VMAT) into synaptic vesicles for reuse or oxidized by monoamine oxidase (MAO) for degradation. **(B)** Reserpine's effect was first discovered based on its ability to lower serotonin levels. In this experiment, the level of serotonin released in the intestine (serotonin is used as a major neurotransmitter in the enteric nervous system) was found to decrease progressively after rabbits received increasing reserpine doses. Reserpine exerts this effect by inhibiting VMAT. (Adapted from Pletscher A, Shore PA, & Brodie BB [1955] *Science* 122:374–375.)

Figure 12-24 Competitive binding assay for testing drug action. (A) Schematic illustration. A fixed amount of radioactively labeled neurotransmitter (or known receptor agonist) and variable amounts of drugs are used for competitive binding (a 1:1 ratio of the drug and agonist is illustrated for simplicity). Retention of a large fraction of radioactivity by the receptors indicates low affinity of the drug to the receptor (left), whereas retention of a small fraction of radioactivity indicates that the drug outcompetes the neurotransmitter or agonist and binds the receptor with higher affinity (right). **(B)** The efficacy of various antipsychotic drugs, as measured by the amount required for effective treatment, correlates well with their ability to compete with radioactively labeled dopamine (blue data points) or haloperidol (itself an antipsychotic drug; red data points) for binding to dopamine receptors from brain extract. The diagonal line represents a 1:1 relation between the blocking molarity and the clinical dose. (Adapted from Seeman P, Chau-wong M, Tedesco J, et al. [1975] *Proc Natl Acad Sci U S A* 72:4376–4380.)

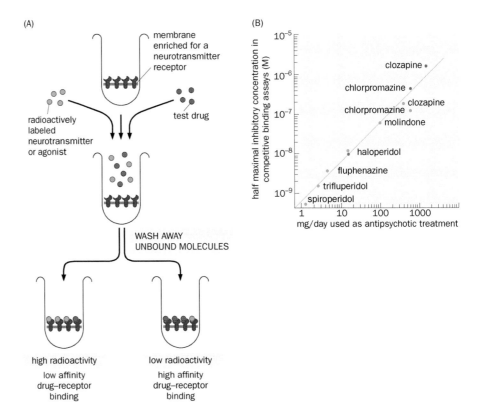

Chlorpromazine acts differently. In the 1970s, competitive binding assays were established to test drug effectiveness: a given drug would compete against a particular neurotransmitter for binding to brain membrane extracts presumed to contain high levels of neurotransmitter receptors (**Figure 12-24**A). When chlorpromazine and other antipsychotic drugs were tested in such competitive binding assays against different neurotransmitters, it was found that a drug's affinity for dopamine receptors (but not for serotonin or adrenergic receptors) correlated well with its effectiveness as an antipsychotic (Figure 12-24B). Subsequent work suggested that blockade of a specific subtype, the D_2 dopamine receptor, correlated best with antipsychotic effects. Unfortunately, blocking the D_2 dopamine receptor also leads to PD-like symptoms. New generations of antipsychotic drugs with milder side effects on movement control have been developed, although all of these have significant metabolic side effects.

The fact that reserpine and chlorpromazine both reduce dopamine function through distinct mechanisms suggested that dopamine system abnormalities contribute to the positive symptoms of schizophrenia. This hypothesis was further supported by cases of drug-induced psychosis. Two commonly abused drugs, cocaine and amphetamine, are known to induce paranoid delusions and hallucinations similar to the positive symptoms of schizophrenia. These **psychostimulants** produce transient euphoria and suppress fatigue but are potently addictive because they modulate the brain's reward system. As we will learn in Section 12.18, both cocaine and amphetamine increase dopamine levels. As in schizophrenia, the positive symptoms induced by these psychostimulants can be effectively treated with antipsychotic drugs that reduce dopamine receptor function. Additional support for the involvement of dopamine in schizophrenia came from PET imaging studies that correlated acute psychotic states with increased dopamine levels.

Current antipsychotic drugs are not effective at treating about a third of schizophrenia patients, suggesting considerable heterogeneity in the disorder. For those who respond, antipsychotic drugs are effective only at reducing positive symptoms; they have no effect on negative symptoms or cognitive impairment. This is likely because schizophrenia alters more than just the dopamine system. Indeed, drugs that act as NMDA receptor antagonists, such as phencyclidine (PCP) and ketamine, also induce psychosis resembling schizophrenia, with associated negative symptoms, suggesting that a reduction in NMDA receptor function also contributes to schizophrenia.

Although informative, studies investigating drug action have not revealed the root causes of schizophrenia. Recent structural magnetic resonance imaging studies indicate that schizophrenia is associated with significant thinning of the cerebral cortex. The most affected cortical areas include the **prefrontal cortex**, an executive control center that integrates multisensory information, processes working memory, and performs complex executive functions such as goal selection and decision making. Cortical thinning is suspected to be due to excess synaptic pruning normally associated with cortical development (Section 7.14), suggesting a neurodevelopmental origin of schizophrenia. Indeed, prior to their first psychotic episodes, schizophrenia patients usually already exhibit considerable social, mood, and cognitive impairments collectively known as schizophrenia prodrome. As will be discussed in Section 12.19, schizophrenia has a strong genetic component. Identifying genetic factors and studying their mechanisms of action may shed more light on the causes of this calamitous disorder and thereby suggest more effective therapeutic strategies.

12.16 Mood disorders have been treated by manipulating monoamine neurotransmitter metabolism

We all experience happiness and sadness in our lives. However, a sizable fraction of the general population suffers from mood disorders that can take over their lives. Mood disorders fall into two major categories: bipolar disorder and major depression. Individuals with **bipolar disorder** have episodes of mania and depression interwoven with euthymia (a normal, tranquil state of mind or mood). During the manic phase, patients often experience elevated feelings of grandiosity and irritability and have a decreased need for sleep; by contrast, during the depressive phase, they feel sad, empty, and worthless, and experience physical and mental slowness. The artist Vincent van Gogh (Figure 12-25) is one of many historical figures suspected to have suffered from bipolar disorder, which has a lifetime prevalence of 1%. **Major depression** has only the depressive phase (and is thus also referred to as unipolar depression) and is more common than bipolar disorder, with a lifetime prevalence of more than 5%. Both of these mood disorders can be life threatening, as they are implicated in a large fraction of suicides.

Like the first antipsychotics, the first antidepressant was discovered serendipitously in the 1950s. **Iproniazid** (Figure 12-26A) was originally developed for treating tuberculosis. Physicians reported that iproniazid-treated patients were happier, despite the drug's ineffectiveness as a tuberculosis treatment. This, and other clues from animal studies, led to trials of iproniazid for depression, with the finding that the drug significantly improved patients' depressive states. Further studies showed that iproniazid acts by inhibiting monoamine oxidase (Figure 12-23A), hence increasing the concentration of monoamines in presynaptic terminals and synaptic clefts. These findings suggest that elevation of one or more of the monoamine neurotransmitters—serotonin, norepinephrine, or dopamine—may have a therapeutic effect on depression. However, because monoamine oxidase inhibitors indiscriminately increase monoamine levels, they have many side effects. Over the years, they have been replaced by tricyclic antidepressants such as **imipramine**, named for their characteristic three-ring molecular structures (Figure 12-26A).

Imipramine was synthesized as a variant of chlorpromazine in the hope of identifying a more effective antipsychotic. Although ineffective at treating psychosis, imipramine had a pronounced effect on depression. Further studies led to the discovery that imipramine and other tricyclic antidepressants inhibit the plasma membrane monoamine transporters (PMATs; Figure 12-23A) that mediate neurotransmitter reuptake from presynaptic terminals. Originally discovered by studying the action of norepinephrine on the targets of sympathetic nerves, such pump-mediated reuptake turns out to be a general mechanism by which the

Figure 12-25 Self-portrait by Vincent van Gogh. A brilliant artist, van Gogh suffered from illnesses that may have included bipolar disorder—he was prolific during normal and possibly manic phases, but committed suicide at the age of 37, probably during a depressive episode.

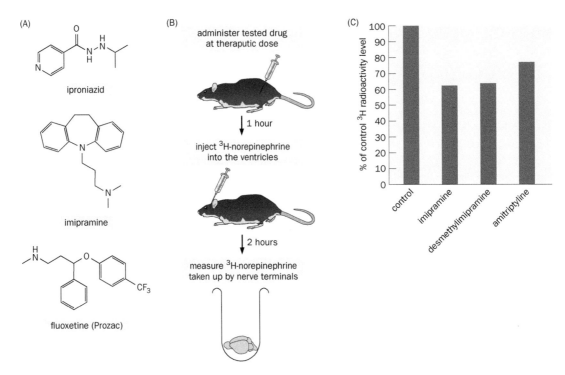

Figure 12-26 Structures and mechanisms of action of representative antidepressants. (A) Resembling the structure of monoamines (Figure 3-16), iproniazid inhibits monoamine oxidase, while imipramine inhibits the plasma membrane transporters for serotonin and norepinephrine, and fluoxetine selectively inhibits the plasma membrane transporter for serotonin. **(B)** Procedure to assay norepinephrine uptake by the brain. First, a drug was administered. Then, radioactive ^{3}H-norepinephrine was injected into cerebral ventricles (because norepinephrine does not cross the blood–brain barrier). ^{3}H-norepinephrine molecules retained in the brain 2 h after injection indicated nerve terminal uptake; that not taken up by nerve terminals was presumably metabolized. **(C)** Compared to a control, application of three clinically effective antidepressants markedly reduced brain uptake of ^{3}H-norepinephrine. In the same assay, clinically ineffective antidepressants did not inhibit norepinephrine uptake (not shown). (B & C, from Glowinski J & Axelrod J [1964] *Nature* 204:1318–1319. With permission from Springer Nature.)

action of some neurotransmitters (particularly monoamines) is terminated. The effect of various drug treatments on reuptake were determined using a quantitative assay that measured how much experimentally administered radioactively labeled norepinephrine brain tissues retained (Figure 12-26B). Antidepressants such as imipramine reduced the level of norepinephrine retained in brain tissues compared with controls and other drugs lacking antidepressant effects (Figure 12-26C), indicating that imipramine inhibits norepinephrine reuptake.

Each monoamine neurotransmitter has its own PMAT encoded by a distinct gene. Inhibiting its reuptake prolongs a neurotransmitter's actions. Imipramine inhibits the PMATs of norepinephrine and serotonin. Drugs developed subsequently, such as **fluoxetine** (brand name Prozac; Figure 12-26A), selectively block serotonin reuptake. **SSRIs (selective serotonin reuptake inhibitors)** are the most widely used antidepressants today. Thus, enhancing the actions of monoamine neurotransmitters, notably serotonin, can significantly relieve depressive states. As discussed in Box 9-1, serotonin (and norepinephrine) neurons cluster in various brainstem nuclei, but their axons project throughout the central nervous system, from the forebrain to the spinal cord, enabling these neurons to modulate many excitatory and inhibitory target neurons, with diverse effects. Thus, a general elevation of serotonin levels, while relieving depression in a subset of patients, also causes undesirable side effects. Identifying the primary target neurons and circuits relevant for mood regulation will be an important future goal; it will enhance our understanding of the neurobiology of mood and facilitate strategies for targeting specific populations of serotonin neurons that modulate relevant circuits so as to achieve more targeted therapeutic effects.

A mystery regarding SSRIs is that it usually takes several weeks to alleviate depression, suggesting a more complex mechanism than simply acutely elevating synaptic serotonin levels. In addition, about one-third of depression patients do not respond to SSRIs. Among alternative strategies for treating depression, ketamine has received the most attention in the past two decades. Originally developed as an anesthetic, low-dose ketamine was later found to have a rapid effect (within hours) of alleviating depressive symptoms of patients, including those resistant to SSRIs. As introduced in the previous section, ketamine is best known as an NMDA receptor antagonist, can induce psychosis, and is highly addictive. Intense research is being carried out to determine the primary molecular targets and brain circuits via which ketamine achieves its anti-depressant effects, with the goal of developing more specific antidepressants with fewer side effects.

12.17 Modulating GABAergic inhibition can alleviate symptoms of anxiety disorders

Anxiety disorders, the most prevalent class of psychiatric disorders, include generalized anxiety, phobias, panic disorder, obsessive–compulsive disorder (OCD), and posttraumatic stress disorder (PTSD). Anxiety is the emotional state of worry in the absence of immediate danger. Generalized anxiety disorder alone has a lifetime prevalence of more than 5% in the general population; patients with this disorder exhibit persistent worries about impending misfortune, often with physical symptoms such as fatigue, muscle tension, and sleep disturbance.

Barbiturates and their derivatives were the first classes of drugs used to treat anxiety disorders. Barbiturates are also potent sedatives, and a more serious concern is that overdose is lethal. Once again, fortuitous discoveries since the 1950s produced a new class of molecules called **benzodiazepines** (Figure 12-27A), which are effective at relieving anxiety. Importantly, while benzodiazepine overdose induces lengthy sleep, the lethal dose is much higher than that of barbiturates. Benzodiazepines have therefore largely replaced barbiturates in treating several anxiety disorders.

Both barbiturates and benzodiazepines exert their anxiolytic (anxiety-reducing) effects by binding the ionotropic GABA$_A$ receptors to enhance GABA transmission (Section 3.17). However, barbiturates, benzodiazepines, and GABA have distinct binding sites on GABA$_A$ receptors. At high concentrations, barbiturates can activate GABA$_A$ receptors independently of GABA, causing hyperpolarization of target neurons via chloride influx through the GABA$_A$ channels. By contrast, benzodiazepines enhance GABA$_A$ receptors' affinities for GABA and other GABA agonists such as barbiturates or alcohol without directly activating the receptors. Thus, in enhancing the action of endogenous GABA, benzodiazepines act as **allosteric agonists** (Figure 12-27B). This accounts for the relative safety of benzodiazepines compared to barbiturates: their maximal effect is limited by the amount of endogenous GABA.

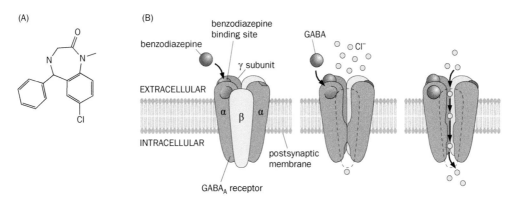

Figure 12-27 Benzodiazepines act as allosteric agonists of the GABA$_A$ receptor. (A) Structure of diazepam (Valium), a benzodiazepine. **(B)** Schematic of benzodiazepine action as an allosteric agonist of the GABA$_A$ receptor. Here, the GABA$_A$ receptor pentamer consists of two α subunits, two β subunits, and one γ subunit, the most commonly occurring subunit composition. Binding of a benzodiazepine to the GABA$_A$ receptor at the interface of the α and γ subunits does not open the channel (left) but enhances the receptor's affinity for GABA. The GABA$_A$ receptor channel is opened by binding of two GABA molecules at the interface of the α and β subunits (middle and right). To visualize the channel opening, the β subunit at the front is omitted in the middle and right panels. (See Zhu S, Noviello CM, Teng J, et al. [2018] *Nature* 559:67–72 for cryo-EM structures of the GABA$_A$ receptor in complex with GABA and a benzodiazepine site antagonist.)

Figure 12-28 Dissociating functions of GABA$_A$ receptor subunits α1 and α2 in promoting sedation and relieving anxiety. **(A)** Left, when the α1 subunit of the GABA$_A$ receptor was rendered incapable of binding benzodiazepines by changing a histidine residue to arginine (H101R), diazepam no longer induced sedation as assayed by locomotor activity, whereas locomotor activity of wild-type animals was reduced by diazepam in a dose-dependent manner. Right, benzodiazepine still had anxiolytic effects on mutant mice, as assayed by the time mice spent in the lit area of an open field. **(B)** When the α2 subunit of the GABA$_A$ receptor was rendered incapable of binding to benzodiazepine, diazepam still induced sedation as it did in wild type (left), while the anxiolytic effects of diazepam were abolished (right). In both panels, locomotor activity was measured by the number of line crossings as mice explored an open field. Anxiety was measured by the amount of time mice spent in areas that were lit versus dark (anxious mice excessively avoid lit areas). (A, from Rudolph U, Crestani F, Benke D, et al. [1999] *Nature* 401:796–800. With permission from Springer Nature. B, from Löw K, Crestani F, Keist R, et al. [2000] *Science* 290: 131–134.)

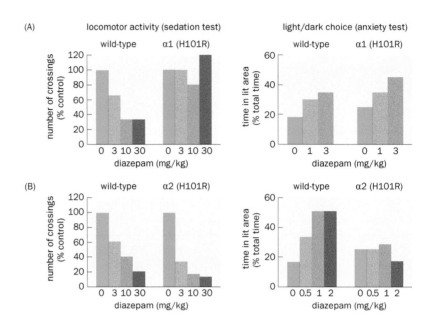

Benzodiazepines have fewer side effects than barbiturates, but they still induce sedation. Is it possible to separate the anxiolytic and sedative effects? Molecular-genetic studies of benzodiazepine action have offered some clues. In humans and mice, six separate genes, which are differentially expressed in the brain, encode the GABA$_A$ receptor subunits α1–α6. Subunits α1, α2, α3, and α5 possess a specific histidine residue (located at position 101 of α1 and α2) that renders them sensitive to benzodiazepines; α4 and α6 have an arginine at the same position, which disrupts benzodiazepine binding. In other words, benzodiazepines affect only brain regions expressing high levels of α1-, α2-, α3- or α5-containing GABA$_A$ receptors. Replacing the histidine (H) with arginine (R) can cause a benzodiazepine-sensitive subunit to become insensitive without affecting GABA$_A$ receptor function. Using genetic knock-in (Section 14.7), researchers produced mice with a H101 → R101 substitution in the α1 subunit. These mice (designated as H101R) were no longer sedated by benzodiazepines but still responded to benzodiazepines' anxiolytic effects. By contrast, benzodiazepines could still sedate mice with the H101R substitution in the α2 subunit but lost their anxiolytic effects in these mice (**Figure 12-28**). These experiments suggested that benzodiazepines promote sedation via α1-containing GABA$_A$ receptors and relieve anxiety via α2-containing GABA$_A$ receptors. These differential effects are likely caused by differential expression of α subunits in brain regions that control the functions of different circuits. Indeed, one of the highest α2-expressing regions is the amygdala, a center that regulates emotion and fear responses (Section 11.23). In principle, drugs that specifically elevate the function of GABA$_A$ receptors containing α2 but not α1 subunits should be more effective and specific anxiolytics than the benzodiazepines currently in use.

GABA is the major inhibitory neurotransmitter in the brain and mediates many diverse physiological functions. In addition to reducing anxiety, drugs that affect the GABAergic system have been used to treat epilepsy (Box 12-4), pain, and sleep problems. Studies of benzodiazepine action illustrate how investigating drug action can uncover a neurotransmitter's diverse functions; such investigation, in turn, helps inform design of drugs better suited for treating specific disorders.

A limitation of benzodiazepines in treating anxiety disorders is that long-term use can lead to both physiological tolerance and addiction. Increasing doses become required to achieve similar effects, and halting treatment results in withdrawal symptoms. SSRIs used to treat depression (Section 12.16) also have anxiolytic effects but are not addictive, and thus are the first-line treatment for most anxiety disorders. The mechanisms by which altering serotonin levels alleviates anxiety are being intensely investigated.

In addition to medication, other effective treatments for anxiety and related disorders include cognitive behavioral therapy, which focuses on symptom management via the guidance of a therapist. The goal is to be able to realize when one is at risk of entering into a maladaptive pattern of behavior or thought in the moment before it happens and to prevent it. As a specific example, PTSD can be treated with prolonged exposure therapy, wherein patients describe in extensive detail a traumatic experience in a safe environment in the presence of a therapist who can help maximize the exposure but stop the experience in case it becomes dangerous. Such therapy takes advantage of the malleability of memory during reconsolidation (Section 11.23).

12.18 Addictive drugs hijack the brain's reward system by enhancing the action of VTA dopamine neurons

Addictive substances, including alcohol from fermentation, opium from poppy plants, and cocaine from coca leaves, have been used by humans for thousands of years. Successful chemical syntheses of active components and new methods for efficient delivery have both contributed to an increasing prevalence of drug abuse, which constitutes a significant problem for humanity spanning many nations and cultures. **Drug addiction** is defined as compulsive drug use despite long-term negative consequences and is associated with loss of self-control and propensity to relapse. What are the neurobiological bases of addiction?

Remarkably, almost all drugs of abuse have one common effect: they increase dopamine concentration at the output targets of **ventral tegmental area** (**VTA**) dopamine neurons, including the VTA itself. Two major output targets of VTA dopamine neurons are the **nucleus accumbens** (**NAc**, a major part of the ventral striatum), best known for processing reward signals (Section 11.24), and the prefrontal cortex, responsible for executive functions such as goal selection and decision making (**Figure 12-29**). VTA dopamine neurons receive inputs from many parts of the brain, including glutamatergic excitatory input from the prefrontal cortex and GABAergic inhibitory input from local and NAc neurons.

Different drugs of abuse enhance dopamine action via distinct mechanisms. For example, nicotine enhances excitatory input onto VTA dopamine neurons by presynaptic excitation—it activates nicotinic ACh receptors (Section 3.13) located on glutamatergic presynaptic terminals, causing increased glutamate release and, hence, greater excitation of dopamine neurons. Nicotine can also excite dopamine neurons directly through nicotinic ACh receptors on the dopamine neurons themselves. By contrast, opioids, benzodiazepines, and cannabinoids act by hyperpolarizing, and thereby inhibiting, local GABAergic neurons in the VTA, causing disinhibition of dopamine neurons. Ethanol is known to boost dopamine concentration at the VTA and NAc, but the exact mechanisms and site(s) of action remain unclear. The psychostimulant drugs cocaine and amphetamine act by enhancing dopamine's effects at presynaptic terminals of dopamine neurons. Cocaine blocks the **plasma membrane dopamine transporter** (**DAT**) for dopamine reuptake, thus increasing the dopamine concentration in the synaptic cleft post-release.

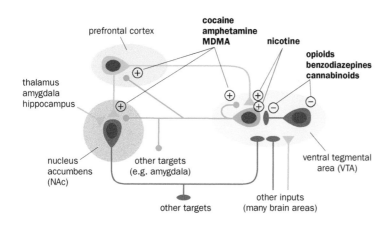

Figure 12-29 Drugs of abuse increase dopamine action at VTA dopamine neuron targets. A highly simplified circuit diagram illustrating connections between the ventral tegmental area (VTA), nucleus accumbens (NAc), and prefrontal cortex. Blue, dopamine neurons; red, GABAergic neurons; green, glutamatergic neurons. Also summarized are the action sites of the most common drugs of abuse, all of which enhance dopamine concentration in target regions. +, enhancement; –, suppression. See text for details. (Based on Lüscher C & Malenka RC [2011] *Neuron* 69:650–633 and Sulzer D [2011] *Neuron* 69:628–649.With permission from Elsevier Inc.)

Amphetamines (including MDMA, or 3,4-methylenedioxy-*N*-methylamphetamine, commonly known as ecstasy) have more complex effects: (1) they reverse the normally unidirectional transport mediated by DAT (from synaptic cleft to presynaptic cytosol), causing presynaptic vesicle-independent release of dopamine into the synaptic cleft; (2) they enhance dopamine biosynthesis; and (3) they inhibit dopamine degradation. All these effects enhance dopamine action (Figure 12-29).

How does enhancement of the dopamine action of VTA neurons lead to addiction? As we learned in Section 11.24, projections from VTA dopamine neurons to the NAc play a critical role in reward-based learning. Specifically, studies in primates and rodents have shown that many VTA dopamine neurons encode reward prediction errors. These error signals are proposed to affect synaptic plasticity in target neurons in the NAc and prefrontal cortex for reinforcement learning. If VTA dopamine neurons signal a reward, the action or behavior that immediately preceded the reward is reinforced through dopamine modulation of downstream circuits (Figure 11-47). Drugs of abuse bypass natural signals that activate these dopamine neurons, thus dissociating the reward system from its natural stimuli. Specifically, by increasing dopamine concentration at dopamine neurons' presynaptic terminals, drug consumption mimics dopamine neuron activation, reinforcing the preceding actions, including drug consumption itself. Thus, addictive drugs hijack the brain's reward system and exploit mechanisms that otherwise regulate learning and motivated behaviors.

What are the cellular and molecular mechanisms underlying the long-lasting behavioral changes in drug addiction? As discussed in Chapter 11, learning is mediated by synaptic plasticity in relevant neural circuits. Drugs of abuse likely act by altering synaptic weights in circuits involving VTA dopamine neurons and their targets. For example, a single *in vivo* exposure to cocaine induced marked enhancement of excitatory input onto VTA dopamine neurons, as measured by an increased ratio of AMPA receptor (AMPAR)-mediated current to NMDA receptor (NMDAR)-mediated current in subsequent whole-cell patch clamp recording in VTA slices *in vitro* (**Figure 12-30**A). This effect is similar to long-term potentiation at hippocampal synapses (Section 11.7). Indeed, the cocaine-induced increase in the AMPAR/NMDAR ratio was NMDAR-dependent and occluded subsequent LTP induced *in vitro*. Other drugs of abuse, including morphine, nicotine, and ethanol, cause similar increases in the AMPAR/NMDAR ratio of VTA dopamine neurons (Figure 12-30B). Addictive drugs can likewise affect excitatory synapses onto spiny projection neurons in the NAc, which receive input from diverse brain regions (Figure 12-29). The VTA-NAc-prefrontal cortex circuits are complex and heterogeneous: for instance, while some VTA dopamine neurons signal reward prediction errors, other dopamine neurons signal aversion, and still other dopamine neurons signal the salience of stimuli (Section 11.24 and Box 11-4). Identifying the specific synaptic connections and circuits that are modulated by drugs of abuse will be crucial in establishing causal relationships between synaptic changes and addictive behaviors.

In addition to the drugs' rewarding effects, one difficulty of treating drug addiction is the extremely unpleasant physical and emotional responses to drug *with-*

Figure 12-30 Exposure to drugs of abuse causes long-lasting enhancement of excitatory input to VTA dopamine neurons. (A) Dopamine neurons in VTA slices were recorded by whole-cell patch clamp 24 h after a single *in vivo* exposure to cocaine to measure the magnitude of excitatory postsynaptic currents conducted by AMPA and NMDA receptors (AMPARs and NMDARs). The total AMPAR- and NMDAR-mediated excitatory postsynaptic current in response to stimulating input axons was measured at +40 mV, relieving the NMDAR Mg²⁺ block (Section 3.15); then an NMDAR antagonist was added to isolate the AMPAR current in response to the same input stimulation. The NMDAR current was calculated by subtracting the AMPAR current from the total current. Compared to naive and saline injection controls, cocaine exposure caused an increase in the AMPAR current and an enhanced AMPAR/NMDAR current ratio, indicative of potentiated synapses. **(B)** The AMPAR/NMDAR current ratio is similarly enhanced following *in vivo* exposure to five addictive substances, but not to nonaddictive drugs such as fluoxetine. (A, adapted from Ungless MA, Whistler JL, Malenka RC, et al. [2001] *Nature* 411: 583–587. With permission from Springer Nature. B, adapted from Saal D, Dong Y, Bonci A, et al. [2003] *Neuron* 37:577–582. With permission from Elsevier Inc.)

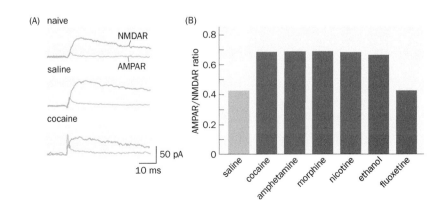

drawal. Interestingly, brain regions involved in the rewarding effects of a drug also contribute to these negative withdrawal symptoms. For example, while enhanced dopamine release in the NAc contributes to the rewarding effects of opioids, pharmacological experiments suggest that dopamine D_2 receptors in the NAc also mediate opioid withdrawal symptoms. Recent circuit analyses revealed that activation of a pathway from the paraventricular nucleus of the thalamus (PVT) to the NAc mediates several morphine withdrawal symptoms. Chronic morphine exposure causes long-term potentiation of the synapses between PVT and NAc dopamine D_2 receptor–expressing spiny projection neurons, and a reduction of this potentiation suppresses morphine withdrawal symptoms. A deep understanding of the neurobiology underlying drug withdrawal is crucial to combating drug abuse.

Finally, knowledge of the neurobiology and pharmacology of addiction has led to effective treatments using partial agonists. A **partial agonist** is a drug that binds to a target receptor but elicits only a partial biological effect (**Figure 12-31**). If the partial agonist binds with higher affinity than a full agonist to overlapping binding sites on the receptor, it can *antagonize* the effect of the full agonist. For example, varenicline was developed as a high-affinity partial agonist to the α2β4 nicotinic acetylcholine receptor, the major isoform expressed in VTA neurons, and has proven effective in promoting smoking cessation. Being a partial agonist, varenicline triggers a small amount of dopamine release, thus relieving the negative withdrawal symptoms of not smoking. At the same time, it diminishes the reinforcing effect of smoking by antagonizing nicotine-mediated dopamine release. Likewise, buprenorphine, a high-affinity partial agonist of the μ opioid receptor, is efficacious in treating opioid addiction.

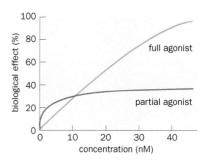

Figure 12-31 Partial agonist for treating drug addiction. A partial agonist activates its receptor to a lesser extent than the full agonist, and thus elicits a smaller biological effect. In the presence of a full agonist, a higher affinity agonist (as depicted here) can antagonize the activity of a full agonist if they compete for binding sites on the receptor. Partial agonists have been used to effectively treat nicotine and opioid addictions.

12.19 Human genetic studies suggest that many genes contribute to psychiatric disorders

As we see from the previous sections, studies of the actions of therapeutic and abuse-related drugs have enriched our understanding of normal brain function. A major limitation of relying on drug action to reveal the pathophysiology of psychiatric disorders, however, is that our understanding is limited to drug targets, which may be related only to the symptoms of disorders rather than their root causes. Twin studies of schizophrenia and mood and anxiety disorders all indicate significant genetic contributions, with a heritability (the proportion of phenotypic differences contributed by genetic differences; see Section 1.1 on how to estimate heritability) of up to 80% for schizophrenia and bipolar disorder. Heritability of major depression and general anxiety disorders is lower, at 30–40%. Family studies also suggest that various psychiatric disorders may share common genetic factors. For example, family members of a schizophrenic patient have an increased chance of developing not only schizophrenia but also bipolar disorder. Likewise, major depression and generalized anxiety disorders often run together in families. These studies implicate strong genetic contributions to these psychiatric disorders. At the same time, nonheritable factors also contribute significantly— these could be environmental factors, epigenetic influences, or *de novo* mutations that occur in the parental germline or during early embryonic development (**Box 12-3**). Because environmental factors are multifaceted and difficult to track, hope has been placed on identifying genes that contribute to psychiatric disorders as a means to gain new insight into their origins and identify new targets for drug development.

So far, a simple Mendelian inheritance pattern has not been identified in any of the psychiatric disorders we discussed thus far (in contrast to Huntington's disease and certain familial forms of Parkinson's and Alzheimer's diseases caused by dominant or recessive mutations in single genes). This suggests that multiple genetic factors contribute to each psychiatric disorder and that each factor in itself only increases susceptibility, as with ApoE *ε4* in Alzheimer's disease. The recent genomic revolution has generated new tools for identifying genetic variations that contribute to complex diseases, including psychiatric disorders. These studies suggest that schizophrenia and bipolar disorder result from modest contributions of

Table 12-2:   Selected candidate genes associated with psychiatric disorders[a]

Protein encoded by candidate gene	Identified on the basis of[b]	Associated with	Also associated with	Physiological functions
Drd2[c]	Genome-wide association study (GWAS)	Schizophrenia	Major depression[g]	Dopamine receptor D$_2$ that couples dopamine binding to G protein signaling
Ca$_v$1.2[c]	GWAS	Schizophrenia	Autism spectrum disorder	Voltage-gated Ca^{2+} channel with large conductance, used in neuronal and synapse-to-nucleus signaling, and cardiac muscle contraction
Neurexin-1[d]	Copy number variation	Schizophrenia	Autism spectrum disorder	Cell-surface protein for synapse development and trans-synaptic signaling
Teneurin-4[e]	GWAS	Bipolar disorder	Schizophrenia	Cell-surface protein for wiring specificity and trans-synaptic signaling
Laminin-α2[f]	Whole-exome sequencing	Schizophrenia		Extracellular matrix protein for cell adhesion, axon growth, and synapse development
Component 4 (C4)	GWAS	Schizophrenia		Component of the complement cascade involved in synapse pruning

[a]This list represents a very small subset of findings from a large body of human genetic studies of psychiatric disorders.

[b]See Box 12-3 for definitions.

[c]Data from Schizophrenia Working Group of the Psychiatric Genomics Consortium (2014) *Nature* 511:421.

[d]Data from Rujescu et al. (2009) *Hum Mol Genet* 18:988.

[e]Data from Psychiatric GWAS Consortium Bipolar Disorder Working Group (2011) *Nat Genet* 43:977.

[f]Data from Xu et al. (2012) *Nat Genet* 44:1365.

[g]Data from Howard et al. (2019) *Nat Neurosci* 22:343.

many genetic variants, with the total number of variants estimated to be in the hundreds. These variations can take the form of point mutations or differences in gene copy numbers; some are inherited, while others are produced *de novo*. The identities of the candidate susceptibility genes (Table 12-2) also suggest that abnormal neuronal signaling and neural development contribute significantly to psychiatric disorders, and that each of these genes may increase susceptibility to multiple disorders.

For example, the *Drd2* locus, which encodes the dopamine receptor D$_2$ and is widely expressed in the brain, including in the spiny projection neurons that constitute the indirect pathway of the basal ganglia (Figure 8-21), was identified in a large-scale **genome-wide association study** (**GWAS**; see Box 12-3 for details) as a risk factor for schizophrenia. As discussed in Section 12.15, the D$_2$-type dopamine receptor is a major target of all effective antipsychotic drugs used today; thus, the GWAS finding provides a satisfying link between recent genomic approaches and decades of drug-based investigation. (Notably, *Drd2* is also associated with major depression.) GWAS of schizophrenia also identified genetic loci encoding other neuronal signaling molecules, including glutamate receptors and voltage-gated Ca^{2+} channels that play important roles in synapse-to-nucleus signaling (Section 3.24). Genes important in neural development have also been associated with schizophrenia and bipolar disorder; these include genetic loci encoding Neurexin-1 and Teneurin-4, transmembrane proteins that play important roles in trans-synaptic signaling and wiring specificity (Sections 7.11 and 7.23 and Box 7-2). A strong link between neural development and psychiatric disorders was further suggested by the observation that some psychiatric disorder susceptibility genes are also associated with neurodevelopmental disorders such as autism spectrum disorder.

Rapid advances in human genetics will undoubtedly uncover many more susceptibility genes. Progressing from genetic variations to mechanistic understanding and rational drug design for treatment, however, poses substantial challenges. As discussed in the first part of this chapter, animal models are instrumental for studying disease mechanisms and testing therapeutic strategies. However, many symptoms of psychiatric disorders, such as hallucination, delusion, and depression, are more difficult to model in animals than neurodegeneration. The small effects of individual susceptibility genes further complicate efforts to create effective animal models. Researchers are developing behavioral (Section 14.29) and physiological assays in animals that mimic specific aspects of psychiatric disorders. Additional model systems, including nonhuman primates and induced pluripotent stem cell–derived organoids (Box 7-1), are being explored. Finally, investigating the physiological and developmental functions of susceptibility genes may ultimately provide new insights into why their disruption contributes to psychiatric disorders and how different susceptibility genes interact with each other and with environmental factors.

We end with an example of how human genetics data can shed new light onto the biology of schizophrenia. Among the >100 loci identified from large-scale GWAS studies, a locus on human chromosome 6 exhibits the strongest association with schizophrenia (**Figure 12-32**A). Analysis of this locus revealed considerable variations in the gene *C4*, encoding component 4 of the complement cascade. These variations—isoforms of the gene (*C4A* and *C4B*, which bind different molecular targets), copy numbers, the presence or absence of an intronic insertion—result in different *C4A* expression levels. Remarkably, schizophrenia risks are proportional to *C4A* expression levels (Figure 12-32B). Moreover, *C4A* expression levels

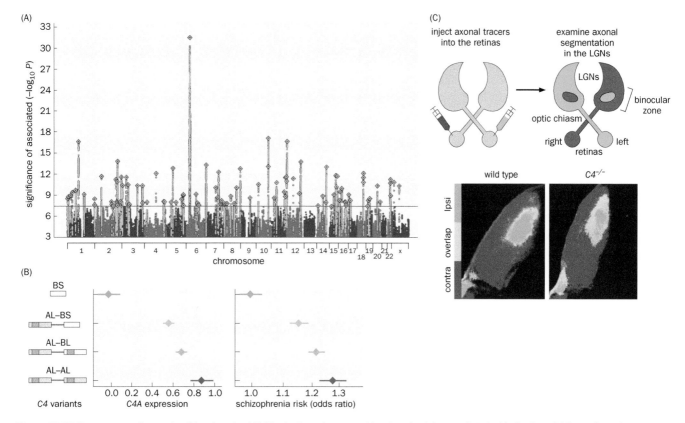

Figure 12-32 Synapse pruning and schizophrenia. (A) Manhattan plot summarizing genome-wide (34,214 patients and 45,604 controls) and family-based (1,235 parent affected–offspring trios) association studies of schizophrenia. The *x* axis is chromosomal position and *y* axis is significance of association. The red line shows the genome-wide significance level ($p = 5 \times 10^{-8}$). Each dot represents a single-nucleotide polymorphism (see Box 12-3). Green, significant association. **(B)** Positive correlation between C4A expression and schizophrenia risk. Left, four common forms of the *C4* gene, located at the chromosome 6 locus with the highest GWAS signal in Panel A. BS and BL represent short and long versions of *C4B*, respectively, and AL represents the long version of *C4A*. Long and short versions differ by an intronic insertion of a retroviral element (orange) that elevates the expression level. Middle, relative *C4A* expression level for the four *C4* forms, estimated from RNA expression in postmortem samples. Right,

schizophrenia risk associated with the four *C4* forms based on analysis of 28,799 patients and 35,986 controls. See Box 12-3 for an explanation of odds ratio. **(C)** *C4* mutant mice exhibit deficient eye-specific segregation of retinal ganglion cell (RGC) axons in the lateral geniculate nucleus (LGN). Top, schematic borrowed from a similar experiment in ferrets (see Figure 5-22). Different anterograde tracers were injected into the left and right retina, and RGC axons were visualized at the LGN. Bottom, while control RGC axons from ipsi- and contralateral eyes show minimal overlap (yellow) in the LGN, RGC axons from two eyes in *C4* mutant mice have substantial overlap, suggesting a deficit in synapse pruning. (A, from Schizophrenia Working Group of the Psychiatric Genomics Consortium [2014] *Nature* 511:421–427. With permission from Springer Nature. B & C, from Sekar A, Bialas AR, de Rivera H, et al. [2016] *Nature* 530:177–183. With permission from Springer Nature.)

measured in postmortem brain samples from schizophrenia patients are significantly higher than those in controls. As we learned in Section 7.13, one function of the complement cascade is to tag synapses for pruning by microglia during development. Indeed, *C4* mutant mice exhibit deficient eye-specific segregation of retinal ganglion cell axons in the lateral geniculate nucleus (Figure 12-32C), a process believed to be mediated by activity-dependent synapse pruning. These data are consistent with the possibility that elevated *C4A* expression increases schizophrenia risk by causing excessive synapse pruning.

Box 12-3: How to collect and interpret human genetics data for brain disorders

Brain disorders caused by single-gene mutations that follow Mendelian inheritance patterns are the simplest to study from a genetics perspective (**Figure 12-33**). **Autosomal dominant** mutations cause phenotypes resulting from toxic gain-of-function effects due to the mutant allele or from loss-of-function effects due to insufficient normal gene products produced from the remaining wild-type allele. **Autosomal recessive** mutations cause phenotypes resulting from loss-of-function effects due to disruption of both alleles. (We have seen both in familial forms of neurodegenerative diseases.) Single-gene mutations can also be **sex-linked** (when the mutant gene is located on the X chromosome). Sex-linked mutations usually affect males more severely than

females, because mutations on a male's single X chromosome exert their effects in all cells (the human Y chromosome carries few genes). In females, by contrast, one of the two X chromosomes is randomly inactivated in each cell since early development (**random X-inactivation**), so that sex-linked mutations are expressed in about half of a female's cells. Red–green color blindness (Section 4.11) is a good example of a sex-linked trait. Genes causing Mendelian disorders can be mapped by pedigree analyses and by molecular markers distributed across the genome (Figure 12-4A).

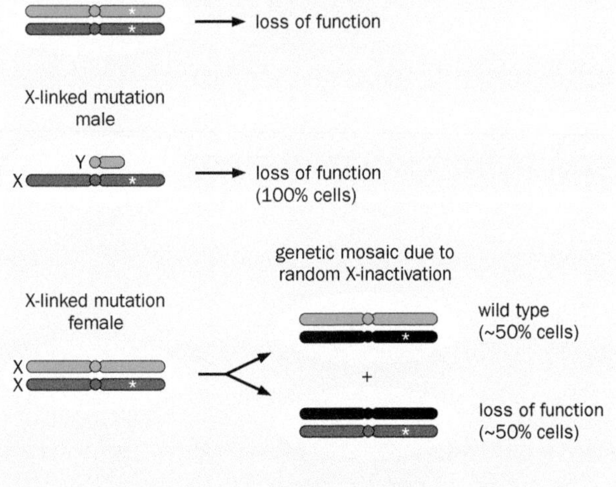

Most brain disorders that are defined by symptoms or pathology do not follow simple Mendelian inheritance patterns. These disorders can in principle be caused by (1) inheritance of multiple genetic variants that interact with each other, (2) ***de novo* mutations** that occur in the parental germline and—with the exception of X-chromosome mutations inherited by a female—affect all of the patient's cells, (3) ***de novo* somatic mutations** that occur in progenitor cells and therefore affect a subset of the patient's cells, (4) environmental factors, and (5) any of the above factors acting in combination. Of these, only factor 1 (and a small fraction of factor 2; see below) contributes to heritability. Therefore, if a genetic disorder such as schizophrenia or bipolar disorder has a high heritability but no clear Mendelian inheritance pattern, then multiple inherited mutations must contribute to the disorder, interacting either with each other or with additional factors.

A conceptually simple way to identify genes contributing to a given disorder is to perform a genome-wide association study (GWAS), taking advantage of **single nucleotide polymorphisms (SNPs)** present throughout the human genome. All individuals have about 3.5 million SNPs when compared to the reference human genome. If a SNP is close to or within a gene whose mutations contribute to a disease, then it should be tightly linked with the disease-contributing mutation in the general population. DNA samples collected from many patients (usually thousands or more) can be compared with those from a similar number of healthy controls (ideally relatives or populations with the same ethnicity and geographic distribution) to identify the SNPs most strongly linked with the disease. The strength of the association can be quantified by parameters such as the **odds ratio**, the probability of having the disease among people with the SNP divided by the probability of having the disease among people without the SNP. Given that most brain disorders have

Figure 12-33 Three types of Mendelian inheritance. Left, genotypes are represented by pairs of homologous chromosomes, with one chromosome inherited from the father (blue) and one from the mother (red), in each cell (yellow oval). * designates a mutation. Right, summary of mutation effects. Black chromosomes indicate inactivated X chromosomes.

Box 12-3: continued

multiple genetic risk factors, the odds ratio is a complex function of both the heterogeneity of the patient population and the penetrance of the linked mutation that contributes to the disease. For schizophrenia and bipolar disorder, disease-associated SNPs have odds ratios between 1.1 and 1.3. By comparison, GWAS studies identified *Apoe ε4* as having an odds ratio of about 3.5 for Alzheimer's disease.

While SNPs have historically been detected by DNA microarray analyses (see Section 14.12 for details), recent advances in sequencing technology (Box 14-2) have made it possible to sequence whole exomes (the roughly 1% of genomic DNA sequences that correspond to exons) or whole genomes of patients and control subjects; these methods offer powerful ways to identify disease-causing DNA variants. Whole-exome and whole-genome sequencing have revealed that *de novo* mutations contribute significantly to many brain disorders. *De novo* mutations usually occur spontaneously in the germline of parents, more commonly in the paternal germline, as spermatogenesis involves many more cell divisions than oogenesis and hence presents more opportunities for DNA replication errors. By definition, *de novo* mutations do not affect the phenotypes of parents and do not contribute to heritability (except in the rare case where the mutations occur early in parental germline development and affect the sperm or eggs inherited by multiple progeny, contributing to sibling similarities). Thus, whole-exome or whole-genome sequencing of patients with a specific disease and their healthy parents (trios) should reveal *de novo* mutations that contribute to the disease. A complication is that *de novo* mutations occur even in healthy individuals, with an incidence of about one gene-disrupting *de novo* mutation per individual; in fact, each of us has about 100 inherited gene-disruption mutations in our genome. As a result, identifying which *de novo* mutations contribute to a given disease involves complex statistical analysis, taking into consideration sequence conservations and possible physiological functions of affected proteins. In general, if *de novo* mutations affect the same gene in more than one patient with the same disease (as was the case with laminin α2; Table 12-2), then the probability that these mutations contribute to the disease increases.

Among *de novo* mutations, **copy number variations (CNVs)** make a major contribution to brain disorders. CNVs are deletions or duplications of chromosomal segments varying in length from 500 base pairs to several million base pairs (Mb) and containing coding sequences ranging from a small fraction of a single gene to many genes. CNVs can also be inherited if carriers bear progeny. Frequently occurring CNVs are associated with repeat elements in the genome that cause errors in DNA recombination during the meiotic cell cycles that produce sperm and eggs. Healthy humans carry an average of about 1,000 polymorphic CNVs; having one or three copies of most genes does not significantly impact health. However, some genes are dosage sensitive, such that losing a copy or gaining an extra copy can contribute to or cause specific disorders. For instance, spontaneous deletion of a 3-Mb segment on chromosome 17 flanked by genomic repeats affects 1 in 15,000–25,000 people, causing a neurodevelopmental disorder called **Smith–Magenis syndrome**, which is characterized by mild to moderate intellectual disability, delayed speech, sleep disturbances, and impulse control and other behavioral problems. Although the common deletion contains >30 genes, losing one copy of a single gene called *Rai1* (retinoic acid induced 1) within the common deletion interval is sufficient to cause most of the symptoms. Remarkably, duplication of this genomic region (which occurs at the same frequency as the common deletion) results in **Potocki–Lupski syndrome**, which is likely caused by an increased dose of *Rai1* and is associated with mild intellectual disability and autistic symptoms. Thus, the proper expression level of *Rai1*, which encodes a chromatin-binding protein that regulates gene expression, is critical for proper brain development and function. As another example, deletions of one copy of part of the gene encoding Neurexin-1 markedly increase the odds of developing schizophrenia and autism (Table 12-2).

NEURODEVELOPMENTAL DISORDERS

Whereas neurodegenerative and psychiatric disorders usually have an adult or adolescent onset, the symptoms of neurodevelopmental disorders first appear in infancy or early childhood. Depending on the types of symptoms, neurodevelopmental disorders are categorized as intellectual disability (ID, previously referred to as mental retardation), autism spectrum disorder (ASD), communication disorders, attention deficit hyperactivity disorder (ADHD), learning disorder, or motor disorders. Recent work suggests that some neurodevelopmental disorders share similar underlying genetic causes with some psychiatric disorders. In the following sections, we start with a general discussion of ID and ASD, two developmental disorders that are significant both in their frequency and their profound effects on patients and their caregivers. We then focus on two specific syndromes featuring symptoms of both ID and ASD, as approaches pioneered by research on these syndromes can apply to studies of other neurodevelopmental disorders.

12.20 Intellectual disability and autism spectrum disorder are caused by mutations in many genes

Intellectual disability is characterized by deficits in general mental abilities such as reasoning, problem solving, planning, abstract thinking, judgment, and learning from experience. ID patients usually have an intelligence quotient (IQ) of 70 or less, which is two standard deviations below the age-matched population mean (Figure 1-2). ID affects 1–3% of the general population.

Genetic factors, including chromosomal abnormalities and monogenic causes, account for a large fraction of ID cases, especially for those with IQs below 50. ID can also be one feature of **syndromic disorders** characterized by defined constellations of behavioral, cognitive, and physical symptoms. For example, Down syndrome results from having an extra copy of chromosome 21 and is the most common genetic form of ID (affecting 1 in 500–1000 births). ID can also be caused by genetic mutations in the absence of recognizable syndromes or global structural abnormalities of the brain; these are called nonsyndromic ID (NS-ID). Because the primary symptom of NS-ID is intellectual impairment, the corresponding genes may function more specifically in processes related to learning and intellectual capabilities and are thus of considerable interest to scientists seeking to understand the biological bases of cognition.

Genetic mapping studies in the past two decades have identified dozens of genes that, when mutated, cause NS-ID; a large fraction reside on the X chromosome. As males have only one copy, mutations on the X chromosome affect all cells (Figure 12-33) and are therefore technically easier to identify than autosomal recessive mutations. Because mutations on the X-chromosome likely account for just a small fraction of NS-ID cases, genetic causes for NS-ID may involve hundreds of genes across the genome. ID-associated genes identified to date encode proteins functioning as transcriptional regulators and cell-adhesion and signaling molecules important for brain wiring, as well as molecules known to regulate synapse development and function. In the following, we discuss one specific example.

A class of proteins often disrupted in ID are involved in Rho GTPase signaling. These proteins transduce extracellular signals to regulate cytoskeletal changes underlying axon growth and guidance, dendrite morphogenesis, and synapse development (Box 5-2). Rho GTPase signaling pathway members associated with NS-ID or syndromic ID include guanine nucleotide exchange factors (GEFs) that promote the active GTP-bound form, GTPase activating proteins (GAPs) that promote the inactive GDP-bound form, and protein kinases downstream of GTPases (**Figure 12-34**A). The first identified X-linked NS-ID gene encodes a protein called oligophrenin1 (Ophn1), which acts as a GAP for Rho GTPases. Ophn1 is widely expressed in the nervous system and is distributed in axons, dendrites, and dendritic spines (Figure 12-34B). RNAi knockdown in cultured rat hippocampal neurons resulted in decreased spine length (Figure 12-34C) and impaired synaptic transmission and synaptic plasticity. *Ophn1* knockout mice exhibited a variety of cognitive deficits, including impaired spatial learning in the Morris water maze (Figure 12-34D; Figure 11-33). Thus, in the case of Ophn1, the cognitive deficits observed in human patients may be caused in part by impaired synaptic structure and function.

Autism spectrum disorder (**ASD**) covers a wide range of symptoms and in total affects >1% of children. ASD patients often show a reduction in, or absence of, sharing of interests and emotions with others, and have difficulty adapting to different environments. They also exhibit restricted, repetitive activity patterns and excessive adherence to routines. About 70% of ASD patients also have ID, but others have normal intelligence, and some exhibit exceptional ability in memory, mathematics, art, or music.

Genetic factors are a predominant cause for ASD. For example, compared with the general population, the relative risk of a child being diagnosed with ASD is >25-fold greater if a sibling is affected. Recent genome-wide association, copy number variation, and whole-exome and whole-genome sequencing studies (Box 12-3) have identified many independent genetic variants associated with ASD.

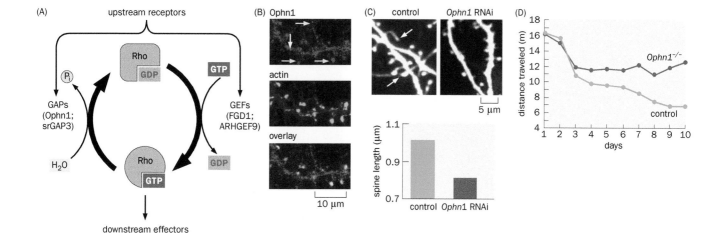

Figure 12-34 Defects in Rho GTPase signaling can cause intellectual disability. (A) Schematic of the Rho GTPase signaling pathway. Mutations in two GTPase activating proteins (GAPs), two guanine nucleotide exchange factors (GEFs), and two downstream kinases are associated with intellectual disability. The protein names corresponding to these mutations are in parentheses. **(B)** Oligophrenin1 (Ophn1), a RhoGAP, is highly concentrated in axons (cyan arrow), dendrites (white arrow), and dendritic spines (yellow arrows) of cultured rat hippocampal neurons. These cultures are doubly stained with antibodies against Ophn1 (red) and actin (green, which is concentrated in dendritic spines). **(C)** Compared with wild-type dendritic spines (arrows), dendritic spines in neurons treated with RNAi against *Ophn1* show reduced length, as quantified below. **(D)** Compared with controls, *Ophn1* mutant mice travel longer distances to reach the hidden platform during daily trials in a Morris water maze. (A, based on Pavlowsky A. Chelly J, & Billuart P [2012] *Mol Psychiatry* 17:663. B & C, from Govek EE, Newey SE, Akerman CJ, et al. [2004] *Nat Neurosci* 7:364–372. With permission from Springer Nature. D, adapted from from Khelfaoui M, Denis C, van Galen E, et al. [2007] *J Neurosci* 27:9439–9450.)

Similar to ID, genes associated with ASD encode proteins that regulate synapse development and synaptic function, transcription, and chromatin structures (see Section 12.24). Most ASD-associated genes are risk factors, and little is known yet about how they affect neural development. However, studies of specific syndromes whose symptoms overlap with ID and ASD can shed light on the underlying neurobiological mechanisms. These syndromes are usually caused by mutations in single genes with complete penetrance, and animal models often recapitulate key symptomatology, making it possible to study pathogenic processes and reveal the underlying mechanisms. Below, we use studies of Rett and fragile-X syndromes to illustrate.

12.21 Rett syndrome is caused by defects in MeCP2, a regulator of global gene expression in postmitotic neurons

First described by Andreas Rett in the 1960s in severely disabled girls who exhibited a common set of symptoms including incessant hand wringing, **Rett syndrome** is a neurodevelopmental disorder that affects 1 in 10,000–15,000 girls during early childhood. Rett patients usually develop normally for the first 6–18 months, achieving milestones such as walking and first words at a normal age. Their development then slows, arrests, and regresses. Patients exhibit social withdrawal, language loss, and other autistic features. The onset of mental deficits is also accompanied by motor symptoms such as hand wringing. The condition subsequently stabilizes and patients usually live to adulthood, albeit with severe and persistent disability.

In 1999, genetic mapping and candidate gene sequencing revealed that Rett syndrome is caused by mutations in an X-linked gene encoding a protein called **methyl-CpG-binding protein 2** (**MeCP2**; Figure 12-35). Loss-of-function *Mecp2* mutations usually lead to prenatal or infant lethality in boys, who have only one X chromosome. Girls with a loss-of-function *Mecp2* mutation are genetic mosaics for MeCP2 function and develop Rett syndrome. Because of random X-chromosome inactivation (Figure 12-33), about half of their cells have defective MeCP2, and the severity of the disorder is influenced by the pattern of random inactivation. Rett syndrome is almost always caused by *de novo* mutations in *Mecp2,* as patients are

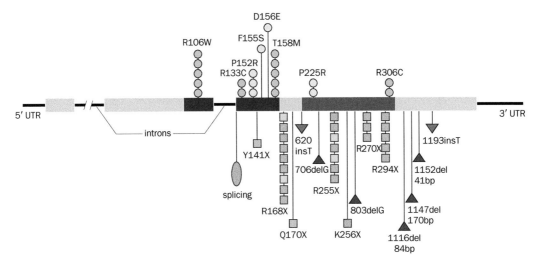

Figure 12-35 Rett syndrome is caused by mutations in the *Mecp2* gene, which encodes methyl-CpG-binding protein 2. Distribution of mutations in the coding region of the *Mecp2* gene (boxed region, interrupted by two introns) identified in Rett patients. Above the gene structure are missense mutations, each of which changes the identity of a single amino acid. Below are a splicing mutation (oval), nonsense mutations (squares), insertions (downward triangles; the inserted nucleotide is followed by "ins"), and deletions (upward triangles).

Mutations observed in multiple patients are represented by stacked symbols at the same position; most of these occur at mutational hot spots. The methyl-binding domain and transcriptional repression domain are colored blue and red, respectively. 5' UTR and 3' UTR, 5' and 3' untranslated regions. (From Amir RE, Van Den Veyver IB, Schultz R, et al. [2000] *Ann Neurol* 47:670–679. With permission from John Wiley & Sons.)

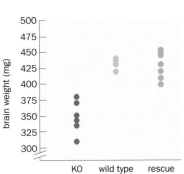

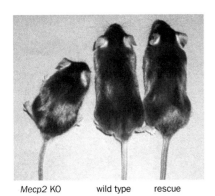

Figure 12-36 MeCP2 acts primarily in postmitotic neurons. Compared with wild-type male mice at 8 weeks, *Mecp2* knockout (KO) males have reduced size (top) and brain weight (bottom). Both phenotypes are rescued by transgenic expression of MeCP2 in postmitotic neurons of KO mice. (From Luikenhuis S, Giacometti E, Beard CF, et al. [2004] *Proc Natl Acad Sci U S A* 101:6033–6039. Copyright the National Academy of Sciences, U.S.A.)

so severely disabled that they rarely have children. Since the original discovery that *Mecp2* mutations underlie Rett syndrome, specific missense mutations in *Mecp2* that presumably have weaker effects than a complete loss-of-function have been associated with sporadic ASD and schizophrenia, highlighting shared molecular pathways between different brain disorders. The proper MeCP2 expression level is important, as duplication of *Mecp2* also causes severe neurodevelopmental defects.

MeCP2 is a nuclear protein that binds methylated DNA, including CpG sites (CpG refers to a cytidine followed by a guanosine in the DNA sequence). DNA methylation, a major form of epigenetic regulation, is usually associated with *gene repression*; for instance, methylation is the primary contributor to random X-chromosome inactivation. In the mammalian genome, CpGs are usually methylated, except where they are present in large clusters called CpG islands, which are often associated with active transcription. Because methylation states affect gene expression, MeCP2 links chromatin structure to gene expression. MeCP2 is most abundantly expressed in the brain. In mice, MeCP2 expression increases greatly during the first 5 weeks of life as the final stages of neural development occur. Biochemical analysis indicated that MeCP2 binds CpG sites along the entire genome, suggesting that MeCP2 acts as a general regulator of chromatin structure and thus affects global gene expression. Indeed, the number of MeCP2 molecules per neuronal nucleus is similar to that of histones, the principal proteins that complex with DNA to form chromatin.

How does loss of a global regulator of chromatin structure cause the neurological deficits characteristic of Rett patients? As discussed earlier in this chapter, animal models can provide important insights into human disease. MeCP2 is present in all vertebrates and is highly conserved in mammals. Indeed, *Mecp2* knockout mice mimic many aspects of Rett syndrome. As in humans, mouse *Mecp2* is located on the X chromosome. *Mecp2* mutant male mice grow normally for the first several weeks. Between 3 and 8 weeks of age, they begin exhibiting motor coordination deficits, impaired growth, and reduced brain weight (**Figure 12-36**). These symptoms progressively worsen, and most mutant males die by 12 weeks of age. Female mice heterozygous for the *Mecp2* mutation, a genetic condition equivalent to that of girls with Rett syndrome, initially develop normally. They begin exhibiting symptoms such as mild motor deficits and inertia when several months old. Importantly, conditional knockout of *Mecp2* only in neurons and glia resulted in phenotypes essentially identical to those of *Mecp2* knockout mice. Conversely, restoring MeCP2 function only in postmitotic neurons rescued many neurological phenotypes and prevented the death of *Mecp2* mutant males (Figure 12-36). These experiments indicate that MeCP2 exerts its function predominantly in postmitotic neurons.

Detailed analyses of the *Mecp2*-deficient mouse model (mostly in males) have revealed a host of defects, including size reductions in the brain, neurons, and dendritic spines, as well as alterations in dendritic morphology, synaptic transmission, and synaptic plasticity. Conditional knockout (Section 14.8) of *Mecp2* in specific neuronal populations indicated that MeCP2 plays important roles in excitatory, inhibitory, modulatory, and peptidergic neurons as well as in glia, each of which contributes to a subset of phenotypes found in mice lacking *Mecp2* in all cells. Deleting *Mecp2* in GABAergic neurons caused the most severe phenotypes, including Rett syndrome features such as repetitive, compulsive behaviors, motor dysfunction, learning deficits, and premature death. Reduction of GABAergic inhibition could also contribute to the seizures (Box 12-4) often associated with Rett syndrome.

12.22 Restoring MeCP2 expression in adulthood reverses symptoms in a mouse model of Rett syndrome

A key question in neurodevelopmental disorders is whether the symptoms are reversible. A given neurodevelopmental disorder caused by early and irreversible deficits in nervous system development presents more challenges for developing therapeutic interventions. Alternatively, when developing and adult nervous systems require a continuous supply of a disrupted gene's product for maturation and function, symptoms may be more easily reversible, given proper treatment. The mouse model for Rett syndrome provided an opportunity to ask whether MeCP2 is continuously required in the adult nervous system and to determine if phenotypes caused by *Mecp2* deficiency can be rescued by restoring MeCP2 expression in adults.

To answer the first question, a temporally controlled knockout scheme featuring CreER (see Section 14.8 for details), a drug-inducible variant of the Cre recombinase, was used to remove the *Mecp2* gene upon drug application in adulthood. Mice that lost *Mecp2* as adults developed symptoms resembling those observed in mice born lacking *Mecp2*, including deficits in motor coordination and premature death (**Figure 12-37**). This experiment indicated that MeCP2 is continuously required in adulthood.

To determine whether later expression of MeCP2 could rescue phenotypes resulting from developmental deficiencies in *Mecp2*, one could, in principle, transgenically express *Mecp2* under the control of a drug-inducible promoter. However, MeCP2 *overexpression* also causes significant neurological defects (echoing the symptoms caused by human *Mecp2* duplication). To restore MeCP2 expression at physiological levels, a conditional transcriptional stop cassette was *knocked in* (Section 14.7) between the promoter and the coding sequence of *Mecp2* such that MeCP2 is normally not expressed from this modified allele. Upon drug-induced CreER activation, the stop cassette is excised by recombination, and transcription is reactivated from the endogenous promoter, such that MeCP2 is expressed at physiological levels (**Figure 12-38**A). Remarkably, Rett-like symptoms and premature death were rescued after reactivation of MeCP2 in young adult males carrying the modified allele (Figure 12-38B). In female mice heterozygous for the modified allele, and therefore genotypically similar to girls with Rett syndrome, deficits in hippocampal long-term potentiation also reverted after MeCP2 reactivation (Figure 12-38C). While these results do not suggest a specific therapy for treating Rett syndrome patients, they do offer hope that, once a therapy is found, Rett syndrome can be treated even after symptom onset.

Studies of Rett syndrome have inspired similar approaches to test reversibility of other neurodevelopmental disorders in mouse models. For example, Smith–Magenis syndrome is caused by haploinsufficiency of *Rai1*, which encodes a

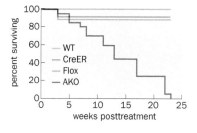

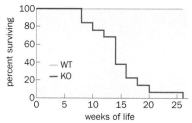

Figure 12-37 MeCP2 is required in adulthood. Top, mice in which the *Mecp2* gene is knocked out in adulthood (AKO, red) die prematurely compared with controls. WT, wild type; CreER, CreER only; Flox, conditional allele only. Bottom, mice born lacking *Mecp2* (KO, brown) die prematurely compared to controls (green). Note the similarity in AKO and KO survival curves from the time point at which *Mecp2* is deleted. (Adapted from McGraw CM, Samaco RC, & Zaghbi HY [2011] *Science* 333:186.)

Figure 12-38 Reversing Rett symptoms by restoring MeCP2 expression in young adult mice. (A) Strategy for conditionally restoring MeCP2 expression. Top, the endogenous *Mecp2* locus. Middle, a transcriptional stop flanked by two *loxP* sites is inserted between the transcriptional start and the coding region, such that no MeCP2 protein is made from this *stop* allele. Bottom, upon tamoxifen (TM) induced CreER excision of the transcriptional stop, MeCP2 is expressed at the endogenous level (see Sections 14.8 and 14.9 for more details of these techniques). **(B)** Male *stop/Cre* mice survived after TM injection in adulthood (red curve) but died prematurely without TM injection (blue curve). **(C)** Restoration of MeCP2 expression in adult females (orange) rescued the long-term potentiation deficits exhibited by *Mecp2* heterozygous females (blue) to wild-type levels (green), as seen in the magnitude of field excitatory postsynaptic potential (fEPSP) increases in response to high-frequency stimulation (HFS). (Adapted from Guy J, Gan J, Selfridge J, et al. [2007] *Science* 315:1143–1147.)

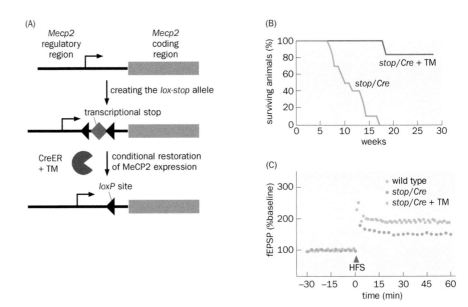

chromatin binding protein (Box 12-3). Restoring *Rai1* expression level in early adolescence (3–4 weeks), but not adulthood (8 weeks), fully rescued a social interaction deficit. Mutations in *Shank3*, encoding a postsynaptic density protein (Section 3.13), contribute to ~1% of ASD cases. Disruption of mouse *Shank3* causes a host of neurodevelopmental and behavioral deficits. Interestingly, adult restoration of *Shank3* function rescued abnormal synaptic protein composition, spine density, and social and grooming behaviors but could not rescue anxiety and motor coordination deficits. Therefore, the reversibility of each disorder, and different symptoms of the same disorder, must be evaluated individually.

12.23 Fragile-X syndrome results from loss of an RNA-binding protein that regulates activity-dependent translation

Fragile-X syndrome (FXS) is a leading cause of inherited intellectual disability (ID), affecting about 1 in 5000 boys. FXS patients have reduced IQ and a significant developmental delay in speech and motor skills. Many patients exhibit autistic features, such as a tendency to avoid eye contact and repetitive, stereotyped behaviors; 20–30% of FXS patients are diagnosed with autism spectrum disorder (ASD) based on behavioral criteria. The name fragile X came from an unusual chromosomal gap on the X chromosome of patients (Figure 12-39). The defective X-linked gene that causes FXS was identified in 1991 and named ***Fmr1*** (fragile-X mental retardation 1). A polymorphic CGG trinucleotide repeat was found in the 5′ untranslated region of the *Fmr1* gene. Healthy individuals have 6 to 54 CGG repeats, but the number expands to >200 in FXS patients. CGG repeats of between 55 and 200 are called premutations, which are at high risk of expansion. FXS is inherited from mothers who carry premutations or full mutations. Boys with the syndrome are most severely affected, while the severity of the syndrome in girls varies widely depending on X-inactivation patterns. Expanded CGG repeats cause extensive methylation near the *Fmr1* promoter and silencing of *Fmr1* expression. Therefore, FXS is caused by loss of the one protein, **FMRP**, encoded by the *Fmr1* gene.

FMRP is an evolutionarily conserved RNA-binding protein. A point mutation affecting a conserved residue in the RNA-binding domain of FMRP causes FXS with symptoms similar to those in patients with silenced *Fmr1* expression, demonstrating the importance of RNA binding for normal FMRP function. FMRP is expressed highly during embryonic development and continuously in all neurons throughout life. Within neurons, FMRP localizes to cytoplasm, axons, dendrites, and postsynaptic compartments and thus can regulate local protein translation (Section 2.2). FMRP is enriched in polyribosomes, where it binds mRNA mole-

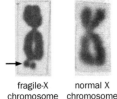

Figure 12-39 Fragile-X chromosome. X chromosomes from the cells of fragile-X syndrome (FXS) patients exhibit a gap near the tip of the long arm (arrow on the left), which is not seen in normal X chromosomes (right). (From Lubs HA [1969] *Am J Hum Genet* 21:231–244. With permission from Elsevier Inc.)

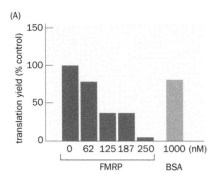

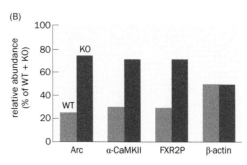

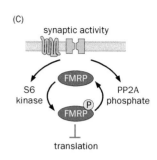

Figure 12-40 Regulation of protein synthesis in dendrites and synapses by fragile-X mental retardation protein (FMRP). (A) In an *in vitro* translation assay, purified FMRP represses translation of total brain mRNA in a dose-dependent manner, while a control protein, bovine serum albumin (BSA), does not. **(B)** Synapse-enriched protein extracts from *Fmr1* knockout mice contain higher amounts of the proteins Arc (a postsynaptic signaling protein), α-CaMKII (the α subunit of the Ca^{2+}/calmodulin-dependent kinase), and FXR2P (fragile-X mental retardation syndrome-related protein 2, encoded by a *Fmr1* paralog) than do control extracts from wild-type mice. By contrast, the level of β-actin protein is unaffected. This suggests that FMRP represses translation of select mRNAs such as those that produce Arc, α-CaMKII, and FXR2P. **(C)** Only phosphorylated FMRP represses translation. S6 kinase and phosphatase 2A (PP2A) phosphorylate and dephosphorylate FMRP, respectively. Synaptic activity regulates FMRP phosphorylation via PP2A and S6 kinase and, hence, local translation. For example, type I metabotropic glutamate receptors activate PP2A more rapidly than they activate S6 kinase, causing transient FMRP dephosphorylation and de-repression of translation. (A, from Li Z, Zhang Y, Ku L, et al. [2001] *Nuc Acid Res* 29:2276–2283. With permission from Oxford University Press. B, from Zalfa F, Giorgi M, Primerano B, et al. [2003] *Cell* 112:317–327. With permission from Elsevier Inc.; C, based on Santoro MR, Bray SM, & Warren ST [2012] *Annu Rev Patho Mec Dis* 7:219–245; see also Bear MF, Huber KM, & Warren ST [2004] *Trends Neurosci* 27:370–377.)

cules and represses translation *in vitro* and *in vivo* (**Figure 12-40**A, B). Normally most FMRP is phosphorylated at a conserved serine residue, and phosphorylated FMRP represses translation. Upon specific signals, such as synaptic activity, FMRP is dephosphorylated, which transiently reduces its translational repression, thereby allowing rapid local translation (Figure 12-40C). This activity-dependent regulation of local translation contributes to synaptic plasticity and learning.

Biochemical studies using cross-linking of RNA bound to FMRP followed by immuno-purification of FMRP identified many specific target mRNAs associated with FMRP. These include microtubule-associated proteins involved in axonal and dendritic transport, presynaptic proteins that regulate synaptic vesicle release, postsynaptic scaffold proteins, and components of the NMDA and metabotropic glutamate receptor (mGluR) signaling pathways for neurotransmitter reception. FMRP also binds to the mRNAs of many ASD-associated proteins, providing a molecular link between FXS and ASD.

There has been intense effort to find a treatment for FXS since the discovery of *Fmr1*. Several strategies have been reported to effectively ameliorate symptoms in *Fmr1* knockout mice. For example, evidence suggests that type I mGluRs oppose the function of FMRP in translational regulation (Figure 12-40C). Application of an inhibitor of type I mGluRs in adult *Fmr1* knockout mice corrected deficits in protein synthesis, dendritic spine density, hippocampal long-term depression, and learning and memory. However, none of these drugs has proven effective in human clinical trials so far. These efforts highlight the challenges of treating neurodevelopmental disorders. These include establishing animal models that resemble human physiology more closely for preclinical studies, designing clinical trials for appropriate developmental stage and duration (in general, drug safety is of paramount concern with long-term application to children), and establishing objective, quantitative, and sensitive criteria for evaluating therapeutic efficacy.

12.24 Synaptic dysfunction is a cellular mechanism underlying many neurodevelopmental and psychiatric disorders

Studies of Rett and fragile-X syndromes have reinforced the theme we introduced in Section 12.20: disruption of synaptic development and function is a cellular mechanism underlying many neurodevelopmental disorders. Further support comes from the identification of other syndromic, nonsyndromic, and sporadic

Figure 12-41 Defects in many proteins involved in synaptic signaling contribute to neurodevelopmental disorders. Mutations in genes encoding proteins bolded in this figure have been implicated either in various syndromes with increased risks of intellectual disability (ID) or autism spectrum disorder (ASD) or in nonsyndromic or sporadic ID/ASD. For example, mutations in synaptic adhesion proteins neurexin and neuroligin, a voltage-gated Ca²⁺ channel, and postsynaptic scaffold protein Shank3 are associated with ASD; mutations in several components of the Rho GTPase signaling pathways cause ID; mutations in translation regulators TSC1 and TSC2 yield ASD symptoms; mutations in small GTPase Ras, its negative regulator NF-1, and effector kinases are associated with learning disability; and mutations in MeCP2, a chromatin-binding protein that is phosphorylated in an activity-dependent manner, cause Rett syndrome, with both ID and ASD features. See also Figure 3-27; Figure 3-41.

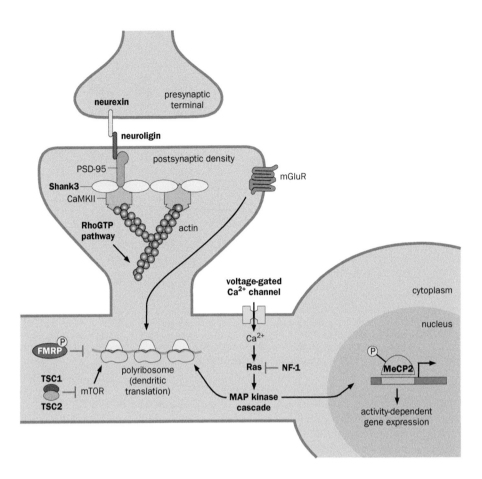

ASD and ID genes that affect different aspects of synaptic signaling (**Figure 12-41**). For example, as we discussed in Chapters 3 and 7, neurexins and neuroligins form trans-synaptic complexes that regulate synapse assembly and organization. Genetic studies have identified mutations in human genes encoding several neurexin and neuroligin isoforms to be associated with ASD. As introduced in Section 12.22, disruption of Shank3, a postsynaptic scaffold protein, has been associated with syndromic and sporadic ASD. Expanding on FMRP's role in regulating synaptic protein translation, a key regulator of protein translation in response to extracellular signals is **mTOR** (<u>m</u>ammalian <u>t</u>arget <u>o</u>f <u>r</u>apamycin), which in turn is negatively regulated by a complex consisting of Tsc1 and Tsc2 (tuberous <u>scl</u>erosis 1 and 2). A large fraction of patients with tuberous sclerosis, which is characterized by nonmalignant tumors in the brain and other organs and results from mutations in *Tsc1* or *Tsc2*, exhibit ASD symptoms, providing more evidence that abnormal protein translation can contribute to ASD.

Many factors involved in postsynaptic signaling, including synapse-to-nucleus signaling (Section 3.22), are also implicated in neurodevelopmental disorders. For example, a gain-of-function mutation in a voltage-gated Ca²⁺ channel, Ca$_V$1.2, causes **Timothy syndrome**, which features cardiac arrhythmia and autistic symptoms. Disruption of components of the Ras/MAP kinase pathway, including the small GTPase Ras, or its downstream MAP kinase cascade, causes Noonan syndrome, which presents multiple developmental defects including learning disability. Disruption of neurofibromin 1 (NF-1), a GTPase activating protein of Ras, causes neurofibromatosis type I, whose symptoms also include learning disability. The Ras–MAP kinase pathway regulates protein translation and is also a key synapse-to-nucleus signaling pathway (Figure 3-41). One nuclear effector of synapse-to-nucleus signaling is MeCP2, which, when mutated, causes Rett syndrome. MeCP2 is phosphorylated at multiple sites in response to neuronal activity, and phosphorylation regulates its gene repression activity.

Defects in synaptic signaling may also underlie many psychiatric disorders such as schizophrenia and bipolar disorder. Indeed, mutations in several genes, including those encoding Ca$_v$1.2, Neurexin-1, and MeCP2, have been associated with both schizophrenia and ASD (Table 12-2). As new genes associated with brain disorders are being discovered at a rapid pace, thanks to recent advances in human genetics (Box 12-3), the overlap between genes associated with psychiatric and neurodevelopmental disorders will likely increase, as will the links between these disorders and synaptic signaling.

12.25 Studies of brain disorders and basic neurobiology inform and advance each other

If synaptic dysfunction is a cellular mechanism for many neurodevelopmental and psychiatric disorders, why do disruptions of different genes (and sometimes different mutations of the same gene) give rise to different disorders? Do these mutations affect all synapses equally, such that the symptoms of these disorders are caused by overall suboptimal synaptic functioning? Or are synapses in specific brain regions with dedicated circuit functions differentially affected in different disorders? We remain far from having satisfactory answers to these questions, and the answers may differ for different disorders. For example, as we learned in Section 12.21, MeCP2 disruption in inhibitory neurons appears to cause the mutation's strongest effects in a mouse model. Excitation–inhibition imbalances have also been suggested to underlie epilepsy (**Box 12-4**), ASD, and schizophrenia. As researchers use more sophisticated tools to dissect the contributions of different neuronal subpopulations, we will certainly find answers to the questions raised here. At the same time, studies focusing on specific diseases may shed light on how the normal brain functions, revealing, for instance, the specific brain regions and circuits crucial for intelligence, social interactions, and other cognitive functions.

Studies of genetically defined brain disorders have also introduced general strategies for treating these disorders (**Figure 12-42**). Identification of defective genes underlying brain disorders leads to the establishment of appropriate animal models. This enables mechanistic studies of the pathogenic process that enrich our understanding of basic neurobiology and at the same time suggest candidate pathways for therapeutic intervention. Development and clinical trials of appropriate drugs may eventually lead to successful therapies. For disorders with multigenic or largely unidentified or nongenetic underlying causes, parts of this discovery-to-treatment path can still apply. Although we do not have effective therapies for most of the disorders described in this chapter, new advances in basic and disease-focused neurobiology research are occurring every day, and breakthrough treatments for disabling brain disorders are anticipated in the coming decades.

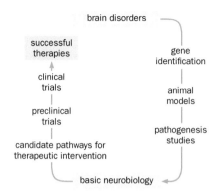

Figure 12-42 A general strategy for understanding and treating brain disorders. The path on the right links brain disorders to underlying basic neurobiology; the path on the left applies knowledge from basic neurobiological research toward development of therapeutic interventions. (Adapted from Zoghbi HY & Bear MF [2012] *Cold Spring Harb Perspect Biol* 4:a009886.)

Box 12-4: Epilepsy is a disorder of neuronal network excitability

We have encountered seizures and epilepsy many times thus far. With a framework for studying brain disorders established, we can now discuss the symptoms, causes, and treatment strategies of seizures and epilepsy. A **seizure** is an episode involving abnormal synchronous firing of large groups of neurons; about 1 in 20 people has at least one seizure in their lifetime. **Epilepsy** is a chronic condition characterized by recurrent seizures, which affects about 1% of the human population. When cortical neurons engage in abnormal synchronous firing, their activity can often be detected on **electroencephalograms** (**EEGs**); EEGs record electrical potential differences between surface electrodes

placed on specific locations on the scalp, which report the collective electrical activity of many nearby cortical neurons underneath the electrodes (**Figure 12-43**).

Seizures are typically categorized as either focal (partial) or generalized. **Focal seizures** are defined by clinical symptoms or EEG changes that indicate initial activation of neurons in a relatively small, discrete region of the brain. Depending on the brain region, symptoms can include temporary loss of sensation, odd sensory experiences, temporary loss of movement control, and confusion. **Generalized seizures** affect multiple bilateral brain regions. In primary generalized

(Continued)

Box 12-4: continued

(A)

FRONT

BACK

(B)

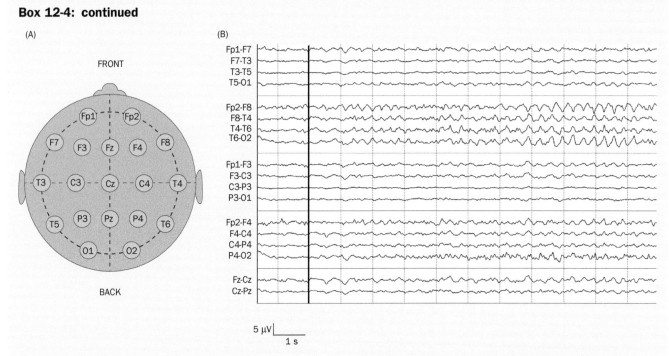

5 µV

1 s

Figure 12-43 **Detecting seizure onset on electroencephalograms.**
(A) Schematic of surface electrode placement on the scalp.
Electrode positions are named according to cortical regions (F, frontal;
T, temporal; P, parietal; O, occipital; C, central—that is, closer to the
midline of the head). **(B)** EEG record of a patient suffering from focal
epilepsy presumed to originate from the right temporal lobe. Each
row is the electrical potential difference between two designated
electrodes. The solid vertical line indicates seizure onset. Before
onset, EEGs are small in amplitude and mostly asynchronous,
reflecting brain activity during normal waking. After seizure onset,
EEG records between multiple pairs of electrodes on the right
hemisphere show a large-amplitude synchronous pattern. Note that
the exact time of onset and location of epileptic activity cannot be
determined via scalp recordings; for these, intracranial electrodes
must be implanted in the presumed areas of seizure activity.
(Courtesy of Dr. Josef Parvizi, Stanford University Medical Center.)

seizures, the entire cortex seems to be activated simultaneously, while a secondary generalized seizure results when a focal seizure spreads to larger areas of the brain. A generalized **absence seizure** (formerly called petit mal) is characterized by a brief lapse in consciousness (~10 seconds or less) and cessation of motor activity without loss of posture. A generalized **tonic–clonic seizure** (previously called grand mal) is associated with loss of consciousness and a predictable sequence of motor activity: patients first stiffen and extend all extremities (tonic phase), then undergo full-body spasms during which muscles alternately flex and relax (clonic phase).

Epilepsy has diverse causes, including head injury, infection, strokes, brain cancers, and brain surgery. Epilepsy can also result from inherited or *de novo* mutations in several dozen identified genes, some of which monogenically cause epilepsy while others only confer risk. Other brain disorders, notably neurodevelopmental disorders like Rett syndrome, can have epilepsy as a symptom. Epilepsy can also be a withdrawal symptom following sudden cessation of drug treatment. Despite its diverse causes, epilepsy features one defining phenotype: abnormal balance between the activities of excitatory and inhibitory neurons (the **E–I balance**) results in hyperactivation of excitatory neurons, which spreads the abnormal excitation across neural networks. This notion is supported by examination of epilepsies caused by defective ion channels (**channelopathies;** Table 12-3).

Given the key roles of voltage- and ligand-gated ion channels in regulating neuronal excitability (Chapters 2 and 3), it is unsurprising that mutations disrupting ion channels can cause abnormal neuronal firing patterns. For instance, voltage-gated K^+ channels repolarize neurons after excitation; reduction of their function can cause abnormal excitation of mutant neurons. Likewise, reduction of $GABA_A$ receptor function can cause epilepsy as neurons fail to receive appropriate inhibition. Let's examine a specific example in more detail: loss of one copy of the gene encoding $Na_v1.1$, a voltage-gated Na^+ channel. This condition, called Dravet syndrome or severe myoclonic epilepsy of infancy (Table 12-3), has been recapitulated in mice: mice with one functional $Na_v1.1$ allele exhibit spontaneous seizures and shortened lifespan. Voltage-gated Na^+ channels underlie action potential generations, and their reduction should therefore inhibit, rather than promote, excitability. However, while most excitatory neurons express multiple genes encoding voltage-gated Na^+ channels, $Na_v1.1$ is highly expressed in GABAergic inhibitory neurons. Thus, a reduction in $Na_v1.1$ activity preferentially reduces the excitability of inhibitory neurons, thereby increasing network excitability. This example illustrates why mutations in different ion channels cause different types of epilepsy: cell types that uniquely express a particular ion channel (and therefore can least compensate for its loss) differ in their expression of each gene implicated in epilepsy. Furthermore, the phenotypic severity of the same mutation often differs between individual patients or mice with differ-

Box 12-4: continued

Table 12-3: **Representative examples of ion channel mutations that cause epilepsy**

Affected protein	Disorder
α_4 subunit of nicotinic ACh receptor	Nocturnal frontal lobe epilepsy[a]
$K_v7.2$ or $K_v7.3$ (voltage-gated K^+ channels)	Benign familial neonatal seizures[a]
α_1 subunit of $GABA_A$ receptor	Juvenile myoclonic epilepsy[b]
$Ca_v2.1$ (voltage-gated Ca^{2+} channel)	Absence epilepsy and episodic ataxia[b]
$Na_v1.1$ (voltage-gated Na^+ channel)	Severe myoclonic epilepsy of infancy[b]

This is only a partial list of known ion channel mutations that cause focal (denoted by [a]) or generalized (denoted by [b]) seizures. All the listed mutations are autosomal dominant, and epilepsy in most cases results from genetic loss-of-function effects.

Data from Lerche H, Shah M, Beck H, et al. (2013) *J Physiol* 591:753–764.

ent genetic backgrounds, revealing that even monogenic mutations are subject to complex interactions with other factors.

Channelopathies can reveal some of the mechanisms that produce seizures, but these mutations account for only a small fraction of epilepsy cases. Although the root causes in other cases are less clear, perturbation of E–I balance may be a common culprit. For example, the sprouting of neuronal process and synapse formation resulting from physical injuries to the brain, such as strokes and surgical removal of brain tissue, may differ for excitatory and inhibitory neurons, thereby perturbing the E–I balance. Another important factor that may contribute to recurrent seizures is seizures themselves: many neurons firing in synchrony can cause significant changes in the circuits involved according to the plasticity rules discussed in Chapter 11. These activity-dependent changes could, for example, decrease the threshold for future seizures. Indeed, excess excitation of glutamatergic neurons, with abnormally high glutamate release and NMDA receptor activation, can drive excessive elevation of intracellular Ca^{2+} concentrations in their post-synaptic target neurons, triggering **excitotoxicity** and neuronal death.

About two-thirds of epilepsy patients are responsive to medication, thanks to the common phenotype of excessive network excitability. These medications include $GABA_A$ receptor agonists (e.g., benzodiazepines; Section 12.17) that boost network inhibition, drugs that enhance voltage-gated Na^+ channel inactivation to curb excitation, and drugs that inhibit voltage-gated Ca^{2+} channels to reduce neuronal burst firing or synaptic transmission efficacy. About one-third of epilepsy patients suffer from intractable seizures and do not respond to medication. Of these patients, some who suffer from focal seizures can be treated with brain surgery. Identification of the seizure focus, usually achieved by intracranial recording and stimulation during surgery, is key. (Indeed, we have learned a great deal about the functions of individual human neurons via this procedure; e.g., Section 1.10.) If the seizure focus regulates non-vital functions and, ideally, is located in the nondominant hemisphere, then surgical removal of the affected brain tissue or severing of its connections can be an effective treatment.

SUMMARY

Each brain disorder, defined by a specific set of symptoms, has a specific pattern of genetic (and sometimes environmental) contributions. Huntington's disease (HD), Rett syndrome, and fragile-X syndrome all result from disruption of single genes. Mutations may follow Mendelian inheritance, as in HD, or occur *de novo,* as in Rett syndrome. More complex disorders, such as Alzheimer's disease (AD), Parkinson's disease (PD), amyotrophic lateral sclerosis (ALS), intellectual disability (ID), and epilepsy, are heterogeneous in their origin. Only a fraction of cases result from mutations in specific genes that follow Mendelian inheritance; most cases are sporadic and have incompletely defined causes, including genetic risk factors, *de novo* mutations, and environmental factors. Even more complex disorders, including the psychiatric disorders we discussed and the nonsyndromic autism spectrum disorder (ASD), are mostly sporadic, as genetic causes with full penetrance have not been identified. Whereas schizophrenia, bipolar disorder, and ASD have strong genetic contributions, environmental factors play a large role in drug addiction, depression, and anxiety disorders. Recent advances in human genetics have identified genetic risk factors and provided biological insights into complex disorders, as exemplified by the engagement of microglia in AD pathogenesis and the involvement of the complement cascade in schizophrenia.

A common pathological feature of neurodegenerative diseases is altered protein homeostasis. AD is characterized by extracellular Aβ deposition and intracellular tau aggregation. Most PD cases involve aggregated α-synuclein. Multiple ALS-causing mutations result in aggregation of distinct mutant proteins. HD and spinocerebellar ataxia are caused by toxic gain-of-function effects of protein aggregation due to expanded polyglutamine repeats. Prion diseases result from propagation of pathogenic PrP[Sc], which converts nonpathogenic PrP[C] to PrP[Sc] aggregates; similar processes may contribute to pathogenesis of other neurodegenerative diseases. The ultimate symptoms of different neurodegenerative diseases reflect the different neuron types affected. AD and prion diseases affect a broad range of neuron types, while ALS preferentially affects motor neurons and PD symptoms primarily result from death of substantia nigra dopamine neurons.

Our current understanding of psychiatric disorders has benefited from studying the actions of drugs serendipitously found to have therapeutic effects. For example, most antipsychotic drugs that reduce schizophrenia's positive symptoms antagonize the dopamine D_2 receptor. The most widely used antidepressants block serotonin reuptake into presynaptic terminals. Enhancing GABAergic inhibition mediated by specific $GABA_A$ receptors can reduce anxiety. As these neurotransmitter systems have broad actions in diverse brain regions, investigating the specific neural circuits that mediate these drug actions and that become dysfunctional in psychiatric disorders is likely key to generating better therapies.

Our understanding of neurodevelopmental disorders has benefited from studies of syndromic disorders. Rett syndrome results from disruption of MeCP2, a global regulator of gene expression that is particularly important in post-mitotic neurons. MeCP2 is required during postnatal development and in adulthood, and reactivation of MeCP2 in adult mice can ameliorate defects caused by developmental disruption of MeCP2. Fragile-X syndrome is caused by disruption of FMRP, a protein that binds the mRNA of many ASD-associated genes and regulates their translation. Recent human genetic studies have identified greater numbers of genes associated with psychiatric and neurodevelopmental disorders. These studies suggest that, despite their diverse symptoms, many disorders share synaptic dysfunction as a common cellular mechanism and potential target for further research and treatment efforts.

Animal models can be used to investigate pathogenic disease mechanisms, trace disease progression, and test therapeutic approaches. Our increasing knowledge of disease mechanisms and technological advances over the past few decades have led to rational drug design, resulting in successful treatment of the causes and symptoms of several brain disorders. These include L-dopa for treating PD and antisense oligonucleotides and gene therapy for treating spinal muscular atrophy. While many challenges remain, our goal of progressively conquering the many brain disorders appears increasingly within reach.

OPEN QUESTIONS

- First and foremost, how can we develop more effective therapies to treat each disorder discussed in this chapter?

- Why do mutations in ubiquitously expressed genes result in degeneration of specific neuron types, thereby causing different neurodegenerative diseases?

- For disorders that impact synapse development and function, are the symptoms caused by synaptic dysfunction in specific brain circuits or by suboptimal synaptic functions across the brain? If specific brain circuits are implicated, which ones are impacted in each disorder?

- How can we effectively deliver therapeutic agents such as antibodies and gene therapy vectors to the adult brain and ideally to affect only certain neuron types?

FURTHER READING

Books and reviews

Abeliovich A & Gitler AD (2016). Defects in trafficking bridge Parkinson's disease pathology and genetics. *Nature* 539:207–216.

Long JM & Holtzman DM (2019). Alzheimer disease: an update on pathobiology and treatment strategies. *Cell* 179:312–339.

Nestler E, Hyman SE, & Malenka RC (2015). Molecular Pharmacology: A Foundation for Clinical Neuroscience, 3rd ed. McGraw-Hill.

Online Mendelian Inheritance in Man. www.omim.org.

Prusiner SB (1991). Molecular biology of prion diseases. *Science* 252: 1515–1522.

Snyder S (1996). Drugs and the Brain. Scientific American Books, Inc.

Sullivan PF & Geschwind DH (2019). Defining the genetic, genomic, cellular, and diagnostic architectures of psychiatric disorders. *Cell* 177:162–183.

Taylor JP, Brown RH Jr., & Cleveland DW (2016). Decoding ALS: from genes to mechanism. *Nature* 539:197–206.

Zoghbi HY & Bear MF (2012). Synaptic dysfunction in neurodevelopmental disorders associated with autism and intellectual disabilities. *Cold Spring Harb Perspect Biol* 4:a009886.

Neurodegenerative disorders

Braak H, Del Tredici K, Rub U, de Vos RA, Jansen Steur EN, & Braak E (2003). Staging of brain pathology related to sporadic Parkinson's disease. *Neurobiol Aging* 24:197–211.

Bueler H, Aguzzi A, Sailer A, Greiner RA, Autenried P, Aguet M, & Weissmann C (1993). Mice devoid of PrP are resistant to scrapie. *Cell* 73:1339–1347.

Finkel RS, Mercuri E, Darras BT, Connolly AM, Kuntz NL, Kirschner J, Chiriboga CA, Saito K, Servais L, Tizzano E, et al. (2017). Nusinersen versus sham control in infantile-onset spinal muscular atrophy. *N Engl J Med* 377:1723–1732.

Goate A, Chartier-Harlin MC, Mullan M, Brown J, Crawford F, Fidani L, Giuffra L, Haynes A, Irving N, James L, et al. (1991). Segregation of a missense mutation in the amyloid precursor protein gene with familial Alzheimer's disease. *Nature* 349:704–706.

The Huntington's Disease Collaborative Research Group (1993) A novel gene containing a trinucleotide repeat that is expanded and unstable on Huntington's disease chromosomes. *Cell* 72:971–983.

Jonsson T, Atwal JK, Steinberg S, Snaedal J, Jonsson PV, Bjornsson S, Stefansson H, Sulem P, Gudbjartsson D, Maloney J, et al. (2012). A mutation in APP protects against Alzheimer's disease and age-related cognitive decline. *Nature* 488:96–99.

Kang J, Lemaire HG, Unterbeck A, Salbaum JM, Masters CL, Grzeschik KH, Multhaup G, Beyreuther K, & Muller-Hill B (1987). The precursor of Alzheimer's disease amyloid A4 protein resembles a cell-surface receptor. *Nature* 325:733–736.

Keren-Shaul H, Spinrad A, Weiner A, Matcovitch-Natan O, Dvir-Szternfeld R, Ulland TK, David E, Baruch K, Lara-Astaiso D, Toth B, et al. (2017). A unique microglia type associated with restricting development of Alzheimer's disease. *Cell* 169:1276–1290.

Langston JW, Ballard P, Tetrud JW, & Irwin I (1983). Chronic Parkinsonism in humans due to a product of meperidine-analog synthesis. *Science* 219:979–980.

Luk KC, Kehm V, Carroll J, Zhang B, O'Brien P, Trojanowski JQ, & Lee VM (2012). Pathological alpha-synuclein transmission initiates Parkinson-like neurodegeneration in nontransgenic mice. *Science* 338:949–953.

Mendell JR, Al-Zaidy S, Shell R, Arnold WD, Rodino-Klapac LR, Prior TW, Lowes L, Alfano L, Berry K, Church K, et al. (2017). Single-dose gene-replacement therapy for spinal muscular strophy. *N Engl J Med* 377:1713–1722.

Neumann M, Sampathu DM, Kwong LK, Truax AC, Micsenyi MC, Chou TT, Bruce J, Schuck T, Grossman M, Clark CM, et al. (2006). Ubiquitinated TDP-43 in frontotemporal lobar degeneration and amyotrophic lateral sclerosis. *Science* 314:130–133.

Roberson ED, Scearce-Levie K, Palop JJ, Yan F, Cheng IH, Wu T, Gerstein H, Yu GQ, & Mucke L (2007). Reducing endogenous tau ameliorates amyloid beta-induced deficits in an Alzheimer's disease mouse model. *Science* 316:750–754.

Shi Y, Yamada K, Liddelow SA, Smith ST, Zhao L, Luo W, Tsai RM, Spina S, Grinberg LT, Rojas JC, et al. (2017). ApoE4 markedly exacerbates tau-mediated neurodegeneration in a mouse model of tauopathy. *Nature* 549:523–527.

Strittmatter WJ, Saunders AM, Schmechel D, Pericak-Vance M, Enghild J, Salvesen GS, & Roses AD (1993). Apolipoprotein E: high-avidity binding to beta-amyloid and increased frequency of type 4 allele in late-onset familial Alzheimer disease. *Proc Natl Acad Sci U S A* 90:1977–1981.

Yeh FL, Wang Y, Tom I, Gonzalez LC, & Sheng M (2016). TREM2 binds to apolipoproteins, Including APOE and CLU/APOJ, and thereby facilitates uptake of amyloid-beta by microglia. *Neuron* 91:328–340.

Psychiatric disorders

Coe JW, Vetelino MG, Bashore CG, Wirtz MC, Brooks PR, Arnold EP, Lebel LA, Fox CB, Sands SB, Davis TI, et al. (2005). In pursuit of alpha4beta2 nicotinic receptor partial agonists for smoking cessation: carbon analogs of (-)-cytisine. *Bioorg Med Chem Lett* 15:2974–2979.

Glowinski J & Axelrod J (1964). Inhibition of uptake of tritiated-noradrenaline in the intact rat brain by imipramine and structurally related compounds. *Nature* 204:1318–1319.

Howard DM, Adams MJ, Clarke TK, Hafferty JD, Gibson J, Shirali M, Coleman JRI, Hagenaars SP, Ward J, Wigmore EM, et al. (2019). Genome-wide meta-analysis of depression identifies 102 independent variants and highlights the importance of the prefrontal brain regions. *Nat Neurosci* 22:343–352.

Rudolph U, Crestani F, Benke D, Brunig I, Benson JA, Fritschy JM, Martin JR, Bluethmann H, & Mohler H (1999). Benzodiazepine actions mediated by specific gamma-aminobutyric acid(A) receptor subtypes. *Nature* 401:796–800.

Saal D, Dong Y, Bonci A, & Malenka RC (2003). Drugs of abuse and stress trigger a common synaptic adaptation in dopamine neurons. *Neuron* 37:577–582.

Schizophrenia Working Group of the Psychiatric Genomics Consortium (2014) Biological insights from 108 schizophrenia-associated genetic loci. *Nature* 511:421–427.

Sekar A, Bialas AR, de Rivera H, Davis A, Hammond TR, Kamitaki N, Tooley K, Presumey J, Baum M, Van Doren V, et al. (2016). Schizophrenia risk from complex variation of complement component 4. *Nature* 530:177–183.

Zhu Y, Wienecke CF, Nachtrab G, & Chen X (2016). A thalamic input to the nucleus accumbens mediates opiate dependence. *Nature* 530:219–222.

Neurodevelopmental disorders

Amir RE, Van den Veyver IB, Wan M, Tran CQ, Francke U, & Zoghbi HY (1999). Rett syndrome is caused by mutations in X-linked MECP2, encoding methyl-CpG-binding protein 2. *Nat Genet* 23:185–188.

Chao HT, Chen H, Samaco RC, Xue M, Chahrour M, Yoo J, Neul JL, Gong S, Lu HC, Heintz N, et al. (2010). Dysfunction in GABA signalling mediates autism-like stereotypies and Rett syndrome phenotypes. *Nature* 468:263–269.

Darnell JC, Van Driesche SJ, Zhang C, Hung KY, Mele A, Fraser CE, Stone EF, Chen C, Fak JJ, Chi SW, et al. (2011). FMRP stalls ribosomal translocation on mRNAs linked to synaptic function and autism. *Cell* 146:247–261.

Dolen G, Osterweil E, Rao BS, Smith GB, Auerbach BD, Chattarji S, & Bear MF (2007). Correction of fragile X syndrome in mice. *Neuron* 56:955–962.

Guy J, Gan J, Selfridge J, Cobb S, & Bird A (2007). Reversal of neurological defects in a mouse model of Rett syndrome. *Science* 315:1143–1147.

Huang WH, Guenthner CJ, Xu J, Nguyen T, Schwarz LA, Wilkinson AW, Gozani O, Chang HY, Shamloo M, & Luo L (2016). Molecular and neural functions of Rai1, the causal gene for Smith-Magenis syndrome. *Neuron* 92:392–406.

Mei Y, Monteiro P, Zhou Y, Kim JA, Gao X, Fu Z, & Feng G (2016). Adult restoration of Shank3 expression rescues selective autistic-like phenotypes. *Nature* 530:481–484.

CHAPTER 13

Evolution of the Nervous System

In the preceding chapters we have mostly asked questions regarding the workings of the nervous system in the present: How do neurons communicate with each other? How do we see and smell? How is neural circuitry established during development and altered by experience? When studying biological systems, we can ask a second type of question: How did it arise?

Evolution has produced wonders. Consider the human eye. As we learned in Chapters 4 and 5, our rods are exquisitely sensitive, able to detect single photons in near darkness, yet also able to discern contrast over a 10^4 range of ambient light levels. Cones extend this range by an additional factor of 10^7, enabling us to see over a 10^{11}-fold range of ambient light levels. Processing of visual signals by dozens of retinal cell types enables information about luminance, contrast, color, and motion to be extracted and transmitted to the brain. These features are products of a developmental process generating not only the exquisite structure of the eye but also the precise network of connections linking retinal neurons to each other and their brain targets. How did these properties arise during evolution? What forces selected for their emergence? How do these properties differ across species and environments, and how have they changed over time?

The eye exemplifies the complex and multifaceted nature of evolutionary investigations. While we can definitively answer many "how does it work" questions through carefully designed experiments, we cannot experiment on the distant past to address "how did it arise" questions. However, studies of evolution can be aided by analyzing the results of numerous past and present-day "experiments" carried out in nature, which have generated the wonderful diversity of life. Furthermore, as biological evolution occurs at the level of DNA, allowing traits to be inherited by future generations, evolutionary history is documented in DNA, which can be extracted from extant and recently extinct life forms. We can infer answers to "how did it arise" questions with increasing levels of confidence by combining analyses of DNA sequences; examination of fossil records of past life forms; comparative studies of anatomical, physiological, and behavioral traits in extant life forms; and incisive simulation experiments. Such investigations deepen our understanding of the biological world by revealing clues about the unitary nature of life and the relationships among living organisms, offer a rationale for our use of model organisms in understanding the human brain, enhance our understanding of how the nervous system works by providing a historical perspective on how changing nervous systems evolve new capabilities, and fulfill our desire to know where we came from.

In this chapter, we focus on selected topics regarding the evolution of neuronal communication, sensory systems, and the structure and development of the nervous system. Our treatment of these topics is not intended to be comprehensive, but instead to encompass examples for which there are sufficient data to construct plausible evolutionary histories, and from which we can learn about general principles of evolution. Before addressing these specific topics, we first introduce several key concepts and approaches in evolutionary analysis that form the foundation for studying the specific problems presented later in the chapter.

GENERAL CONCEPTS AND APPROACHES IN EVOLUTIONARY ANALYSIS

The publication of *On the Origin of Species by Means of Natural Selection* by Charles Darwin in 1859 was a landmark in human history. Darwin's theory of evolution offered a mechanism by which life forms could change over time and provided the first satisfactory explanation for biological diversity without resorting to supernatural causes. Its profound influence on biology cannot be overestimated, as reflected in the epigraph by Theodosius Dobzhansky. In essence, Darwin proposed that (1) species are not static but evolving, (2) evolution is a gradual and continuous process, (3) all living organisms descended from a common ancestor, and (4) **natural selection** is the primary force directing the evolutionary process and its outcomes (**Figure 13-1**).

Evolution requires variation and selection. **Variation**, the presence of differences in traits, must be heritable to contribute to evolution. Darwin was unaware of genetic mechanisms; Gregor Mendel's laws of inheritance, originally published in 1866, were not rediscovered by the scientific community until the beginning of

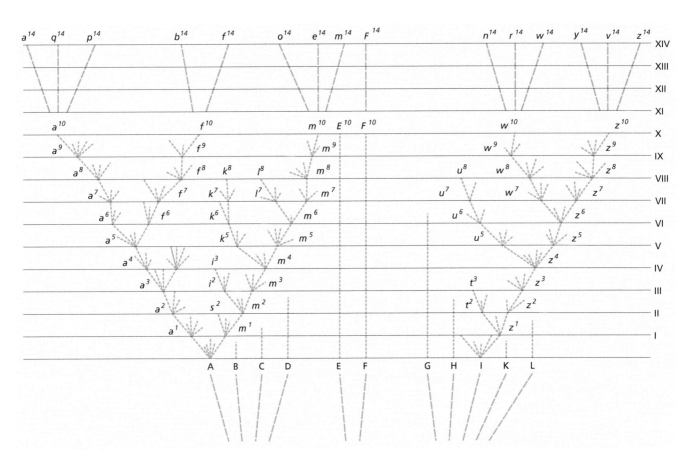

Figure 13-1 Darwin's depiction of evolution. In the sole diagram in *Origin of Species*, letters A to L represent the species of a hypothetical genus at the beginning of a time period. The horizontal distance represents the degree of their resemblance (for example, D shares more similarities with C than it does with E). The vertical distance between an adjacent pair of horizontal lines labeled by Roman numerals represents the span of 1,000 generations (a number chosen arbitrarily by Darwin). The dashed lines extending upward from species represent the species' progeny, and the fan-shaped dashed lines that radiate from many species indicate variations among the progeny. For example, after 1,000 generations, species A, which generates more varieties than other species, has produced two distinct varieties, a^1 and m^1, both of which continue to produce many varieties, perhaps as an inherited trait from species A. Superscripts represent varieties at specific thousands of generations (for example, a^5 represents a variety at the 5,000th generation). Most progeny lines go extinct (dashed lines that terminate) during natural selection; only a small subset continues to produce progeny and new varieties. Species that produce more varieties (such as A and I) are more likely to have surviving progeny lines, but this is not absolute (for example, species F). At the 14,000th generation, 15 varieties (or species, as they are sufficiently far apart from each other) have survived; the traits of these survivors differ more widely than in the founding species, as represented by the increased horizontal spread. (From Darwin [1859] On the Origin of Species by Means of Natural Selection. John Murray.)

the twentieth century. Subsequent advances in our knowledge of genes, mutations, molecular biology, and genomics have illustrated how variations in DNA can generate phenotypic variants. Evolution also requires **selection** of favorable variants. In any population, an individual whose genetic variants confer a higher chance of reproductive success is more likely to pass its versions of those genes to the next generation.

A key approach Darwin employed to reach his conclusion was to compare morphological traits of different organisms shaped by human selective breeding or natural selection. Comparative studies of morphological, physiological, and behavioral traits are still widely used today to explore similarities and differences among living organisms. The revolution in molecular biology and genomics added nucleotide and protein sequences to the list of traits that can be quantitatively compared across different species; these advances have provided a deeper understanding of the evolutionary process.

13.1 Phylogenetic trees relate all living organisms in a historical context

From *E. coli* to elephants, all organisms share the building blocks of life: we string together the same nucleotides to form a genetic blueprint; we assemble the same amino acids into proteins that serve as the major executors of cellular functions; we employ a nearly universal genetic code for translating nucleotide sequences into protein sequences. Coupled with the biochemical similarities shared by all cells in processes ranging from energy metabolism to macromolecule synthesis, these findings provide evidence beyond doubt that *all living organisms descended from a common ancestor*. By comparing sequence similarities and differences in ancient molecules, such as ribosomal RNAs and enzymes involved in basic metabolism, **phylogenetic trees** illustrating the evolutionary relationships between diverse living organisms have been constructed. Assuming a constant mutation rate, **molecular clocks** use the rates of sequence changes, calibrated against fossil records, to estimate the times at which species diverged, that is, the times at which branching occurred on the phylogenetic tree.

Life on Earth started about 4 billion years ago with single-celled **prokaryotes**. Prokaryotes today belong to two large branches: eubacteria and archaea. The first **eukaryotes** (cells with a nuclear membrane separating the nucleus from the cytoplasm) originated from prokaryotes about 2.5 billion years ago, and with eubacteria and archaea form three domains of life. A precondition of nervous system evolution was the emergence of multicellular organisms, which occurred more than a billion years ago in multiple eukaryotic branches, including the one that gave rise to animals (**Figure 13-2**). With multicellularity came differentiated cells with specialized functions, such as sensor cells for detecting external stimuli, effector cells for producing movement, and connecting cells linking sensor and effector cells. The first nervous system featuring such interconnected cells arose before the divergence of **cnidarians** (radially symmetric animals such as hydra, jellyfish, and corals) and **bilaterians** (bilaterally symmetric animals with three germ layers; Section 7.1), including all vertebrates and most invertebrates (see **Box 13-1** for different views on the origin of nervous systems).

Within the bilaterian lineage, increased numbers of peripheral nerve nets and a need for overall coordination led to centralization of neurons, resulting in the prototypes of the central nervous system and the brain. This centralization likely occurred before the divergence of **protostomes** ("mouth first," referring to the mouth appearing first during embryonic development connecting the future digestive system to the outside world) and **deuterostomes** ("mouth second," with the anus appearing first). During the **Cambrian** period (542–488 million years ago), major phyla within the animal kingdom diversified, as evidenced by an abundance of corresponding fossils. However, molecular clock analyses suggest that the roots of bilaterian diversification originated long before, in the pre-Cambrian sea. One branch, the **chordates** (animals with a notochord), gave rise to the vertebrates we know today—fish, amphibians, reptiles, birds, and mammals, with birds emerging as a branch of reptiles. Other branches led to many phyla of

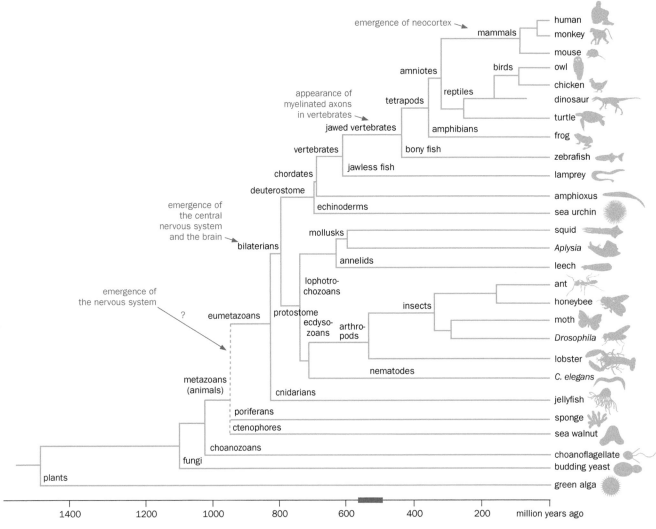

Figure 13-2 Phylogenetic tree of the animal kingdom. Times of divergence were estimated based on molecular clock data at www.timetree.org as of July 2019. Dates (scale at bottom) are mean estimates, as different molecules and analysis methods yield considerable variation. Dates before the Cambrian period (red on the time scale) are rough estimates at best, as fossil records for calibration are scarce. The dashed vertical line indicates uncertainty in the branching pattern (see Box 13-1). Red arrows point to several milestones of nervous system evolution discussed in this chapter.

invertebrates that have been used as model organisms in neurobiological research (Figure 13-2).

The construction of accurate phylogenetic trees not only unifies the animal kingdom, but also provides an important basis for evolutionary analysis of the gain and loss of specific traits, as we discuss in the next section.

Box 13-1: When did nervous systems first emerge?

The nervous system has been thought to first emerge in **eumetazoans**, a taxon that includes the most recent common ancestor of cnidarians and bilaterians (Figure 13-3A). This is because the nervous system is absent from poriferans (sponges), a more distantly related species to eumetazoans (Figure 13-2). However, recent phylogenetic analyses of ctenophores (comb jellies), culminating in the complete genome sequencing of *Mnemiopsis leidyi* and *Pleurobrachia*

bachei (commonly known as sea walnut and sea gooseberry, respectively), suggested that despite their morphological similarities with jellyfish, ctenophores are more distantly related to jellyfish than are sponges. Yet, unlike sponges, comb jellies possess neurons, synapses, and nerve nets. If ctenophores were indeed more distant from eumetazoans than were sponges (note that there is considerable debate about this issue), these findings would suggest that the ner-

Box 13-1: continued

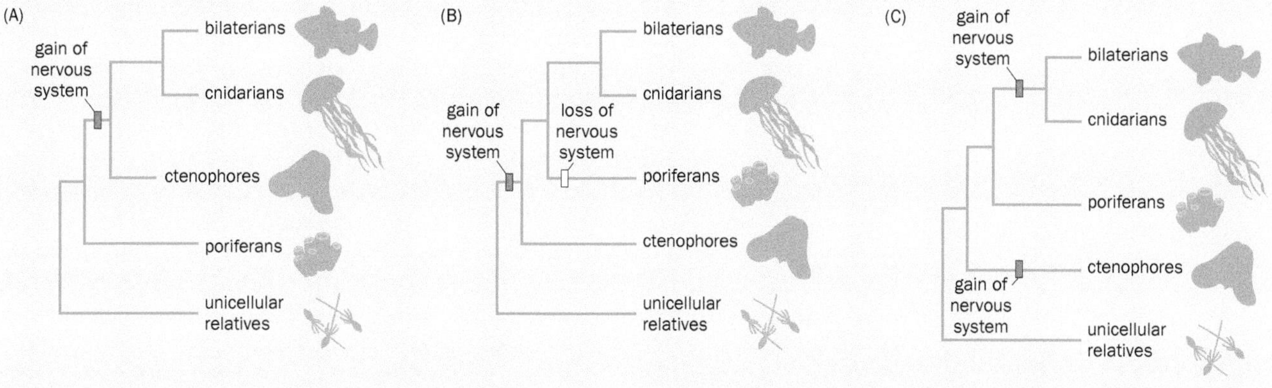

(A)

gain of
nervous
system

bilaterians

cnidarians

ctenophores

poriferans

unicellular
relatives

(B)

gain of
nervous
system

loss of
nervous
system

bilaterians

cnidarians

poriferans

ctenophores

unicellular
relatives

(C)

gain of
nervous
system

bilaterians

cnidarians

poriferans

ctenophores

gain of
nervous
system

unicellular
relatives

Figure 13-3 Three views on nervous system emergence.
(A) According to this phylogenetic tree, cnidarians and bilaterians
are more closely related to ctenophores, which have a nervous
system, than to poriferans, which lack a nervous system. The nervous
system first emerged before the divergence of ctenophores and
cnidarians/bilaterians. **(B & C)** Recent genomic analysis suggests
an alternative phylogenetic tree, in which cnidarians and bilaterians
are more closely related to poriferans than to ctenophores. If this

were the case, then either the nervous system emerged before the
divergence of ctenophores and was subsequently lost in poriferans
(Panel B) or nervous systems emerged independently in ctenophores
and eumetazoans (Panel C). (Based on Ryan JF, Pang K, Schnitzler CE,
et al. [2013] *Science* 342:1336–1344; Moroz LL, Kocot KM,
Citarella MR, et al. (2014) *Nature* 510:109–114; King N & Rokas A
[2017] *Curr Biol* 27:R1081–R1088.)

vous system might have first emerged before poriferans
and eumetazoans diverged and had subsequently been lost
in poriferans (Figure 13-3B). Alternatively, the nervous systems of eumetazoans and ctenophores may have evolved
independently (Figure 13-3C), using common building blocks
that were present in unicellular organisms (see Sections
13.6–13.9 for details).

While resolving these alternatives will require further data
and analyses, these hypotheses draw upon two principles of
evolution: First, loss of traits can be just as significant as gain
of traits in the course of evolution. Second, similar solutions
to a common problem can arise independently in different
branches of the tree of life, a concept called **convergent
evolution**.

13.2 Cladistic analysis distinguishes processes of evolutionary change

As exemplified in Box 13-1, evolutionary processes proceed via both gain and loss
of traits. Let's consider the following hypothetical example: suppose that while
comparing three animals, *a*, *b*, and *c*, we find that trait *T* is present in *a* and *b* but
not in *c*. *T* might be a new trait acquired by *a* and *b*. Alternatively, *T* might have
existed in the last common ancestor of *a*, *b*, and *c*, and been lost from *c* during
subsequent evolution. How do we distinguish between these alternatives?

The phylogenetic relationship between animals *a*, *b*, and *c* can provide useful
clues (Figure 13-4). If we know that the phylogenetic tree follows Figure 13-4A,
then it is more likely that *T* was lost by *c* rather than acquired by *a* and *b*, as the
former scenario requires only one event—the loss of *T* after *b* and *c* diverged,
whereas the latter scenario requires two separate events—the independent acquisition of *T* by the *a* and *b* clades. (A **clade** is a branch of the tree of life, consisting
of an ancestor and all of its descendants.) If we know that the phylogenetic tree
follows Figure 13-4B, then both scenarios are equally probable, since each can be

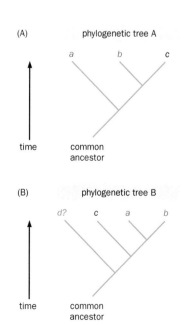

(A)   phylogenetic tree A

a   *b*   *c*

time   common
ancestor

(B)   phylogenetic tree B

d?   *c*   *a*   *b*

time   common
ancestor

Figure 13-4 Illustration of cladistic analysis. Animals with trait *T* (*a* and *b*) are in red; animals
lacking trait *T* (*c*) are in black. The likelihood that *T* is gained in *a* and *b*, or lost in *c*, can be inferred
if we know the phylogenetic relationship between the three animals. **(A)** In this phylogenetic tree,
T is more likely to have been lost in *c* (requiring a single event) than gained in *a* and *b* (requiring
two events). **(B)** In this phylogenetic tree, the outcome is explained equally well by either loss of
T in *c*, or gain of *T* in the common ancestor of *a* and *b*. Analysis of whether *T* is present or absent
in the outgroup *d* can help determine which evolutionary process is more likely.

accounted for by a single event—gain of T by the common ancestor of a and b after it diverged from c, or loss of T in the c clade, assuming here for simplicity that losses and gains are equally likely events. To further distinguish between these possibilities, we can ask whether T is present or not in a more distantly related species d, which is called an **outgroup**. The absence of T in d would favor a gain in the a/b clade, whereas the presence of T in d would favor a loss in the c clade.

This type of evolutionary comparison in the context of phylogenetic relationships is called **cladistic analysis**. The method we just used to deduce whether T was gained by a/b or lost by c is called **maximum parsimony** and is a way to generate phylogenetic predictions by favoring the smallest number of evolutionary changes needed to explain data. As the example in Figure 13-4 illustrates, conclusions about evolutionary changes become more robust when more clades are included in the cladistic analysis. For the simplicity of this discussion, we have assumed phylogenetic trees to be known. In reality, phylogenetic trees are usually *constructed* by methods such as maximum parsimony by comparing many traits (for example, protein and nucleic acid sequences) across different organisms.

Let's apply cladistic analysis to a real-world example: the evolution of gyri (ridges) and sulci (furrows) in the mammalian neocortex. Because the human neocortex is full of gyri and sulci (gyrencephalic), while the neocortex of smaller mammals such as the mouse is smooth (lissencephalic), one might think that gyri and sulci evolved along with the complexity of the primate brain. However, it turns out that all three major clades of mammals—monotremes, marsupials, and placental mammals—have species with gyrencephalic or lissencephalic neocortices (**Figure 13-5**). Thus, gyri and sulci most likely arose in all three branches independently. It is less likely that they were independently lost by subsets of mammals in all three branches, as extant reptiles (an outgroup of mammals) do not have gyri, sulci, or neocortex. One possible explanation for the independent development of gyri and sulci in all three mammalian branches is that the mechanisms underlying gyri and sulci formation are ancestral, and the degree to which they are utilized might be what is under selection. Indeed, the degree of gyri and sulci correlates with brain size. As we will discuss in Section 13.21, gyri and sulci may be byproducts of neuronal number expansion.

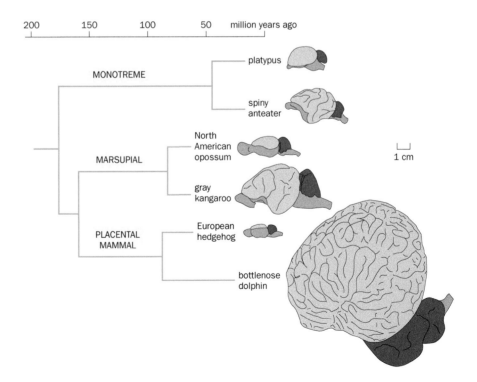

Figure 13-5 Gyri and sulci in the neocortex likely arose independently in the three mammalian clades. Examples of mammals with either smooth (lissencephalic) or folded (gyrencephalic) neocortices (tan); both traits are present in all three mammalian branches. All brains are drawn to the same scale, and large brains tend to have gyri and sulci. The estimated times of divergence are according to www.timetree.org.

13.3 Gene duplication, diversification, loss, and shuffling provide rich substrates for natural selection

Evolutionary changes must occur at the level of DNA to be passed on to future generations. With few exceptions (see Box 14-1 for one in prokaryotes), changes in DNA are *not directed* in favor of particular traits. Instead, **mutations**—insertions, deletions, or alterations of the identity of DNA base pairs—occur randomly. Natural selection tends to increase the frequency of **adaptive** changes, changes that render an individual and its progeny more likely to survive and reproduce in a particular environment. (Note that **adaptation** in the context of evolution has an entirely different meaning than the adaptation in sensory systems we discussed in Chapters 4 and 6.) The increase in frequency of beneficial **alleles** (an allele is a specific version of a gene) through this process is described as **positive selection**, while the decrease in frequency of detrimental alleles is described as **negative selection**. In addition to positive and negative selection, the allele frequency of a gene can change due to random events collectively referred to as **genetic drift**. For example, an allele may not be passed on to progeny because alleles in progeny are a random sample of those in the parents. An allele may be lost when progeny carrying the allele die or fail to reproduce by chance. Such events leading to the loss of one allele would increase the prevalence of other allele(s). Genetic drift has a larger impact on small populations and highlights the stochastic aspect of evolution. Whether via natural selection, genetic drift, or their combination, when the frequency of an allele representing a change becomes 1, the change is **fixed**.

Population geneticists use **fitness** to quantify the ability of individuals to pass their alleles to progeny. The fitness of a specific allele is defined as the ratio of the allele frequency in a population after one generation of selection to the frequency before the selection. A fitness value greater than 1 indicates that the specified allele is, on average, beneficial for the carrier's survival and reproduction *in a given environment*. (Fitness can also be similarly defined with respect to a specific phenotype.) Note that fitness depends upon the environment. For example, sickle-cell anemia is a human genetic disorder caused by homozygosity of a point mutation in the β-globin gene that affects the morphology and function of red blood cells. Despite its detrimental effect when homozygous, the mutant allele can occur at relatively high frequencies in populations living in areas where malaria infection is prevalent, as heterozygous carriers of the mutant allele are more resistant to malaria infection than non-carriers. Thus, in different environments, the mutant β-globin allele confers different fitness levels.

What are the mechanisms of evolutionary change at the level of DNA? In principle, evolutionary changes can be realized by the birth of new genes, loss of old genes, or mutation of existing genes. Mutations can occur in the protein-coding region, which could alter the protein product's function, or in a regulatory region, which could alter the gene expression pattern. We discuss these mechanisms in the rest of this section and the next section.

While evidence in recent years suggests that new protein-coding genes can be produced *de novo* from non-coding sequences in the genome, the vast majority of new genes are produced via gene duplication followed by diversification. Thus, a major step in evolutionary innovation is **gene duplication**. If a gene is essential for survival, the effects of random mutations are more likely to be detrimental than beneficial. If a gene is duplicated, however, the extra copy can undergo changes more freely without compromising the animal's survival. Although most mutations are neutral or detrimental, beneficial changes occasionally arise. Through trial and error across many generations, the duplicated copy of the gene may evolve to carry out the original gene's function more effectively or acquire entirely new functions. The evolution of cone opsin genes that brought trichromatic color vision to Old World monkeys and apes is one such example (Section 4.11). The most common mechanism underlying gene duplication is error in DNA recombination. Unequal crossing over between homologous DNA sequences can produce an extra copy of a chromosomal segment in one daughter cell and a deletion of the same chromosomal segment in another daughter cell (**Figure 13-6**A).

Figure 13-6 Selected mechanisms of gene gain, loss, and shuffle. (A) Top, schematic of unequal crossing over. During meiosis, error in pairing of homologous chromosomes (gray bars) at DNA repeat sites (light and dark blue ovals with similar sequences) causes unequal crossing over (×), producing one germ cell with a duplication of the ancestral gene (yellow) and another with a deletion. Bottom, as each new gene acquires different mutations, the two genes diverge (represented by color changes). **(B)** Exon shuffling. Translocation between two nonhomologous chromosomes (different shades of gray) is illustrated. If the breakpoint of each chromosome is located in introns (cyan), exons can be shuffled between two genes (*a* and *b*). This process separates functional domains that were unified in the old proteins and pairs them with different functional domains to create new proteins (encoded by *a/b* and *b/a*).

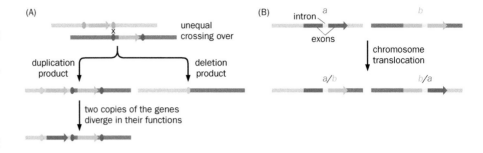

When a gene's function can be better served by other genes, or when the function becomes obsolete to the organism (for example, as a result of environmental changes), the integrity of the gene is no longer selected for. As mutations accumulate, the gene may become a nonfunctional pseudogene. We have seen examples of this in our discussion of genes encoding odorant receptors (Section 6.4). Pseudogenes may eventually be deleted from the genome via mechanisms like those in Figure 13-6A.

Another mechanism for creating new genes is **DNA shuffling**: part or all of the protein-coding sequence of one gene can be fused with that of another gene. The resulting fused gene may be able to perform some of the functions previously performed by two separate genes, or it may acquire novel properties. Conversely, a gene encoding two different domains of a protein can be split into two separate genes, each encoding one domain. Such shuffling can be achieved by chromosomal duplications or translocations that place part or all of a gene into a new genomic context.

Most eukaryotic genes contain introns between protein-coding exons. Translocations between the introns of two genes cause **exon shuffling** (Figure 13-6B), a special case of DNA shuffling. The emergence of repeating modules in ion channels is an example of how genes encoding individual modules can be duplicated and shuffled (see Section 13.6). DNA shuffling can also cause proteins to be expressed under the control of new regulatory elements, thus altering gene expression patterns.

13.4 Altered patterns of gene expression can drive evolutionary change

Recent genomics data have highlighted the importance of altering gene expression patterns in the evolutionary process, in addition to generating new proteins and changing protein-coding sequences. The number of protein-coding genes does not increase considerably as the nervous system becomes more complex. For example, while the sea anemone (a cnidarian), nematode *C. elegans,* and fruit fly *Drosophila melanogaster* have about 18,000, 19,000, and 15,000 protein-coding genes, respectively, the human genome contains only 21,000. (By comparison, the rice genome contains about 50,000 protein-coding genes.) Furthermore, whole-genome comparisons reveal that at least 6% of the genome has been highly conserved between mice and humans since their divergence about 90 million years ago (Figure 13-2), but only about 1% of the human genome consists of protein-coding sequences. Among the known functions of conserved non-protein-coding sequences are ***cis*-regulatory elements** (DNA elements that enhance, suppress, or insulate the expression of genes on the same chromosome) and DNA sequences encoding non-protein-coding RNAs, which also regulate gene expression.

Even though we are far from a complete understanding of the function of the non-protein-coding genome, the data summarized here point to the importance of changes in gene expression patterns during evolution. Indeed, in examples ranging from the wing spot patterns of butterflies to the skeletal structures of three-spined sticklebacks (**Figure 13-7**), researchers have repeatedly identified changes in the *cis*-regulatory elements that regulate the expression patterns of developmental control genes as the underlying cause of phenotypic variations.

As discussed in Box 12-3, human genome sequencing studies revealed that, in comparison to the reference genome, each human contains approximately 3.5 million single-nucleotide polymorphisms, 100 gene disruption mutations, and 1000 polymorphic copy number variations. An estimated 100 new mutations are introduced with each conception. These numbers highlight the abundance of genetic variation in our species. Similarly abundant genetic variation is likely present in other species as well. These genetic variations in copy numbers, protein-coding sequences, and gene regulatory elements provide rich substrates for natural selection.

13.5 Natural selection can act on multiple levels of nervous system organization to enhance fitness

A major question in evolutionary biology is how inherited variation, which occurs at the level of DNA and genes, contributes to adaptation. Although historically there has been considerable debate, scientists now agree that the *object of selection* in most cases is the *individual organism*. An individual better at finding food, avoiding predators, securing mates, and caring for offspring has a better chance of passing its genetic variants onto future generations. In other words, while variation occurs at the level of *genotypes*, natural selection acts on an individual's *collective phenotypes*. (Note that there are notable exceptions to the individual organism being the object of selection. In an ant colony, for instance, the individual worker ants are not the objects of selection, as they do not produce offspring. Nevertheless, their behavior contributes to the well-being of the queen, who produces all the progeny. Thus, the object of selection in an ant colony is likely the set of genes contributing to the collective behavior of the entire colony.)

We defined fitness with respect to individual alleles or phenotypes in Section 13.3. Fitness can also be defined with respect to an individual organism (or genome, which collectively has allelic variants of many different genes). In this context, fitness is defined as the number of second-generation descendants (the "grandchildren") that the *type* of individual with a particular genome can expect to have. Note that this definition of fitness focuses on a type of individual in order to guard against stochastic events that might affect the offspring number of any particular individual. Also, as it is based on the expected number of grandchildren, this definition of fitness removes influences other than the individual's genome (such as, for example, maternal effects, in which an individual's phenotype is affected by its mother's phenotype or genotype rather than its own genome) and takes into account the well-being and fertility of the individual's offspring.

Genes can influence an individual's fitness at multiple levels of nervous system organization—molecules, cells, circuits, behaviors, and developmental processes. Likewise, natural selection can exert its primary effect on any of these levels (**Figure 13-8**): on proteins, such as a sensory receptor that improves an individual's ability to detect predators; on cells, such as a photoreceptor neuron that transduces light signals into electrical activity more efficiently; on circuits, such as a wiring pattern that offers greater flexibility or more efficient computations for

Figure 13-7 Evolution of three-spined sticklebacks. Marine sticklebacks have colonized numerous freshwater lakes and streams since the last Ice Age. Comparison of marine and freshwater sticklebacks provides a rich repertoire of phenotypic differences from which to trace evolutionary changes that have taken place over 10,000 generations as freshwater sticklebacks have adapted to their new environments. Compared to their marine counterparts (top of each pair), different freshwater varieties demonstrate loss of a pelvic hind fin (left pair; arrow indicates the pelvic hind fin), significant reductions in bony armor (middle pair), and lighter skin color (right pair). Different stickleback species can be crossed to produce hybrids, allowing researchers to examine evolutionary changes using molecular-genetic and genomic methods. Each of the changes in the three examples shown has been traced to variations in the *cis*-regulatory elements that regulate transcription of a developmental control gene. (From Kingsley DM [2009] *Sci Am* 300[1]:52–59. With permission from Springer Nature.)

Figure 13-8 Natural selection can exert its effects on multiple levels. Since genes act on proteins, cells, circuits, behaviors, and developmental processes to affect nervous system function, natural selection can exert its primary effects on any of these levels to determine the fitness of individuals. Individuals with greater fitness are more likely to pass their versions of genes to the next generation (red feedback arrow) for a new round of selection. While there is a left-to-right hierarchy from protein to cell to circuit to behavior, with developmental processes largely in parallel, the reciprocal arrows show that changes at higher levels can also affect properties at lower levels.

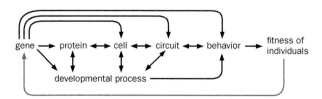

extracting behaviorally relevant information; on behaviors, such as a daily activity rhythm better adapted to the circadian cycle of the environment; and on developmental processes, such as a pattern of cell division that produces more neurons, allowing for more sophisticated computation. A combination of alleles that collectively improves the fitness of an individual will have a better chance of being passed on to the next generation during natural selection (red arrow in Figure 13-8). Numerous cycles of selection occurring simultaneously in many branches of the phylogenetic tree (Figure 13-1; Figure 13-2) during the past billion years have produced myriad nervous systems in animals that occupy diverse niches in the web of life.

Having introduced general concepts and approaches in evolutionary analysis, we now investigate how neuronal communication, sensory systems, and the structure and development of the nervous system have arisen.

EVOLUTION OF NEURONAL COMMUNICATION

What advantage does having a nervous system give an animal? All animals must sense nutrients and danger in the environment and respond accordingly, often by controlled movement. Many single-cell organisms are already capable of that (see examples of bacteria and green algae later in the chapter). The emergence of functionally specialized cells in metazoans enabled development of different cell types, such as sensory neurons to detect environmental signals, motor neurons to control muscle contraction, and muscles to execute movement. These specialized cells enabled better detection of food and predators and allowed more sophisticated behaviors, which facilitated exploration of new environmental niches and provided selective advantages for ctenophores and eumetazoans. Interneurons connecting sensory and motor neurons evolved to better coordinate inputs with outputs and perform simple computations (Box 1-2). The emergence of central nervous systems and brains in bilaterians likely resulted from progressive elaboration of interneuronal circuits in response to an increasing need to integrate sensory input and coordinate movement accompanying the greater complexity of bilaterians' body plans. The elaboration of interneuronal circuits further enabled more sophisticated functions that benefited animals possessing them. These include internal neural representations of the external environment and the body, long-term memories of past experiences, and cognitive systems that take into account the animal's needs and past experiences to generate optimal behavioral responses to environmental signals.

While most of the mechanisms underlying the progressive sophistication of the nervous system remain unknown, an important element must have been the evolution of the speed and sophistication of neuronal communication (Movie 2-1). As a way to investigate how different features of the nervous system arose, we examine in the following sections the origin of the key components that allow signals to propagate within and between neurons.

13.6 Ion channels appeared sequentially to mediate electrical signaling

As we learned in Chapter 2, rapid neuronal communication relies on electrical signaling. Action potentials, the primary means of propagating signals from somata to axon terminals, require the sequential opening of voltage-gated Na^+ and K^+ channels, and neurotransmitter release depends on voltage-gated Ca^{2+} channels at nerve terminals. These voltage-gated channels allow neurons to perform nonlinear computations, such as amplification of synaptic potentials or integration of analog signals into digital signals in spiking neurons. Tracing the origin of voltage-gated ion channels suggests steps by which the building blocks of electrical signaling might have evolved.

Among voltage-sensitive ion channels, K^+ channels are the most ancient, as they are prevalent in prokaryotes (**Figure 13-9**A). Many bacterial species have K^+ channels with two transmembrane domains (2TM). Some also have K^+ channels

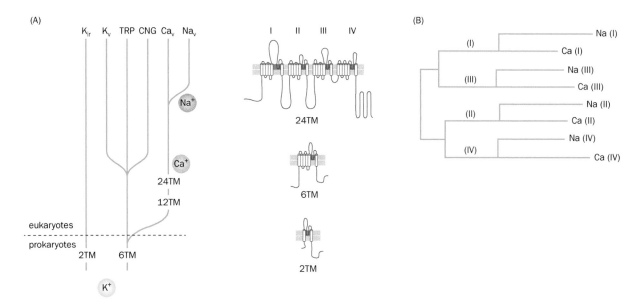

Figure 13-9 Evolutionary path of ion channels inferred from sequence comparison. (A) Of the three voltage-gated cation channels (K_v, Ca_v, and Na_v), K^+ channels are the most ancient. The two major types of K^+ channels—designated 2TM or 6TM according to the number of their transmembrane (TM) domains (see schematic on the right, taken from Figure 2-34)—arose before the split of prokaryotes and eukaryotes. Voltage-gated Ca^{2+} channels diverged from 6TM K^+ channels through two rounds of gene duplication and fusion, yielding a 24TM channel, from which voltage-gated Na^+ channels later derived. TRP (transient receptor potential) and CNG (cyclic nucleotide-gated) channels also derived from 6TM K^+ channels. K_{ir}, inward-rectifier K^+ channel. **(B)** Each of the four repeats of the voltage-gated Na^+ channel is most similar to the corresponding repeat of the voltage-gated Ca^{2+} channel. Furthermore, repeat I is more similar to repeat III than to repeats II and IV. The tree was built based on the most parsimonious amino acid sequence alignment of the transmembrane segments of the rat $Na_v1.1$ and $Ca_v1.1$ channels. The length of each branch reflects the number of amino acid substitutions. (A, from Hille [2001] Ion Channels of Excitable Membranes. With permission from Sinauer. B, from Strong M, Chandy KG, & Gutman GA [1993] *Mol Biol Evol* 10:221–224. See also Moran Y, Barzilai MG, Liebeskind BJ, et al. [2015] *J Exp Biol* 218:515–525.)

with six transmembrane domains (6TM) that resemble the voltage-gated K^+ channels of animals, which have S1–S4 added to the 2TM pore. Extensive **horizontal gene transfer** (gene transfer from one organism to another via mechanisms other than reproduction, such as viral transduction) during the early history of life makes it difficult to determine the origin of K^+ channels, but they likely existed well before the emergence of eukaryotes. K^+ channels function in prokaryotes to create a negative resting membrane potential, a feature shared by eukaryotic cells, and maintain ion gradients across the cell membrane.

Ca^{2+} channels were next to emerge, likely via duplication and divergence of a gene for a 6TM K^+ channel, followed by two rounds of duplication and fusion (Figure 13-6). The resulting 24TM channel comprises four tandem repeats of an ancestral 6TM structure. Single-cell yeast and green algae have similar 24TM Ca^{2+} channels, suggesting that 24TM Ca^{2+} channels likely appeared in early eukaryotes before the split of the plant, fungal, and animal lineages (Figure 13-9A). The appearance of Ca^{2+} channels enabled control of intracellular Ca^{2+} concentrations, which is essential for many eukaryotic intracellular signaling events. Indeed, Ca^{2+}-binding proteins such as calmodulin are prevalent in all eukaryotic branches, suggesting that Ca^{2+} has been used as a signaling ion since ancient times. With the emergence of the nervous system, one of the earliest functions of voltage-gated Ca^{2+} channels might have been control of neurotransmitter release. Together with voltage-gated K^+ channels, voltage-gated Ca^{2+} channels can also amplify electrical signals by producing "Ca^{2+} spikes," likely precursors to Na$^+$-based action potentials. Ca^{2+} spikes are used by some muscle cells, invertebrate (including cnidarian) neurons, and dendrites of large mammalian neurons. These Ca^{2+}-based active properties allow electrical signals to propagate more efficiently than passive spread (Sections 2.8 and 2.9) and enable neurons to perform nonlinear computations.

Voltage-gated Na$^+$ channels evolved most recently, likely via duplication and diversification of a gene encoding a 24TM Ca^{2+} channel. Comparison of amino acid sequences revealed that each of the Na$^+$ channel's repeat segments is more similar to the corresponding repeat of the Ca^{2+} channel than to the three other repeats in its own structure (Figure 13-9B). This suggests that the two duplications of a 6TM channel that gave rise to the 24TM channel occurred before the divergence of Na$^+$ channels from Ca^{2+} channels. Recent genome sequencing revealed homologs of voltage-gated Na$^+$ channels in single-celled choanoflagellates, implying that the duplication event that gave rise to voltage-gated Na$^+$ channels occurred prior to the emergence of the nervous system or even multicellular organisms. However, the pore sequence suggests that the choanoflagellate channel conducts both Na$^+$ and Ca^{2+}, likely reflecting a transitional stage before the emergence of voltage-gated channels that conduct Na$^+$ specifically. Analysis of pore sequences followed by functional studies suggest that channels with Na$^+$ selectivity emerged independently at least twice in animals: once in the bilaterian lineage that gave rise to all voltage-gated Na$^+$ channels in vertebrates and another time in a specific cnidarian clade. The emergence of voltage-gated Na$^+$ channels freed Ca^{2+} from its role in propagating electrical signals. Ca^{2+} could thus be dedicated to regulating many other processes, such as biochemical reactions through Ca^{2+}-dependent kinases and phosphatases and synaptic transmission. Moreover, as cells can tolerate greater fluctuations of Na$^+$ than of Ca^{2+}, action potentials could be produced with larger membrane potential changes, enhancing the signal-to-noise ratio for long-distance propagation of electrical signals.

Interestingly, *C. elegans* has voltage-gated Ca^{2+} channels but no Na$^+$ channels. Given that voltage-gated Na$^+$ channels are highly conserved between squid, insects, and mammals, it is almost certain that the last common bilaterian ancestor (Figure 13-2) already had a voltage-gated Na$^+$ channel, which was subsequently lost in the clade leading to *C. elegans*. This may be because passive properties and voltage-gated Ca^{2+} channels are sufficient for propagating electric signals in small animals like *C. elegans*, and the cost of maintaining voltage-gated Na$^+$ channels may have outweighed the benefit.

Tracing the evolutionary history of ion channels has illustrated several principles we will repeatedly encounter. (1) Gene duplication followed by diversification is a prevalent mechanism for evolution of new functions. (2) Novel functions can be built on more ancient molecules. (3) Evolution involves both gene loss and gene gain, which together shape the genomes of present-day animals.

13.7 Myelination evolved independently in vertebrates and large invertebrates

Larger animals require more rapid propagation of action potentials. Two different solutions have evolved (Section 2.13). The first is to increase the diameter of the axon, as the conduction speed of action potentials is proportional to the square root of the axon diameter. We have seen a striking example in the squid giant axon, whose rapid conduction enables the squid to swiftly escape from danger. Likewise, *Drosophila* and zebrafish respectively possess the giant fiber and Mauthner fiber, both of which are large-diameter axons that connect the brain with motor neurons in the ventral nerve cord or spinal cord to mediate escape reflexes.

The second and more efficient solution is glial wrapping of axons, which increases membrane resistance, reduces membrane capacitance, and enables saltatory propagation to speed up action potential conduction and conserve energy. Axon myelination in vertebrates shares many properties and likely evolved only once in jawed fish, as jawless fish such as lamprey (Figure 13-2) do not have myelinated axons. The appearance of myelination must have been a crucial event in the evolution of nervous systems of large vertebrates. Glial wrapping of axons likely evolved independently in several invertebrate clades, including annelids and arthropods, as the properties and protein compositions of their glial membranes are distinct from each other and from those of vertebrates. Despite independent origins, the morphology of myelinated axons in some invertebrates closely resembles that of myelinated axons in vertebrates (**Figure 13-10**). Thus, myelination provides a striking example of **convergent evolution**, independent evolution of similar features in animals from different clades.

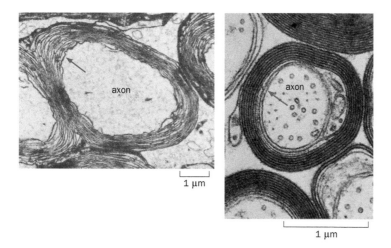

Figure 13-10 Convergent evolution of axon myelination in invertebrates and vertebrates. Electron micrograph of a cross section of a myelinated axon in the nerve cord of a sea prawn (*Palaemonetes vulgaris*) (left) and the spinal cord of a dog (right). Arrows point to the myelin sheaths. (Left, from Heuser JE & Doggenweiler CF [1966] *J Cell Biol* 30:381–403. With permission from Rockefeller University Press. Right, courtesy of Cedric Raine.)

13.8 Synapses likely originated from cell junctions in early metazoans

How did the synapse—the nexus of interneuronal communication—arise? Tracing the evolutionary origins of proteins found in chemical synapses by comparing the genomes of simple eukaryotes has offered interesting insights (**Figure 13-11**).

The chemical synapse is a cell–cell junction between pre- and postsynaptic compartments. Proteins that hold synaptic compartments together likely evolved

(A)

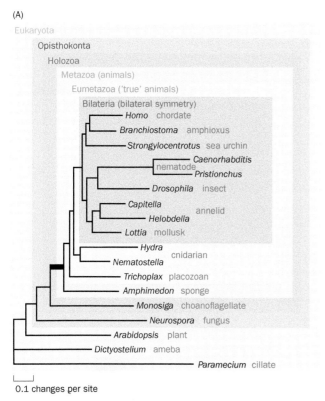

(B)

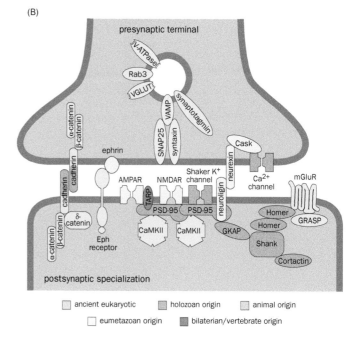

Figure 13-11 The origin of synaptic proteins. (A) Phylogeny of animals based on alignment of 229 conserved genes from the sequenced genomes of 18 species. The bold line represents the metazoan branch. Both genera and common taxon names are given. **(B)** Origin of individual synaptic proteins; the color code matches that of the phylogeny in Panel A. See Figures 3-10 and 3-27 for the organization of presynaptic terminals and postsynaptic specializations, respectively. Note that the extracellular and intracellular domains of cadherins emerged at different times. For proteins we did not introduce in Chapter 3: VGLUT is a vesicular membrane transporter for glutamate; Cask is a presynaptic scaffold protein; Homer, Shank, GKAP, and GRASP are all postsynaptic scaffold proteins; Cortactin is an actin-binding protein; α-, β-, and δ-catenins are intracellular partners of cadherins. (From Srivastava M, Simakov O, Chapman J, et al. [2010] *Nature* 466:720–726. With permission from Springer Nature. See also King N, Westbrook MJ, Young SL, et al. [2008] *Nature* 451:783–788.)

(A)

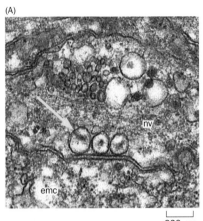

200 nm

(B)

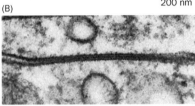

100 nm

Figure 13-12 A cnidarian chemical synapse and gap junction. (A) Electron micrograph of a chemical synapse between a nerve cell (nv) and an epitheliomuscular cell (emc; a cell type with properties of both epithelium and muscle) in *Hydra*. The vesicles docked at the presynaptic membrane (arrow) have larger diameters (~150 nm) than bilaterian synaptic vesicles (~40 nm; Figure 3-3). **(B)** Electron micrograph of a gap junction between epitheliomuscular cells in *Hydra*. (From Chapman JA, Kirkness EF, Simakow O, et al. [2010] *Nature* 464:592–596. With permission from Springer Nature.)

from those that mediate cell adhesion in ancestral multicellular metazoans. For example, members of the cadherin family of homophilic cell adhesion molecules are abundant in the synapses of animals ranging from insects to mammals (Figure 3-10). As they are present in all animals, cadherins predate the emergence of the nervous system. Indeed, proteins with cadherin extracellular domains are found even in single-celled choanoflagellates, suggesting a multistep evolution of these important cell adhesion molecules (Figure 13-11B). Heterophilic adhesion partners such as neurexins and neuroligins, prominent synapse-organizing molecules in both vertebrates and insects, first emerged in eumetazoans, a group that includes the last common ancestor of bilaterians and cnidarians. Indeed, chemical synapses in cnidarians (Figure 13-12A) share morphological features with those in bilaterians. Ephrins and Eph receptors, which play important roles in synapse formation in addition to their axon guidance functions discussed in Section 5.4, first functioned as a pair in eumetazoans, although Eph receptors appeared earlier, presumably as partners for other molecule(s). The addition of heterophilic trans-synaptic interaction molecules may have contributed to the asymmetry and directionality of chemical synapses, which are distinct from the symmetrical cell junctions between epithelial cells. Once trans-synaptic communication molecules in the presynaptic and postsynaptic membranes were no longer identical, their cytoplasmic domains could then interact with distinct sets of scaffold proteins present in the pre- and postsynaptic compartments. Some of these scaffold proteins are ancient, others co-evolved with synapses, and still others were added later as regulation of synaptic transmission became more elaborate (Figure 13-11).

What about electrical synapses, which directly couple neurons via gap junctions (Figure 1-14 and Box 3-5)? The protein classes that mediate most gap junctions in vertebrates, the connexins, appear to be a chordate innovation. But gap junctions are also prevalent in invertebrates, where they are mediated by a distinct class of proteins called innexins (invertebrate connexins; their vertebrate orthologs do not appear to function as gap junctions). Innexins were found in the genomes of cnidarians but not sponges; the earliest functions of these proteins likely involved formation of gap junctions between non-neuronal cells (Figure 13-12B). Thus, electrical synapses, like chemical synapses, also evolved from cell junctions. As nervous systems have become more complex, these two types of synapses have collaborated as independent but complementary systems for interneuronal communication.

13.9 Neurotransmitter release mechanisms were co-opted from the secretory process

Among the most ancient synaptic proteins are those involved in synaptic vesicle exocytosis (Figure 13-11). The v-SNARE and t-SNARE proteins that mediate vesicle fusion (Figure 3-8) are present in all eukaryotes. This is because neurotransmitter exocytosis utilizes a general mechanism fundamental to all eukaryotic cells— fusion of vesicles with intracellular or plasma membranes. In the secretory pathway, newly synthesized proteins destined for the cell surface or extracellular environment enter the endoplasmic reticulum (ER) upon translation and travel via membrane-enclosed compartments through the Golgi apparatus and to the plasma membrane (Figure 2-2). Vesicle budding and fusion are key steps in this secretory pathway. A striking convergence in scientific inquiry occurred in the early 1990s when cell biologists studying the machinery of the general secretory pathway in yeast and mammalian cells and neurobiologists studying neurotransmitter exocytosis in neurons found that they were studying the same protein families (Figure 13-13). In fact, due to the abundance of synaptic vesicles in the brain, neurotransmitter exocytosis has become a model system for cell biologists to study the general mechanisms of vesicle fusion.

Across the synaptic cleft, many proteins enriched in the postsynaptic density also predate the emergence of the nervous system (Figure 13-11), suggesting that they had ancestral functions prior to being recruited for neuronal communication. For example, postsynaptic scaffold proteins such as PSD-95, Homer, and

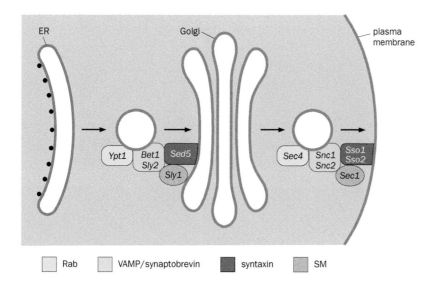

| Rab | VAMP/synaptobrevin | syntaxin | SM |

Figure 13-13 Conservation of membrane fusion machinery in yeast and neurons. This schematic summarizes genes identified by yeast geneticists as required for two different steps of membrane fusion in the secretory pathway: from the endoplasmic reticulum (ER) to the Golgi, and from the Golgi to the plasma membrane. As in neurotransmitter release, each step is regulated by a Rab GTPase and requires a VAMP/synaptobrevin-like v-SNARE, a syntaxin-like t-SNARE, and an SM protein (Figure 3-8; Figure 3-10). The names of yeast genes are in italics; their colors correspond to the protein classes they encode, shown at the bottom. (Adapted from Bennett MK & Scheller RH [1993] *Proc Natl Acad Sci U S A* 90:2559–2563.)

Shank first emerged in single-celled choanoflagellates. G-protein-coupled metabotropic glutamate receptors first arose in sponges. Ionotropic glutamate receptors such as AMPA and NMDA receptors first appeared in eumetazoans, although an ancestral glutamate receptor is found in all animals. Ionotropic glutamate receptors underwent expansion in ctenophores, as glutamate is a major neurotransmitter in nervous systems most distant from ours (Figure 13-2). The ancestral functions of these postsynaptic proteins may have been to sense chemicals in the environment, as will be discussed in the next part of the chapter.

In summary, proteins with specialized functions in neuronal communication were gradually recruited or added during evolution as the nervous system became more complex. Indeed, the assembly of the first nervous systems relied mostly on recruitment of preexisting proteins rather than simultaneous generation of multiple new proteins, echoing the epigraph by François Jacob: "evolution does not produce novelties from scratch."

EVOLUTION OF SENSORY SYSTEMS

Sensory systems, which function to detect food, mates, and predators, are under strong selection throughout an organism's struggle for existence and reproductive success. Indeed, sensory systems provide dramatic examples of how interactions between certain animals and their environments over time can build upon the five common senses discussed in Chapters 4 and 6 to form distinct, extraordinary, and niche-specific senses. Living in swamps and marshes, star-nosed moles are functionally blind, but their snouts are surrounded by mechanosensory appendages (rays) that allow them to capture small prey with incredible speed (Figure 13-14A; Movie 13-1). Rattlesnakes possess extremely sensitive pit organs for detecting infrared radiation (heat), which signals warm-blooded prey nearby (Figure 13-14B); these pit organs employ a TRPA1 channel that is the most sensitive heat-activated channel identified to date. Echolocating bats possess special organs that produce high-frequency ultrasound and detect the resulting echoes to navigate and identify their prey in the dark (Figure 13-14C; Section 6.26). Finally, monarch butterflies integrate diverse information—the direction of the sun, the circadian cycle, and probably the magnetic field (via magnetoreception)—to achieve their remarkable annual migration across North America (Figure 13-14D).

Development of sophisticated sensory systems with greater sensitivity, speed, and resolution should aid animals in survival and reproduction. However, elaborated systems are also expensive to build and maintain. These conflicting forces, together with historical constraints, have shaped the evolution of sensory systems. In the following sections we use chemosensation (Box 13-2) and vision to explore how sensory receptors, neurons, and circuits came about.

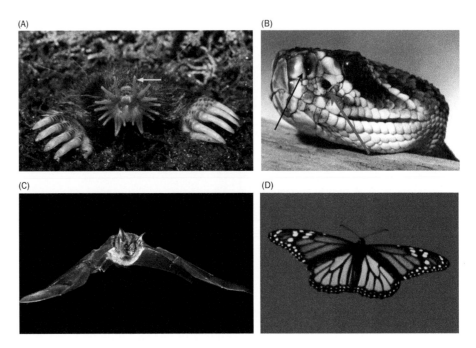

Figure 13-14 **Extraordinary senses. (A)** Star-nosed moles extend 11 pairs of appendages (the arrow points to one appendage) that surround the nose and use them to detect and capture small prey in swamps. **(B)** Rattlesnakes express a highly heat-sensitive TRPA1 channel in nerve terminals innervating the pit organ (red arrow), which is located between the nostril (black arrow) and the eye and is used to sense infrared radiation from warm-blooded prey. **(C)** Bats use echoes of the ultrasounds they emit to navigate and hunt. **(D)** Monarch butterflies fly thousands of miles each fall from the northern United States and Canada to Mexico. To guide their migration, they use a circadian-rhythm-adjusted sun compass and possibly an additional mechanism for detecting the magnetic field. (A, from Catania KC [2012] *Curr Opin Neurobiol* 22:251–258. With permission from Elsevier Inc. B, from Gracheva EO, Ingolia NT, Kelly YM, et al. [2010] *Nature* 464:1006–1011. With permission from Springer Nature. C, courtesy of Brock Fenton. D, from Reppert SM, Gegear RJ, & Merlin C [2010] *Trends Neurosci* 33:399–406. With permission from Elsevier Inc.)

Box 13-2: Chemotaxis: from bacteria to animals

Chemosensation is probably the most ancient sense and is well developed in bacteria. Like all living organisms, bacteria require nutrients. When a capillary tube containing attractants such as amino acids is inserted into a suspension of *E. coli,* the bacteria quickly accumulate near the mouth of the capillary in a process called **chemotaxis** (moving toward or away from a chemical source). How do bacteria achieve this?

Tracing the trajectory of individual bacteria provided important insights. In an isotropic solution (a solution of uniform concentration), bacteria exhibit two kinds of motion: runs of smooth swimming and tumbles that result in reorientation. During a run, bacteria travel a fairly straight path, but they change direction randomly during a tumble, resembling a random walk (**Figure 13-15**A). In a gradient of attractant, bacteria behave differently depending on the direction in which they are moving. When swimming away from attractant, a bacterium behaves in a manner similar to that observed in an isotropic solution, exhibiting frequent tumbles. However, when swimming toward an attractant, its runs are interrupted less frequently by tumbles. This **biased random walk** strategy effectively moves bacteria toward the attractant source. Conversely, as bacteria travel down a gradient of a repellent, they tumble less frequently than in

an isotropic solution; thus, they tend to swim away from the repellent source.

The molecular mechanisms enabling bacteria to sense chemicals and modify their movement have been worked out in detail. In essence, attractant binding to specific transmembrane receptors inhibits phosphorylation of an effector, which in turn decreases flagellar motor activity and thus inhibit tumbles. Repellent binding does exactly the opposite. Remarkably, *E. coli* can sense attractants at minute concentrations of 10 nM (equivalent to 10 molecules dissolved in a volume equaling that of a bacterium) but can also differentiate between a 10^5-fold range of concentrations so as to swim toward an attractant source. Signal amplification and sensory adaptation, two concepts introduced in our study of vision (Sections 4.4 and 4.7), are key to solving these problems. Abundant chemotactic receptors and all components of the chemotactic signaling pathway are clustered at one end of the cell to facilitate signal amplification. Regulation of receptor methylation allows receptors to adapt to different concentrations of an attractant, facilitating approach toward progressively higher concentrations (Figure 13-15B).

Thus, although bacteria are separated from animals by more than 2 billion years of evolution and use different molecular

Box 13-2: continued

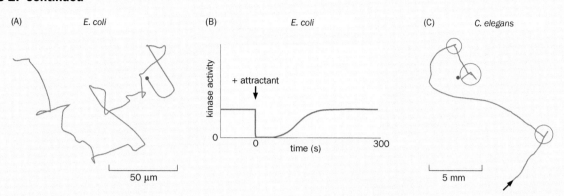

Figure 13-15 Biased random walk in bacterial and *C. elegans* chemotaxis. (A) The trajectory of a wild-type *E. coli* cell in an isotropic solution was recorded for 29.5 s, starting at the dot; linear runs are interrupted by tumbles that change the cell's direction of movement. **(B)** Addition of attractant to *E. coli* causes a rapid decrease in the activity of a kinase called CheA, which phosphorylates an effector to promote tumble. Thus, attractant binding to its receptor inhibits tumble. CheA also phosphorylates an enzyme that removes methylation from the attractant receptor and enhances its sensitivity. (Another enzyme methylates the receptor and reduces its sensitivity.) Thus, high attractant concentrations reduce CheA activity and receptor sensitivity, so that more receptors must be activated to keep CheA in an inhibited state to prevent tumble.

This adaptation allows bacteria to respond continually to increasing attractant concentrations. **(C)** Track (spanning 170 s) of a *C. elegans* moving toward a NaCl spot (dot) on a plate. Arrow indicates the start of the track. Circles indicate pirouettes, and the rest of the track is defined as runs. Note that runs can curve toward the NaCl source. (A, from Berg HC & Brown DA [1972] *Nature* 239:500–504. With permission from Springer Nature. B, from Sourjik V [2004] *Trends Microbiol* 12:569–576. With permission from Elsevier Inc. C, from Lino Y & Yoshida K [2009] *J Neurosci* 29:5370–5380. Copyright ©2009 Society for Neuroscience. See also Pierce-Shimomura JT, Morse TM, & Lockery SR [1999] *J Neurosci* 19:9557–9569.)

circuits, they face the same problems of signal detection, transduction, and response that we encountered in our discussion of animal sensory systems in Chapters 4 and 6. Bacterial chemotaxis solves these problems by employing strategies similar to those used in the early steps of animal sensory systems: signal amplification and sensory adaptation.

Bacteria use a temporal strategy to assess the spatial gradient of a chemical by comparing the concentration at the present time to that at an earlier time. This is because bacteria are too small to use a spatial strategy in which they compare concentrations at two ends of a cell at any given time. With a small body size, the nematode *C. elegans* chemotaxis also consists of runs and turns called pirouettes (Figure 13-15C), similar to the biased random walk of bacterial chemotaxis. However, *C. elegans* can also curve their run trajectories toward an attractant, suggesting that they can detect the direction of chemical signals during runs. This is likely achieved by swinging their heads to detect concentration differences across space. Insect olfactory organs sit at the distal ends of their antennae, allowing sampling of concentration differences across larger distances. This enables them to compare odorant concentrations simultaneously detected by the two antennae to drive their chemotactic behavior. Some mammals (including humans) also use inter-nostril comparisons to help localize an odor source.

13.10 G-protein-coupled receptors (GPCRs) are ancient chemosensory receptors in eukaryotes

As we learned in Chapters 4 and 6, the mammalian sensory receptors for vision, olfaction, itchy chemicals, and sweet, bitter, and umami tastes are all seven-transmembrane G-protein-coupled receptors (GPCRs). When did GPCRs first appear and what was their ancestral function?

GPCRs are not found in prokaryotes but are present in all eukaryotic branches, including in protists, plants, fungi, and animals. Thus, GPCRs likely appeared early in the eukaryotic lineage. The best-studied GPCRs in unicellular organisms are those in the budding yeast *S. cerevisiae,* a model genetic organism and the first eukaryote for which the genome was sequenced (in 1996). Only three GPCRs were found in the budding yeast genome, and studies of these GPCRs have shed light on the general function and mechanisms of action of GPCRs.

Although *S. cerevisiae* can reproduce asexually through budding, they also engage in sexual reproduction through production and reception of peptide pheromones. Haploid budding yeast has two mating types: a and α. a-cells produce

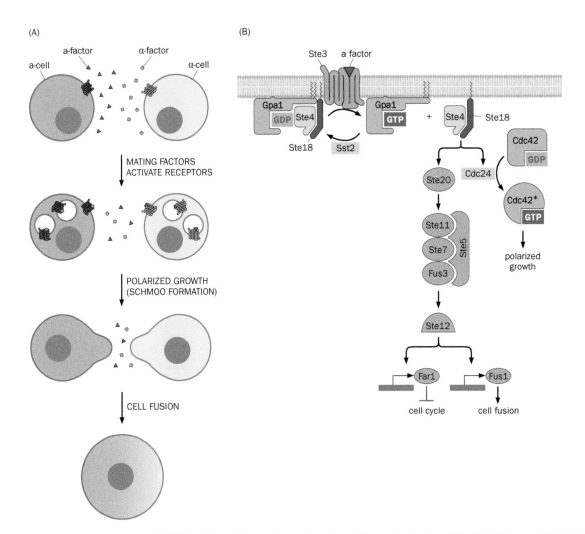

Figure 13-16 G-protein-coupled receptor (GPCR) pathways in yeast mating. (A) Schematic of the mating process of budding yeast. a-cells secrete a-factor, a peptide pheromone, and express a GPCR as the receptor for the α-factor peptide pheromone. α-cells secrete α-factor and express a GPCR for the a-factor. The release of mating factors by cells of opposite mating types triggers receptor activation, internalization of the partner cell's mating factor along with its receptor, and activation of intracellular pathways leading to the polarized growth of a- and α-cells toward each other and subsequent cell fusion. **(B)** Signal transduction pathways. Activation of Ste3 (a-factor receptor) triggers dissociation of the βγ subunits (Ste4/Ste18) from the α subunit (Gpa1) of the trimeric G protein. βγ subunits activate Ste20, a kinase that triggers activation of the MAP kinase cascade, consisting of Ste11, Ste7, and Fus3. Ste5 is a scaffold protein that assists in formation of the MAP kinase complex. The kinase cascade leads to phosphorylation and activation of the Ste12 transcription factor, which turns on many genes, including Far1 (for cell cycle arrest) and Fus1 (for cell fusion). The βγ subunits also locally activate Cdc24, a guanine nucleotide exchange factor for the Rho-family small GTPase Cdc42. Cdc42 regulates polarized growth through its action on the actin cytoskeleton, thereby promoting shmoo formation at the site of the highest concentration of the partner cell's mating factor. The G protein cycle is terminated by GTP hydrolysis of Gpa1-GTP, assisted by the GTPase activating protein Sst2, causing reassociation of βγ subunits with Gpa1-GDP. Pathways similar to those shown here participate in neuronal GPCR signaling, such as the MAP kinase cascade in synapse-to-nucleus signaling (Section 3.24) and small GTPase Cdc42 signaling in growth cone guidance (Box 5-2). (Based on Herskowitz I [1995] *Cell* 80:187–197 and Akrowitz RA [2009] *Cold Spring Harb Perspect Biol* 1:a001958.)

a-factor and α-factor receptor, while α-cells produce α-factor and a-factor receptor. Two cells of opposite mating types release mating factors that activate each other's receptors. Receptor activation leads each cell to extend toward its partner a cytoplasmic projection—called a shmoo for its resemblance to the Al Capp cartoon character—and triggers cell fusion (**Figure 13-16**A). Genetic screens identified many genes in the mating pathway (Figure 13-16B); some of these genes were named *Ste* for their sterile phenotypes. Molecular-genetic analysis revealed that the two most upstream genes in the mating pathway encode receptors for the a-factor and the α-factor; these receptors constitute two of the three yeast GPCRs. Signals are transduced through trimeric G proteins (Section 3.19) and amplified by the MAP kinase cascade (Box 3-4) to induce transcription and cell cycle arrest.

G proteins also activate a separate pathway that directs the polarized growth of shmoos via the Rho-family small GTPase Cdc42 (Figure 13-16B).

The third GPCR in the budding yeast is a sugar receptor that senses extracellular glucose and sucrose and activates a Gα protein distinct from the one used in the mating pathway. The downstream pathway is of a type commonly associated with GPCR signaling, with the Gα protein activating an adenylyl cyclase that produces cAMP and thereby activates PKA (Figure 3-33). GPCRs in other fungi and protists studied thus far all participate in sensing sugars, amino acids, other nutrients, or pheromones.

Given the striking conservation between the molecular components and signaling mechanisms of GPCRs employed in present-day single-cell and multicellular organisms, it is likely that their single-cell common ancestors already employed the same signaling pathways, and the ancestral function of GPCRs in unicellular organisms is chemosensation of nutrients and pheromones. The sensory systems of multicellular organisms essentially inherited these functions. When animals left the predominantly aqueous environment wherein early unicellular organisms resided, GPCRs diversified in function to detect volatile chemicals. GPCRs also expanded for use in intercellular communication, as all metabotropic neurotransmitter receptors are GPCRs.

13.11 Chemosensory receptors in animals are predominantly GPCRs

Following the precedent set in unicellular organisms, most multicellular organisms continue to dedicate most of their GPCRs to sensory systems that detect chemicals. In *C. elegans,* hundreds of predicted GPCRs are expressed in ciliated chemosensory neurons that allow detection of soluble and volatile chemicals in the environment (Figure 6-21). Some mammalian species have evolved more than 1000 odorant receptors alone (Figure 6-9). Mammals that possess a functional accessory olfactory system, such as rodents, also employ up to a few hundred additional GPCRs to detect chemicals produced by conspecifics, predators, and prey (Box 6-1).

Chemoreceptors in the sensory neurons of multicellular organisms must convert binding of a chemical ligand into electrical signals in order to transmit information to the rest of the nervous system. From *C. elegans* to mammals, modulation of two types of channels—cyclic nucleotide-gated (CNG) channels and transient receptor potential (TRP) channels (Box 2-4)—are among the most commonly employed effectors of GPCR activation in chemosensory neurons (**Figure 13-17**). Both CNG and TRP channels evolved from K$^+$ channels in early eukaryotes (Figure 13-9) but are usually nonselective cation channels. Opening of these channels causes net cation influx and depolarization of sensory neurons, which triggers neurotransmitter release from sensory cells and thereby transmits information to second-order neurons. Signals can also be sent by closing these channels, causing hyperpolarization and decreased neurotransmitter release, as is the case in *C. elegans* olfaction (Section 6.10).

13.12 Odors are sensed by ligand-gated ion channels in insects

While GPCRs perform most of the chemical sensing in the animal kingdom, insect olfaction provides an instructive exception. As we learned in Chapter 6, both insect and mammalian olfactory systems use glomeruli to organize axonal input

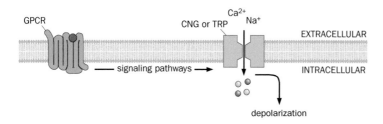

Figure 13-17 Chemosensory GPCRs activate transduction pathways that change the open probabilities of CNG or TRP channels. Most chemosensory GPCRs activate signal transduction pathways that eventually lead to opening of CNG or TRP channels and depolarization of sensory neurons. See Figure 6-4 for a specific example of this pathway in mammalian olfaction.

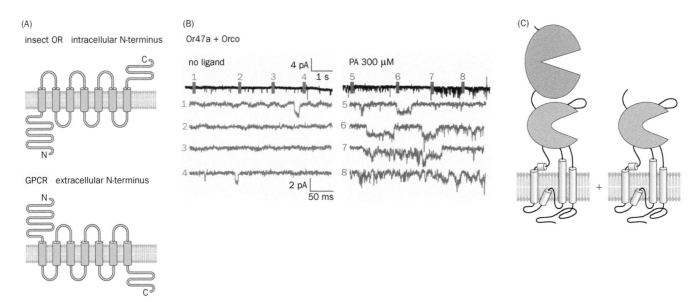

Figure 13-18 Insect odorant receptors are ligand-gated cation channels. (A) Insect ORs have an inverted membrane topology compared to GPCRs. **(B)** Co-expression in *Xenopus* oocytes of *Drosophila* Or47a, a specific OR activated by pentyl acetate (PA), and Orco, a co-receptor, led to inward currents in an excised membrane patch (Box 14-2) in response to PA application (red) compared to control (blue). Traces in the numbered regions are expanded below. **(C)** A subset of *Drosophila* olfactory receptor neurons use heterotetramers of ionotropic receptors as odorant receptors. The common subunit (left) has the same structure as a subunit of an ionotropic glutamate receptor (Figure 3-25); the specific subunit (right) lacks an N-terminal domain. (A, from Benton R, Sachse S, Michnick SW, et al. [2006] *PLoS Biol* 4:e20. B, from Sato K, Pellegrino M, Nakagawa T, et al. [2008] *Nature* 452:1002–1006. With permission from Springer Nature. C, based on Abuin L, Bargeton B, Ulbrich MH, et al. [2011] *Neuron* 69:44–60.)

from olfactory receptor neurons (ORNs) (Figure 6-24). Like the GPCR odorant receptors identified in mammals and *C. elegans,* the first odorant receptors (ORs) identified in *Drosophila* were predicted to contain seven transmembrane domains and were assumed to be GPCRs. However, further studies indicated that, unlike GPCRs, which have an extracellular N-terminus and an intracellular C-terminus, insect ORs have the opposite configuration (**Figure 13-18**A). In addition to expressing a unique OR, each ORN also expresses a co-receptor (Orco) that has the same 7TM topology as other ORs and is highly conserved among insects. Together, ORs and Orco act as ligand-gated cation channels (Figure 13-18B). A recent cryo-EM structure of Orco homomers revealed that the channel consists of four subunits and has an architecture distinct from those of all previously studied ion channels.

A second family of *Drosophila* olfactory receptors belongs to an entirely new class of proteins resembling ionotropic glutamate receptors and are named ionotropic receptors (IRs). IRs function as heterotetramers, composed of a common subunit that retains the architecture of ionotropic glutamate receptor subunits (Figure 3-25) and a specific subunit that resembles an ionotropic glutamate receptor subunit without an N-terminal domain (Figure 13-18C). While the OR family is specific to the insect lineage, IRs are present in a broad range of invertebrates, including nematodes and mollusks. IRs in *Aplysia* and *C. elegans* are also expressed in chemosensory neurons. Thus, IRs may have had an ancestral chemosensory function in all protostomes. It is unclear why insects use ionotropic receptors instead of GPCRs as olfactory receptors; one possibility is that ligand-gated ion channels transmit signals faster than GPCRs.

These findings provided several interesting insights into the evolution of sensory systems. First, the extant olfactory receptor repertoire in *Drosophila* appears to have been acquired in a piecemeal fashion during evolution, with 20% of ORNs utilizing the more ancient IR family and the remaining 80% the more recently evolved OR family. Despite employing distinct receptors, these ORNs cooperate to provide complementary coverage of the chemical world, and their axons project to complementary glomerular targets within an antennal lobe. Second, the identification of chemoreceptors resembling ionotropic glutamate receptors suggests a close link between chemosensation and neurotransmitter reception by postsynaptic neurons—after all, both of these processes transform ligand binding into electrical signals. Third, despite using completely different olfactory receptors, the olfactory systems of flies and mice share strikingly similar organizations, with most ORNs expressing one specific olfactory receptor and ORNs expressing the same receptor projecting their axons to the same glomeruli (Figure 6-24). The glomerular organization of olfactory information processing might have already been present in the last common ancestor of insects and vertebrates (and might have been independently lost in other clades, such as nematodes and mollusks). Alternatively, this shared glomerular organization might be the product of convergent evolution solving a similar problem in insects and mammals. The finding that insect and mammalian olfactory systems employ different kinds of receptors favors the convergent evolution hypothesis.

As discussed in Chapter 6, ion channels are used in mammals for sour, salt, and trigeminal chemosensation (for example, TRPV1 for hot spice). In flies, a family of gustatory receptors (GRs) with structures similar to ORs, and hence likely ion channels, are also receptors for sweet and bitter tastants. GR-expressing gustatory neurons also sense chemicals that regulate courtship behavior (Box 13-3).

Box 13-3: Case studies in the evolution of courtship behavior

Sensory systems play a key role in the evolution of behavior such as courtship. Animals mate with conspecifics in order to propagate their genes but avoid mating with other species, as interspecies mating usually results in progeny that are unviable or infertile. In this box, we discuss two case studies related to the evolution of courtship behavior in *Drosophila*.

As discussed in Chapter 10, chemosensory cues play a central role in regulating the mating of fruit flies. A *Drosophila* male uses its foreleg to tap a potential mating partner so that the gustatory neurons in its foreleg can "taste" cuticular pheromones that indicate the sex and species of its partner. *Drosophila melanogaster* females produce 7,11-HD (heptacosadiene), which promotes courtship behavior by *D. melanogaster* males. However, 7,11-HD inhibits courtship of male *Drosophila simulans*, a species that diverged from *D. melanogaster* 2–3 million years ago. How does the same chemical produce opposite mating behavior in two closely related species? Detailed comparative studies revealed that in both species, 7,11-HD is detected by the same sensory neurons, which relay the signal to homologous excitatory and inhibitory interneurons that converge onto the P1 command neurons for mating behavior (Figure 13-19A; see also Figure 10-9). However, the signals from the excitatory pathway are stronger than those from the inhibitory pathway in *D. melanogaster*, causing P1 activation, which promotes courtship. In *D. simulans*, the inhibitory pathway prevails, resulting in inhibition of courtship.

A different strategy is used by *D. melanogaster* males to avoid mating with *D. simulans* females, which produce the cuticular pheromone 7-T (tricosene). 7-T is detected by a specific group of gustatory neurons in the forelegs of *D. melanogaster* males, the activation of which inhibits courtship (Figure 13-19B). (7-T is also produced by *D. melanogaster* males to inhibit male–male courtship.) It is unknown how the 7-T signal is relayed to the command neurons, or why the signal

does not inhibit *D. simulans* males courting their conspecific females. Even in the better characterized 7,11-HD pathway, we do not yet know what change(s) in DNA underlie the altered excitation–inhibition balance accounting for the opposite behavior elicited by the same pheromone.

Genetic mapping can be used to identify changes in DNA that cause evolutionarily divergent behavior. We use another example related to *Drosophila* courtship to illustrate. As discussed in Chapter 10, *Drosophila* males sing courtship songs to attract females (Figure 13-19C). Different species produce songs with different parameters, which can be used for species discrimination during mate selection. For example, *Drosophila simulans* and *Drosophila mauritiana*, two species that diverged about 240,000 years ago, produce sine songs with carrier frequencies 10-Hz apart (Figure 13-19D). Researchers mapped the genetic locus responsible for sine song frequency to the *Slowpoke* gene, which encodes a Ca^{2+}-activated K^+ channel (Box 2-4). Specifically, insertion of a transposon (retroelement) into an intron of *Slowpoke* in *D. simulans* affected alternative splicing of an adjacent exon. Remarkably, precise removal of the retroelement insertion utilizing CRISPR/cas9-based genome editing (Box 14-1) increased the sine song frequency by 8.7 Hz, indicating that this single genetic change is largely responsible for the altered sine song carrier frequency (Figure 13-19E). (A survey of many strains of *D. simulans* revealed that this transposon insertion is rare within *D. simulans*. Thus, other genes must also have evolved to cause the average difference between *D. simulans* and *D. mauritiana* strains.) *Slowpoke* is widely expressed in the nervous system. Flies with a null mutation exhibit a host of deficits, including an inability to sing the courtship song. Nevertheless, a change in alternative splicing patterns of *Slowpoke* could be exploited by evolution to alter a specific component of the courtship song. Which neurons and circuits are affected by this molecular change remains to be investigated.

(Continued)

Box 13-3: continued

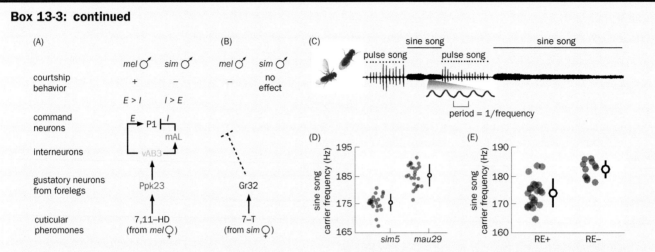

Figure 13-19 Evolutionary changes in *Drosophila* courtship behavior. (A) Female *D. melanogaster* (*mel*) produces 7,11-HD, which promotes courtship of *D. melanogaster* males but inhibits courtship of *D. simulans* (*sim*) males. The 7,11-HD signal is processed by homologous circuits in two species, involving gustatory neurons expressing Ppk23, and excitatory (vAB3) and inhibitory (mAL) interneurons that converge onto the P1 command neurons that promote courtship. Promotion or inhibition of courtship is determined by the balance between the excitatory and inhibitory input onto P1 neurons. **(B)** Female *D. simulans* produce 7-T, which inhibits mating of *D. melanogaster* males via gustatory neurons expressing Gr32, a member of the family of gustatory receptors resembling *Drosophila* ORs. **(C)** During courtship, *Drosophila* males vibrate one wing to produce courtship songs consisting of a pulse song and sine song with a specific carrier frequency. **(D)** Two species, *D. simulans* (*sim5*) and *D. mauritiana* (*mau29*), produce sine songs differing in carrier frequency. Each dot is a measurement from an individual fly; mean ± standard deviation is shown on the right. **(E)** Deleting a retroelement (RE) present in an intron of *sim5* (but not *mau29*) causes the sine song frequency of *sim5* to be similar to that of *mau29*, indicating that this RE insertion contributes to species-specific sine song frequency variation. (A & B, based on Seeholzer LF, Seppo M, Stern DL, et al. [2018] *Nature* 559:564–569 and Pan F, Manoli DS, Ahmed OM, et al. [2013] *Cell* 154:89–102. C–E, from Ding Y, Berocal A, Morita T, et al. [2016] *Nature* 536:329–332. With permission from Springer Nature.)

These examples illustrate some of the challenges of discovering the mechanisms underlying evolutionary changes in behavior. The best chance of acquiring a satisfying understanding is to study species that are sufficiently closely related such that behavioral changes are likely caused by a small number of genetic changes. It is helpful to select a behavior that is robust, can be quantified, and is well studied in terms of molecular and neural mechanisms. One needs also to develop genetic and neurobiological tools in both species to manipulate genes and neurons to establish causal links between changes in DNA, neuronal signaling, and behavior.

13.13 Retinal- and opsin-based light-sensing apparatus evolved independently at least twice

Light sensing, like chemical sensing, is an ancient faculty. Light enables phototrophic archaea, bacteria, algae, and plants to generate chemical energy; directs the movement of a wide range of organisms; and regulates physiology in a circadian fashion in nearly all life forms. In each of these processes, light reacts with a chromophore. The structural change of the chromophore as it absorbs a photon causes an associated protein to change its conformation.

The process in microorganisms that most closely resembles animal vision is **phototaxis**, the ability to move toward or away from a light source. Light sensing in bacterial phototaxis is mediated by **sensory rhodopsins**, which activate membrane-anchored histidine kinases, transducing signals to regulate the flagellar motor in a manner similar to chemotaxis (Box 13-2). Remarkably, bacterial sensory rhodopsins consist of an opsin protein with seven transmembrane helices and a retinal as their chromophore, just like animal rhodopsins (Section 4.3). Do animal vision and bacterial phototaxis share a common evolutionary origin?

Several lines of evidence argue against this hypothesis. First, although they both possess seven transmembrane helices, the opsins in the prokaryotic sensory rhodopsins (type I rhodopsins) and animal rhodopsins (type II rhodopsins) share no sequence similarities and differ in the arrangement of their transmembrane

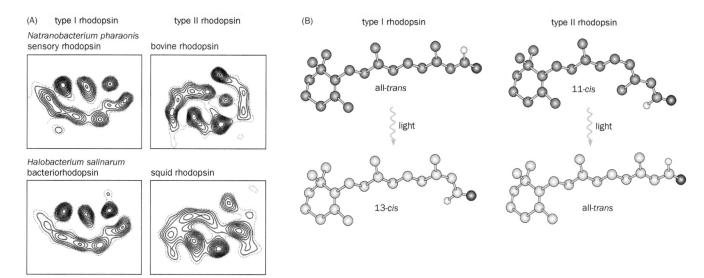

Figure 13-20 Differences between type I and type II rhodopsins.
(A) Electron density maps of the transmembrane helices of two type I rhodopsins from archaea (left) differ from those of type II rhodopsins from animals (right). **(B)** In type I rhodopsins, light absorption triggers isomerization of all-*trans* retinal to 13-*cis* retinal (left). In type II rhodopsins, light absorption triggers isomerization of 11-*cis* retinal to all-*trans* retinal (right). The black circle is from a lysine residue of the opsin (Figure 4-6). (From Spudich JL, Yang CS, Jung KH, et al. [2000] *Ann Rev Cell Dev Biol* 16:365–392. With permission from Annual Reviews.)

helices (**Figure 13-20**A). Second, although both use retinal as their chromophore, photon absorption triggers different retinal isomerization reactions, converting all-*trans* retinal to 13-*cis* retinal in type I rhodopsins and 11-*cis* retinal to all-*trans* retinal in type II rhodopsins (Figure 13-20B). Third, only type II rhodopsins are G-protein-coupled receptors.

Phylogenetic analysis indicates that prokaryotic sensory rhodopsins evolved from more ancient type I rhodopsins that harvest light energy. **Bacteriorhodopsin**, a light-driven proton pump, and **halorhodopsin**, a light-driven chloride pump, are used by some prokaryotes to convert solar energy into chemical energy in the form of ionic gradients across the membrane (Section 2.4). Type I rhodopsins are also present in many eukaryotic microbes, including algae, fungi, and amoeba. For example, the single-cell green alga *Chlamydomonas reinhardtii* has two type I rhodopsins that are enriched in its eyespot (**Figure 13-21**A) and play a role in phototaxis. When expressed in *Xenopus* oocytes or mammalian cells, these proteins function as light-activated cation channels (Figure 13-21B) and were thus named **channelrhodopsins**. Remarkably, as mammalian neurons contain endogenous retinal, expression of *Chlamydomonas* channelrhodopsin-2 (ChR2) in mammalian neurons can cause light-induced depolarization and hence neuronal activation. Halorhodopsin, on the other hand, has been adopted for inactivating neurons, as it pumps Cl⁻ into mammalian neurons in response to light, resulting in hyperpolarization. These light-activated channels and pumps have become powerful tools for manipulating the activity of genetically defined neurons, an approach called **optogenetics**, which we have encountered numerous times in preceding chapters (see Section 14.25 for more details).

Figure 13-21 Light sensing in
Chlamydomonas. (A) An image (left) and a schematic drawing (right) of a *Chlamydomonas* cell. In addition to harvesting light energy for photosynthesis in the chloroplast, this single-cell green alga uses light-sensing pigments in the eyespot to direct flagellar movement for phototaxis. In the latter process, light is detected by two type I sensory rhodopsins, named channelrhodopsin-1 and channelrhodopsin-2 (ChR1 and ChR2). **(B)** Excised patch clamp recording of a piece of plasma membrane from a *Xenopus* oocyte expressing ChR2. In the presence of exogenous all-*trans* retinal, light induces an inward current from the membrane patch containing ChR2. (A, courtesy of Moritz Meyer. B, from Nagel G, Szellas T, Huhn W, et al. [2003] *Proc Natl Acad Sci U S A* 100:13940–13945. Copyright The National Academy of Sciences, U.S.A.)

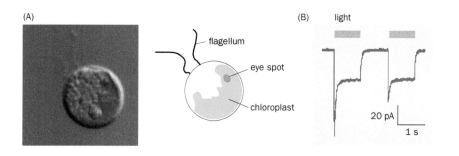

Type II rhodopsins are present in cnidarians and bilaterians but not in sponges, and thus likely first emerged in eumetazoans. The overall structure of type II opsins is much closer to other eukaryotic GPCRs than to type I opsins. Thus, nature has coupled opsins to retinal to form a light-sensing apparatus at least twice independently: once in ancient prokaryotes, generating a light sensor that later spread to eukaryotic microbes, and again during early metazoan evolution, modifying chemosensory GPCRs to produce what would become a critical component in all animals' visual systems.

13.14 Photoreceptor neurons evolved in two parallel paths

Emergence of a light-sensing molecule is just one step in the evolution of animal vision. Rhodopsins must be expressed by appropriate sensory neurons (photoreceptors), and light activation of rhodopsin must be linked to signal transduction and amplification pathways that convert light sensing into electrical signals. Two distinct photoreceptor types are widely used in animals. Vertebrate rods and cones belong to the **ciliary type**, so named because the outer segments, into which the light-sensing opsins are packed, derive from the **primary cilium**. The primary cilium is a short, single, non-motile cilium that projects from the surface of many animal cell types and is used in other contexts as a signaling center. The eyes of most invertebrates, including the compound eyes of flies (Figure 5-34) and the sophisticated eyes of cephalopods, use photoreceptors of the **rhabdomeric type**, in which the apical surface folds into microvilli that house opsins. In addition to these morphological distinctions, the two photoreceptor types also differ in many other respects (**Figure 13-22**).

Although both photoreceptors use G-protein-coupled type II rhodopsins, their opsins belong to two distinct subfamilies based on sequence similarities. Rhabdomeric photoreceptors employ r-opsins, while ciliary photoreceptors employ c-opsins. Their phototransduction pathways also differ. Ciliary photoreceptors use the heterotrimeric G protein transducin to couple photon absorption to activation of phosphodiesterase (PDE), which reduces cGMP levels, causing closure of cGMP-gated channels (Figure 13-22, top; Figure 4-10). Thus, photon absorption leads to photoreceptor *hyperpolarization*. The phototransduction pathway in *Drosophila,* which likely applies to other rhabdomeric photoreceptor neurons, uses G_q, which couples photon absorption to activation of a phospholipase C (PLC) to produce diacylglycerol (DAG) (Figure 3-34), triggering opening of a TRP channel. Thus, photon absorption leads to photoreceptor *depolarization* (Figure 13-22, bottom). How did these two divergent types of photoreceptors arise?

Recent studies revealed that ciliary and rhabdomeric photoreceptors have coexisted in both vertebrate and invertebrate clades (**Figure 13-23**). In fact, opsins in cnidarians such as jellyfish more closely resemble the c-opsins used in vertebrate photoreceptors, and phylogenetic analysis suggests that c-opsin and r-opsin diverged very early, predating the cnidarian–bilaterian split. Remnants of both types of photoreceptors have been found in both the invertebrate and vertebrate branches of bilateria.

Figure 13-22 Ciliary and rhabdomeric photoreceptors. Ciliary photoreceptors and rhabdomeric photoreceptors, which are primarily used in vertebrate rods and cones and in invertebrate eyes, respectively, differ across many features, including morphology, opsin type, signaling pathway, and direction of membrane potential changes. See text for more details. (Adapted from Fernald RD [2006] *Science* 313:1914–1918.)

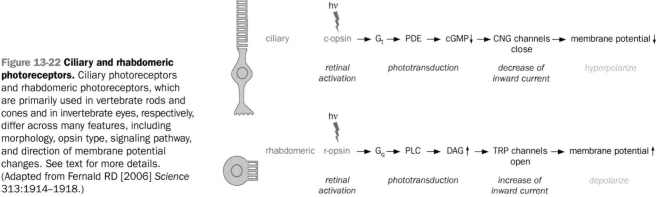

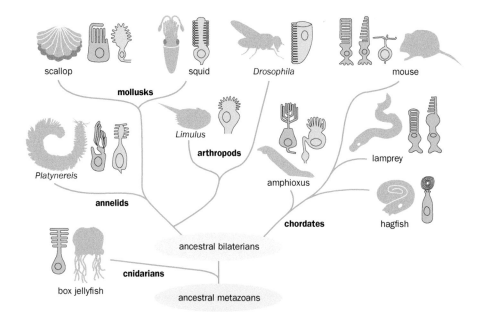

Figure 13-23 Parallel evolution of ciliary and rhabdomeric photoreceptors.
Ciliary (green) and rhabdomeric (blue) photoreceptors coexist in many phyla of animals, indicating an early origin for both types of photoreceptors. (Adapted from Fain GL, Hardie R, & Laughlin SB [2010] *Curr Biol* 20:R114–R124. With permission from Elsevier Inc.)

Although rhabdomeric photoreceptors dominate most invertebrate phyla, ciliary photoreceptors coexist in certain species. In scallops, hyperpolarizing and depolarizing photoreceptors have been found side-by-side in the same retina. Like most invertebrates, the marine annelid *Platynereis* uses r-opsin in its eyes. However, c-opsin has been found in *Platynereis* brain neurons associated with ciliary structures, likely for sensing light intensity for circadian rhythm regulation.

Among chordates, amphioxus (a basal chordate; Figure 13-2) also possesses both rhabdomeric and ciliary photoreceptors. In vertebrates, including humans, remnant rhabdomeric photoreceptors coexist with the dominant ciliary photoreceptors. As discussed in Section 4.19, intrinsically photosensitive retinal ganglion cells (ipRGCs), whose functions include circadian rhythm entrainment and pupil constriction, directly sense light. ipRGCs use melanopsin, an r-opsin, for light detection, and its transduction requires phospholipase C and TRP channels just like rhabdomeric photoreceptors in *Drosophila*. These data suggest that mammalian ipRGCs and invertebrate rhabdomeric photoreceptors derived from a common ancestor.

While it remains unknown why vertebrate vision predominantly utilizes ciliary photoreceptors and invertebrate vision rhabdomeric photoreceptors, the parallel evolution of these two photoreceptor types illustrates that multiple solutions exist for a common problem, with individual organisms using these solutions in different ways based on their evolutionary histories. Indeed, evolutionary analyses of eye morphology suggest that each of the photoreceptor types has been independently integrated into many different types of eyes (Box 13-4). The collaboration of different light-sensing strategies in the same animal also illustrates that natural selection does not always produce a "winner take all" scenario.

Box 13-4: Darwin and the evolution of the eye

The eye has occupied a special place in the history of evolutionary theory. In a chapter titled "Difficulties on Theory" from the first edition of *Origin of Species,* Darwin described the eye as an "organ of extreme perfection." He stated that "To suppose that the eye, with all its inimitable contrivances for adjusting the focus to different distances, for admitting different amounts of light, and for the correction of spherical and chromatic aberration, could have been formed by natural selection, seems, I freely confess, absurd in the highest possible degree." Indeed, attackers of evolutionary theory have often used the eye as an example of the improbability of natural selection. But Darwin went on:

Yet reason tells me, that if numerous gradations from a perfect and complex eye to one very imperfect and simple, each grade being useful to its possessor, can

(Continued)

Box 13-4: continued

be shown to exist; if further, the eye does vary ever so slightly, and the variations be inherited, which is certainly the case; and if any variation or modification in the organ be ever useful to an animal under changing conditions of life, then the difficulty of believing that a perfect and complex eye could be formed by natural selection, though insuperable by our imagination, can hardly be considered real.

In the 160 years since *Origin of Species* was published, studies on the structure, function, and development of the eye have validated Darwin's intuition, revealing the details of eye evolution at multiple levels of organization. In research based primarily on morphological criteria, careful comparative study of photoreceptors and eyes in different organisms concluded that the eye as a photo-sensing apparatus has evolved independently 40 to 65 times across animal phyla. Twenty phylogenetic lines include animals with a regular series of "ever more perfect" eyes among still-living relatives. The "ever more perfect" eye has also been a subject of a simulation experiment with the objective of determining the number of generations required for the evolution of an eye's optical geometry. The study supposed that the driving force is for higher spatial resolution (visual acuity) and that changes take the form of sequential 1% modification steps. For example, doubling the length of a structure would take about 70 1% modification steps (as $1.01^{70} \approx 2$). An estimated 1,829 steps are needed to convert a simple patch of photosensitive epithelium into an eye with a lens focusing onto a curved layer of photosensitive cells, as in the vertebrate retina (**Figure 13-24**). A conservative estimate using inheritance of quantitative traits gives about 360,000 generations, or, if the generation time is about 1 year, less than

half a million years. Considering that both r-opsin-based and c-opsin-based photoreceptors most likely existed before the cnidarian–bilaterian divergence around 1 billion years ago, a half-million-year period is a very short time in which to perfect optical geometry.

Hermann von Helmholtz, a nineteenth-century physicist and authority on vision, recognized that the human eye is not perfect after all. After detailing its numerous defects, such as spherical aberration and the blind spot created by the optic nerve head, Helmholtz stated that "Now it is not too much to say that if an optician wanted to sell me an instrument which had all these defects, I should think myself quite justified in blaming his carelessness in the strongest terms, and giving him back his instrument." However, Helmholtz went on, appreciating Darwin's evolutionary theory: "We have now seen that the eye in itself is not by any means so complete an optical instrument as it first appears: its extraordinary value depends on the way in which we use it: its perfection is practical, not absolute, consisting not in the avoidance of every error, but the fact that all its defects do not prevent its rendering the most important and varied services."

In essence, natural selection does not produce perfect products; it yields products "good enough" to help animals survive and reproduce. This is because natural selection is constrained by evolutionary history and by the competing energy costs and fitness benefits of maintaining complex structures. In a fitness landscape with peaks and troughs, natural selection can bring organisms to only a local peak of fitness from any given starting point; it does not have the foresight to overcome further troughs to reach distant and perhaps higher peaks.

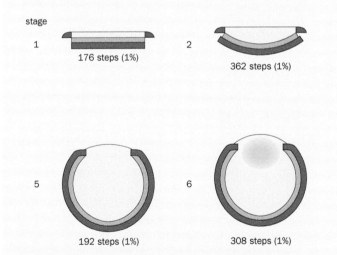
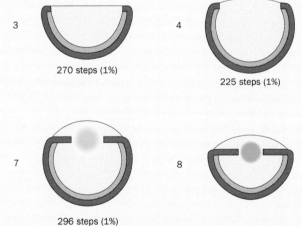

Figure 13-24 Simulating the evolution of optical geometry. The evolution from a patch of light-sensitive cells to an eye with a focused lens is divided arbitrarily into eight stages. Changes between successive stages are divided into 1% modification steps. At stage 1, the initial structure (hereafter called the *retina*) consists of a layer of light-sensitive cells (orange) sandwiched between a transparent protective layer (yellow) and a layer of dark pigment (red). At stages 2 and 3, the retina invaginates, forming a sphere.

At stages 4 and 5, the retina continues growing without changing its radius, causing the retinal pit to deepen and an aperture to form. At stages 6 through 8, a graded-index lens appears with gradual shortening of the focal length; by stage 8, the focal length equals the distance to the retina, producing a sharply focused system. Spatial resolution increases during each of the modification steps. (Adapted from Nilsson DE & Pelger S [1994] *Proc R Soc Lond B* 256:53–58.)

13.15 Diversification of cell types is a crucial step in the evolution of the retina

In environmental niches exposed to light, the extraction of useful features from light signals enables animals to identify behaviorally relevant objects such as food, predators, and mates. This is the fitness benefit of vision. This feature extraction starts downstream from the photoreceptors in retinal circuits. As we learned in Chapter 4, the mammalian retina has dozens of well-characterized neuronal types belonging to five classes. Photoreceptors (rods and cones) provide the input; retinal ganglion cells send the output to the brain; bipolar cells connect photoreceptors to retinal ganglion cells; horizontal and amacrine cells participate in transforming light intensity signals received from the input cells into contrast, motion, and color signals transmitted by the output cells. How did such exquisite circuits combining so many types of neurons come about?

We are far from having a satisfactory answer to this question. A key step must be the creation of different cell types. It is generally believed that as animals evolve sophisticated functions, cells become more and more specialized, such that each increasingly distinct cell type performs a subset of the functions once performed by a multifunctional ancestral cell. Viewed from this perspective, unicellular microbes capable of phototaxis may be considered the most versatile living cells. They are, at the same time, a sensory neuron that detects light, an interneuron that integrates sensory information, a motor neuron that converts the information into instructions for movement, and a muscle cell that produces movement in the form of flagellar beating. However, multifunctionality comes at a cost: each function is constrained by other functions the cell must perform. One way to improve existing functions or add new ones is to duplicate and diversify the cells, in analogy to the duplication and diversification of genes (Figure 13-6A). As cellular diversity increases, each cell can perform a specialized function with greater sophistication, free from the constraint of having to perform other functions.

The origin of the bipolar cell provides an illuminating example in the evolution of vertebrate retinal cell types and circuits. Bipolar cells share many similarities with photoreceptors in their shape, developmental sequence, synaptic organization, and gene expression patterns, suggesting that bipolar cells might be phylogenetic "sister cells" to photoreceptors. Indeed, in hagfish, a basal vertebrate (Figure 13-2), bipolar cells are absent and photoreceptors directly connect to retinal ganglion cells. It is likely that bipolar cells originated from photoreceptors in early vertebrates through division of labor. The functions of light sensing and signal propagation, both of which were once carried out by photoreceptors, were gradually distributed across two different cell types. One became specialized in light sensing and lost its connections to retinal ganglion cells, while the other became specialized in communicating signals from light-sensing cells to retinal ganglion cells and lost the capacity to sense light (**Figure 13-25**).

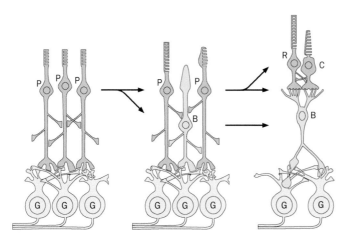

Figure 13-25 A likely scenario for the diversification of retinal cell types. In this hypothetical model, bipolar cells (B) derived from early multifunctional photoreceptors (P) and gradually became specialized in connecting photoreceptors to retinal ganglion cells (G) and lost their light-sensing capability, while photoreceptor cells gradually lost their connection to ganglion cells and became more specialized in light sensing. Likewise, specialized cones (C) and rods (R) probably derived from an earlier photoreceptor that resembled modern cones. (From Arendt D [2008] *Nat Rev Genet* 9:868–882. With permission from Springer Nature.)

Diversification and division of labor may also account for the subdivision within each retinal cell class. For example, rods likely evolved from photoreceptors that resemble modern-day cones during the early evolution of chordates (Figure 13-25). A more recent example of photoreceptor diversification is the elaboration of cones that endowed some primates, including humans, with trichromatic color vision, which we will study in the next two sections.

13.16 Trichromatic color vision in primates originated from variations and duplications of a cone opsin gene

As discussed in Chapter 4, color vision is enabled by the presence of multiple types of cone cells, each expressing an opsin with a distinct spectral sensitivity. Ancestral vertebrates appear to already have had four cone opsin genes, likely a result of two genome duplication events in stem vertebrates after they diverged from amphioxus. This was then followed by duplication of one of the cone opsin genes that produced the gene for the rod opsin in jawed fish after they diverged from jawless fish such as lampreys (Figure 13-2). Many extant vertebrates, including jawed fish, reptiles, and birds, have retained four cone opsins in addition to the rod opsin and are therefore tetrachromatic. However, during early mammalian evolution, genes encoding two cone opsins were lost, likely because early mammals were nocturnal and there was a lack of selection for the multitude of cone opsins. Thus, most mammals are dichromatic, having short-wavelength and longer-wavelength cone opsins encoded by the *S* and *L'* genes, respectively (**Figure 13-26**).

Humans are trichromatic, possessing three types of cones: S, M, and L (Figure 13-26; Section 4.10). The genes encoding M- and L-opsins derived from a recent duplication of the ancestral *L'* gene about 35 million years ago in the ancestor that gave rise to catarrhines (Old World monkeys and apes, including humans). Indeed, *M* and *L* genes in humans and other catarrhines are located next to each other on the X chromosome, suggesting that they derived from an unequal crossing over of the ancestral *L'* gene (Figure 13-6A). This duplication was accompanied by diversification of the *M* and *L* genes). As few as three amino acid changes

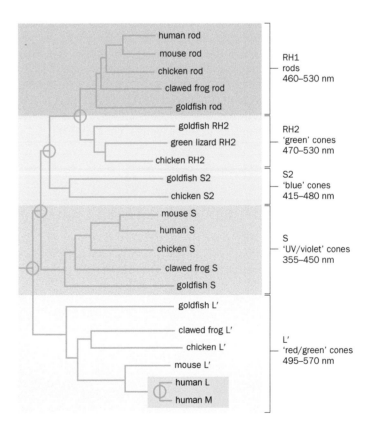

Figure 13-26 Phylogenetic relationships in the vertebrate opsin gene family. Circles indicate gene duplication events. The leftmost three, which gave rise to four cone opsin genes, likely correspond to two rounds of genome duplication in ancestral vertebrates (>500 million years ago). The duplication that gave rise to the rod opsin gene occurred in jawed fish after their divergence from jawless fish. The bottom duplication of the *L'* cone opsin gene occurred more recently (about 35 million years ago). The *Rh2* and *S2* cone opsin genes were lost in the mammalian clade. (From Bowmaker JK [2008] *Vis Res* 48: 2022–2041. With permission from Elsevier Inc.)

in the transmembrane helices of the cone opsin can account for the spectral shift of M- and L-cones, which have maximal absorbances around 530 and 563 nm, respectively (Figure 4-18).

As the acquisition of trichromacy in primates was more recent than most of the adaptations we have discussed thus far, more details of this evolutionary change have been discerned. Studies of New World monkeys, which are indigenous to the Americas, provided particularly interesting insights. Most New World monkeys, like most mammals, have only S and L' genes. However, multiple polymorphic alleles have been found in the L' gene encoding opsins that confer maximal absorbance ranging from 535 nm to 563 nm. Because the L' gene is located on the X chromosome, males have only one copy and are dichromatic. Females homozygous for a particular L' allele are dichromatic as well. However, females heterozygous for two polymorphic L' alleles are *trichromatic* (**Figure 13-27**A), as they express each allele in complementary patches of cones due to random X-inactivation, like the patchy distributions of M- and L-cones in trichromatic humans (Figure 4-17).

Interestingly, one New World monkey species—the howler monkey—has also duplicated the ancestral L' gene independently of Old World monkeys and apes, such that these monkeys are uniformly trichromatic. Sequence comparisons indicate that M and L alleles in howler monkeys are very similar to polymorphic L' alleles in other New World monkeys that lack the duplication event. These data suggest that, at least in the case of howler monkeys, diversification of L' opsins actually preceded the duplication event (Figure 13-27B). Indeed, the three key amino acids that account for the spectral shift of M- and L-cones are identical in howler monkeys and Old World monkeys and apes. This suggests that different alleles of the ancestral L' gene might have already existed before the divergence of New World monkeys from Old World monkeys and apes over 35 million years ago. Furthermore, a process similar to the acquisition of trichromacy in howler monkeys (Figure 13-27B) might account for the acquisition of trichromacy in Old World monkeys and apes as well.

To utilize spectrally distinct opsins for color vision, two additional properties are necessary. First, M- and L-opsins must be expressed in distinct cone cells. This problem was solved in trichromatic female New World monkeys by random X-inactivation, whereby each cone expresses only one allele. In uniformly trichromatic primates, the adjacent M and L genes share the same locus control region (LCR), a *cis*-regulatory DNA element conferring cone-specific opsin expression (Figure 13-27B). In a given cell, the LCR can be paired with only one of the two

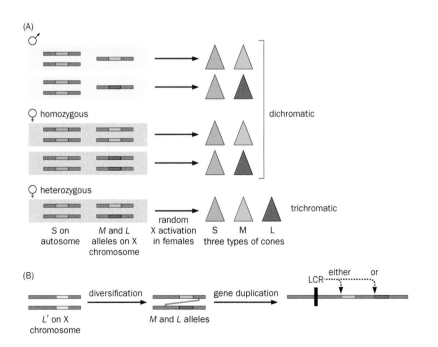

Figure 13-27 Origin of trichromacy in primates. (A) In addition to carrying two autosomal copies of the short-wavelength opsin gene (S), some species of New World monkeys acquired polymorphisms in the *L'* gene, which endows cones with spectral sensitivities at medium (M) and long (L) wavelengths. Males and females homozygous for the *M* or *L* alleles are also dichromatic (top, middle). However, females heterozygous for the *M* and *L* alleles (bottom) are trichromatic, as random X-inactivation yields a retinal mosaic of cones expressing either *M* or *L*. Gray segments represent the chromosomal DNA surrounding the opsin genes. **(B)** A likely sequence of events that gave rise to trichromacy in primates. Mutations in the *L'* gene first produced polymorphisms in spectral sensitivity, as is seen in many New World monkeys (first step). In howler monkeys, a recent duplication caused each X chromosome to harbor both the *M* and *L* alleles, such that males and females are uniformly trichromatic (second step). A similar event might have occurred in an ancestor of Old World monkeys and apes. The same LCR (locus control region) activates expression of either the *M* or *L* opsin gene, but not both, in a given cone cell.

Figure 13-28 **Trichromatic vision helps primates find fruits**. Juvenile white-faced capuchins forage fruit of *Allophylus occidentalis* in Costa Rica. Images simulate capuchin trichromacy **(A)** and dichromacy **(B)**. Field studies reveal that trichromatic capuchin females have faster fruit intake than dichromatic females. (From Melin AD, Chiou K, Walco E, et al. [2017] *Proc Natl Acad Sci U S A* 114:10402–10407. With permission from the National Academy of Science U.S.A.)

(A)   (B)

opsin genes, and this pairing appears to be random, thus giving rise to the random distribution of M- and L-cones (Figure 4-17).

Second, retinal circuitry must be able to extract the difference between signals detected by M- and L-cones. This is accomplished by the midget bipolar cell, which transmits signals from the cones in the fovea with an extremely small receptive field, contacting only a single cone. As a result, the bipolar cell and retinal ganglion cell downstream of the cone can compare a center dominated by a single cone—and therefore a single color—with a surround that samples mostly a random mix of M- and L-cone signals (Figure 4-34).

Trichromacy in primates appears to confer a strong evolutionary advantage. Evidence for this includes the maintenance of multiple polymorphic L' alleles in New World monkeys and the duplication of the L' gene in catarrhines and, independently, in howler monkeys, both of which have since become fixed in the genome. A predominant theory is that the distinction of green and red allows primates—many of which are frugivores—to identify red, orange, and yellow fruits among green leaves, using colors to judge their ripeness (Figure 13-28). In fact, the evolution of trichromacy is closely linked with the evolution of colored fruits. Frugivores use color vision to feed on ripe fruits, while plants bearing colored fruits that can be seen, eaten, and transported by trichromatic animals will enjoy advantages in seed dispersal. As an added dividend, if the seeds are small enough to be ingested, they come to rest in a nutrient-rich environment.

13.17 Introducing an extra cone opsin into dichromatic animals enables superior spectral discrimination

While many evolutionary studies compare traits in different species to infer processes that likely led to these differences, elegant experiments that simulate the evolutionary process can provide strong support for the inferred processes and generate new insights. We end our discussion of sensory system evolution with two such experiments.

Mice are dichromatic, having one S-opsin and one longer-wavelength opsin that we will call M (for medium wavelength, with maximal absorbance near 510 nm). Using the knock-in technique (Section 14.7), a human gene encoding an L-opsin (with maximal absorbance of 556 nm) replaced one of the endogenous mouse *M* alleles (also located on the X chromosome). Female mice heterozygous for the *M* and *L* alleles thus have three opsin genes—the endogenous autosomal *S*, the endogenous *M* on one X chromosome, and the knocked-in *L* on the other X chromosome—which, due to random X-inactivation, results in three types of cones, as in heterozygous female New World monkeys (Figure 13-27A). Female mice carrying both the *M* and *L* alleles indeed have expanded spectral sensitivity compared to mice expressing only the *M* or *L* allele. Do these trichromatic mice possess superior color vision compared to normal, dichromatic mice?

(A)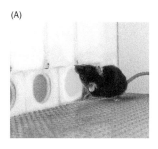

(B)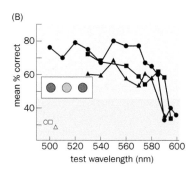

Figure 13-29 **Superior spectral discrimination of trichromatic mice.**
(A) In a color discrimination task, thirsty mice were trained to indicate which port displayed a color different from two other ports in order to receive a water reward. The light wavelength was set at 600 nm for two ports and was variable for the third port. (B) In this task (symbolized by the three color circles), a control dichromatic mouse and two trichromatic mice with imbalanced M:L ratios (open symbols) performed the 500 nm versus 600 nm discrimination test at the chance level (~33%; the dark yellow area indicates chance performance at the 95% confidence level). Three trichromatic mice with balanced M:L ratios (filled symbols) discriminated between wavelengths in the 500–580 nm range and 600 nm significantly above chance. (A, from Jacobs GH & Nathans J [2009] *Sci Am* 300[4]:56–63. With permission from Springer Nature. B, from Jacobs GH, Williams GA, Cahill H, et al. [2007] *Science* 315:1723–1725. With permission from AAAS.)

A spectral discrimination task was designed to have mice report their detection of spectral differences. In a three-alternative forced-choice task, mice were trained to indicate which color was different from the other two to obtain a reward (Figure 13-29A). Colors for two ports were set at 600 nm, and the remaining port varied within a range of 500–600 nm. Control dichromatic mice completely failed at the task, making correct choices at chance level (33%) for discrimination between 500 nm and 600 nm. However, heterozygous female mice with a balanced M:L ratio performed significantly above chance level, even when discriminating between 580 nm and 600 nm (Figure 13-29B), indicating that these genetically engineered trichromatic mice indeed have superior color discrimination capabilities.

In a conceptually similar experiment, viral transduction was used to express human L-opsin in the retina of the genetically dichromatic male adult squirrel monkey, a New World monkey. These monkeys were trained in a three-alternative forced-choice task; to receive a juice reward, they were required to touch a patch of colored dots surrounded by gray dots of varying size and intensity (Figure 13-30A). Even after extensive training, the dichromatic monkeys still could not identify a color patch with a wavelength near 490 nm, a so-called spectral neutral point. However, monkeys that were virally transduced to express human L-opsin could discriminate the color patches from the gray background at all wavelengths tested (Figure 13-30B). Thus, expression of L-opsin in adult retina was sufficient to confer new color discrimination capability to dichromatic squirrel monkeys. This experiment also provided a proof-of-principle for gene therapy to correct color blindness in humans.

Whereas trichromatic primates in principle have had millions of years to evolve their color perception circuits to match their polymorphic cone genes, in the mouse simulation experiment, a new color-processing channel was introduced within a single generation. That these mice could immediately employ the new opsin for spectral discrimination and to inform behavior speaks strongly to the flexibility of the nervous system to adapt to abrupt changes in sensory input. The squirrel monkey experiment further demonstrates that a new spectral channel introduced in adulthood can be utilized almost immediately. How the retina and downstream circuits process the new chromatic contrast information to inform behavior in these genetically modified animals is an interesting subject for further investigation.

Figure 13-30 **Viral expression of a third cone opsin in adult dichromatic squirrel monkeys improved spectral discrimination.**
(A) Color discrimination task. Monkeys were trained to touch a color patch that appeared randomly against a gray background in one of three compartments separated by two light bands in order to receive a juice reward. (B) Performance of a male squirrel monkey before (black line) and after (blue line) viral transduction of a human L-opsin gene. Before viral transduction, the dichromatic monkey could not discriminate color patches with wavelengths near 490 nm, as indicated by very high threshold values (derived from psychometric measurements of the monkey's performance with varying degrees of color saturation). After expressing the human L-opsin, the male squirrel monkey was able to discriminate color patches of all wavelengths tested. Red triangle indicates the threshold for a trichromatic female control. (From Mancuso K, Hauswirth WW, Li Q, et al. [2009] *Nature* 461:784–787. With permission from Springer Nature.)

(A)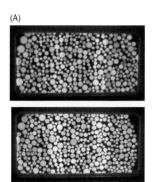

(B)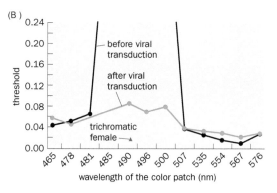

EVOLUTION OF NERVOUS SYSTEM STRUCTURE AND DEVELOPMENT

We now turn to the final part of the chapter on the evolution of nervous system structures. An important angle from which to approach this question is through the study of development. After all, evolution of new structures must occur through changes in developmental processes. In recent decades, advances in developmental biology have revealed a remarkable degree of conservation in developmental mechanisms across the animal kingdom. Such is the case for the establishment of the body plan, our first topic of discussion.

13.18 All bilaterians share a common body plan specified by conserved developmental regulators

Bilaterians, which include all animals with a central nervous system, have two major branches: the protostomes, comprising most invertebrate phyla, and the deuterostomes, comprising all vertebrates (Figure 13-2). Among the substantial differences between protostome and deuterostome body plans is the location of their central nervous systems: whereas the spinal cord is located dorsally in vertebrates, the nerve cords of protostomes are usually located ventrally.

In 1822, Étienne Geoffroy Saint-Hilaire first suggested that vertebrate and invertebrate body plans share similarities; for example, an upside-down lobster somewhat resembles a vertebrate (**Figure 13-31**A). Elucidation of the molecular mechanisms of development in the 1990s validated Geoffroy Saint-Hilaire's hypothesis. Researchers found that the determinants of the dorsal–ventral axes in flies and frogs are homologous. In the fruit fly embryo, the ventral neuroectoderm, which gives rise to the ventral nerve cord, expresses the secreted protein Sog (short gastrulation); Sog antagonizes the dorsally expressed secreted Dpp (Decapentaplegic), which promotes dorsal fate (Figure 13-31B). In the frog embryo, the dorsal structure that includes the neuroectoderm is determined by the dorsally expressed Chordin, a homolog of Sog; Chordin antagonizes Bmp4 (bone morphogenetic protein 4), a homolog of Dpp, which promotes ventral fate (Figure 13-31C). Remarkably, expression of frog Chordin in fly embryos promoted ventral fate, whereas expression of fly Sog in frog embryos promoted dorsal fate. These data demonstrated the functional equivalence of Sog/Chordin and Dpp/Bmp4 in specifying dorsoventral axes in fly and frog embryos. Furthermore, following global dorsoventral axis patterning, homologous transcription factors in the fly and frog expressed at different distances from the ventral and dorsal midlines, respectively, are used to

Figure 13-31 Vertebrates have an inverted dorsoventral axis compared to invertebrates. (A) Étienne Geoffroy Saint-Hilaire's drawing of an upside-down lobster illustrates that after this inversion, the body plan of an invertebrate resembles that of a vertebrate, with the central nervous system (CNS) above the digestive system, including the stomach (S), liver (li), and intestine (in), which are farther above the heart (he) and blood vessels (bl). Muscles (mu) flank the CNS. **(B & C)** Cross sections through embryos of a fly (B) and a frog (C) show that both are patterned along the dorsoventral axis by homologous secreted proteins Dpp/Bmp4 and Sog/Chordin, with Sog promoting CNS on the ventral side of the fly embryo and Chordin promoting CNS on the dorsal side of the frog embryo by antagonizing the activity of Dpp/Bmp4 (red inhibitory signs). At a later stage, the neuroectoderm in each embryo is patterned by the expression of three homologous transcription factors: Msh/Msx, Ind/Gsh2, and Vnd/Nkx2. These transcription factors define the dorsoventral axis within the ventral nerve cord and spinal cord, respectively. Note that in the fly, neural progenitors delaminate from the epidermis to form the nervous system, whereas in the frog, neural progenitors invaginate from the epidermis (Figure 7-2). Thus, the three pairs of transcription factors follow the same (rather than the opposite) dorsoventral orientation. (From De Robertis EM & Sasai Y [1996] *Nature* 380:37–40. With permission from Springer Nature. See also Urbach R & Technau GM [2008] *Adv Exp Med Biol* 682:42–56.)

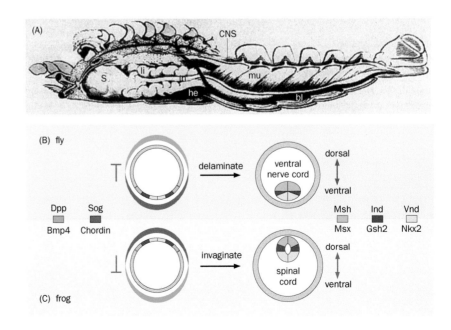

pattern the dorsoventral axes within the fly ventral nerve cord and the vertebrate spinal cord (Figure 13-31B, C, right). These data suggest that the central nervous systems of vertebrates and invertebrates have a common origin within a common body plan. The inversion of the dorsoventral axis in vertebrates from the more basal state of the invertebrates may have occurred during early chordate evolution.

That all bilaterians share a common body plan is further demonstrated by the role of **Hox** **gene** clusters in patterning the anteroposterior body axis. *Hox* genes were originally discovered in *Drosophila* as mutants that exhibit **homeotic transformation**, transformation of one body part into another. In *Antennapedia* mutants, a pair of antennae is transformed into a pair of legs; in *Ultrabithorax* mutants, the thoracic segment that gives rise to the wing is duplicated at the expense of another thoracic segment. Molecular-genetic analysis indicated that the corresponding genes are part of a gene cluster; all members of the cluster encode transcription factors that share a common DNA-binding domain called the **homeobox** (hence the name *Hox* gene cluster). Each *Hox* gene is expressed in specific body segments along the anteroposterior axis, and there is a co-linear relationship between the expression domain in the body and the location of the gene on the chromosome (**Figure 13-32**, top). Remarkably, the *Hox* gene clusters of vertebrates exhibit the

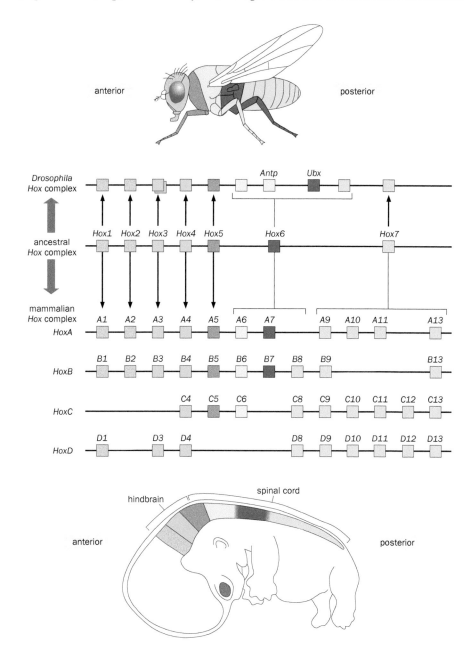

Figure 13-32 *Hox* **gene clusters pattern the anteroposterior body axis.** Each square represents a *Hox* gene. The top row shows the *Hox* gene cluster in *Drosophila*, including *Antennapedia (Antp)* and *Ultrabithorax (Ubx)*. The bottom four rows show the four *Hox* gene clusters in mammals. Colors represent the relationship between individual *Hox* genes and the segment identities they control in their respective organisms (top and bottom). Although *Hox* gene clusters regulate many aspects of vertebrate patterning, only the CNS segments are shown for simplicity. In both fly and mammal, there is a co-linear relationship between the location of a particular *Hox* gene on its chromosome and its expression domain along the anteroposterior axis of the body. Compared to the ancestral *Hox* gene cluster (second row), there is considerable addition and deletion of individual members in both fly and mammalian *Hox* clusters. (Adapted from Alberts et al. [2015] Molecular Biology of the Cell, 6th ed. Garland Science.)

(A)

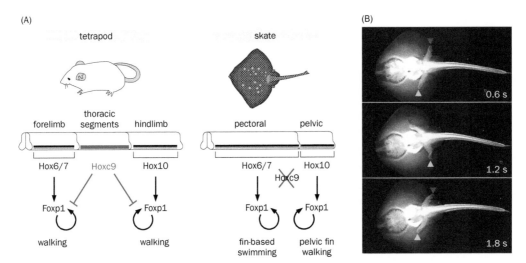

(B)

Figure 13-33 Hox proteins and regulation of vertebrate locomotion.
(A) In tetrapods and skates, Hox6/7 and Hox10 both promote expression of transcription factor Foxp1, which directs motor neurons to innervate limbs for walking in tetrapods (left) and pectoral and pelvic fins for swimming and walking, respectively, in some skates (right). Note that there are no intervening Foxp1⁻ segments in skates because Hoxc9, which is expressed in the intervening thoracic segments in tetrapods and inhibits Foxp1 expression, is absent in skates. **(B)** Time-lapse images of *Leucoraja*, a skate, taken from the bottom of a tank, illustrating its walking mediated by a pair of pelvic fins (red and green arrowheads). (From Jung H, Baek M, D'Elia KP, et al. [2018] *Cell* 172:667–682. With permission from Elsevier Inc.)

same co-linear relationship as in the fly. Notably, vertebrates have four *Hox* clusters that likely resulted from the two rounds of genome duplication during early chordate evolution (Figure 13-32, bottom) that also gave rise to four cone opsin genes, as noted earlier.

Although *Hox* gene expression patterns are established by distinct mechanisms during development in flies and vertebrates, once such patterns are established, *Hox* genes auto-activate their own expression and repress expression of other *Hox* genes so as to maintain and confine their expression within specific body segments. Thus, a loss-of-function mutation in one *Hox* gene can cause misexpression of another *Hox* gene. *Hox* genes also regulate the expression of many downstream genes to confer the identity of individual body segments. This is why mutations in these genes give striking homeotic transformation phenotypes.

We give a specific example of how *Hox* genes in different animals control analogous functions. In tetrapods (a group of animals including amphibians, reptiles, birds, and mammals; Figure 13-2), spinal segments corresponding to the forelimbs and hindlimbs express Hox6/7 and Hox10, respectively. Both *Hox* genes promote the expression of Foxp1, a transcription factor that directs motor neurons in the lateral motor column to innervate limb muscles (Figure 8-9). The intervening thoracic segments express Hoxc9, which inhibits Foxp1 expression, such that motor neurons innervate trunk muscles instead (**Figure 13-33**A, left). Fish undulate their trunk muscles to move. However, certain skate species, such as *Leucoraja*, use pectoral fins to swim and pelvic fins to walk on the seafloor (Figure 13-33B). A recent study suggested that Hox6/7 and Hox10 in *Leucoraja* also promote Foxp1 expression in motor neurons innervating the pectoral and pelvic fins, respectively (Figure 13-33A, right). Thus, motor programs for walking in tetrapods appear to have ancient origins in fish and are controlled by conserved *Hox* genes.

Besides the *Hox* gene cluster, homeobox DNA-binding domains are also present in many other transcription factors; these transcription factors often play evolutionarily conserved roles in regulating development, as illustrated in the next section.

13.19 Eye development is controlled by evolutionarily conserved transcription factors

Developmental genetic analysis in *Drosophila* has revealed a number of transcription factors that initiate the formation of its compound eyes. One such factor is **Eyeless**, mutation of which leads to a complete absence of compound eyes.

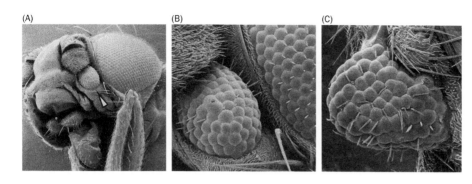

Figure 13-34 Misexpression of *Drosophila Eyeless* or its mouse homolog *Pax6* leads to ectopic eyes. (A & B) Scanning electron micrographs of a fly reveal that misexpression of *Eyeless* in progenitors of the antenna caused an eye to form instead of an antenna (arrowheads in A; magnified in B, with the ectopic eye transformed by the antenna to the left and the normal eye to the right). **(C)** Misexpression of mouse *Pax6* in flies produced an ectopic eye on a leg. (From Halder G, Callaerts P, & Gehring WJ [1995] *Science* 267:1788–1792. With permission from AAAS.)

Remarkably, misexpression of Eyeless in progenitors that give rise to other tissues can cause eye formation in places such as the antenna and underneath the wing (**Figure 13-34**A, B), similar to the homeotic transformations discussed earlier. Several additional genes that exhibit loss- and gain-of-function phenotypes similar to those of *Eyeless* have been identified in subsequent studies. *Eyeless* and these other genes together form a network that acts to direct the formation of the fly eye.

The Eyeless protein belongs to the Pax family of transcription factors; each member contains two DNA-binding domains: a paired box and a homeobox. Eyeless is most similar to mammalian **Pax6**. Interestingly, deletion of *Pax6* in mice results in the absence of eyes. Losing a single copy of *Pax6* in humans causes a partial or complete absence of irises. Losing two copies of *Pax6* in humans prevents eye formation and causes stillbirth. Strikingly, expressing mouse Pax6 in *Eyeless* mutant flies rescues development of normal eyes, and misexpression of mouse Pax6 in flies can produce ectopic eyes (Figure 13-34C) similar to those resulting from misexpression of *Drosophila* Eyeless. These data indicate that *Eyeless* and *Pax6* are evolutionarily conserved genes that regulate eye formation. Indeed, even some jellyfish species have *Pax* genes that are expressed in their eyespots, suggesting an ancient role for Pax genes in eye development.

We learned in Section 13.14 that eyes in vertebrates and most invertebrates have different morphologies and use different opsins and signal transduction pathways. These findings have led to the proposal that vertebrate and invertebrate eyes have multiple origins (Box 13-4). However, research regarding the role of *Pax* genes in eye development seems to suggest that all eyes have a common origin. A parsimonious model that reconciles these different views is that the Pax transcription factors regulated the development of photosensitive cells early in animal evolution, before the cnidarian–bilaterian split, and have retained this function across different animal phyla. These photosensitive cells evolved independently in multiple taxa, giving rise to rhabdomeric photoreceptors (present in invertebrates and a subset of vertebrate retinal ganglion cells) and ciliary photoreceptors (present in vertebrates and photoreceptors and brain neurons in some invertebrates). At the same time, eyes evolved independently with different morphologies that converged on the basis of shared optical principles.

The loss-of-function and, particularly, misexpression phenotypes of *Eyeless/Pax6* and the *Hox* genes illustrate how altering the expression pattern of key developmental regulators can readily produce large structural changes. Because these transcription factors sit at the top of a regulatory hierarchy controlling the expression of many downstream target genes, altering the expression patterns of these transcription factors may represent a key mechanism for evolutionary change (Section 13.4).

13.20 The mammalian neocortex underwent rapid expansion recently

In the remaining sections, we discuss the structure that endows us with complex cognitive functions: the mammalian **neocortex**. The neocortex can be traced back to a homologous but much simpler structure in the dorsal telencephalon of reptiles called the **dorsal cortex**. Unlike the six-layered mammalian neocortex,

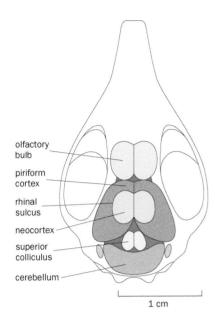

Figure 13-35 Brain organization of an early mammal reconstructed from the fossil record. In this dorsal view of a mammalian skull dating from approximately 85 million years ago, the neocortex is seen at the top, separated by the rhinal sulcus from the piriform cortex. Compared to most extant mammals, the structures in the fossil's brain related to the processing of olfactory information, such as the olfactory bulb and piriform cortex, occupy much larger areas than the neocortex. (Adapted from Kaas [2007] Evolution of Nervous Systems. Elsevier.)

olfactory bulb
piriform cortex
rhinal sulcus
neocortex
superior colliculus
cerebellum

1 cm

the reptilian dorsal cortex has three layers, with excitatory neurons concentrated in the middle layer (layer 2). Recent single-cell RNA-sequencing analysis of turtle and lizard dorsal cortices revealed that reptilian layer 2 excitatory neurons are composed of different transcriptomic clusters, each having distinct mixtures of marker genes found in superficial (2–4) and deeper (5–6) layer excitatory mammalian neocortex neurons (Box 4-4). This suggests that excitatory neuron classes in reptilian dorsal cortex and mammalian neocortex are not homologous; rather, they both diverged from ancestral telencephalon excitatory neurons in their common amniote ancestor (amniotes are a clade of vertebrates including reptiles, birds, and mammals; Figure 13-2). By contrast, GABAergic inhibitory neuronal classes in the reptilian cortex correspond well to analogous classes in the mammalian neocortex, suggesting that they both descended from GABAergic inhibitory neuron classes already present in their amniote ancestor.

Since mammals' divergence from reptiles some 300 million years ago, the mammalian neocortex has expanded both vertically (from 3 to 6 layers) and horizontally, encompassing dozens of areas responsible for the sensory, motor, and association functions we have studied in preceding chapters. Fossil records suggest that much of the expansion might have occurred during the past 65 million years, following the extinction of dinosaurs, when mammals started to expand from mostly small and nocturnal animals into the great diversity seen today. For example, reconstruction of the skull of an early mammal (**Figure 13-35**) suggested that the rhinal sulcus, which separates the dorsally located neocortex from the ventrally located piriform cortex, had a much more dorsal position than in extant mammals, indicating that the neocortex occupied a much smaller fraction of the forebrain in this early mammal. (The piriform cortex, which processes olfactory information, has a three-layered structure and is not part of the neocortex.)

What accounts for the rapid expansion of mammalian neocortex over a relatively short time period during mammalian evolution (**Figure 13-36**)? One factor may be the modular nature of cortical circuit organization (Box 4-4), which enables expansion of cortical areas when mechanisms that produce increased numbers of neurons evolve in tandem. A second factor may be the flexibility and plasticity with which different subregions of the neocortex accommodate diverse subcortical input sources. In the following sections, we discuss possible mechanisms by which the neocortex could expand in size and in the number of specialized areas it contains.

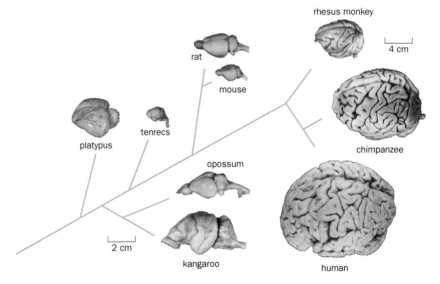

rhesus monkey
4 cm
rat
mouse
platypus
tenrecs
chimpanzee
opossum
kangaroo
human
2 cm

Figure 13-36 Images of selected mammalian brains. Brains are shown from a slanted side view, with rostral toward the left. The 4 cm scale applies to the three primate brains; the 2 cm scale applies to the rest. The phylogenetic relationships of the taxa are drawn in blue. Images are from http://brainmuseum.org. These samples are from the University of Wisconsin and Michigan State Comparative Mammalian Brain Collections, and the National Museum of Health and Medicine, funded by the National Science Foundation and the National Institutes of Health.

13.21 The size of the neocortex can be altered by modifying the mechanisms of neurogenesis

Humans have the greatest neocortex expansion among all mammals—the human neocortex occupies about 75% of the brain's volume. By comparison, the neocortices of the mouse lemur (a small primate) and the rat account for about 40% and 30% of their brain volumes, respectively. As we learned in Section 7.2, excitatory neurons in the neocortex can be produced directly from radial glia in the ventricular zone or via intermediate progenitors that are products of radial glia. At an earlier stage, radial glia are produced by neuroepithelial progenitors through symmetrical expansion (Figure 7-4). In principle, an increase in the number of cell divisions (and a concomitant expansion of the volume of the skull) can produce a larger brain with more cortical neurons.

Several mechanisms have been proposed to increase the number of cell divisions during cortical neurogenesis. The first is to increase the pool of radial glia by lengthening the period during which neuroepithelial progenitors undergo exponential symmetric cell division; each extra division doubles the total neuronal number. The second is to increase the number of cell divisions that yield intermediate progenitors, such that each radial glial cell produces more postmitotic neurons. While there is evidence for both mechanisms, recent imaging studies of human cortical neurogenesis *in vitro* have identified a third mechanism: the developing human neocortex has an expanded subventricular zone, which contains an additional type of radial glia cell called an **outer radial glia** (**oRG**). An oRG can divide asymmetrically to produce another oRG and an intermediate progenitor. The processes of oRGs that extend to the pia surface serve as additional guideposts for radial migration of neurons to the cortical plate, thus expanding the cortical surface for a given unit area of ventricular surface (**Figure 13-37**). oRGs likely derived from ventricular radial glia, reinforcing the importance of cell type diversification in the evolutionary processes discussed in Section 13.15.

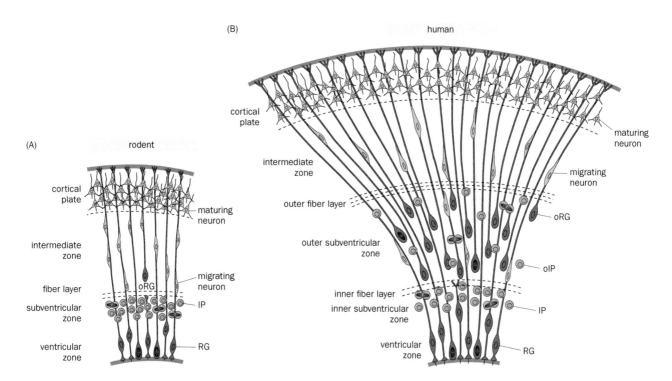

Figure 13-37 Schematic of cortical neurogenesis in rodents and humans. In both rodents and humans, cortical neurogenesis begins in the ventricular zone, where radial glia (RG) produce neurons or intermediate progenitors (IP). IP divisions in the subventricular zone produce more neurons (Figure 7-4). Humans feature a significantly expanded outer subventricular zone, which contains not only outer intermediate progenitors (oIP) but also abundant outer radial glia (oRG) that can produce oIPs and neurons, thus greatly expanding cellular proliferation. oRGs have subsequently been found in rodent brains but are far fewer than ventricular RGs. (Adapted from Lui JH, Hansen DV, & Kriegstein AR [2011] *Cell* 146:18–36. With permission from Elsevier Inc.)

Figure 13-38 Simulation of gyrification of human neocortex during development. Modeling the brain as a soft elastic solid, tangential expansion of the cortical plate relative to the white matter zone underneath (bottom left, magnified from top left, illustrated from a coronal plane) produces gyrification patterns shown on the right, starting from a smooth fetal brain at gestational week (GW) 22. The shape of the GW22 brain was based on a three-dimensional magnetic resonance image at GW22. (From Tallinen T, Chung JY, Rousseau F, et al. [2016] *Nat Physics* 12:588–593. With permission from Springer Nature.)

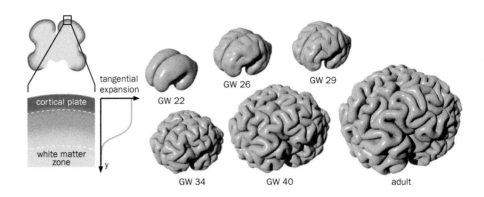

Brains with large neocortices and an increased number of neurons develop gyrencephalic cortices (Figure 13-5). Compared to smooth lissencephalic cortices, gyrencephalic cortices have more surface area per brain volume and can therefore accommodate more cortical processing modules. Another advantage of gyrencephalic cortices is the reduced axon length needed for cortico-cortical connections. How are gyri and sulci produced? Several experiments in mice (which are lissencephalic) suggested that simply increasing the number of cortical neurons, for example, by increasing the size of progenitor pools or inhibiting programmed cell death, is sufficient to produce a cortex with folds that resemble gyri and sulci. Physical modeling of the brain as soft material suggests that due to mechanical instability, tangential expansion of an outer layer relative to an inner layer is sufficient to produce folds with gyri and sulci (**Figure 13-38**). These data and simulations suggest the relative ease with which gyri and sulci could evolve, accounting for their independent appearance multiple times during mammalian evolution (Figure 13-5). A future challenge is to determine which differences in the genomes of different mammals (e.g., mouse versus human) account for the various developmental mechanisms employed, thus ultimately giving rise to species-specific cortical sizes.

13.22 Cortical area specialization can be shaped by input patterns

Mammals with a larger neocortex not only have more neurons but also develop more specialized cortical areas and elaborate connections. For example, the prefrontal cortex in primates is greatly enlarged compared to that in rodents, contributing to their more sophisticated executive functions. Humans have numerous specialized neocortical areas, such as Broca's and Wernicke's areas for language processing (Figure 1-23; Box 13-5). How did these specializations evolve?

There has been an interesting debate about the developmental mechanisms of neocortical area specialization. In one extreme view, radial glia at the ventricular zone are already specified based on their locations with respect to the body axes (Figure 7-3); these progenitors then give rise to prespecified neurons occupying specific cortical areas. In another extreme view, the neocortex is a blank slate during early development; area specializations are determined by the subcortical axonal input they receive. The truth is likely somewhere in between: the gross patterning of motor, sensory, and association cortices is determined by secreted morphogens and intrinsic transcription factors, as we learned in Chapter 7; and input patterns fine-tune cortical areas with respect to their functions. The examples we discuss next provide support for the latter proposal.

As discussed in Chapter 5, visual experience can profoundly affect the connectivity of the primary visual cortex (V1). The following experiment suggested neocortical plasticity at a larger scale after surgery-induced rewiring during development. This experiment involved ablating auditory axonal projections to the thalamus in neonatal ferrets such that the medial geniculate nucleus (MGN), which normally relays auditory input to the primary auditory cortex (A1), no longer received auditory input. At the same time, the superior colliculi were ablated to deprive retinal ganglion cell axons of a major target. In combination, these abla-

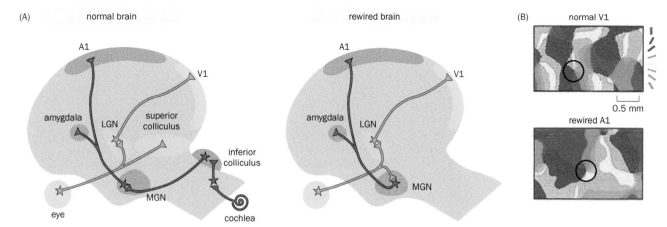

Figure 13-39 Rewiring the auditory cortex to process visual information. (A) Left, schematic of the axonal projections of the visual (blue) and auditory (red) systems in a normal ferret. The lateral geniculate nucleus (LGN) relays retinal input to the primary visual cortex (V1), whereas the nearby medial geniculate nucleus (MGN) relays auditory input from the inferior colliculus to the primary auditory cortex (A1). Right, after bilateral ablation of the superior colliculi and severing of the inferior collicular input to the MGN during early development, some retinal input was channeled into the MGN and relayed to A1. **(B)** A1 neurons in the rewired brain not only respond to visual stimuli but in some cases also exhibit orientation selectivity and a functional organization similar to V1 neurons. Although different in detail, the orientation maps in V1 and rewired A1 display similarities, such as the pinwheel structures. The bars to the right of the top panel assign a color to each orientation; circles highlight the centers of two pinwheels. (A, adapted from Sur M & Rubenstein JLR [2005] *Science* 310:805–810. B, from Sharma J, Angelucci A, & Sur M [2000] *Nature* 404:841–847. With permission from Springer Nature.)

tions created a situation in which some retinal ganglion cell axons were connected to the MGN, enabling A1 to receive visual input (Figure 13-39A). Remarkably, A1 neurons not only responded to visual stimuli but also exhibited orientation selectivity resembling normal V1 neurons. Moreover, the orientation-selective neurons in rewired A1 showed an arrangement reminiscent of the pinwheel maps observed in V1 (Figure 13-39B; Figure 4-45). This experiment suggested that certain cortical properties are not entirely predetermined but can be influenced by the nature of its thalamic input. In principle, then, evolution can act on variations in axonal pathways to shape the properties of the neocortex.

This rewiring example, while striking, represents a highly artificial case. Many examples in nature do support the notion that functional specialization in the neocortex is shaped by the input it receives. For example, bats that use echolocation to identify prey have expanded areas in the auditory cortex specifically tuned to the ultrasonic frequencies they emit (Section 6.26), but nonecholocating bats do not. To analyze a rich array of tactile information, rodents that use whiskers on their snout to explore their environment develop barrels in layer 4 of their somatosensory cortex, with each barrel corresponding to a single whisker (Box 5-3). As nearly blind predators, star-nosed moles rely on special mechanosensory rays surrounding the snout to catch small prey (Figure 13-40A; Movie 13-1); a large fraction of the mole's somatosensory cortex is devoted to representing individual rays (Figure 13-40B). Rays 1 through 10 are used to search for prey, and the 11th pair functions as the equivalent of a "fovea," allowing for rapid fine discrimination

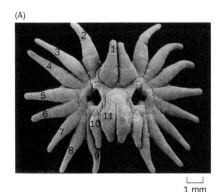

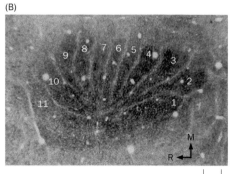

Figure 13-40 Brain representation of mechanosensory rays in the star-nosed mole. (A) Scanning electron micrograph of 11 pairs of rays surrounding the nostrils. **(B)** Horizontal section showing cytochrome oxidase staining of the somatosensory cortex. Each of the 11 bands of dark staining represents a sensory ray on the contralateral side. The 11th band occupies a disproportionally larger cortical area, reflecting the use of the 11th pair of rays in fine discrimination of objects before eating. R, rostral; M, medial. (From Catania KC & Kaas JH [1995] *J Comp Neurol* 351:549–567. With permission from John Wiley & Sons, Inc.)

of potential food objects. Interestingly, cortical representation of the 11th pair of rays is enlarged compared to that of other appendages (Figure 13-40B), just as primate V1 is disproportionally dedicated to representing the retinal fovea (Figure 4-39).

The detailed mechanisms by which these cortical specializations arise are interesting subjects for future exploration. Natural selection can act on intrinsic cortical patterning mechanisms, axonal input, or both. The visual-to-auditory rewiring experiment, along with the superior color discrimination of trichromatic mice (Section 13.17), suggest that the neocortex can modify its functional organization according to changes in input patterns in a single generation, likely as a consequence of its modular circuit design (Box 4-4). This flexibility enables the neocortex to accommodate new input patterns and provides a mechanism by which environmental changes can impact brain structure and organization.

Box 13-5: Brain evolution of *Homo sapiens*

One motive for studying evolution is to know where we came from. Traditionally belonging to the disciplines of paleontology, archaeology, and anthropology, the study of human origins has been greatly aided by the recent genomic revolution. We now have the genome sequences of our own species, the extant great apes, and two extinct archaic humans, Neanderthals and Denisovans, from which we can infer phylogenetic relationships (Figure 13-41A). Furthermore, comparing the genome sequences of humans from different geographic locations—both the present day and the recent past—has revealed a natural history of our own species (Figure 13-41B). *Homo sapiens* originated in Africa between 200,000 and 300,000 years ago. They lived in Africa and the Middle East until about 60,000 years ago, when they started to spread across Eurasia, Oceania, and the Americas. They interbred with Neanderthals and Denisovans, who were living in Europe and Asia before the arrival of *Homo sapiens* but subsequently went extinct. These genomic data also pose interesting questions for brain evolution: what makes us different from our closest extant relative, the chimpanzee? What makes us different from the archaic humans that went extinct?

In principle, the differences between human and chimpanzee genomes—including roughly 35 million single-nucleotide substitutions and 5 million insertions/deletions—should underlie their different anatomy, physiology, and behavior (for example, the human brain contains three times as many neurons as the chimpanzee brain). These are a lot of differences to sift through, most of which reflect genetic drift and do not have phenotypic consequences. Nevertheless, researchers have used these genomic differences to study candidate genes thought to contribute to brain evolution. We discuss two examples. The first regards the *Srgap2* gene, encoding a GTPase activating protein of the Rho-family of small GTPases, which regulate neuronal morphogenesis (Box 5.2). Compared to the genomes of chimpanzee and other mammals, the human genome contains duplications encoding truncated Srgap2 proteins; these truncated proteins are co-expressed with a full-length Srgap2 (encoded by the ancestral gene) and antagonize its function. Expressing truncated Srgap2 in the mouse neocortex delays neuronal migration and dendritic spine maturation, which has been interpreted as mimicking the prolonged development of the human brain. The second example started with identifying human-specific deletions of genomic fragments that are conserved in mammals through to chimpanzees; the rationale is that these human-specific deletions could contain regulatory elements that alter gene expression (Section 13.5). Of 510 such genomic fragments, one maps next to the *Gadd45g* gene, which encodes a negative regulator of cell division. The fragment drives reporter expression in subventricular zones in the embryonic mouse brain, raising the possibility that the loss of this fragment in humans might increase neurogenesis and thus contribute to increased neuronal numbers.

While tantalizing, these studies also expose a limitation in using the mouse model—a relatively distant species with major differences in development, anatomy, and behavior—to infer causality of genome changes that contribute to the difference between chimpanzees and humans. Recent technological advances in brain organoids produced from induced pluripotent stem cells (iPSCs; Box 7-1) allow direct comparisons of neural development in organoids derived from humans and chimpanzees (Figure 13-42). In principle, genome editing tools (Box 14-1) can be applied in either human or chimpanzee iPSCs to re-create genomic differences and test their consequences in brain organoids. While it remains challenging to mimic the late stages of neural development, early steps, including neurogenesis and migration of neocortical neurons, can be recapitulated in brain organoids (Box 7-1), allowing researchers to test genomic differences that may account for the neocortical expansion from chimpanzees to humans. Recent advances in producing genetically modified monkeys can potentially extend studies of candidate genes into later stages of brain development and function (see Section 14.4 for a discussion of ethical considerations).

The genomic differences between modern and archaic humans are much smaller than that between humans and chimpanzees. Only ~30,000 single nucleotide substitutions and ~4,000 small insertions/deletions shared by all ~1,000 analyzed modern human genomes differ from the ancestral versions shared by the Neanderthal, Denisovan, and great ape. Of these, 3,117 fall in regulatory regions, 32 affect splicing sites, and 96 alter amino acids in 87 proteins. Some of these differences presumably account for the evolution of

Box 13-5: continued

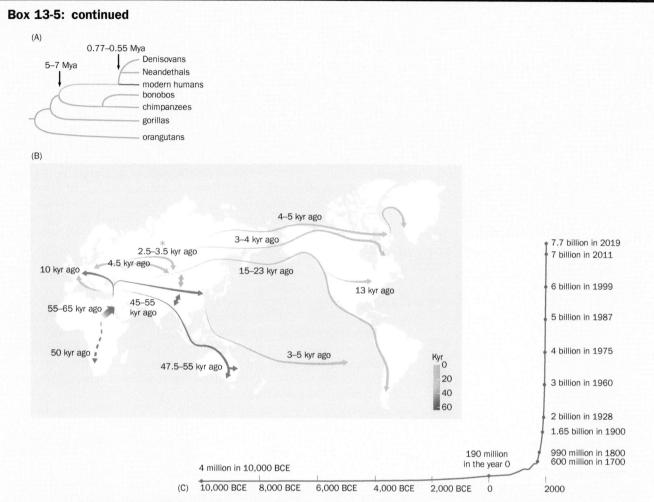

(A)

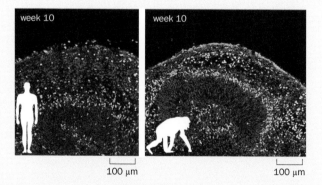

Figure 13-41 Phylogenetic tree, migration, and population growth of *Homo sapiens*. (A) Phylogenetic relationship of modern humans (*Homo sapiens*) to two extinct archaic human species (Denisovans and Neanderthals), as well as extant great apes. Mya, million years ago. **(B)** Major migrations of *Homo sapiens*. Blue star is the location of the Denisova Cave, where Denisovan fossils were found. Kya, thousand years ago. **(C)** *Homo sapiens* population growth over the past 12,000 years. (A & B, based on Pääbo S [2014] *Cell* 157:216–226; Nielsen R, Akey JM, Jakobson M, et al. [2017] *Nature* 541:302–319; Reich D [2018] Who We Are and How We Got Here. Vintage Books. C, adapted from OurWorldinData.org.)

our own species in their exploration and expansion over the world, while archaic human species went extinct (Figure 13-41B). Establishing the causality of these differences in *Homo sapiens* evolution, however, is extremely challenging. This is in part because we know of the physiological and behavioral traits of archaic human species only through indirect inferences from fossils and archaeological artifacts, and in part because chimpanzee brain organoids and genetically modified monkeys are inadequate for modeling cultural differences between different human species.

The evolution of our own species has been greatly affected in the recent past by the development of language, symbolic thinking, transgenerational knowledge transfer, and sophisticated tools made from stone, bronze, and iron. Such advances led to the Industrial Revolution of the 1700s, which was accompanied by rapid population growth (Figure 13-41C). Thus, the evolution of *Homo sapiens* is now increasingly affected by its culture. How cultural changes impact brain evolution will be a fascinating subject of investigation.

Figure 13-42 Brain organoids can be used to compare development of human and chimpanzee neocortices. Sections of 10-week-old brain organoids derived from human (left) and chimpanzee (right) induced pluripotent stem cells immunostained for markers of radial glia (red), intermediate progenitors (green), and subcerebral-projecting (magenta) and callosal-projecting (cyan) neurons. The distributions of these markers mimic *in vivo* distributions. (From Pollen AA, Bhaduri A, Andrews MG, et al. [2019] *Cell* 176:743–756. With permission from Elsevier Inc.)

SUMMARY

Evolution requires two interrelated processes. The first is the production of heritable variation at the level of DNA. Mutations in coding and regulatory regions of genes can alter protein functions and expression patterns, respectively. Gene duplication creates new templates that mutations can alter to generate functions distinct from those of the ancestral gene. Gene loss can eliminate an existing function. DNA shuffling can produce novel protein functions by fusing together domains from separate genes, or novel expression patterns by juxtaposing protein-coding sequences to new regulatory sequences. These events confer populations a rich substrate for evolutionary changes.

The second process is the selection of genetic variants that improve the fitness of individuals. Animals better at finding food and mates, avoiding predators, and caring for their offspring are more likely to pass their variants on to future generations. Numerous iterations of these two processes in all branches of the phylogenetic tree of animals over the last billion years have created a rich diversity of nervous systems. We summarize examples discussed in this chapter in the following 10 general lessons.

1. *Most evolutionary changes are gradual and sequential.* In the propagation of electrical signals, the combined action of voltage-gated K⁺ and Ca²⁺ channels produced active conductances, allowing amplified electrical signals to spread farther than via passive conductance alone. The appearance of voltage-gated Na⁺ channels increased the speed of signal propagation, and the emergence of the myelin sheath further enhanced the speed and reliability of signal propagation over long distances.

2. *New functions often emerge via modification of the coding and regulatory sequences of existing genes rather than acquisition of new genes.* Primitive synapses were built mostly with proteins that existed in animals without a nervous system. The initial neurotransmitter release machinery may have co-opted existing machinery for secretion.

3. *Loss of traits can be just as significant as gain of traits in the course of evolution,* whether due to active selection, absence of selection, genetic drift or other stochastic factors. Early mammals lost two of the four cone genes present in their ancestors, likely due to their nocturnal lifestyle. *C. elegans* lost voltage-gated Na⁺ channels likely due to a lack of need for long-distance electrical signaling in their small bodies.

4. *Useful mechanisms that arose early in the phylogenetic tree are often conserved throughout subsequent evolutionary history.* Molecules used to establish the dorsoventral and anteroposterior body axes emerged in early metazoans and have been conserved in all bilaterians, surviving a dorsoventral body-axis inversion in early chordates.

5. *Useful mechanisms can be extended to solve new problems.* The emergence of G-protein-coupled receptors (GPCRs) in early eukaryotes created an effective way of sensing the external environment and modifying intracellular signaling. While the budding yeast has three GPCRs for detecting mating pheromones and nutrients, most animals have many hundreds of GPCRs for sensing not only environmental chemicals (conservation, as in point 4) but also light and for detecting neurotransmitters, neuropeptides, and hormones for intercellular communication.

6. *Animals in different clades can independently evolve similar solutions to common problems.* Such convergent evolution can occur at the levels of molecules, cells, circuits, or strategies. Retinal-based light sensing has been adopted independently by prokaryotes and multicellular eukaryotes. Vertebrates and invertebrates independently acquired myelination. The glomerular organization of the olfactory system appears to have emerged independently in insects and vertebrates. Sensory systems in animals employ signal amplification and adaptation much like bacterial chemotaxis.

7. *Multiple solutions can evolve to solve the same problem and can coexist to play complementary functions.* Electrical and chemical synapses have coexisted since the beginning of the nervous system as two complementary means of intercellular communication. The prototypes of rhabdomeric and ciliary photoreceptors were likely present before the cnidarian–bilaterian split and have coexisted in both vertebrates and invertebrates. Two distinct families of olfactory receptors cooperate in the *Drosophila* olfactory system.

8. *Evolutionary changes are constrained by phylogenetic history.* Rhabdomeric and ciliary photoreceptors dominate vision in invertebrates and vertebrates, respectively, probably because their ancestors adopted one type, rather than the other, by chance. Insects expanded families of ionotropic receptors to detect odorants because their ancestors did not use GPCRs to detect odorants. Primates reacquired trichromatic color vision from dichromatic ancestors because early mammals had lost two of the four cone opsin genes present in other vertebrates.

9. *Diversification of cell types is an important step in the evolution of complex nervous systems.* Diversification of photoreceptors and bipolar cells in the vertebrate retina allowed dedication of photoreceptors to sensing light and bipolar cells to processing signals. Diversification of rods and cones enabled each type to specialize, with different sensitivities, speeds, and dynamic ranges, thus expanding the collective capacity to detect visual signals. Diversification of radial glia facilitated neurogenesis and cortical expansion.

10. *Flexibility of neuronal circuits is instrumental to the evolution of complex nervous systems.* Retinal circuits in the primate fovea are capable of extracting new color information as soon as a new cone with different spectral sensitivity appears. The modular nature of neocortical circuits and their ability to be patterned by input pathways may have contributed to their recent rapid expansion.

While most of these principles apply to the evolution of all biological systems, the evolution of the nervous system has provided striking examples of natural selection in action. Studying "how did the brain arise" deepens our understanding of "how does the brain work" by providing a historical perspective and by considering brain function in the context of an interconnected web of life. We are living at an exciting time for exploring these rich and complex relations.

OPEN QUESTIONS

- When did the nervous system(s) first emerge?
- How did the centralized nervous systems first emerge?
- How can natural selection, which acts on individuals' numerous collective phenotypes, ensure that beneficial alleles of each specific gene are enriched in future generations?
- How do new neuron types emerge? How are evolutionary changes in different neuron types of a circuit coordinated?
- What properties endow neuronal circuits with flexibility, allowing them to accommodate and take advantage of new evolutionary changes?

FURTHER READING

General concepts and approaches
Darwin C (1859) On the Origin of Species by Means of Natural Selection. John Murray.

Jacob F (1977) Evolution and tinkering. *Science* 196:1161-1166.

Kingsley DM (2009) From atoms to traits. *Sci Am* 300(1):52-59.

Mayr E (1997) The objects of selection. *Proc Natl Acad Sci U S A* 94:2091-2094.

Woese CR, Kandler O, & Wheelis ML (1990) Towards a natural system of organisms: proposal for the domains Archaea, Bacteria, and Eucarya. *Proc Natl Acad Sci U S A* 87:4576-4579.

Origin of the nervous system and evolution of neuronal communication

Bennett MK & Scheller RH (1993). The molecular machinery for secretion is conserved from yeast to neurons. *Proc Natl Acad Sci U S A* 90:2559-2563.

Hartline DK & Colman DR (2007). Rapid conduction and the evolution of giant axons and myelinated fibers. *Curr Biol* 17:R29-R35.

King N & Rokas A (2017). Embracing uncertainty in reconstructing early animal evolution. *Curr Biol* 27:R1081-R1088.

Moran Y, Barzilai MG, Liebeskind BJ, & Zakon HH (2015). Evolution of voltage-gated ion channels at the emergence of Metazoa. *J Exp Biol* 218:515-525.

Moroz LL, Kocot KM, Citarella MR, Dosung S, Norekian TP, Povolotskaya IS, Grigorenko AP, Dailey C, Berezikov E, Buckley KM, et al. (2014). The ctenophore genome and the evolutionary origins of neural systems. *Nature* 510:109-114.

Srivastava M, Simakov O, Chapman J, Fahey B, Gauthier ME, Mitros T, Richards GS, Conaco C, Dacre M, Hellsten U, et al. (2010). The *Amphimedon queenslandica* genome and the evolution of animal complexity. *Nature* 466:720-726.

Evolution of sensory systems and behavior

Arendt D (2008). The evolution of cell types in animals: emerging principles from molecular studies. *Nat Rev Genet* 9:868-882.

Berg HC & Brown DA (1972). Chemotaxis in *Escherichia coli* analysed by three-dimensional tracking. *Nature* 239:500-504.

Collin SP, Knight MA, Davies WL, Potter IC, Hunt DM, & Trezise AE (2003). Ancient colour vision: multiple opsin genes in the ancestral vertebrates. *Curr Biol* 13:R864-865.

Ding Y, Berrocal A, Morita T, Longden KD, & Stern DL (2016). Natural courtship song variation caused by an intronic retroelement in an ion channel gene. *Nature* 536:329-332.

Fernald RD (2006). Casting a genetic light on the evolution of eyes. *Science* 313:1914-1918.

Jacobs GH, Williams GA, Cahill H, & Nathans J (2007). Emergence of novel color vision in mice engineered to express a human cone photopigment. *Science* 315:1723-1725.

Julius D & Nathans J (2012). Signaling by sensory receptors. *Cold Spring Harb Perspect Biol* 4:a005991.

Mancuso K, Hauswirth WW, Li Q, Connor TB, Kuchenbecker JA, Mauck MC, Neitz J & Neitz M (2009). Gene therapy for red-green colour blindness in adult primates. *Nature* 461:784-787.

Melin AD, Chiou KL, Walco ER, Bergstrom ML, Kawamura S, & Fedigan LM (2017). Trichromacy increases fruit intake rates of wild capuchins (*Cebus capucinus imitator*). *Proc Natl Acad Sci U S A* 114:10402-10407.

Nagel G, Szellas T, Huhn W, Kateriya S, Adeishvili N, Berthold P, Ollig D, Hegemann P, & Bamberg E (2003). Channelrhodopsin-2, a directly light-gated cation-selective membrane channel. *Proc Natl Acad Sci U S A* 100:13940-13945.

Nilsson DE & Pelger S (1994). A pessimistic estimate of the time required for an eye to evolve. *Proc Biol Sci* 256:53-58.

Ramdya P & Benton R (2010). Evolving olfactory systems on the fly. *Trends Genet* 26:307-316.

Reppert SM, Gegear RJ, & Merlin C (2010). Navigational mechanisms of migrating monarch butterflies. *Trends Neurosci* 33:399-406.

Salvini-Plawen LV & Mayr E (1977). On the evolution of photoreceptors and eyes. *Evol Biol* 10:207-263.

Seeholzer LF, Seppo M, Stern DL, & Ruta V (2018). Evolution of a central neural circuit underlies *Drosophila* mate preferences. *Nature* 559:564-569.

Spudich JL, Yang CS, Jung KH, & Spudich EN (2000). Retinylidene proteins: structures and functions from archaea to humans. *Annu Rev Cell Dev Biol* 16:365-392.

Evolution of nervous system structure and development

Carroll SB (2005) Endless Forms Most Beautiful. Norton.

Catania KC & Kaas JH (1995). Organization of the somatosensory cortex of the star-nosed mole. *J Comp Neurol* 351:549-567.

Halder G, Callaerts P, & Gehring WJ (1995). Induction of ectopic eyes by targeted expression of the eyeless gene in *Drosophila*. *Science* 267:1788-1792.

Hansen DV, Lui JH, Parker PR, & Kriegstein AR (2010). Neurogenic radial glia in the outer subventricular zone of human neocortex. *Nature* 464:554-561.

Holley SA, Jackson PD, Sasai Y, Lu B, De Robertis EM, Hoffmann FM, & Ferguson EL (1995). A conserved system for dorsal-ventral patterning in insects and vertebrates involving sog and chordin. *Nature* 376:249-253.

Jung H, Baek M, D'Elia KP, Boisvert C, Currie PD, Tay BH, Venkatesh B, Brown SM, Heguy A, Schoppik D, et al. (2018). The ancient origins of neural substrates for land walking. *Cell* 172:667-682 e615.

Scott MP (2000). Development: the natural history of genes. *Cell* 100:27-40.

Sharma J, Angelucci A, & Sur M (2000). Induction of visual orientation modules in auditory cortex. *Nature* 404:841-847.

Tosches MA, Yamawaki TM, Naumann RK, Jacobi AA, Tushev G, & Laurent G (2018). Evolution of pallium, hippocampus, and cortical cell types revealed by single-cell transcriptomics in reptiles. *Science* 360:881-888.

Human Brain Evolution

Charrier C, Joshi K, Coutinho-Budd J, Kim JE, Lambert N, de Marchena J, Jin WL, Vanderhaeghen P, Ghosh A, Sassa T, et al. (2012). Inhibition of SRGAP2 function by its human-specific paralogs induces neoteny during spine maturation. *Cell* 149:923-935.

McLean CY, Reno PL, Pollen AA, Bassan AI, Capellini TD, Guenther C, Indjeian VB, Lim X, Menke DB, Schaar BT, et al. (2011). Human-specific loss of regulatory DNA and the evolution of human-specific traits. *Nature* 471:216-219.

Pääbo S (2014). The human condition–a molecular approach. *Cell* 157:216-226.

Pollen AA, Bhaduri A, Andrews MG, Nowakowski TJ, Meyerson OS, Mostajo-Radji MA, Di Lullo E, Alvarado B, Bedolli M, Dougherty ML, et al. (2019). Establishing cerebral organoids as models of human-specific brain evolution. *Cell* 176:743-756.

CHAPTER 14

Ways of Exploring

Progress in science depends on new techniques, new discoveries, and new ideas, probably in that order.

Sydney Brenner, 1980

We have seen throughout this book how new techniques have led to the discovery of fundamental principles in neurobiology. In this final chapter, we discuss in greater detail some of the key techniques that have advanced our understanding of the nervous system. Studying how these techniques work will enable you to better understand the experiments discussed in this book and to apply these techniques to explore new terrain in neurobiology. We also discuss theory and modeling, as they have become increasingly important to neurobiology. I hope this chapter will also inspire some of you to invent new ways of exploring that will, in turn, bring new discoveries, new ideas, and new principles.

ANIMAL MODELS IN NEUROBIOLOGY RESEARCH

A major goal of neurobiology is to understand how the human brain works. Because the human brain is so complex and because our ability to perform well-controlled experiments in humans is limited for ethical reasons, most neurobiologists use animal models to conduct their research. Virtually every medical advance we have made comes from research done on animals or tissues derived from animals. Many principles identified in animal models apply widely across diverse nervous systems, including the human nervous system. Variations observed in different animal models (Figure 13-2) can be equally informative, as they reveal how the evolution of different nervous systems enabled animals to better adapt to their environmental niches.

What do researchers look for in animal models? As one researcher (William Quinn) put it, an ideal animal for neurobiology research "should have no more than three genes, a generation time of twelve hours, be able to play the cello or at least recite classical Greek, and learn these tasks with a nervous system containing only ten large, differently colored, and therefore easily recognizable neurons." Of course such an ideal animal does not exist, but this statement reflects the qualities that neurobiologists look for in an animal model: a compact genome and short generation time to facilitate gene manipulations and genetic studies; complex brain functions and behaviors that extrapolate results more easily to humans; and large, easily identifiable neurons, the activity of which can be recorded and manipulated individually or together to study the principles of information processing within the neural circuits they constitute.

Before discussing specific techniques, we first review the most commonly used animal models, upon which subsequent discussions of specific techniques are based.

14.1 Some invertebrates provide large, easily identifiable neurons for electrophysiological investigation

Recording electrical signals from individual neurons and manipulating their activity are essential for investigating the mechanisms by which the nervous system functions (Sections 14.20–14.25). The larger the neuron, the more easily researchers can record its activity by placing an electrode inside it. For example, the giant axon of the squid *Loligo* was used to discover the ionic basis of action potentials

(A)

5 cm

(B)

200 μm

Figure 14-1 Invertebrates with large neurons aid neurophysiological investigation. The hawkmoth *Manduca sexta* has been used as a model organism in olfactory and pheromone signaling because it has a superb sense of smell and large neurons, from which it is easy to obtain physiological recordings. **(A)** A nectar-feeding *Manduca*. **(B)** A local interneuron of *Manduca* arborizes its processes in the antennal lobe, the first olfactory processing center in the insect brain. The neuron has been filled with a fluorescent dye via an intracellular recording electrode. Compare the scale here with that of a similar neuron in *Drosophila* (Figure 14-26C). (A, courtesy of John Hildebrand and Charles Hedgcock, R.B.P B, from Reisenman CE, Dacks AM, Hildebrand JG [2011] *J Comp Physiol A* 197:653–665. With permission from Springer.)

(Sections 2.9 and 2.10). *Loligo* also offered giant synapses for intracellular recordings from its presynaptic terminals, which validated the role of Ca²⁺ entry in the control of neurotransmitter release (Section 3.4).

In addition to offering simple preparations for elucidating fundamental principles of neuronal communication, invertebrates have been used to investigate the mechanisms by which neural circuits process and store information. These studies take advantage of the relatively small number of neurons (when compared to vertebrate nervous systems), their large size, and their stereotyped arrangements. These properties enable electrophysiological recording and manipulation of neurons that are individually identifiable, allowing comparisons of experimental results across different animals. For example, studying the *Aplysia* gill-withdrawal reflex enabled the discovery that changes of synaptic connection strengths underlie behavioral habituation and sensitization (Section 11.15), and the lobster stomatogastric ganglion has been used to elucidate the mechanisms of central pattern generation underlying rhythmic movement (Section 8.5). Many other invertebrates, such as snails, leeches, locusts, cockroaches, and moths, have been used to probe the neural bases of sensation and motor control (**Figure 14-1**).

14.2 *Drosophila* and *C. elegans* allow sophisticated genetic manipulations

The two invertebrates we have encountered most often in this book are the fruit fly *Drosophila melanogaster* and the nematode *Caenorhabditis elegans*. These species are popular among neurobiologists because researchers can employ sophisticated genetic tools to manipulate their genes and neural populations with precision (Sections 14.6–14.12). In contrast to the invertebrate models discussed earlier, these animals do not offer researchers the benefits of large neurons; in fact, *C. elegans* and *D. melanogaster* have the smallest neurons of any animal model commonly used in neurobiology. Among animals with nervous systems of similar complexity, neuronal size usually correlates with the size of the animal, which is inversely correlated with its generation time. Model organisms for genetic research have been selected for short generation times—about 10 days for *Drosophila* and 3 days for *C. elegans*; hence their small bodies and neurons.

Drosophila has served as a genetic model organism for over a century. Research first conducted in *Drosophila* laid the foundation for many fundamental concepts in genetics, such as the nature of genes, mutations, chromosomes, and the basis of linkage mapping. *Drosophila* has on the order of 10⁵ neurons, considerably fewer than the mouse (~10⁸ neurons) or human (~10¹¹ neurons), but a number sufficient to mediate sophisticated neural computations and behaviors. Studies in *Drosophila* can also be compared with studies in other insects such as moths, honeybees, ants, locusts, and mosquitoes that act as pollinators or pests in agriculture or vectors of human diseases; insects constitute the most diverse order in the animal kingdom.

Sydney Brenner, the author of this chapter's epigraph, introduced *C. elegans* in the 1960s for the purpose of studying the nervous system and behavior of a simple organism. *C. elegans* has since been used in many fields of biological research, contributing to fundamental discoveries such as programmed cell death and RNA interference. Not only is *C. elegans* well suited for genetic manipulation, its transparent body is also advantageous for developmental and imaging studies. *C. elegans* is the only organism for which the entire **connectome**—the complete set of synaptic connections linking its 302 neurons—has been deciphered using serial electron microscopy (**Figure 14-2**); this advance has guided developmental biology and neural circuit research (e.g., Section 6.10).

14.3 Diverse vertebrates offer technical ease or special faculties

Cold-blooded vertebrates, including fish, amphibians, and reptiles, are useful for producing robust explant preparations in which to study neurobiological problems. Unlike mammalian tissues, these *in vitro* preparations often do not require constant temperature and oxygenation to maintain tissue integrity. As vertebrates,

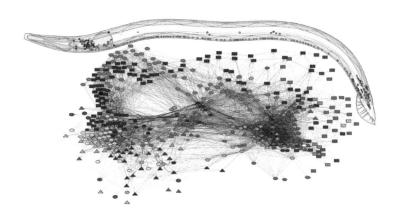

Figure 14-2 **The wiring diagram of *C. elegans*.** Top, side view of a *C. elegans* male with positions of neuronal nuclei depicted. For each bilaterally symmetric pair, only one is depicted. Bottom, schematic of the wiring diagram. Gray lines, chemical synapses; red lines, electrical synapses; triangles, sensory neurons; hexagons, interneurons; circles and ovals, motor neurons; rectangles, muscles. (From Cook SJ, Jarrell TA, Brittin CA, et al. [2019] *Nature* 571:63–71. With permission from Springer Nature. See also White JG, Southgate E, Thomson JN, et al. [1986] *Phil Trans R Soc Lond B* 314:1–340.)

their nervous systems share organizational similarities with the human nervous system that are not found in invertebrate models. Studies in fish and amphibian models have contributed to many fundamental discoveries in neurobiology, such as those regarding wiring specificity in the retinotectal system (Sections 5.1 and 5.2) and the mechanisms of synaptic transmission (Sections 3.1 and 3.2). Zebrafish (*Danio rerio*) have in recent decades become a popular vertebrate model organism due to their transparency in the larval stage, which facilitates developmental and imaging studies (**Figure 14-3**). Whole-brain imaging of Ca^{2+} activity (see Section 14.22) has been performed in behaving zebrafish larvae. Furthermore, the zebrafish's relatively short generation time is well suited for genetic studies.

While some animal models have been selected for technical ease, others have been chosen for their special faculties. One principle of neuroethology is to select a model animal in which the behavior of interest is robustly displayed. For instance, barn owls have been used to study audition because of their superb ability to localize sounds (Sections 1.3, 6.24, and 11.25). Songbirds have been used to study vocalization and learning because they have advanced vocal communication systems and an elaborate song-learning process (Box 10-1). Uncovering the neural mechanisms underlying a particular behavior in a species well suited to its study can benefit researchers who study the same behavior in other animals.

14.4 Mice, rats, and nonhuman primates are important models in mammalian neurobiology research

Among mammalian species, rats and mice have been the predominant animal models for many branches of biology, including neurobiology. A major advantage of mice is the ease of genetic manipulation; a large number of transgenic and knockout mice are available for analyzing gene function and for genetic access to neuronal populations for activity recording and manipulation (Sections 14.6–14.12). Rats have been used as models in neurobiology research longer than mice. Many behavioral paradigms such as operant conditioning (Figure 11-22) were first developed in rats. Genetic tools first developed in mice are now being extended

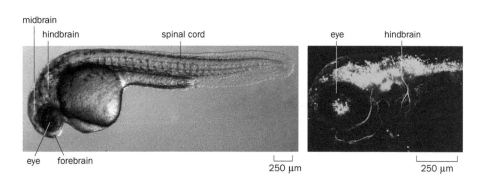

Figure 14-3 **The transparency of zebrafish larvae facilitates developmental and imaging studies.** Left, differential interference contrast microscopic image of a live zebrafish embryo at 36 h postfertilization. Major nervous system structures are indicated. Right, GFP (green) expression in neurons of a live zebrafish at 3 days postfertilization. (Left, from Schier AF & Talbot WS [2005] *Ann Rev Genet* 39:561–613. With permission from Annual Reviews. Right, courtesy of Thomas Glenn & William Talbot.)

to rats, and many physiological and behavioral paradigms first used in rats have been adapted to mice.

In addition to studying intact animals, reduced preparations from mice and rats have been widely used in neurobiology research. For instance, neurons can be dissociated and cultured *in vitro* to study a wide range of topics, such as the development of neuronal polarity (Figure 7-17) and the molecular mechanisms of synapse formation (Figure 7-26). Acute or cultured brain slices have been widely used to study neuronal connectivity (Figure 3-49; Figure 4-46), electrical signaling (Figure 3-44), synaptic transmission (Figure 3-23), and synaptic plasticity (Figure 11-10). These *in vitro* preparations allow for easy experimental manipulations, such as performance of patch clamp recordings of multiple neurons and simultaneous control of the extracellular environment.

Compared with mice and rats, nonhuman primates, such as macaque monkeys, have brain structures (Figure 13-36), gene expression patterns, and physiology more similar to those of humans; likewise, their cognitive abilities are superior to those of rodents. Many sophisticated psychophysical and cognitive tasks, such as decision making (Figure 4-55), were first developed in primate models. The visual systems of trichromatic Old World monkeys and apes are very similar to our own (Figure 4-16). Nonhuman primates are also valuable models for human disease studies and drug testing, because, among all animals, their physiology is more similar to that of humans. In additional to the widely used macaques, the common marmoset—a small New World primate—has recently emerged as a model organism because of its short generation time, sophisticated social behavior, and smooth neocortex (allowing for easier optical imaging of neuronal activity than the gyrencephalic macaque cortex).

When working with animals, researchers are obligated to follow rigorous ethical practices; these include replacing animals with nonanimal systems whenever possible, using the smallest number of animals necessary to obtain the desired information, and using all available methods to minimize pain and distress to animals being used in research. These practices apply particularly to vertebrates, whose proper use is regulated by governments and research institutions.

14.5 Human studies are facilitated by a long history of medicine and by the recent genomic revolution

Medicine has provided many case studies in human neurobiology and neuropathology, thus contributing uniquely to our understanding of the nervous system. Lesions due to injury in patients provided evidence for the existence of language centers in the human brain (Figure 1-23). Electrophysiological recordings of epilepsy patients elucidated the topographic organization of the sensory and motor cortices (Figure 1-25). Studies of amnesic patients revealed the existence of multiple functionally distinct memory systems and their locations in the brain (Section 11.1). Likewise, experimental psychology using healthy human subjects has contributed substantially to our understanding of perception (Figure 4-3), cognition, and behavior. Functional brain imaging studies have greatly improved our understanding of normal human brain organization (Figure 1-24; Figure 11-38) and can be used to monitor disease progression (Figure 12-11) and therapeutic efficacy (Figure 12-21). Genetic variation among humans has helped researchers identify genes essential for basic neurobiological processes (Section 6.17). Mutations that cause brain disorders have made important contributions to our knowledge of normal nervous system development and function (Chapter 12). The development of induced pluripotent stem cell and brain organoid technologies (Box 7-1) has opened new horizons for studying basic and disease-related neurobiology.

With the sequencing of the human genome now complete and the cost of sequencing individual genomes becoming readily affordable (Box 14-2), we can anticipate a wealth of data correlating genetic variants with phenotypes ranging from brain disorders to personality traits. These data provide entry points to many fascinating areas of neurobiological investigation.

GENETIC AND MOLECULAR TECHNIQUES

Genes, the basic functional units of the genome, encode the RNAs and proteins that execute all cellular functions. Many biological processes can be viewed as the consequences of a series of actions by individual genes. Thus, by manipulating individual genes, one can dissect complex biological processes into discrete steps. This genetic approach has made fundamental contributions to all branches of biology, including neurobiology. While gene-centric approaches have been used more widely in molecular and cellular neuroscience, genetic frameworks have also been used to study individual neuronal cell types and have thus become instrumental to investigation of problems in circuit, systems, and behavioral neuroscience.

The most fundamental genetic manipulation is disruption of the function of a gene—that is, creation of a **loss-of-function mutation** in a gene of interest without affecting other genes in the genome. Researchers have taken two general approaches to link a gene with its function inferred from loss-of-function phenotypes: forward genetics, which traces an observed phenotype to a gene, and reverse genetics, which follows a gene to its associated phenotype (**Figure 14-4**).

14.6 Forward genetics uses random mutagenesis to identify genes that control complex biological processes

The **forward genetic screen** has provided key insights into many complex biological processes, from cell division and protein secretion to development of multicellular organisms. Forward genetic screens employ **random mutagenesis** to identify genes involved in a biological process of interest. Suppose that several unknown genes play essential roles in the process of interest. Researchers use chemical mutagens, radiation, or transposon insertion (insertion of a transposable DNA element that disrupts the function of a gene) to mutagenize a population of animals, such that each treated animal carries a different set of random mutations in a small number of genes or a single gene. Researchers then screen for mutations that disrupt the biological process of interest based on phenotypes exhibited by the offspring of the mutagenized animals (**Figure 14-5**).

The mutated gene that causes the phenotype can be traced using a variety of molecular-genetic methods, depending on the nature of the mutagen. Mutations caused by transposon insertions can readily be mapped by identifying the DNA sequences neighboring the insertion sites. Mutations induced by chemicals or radiation can be mapped by molecular-genetic procedures such as **positional cloning**. In this strategy, many meiotic recombinant chromosomes are produced,

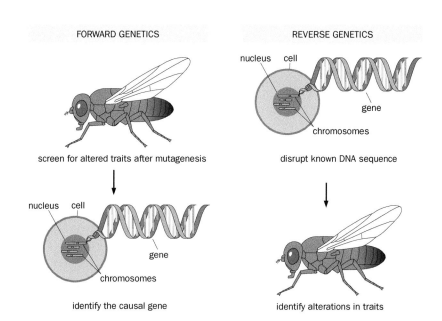

FORWARD GENETICS

screen for altered traits after mutagenesis

identify the causal gene

REVERSE GENETICS

disrupt known DNA sequence

identify alterations in traits

Figure 14-4 Forward and reverse genetics. In forward genetics, researchers start by observing a phenotype and then identify the gene, the alternation of which causes the phenotype. In reverse genetics, researchers start with a gene of interest and disrupt its function to examine the phenotypic consequences.

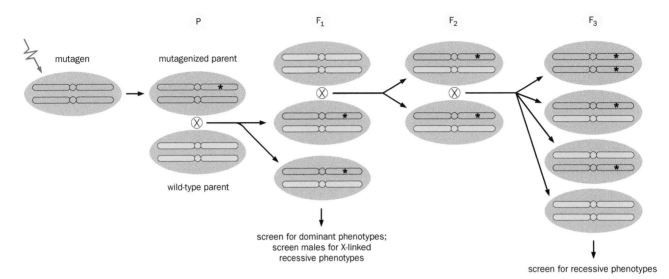

P   F₁   F₂   F₃

Figure 14-5 Forward genetic screens for identifying single-gene mutations that cause specific phenotypes. After mutagen treatment, the mutagenized individual is crossed with a wild type individual (P, for parents); the mutation is indicated by * on the chromosome. Individual progeny from the next generation (F₁, for first filial generation) can be screened directly for mutations that exhibit dominant phenotypes (bottom), or crossed with wild-type individuals to produce a larger population of progeny (F₂) heterozygous for the mutation, from which homozygous mutant progeny (F₃) can be bred and screened for recessive phenotypes (top). Because males have only one X chromosome, they can be screened for X-linked recessive traits in F₁ instead of F₃. Note that for simplicity only the subset of progeny from each cross relevant to the propagation of the mutagenized chromosome is drawn out in the F₁ and F₂ generations.

and the linkage between the mutant phenotype and genetic or molecular markers with known positions in the genome is used to identify where the mutated gene resides. The closer the mutation is to a particular marker, the less frequently the mutation and the marker will be separated following recombination events. The causal gene can be validated by identifying the disruptive mutation in the candidate gene and by rescue of the mutant phenotype via a wild-type **transgene**. (A transgene is an engineered gene that is introduced into an organism; see Section 14.9 for more details). With the development of high-throughput genome sequencing (Box 14-2), researchers can also compare the whole-genome sequences of mutants and wild-type controls to identify the causal genes.

Forward genetic screens are particularly powerful when studying processes for which the cellular and molecular pathways are poorly understood. Researchers rely on mutant phenotypes to identify genes involved in a particular biological process without bias or foreknowledge as to what kinds of genes are involved. The identification of the *Drosophila Period* gene and mouse *Clock* gene provide striking examples of how forward genetic screens have led to our current understanding of the molecular mechanisms controlling circadian rhythms (Figure 9-21). The same procedures used to identify mutant phenotypes and their causal genes resulting from mutagenesis can also be applied to mutations that arise spontaneously, as in the case of shaking flies (Section 2.15), obese mice (Section 9.6), narcoleptic dogs (Section 9.15), and inherited human disorders (Chapter 12).

14.7 Reverse genetics disrupts predesignated genes to assess their functions

We now discuss **reverse genetics**, a term that encompasses strategies for disrupting predesignated genes (Figure 14-4, right). Many molecular components of the nervous system were identified by means other than mutant phenotypes arising from forward genetic screens or spontaneous mutations. For example, genes encoding Na⁺ channels, synaptotagmins, and ephrins were first identified by biochemical purification of their protein products enriched in the electric organ, presynaptic terminal, and developing tectum, respectively (Sections 2.15, 3.6, 5.4, and 14.14). The TRP channels for sensing temperature were identified from expression cloning (Section 6.29). Most ion channels and neurotransmitter receptors were first identified based on their sequence homology with known proteins with similar functions. With genome sequences completed for most model organisms, researchers can search databases to identify candidate genes that might perform certain functions based on expression patterns and predicted protein sequences. A key approach for testing the function of a candidate gene in a biological process is to create loss-of-function mutations and examine the phenotypes of the resulting mutants.

The most widely used method for disrupting a specific gene of interest is via **homologous recombination**, in which a piece of endogenous DNA essential for the function of a gene is replaced by a piece of engineered DNA, the ends of which have sequences identical (hence the term *homologous*) to the endogenous DNA. Homologous recombination is an intrinsic process essential for meiosis in germline cells; it also occurs efficiently in other cell types, including embryonic stem (ES) cells. In multicellular animals, homologous recombination-based gene disruption, known as gene **knockout**, was first developed in mice (**Figure 14-6**). The first step is to engineer a DNA construct carrying a drug-resistance gene flanked on both sides by pieces of DNA (homology arms) derived from the endogenous gene of interest. This construct is then introduced into ES cells, where recombination at both homology arms causes the replacement of an essential part of the gene of interest with the drug-resistance gene (Figure 14-6A). ES cell clones carrying the knockout allele are identified based on their drug resistance, expanded, and injected into blastocyst-stage host embryos. These embryos are then implanted into surrogate mothers, where they develop, producing chimeric pups in which a fraction of cells derives from the *in vitro* engineered ES cells. (A chimera contains some cells with the genotypes of injected ES cells and others with the genotype of the host embryo.) If germline cells in the chimera are derived from injected ES cells, subsequent breeding with wild-type animals can generate

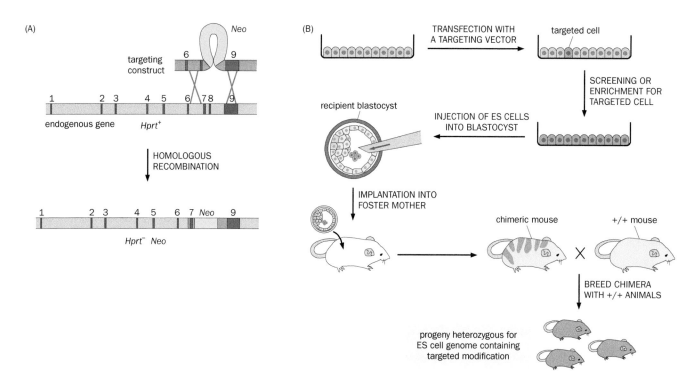

Figure 14-6 Gene knockout in mice. (A) Homologous recombination in ES cells causes the replacement of an essential part of a target gene with a marker gene encoding drug resistance (in this case *Neo*, for resistance to neomycin). Here, *Neo* replaced the sequences corresponding to exon 8 of the *Hprt* gene, which encodes an enzyme for nucleotide biogenesis. Homologous recombination at two crosses results in deletion of the DNA segment corresponding to exon 8. The resulting recombinant chromosome is deficient for *Hprt* and confers neomycin resistance. Light gray and light red: homologous introns of the *Hprt* gene; dark gray and dark red, homologous exons of the *Hprt* gene. **(B)** Modified ES cells can be used to create knockout mice by following the steps in this flow chart. Modified ES cells and their derivatives are red, while cells of the host blastocyst are blue. Note that the final product in the scheme is heterozygous for the modified ES cell genome but appears red because the marker of the modified ES cell (coat color) is dominant. (Adapted from Capecchi MR [1989] *Science* 244:1288–1292.)

offspring in which *all* cells carry the knockout allele (Figure 14-6B), and further breeding of the offspring yields mice homozygous for the knockout allele.

Since the knockout procedure was established in the 1980s, many variations and extensions have made this technique more versatile. For instance, instead of disrupting a gene, single nucleotide changes can be made to test the contribution of specific amino acid residues to protein function *in vivo* (Figure 3-9B); inserting engineered constructs into predetermined genomic loci is also possible. Both procedures are called **knock-in**. Among its many uses, knock-in mice can express a marker gene in a spatiotemporal pattern defined by an endogenous gene's *cis*-regulatory elements (Figure 6-12); we will discuss many applications of this technology in subsequent sections.

Homologous recombination techniques have been used to disrupt genes in flies and rats, in addition to mice. The rate-limiting step is screening for rare recombination events; this has been facilitated in mice and rats by developing ES cell culture, so that such screens can be performed *in vitro*. In *Drosophila,* the homologous recombination procedure has been sufficiently streamlined to permit screening recombination events directly *in vivo*. For most model organisms, however, techniques for gene disruption via homologous recombination have not been established. The recent development of genome editing tools has enabled genetic manipulations, such as the production of knockout and knock-in animals, in species other than the traditional genetic model organisms (**Box 14-1**).

An alternative to gene knockout for examining loss-of-function phenotypes is **RNA interference** (**RNAi**). RNAi takes advantage of the naturally occurring process of degrading double-stranded RNAs (dsRNAs) for defense against viral invaders. This process utilizes a cascade of RNA-processing enzymes for the production of endogenous **microRNAs**, which are short, noncoding RNAs (21–26 nucleotides) used to regulate gene expression by triggering the degradation and inhibiting the translation of mRNAs with complementary sequences. In RNAi-mediated experimental gene silencing, dsRNAs can be produced by base pairing of sense and antisense transcripts with sequences corresponding to a target gene of interest (**Figure 14-7**A). Alternatively, researchers can express a transgene encoding the region homologous to the target gene of interest in an inverted repeat (a sequence followed by its reverse complement). Because the two halves of the repeat can base pair with each other, the transgene's RNA product folds into a hairpin, forming a dsRNA substrate for further processing (Figure 14-7B). Both approaches utilize the cell's microRNA-production machinery, which cleaves dsRNA to produce **siRNA**—double-stranded short interfering RNA, with a length of 21–26 nucleotides. The siRNA causes the degradation or translation inhibition of the target mRNA through base pairing.

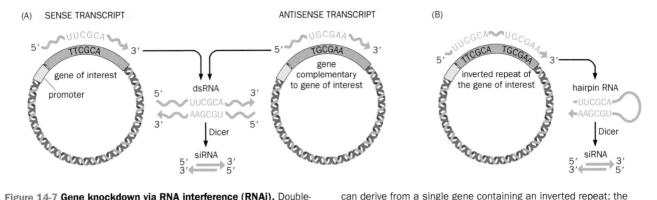

Figure 14-7 Gene knockdown via RNA interference (RNAi). Double-stranded small interfering RNAs (siRNAs) cause degradation and translation inhibition of target mRNAs bearing the same sequence and therefore can be used to knockdown the expression of an endogenous gene. **(A)** siRNAs can derive from two genes encoding sense and antisense RNA transcripts from the same DNA sequence. **(B)** siRNAs can derive from a single gene containing an inverted repeat; the transcript folds back onto itself and the complementary sequences base pair to produce a hairpin. In both cases, double-stranded RNA molecules are cleaved and processed by enzymes (such as Dicer) in the microRNA-processing pathway to produce siRNA. For the original discovery of RNAi in *C. elegans*, see Fire et al. (1998) *Nature* 391:806.

Since inhibition of gene expression by RNAi may be incomplete, this procedure is referred to as gene *knockdown* rather than knockout. RNAi also has the potential for off-target effects due to unintended targeting of similar sequences. Proper controls, such as the use of multiple and nonoverlapping target sequences or rescue of RNAi phenotypes by expression of an RNAi-resistant transgene (a transgene that does not contain sequences complementary to those of the dsRNA), are necessary. The advantage of RNAi over gene knockout is its relative ease and potential for high-throughput screening; this has enabled RNAi to be employed not only in reverse genetics but also in forward genetic screens (Section 14.9). Candidate genes identified via RNAi screens are often subsequently validated by gene knockout.

Box 14-1: Genome editing with the CRISPR-Cas9 system

Genome editing refers to the process of altering the genome at predetermined loci, whether by deleting a piece of endogenous DNA, inserting a piece of foreign DNA, or changing specific base pairs. The knockout and knock-in procedures discussed in Section 14.7 are genome editing strategies that employ homology arms to guide alterations using the homologous recombination process intrinsic to germline and embryonic stem cells. An alternative strategy is to induce, at a genomic locus of interest, double-strand DNA breaks that activate endogenous DNA repair systems, and, in so doing, introduce sequence alterations. In genome editing, double-strand breaks are typically induced by DNA-sequence-specific targeting of exogenous nucleases such as zinc finger nucleases (ZFNs), transcription activator-like effector nucleases (TALENs), or the most recently developed and most versatile system for this approach: the CRISPR-Cas9 system.

Discovered in the first decade of the 2000s, **CRISPR** (clustered regularly interspaced short palindromic repeat) is an adaptive immune system present in many bacteria and archaea. CRISPR is a genomic locus containing repetitive DNA elements derived from the genomes of invading pathogens such as viruses or plasmids. These DNA repeats are then made into small RNA molecules that guide nucleases to degrade the genomes of the invading pathogens through sequence-specific base pairing. Thus, bacteria previously exposed to a pathogen can rapidly defend against future infection by the same pathogen. Since the modification occurs at the level of genomic DNA, this anti-pathogen trait is inherited by progeny; this constitutes a rare case in which directed changes in DNA sequences can contribute to natural selection (Section 13.3).

Although several variants of the CRISPR system exist, the type II system present in bacteria such as *Streptococcus pyogenes* utilizes a single protein called **Cas9** (CRISPR-associated 9) with two nuclease domains that cut both DNA strands to produce a double-strand break (Figure 14-8). Cas9 is brought to a specific sequence on the target DNA via a **guide RNA**, which contains a sequence that base pairs with the DNA target (part of the guide RNA is normally transcribed from the CRISPR locus). Double-strand breaks created by the CRISPR-Cas9 system can be repaired by the **nonhomologous end joining** system in the absence of homologous

DNA sequences serving as a template. Such repairs usually introduce small deletions or duplications at the break point; if the breaks occur in the coding sequence, repairs have a two-thirds chance of creating a frame-shift mutation that disrupts the protein-coding sequence after the breakpoint. Double-strand breaks can also be repaired by homologous recombination, which utilizes donor DNA that shares sequence identity on both sides of the break as a template; homologous recombination-based repair can produce arbitrary changes to the DNA sequence, from single-base-pair changes to insertions of a transgene, at predetermined genomic loci (Figure 14-8).

The CRISPR-Cas9 system has been shown to target double-strand DNA breaks to specific DNA sequences in human cell lines, induced pluripotent stem cells (Box 7-1), and germline cells of a growing list of organisms including *C. elegans, Drosophila,* zebrafish, mouse, and monkey. The efficiency is remarkably high, such that multiple guide RNAs can be injected into the same pronuclei of fertilized mouse eggs (zygotes) to create mutations in both copies of multiple genes simultaneously, without requiring ES cell culture, transfection, screening, and injection into blastocysts (Figure 14-6). Cas9 can be expressed from a transgene; alternatively, Cas9 mRNA or proteins can be co-injected with guide RNAs. The CRISPR-Cas9 system has also been used to create large deletions between sequences targeted by two guide RNAs and to insert transgenes such as fluorescent proteins. One limitation is the potential for off-target effects due to the presence of sequences elsewhere in the genome similar to the intended target, although multiple techniques have been developed to minimize off-target effects. Another limitation of homologous recombination-mediated repair is that homologous recombination does not occur efficiently in postmitotic cells including neurons.

The CRISPR-Cas9 system has been extended in many new directions. The high efficiency of CRISPR-Cas9-mediated genome editing has enabled genome-wide screens using guide RNA libraries in cell-based assays. Furthermore, nuclease-impaired Cas9 variants can be fused to a transcription activator or repressor, and these fusion proteins can target specific DNA sequences recognized by guide RNAs to activate or repress corresponding genes. In a process called base editing,

(*Continued*)

Box 14-1: continued

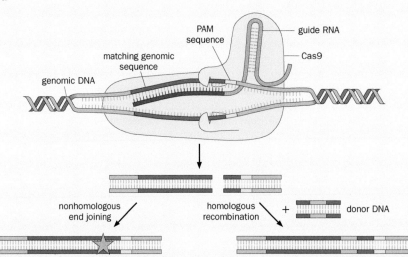

Figure 14-8 The CRISPR/Cas9 system for genome editing. Any eukaryotic DNA containing a PAM sequence (protospacer-adjacent motif, which is usually two or three nucleotides and thus occurs frequently) can be targeted by the CRISPR-Cas9 system. A guide RNA containing sequences complementary to a piece of DNA from the target gene of interest brings the Cas9 enzyme to the target site on the chromosome through DNA–RNA base pairing (purple and red). The two nuclease domains of Cas9 create a double-strand break in the genomic DNA. This break can be repaired by nonhomologous end joining, through which small deletions or insertions may be created at the repair site (indicated by the star). The break can also be repaired by homologous recombination utilizing donor DNA as a template, through which specific modifications, such as insertion of a transgene (green), can result. (Adapted from Charpentier E & Doudna JA [2013] *Nature* 495:50–51. With permission from Springer Nature. See also Ran FA, Hsu PD, Wright J, et al. [2013] *Nat Protocol* 8:2281–2308.)

nuclease-impaired Cas9 is fused to an enzyme that modifies specific nucleotides (for example, causing C → T substitutions) without generating double-strand breaks. While greatly accelerating basic research and therapeutic applications (for instance, treating human genetic diseases via gene ther-apy; Box 12-2), the ease of editing genomes (including those of humans) has raised considerable ethnical issues and calls for societal regulation of genome editing in human germline cells.

14.8 Gene disruption can be spatially and temporally controlled

In multicellular organisms, determining in which cells and at what times the function of a gene is required for a biological process can provide valuable information about the gene's mechanism of action. The general procedures discussed for generating loss-of-function mutations by random mutagenesis or gene targeting require breeding for homozygous mutants. As the gene is disrupted in all cells, these methods cannot pinpoint when and where the gene of interest contributes to developmental, cellular, and circuit functions or animal behavior. Fortunately, multiple powerful methods have been developed to address such questions.

A key extension of the mouse knockout technique (Figure 14-6) is the production of **conditional knockout** mice, first developed using the bacteriophage Cre/*loxP* system. **Cre recombinase** is a bacteriophage-derived enzyme that catalyzes recombination of two sequence-specific DNA elements called ***loxP*** sites. When two *loxP* sites are in the same orientation, a recombination event will delete the intervening sequence; when two are in opposite orientations, a recombination event will invert the intervening sequence. In conditional knockouts, two *loxP* sites are inserted in the same orientation by homologous recombination into introns flanking essential exon(s) of a gene. An allele in which essential exons are located between two *loxP* sites is called a "floxed" allele (for flanked by *loxPs*). In the absence of Cre-mediated recombination, these *loxP*-containing introns are spliced out of RNA transcripts and do not affect gene expression. The gene can be knocked out—the floxed exon(s) excised—only in cells in which Cre has been active (**Figure 14-9**). Researchers have generated hundreds of transgenic Cre

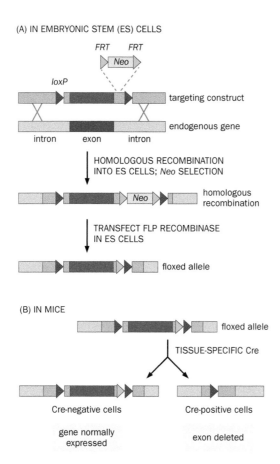

(A) IN EMBRYONIC STEM (ES) CELLS

FRT   FRT

Neo

loxP

targeting construct

endogenous gene

intron   exon   intron

HOMOLOGOUS RECOMBINATION
INTO ES CELLS; Neo SELECTION

Neo   homologous
recombination

TRANSFECT FLP RECOMBINASE
IN ES CELLS

floxed allele

(B) IN MICE

floxed allele

TISSUE-SPECIFIC Cre

Cre-negative cells   Cre-positive cells

gene normally
expressed   exon deleted

Figure 14-9 Conditional knockout in mice.
(A) Floxed allele production in ES cells. In the targeting construct, a pair of *loxP* sites is inserted into two introns flanking an essential exon of interest. In addition, the *Neo* gene, flanked by a pair of *FRT* sites, is inserted into one of the introns. Recombination at crosses between homologous sequences produces the desired recombinant after neomycin selection. Subsequent transient expression of FLP recombinase induces recombination between the two *FRT* sites, thus removing the neomycin resistance gene to produce the floxed allele. Because the two *loxP* sites and one *FRT* site are all located in the intron, the floxed allele does not affect the expression of the target gene of interest. **(B)** In mice bearing the floxed allele and a transgene expressing Cre recombinase, cells that have not expressed Cre are unaffected, but the essential exon is removed by Cre/*loxP*-mediated recombination in cells in which Cre has been active, thus generating conditional knockout of the gene of interest.

lines with different spatiotemporal Cre expression patterns, such that gene deletion can be achieved in specific cell types and only after Cre is expressed. In addition to the Cre/*loxP* system, other site-specific recombinase systems are used. For instance, the yeast **FLP recombinase** mediates recombination of two ***FRT*** (FLP recognition target) sites through a mechanism analogous to Cre-mediated recombination of *loxP* sites.

An important extension of conditional knockout technology was the engineering of Cre so that its activity could be temporally controlled by drug application. Temporal control usually occurs by regulating translocation of Cre into the nucleus, where recombination takes place. As discussed in Section 10.10, the estrogen receptor (ER) normally remains in the cytoplasm but translocates into the nucleus in the presence of estradiol. **CreER** is a fusion of the Cre recombinase and the portion of the ER responsible for cytoplasmic retention. Like the endogenous ER, CreER remains in the cytoplasm when unbound by its ligand but translocates into the nucleus in the presence of the estrogen analog **tamoxifen** (see Figure 10-19; Figure 14-14A). Thus, the CreER/tamoxifen system allows temporal control of recombination of floxed alleles, and hence control of the time at which an endogenous gene is deleted.

Another method for spatiotemporal control of gene is to create genetic mosaics using **mitotic recombination**. In this procedure, DNA recombination occurs between two homologous parental chromosomes in a somatic cell, such that one of the daughter cells can be homozygous for part of one parental chromosome. If an animal is heterozygous for a recessive mutation in a gene of interest and is thus phenotypically normal, one daughter cell (and all its descendants) can be made homozygous for the mutation and thus phenotypically mutant (**Figure 14-10**); the resulting animal is called a **genetic mosaic**—that is, an animal containing cells of more than one genotype.

If cells of distinct genotypes can be differentially labeled, then phenotypic analysis of such mosaic animals can provide information about whether the gene

Figure 14-10 Mitotic recombination can create genetic mosaics. A pair of homologous chromosomes is shown. The parent cell is heterozygous for a recessive mutation (*) in the gene of interest. If DNA recombination occurs between the homologous chromosomes (red cross) following DNA replication, chromosomal segregation in the subsequent cell division can create daughter cells homozygous for either the mutant (left) or wild-type (right) alleles.

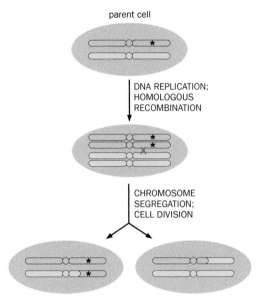

of interest acts **cell autonomously** (only within the cell that produces the gene product) or **nonautonomously** (on cells that do not produce the product) to regulate a given biological process. We have seen examples of genes that act either way in our studies of cell fate determination (Figure 5-35), wiring specificity (Figure 5-28; Figure 5-37; Figure 7-44), and mating behavior (Figure 10-9). The rates of mitotic recombination are naturally very low, but can be markedly enhanced by X-ray irradiation or by introducing into the genome a recombinase and its recognition sites, such as the Cre/*loxP* or FLP/*FRT* systems discussed earlier. When two recombinase recognition sites are at identical locations on two homologous chromosomes in the same orientation, a recombination event between the two sites produces recombination of the two chromosomes (see Figure 14-26 for an illustration). Site-directed recombinases also enable spatiotemporal control of mitotic recombination via the control of recombinase expression. Because the efficiency of recombination between two chromosomes (as in mitotic recombination) is usually lower than within the same chromosome (as in conditional knockouts), mitotic recombination-based genetic mosaics usually produce more sparse knockout cells, which are particularly useful for determining cell-autonomy of gene action. These methods can be used to label single cells, trace cell lineages, and access specific neuronal populations for genetic manipulation (Section 14.16).

While these procedures are most applicable to genetic model organisms such as mice and *Drosophila*, RNAi-mediated knockdown (Section 14.7) and CRISPR-Cas9-mediated genome editing (Box 14-1) can be used for conditional gene disruption in any organism conducive to transgene expression. Transgenes expressing inverted repeats to produce RNAi effects against a specific gene of interest can be expressed from cell-type-specific promoters (discussed in detail in Section 14.9), such that gene disruption is restricted to specific cell types. Likewise, transgenes expressing Cas9 (or guide RNAs or both) to disrupt a specific gene can also be driven from cell-type-specific promoters. Finally, viral transduction and other transient methods (Section 14.10) can be used to deliver transgenes that mediate RNAi or CRISPR-Cas9-based gene editing to specific tissues through targeted injections.

14.9 Transgene expression can be controlled in both space and time in transgenic animals

The ability to introduce engineered DNA sequences into an organism—to create a **transgenic organism**—has revolutionized biology. Transgenic animals serve two broad purposes in neurobiology research. The first is to examine the function of an endogenous gene *in vivo*. For instance, expression of a transgene encoding a wild-type protein can be used to rescue a loss-of-function mutant phenotype and thereby confirm a causal relationship between the disruption of a gene and a given phenotype. Expression in defined spatiotemporal patterns of a transgenic hairpin construct that produces RNAi effects against a specific mRNA (Figure 14-7B) can be used to assess gene knockdown phenotypes. The gene of interest can also be misexpressed at different levels or in different spatiotemporal patterns to assess **gain-of-function** effects. Finally, transgenes with specific modifications can be used *in vivo* in both gain- and loss-of-function contexts to assess the structure–function relationships of the gene of interest (i.e., which specific domains or amino acids are required for a given function).

The second broad application of transgenes is to express molecular tools, such as cell markers for visualizing neuronal morphologies and projection patterns, Ca^{2+} or voltage indicators for recording neuronal activity, and effectors such as light- or chemical-activated channels for silencing or activating neuronal activity (Sections 14.16, 14.18, and 14.21–14.25).

In both applications, researchers must control the transgene's spatiotemporal expression pattern. A protein-coding gene usually consists of enhancer/promoter elements that direct the spatiotemporal expression pattern of a transcription unit consisting of a 5′-untranslated region (UTR), a coding sequence that dictates the production of a specific protein, and a 3′-UTR including a poly-adenylation (poly-A) signal to regulate mRNA stability and nuclear export (Figure 2-2). Various mechanisms can be employed to mimic the expression pattern of an endogenous gene or create an artificial expression pattern. One simple method for driving transgene expression in a particular pattern is to use the DNA sequence 5′ to the coding sequence of an endogenous gene whose expression pattern is being mimicked (**Figure 14-11**A).

Transgenic animals are usually created by injecting DNA into early embryos (for example, pronuclei in single-celled mammalian embryos); the injected transgene integrates randomly into a host chromosome(s). Expression of randomly integrated transgenes can be subject to the influence of endogenous regulatory sequences near the integration sites, causing variable transgene expression patterns in different transgenic lines. Site-specific integration of transgenes, in which an integrase catalyzes insertion of a transgene into a predetermined genomic locus via DNA recombination, offers greater consistency (Figure 14-11B). The most faithful mimicry of endogenous gene expression patterns is achieved by inserting the transgene into the genomic locus of the endogenous gene via knock-in (Section 14.7; Figure 14-11C). Historically, knock-ins were achievable in only a few organisms, including flies and mice. However, the genome-editing tools discussed in Box 14-1 allow knock-ins in any organism in which transgenesis is possible.

In a strategy called **binary expression**, the regulatory and protein-coding components of a gene of interest can also be expressed separately via two transgenes, a **driver transgene** and a **responder transgene**. For instance, regulatory elements drive expression of the yeast transcription factor **GAL4** in a driver transgene, while in a responder transgene, a **UAS** (upstream activating sequence by GAL4) controls the expression of the coding sequence of a gene. When the two transgenes are present in the same animal, the gene of interest will be expressed in the specific cells expressing GAL4 (**Figure 14-12**A). This GAL4/*UAS* binary expression system is widely used in *Drosophila* (e.g., Figure 10-6) and zebrafish. A binary expression

Figure 14-11 Methods for regulating transgene expression patterns. (A) The transcription unit of interest (*T*, brown) is placed under the control of an enhancer/promoter *(E/P)* element, and the resulting transgene is randomly integrated into a host chromosome (gray). **(B)** Transcription unit *T*, under the control of an enhancer/promoter, is integrated at a predetermined locus in the host genome by expressing a bacteriophage integrase that catalyzes an irreversible recombination between the *attP* and *attB* sites. **(C)** *T* is knocked in at the endogenous locus of the gene whose expression pattern is being mimicked (Section 14.7). (Adapted from Luo L, Callaway EM, & Svoboda K [2008] *Neuron* 57:634–660. With permission from Elsevier Inc.)

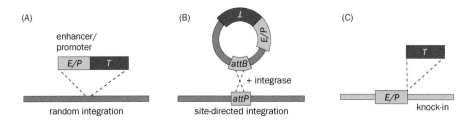

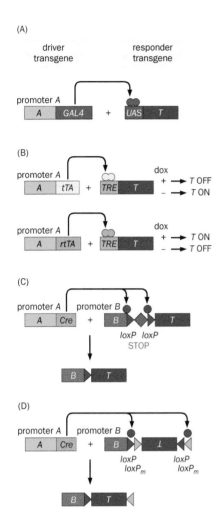

(A)

driver transgene   responder transgene

promoter A

| A | GAL4 | + | UAS | T |

(B)

promoter A

| A | tTA | + | TRE | T | dox + → T OFF
− → T ON |

promoter A

| A | rtTA | + | TRE | T | dox + → T ON
− → T OFF |

(C)

promoter A | promoter B

| A | Cre | + | B | | T |

loxP   loxP
STOP

| B | T |

(D)

promoter A | promoter B

| A | Cre | + | B | T | |

loxP   loxP
loxP_m   loxP_m

| B | T |

Figure 14-12 Binary expression systems. Circles indicate protein products from the left (driver) transgene that act on the right (responder) transgene to regulate its expression. **(A)** Transcription unit T is expressed indirectly under the control of promoter A through the GAL4/UAS system via two transgenes. **(B)** T is expressed indirectly under the control of promoter A through the tTA/TRE or rtTA/TRE systems; exogenous application of doxycycline (dox) confers temporal control. tTA is active only in the absence of dox, while rtTA is active only in the presence of dox. **(C)** Binary expression can be achieved by Cre recombinase excising transcription or translation stop signals (diamond) between two *loxP* sites (triangles), enabling expression of T. Often a ubiquitous promoter is chosen as promoter B, as specificity is already provided by the promoter controlling Cre expression. **(D)** A variation of the system in Panel C, in which T is inverted with respect to promoter B and is flanked by a pair of incompatible *loxP* sites (recombination can occur between two *loxP*s or two *loxP_m*s, but not between a *loxP* and a *loxP_m*) in an inverted configuration. Cre-mediated recombination between either the *loxP* pair or the *loxP_m* pair causes an inversion; an additional recombination between two tandem *loxP_m*s or *loxP*s would delete the intervening *loxP* or *loxP_m* (not shown), producing the final product. (Adapted from Luo L, Callaway EM, & Svoboda K [2008] *Neuron* 57:634–660. With permission from Elsevier Inc.)

system often used in mice consists of the transcription factor **tTA** (tetracycline-regulated trans-activator) as a driver and its binding sequence ***TRE*** (tetracycline response element) as a responder. A gene under the control of a *TRE* is activated only in cells expressing tTA, which is typically driven by selected regulatory elements (Figure 14-12B). In addition, this system can be regulated using a drug: tTA activates *TRE* only in the *absence* of tetracycline, while reverse tTA (**rtTA**), a variant of tTA, activates *TRE* only in the *presence* of tetracycline. Tetracycline and its analog **doxycycline** are small molecules that readily diffuse across cell membranes and the blood–brain barrier, and drug application (often via food or water) provides temporal control of transgene expression.

Another binary expression system widely used in mice is the Cre/*loxP* system described earlier. A gene of interest can be placed after transcriptional and/or translational stop sequences flanked by *loxP* sites (called a *loxP-stop-loxP* sequence) following a ubiquitous promoter (a promoter strongly active in a wide range of cells, tissues, and developmental stages). Only in Cre-expressing cells will *stop* be excised by recombination so that the transgene can be expressed under the control of the ubiquitous promoter (Figure 14-12C). A variant of this strategy places the gene of interest in the opposite orientation as the promoter, and Cre-mediated recombination inverts the transcription unit, allowing its expression from the promoter (Figure 14-12D). This strategy is widely used in viral vectors that can accommodate only limited DNA lengths (discussed later). As discussed in Section 14.8, CreER can be used instead of Cre to enable temporal control of transgene expression in this system.

The binary systems illustrated in Figure 14-12 are more flexible and versatile than the single transgene expression systems shown in Figure 14-11. For instance, a transgenic animal carrying a responder transgene can be crossed to a series of transgenic animals carrying different driver transgenes, resulting in offspring expressing the responder transgene in different spatiotemporal patterns. Likewise, a transgenic driver animal can be crossed to many animals carrying different responder transgenes. As a specific example, *UAS*-hairpin RNA transgenes (Figure 14-7B) have been generated for almost all of the approximately 15,000 protein-coding genes in the *Drosophila* genome, such that researchers can use specific GAL4 drivers to knockdown genes one at a time in cell types of interest. This has facilitated unbiased genetic screening to identify genes involved in diverse biological processes.

14.10 Transgene expression can also be achieved by viral transduction and other transient methods

Section 14.9 discussed methods of transgene expression that rely on integration of transgenes into the germline to produce transgenic animals. Using these methods, the same expression pattern can be reproduced in different animals across multiple generations. Other methods are used to express transgenes transiently

Table 14-1: **Properties of commonly used viral vectors for gene expression in the nervous system**

Property	Adeno-associated virus (AAV)	Lentivirus	Herpes simplex virus (HSV)
Genetic material	Single-strand DNA	RNA	Double-strand DNA
Capacity	~ 5 kilobases	~ 8 kilobases	~150 kilobases
Speed of expression	Weeks	Days to weeks	Days
Duration of expression	Years	Years	Weeks to months
Tropism (cell types susceptible to viral transduction)	From broad to highly preferential depending on the serotype	Usually pseudo-typed[a] with coat proteins from other viruses for broad tropism	Broad tropism for neurons

[a]To pseudo-type a virus, the gene encoding the endogenous coat protein is deleted, and viruses are assembled in cell lines co-expressing genes encoding coat proteins from other viruses, resulting in production of viral particles coated with the co-expressed coat proteins.

From Luo L, Callaway EM, Svoboda K (2008) *Neuron* 57:634–660.

in somatic cells. Transient methods have the drawback that expression levels and patterns may differ from animal to animal. However, they are simpler and faster, especially in animals with long generation times or for which germline transgenesis has not been established.

One transient method involves directly injecting DNA or mRNA encoding genes into large cells, such as those in early embryos. Incorporation of DNA into a host cell genome confers all progeny derived from the cell the potential to express the transgene, whereas mRNA injection is usually limited to studying early development (as mRNAs are diluted with cell division and degraded over time). Another transient method is **electroporation**, wherein DNA containing the transgene is introduced into cells in a specific brain region of a host animal by placing a micropipette containing the DNA near the cells of interest and applying electrical currents to facilitate the transfer of negatively charged DNA molecules into the cells.

A transient method of transgene expression widely used in neurobiology, especially in mammals, is **viral transduction**. Transgenes can be expressed from viral vectors used to produce viruses, which are usually delivered to a specific region of interest by **stereotactic injection** (injection via a device positioned precisely in a three-dimensional coordinate system to target substances such as viruses to a small region). The most commonly used viruses in neurobiology include **adeno-associated virus (AAV)**, **lentivirus**, and **herpes simplex virus (HSV)**, each of which is best suited for specific purposes based on its characteristic properties (**Table 14-1**). These viral vectors have been engineered to minimize the deleterious effects of transduction and to allow spatiotemporal control of transgene expression using strategies similar to those described, such as Cre-dependent expression (Figure 14-12D). Viral vectors are the predominant vehicles used to alter gene expression in humans for the purpose of **gene therapy**, the use of DNA as a therapeutic agent for treating disease (Box 12-2).

14.11 Genetic access to specific neuronal populations facilitates functional circuit dissection

In complex nervous systems, a cell type, rather than an individual cell, is an important unit of neural circuit organization. The ability to monitor and manipulate neuronal activity of specific cell types is therefore critical for neural circuit analysis. Thus, establishing genetic access to specific cell types for recording, silencing, or activation has become a fundamental experimental approach (Sections 14.20–14.25). The most common strategy employed to gain genetic access to a specific cell type is to identify genes expressed specifically in that cell type and then use *cis*-regulatory elements (enhancers and promoters) from those genes to drive responder transgenes in that cell type. For example, the *cis*-regulatory elements of odorant receptors allow genetic access to specific types of olfactory

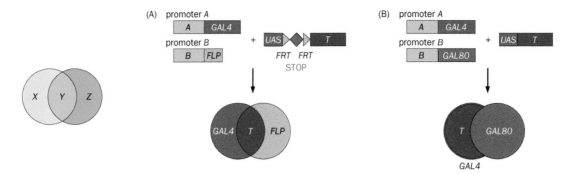

Figure 14-13 Refining transgene expression by intersectional methods. In both examples, promoter *A* drives gene expression in cell populations *X* and *Y*, while promoter *B* drives gene expression in populations *Y* and *Z*. **(A)** An AND logic gate strategy in which target gene *T* is expressed only in population *Y*, utilizing a combination of the GAL4/*UAS* binary expression system and the FLP/*FRT* recombination system. *T* will be expressed only in cells in which GAL4 drives *UAS* expression *and* FLP removes the stop signal. **(B)** A NOT logic gate strategy in which *T* is expressed only in population *X*. GAL80 inhibits GAL4; in cells expressing both GAL80 and GAL4, transcription from *UAS* is repressed. (From Luo L, Callaway EM, & Svoboda K [2008] *Neuron* 57:634–660. With permission from Elsevier Inc.)

receptor neurons (Figure 6-12; Figure 6-25); likewise, the *cis*-regulatory elements of *Fruitless* permit genetic access to neurons expressing *Fruitless* (Figure 10-6).

As the expression pattern of *Fruitless* illustrates, a given gene may be expressed in many cell types (Sections 10.5). Likewise, for most neuronal cell types, it has not been possible to identify endogenous genes expressed exclusively by those cells. Thus, additional methods have been employed to identify regulatory elements or binary system drivers expressed in specific neuronal subpopulations. One approach is to use only a fraction of an endogenous enhancer's elements to drive transgene expression; this is effective when distinct, separable regulatory elements control expression in different cell types. Another approach relies on **intersectional methods**: if promoter *A* drives gene expression in cell types *X* and *Y*, and promoter *B* drives gene expression in cell types *Y* and *Z,* then one can create an AND logic gate for expression only in *Y* (**Figure 14-13**A) or a NOT logic gate for expression only in *X* or *Z* (Figure 14-13B). Indeed, the dissection of the *Fruitless* circuit for mating behavior extensively utilized these intersectional approaches. Other methods access specific neuronal populations based on cell lineage or the timing of neurogenesis; this is effective when specific neuronal populations are born within a specific developmental window and/or arise from a common ancestor (Section 14.16). Yet other methods access specific neuronal populations based on their axonal projection patterns.

Neurons of the same genetically defined cell type may nonetheless exhibit heterogeneity in physiological response properties. This could be because their connectivity was determined by experience or stochasticity during development (see Chapters 5 and 7). Genetic access to neurons based on their activity patterns thus provides an orthogonal approach to cell-type-based methods. The most widely used activity-based approach utilizes immediate early genes, such as *Fos*, whose transcription is rapidly and transiently activated by neuronal depolarization (Figure 3-40). For example, transgenic mice have been made in which the coding sequence of *CreER* is knocked into the *Fos* locus such that its transcription is induced by neuronal activity (**Figure 14-14**A). Only in the presence of both activity *and* tamoxifen is CreER produced *and* translocated into the nucleus to catalyze recombination. Thus, researchers can "TRAP" neurons activated by a specific experience during the taxomifen-active period to permanently express any Cre reporter (Figure 14-14B). This approach was used to identify the cellular correlates of the thirst drive (Figure 9-18) and a cellular memory engram (Figure 11-37; in this case tTA, rather than CreER, was expressed from the *Fos* promoter). A limitation of current activity-based methods is that the drug-active period is at least a few hours; all active neurons during that period, whether related to the experience of interest or not, are potentially TRAPed. Reducing this period, ideally to minutes or seconds, will grant more specific access to neurons activated by an experience of interest.

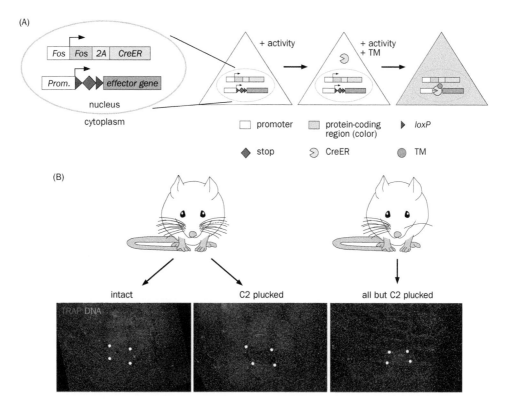

Figure 14-14 Genetic access to neurons based on activity patterns.
(A) Schematic of TRAP (targeted recombination in active populations). The coding region of *CreER* followed by *2A* is knocked into the *Fos* locus. White boxes are promoters (Prom.), including the *Fos* promoter. Note that *2A* produces a self-cleaving peptide, resulting in each mRNA producing two separate proteins, Fos and CreER. Active neurons will thus produce CreER, but CreER will enter the nucleus and catalyze recombination only when tamoxifen (TM) is present. **(B)** Proof-of-principle of the TRAP strategy. Experimental mice have two transgenes: *Fos-CreER* and a tdTomato Cre reporter. Two days after plucking specific whisker(s), mice were injected with tamoxifen and returned to their home cage with toys to stimulate whisker exploration. One week later, mice were sacrificed for histological examination. The barrel cortex corresponding to the C2 whisker (outlined by white dots) lacks recombination (has no tdTomato signal) in a mouse in which the C2 whisker was plucked, while the C2 barrel has the highest signal in a mouse in which all but the C2 whisker were plucked. See Box 5-3 for the organization of the whisker-barrel system. (From DeNardo LA, Liu CD, Allen WE, et al. [2019] *Nat Neurol* 22:460–469. With permission from Springer Nature. Guenthner CJ, Miyamichi K, Yang HH, et al. [2013] *Neuron* 78:773–784. With permission from Elsevier Inc. See DeNardo LA, Luo L [2017] *Curr Opin Neurobiol* 45:121–128 for a summary of other activity-based methods.)

14.12 Multiple, complementary techniques can be used to determine gene expression patterns

In previous sections, we have repeatedly referenced recapitulating patterns of endogenous gene expression. How do we determine gene expression patterns in the first place? Many techniques have been developed for these purposes, allowing characterization of the expression patterns of one or many genes, at the level of mRNA or protein, and with varying levels of quantitative precision and spatial resolution.

The first technique involves isolating mRNAs or proteins from specific tissues, using gel electrophoresis to separate each tissue's mixture of mRNAs or proteins according to physical properties such as molecular weight or net electrical charge, and transferring the separated contents of the gel onto a nylon or nitrocellulose membrane. A labeled gene-specific nucleic acid probe can be hybridized to membrane-bound mRNAs to produce a **northern blot** (Figure 6-8B), while probing of membrane-bound proteins to generate a **western blot** is based on antigen recognition by labeled antibody probes. (These names were inspired by the **Southern blot** developed by Edwin Southern, in which membrane-immobilized DNA hybridizes with sequence-specific DNA probes.) Northern and western blots provide information about the molecular weights and relative abundance of specific mRNAs or proteins, respectively, within a sample, but their spatial resolution is

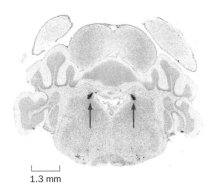

1.3 mm

Figure 14-15 Determining gene expression patterns by *in situ* hybridization. The expression pattern of dopamine β-hydroxylase, an enzyme that converts dopamine to norepinephrine, in a coronal section of the mouse brain determined by hybridizing the brain section with a probe that specifically recognizes mRNAs for dopamine β-hydroxylase. Arrows point to the bilateral locus coeruleus, where most norepinephrine neurons in the brain reside (Box 9-1). Expression patterns of all mouse genes in the brain have been determined systematically by *in situ* hybridization. (From Lein ES, Hawrylycz MJ, Ao N, et al. [2007] *Nature* 445:168–176. With permission from Springer Nature. See also http://mouse.brain-map.org.)

limited by how cleanly the tissue/anatomical region of interest can be separated by dissection.

To characterize mRNA distribution in intact tissue, fixed tissue sections can be hybridized with gene-specific probes, a procedure called ***in situ* hybridization**, which we have encountered many times. *In situ* hybridization has been applied systematically to the adult mouse brain to map the expression patterns of all ~20,000 genes in the mouse genome (**Figure 14-15**), creating a valuable gene expression database. Protein distribution can likewise be determined by a technique called **immunostaining**, which we have already seen many times. In this approach, intact, fixed tissues are incubated with **primary antibodies** that bind to specific proteins, followed by **secondary antibodies** that selectively recognize primary antibodies made by specific animal species. Secondary antibodies can be tagged with an enzyme to produce a color substrate (Figure 7-13), or with any of several differently colored fluorescent probes to allow simultaneous visualization of multiple proteins (Figure 2-26B). In contrast to northern and western blotting, *in situ* hybridization and immunostaining provide cellular and subcellular resolution.

The complete sequencing of whole genomes in the past two decades (**Box 14-2**) ushered in new methods that allow quantitative determination of gene expression at the scale of the entire **transcriptome** (the collection of all expressed mRNAs) in specific tissues or cell types. With **DNA microarray** technology, oligonucleotides or gene-specific probes are synthesized *in situ* photolithographically or individually immobilized onto specific spots on a solid substrate; these spots can be arranged densely such that probes corresponding to expressed sequences from the entire genome can be packed onto a single array (which can contain more than a million probes per cm²). mRNA mixtures from a specific tissue can be labeled and hybridized to the array, and expression levels of all genes can be read out from individual spots via their fluorescent signal intensities, which are approximately proportional to the RNA quantities. An alternative to DNA microarrays is RNA sequencing (**RNA-seq**), in which mRNA molecules from a given tissue, after being converted into complementary DNA (cDNA), are simply sequenced one by one using massively parallel sequencing platforms (Box 14-2). Each mRNA molecule yields a single "read" of its partial nucleotide sequence that is sufficiently long to identify the gene from which the mRNA was transcribed. The entire set of reads provides information about not only which genes are expressed but also the numbers of any given mRNA species present in the sample.

Gene expression profiling methods such as microarrays and RNA-seq are more informative and powerful when they can be applied to highly purified populations of specific cell types. Many methods for purifying cell types of interest have been developed. For instance, as an extension of traditional physical dissection, laser-capture microdissection allows histologically or fluorescently labeled cells from fixed tissue sections to be cut out with a laser beam for mRNA extraction. Other methods of purifying specific cells include sorting dissociated cells based on fluorescence or cell-surface markers, purifying mRNAs specifically from cells expressing transgenes encoding a tagged poly-A binding protein or ribosomal subunit, or using a micropipette to aspirate the cytosols of genetically labeled fluorescent cells in acute tissue preparations. Recent technical advances have enabled researchers to profile the mRNAs from large numbers of single cells, which we focus on in the next section.

Box 14-2: DNA sequencing and the genome revolution

Along with recombinant DNA technology, DNA sequencing has transformed modern biology. First developed in the 1970s, DNA sequencing technology has seen rapid growth, thanks in large part to the Human Genome Project initiated in the late 1980s. In the subsequent decades, and in particular with the introduction of massively parallel sequencing platforms in the first decade of the 2000s, the cost of sequencing has fallen dramatically while its speed has increased many orders of magnitude (Figure 14-16).

In parallel with human genome sequencing, the first drafts of which were completed in 2001, the whole genomes of many organisms on the tree of life have been sequenced. The impact of these data and knowledge on research has been enormous. For example, when researchers identified a gene of interest in a model organism in the 1980s or 1990s, it often took months to years to determine how many similar genes might exist in the same organism or in other organisms (Section 4.11). These questions can now be answered definitively in minutes by searching genome sequence data-

bases. As we learned in Chapter 13, comparative genomics in different species has provided insight into how individual genes arise during evolution and how different organisms are related to each other.

Likewise, comparing the genomes of different individuals within the same species can reveal genetic contributions to individuality. Comparative human genomics has already greatly expanded our understanding of the genetic bases of diseases, including brain disorders that are inherited or caused by *de novo* mutations (Box 12-3). It has also launched a new era of personalized medicine, in which treatment strategies are customized based on genetic etiology rather than symptoms; specific treatments may be more successful with patients who share genetic etiologies in addition to, or instead of, similar symptoms.

Besides genome sequencing, DNA sequencing technology has many other applications, including massively parallel RNA sequencing to determine gene expression patterns in single cells (Section 14.13).

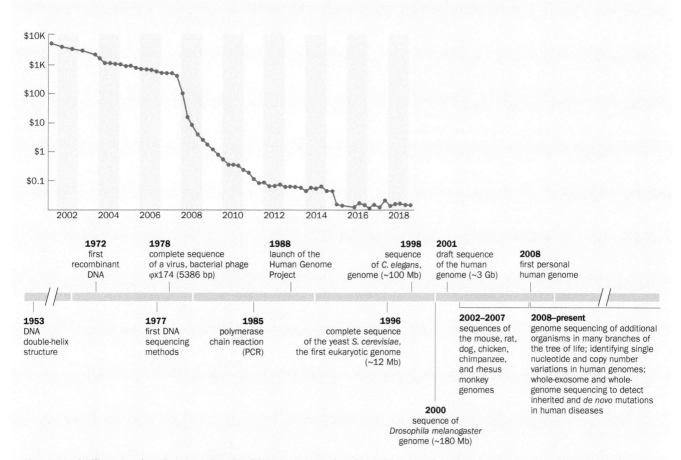

Figure 14-16 **Milestones for advances related to DNA and sequencing.** Top, illustration of the dramatic decrease in sequencing costs (per raw megabase or Mb of DNA), accelerated by the development of massively parallel sequencing platforms in the mid-2000s. Bottom, major milestones. (Top, from the National Human Genome Research Institute; see update at www.genome.gov/sequencingcosts.)

14.13 Single-cell RNA-sequencing can powerfully interrogate transcriptomes and classify cell types

The combination of high-fidelity amplification by polymerase chain reaction and high-sensitivity sequencing technology has enabled reliable sequencing of RNAs from single cells. **Single-cell RNA-seq** provides quantitative information about RNA levels at the scale of the entire transcriptomes of individual cells, the fundamental units of multicellular organisms. In a typical experiment, cells are dissociated from tissue into a suspension. Individual cells are then sorted by fluorescence-activated cell sorting (FACS) or a microfluidic device into individual wells of a 96- or 384-well plate for parallel cell lysis, cDNA library preparation, barcoding, and amplification, followed by high-throughput sequencing (**Figure 14-17**A). Alternatively, dissociated cells are mixed with barcode-containing droplets in a way to ensure that most droplets contain single cells, so that subsequent steps can be performed in bulk (Figure 14-17B), reducing costs and boosting throughput at the expense of sequencing depth compared to plate-based methods. In tissues with cells that are not easily dissociated (such as formalin-fixed postmortem human tissues), individual nuclei can nevertheless be dissociated for single-nucleus RNA-seq. Sequence data are mapped onto the genome and statistical methods are used to cluster cells based on their transcriptomic similarities. Single-cell RNA-seq is a rapidly evolving technology; new methods and data analysis tools that enhance sensitivity and reliability and reduce costs are regularly introduced.

To illustrate the use of this methodology, we discuss a study that aimed to broadly survey cell types in the mammalian nervous system. In the first step, tissues containing different components of the mouse nervous system, such as the cerebral cortex, midbrain, and spinal cord, were dissected out and dissociated into individual cells (**Figure 14-18**A). Droplet-based sequencing was then used to obtain RNA-seq data from 160,796 individual cells collected from these tissues. The data filled a matrix of 160,796 cells × ~20,000 genes (the number of mouse protein-coding genes) wherein each entry specified the number of "reads" (corresponding to the mRNA level) of a particular gene in a particular cell. The high-dimensional gene expression data were processed by dimensionality reduction

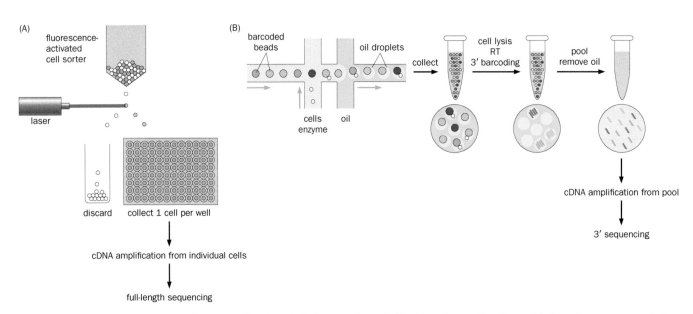

Figure 14-17 Single-cell RNA-seq platforms. (A) Plate-based single-cell RNA-seq. Cells of interest are fluorescently labeled, and selected cells are sorted into individual wells of a 96-well plate. Cell lysis, reverse transcription, and cDNA amplification for next-generation sequencing are performed individually. **(B)** Droplet-based single-cell RNA-seq. Cells are mixed with barcoded beads in oil droplets in a microfluidic chamber. To minimize droplets containing multiple cells, barcoded beads are in excess such that many droplets contain only barcoded beads without cells. After cell lysis and reverse transcription (RT), cDNA from each cell is tagged at the 3′ end by a unique barcode, and subsequent steps are performed on pooled cells. Compared to plate-based sequencing, droplet-based sequencing costs much less per cell and can be performed with higher throughput, but it can sequence only from the 3′ ends of cDNAs containing barcodes, and thus cell identity information. In addition to the two platforms illustrated here, many other variations have been developed.

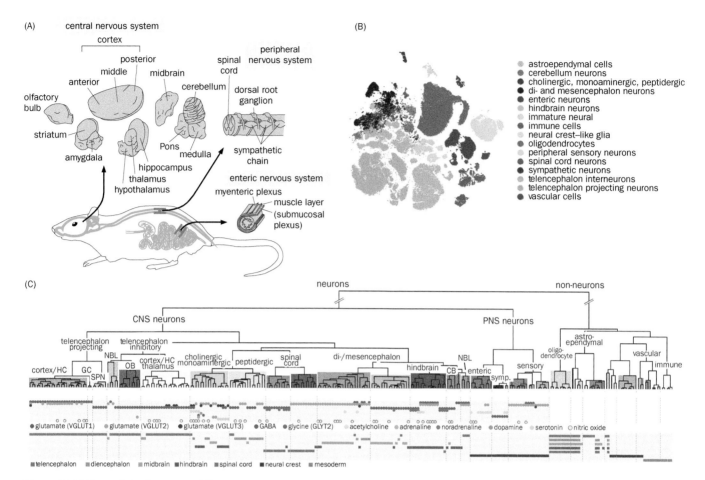

Figure 14-18 Transcriptomic survey of the mouse nervous system.
(A) Illustration of sampling strategy. The mouse brain was divided into several anatomical regions for cell dissociation. Additional samples were taken from the spinal cord, dorsal root ganglia, sympathetic ganglia, and enteric nervous system. **(B)** Visualization of single-cell transcriptomic data by two-dimensional tSNE plot. Each cell is a dot, and cells are color coded by classes indicated on the right. **(C)** Dendrogram of hierarchical organization of cells of the mouse nervous system. Main branches are annotated with labels and colors. Regions and

neurotransmitter properties of each of the 256 transcriptomic types (final branches of the dendrogram) are indicated below with color codes. CB, cerebellum; GC, granule cell; HC, hippocampus; SPN, spiny projection neuron of the striatum; NBL, neuroblast; OB, olfactory bulb; Symp, sympathetic. VGLUT1–3 and GLYT2 are glutamate and glycine transporters, respectively, that serve as cell-type-specific markers. (From Zeisel A, Hochgerner H, Lönnerberg P, et al. [2018] *Cell* 174: 999–1014. With permission from Elsevier Inc.)

techniques such as principal component analysis and visualized on a two-dimensional t-distributed stochastic neighbor embedding (tSNE) plot (Figure 14-18B; tSNE is another dimensionality reduction technique from machine learning that models similar high-dimensional objects with nearby points into two- or three-dimensional space; see Section 14.33 for a general discussion of machine learning). Known marker genes in specific cell classes allowed researchers to determine which clusters corresponded to which cell classes (color coding in Figure 14-18B). Additional statistical methods were used to build a dendrogram, in which the distance between two clusters reflects differences in their gene expression patterns (Figure 14-18C). Thus, the first major division of CNS cells was neurons versus non-neurons; the first major division among neurons was central versus peripheral nervous system (CNS versus PNS) neurons; CNS neurons were further divided by a combination of brain regions and neurotransmitter types.

This survey likely underestimates cell types in many brain regions because of limited cells and reads dedicated to each region. In addition, studies have shown that some cell types are more difficult to dissociate into single cells, thus potentially biasing the abundance of different cell types surveyed via dissociation. Furthermore, distinct cell types with considerable transcriptomic differences during development (for establishing their wiring specificity) may have more similar transcriptomes in the adult nervous system. Nevertheless, this example illustrates

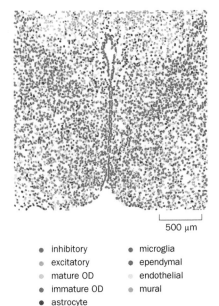

500 μm

- inhibitory
- excitatory
- mature OD
- immature OD
- astrocyte

- microglia
- ependymal
- endothelial
- mural

Figure 14-19 Spatial distribution of different cell classes in the preoptic area of the hypothalamus. Distribution of all cells in a 10 μm section (color coded; cell classes below). Cell classes were determined by expression of 155 preselected genes imaged using a combination of a spatial transcriptomic method called MERFISH (multiplexed error-robust fluorescent *in situ* hybridization) and conventional two-color fluorescence *in situ* hybridization. OD, oligodendrocyte. (From Moffitt JR, Bambah-Mukku D, Eichhorn SW, et al. [2018] *Science* 362:eaau5324.)

the power of single-cell RNA-seq in classifying cell types and obtaining comprehensive data on gene expression patterns.

Neuronal cell types were traditionally defined by a collection of parameters, including cell body location, dendritic morphology, axonal projections, physiological characteristics, gene expression patterns (especially those related to neurotransmitter synthesis and transport), and function. With the exception of a few neural tissues, such as the retina, what constitutes a cell type in the mammalian central nervous system is still mostly under debate. Single-cell RNA-seq provides rich high-dimensional data (each cell can be considered a point in a ~20,000-dimensional space in which each axis represents the expression level of one gene) and can thus contribute to defining cell types. Indeed, in neural circuits where cell types have been analyzed in detail using anatomical and neurophysiological methods, the correspondence between these measurements and transcriptomic clusters has generally been excellent (Figure 4-27). Thus, single-cell transcriptomes have already had a major impact on defining neuronal types in complex brain regions. Furthermore, comparing single-cell transcriptomes of analogous cell types in different organisms will shed light on the evolution of cell types.

While rich in quantitative gene expression data, single-cell RNA-seq does not provide the precise spatial locations of the sequenced cells. Thus, it is often followed by *in situ* hybridization analysis of cluster-specific genes to map clusters to their precise spatial locations. New spatial transcriptomic methods have also been developed to determine the expression patterns of hundreds to thousands of genes in a piece of tissue; some of these have achieved single-cell resolution (**Figure 14-19**). These methods can in principle be combined with measurements of axonal projection patterns and *in vivo* physiological response properties of individual neurons so that researchers can integrate the anatomy, physiology, and gene expression patterns of the same cell.

14.14 Biochemical and proteomic approaches can identify key proteins and their mechanisms of action

Proteins are the ultimate executors of almost all biological functions. Identifying key proteins and studying their mechanisms of action can provide new insight into neurobiological processes, as we have seen throughout the textbook. For example, biochemical identification of SNARE proteins enriched in synaptic vesicles and presynaptic terminals, followed by studies of how these proteins work together, revealed the mechanisms of vesicle fusion and neurotransmitter release (Section 3.5). Biochemical purification of rhodopsin and transducin from bovine retina elucidated the mechanisms of phototransduction (Sections 4.3 and 4.4). Biochemical purification of netrins by following activity in axon outgrowth and turning assays shed light on axon guidance mechanisms (Box 5-1).

In biochemical approaches, researchers often start with a functional assay and a source of tissue enriched for activity in that assay. Through biochemical fractionation of proteins from the tissue, following activity in the functional assay, researchers can pinpoint a single protein or protein complex responsible for the activity. Once the protein is identified, researchers can identify the corresponding gene, predict its primary structure, determine its three-dimensional structure, perform structure–function analysis (for instance, disrupt a specific domain or key amino acid residue and examine the effects on the activity), and identify its interacting partners, thus gaining insight into its mechanisms of action. Researchers can also use reverse genetic techniques (Section 14.7) to validate the function of the protein *in vivo*.

Classic biochemical approaches study one protein (or protein complex) at a time, just like classic genetic approaches study one gene at a time. While these time-honored approaches are still a major force in biochemistry and genetics today, they can be complemented by proteomic and genomic approaches. With genome sequences for all major model organisms (Box 14-2), we can predict all proteins produced in these organisms with high accuracy. This knowledge, along with advances in protein analysis techniques, in particular **mass spectrometry** (MS), has enabled researchers to ask questions at the scale of the entire **proteome**

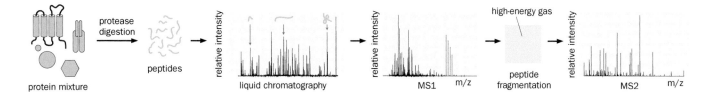

(the collection of all proteins in a sample). Mass spectrometry is an analytical technique that can precisely determine the mass-to-charge ratio of electrically charged molecules (ions). Specifically, it can be used to determine the identities of individual proteins in a protein mixture. In a typical experiment called LC-MS/MS (liquid chromatography followed by *tandem* mass spectrometry; **Figure 14-20**), proteins isolated via a specific procedure are first digested by a protease into peptides. Peptides are then fractionated by liquid chromatography based on their physical and chemical properties. Peptide fractions sequentially enter the first mass spectrometer, where they are ionized and separated further based on their mass-to-charge ratio. Individual peptides, called precursor ions, then sequentially enter a colliding chamber where each peptide collides with high-energy gas, resulting in fragmentation into *random* smaller peptides. These smaller peptides then enter a second mass spectrometer where the mass-to-charge ratio for each peptide is precisely measured. The molecular weight information obtained from this last step, when combined with the predicted proteome of that organism, is often sufficient to determine the amino acid sequence, and hence the identity, of the protein of origin for that peptide. Additional methods have been devised to compare the quantity of the same protein from different conditions (for example, different tissue sources, developmental stages, or particular experimental conditions) or to analyze post-translational protein modifications.

This LC-MS/MS procedure enables researchers to determine the proteome of a given protein mixture, whether isolated by enrichment for a specific protein complex, subcellular compartment, cell type, or tissue at a specific developmental stage. We use a specific example to illustrate: the characterization of synaptic cleft proteomes by a technique called **proximity labeling**, which labels proteins (or mRNAs) based on their proximity to a specific protein linked to an enzyme to catalyze the labeling reaction. In this example, a transmembrane protein targeted to the postsynaptic density is linked to horseradish peroxidase (HRP) in its extracellular domain, so that HRP is enriched in the synaptic cleft (**Figure 14-21**A). In

Figure 14-20 Using mass spectrometry to determine proteome composition. Schematic of LC-MS/MS. The last two graphs are called mass spectra. m/z, mass-to-charge ratio. Individual peaks coming off of liquid chromatography usually still contain a mixture of peptides. They are further separated in the first mass spectrometer (MS1). Individual peptides coming off of MS1 are randomly fragmented into smaller peptides and their m/z are determined in MS2. See text for details. (Adapted from Hosp F & Mann M [2017] *Neuron* 96:558–571. With permission from Elsevier Inc.)

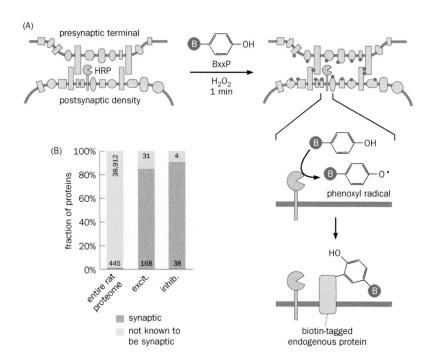

Figure 14-21 Using proximity labeling to discover synaptic cleft proteomes. (A) Schematic of experimental design. Horseradish peroxidase (HRP) is fused to the extracellular portions of transmembrane proteins localized to excitatory or inhibitory postsynaptic sites. When supplied with BxxP and H_2O_2, HRP catalyzes production of phenoxyl radicals, which attach biotin (red) covalently to nearby proteins. After cell lysis, protein extracts are incubated with streptavidin beads to purify biotinylated proteins, which are then subjected to LC-MS/MS analysis. **(B)** Among 199 excitatory (column 2) and 42 inhibitory (column 3) synaptic cleft proteins identified in this study, most match proteins previously described to be associated with synapses; some new proteins were experimentally validated to be present in the synaptic cleft. (From Loh KH, Stawski PS, Draycott AS, et al. [2016] *Cell* 166:1295–1307. with permission from Elsevier Inc.)

the presence of H_2O_2 and BxxP, a biotin-containing peroxidase substrate, HRP produces a phenoxyl radical that can covalently link the biotin moiety to nearby proteins. Because the phenoxyl radical is extremely short lived, only proteins in the vicinity of HRP can be tagged with biotin. Biotinylated proteins can then be purified via their strong affinity to avidin and subjected to LC-MS/MS. As BxxP is membrane impermeable, only proteins exposed to the extracellular environment can be biotinylated. In this study, researchers enriched proteins based on their proximity to excitatory and inhibitory postsynaptic transmembrane proteins relative to proteins identified by their proximity to a ubiquitously distributed transmembrane protein in cultured neurons. The deduced excitatory and inhibitory synaptic cleft proteomes recapitulated existing knowledge and created new knowledge of this important site of neuronal communication (Figure 14-21B).

ANATOMICAL TECHNIQUES

To comprehend how the nervous system operates, we must understand its structure at different levels. In the following sections, we examine the major anatomical techniques that have advanced our knowledge of nervous system structures. We begin with histological methods that have provided overviews of nervous system organization. We then review techniques for visualizing individual neurons, the building blocks of the nervous system. We probe further into the fine structures of individual neurons. Last, we study methods that reveal how neurons connect with each other.

14.15 Histological analyses reveal the gross organization of the nervous system

The anatomical organization of the nervous system is typically examined in **histological sections**; frozen or chemically fixed tissues are sliced into sections using microtomes, with the thickness of the slices ranging from several micrometers to several hundred micrometers, so that the sectioned tissues can be examined with a light microscope. As discussed in Chapter 1, the three common section planes are coronal, sagittal, and horizontal, which are perpendicular to the anterior-posterior (rostral–caudal), medial–lateral, and dorsal–ventral axes of the body, respectively (Figure 1-8C).

Histological sections are typically stained to create contrast and highlight specific structures for microscopic examination. Starting in the nineteenth century, long before molecular techniques such as *in situ* hybridization and immunostaining became available (Section 14.12), histologists invented staining methods to label cell bodies, axon fibers, or myelin sheaths; this early work revealed the overall organization of gray and white matter in the CNS, as well as subdivisions within gray matter. A widely used staining method for cell bodies is the **Nissl stain**, which utilizes positively charged dyes such as cresyl violet that bind negatively charged RNA molecules, thereby highlighting the rough endoplasmic reticulum enriched for ribosomal RNAs. When applied to brain sections, Nissl stains provide a comprehensive overview of the density, size, and distribution of neurons and glia. Nissl stains are commonly used to construct brain atlases and as a counterstain for *in situ* hybridization, immunostaining, and other anatomical methods. Other staining methods selectively label cell nuclei using dyes that bind DNA. Nissl and nuclear stains enabled many discoveries regarding nervous system organization, such as the layering of the lateral geniculate nucleus and neocortex (Figure 4-38; Figure 4-47A), the whisker-barrel pattern in the rodent somatosensory cortex (Figure 5-26), and sexual dimorphism in certain mammalian brain nuclei (Figure 10-17; Figure 10-23). Indeed, using cytoarchitectonics, an approach based on differences in the laminar (layered) organization of cortical neurons as well as the density and thickness of each layer (**Figure 14-22**A), early-twentieth-century histologists divided the cerebral cortex into distinct areas (Figure 14-22B); these divisions are still used today.

(A)

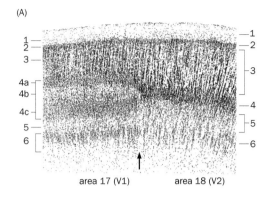

area 17 (V1)   area 18 (V2)

(B)   lateral surface

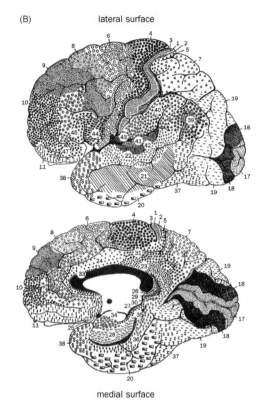

medial surface

Figure 14-22 Nissl stain and cortical area divisions. (A) Nissl stain of a section across the border of the primary (area 17, or V1) and secondary (area 18, or V2) human visual cortices. Layer 4 of V1 is characterized by the presence of three sublayers (4a, 4b, 4c), whereas V2 has a single layer 4. The arrow indicates the border of V1 and V2. **(B)** The type of histological stain shown in Panel A has been used to visualize cytoarchitectonic differences that define distinct areas within the cerebral cortex; for example, the human cerebral cortex is divided here into 50 areas, each represented by a specific symbol. (Adapted from Brodmann K [1909] Vergleichende Localisationslehre der Grosshirnrinde in ihren Prinzipien dargestellt auf Grund des Zellenbaues. Barth: Leipzig.)

Histological sections allow for penetration of staining reagents throughout the sectioned tissue and high-resolution optical imaging of neural structures within the sections. However, deciphering large-scale anatomical organization, such as axonal projections from one brain region to another (Section 14.18), requires reconstruction of three-dimensional volumes from individual two-dimensional sections. By contrast, **whole-mount** preparations enable investigators to examine large pieces of intact nervous tissue, whether in a dissected specimen or an intact organism. High-resolution fluorescence imaging of whole-mount tissues can be obtained via **confocal fluorescence microscopy** (confocal microscopy in short; *confocal* means "having a common focus"). In confocal microscopy, a laser beam is focused on a spot with a volume on the order of one cubic micrometer and a pinhole is employed near the detector to ensure that fluorescence is collected from only this spot (**Figure 14-23**A). By scanning the laser across a plane to image many focal spots, confocal microscopy produces thin optical sections of thick tissues, with the thickness of an optical section ranging from a fraction of a micrometer to several micrometers. Three-dimensional structures can be reconstructed from a series of such optical sections obtained at consecutive *x–y* planes along the *z* axis (the axis perpendicular to the imaging plane). Many images shown in this book are the products of confocal microscopy of whole-mount brains of small animals like *Drosophila* (e.g., Figure 6-30). Confocal microscopy is also commonly used to obtain thin optical sections from thicker physically sectioned tissues.

In the past, the use of whole-mount preparations has typically been restricted to tissue less than a few hundred micrometers thick, limited by penetration of staining reagents and the opacity of tissues due to light scattering. Common whole-mount preparations have included intact *C. elegans*; dissected *Drosophila* brain

Figure 14-23 Confocal and light-sheet fluorescence microscopy. These schematics are simplified to highlight the unique features of each system. **(A)** In confocal microscopy, a small pinhole is placed in front of a detector such that fluorescence emission from only the focal plane (solid lines and arrows), but not from out-of-focus planes (dashed lines and open arrows), is collected by the detector. The dichroic mirror reflects short-wavelength excitation light (blue) but transmits long-wavelength emission light (green). In a typical imaging experiment, the laser beam scans across different x–y positions on the focal plane such that the detector can reconstruct a two-dimensional image for that focal plane. Then the focal plane is adjusted along the z axis to reconstruct two-dimensional images of other focal planes, eventually producing a three-dimensional confocal stack representing the three-dimensional volume of the specimen. For details, see Conchello JA & Lichtman JW (2005) *Nat Methods* 2:920–931. **(B)** In light-sheet microscopy, the laser illuminates the specimen from the side and, through a cylindrical lens, produces a thin sheet of excitation light at the focal plane of the objective lens. Only tissue in the focal plane is exposed to excitation light, and all fluorescence emission in the focal plane is collected simultaneously by a detector (camera) to form a two-dimensional image without scanning. By systematically moving the focal plane and excitation beam along the z axis, a three-dimensional stack can be produced to represent the three-dimensional volume of the specimen. For details, see Keller PJ & Dodt HU (2012) *Curr Opin Neurobiol* 22:138–143.

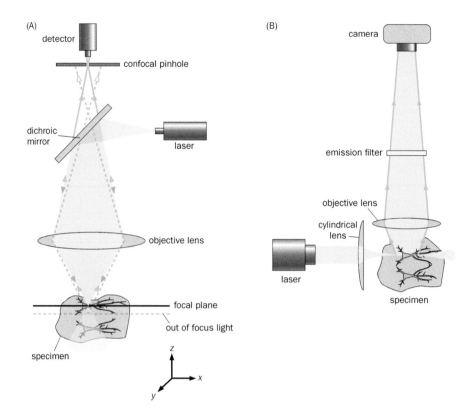

and ventral nerve cord; and nervous systems from organisms at early developmental stages, such as zebrafish larvae and mouse embryos. More recently, development of a number of tissue-clearing methods has enabled high-resolution fluorescence imaging of pieces of intact tissue up to several millimeters in each dimension (e.g., **Figure 14-24**; **Movie 14-1**). **Light-sheet fluorescence microscopy**, which illuminates only the focal plane with a thin sheet of excitation light from the side (Figure 14-23B), offers more rapid imaging (as no scanning is required) and less photobleaching of fluorescent probes (as out-of-focus planes are not illuminated) compared to confocal microscopy. Due to these properties, light-sheet microscopy is particularly advantageous for fluorescence imaging of large blocks of tissue. As discussed later in this chapter, these whole-mount imaging methods can facilitate the study of many aspects of nervous system organization.

14.16 Visualizing individual neurons opens new vistas in understanding the nervous system

As we learned in Chapter 1, development of the Golgi stain for visualizing the morphology of individual neurons was a milestone in neurobiology. This method was used by Ramón y Cajal and his contemporaries to establish individual neurons as the building blocks of the nervous system and provided important insight into how information flows within and between individual neurons. Methods for

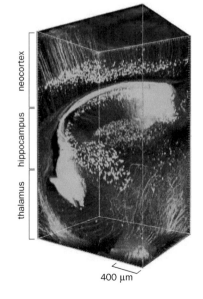

Figure 14-24 CLARITY-based tissue clearing for fluorescence imaging. In this method, intact tissue is fixed in the presence of hydrogel monomers that covalently link DNA, RNA, and proteins into a mesh during subsequent polymerization. Lipids, the major source of opacity in fluorescence imaging, are not covalently linked and are removed during the subsequent clearing process by detergent. The resulting tissue is nearly transparent, enabling fluorescence imaging across several millimeters in each dimension. Shown here is a piece of tissue from a *Thy1-Gfp* transgenic mouse (see Figure 14-25A for details) imaged via confocal microscopy, showing labeled neurons and their processes across the neocortex, hippocampus, and thalamus. (From Chung K, Wallace J, Kim SY, et al. [2013] *Nature* 497:332–337. With permission from Springer Nature.)

visualizing individual neurons are still important in modern neurobiology, as they allow researchers to correlate the morphology and projection patterns of individual neurons with their cell identity, development, and physiological properties.

Due to its simplicity, Golgi stain is still used today to characterize the neuronal composition in a given region of the nervous system and to analyze the effects of gene mutations on neuronal morphology. However, Golgi stain has a number of limitations: it does not reliably stain long-distance axonal projections and fine terminal processes; it cannot be used to visualize neurons in live brains; and it cannot selectively stain specific cell types due to its random nature. A number of methods that overcome these limitations have been developed.

Individual neurons can be labeled by dye injection during intracellular recording (Section 14.21). Injection of small molecules allows neurons to be visualized either in live brains as neurons are being recorded or by *post hoc* staining. Dye-filling methods allow a cell's electrophysiological properties to be correlated with its morphology. They also permit tracing of long-distance projections of individual neurons, as no other neurons are labeled to obscure such tracing (Figure 4-47B).

The development of **green fluorescent protein** (**GFP**) from jellyfish for live cell imaging has revolutionized many fields of biology. Following the initial introduction of GFP as a marker, fluorescent proteins of other colors were both discovered and engineered. In conjunction with genetic methods, these fluorescent proteins enable visualization of singly or sparsely labeled neurons in live tissue. Sparse labeling is necessary for tracing and assigning traced processes to individual neurons across long distances. The simplest genetic strategy for visualizing single neurons is one using a promoter to drive fluorescent protein expression in a sufficiently sparse population of neurons such that labeled cells and their processes are unlikely to come into proximity to one another. While such specific promoters are rare in the central nervous system, integration of transgenes into random locations in the genome can result in the fluorescent reporter being expressed in populations of well-separated neurons (**Figure 14-25**A). Along with

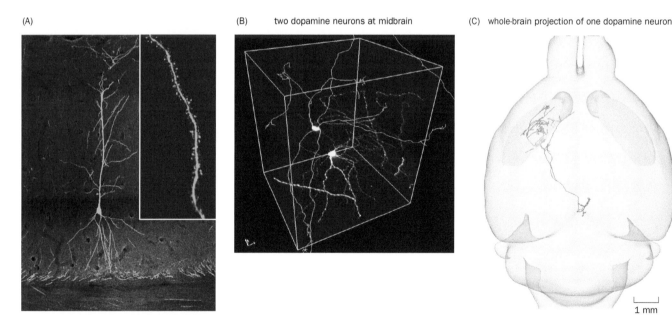

(A)

(B)   two dopamine neurons at midbrain

(C)   whole-brain projection of one dopamine neuron

1 mm

Figure 14-25 Visualizing green fluorescent protein- (GFP-) expressing individual neurons. (A) In *Thy1-Gfp* mice, GFP expression is driven by the promoter of the *Thy1* gene, which is preferentially expressed in excitatory neurons. Due to random transgene integration (Figure 14-11A), this particular transgenic line labels a very small subset of hippocampal pyramidal neurons, allowing visualization of their dendritic trees and spines (inset). **(B)** Two midbrain dopamine neurons are visualized by sparsely expressed GFP using a combination of several binary expression strategies described in Figure 14-12. This image shows cell bodies, dendrites, and local axons in a 300 × 270 × 250 μm volume (outlined). **(C)** Whole-brain projection of the top left neuron in

Panel B, obtained using a technique called fluorescence micro-optical sectioning tomography, which reconstructs high-resolution images on consecutive sections while the brain is being sectioned. In this horizontal view, axons and dendrites of the dopamine neuron are indicated by blue and magenta, respectively. Yellow and green, respectively, outline the dorsal striatum and nucleus accumbens, two major targets of midbrain dopamine neurons. (A, from Feng G, Mellor RH, Bernstein M, et al. [2000] *Neuron* 28:41–51. With permission from Elsevier Inc. B from Lin R, Wang R, Yuan, J, et al. [2018] *Nat Methods* 15:1033–1036. With permission from Springer Nature.)

two-photon microscopy (Section 14.22), such sparse labeling has allowed visualization of individual neurons in live animals over long periods of time for tracking the dynamics and stability of dendritic branches and spines (Figure 11-49). Individual neurons can also be labeled via sparse activation of a recombinase or expression of a reporter in specific cell types via a combination of viral transduction and transgenesis. Axonal projections of sparsely labeled neurons can then be traced across the entire brain (Figure 14-25B, C).

Individual neurons can also be labeled via genetic mosaic methods based on mitotic recombination between the homologous chromosomes of somatic cells (Figure 14-10). The **MARCM** (mosaic analysis with a repressible cell marker; **Figure 14-26**) method has been widely used in *Drosophila* to label individual neurons (Figure 6-30) or groups of neurons that share the same lineage (Figure 7-43). In addition to allowing the labeling of individual neurons, MARCM enables simultaneous deletion of endogenous genes or expression of transgenes in the labeled neurons. An analogous mitotic recombination-based method called MADM (mosaic analysis with double markers) achieves similar purposes in mice. Labeling can be restricted to specific cell types using tissue-specific promoters to drive recombinase expression and can be produced at desired frequencies by controlling the expression level or activity of the recombinase. Such genetic mosaic methods have been used not only to label individual neurons of a specific type but also to examine the relationship between lineage and wiring properties (Figure 7-43) and to determine the cell-autonomous functions of various genes in neuronal morphogenesis and wiring specificity (Figure 5-28; Figure 5-37; Figure 7-44).

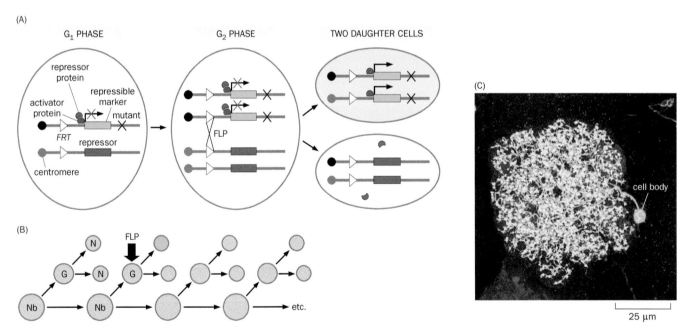

Figure 14-26 Single neuron labeling and genetic manipulation via MARCM. (A) Schematic. A cell marker gene (green bar) is under the control of a repressible promoter: it is activated by the activator protein (blue blob) only in the *absence* of a repressor protein (red blob). The schematic shows a pair of homologous chromosomes in a cell at three cell-cycle stages: before DNA replication (G_1 phase), after DNA replication (G_2 phase), and after cell division (with two daughter cells). The transgene expressing the repressor protein (red bar) is located on the chromosome *in trans* to a mutation (×). FLP/*FRT*-mediated interchromosomal recombination (the cross in the middle panel) followed by cell division results in the loss of the repressor transgene in one of the two daughter cells, thus allowing the marker to be expressed in only that cell, which is also homozygous mutant for a gene of interest. In the original version, the marker is expressed by the GAL4/*UAS* binary expression system and the repressor is GAL80 (Figure 14-12; Figure 14-13). Other repressible binary expression systems have since been developed. **(B)** A common cell division pattern that produces neurons in the insect CNS. The neuroblast (Nb) undergoes asymmetric division to produce another neuroblast and a ganglion mother cell (G), which divides once more to produce two post-mitotic neurons (N). If FLP/*FRT*-mediated recombination is induced in G, then a single neuron is labeled. **(C)** An example of a MARCM-labeled single *Drosophila* olfactory local interneuron that ramifies its processes throughout the entire antennal lobe (Figure 6-24). The cell is labeled in green with a membrane-targeted GFP for visualizing neuronal morphology, and in red with an epitope-tagged synaptotagmin for labeling its presynaptic terminals, which appear yellow because synaptotagmin is expressed within the GFP-labeled cell. Blue is neuropil staining. (A & B, from Lee T & Luo L [1999] *Neuron* 22:451–461. With permission from Elsevier Inc. C from Chou YH, Spletter ML, Yaksi E, et al. [2010] *Nat Neurosci* 13:439–449. With permission from Springer Nature.)

14.17 Fine structure studies can reveal key facets of molecular organization within neurons

Most observations about nervous system structure have been made using light microscopy, which can resolve structures as small as 200 nm, a value defined by the diffraction limit of visible light. This resolution is sufficient for visualizing neuronal cell bodies (which have diameters ranging from 3 μm for small neurons in *Drosophila* and *C. elegans* to 10–20 μm for typical vertebrate neurons to 1 mm for giant neurons in *Aplysia*) as well as most dendritic and axonal processes (which are hundreds of nanometers to a few micrometers in diameter). However, resolving fine subcellular structures of neurons and glia, including individual synapses in densely packed tissues, requires higher-resolution techniques.

Electron microscopy (**EM**) has served neurobiology well since its inception in the 1950s. EM enabled the discovery of the synaptic cleft, observation of synaptic vesicle fusion with presynaptic membranes, and the finding that glial membranes wrap axons (Figure 3-3; Figure 3-4; Figure 2-26). It has also been used to study many aspects of neuronal cell biology, including cytoskeletal organization, intracellular trafficking (Figure 2-6), membrane compartment organization, and pre- and postsynaptic terminal structures. Most EM images in this book were taken using **transmission electron microscopy**, in which high-voltage electron beams transmit through ultra-thin sections of biological specimens (typically <100 nm) to create images. **Scanning electron microscopy** produces images by scanning the surface of a biological specimen, collecting information regarding the interactions of the electron beam with the surface area (Figure 13-34).

As discussed in Section 14.12, light microscopy can be combined with antibody staining to study the tissue-level and subcellular distributions of individual proteins. Likewise, EM can be combined with antibody staining in a procedure called **immuno-EM** to visualize individual proteins at an ultrastructural level. This method can provide important clues about the actions of individual molecules (**Figure 14-27**). High-resolution EM, notably cryo-EM, has been widely used to analyze the atomic structures of proteins (Figure 2-33).

While EM has become the gold standard for ultrastructural analysis, recent development of a number of **super-resolution fluorescence microscopy** techniques has provided impressive views of the fine structures of neurons at resolutions beyond the diffraction limit of conventional light microscopy. In variants called **PALM** (<u>p</u>hoto<u>a</u>ctivated <u>l</u>ocalization <u>m</u>icroscopy) and **STORM** (<u>s</u>tochastic <u>o</u>ptical <u>r</u>econstruction <u>m</u>icroscopy), spatial precision is achieved by photoactivating a random small subset of photo-switchable fluorophores at any one time, such that the position of each fluorophore can be localized to a precision much finer than the diffraction limit of visible light. The fluorophores are then deactivated (bleached), and a different subset of fluorophores is photoactivated to allow a second round of localization. Repeated rounds of localization enable the reconstruction of the entire imaging field, but at a resolution <20 nm in the x–y plane and <50 nm in the z axis in brain sections, far superior to the resolution of conventional light microscopy (**Figure 14-28**A). As a result, for example, STORM can be used to determine the distances of different synaptic molecules from the synaptic cleft and the orientations of certain synaptic proteins with respect to the synaptic cleft (Figure 14-28B, C). Similarly, application of **STED** (<u>st</u>imulated <u>e</u>mission <u>d</u>epletion microscopy), which achieves super-resolution by depleting fluorescence from surrounding regions while leaving a central focal spot of the sample active to emit fluorescence, enabled reconstruction of the spatial distribution of synaptic molecules at the fly neuromuscular junction (Figure 3-11). Compared with immuno-EM for localizing specific molecules with fine resolution, super-resolution fluorescence microscopy is relatively easy to perform and especially well-suited for imaging multiple proteins differentially labeled in the same sample. Super-resolution microscopy can also be used for live imaging to study protein dynamics, whereas EM is limited to fixed tissues.

Another variation of super-resolution microscopy is expansion microscopy. In this technique, tissue is first incubated with swellable polymers to covalently link specific labels to the polymers. The tissue is then allowed to physically expand

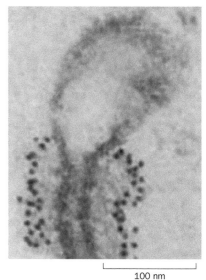

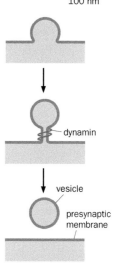

Figure 14-27 Immuno-EM localization of dynamin. Dynamin molecules coat tubular membrane invaginations, visualized here using gold particles in an *in vitro* preparation of nerve terminal membrane. (In this procedure, primary anti-dynamin antibodies bind dynamin, and gold particle-conjugated secondary antibodies bind the primary antibodies.) This suggests a function of dynamin in fission of synaptic vesicles from the plasma membrane during endocytosis, as shown in the schematic below (red, dynamin localization). This role for dynamin is consistent with the protein's loss-of-function phenotypes (Figure 3-14). (From Takei K, McPherson PS, Schmid S, et al. [1995] *Nature* 374:186–190. With permission from Springer Nature.)

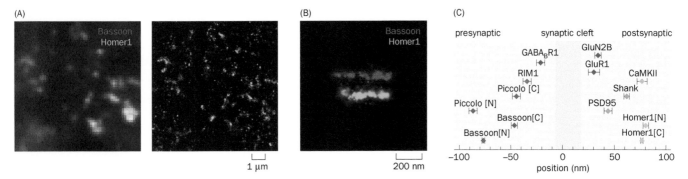

(A)

(B)

(C)

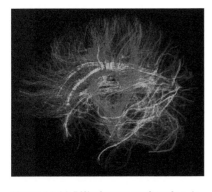

Figure 14-28 Application of super-resolution fluorescence microscopy to map synaptic protein organization. (A) Double labeling of the presynaptic scaffold protein Bassoon (red) and the postsynaptic scaffold protein Homer1 (green) in a section of mouse olfactory bulb glomerular layer, imaged using confocal fluorescence microscopy (left) and stochastic optical reconstruction microscopy (STORM; right, with identical field of view as left). While red and green signals overlap considerably and appear fuzzy at this magnification in the confocal image, they appear distinct in the STORM image. **(B)** High-magnification view of a STORM image, which clearly resolves the distributions of

Bassoon and Homer1 across the synaptic cleft. **(C)** Estimates of the distributions of different synaptic proteins relative to the synaptic cleft, based on the distributions of antibodies against these proteins relative to each other, as visualized via STORM. Colored dots are average positions of individual proteins with respect to the synaptic cleft and vertical bars are standard deviations of the means. To determine the orientation of the molecules with respect to the synaptic cleft, different antibodies were used against N-termini [N] and C-termini [C]. (From Dani A, Huang B, Bergan J, et al. [2010] *Neuron* 68:843–856. With permission from Elsevier Inc.)

isotropically (to the same degree in all directions) via polymer swelling, such that labels located within the diffraction limit are now further apart and can be readily resolved via conventional light microscopy. We will see an example of its application in the next section.

Conventional light microscopy, super-resolution fluorescence microscopy, expansion microscopy, and EM can be used to systematically study the locations of individual molecules with respect to each other and to different subcellular neuronal compartments. Such investigation can in principle provide a realistic model of single neurons at the level of molecular complexes, from the soma to the axonal and dendritic terminals. These molecular neuroanatomical techniques can also probe how protein complexes change with neuronal activity and reveal what goes wrong in brain disorders.

14.18 Mapping neuronal projections allows tracking of information flow across brain regions

While fine structural analyses elucidate how molecular complexes are involved in the functioning of individual neurons, deciphering a nervous system's wiring diagram—the representation of how individual neurons connect with each other to form a complex nervous system—poses a great challenge in neurobiology today. All neurons can be classified into one of two categories: **projection neurons** send information from one region of the nervous system to another, whereas **local neurons** (also referred to as **interneurons**; Section 1.9) confine their axons within a given region. Wiring diagrams are being constructed at different scales with varying resolutions. In the following we discuss methods used to map long-distance connections of projection neurons. In the next section, we study how projection neurons and local neurons form synaptic connections within a given region.

At a global scale, **diffusion tensor imaging** (**DTI**) is a magnetic resonance imaging technique that allows noninvasive imaging of fiber bundles that connect different structures. Its premise is that water diffusion occurs almost equally in all directions in gray matter, but primarily along axonal paths in white matter. By acquiring a series of images, each sensitive to diffusion in a specific direction, DTI can reveal the motion of water at any given volume in the white matter; this is used to estimate the trajectories of the axonal fibers passing through that volume. Information obtained across the white matter can then be used to reconstruct flow lines approximating the trajectories of axon bundles (**Figure 14-29**). The res-

Figure 14-29 Diffusion tensor imaging. In this sagittal view of the human brain, axon bundles running mostly along the medial–lateral axis are colored in red, those running mostly along the anterior–posterior axis are colored in green, and those running through the brainstem are colored in blue. (Courtesy of the Laboratory of Neuro Imaging and Martinos Center for Biomedical Imaging, Consortium of the Human Connectome Project, www.human connectomeproject.org.)

olution of DTI is on the scale of a few millimeters, so it can depict only major white matter pathways in large brains such as the human brain; it is less effective at resolving trajectories of axons within gray matter. Still, DTI has already been used to map gross brain connectivity in healthy subjects and abnormalities in patients with neurological disorders.

A widely used method for determining connections between different brain regions in experimental animals is to follow the trajectory of tracers stereotactically injected at specific regions. **Anterograde tracers** are taken up primarily by neuronal cell bodies and dendrites and travel down axons to label their projection sites. Classical anterograde tracers include radioactively labeled amino acids (which can also be released at the axon terminal and taken up by postsynaptic neurons; see Figure 5-18) and phytohemagglutinin (PHA-L), a lectin from the red kidney bean *Phaseolus vulgaris* (Figure 14-30A). By contrast, **retrograde tracers** are mostly taken up by axon terminals and transported back to cell bodies. Classical retrograde tracers include horseradish peroxidase and cholera toxin subunit b (CTb; Figure 14-30A). Whether a tracer travels in an anterograde or retrograde fashion is mostly determined empirically. Retrograde tracers likely bind receptors that are selectively enriched in axon terminals, are taken up by the axon terminals via endocytosis, and utilize endogenous retrograde axonal transport systems to reach cell bodies (Section 2.3). Anterograde tracers are likely taken up either selectively by receptors in the cell body or nonselectively, with the much greater cell body volume favoring greater uptake by cell bodies than by axon terminals. Anterograde and retrograde tracing are often performed in conjunction with *in situ* hybridization, immunostaining, or Nissl stain; the combination of location, axonal projection, and gene/protein expression have been used to define neuronal types. Much of our knowledge about the connections between different regions in the mammalian brain has been obtained through experiments utilizing such tracers.

A limitation of classical tracers is that all cells at the injection site take up the tracer; therefore, the projection patterns revealed by these methods represent contributions from a combination of different cell types. Much data indicate that different cell types within the same region can have distinct projection patterns.

(A)

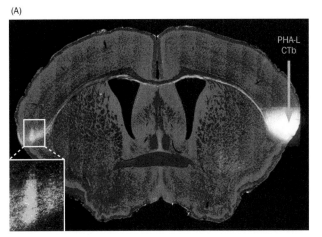

(B)

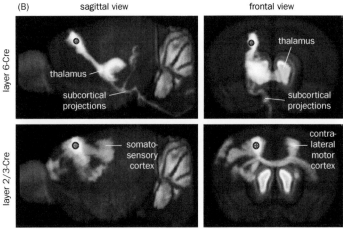

Figure 14-30 Examples of methods for tracing long-distance neuronal projections. (A) A mixture of phytohemagglutinin (PHA-L, green; an anterograde tracer) and cholera toxin subunit b (CTb, magenta; a retrograde tracer) was injected into the right insular cortex (arrow) of a mouse brain. This coronal section is stained in blue with a fluorescent Nissl stain that labels all cell bodies. PHA-L-labeled axons project to the left (contralateral) insular cortex, where CTb-labeled cell bodies (magenta dots in inset) represent retrogradely labeled insular cortical cells in the left hemisphere that project to the injection site. Green and magenta areas elsewhere represent projection sites from or to the right insular cortical neurons, respectively. **(B)** Adeno-associated virus encoding Cre-dependent GFP was injected into the motor cortices of mice expressing Cre recombinase in layer 6 or layer 2/3 neurons. The projection patterns of layer 6 (top) and layer 2/3 (bottom) cortical neurons are shown in sagittal views at left and frontal (coronal) views at right. Circles represent the injection sites. Cre-expressing layer 6 neurons project primarily to the ipsilateral thalamus (with some contralateral thalamic projections) and subcortical regions. Cre-expressing layer 2/3 neurons project mostly within the neocortex. (A, courtesy of Hongwei Dong. See also www.mouseconnectome.org and Zingg B, Hintiryan H, Gou L, et al. [2014] *Cell* 156:1096–1111. B, courtesy of Hongkui Zeng. See also http://connectivity.brain-map.org and Oh SW, Harris JA, Ng L, et al. [2014] *Nature* 508:207–214.)

For instance, many types (~45 in the mouse) of retinal ganglion cells with projections to distinct brain targets (for example, only ipRGCs project to the suprachiasmatic nucleus; Section 4.19) intermingle in the retina; in addition, the types that do project to the same target region can have distinct connection patterns within a target region. Tracing the connections of neurons not only from a particular location but also of a particular cell type thus provides higher resolution in mapping projection patterns. In the mouse, cell-type-specific anterograde tracing can be implemented by injecting AAVs (Table 14-1) encoding markers in a Cre-dependent manner into transgenic lines expressing Cre recombinase in specific cell types (Figure 14-30B). Since many cell-type-specific Cre mouse lines are available, axonal projections from different cell types within a given region can be studied independently.

Projection patterns of individual neurons can be further deciphered by a variety of strategies. For example, a genetic strategy called "brainbow" was developed to label individual neurons with different colors in mice using stochastic Cre-

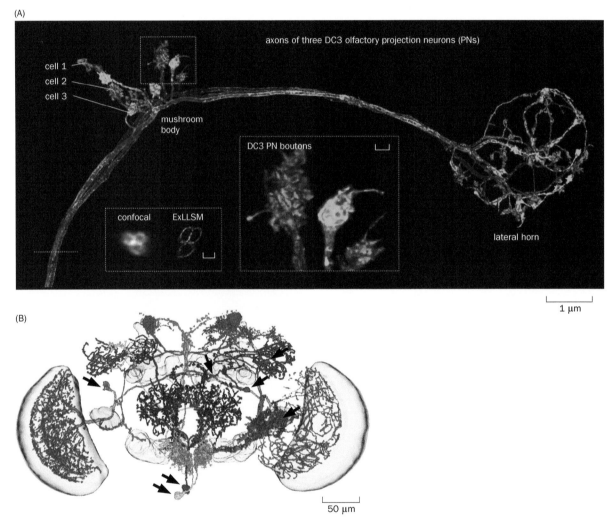

Figure 14-31 **Resolving multiple labeled neurons in the same brain.** (A) Axonal projections of three *Drosophila* olfactory projection neurons expressing a marker driven by the GAL4/*UAS* system can be individually resolved using ExLLSM, a combination of <u>ex</u>pansion microscopy and <u>l</u>attice <u>l</u>ight-<u>s</u>heet <u>m</u>icroscopy, and differentially colored *post hoc*. Their presynaptic boutons in the mushroom body (magnified in the right inset) and lateral horn are readily resolved. The left inset shows a side-by-side comparison of imaging of a cross section of the axon bundle at the position of the dotted line using confocal microscopy and ExLLSM (scaled back to the original size). All scale bars are 1 μm corresponding to preexpansion. (B) Seven neurons, one from each of seven *Drosophila* brains, were individually labeled via the MARCM method using seven different GAL4 lines (representing six different neurotransmitter types); the neurons' projection patterns were imaged in whole mount using confocal microscopy. The seven brains were then individually registered to the same standard brain using a presynaptic marker counterstain (not shown), and the transformation used to register the counterstain was applied to individually labeled neurons, yielding the image. Arrows indicate the seven cell bodies. (A, from Gao R, Asano SM, Upadhyayula S, et al. [2019] *Science* 363:eaau8302. With permission from AAAS. B, from Chiang AS, Lin CY, Chuang CC, et al. [2011] *Curr Biol* 21:1–11. With permission from Elsevier Inc.)

dependent expression of different levels of three fluorescent proteins; the labeling of many neurons in different colors can allow multiple neurons to be visualized and traced within the same brain (Figure 1-12C). Similar strategies have also been developed in other organisms, including *Drosophila* and zebrafish. Alternatively, combining expansion microscopy and light-sheet microscopy, researchers could resolve individual axonal projections when multiple axons were stained in the same color (**Figure 14-31**A).

An alternative strategy to resolving axonal projections of multiple neurons in the same brain is to generate many singly labeled neurons, each in different brains, and then register the images by aligning them to a common standard brain using a counterstain, such that the projection patterns of individual neurons can be compared (Figure 14-31B). While its success depends on the accuracy of image registration, this strategy has the advantage that individual neurons can be labeled using markers driven from different cell-type-specific regulatory elements. The relationships between different cell types, including potential pre- and postsynaptic partners, can thus be examined. However, proximity of the axons of neuron A to the dendrites of neuron B is necessary but not sufficient for them to be synaptic partners. To reconstruct wiring diagrams with synaptic resolution, it is necessary to use the higher resolution anatomical methods discussed in the next section and the physiological methods discussed in Section 14.26.

14.19 Mapping synaptic connections reveals neural circuitry

The most reliable anatomical method for determining whether or not two neurons form synaptic connections is EM, as synaptic connections can be directly visualized by EM. In principle, reconstructions of serial EM sections can establish neuronal wiring diagrams at the synaptic level (also called a **connectome**); this has been completed for the entire *C. elegans* nervous system (Figure 14-2). **Serial EM reconstruction** of a connectome involves making thin sections (50 nm or less) of the tissue of interest, imaging the sections, segmenting individual images into cell bodies and axonal and dendritic profiles belonging to specific neurons, aligning consecutive images to create a three-dimensional volume, and reconstructing individual segments across image stacks to create three-dimensional volumes of individual neurons (**Figure 14-32**; Movie 14-2). Synaptic connections between different profiles in EM sections can then be assigned to specific neurons. For larger nervous systems this procedure is not only labor intensive but also demands tremendous technical precision at each step, as even a very small error rate in a single section may propagate through the tracing of axonal and dendritic profiles across many thousands of sections and compromise the quality of the reconstructed connectome.

Intensive effort is currently focused on tackling these challenges. For example, in serial block-face scanning electron microscopy, a microtome is integrated with a scanning electron microscope; the cut face of a tissue block is imaged, and then each imaged section is sliced away to allow the next section in the tissue block to be imaged. Because successive images are acquired from the same tissue block, images are automatically aligned. A variant of this method is focused ion beam scanning electron microscopy (FIB-SEM), in which tissue is sequentially sectioned away using a focused ion beam instead of a diamond knife, achieving z-resolution of 10 nm or less at the expense of reduced speed. These and other methods, including traditional transmission EM, have enabled reconstruction of synaptic connections of many hundreds of cells in the retina and primary visual cortex of mice as well as in the fly optic medulla (Figure 14-31) and, most recently, the entire adult fly and larval zebrafish brains. These studies have yielded new insights into biological problems, such as direction-selective responses of visual neurons in the mammalian retina and fly optic lobe (Section 4.16 and Box 4-3).

Serial EM is currently the only technique that can reconstruct the entire synaptic connections of all neurons involved in a circuit. At the same time, it provides detailed structural information about other intracellular organelles in all cells. It has been applied to small nervous systems or a local region in the mammalian nervous system. However, it is highly technically demanding to apply serial EM to

Figure 14-32 Constructing a wiring diagram via serial electron microscopy. **(A)** A representative EM micrograph of the *Drosophila* medulla in the optic lobe (Box 4-3). **(B)** High-magnification view of the box in Panel A, showing a presynaptic terminal (red arrow) in contact with four postsynaptic profiles (blue arrowheads). Another profile (green dot) is in contact with the presynaptic terminal but lacks a postsynaptic density in this and adjacent sections, so it was not scored as a synaptic partner. **(C)** The micrograph in Panel A is segmented into profiles of neurites (axons and dendrites) by different colors. **(D)** Neurites are reconstructed by linking profiles in thousands of consecutive sections (of which three are shown here) to construct the three-dimensional object shown on the right. See also Movie 14-2. (From Takemura S, Bharioke A, Lu Z, et al. [2013] *Nature* 500:175–181. With permission from Springer Nature.)

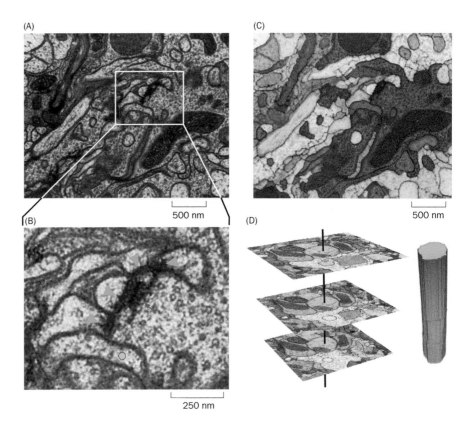

reconstruct the wiring diagram of the entire brain of a mouse, let alone larger mammals. An important challenge in mammalian nervous systems is to distinguish functional motifs versus stochastic and experience-dependent variations in the wiring diagram (see Figure 7-30 for an example of stochasticity in the neuromuscular connectome). Other methods utilizing genetic and viral tracing have been invented to identify synaptic connections between neurons separated by large distances. A powerful strategy for discovering both local and long-distance connections is **trans-synaptic tracing**. An ideal trans-synaptic tracer should have the following properties: if the tracer is expressed in a given neuron (a **starter cell**), a retrograde tracer should be received by only the complete set of neurons presynaptic to the starter cell, while an anterograde tracer should be received by only the complete set of neurons postsynaptic to it. Neurons, axons, and dendrites near the starter cell that do not form synaptic connections with the starter cell should not receive the tracer.

Trans-synaptic tracing strategies are still being developed and refined. The most efficient tracers thus far derive from neurotropic viruses such as **rabies virus** and herpes simplex virus, which spread within the nervous systems of their hosts by crossing synapses. Usually, once a virus infects a neuron, it spreads not only to that neuron's direct synaptic partners but also to the synaptic partners of subsequently infected neurons, making it difficult to determine whether two neurons are connected directly or through intermediate neurons. A strategy has been developed to prevent the retrograde rabies virus from crossing multiple synapses, thereby restricting infection to cells directly presynaptic to starter cells (**Figure 14-33**). To achieve this, the viral gene encoding the glycoprotein (a coat protein of the virus), which is essential for viral entry into a host cell, is replaced with a gene encoding GFP. This mutant rabies virus is pseudo-typed (Table 14-1) with an envelope protein (EnvA) from an avian virus that cannot transduce mammalian cells. Researchers can turn any neuron into a starter cell by expressing two transgenes (Figure 14-33A): one encoding TVA (the receptor for EnvA), so that the neuron can be transduced by pseudo-typed rabies viruses, and the other encoding the rabies glycoprotein to complement the glycoprotein gene deleted from the mutant rabies genome and enable the neuron to produce functional rabies viruses, which

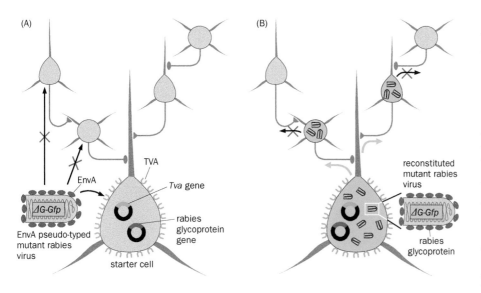

Figure 14-33 Retrograde monosynaptic tracing via modified rabies virus. (A) The gene encoding the rabies glycoprotein (G) is replaced by a gene encoding GFP in the rabies genome, such that this mutant rabies virus (ΔG-*Gfp*, bottom left) can no longer recognize and transduce normal mammalian neurons (red cross). This virus was assembled in a cell line that supplies the EnvA coat protein (blue) from an avian virus, such that the pseudo-typed rabies virus can transduce mammalian neurons (starter cells) expressing the EnvA receptor TVA (cyan). A transgene supplying rabies glycoprotein is also expressed in the starter cells. Circles inside the starter cell symbolize these transgenes. **(B)** When mutant rabies virus enters the starter cell via the EnvA–TVA interaction, G produced by the starter cell complements the G deficiency, producing ΔG-*Gfp* rabies virus with the endogenous rabies glycoprotein (bottom right), which can spread to the starter cell's presynaptic partners. Because the presynaptic partners do not express G, ΔG-*Gfp* rabies virus cannot spread farther. Both the starter cell and its presynaptic partners are labeled by GFP; an additional marker is typically introduced into starter cells (not shown) to distinguish them from their presynaptic partners. (From Wickersham IR, Lyon DC, Barnard RJO, et al. [2007] *Neuron* 53:639–647. With permission from Elsevier Inc.)

spread to the starter cell's presynaptic partners. However, since the presynaptic partners do not express rabies glycoproteins, mutant rabies viruses cannot spread further (Figure 14-33B). While rabies-transduced cells become unhealthy and die within weeks, trans-synaptic spread is sufficiently fast such that researchers can terminate the experiment and examine histological samples with intact starter cells and their presynaptic partners. This monosynaptic strategy for tracing the presynaptic partners of specific starter cell types has been applied to map synaptic connections in many parts of the mammalian nervous system (Figure 4-40; Figure 6-18; Figure 8-10).

In summary, many methods have been developed to construct wiring diagrams of the nervous system at different scales and resolutions. These methods range from noninvasive human brain imaging to the complete reconstruction of all synaptic connections in *C. elegans*. However, many technical challenges must be overcome to comprehensively map the synaptic connections of larger neural systems. Mapping of electrical synapses poses an additional challenge, as they are not as readily identifiable as chemical synapses, even in electron micrographs. Furthermore, a wiring diagram based on anatomical connections alone is just a first step toward understanding a neural circuit. To decipher how a neural circuit operates, researchers must also assess whether synapses are excitatory or inhibitory, how strong they are, and how they are influenced by the actions of modulatory neurotransmitters and neuropeptides (e.g., Sections 6.10 and 8.5). Understanding neural circuit function requires research tools that can measure and manipulate the activity of neurons in the wiring diagram in the context of animal behavior; we will discuss these methods in the next part of the chapter.

RECORDING AND MANIPULATING NEURONAL ACTIVITY

Signals in the nervous system spread predominantly by membrane potential changes. Thus, the ability to record membrane potential as a means of measuring neuronal activity is critical to our understanding of nervous system function. In this part of the chapter, we discuss the principal methods for recording neuronal activity.

While observations and measurements are the foundations of discovery, well-designed perturbation experiments are necessary to elucidate the underlying mechanisms. We will also discuss the loss- and gain-of-function approaches that researchers have employed to silence and activate neurons of interest and thereby identify the roles specific neurons and neuronal populations play in the function of neural circuits and animal behavior.

Figure 14-34 Three principal methods for electrophysiological recording. (A) In extracellular recording, the tip of an electrode is sufficiently close to the neuronal cell body such that when the neuron spikes, the electrode can detect the extracellular voltage changes. **(B)** In intracellular recording, a sharp electrode penetrates into the cell to measure intracellular membrane potential directly. **(C)** In whole-cell recording, a patch electrode forms a tight seal with the plasma membrane to measure intracellular membrane potential (see Figure 14-40 for more details).

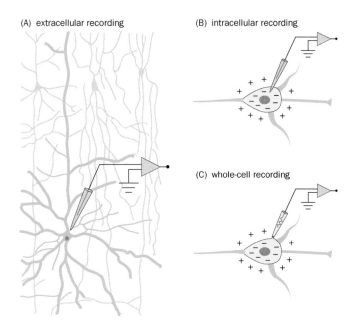

(A) extracellular recording (B) intracellular recording

(C) whole-cell recording

14.20 Extracellular recordings can reveal the firing patterns of individual neurons or ensembles

Three principal electrophysiological methods have been employed to record neuronal activity. In **extracellular recording**, an electrode, often made of metal probe insulated everywhere except at its tip, is placed near the outside of a neuronal cell body to record voltage changes when the neuron fires an action potential (**Figure 14-34**A). In **intracellular recording**, a sharp electrode, usually made of glass with a very fine, open, solution-filled tip, penetrates into the cell to directly record the intracellular membrane potential (Figure 14-34B). **Whole-cell recording** (abbreviated for whole-cell patch clamp recording) is a special form of intracellular recording in which the glass electrode forms a tight seal with the plasma membrane of the recorded cell (Figure 14-34C). We discuss extracellular recording in this section, and intracellular and whole-cell patch recording in the following section.

Although the simplest form of electrophysiology, extracellular recording has proven to be a powerful method. When a cell fires an action potential, the ionic flow creates voltage changes not only inside the cell but also in its immediate surroundings. When an electrode tip is close to a neuronal soma (Figure 14-34A), the signal picked up by the electrode predominantly reflects the spiking activity of this single nearby neuron (**Figure 14-35**A). Action potentials with a specific amplitude and waveform as detected in extracellular recordings define a "unit"; thus, extracellular recording aimed at detecting the firing patterns of individual neurons is also called **single-unit recording**. Sometimes an extracellular electrode can detect and distinguish action potentials from multiple neurons due to the distinct amplitudes or waveforms of action potentials from neurons at varied distances

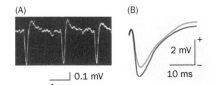

(A)

(B)

$\longrightarrow$ 0.1 mV
1 ms

2 mV
10 ms

Figure 14-35 Examples of single-unit and field potential recordings. (A) Three spikes of a cat retinal ganglion cell recorded by an extracellular electrode in response to light stimulation, from the study that described the receptive fields of retinal ganglion cells (Figure 4-20). **(B)** Two superimposed traces of local field excitatory postsynaptic potentials (fEPSPs) in the dendritic layer of the dentate gyrus in response to electrical stimulation of axons in the perforant path before (blue) and after (magenta) high-frequency stimulation, from the study that discovered long-term potentiation (Figure 11-7). Note that in both examples, the voltages recorded by extracellular electrodes become more negative as neurons are activated. This is because positive ions flow into the cell during the rising phase of the action potential (Panel A) or when postsynaptic cells are depolarized (Panel B). Note also that the time course of the spike in Panel A is much faster than that of the fEPSP in Panel B. (A, adapted from Kuffler SW [1953] *J Neurophysiol* 16:37–68. B, adapted from Bliss TVP & Lømo T [1973] *J Physiol* 232:331–356.)

from the electrode tip. This effect is amplified by the use of **tetrodes**, extracellular electrodes containing four wires enabling four independent recordings of spiking activity of neurons near the electrode. Firing patterns of up to 20 neurons may be resolved by tetrode recording. Single-unit extracellular recording of neuronal activity has been the key method for making many conceptual advances in neurobiology. As we learned in Chapter 4, for example, extracellular recording at successive stages in the visual processing pathway revealed how the brain's representation of a visual scene is transformed as the signals travel from the retina to the visual cortex. Extracellular recording is still the predominant method used for recording neuronal activity *in vivo* today.

Extracellular recording can also be used to measure **local field potentials**—local potential variations measured relative to a distal ground. Local field potentials can be measured by an electrode for single-unit extracellular recording, but high-frequency signals such as those produced by action potentials are removed by a low-pass filter. The remaining low-frequency signals reflect the collective dendritic and synaptic activity of many neurons near the electrode tip. We have seen the application of field potential recording to studies of hippocampal long-term potentiation *in vivo* (Figure 14-35B; Figure 11-7) and in acute brain slices *in vitro* (Figure 11-9).

To explore how groups of neurons together encode and process information, **multielectrode arrays** that record many neurons at the same time were developed. These may consist of many independent electrodes arranged horizontally in grids (**Figure 14-36**A) or vertically at different depths (Figure 14-36B). Multielectrode arrays can record tens to hundreds of neurons simultaneously *in vitro,* such as in brain slices or explants; for example, recording simultaneously from many retinal ganglion cells led to the discovery and characterization of retinal waves (Figure 5-21A). Multielectrode arrays can also be implanted *in vivo* to record the activity of many neurons in awake, behaving animals or humans (Figure 8-40; Figure 8-41).

Electroencephalography (EEG) is essentially a noninvasive field potential recording method that usually involves multielectrode arrays. In a typical EEG setting, the electrodes are attached to the surface of the scalp. As the distance between EEG electrodes and the recorded neurons is greater than in conventional extracellular recording, EEG can detect only the synchronized activity of tens of thousands of neurons or more. EEG enabled the discovery of brain waves of various frequencies and is an extremely useful tool for assessing brain states such as different forms of sleep (Figure 9-28A) and epileptic conditions (Figure 12-43).

Extracellular recordings are typically performed blind to the type of neuron being recorded. They have a bias toward detecting more common neuron types, larger neurons, and more active neurons, because extracellular recording procedures often involve the researcher shifting electrode positions until a spike is detected. Spike waveforms and firing properties from extracellular recordings have been used to classify cell types, but different cell types may have overlapping properties in most brain regions, making such classification ambiguous. This limitation has been addressed by a phototagging method, wherein single-unit recording is performed in a tissue in which neurons of specific types can be stimulated using optogenetics (Section 14.25). The cell types of the recorded units are then classified based on whether or not they respond to optical stimulation (**Figure 14-37**). A caveat of this phototagging method is that spikes can also be driven

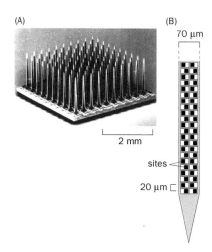

Figure 14-36 Multielectrode arrays.
(A) Multielectrode arrays like this 10 × 10 silicon-based prototype have been widely used in neural prosthetics for recordings of cortical neurons (Section 8.15).
(B) Schematic of the lower portion of a silicon-based Neuropixel probe, which has 384 recording channels that can be programmed to address any of the 960 sites (black squares) along the probe. (A, from Campbell PK, Jones KE, Huber RJ, et al. [1991] *IEEE Trans Biomed Eng* 38:758–768. With permission from IEEE. B, from Jun JJ, Steinmetz NA, Siegle JH, et al. [2017] *Nature* 551:232–236. With permission from Springer Nature.)

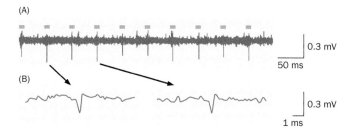

Figure 14-37 Phototagging to identify cell types in extracellular recordings. Extracellular recording of ventral tegmental area neurons was performed in a transgenic mouse that expressed Cre in dopamine neurons and was transduced with an AAV encoding a Cre-dependent channelrhodopsin (ChR2). **(A)** Each photostimulus (cyan bar) resulted in a spike of the recorded neuron. **(B)** Two spikes in Panel A are temporally magnified. These results are consistent with recording of a dopamine neuron expressing ChR2 and producing action potentials in response to depolarization caused by photostimuli. (Adapted from Cohen JY, Haesler S, Vong L, et al. [2012] *Nature* 482:85–88. With permission from Springer Nature.)

indirectly in non-opsin-expressing cells: extremely short latencies between light onset and spiking (~1 ms), or certainty that there is no local excitation, are required for confident cell identification.

14.21 Intracellular and whole-cell recordings can measure synaptic input as well as firing patterns

Compared to extracellular recordings, intracellular recordings with a sharp electrode (Figure 14-33B) offer much higher sensitivity and signal-to-noise ratio for detecting electrical signals. Intracellular recordings can detect not only firing patterns but also subthreshold membrane potential changes resulting from excitatory, inhibitory, or gap junctional input received by the recorded neuron. Many discoveries about neuronal communication we discussed in Chapters 2 and 3, including the ionic basis of the action potential and the mechanisms of synaptic transmission, were made using intracellular recording methods. However, intracellular recording with a sharp electrode can be applied only to large cells, such as muscle cells or giant neurons.

Whole-cell recording (Figure 14-34C) is a widely used variation of the patch clamp method (**Box 14-3**), as it can be used for intracellular recording of small neurons. In this technique, the interior of the electrode forms a continuous compartment with the cytoplasm, allowing the electrode to measure the membrane potential with high sensitivity, comparable to intracellular recording with a sharp electrode. The patch electrode affords greater access to the cell—the tip diameter is usually 1 μm or so for a patch electrode, an order of magnitude greater than the diameter of an intracellular sharp electrode. As a result, whole-cell patch electrodes not only can record the activity of a neuron but can also pass currents that change a neuron's membrane potential, effectively achieving **voltage clamp** (Figure 2-21). Because of this, inhibitory input to a neuron can be recorded in isolation when the neuron is voltage clamped at the reversal potential of the glutamate receptors, such that excitatory input does not make a contribution (**Figure 14-38A**). Likewise, excitatory input to a neuron can be recorded without interference from inhibitory input when the neuron is voltage clamped at the Cl⁻ reversal potential (Figure 14-38B). These techniques are often combined with pharmacological blockers of specific receptors to dissociate excitatory and inhibitory input to specific neurons. Whereas voltage clamp is used to measure current flow across the neuronal membrane, whole-cell recording can also be in a **current clamp** mode, which is used to measure membrane potential changes while holding the current at a set level (Figure 3-23).

Because the solution that fills the patch electrode is continuous with the cell's cytoplasm, macromolecules such as dyes or fluorescent markers can be included in the patch electrode solution to label the recorded cell (**Figure 14-39**); signaling molecules can also be included to alter the properties of the recorded cell (Figure 11-12B). Dye fill during whole-cell recording (or intracellular recording with a

Figure 14-38 Using voltage clamp to dissociate inhibitory and excitatory input. (A) When the voltage in a whole-cell recording is clamped at 0 mV, at the reversal potential of glutamate receptors (Section 3.15), cation influx and efflux balance out even if glutamate receptor channels are open upon glutamate binding. The measured current is mostly contributed by Cl⁻ flow through the GABA$_A$ receptor in response to GABA, thereby reflecting inhibitory input. **(B)** When the voltage is clamped at –65 mV, around the reversal potential of GABA$_A$ and glycine receptors (that is, the equilibrium potential of Cl⁻; Section 3.18), the measured current reflects mostly excitatory input from glutamate receptor channels. V_{CMD}, command voltage; V_p, voltage of the patch electrode.

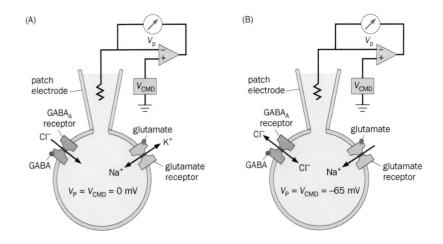

Figure 14-39 Dye fill during whole-cell recording from the dendrite and cell body. In this
example, two patch electrodes are used for whole-cell recording at the cell body (bottom) and
apical dendrite (top) of a cortical pyramidal cell in a rat brain slice. Dual patching of the cell body
and dendrite of the same neuron is confirmed by the mixing of two different fluorescent dyes,
one from each electrode. This dual-patch approach has been used to study synaptic potential
propagation from the dendrite to the cell body and action potential backpropagation from the cell
body to the dendrite. (From Stuart GJ & Sakmann B [1994] *Nature* 367:69–72. With permission
from Springer Nature.)

sharp electrode) can be used to reveal the location, morphology, and projection
pattern of a recorded neuron (Figure 14-1). Such information is highly valuable
in correlating neuronal structure and function (Figure 4-47).

Another important advantage of intracellular recording is that it allows a
genetically defined cell population to be targeted for recording via simultaneous
optical imaging. A given brain region almost always contains a mix of different
neuronal types with varying densities. As discussed in the previous section, blind
recording cannot distinguish unequivocally between different cell types, and it
may prevent rare cell types from being recorded. The job of an electrophysiolo-
gist is made easier if the type of neuron targeted for recording is prelabeled with
fluorescent protein using genetic strategies. For example, recording specific pairs
of genetically labeled pre- and postsynaptic partner neurons in the *Drosophila*
olfactory system enabled investigators to examine how olfactory representations
are transformed as signals travel from presynaptic olfactory receptor neurons to
postsynaptic projection neurons (Section 6.12). With advances in genetic technology for accessing specific neuronal types (Section 14.11), such targeted electrophysiology experiments have become increasingly powerful at revealing how
neural circuits process information.

Box 14-3: Patch clamp recordings can serve many purposes

Patch clamp recording requires formation of a very tight
seal (with a resistance in the range of gigaohms, or 10^9 ohms)
between a glass patch electrode, also called a patch pipette,
and the plasma membrane of a target neuron (**Figure
14-40A**). Due to this high resistance, the very small currents
that pass through individual ion channels in the membrane
underneath the patch electrode can be recorded. Indeed,
the measurement of single-channel conductance in a defined
extracellular environment (the internal solution of the patch
electrode) in this **cell-attached recording** mode (Figure
14-40B) was the first application of the patch clamp technique
(Figure 2-30). Patch clamp can also be used in a number of

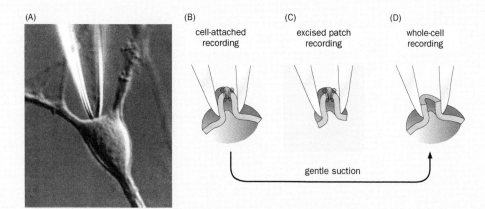

Figure 14-40 The multifunctionality of patch clamp recording.
(A) Photomicrograph of a patch electrode in contact with the plasma
membrane of a neuron in culture. **(B–D)** Three patch clamp modes
are schematized. In a cell-attached patch (Panel B), the patch
electrode forms a tight seal with a neuron's plasma membrane,
allowing measurement of ion flow through a single channel in the
patch of membrane underneath the electrode. In an excised patch
(Panel C), the piece of membrane underneath the electrode is
excised from the cell and can be placed in defined media to study
the properties of the channel. In whole-cell recording (Panel D), the
membrane patch under the electrode is ruptured, such that the
interiors of the electrode and the recorded cell become a single
compartment. (From Neher E & Sakmann B [1992] *Sci Am* 266[3]:
44–51. With permission from Springer Nature.)

(Continued)

Box 14-3: continued

additional modes. For example, the membrane patch underneath the electrode can be excised (Figure 14-40C). The excised patch can be placed in a defined solution so that both the extracellular and intracellular environments of the ion channels in the patch can be controlled. **Excised patch recording** is widely used to study the biophysical and biochemical properties of ion channels (Figure 4-9). In the **whole-cell recording** mode, gentle suction applied to a cell-attached patch ruptures the membrane underneath the electrode, such that the interiors of the electrode and the cytoplasm of the recorded cell form a single compartment (Figure 14-40D). Whole-cell recording is one of the most widely used techniques in measuring neuronal activity in brain slices.

As the electrode internal solution and cytoplasm are continuous in whole-cell recording, cellular contents may be diluted with solution from the patch electrode during prolonged recording. A procedure called perforated patch can be used to minimize this issue; in this method, the membrane between the patch electrode and the cell is not fully ruptured as in whole-cell mode, but the internal solution of the patch electrode contains chemicals that make small holes in the underlying neuronal membrane, such that the patch electrode can record the current and voltage of the cell with minimal macromolecular exchange. In another variation called **loose-patch recording**, a patch electrode is placed against the cell membrane without forming a gigaohm seal. Recording in loose-patch mode thus does not affect cellular content. However, this approach does not provide sufficient sensitivity for recording subthreshold activity; it can record only action potentials. Loose-patch recording is commonly used in *in vivo* recordings. Compared with conventional extracellular recording, it does not have a bias toward active neurons, as the criterion it uses to assess whether the electrode is approaching a cell is a sudden increase in resistance, rather than detection of action potentials. Furthermore, DNA can be added to the internal solution of the patch electrode and introduced specifically into the recorded cell by electroporation after recording. One application of this single-cell electroporation technique is to introduce DNA encoding a fluorescent protein so that the same cell can be located and recorded at a later time.

14.22 Optical imaging allows simultaneous measurement of the activity of many neurons

To appreciate a symphony, it is not enough to hear one instrument at a time; the listener must be able to hear all of the orchestra's musical instruments simultaneously. Likewise, a deep understanding of how neural circuits encode and process information requires that researchers capture the simultaneous activity of many (ideally all) neurons in the circuit. Even with multielectrode arrays, investigators can record only hundreds of neurons at a time. In addition, the spacing of recorded neurons is constrained by the spacing of the electrodes. The only technology currently available for recording the activity of all neurons within a region at cellular resolution is **optical imaging**, which uses changes in fluorescence or other optical properties as indicators of neuronal activity.

Since neurons communicate by membrane potential changes, the ideal indicator would report voltage changes directly. Indeed, many variants of **voltage indicators**, which change fluorescence intensity or other optical properties in response to membrane potential changes, have been developed and begun to be used for *in vivo* recording (Figure 4-31). Other optical sensors have been developed to enable imaging of neurotransmitter release and receptor conformation changes resulting from neurotransmitter binding as proxies for neuronal activity. Due to their superb sensitivity and signal-to-noise ratio, the most widely used sensors of neuronal activity are Ca^{2+} **indicators**, which translate changes in intracellular Ca^{2+} concentration ($[Ca^{2+}]_i$) into changes in fluorescence. A rise in $[Ca^{2+}]_i$ usually accompanies neuronal activation due to activation of postsynaptic neurotransmitter receptors permeable to Ca^{2+} and the opening of voltage-gated Ca^{2+} channels in response to depolarization in both cell bodies and presynaptic terminals.

Some Ca^{2+} indicators of neuronal activity are made from synthetic chemicals, while others are protein based. Chemical indicators typically link a Ca^{2+}-chelating moiety with a fluorophore. For example, binding of Ca^{2+} to the chemical indicator **fura-2** shifts the wavelength of maximal fluorescence excitation about 30 nm shorter (**Figure 14-41**A). Thus, the ratio of fluorescence intensity measured at the excitation wavelengths of 350 nm and 380 nm can be used as a sensitive measure

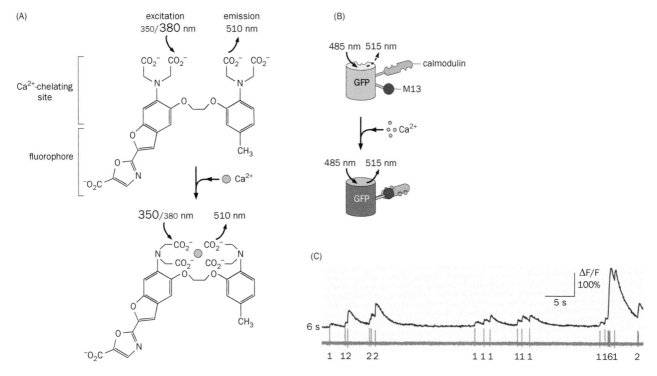

Figure 14-41 Chemical and genetically encoded Ca²⁺ indicators.
(A) Fura-2 is a chemical Ca²⁺ indicator consisting of a fluorophore fused to a Ca²⁺-chelating site from the Ca²⁺ buffer EGTA. When [Ca²⁺]ᵢ is low, excitation at 380 nm produces stronger fluorescence emission than excitation at 350 nm (symbolized by the size of the numbers); when [Ca²⁺]ᵢ is high, the converse is true. Thus, ratiometric imaging at 350/380 serves as a sensitive reporter of [Ca²⁺]ᵢ. **(B)** Design principle of GCaMP. A permuted GFP is restored to its native three-dimensional structure with an associated increase in fluorescence after Ca²⁺-triggered binding of M13 and calmodulin, which are at the ends of GCaMP. The fluorescence intensity of GCaMP can thus be a readout of [Ca²⁺]ᵢ. **(C)** Single action potentials reliably induce fluorescence changes of GCaMP6 in mouse visual cortical neurons *in vivo*, as measured by simultaneous loose-patch recording to identify action potentials (bottom) and GCaMP6 fluorescence intensity changes (top). ΔF/F is the ratio of fluorescence intensity change (ΔF) over basal fluorescence intensity (F). Note that when action potentials occur in rapid succession (indicated by numbers below), individual action potentials cannot be resolved as individual peaks by imaging. (A & B, after Grienberger C & Konnerth A [2011] *Neuron* 73:862–885. With permission from Elsevier Inc. C, after Chen TW, Wardill TJ, Sun Y, et al. [2013] *Nature* 499:295–300. With permission from Springer Nature. See also Grynkiewicz G, Poenie M, & Tsien RY [1985] *J Biol Chem* 260:3440–3450; Nakai J, Ohkura M, & Imoto K [2001] *Nat Biotech* 19:137–141.)

of [Ca²⁺]ᵢ. Protein-based Ca²⁺ indicators are also called **genetically encoded Ca²⁺ indicators**, as they can be expressed transgenically in specific cell types. The most widely used genetically encoded Ca²⁺ indicators are a series of engineered proteins called **GCaMPs**, which report a rise in [Ca²⁺]ᵢ by increasing fluorescence intensity. In GCaMP, the positions of two halves of the green fluorescent protein GFP are switched, and the swapped halves are linked by calmodulin and its target peptide M13. Ca²⁺-triggered binding of calmodulin to M13 restores GFP's original conformation, hence increasing its fluorescence intensity (Figure 14-41B).

Chemical indicators have offered superior sensitivity to genetically encoded indicators in the past and have been widely used in experiments described in this book (e.g., Figure 4-45C). However, protein engineering has generated genetically encoded Ca²⁺ indicators, such as GCaMP6, that can detect single action potentials *in vivo* more reliably than chemical indicators (Figure 14-41C), achieving a milestone in optical imaging. Moreover, expression of genetically encoded Ca²⁺ indicators can be specified by cell type and can enable repeated imaging of the same neuron over the course of months, which is not possible with chemical indicators. These developments have facilitated many recent discoveries and will continue to enable new discoveries.

In addition to activity indicators, optical imaging requires suitable microscopes for visualizing fluorescence changes in live tissues with high spatial resolution, high sensitivity, and minimal photodamage to imaged tissues. Conventional fluorescence microscopes have poor resolution along the *z* axis and are best used with relatively thin tissues such as retinal explants (Figure 5-21B). As described

Figure 14-42 Laser-scanning two-photon microscopy. The design of a two-photon microscope is similar to that of a confocal microscope (Figure 14-22A), except that there is no pinhole at the detector. **(A)** Simultaneous absorption of two long-wavelength (lower energy, red) photons brings a fluorescence indicator molecule to the excited state, from which it subsequently relaxes while emitting fluorescence (green). **(B)** Schematic of a microscope objective on top of a brain tissue sample (cyan) focusing on a portion of a dendrite. The near-simultaneous absorption of two infrared photons (the left and middle photon paths) excites fluorophores at the focal plane; the excitation light outside the focal plane due to scattering (the photon path on the right) is insufficiently intense for two-photon excitation, so excitation is restricted to the imaging spot at the focal plane (highlighted in yellow). **(C)** All emitted photons (green paths) collected by the objective, including those resulting from scattering (indicated by *), contribute to the fluorescence signal from the imaging spot. (From Svoboda K & Yasuda R [2006] *Neuron* 50:823–839. With permission from Elsevier Inc. See also Denk W, Strickler JH, & Webb WW [1990] *Science* 248:73–76.)

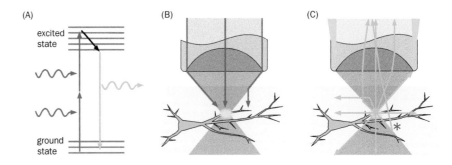

earlier, laser-scanning confocal microscopes have high resolution along the *z* axis (Figure 14-23A). However, tissue above or below the imaging plane is nevertheless exposed to laser excitation during scanning, causing photobleaching of the fluorescence and photodamage via tissue heating. In addition, light scattering in neural tissue limits the depth of effective imaging by confocal microscopy to about 100 μm from the sample surface.

The most widely used method for *in vivo* optical imaging is **laser-scanning two-photon microscopy**, which relies on simultaneous absorption of two long-wavelength photons to excite a fluorophore (**Figure 14-42A**). Only at the focal plane is the density of photons high enough to cause substantial fluorescence emission, so photobleaching is limited to the focal plane (Figure 14-42B). The longer excitation wavelength allows light to penetrate more deeply into neural tissue. Furthermore, because fluorescence excitation is limited to the focal plane, any emission photons collected by the objectives, including those resulting from scattering, can be used without a pinhole (Figure 14-42C; compared with Figure 14-23A). As such, two-photon microscopes collect emitted photons more efficiently than confocal microscopes and require less excitation energy, resulting in less tissue damage (and less photobleaching even at the focal plane). Many optical imaging studies discussed in this book utilized two-photon microscopy, including Ca^{2+} imaging of neuronal activity (Figure 4-45C) and structural imaging of neuronal morphology (Figure 11-49).

Even two-photon microscopes can only effectively image tissue under 1 mm from the sample surface. This restriction prevents imaging of most of the nervous system of even small mammals such as mice. Additional imaging methods have been developed to overcome these limitations. For example, gradient refractive index (GRIN) lens-based fluorescence endoscopy allows imaging of deeper tissues (see **Box 14-3** for an example). One limitation is that insertion of the lens (typically with a diameter of 500–1000 μm) causes substantial damage to the tissue above the imaging area. An alternative method is **fiber photometry**, which uses a thinner fiber (100–200 μm) to measure bulk fluorescence activity from specific neuronal populations expressing genetically encoded activity indicators. Fiber photometry produces less damage to the overlaying tissue but cannot resolve activity from individual neurons. It is nevertheless powerful when used in conjunction with genetic targeting of specific neuronal types (e.g., Figure 8-24; Figure 9-12).

In addition to fluorescence-based imaging, other imaging techniques such as **intrinsic signal imaging** (Figure 4-45B; Figure 6-13) and **functional magnetic resonance imaging (fMRI)** (Figure 1-24; Figure 4-54; Figure 11-38) also report neuronal activity. These techniques use blood flow near excited neurons as an indicator of neuronal activity, as increased neuronal activity is associated with changes in blood flow and oxygenation. These imaging methods have poorer spatial and temporal resolution than fluorescence imaging of individual neurons. For instance, the current resolution of fMRI is about 2 mm in the linear dimension; 8 mm^3 of tissue contains hundreds of thousands of neurons. Nevertheless, an important advantage of fMRI is its noninvasiveness, making it a favored tool for imaging neuronal activity in the human brain.

Box 14-4: From *in vitro* preparations to awake, behaving animals: a comparison of recording methods

What method should researchers choose for recording neuronal activity? The answer depends on the biological question the experimenter intends to address and the preparation used to address the question. In this box, we compare the pros and cons of the recording methods discussed in Sections 14.20–14.22 in the context of different experimental preparations: from neuronal cell culture and tissue explants *in vitro* to anesthetized animals and awake, behaving animals *in vivo* (Table 14-2).

Electrode-based recording methods directly measure membrane potentials and hence have superb sensitivity and temporal resolution. They are widely used in reduced preparations such as cultured neurons or *in vitro* explants such as brain slices, providing insight into the properties of individual neurons and helping researchers investigate synaptic transmission and plasticity as assayed from single neurons or a group of neurons in aggregate (in the case of local field potentials). Reduced preparations also offer easier access to target neurons, permitting greater use of intracellular recording methods such as whole-cell recording. Intracellular techniques feature superb sensitivity, allowing detection of subthreshold membrane potential changes; they also support an array of approaches through which researchers can manipulate the recorded neurons by injecting currents or molecules. Whole-cell recording has been the predominant

method for studies using slice preparations of the mammalian brain. Optical imaging methods can be employed in reduced preparations if recording the activity of many individual neurons is desired (for instance, to observe retinal waves) or when subcellular resolution of neuronal activity is needed.

Recording from *in vivo* preparations poses more challenges regardless of the method because intact systems have greater inherent complexity, the neurons to be recorded are harder to access, and the animals must be kept alive (and, in many cases, awake and behaving) during the recording process. However, obtaining recordings from living animals, including from awake, behaving animals, is essential for addressing many questions in neurobiology related to perception, cognition, and the neural basis of behavior (e.g., Sections 4.28, 8.12–8.15 and Box 11-2). Recording with extracellular electrodes has historically been the predominant method *in vivo*, particularly in awake, behaving animals, due to its overall superior ability to maintain stable recordings (to record from the same cells over long periods of time), penetrate deep into tissues, and simultaneously record from many neurons (Table 14-2). For example, the discovery of place cells in the hippocampus and grid cells in the entorhinal cortex relied on extracellular recordings in freely moving rodents (Box 11-2). Intracellular recordings are difficult to

Table 14-2: Comparison of electrophysiological and optical imaging methods for recording neuronal activity

Property	Electrophysiology		Optical imaging with Ca²⁺ indicators[a]
	Extracellular recording	**Intracellular recording**	
Sensitivity to electrical signals	Spikes and local field potentials	Spikes and subthreshold activity	Generally less sensitive[b]
Spatial resolution	Cellular to network	Cellular to subcellular[c]	Cellular and subcellular
Temporal resolution	<1 millisecond	<1 millisecond	10s to 100s of milliseconds
Number of neurons recorded simultaneously	Up to hundreds	At most several	Thousands or more
Stability during movement	Relatively stable	Poor	Poor[d]
Depth of recording	Any depth	Easier superficially	Limited[e]
Repeated recording	Days to weeks	Limited to one session of 10s of minutes	Hours with chemical indicators; months with protein indicators
Cell-type-specific recording	Poor[f]	Poor[g]	Excellent with protein indicators
Biases	Active cells; large cells; common cell types	Large cells	Cells that take up or express indicators well

[a]Most other indicators have similar properties. However, voltage indicators have temporal resolutions that approach those of electrophysiology. Conversely, the rapid dynamics of voltage indicators means the number of neurons that can be recorded "simultaneously" is limited by the scanning speed of laser-scanning fluorescent microscopy.

[b]GCaMP6 can detect single action potentials but cannot resolve high-frequency spikes (Figure 14-41C). In general, Ca²⁺ transients are an indirect measurement of spikes and may include other signals.

[c]Whole-cell patch recording can be applied to large dendrites (Figure 14-39).

[d]Can be greatly improved by head fixation and motion correction algorithms in image analysis.

[e]At <10 μm with conventional fluorescence microscopy for cellular resolution; ~100 μm with confocal microscopy; <1 mm with two-photon microscopy. GRIN lenses can extend the depth of imaging (Figure 14-43C).

[f]Achievable via phototagging (Figure 14-37).

[g]Achievable by expressing fluorescent proteins in a specific cell type and targeting fluorescent cells.

(Continued)

Box 14-4: continued

maintain when animals are awake and moving, because physical movement can shift the electrode. Likewise, optical imaging requires stability and is sensitive to movement. However, intracellular recording offers superior sensitivity, and optical imaging allows recording of many neurons at once and over long periods of time. Thus, researchers have developed methods to improve stability for *in vivo* recordings.

One way to provide the mechanical stability essential for both intracellular recording and optical imaging is to restrain head movement with respect to the recording equipment, such as a microscope or a micromanipulator that holds an electrode. Such head-fixed preparation was originally developed for extracellular recording of neurons in the visual cortex of awake, behaving monkeys; if the monkey cannot move its head and is trained to fixate, then a stimulus on the screen always falls on the same spot of the retina (Figure 4-55). Head-fixed mice can, for example, be trained to associate an odor with a water reward and to lick to fetch the reward while cortical neurons are being optically imaged (**Figure 14-43**A). A more sophisticated preparation involves virtual reality feedback. For example, a head-fixed mouse navigates on a spherical treadmill made of a ball floating on air; the mouse's movement is used to adjust the visual scene as if the animal were navigating a real environment (Figure 14-43B; **Movie 14-3**). The mouse can be trained to run on linear tracks or make turning choices in virtual reality while its neuron(s) are being recorded with a whole-cell patch electrode or two-photon microscope. An alternative to head-

fixed preparations is to attach a miniature fluorescence microscope to the head of a freely moving mouse, enabling recording of neuronal activity while the animal explores its environment (Figure 14-43C and **Movie 14-4**). The use of fluorescence endoscopy in conjunction with such a miniature microscope has allowed imaging of neuronal activity in many deep brain regions during behavior, such as hypothalamic nuclei while animals engage in mating (Figure 10-32).

Recent advances in optical imaging have enabled researchers to record the simultaneous activity of a majority of the neurons in *C. elegans* and the larval zebrafish brain (up to 100,000 neurons) with cellular resolution, taking advantage of their transparency. Using large cranial windows and advanced optical engineering, simultaneous imaging of tens of thousands of neocortical neurons in the mouse, or two regions such as the neocortex and cerebellum, has become possible. These technical advances have enabled researchers to trace information flow across the brain during behavior and study how different brain regions interact with each other.

Still, in the mammalian brain, optical imaging can provide information on the activity of only a small fraction of the entire nervous system. A strategy that in principle provides information about the activity of neurons throughout the nervous system is accessing the expression patterns of immediate early genes, which are activated when an animal experiences a given sensory stimulus or behavioral episode (Section 3.23). Important limitations of this strategy include

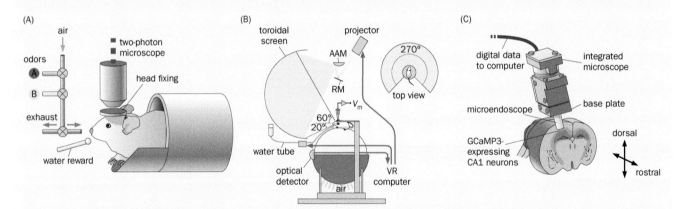

Figure 14-43 Three strategies for recording neurons in behaving mice. (A) In this head-fixed preparation, a metal plate is surgically attached to the mouse's head. During a two-photon imaging experiment, the head plate is mounted onto the microscope to prevent movement of the mouse's head relative to the objective, thus stabilizing the imaging field. A head-fixed thirsty mouse can be trained to lick only when odor A but not odor B is presented to receive a water reward. The motor cortical area controlling tongue extension can be imaged during the learning process. **(B)** In this virtual reality (VR) preparation, a head-fixed mouse is placed on a ball floating on air while a projector presents visual stimuli. The motion of the mouse causes the ball to move; the movement vector is measured by an optical detector and fed into a VR computer to control the projector such that the mouse's movement changes the scene in the projector as if it were moving in the real world (Movie

14-3). Mice could be trained to run along linear tracks in this setting while their hippocampal place cells were subjected to whole-cell recordings of membrane potentials (V_m) or two-photon imaging (not shown). AAM, angular amplification mirror; RM, reflecting mirror. **(C)** A miniature fluorescence microscope weighing 1.9 g can be attached to the head of a freely moving mouse. With an embedded GRIN lens, this miniature microscope could image over a period of more than a month the place fields of hundreds of CA1 pyramidal cells expressing a genetically encoded Ca^{2+} indicator (Movie 14-4). (A, from Komiyama T, Sato TR, O'Connor DH, et al. [2010] *Nature* 464:1182–1186. B, from Harvey CD, Collman F, Dombeck DA, et al. [2009] *Nature* 461:941–946. C, from Ziv Y, Burns LD, Cocker ED, et al. [2013] *Nat Neurosci* 16:264–266. With permission from Springer Nature.)

Box 14-4: continued

its being done *postmortem* in fixed brain tissue, its slow temporal resolution (transcription operates on a time scale of minutes or more, while neuronal activity operates on a time scale of milliseconds), and its indirectness—it is unclear what kinds of activity patterns trigger immediate early gene expression, and this relationship may differ in different cell types. Nevertheless, with proper design and controls, strategies utilizing immediate early gene expression have been successful in many studies, from identifying sensory receptors for specific stimuli (Figure 6-19B) to allowing genetic access to thirst drive neurons (Figure 9-18) and memory traces (Figure 11-37).

14.23 Neuronal inactivation can reveal which neurons are essential for circuit function and behavior

Inactivation of specific neurons or neuronal populations can reveal their necessity in normal nervous system function. The crudest method of inactivating neurons is a lesion, which removes or destroys a chunk of nervous tissue, either by accident or by design. For instance, lesions in the aphasic patients of Broca and Wernicke (Section 1.10) and in amnesic patients (Section 11.1) helped pinpoint brain regions important for speech and explicit memory in humans. Lesions in animal models enable systematic examination of the requirement of specific regions for brain function and behavior (Figure 11-41). Whereas lesions cause permanent damage, researchers can also transiently inactivate specific brain regions by injecting pharmacological agents such as the $GABA_A$ receptor agonist muscimol (Box 3-2) to enhance inhibition, glutamate receptor antagonists to reduce excitation, or Na^+ channel inhibitors to block action potentials (Figure 11-39B). Experiments using lesions and drugs can assess the functions of specific brain regions but cannot differentiate the roles of specific neuronal types within a region.

In invertebrates, many neurons are individually identifiable and make unique contributions to circuit function and animal behavior. For example, individual neurons can be identified in *C. elegans* and subsequently ablated using a high-intensity laser; this approach has been used to identify neurons involved in many specific functions, such as detection of volatile chemicals (Section 6.9). In animals with large neurons, injecting hyperpolarizing current with an intracellular electrode can transiently inactivate a neuron, allowing assessment of its contribution to circuit function or animal behavior (Section 8.5).

In vertebrates, specific neural functions are often carried out by populations of neurons with similar anatomical and physiological properties. It is therefore more difficult to assess neuronal function by inactivating individual neurons. Genetic approaches can obviate this limitation. Since populations of neurons with similar functions often share similar gene expression patterns, researchers can use genetic methods (Sections 14.9–14.11) to express in a neuronal population an effector transgene capable of silencing most or all of these neurons. Diverse approaches for inactivating neurons have been developed. For example, the gene encoding tetanus toxin (Box 3-2), which cleaves synaptobrevin, a SNARE protein essential for neurotransmitter release, has been used to block synaptic transmission from target neurons. Another widely used method for reducing neuronal action potential firing is to overexpress Kir2.1, an inward rectifier K^+ channel (Box 2-4); as the K^+ equilibrium potential is always more hyperpolarized than the resting potential, increasing K^+ conductance causes hyperpolarization of target neurons, making it more difficult for them to reach the firing threshold.

Killing or long-term silencing of neurons may induce compensatory changes in neural circuits. Therefore, a particularly informative method for inactivating neurons to assess their normal function is to silence them acutely and reversibly. Although effectors such as tetanus toxin and Kir2.1 can be temporally regulated at the transcriptional level when they are transgenically expressed (Section 14.9), such regulation is usually slow (on the scale of hours to days). Expression of a temperature-sensitive mutant of the Shibire protein (Shi^{ts}), which reversibly blocks synaptic vesicle recycling only at high temperatures (Figure 3-14), has been a powerful tool in fruit flies to inactivate target neurons transiently (within minutes of a temperature shift; Figure 10-8). However, the Shi^ts strategy cannot be used in mammals, which maintain a constant body temperature.

Chemicals can be used to selectively silence neurons expressing the corresponding receptors; this **chemogenetic** approach has been increasingly used in mammals. We discuss two examples here, both of which utilize chemicals that can cross the blood–brain barrier and have rapid onset and metabolism. The first

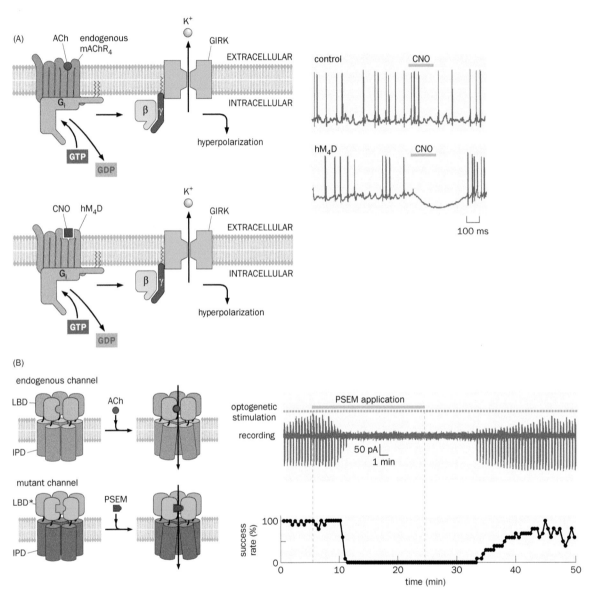

Figure 14-44 Chemogenetic approaches for silencing neuronal activity. (A) Top left, binding of acetylcholine (ACh) to the endogenous metabotropic ACh receptor (mAChR$_4$) is coupled to a G$_i$ protein that activates an inward rectifier K$^+$ channel (GIRK), causing neuronal hyperpolarization. Bottom left, a mutant mAhR$_4$ (hM$_4$D) no longer binds ACh but binds clozapine-N-oxide (CNO) with high affinity; CNO binding to hM$_4$D triggers neuronal hyperpolarization. Right, voltage traces showing that CNO application (horizontal bar) induced hyperpolarization and inhibited spontaneous firing of an hM$_4$D-expressing cultured hippocampal neuron (lower trace) but had no effect on a normal neuron (upper trace). **(B)** Left, whereas the normal ligand-binding domain (LBD) of the nicotinic ACh receptor (nAChR) binds ACh but not PSEM, the mutant LBD (LBD*) binds PSEM but not ACh. When fused to ion pore domains (IPD) of different channels, LBD* can confer PSEM-based gating of different ion conductances. Top right, cell-attached recording of a hypothalamus AgRP neuron expressing channelrhodopsin (ChR2) and an LBD* fused to the IPD of a glycine receptor in a slice preparation, showing that PSEM application inhibits optogenetically induced neuronal firing. Blue, photostimulation period (each rectangle represents bursts of 10 light pulses within 1 s every 30 s); yellow bar, PSEM application period. Bottom right, quantification of the success rate of action potentials induced by optogenetic stimulation. (A, from Armbruster BN, Li X, Pausch MH, et al. [2007] *Proc Natl Acad Sci U S A* 104:5163–5168. Copyright The National Academy of Sciences, U.S.A. For an update, see Nagai et al. [2020] *bioRxiv* https://doi.org/10.1101/854513. B, adapted from Magnus CJ, Lee PH, Atasoy D, et al. [2011] *Science* 333:1292–1296. For an update, see Magnus et al. [2019] *Science* 364:eaav5282.)

example, named DREADD (for designer receptors exclusively activated by a designer drug), uses a mutant metabotropic acetylcholine (ACh) receptor called hM₄D, which binds a chemical, CNO (clozapine-*N*-oxide), but not endogenous ACh. CNO does not have endogenous high-affinity targets, but binding of CNO to hM₄D leads to hyperpolarization of neurons through opening of an inward rectifier K⁺ channel, resulting in effective silencing of hM₄D-expressing neurons (Figure 14-44A). The second approach uses a mutant ligand-binding domain (LBD) of the ionotropic ACh receptor that binds a chemical called PSEM (pharmacologically selective effector molecule) but not endogenous ACh. When this mutant LBD is fused to the ion pore domain of a glycine receptor, the hybrid receptor becomes a PSEM-gated Cl⁻ channel, which can silence neurons in response to PSEM application (Figure 14-44B).

14.24 Neuronal activation can establish sufficiency of neuronal activity in circuit function and behavior

Electrical stimulation can be used to mimic neuronal activation. Indeed, the first discovery that the nervous system communicates using electrical signals was based on the observation that muscles contracted when nerves were electrically stimulated (Section 1.8). Many fundamental discoveries in neuronal communication, such as synaptic transmission and long-term potentiation, involved experimental stimulation of nerve fibers (Figure 3-1; Figure 11-7). Experiments using electrical stimulation also led to the discovery of the sensory and motor homunculi in the human brain (Section 1.11), suggested brain regions related to reward processing (Section 11.24), and helped establish causal relationships between neuronal activity and visual perception (Section 4.28). Electrical stimulation in humans can also be used to treat brain disorders such as Parkinson's disease (Section 12.13).

In most *in vivo* preparations, electrical stimulation has been delivered using an extracellular electrode. In principle, the strength and frequency of electrical stimulation can be controlled to match the endogenous firing patterns of neurons. However, in the complex milieu of the central nervous system, where different types of neurons and passing axons intermingle, it is difficult to control what types of neurons a stimulating electrode activates. Indeed, electrical stimulation usually activates a mixture of excitatory neurons, inhibitory neurons, and axons passing by the electrode tip, making it difficult to assign the effects of stimulation to the activity of specific types of neurons. The recent development of genetically encoded effectors that can activate neurons and be targeted to specific types (Sections 14.9–14.11) has begun to overcome this limitation. These effectors activate target neurons in response to experimentally applied triggers such as heat, light, or chemicals. Ideally, the effector does not act in the absence of the trigger, and the trigger has no impact on the systems being investigated in the absence of the effector. If so, activation can be restricted to the neurons expressing the effector molecule at the time of trigger application.

For example, an effective way to activate neurons in the fruit fly is to express a transgene encoding a temperature-gated *Drosophila* TRPA1 channel, which causes depolarization of dTRPA1-expressing neurons in response to heat (Figure 14-45). By simply changing the temperature of the environment, researchers can investigate the behavioral consequences of selectively activating specific neurons in freely moving flies (Figure 10-8B). This method is simple and noninvasive, but it can be employed only if the behavior being investigated is insensitive to temperature, and it cannot be used in animals that maintain a constant body temperature, such as mammals. The temporal resolution of neuronal activation is limited by how fast temperature changes can occur (usually on the order of seconds to minutes).

Neurons can also be activated by chemogenetic approaches—indeed, both of the approaches discussed in Figure 14-44 can also be adapted to activate neurons rather than silence them. For example, by creating CNO-binding mutations in a metabotropic AChR that couples to G$_s$ instead of G$_i$, CNO application results in an increase in intracellular cAMP, which can depolarize neurons through several mechanisms, including opening of cyclic nucleotide-gated channels (Figure 6-4).

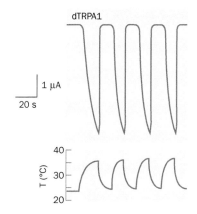

Figure 14-45 Turning heat into depolarization. Expression of a *Drosophila* TRPA1 (dTRPA1) channel in *Xenopus* oocytes results in heat-induced depolarization. In response to heat pulses (bottom), inward (depolarizing) currents are induced in *Xenopus* oocytes voltage-clamped at –60 mV (top). (From Hamada FN, Rosenzweig M, Kang K, et al. [2008] *Nature* 454:217–220. With permission from Springer Nature.)

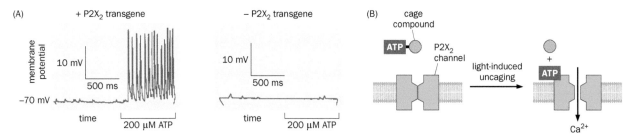

Figure 14-46 Neuronal activation by supplying ATP to neurons expressing an ATP-gated channel. (A) Intracellular recording of *Drosophila* larval muscle in response to ATP application in a dissected neuromuscular preparation. Left, when motor neurons express a P2X$_2$ transgene, ATP application depolarizes the motor neuron, evoking neurotransmitter release and end-plate potentials in muscles. Right, in the absence of the P2X$_2$ transgene, ATP application does not induce end-plate potentials in muscles. **(B)** Schematic of neuronal activation in behaving flies. Flies were injected with caged ATP, which does not bind to P2X$_2$ channels. Light triggers uncaging of ATP, which binds and activates the P2X$_2$ channel, causing Ca^{2+} influx and activation of P2X$_2$-expressing neurons. (From Lima SQ & Miesenböck G [2005] *Cell* 121:141–152. With permission from Elsevier Inc.)

By fusing a mutant LBD to the ion pore domain of a cation channel, PSEM application can likewise cause neuronal activation.

Neurons can also be activated by combinations of chemicals and light. For example, a mammalian ATP-gated P2X$_2$ channel has been used in *Drosophila* to activate neurons. *Drosophila* has neither ATP-gated channels nor sufficient extracellular ATP to activate mammalian P2X$_2$ channels expressed transgenically. Thus, application of exogenous ATP can trigger activation of *Drosophila* neurons expressing transgenic P2X$_2$ (**Figure 14-46**A). The temporal resolution of this approach is limited by the slow time course of ATP application and clearance. To achieve fast activation of neurons in behaving animals, caged ATP (ATP modified chemically so it cannot activate its receptor) can be injected into the central nervous systems of transgenic flies expressing P2X$_2$. A flash of light uncages ATP and thus activates P2X$_2$-expressing neurons (Figure 14-46B). This strategy has been used to effectively induce behaviors such as singing courtship songs (Figure 10-11), and such experiments have helped establish causal relationships between the activity of specific types of neurons and the behaviors they control.

14.25 Optogenetics allows control of the activity of genetically targeted neurons with millisecond precision

Broadly speaking, **optogenetics** is an approach for altering neuronal activity by using light to activate an effector genetically targeted to specific neurons. Thus, it includes the strategy discussed in Figure 14-46: using light to uncage ATP and activate neurons expressing an ATP-gated channel. In most cases, optogenetics refers to the use of microbial opsins as effectors because of their simplicity, effectiveness, and general applicability. Since the term was first introduced in 2006, optogenetics has had a transformative impact in many areas of neuroscience, especially on how researchers evaluate the function of neural circuits and their roles in behavior.

The most widely used optogenetic effector for neuronal activation is **channelrhodopsin-2 (ChR2)**, a protein first found in the green algae *Chlamydomonas reinhardtii*. As we discussed in Section 13.13, ChR2 is a seven-transmembrane type I rhodopsin; instead of coupling to G proteins, it is a cation channel that opens in response to blue light stimulation (**Figure 14-47**A; Figure 13-21B). Although ChR2 requires all-*trans* retinal as a cofactor, mammalian neurons have sufficient endogenous all-*trans* retinal to support ChR2 function. Thus, expression of ChR2 alone is sufficient to allow robust depolarization and firing of mammalian neurons in response to light. Indeed, pulses of blue-light stimulation can precisely direct the firing of ChR2-expressing neurons at rates as high as 30 Hz (Figure 14-47A, bottom), achieving control of target neuron activity with millisecond precision. In nervous systems lacking sufficient all-*trans* retinal (such as those of *Drosophila* and *C. elegans*), supplementing retinal in food is often sufficient for ChR2 function. ChR2 variants with enhanced expression, photocurrent,

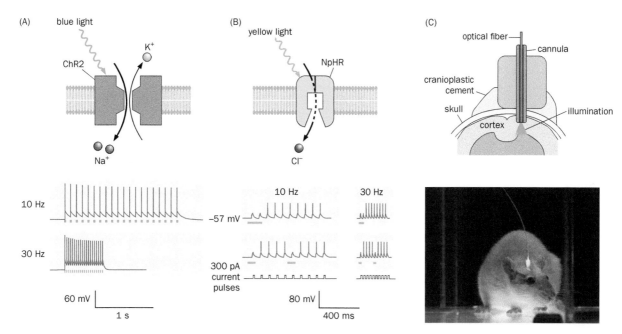

Figure 14-47 Optogenetics for precise temporal control of neuronal activity. (A) Top, channelrhopsin-2 (ChR2) from green algae is a cation channel gated by blue light that causes neuronal depolarization from the resting potential because it allows more Na^+ influx than K^+ efflux. Bottom, voltage traces show that in culture the firing of a hippocampal neuron expressing ChR2 can be precisely controlled by 10 ms blue-light pulses (blue dashes at the bottom) at up to 30 Hz. **(B)** Top, halorhodopsin (NpHR) from archaea is a Cl^- pump activated by yellow light. Bottom, voltage traces show that yellow-light pulses (yellow dashes at the bottom) can cancel with millisecond precision the action potentials produced by depolarizing current pulses in cultured neurons.

(C) Optogenetics *in vivo*. A cannula is surgically introduced into a brain region of interest that expresses ChR2 or NpHR in specific neuronal types. During the experiment, an optical fiber delivering light is introduced into the cannula, so that light-induced behavior can be observed in a freely moving animal (bottom; Movie 14-5). (A, from Boyden ES, Zhang F, Bamberg E, et al. [2005] *Nat Neurosci* 8:1263–1268. B, from Zhang F, Wang LP, Brauner M, et al. [2007] *Nature* 446:633–639. C, schematic from Zhang F, Aravanis AM, Adamantidis A, et al. [2007] *Nat Rev Neurosci* 8:577–581. With permission from Springer Nature. Image courtesy of Karl Deisseroth.)

and faster or slower kinetics have been developed via *in vitro* mutagenesis. Microbial opsins sensitive to different excitation wavelengths have also been developed for optogenetic activation of neurons.

Optogenetics can also be used to reversibly inactivate neurons with high temporal precision. For this purpose, a type I rhodopsin from archaea called **halorhodopsin**, a yellow-light-activated inward Cl^- pump (Figure 14-47B), can be expressed in neurons. Yellow-light stimulation hyperpolarizes neurons expressing halorhodopsin, making it more difficult for these neurons to spike in response to depolarizing signals. In cultured neurons expressing halorhodopsin, short pulses of yellow light were sufficient to block with high temporal precision spiking induced by depolarizing current pulses (Figure 14-47B, bottom). Many new inhibitory opsins have been developed for light-induced neuronal silencing, including **Archaerhodopsin** (Arch), a light-driven outward proton pump from archaea, as well as engineered and naturally occurring anion channelrhodopsins (ACRs) with higher light sensitivity than inhibitory pumps.

To achieve optogenetic manipulation of neurons in behaving animals, fiber optics–based systems have been developed to provide laser pulses in specific brain regions expressing optogenetic effectors via viral transduction or transgenic animals (Figure 14-47C). Optical fibers are usually ~200 µm in diameter, thin enough to limit physical damage but still able to effectively excite ChR2-expressing neurons in approximately 1 mm^3 at the fiber's tip. Thus, the behavior of freely moving animals can be assayed while specific neuronal populations in specific brain regions are activated or silenced (Movie 14-5).

Because optogenetic methods manipulate neuronal activity at the same time scale as fast neuronal communication (with spikes in the millisecond range), they are well suited for probing neuronal signaling and computation. Still, a limitation of the optogenetic approach is that for neuronal inactivation to achieve a biological

Figure 14-48 Combining activity imaging and optogenetic manipulation in the same preparation. (A) The primary visual cortex (V1) of mice was transduced with AAV encoding GCaMP6 for two-photon Ca^{2+} imaging and ChRmine (a red-shifted channelrhodopsin) for optogenetic activation. In a head-fixed preparation, responses of V1 neurons to vertical and horizontal gratings were recorded to identify cell ensembles activated by vertical (green) and horizontal (red) visual stimuli, respectively. **(B)** In mice trained to distinguish vertical and horizontal gratings in a go/no-go task (see Section 14.28), presentation of vertical but not horizontal gratings elicited licking in thirsty mice. Optogenetic stimulation of vertically or horizontally tuned neurons in the absence of visual stimuli could elicit appropriate responses; by contrast, optogenetic stimulation of a similar number of randomly tuned neurons could not. Each symbol represents one session of one mouse with >10 trials. Optogenetic and visual stimuli were randomly interleaved; an auditory signal was given 1 s prior to the onset of visual or optogenetic stimuli. (From Marshel JH, Kim YS, Machado TA, et al. [2019] *Science* 365:eaaw5202. With permission from AAAS. See also Carrillo-Reid L, Han S, Yang W, et al. [2019] *Cell* 178:447–457.)

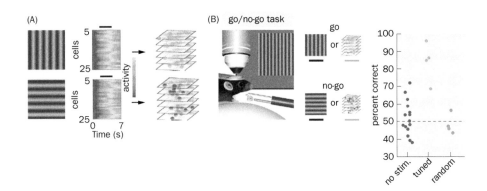

effect, a large fraction of the neurons of interest must be accessible to light delivered by an optical fiber. Optical fiber implantation is also associated with physical damage. By comparison, the chemogenetic approaches discussed earlier, while not as temporally precise (usually taking effects within 10s of minutes after drug application), can affect spatially dispersed cellular populations expressing the relevant receptor. They are minimally invasive if the chemicals can cross the blood-brain barrier and have no side effects in the absence of the effector. Thus, these methods complement each other.

A common limitation of optogenetics and chemogenetics has been that activity of all neurons expressing the effector and exposed to light or chemicals are *simultaneously* activated or silenced, regardless of their natural activity patterns. One way to overcome this limitation is to use immediate early gene promoters to genetically target neurons with specific activity patterns for expression of effectors (Figure 14-14). A more temporally precise approach for optogenetics is to record neuronal activity first, then target a specific subset of neurons in the same tissue volume according to their activity patterns for holographic optogenetic manipulation. In the following example, researchers first used two-photon Ca^{2+} imaging of the mouse primary visual cortex (V1) to identify individual neurons tuned to visual stimuli of vertical or horizontal gratings **(Figure 14-48A)**. Subsequent optogenetic stimulation of vertically or horizontally tuned cells, but not a similar number of random cells in V1, could elicit the same behavioral response as presenting visual stimuli (Figure 14-48B), suggesting that optogenetic activation of orientation-tuned cells might mimic perception of visual stimuli of corresponding orientation.

In summary, classical methods for activating or silencing neurons, including electrical stimulation, lesion, and pharmacology, have now been supplemented by a variety of genetically encoded effectors for activating or inactivating specific neuronal types with light, heat, or chemicals. Coupled with advances in targeting these manipulations to specific cellular populations, these approaches are making a significant impact on our understanding of how neural circuits operate and control behavior.

14.26 Synaptic connections can be mapped by physiological and optogenetic methods

Having studied methods for recording and manipulating neuronal activity, we now return to the subject of mapping neuronal connections discussed in Section 14.19 to study how physiological methods can be employed to address this important problem.

A widely used method for determining whether two neurons are directly connected is to place an electrode in each of the neurons and record from both neurons while manipulating the activity of one (**Figure 14-49**). If injecting a depolarizing current into neuron *A* to produce an action potential causes depolarization of neuron *B* within the time frame of a monosynaptic connection (usually a few milliseconds), we can conclude that neuron *A* forms excitatory synapses directly onto neuron *B* (Figure 14-49A). If firing of neuron *A* causes hyperpolarization of neuron *B* within a few milliseconds, then neuron *A* forms inhibitory synapses onto neuron *B* (Figure 14-49B). Likewise, stimulating neuron *B* and recording from neuron *A* can reveal whether these neurons are connected by *B* → *A* chemical synapses. If hyperpolarization of neuron *A* causes hyperpolarization of neuron *B*, and vice versa, then these two neurons are connected by electrical synapses (Figure 14-49C), as chemical synapses do not transmit hyperpolarizing signals (Box 3-5). Paired recordings have been performed extensively in invertebrates with large neurons using intracellular electrodes (Figure 8-13) and mammalian brain slices using whole-cell patch recording (Figure 4-48). Such tests provide definitive evidence that two neurons are functionally connected and can reveal the type (excitatory, inhibitory, or electrical) and strength of the connection, but are labor intensive and unsuitable for mapping long-range connectivity.

Higher throughput connectivity mapping methods have been developed in mammalian brain slices. One such method utilizes laser uncaging of neurotransmitters. A brain slice can be placed in media containing caged glutamate (glutamate modified chemically so it cannot activate its receptors). Focal laser stimulation leads to uncaging (removal of the chemical modification) and thus local release of glutamate, causing neurons near the laser stimulation site to fire action potentials. If one or several neurons near the laser stimulation site form monosynaptic excitatory connections with a postsynaptic neuron that is being recorded using intracellular or whole-cell recording, then laser uncaging will produce excitatory postsynaptic potentials in the recorded neuron. After the laser scans through a defined area of the brain slice, researchers can create a two-dimensional map of the excitatory neurons in the area that connect with the target neuron being electrically recorded (**Figure 14-50**A).

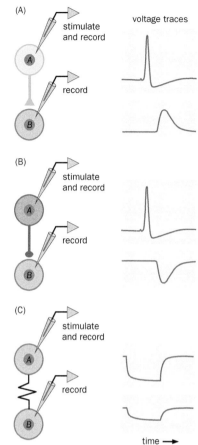

Figure 14-49 Mapping neuronal connections with paired recordings.
(A) Depolarizing an excitatory neuron *A* causes it to fire an action potential that elicits a depolarizing membrane potential in its postsynaptic partner neuron *B*.
(B) Depolarizing an inhibitory neuron *A* causes it to fire an action potential that elicits a hyperpolarizing membrane potential in its postsynaptic partner neuron *B*. **(C)** Hyperpolarizing neuron *A* causes hyperpolarization of neuron *B* via electrical synapses.

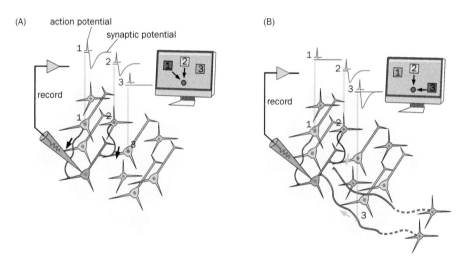

Figure 14-50 Mapping neuronal connections with optical and electrophysiological methods. Arrows indicate the direction of signal flow. For simplicity, axons of neurons that do not form synapses with the recorded neurons are not drawn. **(A)** Glutamate uncaging. The brain slice is incubated with caged glutamate. When a scanning laser (blue) reaches a neuron presynaptic to the recorded neuron, local release of glutamate causes this neuron to fire action potentials, which produce excitatory postsynaptic potentials in the recorded neuron (the *y* axes of the synaptic potential and action potential are not on the same scale). A two-dimensional input map can be produced after the laser systematically scans through the entire slice. In the illustration, neurons 1 and 2 are presynaptic to the recorded neuron, but neuron 3 is not. **(B)** ChR2-assisted circuit mapping. A defined subpopulation of neurons (green) express channelrhodopsin (ChR2), causing them to fire action potentials in response to blue light stimulation. In this scheme, stimulating neuron 1 does not activate the recorded neuron, as neuron 1 does not express ChR2. Stimulating neuron 2 activates the recorded neuron because it expresses ChR2 and synapses onto the recorded neuron. Note that this method can also be used to map connections between ChR2-expressing presynaptic neurons whose cell bodies lie outside the slice (bottom right), because stimulating their axons (3) and terminals (not shown) can often elicit synaptic responses in postsynaptic neurons. Although drawn in the same schematic, the axon-stimulation experiment is performed separately when ChR2 is expressed only in a defined neuronal population outside the slice. (From Luo L, Callaway EM, & Svoboda K [2008] *Neuron* 57:634–660. With permission from Elsevier Inc. The original methods were described in Callaway EM & Katz LC [1993] *Proc Natl Acad Sci U S A* 90:7661–7665 and Petreanu L, Huber D, Sobczyk et al. [2007] *Nat Neurosci* 10:663–668.)

A more versatile approach to this type of experiment replaces laser uncaging of a neurotransmitter with photoactivation of ChR2-expressing neurons (Figure 14-50B) in a method called CRACM (ChR2-assisted circuit mapping). The advantage of CRACM is that the type of presynaptic neurons expressing ChR2 can be genetically defined. Moreover, unlike paired recordings and laser uncaging, which are limited to mapping connections within a brain slice, CRACM can also map long-range connections. If the only source of ChR2 expression is from a defined neuronal population outside the brain slice, photoactivation of ChR2 molecules in their axons and terminals is often sufficient to cause neurotransmitter release at synaptic terminals, which can be detected by recording of postsynaptic neurons located within the slice.

None of these physiological mapping methods is easily applied *in vivo,* at least in the complex mammalian brain. These physiological methods complement anatomical methods such as serial EM reconstruction and trans-synaptic tracing (Section 14.19) for mapping synaptic connections.

BEHAVIORAL ANALYSIS

A major goal of neurobiology is to understand how behaviors arise from the molecular and cellular properties of neurons, the wiring specificity of neural circuits established during development and modified by experience, and the spatiotemporal patterns of neuronal activity at the times when behaviors occur. At the same time, insightful and quantitative analysis of animal behavior is instrumental for studying most neurobiological problems, from perception and action to emotion and cognition. Whereas human studies can utilize self-report, animal studies must rely on observation and measurement of behavior to infer what animals sense, feel, learn, and understand. To analogize Theodosius Dobzhansky's famous quotation about evolution (Chapter 13 epigraph), *nothing in the nervous system makes sense except in the light of behavior.*

In neurobiology research, behavioral analysis serves three broad purposes. First, it aims to explain the behavior itself (see Chapter 10): what is the function of the behavior in animal survival and reproduction, what external factors influence it, what are its constituent motor actions, and what is the underlying neural basis? Second, behavioral analysis is used as a quantitative readout of the functions of brain regions, circuits, and neurons in specific neurobiological processes such as perception (Figure 4-55) and memory (Figure 11-33). Third, behavioral analysis is used to test the effects of manipulating specific genes (Figure 11-34) and to assess animal models of human nervous system disorders (Figure 12-6). Given the extensive links between genes, neurons, circuits, and behaviors (Figure 11-6), these purposes overlap considerably, and we have seen these links throughout earlier chapters. In the following sections, we first highlight two general approaches in behavioral analysis relevant to all of these purposes and then discuss behavioral assays commonly used to assess the functions of genes, neurons, and circuits and to model human brain disorders.

14.27 Studying animal behavior in natural environments can reveal animals' behavioral repertoires and their adaptive values

From an evolutionary perspective, behaviors are products of natural selection that allow animals to interact with their environments in ways that improve their likelihood of survival and reproduction. Thus, an influential approach has been to study behavior in the natural environment: this **neuroethology** approach (Section 1.2) can reveal an animal's natural behavioral repertoire (the behaviors an animal typically exhibits), the relationships between different behaviors (whether one behavior precedes or follows another in a sequence, or whether two behaviors are mutually exclusive in their occurrence), and their adaptive values.

The principal methods of neuroethology include observation and measurement in carefully designed field studies. We use the study of honeybee dancing to

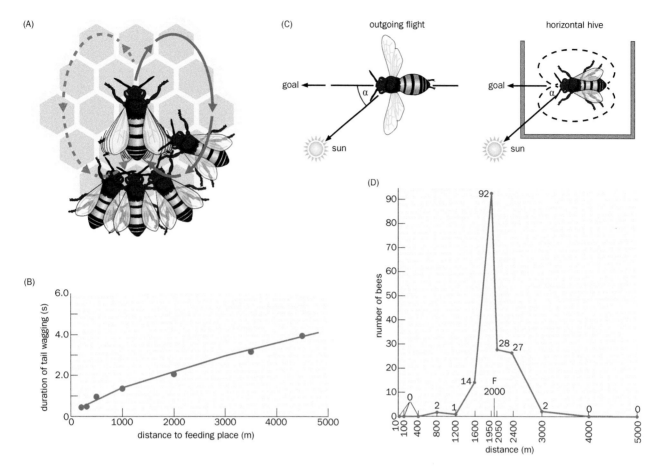

Figure 14-51 Behavioral analysis of honeybee foraging. (A) Illustration of the figure-8 dance of a forager bee in an observation hive. The forager alternates between the right (solid red trajectory) and left halves (dashed gray trajectory) of the figure 8. The four forager recruits take in the information by moving with the forager and maintaining close contact with her, particularly during the straight part of the dance, when the forager exhibits tail wagging. **(B)** The relationship between the duration of tail wagging, measured from film recordings, and the foraging distance signaled by the dancing bee. **(C)** The angle (α) of the

outgoing flight with respect to the sun (left) is signaled by the angle at which the figure-8 dance orients with respect to the sun (right). **(D)** In this field experiment, foragers were fed at a scented plate (F) 2000 m away from the hive. Afterward, similarly scented plates without food were placed at many distances (numbers on the x axis) and the visits of forager recruits were quantified; the y axis values adjacent to the data points reveal that the forager recruits preferred distances of around 2000 m. (Adapted from von Frisch K [1974] *Science* 185: 663–668.)

illustrate. Honeybees are social insects that can perform sophisticated behavioral tasks (Figure 11-23). They are also expert nectar collectors and pollinators. Once forager bees find a good source of nectar, sometimes kilometers from their hive, they communicate to their fellow bees (forager recruits) to direct them to the same place. How do bees achieve this? Researchers have set up observation hives with glass windows so they can observe the behaviors of foragers in the hive environment. Once foragers locate a good source of nectar, they return to the hive and perform dances to convey information to forager recruits. When the source of the nectar is more than 50 m from the hive, foragers typically perform a tail-wagging dance following a trajectory resembling the Arabic numeral 8 (**Figure 14-51**A). By placing scented feeding bowls at different distances and in different directions from the hives and measuring the dances of bees and the subsequent foraging of forager recruits, researchers deduced the following set of rules: the richness of the nectar is indicated by the vigor of the dance; the distance to the nectar source is signaled by the duration of tail wagging, which takes place as the forager bee dances in a straight path between the 8's two circular halves (Figure 14-51B); the orientation of the figure-8 dance in the hive signals the direction of the outgoing flight with respect to the sun (Figure 14-51C); finally, the scent that foragers carry informs forager recruits about the kind of nectar they should look for once they arrive in the vicinity of the nectar source. Experiments using foraging behavior as

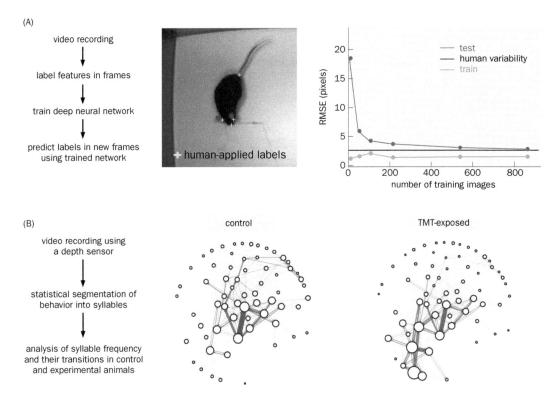

Figure 14-52 Analysis of behavior using machine learning. (A) Left, flow chart of DeepLabCut, a deep neural network-based (Box 14-6) supervised learning procedure for pose estimation. Middle, an example of labeling of the snout, ears, and tail base of a mouse. Right, root mean square error (RMSE) of training and testing, compared to error due to variability in human labeling. The error in "test" (predicting locations of labels from video frames) approaches human variability after training the neural network with 100 labeled images. **(B)** Left, flow chart of MoSeq (motion sequencing), an unsupervised learning procedure that segments time-varying behavior into a sequence of syllables. Middle and right, behavioral state maps generated by MoSeq depicting behavioral syllables (circles, diameter proportional to usage) and transitions (lines, thickness proportional to probability) for a control mouse and a mouse exposed to TMT, a major component of fox urine, which mice innately avoid. The syllables at the bottom left, corresponding to various freezing-like behaviors, occur more frequently in TMT-exposed animals. (A, from Mathis A, Mamidanna P, Cury KM, et al. [2018] *Nat Neurosci* 21:1281–1289. With permission from Springer Nature. B, from Wiltschko AB, Johnson MJ, Iurilli G, et al. [2015] *Neuron* 88:1121–1135. With permission from Elsevier Inc.)

readouts validated some of these rules (Figure 14-51D). The adaptive value of efficiently locating nectar is obvious for the bee colonies, and efficient nectar foraging is also beneficial to the plants that produce the nectar and are pollinated by the bees.

The spirit of neuroethology can be extended to the laboratory, where animal behaviors can be observed, recorded, and quantitatively measured in settings resembling natural environments but offering greater technical ease than field studies. For instance, complete recording of *Drosophila* mating behavior in a laboratory setting (Figure 10-2 and Movie 10-1) enabled this behavior to be dissected into discrete components. Quantitative plots of different behaviors exhibited over time can be used to compare individuals receiving different experimental treatments and thereby to study the neural mechanisms underlying these behaviors (Figure 10-10B). Development of high-speed video recording and automated video analysis have enhanced the sensitivity and throughput of behavioral observations and measurements. For example, supervised and unsupervised machine learning methods (see Section 14.33 for details) have been applied to analyze animal poses (**Figure 14-52**A) and motion sequences (Figure 14-52B; Figure 8-24D, E), respectively.

14.28 Studying behaviors in highly controlled conditions facilitates investigation of their neural bases

Behavior is influenced by many factors: external stimuli, internal drives and brain states, and the individual animal's genetic makeup and life experience. Thus, another influential approach, which at face value might seem to be the polar opposite of neuroethology, is to study behavior under as much experimental control as possible so that factors that might influence behavior can be individually varied to tease apart their contributions to the behavior. To achieve this, researchers carry out behavioral studies in inbred animal strains to decrease genetic variability—all individuals within an inbred strain are essentially genetically identical—and use animals of the same sex and age reared under the same conditions to decrease variability in experience. Behaviors can be performed in a fixed apparatus to reduce the variability of external factors, and standardized conditions can be implemented to control internal factors, such as the circadian cycle and the time when the animal last ate or drank.

This approach has yielded many insights. The discovery of classical and operant conditioning and the identification of factors affecting those learning processes were made using this approach (Section 11.14). Operant conditioning has been particularly influential in studying internal factors such as drives. Indeed, behavioral paradigms are often designed to take advantage of internal drives. For instance, in two commonly employed paradigms, thirsty animals are motivated by a potential water reward to choose to perform an action, such as pressing a lever, or do nothing (the go/no-go paradigm), or to choose one of the two actions (the two-alternative forced-choice paradigm), such as a saccade toward one of two alterative targets. These tasks can be used to study sensory perception, decision making, motor execution, and working memory (Figure 4-55; Figure 6-75; Figure 8-31; Figure 8-37). The combination of behavioral analysis with methods for recording and manipulating the activity of relevant neurons (Box 14-4) has enabled researchers to establish causal links between behavior, neuronal activity, and circuit function.

The two general approaches outlined above are complementary, and many behavioral paradigms incorporate the merits of both. The choice of which behaviors to study under highly controlled experimental conditions and the design of the assays used to study these behaviors should be based on an understanding of the behavioral repertoire of animals in their natural environment. For instance, the Morris water maze (Figure 11-33) takes advantage of the ability of rodents to use external landmarks to navigate and their preference to not swim.

The **closed-loop** design of some behavioral experiments, in which the behavior of an animal changes the environmental stimuli that induce the behavior in the first place, can allow animals to explore naturalistic virtual environments in highly controlled settings. For instance, an influential closed-loop preparation was developed while studying visual control of the flight behavior of houseflies. In a cylinder that delivers panoramic visual stimuli, a test fly is suspended from a torque compensator, which measures the rotation of the fly as it attempts to orient toward visual stimuli and sends a signal to control the rotation of the panorama cylinder (**Figure 14-53**). In this way, the fly's intended movement alters the visual scene as if the fly were actually moving through the scene. This design allows researchers to study in an immobilized fly some of the flight behaviors in more naturalistic settings, such as object tracking, and at the same time makes possible precise delivery of visual stimuli and the quantitative measurement of the fly's behavioral output. This setup also enables researchers to investigate the underlying neural basis of behavior by performing electrophysiological recordings of neurons in the fly brain. Conceptually similar approaches have also been developed for other animals. For example, navigation of head-fixed mice through complex virtual reality environments has facilitated investigation of the neuronal activity associated with spatial navigation using optical imaging and whole-cell recording methods that cannot easily be applied to freely moving mice (Figure 14-43B).

Developing behavioral paradigms for controlled settings also enables researchers to probe animals' capabilities to learn skills they are unlikely to have encountered in their natural environments. Lever pressing for food or drink (Figure 11-22) is one such example. Studying *unnatural* behaviors allow researchers to investigate the neural bases of *generalized learning* with minimal influence from innate preprograms.

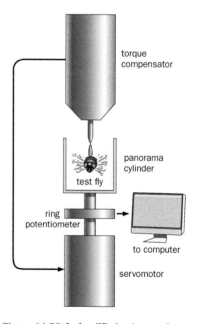

Figure 14-53 A simplified scheme of a closed-loop design for studying the flight behavior of flies. A housefly is suspended from a torque compensator. Signals from the torque compensator provide a quantitative measure of the fly's intended behavioral output. These signals are sent to a servomotor, which controls the rotation of the panorama cylinder through a ring potentiometer. Thus, the fly's intended behavioral output controls its visual environment. (Adapted from Reichardt W & Poggio T [1976] *Q Rev Biophy* 3:311–375.)

In summary, by combining insights derived from neuroethological approaches with the merits of experimental control, researchers can develop sophisticated behavioral paradigms that enable quantitative behavior analysis while recording and manipulating neuronal activity. Thanks to the techniques described in previous sections, these approaches are becoming increasingly powerful for dissecting the neural basis of complex behaviors.

14.29 Behavioral assays can reveal gene and neuronal function and can model human brain disorders

In the final section on behavioral analysis, we highlight behavioral assays commonly used to assess phenotypes associated with disruption of specific genes or activity of specific neuronal populations. These assays are also used to study animal models of human brain disorders and the effects of pharmacological interventions. We focus on assays designed for rats and mice, as they are the most widely used mammalian models (Section 14.4). However, as we discussed in Chapter 12, results obtained from disease models in rodents often do not translate well to effective therapies for human brain disorders, likely due to differences in the physiology and cognitive capacities of rodents and primates. New behavioral paradigms, including those involving nonhuman primates, are sorely needed.

Behavioral assays can be used to assess general sensorimotor functions. One could simply record an animal's behavior in its home cage with a video camera. Such recordings allow researchers to assess general motor activity, eating, drinking, sleeping, and circadian rhythms in a laboratory housing environment with minimal handling. Another assay is to place the animal in an open field—essentially a box with walls but no top—and videotape, analyze, and quantify its movement within a given time period. The recent introduction of machine learning techniques for analyzing and quantifying such natural behaviors (Figure 14-52) will expand the utility of such simple behavioral paradigms. Motor coordination can also be tested by directly recording an animal's locomotion patterns (Movie 8-4) or measuring the length of time an animal can remain on a rotating rod (**Figure 14-54**). These assays evaluate the basic functions of the nervous system. Specific assays can be used to examine specific sensory functions; for example, the hot plate assay tests temperature and pain perception by measuring the time it takes until an animal to flick its tail or lick a hind paw placed on a hot plate as its temperature rises.

Behavioral assays have also been designed to assess aspects of animal cognition, such as learning and memory. For example, the Morris water maze and fear conditioning assays (Figure 11-33; Figure 11-41) are widely used to assess the roles of the hippocampus and amygdala in learning and memory. The radial arm maze tests spatial and working memory; it consists of a number of arms with food pellets located at the ends of the arms (**Figure 14-55**A) and is placed in a room in which numerous visual cues are scattered outside the maze. After becoming habituated to the apparatus and the room, a food-restricted rat placed at the center of the maze efficiently visits all arms to collect food, with minimal revisiting of arms from which food had already been collected. Behavioral experiments indicate that rats rely mostly on extra-maze cues to accomplish the task. By recording the trajectories of animals visiting different arms in the maze, researchers can assess the functions of the hippocampus, which has an essential role in spatial memory, and the prefrontal cortex, which has been implicated in working memory.

Establishing animal models of human psychiatric disorders is challenging, as it is difficult to assess whether a mouse or a rat has a condition resembling schizophrenia, depression, anxiety, or autism. Nevertheless, researchers have devised behavioral paradigms to measure phenotypes that could reflect some aspects of these disorders. For instance, one measure of anxiety-like states is based on the open field test mentioned earlier. When mice are placed in a well-lit open field, their natural tendency is to stay at its periphery, likely because mice are more susceptible to predation when they are exposed than when they are hidden. Normal

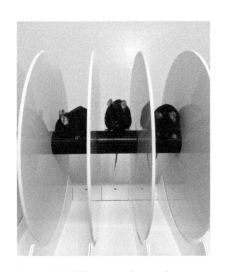

Figure 14-54 The rotarod assay for testing motor coordination and learning. At the beginning of the experiment, mice are placed on a stationary rod. The rod is connected to a motor with an adjustable speed. Typical experiments include running the motor at constant (e.g., 10 revolutions per minute, or rpm) or increasing speeds (e.g., from 5 to 30 rpm), and measuring the amount of time each mouse remains on the rod. In addition to testing motor coordination, the rotarod can also test motor skill learning. (Courtesy of Mehrdad Shamloo.)

(A)

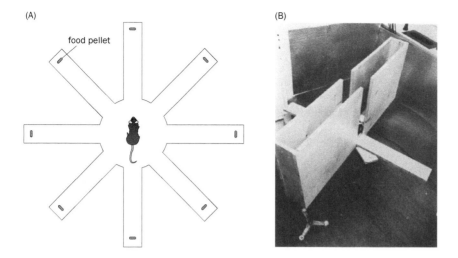

food pellet

(B)

Figure 14-55 Mazes used to test memory and anxiety. (A) Top view diagram of a radial arm maze for testing spatial and working memory. Food pellets are located at the ends of the arms. A food-restricted rat placed in the center of the maze will visit each arm to consume the pellets with minimal repetition, utilizing extra-maze cues (not shown) for navigation. **(B)** Photograph of an elevated plus-maze with a rat located at the intersection, between the open and closed arms. The fractions of time spent in the closed and open arms can be used to quantify anxiety in rats or mice and can be modulated by anxiolytic and anxiogenic drugs. (A, from Olton DS & Samuelson RJ [1976] *J Exp Psychol Animal Behav Proc* 2:97–116. With permission from the American Psychological Association. B, from Pellow S & File SE [1986] *Pharmacol Biochem Behav* 24:525–529. With permission from Elsevier Inc.)

mice venture into the center occasionally. Mice that visit the center less frequently are considered to have a high anxiety level. Another commonly used anxiety assay is the elevated plus-maze (Figure 14-55B). In this assay, a mouse (or rat) is placed onto a four-arm maze that is elevated from the floor, with two arms covered on the sides with walls and the other two arms exposed. Although mice prefer the closed arms, normal mice also spend some time exploring the open arms. Putatively anxious mice spend less time in the open arms. Administering anxiolytic drugs such as benzodiazepines (Section 12.17) can increase the tendency of animals to enter the center of an open field or the open arms of an elevated plus-maze, while anxiogenic drugs do the opposite, suggesting that these behavioral assays test conditions that may be related to anxiety in humans.

Behavior is sensitive to many factors, so it cannot be overemphasized that in performing the assays described here, care must be taken to control the experimental conditions (Section 14.28) to minimize unintended influences on behavior by factors not being tested. Proper control groups should always accompany the experimental group. Deficits in complex assays can have multiple interpretations. For instance, if an animal exhibits a deficit in auditory fear conditioning, the animal might have a problem hearing, sensing electrical shocks, or associating these two events (i.e., learning). Control experiments that test hearing and shock sensitivity are necessary to determine whether or not the deficit is related to learning. The interpretation of behavioral results also depends on the nature of the experimental perturbation. For instance, if the experimental manipulation is to knock out a gene of interest or apply a systemic pharmacological agent, then in principle the entire body could be affected and could contribute to the observed behavioral phenotypes. If the experimental manipulation is conditional knockout of a gene in specific neuronal types or activation or inactivation of specific brain regions or neuronal populations, then the behavioral phenotypes reflect alterations of the manipulated brain regions or neuronal populations.

THEORY AND MODELING

Theoretical and computational approaches are increasingly becoming an integral part of neurobiology. Theory and modeling can test whether our explanations are plausible and make quantitative predictions to guide further experiments. We have seen ample examples of such contributions in early chapters, such as in probabilistic analyses of neurotransmitter release (Box 3-1) and synaptic weight matrices modified by learning (Section 11.2). In the final part of this chapter, we discuss selected neurobiology topics from theoretical perspectives, spanning biophysical properties of neurons that lead to action potentials, information transmission mediated by action potential firing patterns, circuit architectures underlying specific computations, and learning algorithms. A deep understanding of some of these topics requires a command of linear algebra, differential equation, probability theory, and other areas of mathematics. The following sections are simplified so students without these backgrounds can also appreciate the value of theoretical approaches.

14.30 How do electrical properties of neuronal membranes produce action potentials?

Action potentials are a fundamental means by which signals from cell bodies propagate to axon terminals. In Section 2.11, we discussed how changes in voltage-dependent Na^+ and K^+ conductances, measured in voltage clamp experiments in the squid giant axon, *qualitatively* explain different phases of the action potential (Figure 2-24). Can they also *quantitatively* account for the form, amplitude, and other properties of action potentials? In this section, we discuss how Hodgkin and Huxley achieved this feat.

Based on the electrical circuit model of neuronal membranes (Figure 2-15), the total ionic current across the membrane can be expressed as

$$I = I_C + I_K + I_{Na} + I_l = C_m dV_m/dt + g_K(V_m - E_K) + g_{Na}(V_m - E_{Na}) + \bar{g}_l(V_m - E_l),$$

where I_C is the capacitive current (C_m is membrane capacitance, dV_m/dt is the change in membrane potential over time, and their product equals the capacitive current; Box 2-2); I_K and I_{Na} are K^+ and Na^+ currents, respectively, which can be expressed as the product of conductance and driving force, or the difference between the membrane potential V_m and the equilibrium potential (Section 2.7); and I_l is the leak current, which includes Cl^- and other ions whose conductances are assumed to be constant ($\bar{g}_l$; an overline indicates the value is a constant). Hodgkin and Huxley defined the resting potential E_r as 0 mV and introduced V as the membrane potential displaced from the resting potential, or $V = V_m - E_r$. Thus, the above equation can be written as

$$I = C_m dV/dt + g_K(V - V_K) + g_{Na}(V - V_{Na}) + \bar{g}_l(V - V_l), \tag{1}$$

where $V_K = E_K - E_r$ and $V_{Na} = E_{Na} - E_r$, both of which are constant. Unlike voltage clamp conditions, in which the membrane potential is held constant ($dV/dt = 0$), during an action potential, the membrane potential changes rapidly over time, as do K^+ and Na^+ conductances (i.e., g_K and g_{Na} have their own dynamics, which in turn depend on V). Therefore, equation (1) is a complex differential equation with many dependent variables.

The strategy Hodgkin and Huxley took to tackle this complexity was to first determine how the dynamics of g_K and g_{Na} are driven by V. They could then solve a combined set of differential equations for V and the conductances g_K and g_{Na}. They realized that they knew too little about the nature of K^+ and Na^+ conductances to determine them from physical principles. However, their voltage clamp experiments provided empirical data on how g_K and g_{Na} changed over time at different V values (Figure 2-23). Using these data, they could find equations that best fit the series of g_K and g_{Na} curves determined experimentally. Specifically, they found that the following expressions gave the best quantitative fit to the data:

$$g_K = \bar{g}_K n^4$$
$$dn/dt = \alpha_n(1 - n) + \beta_n n,$$

where $\bar{g}_K$ is a constant, α_n and β_n are rate constants that vary with V but not with time, and n is a dimensionless gating variable between 0 and 1. Phenomenologically, these equations correspond to the assumption that the K^+ conductance has 4 gates, all of which must be simultaneously open to allow K^+ to cross the membrane. n represents the probability that any one of these 4 gates is open, and so n^4 is the probability that all four gates are open, assuming they are independent. Additionally, each gate closes at a rate β_n and opens at a rate α_n. In turn, these rates are determined by the membrane voltage V. Hodgkin and Huxley identified the following equations for α_n and β_n that best fit the experimental data:

$$\alpha_n = 0.01(V+10)\bigg/\left[\exp\frac{V+10}{10}-1\right],$$

$$\beta_n = 0.125\,\exp\,(V/80).$$

Fitting the g_{Na} curves was more complex, as in addition to depolarization, there is also an inactivation component to the Na$^+$ conductance (Figure 2-23). Hodgkin and Huxley found that the following equations best fit the experimental data:

$$g_{Na} = m^3 h \bar{g}_{Na},$$

$$\frac{dm}{dt} = \alpha_m(1-m) - \beta_m m,$$

$$\frac{dh}{dt} = \alpha_h(1-h) - \beta_h h,$$

where $\bar{g}_{Na}$ is a constant; α_m, β_m, α_h, and β_h are rate constants for individual gate opening and closing that vary with V but not with time, as before; m and h are dimensionless variables between 0 and 1. m and h can be interpreted as the probabilities of an individual gate being open. Phenomenologically, for the membrane to be permeable to Na$^+$, these equations assume that 3 m gates and 1 h gate must all be open. Again, curve fitting allowed Hodgkin and Huxley to write equations for α_m, β_m, α_h, and β_h as functions of V, just as with α_n and β_n earlier.

The next step was to use these equations derived from voltage clamp data to solve equation (1). In a steady propagating action potential, the membrane current I can also be expressed as

$$I = (a/2R\theta^2)d^2V/dt^2,$$

where a is the radius of the axon fiber, R is the specific resistance of the axon fiber (the total intracellular cable resistance divided by cable length and multiplied by the cable's cross-sectional area), and θ is the action potential conduction speed. Placing that to the left of equation (1) allowed Hodgkin and Huxley to solve the differential equation by numerical methods. Figure 14-56 shows their calculated action potential compared with the action potential measured under a similar condition. They match remarkably well, considering that all the parameters used to calculate the action potential were derived from experimental data obtained

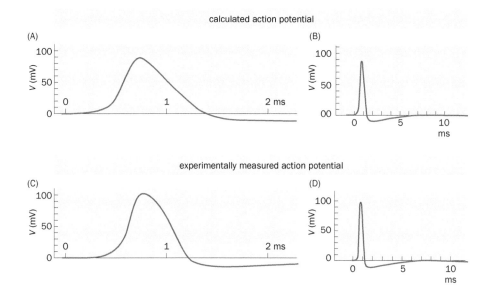

Figure 14-56 **Comparisons of simulated and experimentally measured action potentials**. **(A & B)** A simulated action potential derived from the numerical solutions to the differential equation discussed in the text, displayed at higher (Panel A) and lower (Panel B) time magnifications. **(C & D)** An experimentally recorded action potential similarly displayed at higher (Panel C) and lower (Panel D) time magnifications. Note that Hodgkin and Huxley defined 0 mV as the resting potential. (From Hodgkin AL & Huxley AF [1952] *J Physiol* 117:500–544. With permission from John Wiley & Sons.)

under voltage clamp conditions. In addition, the calculated conduction speed θ matched well the experimentally measured speed (18.8 m/s versus. 21.2 m/s), and so did several other parameters, such as time of inactivation, refractory period, and voltage changes in response to subthreshold stimuli. Thus, a combination of experiments and theory demonstrated that voltage-dependent changes in Na⁺ and K⁺ conductances are sufficient to account for *quantitative* properties of action potentials.

While the Hodgkin–Huxley model applies to action potentials across the animal kingdom, there are interesting variations. For example, the use of K⁺ channels with faster or slower repolarization kinetics contributes to the very short refractory period (~1 ms) of fast spiking neurons and very long interspike intervals in cardiac muscles (~1000 ms). In some cases, such as crustacean muscles or dendrites of certain mammalian neurons, voltage-gated Ca²⁺ channels replace voltage-gated Na⁺ channels in the depolarization phase of action potentials or dendritic spikes. These variations can be similarly modeled using the approach discussed here.

14.31 How do action potential firing patterns transmit information?

Information in the nervous system is conveyed from neuron to neuron mostly by action potential firing patterns, also called **spike trains**. How do spike trains transmit information? This question can be broken down into two related problems: (1) How are stimuli represented by spike trains of a neuron or a neuronal ensemble? This is the problem of **encoding**. (2) What do spike trains of a neuron or an ensemble tell us about the stimuli that induce them? This is the problem of **decoding**, which organisms need to solve in order to extract useful information from neuronal firing patterns.

It is important to emphasize at the outset the *probabilistic nature* of spikes. As we saw in numerous examples (e.g., Figure 6-26; Figure 6-49), identical stimuli do not produce identical spike trains. The source of variability in spike trains unexplained by the stimuli is commonly regarded as **noise**. Such noise can arise due to the probabilistic nature of some of the elementary neuronal communication steps, such as opening and closing of ion channels (Figure 2-29) and fusion of synaptic vesicles with presynaptic membranes (Box 3-1); such noise could also reflect interesting and important variations in the animal's internal state, such as attention or motivation. More generally, in any situation in which a single neuron receives input not only representing the stimuli we are interested in but also a multitude of other unrecorded neurons, we should expect substantial trial-to-trial variability in that neuron's responses, even to identical stimuli. Such noise can be unique to an individual neuron or shared among a group of neurons—for instance, if the group of neurons receives stimulus-unrelated input from a common source. Thus, when analyzing neural encoding and decoding, we always need to take noise into consideration.

Encoding and decoding are two sides of the same coin. We can see this most clearly in neurons in a sensory system, where we are interested in the relationship between a time-varying external stimulus, $s(t)$, and a neuron's spike train described by the times each spike occurs, $t_1, t_2, \ldots, t_N$, abbreviated as $\{t_i\}$. [A spike train can also be expressed as the more familiar time-varying **firing rate (spike rate)**, $r(t)$, which is the mean spike count per unit time.] If $s(t)$ is drawn from a set of discrete stimuli, encoding describes the **conditional probability** of observing different possible spike trains given a particular stimulus, or $P[\{t_i\}|s(t)]$. Conversely, decoding the stimulus requires determining the conditional probability of the different possible stimuli given a particular observed spike train, or $P[s(t)|\{t_i\}]$. These two conditional probabilities are linked by **Bayes' rule**:

$$P[s(t)|\{t_i\}] = P[\{t_i\}|s(t)] \times P[s(t)]/P[\{t_i\}],$$

where $P[s(t)]$ is the probability distribution of stimuli, which is determined by the experimenter (in a controlled experiment) or by the natural statistics of stimuli (in real life); $P[\{t_i\}]$ is the probability distribution over the set of all possible spike trains $\{t_i\}$, and is a property of the neuron. Note that Bayes' rule treats neural computations as statistical inference problems; because sensory stimuli can be ambiguous and there is always noise, as discussed earlier, the brain cannot truly know the state of the world with certainty. Instead, the brain can only attempt to represent the distribution of probable states of the world.

How can we decode stimuli, given observation of a neuron's spike train? One approach is to identify the stimulus most likely to have produced the observed spike train using Bayes' rule—to produce a "Bayes-optimal decoder." In special cases, it is possible to write explicit equations to estimate the stimulus in terms of neural spiking. In many cases, however, mathematical complexity leads researchers instead to rely on the best *linear* approximation of a Bayes-optimal decoder. Linear decoders are attractive because they are simple and can be estimated without large amounts of experimental data, which is often unavailable. A linear decoder computes the weighted sum of the current and past activity of a neuron or group of neurons that best approximates the stimulus. In the following, we use two examples in the visual system to study how researchers linearly decode $s(t)$ given $\{t_i\}$ in a single neuron or a neuronal population.

The first example involves the blowfly H1 neuron in the lobula plate of the optic lobe (**Figure 14-57A**; see Figure 4-31A for the anatomical organization of the insect visual system). The H1 neuron is specialized to detect global horizontal rotations of the visual world. These horizontal rotations occur whenever the fly turns—for example, a rightward rotation of the visual world always accompanies a leftward turn. During flight, this type of visual stimulus can indicate that the fly has been blown off course and needs to perform a compensatory turn to return to its intended trajectory. Researchers have performed extensive extracellular recordings of H1 during motion stimuli in a tethered-fly preparation (Figure 14-53). To reconstruct a time-varying stimulus $s(t)$ from H1's spike train $\{t_i\}$, we look to apply a linear filter, F, to the neuron's spiking to recover the motion stimulus—that is, to produce a weighted sum of the neuron's recent spiking that best approximates $s(t)$:

$$s_{est}(\tau) = \sum_i F_i \cdot t_{\tau-i}.$$

To determine F, the experimenter delivered a battery of random motion stimuli and recorded the resulting spike trains. The Bayes-optimal linear filter turned out

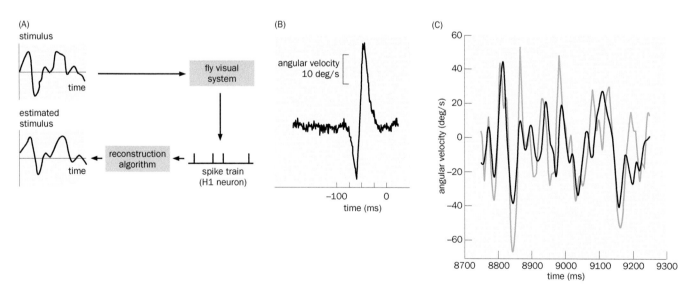

Figure 14-57 Decoding stimuli from spike trains of H1 neurons in the blowfly optic lobe. (A) Schematic of the decoding process. The reconstruction algorithm converts the spike train input into an estimate of the stimulus. **(B)** Linear filter determined based on the best fit of $s_{est}(t)$ to the neuron's response to a previously delivered stimulus sequence s(t). Positive and negative angular velocities correspond to rightward and leftward movements of the visual pattern. Note that stimuli within ~30 ms prior to the spike do not contribute to H1 spiking because of a delay in signal transmission between photoreceptors and the H1 neuron. **(C)** The linear filter in Panel B applied to a novel stimulus (cyan curve) for reconstruction of the estimated stimulus (black curve). (Adapted from Bialek W, Rieke F, de Ryter RR, et al. [1991] *Science* 252:1854–1857. With permission from AAAS. See also Rieke et al. [1997] Spikes: Exploring the Neural Code. MIT Press.)

to simply be the average motion stimulus preceding a neuron's spike. Figure 14-57B shows the shape of F as a function of time for the H1 neuron. We can see that the filter integrates over a time interval of ~40 ms, which is the same order as a fly's fastest behavioral response to angular velocity change. To test the accuracy of this procedure, the filter was applied across time to a novel random stimulus (Figure 14-57C). The good fit indicated that the simple linear decoder can accurately estimate the time-varying $s(t)$.

In most cases, in particular in the vertebrate nervous system, information about a stimulus is conveyed by multiple neurons. A *population decoder* that incorporates contributions from many neurons often produces a better estimate of the stimulus than a single neuron decoder. In the second example, researchers simultaneously recorded the activity of a population of retinal ganglion cells (RGCs) from salamander retinal explant in response to spatially uniform gray stimuli with varying intensity over time. As described in the previous example, the goal was to determine a linear filter for each neuron in the population. The filters were then applied to each cell, as above, and the results were also summed *across* cells:

$$s_{est}(\tau) = \sum_k \sum_i F_i^k \cdot t_{\tau-i}^k,$$

where k indexes across neurons to incorporate contributions from each cell. Here the linear filters for each neuron are not simply computed independently. Instead, this is a multiple linear regression, where the contribution of one cell depends on its *correlation* with the contributions of other cells. Thus, the contribution of an individual neuron will vary depending on the contributions of other neurons included in the decoder.

Figure 14-58 illustrates the simplest example of a two-cell decoder. By varying the number and type (ON versus OFF; Section 4.11) of RGCs included in the stimulus decoder, researchers could learn about how information regarding the stimulus (time-varying gray intensity) is transmitted by an RGC population. For example, a decoder made up of two OFF-RGCs did not improve decoding substantially compared to a single OFF-RGC decoder, whereas a decoder composed of an ON- and an OFF-RGC each did, implying that RGCs of the same type encode largely redundant information, whereas RGCs of different types encode distinct information.

We have focused on filters that predict a stimulus based on a spike train (decoding), but we can use a similar approach to compute filters that predict spike train responses to a stimulus (encoding). While our examples have focused on visual signals, the concepts of encoding and decoding are widely used to analyze neural signals across the nervous system. Many sophisticated encoding and decoding techniques have been developed to analyze increasing amounts of data thanks to recent advances in recording techniques (Sections 14.20–14.22 and Box 14-4). For example, because the biophysical processes that generate action potentials are not actually linear functions of the synaptic inputs to the neuron, many models that use linear filters like those described here also subsequently apply a non-

Figure 14-58 Decoding visual stimuli from spike trains of two simultaneously recorded retinal ganglion cells.
(A) Illustration of the decoding procedure. Stimuli reconstructed from each cell (based on individual spikes and filters in the boxes) are combined to produce the final reconstructed stimulus in Panel B.
(B) Top, reconstructed stimulus (black trace) is superimposed on the actual stimulus (blue trace). Bottom, spike trains of two cells in response to the stimulus. Cell A is an OFF-RGC; cell B is an ON-RGC. The y axis represents light intensity for both the filters in Panel A and traces in Panel B. (From Warland DK, Reinagel P, & Meister M [1997] *J Neurophysiol* 78: 2336–2350. With permission from the American Physiological Society.)

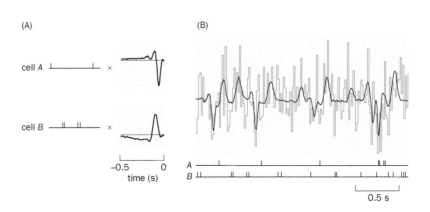

linear sigmoid function to the resulting signal to simulate the nonlinearity of spike generation. In addition, many special cases have been identified wherein it is possible to directly compute the Bayes-optimal decoder, which can be more accurate than the linear decoders. Finally, it is important to note that while we have focused on reconstructing stimuli, in many cases animals do not need to fully reconstruct stimuli from the spike train to extract behaviorally relevant information. We discuss one special case of decoding in **Box 14-5**: binary classification.

Box 14-5: Binary classification: a special case of decoding

Binary classification is the task of classifying observed events into one of two categories, with a contingency table shown in **Figure 14-59**A. We use the binary decision that a monkey makes in the random dot display assay (Figure 4-55) as an example to work through the table. True conditions refer to motion of the dots; decoded conditions refer to the monkey's choice on each trial, as indicated by a rapid eye movement (a saccade) in one of two directions. We designate "positive" to be trials in which a fraction of dots coherently moves in the preferred direction of a recorded middle temporal visual area (MT) neuron. Then, true positives are the trials in which both the motion of the dots and the monkey's saccade are in the neuron's preferred direction. False positives are the trials in which motion is in the opposite (null) direction but the monkey saccades to the preferred direction. False negatives are the trials in which motion is in the preferred direction but the monkey saccades to the null direction. True negatives are the trials in which both the motion of the dots and the monkey's saccade are in the null direction. We can determine the rates of each trial type according to the formulas at the bottom of the table.

Figure 14-59B depicts idealized data recorded from a single MT neuron. The right and left curves represent the normal distributions of spike count in response to dots moving (with a specific coherence) in the preferred and null directions, respectively. With respect to this MT neuron, the decoding problem becomes: how well could a downstream mechanism decode the actual direction of motion on a single trial, based on the spike count recorded on that trial from this MT neuron alone? The downstream mechanism must apply a decision criterion (a threshold) to the spike counts obtained on single trials to decode preferred versus null directions. The four probabilities of the binary classification outcomes are depicted as four denoted areas in Figure 14-59B. As

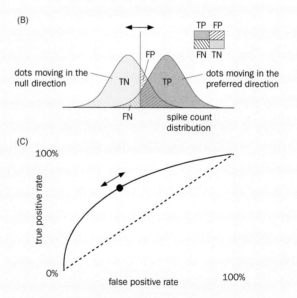

(A)		true condition	
		true condition positive	true condition negative
decoded condition	decoded condition positive	true positive (TP) *true preferred*	false positive (FP; type I error) *false preferred*
	decoded condition negative	false negative (FN; type II error) *false null*	true negative (TN) *true null*
formulas		true positive rate = TP / (TP + FN) false negative rate = FN / (TP + FN)	false positive rate = FP / (FP + TN) true negative rate = TN / (FP + FN)

Figure 14-59 Binary classification. (A) Contingency table of binary classification. Text in red refers to the specific example of a motion discrimination task. The bottom row lists the formulas for calculating true/false positive/negative rates. **(B)** Graphic illustration of two normal distributions representing spike count distributions of an MT neuron when the monkey was viewing a fraction of dots moving coherently in the preferred direction (right) or null direction (left). The vertical line represents the decision threshold. Areas representing true positive (TP), false positive (FP), false negative (FN), and true negative (TN) are color coded. **(C)** A receiver operating characteristic (ROC) curve displaying the true positive rate (*y* axis) against the false positive rate (*x* axis), corresponding to Panel B. As the decision threshold moves from right to left in Panel B, the dot traces out an ROC curve from left to right. The area under the ROC curve estimates the probability of the monkey making the correct choice based on the spike count of this single neuron.

(Continued)

Box 14-5: continued

the threshold moves from right to left, the true positive rate increases, followed by an increasing false positive rate. The **receiver operating characteristic (ROC)** curve plots the true positive rate against the false positive rate as the threshold systematically changes (Figure 14-59C). One can use the area under the ROC curve to estimate the probability of the monkey making the correct choice (these have been shown to be equivalent in a two-alternative forced-choice task, as in the case discussed here). If the two distributions were identical, true and false positive rates would exhibit identical changes as the threshold moves from right to left, and the ROC curve would thus be a diagonal line with the area under it equaling 0.5—that is, the probability of making the correct choice would be 50%, or exactly at the chance level. If the two distributions were well separated, then as the threshold moves from right to left, the true positive rate would increase to almost 1 before the false positive rate would start to increase, so the ROC curve would hug the top and left edges, making the area under the curve close to 1, and the monkey would almost always make the correct choice. If the two populations overlap, as is the example in Figure 14-59B, then the ROC curve would be in the top left half of the square, with the area under the curve between 0.5 and 1 (Figure 14-59C), and the monkey would make the correct choice more than half of the time.

Let's now examine data from actual experiments. **Figure 14-60**A shows single-unit recording data of an MT neuron while an alert monkey was viewing random dot display; the fraction of moving dots with coherent movement was systematically varied in either the cell's preferred direction (blue histogram) or the null direction (red histogram). These two histograms are obviously easily separable when 12.8% of the dots move in the same direction but become more and more similar as coherence decreases. Using area under the ROC curve (Figure 14-60B), we find that at 0.8% coherence, the fraction of correct choices is barely above 0.5

(chance level, corresponding to the area under the diagonal line), whereas at 12.8% it is nearly 1 (always correct). One can thus plot a "neurometric curve" depicting the performance of this neuron in decoding motion direction at different coherence levels (Figure 14-60C, purple curve). This can be compared with the psychometric curve we introduced in Section 4.28 (Figure 14-60C, orange curve), which depicts the behavioral performance of the monkey. Remarkably, the neurometric and psychometric curves coincide, meaning that the monkey could make correct behavioral choices based on the firing pattern of this single neuron alone. This implies that the actual decoding mechanism in the brain may be similar to the simple decision thresholding mechanism used in the calculation. Indeed, in certain experiments the neuron outperformed the monkey's behavior (i.e., the neurometric curve was to the left of the psychometric curve).

These experiments raised an interesting question: why can't the monkey perform better than the neuron, given that it can reduce noise by pooling signals from many neurons (Section 1.12)? Further experiments and theoretical analyses indicate that MT neurons with similar motion direction tuning also have correlated noise, and that when noise correlations reach about 10%, the advantage of pooling (by increasing the signal-to-noise ratio) saturates at about 100 neurons. In reality, a presumed decision neuron downstream of MT, like most cortical neurons, receives inputs from hundreds of neurons and requires co-activation of many of them to reach its own firing threshold. Not all MT neurons that provide input to this decision neuron may have neurometric curves as good as the best neuron (as the direction of coherent motion in these experiments were designed to best match the preferred motion direction of the recorded neuron; Section 4.28). Thus, when studying decoding, we must take into consideration *how neural codes are read by real neurons* in the brain.

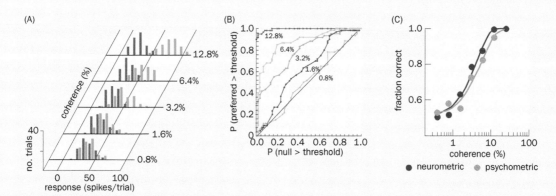

Figure 14-60 Decoding motion direction based on firing patterns of a single MT neuron. (A) A series of histograms of an MT neuron's number of spikes per trial in response to moving dot stimuli with different coherence (percent of dots moving in the same direction) in the preferred (blue histogram) or null (red histogram) directions. **(B)** ROC curves for the five pairs of preferred–null response distributions illustrated in Panel A. **(C)** A neurometric function describing the sensitivity of the MT neuron to motion signals of increasing coherence superimposed on a psychometric function of the monkey's behavior in this experiment. (From Britten KH, Shadlen MN, Newsome WT, et al. [1992] *J Neurosci* 12:4745–4765. Copyright ©1992 Society for Neuroscience.)

14.32 How does neural circuit architecture enable computation?

We have seen many examples of how neural circuits perform specific tasks: extracting contrast information from light intensity (Section 4.14), identifying sound location from a time difference (Section 6.24), coordinating rhythmic movements using central pattern generators (Section 8.5), and transforming receptive fields of input neurons (Section 4.23). In this section, we discuss how neural circuits perform specific tasks from a more abstract and theoretical perspective. Let's start with an idealized neuron with two states: firing (1) and not firing (0); and a simple integration rule: the neuron crosses its firing threshold when two excitatory inputs are simultaneously active, but any simultaneously active inhibitory input would prevent the neuron from firing. Connections between such idealized neurons can perform OR, AND, NOT, XOR (**Figure 14-61**), and all other Boolean logic gates for two inputs. As Boolean logic gates (named after the nineteenth-century mathematician George Boole) are the basis of all operations in modern digital computers, we can state that circuits consisting of these simple neurons can perform complex computations.

Real neurons and neuronal circuits are much more complex than the idealized neuron and integration rule described here. As we learned in Chapter 3, a typical mammalian neuron receives thousands of synaptic inputs, each with different strengths. The contribution of synaptic inputs to the firing threshold is subject to attenuation due to the distance between the input site and action potential initiation site, as well as nonlinear amplification via co-activation of nearby synapses on the same dendritic branch or interaction with back-propagating action potentials. To quantitatively model complex neuronal networks, however, theoretical neurobiologists often make simplifying assumptions about the properties of individual neurons. For example, in a linear threshold model neuron, the activity (which has two states: 1, firing; and 0, not firing) is determined by whether the weighted sum of its synaptic inputs exceeds a threshold θ:

$$y = H\{\textstyle\sum_{i=1}^{N} W_i \cdot x_i - \theta\}, \tag{2}$$

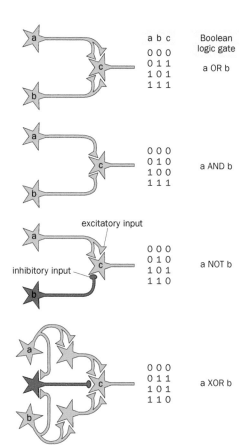

a	b	c	Boolean logic gate
0	0	0	
0	1	1	
1	0	1	a OR b
1	1	1	

a	b	c	
0	0	0	
0	1	0	
1	0	0	a AND b
1	1	1	

excitatory input

inhibitory input

a	b	c	
0	0	0	
0	1	0	
1	0	1	a NOT b
1	1	0	

a	b	c	
0	0	0	
0	1	1	
1	0	1	a XOR b
1	1	0	

Figure 14-61 Boolean logic can be computed by simple neurons and circuits. Specific connection diagrams (left) can lead to specific logic gate operations (right) following a simple integration rule: each neuron fires if it receives two or more excitatory inputs and no inhibitory inputs at any given time. Note that in the XOR gate, two intermediate excitatory neurons are needed so that excitatory and inhibitory inputs reach neuron c at the same time. (Adapted from McCulloch WS & Pitts W [1943] *Bulletin Math Biophys* 5:115–133.)

Figure 14-62 A linear threshold neuron and a perceptron. (A) Schematic of a linear threshold neuron y, which fires when the weighed sum of its input $(\sum_{i=1}^{N} W_i \cdot x_i)$ exceeds a threshold θ. **(B)** Schematic of a simple perceptron. Information flows from an input layer through initially randomly weighted connections to an association area with two neurons and then to a response area with two neurons, each connected to one association neuron. The circuit can be trained to serve as a binary classifier (to detect, for example, whether an image presented by the input layer is or is not a cat) by viewing labeled images and changing synaptic weights (W_{ij}) based on the correspondence between the labels and the circuit's responses. (Based on Rosenblatt F [1958] *Psychol Rev* 65: 386–408.)

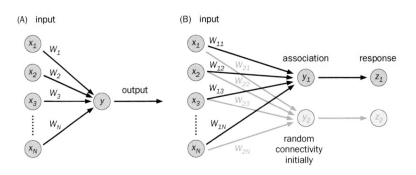

where y is the activity of a neuron receiving N inputs, x_1–x_N (each of which can be 1 for firing or 0 for not firing); W_i is the strength (weight) of the synapse between the neuron and its presynaptic partner i; and H is a step function defined by $H(\sigma) = 1$ when $\sigma \geq 0$ and $H(\sigma) = 0$ when $\sigma < 0$ (**Figure 14-62A**). Note that the vector of weights W is essentially a linear filter (Section 14.31) that describes how presynaptic activity is transformed into postsynaptic activation drive.

An influential early model neural network, the **perceptron,** organizes linear threshold neurons in a feedforward architecture (Figure 14-62B) inspired by the organization of the visual system. In a perceptron, neurons are distributed in specific layers. A given layer receives inputs from the previous layer and sends output to the next layer. Frank Rosenblatt, who originally proposed the perceptron, also posited a learning rule and showed that classification of visual objects can be learned by viewing examples and their correct classifications. Although this was considered a major breakthrough when first proposed, enthusiasm was tempered by subsequent demonstrations that single-layer perceptrons have limits in performing nonlinear classification. However, multilayer perceptrons are not subject to such limits and have become core models underlying the recent resurgence of deep learning, which we will study in Section 14.33 and Box 14-6.

We learned in Chapter 4 that information processing in the visual cortex involves not only feedforward connections but also extensive recurrent and feedback connections. This requires a more complex circuit model than the feedforward architecture of a perceptron. Models with recurrent and feedback connections also exhibit complex temporal dynamics. For example, in a network of N linear threshold neurons, the activity of each neuron *at a given time* is determined by whether the weighted sum of its synaptic inputs exceeds a threshold at an immediately preceding time:

$$x_i(t) = H\{\textstyle\sum_{j=1}^{N} W_{ij} x_j(t-1) - \theta_i\}, \tag{3}$$

where $x_i(t)$ is the activity of a given neuron i at time t, θ_i is its firing threshold, H is a step function defined as in equation (2), and W_{ij} is the synaptic weight between neuron i and its presynaptic partner j. For inhibitory input, $W_{ij} < 0$. If two neurons are not connected, then $W_{ij} = 0$.

Quantitative analysis involving recurrent connections are complex. John Hopfield described a special recurrent network (known as a **Hopfield network**) wherein $W_{ii} = 0$ (i.e., neurons do not form synapses onto themselves) and $W_{ij} = W_{ji}$ (the weights of synaptic connections between any two neurons are symmetrical), which simplifies the computation. In this network, an "energy" function

$$E = -\frac{1}{2}\textstyle\sum_{i,j} W_{ij} x_i x_j + \sum_i \theta_i x_i$$

will either stay the same or decrease as neurons in the network update their states according to equation (3). (Note that this abstract energy function differs from the concept of energy in physics, and the term is used here analogously.) Thus, energy will converge to local minima over time (**Figure 14-63**). These low-energy states are called **attractors**. The Hopfield network was originally conceived of as a way to store associative memory wherein the activity pattern at each attractor corresponds to a specific memory. Activity patterns that diverge slightly from an attractor are guaranteed to move toward the nearest attractor, analogously to recall of

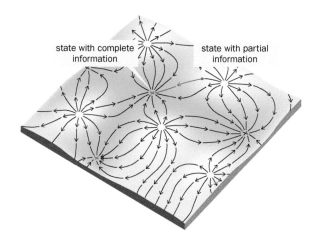

Figure 14-63 Schematic of a Hopfield network. In this energy landscape, in which darker and lighter colors represent valleys and peaks, respectively, each point represents an activity state of a neural circuit (Figure 8-35). Dynamics in the circuit over time can be envisioned as following specific trajectories in this energy landscape. The Hopfield recurrent network guarantees that over time, neural states will reach local minima (marked by crosses). This property can be used for pattern completion in associative memory. If a circuit starts out with partial information, it follows a downhill path (red arrow) to the nearest valley, which contains the complete information (full memory). (From Tank DW & Hopfield JJ [1986] *Sci Am* 257[6]:104–114. With permission from Springer Nature.)

full memory with partial information (a process called pattern completion). The concept of an attractor in an energy landscape has since been widely used to describe dynamics of neural networks beyond associative memory and Hopfield networks.

In our final example, researchers treated excitatory and inhibitory neurons separately in a recurrent model and gained interesting insights after a few simplifications. Suppose that N_E excitatory and N_I inhibitory neurons in a network are reciprocally connected, and additionally receive N_O external inputs (**Figure 14-64**A). Suppose further that each two-state neuron receives K excitatory and K inhibitory inputs ($K \ll N$) through random connections but requires only $\sqrt{K}$ excitatory inputs to cross its firing threshold. All neurons are modeled as binary units. In this network, if a neuron receives strong excitatory input, it also receives strong inhibitory input; each neuron fires when its total excitatory input exceeds its total inhibitory input by a specific amount—this firing pattern resembles what has been observed in cortical neurons. Computational analyses revealed that this network has two remarkable properties. First, even though each neuron is a nonlinear unit—converting analog input signals into digital output signals—network activity (the fraction of neurons firing) varies linearly with the strength of the input (Figure 14-64B). Second, if there is a change in input, the response time for network activity changes more rapidly than in an unbalanced network (a network in which inhibitory and excitatory inputs are not generally equal; Figure 14-64C). Balanced networks respond more quickly because all neurons tend to be near firing threshold at all times.

These examples illustrate some general approaches in theoretical and computational neurobiology. Modeling requires certain simplifying assumptions to

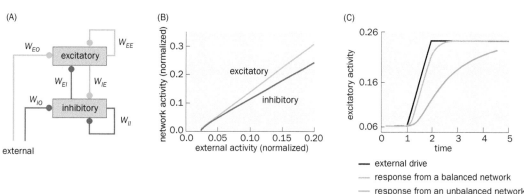

Figure 14-64 A circuit with balanced excitation and inhibition.
(A) Circuit architecture. W_{ij} represents synaptic weight vectors between external input (O), excitatory neurons (E), and inhibitory neurons (I). **(B)** The mean activity of excitatory and inhibitory neurons is a linear function of the amount of external input. **(C)** The response of a balanced network tracked external drive more closely than that of an unbalanced network. (From van Vreeswijk C & Sompolinsky H (1996) *Science* 274:1724–1726. With permission from AAAS.)

make quantitative analysis feasible and to extract the essence of the problem without being overwhelmed by the complexity of the real world. Simplification is often necessary because we are simply ignorant of some details of a biological system. Even when details are known, a modeler may choose to omit those because simpler models are easier to analyze mathematically and easier to understand. That said, one needs to keep in mind the limitations these simplifying assumptions bring. For example, none of the given examples takes into account nonlinear properties of dendritic integration. The Hopfield network further assumes symmetrical synaptic connections, which is far from realistic. However, even when $W_{ij} \neq W_{ji}$, which makes the problem more analytically complex, simulations indicate that neural network activity also moves toward attractor-like states if initial activity patterns do not deviate too much from one of the attractors. Thus, one strategy for dealing with the complexity of reality is to start with simplifying assumptions to discern important properties, and then to test whether these derived properties persist under more complex, realistic conditions.

14.33 How do different algorithms enable biological learning?

In Chapter 11, we studied learning from the perspective of modifications of the strengths (weights) of synapses between neurons in specific circuits. Here we discuss learning from the perspective of algorithms and their implementation, drawing comparisons with machine learning. There are many parallels between biological learning and machine learning; both leverage experience to improve task performance. Indeed, machine learning has been increasingly used in recent years for data analysis in neurobiology, from neural and behavioral recording to serial EM reconstruction.

Algorithms in machine learning can be broadly categorized as unsupervised or supervised. In **unsupervised learning**, the task is to experience a data set and draw inferences about its structure. **Cluster analysis**, which divides a data set into subsets based on similarities in their key feature(s), is a form of unsupervised learning; examples include constructing a phylogenetic tree of proteins based on similarities in their amino acid sequences (Figure 2-34) and visualizing transcriptomic clusters based on single-cell RNA-seq data using tSNE plots (Figure 14-18). **Principal component analysis** as a means of transforming and visualizing high-dimensional data in fewer dimensions is also unsupervised learning; we have seen this in our studies of population coding in the olfactory and motor systems (Figure 6-28; Figure 8-35). In **supervised learning**, the task is to identify a function for mapping an input to an output after experiencing example input–output pairs (training sets). The output during training serves as an instructor to modify the parameters of the function to better match input with output in the future. Examples of supervised learning include linear regression (where output is a value) and object classification (where output is a category label); in both cases, training requires a set of input–output pairs. **Reinforcement learning** is a third type of machine learning somewhat in between unsupervised and supervised learning. In reinforcement learning, an "agent" interacts with an environment by repeatedly performing an action and receiving a reward as a consequence of its action (**Figure 14-65**). In other words, instead of learning from instructed pairs of input and desired output, as in supervised learning, or with no instruction, as in unsupervised learning, an agent receives an indirect instruction from the environment that more closely resembles what animals encounter in their natural environments.

In Section 11.24, we saw a prime example of reinforcement learning in neurobiology: reward-based learning using midbrain dopamine neuron activity as a reward prediction error. In the following we use examples from the visual system to illustrate unsupervised and supervised learning.

Much biological learning is unsupervised—weight changes in synaptic connections resulting from activity or experience without specific instructions. We discuss here a familiar example: development of ocular dominance as a result of correlated activity in nearby neurons in the same eye (Sections 5.7–5.12). Suppose that two lateral geniculate nucleus (LGN) layers, one for each eye, and layer 4 of

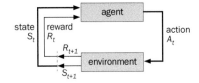

Figure 14-65 Schematic of reinforcement learning. The *agent* refers to the learner and decision maker. Everything else it interacts with constitutes its *environment*. These interact continually: the agent selects an *action*; the environment responds by changing its *state* and providing a *reward*; the agent selects actions to maximize reward over time. The schematic above corresponds to a Markov decision process in which the properties of S_{t+1} and R_{t+1} depend only on the immediate preceding state and action, S_t and A_t, and not on any previous agent–environment interactions. (From Sutton R & Batto A [2018] Reinforcement Learning, 2nd ed., MIT Press.)

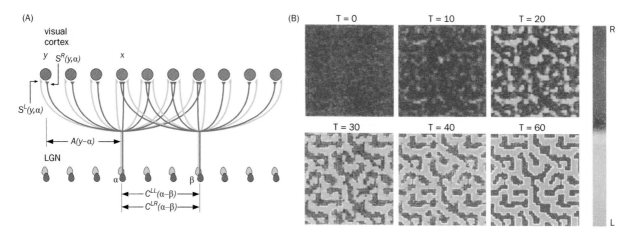

Figure 14-66 Emergence of ocular dominance columns as a result of correlated input and Hebbian plasticity. (A) Circuit diagram. Each LGN neuron, representing left (green) and right (red) eyes at a specific position in the LGN grid (bottom), is connected to 7 × 7 neurons (gray) in the layer 4 grid (top). Synaptic weights from left- and right-eye-representing LGN neurons at position α to the layer 4 neuron at cortical position y are determined by the functions $S^L(y,\alpha)$ and $S^R(y,\alpha)$, respectively. See text for the rest of the notations. **(B)** Layer 4 map of the difference between the total synaptic weights from the LGN representing left versus right eyes, color coded according to the index on the right. At $t = 0$, each layer 4 neuron receives similar inputs from the left- and right-eye-representing LGN, and hence the weight difference is near 0 (blue). As time progresses, the map gradually evolves into cells activated by mostly right- (red) or left-eye-representing (green) LGN. (From Miller KD, Keller JB, & Stryker MP [1989] *Science* 245: 605–615. With permission from AAAS.)

primary visual cortex (V1), are each represented by a 25 × 25 grid of cells. Each LGN cell starts with an axonal arbor connecting with a subgrid of 7 × 7 layer 4 cells following retinotopy (**Figure 14-66**A). We will see that, given correlated input from nearby LGN neurons, unsupervised learning based on Hebb's rule can produce connectivity patterns in these grids that resemble ocular dominance columns.

According to Hebb's rule, synaptic weights are enhanced by correlated firing of pre- and postsynaptic neurons (Section 5.12). At individual synapses this can be expressed as $\Delta S = [(post)(pre) - (decay)] \Delta t$, where ΔS is the change in synaptic weights over a small time interval Δt, *post* and *pre* are functions of the postsynaptic and presynaptic activities, respectively, and *decay* represents homeostatic and other processes that may adjust weights independently of pre/post correlation. Since postsynaptic activity is also a function of presynaptic activity and the synaptic weights of all connected neurons, we can express the changes in synaptic weights over time between a given layer 4 cell at position x and the left-eye-representing LGN cell at position α (Figure 14-66A) as:

$$\frac{dS^L(x,\alpha,t)}{dt} = \lambda A(x - \alpha)\sum_{y,\beta}I(x - y)[C^{LL}(\alpha - \beta)S^L(y,\beta,t)$$

$$+ C^{LR}(\alpha - \beta)S^R(y,\beta,t)] - \gamma S^L(x,\alpha,t) - \varepsilon A(x - \alpha),$$

where β and y are the positions of other LGN and cortical cells, respectively; C^{LL} and C^{LR} are functions representing the correlations between the left-eye-representing LGN cell at one position (α) and LGN cells at other positions (β) representing the left and right eyes, respectively; the cortical interaction function $I(x - y)$ represents the influence of geniculate excitation of the cortical cell at y on cell x; A represents the arbor function of LGN cells; λ is a learning rate; and γ and ε are decay constants. Replacing L with R, one can get an equivalent equation describing $S^R(x,\alpha,t)$. When the initial values of S^L and S^R are given at $t = 0$, these differential equations determine the values of S^L and S^R at subsequent times.

Suppose that at $t = 0$, the system starts with random synaptic weights with intermediate strengths. Thus, each layer 4 neuron is connected with both left- and right-eye LGN layers about equally on average, with little L/R preference (Figure 14-66B, top left panel). This reflects a developmental stage before eye-specific segregation. We now specify the correlation functions such that correlations of neurons in the same LGN layer vary as a normal distribution of physical separation (highest at shortest distance) and correlations of neurons in different LGN layers are weaker on average. Computer simulations following the differential equations revealed that eye-specific segregation develops gradually over time due to modification of synaptic weights (Figure 14-66B). Eventually, the LGN → layer 4 synaptic weight map becomes strikingly similar to the ocular dominance column patterns in cat and primate V1 (Figure 5-18).

By systematically modifying the parameters of the correlation, cortical interaction, and arbor functions, researchers can test how these properties affect the final outcome of the ocular dominance columns. For instance, correlations over a broader area among neurons in the same LGN layer make layer 4 cells more purely monocular, resembling primate V1, whereas narrower correlations produce more binocular cells at the patch border, resembling cat V1. It is important to note that a successful simulation does not *prove* that normal development follows the model exactly. Indeed, other models based on different assumptions can reproduce similar ocular dominance or orientation columns in V1 (Figure 4-45). Nevertheless, this example highlights the power of unsupervised learning: structured input (here, the correlation structure of neural activity) in combination with a simple learning rule can sculpt developing neural circuits to produce seemingly complex but functionally useful organizational patterns.

In our second example, we turn to object recognition in the inferior temporal (IT) cortex in the ventral stream of the primate visual cortex (Figure 4-51). We learned in Section 4.27 that neurons in this part of the higher-order visual cortex are activated by specific objects such as faces. However, little is known about how receptive field properties of inferior temporal cortex neurons arise. Here, supervised machine learning was used to shed light on this problem. Object recognition has been a long-standing problem in computer vision and has experienced great progress in recent years via deep learning using convolutional neural networks (CNNs; **Box 14-6**). CNNs use supervised learning to train weights of connections in artificial neural networks with the goal of improving performance of tasks such as object categorization. Can CNNs help us understand visual cortex?

To address this question, researchers presented photographs of complex sets of objects from natural scenes to monkeys while recording neuronal activity in IT as well as V4, an intermediate station between V1 and IT, using multielectrode arrays. The same sets of photographs were also used to train many CNN variants with different parameters. Furthermore, these CNNs were also used to predict the firing properties of IT for comparison with the actual recording data. A strong correlation between a given CNN's performance in object categorization and in predicting IT firing was found (**Figure 14-67**, left), even though the CNNs were not trained on IT firing data at all. When researchers modified the design of the CNNs to further optimize the accuracy of object categorization, the accuracy of predicting IT firing properties also increased further (Figure 14-67, right), reaching a performance level twice as high as previous best models. Interestingly, while the final layer of the CNN best predicted IT firing patterns, an intermediate layer of the CNN best predicted V4 firing patterns. While there are substantial differences between artificial deep neural networks and neural circuits in the brain (Box 14-6), these findings nevertheless suggest potential algorithmic similarities worth further exploration.

Figure 14-67 Correlation of task performance and IT predictability for different neural network models. Each dot represents a variant CNN model for an object categorization task, with the *x* axis representing performance accuracy. The same CNN model is also evaluated for its ability to predict IT firing patterns, with the *y* axis representing the variance of IT firing explained by the model. The blue dots indicate CNN models produced by systematically varying parameters such as layer number, kernel size, and pooling size (see Box 14-6). The red dots represent custom CNN models designed to further optimize categorization performance. In both cases, there is a strong positive correlation between categorization performance and predictability of IT firing. (From Yamins DLK, Hong H, Cadieu CF, et al. [2014] *Proc Natl Acad Sci U S A* 111:8619–8624.)

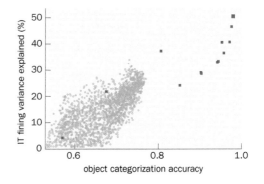

Box 14-6: Deep learning and convolutional neural networks

Due to the revolution over the past decade in machine learning, giving birth to a new discipline called **deep learning**, computers can now tackle many real-world problems with remarkable success. In their simplest forms, artificial neural networks for deep learning closely resemble multilayer perceptrons in architecture, with information flowing through neurons arranged in layers in a feedforward manner (Figure 14-62B). A key difference is the number of layers. While in the shallow perceptron inputs are connected to outputs directly or with one intermediate layer, deep learning uses neural networks with several to a few dozen intermediate layers between input and output (hence the word *deep* in deep learning). The weights of connections are adjusted by training the network with labeled targets as desired output, so deep learning is a supervised learning procedure. Weight adjustments utilize a statistical method called *stochastic gradient descent* that iteratively minimizes the error starting from the output layer backwards (a procedure called *backpropagation*). A large number of technical advances have contributed to recent breakthroughs in deep learning. These include improvements in network architecture, larger training data sets, and the greatly increased processing power of modern computers.

We use a specific example to illustrate the architecture of a deep **convolutional neural network (CNN)** from a land-mark study (Figure 14-68A). This CNN was built for image recognition, an important task in artificial intelligence. Specifically, it took images as input and classified each into 1 of 1,000 categories (Figure 14-68B), a nontrivial task even for humans. After being trained with 1.2 million images, this CNN was tested on 150,000 images that were not in the training set, and achieved ~85% classification accuracy, a marked improvement over the previous best. How does it work? In this box, we first introduce key features of this CNN, many of which were inspired by features of the mammalian visual system. We then compare CNNs to biological circuits.

We start with the input layer (layer I), which comprises 224 × 224 × 3 neurons taking in images of 224 × 224 pixels in three (RGB) colors. It feeds information into layer II. Each recipient neuron in layer II is connected only to a much smaller matrix of 11 × 11 neurons in layer I, and derives its value by linearly combining the values of these 11 × 11 layer I neurons with specific weights and biases (corresponding to **w** and **b** in the linear function $y = \mathbf{w} \cdot x + \mathbf{b}$). The 11 × 11 matrix of weights and biases is termed a *kernel* (or *filter*, analogous to the linear filter we discussed in Section 14.31). Two adjacent neurons in layer II receive inputs from two 11 × 11 matrices in layer I neurons whose positions are shifted in the same direction by a specific distance called a

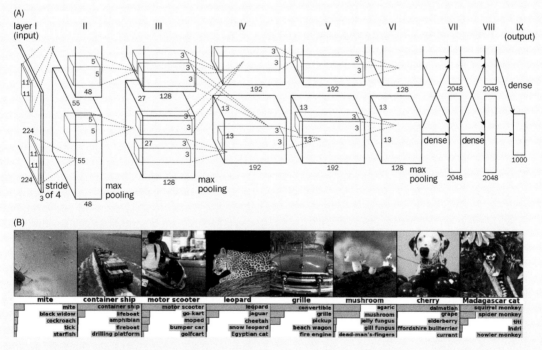

Figure 14-68 A deep convolutional neural network (CNN) for image categorization. (A) Network architecture. This network of 650,000 neurons is made up of five convolutional layers (II–VI), followed by three fully connected layers including an output layer. The top (a duplicate of the bottom, not fully drawn) and bottom halves are only occasionally connected (e.g., between layers III and IV). See text for details. **(B)** Top, example test images with the correct labels underneath. Bottom, top five labels generated by the CNN, with the probabilities of each indicated by the lengths of the horizontal bars. The four images on the left were assigned with correct labels (pink bars) by the CNN. Errors (the four images on the right) are either due to image ambiguity (e.g., dalmatian versus cherry) or close resemblance among the objects (e.g., agaric versus mushroom). (From Krizhevsky A, Sutskever I, & Hinton GE [2012] In *NIPS'2012*.)

(Continued)

Box 14-6: continued

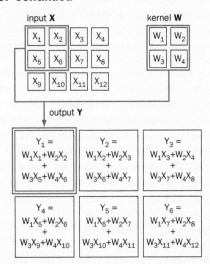

input **X**   kernel **W**

output **Y**

$$Y_1 = W_1X_1 + W_2X_2 + W_3X_5 + W_4X_6$$

$$Y_2 = W_1X_2 + W_2X_3 + W_3X_6 + W_4X_7$$

$$Y_3 = W_1X_3 + W_2X_4 + W_3X_7 + W_4X_8$$

$$Y_4 = W_1X_5 + W_2X_6 + W_3X_9 + W_4X_{10}$$

$$Y_5 = W_1X_6 + W_2X_7 + W_3X_{10} + W_4X_{11}$$

$$Y_6 = W_1X_7 + W_2X_8 + W_3X_{11} + W_4X_{12}$$

Figure 14-69 Illustration of input transformation in a simple convolutional network. In this example, input is a 4 × 3 matrix of neurons with values of X_i. The kernel is a 2 × 2 matrix with specified weights W_n (for simplicity, biases are omitted). Stride length is 1 neuron. Thus, the value Y_j of each neuron in the 3 × 2 output matrix is a dot product of the kernel and a 2 × 2 field in the input matrix (as illustrated for Y_1 in red). This 2 × 2 field shifts across columns and rows one column or row at a time, yielding Y_2 to Y_6.

stride (here the stride length is 4 neurons). Thus, information in the 224 × 224 matrix in layer I is *topographically* represented by a 55 × 55 matrix in layer II, preserving the two-dimensional spatial relationship of the images. **Figure 14-69** illustrates this operation in detail in a simpler example. This operation is analogous to a mathematical function called *convolution*—hence the name *convolutional neural network*. Note that all layer II neurons in this 55 × 55 matrix share the same kernel (weights and biases), and thus extract the same feature of the input data (e.g., vertical lines). This 55 × 55 matrix is therefore termed a *feature map*. The shared weights and biases in a feature map, as well as the topographic representation discussed earlier, confer *translation invariance* to the input images: move a dog across a background, and it is still a dog.

Instead of a single matrix, layer II actually consists of 2 × 48 parallel matrices (Figure 14-68A). Each matrix has its own weights and biases for transforming input from layer I neurons and can thus represent different features of the image. Two procedures are typically performed on the values of layer II neurons (*y* in the discussion above) before sending their output to layer III. The first is a nonlinear function called *ReLU* (rectified linear unit), which maintains positive values but sets negative values to zero (*y* can be negative because weights and biases can be positive or negative). The second procedure is called *max pooling*: the output value of each layer II neuron is compared to its immediate neighbors within a specified region (here a 3 × 3 region) and the maximum is taken as the output. Pooling provides some translational stability—the image appears roughly the same with small changes in its two-dimensional position. After

passing through five convolutional layers, information is then relayed through three fully connected layers. Output layer IX contains 1,000 neurons, each representing an image category.

The similarities between this CNN and the mammalian visual system (Chapter 4) should be evident from the description. The multilayered architecture of the deep neural network resembles that of the visual system, from the retina to LGN, V1 simple and complex cells, and higher-order visual cortical areas (Figure 4-51). The convolutional design mimics retinotopy, where nearby neurons from each layer receive input from nearby neurons of the previous layer. This makes sense for a program intended to process visual information and drastically reduces complexity in computing and data storage. (A fully connected 224 × 224 matrix and 55 × 55 matrix between layer I and one feature map in layer II requires 224 × 224 × 55 × 55 or $1.5 × 10^8$ parameters for weights; by comparison, connecting a 224 × 224 matrix with each feature map in layer II of the CNN requires only 11 × 11 or 121 shared weights, 6 orders of magnitude lower.) The parallel processing in different feature maps resembles the parallel processing of form, color, and motion by different types of retinal ganglion cells or the dorsal and ventral streams in the higher-order visual cortex. Thus, the design of CNNs, and deep learning in general, took inspiration from findings in neurobiology.

A key difference between deep neural networks and nervous systems is in how synaptic weights are determined. Weights and biases in CNNs start with random values and are adjusted through backpropagation of error minimization after a large number of supervised learning sessions. The coarse connections of the visual system are genetically specified via evolutionary selection; fine-scale connections and synaptic weights are established by a combination of genetic specification, stochasticity in developmental process, spontaneous activity, and experience-dependent plasticity, the exact proportions of which remain unknown (see Chapters 5 and 7). So far there is no evidence for the use of mechanisms resembling gradient descent or backpropagation in the nervous system. Nevertheless, as we see from the example in Figure 14-67, CNNs optimized for object recognition can predict firing rates of neurons in the visual cortex specialized for object recognition as well as intermediate-level neurons. Such "convergent evolution" between computer and nervous systems suggests potential algorithmic similarities between artificial and biological intelligence systems.

The success of image recognition in this example has inspired computer scientists to design variations of deep neural networks to solve problems ranging from speech recognition and natural language processing to autonomous driving. In addition to the feedforward architecture in Figure 14-68, researchers have also developed recurrent networks with lateral and feedback connections, as in the nervous system (Figure 4-51); these are particularly useful for modeling sequences like those in speech. Still, much of deep learning is empirical. For example, the parameters that go into the design of Figure 14-68A—the number of layers, the size of

Box 14-6: continued

the kernel, the length of the stride, the number of feature maps, whether or not to perform pooling within a convolutional layer—are very much selected from experience without principled theories guiding such choices. A deeper understanding of how deep neural networks work may help improve problem solving in computer science and contribute to our understanding of how nervous systems work.

Current deep learning methods mostly utilize architectural designs inspired by the visual cortex similar to that in Figure 14-68A. Fundamentally different circuit architectures, such as those of the hippocampus (Figure 11-5), basal ganglia (Figure 8-21), and cerebellum (Figure 8-26), may enable distinct computations to solve different kinds of problems. Combining use of these architectures may help achieve general-purpose artificial intelligence.

SUMMARY AND PERSPECTIVES

Neurobiology research is conducted in diverse animal models. These animal models were chosen for the technical ease with which they can be studied, the special faculties they exhibit, and their resemblance to humans. Recent revolutions in genomics and noninvasive imaging have made humans an increasingly attractive organism to study. While many molecular and cellular processes are highly conserved across the nervous systems of diverse animals, we do not yet know the extent to which information processing at the circuit and systems levels follows general principles that apply across nervous systems of varying complexity. Regardless, studies in diverse animal models enrich our understanding of the diversity of life and the evolution of nervous systems.

Genes are fundamental units in many neurobiological processes. The two most widely used molecular-genetic manipulations in neurobiology are the disruption of endogenous genes and the expression of transgenes. Gene disruption through forward and reverse genetics, as well as transgene expression, can be performed under sophisticated spatiotemporal control and can reveal mechanisms of gene action in neurobiological processes. Transgene expression also provides genetic access to specific neuronal populations for investigating their anatomical organization, physiological properties, and the functional consequences of activity manipulations. Recent advances in single-cell transcriptomics have provided rich data for classifying cells and genetically accessing specific neuronal types. Advances in genome editing have enabled circuit analysis strategies to be applied to diverse animal models.

Classic anatomical methods such as cell staining, axon tracing, and single-cell labeling have provided foundations for our present understanding of nervous system organization. Deeper appreciations of this organization requires building more connections between molecules, neurons, and the brain. One frontier is the fine-structural analysis of how molecules form complexes in different neuronal compartments and the dynamics of these molecular complexes; this will deepen our understanding of how individual neurons function. Another frontier is determining wiring diagrams for complex nervous systems, ultimately at the resolution of individual synapses; this will serve as a blueprint for deciphering information processing principles in neural circuits.

Extracellular, intracellular, and patch recordings of electrical activity have contributed fundamentally to our understanding of information flow within individual neurons, across synapses, in small circuits, and over large networks. These electrophysiological methods have more recently been supplemented by optical imaging, which enables simultaneous recording of many neurons of specific cell types using genetically encoded activity indicators. A crucial approach linking neuronal activity, circuit function, and behavior is the manipulation of neuronal activity with precise spatiotemporal control. Classic lesion, pharmacology, and electrical stimulation methods have been supplemented with sophisticated control of neuronal activity by light, heat, and chemicals. New tools developed in recent years, in particular the ability to control neuronal activity with light, have made it possible to activate and inactivate genetically defined neuronal populations at spatiotemporal scales matching those of neuronal signaling.

Future challenges include to combine the sensitivity and temporal resolution of electrophysiology with the breadth, cell-type specificity, and duration of optical imaging; to expand the methods available for recording neuronal activity in behaving animals across multiple brain regions; to integrate activity recording and manipulation with high spatiotemporal precision; and to develop conceptual frameworks for converting rich data sets into an understanding of the principles of neural circuit operation and the neural basis of behavior. Theory and modeling help test our understanding of specific phenomena in the nervous system, from the ionic basis of the action potential to the neural circuit architectures underlying information processing and learning. As new experimental techniques allow researchers to obtain data with ever-increasing throughput and sophistication, theory and modeling will play an increasingly important role in guiding experiments, analyzing rich data sets, and distilling generalizable principles.

Ultimately, the combination of methods we have studied in this chapter—from deleting and misexpressing specific genes in defined cell types to measuring, activating, and silencing specific neuronal populations with high spatiotemporal precision, and quantitative analysis of animal behavior—will help establish rich connections between genes, neurons, circuits, and animal behaviors. These connections will deepen our understanding of the nervous system in health and disease. With the rapid pace of technology developments, neurobiology research has never witnessed a more exciting time.

OPEN QUESTIONS

- With the advent of genome editing technologies, it is easier than ever before to develop new model organisms to address new neurobiological questions. What new principles can be discovered by exploring new model organisms?

- Can we reach a consensus neuronal cell type classification scheme in complex brain regions?

- Can we build realistic models of neuronal compartments such as the synapse based on structures of constituent proteins and their interactions?

- To what extent do synaptic connectomes reflect consistent themes in all individuals within a species versus variations that differ from individual to individual?

- How can we develop animal models that more faithfully capture key features of human psychiatric disorders?

- What general principles of information processing at the circuit and systems levels apply across nervous systems of varying complexity?

FURTHER READING

Reviews

Capecchi MR (1989). Altering the genome by homologous recombination. *Science* 244:1288-1292.

Datta SR, Anderson DJ, Branson K, Perona P, & Leifer A (2019). Computational neuroethology: a call to action. *Neuron* 104:11-24.

Jorgenson LA, Newsome WT, Anderson DJ, Bargmann CI, Brown EN, Deisseroth K, Donoghue JP, Hudson KL, Ling GS, MacLeish PR, et al. (2015). The BRAIN Initiative: developing technology to catalyse neuroscience discovery. *Philos Trans R Soc Lond B Biol Sci* 370. See also www.braininitiative.nih.gov/2025/BRAIN2025.pdf for the 2015 report, and https://braininitiative.nih.gov/strategic-planning/acd-working-groups/brain-initiative-20-cells-circuits-toward-cures for the 2019 update.

Kim CK, Adhikari A, & Deisseroth K (2017). Integration of optogenetics with complementary methodologies in systems neuroscience. *Nat Rev Neurosci* 18:222-235.

Lin MZ & Schnitzer MJ (2016). Genetically encoded indicators of neuronal activity. *Nat Neurosci* 19:1142-1153.

Luo L, Callaway EM, & Svoboda K (2018). Genetic dissection of neural circuits: a decade of progress. *Neuron* 98:256-281.

Neher E & Sakmann B (1992). The patch clamp technique. *Sci Am* 266(3): 44-51.

Zeng H & Sanes JR (2017). Neuronal cell-type classification: challenges, opportunities and the path forward. *Nat Rev Neurosci* 18:530-546.

Molecular, genetic, and anatomical methods

Brand AH & Perrimon N (1993). Targeted gene expression as a means of altering cell fates and generating dominant phenotypes. *Development* 118:401–415.

Cong L, Ran FA, Cox D, Lin S, Barretto R, Habib N, Hsu PD, Wu X, Jiang W, Marraffini LA, et al. (2013). Multiplex genome engineering using CRISPR/Cas systems. *Science* 339:819–823.

Dani A, Huang B, Bergan J, Dulac C, & Zhuang X (2010). Superresolution imaging of chemical synapses in the brain. *Neuron* 68:843–856.

DeNardo LA, Liu CD, Allen WE, Adams EL, Friedmann D, Fu L, Guenthner CJ, Tessier-Lavigne M, & Luo L (2019). Temporal evolution of cortical ensembles promoting remote memory retrieval. *Nat Neurosci* 22:460–469.

Denk W & Horstmann H (2004). Serial block-face scanning electron microscopy to reconstruct three,dimensional tissue nanostructure. *PLoS Biol* 2:e329.

Feil R, Wagner J, Metzger D, & Chambon P (1997). Regulation of Cre recombinase activity by mutated estrogen receptor ligand-binding domains. *Biochem Biophys Res Commun* 237:752–757.

Feng G, Mellor RH, Bernstein M, Keller-Peck C, Nguyen QT, Wallace M, Nerbonne JM, Lichtman JW & Sanes JR (2000). Imaging neuronal subsets in transgenic mice expressing multiple spectral variants of GFP. *Neuron* 28:41–51.

Fire A, Xu S, Montgomery MK, Kostas SA, Driver SE, & Mello CC (1998). Potent and specific genetic interference by double-stranded RNA in *Caenorhabditis elegans*. *Nature* 391:806–811.

Fodor SP, Read JL, Pirrung MC, Stryer L, Lu AT, & Solas D (1991). Light-directed, spatially addressable parallel chemical synthesis. *Science* 251:767–773.

Gao R, Asano SM, Upadhyayula S, Pisarev I, Milkie DE, Liu TL, Singh V, Graves A, Huynh GH, Zhao Y, et al. (2019). Cortical column and whole-brain imaging with molecular contrast and nanoscale resolution. *Science* 363:aau8302.

Golic KG & Lindquist S (1989). The FLP recombinase of yeast catalyzes site-specific recombination in the *Drosophila* genome. *Cell* 59:499–509.

Gordon JW, Scangos GA, Plotkin DJ, Barbosa JA, & Ruddle FH (1980). Genetic transformation of mouse embryos by microinjection of purified DNA. *Proc Natl Acad Sci U S A* 77:7380–7384.

Gossen M & Bujard H (1992). Tight control of gene expression in mammalian cells by tetracycline-responsive promoters. *Proc Natl Acad Sci U S A* 89:5547–5551.

Jenett A, Rubin GM, Ngo TT, Shepherd D, Murphy C, Dionne H, Pfeiffer BD, Cavallaro A, Hall D, Jeter J, et al. (2012). A GAL4-driver line resource for *Drosophila* neurobiology. *Cell Rep* 2:991–1001.

Jinek M, Chylinski K, Fonfara I, Hauer M, Doudna JA, & Charpentier E (2012). A programmable dual-RNA-guided DNA endonuclease in adaptive bacterial immunity. *Science* 337:816–821.

Lander ES, Linton LM, Birren B, Nusbaum C, Zody MC, Baldwin J, Devon K, Dewar K, Doyle M, FitzHugh W, et al. (2001). Initial sequencing and analysis of the human genome. *Nature* 409:860–921.

Lee T & Luo L (1999). Mosaic analysis with a repressible cell marker for studies of gene function in neuronal morphogenesis. *Neuron* 22:451–461.

Loh KH, Stawski PS, Draycott AS, Udeshi ND, Lehrman EK, Wilton DK, Svinkina T, Deerinck TJ, Ellisman MH, Stevens B, et al. (2016). Proteomic analysis of unbounded cellular compartments: synaptic clefts. *Cell* 166:1295–1307.

Macosko EZ, Basu A, Satija R, Nemesh J, Shekhar K, Goldman M, Tirosh I, Bialas AR, Kamitaki N, Martersteck EM, et al. (2015). Highly parallel genome-wide expression profiling of individual cells using nanoliter droplets. *Cell* 161:1202–1214.

Renier N, Adams EL, Kirst C, Wu Z, Azevedo R, Kohl J, Autry AE, Kadiri L, Umadevi Venkataraju K, Zhou Y, et al. (2016). Mapping of brain activity by automated volume analysis of immediate early genes. *Cell* 165:1789–1802.

Sanger F, Nicklen S, & Coulson AR (1977). DNA sequencing with chain-terminating inhibitors. *Proc Natl Acad Sci U S A* 74:5463–5467.

White JG, Southgate E, Thomson JN, & Brenner S (1986). The structure of the nervous system of the nematode *Caenorhabditis elegans*. *Philos Trans R Soc Lond B Biol Sci* 314:1–340.

Wickersham IR, Lyon DC, Barnard RJ, Mori T, Finke S, Conzelmann KK, Young JA, & Callaway EM (2007). Monosynaptic restriction of transsynaptic tracing from single, genetically targeted neurons. *Neuron* 53:639–647.

Xu CS, Januszewski M, Lu Z, Takemura S, Hayworth KJ, Huang G, Shinomiya K, Maitin-Shepard J, Ackerman D, Berg S, et al. (2020). A connectome of the adult *Drosophila* central brain. *bioRxiv* https://doi.org/10.1101/2020.01.21.911859

Zhou Y, Sharma J, Ke Q, Landman R, Yuan J, Chen H, Hayden DS, Fisher JW 3rd, Jiang M, Menegas W, et al. (2019). Atypical behaviour and connectivity in SHANK3-mutant macaques. *Nature* 570:326–331.

Recording and manipulating neuronal activity, and behavioral analysis

Abdelfattah AS, Kawashima T, Singh A, Novak O, Liu H, Shuai Y, Huang YC, Campagnola L, Seeman SC, Yu J, et al. (2019). Bright and photostable chemigenetic indicators for extended in vivo voltage imaging. *Science* 365:699–704.

Ahrens MB, Li JM, Orger MB, Robson DN, Schier AF, Engert F, & Portugues R (2012). Brain-wide neuronal dynamics during motor adaptation in zebrafish. *Nature* 485:471–477.

Aravanis AM, Wang LP, Zhang F, Meltzer LA, Mogri MZ, Schneider MB, & Deisseroth K (2007). An optical neural interface: in vivo control of rodent motor cortex with integrated fiberoptic and optogenetic technology. *J Neural Eng* 4:S143–156.

Armbruster BN, Li X, Pausch MH, Herlitze S, & Roth BL (2007). Evolving the lock to fit the key to create a family of G protein-coupled receptors potently activated by an inert ligand. *Proc Natl Acad Sci U S A* 104:5163–5168.

Boyden ES, Zhang F, Bamberg E, Nagel G, & Deisseroth K (2005). Millisecond-timescale, genetically targeted optical control of neural activity. *Nat Neurosci* 8:1263–1268.

Chen TW, Wardill TJ, Sun Y, Pulver SR, Renninger SL, Baohan A, Schreiter ER, Kerr RA, Orger MB, Jayaraman V, et al. (2013). Ultrasensitive fluorescent proteins for imaging neuronal activity. *Nature* 499:295–300.

Cohen JY, Haesler S, Vong L, Lowell BB, & Uchida N (2012). Neuron-type-specific signals for reward and punishment in the ventral tegmental area. *Nature* 482:85–88.

Denk W, Strickler JH, & Webb WW (1990). Two-photon laser scanning fluorescence microscopy. *Science* 248:73–76.

Grynkiewicz G, Poenie M, & Tsien RY (1985). A new generation of Ca^{2+} indicators with greatly improved fluorescence properties. *J Biol Chem* 260:3440–3450.

Hamada FN, Rosenzweig M, Kang K, Pulver SR, Ghezzi A, Jegla TJ, & Garrity PA (2008). An internal thermal sensor controlling temperature preference in *Drosophila*. *Nature* 454:217–220.

Hamill OP, Marty A, Neher E, Sakmann B, & Sigworth FJ (1981). Improved patch-clamp techniques for high-resolution current recording from cells and cell-free membrane patches. *Pflugers Arch* 391:85–100.

Harvey CD, Collman F, Dombeck DA, & Tank DW (2009). Intracellular dynamics of hippocampal place cells during virtual navigation. *Nature* 461:941–946.

Johns DC, Marx R, Mains RE, O'Rourke B, & Marban E (1999). Inducible genetic suppression of neuronal excitability. *J Neurosci* 19:1691–1697.

Jun JJ, Steinmetz NA, Siegle JH, Denman DJ, Bauza M, Barbarits B, Lee AK, Anastassiou CA, Andrei A, Aydin C, et al. (2017). Fully integrated silicon probes for high-density recording of neural activity. *Nature* 551:232–236.

Kitamoto T (2001). Conditional modification of behavior in *Drosophila* by targeted expression of a temperature-sensitive shibire allele in defined neurons. *J Neurobiol* 47:81–92.

Lima SQ & Miesenböck G (2005). Remote control of behavior through genetically targeted photostimulation of neurons. *Cell* 121:141–152.

Magnus CJ, Lee PH, Bonaventura J, Zemla R, Gomez JL, Ramirez MH, Hu X, Galvan A, Basu J, Michaelides M, et al. (2019). Ultrapotent chemogenetics for research and potential clinical applications. *Science* 364:aav5282.

Marshel JH, Kim YS, Machado TA, Quirin S, Benson B, Kadmon J, Raja C, Chibukhchyan A, Ramakrishnan C, Inoue M, et al. (2019). Cortical layer-specific critical dynamics triggering perception. *Science* 365:aaw5202.

Mathis A, Mamidanna P, Cury KM, Abe T, Murthy VN, Mathis MW, & Bethge M (2018). DeepLabCut: markerless pose estimation of user-defined body parts with deep learning. *Nat Neurosci* 21:1281–1289.

Petreanu L, Huber D, Sobczyk A, & Svoboda K (2007). Channelrhodopsin-2-assisted circuit mapping of long-range callosal projections. *Nat Neurosci* 10:663–668.

Sweeney ST, Broadie K, Keane J, Niemann H, & O'Kane CJ (1995). Targeted expression of tetanus toxin light chain in *Drosophila* specifically eliminates synaptic transmission and causes behavioral defects. *Neuron* 14:341–351.

Zhang F, Wang LP, Brauner M, Liewald JF, Kay K, Watzke N, Wood PG, Bamberg E, Nagel G, Gottschalk A, et al. (2007). Multimodal fast optical interrogation of neural circuitry. *Nature* 446:633–639.

Ziv Y, Burns LD, Cocker ED, Hamel EO, Ghosh KK, Kitch LJ, El Gamal A, & Schnitzer MJ (2013). Long-term dynamics of CA1 hippocampal place codes. *Nat Neurosci* 16:264–266.

Theory and modeling

Dayan P & Abbott LF (2001). Theoretical Neuroscience: Computational and Methematical Modeling of Neural Systems. MIT press.

Hodgkin AL & Huxley AF (1952). A quantitative description of membrane current and its application to conduction and excitation in nerve. *J Physiol* 117:500–544.

Hopfield JJ (1982). Neural networks and physical systems with emergent collective computational abilities. *Proc Natl Acad Sci U S A* 79:2554–2558.

Krizhevsky A, Sutskever I, & Hinton G (2012). ImageNet classification with deep convolutional neural networks. In *NIPS'2012.*

LeCun Y, Bengio Y, & Hinton G (2015). Deep learning. *Nature* 521:436–444.

McCulloch WS & Pitts W (1943). A logical calculus of the ideas immanent in nervous activity. *Bull Math Biophys* 5:115–133.

Miller KD, Keller JB, & Stryker MP (1989). Ocular dominance column development: analysis and simulation. *Science* 245:605–615.

Pillow JW, Shlens J, Paninski L, Sher A, Litke AM, Chichilnisky EJ, & Simoncelli EP (2008). Spatio-temporal correlations and visual signalling in a complete neuronal population. *Nature* 454:995–999.

Rieke F, Warland D, de Ruyter van Stevenick RR, & Bialek W (1999). Spikes: Exploring the Neural Code. MIT press.

Rosenblatt F (1958). The perceptron: a probabilistic model for information storage and organization in the brain. *Psychol Rev* 65:386–408.

Shadlen MN & Newsome WT (1998). The variable discharge of cortical neurons: implications for connectivity, computation, and information coding. *J Neurosci* 18:3870–3896.

van Vreeswijk C & Sompolinsky H (1996). Chaos in neuronal networks with balanced excitatory and inhibitory activity. *Science* 274:1724–1726.

Yamins DL, Hong H, Cadieu CF, Solomon EA, Seibert D, & DiCarlo JJ (2014). Performance-optimized hierarchical models predict neural responses in higher visual cortex. *Proc Natl Acad Sci U S A* 111:8619–8624.

Glossary

absence seizure A seizure characterized by a brief lapse of consciousness (of about 10 seconds or less) and a cessation of motor activity without loss of posture.

accessory olfactory bulb A brain region adjacent to the olfactory bulb; it is the axonal projection target of sensory neurons from the vomeronasal organ. (Figure 6-19)

accessory olfactory system (vomeronasal system) An anatomically and biochemically distinct system from the main olfactory system; it detects and analyzes nonvolatile chemicals and peptides such as pheromones and cues from predators. (Figure 6-19)

acetylcholine (ACh) The first discovered neurotransmitter; it is used by vertebrate motor neurons at the neuromuscular junction and autonomic nervous system. It is also used in the CNS as an excitatory or modulatory neurotransmitter. In some invertebrates, such as *Drosophila,* it is the major excitatory neurotransmitter in the CNS. (Figure 3-1; Table 3-2)

acetylcholine receptor (AChR) Receptor for the neurotransmitter acetylcholine. Nicotinic AChRs (nAChRs) are nonselective cation channels; they are the postsynaptic receptor at the vertebrate neuromuscular junction and function as excitatory receptors at some CNS synapses. Metabotropic AChRs (muscarinic AChRs or mAChRs) are G-protein-coupled receptors that play a modulatory role. (Figure 3-20 for nAChR)

acetylcholinesterase An enzyme that degrades acetylcholine, enriched in the cholinergic synaptic cleft.

acquisition (of memory) The initial formation of a memory as a consequence of experience and learning.

action potential An elementary unit of nerve impulses that axons use to convey information across long distances. It is all-or-none, regenerative, and propagates unidirectionally in an axon. It is also called a spike. (Figure 2-18; Figure 2-19)

active electrical property A membrane property due to voltage-dependent changes in ion conductances. It can reduce or eliminate attenuation of electrical signals across a distance that occurs due to passive electrical properties.

active transport Movement of a solute across a membrane against its electrochemical gradient via a transporter that uses external energy, such as ATP hydrolysis, light, or movement of another solute down its electrochemical gradient. (Figure 2-10)

active zone An electron-dense region of the presynaptic terminal containing clusters of synaptic vesicles docked at the presynaptic membrane, ready for release. (Figure 3-3; Figure 3-10)

activity-dependent transcription The process by which neuronal activity regulates gene expression.

adaptation (in evolution) Genetic or phenotypic changes that render an individual and its progeny more likely to survive and reproduce in a particular environment.

adaptation (in sensory systems) An adjustment of a system's sensitivity according to background levels of sensory input.

adaptive (in evolution) Adjective for adaptation (in evolution).

adeno-associated virus (AAV) A DNA virus widely used to deliver transgenes into postmitotic neurons. It has a capacity to carry up to about 5 kb of foreign DNA. (Table 14-1)

adenylate cyclase A membrane-associated enzyme that synthesizes cyclic AMP (cAMP) from ATP. (Figure 3-33)

adult neurogenesis The production of new neurons in the adult brain. This occurs in song-production nuclei in the songbird and in limited regions of the mammalian brain.

afferent An axon that projects from peripheral tissue into the CNS. It can also be generalized to describe an input axon to a particular neural center within the CNS.

agonist A molecule that activates a biological process by interacting with a receptor, often mimicking the action of an endogenous molecule.

agrin A protein secreted by motor neurons that induces aggregation of acetylcholine receptors in the skeletal muscle. (Figure 7-25)

AgRP neuron A neuron in the hypothalamic arcuate nucleus that releases the orexigenic peptides agouti-related protein (AgRP) and neuropeptide Y. (Figure 9-14)

AII amacrine cell A type of amacrine cell that links rod bipolars to the pathways that process cone signals. (Figure 4-35)

AKAP (A kinase anchoring protein) An anchoring protein associated with protein kinase A.

allele A specific version of a gene.

allelic exclusion A phenomenon in which mRNAs of a gene are transcribed exclusively from one chromosome of a homologous pair. *See also* **allele**.

allodynia A phenomenon wherein gentle touch or innocuous temperature causes pain when applied to inflamed or injured tissue. (Figure 6-72)

all-or-none Having the property of being binary in occurrence. It applies to action potentials, which have the same amplitude and waveform regardless of the strength of the inducing stimulus, as long as the stimulus is above threshold.

allosteric agonist A molecule that facilitates binding of an endogenous ligand to its receptor. An allosteric agonist binds to a site on a receptor that is different from the site that binds the endogenous ligand.

Alzheimer's disease (AD) A neurodegenerative disorder prevalent in the aging population; it is defined by the combined presence of abundant amyloid plaques and neurofibrillary tangles in postmortem brains, with symptoms including gradual loss of memory, impaired cognitive and intellectual capabilities, and reduced ability to cope with daily life.

amacrine cell An inhibitory retinal neuron whose actions influence the signals transmitted from bipolar cells to retinal ganglion cells. (Figure 4-25)

AMPA receptor A glutamate-gated ion channel that conducts mostly Na^+ and K^+ and can be selectively activated by the drug AMPA (2-amino-3-hydroxy-5-methylisoxazol-4-propanoic acid). It is a usually heterotetramer containing two or more kinds of subunits (GluA1, GluA2, GluA3, and GluA4) encoded by four genes. (Figure 3-24; Figure 3-25)

amygdala An almond-shaped structure underneath the temporal lobe best studied for its role in processing emotion-related information. (Figure 1-8; Figure 11-42)

amyloid β (Aβ) hypothesis   The hypothesis that an increase in amyloid β (Aβ) protein production or accumulation is a common cause of Alzheimer's disease.

amyloid plaque   An extracellular deposit consisting primarily of aggregates of amyloid β protein. (Figure 12-2)

amyloid precursor protein (APP)   A single-pass transmembrane protein from which the amyloid β protein is derived by proteolytic processing. (Figure 12-3)

amyloid β protein (Aβ)   A major component of amyloid plaques in Alzheimer's disease; it is a 39- to 43-amino-acid peptide with a strong tendency to form aggregates rich in β-pleated sheets. (Figure 12-4)

amyotrophic lateral sclerosis (ALS)   A rapidly progressing motor neuron disease that is usually terminal within a few years after symptoms emerge. It is also known as Lou Gehrig's disease.

analog signaling   Signaling that uses continuous values to represent information.

androgen   A male sex hormone, such as testosterone.

androgen receptor   A cytosolic protein that, upon binding of an androgen such as testosterone, translocates to the nucleus, where it acts as a transcription factor. (Figure 10-19)

anions   Negatively charged ions such as Cl$^-$.

anosmic   Unable to perceive odors.

antagonist   A molecule that counters the action of an endogenous molecule.

antagonistic muscles   Muscles that perform opposite actions, such as an extensor and a flexor that control the same joint. (Figure 8-8)

antennal lobe   The first olfactory processing center in the insect brain. (Figure 6-24)

anterior cingulate cortex (ACC)   A neocortical area located near the midline of the frontal lobe. It has extensive connections with the hippocampus and is implicated in long-term memory storage.

anterior pituitary   *See* **pituitary**.

anterior–posterior   *See* **rostral–caudal**.

anterograde   From the cell body to the axon terminal.

anterograde tracer   A molecule used to trace axonal connections; it is taken up primarily by neuronal cell bodies and dendrites and travels down axons to label projection sites. (Figure 14-30)

anterolateral column pathway   An axonal pathway from the spinal cord to the brainstem; it consists of axons from lamina I dorsal horn projection neurons on the contralateral side of the spinal cord. It mainly relays pain, itch, and temperature signals to the brain. (Figure 6-71)

anteroventral periventricular nucleus (AVPV)   A hypothalamic nucleus in the preoptic area that plays a pivotal role in regulating the female ovulatory cycle.

antidromic spike   An action potential that propagates from the axon terminal to the cell body in artificial situations in which experimenters electrically stimulate an axon or its terminal.

antiporter   A coupled transporter that moves two or more solutes in opposite directions. Also called an **exchanger**. (Figure 2-10)

antisense oligonucleotide (ASO)   Short nucleotide that binds to an endogenous target RNA with a complementary sequence, leading to functional inhibition or degradation of the target RNA.

anxiety disorders   A group of psychiatric disorders that includes generalized anxiety disorders (characterized by persistent worries about impending misfortunes), phobias and panic disorders (characterized by irrational fears), and obsessive–compulsive disorder.

AP5 (2-amino-5-phosphonovaleric acid)   A widely used selective NMDA receptor antagonist.

apolipoprotein E (ApoE)   A high-density lipoprotein in the brain involved in lipid transport and metabolism. A specific polymorphic isoform (ε4) is a major risk factor for Alzheimer's disease.

apoptosis   A form of cell death in which a cell kills itself by initiating a cell-death program.

Arc   A cytoskeletal protein present at the postsynaptic density that regulates trafficking of glutamate receptors. It is a product of the immediate early gene *Arc*.

archaerhodopsin   A light-activated outward proton pump in archaea; it can be used to silence neuronal activity in a heterologous system by light. *See also* **optogenetics**.

arcuate nucleus   A ventromedial hypothalamic nucleus that regulates food intake and energy expenditure. (Figure 9-6)

area X   A basal ganglia structure in the songbird essential for song learning. (Figure 10-16)

aromatase   An intracellular enzyme that converts testosterone to estradiol. (Figure 10-19)

arrestin   A protein that binds to a phosphorylated G-protein-coupled receptor and competes with the receptor's binding to Gα proteins. It can therefore terminate GPCR signaling. It can also transduce signals of its own.

ascending arousal system   A neural system consisting of parallel projections from the brainstem and hypothalamus to the forebrain that are essential for maintaining wakefulness. It includes cholinergic projections from the tegmental nuclei, norepinephrine projections from the locus coeruleus, serotonin projections from the raphe nuclei, histamine projections from the tuberomammillary nucleus, and hypocretin projections from the lateral hypothalamus. (Figure 9-29)

association cortex   Cortical areas that integrate information from multiple sensory areas and link sensory systems to motor output.

associative learning   A type of learning involving formation of an association between two events, such as formation of an association between an unconditioned stimulus and a conditioned stimulus in **classical conditioning** or formation of an association between a behavior and a reinforcer in **operant conditioning**.

associativity (of LTP)   A property of long-term potentiation (LTP) whereby activation of a synapse that alone would be too weak to produce LTP can nonetheless lead to LTP if it coincides with the strong, LTP-inducing activation of a different synapse onto the same postsynaptic cell. (Figure 11-8)

astrocyte   A glial cell present in gray matter. It plays many roles, including in synaptic development and function. (Figure 1-9)

asymmetric cell division   A cell division in which the two daughter cells are of different types from birth.

ataxia   An abnormality in coordinated muscle contraction and movement.

attention   A cognitive function in which a subset of sensory information is subjected to more processing at the expense of other information.

attractant   A molecular cue that guides axons toward its source. (Figure 5-11)

attractor   A state in a Hopfield network in which the "energy" function is minimized. It also refers to a stable state in a generic neural network to which neural dynamics converge. (Figure 14-63)

auditory cortex   The part of the cerebral cortex that analyzes auditory signals. It is located in the temporal lobe. (Figure 1-23)

auditory fear conditioning   A classical conditioning procedure in which aversive, fear-inducing stimuli, such as electric shocks, are paired with sound stimuli during training; animals subsequently exhibit fear responses, such as freezing, in response to sound stimuli alone.

auditory nerve   A bundle of axons from spiral ganglion neurons that transmits auditory information to the brainstem. It also contains efferents from the brainstem that synapse primarily onto outer hair cells. (Figure 6-47)

auricle   The external part of the mammalian ear; it collects and focuses sound through the ear canal on the ear drum.

autism spectrum disorder (ASD)   A class of neurodevelopmental disorders characterized by deficits in reciprocal social interactions. Patients also exhibit restricted interests and repetitive behaviors.

autocrine   Of or related to a form of signaling in which a recipient cell receives a signal produced by itself.

autonomic nervous system   Collection of parts of the nervous system that regulate the function of internal organs, including the contraction of smooth and cardiac muscles, as well as secretion and excretion from glands.

autophagy   An ordered cellular process that removes unnecessary or damaged cellular components through the lysosomal pathway.

autosomal dominant   Of a mutation, having a Mendelian inheritance pattern in which mutation of only one allele of a gene located on an autosome is sufficient to produce a phenotype. It can result from a toxic gain-of-function effect of the mutant allele or a loss-of-function effect due to an insufficient amount of the normal gene product being produced by the remaining wild-type allele. (Figure 12-33)

autosomal recessive   Of a mutation, having a Mendelian inheritance pattern in which mutation of both alleles of a gene located on an autosome is required to produce a phenotype. It usually results from a loss-of-function effect of the mutation. (Figure 12-33)

autosome   A non-sex chromosome.

axon   A long, thin process of a neuron; it often extends far beyond the soma and propagates and transmits signals to other neurons or muscle at its presynaptic terminals. (Figure 1-9)

axon guidance molecules   Extracellular cues and cell surface receptors that guide axons along their paths toward the appropriate targets. (Figure 5-11)

axon hillock   *See* **axon initial segment**

axon initial segment (axon hillock)   The segment of the axon closest to the neuronal cell body; it is usually the site of action potential initiation.

axon myelination   The process in which glial cells wrap their cytoplasmic extensions around axons to increase conduction velocity. (Figure 2-27)

bacteriorhodopsin   A light-driven proton pump in archaea.

ball-and-chain   A model of voltage-gated channel inactivation in which a cytoplasmic portion of the channel protein ("ball"), connected to the rest of the channel by a polypeptide chain, blocks the channel pore after the ion channel opens. (Figure 2-31)

barrel   A discrete anatomical unit in layer 4 of the rodent primary somatosensory cortex that represents a whisker. The cortical region containing barrels for all whiskers is called the barrel cortex. The corresponding discrete units in the brainstem and thalamus are called barrelettes and barreloids, respectively. (Figure 5-26)

barrel cortex   *See* **barrel**.

basal ganglia   A collection of nuclei underneath the cerebral cortex; it includes the striatum, globus pallidus, subthalamic nucleus, and substantia nigra and is essential for motor initiation and control, habit formation, and reward-based learning. (Figure 1-8; Figure 8-21)

basilar membrane   An elastic membrane at the base of hair cells in the cochlea. (Figure 6-43)

basket cell   A type of GABAergic neuron; it wraps its axon terminals around the cell bodies of pyramidal cells in the cerebral cortex and Purkinje cells in the cerebellar cortex. (Figure 1-15; Figure 3-46)

basolateral amygdala   A brain region consisting of lateral and basal amygdala. It receives input from the thalamus, cortex, and hippocampus and sends output to the central amygdala and other brain regions. It is involved in regulating emotion-related behavior. (Figure 11-42)

battery   An electrical element that maintains a constant voltage, or electrical potential difference, across its two terminals and that can thus serve as an energy source. (Figure 2-13)

Bayes' rule   The relationship between conditional probabilities that event A occurs given B is true and event B occurs given A is true, expressed as: $P(A|B) = P(B|A) \times P(A) / P(B)$. *See also* **conditional probability**.

BDNF (<u>b</u>rain-<u>d</u>erived <u>n</u>eurotrophic <u>f</u>actor)   *See* **neurotrophins** and **Trk receptors**.

bed nucleus of the stria terminalis (BNST)   A sexually dimorphic brain region that receives direct input from accessory olfactory bulb mitral cells. Its diverse functions include regulation of male courtship behavior. (Figure 10-26)

benzodiazepines   A class of drugs that act as allosteric agonists of $GABA_A$ receptors. They are widely used to treat anxiety, pain, epilepsy, and sleep problems. (Figure 12-27)

biased random walk   A strategy employed in bacteria chemotaxis. When swimming away from an attractant, bacteria exhibit frequent tumbles (reorientation); when swimming toward an attractant, they tumble less frequently. It is also employed by *C. elegans* for chemotaxis. (Figure 13-15)

bilaterians   Animals that are bilaterally symmetrical and have three germ layers. They include all vertebrates and most invertebrate species alive today. (Figure 13-2)

binary classification   A task in which observed events are classified into one of two classes. (Figure 14-59)

binary expression   Expression of a transgene using a strategy in which the *cis*-regulatory elements and coding sequence are separated into two transgenes. (Figure 14-12)

binocular vision   A form of vision involving integration of inputs from the two eyes that carry information about the same visual field location. It is important for depth perception.

binomial distribution   A discrete probability distribution that describes the frequency (f) with which k events occur in n independent trials, given the probability p that an event occurs in each trial. $f(k; n, p) = [n!/k!(n - k)!]\, p^k (1 - p)^{n-k}$. (Box 3-1)

biomarker   A measurable characteristic indicator of normal biological processes, pathogenic processes, and responses to therapeutic interventions. (Figure 12-11)

bipolar (neuron)   Having two processes leaving the cell body.

bipolar cell (in retina)   An excitatory neuron that transmits information from photoreceptors to retinal ganglion cells and amacrine cells. (Figure 4-22; Figure 4-25)

bipolar disorder   A mood disorder in which patients alternate between manic phases characterized by feelings of grandiosity and tirelessness, and depressive phases characterized by feelings of sadness, emptiness, and worthlessness.

bitter   A taste modality that functions primarily to warn the animal of potential toxic chemicals. It is usually aversive.

blastula   The product of cleavage; it is an early-stage embryo consisting of a hollow ball of thousands of cells. (Figure 7-2)

blood–brain barrier (BBB)   Derived from endothelial cell tight junctions in the blood vessels of the brain, it prevents the exchange of many substances between the blood and brain tissues.

blue-ON bipolar cell   An ON bipolar cell that selectively connects with S-cones; it is activated by short-wavelength light and inhibited by longer-wavelength light. (Figure 4-34)

bone morphogenetic proteins (BMPs)   A family of secreted proteins that act as morphogens to pattern embryonic tissues, such as the tissues along the anterior–posterior axis of the telencephalon and the dorsal–ventral axis of the spinal cord.

border cell   A cell in the entorhinal cortex that fires when an animal is at a specific edge of an arena.

Boss (*Bride of sevenless*)   Originally identified from a mutation in *Drosophila* that lacks photoreceptor R7, it is a gene that acts cell nonautonomously in R8 to specify R7 fate. It encodes a transmembrane ligand for the Sevenless receptor tyrosine kinase. (Figure 5-35)

botulinum toxins   A family of proteases produced by *Clostridium botulinum*. Different isoforms cleave synaptobrevin, syntaxin, or SNAP-25 at distinct sites. (Box 3-2)

bradykinin   A peptide released during inflammation; it binds to specific G-protein-coupled receptors on the peripheral terminals of nociceptive neurons. (Figure 6-73)

brain   The rostral part of the central nervous system located in the head. It is the command center for most nervous system functions. (Figure 1-8)

brain slice   A fresh section of brain tissue (usually about a few hundred micrometers thick) that largely preserves the three-dimensional architecture for physiological studies of neuronal and local circuit properties *in vitro*.

brainstem   A structure made up of the midbrain, pons, and medulla. (Figure 1-8)

Broca's area   An area in the left frontal lobe involved in language production. Patients with lesions in this area have difficulty speaking. (Figure 1-23)

α-bungarotoxin   A snake toxin from the venom of *Bungarus* that is a competitive inhibitor of the nicotinic acetylcholine receptor.

Ca²⁺ indicator   A molecule whose optical properties depend on the intracellular Ca^{2+} concentration. Ca^{2+} indicators are used as optical sensors of neuronal activity. (Figure 14-41)

CA3 → CA1 synapse   The synapse between the axons of the Schaffer collaterals of hippocampal CA3 neurons and the dendrites of CA1 neurons. It is a model synapse for investigating mechanisms of synaptic plasticity.

cable properties   *See* **passive electrical properties**.

cadherin   A Ca^{2+}-dependent homophilic cell-adhesion protein.

calmodulin (CaM)   A Ca^{2+}-binding protein that transduces Ca^{2+} signals to many effectors. (Figure 3-34)

calyx of Held   A giant synapse made by a presynaptic terminal of a ventral cochlear globular bushy cell wrapping around the cell body of a medial nucleus of the trapezoid body principal cell. It contains hundreds of neurotransmitter release sites and is responsible for rapid and reliable transmission of auditory signals in the mammalian brainstem sound localization system. (Figure 6-56)

Cambrian   A geological period between 542 and 488 million years ago when major phyla within the animal kingdom diversified, as evidenced by an abundance of corresponding fossils. (Figure 13-2)

CaM kinase II (CaMKII)   A Ca^{2+}/calmodulin-dependent serine/threonine kinase that is highly enriched in the postsynaptic densities of excitatory synapses and that regulates synaptic plasticity, such as long-term potentiation. (Figure 3-34; Figure 11-11)

cAMP-dependent protein kinase   A serine/threonine kinase made up of two regulatory and two catalytic subunits. Binding of cAMP to the regulatory subunits leads to dissociation of the catalytic subunits, which can then phosphorylate their substrates. It is also called A kinase, protein kinase A, or PKA. (Figure 3-33)

capacitance (*C*)   The ability of a capacitor to store charge; it is defined as $C = Q/V$, where Q is the electric charge stored when the voltage across the capacitor is V.

capacitor   An electrical element composed of two parallel conductors separated by a layer of insulator. It is a charge-storing device. (Figure 2-13)

Capricious   A *Drosophila* transmembrane protein that contains extracellular leucine-rich repeats and instructs wiring specificity of axons and dendrites. (Figure 5-38; Figure 7-44)

cardiac muscle   Muscle that controls heartbeat.

Cas9   A key protein in the type II CRISPR system; it is an RNA-guided endonuclease containing two separate nuclease domains that generate double-strand breaks in DNA complementary to a bound RNA. It is used by bacteria for adaptive immunity and experimentally for genome editing. *See also* **CRISPR** and **guide RNA**. (Figure 14-8)

caspases   Proteases best known for triggering apoptosis, a form of programmed cell death.

castrated male   A male from which the testes have been removed.

catecholamines   A class of chemicals that includes the neurotransmitters dopamine, norepinephrine, and epinephrine. (Figure 12-20)

cations   Positively charged ions such as K^+ and Na^+.

CB1   A G-protein-coupled receptor originally identified as the receptor for cannabinoids from the marijuana plant. It serves as a receptor for endocannabinoids under physiological conditions.

CCK (cholecystokinin)   A neuropeptide produced in the small intestine in response to a rise in fatty acid concentrations. It acts as a satiety signal to inhibit eating. (Figure 9-15)

cDNA library   A collection of cloned complementary DNAs synthesized from mRNA templates prepared from a specific tissue or cell type.

cell adhesion molecule   A cell-surface protein that binds to its partners in opposing cells or the extracellular matrix to facilitate cell–cell or cell–matrix adhesion.

cell assembly   A group of interconnected neurons whose firing patterns collectively encode information, such as the location of an animal in an environment.

cell-attached patch recording (cell-attached recording)   A variant of the patch clamp recording method in which the patch pipette forms a high resistance (gigaohm) seal with the plasma membrane of an intact cell, allowing measurement of ion flow through a small number of channels or a single channel in the patch of membrane underneath the electrode. (Figure 2-29; Figure 14-40)

cell autonomous   Of a gene, acting in the cell that produces the gene product.

cell fate   The outcome of the developmental decision as to what type of cell it is.

cell lineage   The developmental history of a cell, comprising the identities of the progenitors from which a cell was derived.

cell nonautonomous   Of a gene, acting on a cell that does not produce the gene product.

cell-replacement therapy   A treatment strategy in which cells differentiated *in vitro* are transplanted into the body to replace dying cells, such as dopamine neurons in Parkinson's disease.

cell-surface receptor   A membrane protein that binds to extracellular ligands and subsequently sends a signal into the recipient cell. (Figure 3-38)

cell theory   The idea that all living organisms are composed of cells as their basic units.

center–surround (receptive field)   A property of a visual system neuron's receptive field, in which light in the receptive field center and light just outside of the receptive field center are antagonistic. (Figure 4-20)

central amygdala   The output nucleus of the amygdala complex; it receives input from the basolateral amygdala and sends GABAergic output to brainstem nuclei, the autonomic nervous system, the hypothalamus, and neuromodulatory systems to regulate diverse physiological functions, including emotion-related behavior. (Figure 11-42; Figure 11-45)

central dogma   The principle that genetic information flows from DNA to RNA to protein. (Figure 2-2)

central pattern generator (CPG)   A CNS circuit capable of producing rhythmic output for coordinated contraction of different muscles without sensory feedback. (Figure 8-12; Figure 8-13)

cerebellar nuclei   The output nuclei of the cerebellum; they receive input from Purkinje cell axons and the collaterals of mossy and climbing fibers, and send information to diverse regions including the thalamus and brainstem. (Figure 8-26)

cerebellum   A structure located dorsal to the pons and medulla that plays an important role in motor coordination, motor learning, and cognitive functions. (Figure 1-8; Figure 8-26)

cerebral cortex   The outer layer of the neural tissue in the rostral part of the mammalian brain. It is associated with higher functions, including sensory perception, control of voluntary movement, and cognition. (Figure 1-8; Figure 1-23)

CGRP (calcitonin gene-related peptide)   A peptide that promotes inflammation when released by the peripheral terminals of sensory neurons. It is also produced in the CNS, for example in a subset the parabrachial nucleus neurons activated by threat and dietary satiety. (Figure 6-73; Figure 9-15)

chandelier cell   A type of GABAergic neuron in the cerebral cortex that forms synapses onto the axon initial segments of cortical pyramidal cells. (Figure 3-46)

channel   A transmembrane protein or protein complex that forms an aqueous pore, allowing specific solutes to pass directly through when it is open. (Figure 2-8)

channelopathies   Diseases caused by mutations in ion channels.

channelrhodopsin   A member of a class of light-activated cation channels in single-celled green algae used for chemotaxis. *See also* **channelrhodopsin-2 (ChR2)**. (Figure 13-21)

channelrhodopsin-2 (ChR2)   A light-activated cation channel from single-celled green algae; it is widely used to activate neurons in heterologous systems by light. *See also* **optogenetics**. (Figure 13-21; Figure 14-47)

characteristic frequency   The sound frequency to which a given cell in the auditory system is most sensitive. (Figure 6-47)

Charcot–Marie–Tooth (CMT) disease   A PNS demyelinating disease characterized by progressive deficits in sensation and movement that preferentially affects neurons with longer axons.

chemical gradient   A concentration difference of a solute over two sides of a membrane, which contributes to the direction and magnitude of solute movement across the membrane. If the solute is not charged, the chemical gradient alone determines the movement direction: from higher to lower concentration. (Figure 2-9)

chemical synapse   A specialized junction between two neurons or a neuron and a muscle cell where communication between cells occurs via neurotransmitter release and reception. It comprises a presynaptic terminal and a postsynaptic specialization separated by a synaptic cleft. (Figure 1-14; Figure 3-3)

chemoaffinity hypothesis   Proposed by Roger Sperry, it posits that growing axons use cell-surface molecules to determine their paths and connect to appropriate synaptic partners.

chemogenetics   An approach using chemicals to activate or silence neurons expressing receptors specifically engineered to be sensitive to those chemicals. (Figure 14-44)

chemotaxis   Movement toward or away from a chemical source.

chlorpromazine   A first-generation antipsychotic drug; it is an antagonist of the D_2 dopamine receptor.

chordates   Animals with a notochord. (Figure 13-2)

chromophore   The light-absorbing portion of a molecule.

ciliary type   A type of photoreceptor in which opsins are packed into the primary cilium-derived outer segment. (Figure 13-22)

circadian pacemaker neuron   A neuron whose activity in isolation oscillates in a circadian fashion (i.e., with a period of close to 24 hours).

circadian rhythms   Self-sustained oscillations in an organism's behavior, physiology, and biochemistry, with a period of close to 24 hours.

circuit motif   A common configuration of a neural circuit allowing the connection patterns of individual neurons to execute specific functions. (Box 1-2)

cis-regulatory element   A DNA element, such as a transcriptional enhancer, repressor, or insulator, that regulates expression of genes on the same chromosome.

clade   A branch in the tree of life, consisting of an ancestor species and all of its descendant species.

cladistic analysis   The study of the emergence and change of traits of organisms in the context of their phylogenetic relationships.

classical conditioning   A form of learning in which repeated pairing of a conditioned stimulus (CS) with an unconditioned stimulus (US) causes a subject to exhibit a novel conditioned response (CR) to the CS. It is also called Pavlovian conditioning. (Figure 11-21)

Cl⁻ channel   An ion channel that allows selective passage of Cl⁻.

cleavage   A series of rapid cell divisions in early embryogenesis that converts a single large zygote cell into thousands of smaller cells. (Figure 7-2)

climbing fiber   An axon that climbs the dendritic trees of Purkinje cells and originates from a neuron in the inferior olive. (Figure 8-26)

Clock   A gene identified from a forward genetic screen in mice for mutations that produce circadian rhythm deficit; it encodes a transcriptional activator, CLOCK, that positively regulates expression of genes whose products feedback to negatively regulate CLOCK function. Its fly homolog serves a similar function. (Figure 9-21)

clonal analysis A method of analyzing the relationships between cells by lineage; it involves labeling a progenitor such that all of its progeny are also labeled.

closed-loop Of a system, having input to the system that is modified by the output of the system. In the context of a behavioral paradigm, having environmental stimuli that induce a behavior in an animal to also change in response to the animal's behavior.

cluster analysis A form of unsupervised learning that divides a data set into subsets based on similarities in their key feature(s).

cnidarians Radially symmetric animals such as hydra, jellyfish, and corals. (Figure 13-2)

CNS (central nervous system) The brain and spinal cord in vertebrates; the brain and nerve cord in some invertebrates.

cochlea A coiled structure in the inner ear containing fluid-filled chambers and the organ of Corti. (Figure 6-43)

cochlear nuclei Brainstem nuclei at the termination of the auditory nerve, subdivided into the dorsal and ventral cochlear nuclei. (Figure 6-52)

coding space A theoretical space used to describe the activity of a neuronal population. The firing rate of each neuron in the population constitutes one dimension/axis in this space, and the activity state of the entire population is represented as a point in this space. (Figure 6-27)

cognitive learning A theory of learning that emphasizes learning as acquisition of new knowledge rather than just modification of behavior.

coincidence detector (in auditory system) A cell maximally activated by simultaneous auditory signals from the left and right ears. (Figure 6-53)

coincidence detector (in synaptic transmission) A receptor that opens only in response to concurrent neurotransmitter binding *and* postsynaptic depolarization, such as the NMDA receptor.

collateral An axon branch.

color contrast (in vision) The difference in light wavelengths between adjacent spaces.

color-opponent RGC A retinal ganglion cell that differentiates signals from cones with distinct spectral sensitivities. The blue–yellow opponent RGC (in all mammals) differentiates short- and longer-wavelength light signals; the green–red opponent RGC (in trichromatic primates) differentiates two long-wavelength light signals. (Figure 4-34)

Comm (Commissureless) A *Drosophila* protein that acts in the secretory pathway to downregulate cell-surface expression of Robo during midline guidance. (Figure 7-13)

commissural neuron A neuron that projects its axon to the contralateral side of the body. Midline crossing of commissural neurons has been used as a model system for studying axon guidance.

compact myelin Closely packed layers of glial plasma membranes wrapped around axons.

complement cascade Part of an innate immune system that "complements" the ability of antibodies and phagocytic cells to clear pathogens and damaged cells from the organism.

complex cell A functionally defined primary visual cortex neuron type. It has no mutually antagonistic ON and OFF regions and is excited by light bars on a dark background or dark bars on an

illuminated background. The stimulus bars must be in a specific orientation but can fall on any part of the receptive field. (Figure 4-42)

conditional knockout The process of disrupting a gene in a specific spatiotemporal pattern or an animal in which a gene has been disrupted in a specific spatiotemporal pattern. The most common strategy for generating conditional knockouts in mice utilizes Cre/*loxP*-based recombination. It usually involves inserting a pair of *loxP* elements into introns flanking (an) essential exon(s) of a gene of interest. The gene of interest is only disrupted in cells in which Cre has been active. (Figure 14-9)

conditional probability The likelihood that event *A* occurs given that *B* is true, expressed as $P(A|B)$. *See also* **Bayes' rule**.

conditioned response (CR) *See* **classical conditioning**.

conditioned stimulus (CS) *See* **classical conditioning**.

conductance (g) The degree to which an object or substance passes electricity; it is the inverse of resistance: $g = 1/R$.

conductor An object or substance that passes electric current.

cone A cone-shaped photoreceptor in the vertebrate retina; it contributes to high acuity, motion, and color vision. (Figure 4-2)

confocal fluorescence microscopy (confocal microscopy) A fluorescence microscopy technique in which a detector pinhole is used to collect fluorescence emission originating only from a focal spot restricted in all three dimensions. By scanning the laser across a plane to record fluorescence emission from many focal spots, it can produce a thin optical section of whole-mount tissue or a thick tissue section. (Figure 14-23)

connectome A representation of the complete set of synaptic connections among a group of neurons of interest. (Figure 7-30; Figure 14-2)

connexin A protein component of gap junctions in vertebrates. (Figure 3-48)

consolidation (of memory) A step in the process of memory formation that occurs between acquisition and storage, during which a newly acquired memory is stabilized for long-term use.

contextual fear conditioning A learning procedure in which a rodent is subjected to aversive fear-inducing stimuli, such as electric shocks, in a specific environment (context). When subsequently placed in the same context, the animal will exhibit a fear response, such as freezing. It requires both the hippocampus and amygdala.

continuous map A type of neural map in which nearby neurons in the input field connect with nearby neurons in the target field, as exemplified by the relationship between the retina and the tectum. (Figure 7-36)

contralateral Of the other side of the midline. For example, a contralateral axonal projection is an axon that crosses the midline and terminates on the side of the nervous system opposite the soma.

convergent evolution The independent evolution of similar features in animals of different clades.

convergent excitation A circuit motif in which several excitatory neurons synapse onto the same postsynaptic neuron. (Figure 1-20)

convolutional neural network (CNN) A class of deep neural network that exhibits space invariant properties; it was originally developed for image analysis.

cooperativity (of LTP) A property of long-term potentiation (LTP) whereby LTP can be induced at a synapse if the presynaptic cell releases neurotransmitter while the postsynaptic cell is in a depolarized state, even if transmitter release from the presynaptic cell alone (in the absence of postsynaptic depolarization) is insufficient to induce LTP. (Figure 11-8)

copy number variation (CNV) A deletion or duplication of a chromosome segment that can vary in length from 500 base pairs to several megabases and may contain coding sequences that range from a small fraction of a single gene to many genes.

coronal section A section plane perpendicular to the rostral–caudal axis; also called cross or transverse sections. (Figure 1-8)

corpus callosum A structure composed of axon bundles linking the two cerebral hemispheres. (Figure 7-12)

cortical amygdala Part of the olfactory amygdala complex that receives direct mitral cell input. (Figure 6-16)

courtship conditioning The process by which a normal *Drosophila* male learns to reduce his attempts at courtship following repeated rejections by mated females.

CRE (c̲AMP-r̲esponse e̲lement) *See* **CREB.**

CREB (c̲AMP-r̲esponse e̲lement b̲inding protein) A transcription factor that binds the cAMP response element (*CRE*), a DNA *cis*-regulatory element in the promoter regions of target genes. It is a substrate for several kinases, including cAMP-dependent protein kinase. (Figure 3-41)

CreER A fusion of the Cre recombinase with the portion of the estrogen receptor responsible for ligand-dependent nuclear translocation. CreER enters the nucleus only in the presence of tamoxifen, an estrogen analog, and thereby catalyzes recombination in a tamoxifen-dependent manner. (Figure 14-14)

Cre recombinase A bacteriophage-derived enzyme that catalyzes recombination between two sequence-specific DNA elements called *loxP* sites. (Figure 14-9; Figure 14-12)

Creutzfeldt–Jakob disease (CJD) *See* **prion diseases.**

CRISPR Acronym for c̲lustered r̲egularly i̲nterspaced s̲hort p̲alindromic r̲epeat; it is a genomic locus in some bacteria and archaea containing repeated DNA elements derived from the genomes of invading pathogens. It is used by bacteria for adaptive immunity. Components of the CRISPR system are used experimentally for genome editing. *See also* **Cas9** and **guide RNA.** (Figure 14-8)

critical period A sensitive period during development when experience plays an important role in shaping the wiring and function of the nervous system.

cryo-EM (cryogenic electron microscopy) A technique used to determine the atomic structures of macromolecules.

cryptochrome A protein that acts as a negative regulator of circadian gene expression in mice but as a light sensor for entrainment of circadian rhythms in flies. (Figure 9-22)

cued fear conditioning A fear conditioning paradigm in which an electric shock is applied at the end of a cue presented during training. *See also* **auditory fear conditioning.**

curare A plant toxin that is a competitive inhibitor of the nicotinic acetylcholine receptor.

current clamp A mode of whole-cell patch clamp recording used to measure membrane potential changes while holding the current at a set level.

cyclic AMP (cAMP) An intracellular second messenger synthesized from ATP by adenylate cyclase. (Figure 3-33)

cyclic GMP (cGMP) A cyclic nucleotide derived from GTP, one of its functions is to activate the cyclic nucleotide-gated cation channel in vertebrate photoreceptors in the absence of light. (Figure 4-7; Figure 4-8)

cyclic nucleotide-gated (CNG) channels Nonselective cation channels whose gating is regulated by the concentration of a specific intracellular cyclic nucleotide. (Figure 2-34)

DCC/Unc40 Homologous proteins in vertebrates (DCC, for d̲eleted in c̲olon c̲ancer) and *C. elegans* (Unc40) that act as receptors for netrin/Unc6 and mediate attraction in the absence of Unc5. The *Drosophila* homolog is Frazzled. (Figure 5-12)

decoding (in neural information processing) The process of deducing stimuli from spike trains of a neuron or a neuronal ensemble.

deep brain stimulation (DBS) A treatment strategy used for some neurological and psychiatric conditions in which electrodes are surgically implanted to stimulate neurons and axons in specific brain nuclei.

deep learning A class of supervised machine learning techniques that employ a few to a few dozen intermediate layers of neural network between the input and output layers.

delay line A thin axon fiber that carries auditory signals to target neurons at different locations along the axon with different time delays. (Figure 6-53)

Delta A transmembrane ligand that activates the Notch receptor. (Figure 7-9)

demyelinating disease A disease in which damage to the myelin sheath decreases the axonal membrane resistance between nodes of Ranvier, leading to disruption of ion channel organization in the nodal region and reduction in action potential conduction speed.

dendrites Thick, bushy processes of a neuron that receive and integrate synaptic inputs from other neurons. (Figure 1-9)

dendritic spine A small protrusion on a dendrite of some neurons that receives synaptic input from a partner neuron. The thin spine neck creates chemical and electrical compartments for each spine such that they can be modulated independently of neighboring spines. (Figure 1-9; Figure 3-45)

dendritic tiling A phenomenon in which the dendrites of certain neuronal types collectively cover an entire receptive field exactly once so they can sample the field without redundancy. For example, certain types of retinal neurons collectively cover the retina exactly once. Certain types of somatosensory neurons cover the body surface exactly once. (Figure 4-26)

dendrodendritic synapse A synapse between dendritic processes of two neurons. The reciprocal synapses between olfactory bulb granule cell dendrites and mitral cell secondary dendrites were the first discovered example of dendrodendritic synapses. (Figure 6-15)

de novo **mutation** A mutation produced in the parental germline that is present in all of the offspring's cells.

dense-core vesicle An intracellular vesicle containing neuropeptides; they are larger and more electron dense than small molecule neurotransmitter-containing synaptic vesicles.

dentate gyrus The part of the hippocampus that receives external input, consisting of granule cells and their dendrites, as well as axons from the entorhinal cortex. (Figure 11-5)

depolarization A change in the electrical potential inside a cell toward a less negative value.

depolarization-induced suppression of inhibition (DSI) A transient reduction of inhibitory input to a postsynaptic neuron induced by depolarization of the postsynaptic neuron. Originally described in hippocampal CA1 pyramidal neurons, DSI was found to result from endocannabinoid signaling.

depressing synapse A synapse at which successive presynaptic action potentials trigger progressively smaller postsynaptic responses. (Figure 3-15)

deuterostomes Animals in which the anus appears before the mouth during development. They include all vertebrates. *See also* **protostomes.** (Figure 13-2)

developmental axon degeneration   The process by which axons fragment into pieces that are subsequently engulfed by surrounding glia during normal development.

diacylglycerol (DAG)   A lipid second messenger that binds to and activates protein kinase C (PKC). (Figure 3-34)

diffusion tensor imaging (DTI)   A magnetic resonance imaging technique that allows noninvasive imaging of axon bundles in white matter based on the direction of water diffusion in a given volume. (Figure 14-29)

digital signaling   Signaling that uses discrete values (0s and 1s) to represent information.

direction-selective retinal ganglion cell (DSGC)   A retinal ganglion cell whose firing pattern is influenced by the direction of motion of a stimulus. (Figure 4-28)

direct pathway (in basal ganglia)   An axonal projection from a subset of spiny projection neurons that link the striatum directly to the basal ganglia output nuclei, GPi and SNr. (Figure 8-21)

discrete map   A type of neural map in which input or target neurons or their processes are spatially organized into discrete units (such as glomeruli or layers) representing different qualities (such as cell types). (Figure 7-36)

disinhibition   A reduction of the inhibitory output of an inhibitory neuron. (Figure 1-20)

divergent excitation   A circuit motif in which an excitatory neuron synapses onto multiple postsynaptic targets via branched axons. (Figure 1-20)

dizygotic twins   Nonidentical (fraternal) twins who share only 50% of their genes because they originated from two different eggs fertilized by two different sperm.

DNA (deoxyribonucleic acid)   Double-stranded chains of nucleotides consisting of the sugar deoxyribose, a phosphate group, and one of four nitrogenous bases: adenine (A), cytosine (C), guanine (G), or thymidine (T).

DNA microarray   A solid substrate containing up to millions of immobilized spots of different oligonucleotides or gene-specific probes. Labeled nucleic acid samples can be hybridized to a DNA microarray to quantify the abundance of different species of nucleic acid molecules in samples.

DNA shuffling   A process by which part or all of the protein-coding sequence of one gene is fused to that of another gene, usually following chromosomal duplication or translocation. The specific type of DNA shuffling that occurs when translocational breakpoints are within introns of two genes is called exon shuffling. (Figure 13-6)

L-dopa   The intermediate metabolite between tyrosine and dopamine in the catecholamine biosynthetic pathway. (Figure 12-20)

dopamine   A monoamine modulatory neurotransmitter derived from the amino acid tyrosine. (Figure 12-20; Table 3-2)

Doppler effect   A phenomenon wherein the sound frequency detected by an observer increases if the sound-emitting object moves toward the observer and decreases if the sound-emitting object moves away from the observer.

dorsal column pathway   An axonal pathway from the spinal cord to the brainstem; it consists of ascending branches of proprioceptive neurons and Aβ-LTMRs, as well as axons of some dorsal horn projection neurons. (Figure 6-70)

dorsal cortex   The evolutionary precursor to the mammalian neocortex in reptiles. It consists of three thin layers (as opposed to the six-layered structure of the mammalian neocortex).

dorsal horn   The dorsal part of the spinal gray matter devoted to processing somatosensory information. (Figure 6-71)

dorsal horn projection neuron   A neuron located in the dorsal horn of the spinal cord that projects its axon into the brainstem to relay touch signals. (Figure 6-71)

dorsal root   The place where somatosensory axons enter the spinal cord. (Figure 8-6)

dorsal root ganglia (DRG)   Clusters of primary somatosensory neurons located along an axis parallel to the spinal cord used for sensation of the body (as opposed to the face). (Figure 6-63)

dorsal stream   A visual processing pathway from primary visual cortex to parietal cortex; it processes motion and depth signals. It is also called the "where" stream. (Figure 4-51)

dorsal–ventral   Of a body axis, from back (dorsal) to belly (ventral). (Figure 1-8)

Doublesex (Dsx)   A *Drosophila* gene encoding sex-specific transcription factors produced by sex-specific alternative splicing. The Dsx isoform determines sex-specific somatic structures and also regulates sexual behavior. (Figure 10-5)

Down syndrome   A syndrome caused by the presence of an extra copy of chromosome 21. It is the most common form of intellectual disability with an established genetic etiology.

doxycycline (dox)   A tetracycline analog that readily diffuses across cell membranes and the blood–brain barrier; it is widely used for temporal regulation of gene expression through the tTA/rtTA/*TRE* system. (Figure 14-12)

drive   A motive for action to maintain homeostasis.

drive reduction theory   A psychological theory of motivated behavior positing that deviations from homeostasis (such as dehydration) create aversive drives (such as thirst) that motivate animals to perform actions (such as searching for and consuming water) to reduce the aversive drives.

driver transgene   In binary expression systems, it is the transgene that expresses a transcription factor or recombinase under the control of a tissue-specific or temporally regulated promoter. (Figure 14-12)

driving force   The force that pushes an ion into or out of a cell; it equals the difference between the membrane potential of the cell and the equilibrium potential of the ion.

drug addiction   Compulsive drug use that persists despite long-term negative consequences. It is often associated with loss of self-control and propensity to relapse.

Dscam (Down syndrome cell adhesion molecule)   Encoded by a gene on human chromosome 21, it is an evolutionarily conserved cell adhesion protein. Insect DSCAMs exhibit extraordinary molecular diversity due to alternative splicing. (Figure 7-21)

dye coupling   The diffusion of small-molecule dyes from one cell to another through gap junctions. It is used as a criterion to identify the presence of gap junctions between two cells.

dynamical state   A point in a coding space, representing the status of a dynamical system at a given time. *See also* **coding space**. (Figure 8-35)

dynamical system   A physical system whose future state is a function of its current state, its input, and some noise. It can model time-dependent changes of neural states in a coding space. *See also* **coding space**. (Figure 8-35)

dynamic range   In sensory systems, the ratio between the largest and smallest values of a given dimension of sensory stimuli that can be detected and distinguished.

dynamin   A protein localized at the neck of vesicles undergoing fission. It is essential for clathrin-mediated endocytosis of synaptic vesicles and ultrafast endocytosis at the presynaptic membrane. It is encoded by the gene *Shibire* in *Drosophila*. (Figures 3-14 and 14-27)

dynein A minus-end-directed, microtubule-based motor protein. (Figure 2-6)

eardrum A membrane at the intersection of the mammalian outer ear and middle ear whose vibrations are transmitted by the bones in the middle ear to the cochlea in the inner ear. (Figure 6-43)

echolocation The ability of certain species to use echoes of their own ultrasonic sound pulses to locate objects.

ectoderm The outer germ layer, which gives rise to skin and the nervous system. (Figure 7-2)

efference copy A copy of motor commands or execution signals, which can be used to predict the sensory consequences of motor actions.

efferent An axon that projects from the CNS to peripheral targets. It can also be generalized to describe an output axon from a particular neural center within the CNS.

efficacy of synaptic transmission (synaptic efficacy) The strength of a synaptic connection; it is usually measured by the mean magnitude of the postsynaptic response to a defined presynaptic stimulus.

E–I balance The relative strength of synaptic excitation versus synaptic inhibition.

electrical circuit Connected electrical elements containing at least one closed current path.

electrical gradient Electrical potential difference between two sides of a membrane; it promotes movement of a charged solute toward the side with the opposite charge. (Figure 2-9)

electrical synapse A cell–cell junction enriched in gap junction channels. It transmits (usually bidirectionally) both depolarizing and hyperpolarizing signals between the two cells. *See also* **gap junction**. (Figure 1-14)

electrochemical gradient A combination of chemical and electrical gradients; it determines the direction and magnitude of movement of a charged solute across a membrane. (Figure 2-9)

electroencephalography (EEG) A method for recording the electrical potential differences between surface electrodes placed on specific locations of the scalp. It reports the collective electrical activities of many cortical neurons underneath the surface electrodes. The resulting record is called an electroencephalogram, also abbreviated as EEG. (Figure 9-28; Figure 12-43)

electromotility A property of the cochlear outer hair cells whereby hyperpolarization causes the cells to lengthen, and depolarization causes them to shorten, along their long axis. (Figure 6-50)

electron microscopy A microscopic technique that uses beams of electrons to create an image of a specimen. It has much higher resolution than light microscopy and can resolve structures separated by a nanometer or less. *See also* **transmission electron microscopy, scanning electron microscopy**, and **cryo-EM**.

electroporation A procedure in which DNA containing a transgene is introduced into cells by applying electrical current to facilitate the transfer of negatively charged DNA molecules into cells. In animals, this can be achieved by placing a micropipette containing the DNA near the cells of interest and applying electrical current.

embryonic stem (ES) cells Pluripotent cells derived from early embryos that can be propagated indefinitely *in vitro* and give rise to all cell types of an embryo *in vivo*. (Figure 7-6)

encoding (in neural information processing) The process of representing stimuli using spike trains of a neuron or a neuronal ensemble.

endocannabinoids Endogenous cannabinoids, which are lipophilic molecules such as anadamide and 2-arachidonylglycerol. They can be produced in response to a rise in intracellular Ca^{2+} concentrations in certain postsynaptic neurons and diffuse across the synapse to affect presynaptic neurotransmitter release by binding to the CB1 G-protein-coupled receptor.

endocrine Of or related to a form of signaling in which a recipient cell receives a signal produced by a remote source and delivered via systemic circulation.

endocrine system A system consisting of glands that release hormones into the bloodstream to affect hormone receptor-expressing cells remotely.

endocytosis The process by which cells retrieve, via budding of intracellular vesicles from the plasma membrane, fluid and proteins from the extracellular space and transmembrane proteins from cells' plasma membranes. (Figure 2-2)

endoderm The inner germ layer, which gives rise to a variety of tissues, such as the liver, the inner linings of the gut, and the respiratory tract. (Figure 7-2)

endoplasmic reticulum (ER) A network of membrane-enclosed compartments in eukaryotic cells where secreted and transmembrane proteins are synthesized and into which secreted and transmembrane proteins translocate. It also serves as a store for intracellular Ca^{2+}. (Figure 2-2)

end-plate current The current that crosses a muscle cell membrane in response to release of acetylcholine from a presynaptic motor neuron.

end-plate potential (EPP) Depolarization produced in a postsynaptic muscle cell by acetylcholine released from a presynaptic motor neuron in response to an action potential. (Figure 3-1)

endosome A membrane-enclosed organelle produced via endocytosis; it carries newly internalized extracellular materials and transmembrane proteins. (Figure 2-2)

engram Physical substrate for memory; it is also called a memory trace.

enteric nervous system A division of the autonomic nervous system that is associated with the gastrointestinal tract and that regulates digestion rather independently of the rest of the autonomic nervous system.

enteroendocrine cells Mechano- or chemosensitive cells in the gastrointestinal tract that produce hormones; these cells may also directly release neurotransmitters to activate afferent axon fibers of visceral sensory neurons. (Figure 9-6)

entorhinal cortex The part of the temporal cortex overlying the hippocampus. It provides major input to and receives output from the hippocampus. It plays a key role in representing spatial information. (Figure 11-5)

entrainment The process by which a stimulus, such as light, resets the phase of the circadian clock.

Eph receptors Receptor tyrosine kinases that bind ephrins with their extracellular domains. Two Eph receptor subtypes, the EphA and EphB receptors, typically bind ephrin-As and ephrin-Bs, respectively. They can also serve as ligands during reverse signaling. (Figure 5-7; Figure 5-12)

ephrins Cell-surface proteins that usually act as ligands for Eph receptors to mediate repulsive axon guidance. The ephrin family consists of two subfamilies: ephrin-As are attached to the extracellular face of the plasma membrane by GPI, and ephrin-Bs are transmembrane proteins. They can also serve as receptors during reverse signaling. (Figure 5-7; Figure 5-10)

epigenetic modifications   Molecular modifications to DNA and chromatin, such as DNA methylation and various forms of post-translational histone modifications. They do not modify DNA sequences but can alter gene expression.

epilepsy   A medical condition characterized by recurrent seizures. *See also* **seizure**.

epinephrine   A hormone produced primarily by chromaffin cells in the adrenal gland that mediates systemic responses to extreme conditions, such as the systemic responses associated with fright, fight, and flight. It also acts as a modulatory neurotransmitter in a small group of neurons in the brainstem. (Figure 12-20)

epithelial Na⁺ channel (ENaC)   A member of a class of Na⁺ channels involved in Na⁺ reabsorption by epithelial cells; it is also essential in mammals for the taste of low concentrations of salts. Its invertebrate homologs participate in mechanotransduction. (Figure 6-41)

EPSC (excitatory postsynaptic current)   An inward current produced by binding of excitatory neurotransmitters to their receptor(s). (Figure 3-23)

EPSP (excitatory postsynaptic potential)   A transient depolarization of a postsynaptic cell associated with an excitatory postsynaptic current (EPSC). (Figure 3-23)

equilibrium potential   The membrane potential at which there is no net flow of an ion across a membrane, when the electrical and chemical forces are equal in magnitude but opposite in direction.

E_{rev} *See* **reversal potential**.

estradiol   A steroid hormone produced by the ovaries of sexually mature females. It is produced *in vivo* by the action of the enzyme aromatase on testosterone. In conjunction with progesterone, estradiol regulates female sexual behavior and reproduction. (Figure 10-19)

estrogen   A female sex hormone, such as estradiol.

estrogen receptor   A cytosolic protein that upon binding of an estrogen (such as estradiol) translocates to the nucleus, where it acts as a transcription factor. (Figure 10-19)

eukaryote   Organism consisting of cell(s) with a nuclear membrane separating the genetic material from the rest of the cellular components.

eumetazoan   A taxon that includes cnidarians, bilaterians, and the most recent common ancestor of cnidarians and bilaterians. (Figure 13-2)

exchanger   *See* **antiporter**.

excised patch   A patch clamp configuration in which the membrane patch underneath the electrode is excised from the cell and placed in a defined medium. It is often used to study the biophysical and biochemical properties of the ion channel(s) in the membrane patch. (Figure 14-40)

excitability   A property of a neuron that defines how readily it fires action potentials.

excitable cell   A cell that produces action potentials, such as a neuron or a muscle cell. It can also refer to any cell that uses electrical signaling to receive, integrate, propagate, or transmit information.

excitation–contraction coupling   A process by which action potentials in muscle cells lead to muscle contraction. It involves actin/myosin-mediated contraction triggered by a rise in intracellular Ca²⁺ concentration. (Figure 8-5)

excitatory neuron   A neuron that, when activated, depolarizes its postsynaptic target cells and makes them more likely to fire action potentials.

excitatory neurotransmitter   A neurotransmitter that depolarizes postsynaptic target cells and makes them more likely to fire action potentials.

excitotoxicity   Toxicity to neurons caused by excessive stimulation by excitatory neurotransmitters such as glutamate, which results in large or persistent increases in intracellular Ca²⁺ concentration.

exocrine system   A system consisting of glands that excrete fluids, such as sweat or tears, locally through specific ducts.

exocytosis   The process by which intracellular vesicles fuse with the plasma membrane to release secreted proteins into the extracellular space and deliver lipids and transmembrane proteins to the plasma membrane. (Figure 2-2)

exon   The part of a pre-mRNA molecule retained in mRNA after splicing. (Figure 2-2)

exon shuffling   *See* **DNA shuffling**.

explicit memory   A form of memory requiring conscious recall, such as memory for names, facts, and events. It is also called declarative memory. (Figure 11-3)

expression cloning   A strategy for cloning a gene by transfecting cells with pools of cDNAs and using a functional assay to identify the pool containing the cDNA of interest. The assay is repeated with progressively divided pools of cDNAs until a single cDNA is identified. (Figure 6-68)

extensor   A muscle whose contraction increases the angle of a joint. (Figure 8-8)

extinction (in classical conditioning)   A decrease in the conditioned response caused by repeated exposure to the conditioned stimulus without the unconditioned stimulus.

extinction (in operant conditioning)   A decrease in a reinforced action or an increase in a punished action when the action is repeatedly not reinforced or punished, respectively.

extracellular recording   A technique for recording voltage changes, such as action potentials from a single neuron or synaptic activity from a population of neurons, by placing an electrode (often made of a metal probe insulated everywhere except at its tip) at close range to a neuronal cell body or a synapse-rich region. (Figure 14-34)

Eyeless   A *Drosophila* transcription factor belonging to the Pax family contains a homeobox and a paired box; it is required for eye development. Its ectopic expression in other structures, such as the antenna or the wing precursors, can induce ectopic eye formation. *See also* **Pax6**.

facilitating synapse   A synapse at which successive presynaptic action potentials trigger progressively larger postsynaptic responses. (Figure 3-15)

familial advanced sleep phase   An extreme variant of the human circadian system characterized by very early morning waking and an early evening sleep onset.

familial Alzheimer's disease (FAD)   A small subset of Alzheimer's disease cases that follows a Mendelian autosomal dominant inheritance pattern.

fast axonal transport   Intracellular transport at a speed of 50–400 mm per day; cargos subject to fast axonal transport include organelles as well as transmembrane and secreted proteins. (Figure 2-4)

fear conditioning   *See* **contextual fear conditioning** and **auditory fear conditioning**.

feedback excitation   A circuit motif in which an excitatory neuron synapses onto its own presynaptic partner. (Figure 1-20)

feedback inhibition   A circuit motif in which an excitatory neuron both provides output to and receives input from an inhibitory neuron. (Figure 1-20)

feedforward excitation   A circuit motif in which serially connected excitatory neurons propagate information across multiple regions of the nervous system. (Figure 1-20)

feedforward inhibition   A circuit motif in which a postsynaptic neuron receives both direct excitatory input from a presynaptic neuron and disynaptic inhibitory input from the same excitatory neuron via an inhibitory interneuron. (Figure 1-20)

fertilization   The fusion of sperm and egg to create a genetically new organism. (Figure 7-2)

Fezf2   A transcription factor that specifies subcerebral projection neuron identity. (Figure 7-12)

fiber photometry   A technique that uses an optical fiber to record bulk fluorescence activity from a cell population that expresses a fluorescence indicator (e.g., a genetically encoded Ca^{2+} indicator).

fibroblast growth factor (FGF)   A member of a family of secreted growth factors that act as morphogens to pattern early embryos during development.

field excitatory postsynaptic potential (fEPSP)   Excitatory postsynaptic potentials recorded from a population of neurons near the tip of an extracellular electrode. fEPSPs evoked by stimulation of axonal inputs to a population are often used as a measure of the strength of synaptic transmission between the stimulated inputs and neurons near the recording electrode. (Figure 11-7)

filamentous actin (F-actin)   A major cytoskeletal element composed of two parallel helical strands of actin polymers. They are also called microfilaments. (Figure 2-5; Figure 8-3)

filopodia   Thin, protruding processes of a growth cone made of bundled F-actin. (Figure 5-16)

firing rate (spike rate)   Mean spike count per unit time.

fissure   A deep invagination of the cortical surface that separates areas of the cerebral cortex.

fitness   With respect to an allele (or phenotype), fitness is the ratio of the frequency of the allele (or phenotype) in a population after one generation of selection to the frequency of the allele (or phenotype) in the same population before the selection. With respect to an individual, fitness is the number of second-generation descendants the type of individual with a particular genome is expected to have.

fixed (of an allele)   The state of an allele when every member of a population is homozygous for the allele.

fixed action pattern   A concept in neuroethology, it refers to an instinctive sequence of behaviors that is usually invariant and runs to completion once triggered.

flavor   A synthesis of taste and olfaction with additional contribution from the trigeminal somatosensory system for special chemicals, temperature, and texture.

flexor   A muscle whose contraction decreases the angle of a joint. (Figure 8-8)

floor plate   A structure at the ventral midline of the spinal cord. (Figure 5-12; Figure 7-10)

FLP recombinase   A yeast-derived enzyme that catalyzes recombination between two sequence-specific DNA elements called *FRT* (FLP recognition target) sites. (Figure 14-9; Figure 14-26)

fluoxetine   A widely used antidepressant that acts as a selective serotonin reuptake inhibitor. Its brand name is Prozac. *See also* **SSRI**. (Figure 12-26)

Fmr1   *See* **fragile-X syndrome**.

FMRP   *See* **fragile-X syndrome**.

focal seizures   Seizures that affect a relatively small, discrete region of the brain.

follicle-stimulating hormone (FSH)   *See* **gonadotropins**.

forebrain   The rostral-most division of the three divisions of the embryonic brain. It gives rise to the cerebral cortex, basal ganglia, hippocampus, amygdala, thalamus, and hypothalamus. (Figure 1-8; Figure 7-3)

forward genetic screen   A procedure for identifying genes necessary for a biological process. It usually involves (1) inducing mutations in a population of experimental animals so that each animal carries a different set of random mutations in a small number of genes or a single gene, and (2) identifying mutations that disrupt the biological process of interest based on the phenotypes exhibited by the offspring of the mutagenized animals. (Figure 14-5)

forward model (in motor systems)   A model that uses efference copy and expected sensory feedback signals to provide online tuning of motor output.

Fos   An immediate early gene encoding a transcription factor. Its expression is commonly used as an indicator of recently active neurons.

fovea   The central part of the primate retina, which has a high density of cones. (Figure 4-13)

fragile-X syndrome (FXS)   A leading cause of inherited intellectual disability; it is caused by expanded trinucleotide repeats in the 5′ untranslated region of the *Fmr1* gene, which encodes an RNA binding protein called the fragile X mental retardation protein (FMRP).

Frazzled   *See* **DCC/Unc40**.

frequency tuning   The property whereby a cell in the auditory system is best activated by sounds of a particular frequency. It is usually represented as a V-shaped curve on a frequency-intensity plot. (Figure 6-47)

frontal eye field (FEF)   A neocortical area that receives extensive feedforward connections from both the dorsal and the ventral streams of visual information and sends feedback projections to many visual cortical areas. (Figure 4-51)

frontal lobe   One of the four cerebral cortex lobes; it is located at the front of the brain rostral to the central sulcus. (Figure 1-23)

FRT   *See* **FLP recombinase**.

Fruitless (Fru)   A *Drosophila* gene that regulates all aspects of male courtship rituals. The splicing of one of its transcripts is regulated by a hierarchy of sex-determining splicing factors: females express a nonfunctional splice isoform of the protein while males express a functional form of the protein (Fru^M), which acts as a transcription factor. (Figure 10-5)

functional architecture   The physical arrangement of neurons in a brain region based on their functional properties.

functional magnetic resonance imaging (fMRI)   A noninvasive functional brain imaging technique; it monitors signals originating from changes in blood flow that are correlated with local neuronal activity. It is also called BOLD (blood-oxygen-level dependent) fMRI.

fundamental frequency   The frequency of the lowest frequency component of a periodic waveform.

fura-2 A small molecule Ca^{2+} indicator whose optimal excitation wavelength shifts from 380 nm to 350 nm when Ca^{2+} is bound. The ratio of fluorescence intensity measured at excitation wavelengths of 350 nm and 380 nm can be used as a sensitive measure of Ca^{2+} concentration. (Figure 14-41)

fusiform face area A specific area of human temporal cortex preferentially activated by images of human faces.

Gα, Gβ, Gγ *See* **trimeric GTP-binding protein**.

GABA A glutamate derivative that is the predominant inhibitory neurotransmitter in vertebrates and invertebrates. (Figure 3-16; Table 3-2)

GABA$_A$ receptor An ionotropic receptor that is gated by GABA and mediates fast inhibition. (Figure 3-21; Figure 12-27)

GABA$_B$ receptor A metabotropic receptor that is activated by GABA and mediates slow inhibition.

gain control Modulation of the slope of a system's input–out function; it is often used to restrict output to a limited dynamic range.

gain-of-function experiment An experiment in which a specific component is added to a system; it is often used to test whether the added component is sufficient for the system to function in a specific context.

GAL4 A yeast transcription factor that binds to a DNA element called a *UAS* (upstream activation sequence) in the promoter regions of genes to activate transcription of those genes. (Figure 14-12)

ganglion A cluster of neurons located in the peripheral nervous system.

ganglionic eminences Developing ventral telencephalon structures including the medial, caudal, and lateral ganglionic eminences (MGE, CGE, LGE); they are the birthplaces of cortical GABAergic neurons (MGE and CGE), GABAergic interneurons in the basal ganglia and amygdala (MGE and CGE), and olfactory bulb interneurons and most GABAergic projection neurons in the striatum (LGE). (Figure 7-5)

gap junction The morphological correlate of the electrical synapse, which usually contains hundreds of closely clustered channels that bring the plasma membranes of two neighboring cells together and allow passage of ions and small molecules between the two cells. (Figure 3-48)

gastrin-releasing peptide receptor (GRPR) A G-protein-coupled receptor activated by gastrin-releasing peptide and involved in processing itch signals.

gastrula The product of gastrulation; it is an embryo with a three-layered structure consisting of ectoderm, mesoderm, and endoderm. (Figure 7-2)

gastrulation The process by which an embryo is transformed from a ball of cells into a structure with three distinct layers: ectoderm, mesoderm, and endoderm. (Figure 7-2)

GCaMP A GFP-based genetically encoded Ca^{2+} indicator; its fluorescence increases in response to a rise in Ca^{2+} concentration. (Figure 14-41)

gene A segment of DNA that carries the instructions for how and when to make specific RNAs and proteins. (Figure 2-2)

gene duplication A major step in evolutionary innovation. Along with associated diversification, it is the primary means of production of new genes. (Figure 13-6)

generalized seizures Seizures that affect multiple, bilateral brain regions.

gene therapy The use of DNA and/or genome modification to treat diseases caused by genetic defects.

genetically encoded Ca^{2+} indicators Proteins whose fluorescence properties change before and after binding to Ca^{2+}. *See also* **Ca^{2+} indicator**. (Figure 14-41)

genetic drift Changes in allele frequency in small populations due to random events. An allele can be lost because it is not passed from parents to progeny due to random allele sampling or because the progeny carrying the allele die or fail to reproduce by chance. The loss of an allele increases the prevalence of the remaining allele(s).

genetic mosaic An animal containing cells of more than one genotype. (Figure 14-10)

genetic susceptibility locus A genomic locus with variant(s) that increase the probability of carriers developing a trait (such as a disease).

genome editing The general process of altering the genome at predetermined loci, such as by deleting a piece of endogenous DNA, inserting a piece of foreign DNA, or creating a specific base-pair change.

genome-wide association study (GWAS) A strategy for identifying genes associated with a specific trait by comparing genome-wide DNA samples collected from many people with or without the trait; it is also used to identify the single nucleotide polymorphisms most strongly linked with the trait.

ghrelin A neuropeptide produced by stomach-associated glands in response to reduced glucose levels; it acts as a hunger signal to stimulate eating. (Figure 9-15)

G$_i$ (inhibitory G protein) A Gα variant that binds to adenylate cyclase and inhibits its activity.

gill-withdrawal reflex A reflex in the sea slug *Aplysia* in which it withdraws its gill into its mantle shelf when a tactile stimulus is applied to its siphon. It has been used as a model system to investigate the mechanisms underlying simple forms of learning and memory. (Figure 11-24)

glia Nonneuronal cells of the nervous system; they play essential roles in the development and function of the nervous system (Figure 1-9)

glomerulus A discrete, ball-like structure in the vertebrate olfactory bulb or insect antennal lobe where axons of olfactory receptor neuron form synapses with the dendrites of their postsynaptic target neurons. (Figure 6-3; Figure 6-24)

GluN1 *See* **NMDA receptor**.

GluN2 *See* **NMDA receptor**.

glutamate An amino acid that is the predominant excitatory neurotransmitter in vertebrates. (Figure 3-16; Table 3-2)

glutamic acid decarboxylase (GAD) An enzyme that converts glutamate into GABA.

glycine An amino acid that is an inhibitory neurotransmitter released by a subset of brainstem and spinal cord neurons in vertebrates. (Figure 3-16; Table 3-2)

glycine receptor An ionotropic receptor that is gated by glycine and mediates fast inhibition. (Figure 3-21)

Goldman–Hodgkin–Katz (GHK) equation An equation that relates the membrane potential at equilibrium to the membrane permeabilities and concentrations of multiple ions on the two sides of a membrane. A variant of the GHK equation relates the membrane potential at equilibrium to the equilibrium potential and conductance of each ion.

Golgi cell A type of GABAergic neuron in the cerebellar cortex that receives input from granule cells and sends inhibitory output back to granule cells. (Figure 8-26)

Golgi outpost A fragment of the Golgi apparatus located in neuronal dendrites. (Figure 7-19)

Golgi stain A histological staining method; it uses solutions of silver nitrate and potassium dichromate, which react to form a black precipitate (microcrystals of silver chromate) that accumulates stochastically in a small fraction of nerve cells so that these cells, and most or all of their elaborate extensions, can be visualized against unstained tissue.

gonadotropin-releasing hormone (GnRH) A prehormone released by hypothalamic neurons (called GnRH neurons) that stimulates release of gonadotropins by anterior pituitary endocrine cells. (Figure 10-22)

gonadotropins A family of hormones including luteinizing hormone (LH) and follicle-stimulating hormone (FSH). Released by anterior pituitary endocrine cells, these hormones stimulate the maturation of male testes and female ovaries during puberty. In adults, they stimulate the testes to release testosterone and the ovaries to release estradiol. (Figure 10-22)

GPCR (G-protein-coupled receptor) A member of a receptor family with seven transmembrane helices that, upon ligand binding, activate trimeric G proteins, which in turn activate intracellular signaling cascades.

GPe (globus pallidus external segment) An intermediate nucleus in the basal ganglia indirect pathway; it contains GABAergic neurons that project to the GPi, SNr, and STN. (Figure 8-21)

GPi (globus pallidus internal segment) One of the two major output nuclei of the basal ganglia; it contains GABAergic neurons that project to the thalamus and brainstem. (Figure 8-21)

G_q A Gα variant that activates phospholipase C, in turn leading to activation of the inositol-phospholipid signaling pathway. (Figure 3-34)

graded potentials Membrane potentials that can change in continuous values, as opposed to the all-or-none property of action potentials. (Figure 2-18)

granule cells Neurons that are granular in appearance because they are densely packed; there are three prominent types. (1) Cerebellar granule cells are the most numerous type of neuron in the brain; their cell bodies and dendrites reside in the granular layer of the cerebellar cortex, where they receive mossy fiber input, and their axons ascend into the molecular layer, where each bifurcates to become a parallel fiber that sends glutamatergic output to Purkinje cells. (Figure 8-26) (2) Dentate gyrus granule cells in the hippocampus receive input from the entorhinal cortex via the perforant path and send glutamatergic output to CA3 pyramidal neurons. (Figure 1-12; Figure 11-5) (3) Olfactory bulb granule cells constitute a large subtype of olfactory bulb interneurons that receive input from the secondary dendrites of mitral cells and send GABAergic output back to mitral cells. (Figure 6-14)

gray matter The parts of the CNS enriched in neuronal cell bodies, dendrites, axon terminals, and synapses and that appear gray in histological sections.

green fluorescent protein (GFP) A jellyfish protein that emits green fluorescence when excited by blue light. It is widely used as a marker of gene expression and in live imaging.

grid cell A cell in the entorhinal cortex whose activity depends on an animal's location in an arena, with peak firing rate occurring at the apices of an imaginary hexagonal grid superimposed on the arena floor. (Figure 11-32)

growth cone A dynamic structure at the tip of a developing neuronal process; it enables the extension of the process and guides its direction. (Figure 1-13; Figure 5-16)

G_s (stimulatory G protein) A Gα variant that binds to adenylate cyclase and stimulates its activity. (Figure 3-33)

GTPase An enzyme that hydrolyzes GTP, converting it into GDP.

GTPase activating protein (GAP) A protein that switches GTPases off by accelerating the GTPases' endogenous activity, which converts GTP into GDP. (Figure 3-32)

guanine nucleotide exchange factor (GEF) A protein that switches GTPases on by catalyzing the exchange of GDP for GTP. (Figure 3-32)

guanylate cyclase An enzyme that produces cGMP from GTP.

guide RNA In the CRISPR/Cas9 system, an RNA molecule that brings Cas9 to a target DNA sequence, where Cas9 generates a double-strand break. The guide RNA must contain sequences that base-pair with the target DNA. *See also* **CRISPR** and **Cas9**. (Figure 14-8)

gustatory nerve A bundle of axons originating from the basal ends of taste receptor cells, the nerve projects to the nucleus of the solitary tract in the brainstem and relays taste information from the tongue to the brain. (Figure 6-32)

habituation A decrease in the magnitude of responses to stimuli presented repeatedly.

hair cell The primary sensory cell for audition; it converts mechanical stimuli—movement of stereocilia at its apical end—into electrical signals. (Figure 6-43; Figure 6-45)

halorhodopsin A light-activated inward chloride pump in archaea; it can be used to silence neuronal activity in heterologous systems by light. *See also* **optogenetics**. (Figure 14-47)

HCN (hyperpolarization-activated cyclic nucleotide-gated) channels Nonselective cation channels activated by hyperpolarization; their gating is influenced additionally by the concentration of intracellular cyclic nucleotides. (Figure 2-34)

head direction cell A cell that fires when an animal's head is facing a specific direction, regardless of the animal's location in its environment. (Figure 11-32)

Hebbian synapse A synapse whose strength can be enhanced by co-activation of pre- and postsynaptic partners.

Hebb's rule A postulate by Donald Hebb that describes how learning can be transformed into a lasting memory. It states: "When an axon of cell *A* is near enough to excite a cell *B* and repeatedly or persistently takes part in firing it, some growth process or metabolic change takes place in one or both cells such that *A*'s efficiency, as one of the cells firing *B*, is increased." (Figure 5-24)

hedonic value The degree to which something is pleasant or unpleasant, which usually correlates with the degree to which something is potentially beneficial or harmful to an animal.

hemisphere One of the two sides of the brain.

heritability A measure of the contribution of genetic differences to trait differences within a population. It can be measured in twin studies as 2 × (the correlation of the trait between pairs of monozygotic twins – the correlation of the trait between pairs of dizygotic twins).

herpes simplex virus (HSV) A DNA virus used to deliver transgenes into postmitotic neurons. It has the capacity to encode ~150 kb of foreign DNA. (Table 14-1)

heterophilic binding Binding of two different proteins, usually two different membrane proteins expressed from adjacent cells across a cellular junction, such as a synapse.

heterophilic cell adhesion protein A protein that facilitates adhesion between cells via direct binding to a distinct partner from apposing cells. The partner can be a different member of the same protein family or from different protein families.

hindbrain   The caudal-most division of the three divisions of the embryonic brain. It gives rise to the pons, medulla, and cerebellum. (Figure 1-8; Figure 7-3)

hippocampus   A structure underneath the cortical surface of the temporal lobe; it has been most studied for its role in spatial representation and acquisition of explicit memory. (Figure 1-8; Figure 11-5)

histamine   A monoamine modulatory neurotransmitter derived from the amino acid histidine. (Figure 3-16; Table 3-2)

histological sections   Slices of frozen or chemically fixed tissue produced by microtomes, with thicknesses ranging from several to several hundred micrometers. They can be stained using a number of different methods and examined under a light microscope.

homeobox   *See* **homeodomain.**

homeodomain (homeobox)   Originally discovered in proteins whose disruption causes transformation of one body part into another, it is a DNA-binding domain shared by all Hox proteins and many other transcription factors.

homeostasis   The maintenance of a steady state of a physiological parameter—such as blood pressure, body temperature, or nutritional level—by feedback physiological and behavioral responses. (Figure 9-6)

homeostatic synaptic scaling   A plasticity phenomenon with a negative feedback mechanism such that a change of a certain parameter, such as the firing rate of a neuron or the strength of a synapse, triggers a compensatory process such that the parameter returns to the original set point.

homeotic transformation   Transformation of one body part into another, such as the transformation of a pair of antennae into a pair of legs in *Drosophila antennapedia* mutants.

homologous recombination   Exchange of nucleotide sequences between two identical or highly similar DNA molecules; it occurs naturally in certain cells due to its role in specific biological processes, such as in germline cells during meiotic crossing over. It is also used experimentally for genome editing, such as in generation of knockout and knock-in alleles. (Figure 14-6)

homophilic binding   Binding of two identical proteins, usually two membrane proteins expressed from adjacent cells across a cellular junction such as a synapse.

homophilic cell adhesion protein   A protein that facilitates adhesion between cells via direct binding of the same proteins from apposing cells.

Hopfield network   A recurrent neural network in which neurons do not form synapses onto themselves and the synaptic weights of each reciprocal pair are the same. It has the property that network dynamics will converge onto a local minimum of an "energy" function. *See* **attractor.**

horizontal cell   An inhibitory neuron in the vertebrate retina; it regulates signal propagation from photoreceptors to bipolar cells through lateral inhibition. (Figure 4-23)

horizontal gene transfer   Gene transfer from one organism to another through mechanisms other than reproduction, such as via viral transduction.

horizontal sections   A section plane perpendicular to the dorsal–ventral axis. (Figure 1-8)

hormone   A chemical or peptide produced by an organism that circulates in tissue fluids such as blood to affect the physiology of recipient cells at a distance.

***Hox* gene**   A member of a family of evolutionarily conserved genes that are arranged in genomes in clusters and encode homeobox-containing transcription factors. *Hox* genes define the anterior–posterior body axes of most invertebrates and all

vertebrates and also regulate neuronal fate at later developmental stages. (Figure 13-32)

5-HT   *See* **serotonin.**

HTMR (high-threshold mechanoreceptor)   A mechanosensory neuron that senses strong and noxious mechanical stimuli. (Figure 6-64)

huntingtin   *See* **Huntington's disease.**

Huntington's disease (HD)   A dominantly inherited disease that usually strikes patients during midlife; it is characterized initially by depression or mood swings and subsequently by abnormal movement due to degeneration of striatal neurons. It is caused by expanded polyglutamine repeats in the huntingtin protein.

HVC (high vocal center)   A dorsal forebrain nucleus in the songbird essential for song production. (Figure 10-16)

hyperalgesia   An enhanced response to noxious (painful) stimuli. (Figure 6-72)

hyperpolarization   A change in the electrical potential inside a cell toward a more negative value.

hypocretin (orexin)   A neuropeptide expressed by specific lateral hypothalamus neurons; it is important for regulating sleep and eating.

hypogonadotropic hypogonadism   A disorder characterized by delayed, reduced, or absent puberty due to reduced gonadotropin levels.

hypothalamus   A collection of nuclei ventral to the thalamus; it controls many bodily functions, including eating, digestion, metabolic rate, drinking, salt intake, reproduction, body temperature, emergency responses, and circadian rhythms. It executes many of these functions by regulating the autonomic nervous system and neuroendocrine system. (Figure 1-8; Figure 9-6)

identified neuron   A neuron that can be recognized across individuals of the same species due to its stereotyped location, size, and/or shape.

Ig CAM (immunoglobulin cell adhesion molecule)   A cell adhesion molecule that contains immunoglobulin domains on its extracellular side.

imipramine   A tricyclic antidepressant that inhibits the plasma membrane monoamine transporters. (Figure 12-26)

immediate early genes (IEGs)   A class of genes whose transcription is rapidly induced by external stimuli without requiring new protein synthesis.

immuno-EM   A combination of immunostaining and electron microscopy used to visualize the distribution of individual proteins at an ultrastructural level. (Figure 14-27)

immunostaining   A staining method that uses antibodies to visualize the distributions of specific proteins in histological sections or whole-mount tissues.

implicit memory   A form of memory in which previous experience aids in the performance of a task without conscious recall. It is also called nondeclarative memory or procedural memory. (Figure 11-3)

inactivation (of ion channels)   A decrease in ion conductance through a channel after its opening. The ion channel, when inactivated, is in a distinct state from when it is closed.

incentive salience theory   A psychological theory of motivated behavior positing that food and water are inherently rewarding to animals; hunger and thirst amplify their reward value.

inclusion bodies   Intracellular foci into which aggregated proteins are sequestered.

indirect pathway (in basal ganglia) An axonal projection from a subset of spiny projection neurons that terminates in the GPe and STN. (Figure 8-21)

induced pluripotent stem (iPS) cells Pluripotent cells produced experimentally from differentiated cells by a variety of means, such as forced expression of key transcription factors involved in maintaining the pluripotency of embryonic stem cells. (Figure 7-6)

induction A mechanism for determining cell fate in which a cell is born with the same potential to develop into different cell types as its sibling or cousins, and its fate is acquired by receiving external signals (i.e., the cell's fate is "induced" by external cues).

inferior colliculus A midbrain nucleus that integrates auditory signals from brainstem nuclei. It sends auditory output to the thalamus and the nearby superior colliculus/tectum. (Figure 6-52)

inferior olive A nucleus in the medulla containing neurons whose axonal projections to the cerebellum form climbing fibers. (Figure 8-26)

inhibitory neuron A neuron that, when activated, hyperpolarizes its postsynaptic target cells and makes them less likely to fire action potentials.

inhibitory neurotransmitter A neurotransmitter that hyperpolarizes postsynaptic target cells and makes them less likely to fire action potentials.

innate A trait or behavior that is genetically programmed and is thus with an organism from birth rather than acquired by experience.

innate song The song a songbird would sing if raised in acoustic isolation during the sensory stage of song learning.

innexin A protein component of gap junctions in invertebrates.

inositol 1,4,5-triphosphate (IP$_3$) A second messenger that binds to the IP$_3$ receptor on the endoplasmic reticulum (ER) membrane to trigger the release of ER-stored Ca^{2+} into the cytosol. (Figure 3-34)

input specificity (of LTP) A property of long-term potentiation (LTP) whereby LTP occurs only at synapses that have experienced an LTP-inducing stimulus and not at unstimulated synapses on the same postsynaptic neuron. (Figure 11-8)

***in situ* hybridization** A method for determining mRNA distribution in tissues by hybridizing labeled gene-specific nucleic acid probes to fixed histological sections or whole-mount tissues.

instrumental conditioning *See* **operant conditioning**.

insular cortex A part of the cerebral cortex that represents taste, pain, and interoception, among other functions; it is located internal to the part of the somatosensory cortex representing the tongue depicted in the sensory homunculus. (Figure 6-32; Figure 9-3)

insulator An object or substance that does not allow electric current to pass. It is equivalent to a resistor with infinite resistance.

insulin A peptide hormone produced by the pancreas in response to a rise in blood glucose levels after meals; it regulates carbohydrate metabolism throughout the body and regulates food intake through its actions on target neurons in the brain. (Figure 9-15)

intellectual disability A condition characterized by deficits in general mental abilities such as reasoning, problem solving, planning, abstract thinking, judgment, and learning.

interaural level difference (ILD) The level difference of a sound received in the left and right ears, used for sound localization.

interaural time difference (ITD) The difference in the arrival time of a sound at the left and right ears, used for sound localization.

intermediate progenitor A progenitor cell produced by division of a radial glial cell. It divides further to give rise to postmitotic neurons. (Figure 7-4)

interneuron A neuron with its axon confined to the specific CNS region that houses the neuron's cell body; it can also be called a local neuron in this context. It may also refer to any neuron that is not a motor or sensory neuron.

interoception The sense of the state of internal organs.

interoceptive system The sensory system that senses the state of internal organs.

intersectional methods (in genetics) Strategies that use two orthogonal binary expression systems to refine patterns of transgene expression. (Figure 14-13)

interstitial branching The extension of a collateral from the side of a growing process. (Figure 7-20)

intracellular recording A procedure for measuring the membrane potential of a cell using an electrode inserted into or continuous with its cytoplasm. (Figure 14-34)

intracellular vesicle A small, membrane-enclosed organelle in the cytoplasm of a eukaryotic cell. (Figure 2-2)

intrinsically photosensitive retinal ganglion cell (ipRGC) A subclass of retinal ganglion cells that expresses melanopsin and can be directly depolarized by light. (Figure 4-36)

intrinsic properties The electrophysiological properties of a neuron determined by the composition, concentration, subcellular distribution, and biophysical properties of its ion channels.

intrinsic signal imaging A method for measuring neuronal activity based on changes in the optical properties of tissue surrounding active neurons, primarily as a result of changes in blood oxygenation in those regions.

intron The part of an RNA molecule that is removed during splicing. (Figure 2-2)

inward-rectifier K⁺ channels A subfamily of K⁺ channels that preferentially passes inward currents over outward currents; these channels pass current at membrane potentials more hyperpolarized than E_K but allow minimal outward currents at membrane potentials more positive than E_K. (Figure 2-34)

ion channel A specialized transmembrane protein that allows passage of one or more specific ions across the lipid bilayer.

ionotropic receptor A neurotransmitter receptor that functions as a neurotransmitter-gated ion channel to allow rapid (within a few milliseconds) membrane potential changes in response to neurotransmitter binding. (Figure 3-21)

iontophoresis A technique by which ions or charged chemicals are locally applied from a micropipette via a current pulse.

IP$_3$ receptor An IP$_3$-gated Ca^{2+} channel on the ER membrane. (Figure 3-34)

iproniazid The first antidepressant, discovered serendipitously in the 1950s; it is an inhibitor of monoamine oxidase. (Figure 12-26)

IPSC (inhibitory postsynaptic current) An outward current produced by the binding of an inhibitory neurotransmitter to its receptor. The fast component is usually mediated by Cl⁻ influx through the GABA$_A$ or glycine receptors.

ipsilateral Of the same side of the midline. For example, an ipsilateral axonal projection is an axon that does not cross the midline and therefore terminates on the same side of the nervous system as its corresponding soma.

IPSP (inhibitory postsynaptic potential) A transient hyperpolarization of a postsynaptic cell associated with an inhibitory postsynaptic current (IPSC).

***I–V* curve** A graphical representation of the relationship between the current that passes through a piece of ion-channel-containing membrane (I) and the voltage across the membrane (V). (Figure 3-17)

kainate receptor A glutamate-gated ion channel that conducts Na^+ and K^+ and can be selectively activated by the drug kainate (kainic acid).

K^+ channels Ion channels that allow selective passage of K^+; they constitute the most diverse channel family. (Figure 2-34)

Kenyon cells Principal cells of the insect mushroom body, which receive synaptic inputs mainly from olfactory projection neurons and play key roles in olfactory learning and memory. *See also* **mushroom body.**

kinase An enzyme that adds phosphates to proteins.

kinesins A family of microtubule-based motor proteins that are mostly plus-end-directed. (Figure 2-6; Figure 2-7)

kisspeptins A family of neuropeptides encoded by the *Kiss1* gene that play an important role in activating GnRH neurons. (Figure 10-22)

knee-jerk reflex The involuntary forward movement of the lower leg due to contraction of the quadriceps femoris muscle (an extensor) and relaxation of the hamstring muscle (a flexor). (Figure 1-19)

knock-in A variation of the knockout procedure in which a DNA construct—either a transgene or a variant of an endogenous gene—is inserted into a specific chromosomal locus; the procedure can produce changes to endogenous genes as small as a single base pair.

knockout A genetic engineering procedure that disrupts a specific gene. In the mouse, it is traditionally achieved by homologous recombination in embryonic stem cells to create a mutation in the target gene. The resulting mutant mouse is called a knockout mouse for that particular gene. (Figure 14-6)

kuru *See* **prion diseases.**

labeled line (in sensory systems) A dedicated processing channel that carries specific information from the periphery to the brain (Figure 6-33)

lamellipodia A veil-like meshwork in growth cones made of branched F-actin. (Figure 5-16)

lamina (in insect visual system) The first neuropil layer underneath the retina in the insect compound eye. (Figure 4-31; Figure 5-34)

lamina terminalis A set of nuclei in the anterior hypothalamus that are key regulators of water intake; it consists of the subfornical organ (SFO), the organum vasculosum of the lamina terminalis (OVLT), and the median preoptic nucleus (MnPO). (Figure 9-16)

landmark-based strategy A navigational strategy in which animals use external cues to determine their locations.

large dense-core vesicle Intracellular vesicle that is responsible for neuropeptide release; it is larger than synaptic vesicles and contains electron-dense materials.

laser-scanning two-photon microscopy *See* **two-photon microscopy.**

late LTP A long-lasting phase of long-term potentiation (LTP), usually lasting longer than 3 hours and requiring new protein synthesis and likely new gene expression.

lateral geniculate nucleus (LGN) A thalamic nucleus that receives visual input from retinal ganglion cell axons and sends output to the primary visual cortex. (Figure 4-37; Figure 4-38)

lateral horn A second-order olfactory center for odor-mediated innate behavior in the insect brain. It and the mushroom body are the two major output sites for olfactory projection neuron axons. (Figure 6-24; Figure 6-30)

lateral hypothalamus A loosely organized hypothalamic region implicated in regulation of multiple functions, including eating, drinking, and sleep.

lateral inhibition (in cell fate determination) The process by which neighboring cells are prevented from adopting identical fates through cell–cell interactions, such as those mediated by Notch/Delta. (Figure 7-9)

lateral inhibition (in information processing) A circuit motif in which an inhibitory neuron receives excitatory input from one or several parallel streams of excitatory neurons and sends inhibitory output to many or all of the postsynaptic targets of these excitatory neurons. It is widely used in sensory systems. (Figure 1-20)

lateral intraparietal area (LIP) A cortical area in the primate parietal lobe implicated in decision making to move eyes in a particular direction.

law of effect A psychological principle positing that behavior followed by a reward will be repeated, whereas behavior followed by a punishment will diminish.

learning The process by which experience changes the nervous system, enabling animals and humans to acquire new knowledge or skills and modify their behaviors.

length constant (space constant, λ) A key parameter that defines the passive properties of electrical signaling; it is equal to the distance along a neuronal process over which the amplitude of a membrane potential change decays to $1/e$, or about 37% of its original value.

lentivirus A retrovirus that can infect postmitotic neurons. It has a capacity to carry ~8 kb foreign DNA. (Table 14-1)

leptin A hormone secreted by fat tissues that negatively regulates food intake through its actions on specific neurons in the brain. (Figure 9-15)

Lewy bodies Intracellular inclusions that are a defining pathological feature of most forms of Parkinson's disease.

ligand A molecule that binds to its receptor to exert a biological activity.

ligand-gated ion channel A transmembrane protein complex that directly conducts ions in response to the binding of a neurotransmitter or other ligand.

light microscopy The most widely used microscopic technique in biology; it uses beams of visible light (photons) to create an image of a specimen and, with the exception of super-resolution fluorescence microscopy, can only resolve structures greater than 200 nm apart.

light-sheet fluorescence microscopy A fluorescence microscopy technique in which only the focal plane (a single plane in the z-dimension) is illuminated with a thin sheet of a laser beam from the side. All fluorescence emissions in the focal plane are collected simultaneously by a detector. (Figure 14-23)

LKB1 A protein kinase essential for determining axon fate during the establishment of neuronal polarity.

LMAN (lateral magnocellular nucleus of the anterior nidopallium) A forebrain nucleus in the songbird essential for song learning but not song production. (Figure 10-16)

local field potential Electrical potential at an extracellular recording site relative to a distal ground. Usually filtered to remove high-frequency signals, it reflects the collective dendritic and synaptic activities of many neurons near the electrode. (Figure 14-35)

local interneuron (LN) (in the insect olfactory system) A neuron whose processes are restricted to the antennal lobe. (Figure 6-24)

local neuron *See* **interneuron.**

local protein synthesis   Translation of mRNA into protein in a neuron's dendrites and axons rather than in its cell body.

locus coeruleus   A brainstem nucleus consisting of norepinephrine neurons that project widely across the brain. (Figure 9-32)

long-range cue (in axon guidance)   A secreted protein that can act at a distance from its cell of origin. (Figure 5-11)

long-term depression (LTD)   A long-lasting decrease in synaptic efficacy that can be induced experimentally by specific stimulus conditions.

long-term memory   Memory that lasts hours to years. (Figure 11-3)

long-term potentiation (LTP)   A long-lasting enhancement of synaptic efficacy. It can be induced experimentally by a variety of stimuli, such as high-frequency stimulation of input axons. (Figure 11-7)

long-term synaptic plasticity   A change in the efficacy of synaptic transmission that lasts hours to the lifetime of the animal.

loose-patch recording   A technique in which a patch electrode is placed against a cell membrane without forming a gigaohm seal. It can be used to record only spiking activity (not subthreshold activity) but, unlike whole-cell recording, does not affect the intracellular content of the recorded cell.

lordosis   A posture that female rodents assume when sexually aroused. It facilitates sexual intercourse.

loss-of-function experiment   An experiment in which a specific component is disrupted, often used to determine if the missing component is necessary for the system to function.

loss-of-function mutation   A mutation that disrupts the function of a gene.

lower envelope principle   The idea that the limits of psychophysical performance are determined by the sensitivities of the most sensitive individual neurons. (Figure 6-74)

loxP   *See* **Cre recombinase.**

LRP4 (low-density lipoprotein receptor-related protein-4)   Along with MuSK, it is an agrin receptor in muscle. It also signals back to motor axons to trigger presynaptic differentiation in a MuSK-independent manner. (Figure 7-25)

LTMRs (low-threshold mechanoreceptors)   Touch-sensitive somatosensory neurons that innervate hair follicles, specialized epithelial cells, and encapsulated corpuscles in the skin. They respond to vibration, indentation, pressure, and stretch of the skin as well as movement or deflection of hairs. (Figure 6-64)

luminance contrast (in vision)   A difference in light intensity between adjacent spaces.

luteinizing hormone (LH)   *See* **gonadotropins.**

lysosome   A membrane-enclosed organelle that contains enzymes for protein degradation. (Figure 2-2)

M1   *See* **primary motor cortex (M1).**

macular degeneration   A disease that causes photoreceptors in the fovea to die, impairing high-acuity vision.

major depression   A mood disorder characterized by persistent feelings of sadness, emptiness, and worthlessness.

major urinary protein (MUP)   A highly stable protein found in urine that is used by some species to mark an individual's territory for a long duration.

MARCM (mosaic analysis with a repressible cell marker)   A genetic mosaic method in *Drosophila* used to label individual cells or groups of cells that share the same lineage and at the same time to delete an endogenous gene or express a transgene specifically in these labeled cells. (Figure 14-26)

Martinotti cell   A type of GABAergic neuron in the cerebral cortex that forms synapses onto the distal dendrites of pyramidal cells. (Figure 3-46)

massively parallel processing   An information processing method; it utilizes a large number of units to perform a set of coordinated computations in parallel. It is a key feature of the nervous system.

mass spectrometry   An analytic technique that can precisely determine the mass-to-charge ratio of electrically charged molecules; it is widely used to determine the identity of individual proteins in a protein mixture. (Figure 14-20)

maximum parsimony   A means of generating phylogenetic predictions by selecting the interpretation of the experimental data that posits the fewest number of evolutionary changes among all the potential interpretations.

MC4R   A G-protein-coupled receptor activated by α-MSH to reduce food intake. (Figure 9-14)

mechanosensitive channel   An ion channel gated by mechanical force.

mechanosensory neurons   Somatosensory neurons activated by mechanical force and responsible for proprioception, touch, and a subset of nociceptive sensations.

mechanotransduction   The process in sensory cells by which mechanical stimuli are converted into electrical signals.

MeCP2 (methyl-CpG-binding protein 2)   A nuclear protein that binds to DNA at methylated CpG sites (adjacent cytosine and guanine nucleotides). It is highly expressed in developing and adult neurons. *See also* **Rett syndrome.** (Figure 12-35)

medial amygdala   Part of the olfactory amygdala complex that receives direct input from accessory olfactory bulb mitral cells. It is sexually dimorphic and regulates male courtship behavior. (Figure 10-26)

medial geniculate nucleus   A thalamic nucleus that processes and relays auditory signals to the auditory cortex. (Figure 6-52)

medial–lateral   Of a body axis, from midline to side. (Figure 1-8)

medial preoptic area (MPOA)   A sexually dimorphic nucleus in the anterior hypothalamus that regulates male courtship behavior. (Figure 10-23; Figure 10-26)

median preoptic nucleus (MnPO)   Part of the lamina terminalis in the anterior hypothalamus; it is responsible for integrating dehydration signals sensed by the subfornical organ (SFO) and organum vasculosum of the lamina terminalis (OVLT) and broadcasting the signals to downstream neurons. (Figure 9-16)

medulla   The caudal-most part of the brainstem, between the pons and the spinal cord. (Figure 1-8)

medulla (in insect visual systems)   A neuropil that lies beneath the lamina in the insect optic lobe. (Figure 4-31; Figure 5-34)

melanopsin   An opsin expressed by vertebrate intrinsically photosensitive retinal ganglion cells; it is a member of the c-opsin subfamily, whose members are most widely used in invertebrate visual systems.

membrane potential   The electrical potential difference between the inside of a cell and its extracellular environment.

memory   Lasting changes in the brain that retain learned information. It is typically divided into several distinct processes, including acquisition, consolidation, storage, and retrieval.

memory trace   *See* **engram.**

Merkel cell A specialized epithelial cell at the junction of the dermis and epidermis; it is closely associated with the peripheral ending of the slowly adapting type I (SAI) LTMR. (Figure 6-63)

mesencephalic locomotor region (MLR) A midbrain region wherein electrical stimulation evokes locomotor activity.

mesoderm The middle germ layer, which gives rise to the skeletal system, connective tissues, muscle, and the circulatory system. (Figure 7-2)

messenger RNA (mRNA) A mature RNA molecule that has undergone 5′ capping, 3′ polyadenylation, and splicing to remove introns and is exported to the cytoplasm to direct protein synthesis. (Figure 2-2)

metabotropic receptor A neurotransmitter receptor that regulates ion channel conductance indirectly through intracellular signaling cascades, modulating membrane potential over a time scale of tens of milliseconds to seconds. (Figure 3-22)

microglia A glial cell that functions as the resident immune cell of the nervous system. It engulfs damaged cells and debris. (Figure 1-9)

microneurography A neurophysiological technique used to record neuronal activity in the peripheral nerves of awake human subjects.

microRNA A short, noncoding RNA (21–26 nucleotides in length) widely used in eukaryotic organisms to regulate protein production. It triggers the degradation and inhibits the translation of mRNAs with complementary sequences. *See also* **RNA interference**.

microstimulation Delivery of small currents through an extracellular electrode with the goal of activating a limited number of nearby neurons.

microtubule A major cytoskeletal element composed of hollow cylinders of 13 parallel protofilaments made of α- and β-tubulin. (Figure 2-5)

midbrain The rostral-most part of the brainstem. It includes the tectum (superior and inferior colliculus in mammals) dorsally and the tegmentum ventrally. It is also the middle part of the three divisions of the embryonic brain caudal to the forebrain and rostral to the hindbrain. It is also called the mesencephalon. (Figure 1-8; Figure 7-3)

middle temporal visual area (MT) A high-order visual cortical area in the dorsal stream specialized for analyzing motion signals. (Figure 4-51)

midget ganglion cell A type of retinal ganglion cell with a small receptive field used for high-acuity and green–red color vision in primates. (Figure 4-34)

miniature end-plate potential (mEPP) Small depolarization of a muscle cell in response to spontaneous neurotransmitter release from a motor neuron. (Figure 3-2)

mirror neuron Neuron in the premotor or posterior parietal cortex of primates, activated when an animal observes actions by another animal as well as when the animal performs the same actions it has observed.

mitogen-activated protein (MAP) kinase cascade A kinase cascade that acts downstream of the small GTPase Ras and other signaling molecules; it consists of three serine/threonine kinases represented by Raf, Mek, and Erk. (Figure 3-39)

mitotic recombination Exchange of a portion of homologous maternal and paternal chromosome during mitotic cell division; it can create daughter cells homozygous for alleles on portions of the paternal or maternal chromosomes. (Figure 14-10; Figure 14-26)

mitral cell A second-order neuron in the vertebrate olfactory bulb; it receives input from ORNs and sends output to the olfactory cortex. It differs from a tufted cell, also a second-order neuron in the vertebrate olfactory bulb, in its cell body location in the olfactory bulb and axon termination pattern in the olfactory cortex. (Figure 6-16)

MnPO *See* **median preoptic nucleus**.

modulatory neurons Neurons that release modulatory neurotransmitters. They can act on excitatory, inhibitory, and other modulatory neurons to up- or downregulate their excitability or synaptic transmission.

modulatory neurotransmitter (neuromodulator) A neurotransmitter that can bidirectionally change the membrane potential, excitability, or neurotransmitter release of its postsynaptic target neurons. It acts through G-protein-coupled receptors.

molecular clock A technique that utilizes the rates of sequence changes, calibrated against fossil records, to estimate the times at which two species diverged.

monoamine neurotransmitter A neurotransmitter derived from an aromatic amino acid, including serotonin, dopamine, norepinephrine, and histamine.

monoamine oxidase An enzyme that oxidizes dopamine, norepinephrine, and serotonin, leading to their degradation. (Figure 12-23)

monozygotic (identical) twins Twins produced from the same fertilized egg or zygote; they share 100% of their genomes.

morphine The active ingredient of opiates and a powerful opioid receptor agonist.

morphogen A diffusible signaling protein that can cause cells located at different distances from the source to adopt different fates.

Morris water maze A navigation task in which rats and mice learn to locate a hidden platform in a pool of milky water using distant cues in the room.

Mosaic analysis A method for analyzing the cell types in which the function of a gene is required by creating genetic mosaic animals containing both wild-type and mutant cells that are usually differentially marked.

mossy fiber An axon that has elaborate terminal arborizations. The two most prominent types are found in the cerebellum and hippocampus. (1) The cerebellar mossy fiber is an axon that terminates in the granular layer of the cerebellar cortex, where it synapses onto granule cells. It originates from neurons residing in the pons, medulla, and spinal cord. (Figure 8-26) (2) The hippocampal mossy fiber is an axon of a dentate gyrus granule cell that synapses onto CA3 pyramidal neuron dendrites. (Figure 11-5)

motor homunculus A map in the primary motor cortex that corresponds to movement of specific body parts. Nearby areas in the motor cortical areas represent movement control of nearby body parts. (Figure 1-25)

motor neuron A type of neuron that extends dendrites within the CNS and projects its axon out of the CNS to innervate a muscle. (Figure 1-15; Figure 8-9)

motor pool A cluster of motor neurons innervating the same muscle. (Figure 8-6)

motor protein A protein that converts energy from ATP hydrolysis into movement along the cytoskeletal polymers.

motor system The collected parts of the nervous system that control the contraction of skeletal muscles and thereby enable movement and maintain body posture.

motor unit A motor neuron and the set of muscle fibers it innervates. (Figure 8-6)

motor unit size The number of muscle fibers a motor neuron innervates.

mount　A posture that male rodents assume when sexually aroused. It facilitates sexual intercourse.

M pathway　A visual processing pathway from the retina to the visual cortex that originates from retinal ganglion cells with large receptive fields and engages lateral geniculate nucleus cells in the magnocellular layers; it carries information about luminance and has excellent contrast and temporal sensitivity. (Figure 4-51)

MrgprA3　A G-protein-coupled receptor activated by pruritogens such as chloroquine.

α-MSH (α-melanocyte-stimulating hormone)　A neuropeptide released by POMC neurons in the arcuate nucleus that reduces food intake.

mTOR (mammalian target of rapamycin)　A key protein in intracellular signaling pathways that plays an important role in regulating protein translation. (Figure 12-41)

Müller glia　A glial cell in the retina wherein the conversion of all-*trans* retinal to 11-*cis* retinal occurs to assist the recovery process in cones.

multielectrode array　A device used to record the spiking activities of many individual neurons. The electrodes can be arrayed either horizontally or vertically. (Figure 14-36)

multiple sclerosis (MS)　A common adult-onset CNS demyelinating disease characterized by inflammatory plaques in the white matter caused by immune cell attack of myelin.

multipolar (neuron)　Having more than two processes leaving the cell body.

muscarinic AChR　*See* **acetylcholine receptor (AChR)**.

muscimol　A mushroom-derived toxin that is a potent activator of the GABA$_A$ receptor.

muscle fiber　A muscle cell.

muscle spindle　A special apparatus in muscle cells that senses muscles stretch. It has embedded endings of peripheral branches of proprioceptive somatosensory neurons. (Figure 1-19; Figure 6-63)

mushroom body　A second-order olfactory center for odor-mediated learning and memory in the insect brain; it and the lateral horn are the two major output sites for olfactory projection neuron axons. (Figure 6-24; Figure 11-29)

MuSK　A muscle-specific receptor tyrosine kinase; it acts together with LRP4 as an agrin receptor to promote acetylcholine receptor clustering. (Figure 7-25)

mutation　A change in DNA, including insertion, deletion, or alteration of one or more base pairs.

myelin sheath　Cytoplasmic extensions of oligodendrocytes and Schwann cells that wrap around axons with multilayered glial plasma membranes to increase resistance and decrease capacitance for action potential propagation. *See also* **axon myelination**. (Figure 2-26; Figure 2-27)

myofibril　A thread-like longitudinal structure in muscle cells composed of repeating sarcomeres and responsible for muscle contraction. (Figure 8-3)

myosin　An F-actin-based motor protein. (Figure 8-3)

β2 nAChR　A subunit of nicotinic acetylcholine receptors that, among other functions, is essential for cholinergic retinal wave propagation. *See also* **acetylcholine receptor (AChR)**.

Na$^+$-K$^+$ ATPase　A pump that uses energy derived from ATP hydrolysis to pump Na$^+$ out of a cell and K$^+$ into a cell against their respective electrochemical gradients; it helps maintain the Na$^+$ and K$^+$ concentration differences across cellular membranes. (Figure 2-12)

narcolepsy　A disorder characterized by difficulty staying awake during the day, especially following moments of happiness or excitement, caused by a deficiency in the neuropeptidergic hypocretin signaling system or by dysfunction of hypocretin-expressing neurons.

nasal (in retinal maps)　In the direction of the nose.

natural selection　A key mechanism of evolution; it is the process by which genetic variations that confer individuals a better chance of reproductive success become more common in a population over time. (Figure 13-1)

nature (in context of nature versus nurture)　The contribution of genetic inheritance to brain function and behavior.

negative reinforcement　A type of learning in which an action occurs more often because it is followed by removal of a punishment.

negative selection　The process by which an allele that is detrimental to an organism becomes less prevalent in a population.

nematocin　The *C. elegans* ortholog of vertebrate oxytocin and vasopressin. (Figure 10-36)

neocortex　The largest part of the mammalian cerebral cortex; it typically contains six layers and is evolutionarily the newest part of the cerebral cortex.

Nernst equation　An equation that relates the equilibrium potential of an ion to the concentrations of the ion on the two sides of a membrane.

nerve　A discrete bundle of axons in the peripheral nervous system.

nerve growth factor (NGF)　A prototypical neurotrophin; it is a target-derived secreted protein that supports the survival and axon growth of sensory and sympathetic neurons. (Figure 7-34; Figure 7-35)

nerve impulse　The historical name for transient changes in membrane potential that propagate along axons; it is the same as an action potential.

netrin/Unc6　Homologous secreted proteins originally identified by biochemical purification (netrins in vertebrates) and genetic screening (Unc6 in *C. elegans*) in the context of midline guidance; they are widely used axon guidance cues. (Figure 5-12)

neural circuit　An ensemble of interconnected neurons that act together to perform specific functions.

neural crest cells　A special group of cells at the junction of the dorsal neural tube and the overlying epidermal cells. They migrate away from the neural tube to produce diverse cell types, including cells of the peripheral nervous system. (Figure 7-2)

neural plasticity　Changes of the nervous system in response to experience and learning.

neural plate　The layer of ectodermal cells overlaying the notochord that invaginates and gives rise to the neural tube during neurulation. (Figure 7-2)

neural progenitor　A dividing cell that gives rise to neurons and glia. In vertebrates, it is usually located near the ventricle in the developing vertebrate CNS. (Figure 7-4)

neural prosthetic device　A device that can substitute a sensory or motor function that has been disrupted due to injury or disease. For example, population activity of neurons in the motor cortex can be used to control an external device such as a robotic arm or a computer cursor to help patients who suffer from paralysis or motor neuron diseases. (Figure 8-40)

neural tube　A hollow tube surrounded by layers of neuroectodermal cells; it is the embryonic precursor to the vertebrate CNS. (Figure 7-2)

neuraxis   Axis of the CNS. The rostral–caudal neuraxis follows the curvature of the embryonic neural tube; the dorsal–ventral neuraxis is perpendicular to the rostral–caudal neuraxis. (Figure 1-8)

neurexin   A protein on the presynaptic membrane that mediates synaptic adhesion. A major binding partner is neuroligin. (Figure 7-26)

neuroblast   A neuronal progenitor.

neurodegenerative disorders   Disorders characterized by progressive neuronal dysfunction, including loss of synapses, atrophy of dendrites and axons, and death of neurons.

neuroendocrine system   The collected parts of the nervous system that control hormone secretion and thus regulate animals' physiology and behavior in response to sensory stimuli and brain states.

neuroethology   A branch of science that emphasizes the study of animal behavior in the natural environment.

neurofibrillary tangle   An intracellular fibril consisting of an abnormal accumulation of hyperphosphorylated tau, a microtubule-binding protein. (Figure 12-2)

neurofilament   An intermediate filament (a cytoskeletal polymer with a diameter between F-actin and microtubules) in vertebrate neurons; it is concentrated in and provides stability to axons.

neurogenic inflammation   Inflammation triggered by release of neuropeptides such as substance P and calcitonin gene-related peptide from the peripheral terminals of sensory neurons.

neuroligin   A protein on the postsynaptic membrane that mediates synaptic adhesion. A major binding partner is neurexin. (Figure 7-26)

neuromodulator   *See* **modulatory neurotransmitter**.

neuromuscular junction   The synapse between a motor neuron's presynaptic terminal and a skeletal muscle cell. (Figure 3-1)

neuron (nerve cell)   An electrically excitable cell that receives, integrates, propagates, and transmits information as the working unit of the nervous system.

neuronal polarity   The distinction between axons and dendrites.

neuronal process   Cytoplasmic extension of a neuron.

neuron doctrine   The proposition that the individual neuron is the working unit of the nervous system.

neuropeptide   Polypeptide that is a few to a few dozen amino acids in length and acts as a neurotransmitter.

neuropil   Region of neural tissue densely packed with neuronal and glial processes.

neuropilin-1 (Nrp1)   A co-receptor for semaphorins in vertebrates.

neurotransmitter reuptake   The process by which neurotransmitters in the synaptic cleft are transported either into nearby glial cells or back into the presynaptic cytosol and then into synaptic vesicles. (Figure 3-12)

neurotransmitters   Molecules stored in synaptic vesicles (or dense-core vesicles in the case of neuropeptides) in presynaptic terminals, released into the synaptic cleft (triggered by presynaptic depolarization), and activate ionotropic or metabotropic receptors on postsynaptic target cells. (Figure 3-16; Table 3-2)

neurotrophic hypothesis   The idea that the survival of developing neurons requires neurotrophins produced by the neurons' postsynaptic targets.

neurotrophin-3 (NT3)   *See* **neurotrophins**.

neurotrophin-4 (NT4)   *See* **neurotrophins**.

neurotrophins   A family of secreted signaling proteins that regulate the survival, morphology, and physiology of target neurons through binding to specific receptors on those neurons. Mammalian neurotrophins include nerve growth factor (NGF), brain-derived neurotrophic factor (BDNF), neurotrophin-3 (NT3), and neurotrophin-4 (NT4). *See also* **Trk receptors** and **p75NTR**. (Figure 3-39; Figure 7-35)

neurulation   The developmental process in vertebrate embryos leading to formation of the neural tube, which gives rise to the nervous system. (Figure 7-2)

nicotinic AChR   *See* **acetylcholine receptor (AChR)**.

Nissl stain   A stain that labels RNA and thus highlights the rough endoplasmic reticulum in cytoplasm. It uses basic (i.e., positively charged) dyes such as cresyl violet that bind to negatively charged RNA molecules. (Figure 14-22)

NMDA receptor   A glutamate-gated ion channel that conducts Na^+, K^+, and Ca^{2+} and can be activated by the drug NMDA (*N*-methyl-D-aspartate). Its opening requires both binding of glutamate and postsynaptic depolarization. It is a heterotetramer of two GluN1 subunits encoded by a single gene, and two GluN2 subunits, of which there are four variants (GluN2A, GluN2B, GluN2C, GluN2D) each encoded by separate genes. (Figure 3-24)

nociception   The sense of noxious stimuli by the somatosensory system.

nociceptive neuron   A somatosensory neuron that detect noxious stimuli.

nodes of Ranvier   Periodic gaps in axonal myelination, usually 200 μm to 2 mm apart, where the axon surface is exposed to the extracellular ionic environment. They contain high concentrations of voltage-gated Na^+ and K^+ channels that regenerate action potentials. (Figure 2-26)

noise (in neural encoding and decoding)   The source of variability in spike trains unexplained by stimuli.

nonhomologous end joining   An endogenous DNA repair system; it rejoins the two ends of a DNA molecule with a double-strand break. It often creates a small deletion or duplication at the break point as a result of the repair process. (Figure 14-8)

non-spiking neuron   A neuron that uses graded potentials rather than action potentials to transmit information.

norepinephrine   A monoamine neuromodulator derived from dopamine. (Figure 12-20; Table 3-2)

northern blot   A method for determining the amount of a specific RNA in an RNA mixture. RNAs are separated by gel electrophoresis and transferred to a membrane; labeled nucleic acid probes are then hybridized to the membrane to visualize specific RNA molecules that hybridize to the probe. (Figure 6-8)

Notch   A transmembrane receptor widely used in diversifying cell fate during development. Binding of a ligand to Notch triggers proteolytic cleavage of Notch in its transmembrane domain, releasing the Notch intracellular domain, which can then enter the nucleus to regulate gene expression. (Figure 7-9)

notochord   A midline mesodermal structure in vertebrate embryos ventral to the spinal cord that produces secreted cues for patterning the spinal cord. (Figure 7-2; Figure 7-10)

NREM sleep   Non-rapid eye movement (NREM) sleep, or sleep stages other than REM sleep. (Figure 9-28)

nucleus accumbens   The major part of the ventral striatum; it is involved in processing reward signals and receives input from VTA dopamine neurons and the prefrontal cortex, thalamus, hippocampus, and amygdala. (Figure 12-29)

nucleus laminaris (NL) A brainstem nucleus in the barn owl that analyzes interaural time differences; it is analogous to the medial superior olivary nucleus in mammals.

nucleus of the solitary tract (NTS) A nucleus in the brainstem that receives input from the taste system as well as sensory information from internal organs. (Figure 6-32; Figure 9-4)

null direction The direction of stimulus motion that elicits the lowest firing rate of a direction-sensitive visual system neuron.

Numb A *Drosophila* protein that segregates asymmetrically in daughter cells during sensory organ precursor and neuroblast divisions; it is essential for conferring different fates to the two daughter cells of an asymmetric division. (Figure 7-8)

nurture (in context of nature versus nurture) The contribution of environmental factors to brain function and behavior.

occipital lobe One of the four cerebral cortex lobes; it is located at the caudal-most part of the neocortex. (Figure 1-23)

octopamine A neurotransmitter in some invertebrate nervous systems that is chemically similar to norepinephrine in vertebrates.

ocular dominance Preference for receiving and representing visual input from one eye over the other. In the primary visual cortex of some mammals, such as cats and monkeys, cells in the same vertical columns share the same ocular dominance, thus producing ocular dominance columns. (Figure 5-18)

ocular dominance column *See* **ocular dominance**.

odds ratio In genetics, a measure of the effect of a genetic variant on the likelihood of having a particular trait, such as a disease. It is calculated by dividing the probability of having the trait among people with the genetic variant by the probability of having the trait among people without the genetic variant.

odorant A molecule that elicits olfactory perception; it is usually volatile. (Figure 6-6)

odorant receptor A receptor on the surface of olfactory cilia that binds odorants. (Figure 6-8)

OFF bipolar A bipolar cell that expresses ionotropic glutamate receptors and is depolarized by glutamate release from photoreceptors; its membrane potential changes follow the sign of photoreceptors, such that it is hyperpolarized by light. (Figure 4-22)

Ohm's law An equation that relates current (I) to voltage (V) and resistance (R); $I = V/R$.

olfactory bulb The first olfactory processing center in the vertebrate brain. (Figure 6-3; Figure 6-14)

olfactory cilium A dendritic branch of an olfactory receptor neuron enriched in odorant receptors. (Figure 6-3)

olfactory cortex Brain regions that receive direct input from olfactory bulb mitral/tufted cells, including the anterior olfactory nucleus, piriform cortex, olfactory tubercle, cortical amygdala, and entorhinal cortex. (Figure 6-16)

olfactory epithelium The epithelial layer in the nose that houses olfactory receptor neurons. (Figure 6-3)

olfactory processing channel A discrete information-processing unit in the olfactory system consisting of olfactory receptor neurons (ORNs) expressing a given odorant receptor, the glomerular targets of those ORNs, and second-order neurons that send dendrites to the same glomeruli.

olfactory receptor neuron (ORN) A primary sensory neuron in the olfactory system; it converts odorant binding to odorant receptors into electrical signals relayed to the brain via its axon. Also called an olfactory sensory neuron (OSN). (Figure 6-3)

oligodendrocyte A glial cell in the CNS that wraps axons with its cytoplasmic extensions to form a myelin sheath. (Figure 1-9)

ommatidium A repeating unit of the arthropod compound eye. In *Drosophila*, each ommatidium contains eight photoreceptors. (Figure 5-34)

ON bipolar A bipolar cell that expresses metabotropic glutamate receptors and is inhibited by glutamate release from photoreceptors; its membrane potential changes are opposite in sign to those of photoreceptors, such that it is depolarized by light. (Figure 4-22)

open probability The proportion of time an individual ion channel is open and able to conduct current.

operant conditioning (instrumental conditioning) A form of learning in which a subject associates performance of a specific action (e.g., pressing a lever) with a particular outcome, such as delivery of a reinforcer (e.g., food) or punishment (e.g., an electrical shock).

opioid receptors A subfamily of G-protein-coupled receptors that serve as receptors for opioids, including morphine and endogenous opioid neuropeptides. They are widely distributed across the nervous system.

opioids Molecules with similar effects to opiates such as morphine. They include opiates from the opium poppy and endogenous neuropeptides such as enkephalin, endorphin, and dynorphin.

opsin A member of a family of G-protein-coupled receptors expressed in photoreceptors of multicellular organisms; it is associated with retinal and converts photon absorption into activation of a trimeric GTP-binding protein. In microbes, it is a member of a family of light-induced channels or pumps that are not G-protein-coupled receptors. (Figure 4-6; Figure 13-20)

optical imaging An approach that uses changes in fluorescence or other optical properties as indicators of neuronal activity.

optic ataxia A condition in which patients cannot guide their hand toward an object using visual information, even though other aspects of their movement and vision are less affected.

optic chiasm The midline structure where a fraction of retinal ganglion cell axons cross to the side of the brain contralateral to the eye of origin. (Figure 4-37; Figure 5-13)

optic lobe The part of the insect brain that consists of the retina, lamina, medulla, and lobula complex and is used to analyze visual signals. (Figure 5-34)

optic nerve The bundle of retinal ganglion cell axons; it sends visual information from the eye to the brain. (Figure 4-37)

optic tract The bundles of retinal ganglion cell axons distal to the optic chiasm. (Figure 4-37)

optogenetics The set of methods used to manipulate neuronal activity by using light to activate genetically encoded effectors, most commonly microbial opsins (e.g., channelrhodopsin-2, archaerhodopsin, halorhodopsin). (Figure 14-47)

orexin *See* **hypocretin**.

organization–activation model A central principle in endocrinology; it proposes that sex hormones have two different types of effects: organizational effects during development, which configure the brain in a sex-typical manner, and activational effects in adults, which stimulate male- or female-typical sexual behaviors. (Figure 10-20)

organ of Corti An organ in the cochlea consisting of hair cells, the surrounding support cells, and the basilar membrane. (Figure 6-43)

organoids Miniature structures that resemble certain aspects of specific organs, produced from pluripotent stem cells in three-dimensional cultures *in vitro*.

otolith organ   A sensory organ in the vestibular system that senses linear acceleration and stationary head tilts. (Figure 6-59)

outer radial glia (oRG)   A type of radial glia whose cell bodies are located in the subventricular zone. They serve, along with ventricular zone radial glia, as neural progenitors. They are greatly expanded in number in humans compared to mice and likely contribute to increased neuronal production in mammals with large neocortices. *See also* **radial glia**. (Figure 13-37)

outer segment   A cytoplasmic extension of a rod or a cone; it contains a highly specialized photon detection apparatus made of tightly stacked membrane disks enriched in opsins. (Figure 4-2)

outgroup   A group of organisms that is closely related to but falls outside of a set of organisms of interest. It is used as a reference group in determining the phylogenetic relationships among a set of organisms.

ovariectomized female   A female from which the ovaries have been removed.

OVLT (organum vasculosum of the lamina terminalis)   A part of the lamina terminalis in the anterior hypothalamus responsible for sensing dehydration signals from the blood, including osmolarity and angiotensin II levels. (Figure 9-16)

oxytocin   A hormone secreted by hypothalamic neurons in the posterior pituitary and a neuropeptide released by some hypothalamic neurons; it regulates maternal and social behavior.

P1 neurons   A cluster of ~20 Fru+/Dbx+ neurons in the male fly brain that integrates mating-related sensory signals and promote courtship behavior.

p75NTR   A 75 kilodalton neurotrophin receptor that has a low affinity for all neurotrophins and is also a receptor for all proneurotrophins. (Figure 7-35)

pacemaker cell   A cell that can produce rhythmic output in the absence of input.

PALM   *See* **super-resolution fluorescence microscopy**.

parabiosis   The joining of the circulatory systems of two animals such that they have limited exchange of substances in systemic circulation.

parabrachial nucleus   A brainstem nucleus that transmits ascending signals from the visceral sensory and pain somatosensory systems to the thalamus, amygdala, hypothalamus, and brainstem autonomic centers. (Figure 6-71; Figure 9-15)

paracrine   Of or related to a form of signaling in which a recipient cell receives a signal produced by nearby cells.

parallel fiber   The portion of the axon of a cerebellar granule cell that runs in parallel to the pial surface and crosses Purkinje cell dendrites at a right angle. (Figure 8-26)

parasol ganglion cell   A type of retinal ganglion cell in primates with a large receptive field and excellent contrast and temporal sensitivity used for motion vision.

parasympathetic system   A branch of the autonomic nervous system that facilitates energy conservation. Activation of the parasympathetic system decreases heart rate and blood flow, constricts airways in the lung, and stimulates salivation and digestion. (Figure 9-1; Figure 9-2)

paraventricular hypothalamic nucleus (PVH)   A hypothalamic nucleus involved in multiple physiological functions, including release of oxytocin and vasopressin into the bloodstream through axonal projections in the posterior pituitary and descending control of autonomic nervous system functions. (Figure 9-6)

parietal lobe   One of the four cerebral cortex lobes; it is located behind the frontal lobe and above the occipital lobe. (Figure 1-23)

parietal reach region (PRR)   A region in monkey's posterior parietal lobe that is selectively associated with arm reaching.

Parkinson's disease (PD)   A common neurodegenerative disease caused by death of substantia nigra dopamine neurons; it primarily affects movement control, with symptoms including shaking, rigidity, slowness, and difficulty walking.

parthenogenesis   A reproductive strategy in which embryos develop from unfertilized eggs without exchange of genetic materials.

partial agonist   A drug that binds to a target receptor but elicits only a partial biological effect.

passive electrical properties   Membrane properties in the absence of voltage-dependent conductances. Two salient examples: (1) due to membrane capacitance, a sharp change in electrical signal (e.g., a current pulse) becomes more diffuse temporally as the signal travels along a neuronal process; (2) due to membrane conductance, the magnitude of electrical signal becomes attenuated over distance. They are also called cable properties. (Figure 2-16)

passive transport   Movement of a solute across a membrane down its electrochemical gradient via a channel or transporter. (Figure 2-8)

patch clamp recording   An electrophysiological recording technique that utilizes a glass electrode (patch pipette) to form a high-resistance seal with a membrane. It has several variants, including cell-attached patch, excised patch, and whole-cell recordings. (Figure 14-40)

patch pipette   *See* **patch clamp recording**.

path-integration strategy   A navigational strategy in which animals use the speed, duration, and direction of their own movement to calculate their current position with respect to their starting position.

pattern separation   A theoretical neuroscience concept that describes the ability to separate similar neural states, such as firing patterns of neuronal populations, or similar memories.

Pax6   A member of the Pax family of transcription factors; it contains a homeobox and a paired box. It regulates patterning of the cerebral cortex and spinal cord and is required for eye development in mammals. Its *Drosophila* homolog is Eyeless. *See also* **Eyeless**.

pC1 neurons   A cluster of Fru–/Dbx+ neurons in the female fly brain that integrates mating-related sensory signals and promotes courtship receptivity.

PDZ domain   Acronym for a domain shared by PSD-95, Discs large (a *Drosophila* protein implicated in cell proliferation and associated with postsynaptic densities), and ZO-1 (an epithelial tight junction protein). It is a protein–protein interaction domain that binds to a specific protein sequence motif present at the C-terminal end of many transmembrane receptors.

percept   A specifically perceived object or the brain representation of the object.

perceptron   An artificial neural network with a feed-forward architecture; it can be used to discriminate visual objects through changes in connection weights (usually in a single layer) after viewing examples and their correct labels.

perforant path   The pathway taken by axons of neurons in the superficial layer of the entorhinal cortex that project to the hippocampus. (Figure 11-5)

periaqueductal gray (PAG)   A midbrain gray matter structure surrounding the cerebral aqueduct; it serves many functions, including descending control of pain and execution of defensive behaviors such as freezing. (Figure 6-71; Figure 11-45)

periglomerular cell　A member of diverse types of interneurons in the olfactory bulb that receive direct input from olfactory receptor neuron (ORN) axons or apical dendrites of mitral cells and send (mostly inhibitory) output to targets within the same glomerulus or in nearby glomeruli. (Figure 6-14)

Period　A fruit fly gene discovered based on mutations that speed up, slow down, or disrupt circadian rhythms. It encodes a protein that participates in negative regulation of its own transcription, and its mammalian homologs serve a similar function. (Figure 9-21)

peristimulus time histogram (PSTH)　A graph that plots firing rates of neurons as a function of time after stimulus onset.

permeability　The ability of a membrane to conduct specific ions, determined principally by the number of open channels capable of conducting those ions.

perturbation experiment　An experiment in which key parameters in a biological system are altered, usually under the experimenter's control, in order to study the consequences.

pharmacodynamics　The effects of a drug in the body, including intended effects on target molecules and processes as well as unintended side effects.

pharmacokinetics　The effects of the body's biological processes on a drug, including the drug's absorption, distribution, metabolism, and excretion.

phase locking　A property whereby the spikes of auditory neurons occur at a specific phase of each cycle of a sound wave. (Figure 6-49)

phasic　Of a neuronal firing pattern, bursts of action potentials in response to specific stimuli.

pheromone　A substance produced by an individual to elicit a specific reaction from other individuals of the same species.

phosphodiesterase (PDE)　An enzyme that hydrolyzes cyclic AMP (cAMP) to AMP, or cGMP to GMP.

phospholipase C (PLC)　A membrane-associated enzyme that is activated by G_q and cleaves inositol-phospholipids to produce inositol 1,4,5-triphosphate (IP_3) and diacylglycerol (DAG). (Figure 3-34)

photoreceptor　A cell that converts light into electrical signals. (Figure 4-2; Figure 13-22)

phototaxis　Movement toward or away from a light source.

phototransduction　The biochemical reactions triggered by photon absorption. (Figure 4-8)

phrenology　A discipline in the 1800s with the goal of mapping the functions of brain areas by studying the shape and size of bumps and ridges on the skull, which were thought to be correlated with an individual's talents and character traits. (Figure 1-22)

phylogenetic tree　A branching diagram showing the relationships among different organisms; it is constructed based on the similarities and differences of different organisms' traits, such as nucleotide and protein sequences. (Figure 13-2)

picrotoxin　A plant toxin that is a potent blocker of the $GABA_A$ receptor.

Piezo　A mechanosensitive channel whose conductance is affected by mechanical force. (Figure 6-65)

pigment cell　A cell in the pigment epithelium layer of the retina adjacent to the outer segments of photoreceptors; it absorbs scattered photons and converts *all-trans* retinal back into 11-*cis* retinal to assist the recovery process in rods. (Figure 4-2)

piriform cortex　The largest olfactory cortical region; it is a three-layered cortex separated from the more dorsally located neocortex by the rhinal sulcus. (Figure 6-16)

pituitary　The endocrine center of the brain; it is located ventral to the hypothalamus. The posterior pituitary contains axon terminals of hypothalamic neurons that directly release hormones into the bloodstream. The anterior pituitary contains endocrine cells that release hormones into the bloodstream in response to prehormones originating from hypothalamic neurons and transmitted by specialized portal vessels. (Figure 9-6)

placebo effect　In the context of pain perception, the phenomenon whereby the perception of pain can be reduced in some patients by the mistaken belief that they have received a treatment thought to reduce pain.

place cell　A hippocampal cell that fires maximally when the animal is at a particular place in an environment. (Figure 11-31)

place field　The physical location in an environment that elicits maximal firing of a particular place cell. (Figure 11-31)

plasma membrane dopamine transporter (DAT)　*See* **plasma membrane monoamine transporters**.

plasma membrane monoamine transporters (PMATs)　A family of proteins on the presynaptic plasma membrane that transport serotonin (serotonin transporter [SERT]), dopamine (dopamine transporter [DAT]), or norepinephrine (norepinephrine transporter [NET]) from the synaptic cleft into the presynaptic cytosol. *See also* **plasma membrane neurotransmitter transporter**. (Figure 12-23)

plasma membrane neurotransmitter transporter　A transmembrane protein on the presynaptic or glial plasma membrane that transports neurotransmitters from the extracellular space into the cell, usually using energy from the co-transport of Na^+ down its electrochemical gradient. (Figure 3-12)

plexin　A member of a class of proteins that serve as receptors for the axon guidance cue semaphorins.

pluripotent stem cell　A cell that has the potential to develop into all cell types of an embryo.

PNS (peripheral nervous system)　Neural tissue and cells outside the CNS, including the nerves that connect the CNS with the body and internal organs as well as isolated ganglia outside the CNS.

Poisson distribution　A discrete probability distribution in which the frequency (f) that k events occur can be determined by a single parameter λ (the mean frequency of occurrence, which equals the product of n and p in the binomial distribution): $f(k; \lambda) = (\lambda^k / k!) \, e^{-\lambda}$. It is an approximation of the binomial distribution when n is large and p is small. (Box 3-1)

polymerase chain reaction (PCR)　A highly sensitive DNA amplification technique that uses a pair of oligonucleotide primers to amplify the DNA segment between the sequences corresponding to the primers through cycles of DNA replication.

polymodal neuron (in the somatosensory system)　A neuron that responds to stimuli of more than one sensory modality.

polymorphism (in genetics)　DNA sequence variation among individuals of the same species.

POMC neuron　A neuron in the arcuate nucleus that expresses pro-opiomelanocortin (POMC), a precursor protein for multiple peptides, including the anorexigenic peptide α-melanocyte-stimulating hormone (α-MSH). (Figure 9-14)

pons　The middle part of the brainstem caudal to the midbrain and rostral to the medulla. (Figure 1-8)

pontine nuclei　Nuclei located in the basal pons that receive input from the cerebral cortex and send output to the cerebellum.

population vector (in movement control)　The sum of the preferred direction vectors of a population of neurons weighted by the firing rate of each neuron. The preferred direction of a motor system neuron is a vector in a three-dimensional space pointing in the direction toward which movement elicits the highest firing rate of the neuron. (Figure 8-31)

positional cloning   A molecular genetic technique that uses molecular and genetic markers on specific chromosomes to identify a gene that causes a particular phenotype or disease.

positive reinforcement   A type of learning in which an action occurs more often because it is followed by receipt of a reward.

positive selection   The process by which an allele that is beneficial to an organism becomes more prevalent in a population.

positron emission tomography (PET)   A noninvasive three-dimensional imaging technique for measuring the distribution of positron-emitting probes introduced into the body.

posterior pituitary   *See* **pituitary**.

postganglionic neuron   A neuron whose cell body is located in a sympathetic or parasympathetic ganglion in the peripheral nervous system and whose axon innervates effectors such as smooth muscle, cardiac muscle, or glands. (Figure 9-1; Figure 9-2)

postsynaptic density   *See* **postsynaptic specialization**.

postsynaptic specialization   A structure on a postsynaptic target cell adjacent to a presynaptic terminal; it is enriched in neurotransmitter receptors and signaling and scaffold molecules. It is also called a postsynaptic density because it is electron dense in electron microscopic images.

Potocki–Lupski syndrome   A neurodevelopmental disorder characterized by mild intellectual disability and autistic symptoms; it is caused by duplication of a chromosome segment including *Rai1*, reciprocal to the deletion that causes Smith–Magenis syndrome.

power stroke   The process by which myosin and actin filaments move relative to each other; it involves conversion of chemical energy from ATP hydrolysis into mechanical force by the myosin motor. (Figure 8-4)

P pathway   A visual processing pathway from the retina to the visual cortex that originates from retinal ganglion cells with small receptive fields and engages lateral geniculate nucleus cells in the parvocellular layers; it carries information about high-acuity color vision. (Figure 4-51)

pre-Bötzinger complex (preBötC)   A caudal brainstem region that generates inspirational breathing rhythms.

precedence effect   The ability of a first-arriving sound to suppress perception of later-arriving sounds.

preferred direction (in the visual system)   The direction of stimulus motion that elicits the highest firing rate of a direction-sensitive visual system neuron.

prefrontal cortex   A neocortical area in the anterior frontal lobe; it is an executive control center that integrates multisensory information, mediates working memory, and performs complex executive functions such as goal selection and decision making.

preganglionic neuron   A neuron whose cell body is located within the CNS and whose axon synapses onto postganglionic neurons in the sympathetic or parasympathetic ganglia. (Figure 9-1; Figure 9-2)

premotor cortex   Area of motor cortex anterior to the primary motor cortex; its neurons send axons primarily to primary motor cortex.

premotor neuron   A spinal cord or brainstem neuron that is presynaptic to motor neurons and thereby participates directly in controlling the firing of motor neurons. (Figure 8-10)

preparatory activity   Neural activity preceding the onset of motor commands and often predictive of upcoming motor actions; it is

abundant in neurons in the frontal and parietal lobes, in particular the premotor cortex.

presenilin   A member of a family (consisting of presenilin-1 and presenilin-2) of multi-pass transmembrane proteins that function as subunits of the γ-secretase complex. They were originally identified based on mutations that cause familial Alzheimer's disease. (Figure 12-5)

prestin   A protein that mediates electromotility in cochlear outer hair cells.

presynaptic facilitation   The process by which neurotransmitter release from cell *A* onto the presynaptic terminal of cell *B* leads to an increase in neurotransmitter release from cell *B*.

presynaptic inhibition   The process by which neurotransmitter release from cell *A* onto the presynaptic terminal of cell *B* leads to a decrease in neurotransmitter release from cell *B*.

presynaptic terminal   A structure at the end (or along the trunk) of an axon that is specialized for releasing neurotransmitters onto target cells. (Figure 1-9)

pretectum   A brainstem structure that receives retinal ganglion cell axon input and regulates pupil, lens, and eye movement reflexes. (Figure 4-37)

primary antibody   An antibody that selectively recognizes a specific antigen (usually a protein).

primary auditory cortex (A1)   The part of the cerebral cortex that first receives auditory sensory information.

primary cilium   A single short, non-motile cilium that projects from the surface of many animal cell types and is often used as a signaling center.

primary motor cortex (M1)   The part of the cerebral cortex that sends descending axons to spinal cord (and in some species directly to motor neurons) to control muscle contraction. (Figure 1-25)

primary somatosensory cortex   The part of the cerebral cortex that first receives somatosensory information from the body. (Figure 1-25)

primary visual cortex (V1)   The part of the cerebral cortex that first receives visual input from the lateral geniculate nucleus. (Figure 4-37; Figure 4-47)

principal component analysis (PCA)   A statistical method used to reduce the dimensionality of a data set. The axes of the reduced data set are called principal components; their orientations in the nonreduced space are selected to maximize the spread of the data along each principal component—data are most spread along the axis of the first principal component, followed by the axis of the second principal component, and so forth.

prion   <u>pr</u>oteinaceous <u>in</u>fectious particle.

prion diseases   Diseases characterized by propagation across the brain of prion protein (PrP) that adopts a specific conformation (PrP^{Sc}), which aggregates and causes massive neurodegeneration and neuronal death. They include scrapie in sheep and goats, mad cow disease in cows, kuru (a human disease that occurred in certain tribes that observed ritual cannibalism), and Creutzfeldt–Jakob disease (CJD; a human disease in which mutations in the *Prp* gene make PrP^C more prone to adopt the PrP^{Sc} conformation spontaneously). (Figure 12-13)

prion hypothesis   The idea that the infectious agent in scrapie is solely proteinaceous in nature.

progesterone   A steroid hormone that regulates female sexual behavior and reproduction in conjunction with estradiol.

projection neuron   A neuron with an axon that projects outside the CNS region that houses the neuron's cell body. In the insect olfactory system, it is a second-order neuron (PN) that receives input from olfactory receptor neuron (ORN) axons and sends output to higher olfactory centers, analogous to a vertebrate mitral/tufted cell. (Figure 6-24)

prokaryote   A single-cell organism without a nucleus. Prokaryotes are members of one of two domains of life: eubacteria and archaea.

proprioception   The sense of body position and movement.

proprioceptive neurons   Somatosensory neurons with peripheral endings embedded in muscle spindles, tendons, and joints to sense muscle stretch and tension. (Figure 6-63)

prostaglandin   A lipid released during inflammation; it binds to specific G-protein-coupled receptors on the peripheral terminals of nociceptive neurons. (Figure 6-73)

protein   Chain of amino acids with a specific sequence linked by peptide bonds.

protein kinase A (PKA)   *See* **cAMP-dependent protein kinase**.

protein kinase C (PKC)   A serine/threonine kinase with diverse substrates; it is activated by binding of both diacylglycerol and Ca^{2+}. (Figure 3-34)

proteinopathy   A disease caused by altered protein conformations, interactions, and/or homeostasis.

protein phosphatase   An enzyme that removes phosphates from phosphorylated proteins, thus counteracting the actions of kinases.

proteome   The collection of all proteins in a specimen.

protocadherin   A member of a class of cell adhesion molecules in vertebrates whose structures and biochemical properties resemble those of cadherins.

protostomes   Animals in which the mouth appears before the anus during development. They include most invertebrate phyla. *See also* **deuterostomes**. (Figure 13-2)

proximity labeling   A technique for labeling macromolecules (proteins or RNAs) based on their proximity to a specific protein tagged with an enzyme that catalyzes the labeling reaction.

pruriception   The sense of itch.

pruritogen   A chemical that causes the sensation of itch.

PSD-95 (postsynaptic density protein of 95 kilodalton)   A postsynaptic scaffold protein highly enriched at glutamatergic synapses. (Figure 3-27; Figure 7-26)

pseudogene   A gene rendered nonfunctional by stop codons in its coding sequence or by other disrupting mutations. Such disrupting mutation(s) are prevalent in a given species.

psychometric function   The quantitative relationship between a parameter of a physical stimulus and the response or perception of a subject.

psychophysics   An experimental approach for characterizing the relationship between physical stimuli and the sensations or behaviors they elicit.

psychosis   A mental state characterized by hallucinations and/or delusions.

psychostimulant   A drug that transiently produces euphoria and suppresses fatigue.

pump   A transporter that uses external energy, such as ATP hydrolysis or light, to actively move a solute across a membrane against its electrochemical gradient. (Figure 2-10)

Purkinje cell   GABAergic neuron of the cerebellar cortex with a highly branched planar dendritic tree; it receives excitatory input from parallel fibers (axons of cerebellar granule cells) and climbing fibers from inferior olive neurons and sends output to the cerebellar nuclei. (Figure 1-11; Figure 8-26)

pyramidal neuron   A type of glutamatergic neuron that has a pyramid-shaped cell body with an apical dendrite and several basal dendrites that branch further; it is abundant in the mammalian cerebral cortex and hippocampus. (Figure 1-15)

quantal content   The number of synaptic vesicle exocytosis events in response to a single action potential.

quantal hypothesis of neurotransmitter release   The idea that neurotransmitters are released in discrete packages of relatively uniform size.

RA (robust nucleus of the arcopallium)   A dorsal forebrain nucleus in the songbird essential for song production; it functions downstream of the HVC. (Figure 10-16)

Rab   A member of a family of small monomeric GTPases involved in intracellular vesicle trafficking.

rabies virus   A neurotropic RNA virus that spreads within the nervous system of its host naturally by crossing synapses. It has been modified for retrograde trans-synaptic tracing. (Figure 14-33)

radial glia   Progenitor cell in the ventricular zone that extends two radial processes—one to the ventricle and the other to the pial surface of the developing cortex. These radial processes serve as substrates for neuronal migration. (Figure 7-4)

random mutagenesis   *See* **forward genetic screen**.

random X-inactivation   A process in which one of the two X chromosomes in female mammals is randomly inactivated in each cell during early development.

raphe nuclei   Brainstem nuclei enriched in serotonin neurons that project widely across the brain. (Figure 9-32)

Ras   A member of a family of small monomeric GTPases involved in signaling pathways required for cell growth and differentiation.

R-C circuit   A circuit containing both resistors and capacitors. (Figure 2-14)

readily releasable pool   A small subset of synaptic vesicles docked at the active zone and primed by an ATP-dependent process to achieve a high-energy configuration that includes preassembled SNARE complexes.

receiver operating characteristic (ROC)   A curve in binary classification where the rate of true positives is plotted against the rate of false positives as the discriminating threshold is systematically changed.

receptive field   In the visual system, the area of the visual field that influences the activity of a given neuron. In the somatosensory system, the area of the body where stimuli can influence the firing of a neuron. Generally, the region of space from which an appropriate stimulus can influence the activity a given neuron in a sensory system.

receptor   A protein that binds and responds to a specific signaling molecule.

receptor potential   A type of graded potential induced at the peripheral endings of sensory neurons by sensory stimuli.

receptor tyrosine kinase (RTK)   A transmembrane protein with an N-terminal extracellular ligand-binding domain and a C-terminal intracellular tyrosine kinase domain. Upon ligand binding, receptor tyrosine kinases add phosphates to tyrosine residues of target proteins.

recombinase   An enzyme that catalyzes recombination between two sequence-specific DNA elements. *See also* **Cre recombinase** and **Flp recombinase**.

reconsolidation (of memory)   A repeated process of memory consolidation after retrieval of a previously consolidated memory.

recording electrode   An electrode used to measure membrane potential changes.

recovery (photoreceptor) The process by which light-activated photoreceptor cells return to the dark state. (Figure 4-11)

recurrent (cross) excitation A circuit motif in which two parallel excitatory pathways mutually excite each other. (Figure 1-20)

recurrent (cross) inhibition A circuit motif in which two parallel excitatory pathways mutually inhibit each other via inhibitory interneurons. (Figure 1-20)

refractory period A time window after an action potential during which another action potential cannot be initiated. (Figure 2-25)

regenerate (axons) Have the ability to reextend and connect with their synaptic partners after damage.

regenerative (action potentials) Propagating without attenuation in amplitude. (Figure 2-25)

regulator of G protein signaling (RGS) A protein that acts as a GTPase activating protein for a trimeric GTP-binding protein.

reinforcement learning A type of machine learning wherein an agent interacts with an environment by repeatedly performing an action and receiving a reward as a consequence of its action.

release probability The probability that an active zone will release one or more synaptic vesicles following an action potential.

releaser A concept in neuroethology; it refers to the essential features of a stimulus that activate a fixed action pattern.

Remak Schwann cell A type of Schwann cell whose cytoplasm extends between individual unmyelinated axons, forming a Remak bundle. (Figure 2-27)

REM sleep A stage of sleep characterized by rapid eye movement. (Figure 9-28)

repellent A molecular cue that guides axons away from its source. (Figure 5-11)

reserpine A first-generation antipsychotic drug; it is an inhibitor of monoamine oxidase.

resistance (R) The degree to which an object or substance opposes passage of electrical current; it is the inverse of conductance: $R = 1/g$.

resistor An electrical element through which passage of current is limited. Current flow through a resistor produces a voltage difference across its two terminals. (Figure 2-13)

responder transgene In binary expression systems; it is the transgene containing the coding sequence for the protein or RNA of interest, along with binding or recombinase sites for the transcription factor or recombinase, respectively, encoded by the driver transgene. (Figure 14-12)

resting potential The membrane potential of a neuron at rest (i.e., in the absence of action potentials or synaptic input), which is typically between –50 and –80 millivolts relative to the extracellular fluid. (Figure 2-11)

reticular theory The idea that the processes of nerve cells fuse and form a giant net constituting the working unit of the nervous system. It has been mostly disproven (with the possible exception of electrical synapses, which allow limited exchange of ions and small molecules between partner neurons).

retina A layered structure at the back of the vertebrate eye with five major neuronal classes (photoreceptors, horizontal cells, bipolar cells, amacrine cells, and retinal ganglion cells) and support cells. Together, these cells convert light into electrical signals, extract biologically relevant features from the outputs of photoreceptors, and transmit such information to the brain. (Figure 4-2)

retinal A chromophore covalently linked to opsins; it changes its configuration upon photon absorption. (Figure 4-6; Figure 13-20)

retinal ganglion cell (RGC) The output cell class of the retina; it transmits information from the eyes to the brain. (Figure 4-2; Figure 4-25)

retinal wave The spread of spontaneous excitation across the developing retina. (Figure 5-21)

retinotopy The topographical arrangement of cells in the visual pathway according to the position of the retinal ganglion cells that transmit signals to them.

retrieval (of memory) The recall of a memory.

retrograde From the axon terminal to the cell body.

retrograde flow The flow of F-actin from the leading edge of the growth cone to its center powered by myosin motors. It contributes to growth cone dynamics. (Figure 5-16)

retrograde tracer A molecule used to trace axonal connections; it is taken up primarily by axon terminals and transported back to cell bodies. (Figure 14-30)

retrograde trans-synaptic tracing *See* **trans-synaptic tracing**.

Rett syndrome A neurodevelopmental disorder in girls caused by disruption of the X-linked gene encoding methyl-CpG binding protein 2 (MeCP2). Patients usually develop normally for the first 6–18 months. Their development then slows, arrests, and regresses, with severe deficits, including social withdrawal, loss of language, and motor symptoms. *See also* **MeCP2**.

reversal potential (E_{rev}) The membrane potential at which current passing through an ion channel changes direction.

reverse genetics The strategy or process of disrupting a predesignated gene to identify its loss-of-function phenotypes. (Figure 14-4)

reward prediction error A theoretical value representing the difference between a received reward and the predicted reward; it is represented by a population of midbrain dopamine neurons.

rhabdomeric type A type of photoreceptor in which the apical surface folds into microvilli that house opsins. (Figure 13-22)

Rho A member of a family of small monomeric GTPases involved in actin cytoskeleton regulation.

rhodopsin A photosensitive molecule in the rod consisting of opsin covalently attached to the chromophore retinal. (Figure 4-6)

RNA (ribonucleic acid) Chain of ribose-containing nucleotides consisting of the sugar ribose, a phosphate group, and one of four nitrogenous bases: adenine (A), cytosine (C), guanine (G), or uracil (U).

RNA editing Post-transcriptional modification that alters a nucleotide sequence of an RNA transcript after it is synthesized.

RNAi (RNA interference) A genetic technique for knocking down expression of a gene of interest by producing a double-stranded RNA with a sequence corresponding to that of the gene of interest. (Figure 14-7)

RNA-seq A technique in which RNA molecules from a given tissue are sequenced one by one in a massively parallel fashion using next-generation sequencing methods. It is used to obtain information about which genes are expressed and at what level at a transcriptome-wide level.

RNA splicing The process by which introns are removed from RNA molecules. In the case of alternative splicing, a subset of exons is removed as well. (Figure 2-2)

Robo (Roundabout) An axon guidance receptor for the ligand Slit. (Figure 7-13)

rod A rod-shaped photoreceptor in the vertebrate retina; it is a very sensitive photon detector specialized for night vision. (Figure 4-2)

rostral–caudal (anterior–posterior) Of a body axis, from head to tail. (Figure 1-8)

rtTA *See* **tTA.**

ryanodine receptor A Ca²⁺ channel on the ER membrane activated by an increase in intracellular Ca^{2+} concentrations and thus amplifies cytosolic Ca^{2+} signals. It is also activated by the plant-derived agonist ryanodine. (Figure 3-41)

saccade A rapid movement of the eyes between fixation points.

sagittal section A section plane perpendicular to the medial–lateral axis. (Figure 1-8)

saltatory conduction The process by which an action potential in a myelinated axon "jumps" from one node of Ranvier to the next. (Figure 2-26)

salty A taste modality that functions primarily to reveal the salt content of food; it is usually appetitive at low concentrations and aversive at high concentrations.

sarcomere The contractile element of a myofibril composed of overlapping F-actin (thin filaments) and myosin (thick filaments). (Figure 8-3)

sarcoplasmic reticulum A special endoplasmic reticulum derivative that extends throughout muscle cells. Ca^{2+} released from the sarcoplasmic reticulum mediates excitation-contraction coupling. (Figure 8-5)

Satb2 A transcription factor that specifies callosal projection neuron identity. (Figure 7-12)

savings A phenomenon whereby less effort is required for an animal to relearn something it has previously learned and then forgotten.

scanning electron microscopy (SEM) A form of electron microscopy that produces images by scanning the surface of a biological specimen and collecting information regarding the interaction of the electron beam with the surface areas.

Schaffer collateral An axonal branch of a hippocampal CA3 pyramidal neuron that synapses onto CA1 pyramidal neurons. (Figure 11-5)

schizophrenia A psychiatric disorder characterized by positive symptoms (such as hallucinations and delusions), negative symptoms (such as social withdrawal and lack of motivation), and cognitive impairment (such as deficiencies in memory, attention, and executive functions).

Schwann cell A glial cell in the PNS that wraps axons with its cytoplasmic extensions to form myelin sheaths. (Figure 2-27)

scrapie *See* **prion diseases.**

secondary antibody An antibody that selectively recognizes primary antibodies made by specific animal species; it is usually conjugated to a fluorophore or an enzyme that produces a color substrate.

secondary dendrite A mitral cell dendrite that extends laterally; it forms reciprocal synapses with granule cells and other olfactory bulb interneurons to spread signals to different olfactory processing channels. It is distinct from the primary (apical) dendrites of mitral cells, which extend into glomeruli. (Figure 6-14)

α-secretase An extracellular protease that cleaves amyloid precursor protein (APP) in the middle of the amyloid β (Aβ) peptide and prevents production of pathology-associated Aβ. (Figure 12-3)

β-secretase An extracellular protease that cleaves amyloid precursor protein (APP) at the N-terminus of amyloid β (Aβ) to produce, along with γ-secretase, intact Aβ. (Figure 12-3)

γ-secretase An intramembrane protease that cleaves α- or β-secretase-processed amyloid precursor protein (APP) at the C-terminus of Aβ. (Figure 12-3)

secreted protein A protein destined for export from the cell. (Figure 2-2)

seizure An episode involving abnormal synchronous firing of large groups of neurons. (Figure 12-43)

selection (in evolution) The process by which genetic variants that confer higher/lower chances of reproductive success become more/less prevalent in future generations. See also **positive selection** and **negative selection**.

selectivity filter The part of an ion channel pore responsible for discriminating between different ionic species so that only some species pass through the channel. (Figure 2-33)

self-avoidance The process in which different axonal or dendritic branches from the same neuron repel each other to avoid overlap of processes from a single cell.

Sema1A, Sema2A, Sema2B (Semaphorin-1A, -2A, -2B) Axon guidance molecules of the semaphorin family in invertebrates; Sema1A is a transmembrane isoform, while Sema2A and Sema2B are secreted isoforms.

Sema3A (Semaphorin-3A) A secreted axon guidance molecule of the semaphorin family in vertebrates.

Sema3F A secreted axon guidance molecule of the semaphorin family in vertebrates.

semaphorins Evolutionarily conserved, widely used axon guidance cues. They consist of secreted and transmembrane variants and act mostly as repellents. Some transmembrane variants can also act as axon guidance receptors. (Figure 5-11)

semicircular canal A sensory organ in the vestibular system that senses angular acceleration in a specific plane. (Figure 6-59)

sensitization An increase in the magnitude of a response to a stimulus after a different kind of stimulus, often noxious, has been applied.

sensorimotor transformation The process by which sensory information is transformed into motor commands, which often involves transformation of spatial coordinates from those of the sensory system into those of the motor system.

sensory homunculus A map in the primary somatosensory cortex; it corresponds to sensation of specific body parts. Nearby somatosensory cortical areas represent sensation from nearby body surfaces. (Figure 1-25)

sensory neuron A neuron that responds directly to external stimuli, such as light, sound, chemical, thermal, or mechanical stimuli.

sensory rhodopsin A type I rhodopsin used in prokaryotes for phototaxis. (Figure 13-20)

serial electron microscopic (EM) reconstruction A method in which consecutive electron micrographs of thin sections are aligned to produce a three-dimensional volume. (Figure 14-32)

serial processing An information processing method in which processing units are arranged in sequential steps.

serine/threonine kinase An enzyme that adds a phosphate onto specific serine or threonine residues of target proteins.

serotonin A monoamine neurotransmitter derived from the amino acid tryptophan that primarily acts as neuromodulator. It is also called 5-HT for 5-hydroxytryptamine. (Figure 3-16; Table 3-2)

Sevenless Originally identified from a mutation in *Drosophila* lacking photoreceptor R7; it encodes a receptor tyrosine kinase that acts cell autonomously in R7 to specify the R7 fate. (Figure 5-35)

sex chromosome   The chromosome whose presence or number determines the sex of an organism.

sex-linked   Of a mutation, having a Mendelian inheritance pattern characteristic of genes located on a sex chromosome. (Figure 12-33)

sex peptide   In *Drosophila,* a peptide transferred with sperm from males to females during mating. It reduces female receptivity to courtship.

sexually dimorphic   Of a trait, differing between females and males.

SFO   *See* **subfornical organ**.

Shaker   Identified as a mutation in *Drosophila* that causes defects in a fast and transient K$^+$ current in muscles and neurons; it encodes a voltage-gated K$^+$ channel.

sharp-wave ripple   Large amplitude rapid oscillations in local field potential in the hippocampus during sleep or during resting while awake. (Figure 11-40)

short-range cue (in axon guidance)   A cell-surface protein that can exert its guidance effects only when axons contact the cell that produces it. (Figure 5-11)

short-term memory   Memory that lasts seconds to minutes. (Figure 11-3)

short-term synaptic plasticity   A change in the efficacy of synaptic transmission that lasts milliseconds to minutes.

sign   In sensory physiology, the direction in which a neuron's activity or membrane potential is changed by a stimulus (for example, the sign is positive if a neuron is depolarized by a stimulus, and the sign is negative if a neuron is hyperpolarized by a stimulus).

signal transduction   The process by which an extracellular signal is relayed via intracellular pathways to varied effectors to produce specific biological effects.

silent synapse   A glutamatergic synapse containing NMDA but not AMPA receptors on the postsynaptic membrane; it can be activated by presynaptic glutamate release that coincides with postsynaptic depolarization but not by presynaptic glutamate release alone.

simple cell   A functionally defined neuronal type enriched in layer 4 of the primary visual cortex; it is best excited by a bar of light in a specific orientation and has separate ON and OFF regions that, when stimulated together, cancel each other's effects. (Figure 4-41)

single-cell RNA-seq   A technique using high-throughput sequencing methods to identify and quantify all mRNAs expressed in individual cells. *See also* **RNA-seq**. (Figure 14-17)

single channel conductance (γ)   The conductance of a single ion channel when open.

single nucleotide polymorphism (SNP)   A single nucleotide of DNA in the genome that varies between members of a species.

single-unit recording   An extracellular recording of the firing pattern of an individual neuron. *See also* **extracellular recording**. (Figure 14-34)

siRNA (short interfering RNA)   Double-stranded RNA with a length similar to microRNA (21-26 nucleotides); it directs a protein complex to degrade target mRNA through base pairing. *See also* **RNAi**.

size principle   The idea that within a motor pool, motor neurons with smaller motor unit sizes (with smaller axon diameters and cell bodies) fire before motor neurons with larger motor unit sizes during muscle contraction. (Figure 8-7)

Slit   A secreted protein best characterized as a repulsive ligand involved in midline axon guidance in many species, from insects to vertebrates. (Figure 7-13)

slow axonal transport   Intracellular transport at a speed of 0.2–8 mm per day. Cargos subject to slow axonal transport include cytosolic proteins and cytoskeletal components. (Figure 2-4)

small bistratified RGC   A blue–yellow color opponent retinal ganglion cell. *See also* **color-opponent RGC**. (Figure 4-34)

Smith–Magenis syndrome   A neurodevelopmental disorder characterized by mild to moderate intellectual disability, delayed speech, sleep disturbances, impaired impulse control, and other behavioral problems. It is caused by mutations that disrupt the function of one copy of a single gene called *Rai1* (retinoic acid induced 1) or loss of one copy of a chromosome segment including *Rai1*.

smooth muscle   Muscle that controls movement of tissue within the digestive, respiratory, vascular, excretory, and reproductive systems.

SM protein   A protein related to yeast Sec1 and mammalian Munc18; it binds SNAREs and is essential for vesicle fusion. *See also* **SNAREs**.

SNAP-25   A t-SNARE attached to the plasma membrane via lipid modification. *See also* **SNAREs**. (Figure 3-8)

SNAREs (soluble NSF-attachment protein receptors)   Proteins on intracellular vesicles and target membranes that are part of a complex mediating membrane fusion. (Figure 3-8)

SNc (substantia nigra pars compacta)   A midbrain nucleus containing dopamine neurons that project mainly to the dorsal striatum. (Figure 8-21)

SNr (substantia nigra pars reticulata)   One of the two major output nuclei of the basal ganglia; it contains GABAergic neurons projecting to the thalamus, superior colliculus, and brainstem motor control nuclei. (Figure 8-21)

solute   A water-soluble molecule, such as an inorganic ion, nutrient, metabolite, or neurotransmitter.

soma   The cell body of a neuron or any other cell.

somatic mutation   A mutation that occurs in a progenitor cell and thus affects only the cells derived from that progenitor.

somatosensory system   The collected parts of the nervous system that process bodily sensation.

Sonic Hedgehog (Shh)   A morphogen that determines cell fate by regulating expression of specific transcription factors in many developmental contexts. For instance, floor plate–derived Shh is responsible for determining the different fates of neuronal progenitors located at different positions along the dorsal–ventral axis of the ventral spinal cord. It is also used as a midline attractant for commissural axons. (Figure 7-10)

sour   A taste modality that functions primarily to warn animals of potentially spoiled food; it is usually aversive.

Southern blot   A method for determining the amount of a specific DNA in a DNA mixture. DNA molecules are separated by gel electrophoresis and transferred to a membrane; labeled nucleic acid probes are then hybridized to the membrane to visualize specific DNA molecules that hybridize to the probe.

space constant   *See* **length constant**.

spatial integration (in dendrites)   The summation of postsynaptic potentials produced by synchronous activation of synapses located at different spatial locations on the postsynaptic neuron. (Figure 3-43)

spectral sensitivity The relationship between a response (e.g., of a photosensitive cell or molecule) and the wavelength of the stimulus light.

spike *See* **action potential**.

spike rate *See* **firing rate**.

spike-timing-dependent plasticity (STDP) A change of synaptic efficacy induced when pre- and postsynaptic neurons repeatedly fire within a restricted time window: synaptic efficacy is potentiated if the presynaptic neuron fires before the postsynaptic neuron and depressed if the presynaptic neuron fires after the postsynaptic neuron.

spike train action potential firing pattern.

spinal cord The caudal part of the vertebrate CNS enclosed by the vertebral column. (Figure 1-8)

spinal muscular atrophy (SMA) A neurodegenerative disease that cause motor neuron death due to homozygous disruption of the *Smn1* (survival motor neuron 1) gene; it is a leading genetic cause of infant mortality.

spinocerebellar ataxia One of a collection of neurodegenerative diseases that share motor defects such as ataxia and are caused by polyglutamine expansion in a number of proteins. (Table 12-1)

spinocervical tract pathway An axonal pathway from the dorsal spinal cord to the lateral cervical nucleus that relays a subset of touch signals, particularly from hairy skin.

spiny projection neuron The most numerous type of neuron in the striatum; it is a GABAergic neuron that projects either directly or indirectly to the output nuclei of the basal ganglia. It is also called a medium spiny neuron. (Figure 8-21)

spiral ganglion neuron A bipolar neuron whose peripheral axon receives auditory information from a hair cell in the cochlea and whose central axon transmits information to the brainstem as part of the auditory nerve. (Figure 6-47)

spontaneous activity Firing of neurons in the absence of environmental stimuli.

sporadic Of a human disease, occurring in a patient without an identifiable family history of the disease.

***Sry* (*Sex determining region Y*)** A gene located on the Y chromosome in mammals; it encodes a transcription factor that determines testes differentiation and other male-specific characteristics.

SSRI (selective serotonin reuptake inhibitor) An inhibitor of the plasma membrane serotonin transporter; it prolongs the action of serotonin in the synaptic cleft.

starburst amacrine cell (SAC) A class of GABAergic inhibitory neurons in the retina that also release acetylcholine. It is a crucial cell type that shapes the responses of direction-selective retinal ganglion cells and participates in generating retinal waves essential for activity-dependent wiring of the visual system. (Figure 4-29)

starter cell *See* **trans-synaptic tracing**.

STED *See* **super-resolution fluorescence microscopy**.

stereocilium A rigid bundled F-actin-based cylinder located on the apical surface of a hair cell. Stereocilia on the same hair cell are arranged in rows of increasing height like a staircase. (Figure 6-45)

stereotactic injection Injection via a device positioned precisely in a three-dimensional coordinate system to target substances such as viruses to a small region.

stereotyped Invariable among individual animals.

stereotyped axon pruning The pruning of exuberant axons with an invariable outcome.

stereotypy A trait or behavior that is largely invariant in different individual organisms of a species.

stimulating electrode An electrode used to pass current into a neuron, usually with the goal of changing the membrane potential of a neuron or its processes.

stomatogastric ganglion (STG) A crustacean ganglion that controls stomach contraction; it has been used as a model system for studying central pattern generators and rhythmic activity in neuronal circuits. (Figure 8-13)

storage (of memory) A step in between memory acquisition and retrieval, in which a memory is encoded as a persistent representation in the nervous system.

STORM *See* **super-resolution fluorescence microscopy**.

striatum The part of the basal ganglia that receives convergent input from the cerebral cortex and thalamus. Also called the caudate-putamen because in some species, the striatum has two separate regions called the caudate and the putamen. (Figure 8-21)

subfornical organ (SFO) Part of the lamina terminalis in the anterior hypothalamus; it is responsible for sensing dehydration signals from the blood, including osmolarity and angiotensin II levels. (Figure 9-16)

substance P A neuropeptide that promotes inflammation when released by the peripheral terminals of sensory neurons. (Figure 6-73)

substantia nigra A midbrain structure named after the high levels of melanin pigments present in the dopamine neurons of healthy human subjects. *See also* **SNc** and **SNr**. (Figure 12-16)

subthalamic nucleus (STN) An intermediate nucleus in the basal ganglia indirect pathway; it contains glutamatergic neurons that project to the GPi and SNr. These neurons receive GABAergic input from the GPe and glutamatergic input from the cerebral cortex. (Figure 8-21)

subthreshold stimulus A stimulus that is insufficient to cause a neuron to generate an action potential. (Figure 2-18)

subventricular zone A cellular layer next to the ventricular zone in the developing nervous system; it contains intermediate progenitors and some radial glia.

superior colliculus A multilayered midbrain structure in mammals that receives retinal ganglion cell axonal input as well as input from other sensory systems. Among many functions, it regulates head orientation and eye movement. It is analogous to the tectum in nonmammalian vertebrates. (Figure 4-37)

superior olivary nuclei Brainstem nuclei in mammals where auditory signals from the left and right ears first converge. The medial superior olivary nucleus (MSO) analyzes interaural time differences, whereas the lateral superior olivary nucleus (LSO) analyzes interaural level differences. (Figure 6-55)

super-resolution fluorescence microscopy A set of fluorescence microscopy techniques capable of imaging specimens at resolutions below the diffraction limit of light. For example: (1) STED (stimulated emission depletion microscopy) achieves super-resolution by exciting fluorophores in a region of tissue smaller than the diffraction limit by depleting fluorescence in an annulus surrounding a central focal spot. (2) STORM (stochastic optical reconstruction microscopy) and (3) PALM (photoactivated localization microscopy) achieve super-resolution by photoactivating a small random subset of photo-switchable fluorophores at any one time, such that the position of each fluorophore can be localized with precision below the diffraction limit; repeated rounds of imaging and deactivation enable reconstruction of the entire imaging field. (Figure 14-28)

supervised learning A category of machine learning in which the task is to identify a function for mapping an input to an output after experiencing example input–output pairs.

suprachiasmatic nucleus (SCN) A hypothamalic nucleus that is the master regulator of circadian rhythms and light entrainment in mammals. (Figure 9-6; Figure 9-24)

suprathreshold stimulus A stimulus that can cause a neuron to generate an action potential. (Figure 2-18)

sweet A taste modality that functions primarily to detect the sugar content of food; it is usually appetitive.

sympathetic system A branch of the autonomic nervous system that facilitates energy expenditure, such as in the case of an emergency response. Activation of the sympathetic system increases heart rate and blood flow, relaxes airways in the lungs, inhibits salivation and digestion, and stimulates production of the hormone epinephrine (adrenaline) in the adrenal glands. (Figure 9-1; Figure 9-2)

symporter A coupled transporter that moves two or more solutes in the same direction. (Figure 2-10)

synapse A site at which information is transferred from one neuron to another neuron or a muscle cell; it consists of a presynaptic terminal and a postsynaptic specialization separated by a synaptic cleft.

synapse elimination The process by which extra synapses are removed during development. It is best described at the vertebrate neuromuscular junction, where the innervation of muscle cells by multiple motor neurons is refined during early postnatal development so that each muscle cell is innervated by a single motor neuron in adults. (Figure 7-29)

synaptic cleft A 20–100 nm gap that separates the presynaptic terminal of a neuron from its postsynaptic target cell. (Figure 1-14; Figure 3-3)

synaptic efficacy *See* **efficacy of synaptic transmission**.

synaptic failure An event in which an action potential in a presynaptic neuron does not produce a postsynaptic response.

synaptic plasticity The ability to change the efficacy of synaptic transmission, usually in response to experience and neuronal activity.

synaptic potential A graded potential produced at postsynaptic sites in response to neurotransmitter release by presynaptic partners.

synaptic tagging The hypothesis that induction of LTP at a synapse causes production of a "tag" at the synapse and that newly synthesized macromolecules necessary for stabilization of LTP are selectively captured by the tag. The hypothesis explains how the input specificity of LTP is maintained despite the cell-wide distribution of newly synthesized macromolecules required for LTP. (Figure 11-20)

synaptic transmission The process of neurotransmitter release from a presynaptic neuron and neurotransmitter reception by a postsynaptic neuron.

synaptic vesicle A small, membrane-enclosed organelle (typically about 40 nm in diameter) localized at the presynaptic terminal; it is filled with neurotransmitters and, upon stimulation, fuses with the plasma membrane to release neurotransmitters into the synaptic cleft. (Figure 3-4; Figure 3-7)

synaptic weight matrix A network of synapses between ensembles of input neurons and output neurons, where the strength (weight) of each synapse can vary between 0 (no connection) and 1 (maximal strength connection). (Figure 11-4)

synaptobrevin A transmembrane SNARE on the synaptic vesicle (a v-SNARE). It is also called VAMP (vesicle-associated membrane protein). *See also* **SNAREs**. (Figure 3-8)

synaptotagmin A Ca^{2+}-binding transmembrane protein on the synaptic vesicle that serves as a Ca^{2+} sensor to trigger neurotransmitter release.

syndromic disorder A disorder characterized by a defined constellation of behavioral, cognitive, and physical symptoms.

syntaxin A transmembrane SNARE on the target plasma membrane (a t-SNARE). *See also* **SNAREs**. (Figure 3-8)

α-synuclein A protein normally enriched in the presynaptic terminal; it is a major component of Lewy bodies, a defining pathological feature of most forms of Parkinson's disease.

T1R1 A G-protein-coupled receptor and a subunit (along with T1R3) of the mammalian umami taste receptor. (Figure 6-39)

T1R2 A G-protein-coupled receptor and a subunit (along with T1R3) of the mammalian sweet taste receptor. (Figure 6-39)

T1R3 A G-protein-coupled receptor and a shared subunit of the mammalian umami and sweet taste receptors. (Figure 6-39)

T2Rs A family of G-protein-coupled receptors that are the mammalian bitter taste receptors. (Figure 6-39)

tamoxifen *See* **CreER**.

tastant A nonvolatile, hydrophilic molecule in saliva that elicits taste perception.

taste bud A cluster of tens of taste receptor cells, with their apical endings facing the surface of the tongue. (Figure 6-32)

taste pore The collected apical endings of taste receptor cells in a taste bud. (Figure 6-32)

taste receptor cell (TRC) A sensory neuron on the surface of the tongue and oral cavity; it converts tastant binding to taste receptor proteins into electrical signals that are transmitted to the peripheral terminals of the gustatory nerve. (Figure 6-32)

tau A microtubule binding protein highly enriched in axons.

tauopathies Neurodegenerative diseases characterized by the presence of neurofibrillary tangles, which consist of aggregates of hyperphosphorylated tau.

Tbr1 A transcription factor that specifies corticothalamic projection neuron identity. (Figure 7-12)

tectorial membrane A membrane on the apical side of hair cells apposed to the stereocilia. (Figure 6-48)

tectum A midbrain structure analogous to the mammalian superior colliculus; it is the major target of retinal ganglion cells in the brains of amphibians and lower vertebrates. (Figure 5-5)

telencephalon The anterior part of the forebrain, including the olfactory bulb, cerebral cortex, hippocampus, and basal ganglia. (Figure 7-3)

temporal (in retinal map) In the direction of the temples.

temporal integration (in dendrites) The summation of postsynaptic potentials produced by activation of synapses within a finite time window. (Figure 3-43)

temporal lobe One of the four cerebral cortex lobes; it is located at the lateral sides of the brain. (Figure 1-23)

teneurins Evolutionarily conserved cell adhesion proteins that control synaptic partner matching and synaptic signaling.

testosterone A steroid hormone that promotes the development of the male reproductive system (masculinization) and inhibits the development of the female reproductive system (de-feminization). In adults, it stimulates sexual and aggressive behaviors. It also serves as a precursor of estradiol. (Figure 10-19)

tetanus toxin   A protease produced by *Clostridium tetani* that cleaves synaptobrevin, thereby inhibiting neurotransmitter release.

tetraethylammonium (TEA)   A chemical that selectively blocks voltage-gated K$^+$ channels.

tetrode   An extracellular recording electrode containing four wires that enable four independent recordings of spiking activities of neurons nearby the electrode tip. The firing patterns of up to ~20 neurons can be resolved based on their different action potential amplitudes and waveforms.

tetrodotoxin (TTX)   A toxin that potently blocks voltage-gated Na$^+$ channels across animal species and is widely used experimentally to silence neuronal firing; it is produced by symbiotic bacteria in puffer fish, rough-skinned newt, and some octopi. (Figure 2-29)

thalamocortical axons (TCAs)   The axons of thalamic neurons that project to the cortex.

thalamus   A structure situated between the cerebral cortex and midbrain; it relays sensory and other signals to the cerebral cortex through extensive bidirectional connections. (Figure 1-8)

theory of dynamic polarization   The idea that every neuron has (1) a receptive component, the cell body and dendrites; (2) a transmission component, the axon; and (3) an effector component, the axon terminals. According to this theory, originally proposed by Ramón y Cajal, neuronal signals flow from dendrites and cell bodies down the axon to the axon terminals.

thermosensation   The sense of temperature.

thermosensory neuron   A somatosensory neuron that senses temperature.

threshold (of action potential)   The membrane potential above which an action potential is generated. (Figure 2-18)

thrombospondin (TSP)   A member of a family of secreted proteins with diverse functions; it can be produced by astrocytes to stimulate synapse formation.

time constant (τ)   The product of resistance and capacitance in an *R-C* circuit. It is a measure of the rate at which both a capacitor charges or discharges and the voltage across a resistor changes in response to changes in current. In neurons, τ corresponds to the time required for the membrane potential change to reach 63% $(1 - 1/e)$ of its maximal value in response to a sudden change in current flow.

Timeless   A fruit fly gene discovered based on mutations that affect circadian rhythms; it encodes a protein that negatively regulates its own transcription. (Figure 9-22)

Timothy syndrome   A syndrome characterized by cardiac arrhythmia and autistic symptoms; it is caused by mutations in the gene encoding Ca$_V$1.2, a voltage-gated Ca^{2+} channel. (Figure 12-41)

tip link   The connection between adjacent stereocilia; it consists of cadherin-23 on the taller stereocilium and protocadherin-15 on the shorter stereocilium. (Figure 6-46)

tonic   Of a neuronal firing pattern, regularly timed and repetitive.

tonic–clonic seizure   A seizure associated with loss of consciousness and a predictable sequence of motor activity: patients first stiffen and extend all extremities (tonic phase) and then undergo full-body spasms during which muscles alternately flex and relax (clonic phase).

tonotopic map   The ordered arrangement of cells in the auditory system in physical space according to their frequency tuning. The cochlea and multiple brain regions contain tonotopic maps. (Figure 6-47)

topographic map   An ordered representation in the brain of features of either the external world or the animal's interactions with the world. For examples, *see* **retinotopy, sensory homunculus**, and **motor homunculus**.

transcription   The process by which RNA polymerase uses DNA as a template to synthesize RNAs. (Figure 2-2)

transcription factor   A DNA-binding protein that regulates transcription of target genes.

transcription unit   The part of the gene that serves as a template for RNA synthesis. (Figure 2-2)

transcriptome   The collection of all expressed RNAs in a tissue or single cell.

transcytosis   The process by which transmembrane or extracellular proteins are first retrieved by endocytosis in one cellular compartment and then delivered for exocytosis at another cellular compartment.

transducin   A trimeric GTP-binding protein complex that links light-activated rhodopsin (or cone opsin) to phosphodiesterase activation in vertebrate photoreceptors. (Figure 4-8)

transgene   An *in vitro* engineered gene introduced into somatic cells or the germline of an organism.

transgenic organism   An organism containing a transgene, usually in the germline.

translation   The process by which an mRNA is decoded by ribosomes for protein synthesis. (Figure 2-2)

transmembrane AMPA receptor regulatory proteins (TARPs)   Transmembrane proteins associated with the AMPA receptors. They regulate trafficking, postsynaptic density anchoring, and physiological properties of the AMPA receptors. (Figure 3-26)

transmembrane protein   A protein destined to span the lipid bilayer of a membrane. (Figure 2-2)

transmission electron microscopy (TEM)   A form of electron microscopy in which high voltage electron beams transmitted through ultra-thin (typically under 100 nm) sections of biological specimens are used to create images.

transporter   A transmembrane protein or protein complex with two separate gates that open and close sequentially to allow solutes to move from one side of a membrane to the other. (Figure 2-8)

trans-synaptic tracing   A method for labeling the synaptic partners of a given neuron or neuronal population of interest (the starter cell or cells). A retrograde trans-synaptic tracer labels presynaptic partners of starter cells, whereas an anterograde trans-synaptic tracer labels postsynaptic partners of starter cells.

transverse section   See **coronal section**.

transverse tubules (T tubules)   An invagination of the plasma membrane that extends into the muscle cell interior, bringing the plasma membrane close to the sarcoplasmic reticulum, such that depolarization effectively triggers Ca^{2+} release from the sarcoplasmic reticulum throughout the entire large muscle cell. (Figure 8-5)

TRE (tetracycline response element)   The DNA sequence to which tTA or rtTA bind. (Figure 14-12). *See also* **tTA**.

TREM2 (triggering receptor expressed on myeloid cells 2)   A cell-surface receptor expressed in the myeloid lineage, including microglia in the brain. Loss-of-function alleles are major risk factors for late-onset Alzheimer's disease. (Figure 12-9)

trichromat   An organism with three different cones for color vision—the S-, M-, and L-cones.

trigeminal ganglia   Clusters of somatosensory neurons near the brainstem involved in sensation of the face.

trimeric GTP-binding protein (G protein)   A GTP-binding protein complex composed of a Gα, a Gβ, and a Gγ subunit with intrinsic GTPase activity in Gα. It has many variants, which couple different GPCRs to diverse signaling pathways. *See also* **G$_s$, G$_i$,** and **G$_q$.**

Trk receptors   A family of neurotrophin receptors that function as receptor tyrosine kinases, including TrkA, TrkB, and TrkC. (Figure 3-39; Figure 7-35)

TRP channels   Nonselective cation channels that share sequence similarities with the *Drosophila* transient receptor potential (TRP) protein. (Figure 2-34)

TRPM8   A TRP channel activated by menthol and by temperatures <26°C. (Figure 6-68)

TRPV1   A TRP channel activated by capsaicin and by temperatures >43°C. (Figure 6-68)

t-SNARE   A SNARE located on the target membrane, such as syntaxin. *See also* **SNAREs.**

tTA (tetracycline-repressible transcriptional activator)   A bacterial transcription factor widely used in heterologous systems, including transgenic mice, to control transgene expression. It drives expression of target genes whose promoters contain a tetracycline response element (*TRE*), but its activity is repressed by tetracycline or its analog doxycycline. A variant called rtTA (reverse tTA) activates *TRE*-driven transgenes in the presence but not the absence of doxycycline. (Figure 14-12)

tuberomammillary nucleus   A hypothalamic nucleus rich in histamine neurons. (Figure 9-29)

tufted cell   *See* **mitral cell.**

two-photon microscopy   A microscopy technique that utilizes simultaneous absorption of two long-wavelength photons to excite fluorophores. Compared with confocal microscopy, it produces less photodamage because only at the focal plane is the density of photons high enough to cause substantial fluorescence emission. Like confocal microscopy, it relies on laser scanning of imaging spots across a plane to produce an optical section. (Figure 14-42)

type III neuregulin-1 (Nrg1-III)   An axonal cell-surface protein, the expression level of which determines the degree of axon myelination by Schwann cells.

tyrosine hydroxylase   An enzyme that converts L-tyrosine to L-dopa; it is the rate-limiting enzyme in the catecholamine biosynthetic pathway. (Figure 12-20)

tyrosine kinase   An enzyme that adds a phosphate onto specific tyrosine residues of target proteins.

UAS   *See* **GAL4.**

ubiquitin-proteasome system   A protein degradation system present in all eukaryotes.

umami   A taste modality that functions primarily to detect the amino acid content of food; it is usually appetitive.

Unc5   A co-receptor for netrin/Unc6 that acts together with DCC/Unc40 to mediate axon repulsion.

Unc6   *See* **netrin/Unc6.**

Unc40   *See* **DCC/Unc40.**

unconditioned response (UR)   *See* **classical conditioning.**

unconditioned stimulus (US)   *See* **classical conditioning.**

unipolar (neuron)   Having one process leaving the cell body that gives rise to both dendritic and axonal processes. (Figure 1-15)

unsupervised learning   A category of machine learning in which the task is to experience a data set and draw inferences about its structure without instruction.

V1   *See* **primary visual cortex.**

vagus nerve   A cranial nerve in the parasympathetic system that connects the brainstem with various internal organs. (Figure 9-2)

variation (in evolution)   The presence of differences in genes or heritable traits.

vasopressin   A hormone secreted by hypothalamic neurons in the posterior pituitary and a neuropeptide released by hypothalamic neurons; it regulates water balance and social behavior.

V-ATPase   A transmembrane protein on synaptic vesicles that pumps protons (H$^+$) into vesicles against their electrochemical gradient using energy derived from ATP hydrolysis. (Figure 3-12)

ventral horn   The ventral part of the spinal gray matter where motor neurons reside. (Figure 8-6)

ventral nerve cord   An invertebrate CNS structure posterior to the brain; it is analogous to the vertebrate spinal cord. (Figure 7-13)

ventral root   The place where motor axons exit the spinal cord. (Figure 8-6)

ventral stream   A visual processing pathway from primary visual cortex to temporal cortex; it is responsible for analyzing form and color. It is also called the "what" stream. (Figure 4-51)

ventral tegmental area (VTA)   A midbrain nucleus containing dopamine neurons that project mainly to the ventral striatum (nucleus accumbens) and prefrontal cortex. (Figure 8-21; Figure 12-29)

ventricle   A cavity derived from the lumen of the neural tube; it is filled with cerebrospinal fluid. (Figure 7-5)

ventricular zone   A layer of cells adjacent to a ventricle. (Figure 7-4)

ventromedial hypothalamic nucleus (VMH)   A hypothalamic nucleus whose best characterized functions include regulation of female lordosis and male mounting and aggression. (Figure 10-26)

vesicular monoamine transporter (VMAT)   A transmembrane protein on synaptic vesicles that transports dopamine, norepinephrine, and serotonin from the presynaptic cytosol into synaptic vesicles. (Figure 12-23). *See also* **vesicular neurotransmitter transporter.**

vesicular neurotransmitter transporter   A transmembrane protein on synaptic vesicles that transports neurotransmitters from the presynaptic cytosol into vesicles using energy from the transport of protons down their electrochemical gradient. (Figure 3-12)

vestibular ganglion neuron   A bipolar neuron whose peripheral axon receives vestibular information from cells in an otolith organ or a semicircular canal and whose central axon transmits information to the brainstem as part of the vestibular nerve.

vestibular nerve   A collection of axons from vestibular ganglion neurons that transmits vestibular information to the brainstem. (Figure 6-59)

vestibular nuclei   Brainstem nuclei that the vestibular nerve innervates; they also receive input from other sensory systems, such as the somatosensory system. (Figure 6-60)

vestibular system   The collected parts of the nervous system that sense the movement and orientation of the head and use this information to regulate a variety of functions including balance, spatial orientation, coordination of head and eye movements, and perception of self-motion.

vestibulo-ocular reflex (VOR)   A reflexive eye movement that stabilizes images on the retina during head movement by moving the eyes in the direction opposite to the head movement. (Figure 6-61)

viral transduction   The process by which a virus infects a host cell, introducing its genome; it is widely used for transgene expression in somatic cells.

visceral afferent   The peripheral axon of a visceral sensory neuron.

visceral motor neurons   Pre- and postganglionic neurons of the autonomic nervous system.

visceral sensory neuron   A sensory neuron whose peripheral branch innervates an internal organ and whose central branch extends into the spinal cord or brainstem. (Figure 9-3; Figure 9-4)

visual cortex   The part of the cerebral cortex dedicated to analyzing visual information.

visual field   The portion of the external world that can be seen at a given time.

voltage clamp   An experimental technique used to measure the current passing through a membrane while holding (clamping) the membrane potential at a set level. (Figure 2-21)

voltage-gated Ca²⁺ channel   An ion channel that allows selective passage of Ca^{2+} and whose conductance is regulated by the membrane potential. (Figure 2-34)

voltage-gated ion channel   An ion channel whose conductance changes as a function of the membrane potential. (Figure 2-30)

voltage indicator   A molecule whose optical properties change in response to membrane potential changes.

volume transmission   The secretion of neurotransmitters (usually neuromodulators) into the extracellular space outside the confines of morphologically defined synapses, where they can affect multiple nearby cells.

vomeronasal organ (VNO)   A sensory organ located above the roof of the mouth that houses the sensory neurons of the accessory olfactory system. (Figure 6-19)

vomeronasal system   *See* **accessory olfactory system**.

VOR gain   The ratio of rotation of the eyes to the rotation of the head in the vestibulo-ocular reflex. *See also* **vestibulo-ocular reflex (VOR)**.

v-SNARE   A SNARE located on a vesicle, such as synaptobrevin. *See also* **SNAREs**.

Wallerian degeneration   The process by which distal axons are eliminated after they are severed from their cell bodies.

Weber's Law   In sensory perception, the property that the just-noticeable difference between two sensory stimuli is proportional to the magnitude of the stimuli.

Wernicke's area   An area in the left temporal lobe involved in language comprehension. Patients with lesions in this area have difficulty comprehending language. (Figure 1-23)

western blot   A method for determining the amount of a specific protein in a protein mixture. Proteins are separated by gel electrophoresis and transferred to a membrane; labeled antibodies are then used to visualize specific proteins bound by the antibody. It can be used to determine protein expression patterns.

white matter   The parts of the CNS enriched in oligodendrocytes and myelinated axons and that thus appear white in histological sections, due to the high lipid content of myelin.

whole-cell patch clamp recording (whole-cell recording)   A form of intracellular recording in which a glass electrode forms a high-resistance seal with the plasma membrane of the recorded cell. After formation of the seal, the membrane underneath the patch electrode is ruptured, such that the interior of the patch electrode and the cytoplasm form a single compartment. *See also* **patch clamp recording**. (Figure 14-40)

whole-mount   A tissue specimen that has not been sectioned.

Wnts   A family of secreted proteins that act as morphogens to pattern embryonic tissues, such as the tissues along the anterior–posterior axes of vertebrates and *C. elegans*. They can also serve as cues in axon guidance and direct formation of synapses along an axon.

working memory   A form of explicit short-term memory; it maintains, updates, and manipulates information for a short period of time. (Figure 11-3)

zygote   A fertilized egg. (Figure 7-2)

Index

Note: Page numbers and ranges suffixed, B, F, or T indicate that material relevant to the topic appears only in a Box, Figure, or Table on that page. Where a text treatment on the same page is already indexed, non-text material is not always distinguished.

When acronyms or their expansions are used consistently, the preferred form in the text becomes the sole entry; where both appear, acronyms are preferred.